BIOETHICAL and EVOLUTIONARY APPROACHES to MEDICINE and the LAW

W. Noel Keyes

Cover design by ABA Publishing

11 10 09 08 07 5 4 3 2 1

Library of Congress Cataloging-in-Publication Data

Keyes, W. Noel.
Bioethics / by W. Noel Keyes.
p. ; cm.
Includes index.
ISBN-13: 978-1-59031-725-9
1. Medical ethics—United States. 2. Bioethics—United States. I. American Bar Association. Committee on Biotechnology. II. Title. [DNLM: 1. Bioethical Issues—United States. 2. Bioethical Issues—legislation & jurisprudence—United States. 3. Bioethics—United States. WB 60 K445b 2006]

R724.K487 2006
174'.957—dc22 2006036128

Discounts are available for books ordered in bulk. Special consideration is given to state bars, CLE programs, and other bar-related organizations. Inquire at Book Publishing, ABA Publishing, American Bar Association, 321 North Clark Street, Chicago, Illinois 60610.

www.ababooks.org

Acknowledgments

These acknowledgments constitute credit deserved by physicians, as well as two men who were both physicians and attorneys who commented on several portions of preliminary drafts of individual chapters where they had a particular interest.

Philip S. Cafarelli, M.D., J.D., FCLM, president of the American College of Legal Medicine, reviewed Chapters 11 through 13, and Melvin A. Shiffman, M.D., F.I.C.S., a member of the Board of Governors of the college, reviewed Chapters 1 through 4. The American College of Law and Medicine publishes *The Journal of Legal Medicine*, to which both have contributed. The publication offers topics of interest in legal medicine, health law and policy, professional liability, hospital law, food and drug law, medical legal research, and education. Physicians who have expressed interest in discussing parts of their volume also included Jerome S. Tobis, M.D., former chairman of the Ethics Committee at University of California–Irvine; Ronald Koons, M.D., current chairman of that committee; and Linh Dan Nguyen, M.D., a committee member who was assistant clinical professor of obstetrics/gynecology. The entire final draft underwent a considerable number of further revisions, including updating, along with cross-references and work toward a goal of clarification in a significant number of sections.

I also wish to thank Boni Goodman, Cheryl Ponti, and Rosemary Matcell, law students who proofread and made revisions to the manuscript, and Leticia Muiños, Betty Andrighetti, and Anindita Dutta (an attorney) for the final typing of this volume.

Because bioethics is a multidiscipline subject (philosophy, religion, science, the law, etc.) I have placed the citations and other material in the footnotes at the bottom of the pages. This will not only save the reader from the need to flip to the end of the chapter and back, it may also refer to more information on a portion of the particular discipline being discussed in the text.

Contents

Acknowledgments iii

About the Author xv

Introduction xvii

1. The Revolution of Bioethics; The American Republic and the Large Increase in the Financing of Healthcare xvi
2. Bioethics and the Development of Its Standards xix
3. The AMA's and ABA's Ethical Standards, Lawyer Ethical Examinations, the *Restatement*, and HHS xxii
4. Members of Hospital Ethics Committees xxiv
5. Some Differences in Methods of Learning the Practice of Bioethics xxviii
6. Divisions of Bioethical Subjects and the Law Thereon xxxiii
7. The Use of Chapters 1-4 Together with Each of Chapters 5-15 ... xxxvi
8. Episodes of good v. Good and "Twenty-first Century Approaches" xxxviii
   a. good v. Good xxxviii
   b. Twenty-first Century Approaches xlii
9. Developing Internationalization of Bioethical Law in the Twenty-first Century xliii

Table of Abbreviations l

Part I
Sources of Bioethics for Members of Hospitals, Ethics Committees, and Regulators of Healthcare 1

Chapter 1
Drawing Lines in Bioethics: Medicine and the Law 3

1. What Are the Subjects of Bioethics and the Law? 3
2. Environmental Parts of Ethics 6
3. The European Union Convention on Human Rights (ECHR) and Bioethics Normal Medical Practice 9
4. The Partial Transition from Bioethics to the Law—Charles Snow 11
   a) Differences between Bioethics (Referred to the Hospital's Ethics Committee) and Medical Malpractice (Referred to the Hospital's Peer Review Committee); Physician Expert Witnesses 14
   b) The Better Way to Face These Problems 23

Chapter 2

The Creation and Evolution of the Universe and Humankind 27
1. Cosmic Evolution from the Beginning of Time 28
   a) The Big Bang and the Meaning of the Creator of the Universe, Its Age, and Evolution 32
   b) Potential of Progress in Astrobiology, Astrology 41
2. The Importance of the Continuance of Evolution into Bioethics, Medicine, and the Law 46
   a) What Constitutes Life Itself? 46
   b) The Importance of Considering the Meaning of Evolution 54
   c) Evolutionary Advances by the Third Millennium 63
   d) A Key to the Basis of Significant Contingencies (Not Design) in Evolution 66
   e) Law and Biology 68

Chapter 3

Religious Sources, Their Restrictions, and Their Possible Bioethical Standards 73
1. Certain Differences between Religious and Bioethical Standards 74
   a) Monotheism and Its Maintenance in Modern Times 76
   b) Consideration of Religious and Ethical Standards under the U.S. Constitution 80
2. A Brief Look at Several of America's Diverse Religions Together with Their Restrictions on Possible Medical and Ethical Sources 86
3. The "Wall of Separation" and Legal Bans on Teaching Religion in Public Schools 127

Chapter 4

Some Differences and Difficulties with Science and Philosophy in the Search for Bioethical Standards 145
1. Better Reasoning and Performance of Science; Law and Ethics Partially Distinguished 146
2. Some Scientists Are Religious, Fewer Are Atheists, and More Are Agnostics 151
   a) Prayer and Certain Conflicts; Reincarnation 152
   b) Limitations of Science; Risk/Benefit Review by IRBs in Research Clinics; Complementary and Alternative Medicine (CAM) 165
3. A Problem With the Use of Either Reason or Philosophy Alone 176
4. The Allegation of Neutral Principles; Science and the Humanities .. 180
5. The Change in Modern Philosophy and Secular Humanism 185

Chapter 5

Autonomy, Responsibility, and Informed Consent 189
1. Autonomy and Responsibility 190
   a) Animal Anatomy of Autonomy in Nature and the Need for Ecology 190
   b) From Animal Autonomy to Human Autonomy 193
   c) Paternalistic Opposition to Individual Autonomy 200
   d) The Balance Between Autonomy and Responsibility, Majority vs. Minority 207
2. Eugenics Examples in Bioethical Issues 211
   a) The Ethical Need for Increased Individual Responsibility 214
   b) Bioethical Considerations of Religious Diversity and Educational Autonomy 218
3. The Informed Consent: Results from Advances in Autonomy 222
   a) The Golden Rule for Use by Physicians 226
   b) Alleged Reliance on Patients to Ask Questions 228
   c) Informed Consent Concerned with Alternative Treatments in Law and Medicine and the Rise of Consumerism 230
   d) Guidelines for Informed Consent in Medicine and the Law; Mandating Reporting of Violence 233
   e) Informed Consent on Clinical Research under IRBs and Experimental Treatments during Life-Threatening Situations through Randomized Placebo-Controlled Investigations 237
   f) Informed Consent by Emergency Department Physicians, EMTALA, the Triad Terrorist Attacks; Radiation Treatment 242
   g) Nuclear Terrorism 250
   h) Informed Consent v. Ethnic Groups and Prisoners 258
   i) Informed Consent v. Laws Authorizing Patient Access to Obtain Physician Profiles; Medical Students Characterized as "Doctors" 261
   j) Mental Competence to Give Informed Consent 269
   k) Informed Consent by Minors 275

Chapter 6

Ethics, Bioethics and Ethics Committees 279
1. Lack of Fundamental Constitutional Rights to Healthcare and Education, Costs 280
2. Bioethicists: With and Without Religion 285
3. Ethics as Viewed by Some Physicians; Errors and Interests; Lack of Effective Discipline; and Use of Placebos 293
   a) Ethics Viewed by a Survey of Physicians vs. Other Readers (of JAMA) 293

b) Evaluations of Physicians (or Bioethicists) Who Work for Manufacturers of Drugs or Medical Devices 294
c) Peer Review Distinguished from Most Bioethical Committee Matters 308
d) Medical Errors, Lack of Reporting, PSQIA and (Medicare Fraud) 312
e) Doctoring Business with Fraud and Waste; and Use of "Double Standards" for Admission to Medical Schools 322
4. The Vagueness of Bioethical Principles and the Need for an Alternative 325
5. The Teaching of Bioethical Ethics and the Law; Ethical Lapses 328
6. The Role of Biomedical Ethics Committees and the Meaning of Care 331
a) JCAHO's Requirement of a "Mechanism" for Ethical Issues ... 332
b) The Lack of Standards for Membership Credentials 336
c) The Goals of Medicine and of Bioethics Committees; "Clarification" and Other Goals 338
d) Physician Dominance in Hospital Ethics Committee Decision Making 341
e) Hospital and Ethics Committee Records 345
7. Organizational Ethics 348
a) Health Maintenance Organizations (HMOs); ERISA and Rationing 349
b) Fraud in Medicare and Medicaid; Contracting Practices of Healthcare Organizations and Needs of Guidelines of the FAR 356

Part II
The Issues Bioethics Must Continue to Solve During the Twenty-first Century 363

Chapter 7
To Conceive or Not to Conceive—The Ethics of Family Planning and Birth Control 365
1. Births within Marriage, Divorce, Same Sex, and Births Outside of Marriage 366
2. The Great Ethical Problem of Overpopulation 378
a) The Current Increasing Population 378
b) The Ethical Need to Establish a National Goal for Family Planning; President Nixon's Goal 383
c) Immigration Ethics and Population Increase 387

d) Environmental (and Bioethical) Problems Growing Out of Overpopulation . . . 392
3. Ethical Actions That Can and Should Be Taken . . . 396
a) Involuntary Birth Controls . . . 396
(i) Involuntary Sterilization, Castration, and Mental Retardation . . . 398
b) Voluntary Birth Controls . . . 404
(i) Voluntary Sterilization . . . 405
(ii) Contraception; Breast-feeding . . . 408
(iii) Unintended Pregnancies . . . 415
(iv) Emergency Contraception . . . 418

Chapter 8
Infertility, Impotence, and Cloning . . . 421
1. Fertility and Impotence . . . 422
a) Some Causes of Impotence . . . 422
b) Public Subsidizing of Reproductive Assistance and Mandatory Insurance Coverage . . . 428
c) Fertility Clinics and Preimplantation Genetic Diagnosis; Lack of Regulation by Many States; Embryo Detraction . . . 433
d) Sperm Banks and Sperm Problems . . . 446
e) Egg Banks . . . 450
f) Frozen Eggs and Embryos and Agreements to Donate Them, Posthumous Reproduction . . . 453
g) Multiple Births Resulting from Fertility Pills . . . 460
h) Artificial Insemination of Single Mothers . . . 465
i) Surrogacy . . . 468
2. Cloning in the Third Millennium . . . 472
a) Origins of Cloning . . . 472
b) Some Religious Reactions . . . 479
c) Some Political and Legal Statements on Human Cloning and Stem Cell Research . . . 481
d) Some Views of Scientists and Physicians on Human Cloning . . . 488
e) Proposed Limits on Cloning in Ethics and the Law . . . 490

Chapter 9
The Choice of Abortion . . . 497
1. Legality, Privacy, and Substantive Due Process . . . 498
2. Restrictions on Public Funding of Abortions . . . 508
3. Religion and Abortion . . . 512
4. Abortion Pills and Other Devices . . . 517
5. Unwanted Pregnancies, Unmarried Mothers, and Abortion Rates . . . 521

6. Spousal and Parental Consent for Abortions 527
7. Harassing and Bombing Abortion Clinics 532
8. Formerly Mandated C-Sections to "Save" a Viable Fetus 537
9. Genetic Testing and Prenatal Counseling; Gender Selection 541
10. Partial Birth and Post-viability Abortions 549
11. Shortage of Physicians for Abortions 556

Chapter 10
Fetal Abuse and Severely Defective Newborns 561
1. Current Liberty Interests of Most Pregnant Drug Users 561
2. Withholding Medical Treatment from Severely Defective Premature Infants with Low Birth Weight 570
3. Newborn Screening Requirements: Intersex Conditions 577
4. Attempts to Prevent Premature Infants with Very Low Birth Weight and Futile Results in Certain Prenatal Births 581
5. Poor and Expensive Outcomes at Neonatal Intensive Care Units (NICUs) 585
6. Proposal that Parents Share Some Expenses and Time 593
7. Infanticide and Abortions after Viability of Certain Types of Severely Defective Infants 595

Chapter 11
Restrictions on the Sources and Allocation of Organ Transplants 605
1. Restrictions on Sources for Organ Transplant and Embryo Transfer in the Twenty-first Century 606
   a) The Need for Consent to Organ Transplants 613
   b) Religious and Cultural Restrictions on Transplants 618
      i) Embryo Transfer for Stem Cell Research 621
   c) Donations to or by Living Mature Minors and Other Legal Incompetents 633
   d) Sales of Organs 638
   e) Organs from Suicidal Donors and Executed Criminals (Murderers) 643
   f) Organs from Anencephalics 648
   g) Fetal Tissue and Mortality Rates in Hospitals; Age Limits 650
   h) Whole Brain Death (by Neurologic Criteria) and Persistent or Permanent Vegetative State (PPVS) 657
   i) Xenotransplants (from Animals to Humans) 668
2. Allocation of Donated Organs 674
   a) Allocation Priority to the "Sickest" Nationwide; HIV Patients . . 675
   b) Non-direct Donations by Living Individuals 680
   c) Directed Donations by Living Persons 683
   d) Transplants to Patients with End-Stage Renal Disease (ESRD) . . 684

Part III **689**
The Right to Die with Dignity

Chapter 12
The Right to Dignity in the Dying Process **691**
1. The Need for End-of-Life-Care Professionals in the Dying Process . . 692
2. The Refusal to Acknowledge the Right to Die 704
3. Culture, Longevity, and Quality of Life Decisions; Monism and Dualism 707
   a) Increasing Longevity 710
   b) Decreasing Longevity and Quality of Life 719
   c) Overweight, Obesity, and Earlier Deaths 735
4. Attempts to Change from the "Slow Course to Death" to "Varieties of Approaches to Death," Kubler-Ross 745
5. Pain and Its Partial Management 749
   a) Pain 749
   b) Pain Management and Its Limitations 751
6. Costs Incurred Nearing End-of-Life Healthcare 761
   a) Entitlements and Their Limitations; Social Security 763
   b) Inequitable Allocations and Deceiving Healthcare Payers v. HMOs 766
   c) Long-term Home Care and Nursing Homes 773

Chapter 13
Improvements Needed in the Twenty-first Century Right to Die **779**
1. Training for End-of-Life Care 780
   a) Advance Directives, Living Wills 782
2. Surrogates, Durable Powers of Attorney, and Emergency Treatment (EMTLA) 791
3. Withholding and Withdrawal of Treatment; Limitations on Nonbeneficial (Futile) Treatment: DNR vs. CPR 798
   a) Withholding Equals Withdrawal 802
   b) Withholding Nutrition and Hydration 803
   c) Patients in a Persistent or Permanent Vegetative State (PPVS) Without a Living Will or DPA; Subjective Personhood 808
4. The Rights of Mature Minors to Die and Jehovah's Witnesses 813
5. The Choices of Do Not Resuscitate (DNR) vs. Cardiopulmonary Resuscitation (CPR): Legal and Ethical 818
   a) Presumed Consent to Cardiopulmonary Resuscitation (CPR) . . . 820
   b) Best Interests v. Substituted Judgment 826
6. Limitations on Nonbeneficial (Futile) Treatment 828
   a) Treatment Beyond the "Goals of Medicine" 831

b) Treatment Regardless of a Lack of Consensus on "Futility" 833
c) The Unique Direction of Informed Consent 836
d) A Way to "Free" Physicians from Self-Interest or Self-Protection in Connection with Life Support; Anencephalics 838
e) The Authority and Responsibility of Physicians with ICUs and Authorized Executions 843
f) The Current Method of Stopping Futile Treatment of Dying Patients by the AMA and by Texas 851
7. Slow Code 855

Chapter 14
End-of-Life Choices of Terminal Patients 857
1. The Choice of Suicide 858
a) Some Religious Views of Jews and Jehovah's Witnesses 858
b) Various Reasons for Possible Suicides; Taking Risks (Motorcycles) 862
c) Limitations on the Duty to Deny Suicide 868
d) The Option of Knowing That You Can 871
2. The Goals of the Hospice Option 873
a) Assistance and Lack of Hospice Assistance in Dying 879
3. The Choice of Physician-Assisted Suicide; Double Effect; Costs of Capital Punishment Procedures 881
a) Some Religious Views on PAS 889
b) The Choice of Double Effect (or Euthanasia) by the Use (or Abuse) of Lethal Doses of Pain Medication 891
c) The Choice of PAS and/or Double Effect in Oregon 896
4. Limitation in Model Statutes; Mandatory Psychiatric Examination .. 907
5. Progressive Nullification of Anti-PAS Statutes 910
a) Civil Disobedience Actions by Physicians and Nurses 910
b) The Prohibition Analogy (Nullification) 911
c) The Example of Dr. Kevorkian 912
6. Other Choices of Euthanasia 914

Part IV
Bioethics and Future Somatic and Germline Gene Therapy 921

Chapter 15
Bioethics on Genetics Superseding the Human Genome Project 923
1. The Human Genome Project 924
2. Genetically Modified Crops 934

3. Genetic Approaches to Disease; Confidentiality; Non-Disease and Privacy 941
4. The Absence of Ethics and Morals in Patents on Genetically Engineered Living Things 952
5. The Possible Future of Inheritable Genetic Interventions and Changes in Homo Sapiens 960
   a) Clinical Trials on Genetic Research 964
   b) Distinctions Between Somatic and Germline Therapy 969
   c) Somatic Gene Therapy and Some Needs for Mandatory Testing . . 972
   d) The Publicity Needed for Recombinant DNA Research by Public and/or Private Entities 982
   e) Germline Gene Therapy 985
      i) Age Limits 991
      ii) Intelligence Quotient; Millionaires; Brains of Males and Females; and Sizes in Animals 996
      iii) The Importance of Germline Gene Therapy for Space Travel 1008

Chapter 16
Epilogue 1021
I. Twenty-first Century Bioethical Decisions within Traditional Cultural, Religious, or Nontraditional Categories by Federal and State Courts 1033

Glossary 1045

Appendix A:
Environmental Crises at the Turn of the Third Millennium 1077
1. Life's Ecological Needs and Purposes According to Existential Philosophers; Environmental Crises 1078
2. Biomass and the "Lack in Accounting" for the Loss in its "Capital" 1081
3. A Sustainable Environment in Forest and for Animals 1085
4. Social Structures of Humans and Two Other Primates 1093
5. The Oceans and Their Fish and Whales 1094
6. The Human Need for Protection Against Release of Chemicals into the Atmosphere 1097
7. Ozone Depletion 1101
8. Global Warming and the Loss of Biological Diversity 1105
   a) Solar, Moon, and Wind Power vs. Coal, Gas, and Oil 1108
   b) The Hydrogen Engine vs. Electric Cars 1110

Appendix B:
Some Views on Medicine in the Two Chinas and Tibet at the Turn of the Millennium 1113
1. Documentary Video "Beyond The Clouds" and an Interview with an Experienced Doctor 1114
2. Birth Control for Han Chinese Majority vs. Minority 1116
   a) The "Critical Appraisal" of Dr. Ruiping Fan 1117
   b) Some Remarks on Articles Contained in the Volume *Chinese and International Philosophy of Medicine* (1998) 1118
3. Adoptions of Abandoned Girls by Foreigners in Beijing 1120
4. Reactions to a Presentation on American End-of-Life Laws in Beijing 1120
5. Reactions by Lawyers in the "Other China" (Taiwan) 1121
6. Tibet 1123
   a) The Herbal Medicine Hospital in Lhasa, Tibet, a Teaching Hospital 1123

Appendix C:
False Claims Act Settlements and Rulings in Healthcare Organizations 1125

Appendix D:
Millennia 1135

Appendix E:
Genetic Research and Economic Advantage of Abbreviated New Drug Applications 1137

Authors of Books Referenced in Footnotes 1143

Author Index 1153

Index 1159

About the Author

W. Noel Keyes is a graduate of Columbia College, Columbia Law School, and holds a Doctor's degree from the University of Paris, France. He was a Fulbright scholar at the International Institute of Private Law in Rome, Italy, and became a member of the New York, New Mexico, and California bar associations. He served as counsel for regions of the Atomic Energy Commission.

Thereafter, he joined the faculty of the Pepperdine University School of Law, where he taught several law classes, developed its Clinical Law Program, founded the Orange and Los Angeles counties Colleges of Trial Advocacy, and wrote a number of articles and books on the law.

After serving for a decade as a member of the Hospital Ethics Committee of the University of California (at Orange), he was appointed to a bioethics committee whose members consist of physicians of this university's medical school, along with a number of other professionals.

Our whole dignity consists in thought. Let us endeavor, then, to think well: this is the principle of ethics.

Blaise Pascal, *Pensées, II,* 1670

It is time for a new generation of leadership to cope with new problems and new opportunities. For there is a new world to be won.

President John Kennedy

The waters I am entering upon nobody has crossed yet.

Dante Alighieri

Introduction

1. ***The Revolution of Bioethics; The American Republic, and Large Increase in the Financing of Healthcare***
2. ***Bioethics and the Development of Its Standards***
3. ***AMA's and ABA's Ethical Standards and the Restatement***
4. ***Members of Hospital Ethics Committees***
5. ***Some Differences in Methods of Learning the Practice of Bioethics***
6. ***Divisions of Bioethical Subjects and the Law Theron***
7. ***The Use of Chapters 1–4 Together with Each of Chapters 5–15***
8. ***Episodes of:***
   - **a. good vs. Good**
   - **b. Twenty-first Century Approaches**
9. ***The Internationalization of Bioethical Law in the Twenty-first century***

The failure to cross the waters of bioethics now being entered should not be done at our peril today. America's "bioethical" revolution was partially inaugurated in the last third of the twentieth century and, having caused considerable excitement, this revolution continues into the new millennium. Many individuals are seeking to understand the ethical issues in medicine and the law regarding "a new life" or new decisions covering an "end of life" for themselves or for their loved ones. Unfortunately, until the dawn of the third millennium, little attention was given to these matters by academic centers or by medical schools.

Brief portions of this nonstop evolution during the final third of the twentieth century will be summarized; they are explained in further detail in subsequent chapters following the mode of Oliver Wendell Holmes who stated:

> It is revolting to have no better reason for a rule than that it was laid down at the time of Henry IV. It is still more revolting if the grounds upon which it was laid down have vanished long since, and the rule simply persists from blind imitation of the past.[1]

1. The Revolution of Bioethics; The American Republic and the Large Increase in the Financing of Healthcare

For many people, the existence of the revolution in bioethics may best be known by the placing of increased emphasis upon the autonomy of the individual. On May 25, 1787, the Constitutional Convention was convened in Philadelphia, and on September 28, the state delegates voted to send the completed Constitution of the United States to the state legislatures for approval which occurred virtually two years later, on September 25, 1789. The first United States Congress adopted 12 amendments to the Constitution and they were sent to the states for ratification. Ten of them became our Bill of Rights by December 15, 1791. Following the Civil War our constitutional freedoms at the state levels were furthered with the 14th Amendment of the Constitution, by providing that "no State shall make or enforce any law which shall abridge the privileges or immunities of citizens of the United States; nor shall any State deprive any person of life, liberty, or property, *without due process* of law; nor deny to any person within its jurisdiction the equal protection of the laws." (Emphasis added.) In recent years the Supreme Court has held that state law controls application of the laws pertaining to healthcare except with respect to the conditions applicable to the use of funds granted by the Federal government.

From many human as well as legal viewpoints, we first entered this modern era in 1965, when the Supreme Court issued a decision via alleged "Sub-

1. Holmes, *The Path of the Law,* HARV. L. REV. (1887) p. 457

stantive Due Process" so that the word "privacy" would enter into the U.S. Constitution. This was first done in order to strike down a state statute prohibiting the sale and distribution of contraceptives. Justice William Douglas declared that "specific guarantees in the Bill of Rights have penumbras, formed by emanations from those guarantees that help give them life and substance."[2] This broadened the Bill on grounds of fairness (see Chapter 7). Such a substantive due process ruling on a bioethical matter initiated the "bioethical revolution" maintained throughout the last third of the twentieth century. Of equal importance were decisions of state courts which initiated our modern medical and legal concept of "informed consent," significantly increasing a patient's and a client's autonomy (see Chapter 5). Other breakthroughs in bioethics followed by a state top court's granting the unconscious Karen Ann Quilan's father the right to order the disconnection of her mechanical respirator, and the U.S. Supreme Court's declaration that most voluntary abortions were being constitutionally performed. Important changes were also made by state initiatives as well as by elected representatives to state legislatures. For example, prior to the end of the second millennium, a slim majority of the people in Oregon (1.10.9) voted to give its citizens who were in terminal health the choice of physician-assisted suicide (PAS). By the 1990s, most members of the Supreme Court would no longer grant many decisions on the basis of "substantive due process" during that newer era. They first refused to affirm an appeal seeking the existence of the fundamental right of "terminal patients" to exercise PAS under the U.S. Constitution. However, a few monts later, the vote of the Oregon people on a second initiative favoring PAS by an overwhelming (3 to 2) majority was upheld by the Supreme Court through its refusal to review the decision of an appellate court that favored the Oregon statute; that is, "certiorari" was denied (see

---

2. Griswold v. Connecticut, 381 U.S.479, 484 (1965). A penubra is a partial shadow cast by the sun. Nixon's call for a strict constructionist had developed during his 1968 presidential campaign. He wanted a justice who believed the Supreme Court "should interpret the Constitution rather than amend it by judicial fiat," i.e., rather than by use of "substantive due process." DEAN, THE REHNQUIST CHOICE (2001), p. 15. William H. Rehnquist was also appointed the Supreme Court by Nixon in 1971. He was appointed chief justice by President Reagan by the time the Supreme Court had largely stopped making alleged "substantive due process" decisions.

Chapter 14).[3] Unlike many state court judges, federal judges are appointed rather than elected. Further, they serve for life under the Constitution "during good behavior" and unlike the Constitution of other countries (e.g. Mexico and France), they may invalidate certain legislative acts of Congress.

The eighteenth-century philosopher, Immanuel Kant, the original proponent of democratic peace, believed that some democracies had been tyrannical, and he specifically excluded them from his conception of "republican" governments.[4] Similarly, James Madison, who was author of much in that century, claimed that "America was better termed a republic," and Article 3, Section 4 of the Constitution specifically declares that the United States "shall guarantee to every state in this Union a Republican Form of Government." Senator John F. Kennedy essentially did so by publishing his book, *Profiles in Courage*.[5] In this connection, it has been pointed out that although people's respect for politicians is often low, "three bodies are always at the top of their list: the Supreme Court, the armed forces, and the Federal Reserve System. All three have one thing in common: they are insulated from public pressures and operate undemocratically. It would seem that Americans admire these institutions precisely because they lead rather than follow."[6] The first

---

3. A writ of certiorari is an order issued by the U.S. Supreme Court that directs a lower court to transmit records for a case that it will hear on appeal. By far, most cases involve a petition for a writ of certiorari.

4. Zakaria, The Future of Freedom (2003), p. 115. Michael Doyle, in his 1997 book *Ways of War and Peace,* is quoted as stating:

> Kant distrusted unfettered, democratic majoritarianism, and his argument offers no support for a claim that all participatory polities – democracies—should be peaceful, either in general or between fellow democracies. Many participatory polities have been non-liberal. . . . The decisive preference of [the] median voter might well include "ethnic cleansing" against other democratic polities.

Id. at 116. *See also* Snyder and Mansfield, *Democratization and the Danger of War,* International Security 20, no. 1 (Summer 1995).

5. The voters selected us because they had confidence in our judgment and our ability to exercise judgment from a position where we could determine what were their own best interests, as part of the nation's interests. This may mean that *we must on occasion* lead, inform, correct, and *sometimes even ignore public opinion for which we were elected.* (Emphasis added.)

Id. at 168-9.

6. *Id.* at 241.

on this "top list" has had a most significant effect on bioethics until the turn of the twenty-first century.

It is interesting to note that government contract expenditures amount to a total of some $250 billion dollars or more by both the military and all U.S. executive agencies every year. However, another field greatly exceeds this amount. The federal government's expenditures in the field of healthcare grants to Medicare, Medicaid, and research have been overwhelming; and total healthcare costs have amounted to more than $1 trillion per year since the beginning of the third millennium—an amount that is more than four times that expended annually on government contracts.[7] Healthcare expenditures will necessarily continue to involve more consideration of bioethics within all the hospital ethics committees and regulators for the duration of the 21st century. Government contract law and regulation (the Federal Acquisition Regulation [FAR]), is currently inapplicable to grants.

Our whole dignity consists in thought. Let us endeavor, then, to think well: this is the principle of ethics.

Blaise Pascal, *Pensées,* II, 1670

2. Bioethics and the Development of Its Standards

Several physicians who were members of a hospital's bioethical committee on which I served stated, "We consider the use of Medical Ethics as being bioethics." However, "bioethics" differs from most of what is called, and limited to, "medical ethics."[8] We consider use of "medical ethics" as generally being different from "bioethics." Medical Ethics is typically used in connection with malpractice cases. "Bioethics," has also been termed a "subfield of ethics," which "focuses on broader issues of social justice in healthcare among many other things.

---

7. Keyes, Government Contracts under the Federal Acquisition Regulation (3d ed. 2003). Because healthcare organizations have spent billions of dollars in fines for various fraudulent actions within the past few years, I recommend that the Federal Acquisition Regulation (FAR) be assigned as a "guideline" for all healthcare organizations, including Medicare and Medicaid, which together receive over a trillion dollars in grants annually. See Chapter 6.

8. Jonsen, *The Birth of Bioethics*, Hastings Ctr. Rep., Special Supplement (November-December 1993), S1.

Van Rennsselaer Potter, the oncologist who coined the term "bioethics" in 1970, foresaw disasters caused by the world's population explosion.[9] (See Chapter 7) Herein it is indicated that bioethics generally involves what can be called "ultra-medical." (See Chapter 1) Until the turn of the third millennium, most physicians, who had not studied bioethics, refused to acknowledge its difference from normal "medical ethics;" one ethicist stated:

> the doctor is involved in a constant stream of choices of an ethical kind, which are made at the local level of his or her interaction with the patient and which bear on its most minute aspects . . . Ethics is what happens in every interaction between every doctor and every patient."[10]

The second edition of the *Encyclopedia of Bioethics* (Becker and Becker, eds. 1992) consists of five volumes which divide materials into different types: many relate to medicine or healthcare, and others refer to "Mental Health and Behavioral" issues, "fertility," and "death and dying." Subjects of equal importance include "Environment and Ecosystem Conditions," "Science," "Population," "Religious Traditions," and several more areas of nonlegal endeavor. That encyclopedia also set forth the area "Law and Bioethics," which states, "Law's influence on bioethics has been so pronounced as to be unmistakable, yet so pervasive as sometimes to be unnoticed." It also pointed out that "perfect legal solutions may not exist for all bioethical di-

---

9. Whitehouse, Hastings Ctr. Rep. (November-December 2001), "Eventually, he returned to the name 'global bioethics.'"

10. Paul Komesaroff, *From Bioethics to Microethics: Ethical Debate and Clinical Medicine*, in TROUBLED BODIES: CRITICAL PERSPECTIVES ON POSTMODERNISM, MEDICAL ETHICS, AND THE BODY (1995).

> To reconstruct bioethics toward a focus on the relationship of nurse-patient or doctor-patient as an embodied, empathic responsiveness is imperative. Such an ethical responsiveness occurs as we come face to face with our patients, and in so doing we come face to face with ourselves.

Bergum, *Beyond Rights: The Ethical Challenge*, 10 PHENOMENOLOGY & MEDICINE 53-74 (1992). Raymond De Vries recently referred to the "social sciences in bioethics," J.L. MED. & AND ETHICS 279–92 (2003).

lemmas." It states:"Scholars differ on the precise influence the law has had shaping the content, methods, and focus of the interdisciplinary field of bioethics, but all would agree that the influence has been significant." In this connection, the American Society for Bioethics and Humanities (ASBH) which was formed "through a consolidation of three separate bioethics organizations—the Society for Health and Human Values, the Society for Bioethics Consultation, and the American Association of Bioethics (ASBH)" that includes philosophers and scholars in the humanitites who specialize in what has come to be known as the "medical humanities."[11]

The development of bioethical standards will continue as those changes to profoundly influence the lives and deaths of many more Americans during the twenty-first century. Uncertainties about continual changes include their timing, which cannot always be predicted with great precision. Consideration of bioethics, the law, and scientific developments offer a significant amount of the evidence why changes remain in great need of attention, as well as the methods by which they might be accomplished.

This evidence must consider various cultures, religions, and their lack of connection with the sciences. An American philosopher early in the twentieth century stated, "The problem with putting the values that are present in our culture, to use in our cultures, can sometimes be undermined by the conclusions of modern science."[12]

Much of this has taken place since the bioethical revolution began. A need clearly exists for considering methods to achieve the possible alternatives related to both the beginning of a new human life and decisions to be made concerning patients who are nearing the end-of-life. The role of emotions may complicate the justification of decision-making.[13]

---

11. *See* J. L. Med. & Ethics 194 (Summer 2005).

12. McGee, Hastings Ctr. Rep. (March-April 1997), p. 21, *citing* The Construction of the Good in the Philosophy of John Dewey, McDermott ed. (1981), p. 577.

13. I would be especially slow to label the moral sentiments or responses of others as squeamishness, or sentimentality, or irrationality. I would be especiallyaware that graver moral danger arises from a deficit of moral emotion than from emotional excess.

Sidney Callahan, *The Role of Emotion in Ethical Decision Making*, 18 Hastings Ctr. Rep. 9-14 (1988).

3. The AMA's and ABA's Ethical Standards, Lawyer Ethical Examinations, the Restatement, and HHS

The American Medical Association's (AMA) was founded in Philadelphia on May 7, 1847. Its current *Code of Medical Ethics* reflects the application of the Principles of Medical Ethics to more than 175 specific ethical and many bioethical issues in medicine (including healthcare rationing, genetic testing, and withdrawal of lifesustaining treatment). Although the Council on Ethical and Judicial Affairs validates the AMA on its ethics, only courts and legislation determine the law with respect to them. For this reason, each principle is followed by many brief citations of articles and legal decisions that contain a sentence setting forth the conclusion thereof. The Preamble to the AMA's Principles states that they are "*not laws*," but standards of conduct which define the essentials of honorable behavior for a physician,"[14] and notes, "a physician shall respect the law." Its Principle III recognizes "a responsibility to seek changes in those requirements which are contrary to the best interest of the patient." See Chapter 4 regarding the AMA's special views on ethics and alleged "unjust laws."

Many provisions of the AMA *Code of Medical Ethics* (e.g.. the 2003 edition is 398 pages) have been integrated into considerations contained in this volume, along with some ethical principles for lawyers set forth in the American Bar Association's *Model Rules of Professional Conduct* (2003 edition, 196 pages). The rules in each of the current editions are about the same length. How can this be so? My estimate is that about 192 pages in the current edition of the AMA edition of its *Code of Medical Ethics* constitute a large gathering of citations of each rule's legal aspects, whereas, the ABA Model Rules contain no citations. The ABA Center for Professional Responsibility develops and interprets standards and scholarly resources on legal ethics, professional regulation, and client protection programs. Like the AMA, the ABA also brings its substantial resources and expertise to sections and divisions, state and local

---

14. *Also see* Pryzbowski vs. U.S. Healthcare, Inc., 64 F. Supp. 2d 361, 370 which states:

> The plaintiff relied on the ABA Council's Opinion 8.13 to show the defendant had a duty to advocate for her in seeking prompt approval for her surgery. The court, however, stated that the plaintiff's reference to the Opinion failed to establish such a duty because the Code of Medical Ethics does not have the force of law.

bar associations, and affiliated organizations, such as the Association of Professional Responsibility Lawyers, the National Client Protection Organization, and the National Organization of Bar Counsel. The center also conducts evaluation of state lawyers and U.S. judicial disciplinary systems,[15] operates a legal ethics research service, ETHIC Search, which helps inquirers resolve questions and manages the National Lawyer Regulatory Data Bank, which records public disciplinary actions against lawyers; and

Many doctors are actually becoming more interested in certain aspects of the law than most lawyers are with medicine. Nevertheless, in 2002, for the first time, the ABA modified its *Model Rules of Professional Conduct* by inserting the key term "informed consent" from the AMA's Medical Code into about 38 of the its rules. Since the dawn of the third millennium, there has existed a significantly greater mutual interest in these two fields. The price inflation changes between 1990 and 2004 has shown Consumer Price Index (CPI) increases of 152.4 percent for hospital services, nursing homes and adult day care; 101.7 percent for legal services; and 82.3 percent for physician services.[16] Nevertheless, only a few others have expressed these sources in publications connected with hospital ethics committees.

The year 2000 also saw the first publication by the American Law Institute's *Restatement* on the Laws Governing Lawyers. Its "Introduction" specifically noted that it addresses "those constraints imposed by law—that is, official norms enforceable through a legal remedy administered by

---

15. A.B.A. J., April 2003, p. 79. The center is online at www.abanet.org/cpr or email cpr@abanet.org. The ABA estimates that there are 1,050,000 licensed lawyers in the United States. A.B.A. J., October 2003, p.8.

> Revisions in 2002 to the ABA Model Rules of Professional Conduct loosened restrictions, but only a few states have accepted them, while some have rejected them and many are still considering them.

A.B.A. J., February 2006, p.31.

16. *The American Institute for Economic Research Report* (Jan. 10, 2005), p. 3.

> In 1978, Federal legislation first explicitly directed the Federal Reserve to conduct monetary policy with a goal that included "stable prices." Yet the CPI has almost tripled since then, suggesting that the purchasing power of the dollar has been cut by almost two-thirds since the Federal Reserve was first directed to pursue "stable prices."

Id.

a court, disciplinary agency, or similar tribunal." The lawyers Multistate Professional Responsibility Examination (MPRE) is a national three-hour multiple choice exam, required for admission to the bars of all but three U.S. jurisdictions.

Finally, bioethical regulations referred to herein have been issued by the many administrations, agencies, and centers of the Department of Health and Human Services (HHS).[17] The *Federal Register*, which issues new rules daily, published a record 75,606 pages in 2002.[18] HHS controls all of its grants for research, the issuance of drugs, etc.; however, it does not directly control research by private industries or laboratories.

Liberty is to the collective body what health is to every individual body. Without health, no pleasure can be tasted by man; without liberty, no happiness can be enjoyed by society.

Thomas Jefferson

There is only one corner of the universe you can be certain of improving . . . and that's your own self.

Aldous Huxley

4. Members of Hospital Ethics Committees

In 1991 the Joint Commission on Accreditation of Healthcare Organizations' (JCAHO's) *Accreditation Manual for Hospitals* referred to ethical con-

---

17. HHS has Administrations for Children and Families and for Aging; its Agencies are for Healthcare Research and Quality, for Toxic Substances and Disease Registry. Its Centers are for Disease Control and Prevention, for Medicare & Medicaid Services, Food and Drug Administration (FDA), Health Resources and Services Administration, Indian Health Service, National Institutes of Health (NIH), Office of Civil Rights (OCR), Program Support Center, and Substance Abuse and Mental Health Services Administration. The HHS has 67,000 employees and a $573-billion budget—bigger than the Pentagon's. L.A. TIMES (December 14, 2004), A12. (The OCR has 10 regional offices each of which covers certain states.)

18. Cres S. Jr., *Cato Inst.*, ORANGE COUNTY REGISTER, July 12, 2003, Com. 6. "Only five agencies are responsible for more than half this torrent: the Environmental Protection Agency and the Departments of Transportation, Treasury, Agriculture and Interior." *Id.*

cerns and ethics committees. It required that organizations have "in place a mechanism(s) for the consideration of ethical issues arising in the care of patients and to provide education to caregivers and patients on ethical issues in healthcare." As of 2002, some 4,960 hospitals have been accredited and the JACHCO requires them to face their bioethical needs. It also accredits thousands of nursing homes and home healthcare agencies (HHAs). JACHCO profoundly affected the growth and structure of the credentialing processing in the early 1980s. Although their standards do not consist of law, they are brought into the bylaws of hospitals and may become contractually binding.[19] Generally, lay members are considered to be "persons not employed by the institution, but not necessarily persons who represent the views of the community in which the institution is located." In 1993, the American Hospital Association estimated that only 60 percent of the hospitals in the United States had ethics committees, along with 10 percent of nursing homes. Yet, at that time there were "more than 4,000 institutional ethics committees in the United States." Ross et al., in "Healthcare Ethics Committees: The Next Generation,"[20] stated, "The committees themselves had between 10 and 25 members. Most committees shared basic values. They described themselves as interdisciplinary both in makeup and approach. Ethics committees for nursing homes are fast becoming a reality. Growth is certain to accelerate in the next five years" (p. 165). "It is not unlikely that in the very near future, all hospitals and many long-term care facilities will have ethics committees" (p. ix). In 2003 JCAHO published the *Comprehensive Accreditation Manual for Home Health Care*; therein are set forth its accreditation standards.[21]

In 1994, the AMA issued its current guidelines to aid in the establishment and functioning of ethics committees in hospitals and other healthcare institutions that are required to form such committees. There are successful committees with 10 members and other committees with 30 or more members. Some committees have had virtually the same members from their be-

---

19. *See* Jost, *The Joint Commission on Accreditation of Hospitals: Private Regulation of Healthcare and the Public Interest*, 24 B.C. L. Rev. 835, 842-45, and 873 (1983) and Am. J. L. & Med. (1996), p. 177.

20. Am. Hosp. Ass'n (AHA 1993), p. 4.

21. Joint Comm'n on Accreditation of Healthcare Orgs., Press Release, JCAHO, Joint Commission Publishes 2003 Home Care Standards Manuals (Nov. 4, 2002).

ginning; others rotate members like clockwork.[22] The functions should be "educational and advisory in purpose and assist in resolving unusual, complicated ethical problems involving issues that affect the care and treatment of patients." The committee's recommendations (using the *Code of Medical Ethics* of the AMA for some of its "guidance") must "receive serious consideration by decision-makers." Their members should be selected "on the basis of their concern for the welfare of the sick and infirm, their interest in ethical matters, and their reputation in the community" (see Chapter 6). In healthcare institutions "operated by religious orders, the recommendations of the ethics committee may be anticipated to be consistent with published religious tenets and principles" (see Chapter 3).[23]

Ethics Committee membership prefers those with "background training in such fields as philosophy, religion, medicine, and law." Ross et al., within The American Hospital Association (AHA), pointed out that "Using a legal format may make dominance of legal (rather than ethical) factors inevitable, but law and ethics are peculiarly intertwined in some areas of bioethics" (p. 93). However, it was also noted that the background of an attorney must be considered:

> because the hospital's attorney has a professional responsibility to protect the interests of the institution rather than those of the patients (even though those interests often may be identical), his or her being a member of the ethics committee could result in a potential conflict of interest.[24]

These subjects are covered in the chapters of this volume. In general, for patients and staff, informed consent may be presumed for ethics consultation (see Chapter 5) and some ethics committees may handle "administrative or organizational ethics issues" (see Chapter 6).

As stated in the AHA publication:

---

22. AMA Code of Medical Ethics Section 9.11 (2003 Ed.).

23. It was based upon the report *Guidelines for Ethics Committees in Health Care Institutions*, adopted December 1984; *see* JAMA 1985, 253L 2698-99.

24. AMA Code of Medical Ethics Section 9.11 (2003).

> Ethics committees are often criticized for not being knowledgeable enough to carry out their mission. The fear of inadequate expertise troubles committee members as well. They worry about whether they know enough and about what it is that they should know more. They are sure they do not know enough ethics, and they doubt that they know enough law.[25]

The resolution for the criticism of these ethics committees and dissolution of the fear of committee members regarding their "inadequate expertise troubles" constitute the goals of this volume (See Chapter 6).

We are hardly born human; we are born ridiculous and malodorous animals; we become human, we have humanity thrust upon us through the hundred channels whereby the past pours down into the present that mental and cultural inheritance whose preservation, accumulation, and transmission place mankind today, with all its defectives and illiterates, on a higher plane than any generation has ever reached before.

Will Durand, *The Greatest Minds, an Ideal* (2002), p. 103.

The superior man is broad-minded toward all and not a partisan; the inferior man is a partisan, but not broad-minded toward all. The superior man is liberal towards others' opinions, but does not completely agree with them; the inferior man completely agrees with others' opinions, but is not liberal toward them.

Confucius (Translated by Yutang),
The Wisdom of Confucius (1938), p. 190.

25. AHA Manual on Healthcare Ethics Committees, p. 134. The Encyclopedia of Bioethics (1995), "Law and Bioethics," p. 1333 stated that "institutional ethics committees usually include at least one lawyer, who can provide analytic abilities as well as expertise on statutory, regulatory, and case law."

5. Some Differences in Methods of Learning the Practice of Bioethics

Until the turn of the third millennium, most physicians still clung to the teaching belief that moral principles such as beneficence, nonmalfeasance, justice, along with some autonomy among patients, may generally provide answers to individual cases in the field of bioethics. However, as stated in the AHA 1993 manual on Healthcare Ethics Committees, "They hope for something that does not exist." In criticizing an updated volume on these principles, it was pointed out:

The method came under fire for being a theory that failed to carry out the tasks that theories typically set for themselves, such as providing a decision procedure for ambiguous and difficult cases. These criticisms impaled [the authors] on their own language.[26]

Stephen Toulmin has also shown that, "The central concerns of philosophers had become so abstract and general that they had, in effect, lost all touch with the concrete and particular issues that arise in actual practice, whether in medicine or elsewhere."[27] This had also been noted in a report on a volume that had been used at medical schools.[28] As Plato stated "time will change and even reverse your present opinions." This may be expected to be seen in bioethics during the twenty-first century.

Hospital ethics committees have continued to be dominated to a large extent by physicians, most of whom had received little training in the field of bioethics. For many, extra training was unnecessary because it was thought

---

26. Ross et al., Healthcare Ethics Com. (AtlA 1993), p. 93.

27. *See also* AMA Code of Medical Ethics Section 9.115(2003).

28. Kuczewshi, *Childress's Greatest Hits*, Hastings Ctr. Rep. (July-August 1998), p. 43.

> The widespread use of Beauchamp's and Childress's *The Principles of Bioethics* (now in its third edition) has created the illusion that if only one could master the principles and some unspecified accompanying method of analysis, the solutions to difficult treatment decisions would be more obvious. In fact, formal ethical analysis and the principles of ethics have been far less helpful to bioethics than many originally expected.

Ross, et al., Healthcare Ethics Committees: The Next Generation (AHA 1993), Chapter 3.

to be a part of "medical ethics." The Preamble to the American Medical Association's (AMA's) *Code of Ethics* states, "As a member of the profession, *a physician must recognize responsibility* not only to patients, *but also to society*." (Emphasis added.)

As noted by the Midwest Bioethics Center (citing this portion of the AMA Ethics Code), "A more complete description of the relational matrix [of bioethics] includes family, friends, the legal system, and regulatory agencies."[29] In a recent biography of the philosopher Martin Heidegger, it was stated, "Philosophy in about 1900 was in deep trouble" (p. 26). Heidegger had stated "science has become our destiny while at the same time leaving us with the unanswered question of its meaning." It was noted that Weber had quoted Tolstoy's view on science and of the path to true happiness: "It is pointless, because it gives no answer to the only question important to us—what are we to do? How should we live?" Heidegger wrote in *Being and Time* in the late 1920s that, "The meaning is Time, but Time 'gives' no meaning."[30] Members of hospital ethics committees also need to have knowledge of restrictions contained in many religions, the law, and other disciplines in order to make recommendations applicable to the individual cases being considered. In 2002, the Institute of Medicine identified the law as a major area of study in modern public health training and many medical centers have begun to offer law courses for medical professionals.[31] Upon the arrival of the twenty-first century, more physicians actually admitted their lack of "ultra-medical" training and recognized their need to collaborate with those who are into such disciplines.

In the middle of the nineteenth century Joseph Henry, Secretary of the National Academy of Science wrote:

> The *increase of knowledge is* much more difficult and in reference to the bearing of this institution on the character of our country and the

29. Toulin, *How Medicine Saved the Life of Ethics*, PERSPECTIVES IN BIOLOGY AND MEDICINE 25 (1982), p. 736.

30. JAMES, THE WILL TO BELIEVE AND OTHER ESSAYS IN POPULAR PHILOSOPHY, Harvard University Press (1979), p. 79.

31. MIDWEST BIOETHICS CENTER (2000), DIFFICULT PROVIDER-PATIENT RELATIONSHIP, p. 53.

> welfare of mankind, *much more important than the diffusion of knowledge.*[32] (Emphasis added.)

The twentieth century adopted an entirely different view, inasmuch as the diffusion of knowledge helps persons who receive it to make better judgments. In the twenty-first century, bioethicists became cognizant of the need for information derived from more fields to be able to help a hospital's ethics committee. These include some knowledge of other sciences (e.g., anthropology, astronomy, biology), the humanities, (philosophy, religion), and governmental rules (the law).[33] Writers in this interdisciplinary field are like historians who must continually face questions involving agriculture, geography, economics, and many other disciplines; they learn and also consult with appropriate outside specialists. In January 2003, bioethicist Arthur Caplan argued that:

> in order to practice bioethics you must know the law; you need to know the medical practice standards . . . you need to have some idea of the moral principles both religious and secular that you can depend upon to make an argument [and] you also need some sense of the cultural, social, and economic setting within which the medical care is being given.[34]

As H. G. Wells wrote near the beginning wrote of the twentieth century, "Human history becomes more and more a race between education and catastrophe."[35] Today, many of the most popular journals in science and medicine (including the *Journal of the American Medical Associton, The New England Journal of Medicine, Nature,* and *Science,*) authorize reporters to preview their articles prior to publication provided such reporters do not ac-

32. WHO WILL KEEP THE PUBLIC HEALTHY? EDUCATING PUBLIC HEALTH PROFESSIONALS FOR THE 21ST CENTURY, Gebbiek, Rosenstock & Hernandez eds., and the Committee on Educating Public Health Professionals for the 21st century, National Academy Press, 2002.

33. Conway, THE SMITHSONIAN (1995), p. 46.

34. http://www.bioethics.net/ 1/31/03, and Marshall, *Embargoes: Good, Bad, or "Necessary Evil"?* 282 SCIENCE 860-65.

35. WELLS, THE OUTLINE OF HISTORY (1920), p. 1100.

tually make advance disclosures. Near the beginning of the twentieth century, the *Merck Manual* began as a reference for doctors and later evolved as a brand of medical reference for consumers, along with *Merck's Manual of Health & Aging*, etc. This volume largely uses medical findings, which have relied upon statistics. In *Basic Statistics; A Primer for the Biomedical Sciences* (1977), Olive Jean Dunn, professor of biostatistics, School of Public Health, UCLA, defines "statistical inference" as "the more exciting part of statistics applied to biological sciences and to medicine." They can become "invaluable to an experimenter" (Chapter 1, p. 1). Many believe that the American magazine *Science* and the foreign one *Nature* (which is often considered the British equivalent of *Science*) are among the best of those containing recent statistical material on science. It has also been noted the "an acquaintance with a few basic statistical techniques cannot give full statistical access to research appearing in the *Journal*."[36] In 2005, a comparison of the citation impact across various study designs and between studies published in 1991 and 2006 (which are utilized for medical studies herein) "is commensurate with most proposed hierarchies of evidence."[37]

At the turn of the third millennium, a writer may be attempting to move forward in the manner of former truth-telling prophets, some of whom had limited effect upon shaping other people's behavior. Hence, Cassandra's accurate but unheeded prophecy of the fall of Troy followed her spurning the advances of Apollo. The "fates" spin out the time of our lives in their threads; Hesiod said it is they who "give men all good and bad." Midgley stated that "They do not assume that the sole reality underlying history is a single, vast, hidden process, a process formally simple and accessible only by a single privileged thought pattern."[38]

---

36. Use of statistical analysis in the *New England Journal of Medicine.* In MEDICAL USES OF STATISTICS, 2d ed., pp. 45-47(BAILAR AND MOSTELLER, EDS.), *NEJM* Books (1992); Moore, "Teaching statistics: resources for undergraduate instructors," Mathematical Association of America, American Statistical Association, 2000; Becker, *Lessons from Darwin: Evolutionary Biology Implications for Alzheimer's Disease Research and Patient Care*, CURR. ALZ. RES., (2005), pp.319-26.

37. Patsopoulos, MD, et al., *Relative Citation Impact of Various Study Designs in the Health Sciences JAMA*, May 18, 2005, p.2362. ("They provide an objective measurement of how often scientists use a specific published work.")

38. MIDGLEY, THE ETHICAL PRIMATE (1994), p. 40.

Much of the bioethical-legal revolution of the last third of the twentieth century has its origin in the increasing dispersion of certain concepts: knowledge of the universe is important if we wish to know where and when we really are; some advances in the knowledge of the billions of years of changes in our cells, if we wish to understand our lives; much reality is found by looking at it in a different way from the past.[39] This was what Thomas Kuhn called a "paradigm" change and noted that contending parties will resort to "the techniques of mass persuasion often including force."[40] Long before the bioethical-legal revolution began, Americans have been pragmatists, as described by William James and John Dewey, our philosophers of the early twentieth century. Near the end of that century, Judge Posner looked at them this way:

Pragmatists of my stripe—pro-science, classical-liberal pragmatists—are not rebels against the Enlightenment; nor are we Utopian dreamers. For us, the significance of pragmatism in relation to the Enlightenment is unmasking and challenging the Platonic, traditionalist, and theological vestiges in Enlightenment thinking.[41]

This pragmatic enlightenment is derived from scientific advances; Kant's basic concern in his *Critique of Pure Reason* at the end of the eighteenth century was the distinction between philosophy and science. Nevertheless, at the end of the nineteenth century Harvard Professor William James noted, "We cannot live or think at all without some degree of faith" because there are "cases where a fact cannot come at all unless a preliminary faith exists in its coming." More assessment of the special links between such faith and science is needed in order to successfully continue the new bioethical process during the first century of the third millennium. In the late 1990s a survey conducted by the National Science Foundation found that 86 percent of Americans agreed that "science and technology are making our lives healthier, easier, and more comfortable."

---

39. *See* Nunberg, The Future of the Book (1948), p. 47. Hesoid wrote this 2,700 years ago.

40. Kuhn, Structure of Scientific Revolutions, p. 47.

41. Posner, Overcoming Law (1995), p. 394.

The only fixed rules of history are no rules. This very rulelessness is what makes the study of history so full of surprise and fascination.
Cowley, *What Ifs*, (2001), p. xvii

6. Divisions of Bioethical Subjects and the Law Thereon

Much of bioethics is concerned with fields that had seldom been taught in medical schools (although they are now being taught at more of them). The traditional field of "medical ethics" concerned matters subject "peer review committees" dealing with the possible malpractice of a physician. Bioethical practice requires help from people in fields quite different from those in the field of medicine. Such people may be requested to serve on "ethics committees" which are in hospitals of the United States (see Chapter 6). This volume covers items that may be normally referred to such committees.

The greatest part of the volume concerns the bioethical aspects of medicine and the law being handled by ethics committees, courts, and others near to both sides of the dawn of the third millennium; over 90 percent of cited material has been published within that last six or seven years. Chapters 1 through 4 constitute matters that often require attention prior to most subsequent ones. For example, the recent twentieth century scientific discoveries on the beginning of the universe by the Creator has resulted in many scientists not becoming (or remaining) atheists (see Chapter 2). Until the dawn of the twenty-first century, the scientific development of evolution was not allowed to be taught in many public high schools due to religious views that prevailed in those states (see Chapter 3).

The extent of the transition of bioethics toward the law often requires recognition of the legal "wall of separation" between religions and the government; certain legal changes in autonomy and informed consent broadened the patient's right to obtain information on medical possibilities before consenting to be treated. A considerable number of bioethical rulings have affected the rights of physicians. As a result of the request of the first female appointed to the Supreme Court, "Justice," before the name has replaced "Mr. Justice" on the bronze doorplates on each Justice's chambers; further, the Federal Rules of Civil and Criminal Procedure are now phrased in gender-neutral terms.[42] However, the continued use of "him and/or her" is not required to be repeated in this portion of today's publications.

42. Justice O'Connor, The Majesty of the Law (2003), p. 196.

Many people have attempted to define "ethics" to be the same as "morals," and most of the latter are derived from their particular religion. For example, Section 1.01 of the AMA *Code of Medical Ethics* defines "ethical" as "matters involving (1) *moral* principles or practices and (2) matters of social policy involving issues of *morality* in the practice of medicine." To the extent morals may be drawn from religion, "such people may believe that their ethics cannot be taught in public schools unless they expressly transfer their origin to another field such as philosophy or sociology." The goals of bioethics differ from some of the goals normally applicable to medicine and, of course, morals and ethics are almost entirely separated from science (see Chapter 2).

Our Constitutional First Amendment of 1791 both prohibits Congress from making any "law respecting an establishment of religion" and "the exercise thereof" including virtually all the viewpoints based upon religions. Following the arrival of the third millennium, there exist two different groups of citizens. The first group consists of a majority who are members of traditional Christian churches. They maintain their autonomy (liberty) to follow the duties and many restrictions expressed in the testaments of the Bible or other religious volumes (see Chapter 3). The second group normally has a sufficient scientific background to chose differently from most of the traditional religions (see Chapter 2) in order to maximize their autonomy (see chapter 5) and consider God as the creator of the universe who did not go beyond setting the laws of nature alone. Their majority would generally adopt each of the bioethical approaches approved by the courts during the final third of the twentieth century (such as those pertaining to abortions, distribution of condoms, *in vitro* fertilization, etc.) together with most of those recommended herein for the twenty-first century (see Chapters 5-15). The key constitutional matter to keep in mind is that neither can those whose moral(and ethics) fall within the first category constitutionally enforce all their religiously based duties and restrictions upon those who are within the second category, nor can it be enforced vice versa. At the close of its 2005 session, in two decisions the Supreme Court took a five to four vote in order to (a) follow the Constitution's "wall of separation" between religion and the law and held that it was being violated by two Kentucky courtrooms in which framed copies of the Biblical ten commandments were hung, but (b) authorized a six-foot granite monument containing them to stand on the grounds of the Texas capital.[43] However the second group's nonreligious autonomous views based on science, experience, and logic may normally result in obtaining the courts authorization to exercise their liberty (autonomy) under the Constitution regardless of objections raised by those who fall within the first group.

The drafters of our Constitution separated the powers of the executive, legislative, and judicial branches. As recommended by Montesquieu's most famous book, *The Spirit of Laws* (1748), the "checks and balances" between each of them became part of the United States Constitution. In this connection, in the case *Marbury v. Madison,* (1803) Chief Justice John Marshall wrote, "It is emphatically the province and duty of the judicial department to say what the law is. Those who apply the rule to particular cases must of necessity expound and interpret that rule. If two laws conflict with each other, the courts must decide on the operation of each." In 2005, the refusal of federal courts to undertake a response from the president and attempt by Congress to make them reconsider decisions by state courts in Florida concerning the right to die when in an unconscious state (the Terri Schiavo case, see Chapters 11 and 14) demonstrated the balance of powers between the federal executive, legislature, and judiciary as separate but equal branches of our government.[44]

Indeed there is but little difference between wisdom and medicine, and between all that is sought in wisdom and all that exists in medicine: contempt of money, honesty and modesty, simplicity in dress, respect, sound judgment, the spirit of decision, neatness, thoughtfulness, the knowledge of all that is necessary for life, hate of vice, the negation of superstitious fear of the gods, divine excellence.

Hippocrates

43. ORANGE COUNTY REGISTER, June 28, 2005, News 3. In the 5-4 ruling in the Texas case, Chief Justice William Rehnquist said in the majority opinion: "Simply having religious content or promoting a message consistent with a religious doctrine does not run afoul of the Establishment clause." In a separate opinion, Justice Stephen Breyer concurred, even though he admitted that it was "borderline" case. It was noted "that Court will revisit this issue." *Id.*

44. See ABA President Robert Grey, *Lawyers Must Defend Judges and Juries*, A.B.A. J., June 2005, p.6.

7. The Use of Chapters 1-4 Together with Each of Chapters 5-15

Some readers may often wish to skip Chapters 1-4 and jump directly into the subsequent chapter dealing with a specific bioethical and legal matter they are facing. When we begin on the bioethical aspects of the fields of medicine and the law, we must first take a look at other fields. For example, many believe that we ought to commence with modern astronomy's "beginning time" in the universe with the Big Bang, which occurred almost 14 billion years ago. This is because space yielded hydrogen and helium and during the subsequent nine billion years galaxies were formed, including our Milky Way, in which enormous supernova explosions yielding interstellar products containing all the new heavier elements making up our solar system. Of course, these are crucial to the evolution of life on our earth over the four billion of years after its formation. Only Chapters 5-15 describe bioethics and the law in particular medical areas and set forth suggested changes to be made early in the third millennium. However, it has been my experience that actual discussions at bioethics meetings on specific types of cases contained in Chapters 5-15 will frequently return to one or more of the matters set forth in these earlier chapters (such as evolution, religions, autonomy, and informed consent) and the general functions of hospital ethics committees—only fewer of which may be repeated in the subsequent chapters.

Bioethical issues must face a number of issues in the third millennium. For many of us the greatest difficulty we face today is the ability to rationally understand those developments in bioethics which appear to extend its recommendations and actions that go beyond our traditional cultures and traditional religions. This requires us to consider the restrictions they contain, along with the demands of most people for their rights to autonomy, and our need to also require them to maintain their responsibility to society. Thus we must consider many areas of ethics and the law including, but not limited to, the following fields: ecology, overpopulation and family planning, infertility, cloning, abortion, fetal abuse, organ transplantation, and serious future decisions to be made on the commencement of somatic and germline gene therapy (along with their maintenance during the third millennium as we expand our activity in space). All of this requires us to consider the tying of the facts and details of (a) their differently alleged expressions of precisely what is or is not "Good," in order to reach (b) the approaches to be taken during the 21st century. Accordingly, the chapters are grouped into four parts:

- Part I – Chapters 1-4 concerns aspects of scientific as well as many of the basic religious ethical and legal changes which arise in bioethics.
- Part II – Chapters 5-6 discuss autonomy, responsibility, informed consent, performance of bioethics committees, and administrative handling.
- Part III – Chapters 7-14 discuss many issues which bioethics must continue to face in the twenty-first century. They focus upon both medical and other policies, together with the legality of treatment to certain patients from the beginning of life (see overpopulation and birth control in Chapter 7) all the way to those concerned with the ending of life by such means as physician-assisted suicide, euthanasia, and double effect, in Chapter 14.
- Part IV – Chapter 15 discusses the bioethical genetical issues commencing in the twenty-first century on somatic and germline gene therapy and will continue throughout the third millennium.

The right to know is the right to live. It is fundamental and unconditional in its assumption that knowledge, like life, is a desirable thing.

George Bernard Shaw

Why should the greatest number be preferred? Why not the greatest good of the most intelligent and most highly developed?

Oliver Wendell Holmes, see Cohn Fed. Lawyer (Jan. 2004) p. 28.

[S]everal independent moral appeals are relevant in determining the rightness or wrongness of actions.

Baruch A. Broch, Taking Issue:
Pluralism and Casuistry in Bioethics, (2003) p. 2.

8. Episodes of good v. Good and "Twenty-first Century Approaches"

a. good v. Good

There still exist some difficulties in science, ethics, and the search for bioethical standards. Sections 5 and 6 above discuss the major distinctions between science and ethics, along with the increase in need for their relationship. This gave rise to ethical considerations for the use of "good v. good," rather than either "right v. right" or "right v. wrong" in Chapters 7-15. St. Augustine stated, "Right is right even if no one is doing it. Wrong is wrong even though everybody is doing it." However, Plato stated, "The *good* is not essence but *far exceeds* essence in *dignity and power*." (Emphasis added.)

In 2003, Justice Sandra O'Connor indicated the need "to ensure that the will of the majority does rights of the minority . . . It is the restraints placed on the majority by the Bill of Rights that make us free" (see Chapter 5). This is done in a manner that may often be considered by members of hospital ethics committees.

Beginning with Chapter 5, virtually every section of subsequent chapters end with a statement of a current ethical issue in terms of "good v. Good." These normally rely upon different morals derived from religions, cultural values, or other sources. This issue is followed by recommended changes in the 21st century to be accomplished by legislatures or the courts or healthcare organizations. Some philosophers might argue that it would better be referred to as "right v. right" for such issues because only others more clearly deserve to be characterized as "right v. wrong."[45]

A "right v. wrong" approach would be close to most of the considerations in trial courts by jurors; and the second one, "right v. right," would

---

45. Corbett v. D'Alesandro, 487 So. 2d 368 (Fla. Ct. App. 1986). Another method was suggested in 2004, namely:

> If individual interests are to give way to communal interests in healthy populations, it is important to understand the value of "the common" and "the good". Public health law is intellectually enticing precisely because it is so difficult, involving complex tradeoffs between individual and collective interests.

Gustin, *Health of the People, The Highest Law*, J. L. MED. & ETHICS (Fall 2004), pp. 510-11.

normally be closer to the majority of cases which go before most appellate courts and many of those before legislatures. It represents all the "more difficult or rather subtle areas," e.g., To what extent should confidential matters be disclosed to third parties? Should a physician be able to carry out all decisions made by a patient? This goal appears to divide it from what virtually everyone would agree with matters that are clearly "right or wrong," from other matters. However, I have used "good v. good" for the issues discussed herein where normally there were quite significant numbers of people (in both minorities and majorities) who have not been able to modify their positions on bioethical matters on a truly objective basis.

It is interesting to note that since the 1920s, all college students at Columbia University (and some other colleges) have been required to include a study of "Classics of Our Civilizations" for two years regardless of their chosen profession. For some students this was not fully appreciated until they had passed several years as professionals. However when they would commence publishing articles they often also cited classical authors. Today most bioethicists seek to draw a difference between logic and rhetoric, and concur with the words of Francis Bacon during the Renaissance:

> It appeareth also that Logic differeth from Rhetoric, not only as the fist from the palm, the one close the other at large; but more is this that Logic handleth reason exact and in truth, and Rhetoric handleth it as it is planted in popular opinions and manners.[46]

---

46. Howell, The Works of Francis Bacon (1961) VI, pp. 168-69. In her volume *Logic Made Easy* (2004), Deborah Bennett not only noted Sir Bacon's distinction, but also remarked about how reasoners use counter examples (such as "good v. Good") in order to:

> Formulate a conclusion based on their mental model and search for alternative models in which the tentative conclusion is false (searching for a counterexample). If the counterexample produces an alternative model, the step is repeated, but if the reasoner is unable to produce an alternative model the conclusion is accepted as following from the premises.

pp. 210-211 (citing Johnson-Laird, Philip, and Byrne. 2002. *Conditionals: A Theory of Meaning, Pragmatics, and Inference*, 109 Psychological Rev. 646-78).

Socrates spoke of knowledge as a virtue. According to the philosophers Plato and Xenophon, Socrates spent most of his time in a modest manner questioning people in Athens rather than spouting dogma. It has been noted that the Greek phrase for "truly good" (kalos kagathos) was referred to by Socrates as freedom (together with self-discipline), a certain degree of education, the ability to get on with people, to do good for one's country, and traditional virtues (such as wisdom, justice, self-control), as well as teaching others what is "truly good."[47] In his pre-scientific age, "Socrates' disavowal of knowledge only means that he disclaims absolutely certain knowledge and leaves the door open for development and improvement."[48] The goal of this volume concerns the need to consider certain scientific conclusions of our day, together with the other virtuous qualities of Socrates, in order to reach twenty-first century approaches toward resolutions for the many bioethical issues we are now facing.

A study by Harvard University educational psychologist Carol Gilligan concluded that men lean toward "ethic of justice," and they view morality as a question of rights. However, women lean toward an "ethic of care" or "goodness." Her abortion study found that:

---

47. Xenophon, CONVERSATIONS WITH SOCRATES, Memories (1990), p. 60. "Self-discipline is presented as the foundation of goodness" (p. 61).

> Suppose. . . that it is right for one man to possess a greater portion of property than another, whether as the fruit of his industry, or the inheritance of his ancestors. Justice obliges him to regard this property as a trust, and calls upon him maturely to consider in what manner it may be employed for the increase of liberty, knowledge, and virtue.

WILLIAM GODWIN, ENQUIRY CONCERNING POLITICAL JUSTICE, Isaac Kramnick ed., London, 1976.

48. *Id.* p. 65. Like the biblical Job in the First Testament, Socrates could only rely upon the knowledge available in his pre-scientific age and, like Jesus in the Second Testament, he accepted no payments for his teaching the seeking of wisdom to his pupils:

> Socrates has often been compared with Jesus, but never, as far as I know, in this respect: principle has the power to change the ways in which people think. Both Jesus and Socrates, in their respective ways, used principle to such effect.

Id. p.25, *citing* BRUNO SNELL, THE DISCOVERY OF MIND IN GREEK PHILOSOPHY & LITERATURE (1982)

The sequence of women's moral judgment proceeds from an initial concern with survival to a focus on *goodness* and finally to a reflective understanding of care as the most adequate guide to the resolution of conflicts in human relationships . . . A moral person is one who helps others; *goodness* is service meeting one's obligations and responsibilities to others, if possible without sacrificing oneself (Emphasis added.)

The essayist Michael de Montaigne expressed, "Nothing is so firmly believed as what we least know." Nevertheless, a modern philosopher wrote:

> "*Right v. Right" is at the heart of our toughest choices* . . . The world unfortunately, faces plenty of right-versus-wrong questions. From cheating on taxes to lying under oath, from running red lights to inflating the expense account . . . the world abounds with instances that, however commonplace, are widely understood to be wrong. But *right-versus-wrong choices are very different from right-versus right ones*. The latter reach inward to our most profound and centurial values, setting one against the other in ways that will never be resolved simply by pretending that one is "wrong."
>
> Right-versus-wrong choices, by contrast, offer no such depth. The closer you get to them, the more they begin to smell. Two shorthand terms capture the differences: If *we can call right-versus-right choices "ethical dilemmas," we can reserve the phrase "moral temptations" for the right-versus-wrong ones. . .*
>
> The really tough choices, then, don't center upon right versus wrong . . . They are genuine dilemmas precisely because each side is firmly rooted in one of our basic, core values . . .
>
> It helps us separate right-versus-wrong from right-versus-right. The more we work with *true ethical dilemmas*, the more we *realize that they fall rather naturally into these paradigms* . . . In this way, the litmus of the paradigms helps us spot the difference between ethical dilemmas and moral temptations." (Emphasis added.)[49]

Similarly, Immanuel Kant, the most famous German philosopher at the end of the eighteenth century, wrote in "The Moral Law" the ethical prin-

49. Rushworth Kidder, How Good People Make Tough Choices (1995), p. 17.

ciple that "What is essentially *good* in the action consists in the mental disposition, let the consequences be what they may."[50] (Emphasis added.) A philosopher has stated after the turn of the third millennium that "in case of practical philosophy" (which he indicated had superseded traditional philosophy) ethics "concerns the Good and its opposite in human affairs."[51] In this connection it may also become necessary to consider the remark by Abraham Lincoln that "You may fool all the people some of the time; you can even fool some of the people all the time; but you can't fool all of the people all the time."

b. Twenty-first Century Approaches

John Stuart Mill, a utilitarian philosopher and writer of the famous essay "On Liberty," had referred to "the peculiar training of a citizen." After stating that this should be "habituating them to act from public or semipublic motives," he emphasized the need "to guide their conduct by aims that unite, instead of isolating them from one another."[52] Public and semi-public unifying "21st century approaches" have been set forth at the end of virtually all sections contained in Chapters 5-15 following the data and the reasons described in each section (and assuming an acquaintance with Chapters 1-4). Changes are proposed in particular sections which contain a number of "legislative facts" that might be considered by a rational legislator prior there being enactment into law.[53] During the last third of the twentieth century, judges and legislators began to make certain significant advances in this new field. Two writers have recently indicated that by 2050 much of the "medical law as we have known it for the last twenty years is unlikely to exist.[54] The

50. CAROL GILLIGAN, IN A DIFFERENT VOICE: PSYCHOLOGICAL THEORY AND WOMEN'S DEVELOPMENT (Harvard Press 1982), pp. 2, 105, & 174.

51. KIDDER, HOW GOOD PEOPLE MAKE TOUGH CHOICES (1995), pp. 157-58.

52. Jonas, Hastings Ctr. Rep. (July-August 2002), p. 34.

53. J. S. MILL, ON LIBERTY (1979), p. 108.

54. Brazier & Glover, *Does Medical Law Have a Future?,* Hayton ed., LAW'S FUTURE(S): BRITISH LEGAL DEVELOPMENTS IN THE 21ST CENTURY (2000), p. 372. This "subject born of family law, tort and jurisprudence may be submerged in consumer law and public law, with the criminal justice system becoming ever more prominent." *Id.* p. 388. *See also* Karst, "Legislative Facts in Constitutional Litigation," (1960) Sup. Ct. Rev. 75; DAVIS, ADMINISTRATIVE LAW 3d ed. 91972), pp. 160–172. In addition, "adjudicative facts" are set forth, which are particular facts relating to the parties at trial in a court or a board.

21st century approaches constitute recommendations concerning those bioethical issues which are in dire need of keys toward the future of humankind.

> Zeno, founder of Stoicism; (Plutarch, *On the Fortunes of Alexander.*) "Never say, 'when you answer the question what country do you belong to, that you are an Athenian or a Corinthian, but that you are a citizen of the world.'"
>
> Ex-slave Epictetus (Arrian, *Epicteti Dissertationes 1.9)*

> *Today, in the absence of any firm sense of manners and morals, the law has become the only recognized authority. Just as the state often acts as a surrogate for the dysfunctional family, so the law is the surrogate for a dysfunctional culture and ethos.*
>
> Himmelfarb, *One Nation, Two Cultures* (1999), p. 67

9. Developing Internationalization of Bioethical Law in the Twenty-first Century

The United Nations has the Security Council, comprised of five permanent members (the U.S., Britain, Russia, China, and France) and 10 members elected for two-year terms.[55]

It constitutes the most important part of the UN because its other members are required to abide by its resolutions.[56] Further, under Article VI of our Constitution, "all Treaties made, or which shall be made, under the authority of the United States shall be the supreme law of the land . . ." Although not everyone currently favors the internationalization of law, it has been noted that it "is happening with phenomenal speed and comprehensiveness."[57]

We are witnessing the growth of an international constitutional common law. It is international because national courts have begun to seek guidance from the decisions of the courts of other nations and because of the recent and rapid proliferation of international tribunals applying treaties, conven-

---

55. UN Charter, art. 23.
56. UN Charter, art. 25.
57. *See, e.g.,* Bork, Coercing Virtue (2003), where he wrote "With that development comes law's seemingly inevitable accompaniment: judicial activism." (p. 15.)

tions, and what they choose to call "customary international law." It is constitutional in that courts insist that their rulings control legislatures and that the legislatures obey. It is common law because the courts piece together, case by case, a fabric of law composed of New Class virtues.[58]

At the dawn of the third millennium, Justice Ruth Bader Ginsburg noted that "readiness to look beyond one's own shores has not marked the decisions of the [United States Supreme Court]."[59] Yet, in his *Humanity and Justice in Global Perspective*, Brian Barry called the issues in international research "justice as mutual advantage."[60]

In 1957, the *Treaty of Rome established the European Economic Community* and the "Rome Statute" of 1998, which was signed by President Clinton, established the International Criminal Court (ICC).[61] Although the ICC will not take jurisdiction if national courts do so, President Bush, Clinton's successor, rejected the statute because the deference of the ICC to national judicial systems was not found to be sufficient.[62]

---

58. *Id.*, p. 137.

59. Ginsburg & Merritt, *Affirmation Action: An International Human Rights Dialogue*, 21 CARDOZO L. REV. 253, 281 (1999).

60. NOMOS XXIV, ETHICS, ECONOMICS, AND THE LAW, Pennock and Chapman eds. (1982), 219-52.

61. Rome Statute of the International Criminal Court, July 17, 1998 (UN Doc. A/CONF.183/9)

62. The ICC is to "determine that a case is inadmissible where . . . the case is being investigated or prosecuted by a State which has jurisdiction over it, unless the State is unwilling or unable genuinely to carry out the investigation or prosecution" (Art. 17, 1(a)) or, the case having been investigated, "the State has decided not to prosecute the person concerned, unless the decision resulted from the unwillingness or inability of the State genuinely to prosecute" (Art. 17, 1(b). Furthermore, "no person who has been tried for conduct also proscribed [as a crime against humanity] shall be tried by the Court with respect to the same conduct unless the proceedings in the other court . . . were for the purpose of shielding the person concerned from criminal responsibility for the crimes within the jurisdiction of the Court; or . . . otherwise were not conducted independently or impartially in accordance with the norms of due process recognized by international law and were conducted in a manner which, in the circumstances, was inconsistent with an intent to bring the person concerned to justice." (art. 30, 3(a)-(b).

BORK, COERCING VIRTUE (2003), p. 36.

The International Law Commission that produced the original draft treaty said, "Let us think about ways in which new developments in substantive law and even new crimes can be brought within the jurisdiction of the Court as time passes and the law progresses."[63] Nevertheless, it has been noted that "international law 'progresses' and new law, especially, human rights law, is created."[64] The Council for International Organizations of Medical Sciences (CIOMS) called for the guidance of ethical review committees and issued some of them.[65] The United Nations Educational, Scientific and Cultural Organization (UNESCO 1997) provided: the "Universal Declaration on the Human Genome and Human Rights" along with its prospective guidance on genetic research.[66]

---

63. *Id.* p. 37, *citing* U.S. Senate Committee on Foreign Relations, *Is a U.N. International Criminal Court in the U.S. National Interest?: Hearing Before the Subcommittee on International Operations of the Committee on Foreign Relations*, 105th Cong. (1998).

64. "International Law Academics tend to see themselves as part of an 'invisible college' devoted to world justice." *Id.*; Sohn, *Sources of International Law,* 25 Ga. J. Intl. & Comparative L. (1995-96) 399. *See* Steiner & Alston, International Human Rights in Context: Law, Politics, Morals, 2d ed. (Oxford University Press, 2000); Commission on Human Rights, Executive Summary, UN ESCOR. 59th Sess., UN Doc. E/CN.4/2003/58 (2003) (para. 8); and 30 Am. J. L. & Med. (2004) 283-304.

65. 19:3-3 Law, Medicine & Health Care 247-58, at 255 (Fall-Winter 1991). The CIOMS guideline states:

> Intimidation in any form invalidates informed consent. Prospective subjects who are patients often depend for medical care upon the physician/researcher, who consequently has a certain credibility in their eyes, and whose influence over them may be considerable, particularly if the study protocol has a therapeutic component. They may fear, for example, that refusal to participate would damage the therapeutic relationship *or result in the withholding of health services.* The physician/investigator must assure them that their decision on whether to participate will not affect the therapeutic relationship or other benefits to which they are entitled . . .

Hastings Ctr. Rep., September-October 2005, p.23, *citing* AFFIRM Investigators, *A Comparison of Rate Control and Rhythm Control in Patients with Atrial Fibrillation*, 1826, and Falk, *Medical Progress: Atrial Fibrillation*, 344 N. Eng. J. Med. 1067-78 (2001).

66. *See* Official UNESCO, *at* http://www.unesco.org/shs/human_rights/hrbc.htm.

The World Trade Organization (WTO) in 2002 established a new Ethics and Health Unit. This unit was expanded to embrace ethics, trade, human rights and health law.[67] Further, a special report was submitted to the UN commission on Human Rights in 2003 stating that the "international human rights law, including the right to health should be consistently and coherently applied across all relevant national and international policy-making processes."[68] In 2003, UNESCO adopted an International Declaration on Human Genetic Data and it has been "in the process of preparing to negotiate a new proposed Declaration on Universal Norms on Bioethics."[69] The World Health Organization holds official observer status on the Council of the WTO and its key committees,[70] "in order to promote health as a legitimate concern on the global trade agenda." In 2003 the Framework Convention on Tobacco Control (FCTC) became the "first convention adopted by the World Health Assembly, the legislative organ of WHO" since it was founded by the UN over half a century earlier.[71] One year after the turn of the third millennium

67. World Health Organization, Ethics and Health Web site *at* http.//www.who.int/ethics/en/. *See also* Hilary, *The Wrong Model: GATS, Trade Liberalization and Children's Right to Health*, SAVE THE CHILDREN 46 (October 2001) (formerly stating that such panels are "on the basis of trade considerations, not public health concerns"). *See also* JAGDISH BHAGWOTI, IN DEFENCE OF GLOBALIZATION (2004), on reputable non-governmental organization (NGOs). It should also be noted that the WTO agreed to abide by the General Agreement on Trade in Services (GATS)—a multi-trade agreement that is applied to all services, including legal services. CAL. BAR J., January 2005, p. 25. Currently, WTO nations are in negotiations to regulate the practice of law between member nations. *Id.*

68. Taylor, *Governing the Globilization of Public Health*, J. L. MED. & ETHICS (Fall 2004), p. 503.

69. DODGSON, LEE & DRAGER, GLOBAL HEALTH GOVERNANCE: A CONCEPTUAL REVIEW (Center on Global Change & Health, London School of Hygiene & Tropical Medicine, 2002).

70. *See* Taylor, *supra* note 58 (at p. 505). WHO has been operating under the conditions of a declining budget in real terms, limiting its autonomy to effect decisions independent of its Member States and compounding pressure on the Organization to institute reforms and implement programmes that are responsive to the demands of key donors. *Id.* p. 506.

71. *See* United Nations Commission for Human Rights, Resolution 2001/71 on Human Rights and Bioethics, April 25, 2001. In 2004 the WHO began intergovernmental negotiations to revise the International Health Regulations, broadening its rules to health threats of international importance and benefit from "small-world

the United Nations Commission for Human Rights adopted resolutions pertaining to human rights and bioethics.[72]

In the 1990s, two legal bioethicists sought for *the establishment of an International Medical Tribunal* so that the medical profession could take a "leading role." They claimed to have done this because:

> It has an *apolitical history*, it has consistently argued for at least some neutrality in *wartime to aid the sick* and wounded, it has a basic *humanitarian purpose* for its existence, *and physician acts intended to destroy human health and life are a unique* betrayal of both societal trust and the profession itself.[73] (Emphasis added.)

However, much medical history has, and continues with "politics" including actions involving "wartime aid to the sick and wounded."[74] Under a number of special circumstances a "physician's acts intended to destroy" a particular human life may be considered the opposite of a "betrayal of both societal trust and the profession itself." (Examples are set forth in Chapters 12–14.) It has been noted that "there is no justification for the creation of a

---

networks" consisting of health professionals, scientists, and non-governmental organizations. WHO Secretariat, *Revision of the International Health Regulations,* Geneva, Switzerland: World Health Organization, Jan. 15, 2004, Publication EB113/3 Rev.1; *see* L.O. Gostin, "The International Health Regulations and Beyond." *See also* Gostin, *International Infectious Disease Law: Revision of the World Health Organization's International Health Regulations*, 291 JAMA 2623-27 (2004).

72. Commission on Human Rights, Executive Summary, UNESCO, 59th Sess., UN Doc. E/CN.4/2003/58 (2003) (para. 8); *see also* 30 Am. J. L. & Med. 283-304 (2004).

73. Annas, Some Choice: Law, Medicine, and the Market 253 (1998). *See also* Annas & Michael A. Grodin, *Medical Ethics and Human Rights: Legacies of Nuremberg,* 3 Hofstra L. & Pol'y Symp. 111, 119 (1999).

74. As stated by Benjamin Meier in *International Criminal Prosecution of Physicians: A Critique of Professors Annas and Grodin's Proposed International Medical Tribunal,* "Few medical decisions are made in a political vacuum, nor should they be." Am. J. L. & Med. 443 (2004). Under the 1949 Geneva Conventions, prisoners of war "are no longer combatants, and military physicians must treat sick prisoners just as they would treat sick or wounded members of their own military units or wounded civilians." Further, they "should have the same right to refuse treatment that an American civilian would have" Annas, Hastings Ctr. Rep., November-December 2004, p. 12. However, two 1977 Additional Protocols to the Geneva Conventions were signed but not ratified by the United States. *Id.* p. 11.

separate criminal tribunal solely for physicians."[75] Further, it would appear that the United States would currently refuse to "provide support for such an International Medical Tribunal" because "it initially voted against and now refuses to ratify the Rome Statute."[76]

In January, 2005, UNESCO's International Bioethics Committee (IBC), recommended a new Universal Declaration on Bioethics and Human Rights, in order "to establish the conformity of bioethics with international human rights law." Its science and technology contributions to healthcare "including for reproductive health" (Art. 13) was objected to by the United States delegation.[77]

Doctors Without Borders/Médecins Sans Frontières (MSF) is an independent, international humanitarian organization that provides emergency medical aid to people affected by armed conflict, epidemics, natural and man-made disasters, and exclusion from health care. MSF currently carries out programs in over 70 countries around the world. However, their New York office recently informed me that as of this time in the twenty-first century their doctors have not yet expressed an interest to discuss bioethical matters.[78]

---

75. To assume that physicians should be held to a different judicial standard, to be judged by shysicians alone and not society as a whole is to follow the flawed logic of medical exceptionalism that led criminal physicians to believe that they were somehow "above the law." *Id.* p. 442.

76. *Id.* p. 447, noting that "On May 6, 2002, the Bush administration took the nearly unprecedented step of withdrawing its signature for the ICC" and "a tribunal would evoke the objections of U.S. medical associations similar to the successful objections voiced by the U.S. military in opposition to the ICC." *Citing* Murphy, *Civil Liability for the Commission of International Crimes as an Alternative to Criminal Prosecution*, HARV. HUMAN RIGHTS J. 1 (1999). Meier concluded with the untested ICC still in its infancy, there is no need for a bold new initiative such as the International Medical Tribunal. Inasmuch as the ICC is able to act toward international criminal prosecutions, it will provide ample opportunity to prosecute physicians within its mandate. (p. 452)

77. Yesley, former staff director of the National Commission for the Protection of Human Subjects, Hastings Ctr. Rep., March-April 2005, p.8. Article 14 calls for sharing the benefits from scientific research and applications. "It is not a law but a declaration." *Id.* IBC's national bioethics committees, which formulate policies for each country, and in many medical centers.

78. The Doctors Without Borders/Médecins Sans Frontières (MSF) office in the United States is located at 333 Seventh Ave., 2nd Floor, New York, NY 10001-5004, and on the Internet at http://www.doctorswithoutborders.org

Finally, throughout the third millennium, it is likely that the UN Treaty on Principles Governing the Activities of States in the Exploration and Use of Outer Space (often called the Outer Space Treaty) will expand in importance. It not only recognizes "the common interest of all mankind in the progress of the exploration and use of outer space," but also states that such exploration:

> shall be carried out for the benefit and in the interests of all countries, irrespective of their degree of economic or scientific development, and shall be the province of all mankind."[79]

---

79. GOLDSMITH, VOYAGE TO THE MILKY WAY (1999) p. 91. "The Outer Space Treaty" has been ratified by all the major spacefaring nations. p. 141. *See also* Chapter 15 in Part IV, "The Future of Bioethics in Somatic and Germline Gene Therapy." See Chapter 15 (Section 5 [e]).

Table of Abbreviations

AAAS	American Association for the Advancement of Science
AANS	American Associations of Neurologic Surgeons
AAP	American Academy of Pediatrics
ABA	American Bar Association
ABMS	American Board of Medical Specialties
ACOG	American College of Obstetricians and Gynecologists
ACP	American College of Physicians
ACS	American Cancer Society
ADA	Americans with Disabilities Act
AHA	American Hospital Association
AHCPR	Agency for Health Care Policy and Research
AIM	Ancestry Informative Markers indicate where one's ancestors came from. Some geneticists have begun searching for such variants. See Leroi, *Mutants*, (2003) p. 339.
Alleles	A variant form of a gene.
ALS	Amytrophic Lateral Sclerosis, also called Lou Gehrig's disease, which progressively destroy neurons of the spinal cord and brain stem resulting in paralysis and death. See Newbery and Abbott, Of mice, men and motor neurons. *Trends in Molecular Medicine* (2002) pp. 88-92.
AMA	American Medical Association
APA	Administrative Procedure Act
APHA	American Public Health Association
ARF	Acute Renal Failure
ART	Assisted Reproductive Technologies
ASBS	American Society for Bariatric Surgery
ASM	American Society for Microbiology
ASRM	American Society for Reproduction Medicine
ATLA	Association of Trial Lawyers of America
BMI	Body Mass Index
BNA	Bureau of National Affairs
BWC	Biological Weapons Convention
CABG	Coronary Artery By-pass Graft

CAM	Complementary and Alternative Medicine
CDC	Centers for Disease Control
Cert.	Certiarary (CERTIORARI)
CIOMS	Council for International Organizations of Medical Sciences, International Guidelines for Ethical Review of Epidemiological Studies (Geneva, 1991); see Law, Medicine & Health Care 19, no. 3-4 (Fall-Winter 1991): 247-58 at 255
COPD	Chronic Obstructive Pulmonary Disease
CPFF	Cost-Plus-Fixed-Fee Contract
CPOE	Comparized Physician Order Entry
CPR	Cardiopulmonary Resuscitation
CREOG	Council on Resident Education of Obstetrics and Gynecology
CT	Computed tomography, a commonly used technology in the evaluation of patients with cancer
DHHS	Department of Health and Human Services
DOD	Department of Defense
DOJ	Department of Justice
DNR	Do Not Resuscitate
EC	Emergency Contraception
EMS	Emergency Medical Services
EMTALA	Emergency Medical Treatment and Active Labor Act
EELC	Education for End-of-Life Care
ELBW	Extremely Low Birth Weight Infants
EPEC	Educate Physicians on End-of-life Care
EPO	European Patent Office (See PTO)
ERISA	Employment Retirement Income Security Act
ERSA	End-Stage Renal Disease
ES	Embryonic Stem cell lines
FAR	Federal Acquisition Regulation
FAA	Federal Aviation Administration
FBI	Federal Bureau of Investigation
FCA	False Claims Act
FDA	Food and Drug Administration
FDCA	Federal Food, Drug, and Cosmetic Act
FEMA	Federal Emergency Management Agency

FECA	Federal Employees' Compensation Act
FOIA	Freedom of Information Act
FRCP	Federal Rules of Civil Procedure
FSH	Follicle Stimulating Hormone
FSS	Federal Supply Schedule
GAO	Government Accounting Office (formerly the General Accounting Office)
GIDA	Genetically Identical Differently Aged
GIFT	Gamete Intra Fallopian Transfer
GLGT	Germ Line Gene Therapy
GMO	Genetic Modified Organism
GSA	General Services Administration
GSC	Proposed U.S. Genetic Science Commission (which was compared to the creation of the Atomic Energy Commission after WWII).
GTSAB	Gene Transfer Safety Assessment Board (WIH)
HAART	Highly Active Antiretroviral Therapy
HBP	High Blood Pressure
HCFA	Health Care Financing Administration
HCQIA	Healthcare Quality Improvement Act
HEDIS	Health Plan Employer Data Information Set
HGP	Human Genome Project
HHA	Home Healthcare Agency
HHS	Department of Health and Human Services
HIPAA	Health Insurance Portability and Accountability Act of 1996
HLA	Human Leukocyte Antigens, a gene type of protein found on the surface of cells which play a role in immune recognition and rejection. A gene donor and the recipient may have matching HLA types.
HMO	Health Maintenance Organization
HUGO	The Human Genome Organization, an international organization of genetic researchers.
ICSI	Intracytoplasmic Sperm Injections of a single sperm into an egg after it is immobilized by crushing its tail. About 48 hours after the eggs are collected, the resulting embryos are transferred to the uterus

IECs	Institutional Ethics Committees
IGBC	International Bioethics Committee
IG/HHS	Inspector General, Department of Health and Human Services
IHGSC	International Human Gene Sequencing Committee
IOM	Institute of Medicine
IV	Intravenous
IVF	In Vitro Fertilization
IQR	Interquartile Range (Normally prognostically variable with division between months) See SEER
IRB	Institutional Review Board
IR&D	Independent Research & Development
IUD	A contraceptive which is an intrauterine device
JAMA	Journal of the American Medical Association
JCAHO	Joint Commission on Accreditation of Healthcare Organizations
L.A. Times	Los Angeles Times
MCO	Managed Care Organization
MHB	Medicare hospice benefit of homecare, hospitalization, or hospice care
MMP–PPA	Medicare and Medicaid Patient and Program Protection Act
MRI	Magnetic Resonance Imaging represents the fields from the synchronous spinning protons in the body aligned by an intense magnetic field, detected by magnetometers, and analyzed by computers. MRI scans can reconstruct excellent patient soft-tissue pictures. Campell, *Earth Magnetism* (2001) pp. 42-43.
MSEHPA	Model State Emergency Health Powers Act
MSPHA	Model State Public Health Act
NBAC	National Bioethics Advisory Committee
NCHS	National Center for Health Statistics
NCI	National Cancer Institute
NCQA	National Committee on Quality Assurance
NCRE	National Clinical Research Enterprise
NEJM	New England Journal of Medicine
NHLBI	National Heart, Lung, and Blood Institute
NHO	National Hospice Organization

NHRA	Nursing Home Reform Act
NICU	Neonatal Intensive Care Unit
NIH	National Institutes of Health—a network of 27 separate institutes that fund medical research
NPDB	National Practitioner Data Bank
OB/GYN	Obstetrics & Gynecology
OCR	Office of Civil Rights (Enforces the HIPAA under the HHS)
O.C. Reg.	Orange County Register
OECD	Organization for Economic Cooperation and Development
OHMO	Office of Health Maintenance Organization (Division of DHHS)
OHR	Office of Human Research, in the Department of Health and Human Services (HHS)
OIG	Office of Inspector General of HHS
PAS	Physician Assisted Suicide
PAS/AE	Physician Assisted Suicide and Active Euthanasia
PCU	Palliative Care Unit
PDMA	Prescription Drug Marketing Act
PET	Positron Emission Tomography
PGD	Pregenetic Diagnosis. A technique with specific and expanding applications for standard clinical practice, including the process by which an embryo can be assessed for lethal genes.
PGDIS	Premarital Genetic Diagnosis International Society founded in 2003 in the United States to promote PGD and workshops on PGD research.
PGH	Preimplantation Genetic Diagnosis
PHS	Public Health Service
PHSA	Public Health Service Act
PKU	Phenylketonuria Mental Retardation due to an inability to metabolize the amino acid phenylalanine
PND	Prenatal Diagnostic Tests
POLST	Physician Orders for Life-Sustaining Treatment
PPVS	Persistent or Permanent Vegetive State
PRO	Peer Review Organization
PTO	U.S. Patent and Trademark Office (See EPO)
PTSD	Post-Traumatic Stress Disorder

PVS	Persistent Vegetative State
RAC	Recombination DNA Advisory Committee (NIH)
RCI	Randomized Clinical Trials
RCT	Randomized Control Trials (in clinical trials)
RASS	Richmond Agitation-Sedation Scale
RICO	Racketeer Influenced and Corrupt Organizations Act
RNAI	RNA Interference
RDNA	Recombinant DNA
SARS	Severe Acute Respiratory Syndrome—a virus
SCHIP	State Children's Health Insurance Program
SCID	Severe Combined Immunodeficiency, an inherited condition
SCNT	A Somatic cell nuclear transfer method of cloning by transferring the nucleus from a donor somatic cell into an enucleated oocyte to produce a cloned embryo
SEER	Surveillance, Epidemiology, and End Results (See IQR)
SGIM	Society of General Internal Medicine
SGT	Somatic Gene Therapy
SIDS	Sudden Infant Death Syndrome
SNF	Skilled nursing facility of a hospital, hospice or nursing home
SNP	Single nucleotide polymorphisms
SOC	Standard of Care
STD	Sexually Transmitted Disease
STI	Sexually Transmitted Infection
SUPPORT	Study to Understand Prognosis and Preferences for Outcomes and Risks of Treatments
TMC	Traditional Chinese Medicine
TSCA	Toxic Substances Control Act
UNOS	United Network for Organ Sharing
USPTO	The U.S. Patent and Trademark Office
USSG	United States Sentencing Guidelines
VA	Department of Veterans Affairs
WHO	World Health Organization
ZIFT	Zygote Interfallopian Transfer

Nothing is so firmly believed as what we least know.
Michael de Montaigne, French Essayist (1533-1592)

Without freedom of thought, there can be no such thing as wisdom.
Benjamin Franklin

Part I

Sources of Bioethics for Members of Hospitals, Ethics Committees, and Regulators of Healthcare

History says, don't hope
On this side of the grave,
But then, once in a lifetime
The longed-for tidal wave
Of justice can rise up,
And hope and history rhyme.

Nobel laureate Seamus Heaney

One generation passeth away and another generation cometh: but the earth abideth forever.

Ecclesiastes 1:1

1

Drawing Lines in Bioethics: Medicine and the Law

1. *What Are the Subjects of Bioethics and the Law?*
2. *Environmental Parts of Ethics*
3. *The European Union Convention on Human Rights (ECHR) and Bioethics Normal Medical Practice*
4. *The Partial Transition from Bioethics to the Law—Charles Snow*
   - (a) Differences between Bioethics (Referred to the Hospital's Ethics Committee) and Medical Malpractice (Referred to the Hospital's Peer Review Committee); Physician Expert Witnesses
   - (b) The Better Way to Face These Problems

1. What Are the Subjects of Bioethics and the Law?

With *bios,* Greek for "life," there is a rather broad definition of bioethics. As shown in the Introduction, when the term was coined in 1970, it referred primarily to ethics in health care. In the third millennium, we must look at related actions to achieve a proper perspective on bioethics in connection with the practice of medicine and the law; we must also draw lines to keep that perspective within limits we can handle. One ethicist stated, "Moral sensitivity is precisely the ability to make distinctions, including some hard ones. We have, then, an important moral interest in discerning that line, however difficult it may be to see at times, because nothing less than our moral character is at stake."[1]

At the close of the twentieth century, it has been noted that the field of bioethics "has not been especially concerned [with] the ethics of health care financing."[2] On bioethics committees where I have served, some doctors liked to emphasize that "costs are not our concern." On the other hand, a majority of them now find that the long-held ideal of completely separating finances from treatment may no longer remain realistic. For example, a Health Maintenance Organization (HMO) has been reported as "the only way to keep health care costs from spiraling out of control." As stated by Dr. Lee:

> I don't think anybody that I know wants to go back to the way it was 30 years ago, when even for a simple procedure everyone was in the hospital for several days. . . . That bubble has been burst. Cost has to be part of the equation. We have a culture in Kaiser Permanente that physicians have to concern themselves with the overall cost of health care. . . . It's a culture which I must say American medicine needs to learn.[3]

Yet by the end of the second millennium, politicians were both attacking and seeking more control over HMOs, in spite of reports that HMOs score high from patients. (See Chapters 6 and 11.) It has been reported that the

---

1. Darff, Matter of Life & Death 189, 197 (1998). The word "bioethics" attributed to Van Rensselaer Potter, a medical researcher at the University of Wisconsin. Bioethicists must work together with others outside their own disciplines.

2. Hast. Cent. Rep., September-October 1999, p. 34, and the *American Heritage Dictionary.* However, it was noted thereafter that:

> Just as our command of chemistry and physics transformed the twentieth century, leading to profound innovations like televisions and transistors, so, too, molecular biology—the science behind the biotech boom—will reshape the twenty-first century. The careful biotech investor can do well—and also do good.

Abate, *The Biotech Investor* (2003) Preface.

3. L.A. Times, Nov. 15, 1999, at 1 (see Chapter 6).

majority of California patients—55 percent—also gave their HMO a satisfaction score of eight or above on a scale of 10.[4] Due to political pressure, medical costs in the third millennium could continue to skyrocket.

Our environment is of prime ethical importance for humanity. The *Encyclopedia of Bioethics* handles religion and many other subjects, including the environment and overpopulation. (Bioethical effects of overpopulation on the environment are considered in Chapter 7.) Other problems must be mentioned first, including the effects of the International Convention on Human Rights and Biomedicine and the effects of the great divide between those knowledgeable in the sciences and those who are not, which could change our approach and end the history of the species *homo sapiens*. Joseph Wood Krutch, a naturalist (and my professor of literature at Columbia University), foresaw the latter as the greatest overall ethical dilemma to be faced in the third millennium:

> Man's ingenuity has outrun his intelligence: He was good enough to survive in a simple, sparsely populated world where he was neither powerful enough nor in sufficiently close contact with his neighbors to do them or himself fatal harm. It is not good enough to manage the more complicated and closely integrated world, which he is, for the first time, powerful enough to destroy.[5]

But the mountain falls and crumbles away, and the rock is removed from its place; the waters wear away the stones; the torrents wash away the soil of the earth; so thou destroyest the hope of man.

Job 14:18-19

The law locks up the man or woman
Who steals the goose from off the common
But lets the greater felon loose
Who steals the common from the goose.

Old English quatrain

---

4. ORANGE COUNTY REGISTER, *3rd millennium*, Sept. 16, 1999, at Bus. 1.
5. KRUTCH, THE MEASURE OF A MAN 25.

2. Environmental Parts of Ethics

As we move into the third millennium, we must commence the transition from a period of human devastation of much on the earth to one when humans more seriously begin to treat our planet in a mutually beneficial manner. Bioethics concerning the life and death of humans depends in part upon our environmental problems, which are discussed briefly.

More than 150 years ago, Henry David Thoreau declared, "In wildness is the preservation of the world." He also said: "A lover of Nature is pre-eminently a lover of man. If I have no friend, what is nature to me?" A decade and a half later, Charles Darwin utilized a holistic approach to discover the relations between the "structure of every organic being and that of all other organic beings with which it comes into competition for food or residence, or from which it has to escape, or on which it preys."[6] Near the turn of the twentieth century, John Muir, founder of the Sierra Club, wrote, "When one tugs at a single thing in nature, he finds it attached to the rest of the world." In the last half of that century, a significant, passionate, famous aviator and environmentalist, Charles Lindberg, wrote in 1965, "I don't want history to record my generation as being responsible for the extermination of any form of life."[7] During the third millennium, insufficient concern for the environment might become the greatest threat to our endurance as a species.

A devastating view has been taken by the writers of a volume entitled *A Moment on Earth:* "When the final book is written on *the human involvement with Earth's biosphere, the most remarkable feature may be how brief that involvement was.*"[8] (Emphasis added.) Our greatest danger is that we

---

6. The Origin of Species, 1959, Burow ed., Penguin (1968).
7. Berg, Lindberg 526 (1998).
8. Westerbrook, A Moment on Earth 691 (1975):

> In early 2003, I received a paper by Bob Berner. For the first time he showed that a rise in atmospheric carbon dioxide would create a drop in the amount of oxygen in the atmosphere. Berner showed that the amount of oxygen in the atmosphere plummeted at the end of the Permian [250 million years ago,] from levels somewhat higher than the 21 percent we enjoy today to levels of 15 percent or less. . . . The world got suddenly hotter, and drier, and all of its creatures—plants, animals, those in the sea, those on land—began to asphyxiate from lack of oxygen and too much carbon dioxide. Only a few survived. If humans had been there we might not have made it.

Ward's *Gordon* 225-26 (2004).

may neither listen enough nor act sufficiently upon such beliefs. For example, in the winter of the year 2000, a study found that temperature increases during the preceding century were mostly due to human activities. This allegedly sparked little reaction,[9] and the administration rejected a global warming treaty.[10] Yet, in the year following the turn of the third millennium, polls showed that more than eight out of every 10 people said they supported environmental goals. "They are not, as some critics claim, elitists, or wealthy people who want to preserve their way of life and block further progress for others. Actually, the label of 'elitist' more readily fits opponents of environmentalism."[11] The President issued Executive Order 13,112 on February 3, 1999, creating an important foreign species policy binding on all federal agencies. A council was created to plan and prevent the introduction, spread, and damage from invasive species. However, under the agreement of the World Trade Organization (WTO), the application of Sanitary and Phytosanitary Standards (SPS), other countries may prevent restrictions on the importation of products and species to continue international trade.[12]

The hopelessness with respect to the environment exhibited by 25 medical students when listening to lectures was demonstrated by Dr. Daniel Goodenough of the Harvard Medical School in 1997:

> Recently, the medical school mounted a series of three seminars on human health and the global environment. The first seminar was overflowing with energetic, aware, and concerned young medical students. Inexplicably, however, the second session was only half full. At the final seminar, only half a dozen students showed up.

When the perplexed Dr. Goodenough asked the students why attendance had fallen off so sharply, their responses were identical: The material was compelling, they said, but it engendered overwhelming personal reactions. The problems were so great—and the ability of the students to affect them so remote—that they could deal with their feelings of frustration and helpless-

---

9. ORANGE COUNTY REGISTER, April 1, 2001.
10. ORANGE COUNTY REGISTER, April 21, 2001.
11. *Id.*
12. VAN DRIESCHE, NATURE OUT OF PLACE 151-52 (2000); BRIGHT, FOREIGN POLICY 50-58 (Fall 1999).

ness and depression only by staying away.[13] Nevertheless, a poll found that 57 percent of other Americans actually agreed that the next 10 years will be the last decade when humans will have a chance to save the earth from environmental catastrophe.[14]

In the 1980s, during the Reagan administration, there was a controversy about a "hit list" of scientific advisors to that agency who were allegedly targeted for exclusion because of their liberal or pro-environment viewpoints.

A 2000 study sadly found:

> The Supreme Court's attitude towards environmental law during the past three decades has generally been marked by apathy. But with the Justices exhibiting increasing signs of skepticism and some hostility, it will not be easy to change the views of the current Justices or to secure the future appointment of Justices who feel different about environmental law.[15]

Ecology deals with the relationship of all organisms with their environment. Bioethics has largely been limited to the human organism, together

---

13. GELBASPAN, THE HEAT IS ON: THE HIGH STAKES OF BATTLE OVER EARTH'S THREATENED CLIMATE (1997).

14. ORANGE COUNTY REGISTER, April 22, 1999, at Metro 8:

> In the 1980s, during the Reagan administration, there was a controversy about a "hit list" of scientific advisors to that agency who were allegedly targeted for exclusion because of their liberal or pro-environment viewpoints.

Steinbrook, "Science, Politics, and Federal Advisory Committees," p. 1456, *citing* Anderson, *Improving Scientific Advice to Government Issues*, SCI. TECHNOL. 2003; 19(3): 34-8 and "EPA's Science Advisory Board Panels: Improved Policies and Procedures Needed," General Accounting Office, June 2001 (GAO-01-536).

15. LAZARUS, RESTORING WHAT'S ENVIRONMENTAL 703, 771-72 (2000):

> Arms control and the banning of weapons of mass destruction, international human rights and disabilities, international labor law and occupational health and safety, environmental law and the control of toxic pollutants, nuclear safety and radiation protection, and fertility and population growth are all intimately related to the domain of international health law.

Tayler, *Governing the Globalization of Public Health*, J. LAW., MED. & ETHICS 502 (Fall 2004).

with other organisms that have a relationship with human beings. However, only a few bioethicists also recognize that in nature, all organisms (not just a few of them) affect life on earth. When nature itself is being considered—together with the changes it makes upon the earth—almost no organisms become subject to complete separation from human beings. Precisely at the turn of the third millennium it was noted that all societies must know and support the bonds that connect people to earth's natural systems if they are to coexist with them.[16] But an insufficient number of people seek a true understanding of an alteration of the earth's environment (see Appendix A).

3. The European Union Convention on Human Rights (ECHR) and Bioethics Normal Medical Practice

In 1996, the Committee of Ministers in the Parliamentary Assembly of the Council of Europe formally adopted the Convention on Human Rights and Fundamental Freedoms—a pan-European convention on issues in bioethics and the law.[17] When ratified by five states, it became effective.[18] More than 50 years ago the members of the United Nations adopted the Universal Declaration of Human Rights. It had less effective medical, social, and economic rights. Article 25 states:

> Everyone has the right to a standard of living adequate for the health and well-being of himself and his family, including food, clothing, housing, medical care, necessary social services, and the right to security in the event of unemployment, sickness, disability, widowhood, old age, or other lack of livelihood in circumstances beyond his control.[19]

---

16. Doubilet, *Coral Complexities*, Forces of Change 145 (2000). A definite need for the establishment of the "World Environment Organization" was expressed by Charnovitz in his article *A World Environment Organization*, 27 Colum. J. Envtl. L. 323-62 (2002).

17. Council of Europe, 1996, Directorate of Legal Affairs, Draft Convention for the Protection of Human Rights and the Dignity of the Human Beings with Regard to the Application of Biology and Medicine, Strabourge.

18. Hast. Cent. Rep., January-February 1997, p. 13.

19. G.A. Res. 217 A (III), U.N. Doc. A/810, 1948, at Art. 25, *available at* http://www.un.org/Overview/rights.html.

The national laws of members of the European Union, including the United Kingdom, Germany, the Netherlands, and France, are subject to the ECHR. The Convention set forth minimum guidelines with regard to the application of biology and medicine (Article 27 based upon the patient's right of self-determination or autonomy). (See Chapter 5, "Autonomy and Responsibility.") Article 5 states:

> An intervention in the health field may only be carried out after the person concerned has given free and informed consent to it.
>
> The person shall beforehand be given appropriate information as the purpose and nature of the intervention as well as on its consequences and risks. The person concerned may freely withdraw consent at any time.

This was the first time an international convention confirmed the bioethical principle of autonomy. In 2002, it was upheld by the European Court of Human Rights, which noted that the right to refuse treatment fell within the privacy guarantees contained in Article 8 of the ECHR.[20]

Some doubts were expressed about the appropriateness of this "consecration of an Anglo-American concept" on the grounds that "we have no reason to believe that expression alone will cover the real interest of the person or affirm the primacy of the human being."[21] Exceptions to autonomy of the individual are many and include those "necessary in a democratic society for the interest of public safety, the prevention of crime, the protection of public health, or the protection of the rights and freedoms of others (Article 26)."[22]

The text of the convention is a mix of law and ethics; as the philosopher Rainer Maria Rilke stated, "Don't be confused by surfaces; in the depths ev-

---

20. Pretty v. United Kingdom, The European Court of Human Rights (Fourth Section), Strasbourg, April 29, 2002 (Application No. 2346/02 at 61).

21. Hennan-Hublet, *Le Projet de Convention de Bioethique*, Hast. Cent. Rep., January-February 1997, p. 20

22. The restrictions contemplated . . . may not be placed on Art. 11, nondiscrimination; Art. 13, interventions on the human genome; Art. 14, nonselection of sex; Art. 16, protection of persons not able to consent to research; Art. 19, the general rule on removal of organs from living donors; Art. 20, the protection of persons not able to consent to organ removal; and Art. 21, the prohibition of financial gain, respectively. Art. 26, para. 2.

erything becomes law." In 2001, some 52 science ministers at UNESCO adopted *Bioethics: International Implications,* stating that bioethics is today a key issue in the protection of human rights and fundamental freedoms. They noted that UNESCO, as a leading agency in bioethics at the international level, should pursue its mission through its International Bioethics Committee (IBC) and its Intergovernmental Bioethics Committee (IGBC).[23] However, it is interesting to note that writers on bioethics who attended a meeting about ethics and genetics had never been to a conference of bioethicists before. As this was being mentioned by one guest to another, almost all agreed they were not bioethicists who were into a multidisciplinary profession:

> Immediately the person next to him sat up straight, as if insulted, and said, "But I'm not a bioethicist." They turned to me. "Don't look at me," I said. "I'm no bioethicist." We asked the same question of the other guests at the table, and while the verdict was not unanimous, many said no, they did not really consider themselves bioethicists either. To an outsider, it must have seemed like an odd response. Many of us present at the table worked in bioethics centers, published in bioethics journals, belonged to bioethics associations and wrote books with the word "bioethics" prominently displayed on the covers.[24]

4. The Partial Transition from Bioethics to the Law—Charles Snow

In the United States, we are undergoing a major transition from bioethics to law due to the vital issues, such as the coming of life and the maintenance of dignity in connection with death, litigated in our courts. Some of these bioethical issues are also addressed by the legislatures. As stated by one lawyer-bioethicist, "American law, not philosophy or medicine, is primarily responsible for the agenda, development, and current state of American Bio-

---

23. http://www.unesco.org/opi/eng/unescopress/2001/01-113e.shtml 10/25/01. The European Union Charter of Fundamental Rights (EUCFR) is a part of the new E.U. Constitutional Treaty. Although it provides for human dignity and Article 2, which safeguards the right to life, these are "non-binding rights." *See* HERVEY & MCHALE, HEALTH LAW AND THE EUROPEAN UNION, Chs. 1, 10.

24. C. Elliott, *Diary: The Ethics of Bioethics*, LONDON REVIEW OF BOOKS 24, No. 23 (2002), 40, *reprinted in* J. LAW., MED. & ETHICS, Summer 2004, p. 287.

ethics."[25] The law will never completely supersede bioethics, as so much of the latter is concerned with voluntary actions of patients, physicians, and health care institutions (although much of this is already inside current law).

A writer at the end of the twentieth century stated that "cultural revolution" really means "great upheaval in consciousness, perception, value systems, and ideology that have affected the way we think of ourselves and our world." The significant problem of perception was expressed by Sir Charles Snow in his Rede Lecture at Cambridge University, "The Two Cultures." In the middle of the twentieth century, he discussed one of the great divisions in our society:

> For constantly I felt I was moving among two groups—comparable in intelligence, identical in race, not grossly different in social origin, earning about the same incomes, who had almost ceased to communicate at all, who in intellectual, moral, and psychological climate had so little in common that instead of going from Burlington House or South Densington to Chelsea, one might have crossed an ocean.[26]

At the end of the twentieth century, a writer broadened Snow's view by stating there are two cultural traditions that are especially important and whose problems remain unresolved. "The first of these is the continued bifurcation between the old linear rational, learned culture centered in the universities, the book-publishing industry . . . on the one side; and popular culture situated in films, television, and rock music on the other. . . . The optimistic forecast in the 1960s that a new synthetic culture would emerge out of the confluence of these two polarities has not been realized."[27] Many high school students not only "hate math," but also cannot do the problems in physics or astronomy.[28]

---

25. ANNAS, STANDARD OF CARE 3 (1993).
26. C.P. SNOW, THE REDE LECTURE, IN TWO CULTURES (1961).
27. *Id.*, p. 509.
28. Only 22% of fourth-graders and 59% of eighth-graders were able to solve this problem about earning money for a class trip: "If Jill earns $2.00 each day on Mondays, Tuesdays, and Wednesdays, and $3.00 each day on Thursdays, Fridays, and Saturdays, how many weeks will it take her to earn $45.00?"

GRANT & MURRAY, TEACHING IN AMERICA 20-21 (1999).

In these subjects, a high school curriculum's first year's study rarely gets beyond what was well known at the end of the nineteenth century.[29] A 1998 study showed 27 percent of U.S. adults believed the sun orbits the earth, not vice versa, and more than half did not realize that the earth circles the sun in one year.[30]

On the brighter side, Harvard biologist Professor Ernst Mayr pointed out at the end of twentieth century that "the sharp break between the 'sciences' and 'nonsciences' does not exist once biology is admitted into the realm of science."[31] If this is so, there should also be no sharp break between the study of biological sciences and bioethics. With respect to the latter, two areas are of significant importance in twenty-first century education. These are evolution (discovered in the midnineteenth century) and the beginning and development of the universe (discovered during the twentieth century). Although these topics were not included in the curriculum of many American high schools, the learning of these subjects furthers bioethical sources and goals (see Chapter 2).

The search for the sources of bioethics and the law mandates several other studies. If this is to be done in a rational manner, life itself must be put into perspective, including reports set forth in local newspapers such as *The Orange County Register* and the *Los Angeles Times*. Looking through standards used to make a judgment upon life requires examination of many of the latest publications on science and the law. This demands that significant attention be paid to the review of periodicals such as *The New England Journal of Medicine, Journal of the American Medical Association (JAMA),* and *Archives of Internal Medicine*, as well as others such as *The Hastings Center Report*, the *American Journal of Law & Medicine,* and numerous opinions of the courts. At the turn of the third millennium, to obtain bioethical literacy, one can sympathize with Alice in *Through the Looking Glass,* who found herself in a high-velocity world where she had to run just to keep in place. This literature is accessible to all readers who desire to understand and attempt to evaluate bio-

29. ORANGE COUNTY REGISTER, Sept. 29, 1999, at News 15.

30. National Science Foundation's SCIENCE & ENGINEERING, *Indicators* 1998, *cited in* SKY & TELESCOPE, September 2000.

31. MAYR, THIS IS BIOLOGY 36-37 (1997). As of 2002, bioethics was being taught in high schools in Australia, Japan, and New Zealand. http://www.biol.tsukuba.ac.jp/~macer/BHS.html, July 27, 2002.

ethical issues early in the twenty-first century, which are both discussed and documented herein. However, bioethics has not become a fully evaluated area for most physicians. A survey at the end of the twentieth century found that physicians would take little time to actively obtain full knowledge of what is medically appropriate according to the legal system.[32]

Bioethics has to stand as one of the most fascinating, and arguably, important intellectual developments in the last third of the twentieth century.

Caplan, *JAMA*, March 3, 1999, p. 849

(a) Differences between Bioethics (Referred to the Hospital's Ethics Committee) and Medical Malpractice (Referred to the Hospital's Peer Review Committee); Physician Expert Witnesses

Bioethics problems generally differ from those finding their legal basis as a "medical malpractice lawsuit." For example, an assault will occur to a patient if he is attached to a hypodevice needle to administer an injection without his consent. A patient's loss of reputation may be attached by the unauthorized release of health information, possibly causing the loss of profit. To patients, these actions can be civil or criminal (see Chapter 6). Furthermore, "negligence" by a professional is "the most common complaint in a malpractice action,"[33] which may also seek punitive damages.[34] According to the American

---

32. LIANG, 3 UNIV. CHI. L. SCH. ROUND TABLE 67 (1996).
33. Between 44,000 and 90,000 Americans die annually due to medical errors. . . .The Inspector General [HHS] found that 84% of HMOs from 1990 to 1999 failed to report even one physician. . . . HMOs and hospitals were required, under the law, to report to the database any disciplinary action that affected a doctor's clinical privileges for at least 30 days.

FRENER, MEDICAL MALPRACTICE LAW 1 (1990). *See also* Lytton, *Using Litigation to Make Public Health Liability, Tobacco, and Gun Policy: Theoretical and Empirical Challeneges in Assessing Product Litigation*, 32 J. LAY MED.& ETHICS (2004).

34. ORANGE COUNTY REGISTER, May 30, 2001, at News 7:

    Punitive damages have generally been available in tort to "punish the defendant for reprehensible conduct and to deter him and others from engaging in similar conduct. *Kemezy v. Peters,* 79 F.3d 33 (7th Cir.

Medical Association, in 18 states physicians and institutional health care providers are having grave difficulties obtaining affordable professional liability insurance. Further, hospitals in several "states that have been hit the hardest have temporarily closed or threatened to close emergency room, obstetrical, or other services."[35] In California, some 6,600 lawyers are currently insured by the bar-approved malpractice insurance program.[36]

Because of the raise in the cost to physicians for malpractice insurance, President Bush proposed a limitation on medical malpractice damages awards in 2003, promoting the "[n]ational adoption of proven standards to make the medical liability system more fair, predictable, and timely."[37] Such medical mistakes occur with relative frequency. A study in 2002 found that nearly all physicians believe the fear of medical malpractice may be a barrier to reporting errors and that greater legal safeguards are needed for reporting systems to be effective. In one study of more than 30,000 randomly selected charts, it was noted that negligence resulted in adverse events in 1 percent of hospital-

---

> 1996). An Ohio court has recently stated that in the absence of state statutes thereon, the courts have a "central role" in determining the distribution of punitive damages.

Dardinger v. Anthem Blue Cross & Blue Shield, 781 N.E.2d 121, 147-48 (Ohio 2002).

35. N. Eng. J. Med., 2281, June 5, 2003.

36. Cal. Bar J. 20 (June 2003): "The Oregon legislature empowered the bar in 1977 to create a mandatory malpractice insurance program. . . 12 states are studying the Oregon model."

> An ABA task force on contingency fees said that only about 2 percent of victims of medical malpractice even file claims. The remaining incidents involving malpractice typically either go undetected by the patients or result in no claim because the potential damages are too small to justify the cost of bringing a case. Moreover, the task force reported in September 2004 that defendants prevail in 75 percent of the cases taken to trial. That compares with a 50 percent success rate in other tort actions.

Giheaut, *The Medical Divide*, A.B.A. J., March 2005, at 41.

37. Press Release, President Calls for Medical Liability Reform (January 16, 2003), *available at* http://www.20030116.html. However, in a review of five states with high malpractice premiums vs. four states with lower rates, the General Accounting Office found no evidence that lawsuits against doctors crimp access to health care.

ization, with a quarter of those resulting in death."[38] Another study found that "90 percent of house officers, of the 114 who responded, reported that they had made errors related to serious adverse events, one-third of which resulted in death."[39]

"Just over half of the house officers reported that they discussed their mistake with their physicians and only one-quarter of them told the patients or families."[40] For example, the University of California was ordered by a court to pay $20 million in a malpractice settlement by a surgeon at its hospital in Irvine. This was due to both his alleged "intentional, despicable, and unprofessional conduct" and the University's action of concealing the existence of a witness.[41]

Normally, a bioethics committee does not become directly involved with a decision based upon how a physician went through the medical treatment learned in medical school; on the contrary, his expertise is largely based upon that which may become the "ultramedical" aspects of a case. Consider this case which has discussed the appropriateness of certain experts: Larry Hall was born prematurely, "blue, flaccid, limp, and with a decreasing heart rate." After making certain efforts, Dr. Anwar concluded that he was dead. Soon, thereafter, the child was "attempting to cry." The doctor made more efforts, but the newborn soon died. The Halls then sued the doctor for malpractice. Dr. Anwar's lawyer presented a videotape of John Paris, a priest and "professor of bioethics," who attempted to provide an "expert opinion" that Dr. Anwar's actions were "within the standard of care." The court stated:

> *Father Paris was not qualified to render opinions about a medical standard of care.* If Father Paris had testified *solely as to whether it was moral or ethical* to resuscitate or terminate resuscitation of the

---

38. Beeman, et al., 112(3), Annals of Int. Med. 221 (1990). A survey titled *Current award trends in personal injury, 2002 edition. Jury Verdict Research*, in 14 Am. Med. News, April 2003, p. 9, found that defendant physicians won 61% of the cases. Accordingly, the payments for expert testimony, which are normally quite large, are only recoverable by plaintiffs in 39% of the malpractice cases.

39. Wu, et al., JAMA 265 (1991).

40. *Id.*; Sisti, *Owning Up: An Erred Physician's Ominous*, Am. J. Bioethics Online, and *Survey on Medical Errors*, Archives of Int. Med., Oct. 27, 2002.

41. L.A. Times, June 21, 1998, at B. Few physicians are really independent contractors in the hospital; they are "de facto" employees.

infant, then his testimony would have been irrelevant to the legal issue of malpractice. *[It] involves the level of care* owed by a similar healthcare provider and *not that owed by an ethicist.*[42] (Emphasis added.)

Most of the cases discussed herein make this distinction. Bioethicists who serve on hospital ethics committees will become aware of malpractice insurance, a growing concern for academic medical centers. The hospitals' peer review committees handle malpractice by the hospitals' physicians. Peer reviews have been long established by health care organizations to scrutinize physicians' professional conduct. The AMA's *Code of Medical Ethics* (2003 Edition) points out, "At least to some extent, each of these types of peer review can be said to impinge upon the absolute professional freedom of physicians. They are, nonetheless, recognized and accepted."[43]

Furthermore, "the composition of committees sitting in judgment of medical students, residents, or fellows should include a significant number of persons at a similar level of training."[44] This contains only a portion of its necessary differences derived from the composition of Ethics Committees.[45] For example, it states that "[p]referably, a majority of the committee should consist of physicians, nurses, and other healthcare providers." See Chapter 6. In California, the Peer Review Committee work may often commence with testimony from the doctor before the Core Committee. This is forwarded to the Medical Executive Committee that votes upon whether to revoke a doctor's

---

42. Hall v. Amwar, 74 So. 2d 41 (Fla. Dist. Ct. App. 2000). See its discussion in Hast. Cent. Rep., July–August 2001, p. 11.

43. Section 9.10. It also prescribes that peer review committees observe its Opinion 9.05, "Due process principles applicable to the physician or medical student whose professional conduct is being reviewed." In connection with this process concerned with scientific work, it has been stated, "We tend to forget that just because peer review reviews scientific work does not mean that it is itself a scientific process." Knoll, *The Communities of Scientists and Journal Peer Review*, 263 JAMA 1331 (1990).

44. Section 9.05 Due Process. To receive accreditation, hospitals must engage in the peer review method of quality assurance. JCAHO COMPREHENSIVE ACCREDITATION MANUAL FOR HOSPITALS, MS.5.11 (2002).

45. Section 9.11 Ethics Committees in Healthcare Institutions. Congress established Peer Review Organizations to monitor hospitals that receive Medicare funds. Peer Review Improvement Act of 1982, Pub. L. No. 97-248, 96 Stat. 381-395.

privileges. The doctor has the right to appeal to a so-called Judicial Review Committee, another group of doctors from this hospital. If this committee upholds the report, the decision is then sent to the Medical Board of California and to the National Practitioner Data Bank (NPDB).[46] The system is confidential and exempt from legal disclosure. Effective January 1, 2003, California's B&P Code § 801(e) amended law requires liability insurance carriers to notify a plaintiff's lawyer that a report to the Medical Board has been made concerning a medical malpractice settlement (over $30,000), judgment or arbitration award (any amount), and an attorney's contingent fee.[47] Under the federal Health Care Quality Improvement Act of 1986 (HCQIA), members of peer review committees were granted immunity from lawsuits filed in response to professional review actions.[48] However, any disciplinary action taken by a review committee must be reported for listing in the National Practitioner Data Bank.[49]

In 1990, the West Virginia University School of Medicine paid $200,000 in malpractice insurance premiums. In 1992, the rate was 10 times that amount, at $2 million. In 2002, the school began paying $7 million in malpractice premiums.[50]

---

46. Orange County Register, May 20, 2003, at Local 4. The patient of a physician whose staff privileges have been revoked might file a medical malpractice lawsuit against him. Salamon, 47 Vill. L. Rev. 643 (2002).

47. Cal. Bar J., March 2003, p. 16. Some 23 states have placed restrictions on attorney contingent fees in medical malpractice actions. For example, the California limit is 40% of the first $50,000 of the damages; 33$^{1}/_{3}$% of the next $50,000; 25% of the next $500,000; and 15% of any amount where the recovery exceeds $600,000. Cal. Bus. & Prof. Code § 6146 (2003). Damages for noneconomic losses cannot exceed $250,000, § 3333.2(b). Mandatory arbitration clauses, which have become common, can be enforced under federal law. Allied-Bruce Terminix Cos. v. Dobson, 512 U.S. 265.

48. 42 U.S.C. § 11111(a)(1).

49. *Id.* § 11133. Typically, physicians disciplined in one state were then relocated in another one. 1986 U.S.C.A.N. (P.L. 99-660) 6384. However, the public is unable to access the information contained in the NPDB. 45 C.F.R. § 60.11(a). *See* Smith, Annotation, *Construction and Application of Health Care Quality Improvement Act of 1986* (42 U.S.C.A. §§ 11101-11152, 121 A.L.R. Fed. 255, § 2 (1994). A failure to report disqualifies the hospital's immunity for three years. 42 U.S.C. § 11133(1)(2000); and 45 C.F.R.§ 60.9(1)(iii)(2003).

50. http://www.aamc.org/newroom/reporter/july02/medicalmalpractice.htm. Proving pain as damages in malpractice "includes virtually any form of conscious suffering, both emotional and physical." Dobbs, Law of Torts 1050 (2000). The stan-

The Association of Trial Lawyers of America (ATLA) contains 90 percent plaintiff litigators in personal injury and malpractice cases, which are generally based upon contingent fees for the winning plaintiffs.[51] Lawyers may receive a basic introduction to scientific reasoning.[52] Only the state of Oregon requires lawyers to carry malpractice insurance.[53] A quarter century ago, California enacted its Medical Injury Compensation Reform Act (MICRA), which both limits contingent attorney fees and caps injury awards at $250,000 for pain and suffering in medical malpractice.[54]

---

dard of care in such case has been whether his conduct was "minimally competent by physicians in the same specialty." Hall v. Hilbun, 466 So. 2d 856, 872-73 (Miss. 1985); Turner v. Temple, 602 So.2d 817 (Miss. 1992); Bahr v. Harper Grace Hosps., 528 N.W.2d 179 (Mich. 1995).

51. The Association of Trial Lawyers of America (ATLA), ranked by *Fortune* magazine as the sixth most powerful lobby group on Capitol Hill, was number one among the legal/lobbyist donors. Ninety percent of its contributions went to Democratic candidates. The expansion of causes of action and damages and the ability to collect enormous contingency fees keep this organization busy in the state houses and on Capitol Hill. In 2001, the U.S. Supreme Court affirmed a decision by the 4th U.S. Circuit Court of Appeals rejecting $120,000 in fees for the lawyer because his work didn't result in an "enforceable judgment." He had represented a group of nursing home residents in whose favor the state of West Virginia later changed the law. Buckhannon Board & Home Care v. W. Va. Dep't of Health and Human Servs., 532 U.S. 598. (In such cases the lawyers sometimes argue for their fee on the basis of "catalyst theory.")

52. Goodstein, *How Science Works*, in REFERENCE MANUAL ON SCIENTIFIC EVIDENCE, 2d ed. 73-75 (Federal Judicial Center, 2000), and Greenland & Robins, *Epidemiology, Justice, and the Probability of Causation*, 40 JURIMETRICS 321-40 (2000). In 1987 the Board of Directors adopted the AANS Position Statement on Testimony in Professional Liability Cases, which stated, "The neurosurgeon should not accept a contingency fee as an expert witness." Pelton, AANS BULLETIN, Spring 2002, at 10.

53. A.B.A. J., January 2003, p. 66. By the year 2002 almost all states had authorized "limited liability partnerships (LLPs) that prevent individual members of a law firm from being held personally liable for the negligence of other partners." A.B.A. J., August 2002, at 46.

54. ORANGE COUNTY REGISTER, July 14, 2002, at News 20. On Sept. 1, 2003, a similar statute took effect in Texas that provides that malpractice plaintiffs can recover no more than $250,000 in noneconomic, or pain-and-suffering, damages at trial. That damage cap can expand to $750,000 if more than one hospital is at fault. LITIGATION, Fall 2003, at 18. Under the MICRA, plaintiffs are required to give a 90-day warning of a claim so that the provider had a chance to settle out of court.

In regard to the expert witnesses in malpractice trials (who differ from bioethicists as noted previously), the Supreme Court recently rendered major opinions that have been of particular interest to both doctors and lawyers. The Court stated "under the [Federal] Rules [of Evidence] that the trial judge must ensure that any and all scientific testimony or evidence admitted is not only relevant, but reliable,[55] that 'abuse of discretion is the appropriate standard' for reviewing a court's decision to admit or exclude expert testimony,[56] but that the trial judge must have considerable leeway in deciding in a particular case how to go about determining whether particular expert testimony is reliable."[57] In 2001, the American Association of Neurological Surgeons (AANS), supported by the American Medical Association (AMA), won the ruling of a U.S. Court of Appeals that a professional society could discipline one of its members on the basis of his testimony in a malpractice case.[58] A suit was brought against

---

55. Daubert v. Merrell Dow Pharm., Inc., 509 U.S. 579, 589, 113 S. Ct. 2786, 1993.

> It was noted that the Court suggested that judges consider several questions, including: (1) is the basis for the opinion testable and has it been adequately tested?; (2) is the error rate associated with the opinion or technique acceptable?; (3) is the research on which the opinion is based published in a peer-reviewed journal?; and (4) is the basis for the opinion generally accepted in the pertinent field?

FAIGMAN, LABORATORY OF JUSTICE 338 (2004).

56. General Elec. Co. v. Joiner, 522 U.S. 136, 139, 118 S. Ct. 512, 1997. The expert must be identified under Rule 26(a)(2)(A) and must provide a written report that complies with Rule 26(a)(2)(B). If a party sees a physician to obtain diagnosis, treatment, and an agreement to render opinions at trial, then the physician has been retained and a report is required. *E.g.*, Thomas v. Consol. Rail Corp., 160 F.R.D. 1, 2 (D. Mass.1996); Widhelm v. Wal-Mart Stores Inc., 162 F.R.D. 591, 594 (D. Neb. 1995). *See* Beane & Karatinos, THE FEDERAL LAWYER, May 2004, at 26.

57. Kumho Tire Co. v. Carmichael, 526 U.S. 137, 152, 199 S. Ct. 1167, 1999. *See also* FED. R. EVID. 104, 702, 703, 403, and the REFERENCE MANUAL ON SCIENTIFIC EVIDENCE, 2d Edition (Federal Judicial Center, 2000).

58. Austin v. Am. Ass'n of Neurological Surgeons, 253 F.3d 967 (7th Cir. 2001), *cert. denied*, 122 S. Ct. 807, 2002. The court ruled that the testimony of Austin, a neurosurgeon, was erroneous and upheld his having been suspended for six months by the American Association of Neurological Surgeons (AANS). An expert who was once retained by the defendant in a personal case was later listed as an expert for the plaintiff. The defense then filed a motion to disqualify both the expert as a witness and the entire law firm of the plaintiff. Although the trial court granted the

the AANS by a suspended neurosurgeon claiming a loss in income. Chief Judge Richard Posner of the Seventh U.S. Circuit Court of Appeals held that the neurosurgeon's annual loss of income as an expert witness is not the "professional body blow that the cases have in mind." Professional self-regulation furthers rather than impedes justice.[59]

In 2002, the Texas State Board of Medical Examiners became the first state medical board to require proficiency through periodic retesting.[60]

The Supreme Court's denial of certiorari (appeal) has resulted in this decision only binding other courts within those states whose jurisdiction is covered by the Seventh Circuit Court of Appeals. However, certain federal and state courts throughout the United States could decide to follow the decision when faced with challenges regarding use of the testimony of experts who have been suspended by other technical organizations. Unlike this case, an important result of "certiorari denial" by the Supreme Court involved a bioethics matter, physician assisted suicide (PAS) (see Chapter 14).

Several medical schools have commenced a dual MD/JD degree pro-

---

motion, the court of appeal reversed, holding that the trial court abused its discretion because there was no evidence that plaintiff's counsel knowingly retained defendant's expert or that the expert intentionally advised both sides. Collins v. State of California, 121 Cal. App. 4th 1112 (18 Cal. Rptr. 3d 112) (2004).

59. Austin v. Am. Ass'n of Neurological Surgeons, 253 F.3d 967 (2001):

> Judges are not experts in any field except law. . . . Judges need the help of professional associations in screening experts . . . and if the association finds in a proceeding that comports with the basic requirements of due process of law that a member of the association gave irresponsible expert testimony, that is a datum that judges, jurors, and lawyers are entitled to weigh heavily.

Id. However, the Coalition and Center for Ethical Medical Testimony (CCEMT) goes "aggressively after plaintiffs' lawyers and expert witnesses who might have stepped over the line." A.B.A. J., August 2004.

60. J. Med. L. Censure & Discipline No. 1, p. 11 (2003). At the annual AMA meeting in 2004, a proposal by a South Carolina surgeon to refuse treatment for attorneys involved in medical malpractice cases was rejected by a committee, saying it would "jeopardize and side-track" the group's efforts to combat high insurance rates and malpractice lawsuits. Orange County Register, June 20, 2004, at News 24. A neurologist said that it should not have been introduced because it seeks to discriminate against a group of people.

gram so that some people now hold both degrees.[61] Several physicians with legal training also provide "consultative services for practicing attorneys, especially in the *medical malpractice* area."[62]

It has been noted that teaching programs in medical ethics were "practically nonexistent in 1970, but by 1990, an "extraordinary diversity" in formal teaching activities existed in almost all medical schools.[63] This remains challenging in the twenty-first century. Ethics committees of most hospitals are served by experts who are "nonphysicians" (e.g., lawyers, biologists, priests, or rabbis), many of whom voluntarily enter the field.

In the twelfth wave of ethics, the method sent to me: 12 communitarians, 11 libertarians, 10 moral rules, 9 stories, 8 ethnographers, 7 sort of feminists, 6 types of virtue, casuistry, 4 just rules, beneficence, nonmalificence, and respect for auto-ho-no-my.

Macklin, "The Twelve Methods of Bioethics," 1997,
sung to the tune of "The Twelve Days of Christmas"

My hypothesis is that the American ethos is strongly tempted to endow various aspects of life with moral meaning in a capricious way.

Albert R. Jonsen, *The Birth of Bioethics*, 1998

The way downward is easy for Avernus.
Black Dis's door stands open night and day.
But to retrace your steps to heaven's air,
There is the trouble; there is the toil.

Virgil, *Aeneid*

---

61. AAMC List of Combined MD/JD Programs, AAMC Web site, *available at* http://www.aamc.org/students/applying/programs/mdjd.htm. "Legal malpractice insurance rates are up across the board for the first time in years, in some cases by 100% or more." A.B.A. J., August 2002, at 26.

62. JAMA 2251 (May 2001). *See also* Rein, *Plaintiffs Lawyer*, A.B.A. J., December 2002, at 88.

63. Forrow & Arnold, Encyclopedia of Bioethics 263 (1995).

(b) The Better Way to Face These Problems

In the mid-1990s, while studying bioethics and other various subjects at Harvard's School of Public Health and working at Brigham and Women's Hospital, one physician wrote to me, stating, "It is true that there seems to be very few physicians in the field of medical ethics."

A hospital's ethics committee, in addition to including physicians, has become the domain of philosophers, lawyers, ministers, priests, and others. These professions do not indicate that bioethicists were among them. Nevertheless, many people still consult physicians with respect to critical problems in bioethics that have legal aspects. This presents a problem: Few physicians consider this new field by continually consulting significant volumes in the field; they find that the help they may give to their patients is limited. A major reason is that some bioethical writers today still cling to the belief that moral principles such as beneficence, nonmalfeasance, justice, and autonomy provide adequate specific answers to individual cases. As Stephen Toulmin stated, "The central concerns of philosophers became so abstract and general that they, in effect, lost all touch with the concrete and particular issues that arise in actual practice, whether in medicine or elsewhere."[64]

Bioethics-oriented committees that deal with actual cases and policies in all U.S. hospitals with 200 beds or more appear to be dominated by physicians, most of whom have received little or no training in the field of bioethics or the law (see Chapter 6). This has been exacerbated by a populace that still relies on the medical profession alone when faced with making certain life-and-death decisions. If people are to achieve the wisdom necessary in the twenty-first century, full cognizance must be taken of the tremendous changes occurring in this recent revolution.

These *revolutionary changes* in bioethics and the law can most be characterized by the increased emphasis now placed upon the autonomy of the individual (see Chapter 5). From human and legal viewpoints, we first entered this modern era with significant historical events: (1) In 1965, the

---

64. Toulmin, *How Medicine Saved the Life of Ethics*, 25 Perspectives in Biology and Medicine 7336-50 (1982). The philosopher Martin Heidegger stated, "Philosophy in about 1900 was in deep trouble." (p. 26).

Supreme Court inserted the word *privacy* into the U.S. Constitution as a "penumbra" (shadow) to the Due Process Clause. This substantive change was made to provide a rationale to strike down a state statute prohibiting the sale or use of contraceptives in Connecticut.[65] (2) The courts initiated the concepts of "autonomy" and modern "informed consent." Shortly thereafter, to a significant extent, this overturned centuries of old and well-used concepts about decision making by physicians, such as abortion and the doctrine of therapeutic privilege. These changes were accomplished by requiring patients to be more fully informed with respect to their autonomous choices (or that of their surrogates) in making decisions concerning proposed treatments, including possible alternative treatments.

One writer noted, "From the destructiveness of complete lies to the destructiveness of unmediated truth took less than three decades at the end of the twentieth century."[66] During that period came a series of important decisions by the courts (and some by elected representatives of the people) significantly altering the then existing course of bioethics and law. For example, precisely 32 years after the Connecticut decision authorizing the distribution of condoms (see Chapter 7), the Supreme Court allowed the people of Oregon to follow their Physician Assisted Suicide Law (see Chapter 14). Oregon remains the sole state in the United States where certain terminally ill patients can be assisted legally by physicians upon request. Physicians may realize that the answers they are seeking to bioethical questions can only be found by taking into consideration law, philosophy, and other disciplines. Medical input from physicians must be available by no-

---

65. Griswold v. Connecticut, 381 U.S. 179 (1965).

66. Cassell, Hast. Cent. Rep., July–August 2000, at 17. In *The Prince*, Machiavelli promises not to fill his book "with pompous phrases or elaborate, magnificent words" or to decorate it "with any form of extrinsic rhetorical embroidery." His most famous sentence may be this one:

> A great many men have imagined states and princedoms such as nobody ever saw or knew in the real world, for there's such a difference between the way we really live and the way we ought to live that the man who neglects the real to study the ideal will learn how to accomplish his ruin, not his salvation.

Machiavelli, The Prince, ed. (Robert M. Adams trans., 1997).

tification to other experts who can apply their disciplines to the individual cases being considered.

Although few physicians fully understand the limits of their training, some are expanding that training and increasing their collaboration with members of other related disciplines that are more pertinent to both bioethics and the law. The many unresolved matters in this field would be of particular concern to everyone during the early part of the third millennium. Such an "ultramedical" background has become a sine qua non. Recent studies, such as the large Project Support (see Chapter 13), have demonstrated the lack of desire by most physicians to seriously consider a patient's Advance Directives on end-of-life matters. More individuals with such knowledge are desperately needed in our society to overcome the errors perpetuated during certain medical practices.

The criticisms of the practice of bioethics and the law must be regarded as follows: The "ultramedical" often requires the expertise of a different group of professionals. Although groups of professionals should be praised for good work done in their respective fields, less praise may be given for counseling on bioethics and the law on the basis of the use made by physicians' subjective viewpoints.

How do we get to a place beyond this? How do we actually consider the nonmedical issues necessary for "ultraphysicians"? We must start with brief discussions of areas of significance for all bioethicists involved with our hospitals, namely, the evolutions of the universe and of ourselves (Chapter 2), patients' religions and directions for possible treatment (Chapter 3), and possible difficulties of looking at both science and ethics in the search for bioethical standards (Chapter 4). These areas constitute matters where hospital patients will demand that attention be given prior to their making a choice among any of the conditions of life set forth in Chapters 5 through 14.

Whatever may have been the case in years gone by, the true use for the imaginative faculty of modern times is to give ultimate verification to facts, to science and to common lives, endowing them with the glows and glories and final illustriousness which belong to every real thing, and to real things only.

Walt Whitman, *A Backward Glance Over Troubled Waters*

We are hardly born human; we are born ridiculous and malodorous animals; we become human, we have humanity thrust upon us through the hundred channels whereby the past pours down into the present that mental and cultural inheritance whose preservation, accumulation, and transmission place mankind today, with all its defectives and illiterates, on a higher plane than any generation has ever reached before.

Will Durand, *The Greatest Minds, an Ideal* (2002), p. 103.

The Creation and Evolution of the Universe and Humankind

1. ***Cosmic Evolution from the Beginning of Time***
   - **(a) The Big Bang and the Meaning of the Creator of the Universe, Its Age, and Evolution**
   - **(b) Potential of Progress in Astrobiology, Astrology**
2. ***The Importance of the Continuance of Evolution into Bioethics, Medicine, and the Law***
   - **(a) What Constitutes Life Itself?**
   - **(b) The Importance of Considering the Meaning of Evolution**
   - **(c) Evolutionary Advances by the Third Millennium**
   - **(d) A Key to the Basis of Significant Contingencies (Not Design) in Evolution**
   - **(e) Law and Biology**

Peace comes within the souls of men
When they realize their oneness with the universe.

J. Kabat-Zinn, 1994

Two things profoundly impress me: The starry heavens above me and the moral law within me.

Immanuel Kant

Now, as God the maker played, He taught the game to Nature,
Whom He created in His image.

Kepler, *The Mystery of the Universe*

It has been said that the highest praise of God consists in the denial of Him by the atheist, who finds creation so perfect that he can dispense with a creator.

Marcel Proust (1921)

Astronomy *enriches our culture, nourishes a scientific outlook in society, and* addresses *important questions about* humanity's place in the universe. It contributes *to areas of immediate practicality, including,* medicine, *and security, and it introduces young people to* quantitative reasoning *and attracts them to scientific and technical careers. (Emphasis added.)*

Washington Charter endorsed by astronomical
societies in Britain, Canada, and Denmark,
Fienberg, *Sky & Telescope,* Sept. 2004, p. 8.

1. Cosmic Evolution from the Beginning of Time

Walt Whitman praised "science, common lives, and real things" in the first epigram. The Biblical Job—of at least two and a half millennia ago—was like all other humans in the prescientific age; he praised God and suffered without knowing why. Early in the twentieth century, the poet Robert Frost brought Job "back to life" so that his questions could be asked again. Many, if not most, ethicists today remain as "Jobists" to some degree.

Since the dawn of the scientific era, many people began to realize that we had come a long way since Biblical times and attained much more knowledge than Job had. Although we recognize that we still have a considerable distance to go, we realize that today's scientific effort may require our recognition of the Creator of the universe, who was mentioned in our Declaration of Independence, as a result of reason. A largely different list of questions has now been raised; our view now is that the universe began with mostly hydrogen (H) and helium (He), whose gravity assembled into stars. This, together with the subsequent development of stars, is quite different from what was explained to Job; it continues to profoundly affect our view of our own evolution and effect upon life. Many of us have considered recent knowledge concerning the necessity of preliminary stellar explosions to provide the ingredients (elements heavier than hydrogen or helium in space) as prerequisites for life.

How we view the possible ending of our life on earth by stellar explosion is of much less importance than other possible astronomical calamity.[1] Although the chances of the human race being wiped out by asteroids or comets may be lower than was previously thought, one set of statistics showed that there is "still probability that the Earth will collide with a space rock before the start of the twenty-second century. While that may seem a remote possibility, it has been estimated as four times higher than the risk of dying in a train crash."[2] An asteroid impact is generally termed an *accretion.* Their average speed is a shocking 70 times that of sound and the accretion may be 20 times its diameter. Following the turn of the third millennium, a 1,000-foot-long asteroid traveled at 68,000 miles per hour as it passed by the earth. Had it hit the earth, its explosion would have released energy thousands of times that of the first atomic bomb drop on Hiroshima.[3]

> A newly discovered asteroid, called 2001 YB5, estimated to be only 300 yards in diameter, zipped by the Earth on January 7th at a relatively close distance, less than twice that of the Moon. There is only one known closer pass of an asteroid, and that will occur in 2027. There is no danger of it hitting Earth, even as far forward in years as astronomers have calculated the position of YB5. An object that size

---

1. *See* KEYES, LIFE, DEATH & THE LAW, Ch. 3 (1995).
2. Bayer, *Asteroid Probability*, http://news.telegraph.co.uk/new/main.jhtml?xml=/news/2001/11/09/wdoom09.xml.
3. ORANGE COUNTY REGISTER, Jan. 7, 2002.

> Near-Earth asteroid Eros is 21 miles (33 kilometers) long and has come within 20 million miles (32 million km) of Earth. While Eros will not collide with Earth anytime soon, the impact of a body this large would place the future of all humanity in jeopardy. Scientists hope any asteroids on impact courses will be discovered decades ahead of time, giving us a chance to divert them.

ASTRONOMY, December 2004, p. 40.

striking Earth would wipe out a medium-sized country. The best estimates are that one such object hits Earth about every 5,000 years.[4]

Thus, it has also been noted *"asteroid collisions* over geologic time scales are nearly certain to have changed the *evolution of life on Earth."*[5] (Emphasis added.)

It has also been argued that comets and meteorites contain amino acids, "the stuff that life is made of." The organics we find in meteorites that we think are similar to the organics we would find in comets contain many amino acids, and amino acids are the building blocks of proteins. So comets probably delivered certain amino acids to earth. By 2008, researchers also expect to complete a 10-year project to locate comets and steroids whose orbits carry them close enough to earth to pose a potential threat.[6] However, the first astronomer to correctly locate the earth in the solar system—which placed bioethics between science and the humanities a half of a millennium ago—correctly entitled his work "On the Revolutions." Nicolaus Copernicus (1473–1543) knowingly published what has been called the "greatest book of the sixteenth century"[7] at the time revolutions were giving birth to the Protestant religions.

---

4. Lyon, Sirius Astronomer, December 2001, p. 7 and February 2002. In March 2004, a small asteroid passed within 26,500 miles of Earth. NASA indicated that such objects pass undetected about this distance roughly once every two years. Orange County Register, Mar. 19, 2004, at News 12. In his volume, *Catastrophe,* Judge Richard A. Posner indicated how "a two-kilometer asteroid collision with Earth might kill a billion people, whereas a 10-kilometer asteroid might extinguish our species. Even a 75-meter rock could release energy equivalent to a 100-megaton nuclear weapon, enough to obliterate a city." *See* Foster, 307 Science, Feb. 25, 2005, at 1205.

5. Sky and Telescope, July 2001, p. 51. In 2005 it was reported:

> If you expect to be alive on April 13, 2029, you can look forward to an asteroid-watching party across three continents like nothing the world has ever seen. Sky & Telescope, May 2005, p. 16. The impact energy in 2029 would be about 850 megatons, 15 times more powerful than the largest hydrogen bomb ever tested. *Id.* at 17.

6. Harwood, Space Odyssey 154 (2000).

7. *De revolutionibus* argued against the common belief that the Earth was solidly fixed in the middle of the universe. Instead, he proposed that the Sun was immovable in the middle, and that the Earth went around it, along with the other planets. In other words, he proposed the arrangement of the solar system much as we know it today. Ginerrich, Senior Astronomer Emeritus, Smithsonian Astrophysical Observatory, The Book Nobody Read 2 (2004).

Today, no scientist has any evidence to support hypotheses that would rule out the Creator. Early in the twentieth century, Edwin Hubble became the actual discoverer of the expanding universe, and by the second half of the twentieth century other scientists stated, "We are all creationalists."[8] In the famous flier Charles A. Lindbergh's *Autobiography of Values,* he wrote, "After my death, the molecules of my being will return to the earth and sky. They came from the stars. I am of the stars." Thus, it has been pointed out that *"Evolution is not limited to biological species:* if other things in the cosmos were not created as they are, but came into being in the course of time, they, too, *must have evolved* . . . in a vast process called *cosmic evolution."*[9] (Emphasis added.) This approach might be taken by a considerably larger proportion of people during the twenty-first century despite their being divided according to Richard Dawkins in his book, *River Out of Eden:*

> Some people are going to get hurt, other people are going to get lucky, and you won't find any rhyme or reason in it, nor any justice. The universe we observe has precisely the properties we should expect if there is, at bottom, no design, no purpose, no evil and no good, nothing but blind, pitiless indifference . . . DNA neither knows nor cares. DNA just is. And we dance to its music.[10]

However, some people still believe that no astronomer has presented sufficient evidence to verify the truth of Dawkin's statement regarding the entire universe.

---

8. Until Edwin Hubble came on the scene, astronomers not only had no way to gauge the age of the universe, they were not sure that anything existed beyond the Milky Way. Then in 1924, Hubble discovered that there was more. By the decade's end, Hubble proved that the universe is expanding—and that its expansion rate can help determine age.

Alverez, *T. Rex and the Creator Doom,* NAT'L GEOGRAPHIC, September 2001, p. 84.

9. LASLO, EVOLUTION, THE GROUND SYNTHESIS 3 (1987). Amateur astronomers, who use the Internet to help the Spacewatch program examine their images for asteroids, found one that approached within 1.2 million miles of Earth on January 23, 2004. It was a fairly close approach, five times the moon's distance. 31 SIRIUS ASTRONOMY, Mar. 2004, p. 20.

10. RICHARD DAWKINS, RIVER OUT OF EDEN: A DARWINIAN VIEW OF LIFE 132-33 (2005) and LASZLO, EVOLUTION, THE GROUND SYNTHESIS 3 (1987). The creation of some elements needed for life is described in Kaler, *Extreme Stars* (2001).

The Big Bang Theory of cosmology makes predictions that are all consistent with observations, and to date, no other theory of cosmology is consistent with all existing observations. This is why nearly all astronomers believe it, and that is a compelling reason for you to believe it, if you choose.

Neil F. Comins, astronomer, *Heavenly Errors* (2001), p. 228

We shall all—philosophers, scientists, and just ordinary people—be able to take part in the discussion of the question why is it that we and the universe exist; if we find the answer *to that, it would be the ultimate triumph of human reason—for then* we would know the mind of God. (Emphasis added.)

Stephen Hawking, *A Brief History of Time*, 1988

a. The Big Bang and the Meaning of the Creator of the Universe, Its Age, and Evolution

Stephen Hawking's words demand that reason be used to ask "Why we and the universe exist?" Why must we consider "where we came from" as much as "where we are going"—*quo vadimus,* the eternal question? To what extent is bioethics religion, reason, or science?[11]

The Declaration of Independence prepared by Thomas Jefferson held "these truths to be self evident, that men are created equal, that they are endowed by *their Creator* with certain *unalienable Rights,"* and it authorized the people to "institute a new government." Our Constitution was instituted a dozen years later, but it failed to mention anything about the Creator.

Toward the end of the twentieth century, Carl Sagan wrote, "In all of the uses of science, it is insufficient to produce only a small, highly competent, well-rewarded priesthood of professionals; *some fundamental understanding of the findings and methods of science must be available on the broadest*

11. In the Foreword of his *A Brief History of Time* (2005), Hawking states that he "maintained and expanded the content of the original book," which "sold about one copy for every 750 men, women, and children on earth." (These books are referenced later in this volume.)

scale." Sagan's "priesthood of professionals" was dedicated to the concept of science; however, it had not become a sufficient attribute to most humans as of the dawn of the third millennium. Science is not a normal source of morals or ethics (see Chapters 3 and 4), although awareness thereof becomes an essential key to our future.

Some conflict between the future of science and religion was recognized in the Middle Ages: Peter Abelard of the University of Paris complained, "How can I teach what I don't understand?" St. Bernard replied, "Believe, don't reason!" Similarly, the Jewish writer Moses Maimonides in those ages dedicated his work, *A Guide for the Perplexed*, to his disciple, Joseph Ibn Akin, who could not reconcile science and faith. When challenged, Maimonides replied, "My theory is not established by demonstrative proof; it is based on the authority of the Bible." Nevertheless, like Thomas Aquinas, Maimonides tried to unite some of Aristotle's logic with his own tradition. Some hospital patients who are seriously ill try to seek understanding more like Abelard did; but they also join both logic and experience in a manner that is beyond how this was done during the Middle Ages. During the Renaissance in the seventeenth century, the French philosopher and scientist Blaise Pascal looked upward and said, "Swallowed up in the infinite expanse of space of which I know nothing, and which knows nothing of me, I am seized with horror."

While attempting to find the sources of bioethics, we must at least commence with astronomy, and examine religion, philosophy, and the law of cultures, asking whether they match reason for standards. Even Charles Darwin in his *Descent of Man* wrote that "the question whether there exists a Creator and Ruler of the universe . . . has been answered in the affirmative by some of the highest intellects." Sagan emphasized that reason cannot be ignored.

The sources of bioethics commence with consideration of the universe at the beginning of time.

> When we look at the largest structures in the cosmic background radiation, we are looking at patterns that were imprinted in the first moments of creation, when these patterns were single quanta—the smallest amount of something (anything), according to quantum theory—far smaller than the smallest subatomic structure ever seen in the laboratory. Whatever special atom-like quantum patterns emerge

> will be a direct sign of the holographic quantization of space-time. For all practical purposes, we could accurately call this a view of the true beginning of time, since we would be penetrating beyond "ordinary" time to some more fundamental, discrete structure from which our apparently continuous time and space have emerged.[12]

Edwin Hubble studied law in Chicago before completing enough courses in physics to become an astronomer. After serving in the army during World War I, he started using the world's largest 100-inch telescope at the Mount Wilson Observatory near Los Angeles, California, and discovered that Andromeda was clearly not part of our galaxy—a matter of great debate until 1924. Only five years later, he extended redshift measurements of galactic Doppler shifts to relate their velocity with distance. The redshifts were indicative of galaxies receding in recession of our expanding universe in 1929. As space expands, the energy in the light from another galaxy dwindles and stretches its wavelengths, causing the color of its radiation to be "redshifted." Velocities that are up to 80 percent of the speed of light are the most distant sources in space known as quasars. Calculating this time in reverse lead to better estimates of the large number of years when the moment of creation ("Big Bang") occurred.[13] As of the turn of the third millennium, using several methods, the universe became known to be between 13 and 14 billion

---

12. Hogan, *Observing the Beginning of Time*, AM. SCI. 420, 427 Sept.-Oct. 2002. *See also* KALER, EXTREME STARS 203 (2001). In 2004 it was noted although some "Biblical literalists insist that at best, the Big Bang cosmology be taught as a theory, on par with but less true than their own Genesis-based ideas, such an approach is nonsense." LINCOLN, UNDERSTANDING THE UNIVERSE 459-60 (2004). The fact is that the Big Bang theory has withstood unparalleled effort to bring it down." *Id.* at 486.

13. BOK & BOK, THE MILKY WAY, p. 327. The Age of the Universe from the Beginning of Time.

"The Hubble constant remained the most sought-after number in astronomy. Its new value was getting closer. After the turn of the third millennium, its 'tension over age' more accurately dropped to between 12 and over 13 billion years." SKY & TELESCOPE, Nov. 2000, p. 28. This magazine took a poll on changing the name "Big Bang," but it found no appropriate rename. *See also* PEIL, THE AGE OF SCIENCE 196 (2001). The Milky Way is 100,000 light-years across. One light-year is about six trillion miles of distance.

years old.[14] In 1931, Albert Einstein visited the Mount Wilson Observatory. Upon finding Hubble's observations to be convincing, he endorsed the expansion and erased the cosmological constant from his equations of general relativity. In 1951, Pope Pius XII delivered an address entitled "The Proofs for the Existence of God in the Light of Modern Natural Science," which included a specific mention of Hubble and his observations. He set forth views that still conflict with those of Southern Baptists and others:

> Present-day science, with one sweeping step back across millions of centuries, has succeeded in bearing witness to that primordial *Fiat lux* uttered at the moment when, along with matter, there burst forth from nothing a sea of light and radiation, while the particles of chemical elements split and formed into millions of galaxies . . . Therefore, there is a Creator. Therefore, God exists![15] (Today, most cosmologists agree that our universe is 13.7 billion years old and is still expanding.[16])

Einstein found that mass becomes energy and vice versa; neither are always conserved. Unlike the atomic energy created on earth by today's reac-

---

> 14. Only Big Bang theory, which is now almost universally accepted by astrophysicists, can satisfactorily explain the mass of observations made by the great telescopes and the results of experiments carried out in particle accelerators and retrace the principal stages in the creation of the universe—*a process that took not six days but 15 billion years!* (Emphasis added.)

LACHIEZE-REY & LUMINET, CELESTIAL TREASURY 255 (2001).

15. SINGH, BIG BANG 360 (2004). Singh called it "the most important and glorious scientific achievement—an archetypal example of the scientific method in action" (p. 469). He added that "[t]he only question that surpasses it in metaphysical magnitude is the biggest question of all: what came before the Big Bang?" *Id.* p. 488. Pope Pius XII concurred with the priest Lemaitre's cosmological solution.

16. The expansion rate is called the Hubble constant (denoted Hc). "The gravitational force holding planets and stars intact and bound together in galaxies are stronger than the outward swelling of space, and so none of these objects expands. Only on . . . such supergalactic scales does the swelling of space drive objects apart." GREENE, THE FABRIC OF THE COSMOS 238 (2004). The current five techniques indicate the Hc equals 72 kilometers per second per megaparsec. (*A megaparsec equals 3.26 million light-years.*) SKY & TELESCOPE, October 2003, p. 33. "The speed of light (c) times the redshift (*z*) gives a velocity. The redshift measures the expansion of space that has taken place while the light from a galaxy is in flight to us." KITSHNAR, THE EXTRAVAGANT UNIVERSE XI (2002).

tors, stars like the sun are stations that squeeze hydrogen (H) mass to both form energy and to become the element helium (with 3.993 units of weight out of four elements of H). Thereafter, helium is squeezed to create carbon, oxygen, silicon, and many more of the elements necessary for life. Our galaxy, the Milky Way, began its own "life" more than 10 billion years ago, roughly twice the age of the sun,[17] which has remained the origin of virtually all the energy that sustains life on earth.

For biological reasons, we know that for humans, up to 25 elements are required to permit our bodies to function.[18] Their source should no longer remain a need about which most people are unaware; to our current knowledge, the earth remains the only home to life in the universe. All its heavier elements are products of former exploding stars (supernovas).[19] They have

---

17. MELIA, THE BLACK HOLE AT THE CENTER OF OUR GALAXY 51 (2003). "The Sun has increased in brightness by about 30 percent in the last 4 billion years." WARD & BROWNLEE, THE LIFE AND DEATH OF THE PLANET EARTH 105 (2002). By weight, the initial universe was composed of about 78 percent hydrogen and 22 percent helium (or atomic abundances of 93 percent hydrogen and 7 percent helium). PRESS & SIEVER, EARTH (1986) and PENROSE, THE EMPEROR'S NEW MIND (1991). Dmitri Mendeleev, at age 35 in 1869, began his arrangement of the then-known 63 elements, in terms of atomic weight, "not atomic number (the quantity of protons in an atom of the element) as we do today." His publication was in the same year as that of Gregor Mendel's initial principles of heredity and one decade after that of Darwin's *Origin of Species.* He was "not aware of about a third of the elements we now know of." Like his contemporaries Darwin and Mendel in their fields, "the crux of Mendeleev's attitude toward what made the periodic system into a 'law' was the role of prediction. It was this capacity for prediction that convinced him of the 'naturalness' of his system of elements." GORDIN, A WELL-ORDERED THING, xix, 27, 32-33 (2004).

18. The 25 elements that are used in the human body are, with atomic numbers, hydrogen (1), boron (5), carbon (6), nitrogen (7), oxygen (8), fluorine (9), sodium (11), magnesium (12), silicon (14), phosphorus (15), sulfur (16), chlorine (17), potassium (19), calcium (20), vanadium (23), chromium (24), manganese (25), iron (26), cobalt (27), copper (29), zinc (30), selenium (34), molybdenum (42), tin (50), and iodine (53).

BARR, MODERN PHYSICS AND ANCIENT FAITH 297 (2003).

19. SKY & TELESCOPE (Cambridge, May 1998):

> The dust and gas from the first supernovae went through another round of condensation and star formation before exploding again to release even heavier elements such as iron, sulfur, and phosphorus into the uni-

been called "the most violent events since the birth of the universe . . . at the distance of the nearest star, they would, for a few seconds, outshine the sun nearly a million-fold and incinerate half the Earth."[20] At the end of the twentieth century, most of their sources were continually observed from the furthest distances in the universe.[21] Although most observed supernovas were not associated with *known* gamma-ray bursts, where a much nearer hypernova, a subclass of supernova, suggested to be the source of gamma rays that might become a very dangerous matter.[22] No supernovas have been seen in our Milky Way galaxy since the ones seen by the astronomer Tycho Brahe in 1572 and the mathematician Johannes Kepler who studied Brahe's material

---

> verse. These condensed once again to form both stars and planets. Our own sun is one of these third generation stars. It seems a bright beginning for life to know that *the stuff of all of us was made in stars and scattered throughout space by supernovae. We are*, quite literally, *stardust*. (Emphasis added)

McFadden, Quantum Evolution 83 (2000).

20. Katz, The Biggest Bangs 1 (2002). "Atomic nuclei known as 'cosmic rays' were also blasted into interstellar space by supernova explosions." Sky & Telescope, January 2004, p. 128. "There are 110 elements whose names are now determined jointly by the International Union of Pure and Applied Chemistry (IUPAC) and the International Union of Pure and Applied Physics (IUPAP). The *mass* of a body is a measure of the quantity of matter it contains. Kinetic energy is what a moving body possesses by virtue of its motion; namely, one-half the product of its mass (the property that manifests itself as inertia) and the square of its velocity. Varieties of energy include heat and electrical, magnetic, chemical and nuclear energy. The conversion factor between mass and energy is the square of the speed of light. A light's visible electromagnetic radiation travels at 186,000 miles per second in a vacuum (or 5.86 x 1,012 miles in one year)."

21. This was done by NASA's Burst and Transient Source Experiment (BATSE), which "made the crucial observations that established that gamma ray bursts come from the far reaches of the universe." *Id.* Italian-Dutch X-ray and gamma-ray-burst satellite (BeppoSAX), launched in 1996, "discovered the X-ray afterglows of gamma-ray bursts and provided the accurate positions that enabled ground-based astronomers to find their visible and radio afterglows." "The redshift of a gamma-ray-burst afterglow is 4.50, held by the afterglow of a burst observed January 31, 2000. This is less than the record redshift of a galaxy or quasar, which is around 6.0." *Id.*

22. *Id.* at 150. It was discovered by M.I. Anderson, et al., Astronomy & Astrophysics, V. 364, L54-61 (2000).

in 1604. However, they have been and continue to be, tracked down in other galaxies. Walter Baade, who came to Mount Wilson in the 1930s, indicated that supernovas could be used as "standard candles" to measure the distances of galaxies because of their standard brilliances. Type I supernovas are derived from explosions of stars like the sun and would end up as "white dwarfs." Type II supernovas come from much more massive stars.[23] They became easily visible when sought by using silicon light detectors called "charge-coupled devices," or CCDs (which were also used by amateurs, including myself). Toward the turn of the third millennium, the Supernova Cosmology Project found from distant stars that the expansion of the universe did not appear to be slowing down. In December of 2003, *Sky & Telescope* reported observations "to confirm dark energy—the mysterious force that is accelerating the expansion of the universe, and that appears to make up some 73 percent of all the matter and energy in a whole new way. . . ." As Heraclitus had said, "A hidden connection is stronger than an obvious one." The stellar explosions occurred, leaving numerous heavy elements in space, a few billion years prior to the formation of the sun, other stars, and planets. In 2004, the first discovery was made of a multiple planet system around a star that is beyond our solar system. The sun's planets included elements (necessary for the evolution of all lives) absorbed during the formation of the earth. Early in the twentieth century, Harvard's astronomer Harlow Shapley said, "We are made of star-stuff." Later, some bioethicists acknowledged that the failure to

23. Type I with mean absolute magnitude 14.3 at maximum and Type II, with an average mean absolute magnitude amounting to -12 at maximum. *Absolute magnitude* is the apparent magnitude a star would have if it were situated at a distance of 10 parsecs, or 32.6 light-years, from the earth. WALLENQUIST, DICTIONARY OF ASTRONOMICAL TERMS.

24. KATZ, p. 192. It is interesting to note that Hans Bethe, who received the Nobel Prize for his explanation of how the sun and other stars are able to pour forth so much energy over long periods of time, referred to a cycle in which carbon and nitrogen atoms (which are also essential for life) act as catalysts for the conversion of four hydrogen nuclei (protons) into helium nucleus: "The very slight mass lost in the process is converted into large amounts of energy, according to Albert Einstein's famous equation, $E = mc^2$." L.A. TIMES, March 8, 2005, at A22. "But deifying Mother Earth fails to appreciate that she shares our finitude and our mortality. She will eventually burn to an ash in the explosion of the dying sun." COX, WHEN JESUS CAME TO HARVARD 289 (2004).

absorb such knowledge may cause some people to fall within the aphorism of the physician Sir William Osler: "The greater the ignorance, the greater the dogmatism."[24]

Background radiation, an "afterglow" derived from the Big Bang, was suggested by George Gamow, who fled from Russia to Europe and later to the United States along with his associate nuclear physicists Ralph Alpher and Robert Herman. In 1948, they found that the sea of light released at the moment of recombination called *cosmic microwave background* radiation (CMB radiation) would prove that the Big Bang really happened.[25] The CMB was actually discovered in the 1960s and resulted in Nobel Prizes for two *Americans.*[26] However, only in the 1990s was it explored by the satellite Cosmic Background Explorer (COBE) to analyze tiny ripples in wave background radiation energy left over from the Big Bang. In space, these microwaves of less than 3° K remain from the original material. On November 18, 1989, NASA launched the *COBE* satellite whose measurements proved that

---

25. *See The Origin of Chemical Elements*, Physical Review J., April 1, 1948:

> Gamow, Alpher and Herman had predicted that, as the universe expanded with time, the wavelength of that primordial light would have been stretched as space itself has been stretched. The light had a wavelength of roughly one-thousandth of a millimeter when it originally emerged from the cosmic fog when the universe was 300,000 years old, but according to the Big Bang model the universe has since expanded by roughly a factor of a thousand. Therefore those light waves should now have a wavelength of roughly 1 millimeter, which would place them in the radio region of the electromagnetic spectrum. *Id.* p. 430.

Singh, Big Bang 333 (2004).

26. Penzias and Wilson received the Nobel Prize for Physics in 1978. In his Nobel lecture Penzias acknowledged the contribution made by Gamow, Alpher, and Herman (who received no portion of the award). *Id.* p. 437.

27. *Id.* p. 462. *See also* Encyclopedia of Bioethics (1995) Intro., p. xxii.

> Soon those galaxies would host generation after generation of exploding stars, and generation after generation of chemical enrichment—the wellspring of those cryptic little boxes that make up the periodic table of the elements. Absent this epic drama, life on Earth—or anywhere else—would simply not exist.

Natural History, December 2004/January 2005, p. 18.

there were genuine variations in the CMB radiation. They were acknowledged to be "evidence that, roughly 300,000 years or so after the moment of creation, there were tiny density variations across the universe at the level of 1 part in 100,000, which grew with time and ultimately resulted in the galaxies that we see today."[27] George Smoot, the project leader, declared, "It's like looking at God." Stephen Hawking exclaimed, "It is the discovery of the century, if not of all time."[28]

What does this have to do with the growth of bioethics? As Ernest Wood pointed out, "The amazing feature of the whole process is the absence of any indication of what may appear at the *next stage of growth.* But the most wondrous fact of all is that everything, *all of life and man himself,* is somehow implicit in the *elements of the universe.*"[29] (Emphasis added.) Toward a natural lead toward our evolution with life and death it was stated:

> None of the superficial contacts between science and myth implies that the ancients possessed an intuitive understanding of our cosmic origins, but that doesn't mean we can't appreciate the bond that the Greeks saw between Creation and Ouranos, the star-studded sky. We now know how most of the atoms constituting the Earth and everything on it, including us, originated in the nuclear furnaces and catastrophic explosions of stars. When I talk about the starry sky, I'm talking about my generation.[30]

Astrobiology is the scientific study of the origin, distribution, evolution, and future life in the universe.

Astrobiology Roadmap Workshop,
Ames Research Center, July 1998

---

28. Ross, Creation and Time 129 (1994). Another NASA mission was launched on June 30, 2001, to further study the afterglow of the Big Bang. Orange County Register, June 30, 2001. "Almost everyone now believes that the universe, and time itself, had a beginning at the Big Bang." Stephen Hawking & Roger Penrose, The Nature of Space and Time 20 (1996).

29. Wood, The Magnificent Frolic 521 (1978).

30. Krupp, Sky & Telescope, September 2005, p. 49.

> *In some remote corner of the universe, poured out and glittering in innumerable solar systems, there once was a star on which clever animals invented knowledge. That was the highest and most mendacious minute of "world history"—yet only a minute. After nature had drawn a few breaths the star grew cold, and the clever animals had to die.*
>
> *One might invent such a fable and still not have illustrated sufficiently how wretched, how shadowy and flighty, how aimless and arbitrary, the human intellect appears in nature.*
>
> Friedrich Nietzsche, *The Nachlass*
> On Truth and Lie in an Extra-Moral Sense (1873), p. 1.

> *Galactic intelligence also looks through the window of the human soul, to see into the very depths of the worlds of matter.*
>
> Ken Carey, *The Third Millennium* (1995), p. 153.

b. Potential of Progress in Astrobiology, Astrology

Until the 29th year in the twentieth century, our ignorance prevailed concerning the creation. By the end of that century, there was scientific confirmation that resulted from an examination of ancient evidence still in the space above us. Although most parts of the universe seem to be indifferent to humankind, some parts may be friendly enough to support life elsewhere.

While with the former Atomic Energy Commission (AEC), I put together a government contract for work needed to make a nuclear energy product (rather than one using solar energy) for the *Viking* spacecraft. This was under conditions to operate upon the first landing on Mars in the late 1970s. After successful landings, it found conditions that appeared to be hostile to life.[31] This was a disappointment to our new "astrobiologists." However, new trips

31. Viking Lander 2 was terminated in April 1980; Lander 1 ended in November 1982.The intriguing feature of this is that if the earth had a similar orbit to that of Mars, with only half the diameter of the earth, it would retain a sufficiently thick atmosphere to have liquid water oceans, and an ideal home for life as we know it. GIBBIN, XTL, EXTRATERRESTRIAL LIFE AND HOW TO FIND IT 3 (2002). *See also* De Gregorio, *Life on Mars?* in SKY & TELESCOPE, February 2004, p. 42. "If organisms can utilize frosts or perhaps tap into the long-proposed underground Martian water reserves, they could potentially survive." *Id.*

were made in the first years of the twenty-first century.[32] (They actually used solar energy at the distance of Mars rather than the nuclear energy as used for the Viking project.)[33] In the third millennium, we shall doubtless try to demonstrate the feasibility of planets circling certain other stars that have a "habitable zone" (HZ) for potential life. More than 100 extrasolar planets going around other nearby stars have been discovered by 2002.[34] Inasmuch as we are a part of the universe, the use of human methods to visit such planets will present bioethical challenges. (See Chapter 15,(5)(e).) We now know of no other place where life exists, or, if it does, whether intelligence reigns.

Since the turn of the twenty-first century, space exploration and commercial use has become of great and similar interest to several cultures.[35] One wealthy man paid $20 million to make the first private voyage into space. In 2000, NASA published its criteria for visitors to the space station. Mother Teresa spoke of its significance by stating, "We realize that what we are accomplishing is a drop in the ocean; but if this drop were not in the ocean, it would be missed." Our recent accomplishments constitute a drop into space; if we were not in space, this would certainly be missed.[36]

In 2002, gravity's speed was confirmed as being identical to that of light; the measurement is one of the last fundamental constants in physics to be

32. Laszlo, Evolution 3 (1987). "Bacteria have been found living encased in lava more than 4,000 feet underground in Hawaii. The implication is that life elsewhere in the universe may live in more harsh conditions than we thought possible." Sirius Asronomer, February 2004, p. 10.

33. Often this may be difficult to operate by landing due to Mars' current coverage by dust storms. Orange County Register, Oct. 12, 2001, at News 34. The European Space Agency has selected teams to design a rover to be launched to Mars in 2009 and drill into the surface to search for evidence of microscopic life. "A team from the United States, Norway, Canada, and South Africa found evidence of the lava-burrowing archaea in 3.5 billion-year-old rock in South Africa." Orange County Register, April 24, 2004, at News 20.

34. Sky & Telescope, June 2002, p. 18. A "continuously habitable zone" has been termed the distance from a star where temperatures will allow liquid water to persist for billions of years. Sky & Telescope, January 2003, p. 26.

35. Tribble's Guide to Space (2000).

36. The criteria are similar to what the government uses in background investigations for positions requiring security clearance made on a case-by-case basis and would require consensus among the participating countries. *See* http://www.nasa.gov.

established and it was noted, "Gravity is not well understood."[37] The heart is accustomed to pumping blood up against gravity. In his *2001 Space Odyssey,* Arthur Clark prescribed rotation of the vehicle (after reaching the acceleration speed) to retain the crew's health against its pull. Thus, in the third millennium many people searching for a meaning of life may wish to reread the words of the deity to Job some centuries prior to the first one:

> Then the Lord answered Job out of the whirlwind:
> Who is this that darkens counsel by words without knowledge?
> Gird up your loins like a man.
> I will question you, and you shall declare to me.
> Where were you when I laid the foundation of the earth?
> Tell me, if you have understanding. (Job 38)

Only today can many of us answer, "We have much more understanding than Job did." We are still searching for more understanding of the many things from the beginning of the universe. In his recent volume *A Brief History of Time,* British physicist Stephen Hawking wrote about the importance of knowing about the creation: "Then we shall all, philosophers, scientists, and just ordinary people, be able to take part in the discussion of the question of why it is that we and the universe exist. If we find the answer to that, it would be the ultimate triumph of human reason—for then we would know the mind of God."

It has been noted that "Although biology is not a Nobel Prize category, it can never be treated as an area of knowledge totally apart from other sci-

---

37. Compared to the other fundamental forces of nature, gravity is 38, 35, and 31 orders of magnitude weaker than the strong nuclear force, electromagnetism, and the weak nuclear force, respectively. . . . Neither general relativity nor quantum mechanics provides a satisfactory explanation for the inherent feebleness of gravity. SKY & TELESCOPE, June 2003, p. 42. Nevertheless, in the following year it was noted:

> For almost two decades, astronomers have been intrigued by what they might learn from *gravitational microlensing*—the distorting and magnifying of a star's light by the gravity of an object passing nearly in front of it. For microlensing to occur, however, the intervening object must pass extremely close to our line of sight to the background star.

SKY & TELESCOPE, July 2004, p. 18.

ences . . . we can easily overlook how seemingly unrelated fields actually complement one another . . . *an example that adequately meets the requirement is astronomy.*"[38] (Emphasis added.) In 2005, Nobel laureate Robert Laughlin wrote, "There is no denying that physicists are culturally the exact opposite of doctors."[39] A new science, astrobiology, which studies the condition of life in the Universe, began in the 1990s. Many writers say that astrobiology forces us to reconsider the life of our planet as but a single example of how life might work, rather than as the only example. "Astrobiology requires us to break the shackles of conventional biology."[40]

Nevertheless, some people actively continue to use *astrology* in their lives. In May of 2000, an alignment of the sun, moon, and the five brightest planets arrived without the disasters predicted by some as doomsday.[41] It was reported that the alignment brought forth mostly "yawns," which in that case appeared to be a lack of "progress."[42] As stated by Nobel laureate Richard Feynman in *The Pleasure of Finding Things Out* (1999):

---

38. Rogers, Biology 1 (1974).

39. Laughlin, A Different Universe 174 (2005). It is also of interest to note that the astronomer Bart Bok assembled a group of 192 well-known scientists, including 19 Nobel Prize winners, to sign a statement condemning the practice of astrology. The astronomer Roger Ravelle wrote, "If some people are comforted by a belief in astrology we should not try to take away their security blanket . . . . If I were to take a strong stand against astrology I would feel it necessary to be equally forceful about the Pope, Norman Vincent Peale and Billy Graham." Levy, Man Who Sold the Milky Way 163 (1992).

40. Ward & Brownlee, Rare Earth XV (2000). "Taking living cells apart reveals that they are composed of the same elements as inanimate matter, held together by the same chemical bonds, interacting by the same laws of physics." G.K. Hunter, *Is Biology Reducible to Chemistry?* Perspect. Biol. Med. 40:130-38 (1996).

41. Orange County Register, May 6, 2000. "Astrology is the oldest of the occult sciences. It is also the origin of science itself. From astrology are derived astronomy, calculation of time, mathematics, medicine, botany, mineralogy, and (by way of alchemy) modern chemistry, among other disciplines. Logarithms were originally devised to simplify the calculations necessary in casting horoscopes; the ray theory of vision—the foundation of modern optics—developed from astrological theories of the effect of stellar rays on the soul. . . . Or, as Ralph Waldo Emerson reputedly remarked, 'Astrology is astronomy brought down to earth and applied to the affairs of men.'" Bobrick, The Fated Sky 5-6 (2005); MacNeice, Astrology 66.

42. Richard Noone's book, *Ice, The Ultimate Disaster* (2000), suggested the alignment's gravitational pull, along with increased solar activity, would trigger massive earthquakes and tidal waves.

I believe that science has remained irrelevant because we wait until somebody asks us questions or until we are invited to give a speech on Einstein's theory to people who don't understand Newtonian mechanics, but we never are invited to give an attack on faith healing, or on astrology—on what is the scientific view of astrology today. (p. 110)

Twenty-first Century Approaches

1. We must recognize that absolutism is not a goal in bioethics and the law; that the true goal normally requires the weighing of what may be a traditional "good" against a "good" derived from knowledge and reason applied to particular factual settings that need to become better known to more people.
2. The achievement of this goal can only be through the attempts to achieve rational analyses of bioethical issues (see Chapters 5 through 15).
3. A prelude to such analyses requires a brief examination of life's evolution and why we need to work toward ascertaining possible bioethical positions for individuals (see Chapters 2 to 4).

To examine the causes of life, we must first have recourse to death.

Mary Shelley

The highest wisdom has but one science—the science of the whole—the science explaining the whole creation, and man's place in it.

Leo Tolstoy

Man in his arrogance thinks himself a great work worthy the interposition of a deity. More humble and I think truer is to consider him created from animals.

Darwin 1848

(11 years before publication of *The Origin of Species*)

Intelligent life on a planet comes of age when it first works out the reason for its own existence.

Richard Dawkins, *The Selfish Gene* (1976), p. 1

2. The Importance of the Continuance of Evolution into Bioethics, Medicine, and the Law

(a) What Constitutes Life Itself?

During the twentieth century, the Howard Hughes Medical Institute (HHMI) for basic biomedical research was founded by Howard Hughes who stated it is "to probe 'the genesis of life itself.'" Following the dawn of the third millennium, we still do not know precisely how life initiated. This remains biology's greatest mystery—particularly with respect to self-replication, an action with which natural selection began (even though Hermann Müller won the Nobel Prize back in 1946 for the discovery that X-rays cause mutations).

We do know what we came from. In 1992, James Watson, Nobel laureate and the codiscoverer of the structure of DNA, indicated that he had accepted Linus Pauling's statement, "We came from chemistry."[43] His codiscoverer, Francis Crick, wrote:

> Everything we know about molecular biology appears to be explainable in a standard chemical way. We also now appreciate that molecular biology is not a trivial aspect of biological systems. It is at the heart of the matter. Almost all aspects of life are engineered at the molecular level, and without understanding molecules, we can only have a very sketchy understanding of life itself. All approaches at a higher level are suspect until confirmed at the molecular level.[44]

It has been found that living systems on earth thermodynamically turn solar radiation into living matter and heat. Photosynthesis has been called "the most important chemical reaction to develop in the history of life on Earth. The energy in photons from the sun is co-opted to run life and power the building of the biochemical stuff we call bodies." Today, "virtually all surface life depends on this." Oxygen is essential to our lives. However, it became a key to our atmosphere only when the earth was half as old—about

---

43. Watson, *A Personal View of the Project*, in THE CODE OF CODES: SCIENTIFIC AND SOCIAL ISSUES IN THE HUMAN GENOME PROJECT, Kevles & Hood eds., 1992, p. 164.

44. *See* CRICK, WHAT MAD PURSUIT, at 61.

2.5 billion years ago—with photosynthesis by cells living upon carbon dioxide (CO_2). "Life created a new world."

After the study of early life began, we entered the field of evolution. In 1859, Charles Darwin proposed it as an alternative to the biblical account in his *Origin of Species*. He called natural selection the "preservation of favorable individual differences and variations, and the destruction of those, which are injurious."[45] Almost a century and a half later, the biologist, Ernst Mayer, professor emeritus at Harvard, wrote *What Evolution Is* (2001), stating that evolution is "the most important concept in biology":

> The evidence for evolution is now quite overwhelming (p. 13). The development of speech, in turn, exerted an enormous selection pressure on an enlargement of the brain, particularly those parts that involved information storage (memory). This enlarged brain made the development of art, literature, mathematics, and science possible (p. 253).
>
> Brain size was increased in H. erectus only slowly for about one million years, but rose in late H. erectus to 800–1,000 cc, finally reaching in H. sapiens an average of 1,350 cc (p. 255). That *evolution* has occurred and takes place all the time *is a fact so overwhelmingly established that it has become irrational to call it a theory*. Evolutionary thinking has enormously enriched almost all other branches of biology. For instance, more than a third of all current publications in molecular biology show how the nature and history of important biological molecules are illuminated by an evolutionary approach.[46] (Emphasis added.)

---

45. Darwin, Origin of Species, in Harvard Classics II (1909) at p. 94. A dozen years later (1871) Darwin (who described himself as agnostic) wrote *The Descent of Man,* concluding "that man with all his noble qualities . . . still bears in his bodily frame the indelible stamp of his lowly origin." James Colbert of Iowa State University "surveyed students in his introductory biology class, asking them if they believe God created humans within the past 10,000 years. Last fall [2005] 32 percent of the 150-member class said they did." He found this unsettling "when one considers that these students are academically among the upper half of high school graduates, and they are students choosing to major in a life science." Science 769 (Feb. 2006).

46. Mayr, What Evolution Is 264 (2001). He did not receive a Nobel Prize for his work because biology is not one of the prize categories. He noted, "Darwin wouldn't have won it, either." Yet, Mayr became "the first scientist to win all three of the major

Although the validity of his concept is acknowledged by the majority of current biologists,[47] in 1996, a strong naysayer, biochemist Michael Behe, actually wrote the opposite: "Molecular evolution is not based on scientific authority. There is no publication in the scientific literature—in prestigious journals, specialty journals, or books—that describes how molecular evolution of any real, complex, biochemical system either did occur or even might have occurred. There are assertions that such evolution occurred, but absolutely none are supported by pertinent experiments or calculations."[48] In his attempt to equate these designs with the origin of biological systems, Behe specifically denied the existence of the journals and books he claimed in the previous paragraph. "Unlike Darwinian evolution, the theory of intelligent design is new to

---

prizes in biology: the Balzan Prize in 1983, the International Prize for Biology in 1994, and the Crafoord Prize in 1999." Maugh II, L.A. TIMES, Feb. 5, 2005, at News 21. He died at 100 years of age in 2005.

47. Although much remains obscure, and will long remain obscure, I can entertain no doubt, after the most deliberate study and dispassionate judgment of which I am capable, that the view which most naturalists until recently entertained, and which I formerly entertained—namely, that each species has been independently created—is erroneous.

DARWIN, ORIGIN OF SPECIES 24. He also wrote:

> The same high mental faculties which first led man to believe in unseen spiritual agencies, then in fetishism, polytheism, and ultimately in monotheism, would infallibly lead him, as long as his reasoning powers remained poorly developed, to various strange superstitions and customs. . . . Yet it is well occasionally to reflect on these superstitions, for they show us what an infinite debt of gratitude we owe to the improvement of our reason, to science, and to our accumulated knowledge.

See KEYNES, DARWIN, HIS DAUGHTER, AND HUMAN EVOLUTION 291 (2002).

48. BEHE, DARWIN'S BLACK BOX 185 (1996). Michael Behe, a biochemist at Lehigh University [9,] contends that the molecular machinery of cells is so complex and interdependent that this is proof of purposeful design. Behe's book, *Darwin's Black Box: The Biochemical Challenge to Evolution*, was chosen as 1997 Book of the Year by the evangelical monthly *Christianity Today*. However, in his book *Quantum Evolution* (1961), the English professor Johnjoe McFadden wrote that biochemist Michael J. Behe of Lehigh University had a solution which is either radical or hopelessly archaic depending on your point of view. . . . [He] proposed that these metabolic pathways could have been put in place only by God. However, I do not believe we need resort to pre-Galilean science to explain their existence. Quantum evolution may instead provide the answer. (pp. 77, 277)

modern science." Behe honestly stated that he was a Catholic with nine children[49] and warned "Men ought not to play God before they learn to be men, and after they have learned to be men they will not play God." My former professor, Joseph Wood Krutch, at Columbia University wrote: "As civilization grows older, it too has more and more facts thrust upon its consciousness and is compelled to abandon one after another, quite as the child does, certain illusions that have been dear to it."[50] In 1999, Kenneth Miller wrote, "No doubt about it—the evolution of biochemical systems, even complex multipart ones—is explicable in terms of evolution. Behe is wrong."[51]

Today, we may agree that men must learn to act responsibly when making ethical and legal choices. In September of 2000, high schools in 19 states received very poor grades on the teaching of evolution in a dispute that put "science against evolution."[52] Until the end of the twentieth century, Kansas, like other states, omitted references to many evolutionary concepts, as well as to the Big Bang theory on the creation of the universe.[53] However, in 2001, a major change was made in Kansas; they authorized the teaching of both of

49. STROBEL, THE CASE FOR A CREATOR 195 (2004).

50. LIVING LEGACIES 37. Professor Barry Hall at Rochester University "demonstrated that two sites just a few bases apart on the same DNA molecule could be subject to widely different mutation rates, depending on whether or not those mutations were adaptive." *Id.* p. 78, *citing* Hall, *Activation of the bgl operon by adaptive mutation*, in MOLECULAR BIOLOGY AND EVOLUTION 1-5 (1998). In his volume *Genetics*, Hall commented: "The selective generation of mutations by unknown means is a class of models that cannot, and should not, be rejected. . . . Spontaneous point mutations occur more often when advantageous than when neutral." (pp. 5-16)

51. KENNETH MILLER, FINDING DARWIN'S GOD 147 (1999). *See also* NATIONAL ACADEMY OF SCIENCES (NAS), SCIENCE AND CREATIONISM: A VIEW FROM . . . (1999), p. 25.

52. ORANGE COUNTY REGISTER, Sept. 27, 2000, at News 13.

53. ORANGE COUNTY REGISTER, Feb. 15, 2001:

> Einstein, in particular, was deeply troubled by the probabilistic character of quantum theory. Physics, he would emphasize again and again, is in the business of determining with certainty what has happened, what is happening, and what will happen in the world around us. Even though he could not say what it was, Einstein wanted to convince everyone that there was a deeper and less bizarre description of the universe yet to be found.

GREENE, THE FABRIC OF THE COSMOS 93 (2005), *citing* ALBRECHT FÖLSING, ALBERT EINSTEIN 208 (1997).

these. But 20 years before in Arkansas, there was a movement in the reverse: in March of 1981, its legislature required its teachers to instruct students to mark "false evidence" or "theory" in the margins next to references to evolution. In 2004, the state school superintendent proposed striking the word *evolution* from Georgia's science curriculum and replacing it with the phrase *biological changes over time* to alleviate pressure on teachers in socially conservative areas where parents object to its teaching.[54] Although former President Jimmy Carter said that it exposed the state to nationwide ridicule, the Atlanta School Board voted unanimously to allow teachers to introduce students to different views about the origins of life.[55] The late Harvard University professor Stephen Jay Gould had stated:

> When I ask myself why evolution, although true by our strongest scientific confidence, has not become generally known or acknowledged in the United States—that is, nearly 150 years after Darwin's publication, and in the most technologically advanced nation on earth—I can only conclude that . . . misreading of his doctrine as doleful, or as subversive to our spiritual hopes and needs, rather than as ethically neutral and intellectually exhilarating, has impeded public acceptance of our best documented biological generality.[56]

Some theistic evolutionists also support a limited amount of science viewed as nature's way of using mutations upon extinct animals in all evolution since the age of dinosaurs as the use of God by such an experimental method. That is, they tended to disagree with Darwinists' use of chance as the only guide toward variation by natural selection. Darwin was not an atheist. He wrote, "My theology is a simple muddle; I cannot look at the universe as the result of blind chance, yet I can see no evidence of beneficent design of any kind in the details." Darwin also wrote: "Nor can I overlook the difficulty from the immense amount of suffering through the world . . . the safest conclusion seems to me that the whole subject is beyond the scope of man's intellect."[57] He was an agnostic and noted, "Not one man in a thousand has accuracy of eye and judgment sufficient to become an eminent breeder" (p. 47).

54. Georgia Department of Education: www.gadoe.org., visited Jan. 30, 2004.
55. http://www.cnn.com/2002/EDUCATION/09/26/creation.evolution/index.html.
56. Orange County Register, March 22, 2001, at News 18.
57. Sir Gavin de Beer, Charles Darwin 268 (1964).

In his *Origin of Species,* Charles Darwin wrote, "I am convinced that Natural Selection has been the most important, but not the exclusive, means of selection." In the late twentieth century, it was found that he was right about the lack of exclusiveness of the natural selection methodology. This became symbiosis: an association between differently named species of plants and animals that persists for a long period that is an evolutionary mechanism quite different from natural selection. A symbiosis involves the union of different genomes, at either the nuclear or the cytoplasmic level (endosymbiosis). In her 1998 volume, *Symbiotic Planet,* professor Lynn Margulis, a member of the National Academy of Science, noted, "This is not the popular idea of the basis of evolutionary change in most textbooks" (p. 33), and concluded, "No evidence exists that we are 'chosen' the unique species for which all others were made" (p. 119). Her overall constitution of evolution is one thing toward which Darwin, in his *Origins of Species* (p. 499), wrote, "That many and serious objections may be advanced against the theory of descent with modification through variation and natural selection, I do not deny." In her own volume, the *Symbiotic Planet,* Margulis wrote more toward which Darwin agreed: "The study of evolution is vast enough to include the cosmos and its starts as well as life, including human life, and our bodies and our technologies. Evolution is simply all of history."

In 1900, the German physicist Max Planck called each *quantum* of light one discrete energy packet (quanta). In 1913, Danish physicist Niels Bohr proposed that like light, electrons were quantized at discrete energy levels in atoms and molecules. After World War II, American physicists developed the theory of quantum electrodynamics (AED), which showed quantum mechanics as *quantum superposition*—a particle in two places at once.[58] By the

---

58. Feynman, Lectures on Physics 37-41 (1963):

> Einstein, in particular, was deeply troubled by the probabilistic character of quantum theory. Physics, he would emphasize again and again, is in the business of determining with certainty what has happened, what is happening, and what will happen in the world around us. Even though he could not say what it was, Einstein wanted to convince everyone that there was a deeper and less bizarre description of the universe yet to be found.

Greene, The Fabric of the Cosmos 93 (2005), *citing* Albrecht Fölsing, Albert Einstein 208-1 (1997).

turn of the third millennium, it was noted how the "chemical changes that lead to mutations also involve the motion of electrons and protons inside the DNA double helix and must be dependent upon quantum measurement," which may be the key ingredient missing in conventional theories to account for the origin of life.[59] Thus, scientists currently expect "that the twenty-first century will see the flowering of quantum biology.[60] At the same time, they would not replace natural selection; but at critical evolutionary junctures, would shift it into the quantum realm.[61]

With the explosion of such diversity in life (sometimes called biology's "Big Bang"), one scientist actually claimed, "Biology is almost unimaginably more complex than physics."[62] Evidence found that the first biological explosion may have begun 3.7 billion years ago with prokaryotic life (bacteria without a nuclei) using the simplest form of reproduction: asexual binary fission. In 1999, the second was noted by stating that there "is convincing evidence for the presence of eukaryotes" [cells with nuclei like ours] "around 2.7 billion years ago."[63] The third has been attached to the "Cambrian Explosion with multicellularity that occurred almost 600 million years ago." Professors Paul Hoffman and Dan Schrag wrote "Snowball Earth," *Scientific*

---

59. McFadden, Quantum Evolution 177, 240 (2000). "The earliest steps in life's emergence took place at the quantum level where primordial chemistry could explore a vast array of multiple states simultaneously. Quantum measurement by a self-replicating biomolecule was the key event that caused life to emerge." (p. 231)

60. *Id.* p. 265 and McFadden & J. Al-Khalili (1999), *A quantum-mechanical model of adaptive mutations*, in 50 Biosystems 203-11.

61. McFadden, Quantum Evolution 270 (2000).

62. E.O. Wilson, Consilience 67 (1998). It has been recently argued that initial life was archaeabacteria, which was separate from prokaryotes:

> The ancestor cannot have been a particular organism, a single organismal lineage. It was communal, a loosely knit, diverse conglomeration of primitive cells that evolved as a unit, and it eventually developed to a stage where it broke into several distinct communities, which in turn became the three primary lines of descent [bacteria, archaea, and eukaryotes].

Woese, *The Universal Ancestor*, 95 Proceedings of the National Academy of Sciences, June 9, 1998, at p. 6854.

63. Orange County Register, Aug. 13, 1999, reporting on findings published in the journal *Science* on that date.

American, January 2000, 68-75, arguing that the "Cambrian Explosion," the biggest change in life, occurred 700 million years ago to freeze the earth, resulting in a profusion of complex life forms (multicellularity) that bacteria survive at the South Pole:

> Suddenly, in the Earth's late middle age, life began frantically procreating, evolving and developing new forms. First came trilobites and ammonites, then dinosaurs and octopuses . . . Most researchers believe that complexity was invented somewhere between 550 and 590 million years ago. That's after more than 3 billion years of simple, single-celled slime.[64]

Outside of the nucleus is our cell's energy-producing part, called mitochondria. It is an asexually reproducing bacterium, the smallest of all living things, which entered into the eukaryotic cells with a nucleus an estimated 1.7 billion years ago.

I asserted—and repeated—that a man has no reason to be ashamed of having an ape for his grandfather.

Thomas Huxley, *Letters* (1900)

This most excellent canopy, the air, look you, this brave o'erhanging firmament, this majestical roof fretted with golden fire . . . What a piece of work is a man! How noble in reason! How infinite in faculties! In form and moving, how express and admirable! In action, how like an angel! In apprehension, how like a god! The beauty of the world! The paragon of animals! *And yet, to me,* what is this quintessence of dust? *(Emphasis added.)*

Shakespeare, *Hamlet* 2.2 (1604)

Nothing in biology makes sense, except in the light of evolution.

Theodosius Dobzhansky

64. Walker, Snowball Earth 202 (2003). "Bacteria survive—somehow—at the South Pole. Other bacteria have shown up beneath glaciers, and even inside solid rock." (p. 197).

(b) The Importance of Considering the Meaning of Evolution

Charles Darwin, the naturalist, lost to William Shakespeare, the humanist, in an English contest in 1999 to determine the most important person of the entire second millennium. We must ask what his "quintessence of dust" is about that the bard's Danish Prince Hamlet spoke in the midst of his personal depression. To what extent should we delve into this profundity after the third millennium has dawned?

To do this, the bioscientist Lynn Margulis stated, "As a culture, we still don't really understand the science of evolution. When science and culture conflict, *culture always wins.*"[65] (Emphasis added.) As also indicated by Professor Paul R. Ehrlich, "Cultural evolution can be much more rapid than genetic evolution. The evolution of that body of extragenetic information—cultural evolution—has been centrally important in making us the unique beasts we are."[66] The only genetic item to match the speed of culture is the "dynamic relationship between *Homo sapiens* and the microorganisms that attack it," such as penicillin. It was once given to virtually everyone, which resulted in the rapid resistance to the use of penicillin as treatment for many dangerous diseases.[67] This demonstrates the rapid speed of evolution pertaining to mammals, which was rarely taught in most schools during the twentieth century.

During the third millennium this could change if it is encouraged by the law; this will make us "freer" (see Chapter 5). It has been noted that although "natural evolution is an immensely powerful process, it operates on geological timescales and yet can also cause [some] changes in a matter of months . . . It can create intricate designs smaller than a pinhead."[68] Microbiologist Richard

---

65. MARGULIS, SYMBIOTIC PLANET 4 (1998). *See also* Margulis, *Microbial Minds,* in FORCES OF CHANGE (2000), at p. 1: "In recent millennia, cultural evolution has shaped human affairs more than biological evolution has." MCNEILL, SOMETHING NEW UNDER THE SUN (2000).

66. EHRLICH, HUMAN NATURES 63 (2000). In 2003, researchers reported proof that the orangutan is a cultured ape, able to learn new living habits and to pass them along to the next generation.

67. *Id.*, p. 281. "Although evolutionary theory is not part of the curriculum in most medical schools, physicians are beginning to give more recognition to the importance of evolution," p. 284.

68. BENTLEY, DIGITAL BIOLOGY 43 (2001).

Lenski, professor at Michigan State University, found populations of the bacterium E. coli. with an evolved new generation every 3.5 hours or so.[69] Serious thought regarding the concept of changes in our species is urgently required, because without understanding evolution, a researcher "has little hope of figuring out how to create new drugs and determine how they should be administered."[70] The Food and Drug Administration expedited approval of new antibiotics, addressing evolutionary issues, by stating, "Anyone you know with AIDS harbors a battle within him or her, because in each HIV-infected person, evolution operates so quickly that *the virus evolves* resistance to new drugs *within months.*"[71] (Emphasis added.)

The tribolic, whose fossil species of three sections I first saw in the Grand Canyon, is one of the earliest animals. It arrived about the Cambrian Age, 500 million years ago, and lasted 300 million years. In his publication of 2000, an expert wrote: "The order of appearance of tribolites is certainly consistent with evolution. Cambrian tribolites have more primitive characteristics than those in the Ordovician [the second oldest age]."[72] It became the beginning of new animals under the process Darwin called "Mutual Selection" by slowly undergoing "many mutations":

> Any being, if it vary however slightly in any manner profitable to itself, under the complex and sometimes varying conditions of life, will have a better chance of surviving, and thus be *naturally selected*

---

69. U.S. News & World Report, July 27, 2002, at 48.

70. Evolution, The Triumph of an Idea 336 (2001). "The same goes for vaccines. As microbes evolve, they become isolated into genetically distinct populations, which create new branches on the evolutionary tree."

71. Polumbi, The Evolution Explosion 10 (2001). "An HIV infection evolves so fast that within two months it has hidden itself from the alerted immune system." (p. 35) It has even been claimed that human evolution of brains "occurred through the spark of advanced development of language as a medium for cultural exchange and innovation." (p. 237)

72. Fory, Tribolite 159 (2000). "The horseshoe 'crab' limitus appears to be the closest living relative of Tribolites." (p. 158)

> In due course, as creatures of many kinds and varying ancestries came to live on the bottom of the seas, the trilobites lost their supremacy. Two hundred and fifty million years ago, their dynasty came to an end. One relation alone survives, the horseshoe crab.

Attenborough, Life on Earth 55 (1999).

> . . . Natural Selection almost inevitably causes much Extinction of the less improved forms of life . . . (p. 23) for *natural selection* acts only by taking advantage of *slight successive variations*; she can never take a great and sudden leap, but must advance by short and sure, though slow steps (p. 206). Nor can organic beings even if they were at any one time perfectly adapted to their conditions of life, have remained so, when their conditions changed, unless they themselves likewise changed; and *no one will dispute that the physical conditions* of each country, as well as the numbers and kinds of its inhabitants, *have undergone many mutations.*[73] (Emphasis added.)

In seeking the actual descent of each animal, it has not always been easy to ascertain the common ancestor from the new structure. The derivation might be from mutational genes (the DNA of cells) or by something epigenic (i.e., beyond the genes).[74] As stated in this third millennium, "By drawing the family trees of genes in modern forms of life and running them backward to the genes at the root of the tree, they may even define the nature of the universal ancestor from which all life sprang."[75] Two important trials were held on controversy over Darwin's *The Origin of Species*. A teacher, named John Scopes, was found guilty in the 1925 "monkey" trial for violating a Tennessee statute prohibiting the teaching of evolution[76] that differed from certain Biblical statements.[77] All expert testimony on evolution was disallowed. Many other states enacted similar statutes. Following the 1958 launching of the Soviet Union's first satellite *Sputnik,* the fear that we as Americans were losing our competitive lead in science developed. It remained a satellite for three months, emitting radio signals for three weeks.

When Governor (later President) Bill Clinton had been defeated in 1980, the Arkansas Legislature enacted Act 590 that was signed into law by his

---

73. DARWIN, ORIGIN OF SPECIES 219 (Colter Press, 1909).

74. WALLS, ICONS OF EVOLUTION (2000). A number of religions believe that God maintains the design.

75. ACHERMAN, CHANCE IN THE HOUSE OF FATE 12 (2001).

76. *See* Scopes v. State, 289 S.W. 363, Tenn. 1927, 131 Tenn. 105, 289 S.W. 363, 1927. Several motion pictures were made thereof.

77. "I made the earth and created mankind upon it." ISAIAH 45:12 "Have you not read that at the beginning God made them male and female?" MATT. 19:4. *See also* GEN. 27.

successor, Governor Frank White, requiring all Arkansas public schools to teach anti-Darwinism alongside the teaching of evolutionary biology. In 1981, a truly "key trial" was held on the mandatory teaching of "scientific creationism" under a statute that emphasized the Bible. No movies have yet been made concerning this trial, where scientists on the witness stand included the late Harvard Professor Stephen Jay Gould, a paleontologist who wrote much on evolution.[78] He noted:

> [N]amed "Scopes II" after the infamous 1925 Dayton, Tennessee, case, the trial brought six of the world's foremost natural scientists and professors of the philosophy of science and theology to Little Rock. Since expert testimony had been disallowed by the judge in Scopes I, it put real scientists on the witness stand in a public trial for the first time in the twentieth century, . . . The Creation-Science act did, however, in White's favor, bring more scholars to the state than his teacher's salaries could have. The problem is, most of these scholars spoke against Frank White and his act.[79]

After hearing many experts on both sides, a federal court in Arkansas held the statute to be unconstitutional.[80] In the late 1980s, the Supreme Court also ruled that the state law of Louisiana requiring teaching in its high schools

---

78. *See* STEPHEN J. GOULD, ROCKS OF AGES: SCIENCE AND RELIGION IN THE FULLNESS OF LIFE 129-31 (1999).

79. HAMILTON, NIGEL BILL CLINTON 380-81 (2003):

> In 1925, when the judge in Scopes I disallowed the testimony of any expert witnesses, only the state's attorney William Jennings Bryan could testify. He had been the defeated Democrat for the Presidency in 1896, 1908, and again for an election during World War I. The anti-evolution statutes remained effective until long after April 12, 1961, when the Soviet cosmonaut Yuri Gagarin became the first man to fly in space.

80. A previous Arkansas statute prohibiting the teaching of evolution in its public schools was struck down by the U.S. Supreme Court. Epperson v. Arkansas, 393 U.S. 97, 1968. The new law tried to avoid the problem with the First Amendment by using the words "scientific creation." However, the federal district court found that the alleged "scientific creation" was not science and held that the statute was unconstitutional. McLean v. Arkansas Bd. of Educ., 529 F. Supp. 1255 (E.D. Ark. 1982).

of such alleged "scientific creationism" was unconstitutional.[81] However, in 1998, a professor of biology at Louisville University, Kentucky, editor of the *American Biology Teacher,* stated that such "Creationists are more powerful than ever. The evolution-creation controversy may not be charged today as it was when Scopes was tried."[82] (He did not mention Scopes II.) As of the dawn of the third millennium, we have experienced rather widespread civil disobedience (called "nullification") under some state laws and cultures, which prohibit a number of bioethical concepts and compromise the teaching of evolution and other scientific concepts.[83] Nevertheless, Harvard biologist Mayr quoted Dobzhansky who stated, "Our basic understanding of the world is now remarkably robust . . . nothing in biology makes sense except in the light of evolution."[84]

The Boy Scout law states, among other things, that a member must be "reverent." A boy named Dale who attained the highest rank (Eagle) became a volunteer assistant of a troop in Oregon when he was a freshman at college. He became an atheist since studying evolution in high school, and "was given

---

81. Wallaw v. Jaffree, 472 U.S. 112, 1987. Statements made by religions against the literal teaching of Biblical (Genesis) creationism include those of The General Assembly of United Presbyterian Churches in the United States (in 1982), the American Jewish Congress, the Central Conference of American Rabbis, the General Convention of the Episcopal Church, the Lexington Alliance of Religious Leaders, the Lutheran World Federation, the Unitarian Universalist Association, and the United Methodist Church. MATSUMURA, VOICES FOR EVOLUTION (1995).

82. ORANGE COUNTY REGISTER, Jan. 13, 1998, at News 8. In 1999, a school board in Kansas voted to exclude evolution from the curriculum and statewide tests in science for high school students (ORANGE COUNTY REGISTER, Aug. 12, 1999, at News 23), whereas a school board in West Virginia defeated a vote "that would lift a ban on teaching the biblical story of creation." ORANGE COUNTY REGISTER, Dec. 18, 1999, at News 17.

83. MAYR, THIS IS BIOLOGY 83 (1997). "Some of the details of how evolution occurs are still being investigated. But scientists continue to debate only the particular mechanisms that result in evolution, not the overall validity of evolution as the explanation of life's history." NAT'L ACAD. OF SCIENCES USA, TEACHING ABOUT EVOLUTION AND THE NATURE OF SCIENCE 4 (1998).

84. *Id.*, p. 47: "It must be said at the outset that no biologist *of any stature* whatsoever denies in general the place of evolutionary theory as the central explanation of the form and origins of living things." *See also* PLOTKIN, DARWIN MACHINES AND THE NATURE OF KNOWLEDGE 38.

a week to find God." It was argued that because the Supreme Court upheld the Boy Scouts' ban on homosexuals in 2000, it could extend their right to exclude atheists.[85] However, while remaining in college, he may learn that Darwin, the founder of evolution, specifically denied that he was an atheist.

In a field as controversial as evolutionary biology, the basic conceptual framework has turned out to be remarkably robust—a key matter for all bioethicists. Attempts "to invalidate Darwinism (and there have been hundreds) have proven unsuccessful, which is also true for most other areas of biology."[86] For example, the 98.6-degree body temperature exists in almost all birds and mammals, each with presumably very different evolutionary histories. However, it has been noted, "There is no simple answer to why our bodies maintain a constant temperature whether we live in the Arctic or the Sahara, why that temperature is 98.6 degrees, or why most mammals and birds have approximately the same temperature."[87]

Mammals have existed for 250 million years, but were dominant for only the last 50 million years, at the beginning of the geological "Age of Mammals."[88] About 20 million years ago, the first primate tailless apes evolved.[89] As noted in Chapter 1, many now approach extinction because of human predation, despite the significant work of Louis Leaky's "three angels" who made friends with them in the wild: Jane Goodall (chimpanzees in Africa), Dian Fossey (gorillas in fAfrica), and Mrs. Biruto Goldikas (orangutans in

---

85. Boy Scouts of America v. Dale, 530 U.S. 640 (2000). The late Chief Justice Rehnquist held for the majority that a group cannot be forced to include someone who violates its members' rights. He wrote: "[I]f the presence of that person affects in a significant way the group's ability to advocate public or private viewpoints . . . Dale's presence in the Boy Scouts would, at the very least, force the organization to send a message, both to the youth members and the world, that the Boy Scouts accepts homosexual conduct as a legitimate form of behavior."

86. *Id.*, MAYR, p. 41.

87. SEGSE, MATTER OF DEGREES xviii and 9 (2002). It states: "According to the physiologist Knut Schmidt-Nielsen, the difference in cooling mechanisms between aquatic and terrestrial animals lies in the relative locations of the insulator and the skin, the organ through which heat is dissipated to the outside." *Id.*, KNUT SCHMIDT-NIELSEN, ANIMAL PHYSIOLOGY, 5th ed., p. 21.

88. WOOD, THE CALL OF DISTANT MAMMOTHS 47 (1997).

89. *Id.*, at 75.

Borneo).[90] Like humans, apes take a longer time to mature. It has been said, "If life is likely to last longer, it is worth taking the risk of growing for a longer time."[91] Peter Singer, the Australian who became a professor at Princeton, wrote *The Great Ape Project,* which would grant the other great apes, as close as possible, civil rights equivalent to those enjoyed by the human great ape.[92] (Certain biologists have argued that the gorilla and chimpanzee should be classified in the genus *Homo.*)[93] "Genetic sequencing in the latter part of the twentieth century proves how similar we are not only to apes, but to frogs, sea urchins, fish, and even to yeast cells. I think one can say the realization of this communality has been the greatest insight of the last twenty years in biology."[94]

Chimpanzees share more than 98.7 percent of their genetic makeup with humans, "which means that they are closer biologically to humans than they are to monkeys."[95] A more recent study in the April 17, 2002, *Science* found that chimpanzees and humans share about 98.7 percent of the same genes and have very similar body tissues. Yet, it has recently been "found that 18 recombination hotspots revealed in the human genome were absent from the chimpanzee genome."[96] Although the chimp's tissue and blood resembled that of humans, the gene activity that develops in the human brain is about

---

90. Karesh, Appointment at the Ends of the World 299 (1999): "Farmers have overrun thousands of acres in Congo's oldest national park, the latest threat to more than half the world's 700 remaining mountain gorillas." Orange County Register, July 16, 2004, at News 27.

91. Hardy, Mother Nature 284 (1999).

92. *See* Dawkins, A Devil's Chaplain (2003): "Ask yourself why this seems so immediately ridiculous; the harder you think, the less ridiculous it seems." (p. 5).

93. Easteal, Collet & Betty, The Mammalian Molecular Clock. The current sequence is *Homo habilis, Homo erectus,* and *Homo sapiens.*

94. Gene, A Matter of Degrees 39 (2002).

95. The Great Apes 13 (1999). "The evolutionary precursors of modern apes and people diverged from ancient monkeys between 29 million and 34.5 million years ago, a new genetic analysis concludes. This evolutionary parting of the ways had previously been placed at between 23 and 25 million years ago." Science News, Jan. 1, 2005, at 13, *citing Proceedings of the National Academy of Sciences,* Dec. 7, 2004.

96. Jorbe, *Where We're Hot, They're Not*, Science, April 1, 2005, at 61 (citing Winckler, et al.).

twice the size of a chimp's.[97] Professor Jared Diamond, an evolutionary biologist at the University of California at Los Angeles (UCLA), classified humans as the "third chimpanzee."[98] Although the apes are our "distant cousins," what makes humans biologically unique is the "possession of a genetically assured potentiality for the acquisition of culture and language."[99]

The great ape family has only four living species: orangutans, gorillas, common chimpanzees, and humans. All but the last are becoming extinct even though it was in apish societies that "we took our first steps toward human intelligence" and violence.[100] It has recently been written:

> Since about 2 million years ago, when the brains of prehumans were literally pint-sized (with a capacity of roughly half a liter), the size of the pre-human and human brain has been growing at an average rate of about one cubic centimeter per 2,000 years. That's a snail's pace by the standards of recorded history, but it's meteoric in evolutionary terms.[101]

---

97. ORANGE COUNTY REGISTER, April 17, 2002.
98. DIAMOND, THE THIRD CHIMPANZEE (1992).

> Humans and chimps are closely related; . . . Chomsky's argument that human language ability is innate makes language part of the bread on top of the similarity sandwich: a specialized human adaptation. But recognizing that chimps and other apes do not share our system of language does not imply that other species might not have their own specialized system of communication—part of the bread on their own sandwiches.

WYNNE, DO ANIMALS THINK? 127 (2004).

99. DOBZHANSKY & BOESIGER, HUMAN CULTURE: A MOMENT IN EVOLUTION (1983). "Sustained syllogistic reasoning, the capacity to state a major premise, then a minor premise, followed by a deductive conclusion, is what our species alone can do." Neurobiologist GAZZANIGA, THE MIND'S PAST 152 (1998).

100. JOLLY, LUCY'S LEGACY 4, 76-7. "A fossil discovery in Africa, a nearly complete skull, dated between 6 million and 7 million years old, has some ape-like features, but the shape of the teeth and lower face suggests human ancestry." ORANGE COUNTY REGISTER, Dec. 20, 2002, at News 30, *citing* the research magazine SCIENCE published that day.

101. DAWKINS, THE SELFISH GENE 13: "We are not alone." In 2005, the bones of a human relative about 3 feet tall were found on the Indonesian island of Flores. It lived just 18,000 years ago, weighed about 55 pounds, but had a "startlingly small brain [with] volume at less than a third of a modern human's. Hobbit had by far the smallest brain of any member of the genus *Homo*. It was small even for a chimpanzee." Morwood et al., NATIONAL GEOGRAPHIC, April 2005, at 9.

The average chimpanzee has a brain of about 330 cubic centimeters; but the average human's brain is about four times larger[102] and with more neural cell division. Although the brains of whales and elephants are still greater than those in humans, they exhibit relationships between factors of intelligence and brain volume.[103]

Nevertheless, the deeper meaning remains unknown to those people who have not yet caught up with the evolution that scientists confirmed. Epistemology deals with what we know and how we know it. In his volume *The Selfish Gene,* Dawkins pointed out, "Philosophy and the subjects known as 'humanities' are still taught almost as if Darwin had never lived."[104] Nevertheless, bioethicists will have to recognize it with respect to many hospital patients. In articles on bioethics, medicine, and the law, I find that discussion of evolution is often ignored. Unlike patients or their surrogates, there may be no desire to discuss knowledge concerning the derivation of lives in the universe of elements and compounds to which they will return. Professor Stephen Gould noted, "I don't think any other ideological revolution in the history of science has so strongly or directly impacted our view of our own meaning and purpose."[104] When discussing evolution, bioethicists should remember that Darwin never completely denied the existence of God.[106]

102. Semendeferi & Damasia (2000).

103. Wickett, Vernon, & Lee (2000), pp. 1095-1122:

> Mammals have larger *cortices* for specialization and a six-layer cortical sheet called the *neocortex*. Charles Darwin had written that "certain higher mental powers" might be "the result of the continued use of a perfect language."

Darwin, The Descent of Man, and Selection in Relation to Sex, 2d ed. (1874).

104. Orange County Register, June 25, 1998, at News 8. However, in 1871, Darwin wrote: "At all times throughout the world tribes have supplanted other tribes; and as morality is one important element in their success, the standard of morality and the number of well-endowed men will thus everywhere tend to rise and increase." Darwin, The Descent of Man, and Selection in Relation to Sex 166, 2 vols. (1871). *See also* Gould, The Structure of Evolutionary Theory (2002) and Wilson, Darwin's Cathedral: Evolution, Religion, and the Nature of Society (2002).

105. *See* Ward, God, Chance and Necessity, One World 93 (1996). Gould argued that his punctuational changes in 5,000 to 10,000 years are as normal in nature as is gradual, linear change.

106. Keynes, Darwin, His Daughter and Human Evolution 308 (2002). Darwin wrote, "I think that generally, and more and more as I grow older, but not always, that agnostic would be the most correct description of my state of mind."

As more individuals are produced than can possibly survive, there must in every case be a struggle for existence, either one individual with another of the same species, or with the individuals of distinct species, or with the physical conditions of life . . . We may feel sure that any variation in the least degree injurious would be rigidly destroyed. This preservation of favorable variation and the rejection of injurious variations, I call "Natural Selection."

Darwin, *Origin of Species* (1859)

(c) Evolutionary Advances by the Third Millennium

In *Science and Religion* (2003, edited by John Haught), the neo-Darwinist Francisco Ayala, professor in the department of ecology and evolutionary biology at the University of California Irvine (UCI), posed the scientific view that "Natural selection does not strive to produce predetermined kinds of organisms, but only organisms that are adapted to their present environments." This depends on which variations happen to be present and turns on the random process of mutation, even though most mutations are disadvantageous. A feature of an organism that increases its reproductive fitness will be selectively favored and such mutations accumulate over the generations. Given enough generations, these mutations will extend to all the members of the population.[107] Professor Ayala's 2003 article, "Intelligent Design: The Original Design," ended with a statement to which scientists agree, namely:

---

107. Science and Religion 21-31 (2000). In *Evolution and Molecular Biology* (1998, published by the Vatican), Francisco Ayala (once a Catholic priest in Spain) cited the 1937 publication of *Genetics and the Origin of Species* by his former associate, Professor Theodosius Dobzhansky, who "advanced a reasonably comprehensive account of the evolutionary process in genetic terms effectively combining Darwinian natural selection and Mendelian genetics." *Id.* at 26. It dealt with mutation as the origin of hereditary variation, the role of chromosomal rearrangements, variation in natural populations, and the concept of species as natural units. Dobzhansky expounded human nature as two dimensions: the biological, shared with the rest of life, and the cultural, exclusive to humans. Encyclopedia of Evolution, Vol. 1 (2002), p. 286. (See also Chapter 15.)

> *Organs exhibit design, but it is not "intelligent design" imposed by God as a supreme engineer,* but the result of natural selection promoting the adaptation of organisms to their environment (p. 18). *Natural Selection must not be understood in the sense of the "absolute creation that traditional Christian theology predicates"* of the divine act of which the universe was brought into being *ex nihilo* (p. 19).[108] (Emphasis added.)

John Haught, a professor of theology at Georgetown University, expressed his belief "that a religiously hierarchical vision can, after all, form a plausible alliance with the scientific atomizing and historical-horizontal understandings of the universe that have apparently leveled the cosmic hierarchy of traditional religions." Nevertheless, he concluded, "There is no obligation on the part of religion to accept the claim that the natural sciences—even with their atomizing and historicizing tendencies—have in the slightest way logically destroyed the discontinuity proper to religious hierarchy."[109]

A reservation of any acceptance of the well-explained method for examing evolutionary fossils was due to a problem admitted at discovery. In *Origin of Species* (1859), Darwin presented this dilemma, namely, the Precambrian fossil record of more than half a billion years ago remained unknown. "Why we do not find rich fossiliferous deposits belonging to these assumed earliest periods before the Cambrian system, I can give no satisfactory answer. The case at present must remain inexplicable; and may be truly urged as a valid argument against the views here entertained." This argument remained sufficient to challenge his views until the *Solution to Darwin's Dilemma* arrived prior to the final decades of the twentieth century. In the volume, *Variation*

---

108. Professor Ayala's 2003 article published by the Center for Theology & Natural Sciences significantly expands that which was published five years earlier by the Vatican. Many scientists are actually "agnostic," which means "One who holds that the ultimate cause (God) and the essential nature of things are unknown or unknowable, or that human knowledge is limited to experience." *(The New Century Dictionary)* In his article, published by the Vatican, Professor Ayala stated, "Therefore, as far as science is concerned, the origin of the universe will remain forever a mystery." (1998, p. 113) As a scientific explanation, this prohibits agnostic scientists from a claim of atheism.

109. *Id.* at 119.

and Evolution in Plants and Microorganisms (published by the National Academy of Science in 2000), William Schoff pointed out that the period from the mid-1960s to the end of the twentieth century culminated "in discovery of the oldest fossils known, petrified cellular microbes nearly 3,500 million years old and more than three-quarters the age of the Earth."[110]

In the more recent millions of years, humans, dogs, whales, and bats, all with similar skeletons, have developed: "Bone by bone, can be observed in every part of the body, including the limbs: yet a person writes, a dog runs, a whale swims, and a bat flies with structures built of the same bones. Scientists call such structures homologous and have concurred that they are best explained by common descent."[111] While species are reproductively isolated from other such groups, the natural population's genetic changes, originating in single individuals, can spread by natural selection to all members of the species. Such changes occurred within a lineage (called *anagenesis*) in humans during the past four million years. One such example is the more than tripled size of the brain.[112] As a result, cultural inheritance for humans—

---

110. *Id., citing* SCHOPF, MICROFOSSILS OF THE EARLY ARCHEAN APEX CHART: NEW EVIDENCE OF THE ANTIQUITY OF LIFE SCIENCE 640-46 (1993) *and* SCHOPF, CRADLE OF LIFE, THE DISCOVERY OF EARTH'S EARLIEST FOSSILS (1999). In 1999, Darwin was being considered as Britain's man of the millennium. However, William Shakespeare was picked as superstar. Professor Colin Blakemore confessed to voting twice for Darwin, stating, "In the end, Darwin will be seen to have told us more about why we are the way we are." L.A. TIMES, Jan. 3, 1999.

> The hypothesis of the molecular clock holds that the number of amino acid (or nucleotide) replacements in any given protein (or DNA) sequence changes linearly with time. Rates used for extrapolation have to be first calibrated by reference to absolute dates drawn from the fossil record. Molecular clock projections have ostensibly pushed back fossil-based dates in many studies. Close approximation to the molecular clock premise should be a necessary condition.

Rodrigues-Telles, Tarrio & Ayala, PNAS, June 11, 2002, pp. 8112-15.

111. AYALA, AN EVOLVING DIALOGUE: THEOLOGICAL & SCIENTIFIC PERSPECTIVES ON EVOLUTION 24 (2001). He claimed to be nonagnostic and received a scientific award from President Bush in 2002.

112. *Id.* at 41.

> Monkeys employ rule-like strategies for promoting the welfare of a group, including maintaining peace, observing boundaries, and sharing food. And they can abide by these rules without necessarily understanding

unlike animals that cannot transmit a "social memory" to any large extent—includes distinct human behavioral traits such as abstract thinking and reasoning, more extensive toolmaking, science, literature and art, ethics and religion, and professions, like law and medicine. While genetic information is encoded in DNA from one generation to the next by means of the sex cells, "cultural inheritance" is based on transmission of information by a teaching-learning process, which is in principle independent of biological parentage.[113] This is expressed by ontology, the science of reality following the investigation of nature.

(d) A Key to the Basis of Significant Contingencies (Not Design) in Evolution

The Cambria explosion some 530 million years ago began with a change of eukaryotic organisms into animals, some of whom were derived over the past half a billion years. However, evidence that the majority of them became extinguished without leaving any ancestors was not correctly determined until the latter part of the twentieth century.

The multicellular fauna of the Burgess Shale in the Rocky Mountains of British Columbia, Canada, was discovered in 1909 by American paleontologist Charles Doolittle Walcott, secretary of the Smithsonian Institution in Washington, D.C. He viewed this fauna as the ancestral versions of the Homo species. The material of Burgess Shale was reexamined by Harry Whittington, along with other paleontologists, experts on fossil arthropods whose writings (between 1975 and 1978) set forth the discovery that most of these soft-bodied fauna were "neither an arthropod, nor anything else ever known be-

---

> them. Humans are a different kind of animal: We can consciously evaluate whether behavior is right or wrong, but we tend to do so depending on the conventions of our society.

Professor Hauser (Harvard University), *Morals, Apes, and Us*, DISCOVER, February 2000.

113. *Ontology Studies* (International Ontology Congress), V. Gomez Pin. ed. No. 1-2, San Sebastian, Spain; Prof. Ayala, *An Evolutionary View of Human Nature: From Biology to Ontology* 292 (2001).

fore."[114] During the 1980s, the same former genera were discovered all over the world[115] and confirmed that they were not ancestors for modern life.[116] As stated by Steven Gould, "If life works more by tracking than by climbing up a ladder of progress, then contingency should reign."[117]

> *The ultimate evolutionary victory, on the theistic hypothesis, does not go to the most ruthless exterminators and most fecund replicators. It will go to beings who learn to cooperate in creating and contemplating values of many sorts, to care for their environment and shape it to greater perfection. It will go to creatures who can found cultures in which scientific understanding, artistic achievement, and religious celebration of being can flourish.*
>
> Keith Ward, *God, Chance & Necessity: One World* (1996)

---

114. GOULD, WONDERFUL LIFE: THE BURGESS SHALE AND THE NATURE HISTORY 172 (1989).

> Paleontology has no Nobel prizes—though I would, without hesitation, award the first to Whittington, Briggs, and Conway Morris as a trio. And, as the old clichés go, you can't fry an egg with your new view of life, or get on the subway, unless you also have a token. (I don't think it even gets you any frequent-flyer miles, though almost everything else does.) *Id.* at 84.

115. P. 224: "The Chinese fauna is half a world away from British Columbia—thus establishing the global nature of the Burgess phenomenon." *Id.* at 226, *citing* Hou Xian-guang & Sun Wei-guo, Discovery of Chengjiang Fauna at Meishucun, Jinning, Yunnan (1988) (in Chinese). ACTA PALAEONTOLOGICA SINICA 27:1–12 and 1987 reports. *Id.* at 226.

116. Life has suffered some astounding mass extinctions since the Cambrian age—the Permian debacle may have wiped out 95 percent or more of all marine species—yet the Burgess phenomenon of explosive disparity never occurred again. Life did rediversify quickly after the Permian extinction, but no new phyla arose. The recolonizers of a depleted earth remained within the strictures of previous anatomical designs. Yet, the early Cambrian and post-Permian worlds were crucially different. Five percent may not be a high rate of survivorship, but no mode of life, no basic ecology, was entirely wiped out by the Permian debacle.

Id. at 229.

117. *Id.* at 300.

Nothing in biology makes sense except in the light of evolution.
Theodosius Dobzhansky, 1971

(e) Law and Biology

Many human systems are derived from the theory of evolution. During the nineteenth century, Karl Marx stated his theory that "Darwin's book is very important and serves me as a basis in natural selection for the class struggle in history."[118] Winston Churchill added, "Men occasionally stumble over the truth, but most of them pick themselves up and hurry off as if nothing had happened." On the judicial side, Justice Oliver Wendell Holmes, Jr., compared theories of the evolution of the common law to evolution as used in biological systems.[119] Furthermore, a late twentieth century lawyer, professor John Beckstrom, advocated initial ascertainment of what biology dictated and then writing that into law.[120]

In 1997, another lawyer (and nonscientist), Yale professor Donald Elliot, wrote, "At most, biology should be followed or accepted as a rebuttable presumption." For him "Law is useful to societies precisely to compensate for those *areas in which biology does not suit us* to live in our current environment."[121] (Emphasis added.) The 1998 Vatican publication of Francisco Ayala (noted previously) also contains this key to the intelligent writing on bioethics and the law: "But once science has had its say, there remains much about reality that is of interest, questions of value and meaning that are forever beyond science's scope" (p. 116). This becomes applicable with respect to the possible dismal future for all humans because of our destruction of the earth's environment (see Chapter 1 and Appendix A), together with a lack of control over the probable geometric increase in population (see Chapter 7). However, as Donald Elliot remarked, "Promise remains to be brought to fruition." He noted, "Legal decision makers, such as judges, are only beginning

118. *Quoted in* RACKELS, CREATED FROM ANIMALS 2 (1990).

119. Holmes, *Law in Science and Science in Law*, 12 HARV. L. REV. 443 (1899). *See also* E. Donald Elliott, *Holmes and Evolution: Legal Process as Artificial Intelligence*, 13 J. LEGAL STUD. 113 (1984), and Elliott, *The Evolutionary Tradition in Jurisprudence*, 85 COLUM. L. REV. 38 (1985).

120. Berkstrom, *Evolutionary Jurisprudence*, 85 Colum. L. Rev. 38, 1986.

121. *See* E. Donald Elliot, *Law and Biology: A New Synthesis?*, 41 ST. LOUIS U. L.J. 595, 606-07 (1997).

to be exposed to law and biology."[122] In the decades after 1987, forensic DNA testing "demonstrated scientific accuracy quickly led jurisdictions to accept expert testimony regarding DNA matches between suspects and crime scene evidence,"[123] and "all fifty states currently have legislation requiring DNA profiles of certain categories of individuals." Although what classes of offenders are to be incorporated varies widely from state to state,[124] this normally constitutes a nonscientific bioethical aspect.[125] Unfortunately, like certain members of hospital ethics committees, judicial decisions often depend upon the testimony of alleged experts, and not on the scientific education of judges who may not have adequate scientific backgrounds (see Chapter 5).

Darwin cited many scientists and noted that several of these writers had said: "I speak of natural selection as an active power or Diety . . . but I mean by Nature, only the aggregate action and product of many laws, and by laws the sequence of events as ascertained by us. With a little familiarity such superficial objections will be forgotten."[126] In a similar manner, the subse-

122. *Id.* at 610.

123. MACKLIN, REGULATION OF BIOETHICS 102 (Spring 2005).

124. *Id.* at 103. *See* participating states, at http://www.fbi.gov/hq/lab/codis/partstates.htm.

> Congress enacted the DNA Analysis Backlog Elimination Act of 2000, 2 U.S.C. § 14135, which authorized the Attorney General to issue grants to the states for the purpose of carrying out, for inclusion in the Combined DNA Index System of the Federal Bureau of Investigation, DNA analyses of samples taken from individuals convicted of a qualifying state offense.

42 U.S.C. § 14135(a)(1).

125. *Id.* at 115. LAFAVE, SEARCH AND SEIZURE § 9.7(b) at 709 (4th ed. 2004); Kaye, *The Constitutionality of DNA Sampling on Arrest*, 10 CORNELL J. L. & PUB. POL'Y 455-509 (2001); Kaye, *Two Fallacies About DNA Data Banks for Law Enforcement*, 67 BROOK. L. REV. 179-202 (2001); Kaye & Smith, *DNA Identification Databases: Legality, Legitimacy, and the Case for Population-wide Coverage,* WIS. L. REV. (2003).

126. ORIGIN OF SPECIES 94-95 (1909). Nevertheless, it has been shown that about 70 percent of high schools omitted the teaching of evolution from the 1920s through the 1960s. LARSON, TRIAL AND ERROR: THE AMERICAN CONTROVERSY OVER CREATION AND EVOLUTION (1989). Even in 2005 the school boards in Odessa, Texas; Frankenmuth, Mich.; and elsewhere voted to assign the Bible class to its high-school curricula. ORANGE COUNTY REGISTER, April 28, 2005, at News 21. *See also* DECKMAN, SCHOOL BOARD BATTLES: THE CHRISTIAN RIGHT IN LOCAL POLITICS 66 (2004).

quent chapters of this volume also discuss the work of many scientists and medical statisticians, 90 percent of whom have written near the opening of the third millennium with approaches recommended for the twenty-first century. It is also believed that future familiarity with these sources makes it at least more probable, and, as Darwin stated, "superficial objections will be forgotten":

> In the *Principia* Newton analyzed the motion of bodies, including projectiles, pendulums, and the planets . . . At a single stroke, Newton severed the link between the physical world and superstition—even the heavens were claimed as part of the rational universe. The *Principia* is generally held to be the greatest and most influential scientific book ever written (although I would opt for *The Origin of Species).* Its publication, more than any other event, marked the end of medieval dogma and launched science into the Age of Reason.[127]

The body of scientists is trained to avoid and organized to resist every form of persuasion but the fact.
J. Bronowski, *Science and Human Values* (Harper, 1965), p. 59

A new scientific truth does not triumph by convincing its opponents and making them see the light, but rather because its opponents eventually die out, and a new generation grows up that is familiar with it.
Max Planck (1858-1947)

Reality is a shimmering presence of infinite planes, aluminous labyrinth of the active now connecting past and future.
Mae-Wan Ho, *The Rainbow and the Worm* (1993)

Twenty-first Century Approaches

1. Teaching something about both the Big Bang and evolution in public high schools in several states largely began at the dawn of the third millennium, which was more than four decades following the launch-

127. Ward & Brownlee, The Life and Death of Planet Earth 120 (2002).

ing of *Sputnik,* the Russian satellite. More consideration should be given by public schools to their meaning in terms of bioethics, medicine, and the law.

2. Many students still face serious problems in their public school curricula. During the twenty-first century, the teaching of biology also should be revised to address the subjects of the Big Bang and scientific evolution, together with their sequences.

It is not the strongest of the species that survive, nor the most intelligent, but the one most responsive to change.

Charles Darwin

Religion illustrates, *perhaps better than any other aspect of human behavior, the way in which* basic biological capabilities *can be built into a vast cultural superstructure, influencing, indeed sometimes virtually* determining, the natures of groups *and individuals. (Emphasis added.)*

Ehrlich, *Human Natures* (2000), p. 221

Science without religion is lame. Religion without science is blind.

Albert Einstein (1879-1955)

I believe that all *of the* great world religions have biological roots, *that religion has such great power because it meets a very deep-seated biological need. (Emphasis added.)*

Dr. Arnold Relman

Religious Sources, Their Restrictions, and Their Possible Bioethical Standards

1. ***Certain Differences between Religious and Bioethical Standards***
   - **(a) Monotheism and Its Maintenance in Modern Times**
   - **(b) Consideration of Religious and Ethical Standards under the U.S. Constitution**
2. ***A Brief Look at Several of America's Diverse Religions Together with Their Restrictions on Possible Medical and Ethical Sources***
3. ***The "Wall of Separation" and Legal Bans on Teaching Religion in Public Schools***

1. Certain Differences between Religious and Bioethical Standards

A brief look into religions, science, reason, and the law is often necessary to focus on potential sources for bioethics on the practice of medicine and the law. As stated in 1993 in Karen Armstrong's *A History of God:* "Indeed, there is a case for arguing that *Homo sapiens* [rational man] is also *Homo religious* [religious man]. Men and women started to worship gods as soon as they became recognizably human; they created religions at the same time as they created works of art" (p. xix).

The seventeenth century English philosopher Thomas Hobbes noted that man possesses essentially a selfish nature and will inevitably bring about anarchy unless he adhered to a "social contract." That is, mankind must submit to a set of laws. In Charles Darwin's *Origin of Species,* he concluded in 1859:

> I see no good reason why the views given in this volume should shock the religious feelings of any one . . . Why, it may be asked, until recently did nearly all the most eminent living naturalists and geologists disbelieve in the mutability of species . . . I look with confidence to the future, to young and rising naturalists, who will be able to view both sides of the question with impartiality. Whoever is led to believe that species are mutable will do good service by conscientiously expressing his conviction; for thus only can the load of prejudice by which this subject is overwhelmed be removed.[1] (p. 520)

In December of 2003, *Sky & Telescope* quoted a statement of the science historian Owen Gingerich at the Harvard-Smithsonian Center for Astrophysics: "A commonsense and satisfying interpretation of our world suggests the designing hand of a super intelligence." However, George Coyne, director of

---

1. A current religious writer has condemned bioethics since the year 2000:

> An absolute base, an immutable foundation for ethical behavior can be found only in God's laws. . . . Biomedical activities in the new millennium years appear to have gone awry. Efforts by upright bioethicists seem ineffectual in filling this empty lacuna, because many bioscientists look upon ethical standards as a farrago of individual choices selected cafeteria-style to suit the project at hand.

FICARRA, BIOETHICS' RISE, DECLINE, AND FALL, Intro. XIX (2002).

the Vatican Observatory, replied, "I disagree. Dragging science in to establish the basis of religious belief on purely rationalistic grounds is an idolatry of science. We sort of latch on to God, especially if we lack sound scientific explanations for our discoveries. I have never sought to prove God's existence through anything like a scientific process or as the logical conclusion of rational thought. Instead, I believe in the God of Creation because doing so enriches my life." The question remains: What uniform standards should be used? The sources for bioethics standards from both religion and the law, or reason, are varied. There appears to be an increase in the tendency of public law toward reason, and in private law toward increased autonomy (see Chapter 5). The effort toward development standards by different beliefs has an accountable history reaching back some 3,000 years.

In view of the need for rationale in the law to govern a diverse society, a means must exist to examine the extent to which religious or other standards become possible for bioethics. The nearest "science" allowing such a study is anthropology.[2] During the past century, anthropologists have worked considerably toward providing a basis for theories of religions by tracing their origins from primitive to more advanced societies. One basic discovery was that societies were often actually using the "best reasons then available" to them. A progression toward a view called "highest reason" would not necessarily be different. For example, as one ancient religious book states, "It is inspired of God and beneficial for teaching, for reproving, for setting things straight, for disciplining in righteousness so that everyone who belongs to God may be proficient, equipped for every good work."[3] A modern book concluded, "Because all the world's religious traditions share the same essential purpose, we must maintain respect and harmony among them."[4] The

2. "Anthropologists compare the facts about various religions in order to attempt to find a theory in much the same manner as scientists proceed. However, results are not conclusive with respect to any one theory." *See* PALS, SEVEN THEORIES OF RELIGION (1996). Clark & Clark's *Measuring the Cosmos* (2004) began with ancient Greek science and cosmology and the first serious measurements that showed "that battles between science and religion as well as the pursuit of research funding are problems that extend all the way back to the Greeks." *See* Birriel, *Struggling to Know the Universe*, SKY & TELESCOPE, February 2004, at 110.

3. 2 *Timothy* 3:16. Polytheism was earlier condemned as an "abomination." *Leviticus* 18:26.

4. W. TEASDALE, THE MYSTIC HEART: DISCOVERING A UNIVERSAL SPIRITUALITY IN THE WORLD'S RELIGIONS (1999); H.G. KOENIG, THE HEALING POWER OF FAITH (1999).

American Psychiatric Association, however, once classified "strong religious belief" as a possible "mental disorder."[5]

As a religious founder, Akhenaten stands at the head of a lineage very different from his predecessors', one represented after him by the Moses of legend, and later by Buddha, Jesus, and Mohammed.

Jan Assman, *The Mind of Egypt*, 2001, p. 216

The Three Monotheistic religious communities we call Judaism, Christianity, and Islam now constitute the single largest block of organized believers on earth.

F. F. Peters, *The Monotheists* (2003)

Religious intolerance was inevitably born with the belief in one God.

Sigmund Freud, *Moses and Monotheism*

(a) Monotheism and Its Maintenance in Modern Times

Monotheism occurred in early Egypt by a very singular pharaoh who believed in "the divine and sole creator of the universe." He developed and "continued in a place unsullied by the old religions."[6] He was "not only the world's first idealist and the world's first individual, but also the world's first monotheist, and the first prophet of internationalism—the most remarkable figure of the Ancient World before the Hebrews."[7] He wrote this hymn containing certain resemblances to the Bible's Genesis:

---

5. AM. PSYCH. ASS'N, DIAGNOSTIC AND STATISTICAL MANUAL OF MENTAL DISORDERS: DSM-IV, 4th ed. (1994).

6. JAMES, TUTANKHAMEN 23 (2000):

> Amenhotep IV had built a new royal capital at Amarna and had tried to abolish the entire religious system, replacing it with sun worship and erasing all other gods' names throughout the country.

FLETCHER, THE SEARCH FOR NEFERTITI 46 (2004).

7. *Id.* Fletcher, at 62 (citing Breasted in ALDRED, 1968, p. 13). "For the history of religion, the Old Testament is significant because it marks the threshold between a Palestinian national cultic religion and what has been called ethical monotheism." With the Old Testament, religion begins to ask about its universally valid truth, and arrives at a radically transcendent view of God, in which religion is inextricably bound up with ethics. LEVIN, THE OLD TESTAMENT 2 (2005).

Splendid you rise in heaven's lightland.
O living Aten [the sun] creator of life!
How many are your deeds,
Though hidden from sight,
O sole God beside whom there is none!
You made the earth as you wished, you alone,
All peoples, herds, and flocks;
You set up every man in his place,
You supply their needs;
Everyone has his food,
His lifetime is counted.
Their tongues differ in speech,
Their characters likewise;
Their skins are distinct,
For you distinguished the people . . .
You are in my heart,
There is no other who knows you,
Only your son, Nefer-khepru-re (Akhenaten), Sole one of Re,
Whom you have taught your ways and might . . .

This hymn (resembling the Bible's Psalm 104) was composed by the pharaoh, Akhenaten (1387–1366 BC), who made the first such religious revolution in recorded history. He embraced a "single god" during the second half of the second millennium BC while the Israelites were still in Egypt. After the Exodus of the Jews, the first commandment, "Thou shall have no other gods before me," appeared, indicating that polytheism was no longer acceptable to them. A reason may have been noted that "Moses was fired with the idea of perpetuating the dying sect of the Aton religion, just as Paul was fired with an ambition to establish Christianity"[8]:

8. Could it have been that when the Egyptians would not accept the Aton religion, Moses determined to give it to the Jews? This is no fanciful concept. Again the comparison can be made with Paul. When the Jews would not have the teachings of Christ, Paul took his gospel to the gentiles—an ironic twist of history. DIMONT, JEWS, GOD, AND HISTORY 39 (1962). "Moses" (the Hebrew "Moshe"), derived from the Egyptian *moser*, means "is born." "Few Christians today realize that in the fifth century, Pope Leo the Great had to tell Church members to stop worshipping the sun. The first ostensibly Christian emperor, Constantine, who converted to the new faith at the beginning of the fourth century, was still worshipping the sun god Helios many years later, as coins and other evidence reveal." HARPUR, THE PAGAN CHRIST 8 (2004).

> The great archaeologist Flinders Petri regards Akhenaten, the creator of monotheism, as "the most original thinker that ever lived in Egypt and one of the greatest idealists of the world. No man appears to have made a greater stride to a new standpoint than he did." His motto was "Living in truth." He referred to his beautiful wife Nefertiti as "Sweet Love" and he maintained family values. Like several subsequent similar revolutions, that of Akhenaten's did not last. The Egyptians reverted to their many gods of former years when the priests forced his successor, Tutankhamon, to abandon the beliefs of Akhenaten.[9]

According to Sigmund Freud, this had an important effect upon Moses:

> The reason for Moses' leaving his country and choosing the Jews as a people in order to realize a new kind of polity based on a new religion and constitution was the failure of the Aton religion. Like Toland at the beginning of the eighteenth century, Freud saw in Moses a princely figure, close to the throne, perhaps the governor of a province such as Goshen, where the Jews settled. Moses was also convinced of the truth of Atonism and too proud to return to orthodoxy. He made the decision not only to emigrate but to found a new nation. For this purpose, he chose the Jews. The situation was favorable for their emigration because the kingdom went through a state of anarchy after Akhenaton's death. Moses led the Jewish tribes out of Egypt, taught them his monotheistic creed (which might have assumed an even more strict and radical form with him than that which Akhenaten had introduced), and gave them laws. (*Id.*, p. 154)

Zoroaster was an ancient Persian prophet who founded the Persians' religion that worshiped the one and only God, Ahura Mazda. Their Bible, the

---

9. The monotheistic revolution of Akhenaten was not only the first but also the most radical and violent eruption of a counterreligion in the history of mankind.

> Immediately after the first publication of the rediscovered inscriptions of Akhenaten, it was realized that he had done something very similar to what memory had ascribed to Moses: he abolished the cults and idols of Egyptian polytheism and established a purely monotheistic worship of a new god of light, whom he called "Aton." Moses was exceedingly important in the land.

Jan Assman, Moses the Egyptian 149 (1997).

Zend-Avesta, or Avesta, and priests believed in Zoroaster's one god, but wanted to sing also to the forces of nature. The Vendidad part of the Avesta gives the laws of the religion. It contains some of the most useful hygienic laws written before the start of modern medicine.[10] The Parsees belong to a Zoroastrian religious group located mainly in Bombay, India. They came to India from Persia in the early years of 700 AD to escape Moslem persecution. The Parsees "believe that the person who does not study is no longer a servant of God." They maintain a school in every temple.[11]

In the postscript to his book, *The Varieties of Religious Experience,* 1902, William James, professor of psychology at Harvard, wrote: "Upholders of the *monotheistic view* will say to such a polytheism (which, by the way, has always been the real religion of common people, and is so still today) that unless there be one all-inclusive God, our guarantee of security is left imperfect. In the Absolute, and *in the Absolute only, all is saved.* If there be different gods, each caring for his part, some portion of some of us might not be covered with divine protection." (Emphasis added.)

> *The belief that,* by cutting ourselves off from the inherited ideas *of our cultures, we can "clean the slate" and make a fresh start,* is as illusory as the hope for *a comprehensive system of theory that is capable of giving us timeless* certainty *and coherence.* (Emphasis added.)
>
> Stephen Toulmin, philosopher of science

---

10. WORLD BOOK ENCYCLOPEDIA. *See also* PARTRIDGE, NEW RELIGIONS (2004); HINTZ, ZOROASTRIANISM 144.

11. *Id.*

> According to Zoroastrian tradition, basic doctrine was rational, anti-ritual and anti-sacrifice, encouraging his followers to come to personal terms with their God. . . . A temple-keeper said "Our basic beliefs are very simple. Choose truth and oppose lies. And always strive for good words, good thoughts and good deeds."

KRIWACZEK, IN SEARCH OF ZOROASTRIAN 229 (2003). Deism required the concept of God by the use of reason, rather than revelation, using the laws of nature (following Newton's *Principia*). David Hume (1711-1776) died the year of Jefferson's Declaration of Independence. Hume wrote his book *Dialogues Concerning Natural Religion* to attack Deism of the Age of Enlightenment on the grounds of human misfortune and fear of death without immortality.

Everyone who is seriously involved in the pursuit of science becomes convinced that a spirit is manifest in the laws of the Universe—a spirit vastly superior to that of man, and one in the face of which we with our modest powers must feel humble. In this way the pursuit of science leads to a religious feeling of a special sort, which is indeed quite different from the religiosity of someone more naive.

Albert Einstein, The Human Side, p. 33

All ambitions are lawful except those which climb upward on the miseries or credulities of mankind.

Joseph Conrad, a British novelist of Polish ancestry

(b) Consideration of Religious and Ethical Standards under the U.S. Constitution

Rule 9.11 of the American Medical Association (AMA) *Code of Medical Ethics* declares that where "particular religious beliefs are to be taken into consideration in the [ethics] committee's recommendations, this fact should be publicized to physicians." It has also been noted that "religion's voice has been prominent in bioethics since the beginning."[12] As stated by Dr. Arnold S. Relman, a longtime editor of the *New England Journal of Medicine,* "All of the great world religions have biological roots."[13]

Others have written:

> Many *people suppose* that human behavior, and any *aspects of human disease* that arise from human nature, are matters *to be dealt with entirely by religion* or sociopolitical action, not by seeking biological causes and remedies.[14] (Emphasis added.)

At the end of the twentieth century, some television evangelists were arguing in favor of the view of religious domination at the time the country was founded. As one historian has shown, however, they "vastly overstated their case." Our U.S. Constitution, for example, is an essentially secular document.[15]

12. Ness & Williams, Why We Get Sick 106 (1995).

13. Park Ridge Center, 1986, editor of the *New England Journal of Medicine* since 1977. Prior thereto he held appointments at Yale, Boston, Pennsylvania, and Harvard Schools of Medicine.

14. Cochran, Hast. Cent. Rept., March-April 2001, p. 43. "In the Bible, God gave some 28 verses for healing, including herbs and food."

No religious group produced a bill of rights. In fact, "much of the thought behind the framing of the American regime was secular or deistic in character—notably, Thomas Jefferson's."[16] The Declaration of Independence he drafted (which preceded the Constitution by a dozen years) invoked the "Creator," which was consistent with the "deism" prevalent in the so-called "Age of Enlightenment." It began by stating that "in the Course of human events, it becomes necessary for one people to . . . assume among the powers of the earth, the equal station to which the *Laws of Nature and of Nature's God* entitle them." Furthermore, "[T]hey are *endowed by their Creator with certain unalienable Rights*." (Emphasis added.) These were references to "Nature" and "its Creator" in a deistic manner, rather than to utilize the essence of traditional religion.[17] Recently, it has been noted that most "Americans in the era of the Revolution were a distinctly unchurched people. The highest estimates for the late eighteenth century make only about 10 percent to 15 percent of the population church members. Americans in 1776 had a long way to go before making themselves strongly Christian or strongly anything else relating to a religious persuasion."[18] Following the American Revolu-

---

15. GLYNN, GOD: THE EVIDENCE 164 (1997); CHERRY, THE DOCTOR WORD 103-72 (1995).

16. Morens, NEJM, April 20, 1999, p. 1222.

17. Glynn, *supra* note 15.

> Deism is a theory about the nature and existence of God. It asserts that God exists and that He created the world. The deist usually proves the existence of God from the order and harmony that exist in the universe. The *deist also tends to reject revelation* as the test of religious truth, *accepting reason instead*. (Emphasis added.)

THE WORLD BOOK ENCYCLOPEDIA.

> Why are the laws of physics what they are? While an unlikely explanation, a deist answer to this question remains tenable.

LINCOLN, UNDERSTANDING THE UNIVERSE 460 (2004).

18. KAMNICK & MOORE, THE GODLESS CONSTITUTION (1996). "Contrary to what modern conservatives claim, America has never been a more religious nation than it is today. Over the past century, church membership has climbed from 25 percent to 65 percent of the American population." GAUSTA ET AL., NEW HISTORICAL ATLAS OF RELIGION IN AMERICA (2001), Fig. 4.16. In 1782 the first printing in English in America of the King James Bible, known as "The Bible of the Revolution," became a fixture in the 19th and 20th century American home. DEAD SEA SCROLLS TO THE FORBIDDEN BOOK—A HISTORY OF THE BIBLE.

tion, both historians and the general public alike noted that "God and Christianity are nowhere to be found in the American Constitution, a reality that infuriated many in 1787, and ratified in 1788 a godless document. Its utter neglect of religion was no oversight; it was apparent to all."[19] Few tried to change this at that time.[20]

After various refinements in committee, Article VI of the U.S. Constitution specifically requires that an oath be taken from federal and state officers to support the U.S. Constitution. It adds, *"No religious test* shall ever be required as a qualification to any office or public trust under the United States."

The antiestablishment clause in the First Amendment (ratified in 1791) was preceded by the debate between Patrick Henry and James Madison over the establishment of the Anglican Church in Virginia and involved religion in the schools. In 1999, this was described by Harvard philosopher John Rawls, as it is still debated (see Chapter 4). Concerning the standards for bioethics he wrote:

> *Henry's argument* for establishment was based on the view that *"Christian knowledge hath a natural tendency to correct the morals* of men, restrain their vices, and preserve the peace of society, which cannot be effected without a competent provision for learned teachers." *Henry did not seem to argue for Christian knowledge as good in itself* but rather as an effective way to achieve basic political values, namely, the good and peaceable conduct of citizens . . . *Madison's objections* to Henry's bill turned largely on *whether religious estab-*

19. *Id.* at 27. However, it has been noted that:

> Many different religious groups—Catholics and Mennonites, Quakers and Jews—established communities and congregations, making the thirteen colonies and the new nation more religiously diverse than any place in Europe. We were already, in John F. Kennedy's phrase, a nation of immigrants.

Baron, The New Americans 2 (2001).

20. Judge William Williams of Connecticut suggested as much in a letter to the *American Mercury*, February 1788, stating: "We the people of the United States, in a firm belief of the being and perfections of one living and true God, the creator and supreme Governor of the world." It was excluded in the final version.

> *lishment was necessary to support orderly civil society. He concluded it was not.* (Emphasis added.)[21]

In his second inaugural address on March 4, 1805, Jefferson stated that he had taken "no occasion to prescribe the religious exercises suited to it: but have left them, as the constitution found them, under the direction and discipline of state or church authorities acknowledged by the several religious societies." Hence, unlike the U.S. Constitution, the preambles of 47 state constitutions refer to "almighty God." However, the nineteenth century Constitutions of two states express "profound reverence for the Supreme Ruler of the *Universe"* and Washington State's Preamble indicates that the people are "grateful to the Supreme Ruler of the *Universe* for our liberties." (Emphasis added.) This is noteworthy because a portion of bioethics falls under state law.

Before and afterward everyone agreed that Jesus was an extraordinary man, regardless of whether or not they belong to a Christian church. For example, a half century following the diests and others who helped to draft the U.S. Constitution, in 1835 David Friedrich Strauss wrote that he was "aware that the essence of Christian faith is perfectly independent of his criticism" and that the cold-blooded way "with which in the course of it, criticism undertakes apparently dangerous operations, be explained solely by the security of the author's conviction that no injury is threatened to the Christian faith." He added, "The boundless store of truth and life which for 18 centuries has been the aliment of humanity, seems irretrievably dissipated; the most sublime leveled with the dust, God divested of his grace, man of his dignity, and the tie between heaven and earth broken."[22]

---

21. RAWLS, THE LAW OF PEOPLES 165 (1999).

22. STRAUSS, THE LIFE OF JESUS, lii. In 1840 Strauss published *The Christian Faith: Its Doctrinal Development and Conflict with Modern Science* and in 1872, *The Old and the New Faith. See also The Life of Jesus Critically Examined,* edited by Peter C. Hodgson, and translated from the fourth German edition in 1972. Similarly, the 19th century Jewish composer Mendelssohn stated, "If a Confucius or Solon were to live among our contemporaries, I could, according to my religion, love and admire the great man without succumbing to the ridiculous desire to convert him. Convert a Confucius or a Solon? What for?" MENDELSSOHN, JERUSALEM AND OTHER JEWISH WRITINGS 113-12 (Jospe trans. and ed.).

The large number of unchurched Protestant people of the last quarter of the eighteenth century changed in both the nineteenth and twentieth centuries. For example, the small earlier amount of Baptist churches rose to very large numbers by the year 1860.[23] The same period also witnessed a dramatic rise in African-American churches.[24] Early in the twentieth century, the three-time Democratic Party candidate for president, William Jennings Bryan, "regarded evolution as a defense of society's weakest members. . . [and] the world was only 6,000 years old." Following World War II, "almost all churches expanded. In the last years of the twentieth century, cooperation between Catholics and Protestants took place."[25] Protestants "have been decisively influenced by the American separation of church and state."[26] (See Section 3 below.) During its eighteenth century revolution, France was first declared a republic in a manner quite different from that in the First Amendment of the U.S. Constitution ratified effective December 15, 1791, stating, "Congress shall make no law respecting an establishment of religion or prohibiting the free exercise thereof" (following the U.S. revolution against the King in England). In a Supreme Court case of the twentieth century, one Justice noted that "while in many contexts the Establishment Clause and the Free Exercise Clause fully complement each other, there are areas in which a doctrinaire reading of the Establishment Clause leads to irreconcilable conflict with the Free Exercise Clause" and another pointed out how "these two clauses may in certain instances overlap" or appear to "conflict."[27] The French Revolution "began as a revolt against a king whom the people could not understand [and like the later Communist Countries] in November 1793 God was formally dethroned."[28]

"In God we trust" was introduced on American currency on April 23, 1864, and on December 28, 1945, Congress officially recognized it; but the

23. Noll, Protestants in America 64 (2000).

24. *Id.* at 74.

25. *Id.* at 113.

26. *Id.* at 136.

27. Abington School Dist. v. Schemp, 374 U.S. 221 (1963), by Justice Potter Stewart, dissenting, in the first quote and by Justice Tom Clark in the second one.

28. Dimont, Jews, God, and History 300 (1962). They could not accept the view in the Bible that humans alone are to be made "in the image of God." (*Gen.* 1:26-27).

word *God* was not placed into the national motto (the Pledge of Allegiance) until 1954. At the turn of the twenty-first century, the display of this motto at public schools in our increasingly diverse society had been criticized.[29] In 2002, it was ruled unconstitutional by the Ninth Circuit Court of Appeals.[30] In 2004, the Supreme Court ruled that an atheist father who challenged the pledge on behalf of his daughter, who recited the words in public school, and was the sole legal representative of her unmarried mother, had no right to sue. Hence, it made no ruling on the constitutionality of the insert.[31] But in the future, were that Court to recognize the deity in the manner expressed in the Declaration of Independence, it would then find that no scientific basis existed on which it might accept the challenge of the Creator as "God" in the current Pledge of Allegiance.

One writer noted that Americans have a kind of "civil religion" that politicians describe on public occasions "as the basis for the moral viability of American society."[32] (See Chapter 4.) This may be part of the growing secularization of many self-described religious people.[33] A comprehensive study in 1994 showed little behavioral difference between those who go to church regularly and those who do not. "On issue after issue, we see that religious teachings make some difference . . . but not a strong difference or one that could readily be anticipated from knowledge of these teachings themselves. Religious leaders want the churches to play a heroic role in our society . . . In reality, religious faith prompts few people in any of these directions."[34] The author of *The Religions of Man* (covering all large religions in the world)

29. Such criticism was directed against a resolution passed by the Colorado State Board of Education encouraging public schools to display "In God We Trust." ORANGE COUNTY REGISTER, July 7, 2000, at News 15.

30. Newdow v. U.S. Congress, 292 F.3d 597 (9th Cir. 2002), 2003 WL 5547.42 (9th Cir., Feb. 28, 2003) (denial of petitions for rehearing).

31. FAIGMAN, LABORATORY OF JUSTICE 318 (2004).

32. ROBERT BELLAH, CIVIL RELIGION IN AMERICA (Daedalus, 1967), *reprinted in* BEYOND BELIEF: ESSAYS ON RELIGION IN A POST TRADITIONAL WORLD 1680-86 (1970). *See also* BELLAH & HAMMOND, VARIETIES OF CIVIL RELIGION (Noble, 1980); ROBERT BELLAH, CIVIL RELIGION AND THE AMERICAN JEREMIAD 88-192 (1982); *and* J.T. Flexner, *Washington after the Revolution V: The Death of a Hero*, AMERICAN HERITAGE 68-74 (1969).

33. *The National Survey of Religious Identification (NSRI)*, p. 274.

34. ROBERT WUHTNOW, GOD AND MAMMON IN AMERICA 6 (Free Press, 1994).

stated, "America has become one of the most traditionless societies in history. As *substitute* she has proposed *reason.*"[35] (Emphasis added.) The AMA *Code of Medical Ethics* (2002-2003) states:

> In denominational health care *institutions* of those *operated by religious orders*, the recommendations of the ethics committee may be anticipated to be *consistent with published religious tenets and principles*. Where particular religious beliefs are to be taken into consideration in the committee's recommendations, this fact *should be publicized to physicians.*[36] *(Emphasis added.)*

> *Congress shall make* no law respecting an establishment of religion *or prohibiting the free exercise thereof. (Emphasis added.)*
>
> First Amendment to the Constitution

> *Let it be simply asked where is the security for property, for reputation, for life, if the sense of religious obligation desert the oaths, which are the instruments of investigation of Courts of Justice? And* let us with caution indulge the supposition that morality can be maintained without religion. *(Emphasis added.)*
>
> President George Washington, "Farewell Address," September 19, 1796

2. A Brief Look at Several of America's Diverse Religions Together with Their Restrictions on Possible Medical and Ethical Sources

The previous epigraphs demonstrate the importance of religion to the first Amendment to the Constitution and to President Washington at the end of the eighteenth century. A bioethicist indicated that the patient's morals

---

35. Huston Smith, The Religions of Man 169 (1958).

36. Rule No. 9.11 (4).

In regard to hospital Ethics Committees, it has been stated, "Medical-moral committees spoke with the authority of the church and the hospital's religious mission, and patients who did not accept those views were free to go to other hospitals." Ross et al., Healthcare Ethics Committee: The Next Generation 3 (AHA, 1993).

and treatment desires might often be expected to be derived from his or her religion; during the first half of the twentieth century, physicians spoke to their patients about religion.[37] Most physicians found this inappropriate during the second half following World War II,[38] however, and they have imposed almost no ethical guidelines thereon since the turn of the twenty-first century.[39] In many circumstances, bioethicists at hospitals must take a patient's particular religion into consideration, as well as the extent to which he or she may actually permit its special rules to become binding on physicians. The word *religion* comes from a Latin root meaning "to tie or bind." Any such tie not only depends on the members of the particular religion, but also on the degree to which its bindings can be observed.

In a 2001 study of America's 208 million adults, 14.1 percent claimed no religion, a rise from 1990 (when there were only 8 percent of such adults). In 2001, among the former members of America's Christian religions, Protestants reported the largest drop (14 percent), Catholics (9 percent), Methodists (7 percent), and others (6 percent).[40] Although the majority of current Americans remain religious, this tendency to drop during the twenty-first century will be of significance to those who are members of hospital ethics committees. Together with immigration, the United States has recently been called "the most religiously diverse nation on earth."[41]

Abraham, a Hebrew who moved in the Middle East over a century following the dawn of the second millennium B.C., became "the patriarch of three

---

37. *See* A WAY OF LIFE AND SELECTED WRITINGS OF SIR WILLIAM OSLER 255 (Dover, 1951).

38. Sloat et al., NEJM (2000), at 1913; Maugane & Wadland, J. FAM. PRACTICE (1991), at 210.

39. Sloat et al., LANCET (1999), at 664; THE BOTTOM LINE OF THE YEAR BOOK 17 (2001). *See* AMA *Code of Medical Ethics* 10.05(3)(c) and 9.11.

40. The 2001 study by the Graduate Center Survey on Religion in America tapped a random sample of more than 50,000 adults over the age of 18. SOUTHERN CAL. CHRISTIAN TIMES, December 2001.

41. DIANA ECK, A NEW RELIGIOUS AMERICA . . . 4 (2001). She added, "The ideal of a Christian America stands in contradiction to the spirit, if not the letter, of America's foundational principle of religious freedom." (p. 46) *See also* Philip Jenkins, *A New Religious America,* FIRST THINGS 125 (August/September 2002), at 25-28. A 1999 British poll shows that only a minority of Britons now believe in the tenets of Christianity. Less than half believe that Jesus was the Son of God. A CONCISE HISTORY OF THE CATHOLIC CHURCH 524 (2004).

monotheistic faiths: Judaism, Christianity, and Islam." Bioethicists, in hospitals, must become aware of their several differences. Unlike most Eastern religions, these faiths require their people to obey the laws that their God is assumed to have rendered. "*Christians* trace their spiritual lineage from Jesus to Abraham, whose obedience, service, and sacrifice prefigured the coming of the Messiah."[42] Statistics show the estimated Christian population in the United States at 225 million in the year 2000, with an estimated growth to 270 million by the year 2025. This country contains the highest number of Christians.[43]

The use of biblical morals and ethics has been studied because of their continued need after the arrival of evolution, along with other sciences and the law, in the nineteenth century.[44] In 1979, the National Book Award–winning *The Gnostic Gospels* by Elaine Pagels (who is now a professor of religion at Princeton University) discussed the more than 50 texts found at Nag

---

42. NATIONAL GEOGRAPHIC, September 2001, at 95 and http://go.compuserve.com/ReligionForum?MSG=2708093.

> Circumcision continued with all Muslim and Jewish males. In accordance with Abraham's discussion, God said "it shall be a sign of the covenant between me and you." *Gen.* 17:11. In the United States, Canada, and England it became less popular in the latter half of the 20th century primarily due to the fact that the foreskin can be retracted away from the glans in more than 90 percent of boys by the age of five.

43. JENKINS, THE NEXT CHRISTENDOM 90 (2002). Based on U.S. government statistics, found respectively in the *Annual Report on International Religious Freedom* (http://www.state.dov/www/global/human rights/irf/ irf_rpt/irf_index.html) and the *CIA World Fact Book* (http://www.odci.gov/cia/publications/factbook/). The U.S. Census Bureau does not collect information on religion. However, in 2002 a study was conducted by the Glenmary Research Center in Nashville, Tennessee. It coordinated the study with 149 participating faiths. ORANGE COUNTY REGISTER, Sept. 18, 2002, at News 17. In the letters of Ignatius, bishop of Antioch from about 113 to 128, we hear for the first time the term *Christianismo* (usually translated "Christianity," but literally "Christianism") set over against *Judaismos* ("Judaism"). WHITE, FROM JESUS TO CHRISTIANITY 346 (2004). He also wrote, "Likewise, let everyone respect the deacons as Jesus Christ; so also, as the bishop is a type of the Father, and the presbyters as the council of God and as the band of the apostles. Without them the church cannot be called church." *Id.* at 371.

44. *See* Chapter 2 and KEYES, LIFE, DEATH, AND THE LAW 140-45 (1995). Bible is derived from byblos, a roll of papyrus. It is also a Phoenician city that produced and exported papyrus more than 2,000 years ago.

Hammadi, Egypt, in 1945 that had been ordered destroyed by church leaders in the fourth century A.D. In 1995, significant analyses were made by scholars writing *The Five Gospels* (including the more recently discovered Gospel of Thomas with 114 of Jesus' sayings).[45] This was done to determine "whether the worldview reflected in the Bible can be carried forward into this scientific age," and if it should be retained "as an article of faith" or for other reasons.[46] In 1999, a survey of data showed that if a person does not accept Jesus Christ as Savior before the age of 14, the likelihood of doing so might diminish.[47] In 2003, the book *Beyond Belief: the Secret Gospel of Tho-*

---

45. The remains of an extraordinary Coptic library, lost for 16 centuries and discovered in 1945 in a ruined tomb near Nag Hamadi, Upper Egypt, has yielded an extensive collection of logia, an anthology of 114 "sayings of Jesus." Preserved by dry sands covering what was apparently a thriving Gnostic community, it is one of the earliest manuscripts related to the *New Testament*.

THE GOSPEL ACCORDING TO THOMAS (Harper & Row, 1959). "Over half of these sayings 'are almost word-for-word parallels to those contained in the recognized (or canonical) Gospels. While it's not certain which came first, most likely the synoptic Gospels (the Gospel of John and the Gospel of Thomas) were all outgrowths from an even earlier body of work known cumulatively as the Q source, a collection of Jesus's sayings and actions probably accrued shortly after his death that likely predate the earliest Gospels by decades.'" Professor Christoph Levin, author of *The Old Testament* (2005), p. 321.

46. Why the Gospel of Thomas didn't make it into the canon, however, isn't difficult to understand when one looks at the entire book; . . . among other things—the concept of rebirth or reincarnation. As such, it is easy to see why parts of it were not included in the canon and, indeed, why they *could not* be." *Id.* Former President Thomas Jefferson "scrutinized the gospels with a similar intent to separate the real teaching of Jesus history from the encrustations of Christian doctrine." THE LIFE AND MORALS OF JESUS OF NAZARETH, EXTRACTED TEXTUALLY FROM THE GOSPELS IN GREEK, LATIN, FRENCH, AND ENGLISH, 1904; and Albert Schweitzer in his *The Quest of the Historical Jesus*, 1906, who contributed in similar fashion. *See also* THE FIVE GOSPELS, 1995, at 6. The "Gospel of Thomas" has a number of sayings parallel to the four gospels now in the Bible; it contains many more unique ones. The scholars used their expertise to help determine the authenticity of the sayings set forth therein. *See* THOMAS, THE NEXT CHRISTENDOM 218.

47. Based on a nationwide representation of more than 4,200 young people and adults, the survey data show that people from ages 5 through 13 have a 32 percent probability of accepting Christ as their savior. Young people from the ages of 14 through 18 have just a 4 percent likelihood of doing so, while adults (ages 19 through

mas, also by Professor Pagels, focused on "the differences between the Gospel of John, one of the four gospels declared official by the early Christian church leaders, and the Gospel of Thomas, one of those rejected as heresy."

What John opposed includes what the Gospel of Thomas teaches—that God's light shines not only in Jesus but, potentially at least, in everyone. Thomas's Gospel encourages the hearer not so much to believe in Jesus, as John requires, as to seek to know God through one's own, divinely given capacity, since all are created in the image of God.[48]

Roman Catholicism, which comes from the Greek *katholike*, "universal," remains this country's second largest specific Christian denomination. The Center for Applied Research in the Apostolate, affiliated with Georgetown University, indicated that 62 million Roman Catholics are in the United States, and between 2 million and 3 million of them are black Catholics. The recent expansion of this group comes from Latin-American immigrants. Most Hispanics acknowledge their religion as Catholic.[49] The Vatican I Council of 1870 initially included the "infallibility of the Pope" when speaking from Peter's chair, meaning incapability of teaching what is false in faith and morals.

The biggest difference between the Eastern Christian Churches and those in the West is that there is no centralized authority in the East to exercise such control. For five centuries after 1100 A.D., the Justinian Code of the Byzantine Empire to be written in Latin became linked to the Roman Catholic Church "to the present day." The "fundamental principles of Roman law in the Justinian Code" constituted "the codification of the canon law, which accorded the pope the position of absolute authority in the church that the

---

death) have only a 6 percent probability of making that choice. The years prior to age 12 are when a majority of children make their decision as to whether or not they will follow Christ. Burna Research Online, *Teens and Adults Have Little Chance of Accepting Christ as Their Savior*, Nov. 15, 1999.

48. In Thomas's Gospel, Jesus performs no physical miracles and others (like the Muhammad) do not refer to him as the messiah. The Bible (from Greek *biblos,* meaning papyrus) was used in ancient bookbinding. The German source "Q" has been claimed to have antedated the Gospels of Matthew and Luke near the middle of the first century, which may have also been the time Joseph's Gospel was written. MACK, THE LOST GOSPEL, and ROBINSON, HOFFMAN, & KLOPPENBORG, THE CRITICAL EDITION OF Q. *See also* HELMUT KOESTER, ANCIENT CHRISTIAN GOSPELS: THEIR HISTORY AND DEVELOPMENT 86 (1990).

49. SO. CALIF. CHRISTIAN TIMES, December 2001, at 3.

Justinian Code gave the emperor."[50] The 1965 Council of Pope John XXIII attempted a new Pentecost to "throw open the windows of the church" while retaining its dogmas. After his Vatican II, masses were sung in English, Spanish, and other languages.[51] Pope John XXIII was followed by a "cold blast" because virtually no other such liberal popes were elected in the twentieth century following Vatican II. Priests now face the congregants (just like Protestant ministers), and their unique confessions are face-to-face rather than in a dark booth.

The Catholic Church renounced the charge that "all" Jews were responsible for Jesus's death by applying that charge only to some of them.[52] In October of 1965, *Nostra Aetate,* a council document stated that "what happened in his [Christ's] passion cannot be blamed upon all the Jews then living, *without distinction,* nor upon the Jews of today." The 2004 motion picture *The Passion of the Christ* accurately expressed the New Testament's description of the aggressive activity of the Jewish officials to obtain the local Roman governor's orders to crucify Jesus. The 35-year-old actor who played that part expected it to be controversial and said: "I think the film has been misunderstood. *It does not condemn an entire race.* There are good people and bad people in any group, and this movie does not blame all Jew-

50. CANTOR, ANTIQUITY 211-19 (2003).

51. ORANGE COUNTY REGISTER, Dec. 16, 1995, at News 12. On November 29, 1964, the U.S. Roman Catholic Church enacted changes in the liturgy, including using English instead of Latin. Pope Pius XII wrote the encyclical *Humani generic,* stating that "if the human body takes its origin from preexisting living matter, the spiritual soul is immediately created by God." MATSUMURA, VOICES FOR EDUCATION 383 (1995). His successor, John Paul II, agreed in 1995; he stated that "unlike evolution, the soul cannot be studied by methods of science." MOORE, FROM GENESIS TO GENETICS 100 (2001).

52. Historically, the Catholic religion has been anti-Semitic. In the last third of the 20th century, the Vatican officially rejected the concept that all of the Jews could have been responsible for the death of Jesus. One writer reported that "John Paul II offers a millennial mea culpa early in the year 2000, and while there was a profound significance in that apology as far as it went, it revealed how far is the distance that must be traveled yet." CARROLL, CONSTANTINE'S SWORD, THE CHURCH AND THE JEWS (2001). This is shown with his setting of dates with explanations, some of which is in Appendix D. However only Catholic priests are required to remain on clerical celibacy and without marriage.

ish people, just as all Italian people should not be blamed for the acts of Mussolini, or all Russian people should not be blamed for the acts of Stalin."[53]

Christianity is virtually the only religion that maintains the concept of an adverse permanent condition called "Hell."[54] "As Christian children (both Protestant and Catholic) during the 1960s, it was largely an unchallenged assumption that both God and the Devil were indeed masculine characters."[55] In the words of Milton's Satan, "the mind is its own place, and can make hell of heaven, a heaven of hell." Today, about one in every five American Catholics, Methodists, Episcopalians, Lutherans, and Presbyterians believes in the devil, and 77 percent of born-again Christians say that human beings are "basically good."[56] In the Gospel of John, 8:43,47, Jesus told the Jews, "You are from your father the devil, and you choose to do your father's desires. . . . Whoever is from God hears the words of God. The reason you do not hear them is that you are not from God." However, these words are seldom repeated in speeches or in church sermons. (For this book my choice was the Revised Standard Version along with advances in biblical scholarship.)

At the close of the twentieth century, new guidelines were issued on exorcism to reflect the Pope's efforts to convince skeptics that the devil exists. These guidelines were the first revision to the exorcism since 1614.

---

53. Orange County Register, Feb. 18, 2004 (Arts 1). In 2003 it was written concerning growing change in the 21st century's American viewpoint:

> This is no place to settle the debate between the evangelical modernists and their anti-modern critics. . . . No one has seen Jesus as inessential. Over the American centuries, some liberals have given up on miracles, the inspiration of the Bible, and (in the case of the "Death of God" theologians of the 1960s) divinity itself. . . . But rather than killing Jesus, these adaptations have only made Him stronger. It almost seems as if the Christians who subtracted this doctrine or that rite were beginning to question their own standing, and in order to convince themselves (and their neighbors) of their bona fides, they bent over backwards to laud and magnify their Savior.

Prothero, American Jesus 155 (2003).

54. In the New Testament, Satan tempts Jesus on his mission (*Matt*. 4:1-11 and *Luke* 13:27, 22:3), until his final defeat in the End Time (*Rev*. 20:2-3, 8, 10).

55. Wray & Mobley, The Birth of Satan, xix (2005).

56. Fox, Jesus in America 293 (2004) (citing *Barna Poll on U.S. Religious Belief—2001*, http://www.adherents.com).

They do not drastically change the old ones, but caution the exorcists to take psychiatric medicine into account. "The existence of the devil isn't an opinion, something to take or leave as you wish," said Cardinal Jorge Medina Estevez, a Vatican official who presented the revisions. "Anyone who says he doesn't exist wouldn't have the fullness of the Catholic faith."[57] Many of today's Catholics appear to go to psychiatrists for cures.

Many Catholics also appear to believe that exorcism (according to Webster, "the conjuration of evil spirits") will cure them of any disease. A "nonreligious" medical problem will be helped by Therapeutic Touch (TT)—a widely used practice alleged by some to actually have a "scientific" basis and for which they are being permitted to go into hospitals. Practitioners of TT claim to treat many medical conditions by using their hands to manipulate a "human energy field" perceptible above the patient's skin. A 1998 study showed unrefuted evidence that the claims of TT are groundless and that further professional use is unjustified.[58] Nevertheless, a more complicated problem derives from a survey showing that more Americans are using "alternative medicine" as opposed to their primary-care doctors. The study concluded that both the alternative and conventional therapies have to undergo rigorous scientific scrutiny.[59]

In 1632, when Galileo Galilei confirmed Nicholaus Copernicus's theory that the earth rotates on its own axis and moves around the sun, he was subjected to the Catholic inquisition and penalty.[60] His confirmation followed his development of the telescope, and its use in analysis of both the phases of the planet Venus and the rotation of four moons around Jupiter. These scientific

---

57. The rite itself remains essentially the same. It involves making the sign of the cross, laying on hands, sprinkling holy water, and ordering the devil to leave the possessed person. ORANGE COUNTY REGISTER, Jan. 27, 1999, at News 11.

58. 279 JAMA 1005-10 (1998).

59. Between 1990 and 1997, visits to alternative practitioners jumped about 47 percent. Over that period, Americans logged about 629 million trips to alternative providers, compared with 386 million visits to primary care doctors.
L.A. TIMES, Nov. 11, 1998, at A3, *citing* JAMA of the same date.

60. His masterpiece, *Concerning the Revolutions of the Celestial Spheres* (1543) was placed on the Vatican Index of prohibited books, where it remained for 200 years. *See also* Galileo's *A Dialogue on the Two Principal Systems of the World* (1632) and *Dialogues on the Two New Sciences* (1638). The latter became the basis for Isaac Newton's physics laws of motion published in 1687.

facts could not be considered consistent with the Old Testament's provisions of Joshua's stopping the rotation of the sun and moon around the earth and other similar biblical passages. In 1992, Pope John Paul II attempted to remove Galileo's writings as being a particular basis for scientific use in misinterpreting the Bible.

Nevertheless, contrary to the U.S. Supreme Court's holdings in the twentieth century, Pope (derived from the Greek *papas,* "father") John Paul II remained firm on the church's strict position against contraception, sterilization, abortion, and artificial conception into the third millennium.[61] The doctrine that prohibits the use of all artificial contraception was enshrined in the 1968 encyclical entitled *Humanae Vitae,* issued by Pope Paul VI and reiterated in the twenty-first century by Pope John Paul II. For example, in 2004 it was held that a California statute that requires employers to include prescription contraceptives in prescription drug coverage under employee benefit plans does not violate the First Amendment.[62] On his death in 2005, Pope John Paul II was succeeded by Cardinal Joseph Ratzinger, who took the name Pope Benedict XVI. At 78 years of age, he became the oldest pope elected since Clement XII in 1730 and the first pope from Germany in more than 400 years.[63] Many Americans considered him too doctrinaire to lead a twenty-first century church. (This is discussed in connection with bioethics and the law in Chapters 7 to 9).

In July of 2000, the Vatican reaffirmed its prohibition on divorced and remarried Roman Catholics receiving communion. It restated that the only

---

61. Orange County Register, Oct. 10, 1998, at News 10. Pope John Paul II repeated this statement in 2002 and added, "Frequently, man lives as if God did not exist, and even puts himself in God's place." Orange County Register, Aug. 19, 2002, at 1. On March 1, 2004, the California Supreme Court set a precedent by ruling (6-1) that a Roman Catholic Church must offer birth-control coverage to its employees even though the church considers contraception a sin. Justice Joyce Werdegar wrote that a "significant majority" of the people who served the charity are not Catholic. Orange County Register, March 2, 2004, at Bus. 6.

62. Catholic Charities of Sacramento Inc. v. Superior Court of Sacramento County, Cal., No. S099822, Mar. 1, 2004. *See also* Church of Lukumi Babalu Aye Inc. v. Hialeah, 508 U.S. 520 (1993).

63. Orange County Register, April 20, 2005, at News 3. The Vatican had put him in charge of doctrinal issues in 1981 and he ruled in favor of papal infallibility and against women priests, use of condoms, homosexuals, sterility, abortion, and more.

exception would be for such couples who refrain from sexual relations. This exception was claimed to be based on "divine law."[64] In 2003, the Vatican's Congregation for the Doctrine of the Faith issued a directive stating that all transsexuals suffer from "mental pathologies," and thus they become ineligible for admission to Roman Catholic religious orders. Furthermore, they should be expelled if they have already entered the priesthood or religious life.[65]

Most American Roman Catholics have ceased to follow recent Vatican dictates regarding such prohibitions and do not look on them as necessarily immoral per se. A survey of American Roman Catholics found that more than half disagreed with the ancient church position barring priests from marriage.[66] However, just before his death on April 2, 2005, Pope John Paul II approved a former Episcopalian priest who converted to Catholicism to become an ordained priest under an exemption to Canon law called the Pastoral Provision.[67] Furthermore, many Catholic priests have undertaken *sexual misconduct*—particularly with children—leading to a significant increase in litigation since the dawn of the third millennium.[68] Pope John Paul II sum-

---

64. The declaration was issued by the Pontifical Council for Legislative Texts, taking into account the views of the Congregation for the Doctrine of the Faith and two other Vatican offices. Orange County Register, July 7, 2000, at News 26.

65. Orange County Register, Feb. 1, 2003, at News 18. The Vatican also launched a global campaign against gay marriages.

66. Study released November 16, 2001, by LeMoyne College and by the research firm Zogby International stating: "More than 60 percent disagreed with church teaching that artificial birth control was morally wrong. Still, most agreed with the church's opposition to abortion and homosexual practices." *Available at* http://www.lemoyne.edu/academics/zogby.htm.

67. His application was "presented by former Cardinal Joseph Ratzinger, now Pope Benedict XVI. Since 1983, about 80 former Episcopal priests in the United States have been ordained as Catholic priests through the provision, according to the Catholic Information Center in Washington, D.C." Orange County Register, Feb. 11, 2006, at News 18.

68. "Sexual misconduct by Roman Catholic priests has cost the church in lost faith and broken lives. . . . Dioceses have sold land and buildings to pay off multimillion-dollar settlements . . . the closing economy and population shifts have forced many dioceses to close parishes and schools." Orange County Register, March 3, 2002, at News 5. Until the 11th century, Catholic priests were allowed to marry. AARP, September/October 2002, at 14. *See also* Southern California Christian Times, May 2002, at 4. In 2005 it was estimated that the costs from sexual predators

moned American cardinals to the Vatican to discuss the misconduct of priests in June of 2002.[69] In December of that year, he accepted the resignation of Cardinal Bernard Law as archbishop of Boston, where the archdiocese had $90 million in insurance to pay victims of approximately 500 alleged incidents of clergy sexual abuse.[70] A survey published in 2003 estimated that about 40 percent of Catholic nuns in the United States experienced either childhood sexual abuse or sexual exploitation within the Catholic community.[71] At 105.4 million, Protestant adults (or other non-Catholic denomina-

---

in the U.S. Roman Catholic priesthood total more than $1 billion. About one-third of the expenses came in the past three years. ORANGE COUNTY REGISTER, June 10, 2005, at News 15.

69. "Upon a return to the U.S., national guidelines adopted at a bishops conference in Dallas require church officials to report any allegation of a minor being abused by clergy. But, while past abusers will be stripped of duties, they will be allowed to remain in the priesthood." ORANGE COUNTY REGISTER, June 17, 2002, at News 10. A judge ruled that a church financial panel had no authority to reject a settlement worth up to $30 million between the Boston Archdiocese and 86 plaintiffs in a sexual-abuse lawsuit. ORANGE COUNTY REGISTER, July 19, 2002, at News 30.

70. L.A. TIMES, Dec. 16, 2002, at A22, and Dec. 17, 2002, at A15.

> The Roman Catholic Archdiocese of Boston agreed to pay $85 million to settle almost 550 lawsuits by people who say they were sexually abused by priests. The settlement is the largest ever by an American diocese to resolve accusations of sexual abuse by members of the clergy. Under the terms of the agreement, victims will each receive $80,000 to $300,000. . . . Parents who filed lawsuits claiming their children were abused will receive $20,000.

ORANGE COUNTY REGISTER, Sept. 10, 2003, at News 21.

71. ORANGE COUNTY REGISTER, Jan. 12, 2003, at News 20. In January 2003, 70 more priests were accused of raping or sexually abusing children. ORANGE COUNTY REGISTER, Jan. 30, 2003, at News 12. A 2004 study conducted by the National Review Board alleged costs of more than $572 million in payments to victims for settlements of priest abuse and psychological treatments. ORANGE COUNTY REGISTER, Feb. 28, 2004, at News 3. Lawyers for alleged victims of such abuse estimate that their clients' claims could cost the Archdiocese of Los Angeles more than $1.5 billion. ORANGE COUNTY REGISTER, Aug. 3, 2004, at News 10. There was also a $100 million settlement between the Catholic Diocese of Orange and alleged sexual-abuse victims: "The diocese will pay half the $100 million, with its eight insurance carriers picking the remainder. With assets in excess of $270 million, diocesan officials said they expect to pay their share without facing bankruptcy, as some other dioceses have." ORANGE COUNTY REGISTER, Jan. 4, 2005, at News 4.

tions) represent a twenty-first century decrease from 60 percent to 52 percent, and thus retain a Christian narrow majority in the United States.[72] Protestant churches began in the sixteenth century with rebellion against the reigning Catholic hierarchy. Unable to find agreement among themselves, they developed different creeds and detailed confessions of faith. Martin Luther, who initiated the Protestant Revolution, even attacked reason viciously.[73] At the same time, Protestants shared ideas as to the priesthood of all believers and the authority of the Scriptures. Most became evangelical about their particular beliefs.[74]

One pious group noticed that in the New Testament, Jesus was immersed by John the Baptist only after he had become a fully mature adult. No biblical authority could be found for infant baptism. This group called themselves Anabaptists and they rebaptized those who joined them. They appeared to be in the forefront with their more rational practices during the Reformation. However, Anabaptists were harshly persecuted for their baptismal beliefs by both Catholics and Protestants (including Lutherans), and thus approached near extinction. During the next century, they reformed and the new "Baptists" established American churches in New England. During the following two centuries they moved westward.

The Southern Baptist Convention was the largest single Protestant group in the United States at the turn of the third millennium.[75] Born in 1846, they split from the Northern Baptist group because the board of the Northern Group refused to appoint a man who owned slaves as a missionary. This group grew continually so that by 1980, just as the field of bioethics was getting started,

---

72. So. Cal. Christian Times, December 2001. Although between 1971 and 1990 the U.S. population increased by 19.9 percent, the Protestant congregations declined by 8.5 per cent. Shibley, Resurgent Evangelicalism in the United States: Mapping Cultural Change Since 1970 (1996).

73. "Reason is the greatest enemy that faith has; it struggles against the divine word, treating with contempt all that emanates from God. The Virgin birth was unreasonable; so the Resurrection; so were the Gospels, the sacraments, the pontifical prerogatives; and the promise of life everlasting." The Evangelical Lutheran Church in America has about 5 million members. In 2004 it was reported that a former Lutheran minister who sexually molested boys brought the total payout in the case to about $69 million. Orange County Register, April 23, 2004, at Local 1.

74. Manchester, A World Lit by Fire 112 (1993).

75. Kelley, Why Conservative Churches Are Growing (1995).

more than 13 million members in 35,000 churches had given more than $2 billion voluntarily.[76] Southern Baptists regard the Bible as containing the written word of God. They believe that the Holy Scripture is the only sufficient, certain, and infallible rule of all saving knowledge, faith, and obedience.

Former President Jimmy Carter's faith figured prominently in his 1976 White House campaign.[77] However, after the turn of the third millennium, he mailed a letter to the Baptists that he was cutting ties to the Southern Baptist Convention after struggling with the "increasingly rigid" creed of the nation's largest Protestant denomination.[78] The recognition of biblical authority is described as "the fundamentalist" position. The Baptists called themselves "the people of the book." During and after the 1970s, concerted efforts were sometimes made to change this view to one called "an inerrant position." Nevertheless, Baptists place less emphasis on written confessions of faith as compared to some segments of the Presbyterian Church, relying on the Westminster Confession of Faith and the Anglican Church in England, with its 39 articles. In 1999, Southern Baptists asked their members to pray that Jews will convert to Christianity during Rosh Hashanah, the Jewish New Year, and Yom Kippur, the Jewish Day of Atonement; the Anti-Defamation League expressed outrage.[79] Although the Baptists believe that death belongs

---

76. Hays & Steely, The Baptist Way of Life 34, 128 (1981).

77. In 1997 television reporter Bill Moyers noted popular political and similar religious people Jesse Jackson, Jesse Helms, former president Bill Clinton, former vice president Al Gore, Pat Robertson, and Jerry Falwell. Kimball, When Religion Becomes Evil 224 (2003).

78. Orange County Register, Oct. 21, 2000, at News 15.
The Southern Baptist Convention has told its overseas missionaries they had until May 5, 2003, to affirm the denomination's revised statement of faith—a document that opposes female pastors and says wives should submit to their husbands—or they could be fired. Nearly all the denomination's 5,500 missionaries already have agreed to support the 2000 Baptist Faith & Message.
L.A. Times, April 17, 2003, at A31.

79. Orange County Register, Sept. 5, 1999, at News 20, and Sept. 29, 1999, at News 6. "St. Paul ensured that they did not even have to undergo the Jewish rite of circumcision." Crocker, Triumph 1 (2001). He said: "Real circumcision is a matter of the heart—it is spiritual and not literal" (*Romans* 2:29). "In the Old Testament, Israel was founded when God intervened in Abraham's sacrifice and the covenant was established between fathers and sons. The covenant came to be symbolized by the circumcision." The Inferno of Dante (1300), Pinsky tr. (1994); Notes: Cantos XXXII-XXXIII, p. 424.

"in the hands of God," they recognize the concern of terminal patients "where the wisdom of God is questioned in continuing life."[80] The *Social Principles Booklet of the General Association of General Baptists*' section on "Termination of Life" states: "We believe life and death belong in thc hands of God. We are thankful to medical science for efforts and accomplishments made in preventing disease and illness and for the great advances in treatment that extend the life and usefulness of those afflicted."

Similarly, in 1980, the General Council of the United Methodist Church adopted the following statement:

> We recognize the agonizing personal and moral decisions faced by the dying, their physicians, their families, and their friends. Therefore, we assert the right of every person to die in dignity, with loving personal care and without efforts to prolong terminal illnesses merely because the technology is available to do so. (But see Chapter 14, Section 3.)

The Anglican Church was originated in 1534 when English King Henry VIII broke from Rome to divorce Catherine of Aragon. When threatened by Pope Clement VII with excommunication, Henry made himself head of the Church of England and confiscated the Catholic churches and monasteries. However, he retained most of the remaining Catholic belief and practice. Queen Elizabeth I adopted 39 Articles of her father's appointed Archbishop of Canterbury and was then excommunicated by Pius V on February 25, 1570. Queen Elizabeth II of the twenty-first century succeeded her. Although Parliament retains authority over civil matters, the Queen is master of the Anglicans. Some 2.3 million of those within the United States belong to the International Anglican Communion of 77 million believers who are in the Protestant Episcopal

---

80. The SOCIAL PRINCIPLES BOOKLET OF THE GENERAL ASSOCIATION OF GENERAL BAPTISTS, *Termination of Life* section states:

> We believe life and death belong in the hands of God. We are thankful to medical science for efforts and accomplishments made in preventing disease and illness and for the great advances in treatment that extend the life and usefulness of those afflicted.

Churches.[81] Many Christian believers rely on the importance of archeological findings. For example, the Pontius Pilate inscription found in 1961 in Caesarea "helps confirm the existence of the controversial Roman governor" who ordered Jesus to be crucified.[82] At the turn of the third millennium, a Gallup poll showed that 80 percent of Americans expressed a wish for spiritual growth whether or not they were "religious" people.[83]

In 2002, an Anglican priest and physicist who won a $1 million religion award stated: "I've tried to take both religion and science seriously and try to see them as complementary to each other and not rivals. The most important thing that they have in common is that both believe there is truth to be sought and found."

Toward the end of the twentieth century, credit has been given to Ronald Reagan, the leader of a republic, who said "God bless you" to Mikhail Gorbachev, the leader of Russia. As a Russian diplomat recalls, "The heretofore impregnable edifice of Communist atheism was being assaulted before their very eyes by a man who had made his name as a hard-line anti-Communist."[84]

Jewish tradition emphasizes belief that they are the "chosen" people under God's covenant with Abraham and his descendants,[85] to worship and remain faithful to him.[86] Unlike Christians, they observe the sabbath on Sat-

---

81. ORANGE COUNTY REGISTER, July 16, 2003, at News 25. "It spread throughout the British Empire and beyond and is now present in 164 countries." ORANGE COUNTY REGISTER, Feb. 26, 2005, at News 20.

82. NEWSWEEK, Aug. 30, 2004, at 35.

83. TIMOTHY GALLUP, JR., THE NEXT AMERICAN SPIRITUALITY: FINDING GOD IN THE TWENTY-FIRST CENTURY (2000).

84. SCHWEIZER, REAGAN'S WAR 272 (2002). "Only by its defeat would the Cold War end, so he chose to force tensions to a decisive conclusion rather than hide them." *Id.* at 283.

85. *Genesis* 12:3 states that God told Abraham, "I will bless you, and whoever curses you I will curse, and in you all the families of the earth shall be blessed." *See also Deut.* 7. When Emperor Hadrian had ordered a ban on circumcision in the early 130s A.D., the Jews, led by Bar Kokhba, rose in revolt, in 134 A.D. Two Roman legions made the Jews "homeless" and created the Jewish Diaspora. BOORSTIN, THE CREATORS 121 (1992). (The year 2004 A.D. is Hebrew 5764.)

86. L.A. TIMES, March 15, 2002. *See also* THE WORLD ENCYCLOPEDIA.

urday. Yet, messianic Jews believe in a messiah, and late in the twentieth century, they recognized the New Testament. The number of Jewish American believers dropped by at least 1 million during the previous decade. However, the 2000–2001 National Jewish Population Survey found that 5.2 million Jews live in the United States, compared with 5.5 million a decade ago.[87] Their median age rose from 37 to 41 in the same period.[88] On May 14, 1948, the independent state of Israel was proclaimed in Tel Aviv. On January 23, 1950, the Israeli Knesset approved a resolution proclaiming Jerusalem the capital of Israel.

The legal, ethical, and ritual writings of the Jewish people are compiled in the Torah (the 613 laws of Moses) and Talmud (the Jewish canonical law). Like the Baptist's ceremony, a boy's bar mitzvah at age 13 recognizes him as a responsible person. (See Chapter 13, Section 4.) Some report that Jewish Americans have undergone their own divisions. A female writer argued in 1997 that much Orthodox Jewish oral education had been traditionally denied to "one-half of its population,"[89] as similarly recommended by apostle

---

87. They had two origins: (1) The Sepharad were originally the Jews of Spain, most of whom were expelled from there in 1492 and spoke Arabic or Castilian. (2) The Ashkenazi, who spoke the German-Hebrew hybrid called Yiddish. Hence the Nazi Holocaust was 90 percent Ashkenazi.

88. So. Cal. Christian Times, December 2001, reported a drop to 2.8 million at the beginning of the third millennium from 3.1 million in 1990. The 1990 report found that 52 percent of Jews marry non-Jews. However, the 2000 study's researchers said it was hard to find people willing to participate. Orange County Register, Nov. 29, 2002, at News 38.

89.
> I was a girl, and the life-line of Jewish learning—Talmud, Mishna, Midrash—was out of my reach. All I got was indoctrination in gender restrictions and a thorough exposure to the great silences—the denial of the past, the suppressed voices, the absence of heroines.

Lerner, Why History Matters 7 (1997). *See also* Aiken, To Be a Jewish Woman 71 (1992). "The fate of American Judaism—whether its adherents will step back from the edge of schism or fall into it—hangs perilously in the balance." Sarna, American Judaism 373 (2004) (citing Freedman, Jew vs. Jew; Debra Cohen, *Are the Jewish People Splitting Apart?*). United Jewish Communities, Press Release, Oct. 8, 2002, The Jewish birthrate dropped to 1.8 children per couple.

Paul in the New Testament regarding the submission of women.[90] This was also done in Southern Baptist churches.

The Christian scripture of John states, "In the beginning was the Word." However, as one Nobel laureate put it, over the centuries, "the word has come to be synonymous with ultimate authority."[91] Many people have personal moral problems translated from biblical passages, for example, those detailing God's commands to kill the Canaanites during the conquest of their land.[92] In this connection, a veteran, recalling the D-Day attack on Omaha Beach, France, during World War II after June 6, 1944, stated: "I never felt like I was killing a man. I just felt like I was killing the enemy. That was our job. I told my men that was our job. We didn't think about it as killing people or we never could've done our jobs. We were trained to kill the enemy and if we didn't kill them, they would've killed us, and we wouldn't be here to talk about it."[93]

The term *killing* was originally mistranslated as being a violation of one of Moses's Commandments by British translators. Although it was corrected to *murder* in its revised editions, many churches continue to use the King James original translation. The death penalty then also followed for both of the parties who are caught in adultery;[94] it was alleviated by the story of Jesus's very different handling of that matter,[95] and changed in our criminal law. At this time, as noted previously, many synagogues also recognize Yeshua as the Messiah (God's "anointed one") of Israel.

---

90. Let your women keep silence in the churches for it is not permitted unto them to speak; but they are commanded to be under obedience. . . . And if they will learn anything, let them ask their husbands at home: for it is a shame for women to speak in the church.

I *Corinthians* 14:34-35. The word *apostle* (*apostolos* in Greek) means "one sent out" or a "missionary."

91. GORDIMAR, LIVING IN HOPE AND HISTORY 195 (1999).

92. *Deuteronomy* 20:10-18; *see also Exodus* 34:11-16; *Deuteronomy* 7:1-6. The authority was also to enslave gentiles: *Leviticus* 25:35-46.

93. ORANGE COUNTY REGISTER, June 7, 2004, at News 4. *See also* FRIEDMAN, THE BIBLE WITH SOURCES REVEALED 319, 334 (2003).

94. *Leviticus* 20:10, reads, "The man who commits adultery with his neighbor's wife will be put to death, he and the woman."

95. "He among you who is sinless—let him cast the first stone at her." "Has no one condemned you?" "No one, sir," answered she. "Nor do I condemn you," said Jesus. "You are free to go. But from now on, avoid this sin." The Gospels of John and of Thomas (that were located in Egypt in 1945).

As predicted by Isaiah 53, messianic Jews were "to whom has the arm of the Lord been revealed. . . . He was despised, and we held him of no account. . . . But he was wounded for our transgressions, crushed for our iniquities." Accordingly, in *The Enduring Paradox* (2000), John Fisher stated: "They speak of themselves as Messianic Jews, call Jesus by his given Hebrew name 'Yeshua,' and visibly demonstrate that a Jew can commit himself to following Yeshua as the Messiah and strengthen—not dilute—his Jewish identity. Messianic congregations or synagogues do not threaten Jewish survival; instead, they enhance it." (p. 8) Some 350,000 messianic Jews exist in the United States; and between 3,000 and 10,000 in Israel.[96] However, the 1950 Israeli Law of Return (which gives every Jew who has not converted to another religion the right to return to the land as a new immigrant entitled to tax breaks) was held by its High Court not to cover a member of a messianic Jewish congregation.[97] They were looked on as those who had professed other religions. For example, it has recently been noted that "In the end the Jews never took wholeheartedly to heaven and hell. It was their Christian and Muslim inheritors who built most enthusiastically on Persian founda-

96. Fisher, The Enduring Paradox 9, 89 (2000). "At Emperor Constantine's Nicaean council in 325 C.E. the Christian day of worship was changed from Saturday to Sunday." *Id.* at 62. Further, the Trinity, "The idea of three persons in the Godhead is well within the concept of Jewish thought, as is taught by the Zohar." *Id.* at 59, *citing* Rabbi Tzvi Nassi, The Great Mystery: How Can Three Be One? (1990). *See also* The Complete Jewish Bible (1998) Introduction:

> Moreover, Jeshua is still a Jew, since he is still alive; and nowhere does Scripture say or suggest that he has stopped being Jewish. His twelve closest followers were Jews. For years all his *talmidim* (disciples) were Jews, eventually numbering "tens of thousands" in Yerushalavim alone. . . . "For the New Testament is in fact a Jewish book—by Jews, mostly about Jews, and for Jews as well as Gentiles." *Acts* 21:20 (p. xxxv) While there are between 100,000 and 500,000 Messianic Jews in the English-speaking countries . . . it is obvious that most of the 13 million to 17 million Jewish people do not accept Yeshua as the Messiah (p. xxxvii).

97. *Id.* at 77 (referring to the case in *Beresford II* as "strengthening the precedent that a Jew who believes in Jesus has 'changed to another religion,' and cannot become a citizen under the Law of Return." That law was specifically amended in 1970 to exclude "ethnic Jews." In his volume *Why The Jews Rejected Jesus* (2005), David Klinghoffer wrote "that Jews have long acknowledged the role played by a few prominent ancestors in the events leading to the Crucifixion." (p. 3)

tions and made the Devil a major symbol of evil and falsehood."[98] Members of many churches (other than Baptists) ignore the prescription in Paul's Letter to the Church at Colossae that "Wives be subject to their husbands, as they should in the Lord." Wisdom, per the Old Testament, as personified by King Solomon's method of dealing with a fight between two women over the identity of a baby, has been modified by today's DNA test, which can now decide that matter in a different, more accurate way.

The hospital's bioethics committees I have attended recognized that the only person to handle a dying orthodox Jew is the rabbi. At one meeting, for example, such specific information was confirmed by a rabbi, after the patient agreed to his position and the bioethics committee concurred. Although a larger percentage of Jews have entered the field of medicine, in America, "Jews, on average, are the least religious of all denominations, with by far the highest mixed-marriage rate and the fewest children when they do marry among themselves."[99] Furthermore, "Intermarriage, low in the early 1960s, rose to 50 percent in the 1990s. Only 28 percent of children in mixed marriages were raised as Jews, with 41 percent raised as non-Jews, and 31 percent with no religion."[100] Israel was founded after World War II when the Germans exterminated men, women, and children in concentration camps; however, "Of these 12 million victims, 7 million were Christians and [below] 5 million were Jews."[101] In Israel and the United States, humanistic

---

98. Kriwaczek, In Search of Zarathustra 198 (2003). In 2005, it was reported on "a subculture that is quite different from the majority of Israelis, about 80 percent of whom define themselves as secular or merely traditional in religious belief." Orange County Register, Aug. 17, 2005, at News 5.

99.
> Do not intermarry with them [non-Israelites], giving your daughters to their sons or taking their daughters for your sons, for that would turn away your children from following me, to serve other gods. Then the anger of the Lord would be kindled against you, and he would destroy you quickly.

Deuteronomy 7:3. *See also* Heynick, Jews and Medicine 557 (2002).

100. Barone, The New Americans 247 (2001). "A study carried out in the summer of 2000 found that 49 percent of Israeli high school students, when asked, said that they 'hate the ultra-Orthodox,' the same percentage that admitted to hating Arabs." Efrom, Real Jews, Secular vs. Ultra Orthodox 7 (2003).

101. Justice Jackson at the Nuremberg trials cited 4,500,000 Jews killed by Germans. . . . Howard M. Sac puts the figure at 4,200,000 to 4,600,000, stating that the figure of 6 million released at the end of the war has since been discounted. The Course of Modern Jewish History 457. Dimont, Jews, God and History 373 (1962).

Jews do not rely only on the orthodox tradition to form ethical judgments; they and others have strong, though unofficial, positions in favor of a right to die. Many may favor both physician-assisted suicide (PAS) and euthanasia in certain cases (see Chapter 14). Israel's right-to-die movement is gaining acceptance, and, according to a recent report, three bills to legalize euthanasia in certain circumstances have been introduced.[102] A number of Jews have left orthodoxy for secularism, at the cost of losing family contact.[103] In 2002, Israel's Supreme Court ruled that the state must officially recognize conversions to Judaism by Reform and Conservative groups in Israel.[104] Nurses in the Christian Science Church give only "practical care" to inpatients (including prayers) and do not necessarily see the reality of the physical symptoms.[105] A parent (not the church) can be found criminally negligent for rely-

---

On November 29, 1947, the U.N. General Assembly passed a resolution calling for the partitioning of the former British-controlled country of Palestine between Arabs and Jews. On January 23, 1950, the Israeli Knesset approved a resolution proclaiming Jerusalem the capital of Israel.

102. HAMER, THE GOD GENE 189 (2004). *See also* Hammer et al., *Human Population Structure and Its Effects on Sampling Y Chromosome Sequence Variation,* GENETICS 1495-509 (2003). As of 2004, DNA evidence showed "that the priestly Y chromosome originated 2,100 to 3,250 years ago, sometime between the Exodus and the destruction of the First Temple—in other words, just about when the Bible says it occurred."

103. Judaism is divided into three subgroups: Orthodox, Conservative, and Reform. "The switch to secular life has been far more painful for others. Israel Segal, an author and prominent television journalist, left his devout home more than 30 years ago. As a result, Segal was banished from the family. 'What my family did to me was a holocaust. I don't have a family. You know, I think I went into the media to show them, to say: "You wanted to exterminate me, to wipe me out. You can't. I live. I exist."'" L.A. TIMES, Dec. 29, 1996, at A16.

104. ORANGE COUNTY REGISTER, Feb. 4, 2002, at News 15. This ruling would force the Interior Ministry to identify those converted by non-Orthodox rabbis as Jews in the "nationality" clause on their ID cards. The Orthodox, meanwhile, vowed to reverse the ruling. Following a six-year legal battle, its Supreme Court granted recognition to non-Orthodox conversions to Judaism partly performed in Israel, which constituted "a sharp blow to the country's Orthodox monopoly over religious affairs." ORANGE COUNTY REGISTER, April 1, 2005, at News 23.

105. The Church's publication states:

> Disease and physical suffering are in no sense caused or permitted by God, and that since they are profoundly alien to His creative purpose, it is

ing exclusively on prayer treatment for a seriously ill child,[106] unless the facts have been found to indicate that the lack of liability prevails.[107] Adults who are of this religion may legally decline Medicaid treatment by physicians pursuant to the doctrine of "informed consent" (see Chapter 5).

The Christian Science Church was established in the second half of the nineteenth century by Mary Baker Eddy (1821–1910), who suffered a severe injury and was healed while reading about Jesus's healing in Matthew 9:1–8. It was expanded both in the United States and 48 countries abroad.[108] Although the word *science* is used, church members follow the proposition that "there is a law of God applicable to healing, and it is a spiritual law instead of material."[109] Mrs. Eddy's 700-page volume entitled *Science and Health* (1875, 1971 Edition) contains her unique interpretations on scriptures, science, and health.[110] Health is a condition of mind, except that "a surgeon"

> wrong to resign oneself to them and right to challenge them. To the Christian Scientist this conviction is rooted in both the *Old* and *New Testaments*. In its fullest implications, such a conviction furnishes the basis on which illness, seen as an aspect of *human alienation from God* can be actively confronted and overcome.

First Church of Christ, Scientist, Freedom and Responsibility: Christian Science Healing for Children 18 (1989). (Emphasis added.)

106. Lundman v. McKown, 530 N.W.2d 807 (Minn. App. 1995); Walker v. Superior Court, 47 Cal. 3d 112 (1988). Christian Scientists generally opposed standard medical care and transplantation.

107. Gayle Quigley v. The First Church of Christ Scientist, et al., 65 Cal. App. 4th 1027 (1998).

108. In 1875 she published the first edition of *Science and Health with Key to the Scriptures*. The church was incorporated in Boston in 1879 and The Massachusetts Metaphysical College in 1872. *The Christian Science Journal* first appeared in 1883. Wilbur, The Life of Mary Baker Eddy 287, Christian Science Publ. Soc. (1907, 1996 Edition). *The Christian Science Monitor*, a daily newspaper, was established in 1908 "to publish the real news of the world in a clean, wholesome manner, devoid of the sensational methods employed by so many newspapers." *Id.* at 361.

109. The World Book Encyclopedia.

110. This volume included Eddy's interpretation of "The Spiritual Peace of Words in the Lord's Prayer: 'And lead us not into temptation, but deliver us from evil' as 'and God leadeth us not into temptation, but delivereth us from sin, disease, and death.'" (p. 17) She pointed out that "in the year 1866, I discovered the Christ Science or divine laws of Life, Truth, and Love, and named my discovery Christian

could give an injection to dull one's pain.[111] In 1996, a federal court held that government payments of Medicare and Medicaid funds to Christian Science nonmedical "healers" violated the constitutional separation of church and state when they teach that prayer is the most effective treatment for illness.[112] (See the following Section 3.)

During a 1996 meeting of members of the American Academy of Family Physicians, it was stated that "seventy-five percent believed that prayers of others could promote a patient's recovery."[113] Unlike Christian Scientists, some other physicians, such as Dr. Cherry Sele, expressed the need for such a promotion after treatment, noting that "at least a third of U.S. medical schools have integrated 'spirituality education' into the required portion of their curricula."[114] Others have indicated that "Doctors don't heal you; they merely assist the body in its own natural process."[115] In 2003, the U.S. Court

---

Science" (p. 197). "I won my way to absolute conclusions through divine revelation, reason, and demonstration." (p. 109) She stated that "leadings of scientific revelation, the *Bible* was my only textbook." (p. 110) "Christian Science is natural, but not physical." The book includes more than 100 pages of reader testimonials.

111. The principle of divine metaphysics is God; the practice of divine metaphysics is the utilization of the power of Truth over error; its rules demonstrate its Science. *Id.* at 111. Health is not a condition of matter, but of mind; nor can the material senses bear reliable testimony on the subject of health. (p. 320) Seized with pain so violent that he could not treat himself mentally—and the Scientists had failed to relieve him—the sufferer could call a surgeon, who would give him a hypodermic injection, then, when the belief of pain was lulled, he could handle his own case mentally. *Id.* (p. 464)

112. Orange County Register, Aug. 9, 1996. In January 1996 the U.S. Supreme Court upheld a $1.5 million award against four Christian Scientists in the case of an 11-year-old Minnesota boy who died of diabetes that had gone medically untreated. *Id.*

113. Bagiella, Powell, & Sloan, *Religion, Spirituality, and Medicine*, 353 The Lancet, No. 9153 (Feb. 20, 1999), at 664.

114. Reginald Cherry, M.D., Healing Prayer 162 (1999). Like Mrs. Eddy, Cherry also cites, "I am the Lord Who heals you" (*Ex.* 15:26). "For I will restore health to you, and I will heal your wounds, says the Lord" (*Jer.* 30:17), and her favorite, the *Assurance of God's Protection* at the beginning of Psalm 91, "You who live in the shelter of the Most High, who abide in the shadow of the Almighty."

115. Shields & Ferrell, Spiritual Survival Guide 116 (2001). Although the Christian Science religion claims to promote health, it can have the opposite effect. Easterbrook, *Faith Healers,* New Republic (July 19, 1999), at 20-29. It has also been

of Appeals for the Ninth Circuit ruled that the allowance of certain payments in the 1997 Balanced Budget Act to Christian Scientists under Medicare and Medicaid for "nonmedical treatment do not violate the First Amendment. That is, religion is not advanced because no one would become a Christian Scientist to obtain admission to a [religious nonmedical health care institution] when the cost of such a conversion would be renunciation of all medical treatment."[116]

A new religion for a new nation begun by members of the Church of Jesus Christ of Latter-day Saints (LDS, known as Mormons) was organized by Joseph Smith, Jr., on April 1, 1830. When he was 33 years old, he claimed to have unearthed tablets on which Moroni, an angel, had etched the reincarnation of Jesus in America about 421 A.D. His translation used some language of the Bible.[117] In 1842, Joseph Smith announced a revival of polygamy and two years later at the funeral of a Mormon man named King Follett, he stated, "As God once was, man is."[118] In June of 1844, a mob in Carthage, Illinois, killed both Joseph Smith and his brother, Hyrum. Brigham

---

noted: "In reality, members have unusually bad health and do not live long lives. For Christian Scientists, healing is seen as a spiritual process, rather than something that can be facilitated through scientific medicine. They are subject to illnesses without recourse to medicines or surgery." BARBER, KINDNESS IN A CRUEL WORLD 322 (2004).

116. Kong v. Scully, 9th Cir., No.02-15057 (Sept. 2, 2003). 14 BNA's MEDICARE REPORT 1014.

117. PROTHERO, AMERICAN JESUS 164 *et seq.* (2003). *The Book of Mormon* describes the religious history of an ancient American civilization whose ancestors were a group of Israelites who left Jerusalem around 600 B.C.E. In the Western hemisphere they split into two groups, the Nephites and the Lamanites. The former were eventually destroyed, leaving the Lamanites, who are the ancestors of today's American Indians. Davis, *Genetic Research*, Hast. Ctr. Rep., July-August 2004, at 42. Grant Palmer, 54, who wrote *An Insider's View of Mormon Origins*, could have been excommunicated because he "suggested Smith wrote it himself, leaning heavily on the King James Bible and personal experiences." ORANGE COUNTY REGISTER, Dec. 13, 2004, at News 14. Nevertheless, the Mormons also pointed to one of the ultimate books in the Old Testament, which indicated that the lost people of Judah, part of ancient Israel, were transferred to another land. *Ezekiel* 15:19, 21 (revised standard version).

118. Voros, Jr., *Was the Book of Mormon Buried with King Follett?,* SUNSTONE 11:2 (March 1987), at 15-18. However, in 1882 Congress outlawed polygamy.

Young (1801–1877) then led the Mormons to Utah, which he called their promised land, a new Israel, and in 1851, adopted a "Word of Wisdom" prohibiting the use of tobacco, alcohol, and tea. In 1877, he initiated the Temple in Salt Lake City. On September 25, 1890, Wilford Woodruff, president of the Church of Jesus Christ of Latter-day Saints, issued a manifesto renouncing polygamy. By then, the LDS church had "become the fifth largest religious group in the United States, with more than 5 million members in its home country and over 10 million worldwide."[119]

The *Unitarian Universalist Association* was published as President Thomas Jefferson's volume called *The Life and Morals of Jesus of Nazareth.* Like the Unitarian movements of today, his group had little interest in most of the Old Testament or the New Testament, which includes the writings of Paul.[120] His work, a *Benevolent Code of Morals* (which was actually published in 1904), was said to "cull" the most sublime code of morals, but tossed out miracles and supernatural references.[121] In 1972, a Unitarian scholar pointed out: "The lack of notable Unitarian Universalist Bible scholars in our time is perhaps a reflection of the widely held liberal philosophy that the Bible has little in the way of unique inspiration that cannot be found in other literary productions."[122] As the Unitarian minister said proudly of his congregation, they are "people of all strands, searching for truth . . . all on their own path." And as a member of the same congregation who was experimenting with different types of spirituality told us, what was helpful in the spiritual life was to follow "whatever seems heart-centered to you."[123] This con-

---

119. PROTHERO, *supra* note 117 at 188. In 1995, the Presbyterian Church (U.S.A.) adopted a "Report on Mormonism" stating that the Latter-Day Saints Church was not "within the historic apostolic tradition of the Christian Church" and insisting that Mormon converts to Presbyterianism be rebaptized. In 2000, the United Methodist General Conference passed a resolution to the same effect. One year later, the Vatican Congregation for the Declaration of the Faith declared that the Roman Catholic Church would require its converts from Mormonism to be rebaptized. *Id.* at 103.

120. ORANGE COUNTY REGISTER, July 22, 2001, at News 16.

121. *Id.*

122. OPPENHEIMAR, KNOCKING ON HEAVEN'S DOOR 33 (2003). *See also* Meyers, *Unitarians, Universalists, and the Bible*, 27 UNITARIAN UNIVERSALIST CHRISTIAN (1972) at 4.

123. HECLAS WOODHEAD, THE SPIRITUAL REVOLUTION 2 (2005).

stitutes an autonomous approach (see Chapter 5). Many, if not most, of the Unitarian churches are largely filled with former adherents to the Jewish faith. In 1999, a writer described a "bleak prognosis for Judaism in North America as more Jews voluntarily abandon the religion and deny future generations exposure to their special heritage."[124] It has some possible comparability with Urantia.[125] The Unitarian Universalist Association adopted by a two-thirds vote the "right to die with dignity." It contained the following:

> BE IT RESOLVED: That Unitarian Universalists advocate the right to self-determination dying, and the release from civil or criminal penalties of those who, under proper safeguards, act to honor the right of terminally ill patients to select the time of their own deaths. (See Chapters 12 to –14.)

Such a statement was reiterated by the Dalai Lama, who, in 1959, fled Tibet, a country where secularism is carried out by Chinese force. He said, "A kind of cultural genocide is in progress in Tibet. And even if losing her independence is acceptable, then still the destruction of our spirituality, of Tibetan Buddhism, is unthinkable."[126] (In 1984, he was awarded the Nobel Peace Prize.) In a rather similar way, the previous Nobel laureate Albert Einstein had stated:

> The religion of the future will be a cosmic religion. It should transcend personal God and avoid dogma and theology. Covering both the natu-

---

124. ORANGE COUNTY REGISTER, March 6, 1999, at News 6. The writer was in the national Jewish Outreach Program called J2K.

125. *The Urantia Book* (UB) was published in 1955; a reference work about the UB is *The Paramony*, which was compiled and privately published in 1986 by Gen. Duane L. Faw, a former co-professor of mine at the law school of Pepperdine University. It consists of 15,000 cross-references that link UB passages to verses in the Bible. *See* GARDNER, URANTIA: THE GREAT CULT MYSTERY, Promether Books.

126. *Quoted in* TUCCI, TO LHASA AND BEYOND 199 (1987). The largest Buddhist facility in California opened in 2002. ORANGE COUNTY REGISTER, Oct. 26, 2002, at News 3. "I call the unique psychological character of modern Tibetans 'inner modernity.' The Western life-purpose is ever greater material productivity, and the Tibetan life-purpose is ever greater spiritual attainment. . . . Tibetan inwardly directed reason put the material world second on its list of priorities." Robert Thurman, Prof. Indo-Tibetan Buddhist Studies, Colum. Univ., N.Y. (p. 20).

> ral and the spiritual, it should be based on a religious sense arising from the experience of all things natural and spiritual as a meaningful unity. Buddhism answers this description . . . If there is any religion that could cope with modern scientific needs it would be Buddhism.[127]

"The basic, most fundamental characteristic of Buddhism is the promise of enlightenment."[128] *Buddha* (as has been characteristic of deism) was the Sanskrit word meaning "to be enlightened" or "awakened."[129] Like Christians, Buddhists place much stress on love; however, because they do not worship Buddha, they are referred to as a nontheistic, or agnostic, and not a dogmatic religion. Tibet inherited India's Buddhism and wrote the *Tibetan Book of the Dead.*[130] A recent writing pointed to the existence of Buddhism as "the decisive period that determines either liberation or rebirth . . . During the experiences of death, we are presented with the opportunity to awaken into the corresponding sphere of enlightenment." Furthermore, there was "an underlying sense of sadness in the fragility and impermanence of life that sometimes becomes overwhelmingly poignant."[131] The "completely different concept of death" is "the fundamental principle of birth and death recur-

---

127. *See* Lama Sura Das, Awakening the Buddha Within, Part One. Gandhi once said, "I claim to be a passionate seeker after truth, which is but another name for God." *Id.* at 11. The word Dharma, frequently used as a synonym for Buddha, means "that which supports or upholds." *Id.* at 25.

128. *Id.* at 13. "Buddhism today is best thought of as an ethical psychological philosophy or nontheistic spiritual practice, needing neither dogma nor belief to be practiced and accomplished. When asked, 'Does God exist or not?' Buddha remained silent. 'Does the universe have a beginning or an end?' Again, the Buddha remained silent." (p. 111) "The Buddha did not introduce the notion of reincarnation." (p. 110)

129. Sanskrit verbal root *budh*, "to fathom a depth, to penetrate to the bottom"; also, "to perceive to know, to come to one's senses, to wake." The Buddha is one awakened to identity not with the body but with the knower of the body, nor with thought but with the knower of thoughts, that is to say, with consciousness.
Joseph Campbell, Myths to Live By 16 (1971).

130. This was not the original title but the name given to it by W.Y. Evans Wentz, editor of its first translation into English in 1927, who showed this title was due to its similarity to the *Egyptian Book of the Dead*, which is also addressed to those who are close to death.

131. Fremantle, Luminous Emptiness 66, 233 (2001).

ring constantly in this life."[132] As noted in Chapters 12 to 14 of this volume, people wish to know of impending death so that they can prepare for it. A Buddhist will state, "As I leave this conditioned body of flesh and blood, I will know it to be a transitory illusion."[133] Someone should read aloud an appropriate text while the family is kept away "so that their grief and anxiety do not cause any disturbance."[134] Transference of life takes place whenever we allow our minds to be turned away from ordinary concerns.[135] The Buddha said: "Do not believe in traditions because they have been handed down for many generations. *Do not believe in anything simply because it is found written in your religious books.* Do not believe in anything merely on the authority of your teachers and elders." (Emphasis added.) The Eightfold Path of Buddha requires knowledge of truth, development of goodwill, consideration of words about others, ethical living, having a job that does not harm others, resistance of evil, self-control, and development of concentration.

Without any supernatural god, Buddhism strives for the salvation of all human beings through internal peace. It exemplifies Jesus's statement, "The kingdom of God is within" (Luke 17:21). They suggest that this should be taught to everyone along with a detachment from cravings of all kinds. During the Cultural Revolution, the Chinese destroyed more than 1,000 monasteries and killed 1 million Tibetans to eradicate most of such teachings. Congressman Frank R. Wolf of Virginia stated, following his 1997 visit to Tibet, "[H]umane progress is not even inching along, and repressed people live under unspeakable brutal conditions in the dim shadows of international awareness."[136] During the twentieth century, the Dalai Lama has been con-

---

132. *Id.* at 7. Buddha was first transmitted orally and committed to writing in the first century B.C.

133. *Id.* at 234. "Dying is transference, which literally means moving from one place to another."

134. CAMPBELL, MYTHS TO LIVE BY 161 (1971), and Fremantle.

135. "It must not be attempted too soon, as that would be tantamount to committing suicide." (p. 239).

> As soon as it is certain that respiration has stopped, the attendant should help the dying person to lie in the sleeping lion posture—lying stretched out on the right side with the head resting on the right hand. Three or four days later, the dying person's attendant is instructed to press the "arteries of sleep" in the neck which will prevent prana [breath] from returning into the right and left nadis.

136. *Quoted in* SCHELL, VIRTUAL TIBET 275 (2000).

demned as a tool of imperialist forces trying to undermine the Chinese communist government even though he speaks of "secular spirituality." But the Dalai Lama never actually condemned the Chinese Communists because "the Buddhist quality of compassion, *karuna,* is equivalent to the Christian of charity, *agape.*"[137] In the year 2000, the Dalai Lama stated: "After 50 years of Chinese rule, Tibet seems no closer to freedom. Although human rights appeals have had little effect thus far, these appeals will no doubt continue to be made; this may eventually help the Tibetans in their fight for self-determination and an end to population transfers. . . ."[138]

Vision is longed for by large portions of present-day humanity, even in America, which has essentially undertaken the materialist pathway. The molecular scientist Mathieu Richard (who became a Buddhist monk) and the philosopher Rovel stated, "Something is still *missing.*"[139] It was noted that "a dialogue has been conducted between Buddhism and science, largely because of an interest in science shown by certain Buddhist thinkers, nota-

---

137. The Dalai Lama has openly stated that Buddhist goals are "the same as those of Western science": to "serve humanity and to make better human beings." *See* DIVORSKY, THE HUMANIST (May/June 2004), at 8. "For me, as a Buddhist . . . he—Jesus—was either a fully enlightened being or a bodhisattva of a very high spiritual realization." DALAI LAMA, THE GOOD HEART: A BUDDHIST PERSPECTIVE ON THE TEACHINGS OF JESUS 96 (1996).

138. Hall, *Chinese Population Transfer in Tibet*, 9 CARDOZO J. INT'L & COMP. L. 173 (2001). The right of the Tibetans to self-determination has never been recognized by the United Nations.

139. RICCARD ROVEL, THE MONK AND THE PHILOSOPHER (1998). In commenting upon Voltaire's praise for the London Stock Market in a country where there are very many and very different value systems, so that they are forced to put up with one another, Auerbach states:

> It is the unexpected contrast of religion and business, in which business is placed higher, practically and morally, than religion. The very device of coupling the two, as though they were forms of human endeavor on the same plane and to be judged from the same viewpoint, is not only an impertinence; it is a specific approach or, if one prefers, an experimental setup, in which religion is ipso facto deprived of what constitutes its essence and its value.

AUERBACH, MINIMIS 354 (1996).

bly the Dalai ["Ocean of Wisdom"] Lama."[140] This acknowledges that "modern science isn't Buddhism's main preoccupation. But there is *interest in the findings of science because Buddhism has long been asking similar questions.*"[141] Einstein stated, "If there is any religion that could respond to the needs of modern science, it would be Buddhism."[142] The Dalai Lama said, "I do not personally advocate seeking a universal religion; I don't think it advisable to do so. And if we proceed too far in drawing these parallels and ignoring the differences, we might end up doing exactly that!"

The word *criticism* comes from the Greek *kritikos,* which means "able to judge." Justice Oliver Wendell Holmes once stated, "The broad-minded see the truth in different religions, the narrow-minded see only the differences." This word tends to have a negative image, rather than an objective judgment, as exhibited by Justice Holmes. The question becomes, to what extent does religion influence the moral behavior of certain modern groups of people?

Religious forms in this country have various views on morality. The Internal Revenue Service will grant a new religion a tax exemption, provided the religion is organized with a sectarian structure.[143] Jehovah's Witnesses, for instance, do not take blood transfusions because of their interpretations of certain portions from the Bible.[144]

---

140. Richard & Thuan, The Quantum and the Lotus 2 (2001).

141. *Id.* at 20. Supreme Court Justice William Douglas found that the United States was not only Christian, but was also "a nation of Buddhists, Confucianists and Taoists." United States v. Seeger, 380 U.S. 163 (1965).

142. *Id.* at 32. However, the Dalai Lama has stated, "Science can operate without spirituality. Spirituality can exist without science. But man, to be complete, needs both."

143. 26 U.S.C. § 501(c)(3). *See* Slye, *Rendering unto Caesar: Defining Religion for Purposes of Administering Religion-Based Tax Exemptions*, 6 Harv. J. L. & Pub. Pol'y at 219, 259-61 (1983).

144. "All fat is the LORD's. It shall be a perpetual statute throughout your generations, in all your settlements: you must not eat any fat or any blood." *Leviticus* 3.17. 1-14. *See also Acts* 15:28-29. However, *Genesis* 9:4 states about blood of the living: "Only you shall not eat flesh with its life, that is, its blood." A study was conducted of a single-center, randomized, double-blind, controlled trial involving children less than one year of age who underwent open-heart surgery, using fresh whole blood or reconstituted blood by bypass-circuit priming. In 2004, it was concluded that the use of fresh whole blood had no advantage over the use of a combination of packed red cells and fresh-frozen plasma during surgery for congenital heart disease. Mou et al., *Fresh Whole Blood versus Reconstituted Blood for Pump Priming in Heart Surgery in Infants*, 3 N. Eng. J. Med. 1635 (Oct. 14, 2004).

Joseph Franklin Rutherford, a lawyer, assumed leadership of the movement; he operated the Watchtower Bible and Tract Society and in 1931 renamed it the Jehovah's Witnesses. Contemporary Witnesses also refuse to participate in military service, and in 1999 their membership worldwide approached 8 million.[145] Death may result in certain cases where a necessary transfusion is denied. The highest proportion of Jehovah Witnesses are female adherents (68 percent).[146] A physician responsible for the death of a patient because of the refusal of a blood transfusion has the ability to defend on such grounds.[147] (See Chapter 14.)

I attest that the only God is Allah, and that Muhammad is His messenger.

Muhammad, in Medina, where he died on June 8, 632

The number of Muslims has increased in the modern world.[148] More than a billion people (one-sixth of humanity) are Muslim. Their belief is that the prophet Muhammad's Koran "is the divine remedy for the distortions that arose in the wake of Islam's earlier prophets—Moses and Jesus."[149] Founded in Egypt in 1928, the Muslim Brotherhood has spread to dozens of countries, promoting a return to so-called "pure" Islamic principles.[150]

---

145. STEIN, COMMUNITIES OF DISSENT 83 (2005).

146. SO. CALIF. CHRISTIAN TIMES, December 2001, at 3. Jehovah's Witnesses also oppose contraception, sterilization, elective abortion, artificial insemination, and genetics/eugenics practices.

147. ORANGE COUNTY REGISTER, June 24, 1999, p. 7. *The Watchtower* of 2004 stated that it is "a Bible-based journal that is devoted to announcing the good news of Jehovah's Kingdom [and] is regularly printed in more languages than any other periodical." (p. 7) In 2004, a Moscow court banned the religious activities of Jehovah's Witnesses. This move critics called a step back for religious freedom. The court thought it to incite hatred or intolerance. L.A. TIMES, Mar. 28, 2004, at A18.

148. SO. CAL. CHRISTIAN TIMES, December 2001.

149. NATIONAL GEOGRAPHIC, January 2002, at 78, and December 2001, at 93. "There are about 1.2 million U.S. residents whose ancestry is solely or partly Arab, less than a half-percent of all Americans. Arabs are nearly twice as likely as the typical U.S. resident to possess a college degree—41 percent to 24 percent. The median income for an Arab family was $52,300, about $2,300 more than the median income for all U.S. families." ORANGE COUNTY REGISTER, March 9, 2005, at News 9.

150. ORANGE COUNTY REGISTER, May 15, 2005, at News 9.

Terrorism by some Islamic believers, which was demonstrated by the destruction of New York's highest buildings on September 11, 2001, and, for example, resulted in a "war" against the former rulers of Afghanistan.[151] Ironically, this was at a time of rapid growth for believers of Islam. Back in 1958, it was written, "In some areas where Islam and Christianity are competing for converts, Islam is growing at rate of ten to one."[152] In 2002, there was a reported 1.2 billion Muslims worldwide, and 6 million in the United States alone.[153] One American scientist ranked the three most influential persons in history in this order: Muhammad, Isaac Newton, and Jesus. Muhammad was the top because he was "the only man in history who was supremely successful on both the religious and secular levels. Within a century after his death, on June 8, 632 A.D, his followers controlled the largest empire in human history."[154] For more than a millennium, both hot and cold wars existed "between Islam and the Christian West, with all the paranoia, wish fulfillment, and sense of unreality implied by that term."[155]

---

151. "The men who attacked America on September 11, 2001, were in effect attacking Islam itself as a religion, a culture, and a civilization. They disfigured and betrayed it, for nowhere in the Islamic texts does it say that you must commit suicide while killing as many people as possible. In Islam, suicide is prohibited and punished in the beyond." JELLOUN, ISLAM EXPLAINED 2 (2002). The Koran states, "Whoever slays a human being, unless it be for murder or for spreading corruption on earth, it shall be as though he had slain all mankind" (5:32). "Do not kill yourselves. . . . Whoever does so, in transgression and wrongfully, we shall roast in a fire" (4:29). ORANGE COUNTY REGISTER, Aug. 7, 2005, at News 16.

152. SMITH, THE RELIGIONS OF MAN 252 (1958).

153. ORANGE COUNTY REGISTER, Nov. 6, 2002, at News 3.

154. *See* PHIPPS, MUHAMMAD AND JESUS 10 (1996). The Islamic numbering of years begins with Mohammed's arrival in Medina in A.D. 622. The calendar is based on the moon's phases. Its year is shorter than the Christian year, which is based on the sun. The moon circles the earth in 29.53 days.

155. FOSS, PEOPLE OF THE FIRST CRUSADE 18 (1997). Mohammed was deemed to be the last of the Prophets to teach the duty and Islam (submission) of God. In 632, he completed his *Koran* or "reading," which is about the size of the *New Testament*. The Islam ideals on punishment resemble retaliation addressed in the *Old Testament*, "an eye for an eye." However, unlike other religions, the *Koran* permits marriage to more than one wife, slavery, and requires prayer five times a day. Saleh al-Sayeri of Saudi Arabia spent much money on weddings and divorce settlements for his 58 wives. He said three of his four current wives have been with him for many years; but "I like to change my fourth wife every year." ORANGE COUNTY REGISTER, Jan. 2, 2005, at News 32.

Like the Catholics, Muslims believe in heaven and hell, and the *Qur'an* refers to Satan. Although most Christians assume that they should be limited to one wife, Muhammad was married to several wives. The Five Pillars of Islam faith include belief in the unity of God and Muhammad as his prophet, prayer, almsgiving, fasting, and pilgrimage. Islam has the potential to expand beyond the current religious followers in America. In Europe, Islam has already become the continent's second largest religion, as it is in France. The Muslim form has greatly appealed to American blacks as well.[156] Islam does not mean peace, but, as Joseph Campbell indicated, it means "submission to their God."[157] There are no miracles and there is no confession in Islam. Although persons preside at prayer and deliver sermons on Friday in the mosque, they do not have the same role as priests or rabbis. Unlike their sources, the prophet Muhammad stated, "From cradle to grave, go in search of knowledge, for whoever longs for knowledge loves God." "Allah is identifiable as the god of Jews and Christians, but as he reveals himself to his messenger."[158] However, "Isa (Jesus) was the penultimate prophet of Islam. He is believed to be *Al-Masih,* the Messiah, and *Kalimat Allah*, the Word of God, but not the Son of God."[159] The prophet

---

156. An extensive study in 1997 found that blacks are significantly more religious than whites: "About 80 percent of blacks and 52 percent of whites said religion was very important." Black adults are most likely to practice as Baptists (57 percent) but . . . have also been known to practice Islam. L.A. Times, April 19, 1997, at B8.

157. Campbell, Reflection on the Art of Living 148 (1991). "This credo gives warriors enormous courage and power: whether they're going to get killed or not get killed, they move in submission to the fate."

158. Leeming, Jealous Gods, Chosen People 121 (2004), *citing* "It has been revealed to me that your god is one god" (*Qur'an* 41:6).

159. *Id.* at 125, *citing Qur'an* 3:40, 4:169, 171. Like the Jews, "Being circumcised is an important rite of passage for any Muslim male" by elders who say a prayer before removing the foreskin with a knife—without anaesthetic. However, it is done later (i.e., between the ages of two and five). Although circumcision causes virtually no medical harm, it offers practically no benefit, either, according to a study made at the end of the 20th century. Orange County Register, Jan. 11, 1999, at News 13.

of Islam wrote, "Do not cheat or commit treachery, nor should you mutilate or kill children, women, or old men."[160]

Until the third millennium, "American Christians had usually followed their government in expressing an absolute and generally uncritical support for the state of Israel."[161] However, students in most Muslim lands are only asked to read the sixth century AD *Qur'an* and not the religious foundational texts from other countries.[162] This is true despite the fact that Islam is also based on Judaism; "Muhammad accepted the Hebrew patriarchs, the prophets, the Hebrew Bible, and monotheism (the belief in one God) as part of his religion."[163] Unlike Muslim schools, the Chapel Hill campus in North Caro-

---

160. However, in Iran a history professor named Aghajari was found guilty of insulting the Prophet Muhammad and questioning the clergy's interpretation of *Islam*. He was sentenced to death in 2002 and only relieved after thousands of university students and some teachers boycotted classes in protest. ORANGE COUNTY REGISTER, Nov. 12, 2002, at News 15.

161. This fact infuriates not just the bulk of the world's Muslims, but also many Third-World Christians. JENKINS, THE NEXT CHRISTENDOM 181 (2002). It was interesting to note in 2004 how the Muslim country Turkey's hopeful junction with the European Union (EU) was delayed when Turkey announced its plans to outlaw adultery. The EU's view is the opposite of Moses's commandment "You shall not commit adultery," *Exodus* 20:14, and expressed developments since Jesus condemned the ancient Jewish "sinful method" of punishing a female who committed adultery. *Matt.* 5:28, *but see John* 8. Accordingly, the Turkish Prime Minister stated that his government was dropping the contentious issue. ORANGE COUNTY REGISTER, Sept. 24, 2002, at News 29.

162. Like the *New Testament*, the *Quran*'s pages were written in a Semitic language and produced by others following Muhammad's death in 632. It also covers more about the *Old Testament*. However, "only 350 verses, or somewhat less than 3 percent of the received quranic text, is legal in content." These verses have been further broken down, with some disagreement on details, into 140 on dogmatic and devotional matters like prayer, fasting, pilgrimage, and the like; 70 on questions of personal status (marriage, divorce, inheritance, etc.); 70 more on commercial transactions (sales, loans, usury); 30 on crimes and punishments; another 30 on justice; and a final 10 on economic matters.

163. THE WORLD BOOK ENCYCLOPEDIA. However, the Muslim scriptures "preach anti-Judaism, seeing the Jewish religion as an inadequate stepping-stone on the path to the true faith. . . . Anti-Semitism is as normal and unexceptional in the Muslim world today as it was in the Europe of the 1920s." JENKINS, THE NEW ANTI-CATHOLICISM 190-91 (2003). *Quran* 22:26 reports that Abraham came to Mecca in search of

lina gave an assignment to 4,200 students to read and discuss the volume *Approaching the Qur'an* (2001). This was to help students understand Islam after the terrorist attacks on buildings in New York and part of the Pentagon in Washington, D.C., on September 11, 2001. Some students and others sought a restraining order against this; however, the request for this order was rejected by a federal judge.[164] The war that commenced in Iraq demonstrated an early split between Sunni and Shi'a Muslims (who had vested more interest in the prophet's family). These, along with other divisions in the Islamic world, appear to some people as partly comparable to the Jewish split with Christians (who later divided among themselves). However, it has been stated that struggles before and after 9/11 may be broadening the boundaries of Islam or making it "denominational."[165] Nevertheless, in the United States, Muslims are divided into two categories, African-American and immigrant, and studies of them were seldom connected,[166] even though "most converts in the mosques were African-American men."[167] "In *Approaching the Qur'an*, which contains substantial and important material on Muhammad's work, Michael Sells states:

> The message of the Qur'an is more explicitly fitted into a prophetic lineage beginning with the creation of Adam, the first prophet in Islam, extending through the stories of Noah, Abraham, Isaac, Jacob, Joseph, Moses, John the Baptist, and Jesus, prophets of the Arab tra-

---

his son, and on that occasion, when Ishmael was grown up, they built the Kaaba together and Abraham instituted the practices of the hajj.

164. Orange County Register, Aug. 16, 2002, at News 7; USA Today, Aug. 20, 2002, at 34. The *Code of Islamic Laws* and subtitle as containing "the criminal and civic laws of Islam directly deduced from the Qur'an" was published in Lahore, Pakistan, in 1997. Pelikan, Interpreting the Bible and the Constitution 17 (2004), *citing* Muhammad Sharif Chaudhry, Code of Islamic Laws (1997).

165. Leonard, Muslims in the United States 142 (2003).

166. "Studies of African-American Muslims stressed black nationalism and racial issues, and studies of immigrants focused on ethnic identity, ethnic entrepreneurship, and gender and generational issues. There was very little, if any, scholarly dialogue across these boundaries." *Id.* at 130.

167. "Between 65 and 70 percent of all mosques agreed 'somewhat' or 'strongly' with the statement that 'America is immoral'; African-American mosques felt that most strongly." *Id.* at 76. "In 2000 in New York State, 32 percent of the African-American prison population was Muslim." *Id.* at 8.

> dition such as Hud and Salih, and ending with Muhammad. On the other hand, most Muslims view the Qur'an as the direct revelation of God to Muhammad. (p. 15)

"Karim" is one of the principle "names of God" in the *Qur'an* [in addition to "Allah" or "God"]. Like creation, the day of reckoning is a boundary moment, an intersection of time and eternity. And as with creation, at the day of reckoning, the spirit is the agent that brings together the eternal and the temporal. (p. 197)

One Muslim scholar and professor of international relations wrote: "Any war against unbelievers, whatever its immediate ground, is morally justified. Only in this sense can one distinguish just and unjust wars in Islamic tradition."[168] However, others emphasize the fading in "the differentiation between Israeli conduct and American foreign policy."[169] In 1988, Salman Rushdie, an Indian expatriate writer, published the novel *The Satanic Verses,* which enraged some orthodox Muslims. The Ayatollah Khomeini of Iran gave this writer a fatwa, or death sentence. However, in 1998, Khomeini's successor revoked the fatwa. During a 2003 interview, Rushdie made a comment regarding "the turbulence" in a "democracy" with freedom of speech—which often represents the manner in which bioethics is discussed.[170] (See Introduction *supra.*) Four years later, a student of Arab politics specified:

> The idea of representation, of elections, of popular suffrage, of political institutions being regulated by laws laid down by a parliamentary assembly, of these laws being guarded and upheld by an inde-

168. Tibi, *War and Peace in Islam, in* ISLAMIC POLITICAL ETHICS: CIVIL SOCIETY, PLURALISM, AND ETHICS 175-93, Sohail H. Hashmi ed. (2002).

169. "Arab public opinion widely believes that the Jewish lobby in the U.S. is the sole determinant of American policy in the region and therefore makes little distinction between U.S. foreign policy and Israeli abuse of the Palestinians." AMANT, THE AGE OF TERROR 48 (2002).

170. I think that democracy, freedom, art, literature—these are not tea parties, you know? These are turbulent, brawling, arguing, abrasive things. I've always seen the work of the imagination and the world of the intellect as being turbulent places. And, you know, out of turbulence come sparks, which are sometimes creative and sometimes not. But without that turbulence, in a calm sea, nothing happens.

COLUM. UNIV. MAGAZINE (Fall 2003).

pendent judiciary, the ideas of the secularity of the state . . . all these are profoundly alien to the Muslim political tradition.[171]

Hinduism of India and elsewhere in the Far East has been called "one of the world's most complex and interesting religions . . . with no one deity worshipped by all" and "with a history of 45 centuries."[172] They believe that animals, as well as human beings, have divine souls. The soul is neither born nor dies; rather it transmigrates from body to body according to a person's deeds during a lifetime. The caste system remains; there are 160 million "Untouchables" in India: "Branded as impure from the moment of birth, one out of six lives and suffers at the bottom of the Hindu caste system."[173] The law of *karma* determines the spiritual quality of the next life based on the actions taken in the previous life. Some 500,000 Hindus live in the United States[174] and they have numerous myths of creation, including life in human, or other, form. The dead are typically cremated and the ashes often scattered in a river. They hope to be released from the cycle of reincarnation and achieve *moksha,* from which the soul never returns. Hindus may be regarded as "one religion consisting of a set of diverse groups." Nevertheless, it has been remarked that "Basic to Hindu ethics is the concept of *dharma,* upholding the order of society and the universe through law, custom, and religion—parallel to the Christian concept of 'natural law.'" That is, their highest goal, *moksha,* or release from rebirth, is realized when all *karmas* (memory traces from previous lives) are removed by the practice of ethical and spiritual living *(yoga).*[175] The population of India exceeded 1 billion at the turn of the third

171. ELIE KEDOURIE, DEMOCRACY AND THE ARAB POLITICAL CULTURE 5 (Washington Institute for Near East Studies, 1992).

172. SULLIVAN, THE A TO Z OF HINDUISM 1 (2002).

173. NATIONAL GEOGRAPHIC, June 2003.

174. Id. at 16. Like the United States, "the Government of India is a secular and democratic institution." (p. 174) Nevertheless, "India's census reports that about 83 percent of its citizens are Hindu." (p. 3) In the United States, Hindus continue to use unique temples. Several of them were constructed just north of Pepperdine University in Malibu, California, on Las Virginas Road.

175. *See* Cromwell Crawford, *Hindu Bioethics for the Twenty-First Century,* JAMA 2759 (June 9, 2004). "In spite of Hindu stress on having a son, Hindu bioethics finds prenatal sex selection to be a perverted use of modern science." *Id.* A demographic study in the United States recorded 1,088 people who claimed to have had past lives. HELEN WAMBACH, RELIVING PAST LIVES: THE EVIDENCE UNDER HYPNOSIS 62-64 (1978).

millennium—the second largest populated country in the world.

With such diversity, how can one religion be used as a standard? Regardless of diversity, many Americans appear to have separated religion from their daily secular activities.[176] In 1998, a theologian and physicist gave a critique of current Christian ecumenical thinking: "The World Council of Churches has more than once issued documents that are simplistic and misleading about issues that required much more balanced and expert consideration." Most important, he expressed the following: "The science and technology community has so far devoted comparatively little attention to ethical matters."[177] In 1953, the Church of Scientology was formally incorporated. "The Internal Revenue Service has had unending conflict with the Church of Scientology, at one point revoking its tax-exempt status. Individuals have repeatedly filed lawsuits against the church, claiming that they were duped out of large sums of money that they paid for auditing." As of 2003, there have been "more than 700 local branch churches affiliated with the Church of Scientology, which has its headquarters in Los Angeles."[178] The Italian novelist Umberto Eco wrote that "Counterculture is an overused term, which, like Resistance, is invariably mentioned so as to reflect well on the person using it. Nobody is against the Resistance, and nobody nowadays would dare suggest that there was anything negative about counterculture . . . *It is religious reform.* It is the heresy of whoever confers a license upon himself and prefigures another church. It is the only cultural manifestation that a dominant culture is unable to acknowledge and accept . . . *offering a model that is capable of sustaining itself.*"[179]

---

176. In his 19th century *Revolt of the Angels*, Anatole France wrote:

> In the century in which we live [the 19th] there are so many modes of belief and of unbelief that future historians will have difficulty in finding their way about. But are we any more successful in disentangling the condition of religious beliefs in the time of Symmachus or of Ambrose? The Episcopalians, the Assembly of God, Latter-Day Saints, American Lutherans, oppose elective abortion, as do Lutherans who also oppose contraception and the use of artificial insemination by a donor other than spouse.

177. POLKINGHORNE, BELIEF IN GOD IN AN AGE OF SCIENCE 92 (1998).

178. STEIN, COMMUNITIES OF DISSENT 129-30 (2003).

179. Eco, *Does Counter-culture Exist?*, APOCALYPSE POSTPONED, Robert Lumley ed. (1994), at 115.

Nevertheless, by the year 2000, "[A]dvances in the rational understanding of the world would inevitably diminish the influence of the last, vexing sphere of irrationality in the human culture: religion."[180] In this connection, in 2005, it was reported that unlike the United States, "Europe is a post-Christian society. Europeans seem to want the church to be available for major life events like christenings, marriages, and funerals ('hatched, matched, and dispatched'), but almost never otherwise darken the door of a parish church."[181] Personal experience with medical ethics committees indicates that membership often includes some physicians and professionals with little or no religious faith.[182]

Diversity in religion, secularism, and philosophy gives rise to the search for a substitute in reason. Goethe's aphorism, "He who is ignorant of foreign languages, knows little of his own,"[183] applies to religion and ethics; it tends to facilitate the attitude of reasonableness for bioethicists. However, in 1944, the Supreme Court held the declaration of religious truth as not being the business of government.[184]

Feeling is the deeper source of religion.

William James

You are a god only insofar as you recognize yourself to be a human being.

Plutarch

---

180. SHERMER, HOW WE BELIEVE: THE SEARCH FOR GOD IN AN AGE OF SCIENCE (2000).

181. ORANGE COUNTY REGISTER, Apr. 21, 2005, at Loc. 9.

182. "It will do no good to cling to our religion in the hope that one day religion or metaphysics will once again be back. They are disappearing forever and that means we can now let go and immerse ourselves in the new world of the secular city." MUZZY, ETHICS AS A RELIGION 145 (1967).

183. JOHANN GOETHE, KUNST AND ALTERTUM (1832). Some 200 religions, or sects of other religions maintained by many cultures are described in the volume *New Religions* (2004), which is a guide edited by Christopher Partridge, Professor of Contemporary Religion at University College, Chester, England (with writings by an international team).

184. United States v. Ballard, 322 U.S. 78, 86-88 (1944).

> *I envy those Chinese schoolboys who were made to memorize every word of Confucius. Never has one man so written his name upon the face and spirit of a people as Confucius has done in China; perhaps the greatest of all historic civilizations is summed up in his name.*
>
> Will Durant,
> *The Greatest Minds and Ideas of all Time* (2002), p. 110

Scientist and Nobel laureate Richard Feynman spoke about the word *atheism* and the moral in the Golden Rule. He stated that "it is possible to doubt the divinity of Christ, and yet to believe firmly that it is a good thing to do unto your neighbor as you would have him do unto you. It is possible to have both these views at the same time; and I would say that I hope you will find that my atheistic scientific colleagues often carry themselves well in society."[185] But the majority of his colleagues were largely agnostic. A sixth-grade student's questions about religion resulted in the description of a "religious feeling of a special sort," according to Albert Einstein.[186] An item in 1987's Public Opinion Poll of the "Political Psychology of Chinese Citizens" found that more than 90 percent were atheists. However, prior to the Communist control of China in

185. FEYNMAN, THE PLEASURE OF FINDING THINGS OUT 251 (1999). Such an atheist was accepted as a teacher at the Simi Valley United Church of Christ. L.A. TIMES, Dec. 11, 2001, at B1. The atheist Stuart Besano, who participated in their Sunday School, stated that he "did so in part because he wanted to 'teach tolerance for people like me.'"

> "The process by which a society becomes increasingly secular (non-religious) has been termed 'secularization.' Although aspects of the *secularization* thesis are debated, looking back over the past couple of centuries it would seem to be overwhelmingly evident that religious beliefs, practices and symbols are gradually being abandoned at all levels of modern society."

Partridge, *The Modern Western World,* NEW RELIGIONS (2004), at 358.

186. Everyone who is seriously involved in the pursuit of science becomes convinced that a spirit is manifest in the laws of the Universe—a spirit vastly superior to that of man, and one in the face of which we with our modest powers must feel humble. In this way the pursuit of science leads to *a religious feeling of a special sort, which is indeed quite different from the religiosity of someone more naïve.* (Emphasis added.)

A. EINSTEIN, HUMAN SIDE 33.

the middle of the twentieth century, Chinese religion was "an amalgam of the 'Three Teachings' (Confucianism, Taoism, Buddhism) and the folk tradition."[187] William James, professor of psychology at Harvard University, wrote in his volume, *The Varieties of Religious Experience,* combining atheism solely with the absence of belief in immortality:

> Religion, in fact, for the great majority of our own race means immortality, and nothing else. God is the producer of immortality; and whoever has doubts of immortality is written down as an atheist without farther trial . . . I know not how to decide. It seems to me that it is eminently a case for facts to testify. Facts, I think, are yet lacking to prove "spirit-return." I consequently leave the matter open, with this brief word to save the reader from a possible perplexity as to why immortality got no mention in the body of this book . . . I feel bound to say that religious experience, as we have studied it, cannot be cited as unequivocally supporting the infinitist belief.[188]

This viewpoint parallels that of many scientists and transcendentalists. As noted in Chapter 1, astronomers have evidence concerning the beginning

---

187. Confucianism has its origins in the oldest school of Chinese thought. Confucius (551-479 B.C.E.) developed a philosophical code of conduct, which was not a religion. He "sought to replace the old religious ideas with moral values" and wrote that everyone should "not do to others what you would not have them do to you." (*Analects* 15.23) PARTRIDGE, NEW RELIGIONS 212-13 (2004). Buddhism provides an elaborate cosmology, a structured priesthood, and a detailed theory of the afterlife. Taoism meets other needs, and offers methods of spiritual and physical healing, a means of commerce with the spirit world, and securing blessings and protection. (p. 9) "At the heart of Taoism is the *Tao te Ching* (the *Classic of the Way and Its Power*). Tradition has it that these teachings were written in the sixth century BCE by Lao-tzu. The text is now thought to be the work of several people and to date from the fourth century BCE; it is likely that Lao-tzu is a mythological figure." OLDSTONE MOORE, TAOISM 35 (2003).

188. JAMES, THE VARIETIES OF RELIGIOUS EXPERIENCE 514-15 (1902). "In the 1960s, the atheist Madalyn Murray O'Hair became notorious by having successfully sued to remove mandatory religious ceremonies from public schools. She and two members of her family were kidnapped and murdered by David R. Waters. In 2003, he died of lung cancer in a Federal prison hospital." ORANGE COUNTY REGISTER, Feb. 5, 2003, at News 18.

of the universe, but no evidence prior to the Big Bang 13.7 billion years ago. Through empirical observation, they can never deny the recognition of the Creator. One scientist stated that the empiricist view "concedes that moral codes are devised to conform to some drives of human nature and to suppress others."[189] For those people, he said, "[E]mpiricism is winning the mind; 'for others,' transcendentalism continues to win the heart." For "those who hunger for both intellectual and religious truth will never acquire both in full measure."[190]

Nevertheless, discussions by members of hospital ethics committees (among each other and patients) can satisfactorily indicate that scientists, "the roving scouts of the empiricist movement," are not immune to the idea of God. It may or may not be necessary to also mention that many would deny a belief in "the personal God who once presided over Western civilization." Yet, they would agree there is "a biologically based human nature, which is relevant to ethics and religion."[191]

---

189. WILSON, CONSILIENCE, THE UNITY OF KNOWLEDGE 256 (1988). It has been noted that many articles in the medical literature and lay press have supported a belief that individuals, including those dying of cancer, can temporarily postpone their death to survive a major holiday or other significant event associated with Christmas, Thanksgiving, or the date of the individual's birthday. However, a large study completed in 2004 "found no evidence, in contrast to previous studies, that cancer patients are able to postpone their deaths to survive significant religious, social, or personal events." Young & Hade, *Holiday, Birthdays, and Postponement of Cancer Death*, JAMA 3012 (Dec. 22/29, 2004).

190. *Id.* at 262. In 1981 President Reagan was shot by a deranged young gunman named John Hinckley, Jr. After leaving the hospital and returning to the White House, he said, "Whatever happens now, I owe my life to God and will try to serve him every way I can." SCHWEIZER, REAGAN'S WAR 135 (2002).

191. Wilson, *supra* note 189 at 263-64.

> It may or may not be true that the great proportion of the world's journalists are not very religious, do not assign religion much importance in world affairs, and do not care enough about religion to study it intensively. It is true that many of them have long and complacently accepted the "secularization thesis," according to which the world is inexorably and inevitably becoming more secular.

NOVAK, THE UNIVERSAL HUNGER FOR LIBERTY 225 (2004).

The more energetic the political activity in the country, the greater the loss to spiritual life.

Aleksandr Solzhenitsyn

Reason obeys itself, and ignorance submits to whatever is dictated to it.

Thomas Paine

Four things belong to a judge: to hear courteously, to answer wisely, to consider soberly, and to decide impartially.

Socrates

3. The "Wall of Separation" and Legal Bans on Teaching Religion in Public Schools

Because the Constitution prohibits Congress from both making any "law respecting an establishment of religion, or prohibiting the free exercise thereof," Congress can neither appropriate funds solely to benefit a religion, or use religion to impose laws on people's lives or public schools.[192] The Supreme Court stated, "The First Amendment has erected a wall between church and state. That wall must be kept high and impregnable. We could not approve the slightest breach."[193] This concept was derived from an 1802 let-

192. Secularism became official in France starting on December 9, 1905, the date when the *separation of church and state became official.* An example: in French public schools, the clergy do not have the right to teach. They can, on the other hand, have their own schools. There are churches, synagogues, and mosques. The state does not get involved in the practice of religion. Turkey was the first Muslim country to become a *secular state* [1922]. (Emphasis added.) JELLOUN, ISLAM EXPLAINED 83 (2002).

193. Everson v. Bd. of Educ. of the Twp. of Ewing, 330 U.S.I. (1947). The "wall of separation" was first mentioned by Thomas Jefferson. The concept derived from John Locke's *A Letter Concerning Toleration*, 1689, which stated:

> I esteem it above all things *necessary to distinguish exactly the business of civil government* from that of religion, and to settle the just bounds that lie between the one and the other. (p. 17) For every church is orthodox to itself; to others, erroneous or heretical. Whatsoever any church believes, it believes to be true; and the contrary thereupon it pronounces to be error. So that the controversy between these churches about the truth of their doctrines, and the purity of their worship, is on both sides equal; *nor is there any judge*, either at Constantinople, or elsewhere upon

ter from Thomas Jefferson to the Danbury Baptists, in which Jefferson, a deist in the "Age of Enlightenment," described "a *wall of separation between church and State*. Believing with you that religion is a matter which lies solely between man and his god . . . their legislature should make no law respecting an establishment of religion."[194] It became applicable to states under the Fourteenth Amendment following the Civil War.

In 1972, the Supreme Court attempted to define the separation: "To have the protection of the Religion Clauses, the claims must be rooted in religious belief . . . philosophical and personal beliefs do not rise to the demands of the Religion Clauses."[195] Nevertheless, the Court ruled that the Old Order Amish was entitled to remove its children from public school prior to high school. "The Amish preserve the teachings of Jacob Amman, who broke with more moderate Mennonites in seventeenth century Switzerland on the issue of shunning—or rejection of those in the religious community who don't keep the faith. They believe the biblical command 'do not be conformed to this

---

earth, *by whose sentence it can be determined.* The decision of that question belongs only to the Supreme Judge of all men. (p. 24). (Emphasis added.)

194. Cord, Separation of Church and State: Historical Fact and Current Fiction 18 (1982). Jefferson was able to admire, respect, and perhaps even love "the first of all Sages" only because he was able to separate the religion of Jesus from the religion of Christianity. Jefferson was not of a sect by himself. Millions of Americans today, Christian and otherwise, harbor similar sentiments. In this sense, Jefferson was a Founding Father not only of the United States of America but also of today's Jesus nation. Prothero, American Jesus 42 (2003).

Jefferson's plan was to open people's minds to deism and secularism, in addition to traditional Christianity, while freeing them from the involuntary civil financial support of religious establishments. Thomas Jefferson, Genius of Liberty 37 (2000).

195. *See* Wisconsin v. Yoder, 406 U.S. 205, 215-16 (1972). In *Epperson v. Arkansas*, 393 U.S. 97 at 106, the Supreme Court stated, "The First Amendment does not permit the State to require that teaching and learning must be tailored to the principles or prohibitions of any religious sect or dogma." In 1963 the Supreme Court struck down a Pennsylvania law that each public school day be started with the reading of at least 10 verses from the Bible. Justice Clark wrote: "[To] withstand the strictures of the [1st Amendment's] Establishment Clause, there must be a secular legislative purpose and a primary effect that neither advances nor inhibits religion." Abington School Dist. v. Schempp, 374 U.S. 203 (1963).

world' requires rejection of higher schooling and modern technology."[196] This special protection, granted to the religious community, has led to some serious health matters.[197] Toward the century's end, courts have upheld the separation in many cases.[198] In 1995, the student government in a California community college voted to quit opening its meeting with the Pledge of Allegiance because of the phrases "under God" and "justice for all." The trustees of the college voted unanimously not to mandate recitation of the pledge before meetings of student and faculty groups on the campus.[199]

In 1980, former Arkansas governor (and later President) Bill Clinton was replaced by a Christian fundamentalist, Governor Frank White. Early in his term, a bill drafted by a San Diego, California, group, referred to "scientific evidences" indicating the "sudden creation of life from nothing," and "the insufficiency of mutation and natural selection in bringing about development from a single organism." This bill also addressed the "separate ancestry of man and ape," as well as a "worldwide flood and recent inception of the earth and living things."[200] This bill, signed by Governor White, was also

---

196. ORANGE COUNTY REGISTER, Nov. 5, 2005, at News 28.

197. Their high number of metabolic disorders made them the subject of genetic research and reason. *The Amish: A Naturalistic Laboratory for Epidemiologic and Genetic Research*, AM. J. PSY. 140, No. 1 (1983), at 75-76; Byers et al., *Bias Crime Motivation: A Study of Hate Crime and Offender Neutralization Techniques Used Against the Amish*, J. CONTEMP. CRIM. JUSTICE 15, No. 1 (1999): 78-96. This group is exempt from newborn screenings, KY. REV. STAT. ANN. § 214.155 (Michie 2000), and immunizations as well, FLA. STAT. ANN. § 2012–71–14 (2001).

198. In 1999 an appellate court held that allowing high school students to choose whether to deliver benedictions at graduations and football games without requiring that prayers be nonsectarian violated the First Amendment. Doe v. Santa Fe Indep. Sch. Dist., 171 F.3d 1013, 5th Cir. 1999. *See also* Coles v. Cleveland Bd. of Educ., 183 F.3d 538 (6th Cir. 1999), Lee v. Weisman, 505 U.S. 577, 1992 ("legislative prayers"), and Marsh v. Chambers, 463 U.S. 783, 1983.

199. ORANGE COUNTY REGISTER, Dec. 5, 1995.

200. HAYS & STEELY, THE BAPTIST WAY OF LIFE 171 (1981).

201. *Id.* at 46. Orthodox creation science embodies four basic tenets:

(1) that a divine Creator created the world from nothing ("ex nihilo");
(2) that the Creator fashioned distinct "kinds" (or "types") of plants and animals that cannot give rise to new "kinds";
(3) that a "worldwide flood" or "Deluge" formed fossils and other paleontological and geological phenomena; and

enacted in several states and required that "balanced treatment" be given to "Scientific Creationism" (see Chapter 2), along with science in classrooms and textbooks of public schools to the extent that they "deal in any way with the subject of the origin of man, life, the earth, or the universe."[201] Such an approach contradicted the words of the Jewish writer Maimonides, who, almost 1,000 years ago, wrote his guide for the perplexed: "First, the account given in Scripture of the Creation is not, as is generally believed, intended to be in all its parts literal. It is therefore right to abstain and refrain from examining this subject superficially and unscientifically."[202]

The enactment of the statute in Arkansas sparked the most significant and important legal trial on religion and the law in the history of the United States. The statute's term, *Scientific Creationism,* fell when used in the biblical sense. The Federal District Court for Eastern Arkansas found Creationism not to be science, and that it constituted an attempt to advance a particular religious viewpoint through legislation.[203] The trial on the merits was the first one to consider the testimony of a number of scientific as well as religious experts.[204] The testimony introduced by several scientists was consis-

---

> (4) that the universe had a "relatively" recent inception (within the past 10,000 years).

FAIGMAN, LABORATORY OF JUSTICE 320 (2004).

202.
> We must blame the practice of some ignorant preachers and expanders of the Bible, who think that wisdom consists in knowing the explanation of words, and that greater perfection is attained by employing more words and longer speech. It is, however, right that we should examine the Scriptural texts by the intellect, after having acquired knowledge of demonstrative science, and of the true hidden meaning of prophecies.

MAIMONIDES, GUIDE FOR THE PERPLEXED 231.

203. McLean v. Arkansas Bd. of Educ., 529 F. Supp. 1255, E.D. Ark. 1982 (mem. op.). Judge Overton found that the definition of creation science "has as its unmentioned reference the first 11 chapters of the Book of Genesis." *Id.* at 1264-65.

204. Some theological witnesses were against legislation specifying what religious theories ought to be included in a public school's curricula. Although the defendants were a fundamentalist group, they were actually supported by a number of people with alleged scientific credentials. The court emphasized, "If they start with a conclusion and refuse to change it regardless of the evidence developed during the course of the investigation . . . creation science has no scientific merit or educational value as science." *Id.* at 1269, 1272.

tent with the following statement by Nobel laureate Steven Weinberg: "Our discovery of the connected and convergent pattern of scientific explanations has done the very great service of teaching us that *there is no room in nature* for astrology or telekinesis or [religious] *creationism* or other superstitions."[205] (Emphasis added.) Five years later, the U.S. Supreme Court declared unconstitutional a statute enacted in Louisiana identical to that enacted in Arkansas. However, in Louisiana there was no trial on the merits or introduction of any scientific evidence. Thus, the court ruled on the ground that the statute required an unconstitutional "enlightenment with religion," stating "The preeminent purpose of the Louisiana legislature was clearly to advance the religious viewpoint that a supernatural being created humankind."[206] The National Academy of Science (NAS) brief declared that "Creation science is not science" and relies on "supernatural means inaccessible to human understanding."[207] In a subsequent case, the Court held that the Equal Access Act, enacted by Congress in 1984, did not violate the establishment clause by authorizing a student-led bible club to meet at a public school.[208] In June of

---

205. Nobel Laureate STEVEN WEINBERG, DREAMS OF A FINAL THEORY 50 (1993).

206. Edward v. Aquillard, 482 U.S. 578, at pp. 591 and 597 (1987). The Court held the Louisiana Balanced Treatment for Creation-Science and Evolution-Science in Public School of Law Institution Act unconstitutional. Justice Brennan found that the primary purpose of Louisiana's act was to advance religion. In 2005, a federal judge ordered school officials in Atlanta, Georgia, to remove stickers they had placed in biology textbooks stating that "evolution is a theory, not a fact," stating: "The sticker communicates to those who oppose evolution for religious reasons that they are favored members of the political community, while the sticker sends a message to those who believe in evolution that they are political outsiders. [The schoolchildren] are likely to view the message on the sticker as a union of church and state." L.A. TIMES, Jan. 14, 2005, at A11.

207. Brief of the National Academies of Science, amicus curiae, in *Edwards v. Aguillard, id.*, stating: "Thus, inferences from paleontology, comparative anatomy and other disciplines as to the evolutionary history of organisms can be tested by examining the sequences of nucleotides in the DNA or the sequences of amino acids in proteins. The potential of such tests is overwhelming. Each of the thousands of genes and proteins provides an independent test of evolutionary history."

208. Westside Cmty. Bd. of Educ. v. Mergens, 496 U.S. 226 (1990). In *Agostini v. Felton*, 521 U.S. 203 (1997), by a 5:4 majority, the Court held that public funds could be provided to support programs that directly aid students of private religious schools and clearly promote secular purposes that contribute to the public good. The estab-

2000, the Court struck down a policy that required elementary and high school teachers to read a disclaimer before *teaching evolution*. This disclaimer required public school teachers to tell their students that the "[S]cientific theory of evolution . . . was not intended to influence or dissuade the Biblical version of creation."[209] The same day it held that the Establishment Clause bars a public school district's use of a student vote to authorize prayers at high school football games.[210] However, the Supreme Court later upheld a state law authorizing *a minute of silence in the public school classroom.*[211] In the public schools of the United States, morals cannot be taught "as religion." "[P]rayer doesn't belong in schools, but there is nothing wrong with a quiet pause. . . . Those who oppose the moment of silence because it might be seen as a victory for prayer are narrow-minded."[212] Some county and city councils started to comply with an appellate court's ruling that a minister cannot say "in the name of Jesus Christ" during an invocation at a public council meeting; however, God could be referred to without mentioning any specific religious figure promoting a particular religion, thus challenging constitutionality.[213] Several appellate courts have ruled that clergy members may be excluded

---

lishment clause is not violated where the religious mission of the private school is not advanced.

209. Orange County Register, June 20, 2000, at News 14.

210. Santa Fe Indep. Sch. Dist. v. Doe, 530 U.S. 290 (2000). The Court found that the "degree of school involvement" in a religious exercise was comparable to that in *Lee v. Weisman*, 505 U.S. 577, 1982, which held that a public middle-school prayer delivered by a rabbi chosen by school officials violated the establishment clause.

211. Orange County Register, Oct. 30, 2001, at Local 6. The Attorney General wrote: "The act does not require students to do anything or say anything or hear anything." In 2002, the U.S. Court of Appeals for the Ninth Circuit held that the Equal Access Act (EAA), 20 U.S.C. § 4071, requires that religiously oriented student activities and other extracurricular activities be allowed under the same terms and conditions. Prince v. Jacoby, 9th Cir., No. 99-35490, Sept. 9, 2002.

212. Balzer, L.A. Times, Dec. 16, 2001, at M5.

213. Orange County Register, Jan. 21, 2003, at Local 6. In 2004 the Supreme Court held that Washington state's exclusion of students pursuing a theology degree from a scholarship program does not violate the Free Exercise Clause of the First Amendment. The state's Constitution prohibited providing funds to students to pursue degrees that are "devotional in nature or designed to induce religious faith." *Locke v. Davey,* Feb. 25, 2004; The Federal Lawyer, May 2004, at 50.

from juries because they tend to be too sympathetic to criminal defendants.[214] In 2002, a U.S. district court held that the chief justice of the Alabama Supreme Court violated the First Amendment of the Constitution when he placed a massive stone carving of the Ten Commandments from the King James translation of the Bible in the rotunda of the State Supreme Court building.[215] The following year, a federal court ordered the removal of large granite tablets (a 5,280-pound granite monument) that contained the Ten Commandments from three school buildings.[216] Later that year, a federal judge

214. "The 9th U.S. Circuit Court of Appeals in San Francisco, for instance, held that a prosecutor 'expressed a legitimate concern that ministers are uniquely forgiving,' and the 5th U.S. Circuit Court of Appeals in New Orleans accepted the rationale that a minister could be excluded because 'perhaps she should have a higher threshold of reasonable doubt.'" ORANGE COUNTY REGISTER, Oct. 13, 2002, at News 14.

215. Glassroth v. Moore, 347 F.3d 016 C.A. 11 (Ala. 2003). *See also* the 5:4 decision of the Supreme Court in *McCreary County, Ky. v. American Civil Liberties Union of Ky.,* 125 S. Ct. 2722, 162 L. Ed. 2d 729 (2005). Justice O'Connor sided with Justices Ginsburg, Stevens, and Souter, who wrote the Court's opinion, in opposing any connection between the government and religion. A Jewish Harvard Law professor wrote:

> Rape is horribly wrong even though the men who wrote the Bible did not think it was wrong enough to include in the Ten Commandments. They did, however, include voluntary adultery, but only if it involved a married *woman!* We can improve on the Ten Commandments, because we have much more human experience on which to base our rules than did the men who wrote the Bible, just as the authors of the Ten Commandments improved on the Code of Hammurabi and earlier laws.

DERSHOWITZ, RIGHTS FROM WRONGS 36 (2004).

216. ORANGE COUNTY REGISTER, June 10, 2003, at News 10. Roy Moore, Chief Justice of the Alabama Supreme Court, who had installed a large Ten Commandments monument, was held to have had the unlawful purpose of an endorsement of religion. Glassroth v. Moore, 347 F.3d C.A. 11 (Ala. 2003). That court noted that Moore "testified candidly that his purpose in placing the monument in the Judicial Building was to acknowledge the law and sovereignty of the God of the Holy Scriptures, and that it was intended to acknowledge 'God's overruling power over the affairs of men.'" Moore burned a copy of the U.S. District Court's order for the monument's removal. The Supreme Court upheld on appeal (ORANGE COUNTY REGISTER, Aug. 21, 2003) and Moore was suspended for his refusal to obey a federal court order. ORANGE COUNTY REGISTER, Aug. 23, 2003, at News 27.

dismissed a lawsuit by three residents seeking to return a 5,300-pound Ten Commandments monument to the lobby of the Alabama Judicial building.[217] Most subsequent courts so ruled under quite comparable sets of facts.[218] However, the U.S. Court of Appeals for the Fifth Circuit ruled to the contrary with respect to a granite monument with the Ten Commandments in a large park with 17 other monuments and 21 historical markers of the Texas state capitol,[219] which was confirmed 5:4 by the U.S. Supreme Court.[220]

In 1993, Congress enacted the Religious Freedom Restoration Act (RFRA),[221] which required the government to demonstrate a compelling interest to justify intrusion into any religious practice. In 1997, the RFRA was deemed unconstitutional as applied to state and local governments. It was held to violate the separation of church and state by granting favors to the religious.[222] Justice Sandra O'Connor wrote in her volume published in 2003

---

217. The judge stated: "The empty space or 'nothingness' in the rotunda of the Judicial Building is neither an endorsement of 'nontheistic belief' nor a sign of disrespect for Christianity or any other religion." ORANGE COUNTY REGISTER, Sept. 5, 2003, at News 28.

218. In 2004, the U.S. District Court for the Western District of Wisconsin held that the city of LaCrosse violated the establishment clause when it sold a Ten Commandments monument and the tiny plot of public parkland it sits on to the monument's donor. Mercier v. LaCrosse, Wis., W.D. Wis., No. 02-C-376-C, Feb. 3, 2004. The same year the Eighth Circuit Appeals Court held that a Nebraska town having a Ten Commandments monument in a public park was an unconstitutional establishment of religion. NATURAL LAW J., March 15, 2004, etc. *See* Am. Civil Liberties Union of Ky. v. Creary County, 6th Cir., No. 01-5935, Dec. 18, 2003; Wynne v. Great Falls, S.C., 4th Cir., No. 03-2069, July 22, 2004; Am. Jewish Cong. v. Corp. of Nat'l and Cmty. Serv., D.D.C., No 02-1948 (GK), July 2, 2004; and Am. Civil Liberties Union of Ohio Found., Inc. v. Ashbrook, 6th Cir., No. 02-3667, July 14, 2004.

219. Van Orden v. Perry, 351 F.3d 173 (2003).

220. 125 S. Ct. 2854, 172 L. Ed. 2d 607 (2005).

221. U.S.C. § 2000(bb)-(bb)(4) (1994). The Supreme Court in City of Boerne, Texas v. Flores, 117 S. Ct. 2157 (1997).

222. In 1991 a number of groups developed legislative plans. One reason was exhibited when the Supreme Court applied the "compelling interest" test to permit Oregon's ban on peyote to stand. The bill was drafted to let the government proceed without this test. The Religious Freedom Restoration Act (RFRA) passed and was signed by President Clinton in 1993. Within a year there were statutes protecting Indian tribes from being penalized for using or transporting peyote. However, in 1997 the Supreme Court (6:3) rejected the statute, holding that Congress cannot expand or create rights pursuant to section.

that in a case "decided in 1803, the Supreme Court declared that it had the authority to declare an act of Congress void if, in the opinion of the Justices, the act violated the Constitution. That principle has survived to this day."[223] These decisions were made by judges who professed diverse religious affiliations.[224] The separation cannot provide any significant behavioral standards for the actions of all citizens independent of the limited respect for other beliefs owing to that separation. In their connection it has been noted in contrast to the United States, Israel does not abide the principle of separation of church and state, and it never did.[225]

This has been the general rule since 1878 when the Supreme Court invoked the belief/action distinction to deny the claim by some Mormons that polygamy was essential to their religion and thus protected by the free-exercise clause in the First Amendment of the Constitution.[226] However, in 2000, the Supreme Court majority (6:3) upheld a Louisiana law that authorized the state to lend computers and schoolbooks to parochial schools.[227] Justice Thomas wrote the majority judgment, agreeing that it showed the "degree to

---

223. O'Connor, The Majesty of the Law 242 (2003), *citing* Marbury v. Madison, I Cranch 137 (1803).

224. Justices Antonin Scalia, Clarence Thomas, and Anthony M. Kennedy are Catholics; Justices Ruth Bader Ginsburg and Stephen G. Breyer are Jews. Justice David H. Souter belongs to a Protestant denomination, as did Justice Sandra Day O'Connor (retired) and former Chief Justice William H. Rehnquist (deceased). Orange County Register, Jan. 25, 1997.

225. Efron, Real Jews, Secular vs. Ultra-Orthodox 223 (2003). The notion that the state funds religious institutions and that religious institutions administer state services remains embedded deeply in the structure of Israeli politics. The principle of "providing for religious needs" has been variously interpreted over the years, and it has given rise to mountains of complicated legislation, much of it the outcome of Byzantine compromises delicately crafted to resolve political exigencies now long forgotten. (p. 224) In 2005 it was also reported that: "The scandals hitting the Greek Orthodox Church have renewed calls for a separation of church and state, which would require a constitutional change in the country. Almost 65 percent of the public supports such a reform." Orange County Register, Feb. 19, 2005, at News 36; Sutton v. United Airlines, No. 97-1943, 1996 WL 588917 (1996), pp. 1-7.

226. Laws are made for the government of actions, and while they may not interfere with mere religious belief and opinions, they may with practices. Reynolds v. United States (1878).

227. Mitchell v. Helms, 530 U.S. 793 (2000).

which our Establishment Clause jurisdiction has shifted." The dissenters, Justices Souter, Stevens, and Ginsburg, argued that "[R]eligious schools could be blessed with government funding as massive as expenditures made for the benefit of their public school counterparts, and religious missions would thrive on public money." Justice Thomas called that statement a "hostility to aid to pervasively sectarian schools [which] has shameful pedigree that we do not hesitate to disavow."

The same majority of Justices authorized a group to meet in a public school at the end of a week for religious classes.[228] Justice Thomas stated, "We disagree that something that is quintessentially religious or decidedly religious in nature cannot also be characterized properly as the teaching of morals and character development from a particular viewpoint." Justices Souter and Ginsburg, in dissent, noted that "[T]he majority avoids reality only by resorting to the bland and general characterization of Good News acting as teaching of morals and character from a religious standpoint." In 2002, the Supreme Court upheld an Ohio program permitting parents to redeem publicly funded education vouchers at sectarian religious schools. The Court stated that the program involved the "genuine" or "independent" private choices of individuals that acted as an intervening cause or "circuit breaker" between church and state.[229] The state legislators came up with the Pilot Project Scholarship Program version of school vouchers, tuition coupons provided for by government funds used at private schools. They are the

228. Good News Club v. Milford Central School, 533 U.S. 98 (2001).

229. Zelman v. Simmons-Harris, 536 U.S. 639 (2002) at 644. The Supreme Court decided it is constitutional for public funds to be used to pay tuition to religious schools under certain conditions. The following year, in the cross-burning case, that court allowed it to be outlawed because a state may "choose to prohibit only those forms of intimidation that are most likely to inspire fear of *bodily* harm." Virginia v. Black, 123 S. Ct. 1536 (2003).

> Since over three-quarters of private school students now attend such schools, this decision could permit a substantial expansion of broad choice plans. It will certainly make it easier for groups of parents united by ideology, religion, or values to assert their right to control the education of their children and have their tax money support the schools that their groups favor.

Hochschild & Scovronick, The American Dream and the Public Schools 124 (2003).

option for families of lower-income resources to send their children to private schools. The funds would otherwise remain with public schools should parents decide to keep their children in public schools. Most private schools are religious schools. The U.S. Supreme Court held that although religious schools did appear to benefit more from school vouchers than the nonreligious private schools, parents had the option of choosing public schools, magnet schools, or nonreligious private schools. Their choice was independent and free of governmental coercion.[230]

In December of 2002, President Bush signed two executive orders to ensure that faith-based groups are not barred from participating in federal programs, including contracts and grants, because of their religious nature.[231]

> *The great writers to whom the world owes what liberty it possesses have most asserted freedom of conscience as an indefeasible right, and denied absolutely that a human being is accountable to another for his religious belief.*
>
> John Stuart Mill, *On Liberty*

At the dawn of the third millennium, it is of interest to reflect on the scientific age in the seventeenth through the nineteenth centuries. The prevalent belief was that "The fear of some divine and supreme power keeps men in obedience." This belief continued through most of the nineteenth century. In the 1830s, Alexis de Touqueville reported:

> The court of common pleas of Chester County (New York) rejected a witness who declared his disbelief in the existence of God. The pre-

230. In *Zelman v. Simmons-Harris*, 536 U.S. 639 (2002), Justices Rehnquist and O'Connor indicated that the voucher program was one of "true private choice." *But see* Ryan & Heise, *The Political Economy of School Choice*, 111 YALE L.J. 2043 (2002), p. 2079 (who stated, "Every proposal to provide for vouchers on a large scale has failed.").

231. E.O. 13,279, "Equal Protection of the Laws for Faith-Based and Community Organizations," permits faith-based organizations to receive federal grants and other financial assistance if they separate their religious activities from social service programs and do not condition receipt of services on holding a particular belief. E.O. 13,280 directs the establishment of new faith-based and community initiative offices at the Department of Agriculture and at the Agency for International Development. U.S. LAW WEEK, Dec. 24, 2002, at 2411.

> siding judge remarked that there was a man living who did not believe in the existence of God; that this belief constituted the sanction of all testimony in a court of justice; and that he knew of no cause in a Christian country, where a witness had been permitted to testify without such belief.[232]

In the 1990s, one writer found that the prohibition against preferring one religion over another has "an underlying predicate . . . that religion does not matter, at least not in the public domain [because] equality implies that religions should be insulated from normative evaluation." Secondly, religious reasons should not be "the basis for a political decision." Fundamentalists, for example, violate "a core tenet of our democratic system that . . . legal policies should be formulated on the basis of a dialogic decision-making process, a process requiring an openness of mind that fundamentalism does not allow."[233]

At the same time, the late Carl Sagan reported, "Our technology has reached astonishing proportions for which, in our hearts, we are inadequately prepared, mentally or emotionally."[234] Perhaps he was attempting to update Thomas Gray's eighteenth century poem, "Elegy Written in a Country Courtyard," in which he wrote the famous expression, "Where ignorance is bliss, 'tis folly to be wise." The word *religion* means to religate, or "bind together" again. Perhaps this will be what scientists desire to see accomplished by bioethicists. Sagan's prognostication was "Whether we will acquire the understanding and wisdom necessary to come to grips with the scientific revelations of the twentieth century will be the *most profound challenge* of the twenty-first."[235]

An *alternative way of coping,* for many people, is to split their minds like schizophrenics and place all the information received from evangelicals

---

232. Robert Burton (1577-1640). Conkle, *Differing Religions, Different Politics: Evaluating the Role of Competing Religious Traditions in American Politics and Law*, J. L. & Religion 10 (1993-1994).

233. Tocqueville, Democracy in America, *quoted in* Hewitt, Searching for God in America 232 (1996). The condition of belief in God (always referred to as a male) in order to qualify as a witness in court was not dropped until well into the 20th century. However, in the 1840s Reverend Theodore Parker of Boston not only made all his prayers to "Our Father and Our Mother God" but was "the first clergyman to invite women to preach from his pulpit." *Id.*

234. Carl Sagan, Billions and Billions 33 (1997).

235. *Id.* at 256.

(and others) in one hemisphere and to locate their mode of living in the other one. As for the latter, the American Psychiatric Association's guidelines stated in 1997 that doctor visits and prescriptions are not enough to treat them.[236] (Occasions to handle this problem by bioethicists will be present in subsequent chapters.) The astrophysicist Sagan had noted:

> *Those who cannot bear the burden of science* are free to *ignore its precepts.* But we cannot have science in bits and pieces, applying it where we feel safe and ignoring it where we feel threatened—again, because we are not wise enough to do so. *Except by sealing the brain* off into separate airtight compartments, *how is it possible* to fly in airplanes, listen to the radio, or take antibiotics *while holding that the Earth is* around *10,000 years old* or that all Sagittarians are gregarious and affable.[237] (Emphasis added.)

In 1890, some 34 percent of the population belonged to an organized church; this rose to 63 percent in 1990. In 1989, George Gallup, Jr., and James Castelli noted a basic difference in the United States between fundamentalist and evangelical churches.[238] In 2000, a Gallup poll alleged that about 80 percent of Americans were Christians, 11 percent had no religion or an undesignated faith, and the remainder included Jews, Muslims, and Buddhists and other increasing religions.[239] In 2004, another poll showed

---

236. Orange County Register, April 1, 1997, at News 7.

237. Sagan, The Demon-Haunted World 297 (1993). In this connection, Will Durant, historian, cited the year 4241 B.C., stating:

> This date alone, the earliest definite date in history, is sufficient to cause some disturbance to fiercely orthodox souls who believe, as did Bishop Usher, that the world was created in 4004 B.C. To accept the testimony of Egyptologists that a calendar existed on the Lower Nile 237 years before the creation of the world might serve as a fertilizing shock to any virgin mind.

Durant, The Greatest Minds and Ideas 108 (2002).

238. Gallup & Castelli, The People's Religion: American Faith in the '90s (1989). More than 90 percent answered "yes" to the question "Do you believe in God?," which was much more than in Europe.

239. Dallas Morning News, June 2, 2001, at 3G. In a 2004 Gallup Poll, six in 10 adults said religion is "very important" in their lives. It was also reported that during the 20th century "church membership has climbed from 25 percent to 65 percent of

that America's Protestants dropped 11 percent (from 63 percent to 52 percent), while the Roman Catholic population has remained relatively stable at about 25 percent of the U.S. population (because of the large recent immigration from Latin America). Thus, there was then a total of 77 percent Christians here.[240] However, the number of current Christians may not be compared to those in France: "[W]hile four-fifths [80 percent] of the French have been baptized, only 17.5 percent are counted as practicing."[241] A serious question arises regarding the extent to which Americans might arrive at the difference from this practice in France. As written by a Christian professor, Diana Eck, at Harvard,[242] America "has become the world's most religiously diverse nation." She claimed to have seen some millions of Muslims, Buddhists, and others expanding here. Yet, she stated that in the United States, "[S]ociety gives immigrants a much yearned for freedom, and for some, this [is] freedom from religion."[243] As noted, "A chemistry of small

---

the American population." GAUSTAD, BARLOW & DISHNO, NEW HISTORICAL ATLAS OF RELIGION IN AMERICA (2001).

240. The study was made by the National Opinion Research Center at the University of Chicago. The Protestants said they were members of a Protestant denomination, such as Episcopal Church or Southern Baptist Convention. The category included members of the Church of Jesus Christ of Latter-day Saints and members of independent Protestant churches. Islam, Orthodox Christianity or Eastern faiths increased from 3 percent to 7 percent, while the share of people who said they were Jewish remained stable at just under 2 percent.
ORANGE COUNTY REGISTER, July 21, 2004, at News 13.

241. FENBY, FRANCE ON THE BRINK 21 (1998).

242. ECK, A NEW RELIGIOUS AMERICA (2001).

243. *Id. at* 337. In subsequent pages (p. 362) she recited how rocks are thrown into mosques and other damages are done to other new religions in the same manner as is done to abortion clinics by some members of Christian churches. After the turn of the millennium, Keith Ward, professor of divinity and Anglican priest at the University of Oxford, wrote *God, a Guide for the Perplexed* (2000). This was a copy of the volume by a Jew named Moses Maimonides over 800 years ago. He indicated that "no product of human imagination can be adequate to God's reality" (p.2) and added, "Spirituality, the cultivation of exalted states of consciousness, may thrive in the borderlands of our culture, but religion, the organized official cult of worship of God, is dying. This is most obvious in Europe, but in America, too, while popular religion remains strong, there is a widespread intellectual revulsion against organized religion."

numbers, a *quantum chemistry,* is needed to account for life's emergence."[244] This development must remain within the memory of bioethicists, whether or not they themselves remain religious persons.[245] This could be significant, based on 2002 arguments by R. P. Sloan and E. Bagiella of Columbia University that none of certain previous studies "provided a sound scientific basis for concluding that religion improves health."[246] One of their cited studies was the *Handbook of Religion and Health,*[247] in which "Only four of the 39 studies cited provided convincing evidence for an effect of religious observance on health." In their own review that focused on all the 266 English-language abstracts on religion included in Medline for the year 2000, "[O]nly 17 percent provided evidence relevant to potential health benefits of religion, suggesting that much of this literature is not helpful in deciding whether religion affects health."[248]

Nevertheless, in 2004, Dr. Laura Schlessinger (whose mother was a "nice Catholic girl from Italy" and whose father was "a nice Jewish boy from Brooklyn"—neither of whom believed in God) finally "found God, converted to Judaism," and broadcast into more than 450 radio stations with lectures and books, concluding that "The notion of God is really, fundamentally, all we have to truly lead us to be good or else we make our own decisions and we become, individually, our own Gods."[249] But Immanuel Kant's *Categorical*

---

244. McFadden, Quantum Evolution 221 (2000). For many scientists their agnosticism is due to the fact that "the world is thus divided between an invisible quantum domain and the familiar realm that we perceive, obeying classical laws" (p. 164). *See also* J. A. Wheeler, *How come the quantum?* and *How come existence?*, Ferris, Best American Science Writing 41 (2001).

245. *See also* Jenkins, The Next Christiandom 161 (2002), who stated: "In their own way, secular, liberal Americans have a distinctly apocalyptic view [prophetic] of the future, with a millenarian expectation of the uprooting of organized religion. And, although the United States remains a far more religious nation than Europe, North American elites are quite as secular as their European counterparts."

246. Sloan & Bagiella, *Claims about Religious Involvement and Health Outcomes*, 24 Annals of Behavioral Medicine 14-21 (2002).

247. Luskin, *Religious and Spiritual Factors*, in Koenig, McCullough & Larson, Handbook of Religion and Health, at 8-15.

248. Barber, Kindness in a Cruel World 326 (2004).

249. Schlessinger, *How Could You Do That? The Abdication of Character, Courage, and Conscience* (1996), and Michael Shermer, the writer of *The Science of Good and Evil* (2004), who "became a born-again Christian" and attended "Pepperdine

Imperative had indicated that to judge the rightness or wrongness of an action, "Act only on that maxim through which you can at the same time will that it should become a universal law."[250]

A 2001 book the title of which begins with the words *Playing God* indicates that "the bioethics profession" has successfully competed with other professions.[251] The findings include debates concerning "formal rationality," "displaced theology," and "philosophy."[252] The Parents Television Council watched every hour of prime time on the broadcast networks during 2003–2004, judged 22 percent of the mentioned religions positive, 24 percent negative, and the rest neutral.[253] In his 2004 volume *Laboratory of Justice,* David Faigman concluded, "The scientific revolution is everywhere; it cannot be ignored with impunity. If the Constitution is to 'endure forever,' its guardians will have to read it in light of the science of today and be prepared to incorporate the discoveries of tomorrow."[254] In this connection, the statutory "oath or affirmation that a federal justice or judge must take before perform-

---

University, an institution affiliated with the Church of Christ" in Malibu, California, to study theology. He also studied "evolutionary biology and experimental psychology in a graduate program at California State University, Fullerton." (pp. 161-62)

250. KANT, FUNDAMENTAL PRINCIPLES OF THE METAPHYSICS OF MORALS (1785), Chicago Press (1952).

251. EVANS, PLAYING GOD? HUMAN GENETIC ENGINEERING AND THE RATIONALIZATION OF PUBLIC BIOETHICAL DEBATE (2001).

252. Among his findings are:

1) The scientific community has moved from substance to formal rationality;
2) Bioethics has displaced theology in the debate and continues its ascendancy; indeed many influential scientists are now part of the bioethics community;
3) Theological discussions have precipitously declined and have been deliberately marginalized by the bioethics profession and by analytic philosophy.

253. ORANGE COUNTY REGISTER, Dec. 20, 2004, at Life 4. Negative words often used when referring to certain religions include *cult* (derived from the Latin *colere*, means "to tend or till"), and *sect*, often used for alternative religions (which is derived from the Latin verb *sequi*, meaning "to follow"). *See* STEIN, COMMUNITIES OF DISSENT 4-5 (2003).

254. SCHLESSINGER, HOW COULD YOU DO THAT: THE ABDICATION OF CHARACTER, COURAGE, AND CONSCIENCE 364 (1996).

ing refers to 'God' in a manner sufficient to reference the Creator of the Universe.[255] Furthermore, unlike legislators, their recusal is demanded if their bias may reasonably be questioned."[256]

255. "I, _____, do solemnly swear (or affirm) that I will administer justice without respect to persons, and do equal right to the poor and to the rich, and that I will faithfully and impartially discharge and perform all the duties incumbent upon me as _____ under the Constitution and laws of the United States, so help me God." 28 U.S.C. § 453.

256. "Any justice, judge, or magistrate [magistrate judge] of the United States shall disqualify himself in any proceeding in which his impartiality might reasonably be questioned." 28 U.S.C. § 455(a) (2004). *See Our Continued Need for Coordination of the 18th Century's "Age of Enlightenment" with the 21st Century's "Ages of Modern Science and Bioethics."* Keyes, 27 Whittier L. Rev. 951-84 (Summer 2006).

And new Philosophy calls all in doubt,
The element of fire is quite put out;
The Sun is lost, and th'erth, and no man' wit
Can well direct him where to look for it.
And freely men confess that this world's spent,
When in the Planets and the Firmament
They seek so many new; then see that this
Is crumbles out again to his Atomies.
Tis all in pieces, all coherence gone;
All just supply, and all Relation.

John Donne, English cleric and poet (1611)

Robert Proctor and others speak of physicians as scientists. They are not. They are professionals building part of their knowledge on science, part on experience, and part on their own good judgment. Today when patients are prepared for surgery, they rightfully sign a statement that indicates the acknowledgment that medicine is not an exact science. That does not diminish that great profession. It only means that ethics—as in all applied professions—become indispensable.

Gisela Konopka

To know is one thing, merely to believe one knows is another.
To know is science, but merely to believe one knows, is ignorance.

Hippocrates, c. 400 B.C.

4 Some Differences and Difficulties with Science and Philosophy in the Search for Bioethical Standards

1. ***Better Reasoning and the Performance of Science; Law and Ethics Partially Distinguished***
2. ***Some Scientists Are Religious, Fewer Are Atheists, and More Are Agnostics***
   - **A. Prayer and Certain Conflicts; Reincarnation**
   - **B. Limitations of Science; Risk/Benefit Review by IRBs in Research Clinics; Complementary and Alternative Medicine (CAM)**

3. A Problem with the Use of Either Reason or Philosophy Alone

4. The Allegation of Neutral Principles; Science and the Humanities

5. The Change in Modern Philosophy and Secular Humanism

We've moved from being a "results tomorrow" to a "results now" mentality. Some of that is being driven by elected leaders who want their results now *. . . Some of that is being driven by urgency and a better understanding about the importance of public health's role in protecting the American people. But our processes have fundamentally changed. We've gone from science and politics, politics with a small "p," to science and Politics, Politics with a capital "P."*

Benjamin, American Public Health Association, *Journal of Law, Medicine, and Ethics* (Fall 2004), p. 13.

The unexamined life is not worth living.

Socrates

1. Better Reasoning and Performance of Science; Law and Ethics Partially Distinguished

Many people will agree to some extent with John Donne's seventeenth century observation in the epigraph that "all coherence [is] gone" and "all [is] relation." Barbara Tuchman, the historian, characterized the twentieth century as "an age of disruption" due to the exceptional pervasiveness of "public immorality making itself so obvious to the average citizen."[1] A writer of "Ethics as a Religion" has noted that several such organizations have existed since the nineteenth century, but stated, "The Ethical Societies have no creed. Their only requirement for membership is devotion to ethics as the supreme end of life."[2]

---

1. Barbara Tuchman, The Guns of August (1993).

2. Muzzey, Ethics as a Religion (1999). *See also* Robert Proctor, Racial Hygiene Under the Nazis (1988): "Medical anti-Semitism represented in this sense not just a desire to gain jobs of Jews but also a vehicle for launching a more general attack on science and its place in society." (p. 162) "[Professor] Kotschan conceded . . . those who say 'I am a doctor, a scientist.'" (p. 165)

To a greater extent, more people now agree with Abbe de Condillas's eighteenth century observation that "the sciences have made progress," and that scientists, "reason better." As Einstein remarked, science is "the most precious thing we have." Justice Holmes emphasized life's experiences over moral and political policy in determining legal rules. Konopka's epigraph notes that physicians are not scientists (with some exceptions). Like many other professionals, they build "their knowledge on science, part on experience," and part on their generally "good judgment." As in all applied professions, "ethics becomes indispensable." At the turn of the third millennium, this was confirmed by Stephen Breyer, Associate Justice of the Supreme Court of the United States, who wrote:

> In this age of science, science should expect to find a warm welcome, perhaps a permanent home, in our courtrooms. The reason is a simple one. The legal disputes before us increasingly involve the principles and tools of science. Proper resolution of those disputes matters not just to the litigants, but also to the general public—those who live in our technologically complex society and whom the law must serve.[3]

A number of my nonscientifically-minded friends and acquaintances actually believe that most of what is considered a scientific fact today may later be deemed incorrect. Such a statement might have been considered quite accurate prior to the scientific age—and even for several years after it had begun. Joseph Campbell's conclusion that Biblical science "was the best science of 1000 B.C." may be illustrated by Joshua's statement, which is consistent with having the sun go around the earth, and to extend the hours of daylight as being necessary to stop the sun from orbiting the earth, and not to stop the earth from rotating:

---

3. Epigraph: Stephen Breyer, Introduction, REFERENCE MANUAL ON SCIENTIFIC EVIDENCE, second edition (Federal Judicial Center, 2000). He had also written, "It is equally true that the law itself increasingly requires access to sound science. This need arises because society is becoming more dependent for its well-being on scientifically complex technology, so, to an increasing degree, this technology underlies legal issues of importance to all of us." Breyer, *The interdependence of science and law*, p. 537. *See also* Judson, in *The Great Betrayal* (2004) (p. 385), who noted that "Breyer's formulation here makes the relationship between science and law seem reciprocal."

> He [Joshua] said in the sight of Israel, sun, stand thou still upon Gebeon; and thou, Moon, in the valley of Aijalon.
> And the sun stood still, and Moon stopped, until the people had avenged themselves upon their enemies.[4]

This viewpoint no longer stands true because virtually no *generally accepted newer scientific developments* of the recent past have been overturned. Early in the seventeenth century Francis Bacon, former Lord Chancellor of England, established principles governing the seeking of the natural world by scientific means. In 1620 his *Novum organum* stated something that would be used by Charles Darwin over three centuries later, namely, "For once a nature has been observed in its variations, and the reason for it has been made clear, it will be an easy matter to bring that nature by art to the point it reached by chance"[5] According to Sir Karl Popper, science is "the only human activity in which errors are systemically criticized and . . . in time corrected."[6] As Nobel Laureate physicist Steven Weinberg wrote in 1992: "One can imagine a category of *experiments that refute well-accepted theories* that have become part of the standard consensus of physics. Under this category *I can find no examples.*"[7] (Emphasis added.) We can generally assume that scientific development will continue throughout the third millennium without overturning any

---

4. *See Joshua* 10:12-13.

5. Francis Bacon, The New Organon, Jardine & Silverthorne eds. (2000). In the 17th century, Descartes also drew science and ethics together with philosophy as a tree "whose roots are metaphysics, whose trunk is physics, and whose branches . . . are all the other sciences . . . medicine, mechanics, and morals." Descartes, preface to *Principles of Philosophy, in* The Philosophical Works of Descartes 1:211 (Haldane & Ross trans.).

6. Popper, The Logic of Scientific Discovery (1968).

7. Steven Weinberg, Dreams of a Final Theory 129 (1992). Max Planck, who became the first developer of the first element of quantum physics in 1900, later stated:

> An important scientific innovation rarely makes its way by gradually winning over and converting its opponents . . . what does happen is that its opponents gradually die out and that the growing generation is familiar with the idea from the beginning.

Niels Bohr, the most active developer of that particular field, stated, "Those who are not shocked when they first come across quantum theory cannot possibly have understood it."

well-accepted scientific theories. Ancients stated, "This too shall pass," and that "the only constant is change." This does not apply to scientific theories that are part of a "standard consensus of scientists." Nevertheless the Old Testament states: "Thus says the Lord: If the Heavens above can be measured, and the foundations of the earth below can be explored, then I will reject all the offspring of Israel because of all they have done."[8]

In 2004, Sir Karl Popper's work was compared with that of Thomas Kuhn, *The Structure of Scientific Revolutions.* Although Kuhn's work was called "the most influential book on the nature of science in the second half of the twentieth century—and arguably, the entire twentieth century,"[9] an author (whose book discussed both of these men) has pointed out that "most actual science—what Kuhn calls 'normal science'—consists of little more than the technical work of fleshing out the paradigm's blueprint" (see Chapter 2). He concluded that "Kuhn was indeed authoritarian and Popper libertarian in their attitudes to science." (p. 9) He emphasized that "[S]cientists should be always trying to falsify their theories, just as people should be always *invited* to find fault in their governments and consider alternatives—and not simply wait until the government can no longer hide its mistakes. This notoriously led Popper and his students to be equal opportunity fault-finders across the natural and social sciences." (p. 27) Social sciences often embrace morals and ethics—which are not covered by science. Nevertheless, disputes raise the question by many religious people and others about the extent to which potential theories of all social subjects, including most of bioethics, should be falsifiable. Although in practice the doctrine may often fall short, in some instances, its goal normally remains a useful principle.

The British philosopher Isaiah Berlin reiterated the general understanding that the "goals of the Enlightenment were to rationalize human thought and human society." He also considered counter-enlightenment thinkers who see knowledge of facts, logical truths, or knowledge of how to do things not as an end, but only as a beginning.[10] What we do with our knowledge of facts depends on our reason. George Bernard Shaw once observed,

8. *Jeremiah* 31:37. *Also see* MURRAY, HUMAN ACCOMPLISHMENT 440 (2003).

9. *See* FULLER, KUHN V. POPPER (2004). Neither Kuhn nor Popper were religious men.

10. Berlin, *The Counter-Enlightenment, in* AGAINST THE CURRENT 1 (1980).

"[A]ll progress depends on the unreasonable man" who pushes his autonomy by attempting to adapt some part of the world in a different manner. Therefore, even though we wish to live in a time of progress, some people may declare this to be unreasonable.

In the *AMA Code of Medical Ethics* (2003 edition), Section 1.02 notes that "ethical values and legal principles are usually closely related, but ethical obligations typically exceed legal duties. In some cases, the law mandates unethical conduct" and physicians "should work to change the law (which activity is discussed herein)." However, it then specifies, "the fact that a physician charged with allegedly illegal conduct is acquitted or exonerated in civil or criminal proceedings does not necessarily mean that the physician acted ethically." Thus, the American Medical Association correctly indicates that although ethics and the law are "closely related," this does not make them identical. Further, the lack of success in the legal prosecution of a physician does not necessarily enable him to escape from a change in his hospital position following an ethical violation. Further, legal principles, generally, are *deontological* and depend upon the act itself in a manner that is becoming more necessary to health care than the mere expression of ethical principles.

Most progress is recognized as such in the legal world. However, many people continually procrastinate their thought on progress regarding most matters of life and death, which really influence the bioethical world. In many instances, they consider such issues only when death is imminent upon themselves or a loved one.[11] These matters ought to be considered as a part of living in the universe.

All the seemingly arbitrary and unrelated constants in physics have one strange thing in common—these are precisely the values you need if you want to have a universe capable of producing life.

Glynn, "The Anthropic Principle in Cosmology," or "The Making and Unmaking of an Atheist," in *God: The Evidence* (1997)

It is thoroughly necessary to be convinced of God's existence, it is not quite so necessary that one should demonstrate it.

Immanuel Kant

11. *See, e.g.,* HANDY, THE AGE OF UNREASON 257 (1990) (see Chapters 12 to 14.)

It seems to me that there is no scientific evidence bearing on the Golden Rule. It seems to me that there is something different.

Richard Feynman, Nobel laureate

If a nation expects to be both ignorant and free, it expects what never was and never will be.

Thomas Jefferson

---

2. Some Scientists Are Religious, Fewer Are Atheists, and More Are Agnostics

An American Religious Identification Survey found that about 13 percent of the American population identify themselves as atheists (a total of about 30 million people) and claimed that about 58 percent of scientists had believed in a god who communicates with humankind and to whom one may pray "in expectation of receiving an answer."[12] However, a detailed (and more accurate 1997 survey) of scientists repeated a 1916 survey verbatim and found only a minority of about 40 percent of responding biologists, physicists, and mathematicians who indicated that they so believed. In 2001 a physicist stated:

> Indeed, from childhood on, we often have competing, incompatible beliefs that live side by side in our minds. For example, I know devoutly religious scientists who spend their professional careers developing scientific theories that are antithetical to the teachings of their faiths.[13]

The Nobel laureate physicist Richard Feynman, following his comment upon the Golden Rule (in the epigraph and Chapter 5, Section 2a), also noted, "It is possible to have these views at the same time; and I would say that I hope you will find that my atheistic scientific colleagues often carry themselves well in society."[14] Would these scientific colleagues in a highly consistent moral society, mostly carried by society's religious people and the law, do so principally by experience and reason alone?

12. Southern California Christian Times, September 2002, at 14.

13. Comins, Heavenly Errors 228 (2001).

14. Feynman, The Pleasure of Finding Things Out 251 (1999). As the Nobel laureate put it when his wife, Arlene, died of infection: "Who was there to be mad at? I couldn't get mad at God because I don't believe in God. And you can't get mad at some bacteria, can you?" Mlodinow, Feynman's Rainbow 159 (2003).

The diagnosis, against which I inwardly rebelled, was that I would have to rely on vitamins and medication forever. But then, when I became acquainted with Christian Science, I was able to maintain my health by trusting in prayer, instead of depending on taking medication.

Christian Science Sentinel, December 31, 2001, p. 28
(see Chapter 3)

Every religion is welcome in our country. All are practiced here. Many of our citizens profess no religion at all. Our country has never had an official faith.

President George W. Bush, National Prayer Breakfast,
February 7, 2002

Science and religion should work together to abolish the gross inequalities that prevail in the modern world. That is my vision, and it is the same vision that inspired Francis Bacon 400 years ago, when he prayed *that through science God would "endow the human family with new mercies." (Emphasis added.)*

Lighthood, physicist, in Ferris, *Best American Science Writer*
(2001), p. 315

When I pray, *coincidences happen.*
I stop praying, *coincidences stop. (Emphasis added.)*

William Temple, Archbishop of Canterbury

(a) Prayer and Certain Conflicts; Reincarnation

Each of these epigraphs expresses recent views on prayer by religious persons, a politician, and a physicist.

In the twentieth century another view was broadly expressed initially by Luther Burbank, the horticulturist who developed the Burbank potato, with which he had previously helped to combat "the blight epidemic" in Ireland. Thereafter he moved from Massachusetts westward to establish a nursery and experimental farm in Santa Rosa, California. There Burbank expressed his doubts about the traditional God and about an afterlife: "I must believe that rather than the survival of all, we must look for survival only in the spirit of the good we have done in passing through. This is as feasible and credible as Henry Ford's own practice of discarding the old models of his automobile. Once obsolete, an automobile is thrown to the scrap heap. Once here and gone, the human life has likewise served its purpose. If it has been a good life it has been sufficient. There is no need for another." On January 22, 1926, the front-page story appeared in the *San Francisco Bulletin* under the headline "I'm an Infidel, Declares Burbank, Casting Doubt on Soul Immortality Theory." The article was reprinted around the world, creating shock waves, and Burbank was inundated with hate mail. "In a desperate effort to make people understand, he attempted to reply to all the letters. But the physical task of doing so, especially amid all the harassment, was so overwhelming for the seventy-seven-year-old that Burbank took ill and died."[15]

In 1957, Pope Pius XII like his predecessors who were involved theologically with prayer and afterlife, noted that "Life, health, all temporal activities are in fact subordinated to spiritual ends."[16] Following the turn of the third millennium, Pope John Paul II wrote *Evangelium Vitae,* stating, "it is precisely [our] supernatural calling that lights the *relative character* of each individual's earthly life. After all, life on earth is not an 'ultimate' but a 'penultimate' reality."[17]

At the same time, the Board of Regents of the State of New York prescribed the use of prayer for use in its public schools, which read "Almighty God, we acknowledge our dependence upon Thee, and we beg Thy blessings upon us, our parents, our teachers, and our country." On June 25, 1962, the

---

15. Napoleon Hill, The Law of Success; The Principles of Personal Integrity 202-03 (2003).

16. Pope Pius XII, The Prolongation of Life (Nov. 24, 1957), *in* Pope Speaks No. 4 (1958) at 395-96. However, a modern writer concluded that "[r]eincarnation is the mechanism through which we may live the very life we've always wanted . . . Yes, but isn't that the problem? the materialist might ask. Couldn't our desire for immortality—for wanting to find a way to realize those dreams or relive a golden past—cause us to 'invent' reincarnation? In effect, could our desire to live forever be the very thing that allows us to deceive ourselves into imagining we live on through multiple incarnations?" Danelek, Mystery of Reincarnation 268 (2005).

17. Pope John Paul II, Evangelium Vitae, *Introduction*, Section 2. *See* Repenshek & Sloser, Hast. Ctr. Rep., November-December 2004, at 13.

Supreme Court held under the First Amendment that prayer was unconstitutional, stating "it is not part of the business of the government to compose official prayers for any group of the American people to recite as part of a religious program carried on by the government."[18] This decision was followed by similar rulings of that Court in subsequent years.[19]

"Praying" is derived from the Latin *precari,* which means to beseech or to beg. In 2004, Donald Spoto wrote: "Any attempt to speak of prayer in the twenty-first century is likely to lead to a confrontation with a cluster of objections—the two most notable being that prayer is irrelevant in a sophisticated age of science and technology; and that prayer is primarily a solipsistic dialogue with the self [i.e., that the self is the key existent thing]. Both rest on a belief that prayer is by its nature psychologically suspect." Although he asked "only of readers that they bring to this book the openness with which I myself have tried to approach this subject, others have warned, there is a danger of infusing prayer with a self-centered, magical spirit that has nothing to do with the search for God; . . ." Authentic prayer does not aim to become a comforting form of self-expression; it is about reaching within and beyond the imagined self to a greater purpose and power."[20]

Another view appears in *A World Treasury of Prayer* whose introduction asked the question: "*Does God really need to be told by us that he created the world and redeemed us and that we are miserable sinners*? Surely, He knew all this already. *Does He demand that we thank Him, praise Him, and plead for mercy?* There is something slightly repellent in this notion, as it suggests a despotic deity who demands endless sycophan-

---

18. Engel v. Vitale, 370 U.S. 421 (1962).

19. For example, in *Abington School District v. Schempp,* 374 U.S. 203 (1963), the Court so held with respect to an eastern Pennsylvania school district that required that ten verses of the Bible be read at the opening of each school day (although students could be excused from the classroom during the readings). *See also* Wallace v. Jaffree, 472 U.S. 38 (1985) (which involved an Alabama statute that authorized teachers to conduct regular prayer services in the classroom).

20. Spoto (a Catholic), *Introduction,* in Silence, Why We Pray (2004). Thereafter he proceeded with discussions of God in most of the next 200 pages (citing the Bible and other sources).

tic obeisance from his worshippers. *All talk of and to God stumbles* under great difficulties." (Emphasis added.)[21]

For some, the answer to a question regarding the meaning of prayer has been that they are in some profound sense talking to themselves. "Prayers, such as that of Saint Patrick, attempt to invoke benign power and strength that will enable us finally to lighten the darkness in the depths of the self."[22] It has been found that "Extemporaneous prayer is another characteristic of Protestantism." As written in 2002: "Although more liturgical Protestants, such as Lutherans and Episcopalians, follow prescribed prayers, low-church Protestants traditionally have preferred extemporary or spontaneous prayer (the Puritans called it "ejaculatory prayer")."[23] However, Einstein's doctrine of strict causality "implies that after the initial act of creation God does absolutely nothing in relation to the world—hardly an attractive concept of the Divine!"[24]

21. *See* ERLICH, HUMAN NATURE 256 (2000), and Byrd, *Positive Therapeutic Effects of Intercessory Prayer in a Coronary Care Unit Population*, 81 SO. MED. J. 826-29 (1988) ("intercessory prayer to a Judeo-Christian God has a beneficial effect in patients admitted to a coronary care unit"). *But see* Walker, et al., *Intercessory Prayer in the Treatment of Alcohol Abuse and Dependence: A Pilot Investigation,* 3-6 ALTERNATIVE THERAPY & HEALTH MED. (1997) at 79-26. Those who were prayed for in order to improve the condition of certain individuals subject to alcohol addiction were no more likely to recover than those who were not.

22. CRAUGHWELL, EVERY EYE BEHOLDS YOU: A WORLD TREASURY OF PRAYER XIV (1998). The Muslims under Mahamet of the seventh century call for prayer five times a day, whereas his recognized Jewish predecessors were only told to pray a couple of times a day.

23. BALMER & WINNER, PROTESTANTISM IN AMERICA 96 (2002).

> General Patton: "Chaplain, I want you to publish a *prayer* for good weather to help our soldiers fight." Chaplain: "Yes, Sir, but it's not customary for my profession to pray for clear weather to kill fellow men." The prayer was offered and printed in a Christmas greeting to the troops stating . . . We humbly beseech thee . . . to restrain these immoderate rains . . . we may advance from victory to victory . . . The day after this prayer the weather cleared. General Patton invited the chaplain, pinned on a bronze medal, and said, "You are the most popular man at this headquarters . . . and we got back to killing Germans."

SEARCHING FOR GOD IN AMERICA (p. 354).

24. MALIN, NATURE LOVES TO HIDE 12 (2001). Nevertheless, *Religious leader: President George W. Bush begins cabinet meetings with prayers*, NATURE, Feb. 2, 2006, at 535.

Studies on effects of prayer on health in 1997 produced opposing and conflicting results. This might be expected due to the difficulty in quantifying the amount of their health effects upon others;[25] some physicians indicated that going to church promotes health [26] and some asserted, "the medicine of the future is going to be prayer and Prozac."[27] However in 2000, several doctors wrote in the *New England Journal of Medicine*: "The evidence is generally weak and unconvincing. There is no convincing evidence that other religious activities are associated with improved health" and there has been major criticism about the federal government's recent contribution of $2.3 million in financing prayer.[28]

---

25. *See* Hast. Ctr. Rep., May-June 2000, at 41-19.

26. Sides, *The Calibration of Belief*, N.Y. Times Magazine, Dec. 7, 1997, at 92-95.

27. Larson D.B. Levin & C.M. Puchalski, *Religion and Spirituality in Medicine: Research and Education*, 278 JAMA 792-93 (1997). However, one study showed falsity:

> In 2001 the *Journal of Reproductive Medicine* published a study by three Columbia University researchers claiming that prayer for women undergoing in vitro fertilization resulted in a pregnancy rate of 50 percent, double that of women who did not receive prayer . . . A clinical professor of gynecology and obstetrics named Bruce Flamm not only found numerous methodological errors in the experiment but also discovered that one of the study's authors, Daniel Wirth, a.k.a. John Wayne Truelove, is not an M.D. but an M.S. in parapsychology who has since been indicted on felony charges for mail fraud and theft, to which he has pled guilty. The other two authors have refused to comment, and after three years of inquiries from Flamm, the journal removed the study from its Web site, and Columbia University launched an investigation.

Shermer, Scientific American, November 2004, at 34.

28. Orange County Register, Oct. 20, 2004, at News 23. In one continuing study financed by the National Institutes of Health called "Healing of Wounds," doctors at California Pacific Medical Center in San Francisco inflict a tiny stab wound on the abdomens of women receiving breast reconstruction surgery with their consent, then determine whether the "focused intention" of a variety of healers speeds the wound's healing. *Id.* at 26.

With or without a theological meaning, an overwhelming portion of the general public claims a belief in God's current activities that go beyond those in Nature, and include one's future "above or down below."[29] In part, this may be accounted for because many have a limited understanding of particular areas of science—that is, physical scientists looking at biological sciences and vice versa. A study found that certain "biologists are the most skeptical, with 95 percent of our respondents evincing belief in atheism or agnosticism."[30] Nevertheless Professor Francisco Ayala (who recently received the highest American award in science and is at the University of California, Irvine) concluded that, "Darwin's theory of evolution and explanation of design does not include or exclude considerations of divine action in the world any more than astronomy, geology, physics, or chemistry do":

> The creation or origin of the universe involves a transition from nothing into being. But a transition can only be scientifically investigated if we have some knowledge about the states or entities on both sides of the boundary. Nothingness, however, is not a subject for scientific investigation or understanding. Therefore, as far as science is concerned, the *origin of the universe will remain forever a mystery.*[31] (Emphasis added.)

Historically, Alan Segal, Professor of Jewish studies at Barnard College, Columbia University, pointed out that during "the First Temple Period in Israel the Hebrew culture foresaw no significant afterlife for the dead, that the covenant had nothing to say about the afterlife except to warn against

29. Orange County Register, Apr. 3, 1997, at News 8; Larson & Whitman, *Scientists and Religion in America*, SCIENTIFIC AMERICAN, September 1999, at 89.

30. Larson & Whitman, SCIENTIFIC AMERICAN, September 1999, at 90. "Today the higher the educational attainment, or the higher the scores earned on intelligence or achievement tests, the less likely are individuals to be Christians." *Id.*

31. AYALA, EVOLUTION AND MOLECULAR BIOLOGY 102, 113 (1998). Prior to his science career, he was ordained a Dominican priest in Spain. He had earlier stated that Darwin's "greatest accomplishment" was to show that "living beings can be explained as the result of a natural process, natural selection, without any need to resort to a Creator or other external agent." CREATIVE EVOLUTION?! 4-5, Campbell & Schopf eds. (1994).

believing that another god could supply one. That belief would make the Hebrews absolutely unique among world cultures and especially strange in the ancient Near East, where elaborate ideas about postmortem existence and even more elaborate rituals were everywhere part of literature, myth, and social life."[32] He also indicated how, during his last supper, Jesus said, "This is the bread that came down from heaven, not like that which your ancestors ate, and they died. But the one who eats this bread will live forever."[33] However, he then referred to additions to the Pauline version (I *Cor.* 11:23-32) that "modifies the literal implication of the Gospels and emphasizes that the actions are set in a liturgy, an imaginative act of reenactment leaving open that the statement is itself a spiritual object of remembrance rather than a literal, physical action."[34] Professor Segal's 866-page study of religions set forth his views of afterlife without expressing knowledge of the science of evolution. Yet, he states, "People who live in a scientific system do not need to interpret their ecstatic experiences as proof of divine providence. Thinking that humans will always find new and important revelations is no salvation."[35] In the Middle Ages much was written on one's future life above or below. In the *Inferno,* written by Dante 700 years ago, he wrote, "bearing the bodies that they had, All shall be clothed. Here Epicurus lies With all his followers, who call the soul dead."[36] Robert Pinski, who translated Dante in 1994, noted that "[t]he Greek philosopher Epicurus [341-270 B.C.] did not believe in the immortality of the soul and taught that 'pleasure'—that is, the absence of pain—was the greatest good. To the medieval Christian mind, this represented the ultimate heresy."[37] The creature Satan, whom Dante also

32. Segal, Life After Death 123 (2004). He also refers to the Second Temple of Judaism with the *Arrival of the Notion of Resurrection*, citing *Ezek.* 37:1-10.

33. Bible, NRV, *John* 6:59.

34. Segal, *supra* at 443. He stated, "This book is a personal view." (p. 698)

35. *Id.* at 712. "Ideas about the afterlife do not exist in a vacuum; they have no life of their own. They do not spread out of inexorable logic or good sense or superior doctrine. . . . Their fortunes are closely tied to the fortunes of their earthly believers. The fact that it takes a close historical analysis of texts that are usually studied by religious specialists has kept the secret from us." (p. 699).

36. Cantos, The Inferno of Dante (1300) (Pinsky trans., 1994).

37. *Id.* at 392.

> Before he was exiled from the city of Florence on a trumped-up charge of political corruption, Dante seemed to be at the summit of his ca-

calls by the names Lucifer, Beelzebub, and Dis, "was an angel of enormous *beauty* before he rebelled against God and was cast down from Heaven."[38] Although Christ is never directly named in the *Inferno,* where he traveled with Virgil, Dante wrote:

> In the faith that conquers every error, "Did ever
> Anyone go forth from here—by his own good
> Or Perhaps another's—to join the blessed, after?"
> He understood my covert meaning, and said,
> "I was new to this condition, when I beheld
> A Mighty One who descended here, arrayed with a crown of victory.
> And Israel's sire and children, and Rachel for whom
> He labored so long, and many others—and His
> Coming here made them blessed, and rescued them.
> Know this: no human soul was saved, till these."

This so-called "Harrowing of Hell" became "official [Catholic] Church dogma in 1215." In 2004, it was noted that "Amid hundreds of new unfamiliar religions, the occasion of one that fails to live up to expectations causes the status of all to be called into question . . . Almost by definition, new religions are dissenters, offering different paths from those that dominate in society."[39]

> reer. . . . Unlike Plato's allegories, however, Dante's journey is autobiographical. Several passages are explicitly so and serve to reinforce the sense in which the descent into Hell may be thought of as a descent into Dante's past.

Id. Freccero, Foreword xiv.

38. *Id.* at 426. Dante was told of the circled shape of Hell by Virgil in Canto XI.

39. Gordon Melton, New Religions, Foreword (2004). As regards "sects," Christopher Partridge pointed out how they "tend to be revivalist, enthusiastic and 'fundamentalist,' and can often be identified by their understandings of truth and authority and by their notions of separateness and distinctiveness as sacred communities. . . . Consequently, it is separated from both 'the world' (wider society) and also the dominant orthodoxy." (p. 18)

All humanity constitutes a part of the universe, as indicated by scientists. The late Carl Sagan commented upon physics evidence, noting that we live in a universe in which much can be "reduced" to a small number of laws of nature. Sagan then asked why should a few simple laws of nature explain so much and hold sway throughout this vast universe? Isn't this just what you might expect from a creator of the universe? (See Chapter 2.) Scientists can never go behind creation (i.e., before time), and hence more modern scientists become agnostics rather than atheists.

However, toward the end of the twentieth century, a Nobel laureate expressed this mystery of human existence and capacity.[40] Similarly, the Supreme Court opinion in 1965 stated that "*an atheist's system of beliefs*" might have "a place in the life of its possessor *parallel to* that filled by the *orthodox* belief in God."[41] However, the court specifically attempted neutrality in 1992 in its "mystery passage" while affirming the constitutional right to abortion:

> *Our Constitution* takes no sides in these ancient disputes about life's meaning. But it *does protect people's right to die as well as live*, so far as possible, in the light of their own intensely *personal convictions about "the mystery of human life."*[42] (Emphasis added.)

Scientific beliefs may be parallel, but are normally quite different. Almost half of the scientists would limit God's activities as those occurring

---

40. What, then, is the meaning of it all? What can we say to dispel the mystery of existence? If we take everything into account, not only what the ancients knew, but all of what we know today that they didn't know, then I think that we must frankly admit that we do not know.

FEYNMAN, THE PLEASURE OF FINDING THINGS OUT 148 (1979).

41. United States v. Seeger, 380 U.S. 163, 166 (1965).

42. Casey v. Planned Parenthood, 112 S. Ct. 2741 (1992). At the turn of the third millennium, the Supreme Court found that a public school's speech, including one religion's prayer, was unconstitutional. Justice Stevens wrote that the prohibition of the establishment of a religion in the First Amendment was designed "to remove debate over this kind of issue from governmental supervision and control." Santa Fe Indep. Sch. Dist. v. Doe, 530 U.S. 290 (2000).

Dworkin's *Life's Dominion: An Argument About Abortion, Euthanasia, and Freedom* (1993) expanded this by stating, "An atheist may have convictions about the point or meaning of human life that are just as pervasive, just as foundational to his moral personality, as those of a devout Catholic, Jew, Muslim, Hindu, or Buddhist."

with the Big Bang to the alleged particular "laws of nature."[43] Although in 1863 Lincoln at Gettysburg stated, "That this nation, *under God* shall have a new birth of freedom," the deity was actually absent from the original Pledge of Allegiance of the 1890s; it was not inserted into that official motto until the 1950s during President Dwight Eisenhower's administration—which is currently being challenged in the 21st century. Many deists and agnostic scientists have actually been called atheists by the general populace solely because of their lack of belief in a purposeful god who can actually receive prayers pertaining to morals and ethics and then intervene meaningfully in human life.[44]

In America, unlike Europe or Asia, being called either an "atheist," secular, or "agnostic" remained "politically incorrect" throughout the twentieth century and the beginning of the twenty-first. For example, in 1988 a reporter asked Vice President George Bush, "Surely you recognize the equal citizenship and patriotism of Americans who are atheists?" Bush replied: "No, I don't know that atheists should be considered as citizens, nor should they be considered patriots. This is one nation under God." To this reply Michael Shermer, the author of *The Science of Good or Evil* (2004), indicated: "This conclusion, however, is not borne out in the data from the scientific study of religion and morality. While individual religious believers may be exceptionally moral and tolerant people, and while religion may inspire in some individuals extraordinary morality and tolerance, religion does not necessarily foster these desirable traits."[45] Such an appellation has also been considered a virtual equivalent

---

43. It is interesting to note that the Greeks, who began serious secular thinking:

> Did not believe that the gods created the universe. It was the other way about: the universe created the gods. Before there were gods heaven and earth had been formed. The Greeks made their gods in their own image. That had not entered the mind of man before. Until then, gods had had semblance of reality. They were unlike all living things.

HAMILTON MYTHOLOGY 24 (1998/1969).

44. Harris, Ph.D., *Archives of Internal Medicine*, found that 11 percent of those who were the object of other people's prayers during their hospital stay had fewer complications.

45. *See* p. 236. Roger L. Shinn, *Union Theological Seminary,* noted in the *Encyclopedia Americana* that "[m]ilitant atheism is now less common than agnosticism, which assumes a smaller burden of proof."

to being called an anarchist or a communist.[46] On the other hand, the editor of *Nature* has written, "It may not be long before the practice of religion must be regarded as anti-science."[47] However, certain Eastern approaches are rapidly entering America regardless of this anti-scientism approach. For example, it has been pointed out that "the Buddha did not introduce the notion of reincarnation."[48] Further, Buddhism today is best thought of as an ethical psychological philosophy or nontheistic spiritual practice, needing neither dogma nor belief to be practiced and accomplished.[49] Thus the Dalai Lama said:

> *We need* some kind of *spirituality, but not necessarily religious faith.* This means basic human good quality such as human affection for one another. You see, *from birth, we have that.* Molecules of our body

---

46. CARL SAGAN, CONTACT (1985). In this science fiction novel, a future President of the United States turned down the heroine for the greatest assignment of mankind, stating: "Some choice! The one's an atheist and the other thinks he's from Vega already. Why do we have to send scientists? Why can't we send somebody . . . normal?" This separation of so-called normal thoughts and scientific ones actually appears to be growing.

47. Maddox, *Defending Science Against Anti-Science,* NATURE at 368, 185 (1994).

> The 12th graders who took the 2000 National Assessment of Educational Progress Scores . . . show that only one in five high school seniors has a grasp of science. On average, they were three points lower than those taking the test in 1996.

ORANGE COUNTY REGISTER, Nov. 21, 2001, at News 22.

48. LAMA SUNY DAS, AWAKENING THE BUDDHA WITHIN 110 (1997).

49. *Id.* at 111. "When asked, 'Does God exist or not?' Buddha remained silent. 'Does the universe have a beginning or an end?' Again, the Buddha remained silent. He did not feel that speculating about such questions greatly facilitates our progress toward freedom and peace." But, when he was on the cross, Jesus cried with a loud voice, Eli, Eli, lama sabachthani? That is to say, My God, my God, why has thou forsaken me?—*Matthew* 27:46. In his volume *When Jesus Came to Harvard* (2004), Professor Cox stated:

> If Christian faith teaches that in this man from Galilee God shared the most racking pain and the most overwhelming sorrow human beings can know, including death itself, then he must also have felt abandoned by God. . . . It merely deepens the mystery of just how God was present in Jesus, and how God continues to suffer the grief and heartache of human existence. All the doctrines and theories that have been invented for nearly two thousand years have not even come close to explaining how such a thing is possible, and none ever will.

> go very well with peace of mind and peace of mind comes, as I said before, with compassion, gives us inner strength. As a result, peace of mind will develop. So therefore, to the nonbeliever, or ordinary people who have no special interest toward religion: Yes, *okay, live as a nonbeliever*, but be a kind person, a warm-hearted person.[50]

Patients whose views must be considered by all members of hospital ethics committees include those described in Chapter 3. These appear to fall into one of several categories: First are the monotheists; they constitute about two billion people who are either Christians, Jews, or Muslims. The second group, like the Hindus, constitute at least one billion people; they are multitheists whose gods are set forth in their songs (Vedas). Third are the Buddhists, millions of people, who have essentially declined a belief in god(s). The fourth are the physicists, evolutionists, and other scientifically oriented people (described in Chapter 2) who have found a lack of evidence for the possibility of an action resulting solely from a prayer; many of these people remain agnostics.[51]

Finally, a fifth group are like the one billion Han Chinese whose overwhelming majority have believed in philosophers and rational moralists, such as Confucius rather than in one or more gods, for more than two and a half

---

50. Initially, it is possible for one individual to be both Buddhist and Christian in the sense that one obtains nourishment from the teachings of his masters. But as you go deeper in religious life, then it may become more difficult. In Tibetan Buddhism, for example, it is very difficult to accept the concept of creator. Of course, God in the sense of infinite love and God's presence everywhere, that's OK. The Buddhist can accept that God of ultimate reality or infinite compassion. But, if all our existence depends on the creator, it is difficult to accept that idea [i.e., that he was the intelligent designer of humans]. Dalai Lama, L.A. Times, June 7, 1997, at B13. Space scientist Carl Sagan stated, "Religions are often the state-protected nurseries of pseudoscience, although there's no reason why religions have to play that role. In a way, it's an artifact from times long gone." The Demon-Haunted World 15 (1995). *See also* Dalai Lama, in Hewitt, Searching for God in America (1996).

51. Thomas Henry Huxley coined an *agnostic* as: "one who holds that the existence of anything beyond and behind material phenomena is unknown and so far as can be judged unknowable, and especially that a First Cause and an unseen world are subjects of which we know nothing." Huxley, Evolution and Ethics 238 (1894).

millennia. Elsewhere it has recently been asked, "If practical ethics has never been simply the application of moral principle to concrete problems, then what is it?" The answer: "A more plausible position describes it as *moral reasoning* about practical questions: another kind of application, but considerably more open than the 'application of principles' model."[52] In *Out of My Later Years*, Albert Einstein stated:

> The main source of the present-day *conflicts between the spheres of religion and of science lies in this concept of a personal God.* In their struggle for the ethical good, teachers of religion must have the stature to give up the doctrine of a personal God, that is give up that source of fear and hope which *in the past placed such vast power in the hands of priests.* In their labors they will have to avail themselves of those forces which are capable of cultivating the Good, the True, and Beautiful in humanity itself. The further the spiritual *evolution of mankind advances*, the more certain it seems to me that the path to genuine religiosity does not lie through the fear of life, and the fear of death, and blind faith, but *through striving after rational knowledge.* (Emphasis added.)

According to a study by the National Science Foundation, 70 percent of Americans do not understand science. Here's the sad part: 30 percent don't even know what percent means.

Orange County Register, June 16, 2002, News 14

In any conflict arising between scientific and theological conclusions, the science is taken to be defective, incomplete, or inadequate or at least suspect, for the bible is seen to be free from any error and is the final authority in all matters concerning faith.

Richard Carlson, *Science and Christianity* (2001)

52. ANDRE, BIOETHICS AS A PRACTICE 66 (2004) 66. In *The Coming of the Golden Age: A View of the End of Progress* (1997), biologist Gunther Stent argued that progress in science and technology must soon end. But John Horgan's subsequent *The End of Science: Facing the Limits of Knowledge in the Twilight of the Scientific Age* (1997) derided this view. *See* MURRAY, HUMAN ACCOMPLISHMENT 438-39 (2003). As Paul Erlich pointed out, "Our challenge is to learn to deal sensibly with both nature and human natures," in his *Human Natures* (2003), at 330.

> *On the other hand, I don't believe that a real conflict with science will arise in the ethical aspects, because I believe that* moral questions are outside of the scientific realm. (Emphasis added.)
>
> Richard P. Feynman, Nobel laureate

(b) Limitations of Science; Risk/Benefit Review by IRBs in Research Clinics; Complementary and Alternative Medicine (CAM)

In the last half of the twentieth century, a historian closed his work with the words, "whether true or not, men have always believed in unscientific concepts, and these beliefs often are the real facts which shape their destiny." Another such author wrote, "I merely wish to indicate that the science-built world of engines and laboratories in which man lives has grown up apart from man himself. Science has not instructed man—it has only implemented him." One modern European philosopher stated, "Science is to be treated as one tradition among many, not as a standard for judging what is, what is not, what can, and what cannot be accepted."[53] Such divergent statements would appear to leave science only as a background for bioethics and the law to develop. Albert Einstein reacted differently, stating, "One thing I have learned in a long life: that all our science, measured against reality, is primitive and childlike—and yet it is the most precious thing we have." But he added, "It has become appallingly obvious that our technology has exceeded our humanity."

Herbert Marcuse claimed, "[T]he mathematical character of modern science determines the range and size of its creativity, and leaves the nonquantifiable qualities of humanitas outside the domain of exact science."[54] He and others argued that the essence of science was exploitation of nature that gave rise to a repressive technological society. "That Marcuse and the other guns of the period knew little about science,"[55] they shouted, as did Feyerbend in 1978: "What's so great about science? What makes modern

---

53. FEYERBEND, FAREWELL TO REASON 390 (1990). Similarly, it has been noted how, under the anthropic principle, the universe must be suitable for life, and we would not wonder about it. Dr. Steven Weinberg, a Nobel laureate from the University of Texas, referred to this as "a guess about the future shape of science." N.Y. TIMES, Oct. 28, 2003, at 1.

54. *See* Sylvan S. Schweber, *Some Reflections on Big Science and High Energy Physics in the United States*, RSS 2 pp. 127-89 (1994).

55. KRAGH, QUANTUM GENERATIONS 442 (1999).

science preferable to the science of the Aristotelians, or to the cosmology of the Hopi?'"[56] They even affected physicists in their "non-physical moments." Alfred North Whitehead, a Fellow of Trinity College at the University of Cambridge, England, and professor of philosophy at Harvard University, was also a mathematician. He limited physical science as only dealing "with half the evidence provided by human experience," stating:

> Science can find no individual enjoyment in nature; science can find no aim in nature; science can find no creativity in nature; it finds mere rules of succession. These negations are true of natural science. They are inherent in its methodology. The reason for this blindness of physical science lies in the fact that such science only deals with *half the evidence* provided by human experience.[57]

Similarly, a writer has claimed that "a perfect example of that is Harvard University . . . Today this prestigious institution is for the most part squarely opposed to the furtherance of Christ's Kingdom. Certainly, we need to put checks and balances into place in our institutions to try to prevent that kind of apostasy."[58] The professor who taught about Christian religion and moral life there pointed out that Harvard:

> . . . was originally founded—in 1626 as a school for the training of ministers, has long since shed its religious identity. For nearly a century now it has been a modern research university, often thought of as quite rich, somewhat snooty, and famously secular. Jesus, on the other hand, who lived two thousand years ago, is associated in most people's

56. Paul Feyerbend, Science in a Free Society (London: New Left Books 1978).

57. Whitehead, Modes of Thought 154. However, some science was found in Buddhism:

> Fritjof Capra, an American particle theorist, believed that he had discovered a deep connection between modern quantum theory and forms of Oriental mysticism, such as Zen Buddhism. His *The Tao of Physics*, first published in 1975, became immensely popular. According to Capra, the insights of quantum physics were the very same as those reached much earlier by Eastern mystics through meditation and intuition.

58. Dr. J. Kennedy, What If Jesus Had Never Been Born? 242 (1994).

minds with simplicity, humility and the spiritual life. He is the central figure in the world's largest religion.[59]

That professor also ran a retrial of Jesus; playing the part of Pontius Pilate, he used his Jewish colleague Alan Dershowitz, at Harvard Law School, as the defense counsel, "and Allen Callahan, on the Divinity School faculty, to serve as the prosecuting attorney." All the students served as his "advisory council" [jury]. He reported, "The verdict was very close, with the actual decision to condemn Jesus winning out by a few votes."[60]

Most humanists claim authority on their own; they have done this largely independent of science. The natural rights to autonomy cannot answer most ethical assumptions solely by scientific means. Charles Darwin acknowledged that noble moral attributes were "by far the most important of all the differences between man and the lower animals."[61] In his *The Descent of Man* (1871), Darwin found harmonious cooperation of the members of a social group, based on altruism, a propensity favored by natural selection.[62] However, he wrote only on science and avoided religion.[63] In the year 2001, Harvard professor Ernst Mayer called a section of his work *The Evolution of Human Ethics.* He noted that evidence of "genuine ethics can be developed only by

59. COX, WHEN JESUS CAME TO HARVARD 1 (2004). While discussing the Resurrection he added:

> "Jesus Christ is risen" means Jesus lives in the lives and actions of those who once followed Him, and still do. "The cause of Jesus," as some of them put it, "moves forward," and this is what the disciples were straining to say in their contradictory descriptions of their encounters with their crucified friend. *Id.* at 282.

60. He "later learned that the 'swing votes' in the council were those of the more conservative Christians who, although they would love to have exonerated Jesus, felt that to do so would somehow contradict the Bible." *Id.* at 227.

61. KEYNES, DARWIN, HIS DAUGHTER AND HUMAN EVOLUTION 289 (2002).

62. *See* FERRIS, BEST OF AMERICAN SCIENCE WRITING 134 (2001).

63. Though I am strong advocate for free thought on all subjects, yet it appears to me (whether rightly or wrongly) that direct arguments against Christianity and theism produce hardly any effect on the public; and freedom of thought is best promoted by the gradual illumination of men's minds, which follows from the advance of science. It has therefore, been always my object to avoid writing on religion, and I have confined myself to science.

Darwin (1880).

adding such global altruism to the 'selfish' altruism of the social group." He pointed to the "chosen group" in the Old Testament[64] in order to "see how consistently a difference is made between behavior toward one's own group and behavior to any outsiders."

This is in contrast to ethics promoted in the New Testament; Jesus's parable of the altruism of the Good Samaritan was a striking departure from a former custom. Altruism toward strangers is a behavior not supported by natural selection. As Mayer stated, "We are not born with a feeling of altruism toward outsiders, but acquire it through cultural learning. It requires the redirecting of our inborn altruistic tendencies toward a new target: outsiders . . . They acquire true ethics (including to outsiders) by learning."[65]

Children are often born with at least a running start on the path to moral development. A number of inborn responses predispose them to act in some ethical ways. For example, empathy (the capacity to experience another person's pleasure or pain vicariously) is part of our native endowment as humans. But children who learn too little about ethical foundation often cause great harm to themselves and others. Unless our children are taught the importance of ethics and morals, and this is practiced, there is no guarantee they will acknowledge use of such things daily. In *The Death of Ethics in America* (1988), Thomas wrote: "Our culture promotes greed, and so greed we get."

At the end of the twentieth century it was pointed out, "*Psychologists do not have* definitive *answers* to these questions, and often their studies seem

64. *See Exodus* 34.10 and *Deuteronomy* 7.

65. MAYER, WHAT EVOLUTION IS 256-58 (2001).

> The discrimination against outsiders, which is perhaps the major reason for the resistance to a worldwide acceptance of a broadly conceived human ethics, is gradually being overcome by some basic social principles, such as equality, democracy, tolerance, and human rights. Moral education has been practiced very successfully by several of the world's great religions . . . even though we have failed so often to follow them. *Id.* at 260.

My neighbor, an American engineer, noted that Professor Mayer's connection of human ethics with altruism was "too lofty, since 'doing good' is for someone else. Human ethics should be definable on an individual basis also."

merely to confirm parents' observations and intuition."[66] (Emphasis added.) Some have argued that many people go through various stages of moral judgment beginning with punishment and reward. "Few if any people reach the most advanced stage, in which their moral view is based purely on abstract principles."[67] One study of Indian and other Americans demonstrated that Indian adults generalized their values to a broad range of social conditions. Toward American children, it was said that moral standards should apply to everyone always. American adults "modified values in the face of changing circumstances." In short, "the Indians began life as relativists and ended up as universalists, whereas the Americans went precisely the other way."[68] This characteristic of most Americans with respect to *values in our society* was described by the Chief Justice of the Supreme Court in the middle of the twentieth century:

> Nothing is more certain in modern society than the principle that *there are no absolutes*, that a name, a phrase, *a standard has meaning only when associated with the considerations which gave birth to* the *nomenclature* . . . To those who would paralyze our Government in the face of impending threat by encasing it in a semantic straitjacket we must reply that *all concepts are relative.*[69] (Emphasis added.)

Scientists appear to be closer to obtaining the absolutes in nature than are many moralists who try to determine human values.

Some volumes have been written describing building blocks of morality recognizable in other animals. It has been found that apes, such as gorillas and chimpanzees, have many of the moral and altruistic behavioral tendencies displayed by human children. Because animals cannot be held legally responsible, they cannot be given "rights." In the middle of the 19th century, Ralph Waldo Emerson indicated, "Human beings are the only animals that have developed religions, ethics, moral codes, and mutually agreed-on norms of conduct." A century and one half later, the Dutch ethologist Franz de Waal wrote, "To classify the chimpanzee's *behavior* as *based on instinct* and the

66. Daman, SCIENTIFIC AMERICAN, August 1999, at 73.
67. *Id.* at 74.
68. *Id.* at 76.
69. Dennis v. United States, 341 U.S. 494, 508 (1951).

person's behavior as proof of moral decency is misleading, and probably incorrect: *I hesitate to call the members of any species other than our own 'moral beings,' yet* I also believe that many of the sentiments and *cognitive abilities underlying human morality antedate the appearance of our species* on this planet."[70] (Emphasis added.)

With altruism, someone performs an act for the well-being or good of another in his group. Although it has been alleged to exist in a number of animals, several evolutionary biologists were able to show such altruism as actually selfish; it is in the genes for a family. In both animals and humans, a generally accepted view is that altruistic actions toward unrelated individuals are motivated by an understanding of reciprocation.[71] In his volume *On Human Nature*, Professor Edward O. Wilson of Harvard contended, "Human emotional responses and the more general ethical practices based on them have been programmed to a substantial degree by natural selection over thousands of generations." These include something humans have in common with the great apes.[72] To a great extent, change in human history is beyond prediction. As Karl Popper stated, "The belief in historical destiny is sheer superstition . . . There can be no prediction of the course of human history by scientific or any other rational methods."[73] Nevertheless, in most, but far from all, members of the species Homo sapiens, moral values of a modern society are derived from reason and learning, not instinct.

70. Franz de Waal, The Mammal in the Mirror 210 (2000).

71. *See, e.g.,* Dawkins, The Blind Watchmaker (1996); The Selfish Gene (1976); Dugatkin, Cooperation among Animals: An Evolutionary Perspective (1997); Feynman, The Pleasure of Finding Things Out 251 (1999).

72. Wilson, On Human Nature 6 (1978).

73. Karl Popper, The Poverty of Historicism v (1957). Similarly, it has been pointed out, "There are certainly no equations for history. Indeed, while the physical sciences reveal numerous regularities that can be captured in immutable scientific laws, this does not seem to be the case in the social world where emotional and unpredictable humans take center stage." Duchanan, Nexus 11-12 (2002). Nevertheless, he cited a mathematician who discovered special uses with humans of dots connected by lines to produce graphs for an organizational principle. Duncan J. Watts & Steven P. Strogatz, *Collective Dynamics of 'Small-World' Networks*, Nature 393, 440-42 (1998).

Institutional review boards (IRBs) have been used to evaluate research proposals to ascertain if their "risks to subjects [volunteers] are reasonable in relation to anticipated benefits, if any, to subjects, and the importance of knowledge that may reasonably be expected to result."[74]

This risk-benefit relationship has been called a Common Rule's "clinical equipoise," which others have indicated "exists when the community of expert practitioners is uncertain as to the relative merits—i.e., the relative balance of benefits and harms—of standard versus experimental therapy."[75] The National Bioethics Advisory Commission (NBAC) has recommended that IRBs "must also be able to put those risks and benefits into a social context in order to judge how ordinary citizens would value them."[76] Further, in 2004, the Department of Health and Human Services recommended that research physicians "consider whether specific financial relationships create financial interests in

---

74. 45 C.F.R. § 46.111(2). The IRBs are to use "procedures which are consistent with sound research design and which do not unnecessarily expose subjects to risk" [45 C.F.R. § 46.111a(1)] and to not "consider possible long-range effects of applying knowledge gained in the research (for example, the possible effects of the research on public policy) as among those research risks that fall within the purview of its responsibility" [45 C.F.R. § 46.111(2)].

75. Miller & Weijer, *Moral Solutions in Assessing Research Risk, in* IRB: A REVIEW OF HUMAN SUBJECTS RESEARCH 22(5):6-10 (2000) (where they indicated the variety between the subject's ages, diseases, etc.). *See also* their *Rehabilitating Equipoise*, 13 KENNEDY INST. ETHICS J. (2003) (where they recognized the clinical experts who were members of the IRBs). Physicians in private practice who participate in industrial research had risen by 300 percent by the year 2000, and the 20 percent of clinical trials funded by industry outside of academia had grown to more than 60 percent. HHS OFFICE OF INSPECTOR GENERAL, RECRUITING HUMAN SUBJECTS: PRESSURES IN INDUSTRY-SPONSORED RESEARCH (June 2000).

76. ETHICAL AND POLICY ISSUES IN RESEARCH INVOLVING HUMAN PARTICIPANTS, VOL. I: REPORT AND RECOMMENDATIONS OF NBAC (2001). In this connection:

> Five trade groups representing pharmaceutical companies worldwide are urging members to release more information about clinical trials. . . . The companies have been under pressure since revelations that they kept trial data for antidepressants and other drugs secret.

Critics such as Drummond Rennie, deputy editor of the *Journal of the American Medical Association*, aren't optimistic. "Marketing forces and self-interest . . . are going to win out every time over the ethics of doing the right thing." SCIENCE, Jan. 14, 2005, at 189.

research studies that may adversely affect the rights and welfare of subjects."[77] In regard to more than 2,900 responding IRB medical faculty members to a survey, almost half (47 percent) had been consultants for industry.[78]

In this connection, the World Health Organization has decided to issue a comprehensive *WHO Health Care Research Handbook*, which is intended to be an educational tool that could facilitate implementation of Good Clinical Practice in clinical trials. The UNDP/World Bank/WHO Special Programme for Research and Training in Tropical Diseases has also formulated *Operational Guidelines for Ethics Committees that Review Biomedical Research.*[79] The International Committee of Medical Journal Editors (ICMJE) seeks to ensure public access to all "clinically directive" trials that it defined as "Any research project that prospectively assigns human subjects to intervention and comparison groups to study the cause-and-effect relationship between a medical intervention and a health outcome."[80] Computerized clinical decision support systems (CDSSs) which have been designed to improve clinical decision making have been assessed. Although "many CDSSs improve practitioner performance; to date, the effects on patient outcomes remain understudied and, when studied, inconsistent."[81]

---

77. HHS, *Financial Relationships and Interests in Research Involving Human Subjects: Guidance for Human Subject Protection* (2004), 69 Fed. Reg. 26,393 at 26,394, titled "Guidance Document."

78. Campbell et al., *Characteristics of Medical School Faculty Members Serving on Institutional Review Boards: Results of a National Survey*, 78 Academic Med. 833-34 (2003).

79. Fluss, *The Evolution of Research Ethics: The Current International Configuration*, J.L. & Ethics 598 (Fall 2004) (citing WHO document TDR/PRD/Ethics/2000.1).

80. Starting on July 1, 2005, it would consider a trial for publication only if it has been registered before the enrollment of the first patient. N. Eng. J. Med. 2436 (June 9, 2005). The nonmandatory registry is administered by the National Library of Medicine under section 113 of the Information Program on Clinical Trials for Serious or Life-Threatening Diseases, Food and Drug Administration Modernization Act of 1997, Pub. L. 105-115. Clinical limitations of medical research with guidance by many medical practitioners when serving patients scientifically. Guyatt et al., *Evidence-based medicine: principles for applying the users' guides to patient care*, JAMA 1290-96 (2000).

81. Garg et al., *Effects of Computerized Clinical Decision Support Systems on Practitioner Performance and Patient Outcomes*, JAMA 1223 (March 9, 2005). *See also* Wears & Berg, Computer Technology and Clinical Work, *id.* at 1261.

An official commission was established by the President of the United States at the turn of the third millennium in order to maximize "the benefits to Americans of complementary and alternative medicine" (CAM),[82] which is supported by his successor.[83] Its 2002 report indicated that CAM approaches "are already in widespread use and those that have the greatest potential for addressing the nation's most serious healthcare problems" such as "biofeedback, meditation, guided imagery, art therapy, music therapy" appear to be effective but may not be profitable to private investors. It also observed that "access is further compounded by lack of scientific evidence regarding safety and effectiveness of many CAM practices and products."[84] In 1998 it was reported that there is *no evidence* that most alternative medicine works according to adequately controlled outcome research."[85] Dr. Lawrence Schneiderman, a major critic, referred to complementary and alternative medicine (CAM) as "a collective romantic fantasy."

---

82. Exec. Order No. 13,147, 65 Fed. Reg. 13,233 (March 7, 2000). Government funding for CAM research is administered by the National Center for Complementary and Alternative Medicine at the National Institutes of Health. In 2000, 73 percent of all U.S. medical schools included CAM in required courses and 51 percent offered it as a separate elective. These numbers suggest that physicians who do not become familiar with CAM and possibly include it in their practices will soon be a distinct and possibly marginalized minority. Some studies have shown that 80 percent of cancer patients use CAM in addition to conventionally prescribed therapies.

83. Boozang, *National Policy on CAM: The White House Commission Report*, 31 J. L., MED. & ETHICS (2003) at 251. The Commission's final report ("Commission Report"), issued in March 2002, similarly skirts the fundamental question of whether evidence exists that CAM interventions are safe or offer sufficient benefit to justify their proliferation. Admittedly, answering this question is extremely problematic because of the wide range of therapies that can be included within the CAM rubric. Nonetheless, a commission established to guide the nation on how to proceed in resolving the myriad public policy questions raised by CAM's popularity must at least first identify what comprises CAM, and address whether—so identified—CAM is safe and effective. *Id. See also* Exec. Order No.13,147, 65 Fed. Reg. 13,233 (March 7, 2000).

84. *Id.* at 53. "The Commission suggested that Congress 'direct the Center for Medicare and Medicaid Services to develop a demonstration project to study evidence-based CAM interventions as part of comprehensive care of persons with chronic disease in both the Medicare and Medicaid programs.'" *Id.* at 254.

85. Boozang, *Western Medicine Opens the Door to Alternative Medicine*, AM. J. L. & MED. (1998) at 185.

Programs in alternative medicine have been established in most medical schools, including Harvard Medical School, which, while steadfastly refusing to lower its standards and provide a residency training program in family medicine, courageously accepted a $10 million grant from the Bernard Osher Foundation to endow a program in alternative medicine. At the inauguration, the benefactor is reported to have declared: "Integrative medicine's time has come. It is here because of patient demand, and it has come about only because of patient demand."[86]

Another critic, Dr. Arnold Relman, asserted that "science denies religion and that is what distresses advocates of CAM because the CAM view has a spiritual foundation."[87] Dr. Sugarman stated, "[T]he interaction of herbal remedies with anesthetic agents and cardiovascular drugs should be described and discouraged."[88] On the other hand, Professor Morreim, Ph.D., concluded that our anecdotal reports on "the actual clinical practice of medicine cannot realistically, faithfully claim to be the scientific enterprise that is presupposed when CAM modalities are criticized as being 'unscientific.'"[89] In 2003, a study of large clinical trials found that as the criterion standard for making treatment decisions, nonpublication of the results of such trials can lead to bias in the literature and contribute to inappropriate medical decisions. They found that "Bias against publishing nonsignificant results is a problem even for large randomized trials. Nonpublication breaks the contract that investigators make with trial participants, funding agencies, and ethics boards."[90]

86. Schneiderman, *The (Alternative) Medicalization of Life*, 31 J. L., Med. & Ethics (2003) at 194, *citing* Lambert, *The New Ancient Trend in Medicine*, Harv. Magazine (March-April 2002) at 46.

87. Hufford, *id.* at 199.

88. *Id.* at 248, *citing* Lyons, *Herbal Medicines and Possible Anesthesia Interactions*, 70 AANA Journal (2002) at 47-51; G. Valli & E.G.V. Giardina, *Benefits, Adverse Effects and Drug Interactions of Herbal Therapies with Cardiovascular Effect*, 39 J. Am. Coll. Cardiology (2002) at 1083-95.

89. *Id.* at 226. "The actual process of daily clinical care often is, and in may ways must inevitably be, rather far removed from those scientific roots . . . to 'hold' both to 'the same' standards appears to bode far worse for medicine than for CAM." (p. 228)

90. Krzyzanonska, Pintilie & Tanneck, JAMA 495 (July 23, 2003).

Ruiping Fan, a Hong Kong scholar, pointed out how traditional Chinese medicine hospitals also use "double diagnosis" and "double therapy."[91] In 2001 it was more tolerantly noted: "In recent decades orthodox medicine's successful focus on specific disease interventions has meant relative neglect of self-healing and holism, and from this shadow complementary medicine has emerged with its counterpointing biases. The gap between them is, however, narrowing with the emerging view, backed by the study of placebo and psychoneuroimmunology, that to ignore whole person factors is unscientific and less successful."[92] Having visited China during the past few years, I discussed some of the Chinese standards with an experienced Chinese physician and those of a Hong Kong scholar, which are set forth in Appendix B.

However, as described throughout the remainder of this volume, moral discipline is only indirectly derived from science. One scientist on the subject of evolution of species stated: "There are matters of value and meaning that are outside science's scope."[93] Rita Charon, who is a professor of clinical medicine at Columbia University's College of Physicians and Surgeons, began a program in narrative medicine for students, doctors, and staff "to encourage them all to lay down their stethoscopes and listen to their patients' stories. The point is to improve the ordinary practice of medicine."[94]

Nevertheless, science's combination with logic and reason constitute the field of bioethics in medicine and the law. A rabbi claimed, "Judaism has gone further than most other religious or secular systems of ethics in trying to deal with morality in legal terms."[95] However, Chapter 2 pursued the views

---

91. *Id.*, at 213. *See* Appendix B.

92. Reilly, *Enhancing Human Healing*, BRITISH MED. J. (2001) at 120.

93. VATICAN OBSERVATORY PUBLICATIONS, EVOLUTIONARY AND MOLECULAR BIOLOGY, SCIENTIFIC PERSPECTIVES ON DIVINE ACTION (1998), Ayala (p. 115), *citing* UPDIKE, TOWARD THE END OF TIME (1995).

94. I once learned from a friend: "In the matter of the value and meaning of the universe, science has all the answers, except the interesting ones." Taking care of patients with cancer intensifies the emotional demands. The doctors, nurses, and social workers who care for these patients are even more challenged than the rest of us with a sense of defeat, loss, anger, and rage at this disease that still is not being conquered.

Quoted by Pollak in THE MAGAZINE OF COLUMBIA UNIVERSITY, Fall 2003, at 45.

95. DORFF, MATTERS OF LIFE AND DEATH 397 (1998).

of other religions regarding the distinction of reason and experience and more recent scientific matters.

But after observation and analysis, when you find anything agrees with reason, and is conducive to the good and benefit of one and all, then accept it and live up to it.

The Buddha

There remains the danger that intellectuals will take the position that their views are absolutely best, instead of just better for them. Such an attitude lacks magnanimity. At the same time, the jurors who voted to condemn Socrates failed to have the patience and tolerance which men in a free society must have if they are to enjoy the fruits of diversity. Here was a fatal misunderstanding for Socrates, the Athenians, and others since.

Finley Hooper, *Greek Realities* (1967), p. 354.

The safest general characterization of the European philosophical tradition is that it consists of a series of footnotes to Plato.

A. N. Whitehead, *Process and Reality*, p. 39.

Homo sapiens (literally, the intelligent human being) have lived to see the time when his or her existence increasingly depends on his or her intelligence.

Erwin Laszlo, a founder of the Club of Rome.

3. A Problem With the Use of Either Reason or Philosophy Alone

In his volume on ethics called *Nicomachean Ethics*, Aristotle emphasized the significance of the fact that all men seek happiness. For him, the highest degree of happiness came with the use of his mind to contemplate. The Greek word *ethos* means a habitual mode of conduct, which normally is addressing life's goals. Aristotle's goal of reflection has been carried by many people through religious convictions, but elsewhere by others. The second century A.D. Roman Emperor Marcus Aurelius was largely (but not totally) a stoical guide.[96] His meditations were frequently cited "good" in a manner close to that of many modern Americans, but unlike that of youths who favor Hollywood's different methods emphasizing happiness by sexual and other means. He stated, "[N]o difference lies here between the logos of rationality and that of justice."[97] In 1876, the first Ethical Culture Society in New York City, as well as the first British Society, was founded. Early in the twentieth century, the American philosopher John Dewey wrote, "[T]he standing temptation of philosophy, as its course abundantly demonstrates, is to regard the results of reflection as having, in and of themselves, a reality superior to that of the material of any other mode of experience."[98] Nevertheless, in 1967 in answer to the question of benefits humanity received from philosophers, Bertrand Russell replied: "I don't know whether humanity has received much *benefit from philosophers*. They haven't, in the first place, made any discoveries . . . I suppose they've enlarged people's imagination in some ways, and given them a capacity to think about the universe as a whole; but *nothing comparable to* what the *men of science* have done."[99] (Emphasis added.) This was followed by Theodore Dreiser's twentieth century comment that "[O]ur civilization is still in the middle stage and 'not yet wholly guided by reason.'" Some TV evangelists and their followers appear to believe with Chesterton that all atheists and agnostics "will believe in anything." Before Russell, Immanuel Kant produced his great *Critique of Pure Reason* that

96. Marcus Aurelius Antoninus was born in 121 A.D., became the Roman Emperor in 161, and ruled for almost 20 years. *See* HAYS, MEDITATIONS, A TRANSLATION OF MARCUS AURELIUS, 2002. His introduction included the statement, "We do not live in Marcus's world, but it is not as remote from us as we sometimes imagine."

97. MARCUS AURELIUS BV11, 1.

98. CAMPBELL, UNDERSTANDING JOHN DEWEY 74 (1995).

99. The first German to take notice of Hume was Immanuel Kant, who had been content, up to the age of about 45, with the dogmatic tradition derived from Leibniz. Then, as he says himself, Hume "awakened him from his dogmatic slumbers." After meditating for 12 years, he produced his great work, the *Critique of Pure Reason;* seven years later, at the age of 64, he produced the *Critique of Practical Reason*, in which he resumed his dogmatic slumbers after nearly 20 years of uncomfortable wakefulness.

BERTRAND RUSSELL,UNPOPULAR ESSAYS 51 (1950).

denied the possibility of accumulating knowledge by reason alone, a work that tended to doom the dogmatism in the medieval age of scholasticism.[100] Nevertheless, students of the law will normally agree with Britain's Lord Coke: "Reason is the life of the law; nay, the common law itself." In speaking of the country's founders, Abraham Lincoln stressed the need for "unimpassioned reason" saying:

> The pillars of the *temple of liberty* . . . must fall, unless we, their descendants, supply their places with other pillars, hewn from the solid quarry of sober reason. *Passion* has helped us; but *can* do so no more. It will in future *be our enemy*. *Reason*, cold, calculating, unimpassioned reason, *must furnish* all the materials for *our future support* and defense.[101] (Emphasis added.)

We have seen the recrudescence of a strong belief in the value of reason with the recent field of bioethics. It represents a culture that will permit science and humanity to continue together.

At the dawn of the United States as a country, Benjamin Franklin pleaded that people should on occasion "doubt a little of [their] own infallibility."[102] Thomas Aquinas wrote that the law "is a rule and measure of acts, whereby man is induced to act or is restrained from acting," and that the law "needs to be in accord with some rule of reason."[103] Lawmakers, whether they are elected representatives of the people or state judges at courts of law, should support public law by joining together adequate facts with reason. With the separation of church and state in our diverse society, the reason for such support must be a secular one—although it may also be consistent with some religious ethics.[104] Judge Cardozo called it "right reason," although he was unable to define "right" except by his individual judicial opinions themselves. Oliver Wendell Holmes made a nineteenth century analysis which actually remains at least partially effective in the twenty-first century: "The life of the law has not been logic; it has been experience. The felt necessities of the

---

100. *See* MOOREHEAD, BERTRAND RUSSELL, A LIFE (1993).

101. Lincoln's 1838 address to the Young Men's Lyceum of Springfield.

102. *See* LEARNED HAND, THE BILL OF RIGHTS 75 (1958).

103. ST. THOMAS AQUINAS, SUMMA THEOLOGICA I-II Q. 990-108.

104. RELIGION IN THE PUBLIC SQUARE: THE PLACE OF RELIGIOUS CONVICTIONS IN POLITICAL DEBATE 25-29 (1997).

time, the prevalent moral and political theories, intuitions of public policy, avowed or unconscious, even the prejudices which judges share with their fellow men, have had a good deal more to do than the syllogism in determining the rules by which men should be governed."[105]

Legal opinions today seldom rule on the bases of maxims of law (or even recite them), although they may sometimes be used as "window dressing" by lesser judges. At the beginning of the twentieth century, President Theodore Roosevelt called ours "a government of law, not men." Others would amend this to read, "a government of laws made by representatives of men" (and since the 19th Amendment was ratified in 1920 would add "and women"). However, due to the field of bioethics, emphasizing an individual's autonomy over his or her own body, many of those rights are being recognized by the courts. But, as will be discussed in Chapter 5, et seq., a number of these rights are limited in certain areas.

> *There are no neutral principles, only so-called principles that are already informed by the substantive content to which they are rhetorically opposed.*
>
> Fish, *The Trouble with Principles* (1999), p. 4

> *I liked the piece by Scott Lively in the August edition. Yes, to* NOT teach religion is TO teach atheism—it cannot be a matter of teaching neither, or "neutrality," *since, as he says, if you neglect teaching about God, you convey atheism* . . . Ethics-neutral rule means a corrupt society. (Emphasis added.)
>
> Roger D. Crane, *Southern California Christian Times*, September 2002

> *In the world of values,* nature in itself is neutral . . . *It is we who created value and our desires which confer value . . . It is for us to determine the good life, not for nature—not even for nature personified as God.* (Emphasis added.)
>
> Bertrand Russell, 1957, pp. 55-56

---

105. The Common Law 1 (1887). As stated in *Ecclesiastes,* which made the unusual statement, "For in much wisdom is much vexation, and those who increase knowledge increase sorrow" (*Eccl.* 1:12-18) (which also represents what happened to Socrates, as set forth in the epigraph on page 184).

> *The truth is that* medicine, *professedly founded on observation*, is as sensitive to outside influences, political, religious, philosophical, *imaginative*, as is the barometer *to the changes of atmospheric density.* (Emphasis added.)
>
> Dr. Oliver Wendell Holmes, Medical Essays

4. The Allegation of Neutral Principles; Science and the Humanities

The epigraphs of this section illustrate what appears to be the major diversity of viewpoints on any need for uniformity regarding the principle of neutrality. Although it may seem hard to believe, in the twentieth century, an attempt was made by Professor Herbert Wechsler of the Columbia University School of Law to express a theory of "neutral principles." It has been cited in hundreds of law review articles, many of which expressed its analytical shortcomings.[106] As Judge Richard Posner, of the Court of Appeals for the Seventh Circuit, noted:

> Wechsler's article does not explain, or adequately define, or indicate the provenance of, or defend, his central concept of "neutral principles." . . . It is not easy to imagine an article with the self-conscious authority of Wechsler's being written today. . . . All he means, apparently, is that decisions should be *principled.* And all that that *seems to mean* to him is that judges should *avoid grounds* of decision that would require them to engage with the messy world of *empirical reality.*[107] (Emphasis added.)

---

106. *See* Gary Peller, *Neutral Principles in the 1950s*, 21 J. L. Reform 561 (1988).

107. Posner, Overcoming Law 74 (1995). In 2002, he went so far as to write a volume that includes the statement:

> Moral theory has nothing for law . . . even if moral theorizing can provide a usable basis for some moral judgments, it . . . is not something judges are or can be made comfortable with or good at; it is socially divisive; and it does not mesh with the issues in legal cases. . . . If I am right that there is no necessary or organic connection between law and morality, then judges need not take sides on moral questions.

Posner, *The Problematics of Moral and Legal Theory* (2002).

Today, it would appear that we should recognize that the overuse of broad, general statements can be of little help in deciding cases; nevertheless, some may be of use. For example, Beauchamp and Childress stated that information provided to patients must be free from the "entrenched values and goals of medical professionals."[108] That is, a physician's own values should not always influence his or her own treatment decisions, which are subject to final decision making upon receipt of informed consent from his or her patients. Many physicians assume their patients have become incompetent when they disagree with the professional's recommendation and forgo treatment.[109] In the year 2000, a test was administered that studied the "moral reasoning" of residents in internal medicine, family medicine, and surgery. It was found that the scores of such residents thereon did not increase during their training and that their average scores were lower than those of many other professions, including law students.[110] This is similar to the study of economics

108. BEAUCHAMP & J.F. CHILDRESS, PRINCIPLES OF BIOMEDICAL ETHICS 88 (3d edition 1989).

109. Wolff, Note, *Determining Patient Competency in Treatment Refusal Cases*, 24 GA. L. REV. 733, 750 (1990). Some people have reached a similar adverse conclusion of the U.S. Supreme Court with regard to a state statute that authorized its courts to allow young children to visit their grandparents to serve "the best interests" of the children, without regard to the wishes of the parents. A plurality of the Court struck down the statute, holding that it "unconstitutionally interferes with the fundamental rights of parents to rear their children," and authorized "any person" to petition for visitation, Troxel v. Granville, 530 U.S. 57 (2000), even though the Supreme Court had actually acknowledged that "[t]he composition of families varies greatly from household to household. . . . For example, in 1998, approximately 4 million children—or 5.6 percent of all children under age 18—lived in the household of their grandparents."

110. The Defining Issues Test (DTI) posed six moral dilemmas to the test-takers and also asked them to prioritize a list of 12 items that may have an impact on their decisions. This test is accepted as a good indicator of a person's level of moral reasoning. Previous studies have observed similar trends in medical school students.

PENN STATE NEWS, July 13, 2000, *available at* http://www.psu.edu/ur/2000/hmcmoral.html.

concerning which it has been stated, economists are struggling with the data. Inflation picked up without the pickup in employment and is still having strong productivity gains. It's a reminder to all of us that *economics is more an art than a science.*[111]

Some general statements prove much more difficult to elude. In his 1739 volume, *A Treatise of Human Nature*, the British philosopher David Hume postulated the "naturalistic fallacy": as previously noted, mere knowledge of scientific facts (or the "is") cannot determine what should (the "ought") be done in ethics or morals.[112] This cannot be totally true in all ordinary instances of daily living. If you are about to be run over by a car (fact), you "should" jump out of the way—assuming that you wish to survive (invariably a good assumption for virtually all humans). Even Maimonides, in his almost 900-year-old *Guide for the Perplexed,* stated, "I therefore maintain that the law, though not a product of Nature, is nevertheless not entirely foreign to Nature."[113] But the postulate of the fallacy applies in other areas, such as bioethical matters that require reflection on conflicting values at a more leisurely pace, or conflicting values concerning a severely suffering terminal cancer patient, whether to prolong his life or inject him with painkillers that may cause him to die sooner (see Chapters 12 and 14).

One of the first to reflect on the issue of prolonged life was August Compte. In the face of what he termed a "meaningless" universe, Compte proposed a "religion of humanity" to replace old moral creeds. Some academics might

---

111. Ethan Harris, co-chief U.S. economist at Lehman Bros. in New York, ORANGE COUNTY REGISTER, July 18, 2004, at Money 3.

112. HUME, A TREATISE OF HUMAN NATURE 469-70 (Vol. 3 1988).

> I am surpriz'd to find, that instead of the usual copulations of propositions, *is*, and *is not*, I meet with no proposition that is not connected with an *ought*, or an *ought not*. This change is imperceptible; but is, however, of the last consequence. For as this *ought* or *ought not*, expresses some new relation or affirmation, 'tis necessary that it should be observ'd and explan'd; and at the same time that a reason should be given, for what seems altogether inconceivable, how this new relation can be a deduction from others, which are entirely different from it.

113. *See* p. 233.

desire to continue those creeds.[114] However, this could lead to the misuse of experts on hospital bioethics committees, which are generally comprised of physicians, plus community representatives (see Chapter 6). The American Hospital Association (AHA) 1993 volume on Healthcare Committees agreed that "the source of their moral authority has never been clearly delineated." That does not mean that "committees can have no moral authority or role in creating a moral community. However, the nature of their moral authority is not a function of their ethics expertise and thus is not well connected to case consultation insofar as case review appears to require ethics expertise."[115] With appropriate education, this expertise will expand in ethics committees of the twenty-first century.

Without abandoning religion, humanity began in the Mideast with an apple from the Tree of Knowledge, eaten by Adam and Eve, and in Greece, with the fire that was stolen by Prometheus. For these acts, they were made to suffer by their deities. Nevertheless, many humans have felt that this knowledge was their own "divine" right, even though a large percentage of people ignored the basic sciences. For example, a Gallup poll reported that of the people who were asked whether the earth revolves around the sun or the sun revolves around the earth, 18 percent of Mayans chose the later arrangement.[116] Even the ancient Mayans believed that "learning on the one hand, and ecstatic spiritual experience on the other, were the two complementary ways that human beings had painstakingly developed to reclaim their original divine sight."[117]

We have not yet found a uniform way to combine the perspective of an individual with an objective view of that world. As a Nobel Laureate scientist confessed, "I didn't have time to learn and I didn't have much patience with what's called the humanities, even though in the university there were humanities that you had to take. I tried my best to avoid somehow learning anything and working at it. I have a limited intelligence and I use it in a

114. AUGUST COMTE, INTRODUCTION TO POSITIVE PHILOSOPHY (Ferre ed. 1988), and MANUEL, THE PROPHETS OF PARIS (1962).

115. ROSS ET AL., HEALTHCARE ETHICS COMMITTEES: IN NEXT GENERATION 183 (1993). "It is possible that future public policy will provide stronger functional (if not moral) authority for ethics committees and their role in case consultation."

116. PARADE MAGAZINE, Dec. 26, 1999, at 18-19.

117. GILLETE, THE SHAMAN'S SECRET 18-19 (1997).

particular direction." Furthermore, "Social science is an example of a science which is not a science."[118] Accordingly, one modern philosopher stated, "It is the most fundamental issue about morality, knowledge, freedom, the self, and the relation of mind to the physical world."[119] In a similar vein, a Harvard University scientist in 1998 wrote a volume attempting to link the sciences and humanities: "The greatest enterprise of the mind has always been and always will be the attempted *linkage of the sciences and humanities*. The ongoing *fragmentation of knowledge and resulting chaos in philosophy* are not reflections of the real world but artifacts of scholarship."[120] (Emphasis added.) One of his critics exclaimed, "Off to such a start, I approached the rest of the book with considerable trepidation."[121] But was he actually wrong about the chaotic philosophy?[122] David Hume's naturalistic fallacy remains with us, although it is not necessarily totally applicable, as we shall see in subsequent chapters.

There is only one good, knowledge,
And one evil, ignorance.

Socrates

---

118. FEYNMAN, THE PLEASURE OF FINDING THINGS OUT 2-3 (1999).

119. "Instead of a unified world-view, we get the interplay of these two uneasily related types of conception, and the essentially incomplete effort to reconcile them." Thomas Nagel, *The View from Nowhere* 43-44 (1986).

120. E.O. WILSON, CONSILIENCE, THE UNITY OF KNOWLEDGE 238 *et seq.* (1998).

121. SCIENTIFIC AMERICAN, June 1998, at 97. "I do not share his optimism that this unity may be our salvation. I fear the inflexibility of social short term." *Id.* at 98.

122. A university president was quoted by John Berrow and Frank Tipler in their book, *The Anthropic Cosmological Principle,* stating: "Why is it that you physicists require so much expensive equipment? Now, the Department of Mathematics requires nothing but money for paper, pencils, and waste baskets. The Department of Philosophy is better still. It doesn't even ask for waste paper baskets."

> *The only way in which a human being can make some approach to knowing the whole of a subject is by hearing what can be said about it by persons of every variety of opinion and studying all modes in which it can be looked at by every character of mind. No wise man ever acquired his wisdom in any mode but this.*
>
> John Stuart Mill

5. The Change in Modern Philosophy and Secular Humanism

Professor Hans Jonas, a classical philosopher throughout most of the twentieth century, explained the major change following the turn of the third millennium: "I now say my goal to be philosophy of organic matter or a philosophical biology. This became my postwar [WWII] pursuit." He noted, "The natural sciences bring to light relevant for the concept of Being, which is, after all, philosophy's concern." These include biology, which "revealed to me the wonders of life, its evolution, its abundance of forms, modes of functioning, etc.," and would enable him to "now tackle my philosophical project . . . I now saw my goal to be a philosophy of organic matter or *a philosophical biology.*" After all, natural sciences:

> . . . have their eminently practical side in the form of technology, which increasingly shapes lives of us all . . . The guiding principle of my interpretation became the concept of freedom . . . It introduces the dimension of the *ethical* that *extends beyond the concept of Being to that of obligation*. . . . This awareness, along with that of many others, finally forced me to turn from the theoretical to practical reason—that is, to *ethics*, which became the *central interest and theme* of the last stage of my intellectual journey. . . . [This practical philosophy] was my final personal experience in the course of my intellectual career.[123] (Emphasis added.)

This outline of the change by a traditional professional philosopher into a modern one—particularly with respect to ethics—reflects a major change

123. Jonas, Hast. Ctr. Rep., July-August 2002, at 32-34.

in philosophy applicable toward bioethics during the launch of the third millennium.

The modern philosophical view of "secular humanism" may be described as an attempt to understand humanity by viewing moral theory with some limitations with respect to the supernatural or the metaphysical characteristics of Western religions. The Supreme Court in 1961 wrote an opinion, which included such "religions" as Buddhism, Taoism, Ethical Culture, Secular Humanism, and others even though none specifically worshipped a supernatural god.[124] The latter are directly opposed to the central concept of the Judeo-Christian-Muslim views of religions.[125] Nevertheless, the courts were able to find that under our Constitution, a secular purpose supported the use in public schools of textbooks promoting secular humanism and this philosophy was not seeking the establishment of any religion.[126] An appellate court decided on June 19, 2000, that a school district could not permit student-led prayer before high school football games. This was because the degree of school involvement made it clear that "the pre-game prayers bear the imprint of the State and thus put school-age children who objected in an untenable position."[127] As pointed out, "unlike the Bible, the Constitution is for most Americans neither God-breathed, inerrant, nor above human alteration."[128] It has been claimed that some humanists overemphasize reliance on technology and the power of reason. In spite of their small numbers, they have been called "moral strangers" because "they have different views regarding the morality of a particular endeavor, such as euthanasia, surrogate motherhood, or justice and health care."[129]

---

124. Torcaso v. Watkins, 367 U.S. 488, 495 n.11 (1961). *See also* Smith v. Bd. of School Com'rs of Mobile County, 655 F. Supp. 939 (S.D. Ala. 1987).

125. Melnick, *Secularism in the Law: The Religion of Secular Humanism*, Ohio N.U. L. Rev. 329, 329-57 (11th Cir. 1987).

126. Smith v. Bd. of School Comm'rs of Mobile County, 827 F.2d 684, 693 (11th Cir. 1987).

127. Santa Fe Indep. School District v. Doe, Docket No. 99-62, *aff'd,* 5th Cir., June 19, 2000 (1999), 168 F.3d 806.

128. Crapan Zano, Serving the Word 2000 223.

129. The American Humanist Association has only a few thousand members and other related groups have even less. Kurtz, Defense of Secular Humanism 189 (1983); Englehardt, Bioethics and Secular Humanism 2-3 (1991). *See also* Dawkins, River Out of Eden 132-35 (1995).

How can bioethics and the law, which applies to all citizens regardless of their religious and other philosophical views, provide an effective approach to such divergent viewpoints? To a large extent, answers began to be provided by courts and other institutions toward the end of the twentieth century; and suggestions for new approaches in the twenty-first century will commence in the next chapter.

What is liberty? We say of a boat skimming the water with light foot, "how free she runs," when we mean, how perfectly she is adjusted to the force of the wind, how perfectly she obeys the great breath of the heavens that fills her sails . . . She is free only when you have let her fall off again, and have recovered once more *her nice adjustment to the forces she must obey and cannot defy. (Emphasis added.)*

Woodrow Wilson, 28th President of the United States

Subjugation to appetite alone is slavery while obedience to a law one has prescribed for oneself is liberty.

Rousseau, "On the Social Contract"

Autonomy, Responsibility, and Informed Consent

1. **Autonomy and Responsibility**
   - a) ***Animal Anatomy of Autonomy in Nature and the Need for Ecology***
   - b) ***From Animal Autonomy to Human Autonomy***
   - c) ***Paternalistic Opposition to Individual Autonomy***
   - d) ***The Balance between Autonomy and Responsibility, Majority vs. Minority***
2. **Eugenics Examples in Bioethical Issues**
   - a) ***The Ethical Need for Increased Individual Responsibility***
   - b) ***Bioethical Considerations of Religious Diversity and Educational Autonomy***
3. **The Informed Consent: Results from Advances in Autonomy**
   - a) ***The Golden Rule for Use by Physicians***
   - b) ***Alleged Reliance on Patients to Ask Questions***
   - c) ***Informed Consent Concerned with Alternative Treatments in Law and Medicine and the Rise of Consumerism***

d) Guidelines for Informed Consent in Medicine and the Law; Mandatory Reporting of Violence
e) Informed Consent on Clinical Research under IRBs and Experimental Treatments during Life-Threatening Situations through Randomized Placebo-Controlled Investigations
f) Informed Consent by Emergency Department Physicians, EMTALA, the Triad Terrorist Attacks; Radiation Treatment
g) Nuclear Terrorism
h) Informed Consent v. Ethnic Groups and Prisoners
i) Informed Consent v. Laws Authorizing Patient Access to Obtain Physician Profiles; Medical Students Characterized as "Doctors"
j) Mental Competence to Give Informed Consent
k) Informed Consent by Minors

Whatever nature has in store for mankind, unpleasant as it may be, men must accept, for ignorance is never better than knowledge.

Enrico Fermi

1. Autonomy and Responsibility

a) Animal Anatomy of Autonomy in Nature and the Need for Ecology

As noted in Chapter 1, ecology is an important part of bioethics. Pacific salmon swim upstream to die in the area where they spawned and in order that their bodies will provide phosphorus for the diatoms that, during the next season, serve as food for their salmon fry. Humans live individual lives in our great sea of air where we breed, care for offspring, and thereafter struggle toward the end seeking dignity. (See Chapters 12 through 14.) Men and animals have certain things in common: Liberty, as Rousseau wrote, is "obedience to a law one has prescribed for oneself." For humans, liberty resembles that of great land mammals; elephants, like humans, have virtually no outside natural predators. For those who live closely, they may inspire affection. In his 1994 volume, *Last of the Free*, Gareth Patterson states:

> Elephants have many parallels with man. They, like us, are social animals that live in cohesive family groups. The females become sexu-

ally mature at about the same age as a young woman does. They hold great attachment and love for their close ones. They shed saltwater tears when traumatized or imprisoned. They, like us, understand death and mourn the loss of those who die.[1]

A passion for the liberty of such animals is derived from a greater source than mere affection; it is a moral source. In the last half of the twentieth century, this was especially exhibited by two famous women and an intrepid man; Jane Goodall with regard to chimpanzees (our closest living animal cousins), Dian Fossey with regard to gorillas (our next closest living animal cousins), and by George Adamson with regard to lions (a much more distant mammalian cousin). Jane Goodall specifically noted that a large part of the scientific interest of chimps was their extreme individuality.[2] George Adamson, with his wife, Joy, wrote a best-selling trilogy, *Born Free*, *Living Free*, and *Forever Free*.

During a period of almost 30 years, Adamson lived with lions, supporting their freedom. Why he did this has been described with singular emphasis on the same moral issue facing mankind. He was often criticized by conservationists for his work, which they termed as "unscientific" and "of no value to conservation." His critics missed the fundamental point, lacking Adamson's wisdom and vision. *His work was* never intended to be a scientific quest, but a *work the essence of which was a strong moral issue*. He enabled lions to have the fundamental right to be free. Freedom, whether in the guise of the abolition of slavery in our past or the pursuance of human rights today, lies in the very fabric of man's being. Long before the days of animal rights activism, Adamson believed that *"freedom* was *essential to all life*, and he believed in a kinship with all life."(Emphasis added.) Adamson once wrote:

> A lion is not a lion if it is only free to eat, to sleep, and to copulate. It deserves to be free, to hunt and to choose its own prey, to look for and find its own mate, to fight for and hold its own territory, to die where it was born—in the wild. It should have the same rights as we have.[3]

Like our 28th president's statement in the epigraph, one is free only when he is allowed to "fall off again" and adjust to the natural "forces she must obey and cannot defy."

1. Patterson, Last of the Free 128 (1994).
2. Goodall, The Chimpanzees of Gombe: Patterns of Behavior (1986).
3. Forever Free 5.

For all their efforts, George Adamson and Dian Fossey were both killed by local African people with different ideas. Fossey was shot by a gorilla poacher. Adamson was shot in the back in the Kora region of Kenya. However, his efforts were not in vain. The Kora National Reserve, "the land he had doggedly defended for nineteen years," was made a national park and the areas with wildlife were to be offered greater protection.[4]

good v. Good

Evolution, if it's true, is a thing of the past and not repeatable. We should ignore most animals and only look to our own future.

v.

Although evolution does not repeat itself in any precise form, life continues, and will always be with us. We should learn what we can from our "nearest relatives in the wild."

Twenty-first Century Approach

In order to better appreciate human bioethical problems, greater effort is needed in promulgating an understanding of the life of most wild animals, which is a living part of the universe that requires our protection from extinction and human exploitation.

Central convictions of theological anthropology in biblical traditions, such that human beings are created in the image of God, *support the notion of intrinsic* human dignity and respect for personal choice conveyed in bioethics discourse by the principle of autonomy. *(Emphasis added.)*

Campbell & Callahan[5]

It is the gods who put this fire in our minds, or is it that each man's relentless longing becomes a god to him?

Virgil, *Aeneid*

---

4. *Id.* at 6. After the Kora ambush incident, Kenya's President Daniel Arap Moi proclaimed a shoot-on-sight policy. Any unauthorized armed person resisting arrest in a wildlife area would be shot. In the month ahead, 70 "shifta" in total—all of Somalia origin—were killed, and as a result of this, a calm settled in the wildlife parks of Kenya. *Id.* at 4.

5. *Theology, Religious Traditions and Bioethics,* Supplement to Hast. Cent. Rep., Campbell & Callahan eds., July-August 1990, at 9.

O airborne voice! Long since, severely clear, A cry like thine in mine own heart I hear: "Resolve to be thyself; and know that he, Who finds himself, loses his misery!

Matthew Arnold

I find it astounding that the word "nature" (or the phrase "natural world"), *in the vast majority of cases*, still means the world of the non-human. *No complex relation of the human and non-human can be worked out until* this meaning of "nature" *is admitted to be* inaccurate. (Emphasis added.)

Moran, *The Grammar of Responsibility* (1996) p. 143

Ignorance more frequently begets confidence than does knowledge: it is those who know little, *and not those who know much,* who so positively assert *that this or* that problem will never be solved by science. (Emphasis added.)

Darwin, *The Descent of Man* (Introduction)

b) From Animal Autonomy to Human Autonomy

Anthropology links to autonomy inasmuch as a primitive man was closer to animals and nature than we are today.[6]

It also mentions "respect for personal choice," which reiterates Virgil's "the fire in our minds" as well as "each man's relentless longing becomes a god to him." This is done so that man may "find himself" and as in Mathew Arnold's epigraph "loses his misery" and to express goals that will reoccur in this volume. As Frederick Nietzsche stated, "Man is in transition." It has also been suggested that we should:

6. The anthropologist Alfred Kroeber wrote:

   Man, to every anthropologist, is an animal in the given world of nature; that and nothing more—not an animal with a soul or destiny or anything else attached to him beforehand, but an animal to be compared, as to structure, and as to function, with other animals; and with the unmistakable conviction that any special traits and qualities which may ultimately be assigned to him are to eventuate from inquiry instead of being supposed.

Quoted in Evans, *The Death of Liberalism*, by Dorothy Buckton.

> Edit and interpret the conclusions of modern science as tenderly as we like, it is still quite impossible for us to regard man as the child of God for whom the earth was created as a temporary habitation. Rather we must regard him as little more than a chance deposit on the surface of the world, carelessly thrown up between the two ice ages by the same forces that rust iron and ripen corn . . . Man is but a foundling in the cosmos, abandoned by the forces that created him . . . It has taken eight centuries to replace the conception of existence as a divinely composed and purposeful drama with . . , a blindly running flux of disintegrating energy.[7]

This might be considered a heresy for people of a fundamentalist persuasion. However, the Greek root of persuasion is "to think for one's self," and Cicero said, "No one can give you better advice than yourself." It might be similar to those who treat the Scriptures as their guide in an allegorical fashion. Others may adopt what Pericles of Athens stated: "Make up your minds that happiness depends on being free, and freedom depends on being courageous." Each of these ancient statements is most applicable to the bioethical revolution, as it commenced in the last third of the twentieth century.

The eighteenth century philosopher and novelist Jean-Jacques Rousseau pressed us toward autonomy by emphasizing the primitive, precivilized man and blamed society for causing him to decline with resulting corruption. In 1753, he wrote "I dared to strip man's nature naked, to follow the evolution of those times and things which have disfigured him; I compared man as he made himself with man as nature made him, and I discovered that his supposed improvement had generated all his miseries."[8]

Early in the twentieth century, the philosopher John Dewey wrote that the purpose of "all social institution [is] to set free and to develop the capaci-

---

7. Becker, The Heavenly City of the Eighteenth Century, *quoted in* Evans, The Theme is Freedom 70 (1994).

8. Cranston, The Solitary Self 1990 (1997). "Rousseau asks how any Christian can believe God to be so unjust as to punish us for sins we have not committed." *Id.* at 49. He presaged democracy in *The Social Contract* where he "insisted that laws cannot be binding on anyone who has not cast his vote on them in person or through freely chosen representative." *Id.* at 179. Most famously in that work he wrote, "Man is born free, yet everywhere he is in chains." However, it was a sentence later revised by Karl Marx to serve his very different approach; namely, Communism.

ties of human individuals without respect to race, sex, class or economic status," to "educate every individual into the full stature of his possibility," and to bring about "the all-around growth of every member of society."[9] By the end of that century the philosopher Dworkin assumed the existence of a certain amount of education; he set forth the concept of "ethical individualism," which insists that "we each have responsibility for making as much of a success of our lives as we can, and that this responsibility is personal, in the sense that each of us must make up his own mind, as a matter of felt personal conviction, about what a successful life would be."[10]

Ronald Beiner, a critic of such liberalism wrote, "The central purpose of a society, understood as a moral community, is not [solely] the maximization of autonomy, or protection of the broadest scope for the design of self-elected plans of life, but the cultivation of virtue, interpreted as excellence or as a variety of excellencies, moral and intellectual."[11]

Beiner failed to answer the question concerning precisely what constitutes virtue or excellence. For us modernists, why could not our approach be a maximization of autonomy whose limits are properly defined? As one late twentieth century commentator explained: Humans are capable of wanting to improve their preferences and goals. These are not simple, monolithic wants. Rather they are based on a fabric of life experiences and values that together draw for the individual the image of a person that individual wants to be.[12]

All humans differ. In his poem "The Road Not Taken," the twentieth century poet Robert Frost begins, "Two roads diverged in a yellow wood," and ends with "I took the one less traveled by, And that has made all the difference." An individual may choose differently according to his genetic nature and his experiential background and even his fancy. Einstein himself noted, "When I examine myself and my methods of thought, I come close to the conclusion that the gift of fantasy has meant more to me than my talent for absorbing positive knowledge." We only need to struggle with the contours of "maximization"—a word that implies at least some limits. Richard Niebuhr said, "The self is fundamentally social, in the sense that it is a being

9. Campbell, Understanding John Dewey 185 (1995).

10. Dworkin, Freedom's Law 250 (1996).

11. Beiner, What's the Matter with Liberals? 51-52.

12. Frankfort, *Freedom at will, and the Concept of a Person*, in Freewill, Watson ed. (1982).

which not only knows itself in relation to other selves but exists as self only in relation."[13] Niebuhr looked upon individuals as being "engaged in dialogue" and thought of "all our actions as having this character of being responses, answers, to actions upon us."[14] This may go too far inasmuch as some of our actions are in response to our own thoughts and desires that are independent of others. Although, for Einstein, "Everything that is really great and inspiring is created by the individual who can labor in freedom"; we are all responsible to our community of humankind. This has been true under international law since the United States finally ratified the right to "dignity" set forth in the Universal Declaration of Human Rights (UDHR), which had been adopted by the United Nations General Assembly back in 1948.[15] (It also represents the core of Chapters 12 through 14.)

The great twentieth century novelist and objectivist philosophical proponent of these concepts, Ayn Rand, was prophetic. The heroine of *Atlas Shrugged* was a woman who managed a large industrial organization several decades before such a thing became quite popular. The first two parts of that work constituted Rand's concept of capitalism—a system based on the recognition of individual rights, including property rights. It was the function of government to protect some rights, which had been condemned by certain bureaucrats (who resembled USSR planners in her time) and politicians (who were not producers), as well as some industrialists themselves.[16] Her brand of individualism concerned a movement that eventually helped bring down

13. NIEBUHR, THE RESPONSIBLE SELF 71.

14. *Id.* at 56.

15. *Universal Declaration of Human Rights*, 279 JAMA 469-70 (1998). UNGA resolution 217 A (III); Dec. 10, 1948, U.N. Doc. A/810 at 71 (1948).

16. When you see that trading is done, not by consent, but by compulsion—when you see that in order to produce, you need to obtain permission from men who produce nothing—when you see that money is flowing to those who deal not in goods, but in favors—when you see that men get richer by graft and pull rather than by work, and your laws don't protect you against them, but rather them against you—when you see corruption being rewarded and honesty becoming a self-sacrifice—you may know that your society is doomed.

RAND, ATLAS SHRUGGED 388. A Russian immigrant to the United States in 1926 at age 21, Rand was without religion. After *Atlas Shrugged*, she spent the rest of her life promoting an objectivist philosophy.

the former USSR 32 years later to end the Cold War. Although communism still continues elsewhere following the turn of the third millennium, the Communist party in China began to present a capitalism, which was derided in the Marxist ideology as the representation of an "exploiting class."

Rand's ideal society worshipped "responsibility" for one's self in lieu of demanding only more "rights," for welfare and other handouts. The American philosopher Eric Hoffer wrote, "We join a mass movement to escape individual responsibility, or, in the words of the ardent young Nazi, 'to be free from freedom.'"[17]

Although few would go as far as Rand, we could often use more responsibility today. She demolished the Communist ideal of "to each according to his ability and to each according to his needs" with example after example of how this may destroy incentive and reduce individualism. She also praised fair competition in business (a subject to which I have devoted another volume).[18]

Although her novels pushed the ideal of maximizing human autonomy and reason, she would do so without many limitations. Neither the modern environmental movement nor that of bioethics existed in her day, and some extremes were presented to demonstrate the ideal of a self-actuated person.[19]

---

17. Hoffer, True Believer (1951).

18. The Ayn Rand Institute (founded in 1985) is located in Irvine, Orange County, California. It is the Center for the Advancement of Objectivism. The Ayn Rand Institute, www.aynrand.org. *See* Keyes, Government Contracts (3d Ed. 2003) §§ 6.1-6.5, which only apply to the quarter trillion dollars per year of *government contract* funds. Unfortunately, they do not apply as a guide to those receiving *grants* given to health care organizations (such as those for Medicaid, Medicare, etc.).

19. For centuries, the battle of morality was fought between those who claimed that your life belongs to God and those who claimed that good is self-sacrifice to your neighbors—between those who preached that the good is self-sacrifice for the sake of ghosts in heaven and those who preached that the good is self-sacrifice for the sake of incompetents on earth. And no one came to say that your life belongs to you and that the good is to live it, the question "to be or not to be" is the question "to think or not to think." When I disagree with a rational man, I let reality be our final arbiter; if I am right, he will learn; if I am wrong, I will: one of us will win, but both will profit. . . . The only means of judging right and wrong is the mind. (pp. 940-41)

Atlas Shrugged, from the speech of Ayn Rand's hero.

Rand followed Aristotle in large part by stating that man's "own happiness is the moral purpose of his life, with productive achievement as his noblest activity, and reason as his only absolute." Ethics recognizes that "Reason is man's only proper judge of values and his only proper guide to action. Rationality is man's basic virtue." Following Kant, man "is an end in himself, not a means to the ends of others; he must live for his own sake." Rand has been criticized for calling "Objectivism" a rejection of a form of altruism. Nevertheless, logicians would argue that it might be included within the term "altruism" as it combined inspiration for individualism and capitalism; it was without those controls needed to protect against abuses of the latter. This was demonstrated in the United States during the first few years of the third millennium by a number of high-ranking industrialists who took millions of dollars away from their corporations and its employees without correctly accounting for it.[20] This was not done by her heroes.

Although Laura Schlessinger's mother was a Catholic and her father a Jew, anything went morally for her in the 1960s and she "grew up with no God." But, after several more years, she found "something was missing" from her life. She converted to Judaism, and resigned from the editorial board of the *Skeptic* magazine. In 1996 she composed a volume entitled *How Could You Do That?!: The Abdication of Character, Courage, and Conscience,* in which she wrote, "The notion of God is really, fundamentally, all we have to truly lead us to be good or else we make our own decisions and we become, individually, our own Gods." (p. 9) In his 2004 volume *The Science of Good and Evil,* Michael Shermer, editor of *Skeptic*, wrote, "For Schlessinger, it comes down to this: you'll be busted by Mr. Big if you sin, so don't." (p.152) (As will be observed later, his other basic views were quite different.) The only probable bioethical problem is that which could result regarding her willingness to recognize the rights of others to make different choices in this area.

---

20. *See* Enron, Worldcom, and other scandals at the turn of the 21st century that caused Congress to pass the Sarbanes-Oxley Act of 2002. As both the 19th century's Alexis du Tocqueville and the 20th century historian Henry Steel Commager point out, "When *Americans* are given the choice between the conflicting values, they overwhelmingly *choose individualism over equality.*" (Emphasis added.) JAMONVILLE, COMMAGER 271 (1999). Yet, the rhetoric of many politicians claims to choose the reverse. "The Sarbanes-Oxley Act of 2002 was enacted to improve corporate governing, public auditing, and Securities and Exchange Commission oversight." KEYES, GOVERNMENT CONTRACTS 939 (3d ed. 2003).

good v. Good

We need more emphasis on the rules set forth in the Scripture than emphasis upon the "god within us." Increasing individual autonomy that is causing most people to act in a less ethical manner can only destroy the values and traditions of our society.

v.

The maximization of individual autonomy that includes responsibility permits the full development of each of the people who comprise our society. The goals of freedom are what Americans cherish most.

Twenty-first Century Approach

Because ethics and the law are tending toward the maximization of autonomy together with responsibility towards others, more effort will be needed to combine autonomy with a deeper understanding of what this may mean to a society governed by ethics and the law.

A paternalist government is the greatest conceivable despotism.

Immanuel Kant

Beings who have received the gift of freedom are not content with the enjoyment of comfort granted by others.

Immanuel Kant, *The Quarrel Between the Faculties* (1798)

According to an estimate by Freedom House, as of early 1993, the percentage of human beings in the world who live in free societies had diminished *in the preceding decade, from one-third or so of the global population to approximately one-quarter. At year's end, this group concluded that the situation deteriorated even further, and that* fewer than one-fifth of the world's inhabitants lived in freedom. *(Emphasis added.)*

Evans, *The Theme Is Freedom* (1994), p. 26

Government is not the solution to our problems; government is the problem.

Ronald Reagan in his first inaugural address. Judis & Teixeira, *The Emerging Democratic Majority* (2002), pp. 151-52

It is not the function of our government to keep the citizen from falling into error; it is the function of the citizen to keep the government from falling into error.

Justice Robert H. Jackson (1950)

c) Paternalistic Opposition to Individual Autonomy

According to Confucian philosophy of the fifth century B.C., "a virtuous ruler represented the apex of humankind, he could bring peace and prosperity to the realm." However, this nondemocratic view may also constitute a paternalism whose essence could be the denial of freedom to an individual by those who assume that they know better than he does and can decide or act for him. This enlightenment caused Immanuel Kant to express the epigraph's opposing view in the eighteenth century A.D. In bioethics and the law, under certain circumstances this freedom has been denied either by the government or by physicians.

Although at the middle of the twentieth century Justice Jackson indicated that the duty of men and women in free societies is "to keep the government from falling into error," at the dawn of the twenty-first century great portions of the people in some societies live with limited autonomy. Just three years prior to the twentieth century's end, the Dalai Lama optimistically stated, "Freedom is basic in human nature. It's an instinctive desire. It's there. No force can stop it. That is one of the main reasons why communism has been a failure everywhere."[21]

At the beginning of this chapter, examples were given of the need for humans to protect the autonomy of certain species of "higher" mammals. Darwin once wrote a note to remind himself not to use the terms "higher" and "lower" levels of life. The conjecture that a species of insects might somehow be regarded as exceeding the capacities of mammals for certain purposes was presented in 1998 by the modern entomologist E. O. Wilson; he composed the following "state-of-the-colony speech" for a female termite leader to deliver to the multitude in her attempt to reinforce the "super termite ethical code":

> Ever since our ancestors, the macrotermitine termites, achieved ten-kilogram weight and larger brains during their rapid evolution through

21. Dalai Lama, *quoted in* Chinoy, China Live (1997).

> the late Tertiary Period, and learned to write with pheromonal script, termite scholarship has elevated and refined ethical philosophy. It is now possible to express the imperatives of moral behavior with precision. These imperatives are self-evident and universal. They are the very essence of termitity. They include the love of darkness and of the deep, saprophytic, basidiomycetic penetralia of the soil, the centrality of colony life amidst the richness of war and trade with other colonies, the sanctity of the physiological caste system; the evil of personal rights (the colony is ALL); our deep love for the royal siblings allowed to reproduce; the joy of chemical song; the aesthetic pleasure and deep social satisfaction of eating feces from nest mates' anuses after the shedding of our skins; and the ecstasy of cannibalism and surrender of our own bodies when we are sick or injured (it is more blessed to be eaten than to eat).[22]

Following this amazing speech boasting of everything frightful to most humans today, Wilson proceeded to quote from a 1940 summary of the actual independent histories of "human civilization" in the Old and New World made by the American archaeologist Alfred Kidder, documenting the long tradition of deception:

> In the New World as well as in the Old, priesthood grew and, allying themselves with temporal powers, or becoming rulers in their own right, reared to their gods vast temples adorned with painting and sculpture. The Priests and chiefs provided for themselves elaborate tombs richly stocked for the future life. In political history it is the same. In both hemispheres, groups joined to form tribes; coalitions and conquests brought pre-eminence; empires grew and assumed the paraphernalia of glory.

Similar paternalism continued in the twentieth century. In 1918, at the end of World War I, the famous prizefighter John L. Sullivan died of a malignancy that his doctors had concealed from him. This deception gave rise to the earliest debates over "truth-telling" by physicians. In 1999 nearly half the physicians surveyed admitted to having exaggerated the severity of pa-

22. Wilson, Consilience 148-49 (1998).

tients' conditions to gain additional financial benefits from patients.[23] It has also been remarked that exaggerated and inaccurate diagnoses may frighten patients and compromise their future clinical care.[24] The AMA Code of Medical Ethics (2002–2003) Rule 8.08 states, "Social policy does not accept the *paternalistic view that* the *physician may remain silent* because divulgence might prompt the patient to forego needed therapy." (Emphasis added.)

The paternalism exhibited in E. O. Wilson's "state of the colony" address of the termite leader tends to portray much of the motif as expressed by Kidder on civilizations in the Old and New Worlds, which have been partially tempered in the late twentieth century. The conservative bioethicist Jay Katz wrote that although physicians tolerate patient autonomy as a necessary legal obligation, the medical profession has demonstrated a strong preference for "custody" over "liberty" in patient care decision making.[25] Lawyer domination occurred in the legal profession because a client's desires may be different from those of his or her lawyer, and the fact that they differ in their wants is not itself a sign that the client is acting in a foolish or self-destructive way. Anthony Kronman, Dean, Yale Law School stated:

> It is from the perspective of the client's own interests that his judgment must be assessed. To do this, *a lawyer needs to place himself* in the *client's position* by provisionally accepting his ends and then imaginatively considering the consequences of pursuing them, with the same combination of sympathy and detachment the lawyer would employ if he were deliberating on his own account.[26] (Emphasis added.)

It has also been stated that "When a lawyer and client disagree, so long as the lawyer does not believe that the direction the client wants to go would be [legally wrong], we believe that the lawyer should defer to the client."[27]

---

23. Novack, et al., *Physicians' Attitudes Toward Using Deception to Resolve Difficult Ethical Problems*, 261 JAMA 2980-85 (1989); Kaiser Family Foundation, Survey of physicians and nurses, *available at* http://www.kff.org/199/1503. Accessed Sept. 17, 1999.

24. Morreim, *Gaming the System: Dodging the Rules, Ruling the Dodgers*, 151 Arch Intern. Med. 443-47 (1991). *See* Kinhorn, *Should Doctors Ever Lie on Behalf of Patients?* JAMA, Nov. 3, 1999, at 1674.

25. Katz, The Silent World of Doctor and Patient 2 (1984).

26. Koonman, The Lost Lawyer 129-30 (1993).

27. Cochrane, et al., The Counselor-at-Law: Lexis 187 (1999). "The powerful lawyer may need to work to respect the dignity of the weak client; the weak lawyer may need courage to confront the powerful client." (p. 185)

The paternalistic effects of bureaucrats in the federal and state governments are sprinkled, particularly in the Department of Health and Human Services (HHS) and in certain private organizations. These private organizations include the American Medical Association (AMA), religious institutions, and with private physicians, whose cultural traditions are very much alive today. Most physicians defer to themselves for their own prescriptions.[28]

Early in the 19th century, the philosopher Georg Hegel conceived his "organic theory of the state" as being the true individual. This developed into the "super termite-like ethical codes" of fascist and communist societies.[29]

Since World War II, we have condemned such societies while embracing detailed federal control over most hospitals in the United States through regulation of money received in grants. Some of the questionable AMA ethical policies have been adopted by states. For example, the AMA maintains in detail a total prohibition of any physician participation in capital punishments or regarding a prisoner's competence to be executed, although no physicians are bound to follow any of these rules under state laws that require such participation.[30]

In 1990, the Harvard Medical Practice Study Group (at the School of Public Health) conducted a comprehensive and objective study of medical malpractice in New York State. It found more than 98,000 patients who suffered "adverse events" from medical care, rather than disease, in one year; 27 percent of the adverse effects were caused by the failure of physicians to

---

28. About 75 percent of the urban physicians considered self-prescription ethical . . . [even though] only about half of the physicians considered self-prescription healthy and about 80 percent did not consider self-prescription "necessary." Minn. survey, www.reutershealth.com/archive/e/2001/10/04. Even in the past 30 years there has been a massive decline in deference to traditional authorities. In the modern Western World, not only are people less respectful of traditional authorities, but, as we have seen, increasingly authority is located in the individual, understood primarily as a consumer. *We decide* what is good for us; *we decide* according to our own desires, conscience and reason.

Partridge, *New Religions*, MODERN WESTERN WORD, at 359.

29. STACE, THE PHILOSOPHY OF HEGEL (Dover 1955).

30. Keyes, *The Choice of Participation by Physicians in Capital Punishment,* 22 WHITTIER L. REV. at 809 (2001) (cited in the *AMA Code of Medical Ethics* (2003) Section 2.06, *Capital Punishment*). *See also* Chapters 12 to 14.

meet reasonable standards of care, resulting in 6,895 deaths.[31] The number of such deaths increased substantially during the decade that followed. An analysis of 5,810 hospitals nationally estimated there were 90,000 deaths linked to hospital infections in 2000 caused by the carelessness of doctors and nurses. In 2002, it was estimated that clean-hand policies alone could prevent the deaths of up to 20,000 patients each year.[32] Steps are now being taken to increase recording possible treatment errors in hospitals, only a few of which had resulted in lawsuits. A study has shown that the severity of the patient's disability, not the occurrence of an adverse event due to negligence, is predictive of payments to those who make malpractice claims.[33]

Many religious people in Catholic, Orthodox Jewish, and other religious communities state, "Justice meant nothing less than obedience to divine law."[34] In Israel, unlike in the United States, that philosophy permits a competent patient to be treated against his or her expressed will if the legally constituted hospital ethic's committee is convinced that there is "reason to believe that after receiving the treatment the patient will give . . . retroactive consent."[35] While visiting Israel, I was informed about the lack of success of several attempts to increase a patient's autonomy there.

Cultural controls are no less exacting. A 1995 study was made of differences in the attitudes of elderly subjects from different ethnic groups toward disclosure of the diagnosis and prognosis of a terminal illness toward end-of-life decision making. It concluded that Korean-American and Mexican-American subjects were more likely to hold a family-centered model of medical decision making, rather than the patient autonomy model favored by most of the African-Americans and former Europeans.[36]

31. Breunan et al., *Incidence of adverse events and negligence in hospitalized patients: results of the Harv. Med. Practice Study II*, N. Eng. J. Med. (1991) at 324.

32. Orange County Register, July 21, 2002, at News 8.

33. N. Eng. J. Med. (1996) at 335.

34. Auerbach, Rabbis and Lawyers 160. Unlike the Orthodox, only the Reform movement in Judaism has "made individual autonomy the centerpiece of its ideology and methodology." In the Bible, some Psalms refer to the "Statutes of the Lord" as being both "trustworthy" and "wise and more precious than gold." *Psalm* 19: 9-10. *Psalm* 111:10 states: "The fear of the Lord is the beginning of wisdom; All who follow his precepts [statues] have good understanding."

35. N. Eng. J. Med., March 27, 1997, at 954.

36. Blackhall, et al., *Technicity and Attitudes Toward Patient Autonomy*, JAMA, Sept. 13, 1995, at 82.

By substituting "physicians" for "family," this attitude was aptly expressed in 1847 with the first code of ethics of the American Medical Association:

> Physicians should unite tenderness with *firmness*, and *condescension with authority,* [so] as to inspire the minds of their patients with *gratitude*, respect, and confidence. The *obedience of a patient to* the prescriptions of *his physician* should be prompt and implicit. *He should never permit their own crude* opinions as to their fitness *to influence his attention* to them. (Emphasis added.)

In 2005 it was reported that "The past 30 years [i.e., since 1975] have seen the gradual disappearance of the model of the autocratic, paternalistic physician who extracts consent from patients—a model that has been a concept of patient autonomy, widely translated as superseded by the freedom of choice. But such a view neglects the complexity of the social world that acts on all of us. This complexity has led to increasing questions about what autonomy means in medical settings"[37] (citing a 1998 authority). However, the fact that the mid-19th century AMA approach had not changed was shown by the results of the 1995 Project SUPPORT: The notion of prompt "obedience of a patient to the prescriptions of his physician," and lack of respect for the patient's "crude opinions as to their fitness" remained implicit in the minds of many physicians at the end of the twentieth century. Project SUPPORT enrolled more than 9,000 patients suffering from life-threatening illnesses in five large U.S. teaching hospitals in order to evaluate, over four years, an intervention to improve the dying experience.[38] Although physicians were specifically informed by the nurses[39] of their patients' wishes in

37. N. Eng. J. Med., Jan. 27, 2005, at 329, *citing* C.E. Schneider, The Practice of Autonomy: Patients, Doctors and Medical Decisions (1998).

38. *A Controlled Trial to Improve Care for Seriously Ill Hospitalized Patients*, JAMA, Nov. 22, 1995. For their patients, the majority of these physicians insisted upon their own view that only the autonomy of the caregiver must be respected in order to provide effective care. Clement, Care, Autonomy and Justice 25-26.

39. Hardwig, *SUPPORT and the Invisible Family*, Special Supp., Hast. Ctr. Rep. 25, no. 6 (1995), at S23-S25.

> The SUPPORT intervention asked physicians to take into consideration patient preferences and prognoses. The *physicians failed* to do so. This study strongly suggests that *inappropriate treatment decisions are being made* at the end of life and they are being made *by the wrong decision makers*. (Emphasis added.)

the living wills they had executed, they mostly ignored them. These results were summarized as follows:

> *Patients' wishes* to withhold care *were* often *overlooked.* Among patients who did not want resuscitation attempted, less than one-half of their physicians understood that preference. The findings do not depict gentle peaceful death, but high technology run amok with *poor communication*, inadequate relief of symptoms, and *little respect for patient preferences*.[40] (Emphasis added.)

As noted in the Introduction, this volume has been dedicated to making those changes needed in order that "appropriate treatment decisions" are made by "the right decision makers" as early as convenient in the twenty-first century.[41]

good v. Good

Well-trained physicians should retain their paternalism because they know their patients' needs better than their patients do and could do well without having to undergo other training.

v.

Patient autonomy, under the law, correctly supersedes a physician's paternalism.

Twenty-first Century Approaches

1. A physician's diagnosis and prognosis must be explained to the patient who controls actions upon himself or herself. Project SUPPORT, the largest study of its kind, showed that a current threat to individual

---

40. Lo, Hast. Ctr. Rep., November-December 1995, at S6. These physicians appear to have acted according to the subjectivist view, because there is no such thing as moral knowledge, for *there are no objective moral facts;* moral terms merely reflect our psychological states, our emotions or preferences, and not facts. BERUMEN, DO NO EVIL (2003), at 33.

41. We believe in the dignity, indeed the sacredness, of the individual. Anything that would violate our right to think for ourselves, judge for ourselves, make our own decisions, live our lives as we see fit, is not only morally wrong, it is sacrilegious. Our highest and noblest aspirations, not only for ourselves, but also for those we care about, for our society and for the world, are closely linked to our individualism.

autonomy remains in the medical field. Physicians and health care organizations fell into error to a significant extent by not ending their tendency to promote paternalistic impulse.

2. The renunciation of a denial of the patient's right to choose treatment often remains necessary in order to terminate the continuation of unauthorized treatment of patients in violation of bioethics and the law.

Liberty may be endangered by the abuse of liberty as well as by the abuse of power.

James Madison

Liberty means responsibility. That is why most men dread it.

George Bernard Shaw

Learning about many things does not produce understanding.

Heraclitus

A culture of self-responsibility *is not just the best chance we have to create a decent world. It is* the only chance. (Emphasis added.)

Nathaniel Branden

d) The Balance Between Autonomy and Responsibility, Majority vs. Minority

Freedom, like love, only exists when you can give it to others. This is the essence of the epigraphs of Madison's "abuse of liberty" and Shaw's "Liberty means responsibility." It may have been most literally expressed in Ingersoll's statement, "He must give liberty to others in order to have it himself," a derivative of the Golden Rule when stated in the positive version. The new "Physician's Golden Rule" is described in Section 2.(a) that follows.[42] The French "Declaration of the Rights of Man and of the Citizen" (adopted two years prior to the American Bill of Rights of 1791, as amendments to our Constitution), declared, "Liberty consists of being able to do whatever does not harm another."[43]

---

42. *See also* KEYES, LIFE, DEATH, AND THE LAW (1995), at 206, 607.

43. *Declaration of the Rights of Man and of the Citizen* (Aug. 26, 1789), Art. 4, 10 ENCYCLOPEDIA BRITANNICA 71 (15th ed. 1985).

When James Madison together with his friend Thomas Jefferson put together most of the first Bill of Rights, he informed Congress that they be directed "sometimes against the abuse of the Executive power, sometimes against the Legislative, and in some cases *against the community itself;* or, in other words, against the majority *in favor of the minority.*"[44] (Emphasis added.) The structure of our original and then unique Constitution followed that expressed in Montesquieu's famous French book, *The Spirit of Laws* (1748), showing how freedom could only be created by a government that separated the executive, legislative, and judicial powers. Each of those powers is subject to the Constitution's "checks and balances."

The twentieth century American philosopher John Dewey attempted to express the limits of autonomy by stating: "The democratic idea of freedom is not the right of each individual to do as he pleases even if it were qualified by adding, 'provided he does not interfere with some freedom on the part of others.'" The intent is rather "freedom of mind and of whatever degree of freedom of action and experience is necessary to produce freedom of intelligence."[45] At the end of that century Paul Root Wolpe stated, "indisputably . . . patient autonomy has become the most powerful principle in ethical decision-making in American medicine."[46]

Early in the twenty-first century, Supreme Court Justice Sandra Day O'Connor wrote, "We have placed within our democratic system a mechanism to protect a range of civil and human rights, and to ensure that the will of the majority does rights of the minority. It is the restraints placed on the

---

44. I Annals of Congress 437 (1834).

45. Campbell, Understanding John Dewey (1995), at 170. An even less specific attempt to express a balance was made by a twentieth century Christian theologian:

> The telos [the fate] of man as an individual is determined by the decisions he makes in existence on the basis of the potentialities given to him by destiny. He can waste his potentialities, though not completely, and he can fulfill them, though not totally. Thus, the symbol of ultimate judgment receives a particular seriousness. The exposure of the negative as negative in a person [at the judgment] may not leave much positive for Eternal Life.

Paul Tilich, Systematic Theology, Vol. 3 (1963), at 144-46.

46. Wolpe, *The Triumph of Autonomy in American Bioethics: A Sociological Perspective*, *in* Bioethics and Society: Contructing the Ethical Enterprise, DeVries & Subedi eds. (1998).

majority by the Bill of Rights, that [also] make us free."[47] These express the citizen's goal to achieve autonomy under these rights and the reason to set forth "goods" of a current majority versus "good" of a minority in twenty-first century bioethical approaches set forth at the end of most sections in this volume. This is done in a manner that may also be considered by members of hospital ethics committees and others for the reasons expressed by Madison and O'Connor. Although "freedom of intelligence" must remain an unfettered goal, a single expression may only contain most of the parameters needed to balance autonomy with responsibility. As bioethicist James Englehardt wrote, "Mutual respect is accepted because it is the one way to ground a common moral world for moral strangers without arbitrarily endorsing a particular ranking of values."[48]

An example of this need for reevaluation may be found among proponents of a deaf culture who vigorously oppose the use of cochlear implants for prelingually deafened children, due to seriously invasive treatment. Some of them viewed efforts to "cure" deafness or ameliorate its effects as an immoral means of ending their culture. They claimed a "birthright of silence" for their minority, because this would justify the cultural right to the expression by George Orwell that "some are born more equal than others." However, for many people, this culture was created for political purposes; any deliberate production of such children or limitation on their right to hear violates their right of autonomy, their right to choose.[49] This view has received support in a court that stated: "It is well established that the plaintiff in a personal injury case cannot claim damages for what would otherwise be a permanent injury if the permanency of the injury could have been avoided by submitting to treatment by a physician, including possible surgery, when a reasonable person would do so under the same circumstances."[50]

---

47. O'Connor, The Majesty of the Law 258 (2003). Two bioethicists expressed the contrast between "the predominant orientation in favor of civil liberties and individual autonomy that one finds in bioethics, as opposed to the utilitarian, paternalistic, and communitarian orientation that have marked public health throughout its history." Callahan & Jennings, Am. J. Pub. Health (2002) at 170.

48. Englehardt, Bioethics and Secular Humanism 119 (1991).

49. Tucker, *Deaf Culture, Cochlear Implants, and Effective Disability*, Hast. Ctr. Rep., July-August 1998, at 7-14.

50. Zimmerman v. Ausland, 513 P.2d 1167, 1169 (Or. 1973) (citing McCormick on Damages 136, § 36 (1948)).

The Americans with Disabilities Act (ADA) was broadly interpreted by the Equal Employment Opportunity Commission (EEOC) in a paternalistic manner. In order to qualify for disability, the regulatory determination of whether an individual's physical or mental impairment "substantially limits a major life activity" must be made *without considering the effects of medical treatment* on the individual.[51] On the other hand, several courts refused to hold that a physical impairment constitutes a disability under the ADA where it is correctable by medical treatment,[52] a view which was confirmed by the Supreme Court in 1999 when the Supreme Court, through Justice O'Connor, held that:

> Congress did not intend to bring under the statute's protection all those whose uncorrected conditions amount to disabilities. We hold that determination of whether an individual is disabled should be made with reference to measures that mitigate the individual's impairment, including, in this instance, eyeglasses and contact lenses. [The] ADA's coverage is restricted to only those whose impairments are not mitigated by corrective measures. A "disability" exists only where impairment "substantially limits" a major life activity, not where it "might," "could," or "would" be substantially limiting if mitigating measures were not taken.[53]

Somewhat analogous are the actions of several insurance companies that have added incentives for people to act more responsibly by providing certain agreements for reduced death benefits if the deceased was not wearing a seat belt during a fatal auto accident or if a pregnant employee or family member delayed seeking care.[54]

---

51. 29 C.F.R. § 1630.2(g).

52. In *Coahlan v. H.J. Heinz Co.*, 851 F. Supp. 808, 813 (N.D. Tex. 1994), the court held that "an insulin-dependent diabetic who takes insulin could perform major life activities would therefore not be [substantially] limited [in a major life activity]." Similarly, in *Pancialos v. Prudential Ins.*, 5 A.D. Cases 1825 (E.D. Pa. Oct. 15, 1996), *aff'd*, 118 F.3d 1577 (3d Cir. 1997), the court held that an individual's refusal to have a colostomy to remedy his severe ulcerative colitis meant that his condition was not "substantially limited" so as to fall within the requirements of the ADA's definition.

53. Murphy v. United Parcel Service, 119 S. Ct. 2133 (1999); Sutton v. United Airlines, Albertson's v. Kirkingburg, 527 U.S. 555, 119 S. Ct. 2162, 144 L. Ed. 2d 518 (1999) (the Court stated that the lower court should not have treated monocular vision as itself sufficient to establish a disability).

54. Schwartz, Life Style, Health Status, and Distributive Justice 197.

We know that we did not choose to be born. That we would not choose to die. That we did not choose our parents. That we can do nothing about the passage of time.

Malraux, *The Struggle with Angels* (1941)

What will cosmic ray research cover at the turn of the next century? We cannot tell—perhaps a direct connection will be found with human genetics, for all of us are continually receiving a cosmic radiation dose, even if small.

Friedlander, *A Thin Cosmic Rain* (2000), p. 224

2. Eugenics Examples in Bioethical Issues

Eugenics, the science that deals with the improvement of inborn or hereditary qualities of animals and humans, was originally devised by Francis Galton in his volume *Hereditary Genius: An Inquiry into Its Laws and Consequences* (1869). Later he stated that "The most merciful form of what I ventured to call 'eugenics' would consist in watching for the indications of superior strains or races, and in so favouring them that their progeny shall outnumber and gradually replace that of the old one."[55] It almost came to a grinding halt as a result of Nazi excesses. For a long time following World War II it was politically correct to simply frown upon this entire science. However, the offer of medical assistance to have a child of a type chosen by a couple is a far cry from the Nazi practices and not all eugenics is objectionable.[56] For example, Rabbi Josef Ekstein, an ultra-Orthodox Jew in Brooklyn who, after losing four of his own children to Tay-Sachs disease, founded Dor Yeshorim (Hebrew for "Generation of the Righteous") in 1983.

> When young Orthodox men and women reach a marriageable age, and receive a recommendation from a *shadchan*, or matchmaker, about

---

55. Galton, Inquiries into Human Faculty and Its Development 307 (1883). Upon his death in 1911, his will created the National Eugenics Foundation in London.

56. "From its very beginnings, many eugenicists, including the founder of the eugenics movement, Francis Galton, were opposed to coercion, believing that if people were properly informed they would naturally make the 'right' reproductive decisions. Definitions of eugenics that exclude Galton can hardly be taken seriously." Savulescu, 25 J. Med. Ethics 121-26 (1999).

a potential mate, they make a very important phone call. They (or their parents) call Dor Yeshorim, which retrieves the assigned numbers for each member of the potential couple and checks to see if they are carriers of genetic disease. If they are, they are told that a union is not advisable . . . More than 135,000 people have been tested by Dor Yeshorim offices in the United States, Europe, Canada, and Israel. No children with Tay-Sachs disease have been born to Ashkenazi Jews who have participated in Dor Yeshorim's screening program. Rosner has written about how "genetic manipulation is not considered to be a violation of God's natural law but a legitimate implementation of the biblical mandate to heal."[57]

In our third millennium, genetics will both continue to advance at greater speed and may be inclined to clash with cultural and political activism. Germ-Line interventions may endanger future generations who did not consent to them (see Chapter 15). As Andre Malraux wrote in the epigraph, "no generation in history ever consented to be born of any particular parent, or the time to die."

In addition to Darwin's twins "Natural Selection" and "Sexual Selection," there has been added "artificial selection" to our multifaceted biological thinking. Some have called it "reprogenics" to express a dedication to improving the life of one's offspring by removing those genes that carry susceptibility to diseases and inserting those carrying perceived advantages.[58] Molecular genetic testing involves nucleic acid analyses of DNA.[59] For example, such tests can be used to identify heterozygote (alternative gene) carriers of mutations for autosomal (non-sex-determining chromosome) reces-

---

57. Rosen, *Eugenics Sacred and Profane*, THE NEW ATLANTIS 80-81 (Summer 2003). Premarital Genetic Diagnosis (PGD) as eugenics' "main effect has been to reduce the number of children born with serious or fatal genetic conditions such as Tay-Sachs. But perhaps we should face the fact of a re-emerging eugenics." *Id.* at 88.

58. SILVER, REMAKING EDEN (1997). Similarly, the term "green revolution" was coined over three decades ago to designate progress by increasing the capacity of green plants to use sunlight, water and soil nutrients more efficiently to "see the world." It was supported by public funds. Thus far it has only been a partial and temporary success. Further, it also raised profound ethical and safety issues in biotechnology.

59. Grody & Pyeritz, JAMA (March 3, 1999), at 845.

sive diseases such as cystic fibrosis (CF). Relatives of affected individuals and people of ethnic groups at high risk need information in order to make informed reproductive choices. A 1998 National Institutes of Health Consensus Conference concluded that all people interested in having children should be offered testing for CF heterozygosity (having different alleles with respect to a different character). This approach should also be used for other fatal diseases such as Huntington's.

States have mandatory neonatal screening programs for diseases for which there are known treatments.[60] Liability may follow from a failure to carry out the program.[61] While most states have statutes providing that a person's genetic material may not be tested without informed consent,[62] a half dozen accept some types of research from these laws.[63] Nondirective, but informative, counseling is needed both when contemplating genetic tests and in interpreting their results.

A need exists in bioethics to balance autonomy and responsibility in such a fashion as to minimize the endorsement or ranking of overly paternalistic values. We have begun to choose the latter approach to medical ethics by specifically endorsing the doctrine of informed consent (discussed below). The next millennium may see a further reduction in the endorsement of certain eugenic and paternalistic values that were retained prior to the year 2000.

---

60. *E.g.*, *see* CAL. HEALTH & SAFETY CODE § 125000.

61. Creason v. State Dep't of Health Services, 957 P. 2d 1323 (Cal. 1998). A mandated test for hypothyroidism (diminished thymus activity) was carried out. But the type of blood test was insufficient to detect certain forms of the disease. Although the doctor diagnosed the condition six months after birth, it was too late to treat. In *Safer v. Estate of George T. Pack*, 677 A.2d 1188 (N.J. Super. Ct. App. Div. 1996), the court held that physicians have a duty to warn their patients' children of an abnormal genetic trait where such traits (colon cancer) are both inheritable and treatable.

62. *E.g.*, FLA. STAT. ANN. § 760.40(2)(a) (West 2000); *see also* JENNIFER NORD, NATIONAL CONFERENCE OF STATE LEGISLATURES, GENETIC TESTING: THE INSURANCE INDUSTRY, THE RESEARCH COMMUNITY, AND INDIVIDUAL RIGHTS 1-5 (1997).

63. GA. CODE ANN. § 33-54-6 (1999); LA. REV. STAT. ANN. § 22:213 (D)(4) (West Supp. 1999); NEV. REV. STAT. § 629.151(4) (1999); N.Y. CIVIL RIGHTS LAW § 79-1(4)(a) (McKinney Supp. 1999); OR. REV. STAT. § 659.715(1)-(2) (1993); TEX. REV. CIV. STAT. ANN. art. 9031 § 3(d)(2) (West Supp. 1999). In Minnesota the coroner may remove the brain of any deceased person suspected of suffering from Alzheimer's for the purposes of research. MINN. STAT. ANN. § 145.132 (West 1999).

good v. Good

Because all life is sacred, all eugenics should remain outlawed and testing of pregnant women should be limited to those required to save the lives of both the mother and her fetus.

v.

Tests and other information should be made available to the patient for her decision regarding treatment pertaining to the future of her fetus.

Twenty-first Century Approach

Significantly more effort should be made toward balancing the need for a patient's autonomy in the testing of pregnant women and their fetuses together with the equally important need for bioethical judgment regarding eugenics or paternalism. (Also see Chapter 10.)

Good people do not need laws to tell them to act responsibly, while bad people will find a way around the laws.

Plato

The greatest dangers to liberty lurk in insidious encroachment by men of zeal, well meaning but without understanding.

Justice Louis Brandeis,
dissenting in *Olmstead v. U.S.*, 277 U.S. 438, 479 (1928)

A stable . . . society in equilibrium with the processes of nature cannot allow much freedom or self-assertiveness to the individual.

Donald Worster, *Wealth of Nature*, p. 111

Mankind are greater gainers by suffering each other to live as seems good to themselves, than by compelling each to live as seems good to the rest.

John Stuart Mill

a) The Ethical Need for Increased Individual Responsibility

Early in the twentieth century, Justice Brandeis characterized the "greatest danger" of liberty as the "insidious encroachment by men of zeal" in a government whose "purposes are beneficent." Nevertheless, he also stated that *"The most important political office is that of the private citizen."*

(Emphasis added.) Eric Hoffer spoke of the "neutralization of the independent individual" by men of power that apparently never ends. Later in that century, Donald Worster noted the clash between individual "assertiveness" and nature's "processes." These epigraphs are assertions of the need for one thing, namely "individual responsibility." Statements limited to individual rights normally constitute a judgment by those who assert them without any balance. In his Gospel, Matthew wrote, "With what judgment you judge, you shall be judged, and with what measure you meet, it shall be measured to you again."[64] No rights can exist without duties, and the balance is a judgment.

During post–World War II days, the American Bar Association (ABA) actually proposed to create a section limited to "individual rights." Only following some protests was this changed by adding the words "and responsibilities."[65] In practice, however, the ABA "has done little throughout its history except to deal with rights." A writer in 1996 stated that apparently "the well-intentioned phrasing of the title was flawed from the beginning, that *rights plus responsibilities always seems to equal rights.*"[66] (Emphasis added.) "Responsibility is a process within which things called 'rights' emerge." What needs pairing with rights are duties or obligations.[67]

Early in the twentieth century, John Dewey described the path of democracy as "the road that places the greatest burden of responsibility upon

64. *Matthew* 7, v. 1-2. However, in his *Bioethical Dilemmas, A Jewish Perspective* (1998), Bleich wrote that the U.S. legal system exhibits far greater solicitude for individual autonomy than does Jewish tradition. Nevertheless, the American legal system also recognizes that liberty is a value that must be subordinated when it comes into conflict with a superior value. (p. 148)

65. Oaks, *Rights and Responsibilities*, THE RESPONSIVE COMMUNITY (Winter 1990–91) at L:40.

66. MORAN, A GRAMMAR OF RESPONSIBILITY (1996).

67. *Id.*, p. 78. However, the avoidance of this obligation has been expressed as follows:

> Much of my happiness comes from knowing about, and helping to cause, the happiness of those I love. . . . What will be best for me may . . . largely overlap with what will be best for those I love. But, in some cases, what will be better for me will be worse for those I love. I am self-interested if, in all these cases, I do what will be better for me.

PARFIT, REASONS AND PERSONS (Oxford 1986).

the greatest number of human beings."[68] Later in that century, some politicians appeared to move in the direction of relieving our citizens from Dewey's type of burden.

Professor Richard Epstein of the University of Chicago's Law School, wrote the volume, *Mortal Peril*, which was criticized by Professor Troyen Brennan of the Harvard Medical School and School of Public Health, by stating: "One gets the impression of mean spiritedness that makes Epstein's position a somewhat unattractive political philosophy." He continued:

> Unfortunately, the current law makes it impossible for a hospital to treat drug addicts or alcoholics just once, or even twice, with this stern warning: there is no treatment next time, period no matter what their personal consequences, including death. To the question, "You cannot let them die, can you?" we have to avoid the effective answer, no. To restore long-term stability to the system of emergency care, the answer has to be "yes, we can sometimes." That threat becomes credible only if it is acted on at least once. In support, Epstein notes a Wall Street Journal article in which a patient was noted to have infected her heart valve through use of heroin. She received a new heart valve, and then re-infected the artificial valve, again using heroin. Epstein would not treat this patient again.

Epstein's example comes from the real world and presents a dilemma faced by many physicians. He states, "Indeed, I have several patients along a patient panel that includes numerous intravenous drug abusers and individuals infected with HIV through use of dirty needles, who have required valve replacement as a result of endocarditis from injecting. Should any of them reinfect an artificial valve, I will advocate (I suspect successfully) that the artificial valve be replaced. Hospitalization and professional costs for this episode will likely run over $75,000. The overall majority of physicians would do exactly the same thing. They would do so because they are committed to the individual patient, and respect the principles of compassion that are central to medical ethics."[69]

---

68. Campbell, Understanding John Dewey 180 (1995). Dewey demanded that people be participants in a democracy: "If one is to be a 'spectator' rather than a 'participant,' however, that person will assume the attitude of a man in the prison cell watching the rain out of the window; it is all the same to him." *Id.*

69. *See Brennan, Moral Imperatives Versus Market Solutions: Is Healthcare a Right?*, 65 U. Chi. L. Rev. 345 (1998).

Dr. Brennan attempted to appeal to his own view rather than to society's moral values, which may not be coextensive. Elsewhere in his article, Brennan states, "we have almost no right to health care, and the patchwork of access to health care we do provide comes from a variety of common law rules, state regulation, and Federal programs. We have inherited this patchwork because we have been unable or, more important, unwilling, to redistribute wealth in a way that would provide a right to health care." Brennan's conclusion that all continuous health abusers always have a "right" to full health care, regardless of their personal and repeated demonstration of irresponsibility, showed an absence of a balance with "responsibility." He would distribute considerable wealth to them. His statement that only his view is "central to medical ethics" is subject to considerable bioethical debate. It remains one that will persist well into the third millennium.

By the 1990s, studies showed "smokers annually generate $100 billion in health care expenditures. Smokers average five more visits to physicians per year than nonsmokers, and their average lifetime medical costs exceed those for nonsmokers by $6,000." Similarly, the annual total cost of alcohol abuse has been estimated at over $100 billion.[70] Such conflicts with an individual's autonomy to act irresponsibly raise ethical issues of distributive justice. They not only involve excessive costs to all the other taxpayers, but also issues concerning whether such patients should be allowed to receive organ transplants on the same basis (priority) as other nonsmoking or drug-abusing patients. (See Chapter 11.) A smoker proclaims smoking as one of his rights, and "rights," said Justice Holmes, "mark the limits of interference with individual freedom."[71] Conversely, "responsibility" marks the limit of freedom to interfere with the rights of other individuals. Bioethics often marks the balance between decisions concerning early life as well as later death.

good v. Good

All available resources for treatment should always be provided to patients regardless of cost or of the continual abuse of their health through drugs or other behavior

v.

Priorities should be set forth regarding the treatment of patients who continually and purposely abuse their health as distinguished from those who do not do so.

---

70. HOSPITAL PHYSICIAN, August 1995, at 13.

71. *Quoted in* PHILIP HOWARD, THE DEATH OF COMMON SENSE 122 (1994).

Twenty-first Century Approaches

1 A principal change in the character of bioethical discussions in the twenty-first century must change from (a) assuming the existence of more "rights" by those citizens who repeatedly fail to care for themselves and precipitate a need for others to pay the costs of providing expensive care to (b) balancing rights with a consideration of the "responsibilities" owed by such individual patients to society.

2. A need exists to set forth specific financial priorities and disincentives for such people to continue to act irresponsibly.

The world is so full of a number of things
I'm sure we should all be as happy as kings.

Robert Louis Stevenson, *A Child's Garden of Verses*

The religion of the future will be a cosmic religion. It should transcend personal God and avoid dogma and theology.

Albert Einstein

A society that does not recognize that each individual has values of his own which he is entitled to follow can have no respect for the dignity of the individual and cannot really know freedom.

F.A. Hayes, Nobel laureate

b) Bioethical Considerations of Religious Diversity and Educational Autonomy

The Amish were derived from the Baptists and Mennonites; they were Christian groups that settled in Pennsylvania, Wisconsin, and other states. A controversial decision rendered by the majority of the Supreme Court faced the issue of educational diversity versus educational conformity. An Amish group was held to be exempt from Wisconsin's law requiring that all children attend school until they are either 16 years old or have graduated from high school. In an unusual manner, the Court in 1972 ruled that sending their children to school after the eighth grade would potentially destroy their community because it "Takes them away from their community, physically and emotionally, during the crucial and formative adolescent period . . . In the Amish belief higher learning tends to develop values they reject as influences that alienate man from God."[72] Without education in the Amish culture and religion beyond adolescence, the probability of the diversity in America (as exemplified by that peculiar culture and religion) could be diminished by their not participating in the broader learning made available in public schools outside the Amish community; they appear to be legally banned from learning about other cultures, religions (or about agnosticism vs. atheism). If a child is an end in itself (following Kantian philosophy), the question arises whether forcing a child to learn only the alleged narrower view of a particular religious community, impermissibly diminishes that end. It has been said, "being Amish is precisely not to make one's life choices on the basis of one's own talents and propensities, but to subordinate those individual leanings to the tradition of the group."[73] The Supreme Court in a case involving Jehovah's Witnesses had previously recognized that parents are "free to become martyrs themselves," but stated it does not follow that they "are free, in identical circumstances, to make martyrs of their children before they have reached the age of full and legal discretion when they can make that choice for themselves."[74]

Several years later, some cultural and religious "circumstances" became more important than others according to that court. The majority of American minors are in public schools that are virtually devoid of any teaching about religions and their morality. They do, however, learn very diverse morals from watching TV for an almost equal amount of time as spent in public schools. Although Amish children are now being legally deprived of the opportunity for making moral choices that are available to those outside their culture, learning restrictions respecting them may be modified early in the third millennium. If so, they may be allowed to "Awake to reality, to truth, to things just as they are."[75] In Pennsylvania, at the end of the twentieth century, the parents of Shannon Nixon belonged to the Faith Tabernacle church, one of the sects that believe that illness should be treated spiritually but not medically. When Shannon became ill, her parents prayed and then took her to be anointed at their church. When weakened, she fell into a coma and later died. After an autopsy

72. Wisconsin v. Yoder, 406 U.S. 205, 211-12 (1972).

73. *See* Davis, Hast. Ctr. Rep., March-April 1997, at 11.

74. Prince v. Massachusetts, 321 U.S. 158, 170 (1944).

75. Lama Das, Awakening the Buddha Within (1997), p. 17. The meaning of "lama" in Tibetan is heavy or weighty: A lama is a teacher who embodies the weighty Dharma.

revealed that she had succumbed to diabetic acidosis, a treatable condition, her parents were charged with involuntary manslaughter. When a jury found them guilty, the judge sentenced them to prison for two to five years and fined them $1,000.[76] However, in 1986 Cardinal Ratzinger (now Pope Benedict) wrote that "human beings alone cannot save themselves. Their innate error is precisely that they want to do this by themselves. *We can only be* saved—that is, be *free and true*—when we stop wanting to be God and when *we renounce the madness of autonomy* and self-sufficiency."[77] (Emphasis added.)

Most adolescents not only have a right to medical treatment so as to remain alive and reach maturity, but also to acquire a knowledge, which takes a certain amount of study to be meaningful. If our ignorance is largely dispelled at that stage of life in order to maximize autonomy, this could also increase individuality and diversity. John Dewey held that "the greatest experiment of humanity" is that of "living together in ways in which the life of each of us is at once profitable in the deepest sense of the word, profitable to himself and helpful in the building up of individuality of others."[78] Such building up of "the individuality of others" requires a broadening of their education indoctrinated during adolescence. However, as stated in the volume, *The Ethical Primate*, "There is no single end for human life."[79] Isaiah Berlin emphasized diversity and individuality in his essay "The Hedgehog and the Fox," which was inspired by a fragment from the Greek poet Archilocus: "The fox knows many things, but the hedgehog knows one big thing." Berlin saw two types of individuals: "Those, on one side, who relate everything to a single central vision, and, on the other side, those who pursue many ends, often unrelated and even contradictory."[80] He believed in the plu-

---

76. Pennsylvania v. Nixon, 718 A.2d 311 (Pa. Super. Ct. 1998). The Pennsylvania Commonwealth's Child Protective Services Act (CPSA), 23 PA. CONS. STAT. ANN. § 6303(b)(3), provides that if a county child protection agency determines that a child has not been provided needed medical or surgical care because of seriously held religious beliefs of the child's parents, the child shall not be deemed to be physically or mentally abused. *See* 23 PA. CONS. STAT. ANN. § 6303(b)(3). This statute was held inapplicable because the CPSA "merely exempt parents who treat their children in this manner from characterization as child abusers."

77. CARDINAL RATZINGER (now Pope), IN THE BEGINNING 73 (1986).

78. *Quoted in* PARK, THE FIRE WITHIN THE EVE 176 (1997).

79. MIDGLEY, THE ETHICAL PRIMATE 182 (1994).

80. Berlin, *The Sense of Reality*, STUDIES IN IDEAS AND THEIR HISTORY (1998).

ralistic liberty of the fox while maintaining that a common core of human values laid beneath these many differences.

good v. Good

Parents of children born into a particular culture or religion ought to have the right to withdraw their children following completion of public elementary school in order to preserve their particular religion.

v.

Adolescents have a right to be exposed to other cultures and religions in order to better enable them to make diverse autonomous choices upon attaining the age of majority.

Twenty-first Century Approaches

1. Just as the balance between autonomy and responsibility is equated to ethical living in a democracy, so must the proper balance between autonomy and diversity in high school education become a goal of the twenty-first century.
2. Only by educating maturing teenagers more broadly can they be accorded true freedom to choose medical treatments from the divergent ethical options that may be offered to them.

Who shall decide when Doctors disagree?

Alexander Pope (1732)

What government is the best? That which teaches us to govern ourselves.

Johann Wolfgang von Goethe

A person should be free to do as he likes in his own concerns, but he ought not be free to do as he likes in acting for another, under the pretext that the affairs of the other are his own affairs.

John Stuart Mill, "On Liberty"

Each patient carries his own doctor inside him. They come to us not knowing that truth. We are at our best when we give the doctor who resides within each patient a chance to go to work.

Albert Schweitzer, M.D.

> *The only part of conduct of anyone for which he is amenable to society is that which concerns others; in the part that merely concerns himself, his independence of right is absolute.* Over himself, *his own body and mind,* the individual is sovereign. (Emphasis added.)
>
> J. S. Mill, "Utilitarianism,"
> printed in *Liberty and Representative Government* (1910)

3. The Informed Consent: Results from Advances in Autonomy

"Informed consent" based upon the principle of patient autonomy provides the apex of today's legal decision making for medical treatment. In theory, this began in the late nineteenth century when the Supreme Court stated that a patient had the right to "possession and control of [one's] own person, free from all restraint or interference of others . . ."[81] But, this was not recognized by most physicians during the next century.[82]

In practice, for most Americans this only became a product of autonomy during the last half of the twentieth century; this was when state court decisions began to make informed consent much more meaningful for patients and modifications were made in the AMA Code of Medical Ethics for physicians. Following the dawn of the third millennium, it also entered the American Bar Association (ABA) Model Rules of Professional Conduct and the ALI Restatement (2000): "The Law Governing Lawyers."

---

81. Union Pac. R.R. Co. v. Botsford, 141 U.S. 250, 251 (1891). There is a continuing debate about the appropriateness of obtaining individual informed consent in non-Western Cultures, N. ENG. J. MED., Jan. 31, 1997, at 370.

82. In his history of the doctrine of informed consent, entitled *The Silent World of Doctor and Patient* (1984), Jay Katz indicated that:

> Physicians have always maintained that patients are only in need of caring custody. Doctors felt that in order to accomplish that objective they were obligated to attend their patients' physical and emotional needs and to do so on their own authority, without consulting with their patients about the decisions that needed to be made. Indeed, doctors intuitively believed that such consultations were inimical to good patient care. The idea that patients may also be entitled to liberty, to sharing the burden of decision with their doctors, was never part of the ethos of medicine. (p. 2)

On October 22, 1957, Justice Absalom F. Bray of the California Court of Appeals coined the term *informed consent.* A man with paralyzed legs sued his physician for failing to warn him of the risk of paralysis inherent in the treatment for limping that he had undergone. At the end of his opinion, Justice Bray wrote: "In discussing the element of risk a certain amount of discretion must be employed consistent with the full disclosure of facts necessary to relay *informed consent.*" (Emphasis added.) His opinion marked the beginning of the legal history of those last two words.)[83]

A Lawyer's Duty to Inform and Consult with a Client requires that he "keep a client reasonably informed about the matter and must consult with a client to a reasonable extent concerning decisions to be made by the lawyer. [He or she] must explain a matter to the extent reasonably necessary to permit the client to *make informed decisions* regarding the representation."[84]

Competent adult patients have the right to refuse medical treatment and informed consent is required before physicians can lawfully treat them.[85] This is the standard by which such patients must be treated. A French Code (which is legally binding unlike the AMA Code) also grants the right of competent patients to refuse medical treatment. Article 36 of the Code of Medical Deontology states: "In every case, the consent of the person examined or attended should be sought. When the patient, in a condition to express his will, refuses the investigations or the proposed treatment, the doctor should respect this refusal having informed the patient of all its consequences."[86]

---

83. KATZ, THE SILENT WORLD OF DOCTOR AND PATIENT. When the book appeared in 1984, surgeons almost uniformly attacked Katz's view that patients should have a say in what happens to their bodies. The magnificent vision of the physician as an artist was at odds with the idea of an informed and mature patient. GIGERENZER, CALCULATED RISKS 89 (2002).

84. For example, the Am. L. Inst. (ALI) § 20.

85. Lane v. Candura, 376 N.E.2d 1232 (Mass. App. Ct. 1978); Truman v. Thomas, 611 D.2d 902 (Cal. 1980); Washington v. Glucksberg, 521 U.S. 702 (1997). PROSSER & KEETON ON TORTS 117 (5th ed. 1984). For example, New York's judiciary found that the standard of care requires health care providers to obtain express or implied consent before performing surgery. Walls v. Shreck, 658 N.W.2d 686 (2003).

86. However, unlike in most American states, that Code does not specifically grant a relative or surrogate the power to make binding treatment decisions on behalf of an incompetent patient (Art. 36.3). Their palliative care law of June 9, 1999, also states that "a sick person may refuse to submit to investigation or therapy of whatever kind."

An American court has indicated: "We agree that right to know—to be informed—is a fundamental right personal in the patient and should not be subject to restrictions by medical practices that may be at odds with the patient's information needs. . . . Materiality, therefore, is the cornerstone upon which the physician's duty to disclose is based."[87] Little was known about the extent to which physicians actually fostered patient involvement in knowledgeable decision making, particularly in the routine office practice of both primary care physicians and surgeons. One commentator recently pointed out that with this requirement, "[T]he art of medicine consists in trying to juggle two conceptions: that of the patient as disease-sufferer and that of the patient as autonomous agent." He also noted that the concept of "doctor" has "teacher" as an etymological root.[88] However, at the turn of the third millennium a study actually found that 91 percent of patient-physician interactions failed to meet the requirement of informed decision making; physicians were often incomplete and there was a need for efforts to encourage informed decision making in clinical practice.[89] A number of those particular physicians did not embrace the concept of shared decision and they did poorly in terms of involving patients in their medical decisions; they refused to take time to do so in order "to move on to the next examination room quickly."[90] This economic motive to largely ignore the law has been overlooked by patients and doctors who react accordingly.

In 2002, the words *informed consent* were introduced for the first time into the ABA *Model Rules of Professional Conduct*. It now continues to specify this phrase 37 or more times with amended rules—several of which are similar to those made applicable to physicians under the AMA *Code of*

87. Wheeldof v. Madison, 374 N.W.2d 367, 375 (1985).

88. Scheurich, Hast. Ctr. Rep., March-April 2002, at 17.

89. Braddock et al., *Informed Decision Making in Outpatient Practice*, JAMA, Dec. 22/29, 1999, at 2313.

> In 2003, it was found that Institution Review Boards (IRBs), which are charged with safeguarding potential research subjects with limited literacy, commonly provide text for informed-consent forms that fall short of their own readability standards. Paasche-Orlow, et al., N. Eng. J. Med., Feb. 20, 2003, at 721.

90. Barry, *Involving Patients in Medical Decisions*, JAMA, Dec. 22/29, 1999, at 2357.

Medical Ethics or by the courts. For example, the AMA *Code of Medical Ethics* (2002–2003 Edition) Rule 8.08 "Informed consent" states:

> The patient should make his or her own determination on treatment. The physician's obligation is to present the medical facts accurately to the patient or to the individual responsible for the patient's care and to make recommendations for management in accordance with good medical practice.

And the ABA *Model Rules of Professional Conduct* (2002) Rule 1.01 "Terminology" states:

> "Informed consent" denotes the agreement by a person to a proposed course of conduct after the lawyer has communicated adequate information and explanation about the material risk of and reasonably available alternatives to the proposed course of conduct.[91]

As ye would that men should do to you, do ye also to them likewise.
Luke 6.31, c. first C.E.

What thou avoidest suffering thyself seek not to impose on others.
Epictetus, *Encheiridion*, c. 100

The highest manifestation of life consists in this: that a being governs its own actions. A Thing which is always subject to the direction of another is somewhat of a dead thing.
St. Thomas Aquinas

Desire nothing for yourself which you do not desire for others.
Baruch Spinoza, *Ethica* 4, 1677

To do as one would be done by, and to love one's neighbor as one's self, constitute the ideal perfection of utilitarian morality.
John Stuart Mill, *Utilitarianism,* 1863

---

91. *See also* LAW GOVERNING LAWYERS (2000) § 20, *A Lawyer's Duty to Inform and Consult with a Client.*

It seems to me that there is no scientific evidence bearing on the Golden Rule. It seems to me that this is somehow different.

Feynman, Nobel laureate, *The Pleasure of Finding Things Out* (1999), p. 251

a) The Golden Rule for Use by Physicians

Changes are continuing to be made to a significant extent in the basic direction "from bioethics toward the law" along with the further integration of law into American society. However, a lawyer for a plaintiff in a personal injury case may not request a jury to "Please do unto my client as you would have him do unto you." This is because he might be held to have violated "the Golden Rule prohibition."[92] It is similar to an attorney's asking the jury to place itself in his party's shoes with respect to the issue of liability and damages.[93]

An appellate decision prohibited such rhetoric "as any argument that urges the jurors to consider the case *not objectively* as fair and neutral jurors, *but rather from the subjective personal perspective of a litigant.*"[94] (Emphasis added.)

---

92. Ferris, On the Popularization of Science 159 (2002). "In everything, do to others as you would have them do to you; for this is the law and the prophets" (*Matthew* 7:12).

93. Whitehead vs. Food Max of Miss., Inc., 163 F.3d 265 at 298 (5th Cir. 1998). Assume that a financial officer embezzled money from his company. Now, he certainly would not want to be caught and handed over to the authorities. Therefore, according to the Golden Rule, if the CEO discovered his foul deed, he should refrain from handing him over to the police, for, after all, that is what the embezzler would want.

Berumen, Do No Evil 79 (2003).

94. Carlson, The Federal Lawyer 38 (January 2002). Spray-Rite Service Corp. vs. Monsanto Co., 684 F.2d 1226, 1246 (7th Cir. 1982); *accord* Lovett v. Union Pac. R.R. Co., 201 F.3d 1074 (8th Cir. 2000); Edwards v. City of Philadelphia, 860 F.2d 568, 574 (3d Cir. 1988); Joan W. v. City of Chicago, 771 F.2d 1020 (7th Cir. 1985). However, a Golden Rule argument may be lost if an objection is not made, 728 F.2d 1576 (10th Cir. 1984). *Id.* at 1581. *Accord* Gonzalez v. Volvo Corp. of Am., 752 F.2d 295 (7th Cir. 1985).

The Golden Rule is set forth in both the Old and New Testaments. The latter states: "In everything, do to others as you would have them do to you; for this is the law and the prophets." (*Matt.* 7:12.) It was also written in a negative version and nonreligious manner by Confucius (551-479 B.C.) as "Do not do to others that which we do not want them to do to us."[95]

Recently it was noted:

> The recognition that each of us has an individual identity that we call "self" or "soul" and links us to other human beings exists with indelible certainty across all cultures and continents. It is no coincidence that Jesus and Confucius independently came up with the Golden Rule.[96]

However, Nobel laureate Feynman's reasons for his epigram was that the Golden Rule "must lie outside of science because it is not a question that you can answer only by knowing what happens; you still have to *judge* what happens . . . in a *moral* way." (Emphasis added.) It has been pointed out that although some people "speak of physicians as scientists, they are not. They are professionals building part of their knowledge on science, part on experience, and part on their own good judgment." Feynman added, "I hope you will find that my atheistic scientific colleagues often carry themselves well in society"[97] and that there exists a "consistency between the moral view—or the ethical aspect of religion—and scientific information." As for bioethicists, this supports their view concerning the need to consider both categories of information.

A physician may treat a patient paternally—from his subjective personal perspective—even though he acts as a professional and not as a juror. However, his need to obtain information due to a competent patient's autonomy

---

95. The translation of Lin Yutang, *The Wisdom of Confucius* (1938), p. 185, contained the following:

> What he does not want done unto himself, he does not do unto others. And so, both in the state and in the home, people are satisfied. Tsekung said, "What I do not want others to do unto me, I do not want to do unto them." Confucius said, "There is a central principle that runs through all my teachings."

96. ANDREASEN, BRAVE NEW BRAIN 542 (2001).
97. FEYNMAN, THE PLEASURE OF FINDING THINGS OUT 251 (1999).

will make a physician different; as previously noted, the basic change, from the traditional application of a Golden Rule by a physician's treatment, is accomplished by the replacement of three pronouns: "Do unto your patients as they would have you do unto them."[98] The original version can now only be used by physicians who, in addition, would always adequately inform their patients and do nothing for them without first obtaining their consent.[99]

good v. Good

Physicians should follow the Golden Rule when treating patients.

v.

That rule has been modified to only treat patients who have concurred in a particular treatment.

Twenty-first Century Approach

Following delivery of information about an illness a physician should only treat a competent patient to the extent that a patient consented to the physician's proceeding with the treatment.

b) Alleged Reliance on Patients to Ask Questions

At the end of the twentieth century, the respective responsibilities of both physician and patient were described at the beginning of the American College of Physicians Ethics manual called *Physicians' Responsibilities*:

(1) After patient and physician agree on the problem and the goal of therapy, the physician presents one or more courses of action. If both parties agree, the patient may authorize the physician to initiate a course of action; the physician can then accept that responsibility.

---

98. Today when patients are prepared for surgery, they rightfully sign a statement that indicates the acknowledgment that medicine is not an exact science. That does not diminish that great profession. It only means that ethics—as in all applied professions—become indispensable. I am quoting from Maxwell Anderson: " . . . science is completely impartial. It doesn't give a damn which way to go. It can invent the atom bomb but it can't tell you whether to use it or not."

Konopka Chapter, LESSONS FROM THE HOLOCAUST, p. 18, *citing* M. Anderson (1949), JOAN OF LORRAINE MENASHA, WI, p. 91.

99. KEYES, LIFE, DEATH, AND THE LAW 206 (1995).

(2) The physician is required to provide enough information to allow a patient to make an informed judgment about how to proceed. The physician's presentation should be understandable to the patient, should be unbiased, and should include the physician's recommendation.[100]

These descriptions are uncontroverted as far as they go. However, the same guideline goes further in a different manner:

> The principle and practice of informed consent *rely on the patients to ask questions* when they are uncertain about the information they receive; to think carefully about their choices; and to be forthright with their physicians about their values, concerns, and reservations about a particular recommendation. (Emphasis added.)

Reasonable patients, however, do not have a duty to ask the appropriate questions on a medical subject about which they are unfamiliar. It is the physician's duty to both raise and attempt to answer these questions. The patient must be competent to give consent to both the agreement with their answers and to the treatment. Such competence is virtually always determined by the attending physician himself as is discussed below. Federal law requires skilled nursing facilities to respect the right of a patient to participate in decision making,[101] and the laws of some states require registered nurses to give "the client" the "opportunity to make informed decisions about health care before it is provided."[102] The doctrine shows care and concern for a patient as a rational human being. As the Supreme Court's former Chief Justice Rehnquist recognized: "[T]he informed consent doctrine has become firmly entrenched in American tort law *** [and] [t]he logical corollary of the doctrine of informed consent is that the patient generally possesses the right not to consent, that is, to refuse treatment."[103] Accordingly, more public

100. 128 (7) Ann. Intern. Med. 577, 579 (1998). In 2004 the Institute of Medicine found that 90 million adults have "limited health literacy." It was recommended that health organizations and medical schools should teach health literacy and how to communicate with patients. ORANGE COUNTY REGISTER, April 9, 2004, at News 19.

101. 42 U.S.C. § 1395i-3b.

102. *E.g.,* California State Nursing Practice Act, 1992, para. 1443.5.

103. Cruzan v. Director, Missouri Dept. of Health, 497 U.S. 261, 270, 110 S. Ct. 2841, 2847, 111 L. Ed. 2d 224 (1990). *See also* KUHSE, CARING: NURSES, WOMEN AND ETHICS 174 (1997), 78.1 CEJA, *Ethical Issues*.

support is often needed to improve health literacy and developing ways to communicate and teach clear health information.

c) *Informed Consent Concerned with Alternative Treatments in Law and Medicine and the Rise of Consumerism*

Under Rule 8.08 of the AMA Code of Medical Ethics (2002–2003), the physician "has an ethical obligation to help the patient make choices from among the therapeutic alternatives consistent with good medical practice."[104] This is similar to Rule 1.0 of the ABA Model Rules of Professional Conduct that requires lawyers to explain "the material risk of and reasonably available alternatives to the proposed course of conduct." To deny the disclosure of this information is a denial of informed consent (which might be done for economic reasons). As of the end of the twentieth century it was noted, "Very few reported cases address the physician's failure to disclose treatment alternatives alone."[105] However, in 2003 a study found that nearly one in three doctors report withholding information from patients about useful medical services that aren't covered by their health insurance.[106] In an important early case, a court stated that the right of informed consent "may oblige the physician to advise the patient of the need for or desirability of any alternative treatment *promising greater benefit* than that being pursued."[107] (Emphasis added.) Other courts established "disclosure of an alternative means of therapy pursued by a respectable segment of the medical profession."[108] Physicians

---

104. Code of Medical Ethics, § 8.08, *Informed Consent*. *See also* § 8.13, *Managed Care* (2)(f). Disclosure has been required even where the alternative is more hazardous, *Gemme v. Goldberg*, 626 A.2d 318, 326 (Conn. App. Ct. 1993), or includes pain control programs, *Martin v. Richards*, 531 N.W.2d 70 (Wis. 1995).

105. Iowa L. Rev. 324 (1999).

106. Orange County Register, July 8, 2003, at News 7, *citing* Matthew K. Wynia, director of the Institute for Ethics at the American Medical Association and lead author of the article being published in the journal *Health Affairs*.

107. Canterbury v. Spence, 464 F.2d. 772, 781 (D.C. Cir. 1972).

108. Dunham v. Wriath, 423 F.2d 940, 946 (3d Cir. 1970). *See also* Archer v. Galbraith, 567 P.2d 1155, 1161 (Wash. Ct. App. 1977) (A patient has the right to be informed about an alternative "means of therapy pursued by a respectable segment of the medical profession.").

are also required to obtain an *informed refusal of treatment*.[109] The reason has been declared quite simply: "When there are several appropriate treatment options for a condition . . . [Patients] want to be able to choose for themselves which one to receive."[110] However, a bioethicist once wrote, "It is true, nonetheless, that patients seldom understand everything."[111]

Many studies have shown the prevalence of a failure to give adequate informed consent that may involve serious ethical and legal problems. Three-fourths of physicians reported that they would spend more time disclosing the risks of surgery than of drug therapy.[112] Not all patients were fully informed about student involvement in their care and during the twentieth century many medical schools lacked adequate policies guiding students on how to clarify their role in the patients' care.[113] More than half of 100 women who were in one trial evaluating breast conservation thought they received inadequate information.[114]

> But some physicians would like to influence women's decisions, not only inform them. One technique is to exploit patients' ignorance concerning risk representations. To increase the patient's willingness to participate in therapy, one simply reports the benefits in terms of relative risks (which appear larger) and the costs in terms of absolute risks (which appear smaller). To decrease the patient's willingness, one simply does the opposite.[115]

A study of elderly patients found they were unlikely to give consent, and that surrogates were rarely used when patients lacked decision-making ca-

---

109. Truman v. Thomas, 611 P.2d 902 (Cal. 1980).

110. Ezekiel J. Emanuel & Nancy Neveloff Dubler, *Preserving the Physician-Patient Relationship in the Era of Managed Care*, 273 JAMA 323, 324 (1995).

111. Macklin, Mortal Choices 40 (1987). Nevertheless, it was pointed out that even "*when there is no single, correct answer* to a substantive moral question, *bioethics can still make a contribution*." (Emphasis added.) (p. 221)

112. Williamson, *Explaining the Risks*, Canadian Doctor 46:12 (1980) at 40-42, 44-45, 48.

113. Cohen et al., *Informed Consent Policies Governing Medical Students' Interactions with Patients*, J. Med. Educ. 62:10 (1987) at 789-98.

114. Fallowfield et al., *Psychological Outcomes of Different Treatment Policies in Women with Early Breast Cancer Outside a Clinical Trial*, Brit. Med. J. 301:6752 (1990) at 575-80.

115. Gigerenzer, Calculated Risks 205 (2002).

pacity.[116] Research of nursing home patients found physicians who would often overestimate decision-making capacity.[117]

good v. Good

Many courts have not held physicians liable solely for nondisclosure of alternative treatments to patients on the ground that patients lack the ability to make any judgment in such cases.

v.

Most patients can make a reasonable judgment on the advantages of a treatment upon receiving an adequate disclosure of alternatives.

Twenty-first Century Approaches

1. Physicians should follow the complete informed consent policy by disclosing possible alternative treatments to patients.
2. The disclosure of alternative treatments would often increase the patient's autonomy; however, the breach of many statutes, which state the policy for physicians to communicate alternative treatments to patients, have not usually rendered them liable unless there were simultaneously other significant offenses.
3. Because some laws have not sufficiently increased a patient's autonomy, legislatures in such states should consider modification of their statutes by requiring physicians to communicate significant alternative treatments to patients.

No one is free who is not master of himself.

Claudius

Liberty does not import an absolute right in each person to be, at all times and in all circumstances, wholly freed from restraint. There are manifold restraints to which every person is necessarily subject for the common good.

Justice John Harlan (1905)

---

116. Auerswald et al., *The Informed Consent Process in Older Patients Who Developed Delirium: A Clinical Epidemiologic Study*, 103:5 Am. J. Med. (1997) at 410-18.

117. Fitten, et al., *Assessing Treatment Decision-making Capacity in Elderly Nursing Home Residents*, 38:10 J. Am. Geriatrics Soc. (1990) at 1097-1104.

d) *Guidelines for Informed Consent in Medicine and the Law; Mandating Reporting of Violence*

In order to receive informed consent, medical patients must understand their options, potential risks, and benefits.[118] Several guidelines and consent forms have been issued to aid physicians in obtaining informed consent in general and for particular types of diseases or diagnoses.[119] Texas appears to have the longest consent form, New Hampshire the shortest, and California has no standardized form except for specialized procedures, such as sterilization and hysterectomy.

In 2000, the American Law Institute published a new Restatement called the "Law Governing Lawyers." Rule 20 resembles those applicable to physicians by requiring the lawyer to "keep a client reasonably informed about the matter and consult with a client in a reasonable extent concerning decisions to be made by the lawyer." He must "promptly comply with a client's reasonable requests for information," as well as "explain a matter to the extent reasonably necessary to permit the client to make informed decisions regarding the representation."

A survey was conducted to assess the informed consent process for research with *cognitively impaired* subjects because many believed that more stringent guidelines were necessary. Surrogates were found to usually make decisions consistent with the patients' wishes.[120]

---

118. Drane, *Competency to Give Informed Consent*, JAMA 252, 925-27 (1984); BUCHANAN & BROCK, DECIDING FOR OTHERS: THE ETHICS OF SURROGATE DECISION MAKING 17-53 (1989). These information requirements differ from consent given to police regarding warrantless searches under the 14th Amendment. United States v. Drayton, 536 U.S. 194, 196 (2002).

119. *E.g.*, Miller & Nora, *Written Informed Consent for Electrodiagnostic Testing: Pro & Con, Muscle Nerve* 20, 352-56 (1997), in GENETIC TESTING FOR SUSCEPTIBILITY TO ADULT-ONSET CANCER. The Task Force on informed consent recommends that informed consent for cancer susceptibility testing be an ongoing process of education and counseling in which (1) providers elicit participant, family, and community values and disclose their own, (2) decision making is shared, (3) the style of information disclosure is individualized, and (4) specific content areas are discussed. 277 JAMA 1467-74 (1997); Geller et al., *Genetic Testing for Cancer*, JAMA, May, 14, 1997, at 1,470.

120. Candilis et al., *A Survey of Researchers Using a Consent Policy for Cognitively Impaired Human Research Subjects*, IRB: A REVIEW OF HUMAN SUBJECTS RESEARCH 15, No. 6 (1993) at 1-4.

Problems exist with respect to the process of obtaining informed consent in many cases. One study of costs to patients admitted to an expensive intensive care unit (ICU) found that a large majority of the patients, and a slight majority of the relatives, believed that *fiscal liability information* should *be disclosed* to patients before such treatment.[121] Of 82 informed consent forms submitted to the Institutional Review Board (IRB) at three university-affiliated hospitals within a one-year period, about a quarter of them were judged to be incomplete regarding the federal requirements for disclosure.[122] Several surveys showed patients who viewed the consent form as little more than a formality meant to protect the institutions or the doctors.[123]

The findings suggested that signing forms do not guarantee subjects' understanding. The use of video did not improve comprehension compared to providing only written information.[124] In 2003 a study looked at the model forms drawn up by medical school ethics committees or IRBs. The study found that these templates are written at a tenth-grade reading level or higher. The average American is believed to read at an eighth-grade level. This study suggests a possible reason: The medical school committees assigned to protect research participants are writing forms that are too dense.[125] However, it is noted that in wartime no such understanding by military personnel is re-

---

121. Turner et al., *Physicians' Ethical Responsibilities Under Co-Pay Insurance: Should Potential Fiscal Liability Become Part of Informed Consent?* J. Clinical Ethics 6, No. 1 (1995), at 68-72.

122. White et al., *Informed Consent for Medical Research: Common Discrepancies and Readability*, Acad. Emerg. Med. 3, No. 8 (1996) at 745-50.

123. Daugherty et al., *Informed Consent (IC) in Clinical Research: A Study of Cancer Patient Understanding of Consent Forms and Alternatives of Care in Phase I Clinical Trials* [Meeting Abstract], 16 Am. Soc'y of Clinical Oncology Proc. (1997), at A188; Saw et al., *Informed Consent: An Evaluation of Patients' Understanding an Opinion (With Respect to the Operation of Transurethral Resection of Prostate)*, 87 J. Royal Soc'y Med. (1994), at 143-44. (Patients generally felt the consent form was necessary, although a slight majority believed that its purpose was to protect the doctor.)

124. Westreich, et al., *Patient Knowledge about Electroconvulsive Therapy: Effect of an Informational Video, Convulsive Therapy 11*, No. 1 (1995), at 35-37. 10 U.S.C. § 1107 (2000) (Notice of Use of an Investigational New Drug or a Drug Unapproved for Its Applied Use).

125. *See* http://abcnews.go.com/wire/US/ap20030219_1880.html 2/23/03.

quired to be obtained with respect to the use of unapproved drugs or vaccines where the decision to do so is made by the President of the United States.[126]

As a result of these pervasive problems, it has been suggested that separating the processes of educating patients and obtaining legal consent would underscore an institution's commitment to obtain meaningful informed consent.[127] Although no studies appear to have been made regarding such a separation, in most states, clinicians are required to report to police injuries due to violence. In some states this is called intimate partner violence (IPV) and must be reported even if it was contrary to a patient's consent.[128] The ABA *Model Rules of Professional Conduc*t authorize a lawyer's disclosure of information without the "informed consent" of the client in order "to prevent reasonably certain death or substantial bodily harm."[129]

The National Resource Council has recommended the installation of a moratorium on such laws until there is more research on their advantages and disadvantages. Some such studies showed a high rate of females objected to such reports. A majority of physicians in California thought that the

---

126. 10 U.S.C. § 1107 (2000) (Notice of Use of an Investigational New Drug or a Drug Unapproved for Its Applied Use). *See also* 10 U.S.C. § 980 (2000).

127. Clark et al., Patient Consent in Seven Medical Care Institutions. Some patients will specifically delegate decisional authority to their physicians; Schneider, The Practice of Autonomy: Patients, Doctors, and Medical Decisions (1998). For example, decision-making capacity as part of clinical care can be defined as the ability "to understand and appreciate the nature and consequences of health decisions" and "to formulate and communicate decisions concerning health care." Veterans Health Administration, VHA Handbook 1004.1, Informed Consent for Treatment and Procedures (revised Jan. 29, 2003).

128. Schillinger & Lo, JAMA (1995), at 781. Noncompliance causes clinician to face penalties of up to $1,000 or jail sentences of a few months.

129. Rule 1.6: Confidentiality of Information states:

> (a) A lawyer shall not reveal information relating to the representation of a client unless the client gives *informed consent;* the disclosure is impliedly authorized in order to carry out the representation or the disclosure is permitted by paragraph (b).
>
> (b) A lawyer may reveal information relating to the representation of a client to the extent the lawyer reasonably believes necessary:
>
> (1) to prevent reasonably certain death or substantial bodily harm.

Adopted in 2002. *See* A.B.A. J., December 2002.

requirements could prevent women from seeking medical care. It was suggested that in domestic violence cases, police "should consider options that include consent of patients before wider implementation."[130]

Physicians could provide patients with an entry code to a confidential Web site. Patients may take their time at home to become satisfied and thereafter consent to the procedure by using an "electronic signature."[131]

good v. Good

Consent forms have the great benefit of protecting physicians from omitting materials needed to obtain informed consent. These materials need not include costs

v.

Patients should be protected from having to sign forms alleging that they understand things that they cannot understand. Patients should also have a right to an advance notice of information regarding the estimated cost of treatment.

Twenty-first Century Approaches

1. Consent forms for treatment should not be signed by a patient unless and until the patient has been diagnosed and matters contained therein have been explained to his satisfaction.
2. Special consent forms that patients understand may be needed for diagnostic testing.
3. Consent forms should contain an estimate of the costs of testing for possible treatment.
4. Consideration to improve current procedures should also be given toward further use of the newly developed "informed consent" via the Internet.

The limits are unknown between the possible and the impossible, what is just and what is unjust, legitimate claims and hopes and those which are immoderate.

Emile Durkheim, *Suicide* (1896)

---

130. Rodriquez et al., JAMA, Aug. 1, 2001, at 580.

131. MILW. J. SENTINEL, Sept. 8, 2001. The Medical College of Wisconsin (together with a software company) developed this on this Web-based system. INSTITUTIONS, PATIENT COUNSELING AND HEALTH EDUCATION 4:2 (1982), at 103-10.

e) *Informed Consent on Clinical Research under IRBs and Experimental Treatments during Life-Threatening Situations through Randomized Placebo-Controlled Investigations*

The Nuremberg Code was formulated in August of 1947 by American judges sitting in judgment on Nazi doctors accused of conducting human experiments in concentration camps. Its 10 principles were derived from that trial. The first one set forth informed consent as its core principle. It required that before the acceptance of an affirmative decision by the experimental subject, there should be made known to him the nature, duration, and purpose of the experiment; the method and means by which it is to be conducted; all inconveniences and reasonable hazards to be expected; and the effects upon his health, which may possibly come from his participation in the experiment.[132] That Code has not been officially adopted in its entirety as law by any nation or as ethics by any major medical association. However, its informed consent requirement is not only the law of all states, but federal law,[133] and international law as well.[134] Invasive procedures are often performed urgently in the intensive care unit (ICU), and patients or their proxies may often not be available to provide informed consent. Nevertheless, handouts describing each procedure are available in the ICU waiting area and understanding by the consenter

---

132. N. Eng. J. Med., Nov. 13, 1997, at 1,436. There were no written principles of research in the United States or elsewhere before December of 1946. However, the AMA House of Delegates adopted some principles in December of 1946. In 1963 it was noted that 22 patients had received injections of cancer cells without their knowledge at Brooklyn's Jewish Chronic Disease Hospital. The following year a suit then caused referring to "the hottest public debate on medical ethics since the Nuremberg trials of Nazi physicians." The Board of Regents of the University of the State of New York found this research was "no basis for the exercise of their usual professional judgment." It added that "[z]eal for research must not be carried to the point where it violates the basic rights and immunities of a human person." Zorner, N. Eng. J. Med., Aug. 12, 2004, at 628-29.

133. 45 C.F.R. § 46. 101 (2000). National Institutes of Health, Office for Protection from Research Risks, protection of human subjects.

134. Article 7 of the United Nations International Covenant on Civil and Political Rights (1966). International ethical guidelines for biomedical research involving human subjects. Geneva: Council for International Organizations of Medical Sciences, 1993.

of the procedure is determined by responses on a questionnaire. In this manner health care clinicians have significantly increased the frequency with which consent was obtained without compromising comprehension of the process by the consenter.[135]

The law generally prohibits doctors from giving patients experimental treatments unless the subjects understand the risks and consent to them. A 1996 change in the federal law provided an exception for studies conducted in an emergency setting. Scientists have also been allowed to obtain waivers of that requirement so they can use experimental drugs and devices on patients without consent in life-threatening situations.[136] These waivers have also been endorsed on the grounds that granting desperately ill patients access to new therapies could result in important benefits to future patients.[137] In 2004 a Washington state jury delivered the first major verdict in a case on human experimentation, which was called "a resounding win for a renowned Seattle cancer institution." A jury found that the patients gave informed consent to participate in an unsuccessful clinical trial and that the experiment

135. Davis et al., *Improving the Process of Informed Consent in the Critically Ill*, JAMA, April 16, 2003, at 1963. Private contracts may become unenforceable following the consent of a party where they are unconscionable. RESTATEMENT [SECOND] OF CONTRACTS § 208, *Mistake of Both Parties Makes a Contract Voidable* (§ 152), or *Discharge by Supervening Impracticability* (§ 261).

136. 21 C.F.R. § 50.24 (revised April 1, 2004). The federal regulations require consent of the Institutional Review Board (IRB) and "of the subject or the subject's legally authorized representative." Waivers may be issued when "the human subjects are in a life-threatening situation, available treatments are unproven or unsatisfactory, and the collection of valid scientific evidence is necessary to determine the safety and effectiveness of particular intervention." 21 C.F.R. § 50.24(a)(1). *Also see* 21 C.F.R. § 50.24(a)(3)(ii). The usual IRB review and community consultation are also required. 20 C.F.R. § 50.24 (a)(7)(i). The waiver can be issued so long as "[t]he clinical investigation could not practicably be carried out without the waiver." 21 C.F.R. § 50.24(a)(4).

137. "Without waiver, this work would never be done." Ellenberg, *Informed Consent: Protection or Obstacle? Some Emerging Issues*, 18 CONTROL CLINICAL TRIALS (1997), at 628-36; Trugg et al., N. ENG. J. MED., March 11, 1999, at 806. "If you see informed consent as an absolute value, then none of this makes a difference. The FDA regulations only make sense when you stop seeing the moral world as governed by these types of absolute values." Brody, Hast. Ctr. Rep., January-February 1997, at 8. *See also* Emanuel et al., *What Makes Clinical Research Ethical?*, 283 JAMA (2000), at 2701-17.

was not carried out negligently.[138]

This approach is being challenged on ethical grounds since they are not designed to benefit the patient himself except peripherally. A California statute and a federal statute cover medical experiments.[139] There is also a Common Rule that applies to recipients of federal research funds and institutions to all research conducted there, regardless of funding. The IRB may waive informed consent where "research could not practically be carried out" without it and there is only a "minimum risk" to the subject.[140]

Rule 8.085 of the current *Code of Medical Ethics* (issued in 1997) waives informed consent for research in emergency situations under several conditions. They include: emergency life-threatening treatment on subjects who lack the capacity to give informed consent for participation in the research,

---

138. Wright v. Fred Hutchinson Cancer Research Center, 269 F. Supp. 2d 1286. The plaintiffs who had leukemia and needed bone marrow transplants stated:

> The 12-year experiment, called Protocol 126, tested whether a new drug could prevent graft-versus-host disease, a complication of bone marrow transplants. The drug caused graft failure and almost all participants died. Had the patients undergone conventional treatment, their chances of cure would have ranged from 15 percent to greater than 50 percent, depending on their age and how sick they were.

A.B.A. J., June 2004, at 20.

139. ORANGE COUNTY REGISTER, Sept. 28, 1999, at Metro 7. The federal regulations preempt state or local requirements to protect human subjects. The 1996 Health Insurance Portability and Accountability Act (HIPAA) permits an institutional review board (IRB) to waive the requirement for consent, provided that criteria are met regarding the protection of individual privacy. Tu et al., *Impracticability of Informed Consent in the Registry of the Canadian Stroke Network*, N. ENG. J. MED. (2004), at 1,414, reported how obtaining written informed consent led to selection bias and "[t]he need for legislation on privacy and policies permitting waivers of informed consent for minimal-risk observational research." *Also see* Arthur W. Frank, *Emily's Scars: Surgical Shapings, Technoluxe, and Bioethics*, Hast. Ctr. Rep. (March-April 2004), at 18-29.

140. The Common Rule permits the IRB to waive or limit informed consent if "(1) The research involves no more than minimal risk to the subjects; (2) The waiver or alteration will not adversely affect the rights and welfare of the subjects; (3) The research could not practicably be carried out without the waiver or alteration; and (4) Whenever appropriate, the subjects will be provided with additional pertinent information after participation." Subpart A—Basic HHS Policy for Protection of Human Research Subjects, 45 C.F.R. § 46.101(b)(4).

obtaining surrogate research is unfeasible, and experimental treatment has a realistic probability of benefit equal to or greater than standard care. An Institutional Review Board (IRB) must follow the FDA's regulations governing the exception from Informed Consent Requirements for human subjects who are in a "life-threatening situation" for research through randomized placebo-controlled investigations (that was issued in 1996) is quite similar.[141] In administration of placebo-controlled trials inert substances were disguised as active medication. However, should such a trial involve placebo surgery, it has been argued that it violates "an essential standard for research: the requirement to minimize the risk of harm to subjects."[142]

In 1993 the International Ethical Guidelines for Biomedical Research Involving Human Subjects of the Council for International Organizations of Medical Sciences (CIOMS) issued a guideline stating that "For all biomedical research involving human subjects, the investigator must obtain in the case of an individual who is not capable of giving informed consent, the proxy consent of a properly authorized representative."[143] In 2002 the

---

141. 21 C.F.R. § 50.24 also requires the IRB to find out whether "appropriate animal and other preclinical studies have been conducted" and that "the clinical investigation could not practicably be carried out without the waiver." Macklin, *The Ethical Problems with Sham Surgery in Clinical Research*, 342 N. Eng. J. Med. (1999) at 992-96.

> Placebo-controlled trials have been controversial, especially when patients randomly assigned to receive placebo have forgone effective treatments. . . . No drug should be approved for use in patients unless it is clearly superior to placebo or no treatment. [Further the] difference between the rates of response to the two drugs is likely to be smaller than that between the rates of response to an investigational treatment and placebo.

Emanuel & Miller, N. Eng. J. Med., Sept. 20, 2001, at 915.

142. Horning & Miller, N. Eng. J. Med., July 11, 2002, at 138, *citing* Clark, *Placebo Surgery for Parkinson's Disease: Do the Benefits Outweigh the Risks?*, 30 J. L. & Med. Ethics 2002, at 58-68, *and* Weijer, "I need a placebo like I need a hole in the head." 30 J. L. & Med. Ethics 2002, at 69-72. However, they would require "informed consent and do not support an absolute prohibition."

143. "Before undertaking research in a population or community with limited resources, the sponsor and the investigator must make every effort to ensure that . . . any intervention or product developed, or *knowledge generated*, will be made reasonably available for the benefit of that population or community." (Emphasis added.)

International Bioethics Committee of UNESCO issued a Preliminary Report on the Possibility of Elaborating a Universal Instrument on Bioethics that could "provide unity while recognizing the special challenges posed by the unique histories, cultures, politics, judicial systems, and economic situations of the various countries involved especially in *relation to ensuring the free and informed consent of research subjects.*" (Emphasis added.) The additional need to obtain consent for publication in the United States would appear to be within the definition of "human subject" in a federal rule and this would require approval from an Institutional Review Board (IRB).[144] It has been argued that "What is urgently needed is a set of guidelines that clearly differentiates public health practice from research."[145] In 2003, the Department of Health and Human Services (HHS) issued its Privacy Rule covering protected health information including individually identifiable data that can be used to identify the individual under a "reasonable basis."[146] Nevertheless, there are exceptions which include health authorities for public health purposes.[147]

The requirements for the assent by children in research are "when in the judgment of the IRB the children are capable of providing assent." Otherwise, their own assent may not be necessary. Thus, only children "capable of providing assent" can refuse to give it under the federal regulation.[148]

CIOMS, International Ethical Guidelines for Biomedical Research Involving Human Subjects, 2d ed. *See also* Hast. Ctr. Rep. (May-June 2004), p. 21, and Council of Europe's Recommendation Concerning Medical Research on Human Beings (1990).

144. 45 C.F.R. § 46.102(f). *See also* Dep't of Health and Human Services, 45 C.F.R. Pts. 160 and 164, *and* Doe v. Roe, N.Y.S.2d 668 (1977)

145. Amoroso & Middaugh, *Research vs. Public Health Practice: When Does a Study Require IRB Review?*, 35 Preventive Medicine 250-53 (2003).

146. 45 C.F.R. § 60.103, § 164.501, and §164.501 (requiring an individual's written authorization).

147. 45 C.F.R. § 164.512(a) and (b) (2002). *See also* Rivara, *Research and Human Subjects*, 156 Archives of Pediatric & Adolescent Medicine 641-42 (2002).

148. Federal regulations require an IRB to determine on a study-by-study (and even child-by-child) basis whether assent is appropriate and in what form. 38 U.S.C. § 7331 (2000). If the IRB determines that the capability of some or all of the children is so limited that they cannot reasonably be consulted or that the intervention or procedure involved in the research holds out a prospect of direct benefit that is important to the health or well-being of the children and is available only in the context of

good v. Good

Incompetent emergency subjects may ethically be used for research experiments without their informed consent or that of their family (or surrogate).

v.

Under those conditions such research should be limited to subjects in a life-threatening situation and without placebo surgery.

Twenty-first Century Approach

Where informed consent for experimental research cannot be obtained from an incompetent emergency patient, such research (a) will be made with respect to its possible benefit to otherwise terminal subjects; but (b) will only engage in placebo surgery after obtaining informed consent from the subjects (or surrogates) and meeting the ethical requirements that are appropriate for clinical research.

Doctors keep their scalpels and other instruments handy, for emergencies. Keep your philosophy ready too—ready to understand heaven and earth. In everything you do, even the smallest thing, remember the chain that links them. Nothing earthly succeeds by ignoring heaven, nothing heavenly by ignoring the earth.

The Meditations, by Marcus Aurelius, Bk 3:13

f) Informed Consent by Emergency Department Physicians, EMTALA, the Triad Terrorist Attacks; Radiation Treatment

As noted previously, the U.S. Constitution alone does not specifically grant any right to health care. However, the legislation of Congress has done so in many directions.

The Emergency Medical Treatment and Labor Act (EMTALA) requires emergency departments (EDs) of hospitals that participate in the federal Medicare program to "provide for an appropriate medical screening examination within the capability of the hospital's emergency department, including ancil-

---

the research, the assent of the children is not a necessary condition for proceeding with the research. 45 C.F.R. § 46.408 (a)-(j). California, on the other hand, requires consent by minor subjects age seven or older (CAL. HEALTH & SAFETY CODE § 24173).

lary services routinely available to the emergency department, to determine whether or not an emergency medical condition . . . exists."[149] Members of emergency medical services (EMS) are often "like their fire service colleagues, planned, trained, and responding to contingencies without preplanned coordination with law enforcement, frequently gaining access to the scene only after termination of active police activity."[150] The Act added a length of time to the emergency departments. This was a special provision, which requires these departments to serve all patients who enter. That is, they may arrive without necessarily seeking emergency treatment and the department is required to provide them with an "appropriate medical screening examination . . . to determine whether or not an emergency medical condition exists."[151]

This may be done after a large number of those in emergency have been served by available physicians.[152] Where there is no evidence of discrimina-

---

149. 42 U.S.C.A. § 1395d(e)(1). On Nov. 8, 2002, a study published in the *Annals of Internal Medicine* found that many emergency patients with nonacute ailments suffered no ill effects after being told to come back to help ease crowding and cut costs. ORANGE COUNTY REGISTER, Nov. 5, 2002, at News 11. D.C. KEYES ET AL., MEDICAL RESPONSES TO TERRORISM 385 (2005). In 2003, polls were taken which found that four out of every 10 U.S. citizens stated they "often worry about the chances of a nuclear attack by terrorists." *Two Years Later, the Fear Lingers* (Pew Research Center for the People and the Press, Sept. 4, 2003). *See* ALLISON, NUCLEAR TERRORISM, THE ULTIMATE PREVENTABLE CATASTROPHE (2004).

150. D.C. KEYES ET AL., *supra* n.149 at 385.

151. 42 U.S.C.A. § 1395dd(e)(1)(A) (2004). An emergency is:

> a medical condition manifesting itself by acute symptoms of sufficient severity *(including severe pain)* such that the absence of immediate medical attention could reasonably be expected to result in (*i*) placing the health of the individual (or, with respect to a pregnant woman, the health of the woman or her unborn child) in serious jeopardy, (*ii*) serious impairment to bodily functions, or (*iii*) serious dysfunction of any bodily organ or part.

42 U.S.C.A. § 1395D(E)(1). On Nov. 8, 2002, a study published in the *Annals of Internal Medicine* found that many emergency patients with nonacute ailments suffered no ill effects after being told to come back to help ease crowding and cut costs. ORANGE COUNTY REGISTER, Nov. 5, 2002, at News 11.

152. *See* Baber v. Hospital Corp. of America, 997 F.2d 872 (6th Cir. 1992). (The doctor "did perform a screening examination that was not so substandard as to amount to no examination.") Many physicians have complained that patients who do not need emergency attention should rely on their personal physicians or urgent care

tion in screening, an incompetent conclusion may be reached that no emergency exists, which does not violate that law.[153] Nevertheless, the majority in another Circuit Court actually held the existence to treat an anencephalic (a virtually brainless baby) by exceeding the prevailing standard of medical care and stating: "We recognize the dilemma facing physicians who are requested to provide treatment they consider morally and ethically inappropriate, but we cannot ignore the plain language of the statute . . . The appropriate branch to redress the policy concerns of the Hospital is Congress."[154]

Emergency physicians may prescribe medications that are controlled substances in the schedules of the federal Controlled Substances Act (CSA) that are subject to enforcement under the federal Drug Enforcement Administration (DEA). Hence such physicians often fear that their patients may have no medical need for the drugs.[155] In this connection, the Federation of State Medical Boards issued guidelines on pain relief for medical boards that were adopted by many states.[156] Furthermore, the American College of

---

centers. In some hospitals, ERs simply are not prepared to treat high volumes of non-emergency cases. Research holds out a prospect of direct benefit that is important to the health or well-being of the children and is available only in the context of the research; the assent of the children is not a necessary condition for proceeding with the research. 45 C.F.R. § 46.408(a).

153. Summers v. Baptist Medical Center Arkadelphia, 91 F.3d 1132 (8th Cir. 1996) (en banc). EMTALA established a provider/patient relationship that led to a legal duty of professionally reasonable care. ROSENBLATT & ROSENBAUM, LAW AND THE AMERICAN HEALTH CARE SYSTEM 823-1138 n.1.

154. In the Matter of Baby K, 16 F.3d 590 (4th Cir. 1994); *cert. denied*, 115 S. Ct. 91 (1994). Unfortunately, Congress has failed to modify the statute. *An Anencephalic Child Kept Alive Through the Fourth Circuit's Interpretation of the Emergency Medical Treatment and Active Labor Act: A "Medical Miracle" or a "Living Corpse"?* St. LOUIS U. PUB. HEALTH L. REV. 323-46 (1994). A host of unfavorable writings from many writers yielded no response from Congress: *e.g.*, *The Case of Baby K*, 330 N. ENG. J. MED. 1542-45 (1994); Dzielak, *The Debate About Continued Futile Medical Care*, JOHN MARSHALL L. REV. 28 (1995); and *A Proposed Amendment to the Act After the Matter of Baby K*, VAND. L. REV. 1491-1538 (1995).

155. Rupp & Delaney, *Inadequate Analgesia in Emergency Medicine*, ANNALS OF EMERGENCY MED. 494 (2004), and Ducharme, *Acute Pain and Pain Control: State of the Art*, ANNALS OF EMERGENCY MED. 592-603 (2000).

156. Johnson, *Providing Relief to Those in Pain: A Retrospective on the Scholarship and Impact of the Mayday Project*, 31 J. LAW, MED. & ETHICS 15-20 (2003). Some 23 states have enacted "in tractable pain statutes" in order "to further affirm the

Emergency Physicians (ACEP) issued its "Clinical Policy: Critical Issues for the Initial Evaluation and Management of Patients Presenting with a Chief Complaint of Nontraumatic Acute Abdominal Pain," (2000).[157] The ACEP Code of Ethics states that "[e]mergency physicians should act fairly toward all persons who rely on the ED for unscheduled episodic care. . . . Provision of emergency medical treatment should not be based on gender, age, race, socioeconomic status, or cultural background. No patient should ever be abused, demeaned, or given substandard care."[158]

A federal survey also found that specialist physicians for economic reasons are refusing to be on call—particularly in states with high rates of uninsured patients.[159] However, many hospitals require physicians on call to treat conditions outside their specialty and on call to provide free care. These have been incorporated into their bylaws and sometimes may increase their libility risk. While hospitals may become subject to fines, these are seldom made.[160]

---

importance of treating pain, and to set out some guidance for appropriate regulatory oversight of prescribing practices. In fact, some medical boards have taken disciplinary action against physicians who have neglected their patients in pain." Johnson, *The Social, Professional, and Legal Framework for the Problem of Pain Management in Emergency Medicine*, J. L., MED. & ETHICS 754 (Winter 2005).

157. *Id.* (Johnson 2005) at 744. *Also see* 21 C.F.R. § 50.24 (2005) (covering medical research in emergency medicine).

158. American College of Emergency Physicians, *Code of Ethics for Emergency Physicians* (October 2001), *at* http://www.acep.org/webportal/PracticeResources/PolicyStatements/Ethics/CodeofEthicsforEmergencyPhysicians.htm. *See also* Furrow, *Pain Management and Provider Liability: No More Excuses*, J. L., MED. & ETHICS 29 (2001) *and* Rich, *Physicians' Legal Duty to Relieve Suffering*, 175 WESTERN J. OF MED. (2001) at 151.

159. The survey of emergency-room personnel published this year by the U.S. Department of Health and Human Services showed that 50 percent of the doctors and nurses surveyed nationwide said emergency rooms were having difficulty filling on-call rosters and 12 percent of those surveyed said specialists refuse to take calls.

160. The federal government prohibited hospitals from "dumping"—transferring emergency indigent patients to other hospitals. If a hospital has specialist services during the day, federal law says it must make the same care available at all hours in emergencies, on penalty of losing Medicare funding. Fines of up to $50,000 are authorized but seldom levied against doctors who agree to take emergency calls but don't show up.

ORANGE COUNTY REGISTER, Jan. 5, 2002, at News 20.

Although the AMA *Code of Medical Ethics* states, "Physicians are free to choose whom they will serve," it adds that "the physician should, however, respond to the best of his or her ability in cases of emergency where first aid treatment is essential."[161] Similarly, the ABA *Model Rules of Professional Conduct* (2002 edition) state that: "in an emergency where the health, safety or a financial interest of a person with seriously diminished capacity is threatened with imminent and irreparable harm, a lawyer may take legal action on behalf of such a person even though the person is unable to establish a client-lawyer relationship."[162]

Emergency physicians are *not* generally exempted from the requirements of informed consent for competent patients.[163] Although a court has held that

---

161. Rule 8.11. Rule 9.06 adds, "In cases of accident or sudden illness, as a practical matter, may preclude free choice of a physician, particularly where there is loss of consciousness." Rule 10.015 states, "In some instances the [mutual consent] agreement is implied, such as emergency care."

162. Rule 1.14 comment (9). This may be done only when the person with seriously diminished capacity:

> or another acting in good faith on that person's behalf has consulted with the lawyer. Even in such an emergency, however, the lawyer should not act unless the lawyer reasonably believes that the person has no other lawyer, agent, or other representative available. The lawyer should take legal action on behalf of the disabled person only to the extent reasonably necessary to maintain the status quo or otherwise avoid imminent and irreparable harm.

163. In Higler, *The Blood of Strangers* (1999) at p.113, a chapter entitled "Power" states:

> Just prior to inserting a 7 needle into the belly of a violently objecting and apparently quite competent patient, he said to him, "You need it, Mr. Hyde. We're going to put it in whether you like it or not." When Mr. Hyde again said "Don't put it in, don't put it in, don't do it," the physician wrote, "I just nod to the techs, and they grab his legs and spread them apart. I put two gloves on my right hand, lubricate the index finger, and go inside him before he can think. Oh, oh, oh, he cries, but I've done it and peeled off the glove, and it looks OK. "All right," I said to the techs. They are strong men. "Mr. Hyde," I say, "you need an operation. We're going to take you upstairs to the operating room right now." "Do what you want to," he says.

Id. at 114. Dialogue might not have been written had the physician not been able to end it with the safe words, "Discharged, 5 days later, in good condition."

when a competent patient comes to an emergency department his consent to treatment is somehow implied,[164] consent cannot be implied if the patient actually objects to a treatment or an intervention.[165]

The American College of Emergency Physicians (ACEP) permits acceptance of implied consent from a patient who is unable to understand the necessary information as a result of mental disability or severe pain and anxiety.[166] It has also been found that few emergency center patients knew that they had authority to make treatment decisions or about the purpose of the consent they signed upon admission.[167] As of January of 2005, the state Emergency Medical Assistance Compact (EMAC) was adopted by 47 states, two territories, and the District of Columbia. Thereunder, out-of-state providers may be credentialed by the receiving state to honor the license of the host state.[168]

---

164. Wright v. John Hopkins Health System Corp., 728 A.2d 166 (Md. 1999).

165. Anderson v. St. Francis-St. George Hosp., 614 N.E.2d 841 (Ohio App. Ct. 1992).

166. *See Procedural Sedation and Analgesia in the Emergency Department*, 31 Annals of Emerg. Med. 663-67 (1998); Blackburn & Visser, *Pharmacology of Emergency Pain Management and Conscious Sedation*, 18 Emerg. Med. Clinics of N. Am. 76-101 (2000); Rupp & Delaney, *Inadequate Analgesia in Emergency Medicine*, 43 Annals of Emerg. Med. 494-503 (2004); K.H. Todd, *Emergency Medicine and Pain: a Topography of Influence*, 43 Annals of Emerg. Med. 504 (2004).

167. In Boisaubin et al., *Informed Consent in Emergency Care: Illusion and Reform*, 16 Ann. of Emerg. Med. 62-67 (1987). *See also* Shine v. Vega, 709 N.E.2d 58 (Mass.1999), *citing* Norwood Hosp. v. Munoz, 564 N.E.2d 1017, 1025.

168. The EMAC Model Legislation, *available at* http://www.emacweb.org/EMAC/Model_Legislation.cfm, Article V, provides:

> [w]henever any person holds a license, certificate, or other permit issued by any state party to the compact evidencing the meeting of qualifications for professional, mechanical, or other skills, and when such assistance is requested by the receiving party state, such person shall be deemed licensed, certified, or permitted by the state requesting assistance to render aid involving such skill to meet a declared emergency or disaster, subject to such limitations and conditions as the governor of the requesting state may prescribe by executive order or otherwise.

See also the federal Volunteer Protection Act (VPA), 42 U.S.C. § 14501 *et seq.*

In regard to the use of physical restraints in medical emergencies, 1999 regulations were issued that apply to all patients in hospitals that participate in federal programs for seniors (Medicare)[169] and for the poor (Medicaid). A restraint must be "selected only when other less restrictive measures [are] found to be ineffective to protect the patient or others from harm."[170] This is intended to make their decision to use restraints when the patient is threatening to harm himself or others. In 2003 it was found that "the risk of retention of a foreign body after surgery significantly increases in emergencies, with unplanned changes in procedure, and with higher body-mass index. Case-control analysis of medical-malpractice claims may identify and quantify risk factors for specific types of errors."[171]

Terrorist attacks on September 11, 2001, on the World Trade Center in New York City and the Pentagon in Arlington, Virginia, resulted in almost

---

169. Medicare was established by Title XVIII of the Social Security Act as enacted in 1965 and codified at 42 U.S.C. § 1395 *et seq*. Part A pays for inpatient hospital care and Part B is for physicians (who submit their claims on an HCFA form).

170. (1) The patient has the right to be free from restraints of any form that are not medically necessary or are used as a means of coercion, discipline, convenience, or retaliation by staff.
(2) A restraint can only be used if needed to improve the patient's well-being and less restrictive interventions have been determined to be ineffective.
(3) The use of restraint must be in accordance with the order of a physician or other licensed independent practitioner [and] . . . never written as a standing [or as needed] order. . . .
(4) The condition of the restrained patient must be continually assessed, monitored, and reevaluated.

Fed. Reg. 1998; 64 (127): 3607089. Health Care Financing Administration, Department of Health and Human Services, *Medicare and Medicaid Programs: hospital conditions of participation: patients' rights.*

171. Gawande et al., 348 N. Eng. J. Med., Jan. 16, 2003, at 229. The purpose of the new National Incident Management System (NIMS), a component of the National Response Plan (NRP), is "to enhance the ability of the United States to prepare for and to manage domestic incidents by establishing a single, comprehensive national approach." The Joint Commission on Accreditation of Healthcare Organizations (JCAHO) was adopting an incident management system for community preparedness and disaster management in 2004. Keyes et al., Medical Response to Terrorism 249 (Wetterhall & Collins 2005).

3,000 deaths when three hijacked airliners crashed into them.[172]

It has been pointed out that "Terrorism is an unfortunate reality of our time," and in Emergency Medical Services (EMS) "professionals *and* hospital personnel must be prepared to conduct all aspects of care for victims of terrorism."[173] The Joint Commission on Accreditation of Healthcare Organizations (JCAHO), in the January 2001 Comprehensive Accreditation Manual for Hospitals, "requires hospital emergency management to function as an integrated entity within the scope of the broader community."[174] The compo-

---

172. However, it appeared that the event increased neither the need for all this blood nor a more significant amount of emergency care. This was later provided. NEW ENG. J. MED., Feb. 21, 2002, at 617. The Terrorism Risk Insurance Act of 2002, Pub. L. No. 107-297, 116 Stat. 2322, was enacted to protect and stabilize property ownership . . . limits state regulatory discretion over insurance. Stat. 594, Public Health Security and Bioterrorism Preparedness and Response Act of 2002, Pub. L. No. 107-188, 116 was also enacted to plan and execute public health and medical care during emergencies. The Patriot Act, Pub. L. No. 107-56, 115 Stat. 272 (2001), which was enacted in the wake of the terrorist attacks of September 11, 2002, provides federal authorities "with the appropriate tools required to intercept and obstruct terrorism." *See also* amendments to the Foreign Intelligence Surveillance Act (FISA), 50 U.S.C. §§ 1801-1862.

173. Terrorism has become a reality of life in many parts of the world. News reports frequently remind us of those individuals and groups who are intent on injuring and killing noncombatants in the name of a cause, for personal satisfaction, or for power.

DANIEL C. KEYES ET AL., MEDICAL RESPONSE TO TERRORISM 2 (2005) p. x. On September 28, 2001, the U.N. Security Council passed Resolution 1373, requiring all U.N. members to refrain from giving any kind of support to groups involved in terrorism; prevent those who plan or finance terrorist acts from doing so within their nation's territory; and prevent terrorist movements by controlling passports, preventing counterfeiting or money and passports, etc.

174. *Id.* at 235. The JCAHO has adopted the Federal Emergency Management Agency's (FEMA) four-phase organizational model for emergency management and disaster readiness. The model defines the four phases of emergency management as mitigation, preparedness, response, and recovery. *Id.* The threat of attack, some argued, is only theoretical, and the threat of serious side effects is real. After much consideration, in September 2002 the CDC released a report providing vaccination guidelines to the 50 states as President George Bush pushed a possible attack on Iraq. The guidelines call for each state to be prepared to vaccinate 1 million people within 10 days of an announcement of a smallpox attack.

nents of the Emergency Preparedness and Response Triad are emergency management, health care delivery systems, and public health systems. It has also been called "disaster management" to cover a mass casualty event, or disaster. For example: The smallpox vaccine causes serious side effects in one of every 13,000 people, from severe rashes to brain inflammation.[175]

There are two statutes requiring the disclosure of information on the adverse effects of poisons. One is the Toxic Substances Control Act (TSCA), which requires manufacturers and processors to keep records of "significant adverse reactions to health or the environment . . . alleged to have been caused by the substance or mixture . . . [and must immediately report] information which reasonably supports the conclusion that such substances or mixture presents a substantial risk of injury to health or the environment."[176] The other is the Federal Insecticide, Fungicide, and Rodenticide Act (FIFRA) stating that "If at any time after the re-registration of a pesticide the registrant has additional factual information regarding unreasonable adverse effects on the environment of the pesticide, the registrant shall submit such information to the regulator."[177]

g) Nuclear Terrorism

In December of 1960, the middle of the Cold War, Herman Kahn published his book *On Thermonuclear War* to warn America to prepare for nuclear war, the threat of nuclear holocaust. He divided them. One reader thought it was "a moral tract on mass murder: how to plan it, how to commit it, how to get away with it, how to justify it." Another thanked him, gulping, "All non-

175. Levy & Fischetti, The New Killer Diseases 76-78 (2003). A federal statute makes it a crime for a person who, without authority, "uses, threatens, or attempts or conspires to use, a weapon of mass destruction, including any biological agent, toxin, or vector . . . against any person within the United States." 18 U.S.C. § 2332. In December 2002, President Bush decreed that 500,000 key health workers nationwide would be vaccinated in the coming weeks. Later, 10 million more health workers, firefighters, police, and emergency medical personnel would be vaccinated—condemning some people to serious illness or death.

176. 15 U.S.C. § 2607(e) (2000) and regulations in TSCA § 8(e); Notification of Substantial Risk; Policy Clarification and Reporting Guidance. 68 Fed. Reg. 33,129-30 (June 3, 2003).

177. 7 U.S.C. § 136d(a)(2) (2000) and 40 C.F.R. Pt. 159 (2003).

sense about conventional praise aside, the country—probably the world—owes you a great debt."[178] That particular concept of a potential WW III terminated in 1991 with the change in the former USSR Communist government. Further, seven larger nuclear power countries (United States, Russia, Great Britain, France, China, India, Pakistan) and a smaller one (Israel) are currently seeking to end any increase in their number. The Nuclear Nonproliferation Treaty (NPT) guarantees "the inalienable right of all the Parties to the Treaty to develop research production and use of nuclear energy for peaceful purposes." But this has been said to be "an escape clause that allows nations to legally build the infrastructure for a nuclear weapons program before withdrawing from the NPT [as one of its 188 members] and going the last mile to make nuclear bombs, test those bombs, and declare themselves nuclear weapons states."[179] Although the International Atomic Energy Agency (IAEA) officials serve as accountants for these nuclear reactors to alert other nations if nuclear material is lost or stolen," some withdrawing nations may refuse to admit them any longer.[180] Additional "bad news, however, is that radioactive materials are so widely available in the industrial economies that even a determined effort to deny terrorists access to such material is bound to fail."[181]

In 2004 the United States introduced, and the Security Council later passed, Resolution 1540, which states that all U.N. member nations "must take cooperative action to prevent illicit trafficking in nuclear, chemical, or biological weapons, their means of delivery, and related materials."[182] In 2005, with the first anti-terrorism convention completed since the September 11, 2001, attacks on the United States, the 191-member U.N. General Assembly (only 50 of which are democratic members) approved a nuclear terrorism

178. TABRIZI, THE WORLDS OF HERMAN KAHN 5 (2005).

179. ALLISON, NUCLEAR TERRORISM 158 (2004). Pakistan, India, and Israel are not members of the NPT and have not undertaken any obligations to comply with the treaty. ORANGE COUNTY REGISTER, Aug. 10, 2005, at News 10.

180. *Id.* Allison at 157. On May 24, 2002, President George W. Bush and Russian President Vladimir Putin signed a nuclear arms–reduction treaty in Moscow.

181. *Id.* at 196.

182. ORANGE COUNTY REGISTER, April 14, 2005, at News 25. The FY 2002 Department of Energy (DOE) budgets were $300 million for nuclear physics and $300 million for fusion. *See* http://www.aip.org/enews/fyi/2001/134.html. *See* the Antiballistic Missiles (ABM) and Strategic Arms Reduction (START) treaties.

treaty. It would oblige governments to punish those who illegally possess atomic devices or radioactive materials.[183]

Although other countries have not yet succeeded in joining those which can detonate a nuclear device (except that North Korea and Iran are currently attempting to do so), they may now utilize the much less expensive radiation dispersal devices (RDDs) that cause the spread of radioactive material without being accompanied by any nuclear detonation. Because RDDs may also be used by terrorists (such as the currently active Al Qaeda for religious purposes), more action thereon is being sought by bioethicists.[184] This "nuclear threat is an invisible threat and the biological effects that occur depend on the dose received and the dose rate. To produce *acute radiation syndrome* the radiation must be penetrating, and the whole body must be exposed to the radiation. The expression of these biological effects may be delayed for hours, days, or even years."[185] Hence, many have thought it would seldom be used for military purposes due to the delay of its effect. Nevertheless, in 2001 the International Atomic Energy Agency (IAEA) noted that "almost every nation in the world has the radioactive materials needed to build a dirty bomb." But "more than 100 countries lack adequate controls to prevent the theft of these materials."[186]

---

183. ORANGE COUNTY REGISTER, Aug. 7, 2005, at News 3. For example, in 2005 the DOE cleared the way to move almost 12 million tons of radioactive waste from the banks of the Colorado River. It also began work on the specifics of how to move the 94-foot-tall uranium mining waste pile from its location near Moab, Utah, the cost of which is expected to be more than $400 million. ORANGE COUNTY REGISTER, Sept. 25, 2005, at News 8.

184. "The Radiation Emergency Assistance Center/Training Site (REAC/TS) is a good resource for radiation expertise. This organization, which is part of the U.S. Department of Energy response network, makes available medical experts, radiobiologists, and health physics experts." Edsal & Keyes, *Treatment of Radiation Exposure and Contamination*, in KEYES ET AL, MEDICAL RESPONSE TO TERRORISM 165 (2005).

185. *Id.* Patrick, Edsal & Keyes, *Radiation Dispersal Devices, Dirty Bombs, and Nonconventional Radiation Exposure,* at 167. "A radiation dispersal device (RDD) is essentially the combination of a conventional explosive and radioactive material."

186. Levi & Kelly, *Weapons of Mass Disruption*, SCIENTIFIC AMERICAN, November 2002, at 78. "Federal regulations prohibit radiation workers from receiving more than five rems annually. The U.S. Environmental Protection Agency recommends that a contaminated area be abandoned if decontamination efforts cannot reduce the extra risk of cancer death to about one in 10,000." *Id.* at 79.

In this connection the Defense Against Weapons of Mass Destruction Act of 1996[187] was enacted "to provide training, exercise and expert advice to emergency response personnel and lend equipment to local jurisdictions" (in large cities).[188] Most physicians "have little knowledge about treating radiation injuries because the subject is not included in medical schools."[189] Advice may be obtained from the experts at the Radiation Emergency Accident Center and Training Site located in Oak Ridge, Tennessee, which is "one of the World Health Organization's designated collaboration centers for support and advice in the event of a radiation accident." It has been pointed out how "this requires extensive training. Emergency and hospital personnel need to understand how to protect themselves and affected citizens during a radiological attack and be able to determine rapidly if individuals have been exposed to radiation. Although the effects of a radiological attack are minor compared with those of even a small nuclear weapon, a dirty bomb could have drastic economic and psychological consequences."[190]

In 1975 the Biological Weapons Convention (BWC) became effective when it was ratified by the United States in Geneva.[191] Its restraints on *germ weapons* constituted the only weapon of mass destruction internationally banned at that time. Unlike nuclear weapons, these remained untested in war. Although their preparation and wartime use were banned, there was virtually nothing that could enforce the stop of a secret production of such weapons.[192]

"The BWC's lack of verification and compliance provisions is frequently referred to as its greatest shortcoming in reducing the threat of prolifera-

---

187. The Nunn-Lugar-Domenici Act (Pub. L. 104-201, Sept. 23, 1996).

188. The program was transferred to the Department of Justice (DOJ) in 2000 and again to the Department of Homeland Security (DHS) on March 2, 2003, for those purposes. KEYES ET AL., MEDICAL RESPONSE TO TERRORISM 372 (2005) at 372.

189. *Id.* at 158, Edsal & Keyes, *Treatment of Radiation Exposure.*

190. Levi & Kelly, SCIENTIFIC AMERICAN 81 (2002).

191. Currently 151 nations are parties to the BWC. *See also* U.S.C. § 175. "Prohibitions with respect to biological weapons." (The term "for use as a weapon" does not include their use for "peaceful purposes.")

192. In July 2001 President Bush "withdrew from negotiations for a legally binding verification and compliance Protocol to the Biological Weapons Convention." GUILLEMIN, BIOLOGICAL WEAPONS 168 (2005). *See also* MILLER, ENGLEBERG & BROAD, GERMS: BIOLOGICAL WEAPONS AND AMERICA'S SECRET WAR 203 (2001).

tion."[193] Nevertheless, the Public Health Security and Bioterrorism Preparedness and Response Act of 2002, was summarized as having "mandated increased surveillance of drinking water supplies and food and required secret stockpiles of drugs, vaccines (especially smallpox vaccine), and other emergency medical supplies."[194] This statute defined "accelerated countermeasure research and development," which was described as "research on pathogens of potential use in a bioterrorist attack."[195] Further, "biologists have been asked to look at their research for the harm it might promote and the possible necessity to categorize it as classified or as 'sensitive but unclassified.'"[196] On the other hand, in 2003 the National Research Council (NRC) committee reported, "Given the increased investments in biodefense research in the United States, it is imperative that the United States conduct its legitimate defensive activities in an open and transparent manner."[197] Professor Guillemin concurred from her study of biological weapons, and concluded that government secrecy in the twentieth century "caused a degradation of biology and increased mortal risk to unsuspecting civilian targets." Accordingly, openness "is an advantage that can be too easily lost, in shortsighted homeland security policies and in failed relations with other nations, to our peril and the peril of future generations."[198]

Certain chemicals (which can be used as nonbiological weapons) are covered in a 1991 statute entitled "Control and Elimination of Chemical and Bio-

---

193. GUILLEMIN, BIOLOGICAL WEAPONS 185 (2005). Further, Pharmaceutical Researchers and Manufacturers of America (PhRMA) was vocal in its objections to on-site inspections by international teams as an economic threat. *Id.* at 189. However, Article 1 of the BWC states that acquisition and retention of biological and toxin weapons are forbidden "with the aim to exclude completely and forever the possibility of their use." *Id.* at 192-93.

194. Pub. L. 107-188, and GUILLEMIN, at 182.

195. *Id.* at 183.

196. *Id.* at 200.

197. NRC, BIOTECHNOLOGY RESEARCH IN AN AGE OF TERRORISM 6. The Federal Emergency Management Agency (FEMA) began to focus more on response to natural disasters such as floods, fires, and hurricanes. However, in 1995, Presidential Decision Directive-39 (PDD-39) made FEMA "the primary federal agency responsible for managing the on-scene response to an act of terrorism in the United States." Cain, in KEYES ET AL., MEDICAL RESPONSE TO TERRORISM (2005) at 371.

198. GUILLEMIN, BIOLOGICAL WEAPONS 204 (2005). The Homeland Security Act (2003) affected more than 25 agencies and relocated thousands of government employees.

logical Weapons."[199] Its purposes include both to mandate U.S. sanctions and to encourage international sanctions, against countries that use them who will impose sanctions against companies that aid in their proliferation. In 1997 the Chemical Weapons Convention (CWCA) Article VIII provided for Review Conferences to "take into account any relevant scientific and technological development."[200] Some scientists in microbiology thought that the National Institute of Allergy and Infectious Diseases (NIAID)—which is a part of the National Institutes of Health (NIH)—diverted its funds from nonbiodefense research. However, it was thereafter reported that since the turn of the third millennium-funding for NIAID nonbiodefense research increased by more than 50 percent.[201] Nevertheless, the possible necessity to categorize it as classified or as "sensitive but unclassified" may require that something about a discovery "concerning a pathogen or drug could all be withheld from publication, to the detriment of "the researchers, their institutions."[202]

As noted above, the preservation of the public's health is constitutionally and essentially reserved to the states. A Model State Emergency Health Powers Act (MSEHPA) was drafted at the request of the Centers for Disease Control and Prevention (CDC). It has been adopted by many state legislatures to

199. 22 U.S.C. § 65, Control and Elimination of Chemical and Biological Weapons. The President was specifically required to report on "regional stability by Iran, Iraq, Syria, Libya, and others" to "use chemical or biological weapons." *Id.* § 5001. *See also* 18 U.S.C. § 229, Prohibited Activities (1998 on Chemical Weapons), which exempts both the Armed Forces and a person "in an emergency situation, any otherwise nonculpable person if the person is attempting to destroy or seize the weapon."

200. The Scientific Advisory Board (SAB) for the CWC prepared a report on relevant advancements in science and technology for the First Review Conference in 2003. SCIENCE, August 2005, at 1021. The next conference to consider revisions to the CWC is scheduled for 2008. *Id.*

201. In 2004, the success rates for NIAID biodefense and nonbiodefense research project grant applications were comparable at 25.9 percent (316 grants awarded) and 23.5 percent (845 grants awarded), respectively. . . . security experts repeatedly express concern that future attacks with biological weapons are likely, if not inevitable. Fauci, Director, NIAID, and Zerhouni, Director, NIH, SCIENCE, April 1, 2005, at 49. *See also* Toxic Substances Control Act (TSCA), 15 U.S.C. §§ 2601-2629.

202. GUILLEMIN, BIOLOGICAL WEAPONS 200 (2005).

detect and contain bioterrorism or a naturally occurring disease outbreak.[203] Federal, state, and local authorities will work together to protect the health and safety of its population. The CDC defines "bioterrorism" as the intentional release of viruses, bacteria, or toxins for the purpose of harming or killing civilians.[204] Bioweapons have been used as a "poor man's atom bomb."[205]

In 2005 a code of ethics was proposed stating, inter alia, that "All persons and institutions engaged in any aspect of the life sciences must [refuse] to engage in any research that is intended to facilitate or that has a high probability of being used to facilitate bioterrorism or biowarfare; and . . . Seek to restrict dissemination of dual-use information and knowledge to those who need to know in cases where there are reasonable grounds to believe that the information or knowledge could be readily misused through bioterrorism or biowarfare," etc. Although the authors recognize that "a code will not be sufficient to ensure that science is not misused," they add, "but it can contribute, in conjunction with other measures, to the deterrence of

---

> 203. The Model Act is structured to reflect five basic public health functions to be facilitated by law: (1) *preparedness*, comprehensive planning for a public health emergency; (2) *surveillance*, measures to detect and track public health emergencies; (3) *management of property*, ensuring adequate availability of vaccines, pharmaceuticals, and hospitals, as well as providing power to abate hazards to the public's health; (4) *protection of persons*, powers to compel vaccination, testing, treatment, isolation, and quarantine when clearly necessary; and (5) *communication*, providing clear and authoritative information to the public.

JAMA, Aug. 7, 2002, at 623. As of January 2004, this Act had been enacted in 20 states and the District of Columbia. Visit http://publichealthlaw.net/Resources/Modellaws.htm.

204. Centers for Disease Control and Prevention, U.S. Dep't of Health and Human Services, *The Public Health Response to Biological and Chemical Terrorism, Interim Planning Guidance for State Public Health Officials* (July 2001), *available at* http://wwww.bt.cdc.gov/Documents/PlanningGuidance.PDF.

205. *See* Wheelis, Rozsa & Dando, Deadly Cultures: Biological Weapons Since 1945 (2006), and Encyclopedia of Bioterrorism Defense, Pilch & Zilinskas eds. (2005), whose Preface recognized that "an informed U.S. population protected by knowledgeable, well-trained first responders and medical professionals will be less likely to panic."

bioterrorism and biowarfare."[206] As previously noted, neither the AMA Code of Medical Ethics nor the ABA Model Rules of Professional Conduct constitute the law; however, they both have substantial ethical influence upon physicians and lawyers, respectively.

good v. Good

Emergency physicians should be completely exempt from the doctrine of informed consent in emergencies due to time restrictions on the treatment needed by their competent patients. The exemption should also include the use of experimental drugs and devices on children too young to consent so that future patients may benefit.

v.

All competent patients retain their right to refuse treatment including those in life-threatening emergencies. If the patient is unconscious, consent of a family member or other surrogate needs to be obtained only if time and circumstances permit.

Twenty-first Century Approaches

1. State courts or legislatures in several states should address the question of circumstances permitting waiver of informed consent for emergency purposes by competent adults and mature minors, and the circumstances of waiver for those who are no longer conscious.
2. Such state legislatures should consider enactment of a law recognizing a competent patient's right (a) to refuse medical treatment in emergencies, including those in many life-threatening situations; (b) to avoid having restraints being placed on him by health care facilities unless he threatens to harm himself or others; and (c) to inform patients that refusal of treatment will not be regarded as an unjustified ground for harming oneself.
3. Because most physicians are unable to treat patients who suffer from radiation injuries—many of which could be caused by terrorists and others:

   (a) medical schools in several parts of the country should consider teaching a course on the radiation treatments, or else

206. Somerville & Atlas, *Ethics: A Weapon to Counter Bioterrorism*, 307 SCIENCE, Mar. 25, 2005, at 1881.

(b) physicians in other parts should seek advice on radiation and/or training thereon at the Radiation Emergency Assistance Center/ Training Site (REAC/TS) in Oakridge, Tennessee.

---

h) Informed Consent v. Ethnic Groups and Prisoners

Members of certain ethnic groups appear to disdain the concept of individual informed consent. A 1995 study found that Korean-Americans and Mexican-Americans were more likely to hold a family-centered model of medical decision making, rather than the patient autonomy model favored by European-American and African-American subjects.[207]

A similar study of Navajos found that for them, language does not merely describe reality, language shapes reality; discussing negative information conflicts with the Navajos' views.[208] Because of such religious and cultural rejections of autonomy and informed consent by an individual, some of these patients may indicate their refusal to hear or let a physician proceed. They may also refuse to sign living wills or durable powers of attorney under state law, and as prescribed in the Federal Patient Self-Determination Act. Orthodox Jews and Korean- or Mexican-Americans can "autonomously" request that end-of-life decisions be deferred by such patients to their rabbi or to their families (see Chapter 12). Physicians have been held not liable for failure to disclose in accordance with a patient's specific demand not to be informed.[209] In particular, immigrants of the first and second generation must be examined in order to explore how much their views may have been transformed while residing in their adopted (American) country.

Although prisoners lose many of their rights to freedom, not all of these are forfeited. As noted previously, prisoners are the unique population in the United States who have a fundamental constitutional right to health care. The only question involving informed consent is to what extent can a prisoner refuse medical treatment. The Supreme Court has decided that he has a general lib-

---

207. Blackhall et al., *Attitudes Toward Patient Autonomy*, JAMA, Sept. 13, 1995, at 820.

208. Cárrese & Rhodes, *Western Bioethics on the Navajo Reservation*, JAMA, Sept. 13, 1995, at 1826.

209. Putensen v. Clay Adams Inc., 91 Cal. Rptr. 319, 333 (1970); Cobbs v. Grant, 104 Cal. Rptr. 505, 516 (1972); Arato v. Avedon, 858 P.2d 598, 609 (Cal. 1993).

erty interest to do so. Most prisons have regulations, some of which require medical treatment against the prisoners' will. The court stated that, "When a prison regulation impinges on inmates' constitutional rights, the regulation is valid if it is reasonably related to legitimate penological interests."[210] In the case of a competent but manic-depressive inmate who refused treatment, it was ruled that "the extent of a prisoner's right under the [Due Process] Clause to avoid the unwanted administration of antipsychotic drugs must be defined in the context of the inmate's confinement."[211] It held in 1990 that the drugs may be administered "against his will if he is dangerous to himself or others and the treatment is in the inmate's best medical interest."[212] Two years later, the Court ruled that a competent prisoner who had trouble sleeping and heard voices "in his head" could not be forced to continue taking prescribed antipsychotic drugs against his will while being detained for trial citing the test of dangerousness to himself or others. It stated, "The Fourteenth Amendment affords at least as much protection to persons the State detains for trial."[213]

A State Supreme Court ruled the following year that a quadriplegic prisoner who was completely dependent on medical personnel for assistance could refuse to be fed forcibly. The court noted that "the right to refuse medical treatment is equally 'basic and fundamental' and integral to the concept of informed consent."[214]

"The fact that an individual's decision to forgo medical intervention may cause or hasten death does not qualify the right to make that decision in the first instance."[215] (See Chapters 12 and 13.)

---

210. Washington v. Harper, 454 U.S. 210, 224 (1990), *citing* O'lone v. Estate of Shabazz, 482 U.S. 342, 399 (1981).

211. Washington v. Harper, 494 U.S. 210, 222 (1990).

212. "Where an inmate's mental disability is the root cause of the threat he poses to the inmate population, the state's interest in decreasing the danger to others necessarily encompasses an interest in providing him with medical treatment for his illness." *Id.* at 227.

213. Riggins v. Nevada, 504 U.S. 127, 135 (1992), *citing* Bell v. Wolfish, 441 U.S. 520, 545 (1979). *See also* Turner v. Safley, 482 U.S. 78, 84 (1987) ("Prison walls do not form a barrier separating prison inmates from the protections of the Constitution.").

214. Thor v. Superior Court, 855 P.2d 375, 381 (Cal. 1993).

215. *Id.* at 383. People under these and other conditions have been called "vulnerable" as to their being given special protections and services. Alexander Morawa has

good v. Good

In view of the fact that the modern version of informed consent has been the law in the United States for over the past quarter century, it can be assumed that it is known to all, including minority ethnic groups of all generations; prison populations lose the right to informed consent along with other civil rights.

v.

Many competent citizens (particularly those in minority ethnic groups) often fail to understand that the doctrine permits them to designate family members or church or synagogue authorities for health-related decision-making purposes; prison inmates should only lose their right to informed consent for reasonable penological purposes.

Twenty-first Century Approaches

To the extent they have not done so, states should consider the adoption of consent forms (specialized forms for special diseases and diagnostic purposes) that may be explained to different ethnic groups.

(a) More effort is necessary toward education on the meaning and need for informed consent in public schools and for adults in minority ethnic cultures. Minority cultures should be assured that the choice of deference to family or a religious teacher might constitute an ethical equivalence to an autonomous decision.

(b) Both state and federal prison regulations should require that inmates be properly informed while competent. They may only be forced to

---

indicated that "[T]here is no single approach to definition of vulnerability. In fact, there is no purposeful categorization at all." Morawa, *Vulnerability as a Concept in International Human Rights Law*, 6 J. Int'l Relations & Dev. 139-55 (June 2003). Nevertheless, HHS regulations cover not only children, prisoners, pregnant women, handicapped, and mentally disabled, but also the economically and even the educationally disadvantaged, stating:

> If an IRB regularly reviews research that involves a vulnerable category of subjects, consideration shall be given to the inclusion of one or more IRB members knowledgeable about and experienced in working with these subjects. And when subjects are likely to be vulnerable to coercion or undue influence, additional safeguards must be included in the study to protect the rights and welfare of these subjects.

45 C.F.R. § 46.107[a].

undergo medical treatment or take drugs for reasonable penological reasons, and when competent, they may refuse treatment if their lives may be threatened.

We must not believe the many who say that only free people ought to be educated, but we should rather believe the philosophers who say that only the educated are free.

Epictetus, former Roman slave

Whatsoever I shall see or hear concerning the life of men, in my attendance on the sick, or even apart therefrom, which ought not to be noised abroad, I will keep silence thereon, counting such things to be as sacred secrets.

The Hippocratic Oath

i) Informed Consent v. Laws Authorizing Patient Access to Obtain Physician Profiles; Medical Students Characterized as "Doctors"

Increased publicity growing out of adverse medical results of physicians has given rise to a need for patient information, and not only about the patient's disease and alternative treatments. Patients often need to be informed about a physician's prior actions and background before giving consent to treatment by him. The Joint Commission on Accreditation of Healthcare Organizations (JCAHO) set forth a guideline:

> The patient has the right to know the identity and professional status of individuals providing service to him.... This includes the patient's right to know of the existence of any professional relationship . . . to any . . . educational institutions involved in his care. Participation by patients in clinical training programs should be voluntary [i.e., with informed consent].[216]

No patient can be assumed willing to participate in the training of medical students merely upon being admitted to an academic medical center. But

---

216. Joint Commission on Accreditation of Hospitals, Accreditation Manual for Hospitals, 1985 Edition (Joint Comm'n on Accred. of Hosps. 1984).

when medical students were introduced by other members of the healthcare team as "doctors" in the mid-1990s, only 42 percent were found to have corrected this information with patients[217] and respect for patients' autonomy was reduced.[218] In 2003, the first law of its kind in the nation was enacted to require women patients facing an operation to give consent before undergoing anesthesia for doctors or medical students to perform pelvic exams or other procedures.[219]

At the turn of the twenty-first century, it has been difficult for patients to know if their physicians have questionable pasts. The federal government keeps a list of many sanctioned doctors that includes malpractice payouts and a hospital's own discipline of staff doctors. By law, the National Practitioner Data Bank is currently kept secret, with only hospital administrators and licensing boards given access. Consumer advocates are pressing for disclosure to help patients shop for good doctors.[220] Years ago Walt Rostow indicated that a consumer society is the "highest stage of progress." However, at the end of the twentieth century, less than 1 percent of patients overall knew the success rates of their physicians or hospitals.[221]

---

217. Beatty & Lewis, *When Students Introduce Themselves as Doctors to Patients*, 70 Academic Med. 175-76 (1995).

218. Silverman, *Narrowing the Gap Between the Rhetoric and the Reality of Medical Ethics*, 71 Academic Med. 227-37 (1996). It has been recently noted, "There is no hard evidence to show that individual patients in specific clinical settings clearly benefit from medical students' participation in their care." Mentello, *Comment: Dealing Ethically with an Inevitable Tension*, J. Med. Ethics, Summer 2001, at 119.

219. Orange County Register, Oct.. 3, 2003, at News 14.

220. Federal law requires insurance companies, hospitals, and state and federal health care regulators to report malpractice payments and disciplinary actions against doctors, dentists, and other health care providers. It is a voluntary compliance system—with no auditing by the government—and nearly 60 percent of hospitals have yet to make a report. The government provides a version of the database to the public, but strips out the names of doctors, and dentists and other identifying information.

Orange County Register, June 30, 2000, at News 38. *See also* www.medbd.ca.gov for information on California doctors, and www.citizen.org/hrg/publications/1506.html.

221. Schneider & Epstein, *Use of Public Performance Reports: A Survey of Patients Undergoing Cardiac Surgery*, 279 JAMA 1638-42 (1998).

The American Medical Association has lobbied to prevent Congress from opening the National Practitioner Data Bank. Hence, it has been largely left up to the states to change this situation. In 1998 California enacted a law providing that if a hospital has terminated or revoked a doctor's privileges, the public can find this out. Also, the public may find out whether a physician has paid any arbitration awards or monetary judgments ordered by a judge or jury.[222]

In 1996, enactment of the Physician Profile Act made Massachusetts the first state to publish malpractice histories and hospital disciplinary records.[223] It was enacted in part as a result of the publication of iatrogenic injuries (those caused by doctors or health care institutions) by physicians who had been sued repeatedly for malpractice but never disciplined by the state board. The profiles, which also list education, honors and awards, hospital affiliations, insurance plans, and specialties are sent by fax or are mailed out on the same day they had been requested. The system allows a consumer to request as many as 10 profiles at a time. These profiles are a function of "informed consent" and appear to meet a compelling public

---

222. ORANGE COUNTY REGISTER March 5, 1998, at News 21. "The only public comprehensive list of doctor discipline, including 50 states' records, Medicare and Drug Enforcement Administration sanctions, is Public Citizen's '16,638 Questionable Doctors.'" The group is now posting the information on the Internet and will charge consumers $23.50 for each book of local information; call (202) 588-7780. *Id.* As of January 1, 2003, California's malpractice settlements only discloses occurring thereafter. CAL. BUS. & PROF. CODE § 803.1 (West 2005). Further, such are only required if they occur above a certain frequency and amount. CAL. BUS. & PROF. CODE § 803.1.2027(b)(1)(West 2005).

223. An act providing for Increased Public Access to Data Concerning Physicians, 1996 Mass. Acts ch. 307, § 5, MASS. GEN. LAWS Ch. 112, § 21. Since 1997 the profiles have been disseminated on CD-ROM or over the Internet. Physicians breach their duty of informed consent if undisclosed financial information is in material conflict with a patient's decision regarding treatment alternatives. *See* Moore v. Regent of Univ. of Cal., 793 P.2d 479, 483 (Cal. 1990) ("Indeed, the law already recognizes that a reasonable patient would want to know whether a physician has an economic interest that might affect the physician's professional judgment."). *Id.* at 493.

need.[224] Privacy interests terminate when information becomes a matter of public record.[225]

In 1997, it was reported that after backing out of a multimillion-dollar deal for the AMA to endorse a line of Sunbeam Corporation home-health products, the AMA announced plans for a proposal to create a seal of approval for doctors who meet certain quality standards. Information that the AMA collects on doctors will be sold to interested health plans, which use such data to evaluate physicians. But the AMA has no plans to withhold accreditation from doctors based on poor performance or result. Nor will the public get to see the background on doctors that the AMA collects.[226]

The California Supreme Court ruled that hospitals must turn over records of reviews of a doctor's conduct by other physicians to the State Medical Board. It granted the board the power to subpoena records of peer-review scrutiny of a doctor on a hospital staff in a disciplinary hearing. It also noted that current law shields the records from disclosure during legal proceedings, such as private damage suits.[227] Legal safeguards for preventing discovery of adverse event data are peer review privileges, enacted in every state except New Jersey as of the year 2000.[228]

224. P. Donohue, Note & Comment, *Developing Issues under the Massachusetts Physician Profile Act*, 23 Am. J. L. & Med. 115, 139 (1997), *citing* Watkins v. United States, 354 U.S. 178, 198-200 (1957). In *Whalen v. Roe,* the Supreme Court upheld the constitutionality of a New York statute requiring pharmacists to report to state officials any prescriptions of certain drugs. *See* Whalen v. Roe, 429 U.S. 589, 591 (1977). It was not an impermissible invasion of privacy in light of the state's legitimate interest in the "health of the community." *Id.* at 602; 156 Archives of Intern. Med. 2,565-69 (1996).

225. Mangels v. Pena, 789 F.2d 836, 839 (10th Cir. 1986); Cox Broadcasting Corp. v. Cohn, 420 U.S. 469, 495 (1975); Oden v. Cahill, 398 N.E.2d 1061, 1063 (Ill. App. Ct. 1979); Cleburne v. Cleburne Living Ctr., 473 U.S. 432, 440 (1985).

226. Orange County Register, Nov. 19, 1997, at News 15.8.

227. Orange County Register, Oct. 22, 1996, at News 22. See Arnett v. Dal Cielo, 923 P.2d 1 (Cal. 1996). *See also* Cal. Bus. & Prof. Code § 803.1(b). In 2004 the U.S. Court of Appeals for the Ninth Circuit ruled that California law in Cal. Bus. & Prof. Code § 651(h)(5)(B) that prevents doctors from claiming to be "board certified" in a medical specialty unless the certifying organization meets certain state standards does not violate the First Amendment. Am. Acad. of Pain Mgmt. v. Joseph, 9th Cir. (2004), 353 F.3d 1099.

228. Gosting in JAMA, April 5, 2000, at 1743.

The Illinois Supreme Court held that information gathered outside a peer review process is not covered by the peer review immunity privilege; that privilege is normally attached to information produced by an identifiable hospital peer review committee.[229]

Peer reviews of doctors have been less effective. In 1986 Congress enacted the Health Care Quality Improvement Act (HCQIA) in order to provide peer review of physicians and interstate monitoring of incompetent physicians. It also grants immunity to such participants and does not grant a federal evidentiary privilege to the records and deliberations of the peer review process.[230] Congress also created the National Practitioner Data Bank (NPDB) as a national clearinghouse of information to prevent unqualified physicians, who had their clinical privileges at a hospital limited, from moving on to other hospitals as they had been doing without restraint. Almost all states have granted immunity to the participants in the peer review process and made the deliberations and records of the process privileged from judicial disclosure;[231] the rate

229. Roach v. Springfield Clinic (157 Ill. 2d 29), 623 N.E.2d 246 (1993). The information, recorded and subsequently sent to the committee, is not privileged. *See* Illinois Medical Studies Act 735 ILCS 5/8-2 101.

230. 42 U.S.C. § 11101-52 (1994). HCQIA was amended by the Patient Safety and Quality Improvement Act of 2005, creating a network of databases to analyze patient safety activities and encourage voluntary reporting of medical errors. 42 U.S.C. § 299. *Also see* Mullan et al., *The National Practitioner Data Bank: Report from the First Year*, 268 JAMA 73, 64 (1992). The regulations became effective September 1, 1990: 45 C.F.R. § 60.5 (2000). Hospital peer review boards differ from their ethics committees.

231. Susan O. Scheutzow, *State Medical Peer Review: High Cost but No Benefit—Is It Time for a Change?*, 25 Am. J. L. & Med. 7-8 (1999). The Joint Commission on Accreditation of Healthcare Organizations (JCAHO) also requires hospitals to perform peer review to qualify for accreditation. Current JCAHO policy may cause hospitals to perform peer review when a "sentinel event" has occurred. This is an unexpected occurrence involving death or serious physical or psychological injury, and a "root cause analysis" is to be performed to determine the cause. However, there is no requirement for an independent evaluation as to whether the hospital's response was adequate.

of reporting is exceedingly low.[232] It was reported that without adequate sanctions, there appeared to be little hope for improvement.[233]

Hospitals generally invoke these statutes successfully to protect themselves from patients who seek to recover damages for the hospital's failure to perform peer review properly.[234] There is "an increasing trend among hospitals to invoke the very statutes meant to encourage peer review as a shield from liability stemming from the hospitals' failure to conduct proper peer reviews." It was reported, "No hospital has ever lost its HCQIA immunity because it failed to report information to the NPDB."[235] In the year 2002, California enacted Senate Bill 1950, which established reporting by the medical board malpractice settlements if a physician had three or more malpractice settlements in a 10-year period.

It has been proposed that abolishing immunity statutes would be consistent with the general restrictions on not reporting adverse peer review ac-

---

232. For example, although hospitals reported 31,154 adverse actions, over twice that many disciplinary actions were taken by state licensing boards and reported to the NPDB over the same time period. *Id.* at 15, *citing* Office of Inspector General, U.S. Dept. of Health & Human Servs., Hospital Reporting to the National Practitioner Data Bank (NPDB) (1995), at 4. Its OIG report cited a 1991 Harvard Medical Practice Study that found that 1 percent of all hospitalizations in New York State in one year involved adverse events caused by professional negligence. Projecting this estimate to hospitals nationwide suggests that approximately 80,000 patients are killed each year in the United States by physician negligence. The magnitude of physician negligence indicated by the Harvard study raises the question of whether peer review is occurring at a sufficient rate.

233. *Id.* at 8.

234. Brochner v. Thomas, 795 S.W.2d 215, 217-18 (Tex. Ct. App. 1990). A patient sued her doctor's estate and a medical center, claiming that the doctor was negligent in his diagnosis and treatment and that the medical center failed to discharge its duty to the patient by not controlling the doctor's actions. *Id.* at 215. The plaintiff sought discovery of various files of the Texas State Board of Medical Examiners and the medical center. The court held that the documents were privileged under the Texas Medical Practice Act. *Id.* at 216, 223.

235. OIG Report, at 25. Although HCQIA preempts state laws that provide less immunity than that offered under federal law, "if state law offers more protection, such as providing immunity for all peer review actions, then the higher level of protection would control. . . . Many states do not require peer review information to be kept confidential . . . but provide no sanctions for violations." *Id.* at 35. There are no penalties for failing to report adverse peer review actions. *Id.* at 37.

tions, state licensing boards, and immunities.[236] A few courts have refused to adopt a federal peer review privilege.[237]

It would appear that laws similar to those in force in Massachusetts and California should be enacted by the legislatures of the other states early in the twenty-first century. Such actions would constitute as important a matter for patients as the Health Insurance Portability and Accountability Act of 1996 (HIPAA). This statute requires the development of a central electronic database containing all health records for every patient in the United States.[238] Early in the twenty-first century, doctors will have access to important health information of their patients; but patients cannot access important data concerning their physicians in most states.

At the turn of the millennium "an individual physician's performance is typically compared against normative standards or with other physicians in the medical group or community." However, little is "known about the reliability of commonly used profiling systems." In 1999, a survey was made "to determine the reliability of a set of physician performance measures for diabetes care" and "to examine whether physicians could substantially improve their profiles by preferential patient selection." It was found that physicians' report cards "were unable to detect reliably true practice differences within the three sites studied. Worse yet, it was determined that use of individual physician profiles may foster an environment in which physicians can

236. Immunities such as charitable immunity and the intrafamily immunity once enjoyed between spouses and between parents and children have largely been abandoned. Government and sovereign immunity have also undergone significant restrictions in past years. In abrogating the doctrine of charitable immunity in *Ring v. Thunig,* 143 N.E. 2d 3 (N.Y. 1957), the court provided the following statement about modern immunities that remains the general law: "The rule of non-liability is out of tune with the life about us, at variance with modern-day needs and with concepts of justice and fair dealing. It should be discarded." *Id.* at 9.

237. Burrows v. Redbud Community Hospital, 187 F.R.D. 606 (N.D. Cal. 1998). *See* Webster, *No Peer Review Protection in EMTALA Action*, 7 Health Law Rep. (BNA) No. 1, at 470, March 19, 1998. Actions under the Emergency Medical Treatment and Active Labor Act were brought against a physician for the transfer of a child that resulted in the child's death. The court held that federal law did not provide a privilege and permitted admission of the hospital's peer review records. *See also* Syposs v. United States, 179 F.R.D. 406, 411 (W.D. N.Y. 1998).

238. Health Insurance Portability & Accountability Act of 1996, 110 Stat. 1936.

most easily avoid being penalized by avoiding or deselecting patients with high prior costs, poor adherence, or response to treatments."[239]

good v. Good

Data banks of records on physicians whose hospital privileges have been revoked, or who have been forced to pay awards, settlements, judgments, or conform to peer review decisions, should remain confidential with respect to consumers (patients). This is because hospitals and other organizations are capable of using them without the need for further dissemination of such data.

v.

Often hospital and other organizations that have access to this data have not adequately protected their patients who should be granted the right to review such data themselves prior to consenting to its release to other parties.

Twenty-first Century Approaches

1. The NPDB should be opened to consumers of medical care. In addition, state legislatures should consider enacting statutes normally requiring the disclosure of the revocation of a doctor's hospital privileges and significant arbitration awards, settlements or judgments paid.
2. State legislatures should consider enacting statutes similar to that in Massachusetts, which permits its resident patients to obtain profiles

---

239. Hofer, et al., JAMA (June 9, 1999) at 2095. Analysis of this article stated:

> Proponents of physician profiles and other forms of report cards acknowledge the limitations of these tools but suggest that inaccurate data are better than none at all and that with time the reliability of the information in these reports will improve. The results of the study by Hofer et al. suggest that this may not be the case, at least for primary care physician profiles. Patients are also likely to be disenchanted when learning that the physician profiles are unreliable and only create the pretense of informing them about health care quality. The results reported by Hofer et al. should make it easier for responsible participants in the health care system to agree that physician profiles should not be used as a part of contract negotiations between physicians and health plans and that physician profiles are not and may never be ready for public consumption.

JAMA, June 9, 1999, at 2142.

of their physicians (or residents). This includes significant settlements in malpractice suits, discipline, and potential significant financial conflicts of interest.

3. Physician profiles used by health care providers should be independently audited where necessary to eliminate their unreliability, which was prevalent at the turn of the twenty-first century.
4. Many state legislatures should also consider a revision of immunity and privilege statutes respecting peer review decisions.

> Psychologists' contempt for the individual autonomy *of their own clients can also be seen in their explanations of the lives of everyone else. They explain social problems in psychological terms, often* to the *detriment of solving or ameliorating* them. One result *of the prevalence of these unwarranted assertions and theorizing has been* to weaken people's trust in their own autonomy, *in their own abilities to deal with the problems they confront in life*. (Emphasis added.)
>
> Dawes, *House of Cards* (1994) p. 35

j) Mental Competence to Give Informed Consent

In 1997, the estimated cost of mental illness exceeded $150 billion.[240] Questions arose concerning whether much of the work of psychologists and psychiatrists is scientifically based. This volume largely excludes psychotherapeutic work that depends much on anecdotal and clinical reports, rather than scientific evidence. When certain legal or ethical issues arise, concerning a patient's competency, they are frequently delegated to psychiatrists, although they could normally be adequately handled by a qualified physician. (*Psyche* is from the Greek life, spirit, or the soul; *iatry* is from the greed *iatreia;* an action of healing.) A bioethicist once wrote that "incompetency is not a mental illness, so psychiatrists who are called upon to assess competency must draw on something other than their expertise in diagnosing psychiatric disorders."[241]

Rule 8.08 of the AMA Code of Medical Ethics (2002–2003) states the exception to obtaining informed consent by physicians themselves is per-

---

240. JAMA, June 10, 1996, at 1772.
241. Macklin, Mental Choices 87 (1987).

mitted not only "where the patient is unconscious or otherwise incapable of consenting and harm from failure to treat is imminent," but also "when risk disclosure poses such a serious psychological threat to detriment to the patient as to be medically contraindicated." Rule 8.081 refers to surrogate decision making as well as the utilization of "an ethics committee to aid in identifying a surrogate decision maker or to facilitate sound decision making." The ABA *Model Rule of Professional Conduct* (2003) also covers clients with diminished capacity. Rule 1.14 states:

> (a) When a client's capacity to make adequately considered decisions in connection with a representation is diminished, whether because of minority, mental impairment or for some other reason, the lawyer shall, as far as reasonably possible, maintain a normal client-lawyer relationship with the client.
> (b) In determining the extent of the client's diminished capacity, the lawyer should consider and balance such factors as: the client's ability to articulate reasoning leading to a decision, variability of state of mind and ability to appreciate consequences of decision; the substantive fairness of a decision.

While psychotherapeutic diagnosis and counseling done by professionals may be helpful, it has been found that "the success rate of minimally trained and paraprofessional psychotherapists is not different from the success rate of the highly paid professionals."[242]

A 1993 study of 200 patients in the intensive care unit (ICU) showed that residents and nurses could assess accurately the cognition, judgment, and capacity of patients and subsequently determine their need for a surrogate.[243]

---

242. ORANGE COUNTY REGISTER, April 28, 1998, at News 4.

243. Cohen et al., *Do Clinical and Formal Assessments of the Capacity of Patients in the Intensive Care Unit to Make Decisions Agree?*, 155 ARCH. INTERN. MED. No. 21, at 2,481-85 (1993). (Residents' and nurses' intuitive judgments of decision-making capacity agreed highly with each other and the results of the Mini-Mental State Exam.) The ABA comment on its MODEL RULE OF PROFESSIONAL CONDUCT (2003) 1.14 states: "[7] If a legal representative has not been appointed, the lawyer should consider whether appointment of a guardian ad litem, conservator, or guardian is necessary to protect the client's interest."

In 1996, competency assessments made by lawyers and health professionals, compared with Competency Interview Schedule (CIS) assessment of decision-making capacity was found to match the determinations of the professionals in 85 of the 96 patients referred for electroconvulsive therapy.[244]

In accordance with the American Bar Association's Model Rule 1.14, "when a client's capacity to make adequately considered decisions in connection with a representation is diminished, whether because of minority, mental impairment or for some other reason, the lawyer shall, as far as reasonably possible, maintain a normal client-lawyer relationship with the client." Comment (5) states when he "reasonably believes" that the client has diminished capacity "unless action is taken," and cannot adequately act in the client's own interest, then the lawyer may "take reasonably protective measures deemed necessary." This includes "consulting with family members" and others "that have the ability to protect the client" and "maximizing client capacities and respecting the client's family and social connections."

California's Due Process in Competence Determination Act, which became effective in 1996, defines legal competency to make health care decisions and provides detailed criteria for a legal determination of incompetency to make them.[245] These include, among others, knowing responses to inquiries about treatment, participation in the decision by rational thought processes, and an understanding of minimum basic treatment information such as:

(a) The nature and seriousness of the illness and treatment recommended,
(b) The probable degree and duration of benefits and risks of such treatment, and
(c) The nature, risks, and benefits of reasonable alternative treatments.[246]

244. Bean, et al., *The Assessment of Competence to Make a Treatment Decision: An Empirical Approach*, 41 CAN. J. OF PSYCH. NO. 2, 85-92 (1996).

245. S.B. 730, Due Process in Competency Determination Act § 811 *et seq.*, effective Jan. 1, 1996, amending CAL. PROBATE CODE § 6. Section 814 to the Probate Code reads: "A person who has the capacity to give informed consent to a proposed medical treatment also has the capacity to refuse consent to that treatment."

246. *Id.* The detail of this statute has resulted in its being sometimes called the psychologist's "Full-Employment Act."

These elements of informed consent are what all physicians and surgeons in that populous state are required to do throughout their careers in medicine; there rarely exists need for referral to psychiatrists in order to obtain informed consent for treatment except where special circumstances exist. In the case of a patient's delirium (an acute disorder of attention and cognition),[247] the assessment of competency may become more difficult, particularly in older patients.[248] Nevertheless, it has been found that "A commonly held misconception is that individuals who suffer any mental incapacity are incompetent to make decisions [and] yet *the presence* of a diagnosis of *delirium dementia, or psychosis does not in itself establish incompetence.*"[249] (Emphasis added.) Seeking the consent of the substituted judgment of a surrogate may often be improper. Therefore, surrogate consent should be sought only when the patients clearly have been deemed incompetent or medically unable to participate in the decision-making process (e.g., coma, critical condition).[250] A report by the National Advisory Mental Health Council (an arm of the National Institute of Mental Health) found that during a one-year period more than 5 million (or 3 percent) of American adults had mental disorders of various types.[251] For example: "between 1987 and 1997, there was a marked increase in the proportion of the population who received outpatient treatment for depression. Treatment became characterized by greater involvement of physicians, greater use

247. Inouye, *The Dilemma of Delirium: Clinical and Research Controversies Regarding Delirium in Hospitalized Elderly Medical Patients*, 97 AM. J. MED. 278-88.

248. Auswald et al., *The Informed Consent Process in Older Patients Who Developed Delirium: A Clinical Epidemiologic Study*, AM. J. MED., November 1997, at 410. "Many nursing home patients in this study were found to lack capacity to give informed consent for medical treatment. In addition, nursing home staff failed to recognize that a portion of those patients were incompetent and failed to involve surrogate decision-makers." Barton et al., *Clinicians' Judgment of Capacity of Nursing Home Patients to Give Informed Consent,* 47 PSYCH. SERVS., NO. 9 (1996) at 956-60.

249. *Id.* Auswald, at 411.

250. *Id.*

251. AM. J. PSYCH. (1993). The mental disorders include schizophrenia, schizoaffective disorder, manic-depressive disorder, autism, or a severe form of some other mental ailments, such as major depression, panic disorder, or obsessive-compulsive disorder.

of psychotropic medications, and expanding availability of third-party payment, but fewer outpatient visits and less use of psychotherapy."[252]

Schizophrenia is a chronic and debilitating mental disease with symptoms that generally begin in late adolescence or early adulthood and usually continue throughout life. It represents one of three major forms of mental illness, which affects 2.2 million American adults in their 20s and 30s; it is a genetic disorder that causes delusion, hallucinations, paranoia, and a prevalence of about 10 percent suicides.[253] With the population growth in the early twenty-first century, the number with mental disorders increased to almost 6 million. Although a wide range of drugs is available, they may have significant problems and side effects. Neither psychiatric drugs nor psychotherapy are cure-alls as noted by a current writer;[254] it appears that "psychiatry is in crisis." In their 2001 volume, *Out of its Mind, Psychiatry in Crisis*, Hobson and Leonard, wrote:

> Psychiatric drugs first made medical news in the 1950s, when millions of the mentally ill were still being warehoused in asylums and when Freudian psychoanalysis was coming to hold undisputed sway at most U.S. universities, mental hospitals, private offices, and clinics. Freudians became isolated and dethroned, but unrepentant, while psychiatry wavered about what besides drugs should be stuck into

252. 287 JAMA, 203-09 (2002).

253. ORANGE COUNTY REGISTER, March 19, 2002, at News 4, and Freedman, N. ENG. J. MED., Oct. 30, 2003, at 1738. In 2000 a hospital sought judicial permission to give a paranoid schizophrenic antipsychotic drugs without his "informed consent." The Ohio Supreme Court called "the right to refuse medical treatment" "a fundamental right" and held that the state could not administer drugs unless it showed by "clear and convincing evidence" that a patient "lacks the capacity to give or withhold informed consent." The "clear and convincing evidence" requirement is significantly higher than the normal one ("a preponderance of the evidence"). 736 N.E.2d 10 (Ohio 2000).

254. WHITAKER, MAD IN AMERICA: BAD SCIENCE, BAD MEDICINE, AND ENDURING MISTREATMENT OF THE MENTALLY ILL 242 (2002). Olanzapine is now Eli Lilly's top-selling drug, surpassing even Prozac. There will be no rethinking of the merits of a form of care that is bringing profits to so many. There is no evidence of any budding humility in American psychiatry that might stir the introspection that would be a necessary first step toward reform. At least in the public arena, all we usually hear about are advancements in knowledge and treatment. *Id.* at 291.

> the yawning therapy gap the dethroned Freudians left behind. But the insurance companies and HMOs that gained authority in the 1980s and 1990s didn't waiver. They simply downsized psychiatry, replacing high-paid psychiatrist with low-paid therapists and counselors, and turning psychiatrists they continued to employ into pill-pushers who hardly spoke with patients—to a point where the best medical school graduates no longer wanted to become psychiatrists (p. 4). We shall watch as psychiatry's pendulum swings from the brainless mind of Freud to the mindless brain of biomedicine (p. 12).

Some courts are expanding the psychotherapist-patient privilege to unlicensed mental health workers serving in "employee assistance programs," ranging from substance abuse and mental health to issues concerning financial affairs.[255]

A clinical epidemiologic investigation on informed consent in medical and surgical procedures indicated that this appeared to be a common practice in the late twentieth century. It found "no evidence of documentation of cases where the patients surrender their autonomy to the physician . . . even after particular scrutiny of cases with no consent." Discussions by the investigators of geriatrics throughout the country "documents that these findings are not unique or unusual.[256] However, this did not change the fact that "communication with patients 'verifying that patients can relay their understanding of the information necessary to give informed consent'" should be done by the "attending physician and need not involve a psychiatrist."[257]

---

255. In *Jaffee v. Redmond*, 518 U.S. 1 (1996), the Supreme Court created a privilege for "confidential communications between a licensed psychotherapist and her patients in the course of diagnosis and treatment." (p. 15) The Court extended the privilege to apply to licensed social workers engaged in psychotherapy and did not make explicit the limits of the privilege. In *Olesko v. State Compensation Insurance Fund*, 243 F.3d 1154 (9th Cir. 2001), the court of appeals found that unlicensed counselors who fall within such programs are similarly situated to the licensed professionals. They would both increase access to mental health treatment for employees and often serve as the link between distressed employees and psychotherapy. *Id.* at 1158.

256. *Id.* at 417.

257. A psychiatric consultant is appropriate in cases where decision-making capacity is equivocal, or when there is an underlying mental illness that may affect decision-making capacity. *Id.*

good v. Good

Only psychiatrists have been properly trained to judge the mental competence of patients—particularly those who are about to make decisions concerning their life or death.

v.

Evidence demonstrates that attending physicians are capable of ascertaining the mental competence of their patients in the overwhelming majority of such instances without seeking to obtain a psychiatric evaluation.

Twenty-first Century Approaches

1. Attending physicians, rather than psychologists or psychiatrists, should be the judge of competence to give informed consent to treatments (including those concerning the life or death of their patients) unless such physicians determine that a specific need exists for the aid of other professionals. This determination normally concerns only those patients who demonstrate substantial impairment in their decision-making capacity, or who are unable to participate meaningfully in the communication process with the physician.
2. Counselors who have recently been granted legal access to mental health treatment for employees should be so utilized.

The love of liberty is the love of others; the love of power is the love of ourselves.

William Hazlitt

k) Informed Consent by Minors

Adolescents have traditionally been given little voice in their health care treatment. A comprehensive study of the literature in developmental psychology indicates that there "is little evidence that minors of age 15 and above as a group are any less competent to provide consent than are adults." It concluded that "minors are entitled to have some form of consent or dissent regarding the things that happen to them in the name of assessment, treatment, or other professional activities that have generally been determined unilaterally by adults in the minor's interest."[258] A 1999 report on the data of many developing brains showed a peak at about 12 years of age that continued to grow in the frontal lobes past the age of 20.[259] A number of courts have held that a minor should have the opportunity to explain medical choices to the court, rather than through representation by counsel or by parents.[260] Most state statues provide that consent from a minor for treatment of a venereal or sexually transmitted disease is valid and binding "as if the minor has achieved majority,"[261] although a division remains regarding their ability to maintain it in confidentiality with respect to their parents.[262]

---

258. Grissori & Vierling, *Minors Consent to Treatment: A Developmental Perspective*, 9 PROF. PSYCHOLOGY (August 1978) at 420. *See also* Weithorn & Campbell, *The Competence of Children and Adolescents to Make Informed Treatment Decisions*, 53 CHILD DEVELOPMENT 1589-98 (1982). In 1995, California dropped the age at which a youth can be charged with a murder in an adult court from 16 to 14. Leikin, *Minors Assent or Dissent to Medical Treatment*, 102 J. PED. 172 (1983). *See also* 45 C.F.R. § 46, 408 and L.A. TIMES, Dec. 26, 1996, at B12.

259. Giedd, et al., *Brain Development During Childhood and Adolescence: A Longitudinal MRI Study*, 2 NATURE NEUROSCIENCE No. 10, 861-63 (1999). The following year it was noted: "[The] remodeling of the brain is seen in adolescents of a variety of species. [And] the focus of research is gradually changing, with the recognition that the brain of the adolescent differs markedly from the younger or adult brain, and that some of these differences are found in neural regions implicated in the typical behavioral characteristics of the adolescent." Spear, *Neurobehavioral Changes in Adolescence*, 9 CURRENT DIRECTIONS IN PSYCH. SCIENCE NO. 4 (August 2000). *See also* STAUCH, THE PRIMAL TEEN, at 16 and 101 (2003).

260. *See, e.g., In re* Rena, 705 N.E.2d 1155, 1157-58 (Mass. App. Ct. 1999), which so stated and held that the power of *parents patriae* is strong only when the minor is immature, and thus lacks the capacity to make medical decisions alone. Massachusetts mandates the confidentiality of the minor's medical information unless the minor consents and only allows provider notification to parents when the condition is "so serious that the minor's life or limb is endangered." MASS. GEN. LAWS ANN., Ch. 112, § 12F (2000).

261. 28 AM. J. LAW & MED. 416 (2002).

262. For example, Pennsylvania policy makers relieve physicians of any obligations to inform parents about an adolescent's treatment for substance abuse, and Oklahoma lawmakers prohibit health care professionals from revealing "any information whatsoever." By contrast, Michigan permits disclosure of otherwise confidential information regardless of the minor's express refusal or lack of consent, and California mandates parental involvement, leaving a determination of the appropriateness of that involvement under the circumstances to physicians. *Id.* at 418.

A study of adolescent patients indicated that adolescents regard confidentiality as "the most important characteristic in their decision to access medical care."[263] With respect to affording adolescents legal autonomy it has been stated, "[P]hysicians wield a power that is largely positive and salutary, and this power may result in a loss of patient autonomy if left unchecked or unguarded."[264] The Mature Minor Rule, pertaining in particular to life and death decisions, is discussed in Chapter 13, Section 4.

good v. Good

Minors are not capable of making treatment decisions and should follow the decisions of their parents or surrogates

v.

Mature minors are as capable of making treatment decisions as are mature adults.

Twenty-first Century Approaches

1. The mental competence of normally mature minors aged 15 or above has been recognized in many states and they should be authorized to give informed consent (or refusal) to treatment.
2. Most state legislatures should consider the enactment of a law that would not require parental disclosure without the consent of the mentally mature minor except in the case where there is a possibility of end-of-life involved with the medical treatment of the minor.

---

263. *Id.* at 426. *See also* American Academy of Pediatrics Committee on Bioethics (AAP), *Informed Consent, Parental Permission, and Assent in Pediatric Practice* 95 PEDIATRICS 314-17 (1995); T.O. Erb, S.R. Schulman, & J. Sugarman, *Permission and Assent for Clinical Research in Pediatric Anesthesia*, 94 PEDIATRIC ANESTHESIA 1155-60 (2002); and the Privacy Rule of the Health Insurance Portability and Accountability Act of 1996 (HIPAA) (which gives parents rights to control clinicians' disclosures of their children's health information to third parties. Subject to state laws on clinicians' disclosures to the parents.) (HHS 2002)

264. Daaleman & Larry VandeCreek, *Placing Religion and Spirituality in End of Life Care*, 284 JAMA 2514, 2516 (2000).

Liberty is to the collective body what health is to every individual body. Without health, no pleasure can be tasted by man; without liberty, no happiness can be enjoyed by society.

Thomas Jefferson

Ethics, Bioethics, and Ethics Committees

1. ***Lack of Fundamental Constitutional Rights to Healthcare and Education; Costs***
2. ***Bioethicists: With and Without Religion***
3. ***Ethics as Viewed by Some Physicians; Errors and Interests; Lack of Effective Discipline; and Use of Placebos***
   - **(a) Ethics Viewed by a Survey of Physicians vs. Other Readers of *JAMA***
   - **(b) Evaluations of Physicians (or Bioethicists) Who Work for Manufacturers of Drugs or Medical Devices**
   - **(c) Peer Review Distinguished from Most Bioethical Committee Matters**
   - **(d) Medical Errors, Lack of Reporting, and Medicare Fraud**
   - **(e) Doctoring Business, and Use of "Double Standards" for Admission to Medical Schools**
4. ***The Vagueness of Bioethical Principles and the Need for an Alternative***
5. ***The Teaching of Biomedical Ethics and the Law; Ethical Lapses***
6. ***The Role of Medical Ethics Committees and the Meaning of Care***

(a) **JCAHO's Requirement of a "Mechanism" for Ethical Issues**
(b) **The Lack of Standards of Membership Credentials**
(c) **The Goals of Medicine and of Bioethics Committees: "Clarification" and Other Goals**
(d) **Physician Dominance in Bioethics Committee Decision Making**
(e) **Hospital and Ethics Committee Records**

7. ***Organizational Ethics***
(a) **Health Maintenance Organizations (HMOs); Rationing**
(b) **Fraud in Medicare and Medicaid; Contracting Practices of Healthcare Organizations**

It is not correct to say that every moral obligation is a legal duty; but every legal duty is founded upon a moral obligation.

Lord Chief Justice John Duke Coleridge, L.R. Q.B. 453 (1893), *Regina v. Instan*, 17 Cox Crim. Cases 602 (1893)

Ensure domestic Tranquility, promote the general Welfare, and secure the Blessings of Liberty to ourselves and our Posterity.

Preamble to the Constitution

Congress has not unlimited powers to provide for the general welfare, but only those specifically enumerated.

Thomas Jefferson

1. Lack of Fundamental Constitutional Rights to Healthcare and Education, Costs

The Constitution makes no mention of either public education or healthcare, in spite of the fact that many people believe otherwise. The Supreme Court has so stated.[1] At the dawn of the third millennium, one writer used the term

---

1. Maher v. Roe, 432 U.S. 464, 469 (1977); *United States v. Caroline Products,* 304 U.S. 144 (1938), also distinguished "fundamental" liberties (including freedom of speech and the right to vote) from "nonfundamental" liberties (including property rights and freedom of contract).

"the negative Constitution" to express the absence of a governmental duty to protect the public's health.[2] The Supreme Court set forth one exception: under the Eighth Amendment protection against cruel and unusual punishment, prisoners appear to be the only significant part of the population with a fundamental constitutional right to healthcare.[3]

For others, there are legal obligations of healthcare organizations governed by federal and state legislatures. The 1983 Report of the President's Commission for the Study of Ethical Problems in Medicine and Biomedical and Behavioral Research stated, "*Society has* [only] a *moral obligation* to ensure that everyone has access *to adequate care* without being subject to excessive burdens." (Emphasis added.) The Commission on a Consumer Bill of Rights did not claim that Americans have an inherent duty to buy health insurance if they are sick.[4] The differences in societies were dealt with by Roland Benabou, who stated: "Many countries made the financing of education and health insurance the responsibility of the state; some, notably the United States, have left them in large

---

2. GOSTIN, PUBLIC HEALTH LAW: POWER, DUTY, RESTRAINT (2002). In his book *Health Security for All: Dreams of Universal Health Care in America* (2005), Alan Derickson points out that the United States differs from all the other wealthy nations by not granting access to health care to all of its citizens. This historian actually concluded in the epilogue: "Future attempts to achieve universal health care will have to overcome the defeatism that surrounds this issue. . . . A century of frustrated universalism should . . . not extinguish the dreams of those pursuing one of the essential elements of a human society." *See* Campbell, JAMA 1826 (Oct. 12, 2005).

3.
> Deliberate indifference to the serious medical needs of prisoners constitutes the unnecessary and wanton infliction of pain . . . proscribed by the Eighth Amendment [in its protection against cruel and unusual punishment]. This is true whether the indifference is manifested by prison doctors in their response to the prisoner's needs or by prison guards in intentionally denying or delaying access to medic al care or intentionally interfering with treatment once proscribed.

Estelle v. Gamble, 429 U.S. 97, 104-05 (1976), *quoting* Greg v. Georgia, 428 U.S. 153, 182-83 (1976).

4. ORANGE COUNTY REGISTER, Nov. 20, 1997, at News 6. "In Great Britain . . . the war had paved the way for rapid improvement in the system of public health benefits. The Churchill coalition, in 1944, had proposed a National Health Service based on the twin principles that every citizen in the country had a right to the best medical facilities available and that these services should be free." HITCHCOCK, THE STRUGGLE FOR EUROPE 55 (2002).

part to families, local communities and employers. . . . [Europeans] choose to sacrifice more employment and growth to social insurance than their American counterparts, even though both populations have the same basic preferences."[5] In 2003 it was noted that "[p]redictions beyond 2005 depend on economy, national will, political influences, and global events. Projections are deemed 'stormy.'"[6] In 1998, the bioethics philosopher Daniel Callahan pointed out, "Most strikingly, by the time of the last major national healthcare reform effort, culminating (and collapsing) in 1994, *proponents of universal health care rarely used the language of rights.*"[7] (Emphasis added.)

Congress has established rights to healthcare by deficit spending. For example, Medicare for seniors began to turn up a deficit of $4.2 billion in fiscal year 1996 due to the greatly increased costs of medical treatment.[8] Medicare was established by Congress in 1965 under the Social Security Act in order to

---

5. Benabou, *Unequal Societies: Income Distribution and the Social Contract*, 90 Am. Economic Rev. 96 (2000), at 97 (The "original question of why the social contract differs across countries, and whether these choices are sustainable in the long run, remains an important topic for further research" (p. 119)). *See also* Kawachi & Kennedy, The Health of Nations 4 (2002).

6. JAMA 824 (Aug. 13, 2003), *reviewing* Health and Health Care 2010: The Forecast, the Challenge, 2d ed. (2003), by the Institute for the Future.

7. Callahan, False Hopes 228 (1998). Back in 1798, Supreme Court Justice Samuel Chase, in *Calder v. Bull,* U.S. 388-89, wrote that "[a]n act of the legislature (for I cannot call it a law), contrary to *the great first principles* of the social compact, cannot be considered a rightful exercise of legislative authority." However, by 1905 Justice Oliver Wendell Holmes, Jr., dissenting, in *Lochner v. New York,* 198 U.S. 76, stated that a very restrictive statute proposed "would infringe *fundamental principles* as they have been understood by the traditions of our people and our law," and rulings began to change in the late 1930s after President Franklin Roosevelt attempted to change the Supreme Court's membership.

8. N.Y. Times, Oct. 29, 1996, at A13. Healthcare spending in the United States grew to an estimated $1.7 trillion in 2003—more than $5,800 for every American. Orange County Register, Feb. 12, 2004, at Bus 5. Washington, D.C. and the states split Medicaid costs, with the federal portion at least 50 percent and sometimes more than 70 percent. In 1990 Medicare cost of $110 billion had risen to over $300 billion. The Medicare Board of Trustees predicts it to be bankrupt in 2020, and the Social Security Board of Trustees predicts that Social Security will become so in 2041. Three-quarters of U.S. citizens pay more in Social Security and Medicare taxes than in income taxes, but nobody pays either of them on dividends.

save the elderly (normally 65 years) generally 80 percent of the cost of many aspects of medical care.[9] Medigap policies are provided by private insurers[10] under voluntary certification standards set by the National Association of Insurance Commissioners (NAIC) (made up by state insurance commissioners).[11]

Medical care increasingly leans toward newer technologies. Medical data shows that physicians, "when offered financial incentives to provide too much care to their patients," often do so.[12] Judge Richard Posner noted, "A profession's characteristic modes of thought might have economic causes. Medicine is rich in examples."[13] At the end of the twentieth century, escalating costs (only partly due to technology) have been fueled by a pervasive "entitlement theory of 'rights'" most of which are not regarded as fundamental to our Constitution. As of the turn of the third millennium, healthcare costs exceeded one trillion dollars a year ($4,000 per year for every man, woman and child in the United States). In 2002 the nation's healthcare budget jumped to $1.424 trillion, up from $1.310 trillion in 2000, according to the report by the Centers for Medicare and Medicaid.[14] Most serious analysts appear to agree that priorities similar to those used in Oregon's Medicaid program will have to be set

---

9. Social Security Act of 1965, ch. 7, 79 Stat. 291 (codified as amended at 42 U.S.C. § 1395-1395vv (2000).

10. *Id.* § 1395ss(p).

11. Health Care Financing Administration, 57 Fed. Reg. 37,980-91 (Aug. 21, 1992). Medigap insurers are required to cover "Medicare eligible expenses" *Id.* In 2002, the 9th Circuit Court of Appeals held that the insurer is only obligated to pay at the Medicare rates for such care. Vencor Hosps. v. Blue Cross Blue Shield of R.I., 284 F.3d 1174 (11th Cir. 2002). Under the regulation the Medicare rates are "to the extent recognized as reasonable and medically necessary by Medicare." Health Care Financing Admin., 57 Fed. Reg. 37,980, 37,988 (Aug. 21, 1992).

12. MARC A. RODWIN, MEDICINE, MONEY, AND MORALE: PHYSICIANS' CONFLICTS OF INTERESTS 140 (1993) at 56. Further, data showed that physicians referred patients for more diagnostic tests and laboratory services using durable medical equipment where they had invested in such outside services. *Id.* at 55.

13. POSNER, OVERCOMING LAW 35 (1995), *citing* ACKERKNECHT, A SHORT HISTORY OF MEDICINE 54, 82, 195 (1955).

14. http://member.compuserve.com/news.story, Jan. 17, 2003. The National Health Accounts (NHA) "found that public spending is between 40 and 45 percent of total spending for health care." FOX, J. LAW, MED. & ETHICS (2003), at 612. For states, public spending for health services is at least 60 percent of the total. FOX & FRONSTIN, 19 HEALTH AFFAIRS 271-74 (2000).

early in the twenty-first century to prevent a sizable increase in the national debt and a heavy burden of interest thereon.[15] At this time in history, most politicians in most states appear to abhor even mentioning the "need for rationing"; but the author of Oregon's plan (to extend insurance to thousands of low-income residents by rationing healthcare to recipients) was elected governor of that state. In order to help keep costs down, the Oregon plan rations services by use of a prioritized list of medical conditions and treatments. The state pays for those treatments deemed most important and most likely to be effective.[16]

good v. Good

Healthcare funding, such as "entitlements," is a constitutional right, which should be expanded to fill all needs regardless of the cost.

v.

There is virtually no constitutional right to healthcare without legislation, and there are limits to the share of costs to be borne by the government under congressional legislation rather than by its citizens doing so independently.

Twenty-first Century Approach

Because of the high costs of healthcare in the twenty-first century and the potential dangers inherent in its increases in national expenditures, the government should make it clear that treatments under entitlements may ethically be limited as they become too costly for society to bear, and may require rationing by both the federal and state governments.

In the end it will all come down to a decision of ethics, how we value the natural world in which we have evolved and now—increasingly—how we regard our status as individuals.

E. O. Wilson, *Biodiversity,* p. 16

15. SHENKIN, CURRENT DILEMMAS IN MEDICAL-CARE RATIONING: A PRAGMATIC APPROACH (1996). The author defines the economic, philosophical, and political methods available to accomplish rationing. The word "priorities" replaced the currently "politically incorrect" term "rationing."

16. L.A. TIMES, Jan. 3, 1999, at A 32.

Religions in this country that do not teach what would commonly be considered a belief in the existence of God are Buddhism, Taoism, Ethical Culture, Secular Humanism, and others.

U.S. Supreme Court, *Torcaso v. Watkins* (1961)

My entire argument has contested this radical sundering of nature and value, of blind objective process and subjective purposing or valuing. . . . The sharp modern or secular split between objective nature and subjective values makes intelligible neither science nor the humanities, neither nature nor humanism, neither technology nor democracy.

Gilkey, Theologian, *Nature, Reality, and the Sacred: The Nexus of Science and Religion* (1993), p.111

2. Bioethicists: With and Without Religion

Ethics has been defined as "the branch of philosophy that investigates morality and, in particular, the varieties of thinking by which human conduct is guided and may be appraised."[17] Isaiah Berlin separated science when he equated ethics with politics, stating, "There is no natural science of politics any more than a natural science of ethics."[18] Traditionally, there existed the problem of moral value and distinguishing it from values of other kinds.[19] These rules have been subject to the charge of vagueness by the courts.[20] (This will be discussed in section 4 below.) In law, we are concerned with types of moral responsibility, which sanctions can influence.[21] On rare occasions, a

17. Ethics is especially concerned with the meaning and justification of utterances about the rightness and wrongness of actions, the virtue or vice of the motives that prompt them, the praiseworthiness or blameworthiness of the agents who perform them, and the goodness or badness of the consequences to which they give rise.

The Harper Dictionary of Modern Thought (1977).

18. *See* Berlin, Political Judgment 49.

19. The theory of value in general is called "axiology." A Gallup poll in 2002 showed that 67 percent rated moral values in the U.S. as "getting worse." Orange County Register, Aug. 5, 2004, at Bus. 6.

20. *See* Gentile v. State Bar of Nevada, 501 U.S. 1030, 1048 (1991).

21. *See* Keyes, Life, Death, and the Law: A Sourcebook on Autonomy and Responsibility in Medical Ethics (1995).

law may be subject to the charge of vagueness. If so, that particular law becomes invalid and no sanctions can be applied.[22]

"Bioethics," a relatively new term, is often rendered virtually indistinguishable from morals. However, the term is normally excluded from the morals of any particular ethnic group or religion. Bioethics is more often similar to the law, in that it applies to all citizens. The rules of bioethics are imposed upon people in the medical profession and their patients. The principle of "patient autonomy" allows most religious or moral treatment choices to be made by patients (Chapter 5).

The First Amendment to the Constitution provides that Congress "shall make no law respecting an establishment of religion, or prohibiting the free exercise thereof." Although this is the denial of law prohibiting a religion's free exercise, in the nineteenth century, the Supreme Court banned the Mormon practice of polygamy, stating: "To permit this would be to make the professed doctrines of religious belief superior to the law of the land and in effect to permit every citizen to become a law unto himself."[23] On the other hand, late in the twentieth century the Court actually allowed Amish reli-

22. Bioethics on the Internet is found in the National Library of Medicine (NLM), which offers free Web-based access to its databases. The following Web sites provide free access to NLM databases:

http://www.nlm.nih.gov (MEDLINE search)
http://www.ncbi.nim.nih.gov/PubMed (Access to PubMed)
http://igm.nml.nih.gov (Access to Internet Grateful Med)

23. Reynolds v. United States, 98 U.S.145 (1878). The Court stated:

> Laws are made for the government of actions, and while they cannot interfere with mere religious belief and opinions, they may with practices. Suppose one believed that human sacrifices were a necessary part of religious worship, would it be seriously contended that the civil government under which he lived could not interfere to prevent a sacrifice? Or, if a wife religiously believed it was her duty to burn herself upon the funeral pyre of her dead husband [a former practice in India], would it be beyond the power of the civil government to prevent her carrying her belief into practice? So here, as a law of the organization of society under the exclusive dominion of the United States, it is provided that plural marriages shall not be allowed.

gious practices to ignore a state law pertaining to the attendance of children at public high schools.[24]

In 1892, the original Pledge of Allegiance to the United States was adopted with no reference to God. In 1954, over a half century later, the pledge was amended to add the words "under God" by Congress. Near the turn of the third millennium, in an atheist's lawsuit, the Ninth Circuit Court of Appeals attempted to strike down that version, claiming it violated the First Amendment's separation of church and state.[25] For the reasons set forth in Chapter 2, no scientific basis exists for the denial of a Creator of the Universe that could completely sustain deletion of the reference in the Pledge (as applicable to educated secular and people in the field of medicine).[26] However, the "content analysis" of the form of Hippocratic oaths administered at 147 U.S. and Canadian medical schools showed that only 11 percent of their versions invoked a deity.[27]

Under the Constitution, religion is a matter of choice and has virtually no limits unless and until certain illegal actions are taken in its name. About 80 percent of Americans express a belief in God, although, as the Supreme Court stated (see the epigraph), others do not have "what would commonly be considered a belief in the existence of God." However, when asked whether they believe in a god that actually watches their actions and later judges them, these

---

24. Wisconsin v. Yoder, 406 U.S. 205 (1972). The Court stated, "Only those [governmental] interests of the highest order . . . can overbalance legitimate claims of free exercise."

25. "No regional court has its decisions reversed as often as the 9th Circuit. In 1997, the high court took up 29 cases from the 9th Circuit and reversed 28 of them. Nearly two-thirds of those reversals were unanimous." L.A. TIMES, June 27, 2002, at 1.

26. The scientist is possessed by the sense of universal causation. . . . His religious feeling takes the form of a rapturous amazement at the harmony of natural law, which reveals an intelligence of such superiority that, compared with it, all the systematic thinking and acting of human beings is an utterly insignificant reflection. . . . *I believe in Spinoza's God* who reveals himself in the *harmony of all that exists, but not in a God who concerns himself with the fate and actions of human beings*. (Emphasis added.)

THE QUOTABLE EINSTEIN 147, 151 (1996).

27. Orr, et al., *Use of the Hippocratic Oath: A Review of Twentieth Century Practice and a Content Analysis of Oaths Administered in Medical Schools in the U.S. and Canada in 1993*, 8 J. CLIN. ETHICS 377-88 (1997).

percentage points drop—both among scientists and among people in those "religions" that the Supreme Court enumerated. Because both bioethics and secular law necessarily deal with believers and such nonbelievers, bioethicists sometimes have to attempt a resolution of conflicts between a particular faith and reason. This resolution often seems like an impossible task, because "To tell people whose *faith influences* their *values* that there *is* something *wrong* with those *values, is* to tell them that there is something *wrong* with their faith." (Emphasis added.)[28] Others, however, agree that "[e]thical culture is a creedless religion. The bond of union among its members is a common devotion to the cultivation of moral excellence as the chief duty of humanity."[29] The "Kantian principle of respect for persons is actually bound up in the very idea of morality, either secular or religious. . . . *Morality has* an objective rationale in *complete independence of religion*."[30] It has been written, "The atheist can also draw a moral line around life."[31] The bioethicist Ruth Macklin stated that a "moral system being an ideal of conduct is based on *principles* that *must be accepted by reason*."[32] (Emphasis added.) This is similar to the description of law by St. Thomas Aquinas many centuries ago:

> Since law is a kind of rule and measure, it may be in something in two ways: First, as in that which measures and rules; and since this is proper to reason, it follows that, in this way, *law is in the reason alone*.[33]
>
> *Law is nothing else than* an ordinance of *reason for the common good*, promulgated by him who has the care of the community.[34] (Emphasis added.)

28. CARTER, THE DISSENT OF THE GOVERNED 12 (1998).
29. MUZZEY, ETHICS AS A RELIGION (1951, 1957).
30. NEILSEN, ETHICS WITHOUT GOD 126 (1990):

> For a devout Jew or Christian to give up his God most certainly is important and does take him into the abyss of a spiritual crisis. . . . I have argued in this essay; namely, that if an *erstwhile believer loses his God but can* keep his nerve, *think the matter over*, and thoroughly take it to heart, *life can still be meaningful and* morality can yet *have an objective rationale*. (Emphasis added.)

Id., at 127.

31. APPLEWOOD, BRAVE NEW WORLDS 32 (1998).
32. Macklin, *Moral Progress*, ETHICS 377 (July 1977).
33. SUMMA THEOLOGICA Q. 90, Art. 2.
34. *Id.*, Art. 4.

A common view of bioethics is that morality constitutes an entirely human construct. However, the primatologist Franz de Wall pointed out, "Humans and other animals have been endowed with a capacity for genuine love, sympathy and care." Animals exhibit aspects of their behavior, such as reconciliation, comforting, sharing of food, distress over the pain of others, and disciplining of malefactors, which have provided an evolutionary beginning for morality in humans.[35] The former popular concept of a "house totally divided" between reasoning moralistic humans involving absolutes and unreasoning animals without any morals cannot stand. Aquinas quoted Aristotle: "As the Philosopher says, as man is the *most* noble of animals if he be perfect in virtue, so he is the lowest of all, if he be severed from law and justice."[36]

The current decrease in the influence of certain moral precepts among high school students was demonstrated by a poll conducted in 1998 in which 80 percent admitted having cheated—a 4 percent increase over the previous year and the highest percentage in three decades.[37] Another survey showed that almost half (47 percent) of such students admitted that they stole something from a store. These students identified declining social and moral values as the biggest problem facing their generation at the end of the twentieth century.[38] Many young people consider it permissible to steal merchandise because they feel that they are merely reducing the profits of large corporations. They fail to reflect that theft not only reduces profits, but also increases the price to consumers. They also fail to realize that cheating is less likely to lead to basic grading reforms than to inflict an injustice on other students.[39]

---

35. Franz de Wall, Good Natured, The Origins of Right and Wrong in Humans and Other Animals (1996) (see Chapter 4).

36. Summa Theologica Q. 95: "For man can use his reason to devise means of satisfying his lusts and evil passions, which other animals are unable to do." (Emphasis added.)

37. Orange County Register, Nov. 24, 1998, at News 24.

38. Orange County Register, Oct. 20, 1998, at 1.

39. Bok, Beyond the Ivory Tower 127 (1982): "Poor instruction can harm any class. But it is devastating to a course on ethics, for it confirms the prejudices of students and faculty who suspect that moral reasoning is almost always inconclusive and that classes on moral issues are either a waste of time, even worse, a forum for expounding the private prejudices of the instructor. *Id.* at 133.

To correct this, it has been written: "The first step to reviving an ethos of public administration is to send the philosophers home . . . and to admit that moral education will take place, much as it always has, through examples and even a bit of 'indoctrination' in the virtues of democracy."[40] For example, Model Rule 6.1 of the American Bar Association states: "Every lawyer has a professional responsibility to provide legal services to those unable to pay. A lawyer should aspire to render at least (50) hours of pro bono publico legal services per year."[41]

Some claimed that virtue appeared to be diminishing just prior to the dawn of the third millennium—toward the top.[42] Historians in the twenty-first century may be better able to evaluate such influences.

An ethicist has written, "Morality [meaning a person's principles] is compatible with reason even though it is not required by it. An unprincipled person need not be irrational."[43] Nevertheless, the fields of ethics and law generally describe results from the reason of a community of so-called "principled" persons. It must be asked, if that term embraces ethics, what are its standards? Can there be standards in a very diverse society? How far can the law, which embraces all citizens, both permit them to remain autonomous and still be

---

40. Mark T. Lilla, *Ethos, 'Ethics,' and Public Service*, 63 The Public Interest (Spring 1981), at 17.

41. Rule 6.1: Voluntary Pro Bono Publico Service.

In fulfilling this responsibility, the lawyer should:

b) provide any additional services through:

1) delivery of legal services at no fee or substantially reduced fee to individuals, groups or organizations seeking to secure or protect civil rights, civil liberties or public rights, or charitable, religious, civic, community, governmental and educational organizations in matters in furtherance of their organizational purposes.

42. William Bennett, author of the *Book of Virtues* and former Cabinet secretary, lost $8 million in high-stakes gambling. In 2003 he announced that he was finally ceasing to enter into such activity. Orange County Register, May 6, 2003, at News 7. This might have also been demonstrated by the U.S. President during impeachment proceedings and his fine for perjury in 1999. The federal judge who assessed President Clinton almost $90,000 for contempt of court said that her order is aimed not only at penalizing him for false testimony in the Paula Jones sexual misconduct case, but also to deter others who might consider emulating the president's misconduct.

43. Neilson, Ethics without God 179.

ethical? Finally, can ethics and the law apply to citizens who also recognize that neither ethicists nor legislators may always be right?

This is particularly difficult in the United States because it requires recognition that many politicians will not listen to those who disagree with them until they are forced to do so. This problem is not new. While the Biblical Job, "God's most perfect image," accepted his suffering, he made several attempts to find an explanation for this. It has been written that he only ceased to do so and demanded no further disclosure by reasoning when his God exercised what has been called "raw power."[44] In ancient Greece, about the same time, Heraclitus (552-487 B.C.) renounced his hereditary claim to the kingship of Ephesus (now in Turkey) because like Job, he found that the universe was constantly changing, there was nowhere to escape to, and human nature had no set purpose.[45]

In the early twentieth century, this view was followed by the "cultural relativist" movement, which began among anthropologists. It was regarded as a moral force for tolerance and diversity at Columbia University by Professor Franz Boas and his key students, Ruth Benedict and Margaret Mead. In 1947, the American Anthropological Association asserted, "Man in the twentieth century cannot be circumscribed by the standards of any single culture."[46] Later, however, others spoke out against this view.[47] Although the law concerns many human acts, it makes little attempt to actually direct all such acts. Harvard law professor Lon Fuller observed that "legal morality can be said to be neutral over a wide range of ethical issues. It cannot be neutral in its view of man himself." (See Chapter 4[e].) His approach recognized both autonomy and its limits. Today's philosophers continued the argument of autonomy for "rational persons." The philosopher John Rawls stated that "principles of justice" are those "that free and rational persons, concerned to further their own interests, would accept in

44. MILES, GOD, A BIOGRAPHY 404 (1995).

45. ZELDIN, AN INTIMATE HISTORY OF HUMANITY 233 (1995). This may be analogous to Socrates, who appeared to welcome his fate that would take him out of the world of suffering and change.

46. Washburn, *Cultural Relativism: Human Rights and the AAA*, 89 AM. ANTHROP. (1987).

47. Lane & Rubinstein, *Judging the Other*, Hast. Ctr.. Rep., May-June 1996, at 21, *citing* Geortz, *Anti-Anti-relativism*, 86 Am. Anthrop. 263-78 (1984); RABINOW, HUMANISM AS NIHILISM: THE BRACKETING OF TRUTH AND SERIOUSNESS IN AMERICAN CULTURAL ANTHROPOLOGY, SOCIAL SCIENCE AS MORAL INQUIRY 52-75 (1983).

an initial position of equality as defining the fundamental terms of their association."[48] It is the function of bioethicists to define these "fundamental terms of association" for medical physicians and their patients. The first step toward doing this is to recognize that they will virtually always be evaluating "good against good."[49] An inability to recognize this may disqualify a person from functioning as a bioethicist, the often unsung hero regarding his work on life's tragedy.

good v. Good

Bioethicists should solely follow the "good" religious and moral traditions of the society to which they belong and where they function.

v.

Bioethicists in a diverse society cannot always function as if they were solely bound only by those traditional morals or religious principles adjudged to be "good" by a particular culture or religion.

Twenty-first Century Approach

1. Bioethical reasoning, like much legal reasoning, must continue as rational reasoning, which is not necessarily bound by any particular religion's peculiar set of morals.
2. Bioethicists must also be able to examine and evaluate one "good" against another in a bioethical setting.

The one person whom it is most necessary to reform is yourself.

Ralph Waldo Emerson

If I tell you *that this is the* greatest good *for a human being, to engage every day in arguments about virtue and the other things you have me*

---

48. Rawls, A Theory of Justice 327-28 (1971), n.12 at 11. *See also* David A. J. Richards, *Kantian Ethics and the Harm Principle: A Reply to John Finnis*, 87 Colum. L. Rev. 457, 461 (1987).

49. There is a collision here of what Hegel afterwards called "good with good." It is due not to error, but to some kind of conflict of an unavoidable kind, of loose elements wandering about the earth, of values which cannot be reconciled. What matters is that people should dedicate themselves to these values with all that is in them. If they do that, they are suitable heroes for tragedy.

Berlin, The Roots of Humanticism 13 (1999).

> *talk about,* examining both myself and others, *and if I tell you that the* unexamined life is not worth living *for a human being, you will be even less likely to believe what I am saying. But that is the way it is, gentlemen, as I claim, though it is not easy to convince you of it. (Emphasis added.)*
>
> Socrates, Plato *Apology* 38A

3. Ethics as Viewed by Some Physicians; Errors and Interests; Lack of Effective Discipline; and Use of Placebos

a) *Ethics Viewed by a Survey of Physicians vs. Other Readers (of JAMA)*

In 1998, a survey was conducted by the *Journal of the American Medical Association* (JAMA) in order to assess the extent of agreement on 73 topics between "experts" and readers (the majority of whom were physicians) as most important for publication. It starts us on Socrates' demand from the epigraph. A comparison was made between the differing views of both the readers and the editorial board members, which included the editors of the 10 American Medical Association (AMA) *Archives Journals*, who are considered experts in certain fields. These experts ranked "managed care" (largely economic matters) as #1—well ahead of "ethics" or "virtue," which Socrates (see Epigraph) would have insisted upon. In fact, "ethics" was ranked only #14 by the experts and #31 (over twice as low) by the *Journal*'s readers. Socrates might have noted that physicians were not actually "examining" life—which he stated was a prerequisite for "a life worth living" in the same manner he noted in the epigraph. If so, Socrates may have also remarked that it would not be easy "to convince some of them of it."

Nevertheless, the experts ranked "Death and Dying" #2, whereas the readers of the *Journal* (generally doctors) ranked it #15 (almost eight times as low), possibly because it also involves some non-medical issues. However, the readers ranked "Aging" as #1 and "cancer" as #2—topics that may demand more medical attention.[50] "Scientific integrity" was ranked #34 by the

50. Lundberg et al., *A Comparison of the Opinions of Experts and Readers as to What Topics a General Medical Journal (*JAMA*) Should Address*, JAMA 288-89 (July 1998).

experts and #62 by the readers (almost twice as low, and very near the bottom). "Malpractice liability" was ranked #52 by the experts; yet it ranked a relatively high #11 by the doctor readers (almost five times higher). "Medical legal issues" (issue) was ranked #59 by the experts and a relatively high #28 (more than twice as high) by readers.

Having the same general nature as other humans may account for why most physicians ranked topics such as Health Maintenance Organizations (HMOs), malpractice, and longevity at or near the top of their interests. Ethics and scientific integrity might be expected to rank there also. However, ranking them in the middle or below for other purposes may actually surprise people accustomed to believing that physicians, as a group, are somehow different from groups in other areas. For example, in 2002 the board that disciplines Washington, D.C. physicians for medical misconduct had been rated the worst in the nation in 2001.[51] In California about 80 percent of the complaints filed with the medical board never resulted in a full investigation. Accordingly, in 2003 a new state law requires its medical board to reprioritize cases to protect the public from doctors who may actually injure their patients.[52]

Often [some] have to relearn things that were known by people who retire or move on. . . . It is also recognized that private companies have a disinclination to release information on their own problems.

NOAA/SEC Satellite Working Group, 1999

b) Evaluations of Physicians (or Bioethicists) Who Work for Manufacturers of Drugs or Medical Devices

Patients and physicians in and out of hospitals often carry a special relationship to corporations manufacturing new types of drugs and medical de-

51. In Public Citizens' Health Research Group's ranking of all the states and the District, the District was 51st. Maryland was 43rd, with 1.78 serious actions per 1,000 physicians, and Virginia was 22nd, with 3.56. Wolfe attributed the District's few disciplinary actions to its unwillingness to spend more on enforcement.

http://www.washingtonpost.com/wp-dyn/articles/A38486-2002Sep4.html

52. Orange County Register, Jan. 31, 2003, at News 20.

vices. The latter include valves and stents implanted into hearts, acrylic lenses for eyes, cushions for knees, and spinal cords. There are about $27.8 billion in sales of such medical devices—largely for people older than 55 years of age.[53]

Sales of drugs by their manufacturers are higher, and in 2002 Americans filled 3.34 billion outpatient prescriptions.[54] A study by the Substance Abuse and Mental Health Services Administration found that 19.5 million Americans used illicit drugs in 2002.[55] American consumers have often found that Canadian prices are significantly lower, and the Canadian Patented Medicines Prices Review Board (PMPRB) regulates drug prices; however, in 2002 the PMPRB noted prices in the Canadian pharmaceutical market that were about 1 percent higher than median foreign prices.[56] In the United States, starting in June 2004, Medicare beneficiaries could buy drug discount cards for up to $30 a year, which provide savings of up to 25 percent off the retail price of many drugs. However, as of 2007, people with incomes of $80,000 or more will pay higher Part B premiums for the first

53. L.A. TIMES, Dec. 16, 2001, C.B. "In terms of dollars, the number of people affected, and the political stakes involved, the Medicare prescription-drug bill is the most important health care legislation passed since the enactment of Medicare and Medicaid in 1965." Altman, Ph.D., NEW ENG. J. MED. 9 (Jan. 1, 2004). *See also* Starr, SIMPLE FAIRNESS: ENDING DISCRIMINATION IN HEALTH INSURANCE COVERAGE OF ADDICTION TREATMENT, 111 YALE L.J. 2321 (2002).

54. M.D. CONNEALY, COAST, February 2004, at 180. For example, every year 150,000 people in the United States are newly diagnosed as having epilepsy—a neurological condition characterized by recurrent, unprovoked seizures. In 2004, it was found that "new antiepileptic drugs are well tolerated with few adverse effects, minimal drug interactions, and a broad spectrum of activity." However, a survey of general practitioners revealed that "only 40 percent felt confident in their knowledge of epilepsy and two thirds were unfamiliar with the new antiepileptic drugs." LaRoche & Helmers, *The New Antiepileptic Drugs*, JAMA 605 (Feb. 4, 2004).

55. ORANGE COUNTY REGISTER, Sept. 6, 2003, at News 24.

56. Cohen, *Pushing the Borders: The Moral Dilemma of International Internet Pharmacies*, Hast. Ctr. Rep. (March-April 2004) at 15. The $2 billion in 1980 research and development expenditures by U.S. pharmaceutical companies increased to $33 billion in 2003. Pharmaceutical Research and Manufacturers of America, *PhRMA: 2004 Pharmaceutical Industry Profile* (2004).

time.[57] Spending on prescription drugs is the fastest-growing component of the healthcare budget.[58]

It was recently pointed out that "[r]ising drug costs are estimated at 15 percent to 20 percent annually. The bulk of the more than $14 billion the drug industry spends each year on marketing is directed at physicians. . . . The AMA is an important tool in how drug makers track doctors' prescribing habits, giving pharmaceutical companies valuable insight into which doctors to target for the latest brand-name drugs. . . . The AMA generates more than $20 million in revenue from selling doctors' biographies."[59] It has been noted that the pharmaceutical industry's "physician-targeted marketing budget is $14 billion—more than $30,000 for every practicing physician in the United States." In her book, *The Truth About the Drug Companies: How They Deceive Us, and What to Do About It* (2004) Marcia Angell indicated that the hundreds of millions of dollars spent to develop a new drug "isn't even important, given that return on investment over the past 20 years has been the best for any industry, regardless of how much is spent."[60] In December 2003 the Medicare Modernization Act (MMA) created a federal tax subsidy for money contrib-

57. ORANGE COUNTY REGISTER, Nov. 27, 2003, at News 7. ("Under current law, Medicare's hospital insurance trust fund, which pays for inpatient hospital care, will be exhausted in 2019, seven years earlier than forecast last year.")

58. Heffler et al., *Health spending growth up in 1999, faster growth expected in the future*, 20(2) HEALTH AFF. 192-203 (2002). "Ninety-one percent of U.S. individuals report having seen consumer-oriented drug advertisements, and the pharmaceutical industry spent nearly $2.5 billion on direct-to-consumer advertising (DTCA) in 2000." JAMA 477 (Jan. 22/29, 2003).

59. ORANGE COUNTY REGISTER, June 18, 2001, at News 7. "The pharmaceutical industry is the most research-intensive of U.S. industries that support their research and development with private funds (as distinguished from defense and space contractors)." *The Pharmaceutical Industry—Prices and Progress*, N. ENG. J. MED. 927 (Aug. 26, 2004).

60. Meyer & Seago, book review, JAMA 3107 (June 22/29, 2005). A study demonstrated that drug testing funded by the pharmaceutical industry is four times more likely to favor the sponsor's product than is publicly funded research. Joel Lexchin et al., *Pharmaceutical industry sponsorship and research outcome quality: systematic review*, 326 British Med. J. 1167 (2003); DEWAR, SECOND TREE 353 (2004).

uted to health savings accounts.[61] Corporations and pharmaceutical companies often establish bioethics committees and provide a fee (or salary) for members. This raises a question regarding whether such bioethicists can be "taken seriously if they are on the payroll of the very corporations whose practices they are expected to assess."[62] It has been noted by a clinical psychopharmacologist that those who "are not clinically depressed may be more likely to commit suicide while taking antidepressants. . . . Manufacturers of antidepressants have gone into the business of selling psychiatric illnesses in order to sell psychiatric doses."[63] It remains important that bioethicists distinguish between a consultant "who has been hired for his integrity and one who has been hired because he supports what the company plans to do."[64] "If bioethicists have gained any credibility in the public eye, it rests on the perception that they have no financial interest in the objects of their scrutiny."

A 2003 study reported that "by combining data from articles examining 1,140 studies, we found that industry-sponsored studies were significantly more likely to reach conclusions that were favorable to the sponsor than were non-industry studies."[65]

---

61. Medicare Prescription Drug Improvement and Modernization Act of 2003, Pub. L. No. 108-173, § 1201, 117 Stat. 2066, 2071 (2003); I.R.C. § 223. *Also see* I.R.C. §§ 62(a)(19), 106(d), 223(a)-(b), 2233(e), 3231(e)(11), 3306(b)(18) (West 2005), and Kaplan, *Who's Afraid of Personal Responsibility? Health Savings Accounts and the Future of American Health Care,* 36 McGeorge L. Rev. 535 (2005).

62. Elliot, *Pharma Buys a Conscience*, The American Prospect, Sept. 14-Oct. 6, 2001. It is interesting to note that the U.S. Board of Tea Experts, founded in 1897, was abolished by Congress in 1996. National Geographic.

63. *Id.* at 11. *See* Noah, *Adverse Drug Reactions: Harnessing Experiential Data to Promote Patient Welfare*, 49 Cath. U.L. Rev. 449, 461 (2000): A physician who prescribes a drug to a patient who later has an accident caused by adverse reaction to it owes a duty to third parties to warn the patients of the drug's potential adverse effects. McKenzie v. Hawaii Permanente Medical Group, Inc., No. 23268, 2002 Haw. LEXIS 362 (June 10, 2002).

64. Elliot *supra* n.62: "If bioethicists have gained any credibility in the public eye, it rests on the perception that they have no financial interest in the objects of their scrutiny."

65. Bekelman et al., *Scope and Impact of Financial Conflicts of Interest in Biomedical Research: A Systematic Review,* 289 JAMA 454, 463 (2003). In 2003 the Inspector General of the Department of Health and Human Services released

Pharmaceutical companies are often obliged to test the efficacy and side effects of drugs by means of a double-blind test, where a placebo may be employed. Some volunteers take a new drug while others take a placebo (in Latin, "I will please"), an inactive medication, rather than another existing drug for the illness involved:

> When a new psychiatric drug looks sufficiently promising to be tested in human subjects, it is ordinarily tested against a control drug. If there is no effective treatment for the illness in question, that control is a placebo. Some subjects get the new drug, others get a placebo, and the researchers measure how the two compare. But, if an effective treatment exists for a research subject's illness, the new drug must be tested against it, not against a placebo. The purpose of this rule is to protect subjects from harm. A patient should not get substandard treatment, such as a placebo, simply because he or she has volunteered for a research protocol.[66]

On the other hand, even though the placebo effect against pain has long been dismissed as a psychological rather than physical phenomenon, a recent study demonstrated the contrary. When all participants were told the medicine

---

guidelines for manufacturers to guard against the risk of liability. *OIG compliance program guidance for pharmaceutical manufacturers,* 68(86) Fed. Reg. 23,731-43 (2003).

66. *Id.* at 18. *See also* 21 C.F.R. § 50.24. In this connection, an editor of the *Law Journal Newsletters Bioethic Legal Review* has written, "The blind allegiance to what I call the 'fool's gold standard' lives on. Anyone with even a passing interest in bioethics knows it is unethical to conduct a double-blind placebo controlled trial where standard therapy exists, except under limited circumstances. The exceptions are where: 1) there is no risk of harm if the patient forgoes treatment during the placebo phase, such as in a trial for a drug that seeks to cure hair loss or impotence; 2) the standard therapy carries such severe side effects that patients might choose to avoid it; or 3) the standard therapy is otherwise of questionable efficacy." Alan Milstein, *LSN: Bioethics* 1 (September 2005). He cited Emanuel & Miller, *The Ethics of Placebo-Controlled Clinical Trials—A Middle Ground,* N. Eng. J. Med. 915-19 (Sept. 20, 2001), and Hellman, *Of Mice But Not Men: Problems of the Randomized Clinical Trial,* N. Eng. J. Med. (May 30, 1991).

was coming and the placebo was given, this prompted the brain to release its own natural painkillers.[67]

The AMA *Code of Medical Ethics* states, "[T]here will almost certainly be conditions for which placebo controls cannot be justified."[68] The report on the seemingly successful results are transmitted to the Food and Drug Administration (FDA), which approves it based solely upon the reported evidence. The manufacturer pays for the study and employs physicians who supervise the study—thus relieving the FDA of this financial burden:

> The FDA argues that such research is justified by its scientific merit unless it results in death or permanent disability to subjects. Yet, the FDA is itself deeply compromised by industry money. The industry pays "user fees" to the FDA in order to speed up review of its products—an amount estimated to be more than $300,000 per drug. And while the FDA employs scientific experts to evaluate new drugs, a recent survey found that over half of those experts have a financial conflict of interest because of industry ties. Dr. Richard Horton, editor of Britain's prestigious medical journal *The Lancet*, has called the FDA "the servant of [the drug] industry."[69]

---

67. Orange County Register, Aug. 25, 2005, at News 11, *citing* the Journal of Neuroscience, Aug. 24, 2005. Professor Zubieta at the University of Michigan Health System said that "when someone was told they were receiving a medicine to ease their pain, they then reported feeling less pain. The mind-body connection is quite clear."

68. The current provision 2:075, *AMA Code of Medical Ethics,* issued in 1997 also states:

> Similarly, the use of a placebo control will more easily be justified as the severity and number of negative side effects of standard therapy increase. . . . Additionally, the interim data analysis and monitoring currently in practice will allow researchers to terminate the study because of either positive or negative results, thus protecting patients from remaining on placebo unnecessarily.

69. *See* note 62, Elliot. The Prescription Drug User Fee Act of 1992, 21 U.S.C. §§ 379g, 379h, granting permission to use user fees paid by drug companies to the FDA, was renewed by the Food and Drug Administration Modernization Act of 1997, 21 U.S.C. § 301 (1997). *See* Kulynycn, *Will FDA Relinquish the "Gold Standard" for New Drug Approval? Redefining "Substantial Evi-*

A pharmaceutical company "will make extraordinary efforts to develop a cost-effective critical path to gain an expeditious approval of its candidate compound."[70] The FDA "requires sponsors of drugs, devices, or other biologists seeking to market their products to submit a disclosure statement on financial arrangements." Although the Department of Health and Human Services warned pharmaceutical companies not to offer any financial incentives to doctors to prescribe or recommend particular drugs (or to switch patients from one medicine to another), those standards do not have the force of law.[71]

A study showed that since the tobacco industry reached settlement agreements with all 50 states, "health needs appear to have little effect on the funding of state tobacco-control programs [and] the settlement represents an unrealized opportunity to reduce morbidity and mortality from smoking."[72] In this connection is noted the Supreme Court's (5:4) refusal to uphold the FDA's 1999 rule to restrict the sale and distribution of cigarettes and smoke-

---

dence" in the FDA Modernization Act of 1997, 54 FOOD & DRUG L. J. 127 (1999), and Rebecca Eisenberg's *How Law Directs Biopharmaceutical Research and Development*, 72 FORDHAM L. REV. 477, at 489 (2003). The FDA was the first regulatory federal agency to protect a citizen's health and welfare. *See* Buenker, *Pure Food and Drug Act, in* THE OXFORD COMPANION TO UNITED STATES HISTORY 637-38 (2002) (Boyer ed.).

70. Montaner et al., *Industry-Sponsored Clinical Research: A Double-Edged Sword*, 358 THE LANCET 1893 (Dec. 1, 2001). Nevertheless, it should be noted that "Nobel Prize Winner Paul Berg claims that the ability to conduct research without concern for immediate commercial applications is what lies behind many of the most innovative discoveries, including Berg's own theoretical academic research, which underlies much of the biotechnology revolution, so much heralded by industry." Lemnens, "Restoring Scientific Integrity to the Commercialized Research Scene," J.L., MED. & ETHICS (Winter 2004) at 653 (citing Press & Washburn, *The Kept University*, ATLANTIC MONTHLY (March 2000) at 54).

71. ORANGE COUNTY REGISTER, Oct. 1, 2002, at News 9. "Janet Rehnquist, Inspector General of the Department of Health and Human Services, issued a compliance guide for the drug industry, telling manufacturers they must not offer any financial incentives to doctors, hospitals, insurers, or pharmacists to encourage or reward the prescribing of particular drugs. Such payments have 'a high potential for fraud and abuse.'" ORANGE COUNTY REGISTER, April 28, 2003, at News 7.

72. N. ENG. J. MED. 1080 (Oct. 3, 2002). "The mother's use of alcohol, cigarettes or other centrally acting drugs, legal or illegal, can also contribute to cognitive impairment in the baby." ROSE, THE FUTURE OF THE BRAIN 84 (2005).

less tobacco products to protect children and adolescents. The Court stated, "It is plain that Congress has not given the FDA the authority that it seeks to exercise here."[73] Nevertheless, in 2004 the Centers for Disease Control and Prevention reported a drop from more than 36 percent of high school students who said they were smokers in 1997 to 22 percent in 2003.[74] However, although the world's first treaty devoted entirely to health was designed in 2003 to reduce the estimated 5 million deaths a year caused by smoking, the United States and China (both large tobacco producers) did not sign it.[75] This was the result even though smoking-related diseases kill an estimated 430,700 Americans annually. According to the American Lung Association, smoking currently costs the United States $97.2 billion each year in healthcare costs and lost productivity.[76] A Columbia University survey released on Aug. 18, 2005, showed that 28 percent of middle-school-student respondents reported that drugs are used, kept or sold at their schools, a 47 percent jump since 2002.[77] Even though 11 states allow use of marijuana for medical purposes,[78] the U.S. Supreme Court has held 6-3 that pa-

---

73. In *Food and Drug Administration v. Brown & Williamson Tobacco Corp.*, 529 U.S. 120 (2000), *aff'g* 153 F.3d 155 (4th Cir. 1998), the Supreme Court indicated that: "No matter how "important, conspicuous, and controversial" the issue, and regardless of how likely the public is to hold the Executive Branch politically accountable, . . . an administrative agency's power to regulate in the public interest must always be grounded in a valid grant of authority from Congress." *Brown & Williamson Tobacco Corp.*, 529 U.S. at 161.

74. ORANGE COUNTY REGISTER, June 18, 2004, at News 15. However, it rose 26% in 2003. *Id.*, Sept. 4, 2003, at News 18.

75. ORANGE COUNTY REGISTER, May 22, 2003, at News 20

76. ORANGE COUNTY REGISTER March 9, 2005, at News 19. "Hotels' non-smoking guest rooms have been on the rise, from about 30 percent of inventory in 1988 to 73 percent last year." [2004] ORANGE COUNTY REGISTER, Dec. 6, 2005, at Bus.6

77. ORANGE COUNTY REGISTER, Aug. 19, 2005, at News 9. "Of the teens surveyed, 58 percent said the legality of cigarettes has no effect on their decision to smoke or abstain, and 48 percent said the fact that marijuana is illegal doesn't affect whether they use or don't use the drug."

78. *E.g.*, California's Compassionate Use Act of 1996, which states that physicians shall not be "punished, or denied any right of privilege, for having recommended marijuana to a patient for medical purposes." CAL. HEALTH & SAFETY CODE ANN. § 11362.5(d) (West 2005), and *see* the Comprehensive Drug Abuse Prevention and Control Act of 1970, 21 U.S.C. § 801.

tients could be federally prosecuted for possessing marijuana prescribed by a physician in accordance with state law.[79] The majority opinion, written by Justice Stevens, concluded that Congress had a "rational basis" for believing that federal anti-drug legislation was a "necessary and proper" exercise to regulate interstate commerce. However the dissent on that point was strong.[80]

In 2000, a total of 528 studies provided data on the controversy that exists over the fact that many physicians have regular contact with the pharmaceutical industry and its sales representatives, who spend a large sum of money each year promoting to them by way of gifts, free meals, travel subsidies, sponsored teachings, and symposia. It was concluded that this affected prescribing and professional behavior and should be further addressed at the level of policy and education.[81] The following year the Association of American Medical Colleges (AAMC) recommended that investigators in clinical research at academic medical centers report their financial interests in such companies.[82] The flowing of drug money to academic doctors has been described as follows:

---

79. Gonzalez v. Raich, 125 S. Ct. 2195 (2005). The court stated, "The case is made difficult by respondents' strong arguments that they will suffer irreparable harm because . . . marijuana does have valid therapeutic purposes. The question before us, however, is not whether [the policy] is wise, [but] whether Congress has the power to regulate . . . medicinal substances . . produced and consumed locally."

80. Justice O'Connor, joined by Chief Justice Rehnquist and Justice Thomas, dissented, pointing to Article I, Sections 8.3 and 8.13 of the U.S. Constitution, and adding, "Even if intrastate cultivation and possession of marijuana for one's own medicinal use can properly be characterized as economic, and I question whether it can, it has not been shown that such activity substantially affects interstate commerce. Similarly, it is neither self-evident nor demonstrated that regulating such activity is necessary to the interstate drug control scheme." *Id.* However, Justices Stevens, Souter, Ginsburg, and Breyer voted to uphold this law, and had dissented in all of the recent decisions limiting the scope of Congress's commerce power.

81. Wazana, M.D., *Physicians and the Pharmaceutical Industry; Is a Gift ever Just a Gift?* JAMA 373, Jan. 19, 2000.

82. AAMC, Task Force on Financial Conflicts of Interest in Clinical Research (December 2001). Its 2002 report stated that:

> An institution may have a conflict of interest in human subject research whenever the financial interests of the institution, or of an institutional official acting within his or her authority, might af-

> The ties between clinical researchers and industry include not only grant support, but also a host of other financial arrangements. Researchers also serve as consultants to companies whose products they are studying, join advisory boards and speakers' bureaus, enter into patient and royalty arrangements, agree to be listed authors of articles ghostwritten by interested companies, promote drugs and devices and company-sponsored symposiums, and allow themselves to be plied with expensive gifts and trips to luxurious settings. Many also have equity interest in the companies.[83]

The director of the National Institutes of Health (NIH) issued a conflict-of-interest policy effective August 25, 2005, which banned (a) its scientists from accepting cousulting fees from the biomedical industry or (b) its almost 7,000 senior employees from owning stock in many types of medical companies. Dissenting scientists stated that the proposed policy had "profoundly detrimental

---

> fect—or reasonably appear to affect—institutional processes for the conduct, review, or oversight of human subject research.

Task Force on Financial Conflicts of Interest in Clinical Research, *Protecting Subjects, Preserving Trust, Promoting Progress II: AAMC* (2002). Nevertheless, it should be noted that national research ethics policy generally does not bind private research ethics boards. McDonald et al., The Governance of Health Research Involving Human Subjects (May 2000).

83. Angell, *Is Academic Medicine for Sale?*, 342 N. Eng. J. Med. 1516-18 (May 18, 2000). Marcia Angell, former editor-in-chief of the *New England Journal of Medicine,* also wrote *The Truth about the Drug Companies: How They Deceive Us and What to Do about It* (2004), stating:

> Drug companies now design clinical trials to be carried out by researchers who are little more than hired hands . . . Sponsoring companies keep the data and in multicenter trials, they may not even let the researchers see all of it. They also analyze and interpret the results, and decide what, if anything, should be published. . . . All of this makes a mockery of the traditional role of researchers as independent and impartial scientists. Academic institutions and their faculties vary in how much control they are willing to cede to sponsors, but in general, they cede much more than they should. Contract research organizations and their networks of private doctors cede most of it. (pp. 102-103)

See Hast. Ctr. Rep. (September-October 2004), *Review,* p. 44.

financial impacts" upon them.[84] Under the rules, the top 200 executives will be required to keep the value of their holdings in any single drug company below $15,000.[85]

In 2001, the International Committee of Medical Journal Editors (ICMJE) issued revised guidelines for investigators' participation in the study design, access to data, and control over publication. However, a 2002 study found that academic institutions routinely engage in industry-sponsored research that fails to adhere to the ICMJE guidelines. The 108 medical schools surveyed generally found themselves powerless when the pharmaceutical company–sponsored research found that a drug did not work or was dangerous, then suppressed the result.[86]

---

84. L.A. Times, Feb. 24, 2005, at A14. *See also* Zerhouni, *Policy Proposal for Management of Conflict of Interest*, Sept. 24, 2004, at http://www.nih.gov/about/09240coi_policymemo.htm. In 2004 the Office of Government Ethics had issued a report recommending that the NIH prohibit its staff from consulting with pharmaceutical companies. Kaiser, *Conflict of Interest: Report Suggests NIH Weigh Consulting Ban*, 305 Science 1090 (2004). On biomedical research, $27 billion per year has been sponsored by the NIH, and some $50 billion per year was spent by the private sector on biomedical research and development. Resnik, *The Distribution of Biomedical Research Resources and International Justice*, 4 Developing World Bioethics 42-57 (2004).

85. The National Institutes of Health (NIH) found that 44 of its 1,200 senior scientists appeared to have violated rules governing consulting and that nine might have violated criminal laws. One senior researcher at the National Institute of Mental Health received more than $500,000 in consulting fees from Pfizer while collaborating on his work studying patients with Alzheimer's disease. Orange County Register, Aug. 26, 2005, at News 10; and 307 Science, March 2005, at 1390.

86. N. Eng. J. Med. 1335, Oct. 24, 2002; Orange County Register, Oct. 24, 2002, at News 15. Typically, the companies hire medical-school faculty members to carry out studies.

> What Dr. Gerald Weinstein, the UCI dermatologist, didn't tell them—wasn't required to tell them—was that he owned more the $100,000 worth of stock in Merck, the company developing the drug. He would stand to make money if the FDA approved it. And it did. Finasteride eventually become Propecia—the first and only prescription pill for male pattern baldness. Of the five University of California medical centers, all but UCI require clinical investigators to warn patients about potential conflicts of interest.

Orange County Register, Sept. 12, 1999, p. 1.

This was noted at a time when University of California–Irvine researchers were trying to maximize its partnerships with private industry.[87] It has been written that patients are receiving prescriptions for higher doses of drugs than they needed, and that lower doses would reduce adverse effects. The industry was accused of causing this effect. As a result, it was recommended that an independent national medical safety board be established.[88] The agency (FDA) can only ask—not order—companies to do further safety studies on drugs already approved.[89]

It has been contended that "a therapeutic orientation to clinical trials obscures the ethically significant differences between clinical research and medical care. As a result, it interferes with informed consent and with the development of a concept of professional integrity that is appropriate to clinical research. . . ." It was emphasized that "their ethical significance has not been sufficiently appreciated," along with how clinical trials actually continue to be conceived from a therapeutic perspective oriented around the physician-

87. ORANGE COUNTY REGISTER, Dec. 21, 1998, at News 8. Most new drugs lack innovation, ORANGE COUNTY REGISTER, May 29, 2002, at News 10, and 10 percent to 20 percent are withdrawn after approval. JAMA 2273, May 1, 2002. It has been pointed out "that virtually every pathway of drug metabolism will eventually be found to have genetic variation." Weinshilboum, *Inheritance and Drug Response*, N. ENG. J. MED. 529 (Feb. 6, 2003). *See* Chapter 15, "Bioethics and Genetics."

88. COHEN, OVER DOSE: THE CASE AGAINST THE DRUG COMPANIES (2001). In his 2003 volume *Science in the Private Interest: Has the Lure of Profits Corrupted Biomedical Research?,* Sheldon Krimsky pointed out how "universities have internalized the ethos of academic capitalism." (p.195) "The passion to demonstrate beyond the skepticism of peers that a hypothesis is confirmed is what we expect of our scientists. The passion to make fortunes from the commercialization of knowledge does not in itself advance science, and it is likely to contaminate the pure drive to create knowledge." (p.142)

89. ORANGE COUNTY REGISTER, Oct. 2, 2004, at News 22. The FDA spokeswoman stated: "Since 2002, Vioxx had a warning about increased cardiac risks based on results of Merck's own post-approval study, but the company disputed its own findings and the drug remained on the market. Merck undertook the latest study because less-rigorous experiments indicated Vioxx could prevent recurrence of potentially cancerous colon polyps." *Id.*

patient relationship.[90] "By contrast, in clinical trials, the principal interests of the investigator and the participating patient may diverge."[91] Others have noted that patients may be used as "guinea pigs."[92]

Nevertheless, it remains important to require publicly supported students or athletes to undergo drug testing. Thus the Supreme Court upheld such a drug-testing policy, stating:

> The most significant element in this case is . . . that the Policy was undertaken in furtherance of a public school system, as guardian and tutor for children entrusted to its care. . . . When the government acts as a guardian and tutor, the relevant question is whether the search is one that a reasonable guardian and tutor might undertake.[93]

---

90. Miller & Rosenstein, *The Therapeutic Orientation to Clinical Trials*, N. ENG. J. MED. 1383, April 3, 2003. It was pointed out how the tendency of patient-subjects to confuse their participation in clinical trials with personalized medical care has been called "the therapeutic misconception." Applebaum, et al., *False hopes and best data: consent to research and the therapeutic misconception*, 17(2) Hast. Ctr. Rep. 20-24 (1987).

91. Dickert and Grady suggest that "paying patient-subjects might help dispel the therapeutic misconception." 341 N. ENG. J. MED. 198-203 (1999).

92. Grunberg & Cefalu, *The Integral Role of Clinical Research in Clinical Care*, N. ENG. J. MED. 1386, April 3, 2003. In this connection it is noted that the National Childhood Vaccine Injury Act of 1986, 42 U.S.C. § 300aa-1 (2004), protects the manufacturer by barring victims of a "vaccine-related injury or death" from seeking recovery by a normal lawsuit without first filing their claims in the specialized Vaccine Court. However, in 2004 it was held that where a child was injured by a vaccine, its parents can sue in federal court for the injuries they suffered as a result of the child's medical condition because, in that particular case, a mercury-containing preservative used in the manufacturer's vaccine was not a part of it as defined in the Act. Moss v. Merck & Co., 381 F.3d 501 (5th Cir. 2004). This was also the first appellate decision allowing a child's parents to sue such manufacturers for loss of consortium. *See* Young, *Vaccine Claim Sidesteps 1986 Tort Cap*, 25 NAT'L L. J., Aug. 30, 2004, at 20.

93. Veronica School v. Acton, 515 U.S. 646 (1995). In 2002, a survey by the Harvard School of Public Health found that 44 percent of college students admitted "binge drinking," which was defined as four or more drinks for women or five or more for men. ORANGE COUNTY REGISTER, March 25, 2002, at News 10. The U.S. Air Force Academy stepped up drug testing and put more classroom emphasis on ethics amid its biggest drug scandal. ORANGE COUNTY REGISTER, March 22, 2002, at News 12.

good v. Good

Upon receiving their education at medical schools, students also are beginning to learn about bioethics, which has become a portion of medical ethics courses. While listening to bioethicists thereafter, they often find that their conclusions differ according to the speaker, which may diminish the importance of bioethics for some of them who speak on behalf of a manufacturer of drugs or of medical devices.

v.

Many physicians have learned insufficient information on "Bioethics" to stop them from refusing to separate its meaning from traditional "medical ethics." They may become medical speakers who receive a high income from, and often testify for, a manufacturer of drugs or medical devices and particular experts who primarily rely upon placebo tests given during a limited time period.

Twenty-first Century Approaches

1. Many medical schools that have begun to teach bioethics should be examined in order to ascertain whether or not this teaching largely remains a part of traditional medical ethics—that is, how to avoid a "malpractice" charge for a practice that differs from what they learned to do, rather than "ultramedical" portions of bioethics.
2. The position and integrity of physicians who have been hired by corporations that are manufacturing drugs or medical devices should be determined by independent investigators to correctly ascertain the objectivity of their recommendations with respect to the products of their employers.
3. The recommendations by experts may be excluded where they are involved with the approval of drugs with regard to the basis for the limited term of a placebo test supplied by the manufacturer of the drugs.
4. For certain drugs, reasonable testing in high schools and colleges should be adopted as a policy with regard to publicly supported student athletes.

c) *Peer Review Distinguished from Most Bioethical Committee Matters*

"Peer Review Committees" differ from "Ethical Committees." As noted in Chapter 1, most medical malpractice cases against physicians and their healthcare organizations are not directly connected to bioethics, although there are certain related actions. The federal Peer Review Improvement Act, an amendment to the Social Security Act, states that the peer review organizations (PRO) "shall inform the individual . . . of the organization's final disposition of the complaint."[94] An appellate court has held that the PRO must not only inform the complainant that the review has taken place, but also disclose results of deliberations, even if the healthcare provider objects.[95]

Because such termination actions are taken by their organization, most states enacted peer review committee statutes to protect these organizations from civil liability.[96] They provide an immunity to peer review committee members and generally to those who testify before that committee (unless testimony is known to be false). A discovery privilege extends to terminated physicians, as well as protection regarding the direct use of that committee's evidence by a plaintiff in a malpractice case.[97] The California Medical Board runs a program for addicted physicians. Similarly, in 2002 the California State Bar commenced its attorney diversion program, committing to meetings several times a week and random drug and alcohol testing.[98]

---

94. 42 U.S.C. § 1320c-3(a)(14).

95. Public Citizen Inc. v. Dep't of Health and Human Services, 357 U.S. App. D.C. 1 (2003). The Department of Health and Human Services' (HHS) *Peer Review Organization Manual* had prohibited Peer Review Organizations from disclosing the results of their investigations if this identified a hospital that did not consent.

96. Browning v. Burt, 613 N.E.2d 993, 1007 (Ohio 1993), *cert. denied,* 114 S. Ct. 1054 (1994); Busk & Williger, Fed. Law., January 2002, at 32.

97. "Peer review statutes confer upon peer review organizations the qualities of a black hole; what goes in does not come out, and, unless the information exists in duplicate in the surrounding orbit, nothing that went in is discoverable." Doe v. Unum Life Insur. Co. of America, 891 F. Supp. 607, 610 (N.D. Ga. 1995). *See also* Ohio Rev. Code Ann. § 2305.251.

98. Cal. Bar J., July 2002, at 1.

Under federal law, drug and alcohol addictions are covered by the Americans with Disabilities Act (ADA) of 1990, restricting disability-based discrimination in employment.[99] They are "impairments" in the ADA regulations of thc Equal Employment Opportunity Commission (EEOC).[100] This regulation also prevents employers from discriminating in hiring and in hospitals and other "places of public accommodation."[101]

Congress enacted the Healthcare Quality Improvement Act (HCQIA), requiring that peer review committee findings be reported to a national clearinghouse.[102] A court has pointed to its objective: "The Objective of HCQIA was to facilitate the frank exchange of information among professionals conducting peer review inquiries without the fear of reprisals in civil lawsuits . . . [while] balanc[ing] the chilling effect of litigation on peer review with concerns for protecting physicians improperly subjected to disciplinary action."[103] This goes from an organization's "firing" to "hiring" standards. A hospital hiring a physician must first access the clearinghouse to protect patients from incompetent doctors.[104] It also generally continues a state's immunization from

99. 42 U.S.C. §§ 12101-12213 (1994). Hoffman v. MCI Worldcom Communications, Inc.,178 F. Supp. 2d 152, 155.

100. 28 C.F.R. § 41.31(b)(2001). Although it does not protect *current* users of illegal drugs, it covers them when they are already in recovery. *See also* The Contract with America Advancement Act (DA&A) of 1996 (Pub. L. 104-121), which was enacted to prevent Social Security Administration (SSA) disability claimants who were disabled due to drug and/or alcohol abuse from becoming eligible for benefits. The SSA regulations require claimants to comply with prescribed medical treatment. 20 C.F.R. §§ 404.1530 and 416.930. In DA&A cases, the claimant has the burden to prove that drug addiction or alcoholism is *not material* to the disability. Brown v. Apfel, 192 F.2d 492 (5th Cir. 1999).

101. U.S. Equal Employment Opportunity Commission, Interim Enforcement Guidance on the Application of the Americans with Disabilities Act to Disability-Based Distinctions in Employer-Provided Health Insurance, EEOC Notice 915.002, at 2-3 (June 8, 1993). *See also Addiction Treatment*, YALE L. J. (2002) at 2321.

102. 42 U.S.C. § 11101 *et seq*. (1986).

103. Bryan v. James E. Holmes Regional Medical Center, 33 F.3d 1318, 1322 (11th Cir. 1994), *cert. denied*, 115 S. Ct. 1363 (1995).

104. Congress did not intend to "supplant or modify existing state procedures." Caine v. Hardy, 715 F. Supp. 166, 170 (S.D. Miss. 1989), *rev'd on other grounds*, 905 F.2d 8858 (5th Cir. 1990), *cert. denied*, 112 S. Ct. 1474 (1992). *See also* 42 U.S.C. § 11101.

damages of a hospital's peer review committee whose members are on "professional review bodies" that are undertaking "professional review actions."[105]

They must meet certain requirements based on an objective standard. In 2004, it was noted that "[a]t the turn of the twentieth century, most physicians in the United States were general practitioners; by the turn of the twenty-first century, most were subspecialists."[106] In this connection, most practitioners became certified as members of the American Board of Medical Specialties (ABMS). The phrase "board certification" is a term with an established meaning of a high level of specialized skill and proficiency.[107] If the peer review committee reports this information to the Board of Medical Examiners, the healthcare organization is provided with immunity from civil liability for damages. Should the peer committee fail to report required information, it will lose its immunity for three years. It may also cause patients of the terminated doc-

105. *Id.* 42 U.S.C. § 11151 (11) defines a "professional review body" and § 11151(4)(a)(i) defines a "healthcare entity." Goldsmith v. Harding Hosp., 762 F. Supp. 187, 188-90 (S.D. Ohio 1991).

> A professional review action must be taken (1) in the reasonable belief that the action was in the furtherance of quality healthcare; (2) after a reasonable effort to obtain the facts of the matter; (3) after adequate notice and hearing procedures are afforded to the physician involved or after such other procedures as are fair to the physician under the circumstances; and (4) in the reasonable belief that the action was warranted by the facts known after such reasonable effort to obtain facts and after meeting the requirements of paragraph (3).

42 U.S.C. § 11112(a).

106. M.D. Fincher, *The Road Less Traveled*, N. ENG. J. MED. 630 (Aug. 12, 2004). Since the turn of the 21st century, the American Board of Medical Specialties (ABMS) consists of 24 boards, each of which requires between 3 and 6 years of training in an accredited training program and a passing score on a rigorous cognitive examination. JAMA 1040, Sept. 1, 2004.

107. The 9th U.S. Circuit Court of Appeals upheld a California Medical Board restriction on certain doctors' use of the phrase "board certified" in their advertisements; namely, California Bus. & Prof. Code § 651, which prohibits use of the "board certified" phrase unless the certifying organization is a member of the ABMS and has requirements equivalent to those of the ABMS. Am. Acad. of Pain Mgmt. v. Joseph, 353 F.3d 1099, 1102 (9th Cir., 2004). Although the state medical board refused the academy's request for recognition to use the term "board certified" in advertisements by two doctors on their letterheads, the court held it had not violated their free speech right in the First Amendment.

tor to bring malpractice claims. Clients can sue lawyers if they are not aggressive enough or for continuing to maintain the litigation. However, "a lawyer has no liability for continuing an action where probable cause existed at the time of filing."[108] The *AMA Code of Medical Ethics* (2003 Edition), Rule 9.10 (updated in 1994) states, "Peer review committees act ethically so long as principles of due process (Opinion 9.05, 'Due Process') are observed." Rule 9.05 refers to "a committee that will pass judgment on a peer." It appears that medical errors are not always presented to peer review committees, although when presented they are considered secretly by these committees. For example, in California it was reported:

> By law, all peer-review discussions are secret. Lawyers can't subpoena records of the meetings, and hospitals only have to report their findings to the Medical Board if a doctor's privileges are affected.
>
> Hospitals that do not tell the board about bad doctors can be fined $100,000—up from $10,000 under state law that went into effect in January 2002. The fine increase, meant to encourage hospitals to follow the rules, does not deter them from bending the rules, physicians and Medical Board officials said.[109]

At the dawn of the third millennium, the former general apathy of many physicians toward articles and lectures on bioethics (as distinguished from medical ethics) changed, and they became more prevalent. In order to emphasize the reasons for this, I should like to recite again the 12th verse of a 1997 poem by bioethicist Ruth Macklin from the III World Congress. It is titled, "The Twelve Methods of Bioethics," which is intended to be sung to the tune of "The Twelve Days of Christmas":

> In the twelfth wave of ethics, the method sent to me: 12 communitarians, 11 libertarians, 10 moral rules, 9 stories, 8 ethnographers, 7

108. Swat-Fame v. Goldstein, 101 Cal. App. 4th 613, 627 (2002). Other such matters can be found in Mallen's LEGAL MALPRACTICE (2000), Fifth Edition (West Group).
109. "'You can get around fines for not reporting: just don't do the peer review,' said Dr. Ron Wender, president of the Medical Board's Division of Medical Quality." ORANGE COUNTY REGISTER, Sept. 22, 2002, at News 4.

sort of feminists, 6 types of virtue, 5 casuistry, 4 just rules, beneficence, nonmalfeasance, and respect for auto-ho-no-my.

Some physicians have told me that when they listened to bioethicists, they generally received the impression that there were many opinions among them that depended upon the speaker. However, Socrates' characterization of "virtue" recognized no authority beyond that of an examined life of reason—beyond those alleged to have been presented by the gods of his day. Similarly, St. Thomas Aquinas vehemently defended reasoning, like the "pagan" Aristotle, against his critics.[110] Because the approach of reason with enforceable rules is much closer to that of modern law, it would appear that physicians should be likely to show more interest in those ethical discussions that emphasize the law over the several fields related in the "Twelfth Wave of Ethics."

Power tends to corrupt and absolute power corrupts absolutely.

Lord Acton

d) Medical Errors, Lack of Reporting, PSQIA and (Medicare Fraud)

Problems arise between some medical centers and physicians about peer reviews. An appellate court discussed a case that arose out of a series of misunderstandings and missteps on both sides. Robert J. French, a medical doctor employed by a health maintenance organization, was awarded nearly $7.7 million in compensatory and punitive damages for wrongful termination even though his own evidence showed he was in an "impaired physical and mental state" and "never should have been seeing patients." The plaintiff's award was virtually ordained by the trial court in a directed verdict that told the jury that French's employer intentionally violated California public policy by fail-

110. St. Thomas Aquinas, who taught in the first great Catholic university of medieval Paris, encountered opposition for his interest in the pagan philosopher Aristotle. Aquinas tells Dante in no uncertain terms that a good Christian life requires self-examination through reason, including the philosophical use of reason (Paradiso XIII.115-23). Virtue is derived from the Latin *virtus,* meaning *manliness* or *power*. St. Thomas Aquinas defined it as "a good quality of the mind, by which we live rightly and which no one can use for evil."

ing to provide peer review before termination. French chose to sue for damages rather than avail himself of his administrative and judicial remedies to compel peer review. The appellate court stated, "We reverse." Peer review is mandated to protect doctors and patients. It makes no sense that French should be awarded damages for losing a job he concedes he was not fit to hold. Assuming that occurred here, failure to provide peer review does not sustain a public policy wrongful termination tort.[111]

In 1997, it was noted that "almost (95 percent) medical errors are not reported, and most doctors and administrators have no idea of how serious these errors are."[112]

> In 1999, the National Academy of Sciences reported that medical errors kill 44,000 to 98,000 people a year (exceeding death from highway accidents, breast cancer or AIDS); after this report was released some efforts were made to reduce them.[113] In 2002 a study showed that: "Though substantial proportions of the public and practicing physicians report that they have had personal experience with medical errors, neither group has the sense of urgency expressed by many national organizations."[114] In 2003, the U.S. government's Agency for Healthcare Research and Quality found that medical errors lead to more than $9 billion in extra costs annually.[115]

The problem also spills over into "missed" diagnoses. A study showed that the "error rate in reading radiology images is consistently high (the average error rate is 30 percent), as are lung cancer lesions (26 percent-90 percent) missed at major academic medical centers. . . . They are reported to patients only 25 per-

111. Robert French v. Cigna Healthplans of California, GO14698 (1997), *rev. denied*, 1997 Cal. LEXIS 3927.

112. ETHICAL CURRENTS 1 (St. Joseph Health System), Fall 1997.

113. TO ERR IS HUMAN: BUILDING A SAFER HEALTH SYSTEM, Kohn, Corrigan & Donaldson eds. (Nat'l Academy Press, 1999).

114. N. ENG. J. MED. 1933 (Dec. 12, 2002).

115. "The most serious complication was post-surgery sepsis bloodstream infections, which occurred in 2,592 patients. Sepsis resulted in 11 extra days of hospitalization and $57,727 in extra costs per patient." ORANGE COUNTY REGISTER, Oct. 8, 2003, at News 13.

cent of the time."[116] The report failed to locate specific condemnations of missed diagnoses in the medical ethics codes but concluded that "failure to disclose errors and mistakes constitutes *unethical conduct*." In 2000 *four* different studies were cited showing that "[t]he quality of healthcare is, by many accounts, a serious problem at the turn of the millennium."[117] The following examples concern ethical questions arising out of certain nondisclosures:

> (i) Ben, a 12-year old boy, was brought into the emergency department with severe abdominal pain. A physician examined Ben, determined that he had classic signs of appendicitis and recommended immediate surgery. Ben's parents consented to the surgery and Ben was operated on. During surgery, the surgeon discovered that his appendix appeared normal. Following the surgery, the surgeon informed Ben and his parents that Ben's appendix had been removed and assured them that the surgery had been successful.[118]
>
> (ii) While recovering from minor surgery, 79-year-old Helen Lilly suddenly complained of feeling weak and dizzy as she walked around her hospital room. The nursing staff, suspecting an adverse reaction to a cardiac medication erroneously given earlier, returned Mrs. Lilly to bed, placed her on a cardiac monitor, and watched her blood pressure closely. When the physician was notified about the error, he concluded that telling Mrs. Lilly about the mistake would only cre-

---

116. Berlin, *Reporting the "missed" radiologic diagnoses: medicolegal and ethical consideration*, 183 RADIOLOGY (1994); HEALTHCARE ETHICS LITERATURE REV. No.3, 1994. "The studies found that 'adverse events' occurred in 2.9 percent to 3.7 percent of hospitalizations, with 8.8 percent to 13.6 percent of those mistakes leading to death." ORANGE COUNTY REGISTER, Dec. 10, 1999, at News 10.

117. Research has demonstrated that physicians overuse healthcare services by ordering unnecessary interventions that are costly and place patients at risk; underuse services by failing to provide standard care that would produce favorable outcomes; and misuse services by devising the wrong treatment plan or improperly executing the correct plan.

Gostin, JAMA 1742 (April 5, 2000).

118. "A new study showed the dangers of misdiagnosing appendicitis. The researchers examined data involving 52,969 appendix removals in 1997. About 15 percent of the surgeries were on normal appendixes." ORANGE COUNTY REGISTER, July 15, 2002, at News 7; ARCHIVES OF SURGERY, July 2003.

ate needless worry. By the following morning, the side effects disappeared, and Mrs. Lilly was discharged.[119]

In both cases, the statement of Lantos is applicable: "Lack of disclosure leads to suspicion. If doctors don't tell patients about errors, they appear to have something to hide." He added:

> A breast cancer patient in ICU following a bone marrow transplant developed sepsis (toxin in the blood) and had a cardiorespiratory arrest. CPR was unsuccessful and the patient died. A few days later, during a routine postmortem review, it was discovered that the intravenous fluids provided during resuscitation contained 10 times the normal dose of potassium chloride. This increased dose could have killed a healthy person, and contributed to this woman's death. The doctor asked the ethicists whether he should tell the family about the mistake.[120] The hospital's ethics committee actually "ended up a hung jury," and told the attending physician that either option was acceptable. He didn't tell the family.[121]

Is disclosure not a precondition of trust? Many people regard ethics committees as "defenders of physicians" and hospitals—particularly because most of these committees are dominated by its physicians. Once it was even suggested that such committees be abolished.[122]

In California it has been pointed out that the state medical board that oversees doctors "obviously isn't doing a good job."[123] Dr. Trogen Brennon of the

---

119. Gostin, JAMA, April 5, 2000.

120. Lantos, Do We Still Need Doctors? 129 (1997).

121. *Id.*, Lantos, at 122. There are no federal laws requiring hospitals, pharmacies and clinics to report medication mistakes, even when the patient dies. Only a few states have a mandatory reporting system. Orange County Register, July 21, 1996, at News 20.

122. *See* Katz, *in* The Silent World of Doctor and Patient (1981).

123. "The board rarely investigates complaints, responds slowly and gives lenient penalties . . . the physician-dominated board has disciplined fewer doctors while logging more complaints over the past five years. We see where licensing acts more as a protective shield around like-minded colleagues than an advocate for the public it is supposed to serve." Orange County Register, April 9, 2000, at Local 6.

Brigham and Women's Hospital set forth the principal reason for this medical secrecy and cover-ups of iatrogenic injuries. "No matter how much we might insist that physicians have an ethical duty to report injuries resulting from medical care or to work on their prevention, fear of malpractice litigation drags us back to the status quo." Does this provide justification for such conduct in view of the current standards concerning physicians?[124] In 2002 there was found "a fundamental divergence in the beliefs of doctors and the public" about how care should be improved. Most physicians (86 percent) believed that hospitals' reports of errors should be kept confidential, whereas 62 percent of laypersons believed that these reports should be released publicly. It was explained by way of their personal desires: "Physicians believe that confidentiality will promote openness among colleagues; laypersons favor 'transparency' and the pressure of public accountability."[125]

In 1997, it was remarked that "Discussion of *medical error has not often been* seen as *an arena for ethics committees* in great part because *it is usually referred* to morbidity and mortality and *risk management committees.*"[126] (Emphasis added.) Bioethicist Bernard Lo pointed out how unjust it is "that the physician and institution receive financial reimbursement, yet patients are not compensated for future care and lost wages resulting from a medical error."[127] These are reinforced perceptions of patients and family members of a lack of

124. *In re* Spring, 380 Mass. 629, 637 (1980), the former Massachusetts Supreme Judicial Court stated: "Little need be said about criminal liability: there is precious little precedent, and what there is suggests that a doctor will be protected if he acts on a good faith judgment that is not grievously unreasonable by medical standards."

125. N. Eng. J. Med. 1966 (Dec. 12, 2002). "Dr. Edgar Pierluissi, a young safety researcher, observed 332 hour-long M&M conferences in surgery and internal medicine at UCSF and Stanford, and found that only one in four sessions actually tackled errors head-on; and when they did, the discussion was perfunctory and generally blamed other departments." Wachter & Shojanin (2004), p. 277.

126. Ethical Currents 3 (St. Joseph Healthcare System), Fall 1997. For a list of criteria for implementing clinical safety options, *see* Longo et al., *The long road to patient safety: a status report on patient safety systems*, JAMA 2858-65 (2005).

127. Lo, Resolving Ethical Dilemmas (1995).

caring and collaboration by healthcare professionals.[128] One study of physicians "indicated that they were willing to use deception in several situations."[129] In 2001 a federal judge ordered Medicare officials to disclose findings that a doctor committed medical errors that injured the patient.[130] A Kentucky Veterans Administration (VA) hospital, which had a policy in effect since 1987, called for full disclosure to patients who are injured either accidentally or through medical negligence. That hospital offered fair compensation before the patient or his family realized the error. It was found, "This diminishes the anger and desire for revenge that often motivates patient litigation. Plaintiffs' attorneys become more willing to negotiate a settlement without trying to punish the institution with a big verdict."[131]

Effective July 1, 2001, the Joint Commission on Accreditation of Healthcare Organizations (JCAHO) issued new standards to prevent medical errors and require hospitals to inform patients—and family members—of "unanticipated outcomes" from treatment.[132] In 2003, University of California hospitals be-

128. 154 ARCH. INTERN. MED. 1365-70 (1994). *See also* Witman et al., 156(22) ARCH. INTERN. MED. 2565-69 (1996): "If the physician did not inform them of the mistake, they were twice as likely to want to file a lawsuit upon finding out about the error. When the patient was not informed and the mistake was severe, 60-75 percent of patients chose either to report the physician or file suit."

129. FED. LAW., July 2001, at 27. The most populous state (California) ranked lower than that of 33 other states for doctor discipline. L.A. TIMES, March 29, 1996, at A27A.

130. ORANGE COUNTY REGISTER, July 17, 2001, at News 5. "The lack of an explanation, and of an apology, if appropriate, may be experienced by the patient as extremely punitive and distressing and may be a powerful stimulus to complaint or litigation." C. Vincent, *Understanding and responding to adverse events*, 348 N. ENG. J. MED. 1051-56 (2003).

131. Kraman, ANNALS OF INTERNAL MED., Dec. 21, 1999. The hospital's liability records were compared with those of 38 similar veterans' hospitals from 1990 through 1996.

132. JCAHO R.I. 1.2.2. Patient Rights and Organization Ethics Chapter Standard states:

> Patients and, when appropriate, their families are informed about the outcomes of care, including unanticipated outcomes. The responsible licensed independent practitioner or his or her designee clearly explains the outcome of any treatments or procedures to the patient and, when appropriate, the family, whenever those outcomes differ significantly from the anticipated outcomes.

gan an attempt to reduce medical errors by using the Electronic Event Reporting System. Nevertheless, a nurse or doctor would not immediately face punishment; they are encouraged to fill out a computerized "event report" that records the mistake. The system will not be accessible to patients. Further, rather than reporting them to the state medical board, UC hospitals plan to analyze the computer reports to see if they could make sure that no one repeats the same mistake.[133] In 2002 it was again reported that "[t]he problem of medical errors received increasing attention after the publication of a landmark report from the Institute of Medicine in 1999 that estimated that between 48,000 to 98,000 fatal medical errors occur annually in the nation's hospitals. The system will permit medical personnel to report errors anonymously."[134]

In 2005 Don Berwick, founder and CEO of the Institute for Healthcare Improvement, argued that "the quality of U.S. medicine is far below its potential and lags behind that of many industrialized countries even though U.S. per-capita spending on health is the highest in the world."[135] A study noted both that "medical errors are inevitable" and that their disclosure "to patients is

---

Both JCAHO and the National Committee for Quality Assurance (NCQA) evaluate hospitals, nursing homes, hospice programs, home health agencies, and mental-health organizations.

133. Orange County Register, Nov. 1, 2002, at News 11. In this connection JCAHO's revamp methods include what is known as the "tracer methodology" to follow patients through the course of their care. *Tracer Methodology: How It Can Help You Improve Quality*, Healthcare Benchmarks and Quality Improvement 61-63 (2004). However, it has been reported that hospitals and physicians are paid the same regardless of the safety of the care they deliver. Leatherman et al., *The Business Case for Quality: Case Studies and an Analysis*, Health Affairs 17-30 (2002).

134. L.A. Times, March 28, 2003, B7. "You question your competence but fear being discovered. You know you should confess, but dread the prospect of potential punishment and of the patient's anger." Wu, *Medical error: The second victim: The doctor who makes the mistake needs help too*, 320 British Med. J. 726-27 (2000). *See also* Leape, *Institute of Medicine Medical Error Figures Are Not Exaggerated*, JAMA 95-97 (2000); and McDonald, Weiner & Hui, *Deaths Due to Medical Errors Are Exaggerated in Institute of Medicine Report*, JAMA 93-95 (2000).

135. Orange County Register, April 19, 2005, at Bus.8. He also pointed out that up to 98,000 people die because of medical errors and inappropriate care every year in U.S. hospitals.

desired by patients and recommended by ethicists and professional organizations. Several other studies documented physicians' reluctance to fully disclose errors to patients" that could "diminish patient-physician trust, and increase the likelihood of a lawsuit." It was found that "patients and physicians have notably different perspectives" and concluded that physicians "should strive to meet patients' desires for an apology and for information on the nature, cause, and prevention of errors."[136] The "malpractice crises" tort reformers emphasize the threat that liability crises pose to the cost and availability of medical services, and tort defenders emphasize the importance of liability to medical quality. A 2005 study concluded that although tort reform increased physician supply, "further research is needed to determine whether reform-induced increases in physician supply benefited patients."[137] The Institute of Medicine has recommended establishing mandatory error-reporting systems for hospitals and other health settings. An examination of the opinions and experiences of hospital leaders with state reporting systems disclosed in 2005 that although "hospital leaders generally favor disclosure of patient safety incidents to involved patients, fewer would disclose incidents involving moderate or minor injury to state reporting systems."[138] On July 29, 2005, the president signed the Patient Safety and Quality Improvement Act of 2005 (PSQIA).[139] "Patient safety work product" includes information that is "assembled or developed by a provider for reporting to a patient safety organization and . . . reported to a patient safety organization (PSO)." This law is to reduce some medical errors by encouraging medical-care providers to report mistakes. The information will be reported voluntarily to so-called "patient safety organizations" that would be protected from subpoena by providing criteria for certification of an entity. Thus the error reporting remains volun-

136. "Physicians should disclose the following minimal information about harmful errors regardless of whether the patient asks: (1) an explicit statement that an error occurred; (2) a basic description of what the error was, why the error happened, and how recurrences will be prevented; and (3) an apology. M.D.Gallagher et al., JAMA 1001 (Feb. 26, 2003).

137. Kessler, Sage & Becker, *Impact of Malpractice Reforms on the Supply of Physician Services*, JAMA 2618 (June 1, 2005).

138. Weisman et al., *Error Reporting and Disclosure Systems Views From Hospital Leaders*, JAMA 1359 (March 16, 2005).

139. Pub. L. No. 109-41 (2005), tit. IX, Part C of the Public Health Service Act, 42 U.S.C. § 299 et seq.

tary under the new law, and there are no federal penalties for failing to provide information.[140] Nevertheless, the law provides enforcement authority for both civil monetary penalties of up to $10,000 for violation of the statutory privilege and injunctions against continued violation.[141]

It has also been found that in teaching hospitals at the site of nearly all graduate medical education in the United States, there has been little discussion of how they "could be improved to reduce the likelihood of errors made by residents, who provide much of the direct patient care in such hospitals." It was recommended that residents be allowed "to take personal responsibility for their errors and to discuss them constructively as a means of facilitating collective learning and improving clinical practice."[142] In 2003, the Accreditation Council for Graduate Medical Education (ACGME) commenced enforcing limits on residents' duty hours, with a maximum of 80 hours a week or 30 hours per shift.[143] A study found that intensive-care residents working shorter shifts committed fewer errors.[144] In this connection, it should also be noted that the use of Computerized Physician Order Entry (CPOE) would help "most of the medication errors that occur at the prescribing and

140. PSOs are business associates within the HIPAA. The patient safety organizations will evaluate the data for any trends and recommend steps back to the reporting providers to avoid future mistakes. However, the patients and medical-care providers would not be identified.

141. PSQIA, Section 2. The reporting system is largely voluntary. Penalties for a single act or omission are to be imposed only under either the PSQIA or HIPAA.

142. Volpp & Grande, N. Eng. J. Med. 851 (Feb. 27, 2003). In 2004, a study was made in order "to quantify work hours, sleep, and attentional failures among 20 first-year residents (postgraduate year 1) during a traditional rotation schedule that included extended work shifts and during an intervention schedule that limited scheduled work hours to 16 or fewer consecutive hours." It was concluded that "eliminating interns' extended work shifts in an intensive care unit significantly increased sleep and decreased attentional failures during night work hours." Lockley et al., N. Eng. J. Med. 1829 (Oct. 28, 2004).

143. Weinstein, *Duty Hours for Resident Physicians—Tough Choices for Teaching Hospitals*, N. Eng. J. Med. 1275-78 (2002).

144. Landrigan et al., *Effect of Reducing Interns' Work Hours on Serious Medical Errors in Intensive Care Units*, N. Eng. J. Med. 1838-48 (2004).

order-filling stage—whether due to poor handwriting, sloppy instructions, or bad therapeutic choices" (the wrong medicine or the wrong dose).[145]

good v. Good

Public access reporting of many errors made by physicians and hospitals would seriously affect the image of the profession and may even cause their patients to question it. Hospital ethics committees are best able to handle such matters without disclosing them.

v.

In view of the fact that physician and hospital errors appear to have reached crisis proportions at the turn of the millennium, timely public access in their regions becomes paramount toward improvement of the image of the profession. Hospital committees appear to remain largely dominated by physicians on the disclosure or nondisclosure of such matters (see Subdivision 6(d)).

Twenty-first Century Approach

1. The failure to disclose significant errors constitutes unethical conduct; it is often covered up, which avoids possible sanctions that should be enforced.
2. Patients should be given a statutory right to be informed about such errors as soon as practicable following their occurrence.
3. Hospital ethics committees should be informed about the practice of such errors from the peer review committee (without disclosure of all the details that might be needed to support a possible malpractice case).
4. Teaching hospitals should provide procedures for residents to take personal responsibility for their errors and to discuss with them constructively the uses of the Computerized Physician Order Entry (CPOE) system.

145. Wachter & Shojania, Internal Bleeding 73 (2004). In addition: The CPOE system at Brigham and Women's Hospital in Boston keeps track of a patient's kidney function by monitoring a lab test called creatinine, alerting the doctor to adjust the dose of any of the many medicines that are excreted by the kidney when it detects evidence that the organ is failing. The Department of Veterans Affairs now has a national system—online at every VA hospital. *Id.*

There are some patients whom we cannot help; there are none whom we cannot harm.

Arthur I. Bloomfield, M.D., about 1933

Physician, heal thyself.

Luke 4:23 (King James Version)

e) Doctoring Business with Fraud and Waste; and Use of "Double Standards" for Admission to Medical Schools

In the 1990s, it had been noted that "the distinction between medical and economic concerns has become greatly blurred, and cost considerations have accelerated the transformation of healthcare from a purely professional undertaking to a business enterprise providing professional services."[146] Thereafter, following the turn of the third millennium, it was shown that over $20 billion had been lost due to Medicare fraud, waste, and mistakes—on average, about 10 percent of every dollar spent.[147] A study found that 39 percent of physicians admitted that they "sometimes" manipulate reimbursement rules.[148] Some

---

146. Kenneth Abraham & Paul C. Weiler, *Enterprise Medical Liability and the Evolution of the American Healthcare System*, 108 Harv. L. Rev. 381, 395-96 (1994). The Government Accountability Office (GAO, formerly the General Accounting Office) classified five types of specialty hospitals: cardiac, orthopedic, surgical, women's, and other. Inglehart, *The Emergence of Physician-Owned Specialty Hospitals*, N. Eng. J. Med. 78 (Jan. 6, 2005). Physicians "can perform about twice as many cases in a given time period at specialty hospitals as at community hospitals." And their "incomes had increased by 30 percent as a result of increased productivity." *Id.* at 80. "The GAO also reported that more than 90 percent of the specialty hospitals that have opened since 1990 are for-profit entities, as compared with 20 percent of all general hospitals" and "are much less likely than general hospitals to have emergency departments."

147. Orange County Register, March 10, 2000. "Medicare lost $13.5 billion in 1999. . . . In 1996, the first year the comprehensive Medicare audit was done, overpayments accounted for 14 cents of every dollar spent, or $23 billion. . . . In 1997, 11 cents of every dollar spent on the dollar, or $20 billion, was lost."

148. Wynia et al., *Physician Manipulation of Reimbursement Rules for Patients*, 283 JAMA 1858-65 (2000).

57 percent of all "whistleblower" recoveries were attributable to healthcare fraud cases.[149] These cases were often settled by the healthcare organization. For example, in 2001, a large settlement was reached by HCA, Inc. (formerly Columbia/HCA) that involved the improper billing of $731.4 million for healthcare costs to the federal government and $66 million to several whistleblowers.[150] (See subsection 7.(b) and Appendix C.) In 2004, the Health and Human Services Office of Inspector General (OIG) issued a Draft Supplemental Compliance Program Guidance for Hospitals. It would require that a healthcare organization report to the government within 60 days "credible evidence of misconduct" that "after a reasonable inquiry, [the organization] believes . . . *may* violate criminal, civil, or administrative law."[151]

In 1994-95, 13 medical schools offered combined M.D. and M.B.A. degree programs for work with profit- (or nonprofit-) making organizations.[152] Both for-profit and private not-for-profit dialysis facilities provide the majority of hemodialysis care in the United States. There has been debate about whether the profit status of these facilities influences patient mortality. A study published in 2002 stated, "Hemodialysis care in private not-for-profit centers is associated with a lower risk of mortality compared with care in private for-profit centers."[153] Concurrently, medical school standards were lowered for economic reasons as well as for affirmative action programs for minorities (partially abolished in the late 1990s), some of which have backfired.[154] For example, a black student was

149. Fed. Cont. Rep. (BNA), Oct. 23, 2001, at 421.

150. *In re* Columbia/HCA Healthcare Corp. D.D.C., No. 101-MS-50, order on July 7, 2001. Fed. Cont. Rept. (BNA) 8-14-01. Regulations such as those at 42 C.F.R. pt. 50 F pertain to grants and cooperative agreements. Because they do not make the Federal Acquisition Regulation applicable to grants, such improper billing becomes more likely to cause such large expenditures by the government. *See* Keyes, Government Contracts 3 (3d ed. 2003).

151. 69 Fed. Reg. 32,012, 32,030 (June 8, 2004). In 2003 it was held that an insurance company, a fiscal intermediary that had been knowingly approving false claims for Medicare payment to a home health agency, had immunity under the Medicare statute. 43 U.S.C. § 1395h(i). United States *ex rel.* Sarasola v. Aetna Life Ins. Co., 11th Cir., No. 01-14291 (Jan. 28, 2003).

152. Bazansky et al., *Educational Programs in U.S. Medical Schools*, 274 JAMA 716, 720 (1995).

153. JAMA 2449 (Nov. 20, 2002).

154. Orange County Register, Aug. 18, 1997, at Metro 9.

admitted to the University of California at Davis under its standard-lowering program. He was later found to lack the ability to perform "some of the most basic duties required of a physician." In suspending his license, the administrative law judge found him guilty of gross negligence and incompetence in the treatment of three patients, one of whom died. The judge stated, "Letting him continue to engage in the practice of medicine will endanger the public health, safety and welfare."[155] A study in the *Journal of the American Medical Association* (*JAMA*) found that only 49 percent of black students pass examinations to practice medicine compared to 88 percent of others—a disparity caused by lower admission standards.[156] In most instances, however, disciplinary boards have allowed them to continue the practice of medicine.[157]

good v. Good

It is more important to continue affirmative action programs that do not maintain racially equal standards for admission to medical schools in order to increase minority admissions by all means. Virtually all professions of all kinds have been converted to business enterprises.

v.

Although the use of *double standards* will favor minorities, it also has caused significant damage to the public in an egregious manner in the case of the medical profession. It is unethical to continue business and political interests in such a manner as to interfere with the quality of medical practice.

Twenty-first Century Approach

Admission requirements and lower standards that were maintained by many medical schools need to be significantly upgraded. Disciplinary procedures in many states should be enhanced to protect the public as well as (b) their healthcare businesses from unqualified professionals.

---

155. L.A. Times, Sept. 2, 1997, at A18.

156. *Id.*

157. Orange County Register, March 5, 1998, at News 21. Another study "published in the Journal of the American Medical Association found that 761 physicians had been disciplined by state medical boards for sex-related offenses over an 8-year period. Of those, 75 percent of the cases involved patients, and 30 percent of those entailed sexual intercourse." Orange County Register, June 17, 1998, at News 13.

4. The Vagueness of Bioethical Principles and the Need for an Alternative

In 1998, while writing a volume about the demise of classical education during the 1980s and 1990s, the authors noted, "Of more than one million bachelor's degrees awarded in 1994, only 50 were in the classics."[158] It has been demonstrated that, unlike previous generations, "*Physicians are no longer humanists* and that medicine is no longer a learned profession." Formerly, some advocated "an integrated medical school curriculum, with didactic instruction in medical ethics during the preclinical years, enhanced by patient-centered ethical discussions during the final two years."[159]

With respect to the "Twelve Methods of Bioethics" (recited above), ethical rules are often viewed as too vague to permit decision making. A lesson may be learned by making a comparison with the law, where vagueness constitutes a valid ground to declare a statute invalid. For example, in 1991 the Supreme Court considered the issue of whether a Nevada regulation on attorney speech was "so imprecise that discriminatory enforcement became a real possibility," and could create a "trap for the wary as well as the unwary." Justice Kennedy wrote, "Given this grammatical structure, and absent any clarifying interpretation by the state court, the Rule fails to provide 'fair notice to those to whom [it] is directed.' A lawyer seeking to avail himself of [the Rule's] protection must guess at its contours."[160]

The same rule applies to such statutes or regulations in the field of medicine. For example, laws on fetal tissue research bans that were unconstitutionally vague have been stricken down in three federal circuits. Thus, an Arizona law banning fetal tissue research was held unconstitutionally vague because it failed to properly define the terms "investigation," "experimentation," and "routine." The court stated: "Under the Arizona statute . . . doctors might under-

158. HAUSONT & HEATH, WHO KILLED HOMER? (1998). The authors note that, nevertheless, publications by the few classics teachers have increased.

159. Pelligrino, *Educating the Humanist Physician, An Ancient Ideal Reconsidered*, 227 JAMA 1288-94 (1974). *See also* Radwany & Adelson, *Medical Ethics and Literature*, JAMA, March 27, 1987, *citing* CULVER ET AL., BASIC ETHICS AND THE BEDSIDE 951-56 (1978).

160. Gentile v. State Bar of Nevada, 501 U.S. 1030 (1991). *Id.* at 1048, *quoting* Grayner v. City of Rockford, 408 U.S. 104, 112.

take a procedure involving fetal tissue that they consider to be primarily therapeutic, perhaps, even routine, but under the statute, the state might consider such a procedure illegal. The distinction between experiment and treatment in the use of fetal tissue is indeterminate."[161]

For example, a text (which went through several subsequent editions) that came into general use by medical students emphasized several principles: autonomy (which is new in its modern form), beneficence, nonmalfeasance (from Hippocrates' dictum "First do no harm"), and justice (the law profession), which constitute obligations physicians owe to their patients. These generalities are excellent preludes to the duties of physicians (and bioethicists); however, they may be of diminished value without detailed implementation.[162] Such principles resemble the old "maxims of the law," which are no longer cited except as "window dressing" for judicial decisions. In 1998, a writer on these bioethical works remarked: "As a result, the method came under fire for being a theory that failed to carry out the tasks that theories typically set for themselves, such as providing a decision procedure for ambiguous and difficult cases."[163] These principles resemble the seven cardinal virtues first named by Pope Gregory I in the sixth century: faith, hope, and charity (the Christian ones), and wisdom, justice, courage, and temperance (virtues derived from the Greeks). Bioethicists can look upon and may even be inspired by these goals; but they will also have to look well beyond them to decide specific cases assigned to them and weigh two views of what is "good."

---

161. Forbes v. Napolitano, 247 F.3d 903 (9th Cir. Dec. 29, 2000) at p. 2. *See also* Jane L. v. Bangerter, 61 F.3d 1493 (10th Cir. 1995); Margaret S. v. Edwards, 794 F.2d 994 (5th Cir. 1986); Lifchez v. Hartigan, 914 F.3d 260 (7th Cir. 1990).

162. In the prior edition, the initial chapters, heavily weighted with more than 100 pages of philosophical and ethical theory, intimidated many readers.

JAMA 1582 (March 27, 2002) (reviewing *Principles of Bioethical Ethics).*

163. Kuczewski, *Childress Greatest Hits*, Hast. Ctr. Rep. (July-August 1998), at 43.

good v. Good

Having recently integrated more humanism into their practice, physicians are perfectly capable of implementing mere basic principles of bioethics.

v.

Like the ancient maxims of law (which are rarely cited today) and the cardinal virtues, bioethical principles (which may normally be excellent in themselves) are of little value in decision making without detailing different objective views applicable to the specific cases being considered by a hospital's bioethics committee.

Twenty-first Century Approach

1. In the twenty-first century, the medical profession must begin to recognize that broad principles which began to be promulgated on bioethics toward the end of the twentieth century constitute background (values and inspirational material) that cannot normally resolve difficult ethical or legal cases.
2. Sufficient details in medical and legal standards must be found and used in order to make most bioethical decisions as well as to educate physicians and others who may become members of bioethics committees.

Some professions persist in believing that they can determine, based on objective knowledge and their medical skill, what will benefit a patient and what will not. Clinicians cannot claim expertise in value judgments. (Emphasis added.)

Veatch & Spicer, *Medically Futile Care: The Role of the Physician in Setting Limits*, Am. Law & Med. 15 (1992).

5. The Teaching of Bioethical Ethics and the Law; Ethical Lapses

In the mid-1990s, the Association of American Medical Colleges (AAMC) found that only a few such schools taught bioethics and the law as a required course.[164] This failure in medical schools[165] began to be remedied following the turn of the twenty-first century. However, in the AMA's *Code of Medical Ethics*, Rule 9.011 governing "Ethics Committees in Healthcare Institutions" made no reference to bioethics in connection with their members' continuing education.

The key study to make changes in the nation's medical schools early in the twentieth century was accomplished by Abraham Flexner, who was not medically trained. When he was at the Carnegie Foundation, Pritchett suggested that he conduct an independent and impartial survey of all the medical schools in the United States and Canada. Flexner suspected a case of mistaken identity and pointed out that it was Simon who was the director of the Rockefeller Institute for Medical Research, and the he, Abraham, had no medical training. But Dr. Pritchett replied: "That is precisely what I want. I think these professional schools should be *studied not from the point of view of the practitioner*

---

164. Ass'n Am. Med. Colls. 1994-95, AAMC Curriculum Directory 11, tbl. 6 (1994). The American Association of Law Schools (AALS) Committee on Curriculum and Research surveyed 179 law schools regarding new courses; it found that only 26 of these institutions teach bioethics. Goodman et al., J. L., Med. & Ethics, Summer 2002, at 300. Nevertheless, Rule 1.1, Competence, requires a lawyer to "engage in continuing study and education and comply with all legal education requirements to which the lawyer is subject." Model Rules of Prof'l Conduct (2003 Edition).

165. Parade, June 8, 1997, at 9.

> One of the exceptions may have been the University of New Mexico, which has recently created the following set of "Clinical Ethics Competencies" to guide curricula and resource development throughout its Health Sciences Center. It is expected that medical, nursing, and pharmacy students, upon graduation, will be able to diagnose and manage the ethical dimensions of patient care, institutional demands, and societal realities and agendas in (1) Professional Responsibility, (2) Patients' Rights, (3) Privacy and Confidentiality, (4) Truth Telling, (5) Reproductive Ethics, (6) Distributive Justice, and (7) Research Ethics.

Ethical Currents, Summer 1999, at 9.

but from the standpoint of the educator. . . . This is a layman's job, not a job for the medical man." *He quickly turned to* what was arguably the *greatest scandal in American education, the medical schools.*[166] (Emphasis added.)

Thereafter, a number of major changes were made which generally continued throughout the twentieth century. Nevertheless, a 1998 study of scientific researchers and administrators responsible for academic integrity found many differences in approaches to unethical acts. For example, "neither group was eager to communicate with funding agencies or journal editors" about unethical behavior. Worse yet, 23 percent reported that they had received no training in research ethics, 36 percent had observed some kind of misconduct, and 15 percent would be willing to "select, omit or fabricate data to win a grant or publish a paper."[167] A 2005 study was made of 235 graduates of only three medical schools who were disciplined by one of 40 state medical boards between 1990 and 2003. When predictor variables were made from medical schools, it was found that "disciplinary action among practicing physicians by medical boards was strongly associated with unprofessional behavior in medical school. Students with the strongest association were those who were described as irresponsible or as having diminished ability to improve their behavior. Professionalism should have a central role in medical academics and throughout one's medical career."[168] Another study of 989 biomedical papers published by Massachusetts medical scientists found that although 34 percent of the authors stood to gain financially from their data by holding a patent or as officers or consultants to a biotech firm, none of the authors disclosed their financial interest.[169] This failure to disclose has also been true with respect to many research publications.[170]

166. Heynick, Jews and Medicine (2002).

167. Korenman et al., *Norms for Academic Research Integrity*, JAMA 64 (Jan. 7, 1998).

168. Papadakis et al., *Disciplinary Action by Medical Boards and Prior Behavior in Medical School*, N. Eng. J. Med. 2673 (Dec. 22, 2005).

169. Krimsky et al., *Financial Interests of Authors in Scientific Journals*, Science and Energy Ethics 395-410 (1996).

170. "The integrity of an individual researcher is the central legal and scientific issue when determining future eligibility for government funding. It matters little if the research turns out ultimately to be correct, if the researcher showed himself or herself willing to lie or cheat to arrive there." *Commentary*, JAMA 64 (Jan.7, 1998).

Finally, it is noted that Medicare allocates about $200 million annually so that Quality Improvement Organizations (QIOs) can work with hospitals by providing educational materials, using data collection and feedback to track performance on quality indicators, etc. Although it is the federal government's largest initiative for improving the quality of care, a 2005 study found that "hospitals [that participate] show [more] improvement on quality indicators than hospitals that do not participate."[171]

good v. Good

Physicians should not be examined on bioethics and its law because it is "ultra-medical."

v.

Medical students who also receive something on bioethics will become more likely to recognize when ultra-medical decisions arise and seek bioethical advice.

Twenty-first Century Approach

1. Medical schools should be required to teach and examine, to a reasonable extent, bioethics and related law materials.
2. Such teaching should be integrated into clinical rounds, along with the recognition of problem areas that need referral to other than medical specialists who are members of hospital ethical committees.
3. At least some portions on the nature of bioethics, and the law related thereto, should be made a subject of state medical board examinations.

This is the end I aim at: to acquire knowledge of the Union of the mind with the whole of Nature. . . . *[B]ecause health is no small means to achieving this end, the whole of medicine must be worked out. (Emphasis added.)*

Baruch Spinoza

171. Snyder & Anderson, *Do Quality Improvement Organizations Improve the Quality of Hospital Care for Medicare Beneficiaries?*, JAMA 2900 (June 15, 2005). "Defining hospital participation was one of the greatest challenges in this study. The classification of hospital participation by the QIOs is subject to bias . . . QIOs had the incentive to attribute hospital activities as being related to the QIO. This potential misclassification could lead to biased estimation of the QIOs impact."

> *Ethics is not a luxury or an option. It is essential to our survival. To support that point, let me give you three assertions. Here is the first assertion:* We will not survive the twenty-first century with the ethics of the twentieth century. *(Emphasis added.)*
>
> Rushworth M. Kidder, Journalist

> *In the middle of every difficulty lies opportunity.*
>
> Albert Einstein

6. The Role of Biomedical Ethics Committees and the Meaning of Care

One effect of medical students' involvement in patient care has concerned the lack of adherence to a guideline of the Joint Commission on Accreditation of Healthcare Organizations (JCAHO); it emphasized that a physician cannot assume that a patient is implicitly willing to participate in the training of medical students.[172] As noted in the preceding chapter, in the final years of the twentieth century, investigators found that medical students were introduced by other members of the healthcare team as "doctors" "and only 42 percent corrected the information with patients."[173] Some of them assumed that a teaching hospital relegated such considerations to remain a lesser role.[174]

The Austrian philosopher Martin Heidegger illustrated the structure of dealing with the world of anxiety with care by quoting the wonderful antique "Cure fable" of Hyginus:

172.
> The patient has the right to know the identity and professional status of individuals providing service to him. . . . This includes the patient's right to know of the existence of any professional relationship . . . to any . . . educational institutions involved in his care. Participation by patients in clinical training programs should be voluntary.

JCAHO, ACCREDITATION MANUAL FOR HOSPITALS (1985).

173. Beatty & Lewis, *When Students Introduce Themselves as Doctors to Patients*, 70 ACADEMIC MED. 175-76 (1985).

174. Silverman, *Narrowing the Gap between the Rhetoric and the Reality of Medical Ethics,* 71 ACADEMIC MED. 227-37 (1996). *See also* 12 J. CLINICAL ETHICS, Summer 2001, at 111.

> Once when "Care" was crossing a river, she saw some clay; she thoughtfully took up a piece and began to shape it. While she was meditating on what she had made, Jupiter came by. "Care" asked him to give it spirit, and this he gladly granted. But when she wanted her name bestowed upon it, he forbade this and demanded that it be given his name instead. While "Care" and Jupiter were disputing, Earth arose and desired that her name be conferred on the creature, since she had furnished it with part of her body. They asked Saturn to be their arbiter, and he made the following decision, which seemed a just one: "Since you, Jupiter, have given its spirit, you shall receive that spirit at its death; and since you, Earth, have given its body, you shall receive its body. But since "Care" first shaped this creature, she shall possess it as long as it lives.[175]

In his existential work, *Being and Time*, Heidegger noted that care is a basic characteristic of the human condition. Simply put, life is care and time, whereas care itself is creativity, anxiety, and decision making. These, in turn, generally constitute the essential functions of both physicians and bioethicists. However, they are focused quite differently because the recommendations of bioethics committees transcend the actions of the traditions of physicians (with "diagnoses, prognoses, diet and drug prescriptions"), and surgeons (with "scalpels").

a) *JCAHO's Requirement of a "Mechanism" for Ethical Issues*

At the close of the twentieth century, virtually all healthcare facilities were more or less complying with the standard of the Joint Commission on Accreditation of Healthcare Organizations (JCAHO)[176] that they have a "mecha-

---

175. *Quoted in* MARTIN HEIDEGGER, SAFIANSKI 187 (1998).

176. Joint Commission on Accreditation of Healthcare Organizations (JCAHO) is an independent, not-for-profit organization that has developed professionally based standards and evaluated the compliance of healthcare organizations against these benchmarks. It accredits over 17,000 healthcare organizations and programs in the United States. JCAHO accreditation is used as a substitute for federal certification surveys for Medicare and Medicaid and fulfills licensing requirements in many states. *See* JCAHO, *at* http://www.jcaho.org (official commission Web site).
AM. J. L. & MED. (2003), at 414.

nism" for considering ethical issues.[177]

Although the AMA *Code of Medical Ethics* (2003 Edition) Rule 9.11 states, "Preferably, a majority of the committee should consist of physicians, nurses, and other healthcare providers," fortunately most of them have ethics committees also composed of needed volunteers from non-medical professions or other fields of work. These committees commonly undertake four functions:

1. education of hospital staff and patients about medical ethics,
2. acting as the group designated to discuss medical ethics issues,
3. participating in the development of institutional policy on ethical issues, and
4. participating in individual patient treatment cases and making majority recommendations on their decisions thereon.

Near the end of the twentieth century, 84 percent of American hospitals with at least 200 beds have established such committees. They have been seen as a way to "avoid cumbersome court procedures and unwieldy litigation."[178] Although this has been an "untested assumption,"[179] bioethical committee actions are less costly and give more timely results than do the courts. Even though they only make recommendations, "today's ethics committees . . . often command greater notice and their recommendations may be seen as more

177. *See* JCAHO, 1995 COMPREHENSIVE ACCREDITATION MANUAL FOR HOSPITALS 66 (discussing Standard R.I. 1 for accreditation, which requires a "functioning process to address ethical issues" and states that "patient rights mechanisms may include a variety of implementation strategies: for example, established ethics committees…"). However, in 2005 the Chicago-based JCAHO indicated that it denies accreditation to less than 1 percent of U.S. hospitals that seek it. ORANGE COUNTY REGISTER, Feb. 20, 2005, at News 14.

178. Robin Fretwell Wilson, *Hospital Ethics Committees as the Forum of Last Resort: An Idea Whose Time Has Not Come*, 76 N.C. L. REV. 353, 357 (1998), *quoting* Janet Fleetwood & Stephanie S. Unger, *Institutional Ethics Committees and the Shield of Immunity*, 120 ANNALS INTERNAL MED. 320 (1994).

179. Robin Fretwell Wilson, *Hospital Ethics Committees as the Forum of Last Resort: An Idea Whose Time Has Not Come*, 76 N.C. L. Rev. 353, 360 (1998).

binding."[180] A few states have granted civil and criminal immunity to its members, who generally serve without pay.[181] Healthcare providers often rely on committee advice because its professionals "will seek the comfort and security of the 'safe harbor' created by the grant of immunity."[182]

Some bioethicists have expressed a good deal of pessimism concerning these committees. One private consultant stated, "Many flourished for awhile, but have since languished. Such committees may stir themselves every three years to prepare materials for JCAHO accreditations, but they are often cosmetic most of the time. . . . Static, cosmetic healthcare ethics committees may well be an idea whose time has passed."[183] On the occasion of her retirement, Judith Ross, after spending 22 years in and around hospitals and their ethics committees, wrote that although ethics committees often include "some of the most thoughtful, concerned people in the hospital," in her experience they have not been able "to make a significant difference in practice."[184]

From my experience there has often been a lack of adequate institutional support and "physician dominance" of ethics committees at most healthcare institutions (see (d) below). In addition, institutions sometimes fear that the committee members may be "looking over their shoulders," and wish to free themselves from independent committee oversight in many matters. This may be true in spite of the limitations expressed in the 2003 version of the AMA *Code of Medical Ethics*, which states:

---

180. *Id.* at 362 & 406 n.39 (quoting Karen Ritchie, *When It's Not Really Optional*, Hast. Ctr. Rep., Aug.-Sept. 1988, at 25; *also* Diane E. Hoffmann, *Does Legislating Hospital Ethics Committees Make a Difference? A Study of Hospital Ethics Committees in Maryland, The District of Columbia, and Virginia*, 19 L., MED. & HEALTHCARE 105, 115 (1991).

181. Wilson, *supra* note 178 at 362 & 406 n.39. *See also* ARIZ. REV. ST. ANN. § 36-3231 (West Supp. 1996) (permitting a healthcare provider to consult with and obtain the recommendations of an institutional ethics committee when a surrogate is unavailable). *See also* IOWA REV. STAT. § 663-1.7 (1993) (designating the functions of an ethics committee to include decision making); *also* MD. CODE ANN., HEALTH-GEN. II § 19-374 (1996); there is no mention of immunity for consultants to the committee, ROSS ET AL., HEALTHCARE ETHICS COMMITTEES: THE NEXT GENERATION 140 (1993). *See also* MONT. CODE ANN. § 37-2-201 (1995).

182. Wilson, *supra* note 178 at 362 & 406 n.39.

183. Dugan, 58 ETHICAL CURRENTS 11 (Summer 1999).

184. *Id.*

> Recommendations of the ethics committee should impose no obligation for acceptance on the part of the institution, its governing board, medical staff, attending physician, or other persons. However, it should be expected that the recommendations of a dedicated ethics committee would receive serious consideration by decisions makers.[185]

However, the American Hospital Association's (AHAs) *Handbook for Ethics Committees* (Ross et al., 1986) stated, "One thing is certain, and committee members would do well to keep it in mind: *ethics committees should exist primarily to serve patients and to protect their interests.*"[186] In the 1990s the AHA further discussed the authority of the "future generation" of ethics committees, stating: "Doing ethics seriously means institutional and individual change. Too often, the administration or medical staff ordered up a committee to do education, policy writing, and case review, without any real commitment to any of those activities or without thinking seriously about what an ethics committee doing case consultation really means."

The AHA pointed to bioethics theorists such as Annas and Veatch, who, in the 1980s, appeared to be "very dubious about the abilities of ethics committees to steer a clear course." However, "Moreno, while voicing support for committees generally, expressed a deep concern about the process by which ethics committees make decisions, about their trying to work out values in a health care system that is in flux, and about bureaucratic problems inherent in hospital committees." They were particularly concerned about the drop in case review by ethics committees.[187]

Good manners are a part of good morals; and it is as much our duty as our interest to practice both.

John Hunter

---

185. AMA CODE OF MEDICAL ETHICS § 9.11, Ethics Committees in Healthcare Institutions, which covers "the establishment and functioning of ethics committees in hospitals and other healthcare institutions."

186. *See* Introduction, p. X.

187. ROSS, ET AL., HEALTHCARE ETHICS COMMITTEES: THE NEXT GENERATION 185-86 (1993).

b) The Lack of Standards for Membership Credentials

In the 1990s, there have been proposals concerning standards for bioethics and credentialing of members of bioethics committees. In November 1997, a Task Force on Standards for Bioethics Consultation issued a draft report concluding, "There should be no certification either for individuals who do ethics consulting (as individuals, members of a team, or members of ethics committee) or for programs that provide bioethics education."[188] It did recommend the development of voluntary standards that might be used by ethics committees and consultants with the goal of a gradual change to possible national standards sometime in the twenty-first century.

The task force decided to delay non-voluntary standards due to a lack of consensus among most ethics societies, including the Society for Health and Human Values, the Society for Bioethics Consultation, American Hospital Association, JCAHO, and the Society of Law and Medicine and Ethics. One critic called the proposal for voluntary standards "as toothless a response as one can imagine."[189] A doctor noted that at his hospital's committee, those invited to join represented a variety of disciplines within the medical center: internists, surgeons, family practitioners, pediatricians, gynecologists, a psychiatrist, nurses, legal counsel, an administrator, and a patients' advocate. From time to time, outside consultants and nationally known experts on ethics joined such a group.[190] It would appear that improvements are needed with respect to the ultra-medical issues being considered; Section 9.11 of the AMA *Code of Medical Ethics* (2003) states, "A majority of the committee should consist of physicians, nurses, and other healthcare providers." This appears to eliminate the importance of bioethicists and it needs to be broadened. As stated in the publication of the American Hospital Association (AHA), "Lay members might simply be thought

---

188. "The group characterizes its work as an attempt to provide guidance in some very uncharted waters about the nature and goals of ethics consultation, the core competencies that are needed to conduct it, and how those competencies might be acquired." ETHICAL CURRENTS 2 (Winter 1998).

189. *Id.*

190. GEORGE BURRELL, FINAL CHOICES (1993). The American Hospital Association's (AHA's) *Handbook for Ethics Committees* (1986) stated its "concern here is that the committee not restrict its own best thinking by worrying about the objections of hospital lawyers looking for liability issues" (p. 55).

of as those who are not healthcare professionals (in which case, the nonhospital lawyer and the ethicist could be considered lay members)."[191] In addition, "working in any institution over time places blinders on the employee,"[192] often including physicians and nurses.[193] In 2004 it was stated:

> At a time when 95 percent of those doing ethics consultation in U.S. hospitals have no formal education or training, when there is little to no consensus about how clinical ethics ought to be practiced and by whom, the choice for the future of clinical ethics is a stark one, and never has it been more pressing.[194]

---

191. Ross, et al., *supra* note 187 at 41. In 2000, the Society for Health and Human Values—Society for Bioethics Consultation Task Force on Standards for Bioethics Consultation suggested that "traditional classroom-based approaches may work well for learning about moral reasoning and ethical theory." 133 Annals of Internal Med. at 59-69. However, the following year Thomas May reported that "the vast majority of hospital ethics committee members have *no educational background* in bioethics." May, *The Breadth of Bioethics: Core Areas of Bioethics Education for Hospital Ethics Committees*, 26 J. Med. & Phil. 101-18 (2001).

192. "The ethicist who comes from beyond the hospital walls may be able to broaden the committee members' views because his or her perspective differs from theirs (although, of course, it may have been correspondingly narrowed by some other institution)." *Id.*, at 40.

193. Despite organized medicine's endorsement of ethics committees, many physicians seem unsympathetic or even hostile to them, perhaps because they feel the existence of an ethics committee implies the institution's reluctance to trust them to make treatment decisions without some kind of oversight. Nurses, too, have sometimes been dubious about the value of ethics committees with their long discussions, a process that one nurse described as "agonizing by committee."

Id., Preface.

194. Rubin & Zoloth, *Clinical Ethics and the Road Less Taken: Mapping the Future by Tracking the Past*, J. L., Med. & Ethics 218 (2004). Earlier Laurie Zoloth criticizes the ethicists as those who "describe the cascade of troubling circumstances without halting them, or at least calling out the questions of motives, relationships, and essential duties. . . . It is in part the role of the ethicist to . . . raise the difficult questions that stand just outside of the therapeutic encounter." Zoloth, *Audience and Authority: The Story in Front of the Story*, 7 J. Clin. Ethics 356 (Winter 1996).

Institutional Review Boards (IRBs) under federal research regulations require diversity, such as those in the law for review of a hospital's research.[195] However, not only is this less true of hospital ethics committees, but some such committees may include the hospital's own counsel, who may refuse to risk the institution's management concerns.[196]

Many of the critical decisions in medical ethics during the last quarter of the twentieth century have been made by the courts. Judges decide cases under statutes, but also render decisions comprising what is called "the common law"; these will be considered in all subsequent chapters.

good v. Good

Recommendations of hospital ethics committee members (including legal counsel) are subject to the management of their hospital. It must be recognized that often none may have received training in bioethics.

v.

The key to hospital ethics committees lies in their independence and the training in bioethics of at least one of its members.

Twenty-first Century Approach

1. Hospital ethics committees should render independent recommendations and require members who have received some training in bioethics.
2. Hospital committees that permit attendance by a hospital's own counsel or management should not assume they are bound to accept their votes on the committee's final decisions.

c) The Goals of Medicine and of Bioethics Committees; "Clarification" and Other Goals

In 1996, a group launched by Daniel Callahan, a philosopher who founded the Hastings Center, published a report stating that the four goals of medicine should be:

> [1] *the prevention of disease* and injury and promotion and maintenance of health, [2] the *relief of pain* and suffering caused by maladies, [3] the *care and cure* of those with a malady, and [4] the care of those who

195. 45 C.F.R. § 46.107(a).

196. Wilson, *Hospital Ethics Committees as The Forum of Last Resort: An Idea Whose Time Has Not Come*, 76 N.C. L. Rev. (1998) at 355 & n.5.

cannot be cured: the *avoidance of premature death* and the *pursuit of a peaceful death*. (Emphasis added.)

In bioethics, the focus has been different. The Task Force noted above defined the goals of ethics consultation as being to: "Improve the provision of healthcare and its outcome through the *identification, analysis* and *resolution of ethical issues* as they emerge in consultation regarding particular clinical cases in healthcare institutions." The goals of a bioethics committee include the identification, review, and resolution of ethical issues in particular cases. Neither the existence nor resolution of such issues is expressed as a goal of medicine.[197] The American Hospital Association (AHA) publication has even noted, "Often, physicians have never consciously thought about the goals of medicine and so there may be considerable disagreement as to whether these statements accurately reflect what the physicians intuitively judge to be appropriate."[198] Ethical processes include "gathering relevant data," "clarifying relevant conceptual issues" (often the legal ones), "ensuring that concerned parties have their voices heard," and "assisting concerned individuals in clarifying their own values."[199] Many medical ethics committees go beyond these functions by making specific recommendations in particular cases, providing education of physicians, staff, and others on bioethical matters, and drafting institutional policies regarding medical ethically related matters.

"Heuristics" constitutes a logical approach that increases the chances of finding a reasonable solution but cannot guarantee it because there may be areas that were not explored sufficiently. Some have claimed that building consensus offers such a tool for ethics consultation. Certain medical ethics committees have actually limited themselves to the singular goal of a religion.[200] However, knowledge is required in order to identify those primary

---

197. ACADEMIC PHYSICIAN & SCIENTIST, November/December 1997, at 1.

198. ROSS, ET AL., HEALTHCARE ETHICS COMMITTEES: THE NEXT GENERATION x, 48 (1993). However, the sympathy by physicians and nurses for ethics committees appears to have increased somewhat since the turn of the millennium.

199. I had noted in 1995 that the term "Clarification Committees" might be accurate for the functioning of some committees. Being "process goals," they may seldom offer guidance toward making new bioethical decisions in their process to resolve issues brought before them. KEYES, LIFE, DEATH, AND THE LAW (1995), ch. 5.

200. *See* Cassarett et al., Hast. Cent. Rep., November-December 1998, at 7. The Council for International Organizations of Medical Sciences (CIOMS) called

issues that are legally unresolved, and such knowledge is often lacking. In connection with this, it was emphasized by James M. Buchanan, Nobel laureate in Economic Science, in 2003, that "in a constitutional democracy, persons owe *loyalty to the constitution rather than* to *the government*."[201]

Non-legal ethical expertise may result from the experiences of considering many cases and comparing their similarities or dissimilarities together with some knowledge of philosophical theories and values. Although a philosopher may make moral judgments, he may or may not be approaching the answer. This lead an ethicist to state, "To my knowledge there is no uniform agreement on the goals of clinical consultation."

good v. Good

The goals of medicine presented before ethics committees are holistic with respect to the suffering, care, and cure of patients and hence they enhance all the goals of bioethics.

v.

The goals of bioethics involve the analysis and the objective resolution of the ultra-medical issues concerning them. Until the twenty-first century, most physicians had been inadequately trained in these areas and often tended to resort to their subjective views when making such decisions.

Twenty-first Century Approach

Because of the traditional goals of medicine and the often-different goals of bioethics and the law:

(a) some procedures of healthcare organizations need revision in order to embrace the different goals of ultra-medical decision making; and
(b) persons who have been so trained must more often be utilized for making such recommendations.

---

for the guidance of ethical review committees and issued some of them. *See* 19:3-4 L., MED. & HEALTH CARE 247-58, at 255 (Fall-Winter 1991).

201. IMPRINTS, National Speech Digest of Hillsdale College, March 2000. He added, "I have long argued that on precisely this point, American public attitudes are quite different from those in Europe."

> *As a member of this profession*, a physician must recognize responsibility *not only to patients,* but also to society, to other health professionals, and to self. (Emphasis added.)
>
> From Preamble to the AMA *Code of Medical Ethics* 2002

d) *Physician Dominance in Hospital Ethics Committee Decision Making*

A study of ethics committees published in the year 2000 found that "all had physicians and nurses on the committee (100 percent), virtually all had a social worker (97.5 percent) and a representative of the institution's administration (95 percent) on the committee."[202] Bioethics is interdisciplinary; some committee members are expected to have access to advanced knowledge outside of their own particular disciplines. Formal bioethics education has been lacking for many ethics committee chairs, most ethics committee members, and approximately two-thirds of those individuals designated to perform ethics consultations.[203]

To the extent that this expectation includes knowledge in the humanities, bioethics, religion, and the law, most physicians fail to qualify. Nevertheless, as the committees are constituted, physician members tend to dominate them.[204] As noted above, in large measure this is due to a section of the current AMA *Code of Medical Ethics* (issued in 1997) stating that the "majority of the committee should consists of physicians, nurses, and other healthcare providers." Their appearance is normally needed only to testify regarding facts concerning the case submitted to the committee. As the American Hospital Association's (AHA's) publication stated, "Because hospitals have a hierarchical structure, with physicians at or near the top, physician dominance of ethics committees has consistently been a problem. Even though physicians may numerically dominate the committee, they need not and should not dominate the nature of its work." Other ways that physician dominance can be reduced include those recommended by the AHA ensuring that:

---

202. Hoffman et al., Hast. Ctr. Rep., November-December 1998, at 7.

203. *Id.* at 37.

204. Apparently, in large measure this is due to section 9.11 (issued in 1949) of the current *AMA Code of Medical Ethics. See* § 9.11 (2).

(a) Subcommittees are normally *chaired by members other than physicians.*
(b) The right to vote is extended to all members of the committee (not just physicians, as is required by some medical staff rules).
(c) Meetings begin when a quorum of *members* is present, not a quorum of *physicians.*[205]

As one student of an ethics committee stated, biomedical issues "involve normative decisions in which medical science offers little help."[206] Another member of a committee stated, "We have left the domain of physiology, and entered the realm of values." A third noted, "Because treatment choices are 'individualized,' physicians have no particular ability to determine whether, from the patient's perspective, a treatment's hazards outweigh its benefits."[207] Further, "occasionally *a paternalistic physician* or perhaps an older one parked on a bioethics committee by the medical staff *assumes the chairmanship and destroys the committee*."[208] In 1999, the director of a large center for healthcare ethics wrote, "There are few things more consistent and constant in clinical ethics over the last fifteen to twenty years than the frustrations and complaints of other healthcare professionals over the behavior of physicians. Just in case some haven't noticed: *Physicians hold the ethics franchise* of the healthcare organization."[209] Physicians are devoted to the saying of the Greek poet Pindar, to the effect that "custom rules all." A prominent West Coast expert has written:

---

205. Ross, et al., Healthcare Ethics Committees: The Next Generation 149 (AHA 1993).

206. Wilson, *supra* note 178 at 374 & 406 n.105 (quoting Jerry Avon, *A Physicians' Perspective*, Hast. Ctr. Rep., June 1982, at 11, 12). *See also* Robin Fretwell Wilson, *Hospital Ethics Committees as the Forum of Last Resort: An Idea Whose Time Has Not Come*, 76 N.C. L. Rev. 353, 374 (1998).

207. Michael L. Perlin, *Are Courts Competent to Decide Competency Questions? Stripping the Facade From* United States v. Charters, 38 U. Kan. L. Rev. 957, 970 (1990). It has even been remarked that on "those committees where ethicists serve, they 'are not thought to be very influential, especially by physicians.'" *See* Wilson, *supra* at 385 (quoting Diane E. Hoffman, *Regulating Ethics Committees in Healthcare Institutions - Is It Time?*, 50 Md. L. Rev. 746, 780 (1991)).

208. Hosford, Bioethics Committees: The Health Care Provider's Guide 99 (1986).

209. Blake, 58 Ethical Currents, Summer 1999, at 5.

> Despite a consensus in the ethics committee literature as well as the JCAHO's requirement that the institution's [hospital's] ethics response (whatever form it takes) not be limited to physicians, several committee members reported that *they do not accept cases unless the physician either initiates it* or approves of its going forward. The *explanation* for this practice *was, largely, political.* What seemed to be driving committees in these decisions was their desire to be accepted by physicians. . . . This points to a central—and perhaps *irresolvable—problem for ethics committees.*[210] (Emphasis added.)

This situation prevailed in most hospitals of the United States at the dawn of the twenty-first century. Members of an ethics committee do not direct the institution; however, they should often be considered analogous to members of a board of directors with regard to the particular purposes for which they belong there. For example, a man was appointed by the National Association of Executive Directors as chairman of a blue ribbon commission to encourage "director professionalism" and issue a report to set new standards for directors. In discussing the standards for these members, he stated, "I'd take the person who knows that he or she has a commitment to do a lot on this board, and is prepared to do it *independently of management.* As somebody said, 'The best board member is a pain in the neck.'"[211] However, most *members of "ethics committees,* wholly or in part, may *conceive of their role as serving physicians,*" and the committee as a whole "may be concerned about preserving its place in the institution by *decreasing physician anxiety* about its role and authority."[212] (Emphasis added.)

A 1997 Task Force studying ethics committees designated the skills of committee members to include "knowing the difference between ethical, legal, medical, etc., issues and identifying assumptions that the participants are making." They also include "analysis skills for obtaining relevant legal and ethical facts (e.g., professional codes), clarifying concepts, and critically

---

210. Ethical Currents, March 1995, at 1.

211. Columbia College Today 17 (1998).

212. Robin Fretwell Wilson, *Hospital Ethics Committees as the Forum of Last Resort: An Idea Whose Time Has Not Come*, 76 N.C. L. Rev. 353, 388 (1998).

evaluating and using the relevant knowledge."[213]

Many of these requirements would eliminate some of the participation on ethics committees by most physicians and medical staff. Healthcare facilities have attempted special programs to further the education of such people, but these have not always been effective.[214] Only a few institutions have corrected this and thereby increased their effectiveness.

good v. Good

Bioethics committees should all be chaired by a physician and constituted by a majority of physicians, just as most of them are at this time.

v.

No bioethics committee should be constituted only of a majority of physician decision-makers or chaired by a physician (unless well educated in bioethics) because bioethics is, to a significant extent, an ultra-medical field.

Twenty-first Century Approach

1. In order to increase public confidence beyond the technical medical decisions made on bioethical matters, medical ethics committees should not always consist of a majority of physicians.
2. The committees should not be chaired by physicians unless they are also people who are significantly educated in bioethics.

---

213. Ethical Currents, Winter 1998, at 3. To acquire such skills takes more ustained programs—"fellowships, monitoring programs, extended course work in bio-ethics, and advanced degrees in philosophy, theology, or religious studies with specialization in bio-ethics." *Id.* Legal programs were not included. However, in July 2000, an editorial in the *Annals of Internal Medicine* (p. 56) by Spike and Greenlaw stated: "In our opinion, every ethics consultation service should have at least three members, one who holds an M.D., one who holds a J.D., *and one who should not be employed by the institution's administration or malpractice office.*"

214. One ethics committee on which I have served conducted biweekly or monthly lunchtime meetings for over a decade. Although I found them to be a very effective method of receiving education in this field, very few physicians (including members of the bio-ethics committee) or staff at the facility bothered to attend. This institution even changed its bylaws to delete a requirement that members must attend not less than 50 percent of committee meetings.

3. More programs should be set up to educate physicians and staff on bioethics and the law related thereto.

---

e) Hospital and Ethics Committee Records

The task force noted above made no specific recommendations regarding committee records. Without adequate records, evaluation, and accountability, a committee may not be in accordance with the Health Insurance Portability and Accountability Act (HIPAA) of 1996.[215] As one critic noted, "Every profession says a lot about self-scrutiny, but rarely does anything about it, so the report is typical in that respect."[216] An audit has been suggested that would review the records and outcomes by analyzing them.[217]

On April 14, 2003, extensive privacy regulations became effective under HIPAA requiring that a privacy notice be presented to patients at their first appearance for care and that receipt of the notice be acknowledged and recorded. Physicians and nurses cannot discuss diagnostic information in waiting rooms or public areas. Also, it is separated from matters of payment and treatment records or "healthcare operations" that can be disclosed without any additional records or that can be disclosed without any additional notification or authorization.[218] Patients have the right to inspect and copy their medical records and an authorization to release such records must contain certain information. This includes "a description of the information to be used or disclosed that identifies the information in a specific and meaningful fashion," "the name [of the person or

---

215. Provision 9.115 (5) of the *AMA Code of Medical Ethics* provides that procedural standards "should include establishing who must be involved in the consultation process and how notification, informed consent, confidentiality, and case write-ups will be handled." The Health Insurance Portability and Accountability Act of 1996 (HIPAA), Pub. L. No. 104-191, 110 Stat. 2021 (1996), 42 U.S.C. § 1320d requires safeguards to protect the security and confidentiality of patients' personal health information. U.S. Dep't of Health and Human Services, HHS *Fact Sheet: Protecting the Privacy of Patients' Health Information* (May 9, 2001). *Also see* 45 C.F.R. § 134.502 (b)(1).

216. *Id.* at 9.

217. Hoffman, HEALTHCARE EXECUTIVE, November/December 1995, at 47.

218. Office of Civil Rights, Dep't of Health and Human Services, Standards for privacy of individually identifiable health information: final rules. 67(157) Fed. Reg. 53,182-272 (2002); 45 C.F.R. § 164.501.

entity] to whom the disclosure may be made," "a description of each purpose of the requested use or disclosure," "an expiration date or expiration event," and "the signature of the individual and date."[219] However, a patient has no right to inspect psychotherapy writings that are "the contents of conversation during a private counseling session," information compiled in reasonable anticipation of, or for use in, a civil, criminal, or administrative action or proceeding, and other information. Other reasons for refusal of access include those where "a licensed healthcare professional has determined, in the exercise of professional judgment, that the access requested is reasonably likely to endanger the life or physical safety of the individual or another person," "the protected health information makes reference to another person the access requested is reasonably likely to cause substantial harm to such person," and where "the request for access is made by the individual's personal representative is reasonably likely to cause substantial harm to the individual or another person."[220] The Department of Health and Human Services (HHS) may impose penalties for violation of the regulation.[221]

Normally, confidentiality need not be as important a problem for bioethics committees or its records as some would imagine. Certain documents are given to committees with the patient's consent. The names and addresses of patients have typically been excluded (blacked out) from them. Most privileges belong to the patient, not to the physicians.

Finally, financial support must be forthcoming from the healthcare institution for all ethics committee activities.[222] Except for financial and other sup-

219. *Id.*

220. *Id.*

221. 42 U.S.C. § 1320D-6; Annas, N. Eng. J. Med. 1486 (April 10, 2003); Orange County Register, April 12, 2003, at Bus 1.

222. Dougherty, *The Costs of Commercial Medicine*, Theoretical Medicine 11 (1991):

> The director of a nonprofit state university's health care institution expressed a view following a series of ethical breaches at that institution, some of which happened as a result of the major defaults in management oversight. At a committee meeting he decided that "organizational ethics" was not a consideration of the bioethics committee. He stated that ethics was only a "personal matter" expected of all the employees. Several people who were at that meeting told me how disappointed they were. Yet, he was later transferred to another state institution where he received higher pay.

port, including record keeping and audit, medical ethics committees should operate independent of the healthcare organization.

good v. Good

All types of committees that serve in hospitals should be a part of the organization's hierarchy

v.

Because most matters referred to bioethics committees are also ultra-medical, these committees should essentially act independent of decisional control by the organization's hierarchy.

Twenty-first Century Approach

1. Many bioethics committees should acknowledge that their own medical institution's administration might offer only limited help in normative ethical decision making. Such committees should normally:
   (a) put increased emphasis upon locating members with backgrounds in other fields (including bioethics and applicable law).
   (b) establish a record system with sufficient independence from the hierarchy of the healthcare facility so that potential conflicts of interest between members and the organization's officials (including administrators, corporate counsel, and risk managers) are reduced or eliminated.
2. For their part, healthcare institutions should:
   (a) set standards for record keeping so that ethics committees may be evaluated.
   (b) make arrangements for the conduct of appropriate audits periodically based upon the committee's records to ascertain whether the committee is making independent recommendations and keeping adequate records thereof.
   (c) seek to overcome current dominance on the committees of physicians who have only a very limited bioethical education (including its chairman), in order to broaden consideration by other professionals with expertise in ultra-medical disciplines.

Everyone wants to live at the expense of the state. They forget that the state lives at the expense of everyone.

Frederic Bastiaat

7. Organizational Ethics

The increasing costs of new technology and expensive government entitlement programs have changed the practice of medicine. Healthcare is generally viewed as a commodity and patients are consumers.

Organizational ethics is a subject covered in some detail by the JCAHO standards. These matters may be of considerable interest for bioethicists because there are ethical issues in the business of healthcare delivery that affect individual patient care. Some institutions either set up separate organizational ethics committees or combine such activities with those of their bioethics committees. Unless an organization's ethics committee can separate itself from direction by the institution's management, the latter may tend to cover up its shortcomings.

The AHA has argued that bioethics committees should continue their practice of limited activity in healthcare institutions to prevent these committees from becoming "overburdened with broader institutional dilemmas." However, it is a strong argument only where certain other considerations can be ignored. First, it must be recognized that bioethical matters in healthcare represent a "seamless web," as enforceable matters are within the law. If virtually all ethical affairs are confined to those concerned only with cases involving individual patients admitted into the hospital, many of the committee members tend to disregard other ethical issues peculiar to other institutions. However, JCAHO added "organizational ethics" to its standards for accrediting healthcare institutions.

If a healthcare institution wishes to emphasize ethical standards, the institution must take significant action to sanction those employees who violate these standards. A simple warning, usually given in private, may create an atmosphere conducive to repetition by others if it appeared that the person receiving the warning for his behavior either "got away with it" or that his action had "the tacit approval" of management. Managers must be able to risk a possible lawsuit by disgruntled employees. For these reasons, the 1991 Federal Sentencing Guidelines provide increasing sanctions where there was a lack of investigation and reporting on misdeeds.[223] Care is required

223. Paine, *Managing Organizational Integrity*, HARV. BUS. REV., March/April, 1994, at 106.

for patients even without "allocating guidelines at gatekeeping directives."[224]

Healthcare institutions may have two offices, an office for the bioethics committee and another ethics office (under a manager of the organization). Organizational ethics violations are then defined to be excluded from matters of interest to the bioethics committee. Many managers feel that such an ethical violation is damaging to an institution's image and can or will arise sometime in all large organizations because of unethical employees. The most fatal harm was called "the culture of *cover-up*" by management, which has been a tendency of many organizations.

a) Health Maintenance Organizations (HMOs); ERISA and Rationing

In 1960 the health sector in the United States consumed about 6 percent of GNP. By 1997 nearly 15 percent of GNP was spent on health, "which had become the largest industry in the U.S., employing 1 of every 7 Americans."[225] By 1981, after several large mergers, nearly three quarters of the beds of profit-making multihospital systems were operated by the top three companies.[226] The top three nonprofits operated less than a tenth of the beds in nonprofit systems.[227] At that time, the AMA functioned to "represent doctors in negotiations with hospitals and Health Maintenance Organization (HMOs)" and the efforts to control expenditures for health services also

224. The *Code of Medical Ethics* states:

> Regardless of any allocation guidelines or gatekeeper directive, physicians must advocate for any care they believe will materially benefit their patients. Section 8.13 (2)(b).

225. BASCH, TEXTBOOK OF INTERNATIONAL HEALTH (1999), *citing* D.M. FOX, *Managed care: the third reorganization of health care*, 46 J. AM. GERIATRICS SOC'Y 314-17 (1998).

226. Hospital Corp. of America, Humana, and American Medical Int'l.

227. STARR, THE SOCIAL TRANSFORMATION OF AMERICAN MEDICINE 432 (1982). "New organizations, such as HMOs, reduce the demand for hospital care, and the growing supply of physicians encourages doctors to 'invade' services performed by hospitals to capture a larger share of ancillary profits. Hospitals face a more competitive market, and many may not endure." (p. 436)

stimulated corporate development.[228] The majority of HMOs were being drawn into several large networks run by Kaiser, Blue Cross, INA, and Prudential as more doctors entered group practice, more hospitals became multihospital systems, and more insurance companies became "directly involved in providing medical care through HMOs."[229]

Medical and dental plans were also cut, with growing numbers of workers being shunted into health maintenance organizations (HMOs) that many people found were skilled at erecting obstacles to seeing a doctor or getting needed care. Even those workers who stayed in medical insurance plans that let them pick their doctors found that more and more procedures were excluded from coverage. However, the larger HMOs became "both insurer and care provider—the doctors work for the same company that pays for treatment."[230]

---

228. *Id.* at 427-28.

> Health care in Israel is provided through "Patient Cooperatives" roughly equivalent to health maintenance organizations (HMOs). . . . There are now three large HMOs and about a dozen small ones operating in Israel. In the mid-1990s, the Knesset finally ratified a national health bill that had been debated, in one form or another, since 1948, when Israel was established. The bill guaranteed universal nearly free access to health care, funded by a newly levied health tax. The care itself is provided by existing HMOs, which receive budget allocations based on their membership rolls. Each citizen is free to choose any provider and to move from provider to provider.

ERON, REAL JEWS 163 (2003).

229. *Id.* at 439-40.

> Kaiser Permanente was rated among California's top health plans in recent reviews by the National Committee for Quality Assurance (NCQA)—the nation's leading reviewer of health plan quality. . . . Medical and dental plans were also cut, with growing numbers of workers being shunted into health maintenance organizations (HMOs), which many people found were skilled at erecting obstacles to seeing a doctor or getting needed care. Even those workers who stayed in medical insurance plans that let them pick their doctors found that more and more procedures were excluded from coverage.

JOHNSTON, PERFECTLY LEGAL 53 (2004).

230. Starr, *Simple Fairness: Ending Discrimination in Health Insurance Coverage of Addiction Treatment*, YALE L. J. (2002), at 2321. "This creates cost-cutting incentives that help control insurance premiums, but may compromise the quality of care." *Id.*

In 1990, the Council on Ethical and Judicial Affairs, American Medical Association (AMA) concluded that the welfare of patients must remain the first concern of physicians working in HMOs. However, their managed-care plans often constrain the costs of participating physician practices, which appear to be rising almost 10 percent per year. Physicians are encouraged to make cost-conscious decisions and may be granted bonuses on their limited budget; the proportion of the population enrolled in HMOs increased from 13.5 percent in 1990 to 30 percent by 2000.[231] Because it is unethical to knowingly provide unnecessary care, physicians must assess the seriousness of patients' conditions accurately.[232]

Neither harm nor inadequate outcomes have been fully demonstrated in managed-care arrangements.[233] A 2002 study showed "there are few overall differences in assessments of medical care between enrollees in for-profit and non-profit HMOs."[234] Justice Souter, writing for a unanimous Supreme Court, concluded that the courts should not set standards that define the "unacceptably risky HMO structure," which "would be nothing less than elimination of the for-profit HMO."[235]

However, a physician's obligation to disclose treatment alternatives to patients "includes informing patients of all their treatment options, even those

---

231. Orange County Register, Sept. 10, 2002, based on data from the American Association of Health Plans, a trade association for the managed-care industry; Health, United States, 2001: Urban and Rural Health Chartbook (National Center for Health Statistics, 2001).

232. The American Medical Association (AMA) proposed legislation that would require managed care organizations to establish a medical staff structure, much like that in existence in every hospital in the United States. This proposal includes . . . a medical board composed entirely of participating physicians . . . responsible for periodically reviewing restrictions on services to subscribers.

Ethics in Managed Care—AMA Council on Ethical and Judicial Affairs 330.

233. *Id.* at 333; AMA, Physician Health Plans and Networks Act of 1994.

234. 346 N. Eng. J. Med. 1288-93 (2002). In 2004, a study of the likely effects of financial incentives showed that the rates of use of high-cost operative procedures were not lower among beneficiaries enrolled in for-profit health plans than among those enrolled in not-for-profit health plans. N. Eng. J. Med. 143 (Jan. 8, 2004).

235. Pegram v. Herdrich, 530 U.S. 211, 220 (2000) at 221 and 233. *See also* Aetna Health Inc. v. Juan Davila, 542 U.S. 200, 124 S. Ct. 2488 (2004).

that may not be covered under the terms of the managed care plan. Patients may then determine whether an appeal is appropriate or whether they wish to seek care outside the plan for treatment alternatives that are not covered."[236]

The Federal Trade Commission (FTC) was upheld when it prohibited the AMA from "regulating" or "advising on the ethical propriety of . . . the consideration offered or provided to any physician in any contract with any entity that offers physicians' services to the public."[237] The belief that "the patient's interest always coincides with the physician's interest" has been questioned, and it was recommended that "the objectives, organization, and function of managed care be added to clinical training" at medical schools.[238]

In 2002, a study by Quality Assurance, a Washington-based HMO accrediting group, described "a slow but steady improvement in the quality of care being delivered to Americans."[239] The Employee Retirement Income Security Act of 1974 (ERISA), which regulates employee benefits, was originally held to supersede any and all state laws that relate to any employee benefit plan.[240] However, an Illinois law that gives patients a right to timely review of medical claims by an outside doctor was ruled (5:4) to be closer to a state insurance regulation than a new employee benefit;[241] this decision did not endorse any

236. *Ethics in Managed Care—AMA Council on Ethical and Judicial Affairs*, JAMA 331 (Jan. 25, 1995). "For-profit companies have a major role in the health maintenance organization (HMO) industry. Thanks to the wide for-profit status and the aggressive expansion of large for-profit companies, the market share of for-profit plans increased from one quarter of HMO enrollment in the mid-1980s to approximately two thirds by the late 1990s." *Id.*

237. FTC Order, Docket No. 99064, 248 JAMA 981-82 (1982). Section 9.12 of the *Code of Medical Ethics* states: "The creation of the patient-physician relationship is contractual in nature. Generally, both the physician and the patient are free to enter into or decline the relationship."

238. Coulehan & Williams, Academic Med., June 2001, at 596-604.

239. Orange County Register, Sept. 19, 2002, at News 11.

240. 29 U.S.C. §§ 1001-1461 (2000). Pilot Life Ins. Co. v. Dedeaux, 481 U.S. 41 (1987). A Mississippi dockworker's judgment against an insurance firm for denying him disability benefits was overturned by the Supreme Court's holding that ERISA preempted all such state law claims.

241. Rush Prudential HMO v. Moran, 122 S. Ct. 2151 (2002). An Illinois woman filed suit to compel her HMO to pay for her husband's experimental nerve surgery (which was successful). A.B.A. J., Sept. 2003, at 54. In *Kentucky Ass'n of Health Plans, Inc. v. Miller*, 123 S. Ct. 1471, 1474 (2003), the Supreme Court unanimously held that states' "any willing provider" laws are not preempted by ERISA.

lawsuits against HMOs for denying a particular medical care. State laws apply to individuals who are not in employer-sponsored health plans, and 41 states passed measures allowing consumers to ask for an external review by independent medical experts.[242] A survey indicated that the money spent by insurers and consumers exceeded 10 percent each of the preceding two years,[243] and an HMO survey showed that, as a result, their rates would jump 17 percent on average in 2003.[244] Effective January 1, 2003, California, a state with one of the highest levels of HMOs, prohibited the issuing, amending, or renewing of any contract that gives the insurer the authority to change a material term of the contract "unless the change has first been negotiated and agreed to by the provider and the plan." Also, insurers became unable to compel doctors to take more patients.[245] However, in 2004 the U.S. Supreme Court ruled that patients cannot sue health insurers for refusing to cover doctor-recommended medical care. This decision effectively invalidated a state law that lets patients seek damages when they believe an HMO has improperly denied coverage.[246] A Texas statute indicated that managed-care organizations (such as HMOs) became li-

242. AARP, Feb. 2, 2002, at 15. ERISA's preemption doctrine states that it "shall supersede any and all state laws insofar as they related to any employee benefit plan." 29 U.S.C. § 1144. Nevertheless, 11 states—Arizona, California, Georgia, Maine, New Jersey, North Carolina, Oklahoma, Oregon, Texas, Washington and West Virginia—sought to allow patients to sue their HMOs for damages if care denied to them causes some harm.

243. A study by the Center for Studying Health System Change, a nonpartisan research group in Washington, D.C. USA TODAY, Sept. 25, 2002, at 38.

244. http://www.reutershealth.com/archive/2002/07/01/.html. In 2003 the U.S. Court of Appeals for the Second Circuit held that an HMO could not claim that ERISA preempts an allegedly flawed "mixed eligibility and treatment decision" with respect to a beneficiary's symptoms under the plaintiff's medical malpractice claim. Cicio v. Does, 321 F.3d 83, 106 (2d Cir. 2003).

245. VEATCH, THE BASICS OF BIOETHICS, 2d ed. 127 (2003); Associated Press, Nov. 6, 2002.

246. Aetna Health Inc. v. Davila; CIGNA Healthcare of Texas, Inc. v. Calad, 124 S. Ct. 2488 (2004). This decision essentially negated the laws in 10 states (including California and Texas). ORANGE COUNTY REGISTER, June 22, 2004, at 1. The Court unanimously declared all such claims were preempted by ERISA. In addition, a U.S. Court of Appeals has held that an HMO's fiduciary obligations under ERISA do not automatically require it to disclose physician incentive arrangements to HMO members. Horvath v. Keystone Health Plan East, Inc., 333 F.3d 450 (2003).

able for failure to provide necessary medical treatment for a patient.[247] However, ERISA provides a remedy for a plan participant "to recover benefits due to him under the terms of his plan, to enforce his rights under the terms of the plan."[248] That was interpreted as preempting any claim of denial of benefits under Texas law. HMOs, which are currently facing rising healthcare costs, attempt to control premiums on their insurance.[249]

HMOs are managed-care organizations that compete with other organizations in the healthcare market and "exist to coordinate individual decisions through the price mechanism to yield an acceptable distribution of goods and services." They fall within a basic subject matter of economics and are managing a "cost-conscious physician-administered standard of care—a rationing strategy." An economist by training stated this to be "in direct contrast to the bioethics perspective."[250] This is because "[e]conomists do not expect [a managed care organization] to view the task of determining its [standard of care (SOC)] as one of distributing fixed resources over a fixed population according to principles of justice." It was stated, "There is no easy resolution to this conflict."

The AMA *Code of Medical Ethics* (2003) does not totally agree that a true conflict exists, nor do all bioethicists. The Code points to the fact that under Rule 8.13, "expansion of managed care has brought a variety of changes to medicine for physicians with complex referral restrictions and benefits packages for patients, including new and different reimbursement systems." It then states that managed care plans should "allow physicians to have meaningful input into the plan's development of allocation guidelines." Their obligation to disclose treatment alternatives to patients includes "even those that may not be

---

247. Texas Health Care Liability Act, Tex. Civ. Prac. & Rem. Code §§ 88.001-88.003.

248. ERISA, 29 U.S.C. § 1132(a)(1)(B).

249. Lent, et al., *Health spending rebound continues in 2002*, Health Aff. 147-59 (2004).

250. Mary Ann Baily, *Managed Care Organizations and the Rationing Problem*, 33:1 Hast. Ctr. Rep. 33-42 (2003). The enrolled population, the available resources, and the SOC are actually interdependent, because people enter and exit the organization in response to the SOC and the premium. This interdependence seriously undermines the bioethics view of the MCO as a stable mini-society establishing a stable morally obligatory standard by democratic deliberation.

covered under the terms of the managed care plan," and the financial incentives in the plan "are permissible only if they promote the cost-effective delivery of healthcare and not the withholding of medically necessary care." Assuming that these conditions have been met by the HMOs, their existence appears to have no necessary conflict with bioethical standards. The economist noted that although "business people have ethical obligations," their choice of product characteristics is not considered an ethical issue. Even though "consumers are well-informed about the various [standards of care] available," it was stated that some of those standards "may not meet ethical criteria for adequacy." However, bioethics does not require that all healthcare institutes be adequate to handle all medical problems and all diseases at all times.[251] It appears that the goal of physicians to limit beneficial care has been increasing in medicine and ethics throughout the country, with overall public rationing receiving the most attention in Oregon.[252] In November 2002, Oregon voters overwhelmingly defeated (79 percent to 21 percent) an initiative to make Oregon the first state to offer free tax-financed medical care for everyone.[253] As bioethicist Robert Veatch wrote in 2003, "[S]ome people argue that we do not need to ration healthcare [but] rationing is inevitable."[254]

Self-deceit, this fatal weakness of mankind, is the source of half the disorders of human life.

Adam Smith

---

251. *Id. Also see* Buchanan, *Managed Care: Rationing without Justice, But Not Unjustly*, 23:4 J. HEALTH POLITICS, POL'Y & LAW 617-34 (1998).

252. ABIM Foundation, *Medical Professionalism in the New Millennium: A Physician Charter*, 136:3 ANNALS OF INTERNAL MED. 243-46 (2002); MENZEL, STRONG MEDICINE: THE ETHICAL RATIONING OF HEALTHCARE (Oxford Univ. Press, 1990); UBEL, PRICING LIFE: WHY IT'S TIME FOR HEALTHCARE RATIONING (MIT Press, 2000); Jacobs et al., *The Oregon Health Plan and the political paradox of rationing: What advocates and critics have claimed and what Oregon did*, 24 J. HEALTH POLITICS, POL'Y & LAW 161-80 (1999); Leichter, *Oregon's bold experiment: Whatever happened to rationing?* 24 J. HEALTH POLITICS, POL'Y & LAW 147-60 (1999); and Bodenheimer, *The Oregon Health Plan—Lessons for the nation,* N. ENG. J. MED. 651-55 (1997).

253. Associated Press, Nov. 6, 2002.

254. VEATCH, THE BASICS OF BIOETHICS, 2d ed. (2003).

b) *Fraud in Medicare and Medicaid; Contracting Practices of Healthcare Organizations and Needs of Guidelines of the FAR*

During the final year of the twentieth century, Medicare lost $13.5 billion due to fraud, waste, and mistakes.[255] This loss of 10 percent of every Medicare dollar headed Medicare toward bankruptcy. Whistleblowers have been reporting fraud by healthcare organizations so that they may receive a significant (but smaller) portion of the government's recovery under the False Claims Act (FCA). It must be shown that they were an "original source" in a whistleblower action against a hospital's owner.[256] Under this statute, the U.S. government has recovered almost $600 million in 2002 on about 2,000 cases of healthcare fraud. This also became a reduction in our greatly increasing Medicare costs.[257] The Medicare law provides that no payment may be made for services if the person receiving such services "has no legal obligation to

---

255. ORANGE COUNTY REGISTER, March 18, 2000. That is, Medicare lost more than "the $12.6 billion in losses in 1998, but remains down significantly from losses in the $20 billion range in previous years by the Federal insurance plan for the elderly and disabled."

256. 31 U.S.C.A. §§ 3729-3733 (2002). United States *ex rel.* Dhawan v. New York City Health & Hospitals Corp., 252 F.3d 118 (2d Cir. 2001). A whistleblower may share in a government fraud settlement even when the government has not intervened in his False Claims Act suit. This is because his *qui tam suit constitutes pursuit of "alternate remedy" under* 31 U.S.C. § 3 730(c)(5); United States *ex rel.* Bledsoe v. Community Health Systems Inc., 6th Cir., No. 01-6375 (Sept. 10, 2003). Further, it has been held that the FCA does not preclude federal employees who learn of fraud on the job and have a duty to report it from filing whistleblower suits. United States *ex rel.* Holmes v. Consumer Ins. Group, 10th Cir. (en banc), No. 01-1077 (Feb. 10, 2003); "'Federal employees may be considered "persons" eligible to bring qui tam suits under the FCA, 31 U.S.C. § 3730(b)(1), because the ordinary meaning of the term is broad, and reading it to exclude federal employees would render some specific statutory exclusions superfluous,' Judge Mary Beck Briscoe said in rejecting the government's public policy arguments." F.C.R. (Feb. 8, 2003) at 214.

257. *See* WEST GROUP, GOV'T CONTR. YEAR OF REVIEW 11.4 (2002). Nobel laureate Milton Friedman, in his *Capitalism and Freedom* (1962), stated, "There is one and only one social responsibility of a business, to use its resources and engage in activities designed to increase its profits so long as it stays within the rules of game, which is to say, engages in open and free competition, *without deception or fraud.*" (p. 133) (Emphasis added.)

pay."[258] An appellate court held that the Medicare fraud case, civil penalties, and treble damages awarded under the False Claims Act (FCA) *qui tam* provisions are "not purely remedial" and are subject to review under the Eighth Amendment's excessive fines clause.[259]

Over 550,000 physicians provide Medicare to 39 million disabled or elderly beneficiaries.[260] However, no healthcare institutions are bound by the Federal Acquisition Regulation (FAR) when spending billions of dollars in procurement.(This is a matter covered in another volume[261] and Appendix C) Nevertheless, settlements for a large portion of these dollars have been recovered from both healthcare institutions and drug companies. A few states have somewhat similar false claims laws that are applicable to the healthcare industry.[262]

Procurement contracts are used when the agency's principal purpose is to acquire construction, products, or services for the direct benefit or use of the federal government.[263] Grants and cooperative agreements, on the other

---

258. 42 U.S.C. § 1395y(a)(2). "Medicare, as an example of a 'national' health care financing system, has more than 110,000 pages of rules and regulations that mire professionals in many hours of complicated paperwork." Donald J. Palisano, president of the AMA, "Debating How to Fix Health Insurance." *Letters*, 303 SCIENCE 1467-68 (March 5, 2004).

259. United States v. Mackby, 261 F.3d 821 (9th Cir. 2001).

260. N. ENG. J. MED. 1924 (June 13, 2002).

261. *See* KEYES, *Fraud, Whistleblower, and the False Claims Act,* ch. 3, sub. 3.11, *in* GOVERNMENT CONTRACTS UNDER THE FEDERAL ACQUISTION REGULATION, 3d ed. (2003). In February 2003, the West Group published a summary of qui tam statistics showing recoveries in Health and Human Services of $3,902,704,396 and in Defense Department of $1,396,326,012. *West Group Government Contracts Year in Review Conference* (2003).

262. "Tennessee, California, and Texas also have false claims provisions specific to the healthcare industry, which parallels the growth in interest in Federal false claims actions against healthcare providers." Dauer & Dauer, *Incipient Issues in State and Local False Claims Authority,* 76 F.C.R. 635, June 19, 2001. "According to the *San Francisco Chronicle,* April 18, 2005, those convicted of health care fraud usually get short federal prison sentences—an average of just 14 months for criminal schemes that often total millions of dollars in losses and threaten the Medicare benefits of law-abiding citizens." Stacy, president, We the People Institute.

263. 31 U.S.C. § 6303 (1994). *See also* 31 U.S.C. §§ 6304, 6305 (1994).

hand, are used when the principal purpose is to transfer a thing of value to a state or local government or other recipient in order to promote a broad national interest. In 2002 "the Federal government provided 57 percent of all (federal and state) Medicaid expenditures, which totaled $259 billion—$2 billion more than Medicare expenditures."[264] The federal statute titled the State Children's Health Insurance Program (SCHIP) "provides coverage to some 4.6 million children whose families earn incomes that are too high to allow them to qualify for Medicaid but too low to permit them to afford private health insurance."[265] In 2004 the cost of the program shot up to exceed $300 billion a year.[266] "Medicaid has grown to a projected total of $330 billion in fiscal 2005 (nearly 2.7 percent of GDP). Outlays have increased more than 50 percent since 2000, and Medicaid, the program created as an afterthought, has become even larger than Medicare. The program provides first-dollar coverage of all available medical benefits, the only cost to the patient being a $3 copayment per doctor visit, which is routinely waived if the patient does not have the cash. Because patients never see a bill for the 'free' care they receive, they have no incentive to use medical services efficiently."[267]

A principal difference between a grant and a cooperative agreement is that there has been a greater degree of federal participation in the former than in the latter. The Federal Grant and Cooperative Agreement Act of 1977 encourages competition in the award of grants and cooperative agreements.[268] Only statutes for some grants, such as those awarded by the U.S. Department of Agriculture, specifically require competition.[269] Because grants are not sub-

264. N. Eng. J. Med. 2141 (May 22, 2003).

265. "Medicaid also finances the major gaps in the coverage provided by Medicare to the 6 million persons who are defined as 'dually eligible' for both programs." *Id.* The Medicare Modernization Act of 2003 provides Medicare beneficiaries with a prescription drug benefit that pays 75 percent of a beneficiary's drug costs up to $2,250 a year. After a beneficiary has spent another $3,600, Medicare pays 95 percent of additional drug costs. Hacker & Marmor, *Medicare Reform: Fact, Fiction and Foolishness*, Public Policy and Aging Report 1 (2004); Oberlander, *The Politics of Medicare Reform*, Wash. & Lee L. Rev. (2003) at 1135. (See Chapter 7, 2c.)

266. Orange County Register, June 16, 2004, at News 8.

267. American Institute for Economic Research 78-79 (July 25, 2005).

268. 31 U.S.C. § 6301(3) (1994).

269. 7 U.S.C. § 450i.

ject to the FAR, there is typically neither any bidding nor any source selection evaluation in the manner described in the FAR. Moreover, the recipient of a grant need not even be "responsible" (as described therein).

Healthcare organizations that receive grants for Medicaid, Medicare or research should be "guided by the FAR," particularly those portions mentioned above and many others, including such provisions as are applicable to Environment Cost Accounting Standards, Cost Principles, service contracting, and to subcontracting by prime contractors.[270] Although Department of Health and Human Services' *Application for Continuation Grant Certificate of Costs* in 2001 began to require funded recipients to certify that indirect costs have been calculated in a manner consistent with federal guidelines, this is only a very small portion of those many and much-needed guidelines.[271] In 2003 the White House Office of Management and Budget (OMB) Office of Information and Regulatory Affairs (OIRA) issued a Proposed Bulletin on Peer Review Quality. The bulletin would establish government-wide standards for peer review (that support a major regulatory action or that might affect important public policies or decisions in the private sector "with a possible impact of more than $100 million in any year."[272] Recent attention has been brought to hospital corporations that have violated Security and Exchange Commission (SEC) regulations. (See Appendix C.) For example, the chair of the HealthSouth company was charged by the SEC in 2003 with mastermind-

---

270. *E.g., see* Parts 23, 30, 31, 37, and 44 of the F.A.R. (which are also the chapter numbers in *Government Contracts Under the Federal Acquisition Regulation* (West Group 2003).

271. *See* KEYES, GOVERNMENT CONTRACTS UNDER THE FEDERAL ACQUISITION REGULATION 3d ed. (West 2003).

272. *See* OMB's proposed bulletin on peer review and information quality, Sept. 15, 2003 (68 Fed. Reg. 54,023-29). The administrator of OIRA said, "The goal is fewer lawsuits and a more consistent regulatory environment, which is good for consumers and businesses." 3 N. ENG. J. MED. 104 (Jan. 8, 2004). However government procurement of more than $250 billion is spent annually in public funds, under the FAR and the Financial Accounting Standards Board (FASB), which sets the accounting principles used by publicly traded companies, and the Securities and Exchange Commission, which delegated accounting standards to the FASB, the healthcare industry is supported by "grants" of greater amounts.

ing a scheme to inflate HealthSouth's earnings by at least $1.4 billion since 1999. The company fired its auditor and its board of directors fired the company's chair.[273]

good v. Good

Organizational ethics committees are unnecessary because virtually all employees or agents are responsible for themselves and not for the healthcare organization's ethics; further, there is no need to regulate management procurement using grant funds.

v.

Either the ethics committee or another independent group of the healthcare institute should provide organizational ethics oversight, as well as governmental procurement guidelines for use of taxpayer funds.

Twenty-first Century Approach

1. Healthcare organizations should either (a) refer all ethical matters to their hospital ethics committees or (b) have another independent committee consider certain organizational ethical matters, define the separation, and maintain coordination with its other ethics committee.
2. In view of three key facts, namely:
   (a) As of the beginning of the third millennium, healthcare institutions have been receiving grants amounting to several hundreds of billions of dollars per year of taxpayer money (for Medicare, Medicaid, etc.) with virtually no regulations governing or guiding expenditure of government funds for the procurements made by these institutions;
   (b) Government (taxpayer) funds for federal contracts each year, which are governed by the Federal Acquisition Regulation (FAR), amount to much less than that granted to healthcare institutions; and

273. USA Today, April 1, 2003, at 3B. "On July 30, 2002, the Sarbanes-Oxley Act, 15 U.S.C. § 7201, charged the SEC with promulgating regulations on 'auditor independence.' These were also issued in early 2003 to require top executives to certify that they've disclosed to auditors 'any fraud, whether or not material, that involves management or other employees who have a significant role in the [company's] internal controls.'" A.B.A. J., June 2003, at 48.

(c) Healthcare institutions that have committed frauds have had to pay greater amounts back to the government than other organizations that were subject to the FAR,

the FAR should be prescribed as guidance applicable to healthcare institutions that qualify for federal grants.

Part II

The Issues Bioethics Must Continue to Solve During the Twenty-first Century

We have contented ourselves by assuming that marriage makes sex relations respectable. We have not yet said that it is only beautiful sex relations that can make marriage lovely.

Mary Ware Dennett

Years ago we decided to have only one child; having a small family was, for us, one very effective way to lessen dramatically our "footprint" on the planet.

Ehrlich, *Betrayal of Science and Reason*, 1996, p. 20

The Apostle most explicitly declared that for no cause whatsoever may a woman marry another without being an adulteress if her husband is still alive.

St. Jerome, ca. 342-420

7

To Conceive or Not to Conceive—The Ethics of Family Planning and Birth Control

1. ***Births within Marriage, Divorce, Same Sex, and Births Outside of Marriage***
2. ***The Great Ethical Problem of Overpopulation***
   a) **The Current Increasing Population**
   b) **The Ethical Need to Establish a National Goal for Family Planning; President Nixon's Goal**
   c) **Immigration Ethics and Population Increase**
   d) **Environmental (and Bioethical) Problems Growing Out of Overpopulation**
3. ***Ethical Actions That Can and Should Be Taken***
   a) **Involuntary Birth Controls**
      i. **Involuntary Sterilization, Castration, and Mental Retardation**
   b. **Voluntary Birth Controls**
      i. **Voluntary Sterilization**
      ii. **Contraception; Breast Feeding**
      iii. **Unintended Pregnancies**
      iv. **Emergency Contraception**

A woman shall not wear a man's apparel, nor shall a man put on a woman's garment; for whoever does such things is abhorrent to the Lord your God.

Deuteronomy 22:5

With regard to marriage, it is plain that it is in accordance with reason if the desire of connection is engendered not merely by external form, but by a love of begetting children and wisely educating them; and if, in addition, the love both of the husband and wife has for its cause not external form merely, but chiefly liberty of mind.

Spinoza, *Ethics* (Hafner 1949), "Of Human Bondage" p. 247.

1. Births within Marriage, Divorce, Same Sex, and Births Outside of Marriage

Historically, the purpose of marriage was to have and raise children. At the height of the Middle Ages, marriage was a holy sacrament. The joint interment scandal of the famous medieval unmarried lovers Abelard and Heloise caused them to be twice disinterred and buried separately and later together again.[1] However, in 1520, Martin Luther, the first major Protestant,

---

1. "In 1630 an abbess had had their skeletons exhumed and buried separately, thinking it indecent to have them lying together in deadly embrace in a common grave at the convent of the Paraclete; in 1917 sentimentalists exhumed them again and buried them together." Heinz, *The Last Passage*. 1999, p. 167.

> In ancient Greece Orpheus was said to have incurred women's wrath by rejecting them in favour of love for other males. Was it the case, then, that one cause of mythological family—disruption was the promotion of homosexual over heterosexual tendencies, with the consequent loosening of the marriage bond? The answer is emphatically: no . . . Moreover—an absolutely crucial point—it was assumed that each partner was expected also to take a full part in relationships with women: that is, there was no perceived incompatibility between, on the one hand, participation in the older/younger same-sex liaisons, and, on the other hand, marriage.

Buxton, The Complete World of Greek Mythology (2004), p. 174.

published a manifesto in Latin and German denying that marriage was a sacrament.[2] In the third millennium, the law only requires a "civil marriage" in most countries with exceptions such as Israel.[3]

Unlike with Jews or Islamics, Catholic marriages, if consummated, cannot be dissolved for any cause except death. At the turn of the third millennium, only a fraction of U.S. households with children remain traditional. The percentage of American households made up of married couples with children dropped from 45 percent in the early 1970s to just 26 percent in 1998, and only 56 percent of adults were married, compared with 75 percent in 1972 (when the survey was first taken).[4] In 1967, the Supreme Court held that the right to a civil marriage was fundamental and could be restricted only by the finding of a "compelling state interest."[5] Such an interest was

---

2. Manchester, *A World Lit by Fire*. 1993, p. 167: "[Luther] said that any wife married to an important man should sleep around until she conceived a child, which she could pass off as her husband's. If he objected, she could divorce him." See also Crocker, *Triumph*. 2001, p. 245. "He declared that lust was invincible *concupiscentia invincibilis*.

3. Reform Rabbi Uri Regev, who established the Israel Religious Action Center (IRAC) for non-Orthodox Jews, objected to Israel's marriage laws:

> Because literally hundreds of thousands of Israeli citizens cannot exercise their rights to marry because of religious compulsion. [This marriage] law has no parallels in any democratic country on earth. The only countries with parallel approaches are in fundamentalist Moslem states: Iran, Saudi Arabia, Afghanistan, Somalia—these are the only places you'll find laws like that of Israel, that outlaw or limit alternatives to religious marriages.

Efron, *Real Jews, Secular vs. Ultra-Orthodox* (2003), p. 226.

4. ORANGE COUNTY REGISTER, Nov. 24, 1999, at News 26. "While white families have an 81 percent chance of including a married couple, Hispanics have a 67 percent chance and African Americans a 46 percent chance." *Time*. Nov. 8, 2004, p. 60.

5. Loving v. Virginia, 388 U.S. 1, 12 (1967). The Court stated, "Biological race is no longer a significant category in U.S. marriage law." Nevertheless, South Carolina until 1998 and Alabama until 2000 retained antimiscegenation clauses in their respective state constitutions. Joseph Campbell stated, "[m]arriage is the relationship between two people; you have to think of yourself not as this one, but as these two as one." *Thou Art Thou*. 2001, p. 91.

found under a Utah statute prohibiting persons with HIV infection from marrying, which was actually struck down in 1993.[6]

Males are relatively more interested in their fianceé's fertility, sexual loyalty, and (sometimes) her virginity, whereas females apparently prefer strong, tall men with good genes, who are older than them and have an earning capacity.[7] Both the age chosen before marriage and the divorce rate have been climbing. In 1996, the Census Bureau reported that the typical age for a first marriage was 26.7 for men and 24.5 for women. However, the number of births outside of marriages exceeded 30 percent in both Britain and France.[8] "Analyzing a sampling of 600 children born to unwed mothers, University of Washington economists Shelly Lundberg and Elaina Rose found men are as much as 42 percent more likely to marry the mother of their child when the offspring is a boy."[9]

The annual divorce rate climbed from 4.3 million in the 1970s to 17.4 million by 1994 and continues at a higher rate. It has been suggested that Americans may be less tolerant of marital problems than were earlier gen-

---

6. T.E.P. v. Leavitt, 840 F. Supp. 110, D. Utah 1993.

> According to a 1999 study by the American Bar Association, a third of all women lawyers have never married, as compared to 8 percent of male lawyers, and nearly half of the women lawyers reported being unmarried, compared to 15 percent of the men. Most lawyers score high on the "thinking" scales and low on the "feeling" scales. Another major threat to lawyer marriages is alcoholism—a lawyer's way of dealing with the high stress.

THE NATIONAL L.J., April 12, 2004, p. 42.

7. David Buss, *Behavioral and Brain Sciences*. 12: —49. "The number of married teenagers increased during the 1990s, reversing a decade long decline. As reported by the Census Bureau, 4.5 percent of 15- to 19-year-olds were married in 2000 even though the typical age was 26.8 in 2000, up from 26.1 in 1990. However, the rise during the 1990s was attributed to the influx of immigrants."
ORANGE COUNTY REGISTER, Nov. 9, 2002, at News 27.

8. In Europe "the number of births outside marriage as a percentage of total births was just 7.3 percent in Italy in 1993, compared to Spain's, 10.5 percent; Germany's, 15.4 percent; Britain's, 32 percent; and France's, 34.9 percent." Ginsburg, *Italy and Its Discontents, 1980-2000* (2003), p. 74.

9. ORANGE COUNTY REGISTER, May 11, 2003, at News 25; the May issue of the journal DEMOGRAPHY.

erations.[10] In this connection it is noted that the Supreme Court of one state held that a man may not obtain a divorce on the ground of adultery predicated on his wife's sexual relationship with another woman. That court concluded that the "sexual intercourse" underlying such conduct can occur only between members of the opposite sex.[11] Women who marry at 25 or older have half the breakup rate. At the dawn of the twenty-first century, "40 percent of second marriages for women end in separation or divorce within 10 years."[12] In 1997, divorce became legal in Ireland for the first time in that overwhelmingly Roman Catholic country.[13] However, the Vatican stated that divorced Catholics who remarry should be urged to stop living in a "state of sin," meaning no sexual intercourse in the new relationship,[14] as noted in St. Jerome's epigraph. On other hand, in 1998, the Southern Baptist Convention, the nation's largest Protestant denomination, issued a statement that a

---

10. ORANGE COUNTY REGISTER, March 13, 1996. "Born-again Christians continue to have a higher likelihood of getting divorced than do non-Christians," and "Atheists are less likely to get divorced than are born-again Christians." BARNA, INDEX OF LEADING SPIRITUAL INDICATORS (1996) (cited in SHERMER, THE SCIENCE OF GOOD & EVIL (2004) p. 236.

11. *In re* Blanchflower, 150 N.H. 226 (2003). An Ontario court approved Canada's first same-sex divorce after a judge ruled that the definition of spouses in the Divorce Act is unconstitutional because it stated that only spouses—defined as a man and woman—can divorce. ORANGE COUNTY REGISTER, Sept. 16, 2004, at News 23.

12. ORANGE COUNTY REGISTER, May 21, 2001, at News 5, 25, and May 25, 2001, p. 27.

13. ORANGE COUNTY REGISTER, Feb. 27, 1997, at News 3. On Feb. 27, 1997, divorce became legal in Catholic Ireland. The people in virtually all Latin American countries are Roman Catholics. Also, in 2004, Chile's Congress passed a divorce law, defying the fierce opposition of the Catholic Church. It became the last country in the Americas to legalize divorce and authorizes couples to obtain a divorce only after living apart for at least one year, if both spouses agree, and three years if only one party agrees (unless one spouse proves domestic violence, homosexuality, prostitution, drug addiction or a criminal conviction). Amazingly, some 6,000 annulments were previously approved every year at a cost of $670. ORANGE COUNTY REGISTER, Nov. 19, 2004, at News 30.

14. ORANGE COUNTY REGISTER, Feb. 26, 1997, at News 1. In 2002, Pope John Paul II urged judges and lawyers to refrain from taking on divorce cases, calling the end of a marriage a "festering wound." He also called for more social recognition of "true matrimony" as opposed to other unions such as those between homosexuals. ORANGE COUNTY REGISTER, Jan. 29, 2002, at News 16.

woman should "submit herself graciously" to her husband's leadership.[15] Islamic law permits men to divorce their wives anytime without the approval of a court.[16]

Inbreeding or incest is illegal in most states.[17] Normally, where a boy and girl are brought together before one or the other is 30 months of age and raised in close domestic proximity, "they are devoid of later sexual interest in each other."[18] This is known as the "Westermarck Effect" and it applies to many animals as well. Nevertheless, it occurred in 1997 when a brother and sister who had at least three children together in Milwaukee were sent to prison for incest.[19] In 1879 the Supreme Court ruled that polygamy, as practiced by the Mormons, was "subversive of good order" and not sufficient to shield them from prosecution for maintaining multiple wives.[20] Although *polygamy* is a felony in Utah,[21] it is rarely prosecuted. It has been remarked that polygamists "get around bigamy statutes by legally marrying only one wife; the others are recognized by religious leaders, or simply by the indi-

15. INTERNATIONAL HERALD TRIBUNE, June 11, 1998, p. 3.

16. ORANGE COUNTY REGISTER, Jan. 28, 2000. Since 1997, 36 people have been charged under North Carolina's anticohabitation law; seven have been convicted as unmarried couples who were living together. North Carolina is one of seven U.S. states that still maintain such statutes. ORANGE COUNTY REGISTER, May 12, 2005, at News 14.

17. *E.g.*, Cal. Penal Code § 285 (1999). Persons being within the degrees of consanguinity within which marriages are declared by law to be incestuous and void, who intermarry with each other, or who commit fornication or adultery with each other, are punishable by imprisonment in the state prison. See also Cal. Fam. Code § 2200 (1999). See also Wis. Stat. § 944.06 (1999).

18. Wilson, *Consilience*. 1995, p. 173.

19. ORANGE COUNTY REGISTER, Nov. 12, 1997, at News 14. In 2003, a mother who was convicted of helping her husband impregnate her daughter was released from prison because she showed remorse. Her daughter had testified that at 16 she agreed to be inseminated because her stepfather had threatened her and her mother if she didn't comply. The baby, born in 1999, was put into foster care. The husband was convicted of rape and child endangerment and sent to prison for 20 years. ORANGE COUNTY REGISTER, Feb. 5, 2003, at News 18.

20. Reynolds v. United States, 98 U.S. 145, 1879.

21. *See* Utah Const., Art. III. In April 2000, a man in Utah with 5 wives and 25 children was finally charged with child rape, bigamy, and failing to pay child support. ORANGE COUNTY REGISTER, April 20, 2000, at News.

viduals themselves. Some argue that it is merely a consensual relationship between adults, although many children result."[22]

A federal district court ruled an inmate serving a life term had no right to send his semen in order to artificially inseminate his wife. However, this ruling was actually reversed on appeal. The majority of the appellate court stated that the right to procreate survives even during incarceration,[23] even though a 2000 report showed that "65 percent of inmates released from state prisons committed at least one serious new crime within three years."[24]

The report was prepared for the Bureau of Justice Statistics. The report examined 272,111 former inmates in 15 states during the first three years after their release. Studies dating back to the 1960s have all found that re-arrest rates of prisoners tend to hover at about two-thirds within the first three years after release. These findings have occurred even as the prison philosophy of the day has shifted from rehabilitation to getting tough on crime to deterrence, with seemingly little difference in outcome.

In 1996, Congress enacted the Defense of Marriage Act (DOMA), limiting the recognition for *same sex marriages*. While declaring "Federal neutrality" toward such marriages, it provides that should they occur in one state, no other state is required to recognize them through the "full faith and credit"

---

22. Orange County Register, Jan. 5, 1998. Incest also occurs between a parent and child. For example, in 2002 a father was convicted of repeatedly paying his 12-year-old daughter for sex and impregnating her. She didn't know she was pregnant until she showed up at an emergency room and later gave birth to a boy. DNA evidence was presented at trial and her father was sentenced to 19 years in prison. Orange County Register, June 22, 2002, at Local 7.

23. Orange County Register, Sept. 6, 2001, at News 10. The dissent stated that it cannot "accept the fact that there are certain downsides to being confined in prison." However, a prisoner's bid to procreate via mail was rejected on appeal in California by the Ninth Circuit Court of Appeals (ruling only 6-5).

24. Orange County Register, June 3, 2002, at News 14.

> The report was prepared for the Bureau of Justice Statistics. The report examined 272,111 former inmates in 15 states during the first three years after their release . . . studies dating back to the 1960s have all found that re-arrest rates of prisoners tend to hover at about two-thirds within the first three years after release. These findings have occurred even as the prison philosophy of the day has shifted from rehabilitation to getting tough on crime to deterrence, with seemingly little difference in outcome.

clause (Article IV Section 1 of the Constitution).[25] If any state were to grant *same-sex couples* the right to marry, such couples may be denied legal entitlements that flow from marriage in a sister state, while leaving those rights untouched under forum state law. The Supreme Court ruled that although a court may be guided by the forum state's "public policy" in determining the *law* applicable to a controversy: "The Full Faith and Credit Clause does not compel a state to substitute the statutes of other states for its own statutes dealing with a subject matter concerning which it is competent to legislate."[26] Several states specifically ban same-sex unions.[27] The 2000 U.S. decennial

---

25. Pub. L. 104-199 § 2(a), 110 Stat. 2419 (Sept. 21, 1996). It was passed by the U.S. House of Representatives in 1996 by a vote of 332 to 67, by the Senate 85 to 14, and signed by the President on Sept. 21, 1996. *See also* 28 U.S.C. § 1738 C (1999). In 2005, the 1996 Defense of Marriage Act was upheld by a federal court's dismissal of a lawsuit brought by two Florida women seeking to have their Massachusetts marriage recognized in their home state. ORANGE COUNTY REGISTER, Jan. 20, 2005, at News 14. The so-called mini-DOMAs in some 40 states should be applied to benefits-related statutes. Neil, ABA JOURNAL, September 2005, p. 24.

26. Baker v. General Motors Corp., 522 U.S. 222, 232-3 (1998). In 2004, voters in 11 states approved constitutional amendments to ban same-sex marriage—in most cases by overwhelming margins. ORANGE COUNTY REGISTER, Nov. 3, 2004. In spite of such views, "Dalai Lama's U.S. publisher even asked him to remove the injunctions against homosexuality from his book *Ethics for the New Millennium* for fear that they would offend American readers, and the Dalai Lama acquiesced." FRENCH, TIBET, TIBET (2003), p. 218.

27. *E.g.* Ind. Code Ann. § 31-11-1-1 (West 1999); and ME. REV. STAT. ANN. tit. 19-A, § 701(5) (West 1998) that also states:

> The union of one man and one woman joined in traditional monogamous marriage is of inestimable value to society; the State has a compelling interest to nurture and promote the unique institution of traditional monogamous marriage in the support of harmonious families and the physical and mental health of children; and that the State has the compelling interest in promoting moral values inherent in traditional monogamous marriage.

Id. § 650(1)(A). *See also* Rosengarten v. Downes, 71 Conn. App. 372 (2002), that held that it had no jurisdiction to dissolve the civil union of a resident who entered it with his same-sex partner. The court indicated that the civil union is not a "family relations" matter under Connecticut law and that the Vermont statute is not entitled to full faith and credit. A Georgia appeals court has also ruled that a civil union of two women in Vermont was not equivalent to a marriage for the purposes of interpreting a child-custody agreement entered into in Georgia. Burns v. Burns, 560 S.E.2d 47, 49 (Ga. Ct. App. 2002).

census showed that 0.6 percent of men and 0.5 percent of women 18 years of age and older live together as same-sex unmarried partners; about a quarter of a million children live in households headed by same-sex couples; and that nearly one in five people in same-sex couples is 55 and older.[28]

In 1993 Congress approved the "Don't Ask, Don't Tell" policy for the gays in the military. Over the period from 1994 through 2003, the 9,488 troops who left because of the policy made up only 0.37 percent of all troops discharged during the decade.[29] On July 19, 1993, President Clinton announced a compromise allowing homosexuals to serve in the military, but only if they refrained from all homosexual activity. In 2003, although the Supreme Court struck down a Texas antisodomy law that forbade homosexual sex (6:3),[30] the Court was silent regarding their marriages. In this connection, the historical and cultural homosexual acts of the fifth century B.C. were shown to be unrelated to marriages. This was summarized as follows:

> In ancient Greece, as is well known, a relation of apprenticeship often entailed sexual activity. In many cultures, homosexual acts take

28. Doyle, *Scientific American*. March 2005, p. 28. On April 5, 2005, Kansas became the latest and 18th state to pass a constitutional amendment barring gay marriage. ORANGE COUNTY REGISTER, April 7, 2005, at News 15.

29. ORANGE COUNTY REGISTER, Feb. 25, 2005, at News 8. This was a response to a GAO report. Although some law schools have challenged the military's "don't ask, don't tell" policy on homosexuality, Harvard Law School agreed to go along with that policy in connection with the recruitment access of its students who plan to join the judge advocate offices of the Department of Defense. THE NATIONAL JURIST (November 2005), p. 41.

30. Lawrence v. Texas, 123 S. Ct. 2472 (2003). Justice Kennedy declared that "a State's governing majority has traditionally viewed a particular practice as immoral is not a sufficient reason for upholding a law prohibiting the practice," and therefore that "the Texas statute furthers no legitimate state interest which can justify its intrusion into the individual's personal and private life." Texas was one of four states that banned sodomy among homosexuals; nine states banned the practice for everyone. California had repealed its antisodomy law in 1976. The Supreme Court reversed its 1986 decision Bowers v. Hardwick, 478 U.S. 186, that upheld Georgia's antisodomy law. In the Texas case the two men were having anal sex in private when they were arrested. ORANGE COUNTY REGISTER, June 27, 2003, at News 1.

place willingly, and happily, *between partners who are not "homosexuals,"* and *who are expecting thereby neither to reproduce nor to marry*. They are relevant as examples of the multiple meaning of sexual activity among humans, and as a demonstration of the cultural construction of "the homosexual."[31] (Emphasis added.)

It was also written that the alleged "gay gene—which has never subsequently been found—entered the popular mind as a fact of science."[32] Although in 2002 Chief Justice William Rehnquist acknowledged that "homosexuality has gained greater social acceptance,"[33] in 2003, the potential of such a legal marriage was found to have become opposed by most people.[34] In September of 2003, former California Governor Gray Davis (who was about to be removed from office) signed a law that gave homosexuals most of the commonly exercised legal rights of marriage.[35] In November of 2003,

---

31. MARKS, WHAT IT MEANS TO BE 98 PERCENT CHIMPANZEE (2002), p. 11. A federal court in Boston, Masschusetts, ruled that being called a homosexual is not libelous, stating: "In fact, a finding that such a statement is defamatory requires this court to legitimize the prejudice and bigotry that for too long have plagued the homosexual community." ORANGE COUNTY REGISTER, May 30, 2004, at News 30.

32. *Id.* p. 116. The AMA no longer declares homosexuality an illness.

33. Boy Scouts v. Dale, 530 U.S. 640 (2001). In this case the Boy Scouts, a private organization, was held to possess a constitutionally protected right of "expressive association" that would be violated if it had to grant a leadership role to a homosexual.

34. U.S.A., July 29, 2003, p. 1. "The survey found rising opposition to civil unions that would give gay couples some of the rights of married heterosexuals. They were opposed 57 percent–40 percent." *Id.*

35. ORANGE COUNTY REGISTER, Sept. 20, 2003, p. 1. A partner is not legally called a "spouse" as in Vermont, but he or she has child-custody rights and obligations to pay child support should a couple split up, family student housing and senior-citizen housing eligibility, exemption from estate and gift taxes involving the partner, and a right to financial support, including community property if the relationship ends. The law became effective Jan. 1, 2005. On Aug. 15, 2000, California Proposition 22, the "Limit on Marriage" initiative, passed with 61 percent of the vote. The law states: "Only marriage between a man and a woman is valid or recognized in California." In the state of Vermont the governor signed a bill legalizing "civil unions" for same-sex couples, which went into effect July 1, 2000. BALMER AND WINNER, PROTESTANTISM IN AMERICA (2002), p. 99.

the Massachusetts Supreme Court majority (4:3) became the only state to allow gay marriages. It permitted some 3,000 gay Massachusetts couples to wed. However, Massachusetts voters may change the state constitution to bar gay marriages, but allow same-sex couples to form unions that are partially outside of marriage.[36] For example, California provides domestic-partner rights to (a) adopt, as a stepchild, the child of their partner; (b) make health-care decisions for an incapacitated partners; (c) automatically inherit some of the partner's separate property if the partner dies without a will. As of 2005 the partner has (d) the obligations to pay child support, should the couple split up; (e) the right not to have to testify against a partner in state courts; (f) the right to consent to a partner's autopsy and make organ donations. However, unlike a civil marriage, a *partner* (1) *is not legally a spouse;* (2) receives no Social Security benefits; (3) cannot sponsor a partner who is a foreigner to become an American citizen; (4) has no right to file taxes jointly; and (5) has no legal recognition across state lines.[37] Technically, at least, it may be argued that such criteria might amount to those necessary in order to be excluded from the 2004 statement by President Bush complaining about the Massachusett's decision the previous year, namely, that "Marriage is a sacred institution between a man and a woman. Today's decision of the Massachusetts Supreme Judicial Court violates this important principle. I will work with congressional leaders and others to do what is legally necessary to defend the sanctity of marriage."[38] A professor who teaches constitu-

---

36. ORANGE COUNTY REGISTER, Nov. 19, 2003, at News 3. On Nov. 29, 2004, the U.S. Supreme Court declined to hear the Court's ruling overturning the Massachusetts ruling. ORANGE COUNTY REGISTER, Nov. 30, 2004, at News 13.

37. ORANGE COUNTY REGISTER, Feb. 5, 2004, at News 4. In 2005, the California Supreme Court ruled that a lesbian partner who agreed to raise twins is obligated to pay child support like any parent after a breakup. The court also ruled that a lesbian mother could not break a prebirth agreement with her former partner to share parental rights of their children. ORANGE COUNTY REGISTER, Aug. 23, 2005.

38. *Id.* ORANGE COUNTY REGISTER, Nov. 30, 2004, at News 13. For this reason homosexual sex was frequently condemned in the Bible: *Genesis* 9:1-8, *Leviticus* 18:22, *Romans* 1:26-27, I *Corinthians* 6:9-10, and 1 *s* 1:10. However, in CHRISTIANITY TODAY, Oct. 4, 1999, p. 50, Richard Mouw *et al.* wrote: "If people can demonstrate that they are emotionally and economically committed to one another, then they should have some of the tax benefits in that particular culture that would be given to a married couple."

tional law and who does not believe that the U.S. Constitution would require states to allow same-sex marriage on the basis of "equal protection" of the laws, stated: "We have not reached that point where the discrimination is on the basis of sexual preference . . . You can make the argument that it is rational to *treat marriage and* gay *partnership differently* since one leads to procreation of children and the other does not."[39] Although a worker at the U.S. National Cancer Institute in 1993 claimed to have identified a gene marker common to the gay men among families in which there were at least two admitted homosexual brothers,[40] no such results have been replicated in other family studies.[41]

The *Arizona statutes* that both *prohibit same-sex marriage and define valid marriage as one between man and woman* were held not to violate either the federal or state constitutions.[42] In a similar manner, the U.S. Court of Appeals for the Eleventh Circuit decided that a Florida statute prohibiting the adoption of children by any person who engages in homosexual activity does not violate the U.S. Constitution.[43]

---

39. He was referring to an addition to the California state constitution in 2000 that only marriage between a man and a woman is valid or recognized. (Cal. Fam. Code § 308.5), ORANGE COUNTY REGISTER, Aug. 13, 2004, p. 1. The Netherlands legalized gay marriage in 2001 and same-sex couples also have the right to adopt children. In 2003, they were legalized in Belgium, but without the right to adopt children. In 2005, Spain became the third country to legalize gay marriage (including the right to adopt children) and Canada legalized gay marriage. ORANGE COUNTY REGISTER, July 1, 2005, at News 25.

40. Hamer *et al.*. "A linkage between DNA markers on the X chromosome and male sexual orientation," SCIENCE (1993) pp. 321-37.

41. Rice *et al.* "Male homosexuality: absence of linkage to microsatellite markers at Xq28," *Science* (1999) pp. 665-67.

42. Standhardt v. Superior Court, Maricopa County, Ariz. Ct. App., 206 Ariz. 276 (2003).

43. Lofton v. Sec. of the Dept. of Children and Family Servs, 358 F.3d 804 (11th Cir. 2004); 125 S.Ct. 869 (2005). "A family is not an association of independent people; it is a human commitment designed to make possible the rearing of moral and healthy children. Governments care—or ought to care—about families for this reason, and scarcely for any other." JAMES Q. WILSON, THE MORAL SENSE.

good v. Good

While the commitment of incest, homosexuality, or polygamy deserves prosecution, there should be no illegality for breeding by male prison inmates, for the sending of his sperm to women outside of prison, or of marriages by homosexuals.

v.

The legality of sending sperm to any outside women by such convicted prisoners should be reversed in view of their overwhelming recidivism. The homosexual activity of people is not always illegal; however, the marriages of such individuals have no normal family basis toward their legality.

Twenty-first Century Approach

1. State legislatures should
   (a) enact a law making an increase in population illegal by sending sperm to women outside of the prison where felonious inmates are serving, and
   (b) not transfer all rights and conditions of same-sex "domestic partnerships" into legally identical "heterosexual marriages."

In the United States a large pool of new immigrants keeps the birthrate higher than in any other prosperous country.

Orange County Register, July 19, 1998, News 10

One in every five U.S. births is now to a Latina, the Census Bureau reports.

Orange County Register, February 13, 1998, News 25

No species has ever been able to multiply without limit. There are two biological checks upon rapid increase in number—a high mortality and a low fertility. Unlike other biological organisms, man can choose which of these checks shall be applied, but one of them must be.

George W. Cox

Readings in Conservation Ecology, 1969

2. The Great Ethical Problem of Overpopulation

a) The Current Increasing Population

The turn of the third millennium was also the moment in time when the world's population of human species exceeded 6 billion. In 2003 the U.N. Department of Economic and Social Affairs predicted a population of nine billion by the year 2300.[44] After a detailed analysis the world could support a population of only two billion.[45] The two-thirds of additional population means that we have entered a world crisis at the dawn of the twenty-first century—a status that is little recognized. It has been estimated that there are one hundred times more people on earth than any land animal of comparable size that ever lived.[46] In 1968, the United Nations' population forecasting unit

---

44. Villard and Cook, Infinite Worlds (2005) p. 206. For Catholics it was noted that "[t]he present Pope has ordered his followers not to limit the number of babies they have. If people followed his authority slavishly as he would wish, the results could be terrible famines, diseases and wars, caused by overcrowding." Dawkins, *A Devil's Chaplin* (2003), p. 245. However, in 2005 it was noted that "Muslims have the highest birth rate—three times the rate of non-Muslims—of any demographic group in Europe. 'With current trends,' Professor Bernard Lewis has said, 'Europe will have Muslim majorities in the population by the end of the 21st century. Europe will be part of the Arab west, of the Maghreb [Muslim northwest Africa].'" Time (Nov. 21, 2005) p. 162.

45. Kendall & D. Pimentel, *Constraints on the Expansion of the Global Food Supply*, Ambio 23, 1994, pp. 1998-2005.

> A world population of about two billion—*one-third of the present population—would not be living off progressively smaller plots of land.* To generalize from the example above of pre-modern corn yields, it could raise its food mostly by recycling nutrients rather than unloading fertilizers into the biosphere. It would not need to worry about either the cost of fertilizer or its impact on the environment. Rising populations would not be crowding out farmland, and *farmers would not be burning forests* to create more arable land. That alone would benefit agriculture and help to forestall the climate warming in which the human race is now engaged. And *arid countries would not be running out of water, as they are now.* (Emphasis added.)

Grant, Too Many People, 2000, p. 20.

46. Chivian, 1997, "Global environmental degradation and biodiversity loss: Implications for human health," in Biodiversity and Human Health (F. Grifo and J. Rosenthal, eds.).

reported that global population will peak at 11 billion about the year 2200.[47] Other experts state that 10 billion is an upper limit for sustained population levels;[48] but even that number might result in serious destruction of environmental resources.

The human species, with its domesticated plants and animals and agricultural and industrial practices, is now dramatically altering the conditions of existence for every living species on earth. The human species everywhere has killed most of its large predators and is hard at work eradicating those diseases that might still limit it numbers.

In 2004 the president of the Population Reference Bureau, which is supported by government, foundations, and other grants, stated, "Developing countries in Africa and Asia alone account for about 90 percent of the increase in world population."[49] Yet a senior demographer of the bureau pointed out that "The United States is the only more developed country where we anticipate a significant increase during the first half of the twenty-first century. The U.S. population of 294 million is projected to grow 43 percent and reach 410 million, due mainly to immigration."[50]

Nevertheless, in 2004 the U.S. Court of Appeals for the Ninth Circuit

---

47. Orange County Register, Feb. 22, 1998, at News 33. Even today we should recognize Thomas Malthus's pessimistic original work, *An Essay on the Principle of Population,* first published in 1798. His updated version of 1830 became crucial in Darwin's also original development of evolution.

48. Silver and DeFries, One Earth, One Future, 1992; Cohen, How Many People Can the Earth Support, 1995.

> The human species, with its domesticated plants and animals and agricultural and industrial practices, is now dramatically altering the conditions of existence for every living species on Earth. The Human species everywhere has killed most of its large predators and is hard at work eradicating those diseases that might still limit it numbers.

49. Today, about 5.2 billion live in less-developed countries. That population is projected to grow 55% by 2050 and exceed 8 billion. Their healthcare is improving, their populations include more people of childbearing age, and their birth rates remain high.

The Report of the Population Reference Bureau, Aug. 18, 2004, News 3; Hast. Ctr. Rep., Nov. -Dec. 1998, S16 and S19.

50. "The United States would remain the third most populous nation in 2050, behind India, which would rise to first place, and China, which would slip to second." *Id.* (The Report).

actually ruled that an employer's attempts to discover the immigration status of 23 former employees who were terminated undermined the anti-discrimination goals of Title VII of the 1964 Civil Rights Act.[51] However, this decision is being appealed in view of the U.S. Supreme Court ruling that undocumented workers are ineligible for damages under labor law.[52] Further, the denial of state university admission to illegal aliens is neither preempted by federal immigration law nor violates the Constitution.[53] The Court of Appeals for the Ninth Circuit granted asylum to a Chinese national who illegally married an older woman while he was underage. She became pregnant and forcibly aborted under China's population control policy. When her husband fled to the United States, the Board of Immigration Appeals (BIA) found no connection between China's population control policy and the alien's inability to establish a legal marriage. The Circuit Court stated that BIA's position "leads to absurd and wholly unacceptable results."[54]

Their opposition to the termination of out-of-wedlock pregnancies being population control was actually expressed as a basis for its ruling on this immigrant. (See Appendix A.) However, unfortunately such figures and warnings have been essentially disregarded by the U.N. International Conference on Population and Development. At a meeting held in Cairo, reproductive health was explained as:

> A state of complete physical, mental and social well-being is not merely the absence of disease or infirmity; in all matters relating to the repro-

51. Rivera v. NIBCO Inc., 384 F.3d 822, C.A.9 (Cal. 2004). The employees were being fired because they were not proficient in English.
52. Hoffman Plastic Compounds Inc. v. NLRB, 535 U.S. 137 (2002).
53. Equal Access Education v. Merten, E.D. Va., 305 F. Supp. 2d 585 (2004).
54. Ma v. Ashcroft, 361 F. 3d 553 (9th Cir., 2004).

> In 2003, Justice William H. Rehnquist wrote the majority opinion rejecting the claim made by states that they were immune from having to enforce the Family and Medical Leave Act, [ruled] that the Act was necessary to eliminate the "pervasive sex-role stereotype that caring for family members is women's work." What Justice Rehnquist didn't say was that his daughter was a single working mother with child-care problems of her own, and occasionally she had to recruit the Chief Justice of the U.S. Supreme Court to pick up his granddaughters from school.

Rosenbaum, The Myth of Moral Justice (2004) p. 175.

> ductive system and to its functions and processes. *Reproductive health,* therefore, implies that people be able to have a *satisfying and safe sex life* and that they have the *capability to reproduce and the freedom to decide if, when and how often* to do so. (Emphasis added.)

This U.N. policy attempted to actually confirm the current existence of a right of people to reproduce at will and does not recognize the world's great environmental problem of overpopulation, a problem that concerns the survival of the species Homo sapiens on our planet. This approach must be reviewed early in the third millennium. As stated by Van Rensselaer Potter, the original bioethicist, our current overpopulation constitutes the greatest need for "global bioethics." He transfers this into "A bridge to the future."

Population Action International pointed out that such a fast-growing population will make it more difficult to curb smoke tailpipe emissions, a significant cause of global warming. The Ehrlichs noted, "America's total environmental impact" may one day lead to the "most over-populated nation in the world."[55] In 2003 a report found that 20 percent of American teenagers have had sex before they turn 15 and only a third of their parents are aware of it.[56] Others have written, "The fundamental parameters governing the outlook for humanity's *future* in terms of energy, war, water, food, and population are hopeful."[57] Our comparison with the very different policy in China has been noted at the turn of the third millennium. By the time the Chinese woke up to the need to slow their population growth in 1979, the momentum was so great that the state felt compelled to limit couples to just one child. Even with this policy, the number of Chinese

---

55. ORANGE COUNTY REGISTER, Nov. 4, 1998, at News 6. On Aug. 1, 1790, the first United States Census was completed, showing a population of nearly 4 million people. ORANGE COUNTY REGISTER, Aug. 1, 2004, at News 7.

56. The report was released by the director of the nonprofit National Campaign to Prevent Teen Pregnancy who said, "We can no longer pretend that sex waits until high school," ORANGE COUNTY REGISTER, May 21, 2003, at News 27.

57. MORRISON & TSIPIS, REASON EENOUGH TO HOPE: AMERICA AND THE WORLD OF THE TWENTY-FIRST CENTURY, 1998. In his volume *When Life Nearly Died*, Michael Benton wrote, "and even though the exponential rise in global human population is damped, or slowed down, by famines and wars, the rate continues to go up." (301)

grew (the limited amount) from 989 million in 1979 to more than 1.3 billion near the turn of the third millennium.[58]

In the early 1970s our population control was analyzed here. One consideration is the limitation of any income tax exemption for children—particularly after the birth of the third child and later children. Another was to set up a low rate for childless couples and a high rate for others.[59] In one case an obstetrician testified:

> My policy was with people who were unable to financially support themselves, whether they be on Medicaid or just unable to pay their own bills, if they were having a third child, *to request they voluntarily submit to sterilization following the delivery of the third child.* If they did not wish this as a condition for my care, then I requested that they seek another physician.[60] (Emphasis added.)

For the United States, our controls are extremely limited. In the third millennium's first year a state made the initial decision against the right to reproduce. The Wisconsin Supreme Court upheld the constitutionality of sentencing a man convicted for his refusal to support his nine children. He was sentenced to three years in prison and five years of probation on the condition that he refrain from reproducing during that time.[61] The case goes beyond the

58. Anne and Paul Ehrlich, *Growing Beyond Our Limits* in *Triumph of Discovery*, SCIENTIFIC AMERICAN, 1995, p. 182. In 1995, the Packard Foundation gave $ .3 billion to slow population growth. Malcolm Potts, SCIENTIFIC AMERICAN, January 2000, p. 93. About a sixth of the world's population—nearly 1 billion people—live in slums, and that could double by 2030 if developed nations don't reverse. According to a 2003 U.N. Human Settlements Program's report: "Asia has the largest number of slum dwellers overall, with 554 million, while sub-Saharan Africa has the largest percentage of its urban population living in slums—about 71 percent." ORANGE COUNTY REGISTER, Oct. 6, 2003, at News 15.

59. Rabin, *Population Control through Financial Incentives*, 23ds HAST. L.J. 1358, 1972. *But see* Note, *Legal Analysis and Population Control: The Problem of Coercion*, 84 HARV. L. REV., 1971.

60. Walker v. Pierce, 560 F.2d 600 (4 Cir. 1977). "In 1970 the average number of children in Italy was 2.42, around the norm for the European community; by 1980 it was 1.64; by 1990, 1.30; and by 1993, 1.21, p. 69. In Italy in 1972 the average age of women at the birth of their first child had been 24.9 years; by 1990 it was 29, p. 72." GINSBURG, ITALY AND ITS DISCONTENTS, 1980-2000.

61. Wisconsin v. Oakley, 629 N.W.2d 200, 2001. Of the Supreme Court's seven judges thereon, four were in the majority, while three women justices dissented. (These

facts of the Supreme Court's earlier holding requiring people to demonstrate ability to comply with support obligations before they could obtain a marriage license and not to reproduce.[62] However, neither of these cases established any need for a family in the United States to plan its numbers.

b) *The Ethical Need to Establish a National Goal for Family Planning; President Nixon's Goal*

It is interesting to note that as early as 1969, President Nixon specifically referred to the population problem noting that "by the year 2000, or shortly thereafter, there will be more than 300 million Americans." Accordingly, Nixon set forth a family-planning proposal: "We should *establish as a national goal* the provision of *adequate family planning* services within the next five years to all those who want them but cannot afford them."[63] (Emphasis added.) He named a Commission on Population Growth and the American Future. However, population control actually "provoked grumbling in Congress"[64] and when Nixon fell from power, this recommendation fell with him. No other such attempt has been made to seek the establishment of this national goal of adequate family planning. There has been no U.S. leadership on family-planning adequately to *the third most populous nation in the world*, (after China and India) since that time. In the year 2002, the administration reversed our policy to pay for U.N. family-planning programs overseas even because conservative groups said the programs would tolerate abortions.[65] "Birthrates in developing countries would decrease by about another

---

women feared that the defendant might insist upon abortion for any partner who became pregnant in order to maintain his probation.)

62. Zablocki v. Redhail, 434 U.S. 374 (1978).

63. U.S. President Richard M. Nixon, "Special Message to the U.S. Congress on Problems of Population Growth," July 18, 1969, "Public Pipers of Presidents of the United States," Office of the Federal Register 1.969, p. 521.

64. Kotlowski, Nixon's Civil Rights, 2001, pp. 250-251, and its Footnote 112 *citing* NHC on "Burns to Nixon, Mar. 19, 1969, PNWH, 6!, Fiche to; Wall Street J. Sept. 15, 1969, Nixon to the Vice President *et al.* July 25, 1969; and White House Press Release, March 16, 1970, Box 2 Chester E. Finn Papers. HI."

65. Orange County Register, July 22, 2002, at News 9, and July 25, 2002, at News. In 2003, the Centers for Disease Control and Prevention found that the average age at which American women are having their first child has climbed to an all-time high of 25.1 and that (except for certain ethnic groups) the teen birthrate dropped 30 percent in the past decade to a historic low of 43 births per 1,000 women in 2002. Orange County Register, December, 2003, at News 21.

20 percent, if the demand for family planning were fully met."[66] For example, in Swaziland, Africa, almost one-half of the nation's one million people are estimated to be infected by HIV. AIDS hit Swaziland harder than almost any country in the world. Under the five-year ban, Swazi girls were told to wear a tasseled scarf as a symbolic badge of virginity. But with criticism mounting, the king decided to end the ban a year early. He has 12 wives, 1 bride-to-be, and 27 children.[67]

Immediately following the turn of the third millennium, Pope John Paul II in Italy said couples must "rediscover the culture of love and life, rediscover their mission as parents." Italy has the largest diminution in the population of European countries; the current birth rate will cut Italy's population from 57 million today to 41 million by 2050.[68]

In 1952, India became the first country in the world to actually establish an official government family-planning program in order to achieve zero growth early in the twenty-first century. That first attempt at control ended in disarray in the mid-1970s under the direction of Sanjay Gandhi. In 1990 a spokesperson of India stated, "We have wasted 50 years."[69] The annual population increase in the United States of 3 million people puts more pressure on the world's resources than India's annual increase of 17 million because Americans consume much more food, steel, wood, and energy.[70] The U.N. General Assembly in 1999 began to look at possible curbs on the world's population growth, by adopting proposals giving women greater access to abortions and giving adolescents greater access to sex education. These proposals constituted an advance beyond what was agreed to at the 1994 U.N. population conference in Cairo because it compels governments to take action to increase and improve access to reproductive health care. The Cairo

---

66. Gleichetal, Life Counts (2001), p. 227.

67. Orange County Register, Aug. 21, 2005, at News 13.

68. Orange County Register, Feb. 7, 2000, at News 3. "Italy is a country whose birthrate is one of the lowest in the world. In 2002, a psychologist warned that its highest appeals court's ruling that fathers must carry on supporting adult children until they find a job to their liking could discourage people from having children." http://member.compuserve.com/news/story, April 5, 2002.

69. The Dallas Morning News, May 7, 1999, 50A.

70. Brown, Lester R., *et al.*, 1997. "According to the report by the Population Reference Bureau by 2050, India could surpass China as the country with the world's largest population." State of the World, Sagan, Billions and Billions.

conference had adopted as its main premise that population and developments are linked.[71] As noted in an epigraph above, th*e immigration population has fueled higher birth rates* in the United States. At the turn of the third millennium it was reported that one U.S. resident in 10 is foreign-born. This particular population has grown at a rate four times faster than the rest of the country as a whole according to the Census Bureau.[72] It represented a threefold increase in the past three decades of the twentieth century. More than two-thirds of the nation's foreign-born population lives in just five states—California with 8 million people born abroad, New York, Florida, Texas, and New Jersey. In California, one person in four is a native of a country other than the United States.[73] The largest single group[74] are natives of Mexico who have the nation's highest fertility and immigration rates.[75] In 1998, the

---

71. L.A. TIMES, July 3, 1999, AS. The Islamic definition of birth control was illustrated by Senator Hillary Clinton when she was in Pakistan:

> I spoke to a woman who had ten children, five boys and five girls. She sent her five boys to the secondary school, but her girls had nowhere to go because they couldn't travel to or attend the nearest girls' school. She wanted a secondary school nearby for her girls. She talked very openly about birth control and said that if she had known then what she knew now, she wouldn't have had so many children.

H. CLINTON, LIVING HISTORY, (2003), p. 274. However, in 2004 it was noted by the United Nations that implementing programs agreed on at a 1994 U.N. conference on population and development in Cairo resulted in many Mexican women "making their own free choices about family size." ORANGE COUNTY REGISTER, May 5, 2004, at News 26.

72. ORANGE COUNTY REGISTER, Sept. 17, 1999, at News 12.

73. U.S. Census Bureau.

74. ORANGE COUNTY REGISTER, Feb. 13, 1998, at News 2. "Latina women, a category that includes citizen and immigrant alike, gave birth to nearly one in every five babies in the United States. Births of Hispanic origin, 1989-1995, by the National Center for Health Statistics." *Id.* "On average, according to the report, women of Mexican origin have 3.3 children, compared with 1.8 for non-Latinos." L.A.TIMES, Feb. 13, 1998, A16.

75. "The population of Mexico doubled in 28 years." L.A. TIMES, April 27, 1996, A6. "80 percent of all babies here are born to the poorest 20 percent of the population, most of which lives in the countryside. The church will continue condemning the use of contraceptives to impede natural fertility." L.A. TIMES, Jan. 22, 1999, A10. "By 2050, the United States population will grow to 394 million, some 50 percent more than at present. The fastest rates of growth for minorities are expected for His-

Census Bureau actually dropped from the 2000 questionnaire questions about how many children a person has had for the political reason that they were somehow being too intrusive.[76] In the year 2000, California adopted a law similar to one of the churches' during the Middle Ages, namely, parents who do not want their newborns are able to abandon them at hospitals within 72 hours of birth.[77] However, as stated in a report by the Interaction Council entitled "In Search of a Global Order," "there are no market solutions for such problems as population growth."

At the turn of the third millennium, a volume, *Beyond Six Billion: Forecasting* the *World's Population,* dealt with what was called "the most important issues in the field of demography population forecasts." It reached rational conclusions regarding its great increased projection by the 2050s.[78] Others previously attempted to indicate that there could exist a "transition to bring population growth to a halt by the end of the twenty-first century."[79] Nevertheless, foreign-born residents of the United States rose to 10

---

panic, Asian and Pacific Islander populations." ORANGE COUNTY REGISTER, NOV. 19, 1996, at News 27.

76. ORANGE COUNTY REGISTER, April 1, 1997, News 10. RU-486 is a drug that can end a pregnancy during a period of about 49 days: "By the early 1990s this drug had been approved for clinical use in several countries, but prolife and prochoice forces struggled for more than a decade before the FDA deemed the product safe and effective for U.S. women . . . *We cannot banish values and politics from FDA decision making* but we can insist that they be brought into the open." Dresser, "Plan B: Politics and Values at the FDA, Again," *Hast. Cent. Rep.*, November-December 2004, p. 10.

77. ORANGE COUNTY REGISTER, Sept. 29, 2000. Parents may also reclaim the child within 14 days.

78. The authors are Bongaarts and Bulatao, who wrote for the Committee on Population, Commission on Behavioral and Social Sciences and Education. In its review in JAMA. Feb. 13, 2002, p. 781, regarding "projections of health status" it was noted, "[t]he authors do not discuss this important aspect of population projections."

79. "On that assumption, it showed that population would complete that passage and arrive at 11.5 billion and zero growth by 2200. Alternatively, the model projected population growth at present fertility rates that run, brought the population to 22 billion in 2200 and 28 billion in 2150 and still growing. To the 28 billion population, it showed the present poorest billion people contributing 14 billion descendants." PIEL, THE AGE OF SCIENCE, 2001, p. 441.

percent in 2000.[80] The United States, which is the third most populous country (after China and India), has increased concern for the environmental destruction caused by overpopulation.

The United States has strict immigration policies but lax enforcement, so many people manage to slip illegally over the border. Once here, the illegal immigrants pay dearly in terms of quality of life. Then, periodically, the United States considers granting amnesty to illegal immigrants. This is a crazy system, and we could imagine a better one that could someday handle immigration.

Shiller, *The New Financial Order Risk on the Twenty-first Century* (2003), p. 159.

c) Immigration Ethics and Population Increase

In the United States, the Bureau of Immigration Appeals attempted to follow its ruling that the mere existence of China's population control policies does not qualify an immigrant from that country as a refugee eligible for asylum.[81] However, following a habeas corpus appeal, a district judge held that an alien's expression of views opposing his home country's population control policies might constitute purposes of asylum eligibility under the Immigration and Nationality Act. This would actually appear to qualify for asylum anyone among the 1.3 billion Chinese who may desire to immigrate to the United States in order to avoid its policy of one child per family. Fortunately, this case was appealed and the ruling was reversed.[82] Similarly, a rule that "female cir-

80. Barone, The New Americans, 2002, p. 5.

81. Matter of Chang, Int. Dec. 3107 (BIA 1989). The government's persecution in furtherance of a coercive population control policy that included involuntary sterilization does not constitute "persecution on account of race, religion, nationality, membership in a particular social group, or political opinion" under 8 U.S.C. § 1101(a)-(42)(A), which defines "refugee" as eligible for asylum. Congress set up the Bureau of Immigration in 1894.

82. *See* Guo Chun Di v. Carroll, 842 F. Supp. 858 (E.D. Va. 1994) (*rev'd* Guo Chun Di v. Moscato, 66 F.3d 315 (4 Cir. 1995). An unmarried Chinese woman who underwent a forced pregnancy examination in China before coming to the United States did not qualify for asylum in the United States. Liv. Ashcroft, 9 Cir., No. 00-70157, Dec. 5, 2002. Even the Harvard philosopher John Rawls agreed that "[i]n the absence of a world state, there *must* be boundaries of some kind." Rawls, The Laws of Peoples (1999), p. 39.

cumcision," which is practiced among virtually all Muslim populations in Africa,[83] could qualify all females in those countries for asylum in the United States. In 1997, a Cairo court overturned a year-old ban on health workers performing this operation because it placed undue restriction on doctors. Sheik Yusuf Badry, who led the fight against the ban, stated, "It is our religion . . . For 14 centuries of Islam, our mothers and grandmothers have performed this operation."[84] In 2003 New Zealand introduced pregnancy tests for women of child-bearing age applying for visas to stem the flow of foreigners flying there to give birth using the country's free health-care system and gain New Zealand citizenship for the child.[85] This would appear to be desirable in the United

83. "In 1998 the World Health Organization estimated that more than 135 million girls and women had undergone some form of [female genital mutilation] FGM, and 2 million girls are at risk each year. The practice is most widespread in Africa; in some countries as many as 97 percent of all women have been circumcised." JAMA, Sept. 4, 2002, p. 1131. However, the U.S. Court of Appeals for the Seventh Circuit stayed a mother's deportation because the Board of Immigration Appeals failed to consider threat of female genital mutilation of her young U.S. citizen daughter if she were returned to Nigeria. Nwaokolo v. Immigration and Naturalization Serv., 314 F.3d 303 (7th Cir. 2002).

84. Orange County Register, June 25, 1997, at News 3.

> The ruling had been sought by eight Islamic scholars and doctors, who argued that the health Ministry decree violated religious beliefs and interfered with physicians' prerogatives to perform medical duties. A recent survey of 14,779 women between 15 and 49 in urban and rural areas revealed that 97 percent had undergone the operation. Many Egyptians support circumcision on grounds that it has always been done and helps to keep women's sexual drives "to acceptable and reasonable levels."

L.A. Times, June 25, 1997, A4.

85. Orange County Register, June 18, 2004, at News 30. In this connection it should be recalled that back in 1949 the famous Judge Learned Hand ruled that a man who "in a moment of what may have been unnecessary frankness," admitted that he had had consensual sexual intercourse with unmarried women should not have been denied citizenship. About such morals he stated: "A majority of the votes of those in prisons and brothels, for instance, ought scarcely to outweigh the votes of accredited churchgoers. Nor can we see any reason to suppose that the opinion of clergymen would be a more reliable estimate than our own . . . [We] resort to our own conjecture, fallible as we recognize it to be." Schmidt v. United States, 177 F.2d 450, 451 (2d Cir. 1949).

States because Constitutional Amendment XIV (ratified after the Civil War in the nineteenth century) states: *"All persons born or naturalized in the United States* and subject to the jurisdiction thereof, *are citizens of the United States* and of the State wherein they reside. (Emphasis added.)[86]

As of 2002, the U.S. Census Bureau estimated that a total of 8.7 million illegal aliens resided in the United States. A study at Northeastern University put the total at 11 million.[87] In 2002 the U.S. foreign-born population reached a record high of 32.5 million according to the Census Bureau.[88] The Justice Department Inspector General reported in that year that thousands of foreign-born prison inmates should be deported after completing their sentences, but instead, they are released back into U.S. communities, where some go on to commit more crimes. The Immigration and Naturalization Service is not identifying and properly processing criminal immigrants serving time in jails.[89]

---

86. In Europe, Ireland has recently done away with birthright citizenship following the approval thereof by 80 percent of its voters. In 2004, a federal judge actually refused to deport a pregnant illegal alien based upon the Unborn Victims of Violence Act by so interpreting its "protection" for unborn children as also including saving future "citizenship" for the fetus.

87. IMMIGRATION WATCH, April/May 2003, p. 5. According to the National Center for Health Statistics, Hispanics, as compared to whites, are 30 percent more likely to not have a primary physician, 1.7 times more likely to die of cancer, 1.4 times more likely to die of coronary heart disease, and 4 times more likely to be diagnosed with AIDS. ORANGE COUNTY REGISTER, Nov. 20, 2003, at News 20. Such illegal immigrants cost the United States more than $40 billion a year. Professor Hason, Hillsdale College, November 2003.

88. ORANGE COUNTY REGISTER, March 10, 2002, at News 12. Money sent to Mexico by relatives in the United States reached $8.9 billion in 2001; this was "nearly $24.4 million a day. That was an increase of 35.3 percent from the figure for 2000." ORANGE COUNTY REGISTER, Sept. 16, 2002, at News 14. Health care professionals and hospitals have been forced by former President Clinton's Executive Order 13166 to provide and pay for translators for non-English-speaking patients on demand. The implementation of E.O. 13166 in health care costs $267 million annually.

89. ORANGE COUNTY REGISTER, Oct. 8, 2002, at News 13. On Nov. 24, 2003, the Inter-American Development Bank and the Pew Hispanic Center released their estimate of $30 billion being sent that year by immigrants in the United States to their families in Latin America. ORANGE COUNTY REGISTER, Nov. 25, 2003, at Business 1. The Bank of America indicated that the number of U.S. accounts set up for sending money to Mexico soared 1,500 percent in the first half of 2004. ORANGE COUNTY

At the turn of the millennium, the foreign-born population in the United States soared 60 percent since 1990—the largest proportion of foreign-born residents in 70 years.[90] In 2004 there was also a 60 percent rise in the apprehensions of undocumented immigrants crossing the U.S.-Mexico border.[91] The Federation for American Immigration Reform estimates California spends $7.7 billion a year educating illegal immigrants and their legal siblings, who account for about 15 percent of the state's 6 million students.[92] Because the poverty rate for immigrants is 50 percent higher than that of U.S. citizens, a large expense is incurred upon America's social and welfare services. Unlike the United States, Great Britain is undergoing a tightening of its immigration.[93] (See Appendix B on China's Population Controls.)

REGISTER, Aug. 18, 2004, at Business 3. In December, the ABA pointed out the actions of the Catholic Legal Immigration (CLINIC) members, which grant immigrant legal services to 100,000 low-income immigrants each year. ABA JOURNAL, December 2003, p. 20.

90. ORANGE COUNTY REGISTER, June 5, 2000, at News 13. "No country has ever attempted to incorporate and assimilate 31 million newcomers into its society," Steven Camarata, director, Center for Immigration Studies, Washington, D.C. Hispanic women gave birth to nearly 250,000 babies in 2002 in California; 9.4 percent of the mothers were of Central and South American origin. ORANGE COUNTY REGISTER, Oct. 3, 2003, at News 14.

91. CHRISTIAN SCIENCE MONITOR, May 4, 2004, p. 1. The Californians for Population Stabilization (CAPS) indicated that their current (2004) task was to reverse the endless growth agenda supported by most of our elected officials and based on a "business as usual" attitude that continuing increases in population and consumption can go on indefinitely. We are growing at the rate of 1.74 percent a year, which means a population doubling time of 40 years. By sometime in the next century, our great grandchildren will be living in an America crammed with more than a billion people.

Diana Hull, President CAPS/Keyes, March 31, 2004 (with attachments). "Currently, 11 states issue driver's licenses to undocumented immigrants." WALL STREET JOURNAL (Dec. 6, 2004) p. B1. Mexico's population increased by almost 1.1 million in 2004 to reach an estimated 105,909,000, allowing the country to maintain its place as the 11th largest in the world. ORANGE COUNTY REGISTER, Dec. 29, 2004, at News 23.

92. ORANGE COUNTY REGISTER, May 18, 2005, at News 13. The U.S. Supreme Court ruled in 1982 that the children of illegal immigrants are entitled to a public education under the "equal protection clause" of the 14th Amendment of the Constitution. The case, which originated in Texas, was Plyler v. Doe, *Id.*

93. ORANGE COUNTY REGISTER, May 31, 2002, at News 34. In 2004, an appellate court upheld a Colorado statute that eliminates Medicaid coverage for some legal

With few exceptions, the federal law requires a person to be a U.S. citizen to receive Medicaid benefits. Yet most states accept a signed declaration as proof of U.S. citizenship.[94] However, in 2005 the Medicare Modernization began the allocation of $250 million a year for four years to help hospitals offset the cost of emergency treatment for illegal immigrants.[95] Nevertheless, the U.S. Ninth Circuit Court of Appeals refused to block implementation of Arizona's law that bars illegal immigrants from receiving certain public benefits and makes it a crime for public employees to fail to report undocumented immigrants who seek the benefits.[96]

[P]lainly, nobody will be afraid who believes nothing can happen to him . . . [F]ear is felt by those who believe something is likely to happen to them. People do not believe this when they are, or think they are, in the midst of great prosperity, and are in consequence insolent, contemptuous and reckless . . . [But if] they are to feel the anguish of uncertainty, there must be some faint expectation of escape.

Aristotle, 384-322 B.C., Rhetoric

---

aliens, Soskin v. Reinerston, 10th Cir., No. 03-1162, 1/12/04. The court ruled that the federal 1996 Personal Responsibility and Work Opportunity Reconciliation Act permits states to stop funding Medicaid benefits for aliens who do not meet the definition of "qualified aliens" under the statute. The executive director of Diversity Alliance for a Sustainable America asked: "Are we going to wait until it becomes another China? It'll be too late." SIERRA (November/December 2004), p. 38.

94. Only Montana, New York, New Hampshire, and Texas require applicants to submit documents verifying U.S. citizenship. Oregon is the only state that has conducted an audit to determine how often "non-citizens" gained access to Medicaid. It estimated that it would cost an additional $2 million if 1 percent of the Medicaid rolls are not U.S. citizens. ORANGE COUNTY REGISTER, Aug. 4, 2005, at News 11.

95. "$167 million will be allotted in all 50 states based on their undocumented immigrant population. $83 million will be distributed to the six states with the largest number of illegal immigrant apprehensions. California gets $70.8 million; Texas, $46 million; Arizona, $45 million; New York, $12.25 million; Illinois, $10.3 million; and Florida, $8.7 million." ORANGE COUNTY REGISTER, May 10, 2005, at News 7.

96. L.A. TIMES, Aug. 10, 2005, A24.

d) Environmental (and Bioethical) Problems Growing Out of Overpopulation

In June of 1998, the president of the U.N. General Assembly opened Earth Summit II stating, "We as a species—as a planet—are teetering on the edge, living unsustainably and perpetuating inequity, and may soon pass the point of no return."[97] Amplifying this, a scientist reported:

> Homo sapiens are approaching the limit of its food and water supply. Unlike any species that lived before, it is also changing the world's atmosphere and climate, lowering and polluting water tables, shrinking forests, and spreading deserts. *Most of the stress* originates directly or indirectly *from a handful of industrialized countries.*[98] (Emphasis added.)

Regarding the food problem, it has been noted, "The Green Revolution has been implemented in a manner that has not proved environmentally sustainable. The technology has enhanced soil erosion, polluted groundwater and surface water resources, and increased pesticide use has caused serious public health and environmental problems."[99] In his volume *Who Will Feed China?*. the author pointed out that "Since 1990, there has been no growth in the world grain harvest. Indeed, the 1994 grain harvest is actually

---

97. Zucherman, L.A. Times, March 15, 1998.

98. Wilson, Consilience, 1998, p. 28.

> Unless the birthrate declines, and declines rapidly, this already overcrowded planet will become unlivable for most of humanity. If we don't solve the numbers problem, we won't have the luxury of worrying about mutation accumulation, eugenics, and ethical dilemmas arising from genetic knowledge. Yours may be the last generation in which reproduction is regarded as a right rather than a rationed privilege; this is already happening in China.

Crow, Genetics Yesterday, Today, and Tomorrow, Principles of Genetics (3rd ed. 2003) p. 257.

99. Pimentel, *Constraints on the Expansion of the Global Food Supply*, Ambio, Vol. 23, May 3, 1994, J. Royal Swed. Academy of Sciences.

smaller than that of 1990."[100] He added, "Time is not on our side.[101] The world has waited too long to stabilize population." The twenty-first century will both determine the extent to undergo the additional suffering for which this is true.

Clearly, the earth has finite limits. "No matter what any human law says about individual liberties, *parents cannot continue to have,* on the average, *more children than required to replace themselves*. The finiteness of the Earth guarantees that there are ceilings on human numbers"[102] (Emphasis added.)

If the earth's population were to double yet again before the year 2100 and if the modernizing countries were to consume, on the average, at least a third of the energy used per capita by affluent nations, the global demand for fuel and electricity could quadruple.[103]

---

100. Brown, Who Will Feed China?, p. 12. "Malthus' *Essay on Population* argued that mankind will always produce more offspring than the available resources can support. The inevitable consequences, according to Malthus, are war, famine, poverty and disease, as increasing populations fight for limited resources." McFadden, Quantum Evolution (2001), p. 55.

101. *Id.* p. 141. "The only other major grain exporter is the United States, which can produce and export more grain than it now does—but probably not nearly as much as some people think. In 1994, the United States returned to production all the grain land idled under commodity supply management programs. Even with this land in use and one of the best U.S. harvests in memory, world grain stocks still fell." *Id.* p. 1109.

102. Cohen, How Many People Can the Earth Support?, 1995, p. 16.

> About 3.8 billion people—nearly two-thirds of the world's population—live in food-deficit countries where millions experience hunger and starvation. By 2025, when the world population approaches eight billion people, 48 countries containing three billion people will face severe shortages of fresh water, leading to profound problems with food production.

Salk's *Signals* (Fall 2003), p. 9.

103. The Earth's Biosphere, (2003), Chapter 10. A boy born at the beginning of 2005 was declared China's 1.3 billionth citizen to promote the government's "one child" birth limits. It stated that without the policy, China would have at least 200 million more mouths to feed, straining farm, water, and other resources. Thus, just within the year 2004, China would add two-thirds of the U.S. population without that policy. California Bar J., January 2005, p. 9.

What so affects our liberty will also have ethical consequences. One child per family is limited to a small group in the United States that has doubled in the past 25 years from 9.6 percent in the 1970s to an estimate of about 20 percent in 2001.[104] However, this has had little effect upon the increase of the total population in the United States. The opposite view (i.e., a family view favoring the increase of its children) is strongly held by most others. For example, in 2003 a study by the University of California in Los Angeles (UCLA) showed that Hispanic babies account for more than half of all births in California, the most populous state in the union.[105] In 2003, Census figures showed that they have surged past blacks as the country's largest minority group.[106]

In Africa between 1950 and 1994, the *population of Rwanda* favored by better health care and temporarily improved food supply, more than tripled, from 2.5 million to 8.5 million. "In *1992 the country had the highest growth rate in the world*, an average of eight children for every woman. Although total food production increased dramatically during this period, it was soon overbalanced by population growth. Water was so overdrawn that hydrologists declared Rwanda one of the world's 27 water-scarce countries. It was actually stated that "the teenage *soldiers* of the Hutu and Tutsi then *set out to*

104. This was according to the Census Bureau. ORANGE COUNTY REGISTER, Dec. 10, 2001, Accent (also listing famous personalities of persons with only one child); PICKHARDT, KEYS TO PARENTING THE ONLY CHILD, 1997.

105. "From July to September 2001, there were 138,892 births in California and 69,672, or 50.2 percent, were Hispanic, according to the UCLA Center for the Study of Latino Health and Culture." ORANGE COUNTY REGISTER, Feb. 6, 1003, at News 14.

106. Census Bureau estimates in 2000 showed due to high birth and immigration rates, the Hispanic population more than doubled during the 1990s. The Hispanic population rose 4.7 percent between April 2000 and July 2001. During the same period, the non-Hispanic black population rose about 2 percent. ARMAS, http://member.compuserve.com/news/1/21/03. Recent studies show that while 45 percent of children in California public schools are Latinos, only 2 percent graduate as eligible to attend a California university. CALIFORNIA BAR J., January 2005, p. 9. Nevertheless, the United States actually has withheld money from the United Nation's Population Fund the past three years because the agency supports the Beijing regime's family-planning program.

solve the population problem in the most direct possible way."[107] The massacres that ensued became well known because Rwanda has been called "a microcosm of the world." As the space scientist Carl Sagan wrote in his last work, without "a voluntary halt to world population growth, many of the other approaches to preserving the environment will be nullified."[108] The stark reality of the population problem has been expressed as follows:

> The global population growth rate can fall from its present value of around 1.6 per year to zero or below only by some combination of *fewer births and more deaths. Hardly anybody favors more deaths.*[109] (Emphasis added.)

As so described, although, the prescription of "more deaths" could actually be favored; for obvious reasons, it cannot be regarded as an ethical approach. Fortunately, several ethically correct involuntary and voluntary approaches do exist.

good v. Good

The United States can easily absorb many more immigrants. Further, no special consideration should be given to the fact that these people come from ethnic groups that have the highest birthrates and this should never be considered important to us as a matter of public policy.

v.

In view of our fiscal, environmental, and current bioethical crisis of overpopulation, immediate steps should be taken toward promulgating an awareness of this crisis as a serious one and inform ourselves how it might be avoided at this time.

---

107. It has been estimated that within a period of 100 days 800,000 Tutsis (out of 1,200,000, or two-thirds of the population) were massacred. More than 125,000 people have been charged with genocide. GOUREVITCH, WE WISH TO INFORM YOU THAT TOMORROW WE WILL BE KILLED WITH OUR FAMILIES, 1998. Of course, I recognize that significant cultural differences exist between countries that may cause differences in the responses. *Id. See also* WILSON, CONSILIENCE, 1998, p. 28.

108. SAGAN, BILLIONS AND BILLIONS, 1997, p. 174.

109. *Id.* p. 17.

Twenty-first Century Approach

To avoid the impending crises, a public policy on population growth should be considered early in the twenty-first century together with a natural goal of adequate family planning limited to two infants.

The laws that in many countries on the continent forbid marriage unless the parties can show that they have the means of supporting a family do not exceed the legitimate powers of the State. They are not objectionable as violations of liberty.

J. S. Mill, "On Liberty," Penguin, 1985, p. 197

To check population, one must bring down the birthrate. Birthrate does not fall in conditions of misery. It falls when children are more likely to survive.

Jolly, *Lucy's Legacy* 1999, p. 426

3. Ethical Actions That Can and Should Be Taken

a) Involuntary Birth Controls

It is interesting to note that in the epigraph from his great nineteenth century essay, "On Liberty," John Stuart Mill appears to have anticipated China's current population control policy. What is most important is his *specific remark* that such controls by the state are "not objectionable as violations of liberty." Today, because discussion of such control is "politically incorrect," many Americans probably disagree with Mill. Some of Mill's reasons were that:

> . . . to bring a child into existence without a fair prospect of being able, not only to provide food for its body but instruction and training for its mind, is *a moral crime* both against the unfortunate offspring and *against society,* and that if the parent does not fulfill this obligation, the State ought to see it fulfilled, at the charge, as far as possible, of the parent. (Emphasis added.)

The current rule of "one child per family" only applies to the Han Chinese who constitute 95 percent of the Chinese people. Minorities are sometimes allowed two in the cities and countryside.[110] Although the rule began as a painful imposition, it is unavoidable and deserves the gratitude of the rest of the world. But this important action in China has not been the recipient of gratitude in the United States—even though statistics show that the policy has been generally effective. At a population conference of the United Nations in 1999, Hillary Rodham Clinton actually praised all of those fighting against China's general one-child per family policy.[111] The distance we are from there may also be illustrated by the limited action of the Wisconsin Supreme Court, which merely affirmed the prevention of a particular father who made no effort to support his nine children from having any more.[112] The results of a 20-year American tracking study of couples and their offspring has demonstrated that single "children have higher IQs than their peers with one sibling; only single children remained significantly superior in average vocabulary performance to children in all other family sizes," even those with two.[113] Should this trend continue it would appear that the average Chinese IQ might even surpass that of the average American sometime in the twenty-first century and this potential should receive more attention in the United States.

According to a 1993 survey, less than half of Americans actually agreed that lowering the U.S. birthrate was important for the environment.[114] As noted previously, most politicians have largely been willing to table major aspects of the population problem. During the 1980s and early 1990s, they

110. Komans, China, Hong Kong, Taiwan, Inc., 1997, p. 341. "China can only achieve zero population growth when its total population reaches 1.6 billion. Effective control of population growth is a prerequisite to maintaining high economic growth." Zhang Weiging, Schenzhen Daily, June 3, 1998, p. 6.

111. Orange County Register, Feb. 10, 1999, at News 19. Nevertheless, three days later the conference adopted an international plan to slow global population growth. Orange County Register, Feb. 13, 1999, at News 10.

112. Orange County Register, Nov. 24, 2001, at News 13.

113. Blake, Family Size and Achievement, 1989, p. 651; Zill and Peterson, Learning to do Things without Help, in Laosa and Sigel Ed., Families as Learning Environments, p. 356, and McKibben Maybe One, 1998, p. 28.

114. McKibben, Maybe One, p. 171, *citing* Maher, *Dodging Numbers*, pp. 1-3.

suspended assistance to the United Nation's Fund for Population control (UNFPA) and federal support for family-planning services dropped almost 70 percent since 1980.[115] This was largely because of pressures from certain religious groups. Worldwide, there are about 1 billion each of Muslims, Roman Catholics, as well as Hindus and Buddhists combined[116] who carry significantly different religious beliefs. (See Chapter 3.)

(i) Involuntary Sterilization, Castration, and Mental Retardation

The few involuntary controls, rarely imposed, will have little effect on population increases in the United States during the early twenty-first century. Examples include involuntary sterilization and castration of rapists. Involuntary sterilization under the laws of many states is limited to severe mental retardation for persons who either "would be so incapable of caring for a child that procreation would be inadvisable" or where "pregnancy would present a substantial danger to life or health" of a woman.[117] This trend commenced in 1981 when the New Jersey Supreme Court ruled in favor of the parents of a Down syndrome girl who wanted her to be sterilized in order to prevent their daughter from becoming pregnant as a result of rape or seduction.[118]

---

115. *Id.* p. 172, *citing* Hempel, *Misplaced Blame*. POPULATION PRESS, September-October 1997, p. 1.

116. Johnson, Ed. 1991, INFORMATION PLEASE ALMANAC ATLAS AND YEARBOOK 1992, 45 edition, p. 292. The Muslims have the highest average family size with six children per woman. Although they currently constitute a minority in the United States, they are rapidly increasing in number.

117. *See* 65 Del. Laws 148, § 5712(b)(3)(e). A Federal District Court in North Carolina held that that state's sterilization statutes, which were limited to sexually active mentally retarded persons who were unwilling or incapable of controlling procreation by other contraceptive means and who were found to be likely to procreate a defective child (or who would be unable because of the degree of retardation to be able to care for a child) were not unconstitutional. United States v. State of North Carolina et al., 420 F. Supp. 451 (1976). *See also* Vaughn v. Ruoff, 253 F. 3d 1124, 1130 (8th Cir. 2001).

118. Matter of Grady, 426 A.2d 467 (N.J. 1981). In 2003, survivors from among several thousand North Carolina residents who were involuntarily sterilized by the state for over five decades will be compensated in the form of health care and education. These were the actions of the Eugenics Board of North Carolina, which ordered sterilization of about 7,600 people from 1929 through 1974. ORANGE COUNTY REGISTER, Sept. 29, 2003, at News 12.

Rape and battery are illegal in all the states. The castration of rapists and child abusers has a long history both as prevention and punishment. Therapeutic castrations of criminals began in 1899. Physicians recommended castration to control the criminals' sexual urges and aggressive behavior.[119] In the United States between 100,000 and 500,000 children are sexually molested every year.[120] Early in the twentieth century, castration was looked down upon as not being worthy of "enlightened times."[121] This view is changing. As noted by *Time* magazine near the end of the century: "What changed was the perception of sex offenders as compulsive recidivists." Such offenders have a recidivism rate as high as 65 percent.[122] There are 13 states with laws for the civil commitment of sexual psychopaths who otherwise would be released from prison.[123]

In 1994, seven-year-old Megan Kanka was sexually assaulted and murdered in New Jersey by a man who lived across the street. Megan's family and neighbors were not informed that he was a convicted pedophile. The resulting outrage produced "Megan's Law," which requires notification of the community in which a violent or repeat offender stays during his lifetime.[124] Every other state followed suit.[125] A 1994 federal statute requires

119. Hem & Hursch, *Castration for Sex Offenders: Treatment or Punishment? A Review and Critique of Recent European Literature*, 18 ARCH. SEX. BEHAV. 281, 1979. *See also* DEL. CODE ANN. tit. 16, § 5712(b)(3)(e), 1999.

120. Le Maire at 294. The public should also be made aware of the United Nations recent report of 150 allegations of sexual abuse committed by U.N. peacekeepers stationed in Congo, and more sexual misconduct by its peacekeepers from Nepal, Pakistan, Morocco, Tunisia, South Africa, and Uruguay. Each of them had been presented with the U.N. code of conduct, which forbids "any exchange of money, employment, goods or services for sex." ORANGE COUNTY REGISTER, Dec. 19, 2004, at News 27.

121. Human Sexuality and Its Problems: Second Child Abuse and Neglect: Critical First Steps in Response to a National Emergency. Washington, D.C., Advisory Ed. on Child Abuse and Neglect 1990.

122. TIME, Sept. 9, 1996, p. 60.

123. *See In re* Blodgett, 510 N.W.2d 910 (Minn. 1994) (holding Minnesota's Psychopathic Personality Statute did not violate offender's substantive due process or equal protection rights). *See also* Kansas v. Hendricks, 521 U.S. 346 (1997) (upholding Kansas's Sexually Violent Predator Act).

124. N.J. STAT ANN. § 20:7-1-2c:7-11 (West 2000).

125. L.A.TIMES, April 5, 1996, and ORANGE COUNTY REGISTER, April 11, 1996, at News 13.

that such information be made available to law enforcement agencies and that a specific registrant's information be released to the community at the discretion of local law enforcement officials.[126] However, most states that have similar laws fail to enforce it for thousands of convicted sex offenders.[127] In 1999 a New Jersey court upheld the right of housing associations to refuse occupancy to the most serious offenders listed on the state's Megan's Law registry.[128]

In 1996, Larry Don McQuay, a paroled child molester, said he might kill his next victim unless he was castrated. He then signed a contract with a victims' group that agreed to pay for the surgery. The chair of the Texas board of pardons and parole encouraged him to do so. The state attorney general outlined how he could become the first man on record to be castrated under state auspices.[129] Surgical castration is very effective in reducing the recidivism rates of pedophiles and child molesters. Recidivism rates of 2.2

---

126. 42 U.S.C. § 14071 (2000).

127. *See* Cal. Penal Code § 290 (West 2000).

> California has lost track of more than 33,000 convicted sex offenders, despite a law requiring rapists and child molesters to register each year for inclusion in the Megan's Law database. Failing to register could put high-risk offenders in jail for up to three more years, but most police departments are not enforcing the law . . . More than 77,000 sex offenders were missing from the databases of 32 states. In the other 18 states and the District of Columbia, which are responsible for 133,705 offenders, no trace of thousands of the ex-convicts could be found.

Orange County Register, Feb. 7, 2003, at News 22.

128. "It was held that the association's interest in preserving home values and the safety of the community outweighs the negligible impact the regulation has on a homeowner's ability to sell the property." Mulligan v. Panther Valley Property Owners Assn, No. L-188-99, Nov. 12, 1999. "Sixteen states now exempt real estate licenses from disclosure requirements. Four states—and Vermont specify that agents only have to tell prospective buyers where to go to find the information." ABA Journal, January 2000, p. 24, 337 N.J. Super. 293 (2001). *But see* Paul P. V. Farmer, 80 F. Supp. 2d 320 (D.N.J. 2000).

129. William Winsdale, *et al. Castrating Pedophiles Convicted of Sex Offenses Against Children: New Testament or Old Punishment?*, 51 SMU L. Rev. 349, 1998. *See also* People v. Harris, 543 N.E.2d 859, at 867-68 (Ill. App. Ct. 1989), where the court took judicial notice of the fact that treatment for paraphilias, including surgical castration, is available and successful.

percent for castrates compares with up to 80 percent for noncastrates. Several states provide that a defendant may voluntarily undergo surgical castration.[130] Felix Spector, a Philadelphia doctor who is reported to "have castrated hundreds of men," believes that the surgery can change the habits of sex offenders.[131] Where this has been requested by a prisoner, no cases have been reported challenging the matter on the Eighth Amendment grounds of "cruel or unusual punishment."

Some such offenders have also undergone "chemical castration," a non-surgical treatment with medroxyprogesterone acetate (MPA). This is a drug that has been shown to reduce recidivism in male sex offenders by diminishing sexual impulses if taken in high doses.[132] A court may order its use under state statutes for parolees who do not undergo a "permanent, surgical alternative."[133] Such treatment is not always successful, "mostly because the inhibition of the secretion or action of testosterone is incomplete."[134] For example, in 1999, a convicted rapist who had been considered a success story for going through a "chemical castration" was again sentenced in Richmond, Virginia to 40 years in prison for a sexual attack on a five-year-old girl.[135] The American Civil Liberties Union argued that because it may reduce the offenders' fantasies, sentencing a man to use MPA is a violation of the right to free speech, an argument that seems to be stretching that First Amendment quite a bit.[136] This use of the drug is not considered experimental and is supported by a large volume of medical literature.[137] Under a California statute it must be ordered by a court for twice-convicted offenders in addition to

---

130. *See* FLA. STAT. ANN. § 794.0235 (West 2000), administration of imedroxyprogesterone acetate (MPA) to persons convicted of sexual battery; *see also* Tex. Prob. Code Ann. Art. 37.07 § 3(h); art. 42.12 § 11(f). *See also* Cal. Penal Code § 645(e), West 1997.

131. Garcia and Pemberton, SAN LUIS OBISPO TRIBUNE.

132. Rosler and Witzam, N. ENG. J. MED., Feb. 12, 1998, p. 416.

133. *See, e.g.,* Cal. Penal Code § 645(a) and (e), West 2000.

134. Mcoonaghy *et al.*, *Treatment of Sex Offenders with Imaginal Desensitization and/or Medroxyprotesterone*. 77 ACTA PSYCHIATRICA SCANDINAVIA 199, 203, 1988.

135. ORANGE COUNTY REGISTER, Feb. 4, 1999, at News 11.

136. TIME, Sept. 9, 1996, p. 10.

137. Berlin, *The Paraphilias and Depo-Provera; Some Medical, Ethical and Legal Consideration*, 17 BULL. AM. ACAD. PSYCH. & L., 1989.

any other punishment prescribed by law.[138] Serial child molester Brian DeVries (who molested at least nine young boys) was surgically castrated in 2001. He was released two years later, in Santa Clara County, California, becoming the first graduate of its state-mandated sex offender treatment program. The judge commented to the residents of the city to which he was returning: "The program is elaborate and can certainly assure every resident . . . that Mr. DeVries' placement will not in any way be a risk to them or their families. The public should feel very safe with Mr. DeVries' release."[139]

The California statute does not apply to any offender who volunteers to be surgically castrated;[140] however, there have been few volunteers. Voluntary castration is a normal medical procedure for other purposes. For example, certain men with spreading prostate cancer are five times more likely to survive if they undergo castration either chemically or surgically, or after the prostate is removed.[141]

Two and one-half million Americans (or about 1 percent of the population) are mentally retarded, 85 percent of whom are mildly retarded.[142] The opportunity for exploiting them is very real. Surveys of special educational teachers reveal their belief that 70 percent of the mildly retarded students were having intimate sexual relationships at a younger age than their non-mentally retarded counterparts.[143] In 1996, a Roman Catholic theologian said it is morally acceptable to give birth-control pills to mentally handicapped women who could be "induced or forced into sex," and "it would be administered—without their consent."[144] Rape convictions placed upon men who

---

138. Cal. Penal Code § 645(b)

139. Orange County Register, Aug. 13, 2003, at News 8.

140. *Id.* § 645(e).

141. New Eng. J. Med., Dec. 9, 1999. "The study is the first to show that hormone-blocking treatments sometimes called 'chemical castration' can save the lives of men with spreading prostate cancer, which kills 37,000 men a year in the United States." Orange County Register, Dec. 9, 1999.

142. American Psych. Assoc., Diagnostic and Statistical Manual of Mental Disorders 39, 4th Edition, 1994.

143. Brantlinger, *Teachers' Perceptions of the Sexuality of their Secondary Students with Mild Mental Retardation.* 23 Educ. & Train, in Mental Retardation, 24, 31, 1988.

144. Orange County Register, April 25, 1996, at News 8.

acted upon such women[145] have resulted. Prior to authorizing the sterilization of a mentally incompetent female, the Washington Supreme Court requires the trial court to make a number of findings: The patient must be incapable of making "her own decision about sterilization," "physically capable of procreation . . . under circumstances to result in pregnancy," incapable of caring for a child "even with reasonable assistance," and sterilization must be the least invasive action available. Only the lack of such findings should cause a court in any state to deny authorization to the sterilization.[146]

In 1998, two years after California became the first state to pass a law requiring "chemical castration" of child molesters, the court stated that injections must be approved by a judge. Susan Carpenter McMillan, who drafted the law, said the criminal justice establishment appears to be undermining it.[147] Reports show that some of the convicts stop taking the treatment when released and proceed to commit further sexual assaults. For example:

> Years after a convicted rapist agreed to undergo "chemical castration" as part of his probation, he has pleaded guilty to new sex crimes. Smith, 45, moved to the Richmond area after his release and appeared on CBS-TV's "60 Minutes" in 1984 as a success story for chemical castration. In 1998, Smith was convicted of aggravated sexual battery and two counts of attempted oral and anal sodomy.[148]

Most current legal involuntary controls have only begun to succeed early in the twenty-first century and they should increase later in that century. However, certain steps should be taken now.

---

145. One case involved the alleged rape of a mildly retarded 16-year-old girl by a 27-year-old man named Adkins. Although Adkins was tried and convicted of rape for taking advantage of the victim's mental incapacity to gain her consent to sexual intercourse, the Court of Appeals reversed for failure to prove her mental incapacity beyond a reasonable doubt. Adkins v. Commonwealth, 457 S.E.2d 382 (Va. Ct. App. 1995).

146. Guardianship of Hayes, 608 P.2d 635 (Wash. 1980). State-funded sterilizations are not limited by federal regulation. 42 C.F.R. 461.253.

147. Orange County Register, Nov. 28, 1998, at Metro 9. "The obligation to preserve life and health takes precedence over the prohibition of castration." Dorff, Matter of Life and Death, 1998, p. 12.

148. L.A. Times, Nov. 28, 1998.

good v. Good

No need exists for involuntary birth controls or limitations on the number of children per family because these are fundamental rights subject only to the control by the deity. Discussion of the problems in connection with immigration will be regarded as racist in origin and the castration of sex offenders is a violation of the Eighth Amendment's prohibition on "cruel or unusual" punishments.

v.

Castration or involuntary sterilization of certain mentally defective women and limitations on numbers of children to limit population growth must be considered at this time in order to prepare for its implementation at a critical environmental point of the twenty-first century. Curtailing excessive breeding by certain ethnic groups and submission to castration of sexual abusers are neither cruel nor unusual punishments under the Eighth Amendment to the Constitution.

Twenty-first Century Approach

1. Although a policy on family planning (including considerations for limiting the high number of children per family) was not found politically correct at the turn of the twentieth century, it should be recognized that such a policy is needed to control the adverse effects of overpopulation on the environment in the twenty-first century.
2. It should be recognized that greater control over sex offenders might be achieved by castration (chemical or surgical).
3. Special standards are needed to protect retarded women from sex offenders.
4. Legislation is needed to authorize the use of birth-control pills and the sterilization of severely retarded men and women who are unable to adequately care for any offspring they might have.

---

b) Voluntary Birth Controls

Well into the twentieth century, the Comstock Act prevented distribution of birth-control information. Such action continued to be viewed as obscene and was illegal. In 1929, Mary Ware Dennett, a 50-year-old grandmother, was initially convicted on criminal charges for sending a sex education pamphlet through the U.S. Post Office.[149] Only years later was the contraceptive clause stricken from the obscene list under the act. Methods of contraception are discussed below.

(i) Voluntary Sterilization

Sterilization has been defined as "any surgical or medical procedure intended to render a person permanently unable to procreate."[150] These procedures have recently become the most common form of contraception in the United States and abroad.[151] A woman requests her physician to perform a legally enforceable tubal ligation at the time of delivery of an infant by cutting and tying her fallopian tubes to keep her eggs released by the ovaries from reaching the uterus.[152] In 2002, the Food and Drug Administration (FDA) approved the first nonsurgical method of sterilizing women.[153] Most states

---

149. *But see* U.S. v. Dennett, 39 F.2d 564 (2d Cir. 1930) (reversing conviction of Mary Ware Dennett).

150. *See, e.g.*, DEL. CODE ANN. tit. 16, § 5701(a) (1999).

> It has been written that rabbinic interpretations forbid this. Both traditional and liberal respondents forbid male sterilization on the basis of the rabbinic interpretation and extension of Deuteronomy 23:2 ("No one whose testes are crushed . . . shall be admitted into the congregation of the Lord") or Leviticus 22:24 ("That which is mauled or crushed or torn or cut you shall not offer unto the Lord; nor shall you do this in your land").

DORFF, MATTER OF LIFE & DEATH, 1998, p. 125.

151. MOSHER & BACHRACH, UNDERSTANDING U.S. FERTILITY: CONTINUITY AND CHANGE IN THE NATIONAL SURVEY OF FAMILY GROWTH, 1988-1995, 28.

152. FAIN. PLAN. PERSP. 4.6 (1996) until the 1970s, more than 100,000 feebleminded people were sterilized under "50 state or Federal laws." RIDLEY, GENOME. *See* MEDLAW, 1998, 17:7-11, 2000, p. 290.

153. A tiny device called Essure requires no cutting and only a local anesthetic in a half-hour procedure that promises to block the fallopian tubes as effectively. It will cost about the same as traditional tubal ligation, about $2,000. Sterilization remains the most widely used form of birth control. More than 180 million women worldwide have had the procedure performed, including an estimated 700,000 Americans a year.

ORANGE COUNTY REGISTER, Nov. 5, 2002, at News 11.

authorize the procedure where the competent person gives informed consent in spite of opposition by the large Catholic Church.[154] Because these laws generally deny such treatment to minors,[155] they have no effect upon teenage pregnancies. Federal law precludes the funding of such procedures until age 21[156] despite the fact that at 18 women reach the age of majority for most other purposes.[157] A 1996 study by the Centers for Disease Control and Prevention (CDC) found that from 1 in 250 to 1 in 500 women who had been sterilized actually became pregnant within 10 years. This slightly higher rate (than previously estimated) is due to the fact that sometimes their tubes were not completely blocked by the surgery.[158] A California couple sued Kaiser Permanente for the full cost of raising their three-month-old daughter to age 21, alleging that the hospital officials assured them a tubal ligation would make future pregnancy impossible.[159]

More than 500,000 American women a year undergo tubal ligation. A 1995 Canadian contraception study found that two-thirds of married couples between 35 and 44 choose tubal ligation (women) and vasectomy (men).[160]

---

154. *See* the *Humanae Vitae* issued by Pope Paul VI. "He was equally aware that it did not change Catholic teaching, but only underlined it, and thus should not have been controversial. He was so shocked by the negative reaction generated by *Humanae Vitae* in 1968 that he never wrote another encyclical." Crocker, *Triumph.* 2001, p. 419. However, his successor, Pope John Paul II, the first non-Italian pope in more than 450 years, "reaffirmed *Humanae* in 1993 and two years later condemned abortion in *Evangelium Vitae"* as well. *Id.* p. 421.

155. Crocker, Triumph, 2001, p. 89.

156. Sterilization of a Mentally Competent Individual Age 21 or Older, 42 C.F.R. § 50.203 (2000).

157. Peck v. Califano, 454 F. Supp. 484 (D. Utah 1977); *see also* Voe v. Califafano, 434 F. Supp. 1058 (D. Conn. 1977). As stated by Margaret Sanger 80 years ago, "[b]irth control is woman's problem," Sanger, Woman and the New Race, 100, 1920.

158. L.A. Times, April 29, 1996, B2. "One-third were ectopic pregnancies, in which the fetus grows outside of the uterus and are life threatening unless discovered quickly." *Id.*

159. Orange County Register, July 13, 1995. See also Depenbrok v. Kaiser Foundation Health Plan, Inc., 79 Cal. App. 3d 167, 1978.

160. "The anticipated number of women becoming pregnant after tubal sterilization is 0.5 percent immediately after the operation to 2 percent, 10 years later." Orange County Register, Oct. 16, 1997.

Some women who lost their children in Asia's 2004 tsunami sought surgery to have their sterilizations reversed.[161] Some 86 percent of Health Maintenance Organizations (HMOs) in California provide sterilization.[162] Unfortunately, only on rare occasions do heroin addicts agree to sterilization. In 1997, 28-year-old Shella Horton of Los Angeles (whose five older children were *all born addicted to drugs)* became the first woman to take $200 from an organization in exchange for a tubal ligation.[163] (See Chapter 12.)

A hysterectomy, an excision of the uterus where a developing embryo normally becomes embedded (and the fetus is nourished), is "the most common major surgical procedure performed in the United States for nonobstetric reasons."[164] A quarter of U.S. women undergo elective hysterectomies before menopause. A study of premenopausal women, age 30 to 50 years, with abnormal uterine bleeding who were dissatisfied with medical treatments found that "hysterectomy was superior to expanded medical treatment for improving health-related quality-of-life after six months."[165]

good v. Good

Sterilization should be outlawed because at several locations in the Biblical Scripture it states that humans should "multiply."

v.

Voluntary sterilization should always be available because unwanted children are generally cared for in a manner that is damaging to such offspring and to society as well.

---

161. ORANGE COUNTY REGISTER, June 16, 2005, at News 17 and 22.

162. L.A. TIMES, Feb. 112, 1995, A31. In 2004, a study by the National Center for Health Statistics (NCHS) found that the female sterilization is used by 10.3 million women. ORANGE COUNTY REGISTER, Dec. 11, 2004, at News 25.

163. ORANGE COUNTY REGISTER, Oct. 23, 1997. L.A. TIMES, Nov. 1, 1997, ES. In the adjacent Orange County, hospitals reported 378 drug-baby births in 1994 and 271 in 1995.

164. "In 2000, approximately 633,000 hysterectomies were performed, and U.S. women have an estimated 25 percent risk of having their uterus removed. Ninety percent of hysterectomies are elective and performed before menopause for abnormal uterine bleeding and other nonlife-threatening reasons." Kupperman *et al., Effect of Hysterectomy vs. Medical Treatment on Health-Related Quality of Life and Sexual Functioning*. JAMA, March 24/31, 2004, p. 1447.

165. *Id.*

Twenty-first Century Approach

1. Special efforts should be made to provide sterilization for both men and women who do not desire to become responsible fathers or mothers or cannot do so.
2. State legislatures should consider enacting a statute providing that the availability and advantages of vasectomy be brought to the attention of men, and tubal ligation for women, to those who already have one or more children at the time of entry into a hospital for another successful delivery.
3. Premenopausal women with abnormal uterine bleeding, which has not been improved by medical treatment, should consider electing to have a hysterectomy about six months thereafter.

(ii) Contraception; Breast-feeding

Back in the nineteenth century, a fear of population decline was a concern; in 1871, Catholic leaders in France actually looked upon the defeat and occupation of Paris by German troops of Bismarck, the "iron chancellor," as signs "of divine chastisement" for the "sin of contraception."[166] In the 1890s, novelist Emile Zola wrote his last novel, *Fecondite,* essentially to "promote" his nation's "population growth." Even a century later, in 1996, the Prime Minister of France again urged his people to increase their birthrate, whereas in 2002, about 20,000 Boy and Girl Scouts between the ages of 14 and 18 at the 20th World Scout International Jamboree in Thailand were given free condoms.[167]

---

166. Teitelbaum and Winter, The Fear of Population Decline, 1985, p. 27. In the United States, the Supreme Court's first major decision in bioethics was set forth in Justice William O. Douglas's most famous opinion overruling Connecticut's "anti-contraceptive law" by finding "various guarantees create zones of privacy," including the protection against the denial of substantive "due process of law." Griswold v. Connecticut, 381 U.S. 479 (1965), p. 484. In dissent, Justice Black concluded: "Connecticut's law as applied here is not forbidden by any provision of the Federal Constitution as that Constitution."

167. "French Prime Minister Alain Juppe stated, "The worrisome weakness of the birthrate in our country, which does not even assure a renewal of the population, should jolt our national community." The 1996 population of France was 58 million people." Orange County Register, April 3, 1996. On the contrary, while visiting Salvador de Bahia on Brazil's coast, the Minister of Health, who was Hillary Clinton's guide for a hospital visit, said that despite laws against abortion, "Rich women have

Some studies indicated condoms do not safeguard against human papillomavirus, (HPV), a widespread sexually transmitted disease that may cause genital warts or cervical cancer. President George W. Bush has asked the FDA to modify the current warning on its label to include information about HIV.[168]

The Catholic Church has always prohibited contraception.[169] It has only authorized use of the "rhythm method" based upon the belief that successful conception is likely to occur when intercourse occurs on the day of ovulation or the five days surrounding ovulation. A 1995 study found that a test to predict the time of ovulation does not work for many people.[170] The timing of intercourse did not influence sex selection, and earlier use of contracep-

---

access to contraception if they choose; poor women do not." H. CLINTON, LIVING HISTORY, (2003), p. 314.

168. ORANGE COUNTY REGISTER, April 2, 2004, at News 9. However, several people indicated that they were among those who "would be unlikely to read or heed a condom package label." *Id.* A recent survey revealed that most German men wear the wrong size condoms. ORANGE COUNTY REGISTER, Jan. 2, 2005, at News 26.

169. The desire for legitimate offspring is, in fact, according to the Catholic Church, the only motive that can justify sexual intercourse. If the wife hates sexual intercourse, if she is likely to die of another pregnancy, if the child is likely to be diseased or insane, if there is not enough money to prevent the utmost extreme of misery, that does not prevent the man from being justified in insisting on his conjugal rights, provided only that he hopes to beget a child.

BERTRAND RUSSELL, MARRIAGE & MORALS, 1929, pp. 52-3.

> A former priest, James Drane, who was fired for writing an article that challenged the church's position on contraception, wrote his new book, *More Humane Medicine: A Liberal Catholic Bioethics,* in which he developed "a liberal Catholic natural law perspective for bioethics founded on a holistic, as opposed to purely physical, anthropology."

Panicola, Hast. Cent. Rep., September-October 2004, p. 46.

170. N.Y. TIMES, Dec. 7, 1995, A15; N. ENG. J. MED., Dec. 7, 1995. At least one of the births of 70 percent of women on welfare have been reported as mistimed or unwanted. Center for Health Statistics, Centers for Disease Control, *1995 National Survey of Family Growth.* "Without contraception, 85 out of 100 women would be pregnant in a year." Mauldon, *Providing Subsidies and Incentives for Norplant, Sterilization and Other Contraception: Allowing Economic Theory to Inform Ethical Analysis*. JIKOW, MEDICINE AND ETHICS.

tives did not affect women who later tried to conceive.[171] Anthropologist Margaret Mead pointed out that "Our humanity depends on our relative infertility, upon the long period of human gestation and dependency possible only where there are few children who can be reared long and lovingly."

Another natural means of stretching periods of infertility is the breast feeding of infants. With the chimpanzee, our closest animal relative, the median time from the birth of one offspring to the conception of the next is approximately three and a half to five years, and gorilla birth spacing is between four and six years.[172] With women in underdeveloped countries, it is about three years. However, women who do not breast-feed their infants may recommence ovulation as soon as 27 days after delivery; and short interpregnancy intervals are associated with an increased risk of adverse perinatal outcomes. A 1998 California law[173] allows women to breast feed in public. If other states enacted such laws, they could encourage the practice of breast-feeding by women and help to partially reduce overpopulation. Furthermore, it has been held that certain employers who excluded prescription contraceptives from their health plan for employees violated the federal law.[174]

---

171. DALLAS MORNING NEWS, Oct. 17, 1998, 5-F. It was predicted that in the next century "[s]ex will be practiced for fun, and reproduction we will do in a test tube." *Id.*

> From the point of view of Jewish law, the diaphragm is the most favored form of contraception, for it prevents conception and has little, if any, impact on the woman's health. If the contraceptive pill or implant is not contraindicated by the woman's age or body chemistry, those are usually the next most favored forms of contraception.

DORFF, MATTERS OF LIFE AND DEATH, 1998, p. 123.

172. Watts, 1990. *Ecology of Gorillas and its Relation to Female Transfer in Mountain Gorillas*. INT. J. PRIMATOL. 11:21-46.

173. Cal. Civ. Code § 43.3, West 1999. It has been reported that a woman in New Zealand breast fed her pet puppy because she wanted it to protect her baby daughter as they both grew up. ORANGE COUNTY REGISTER, Jan. 2, 2005, at News 26, "Breast Feeding."

174. Erickson v. Burtell, 2001, THE FEDERAL LAWYER (August 2001, p. 36). It was stated that the employer of a drugstore chain violated Title VII by excluding prescription contraceptives. In 2004, the California Supreme Court ruled that Catholic employers must provide coverage for contraception as part of employment healthcare plans, Catholic Charities of Sacramento v. Dep't of Health Care, No. 04 C.D.O.S. 1737.

In 1951, Carl Djerassi synthesized a molecule, norethindrone, which is similar to progesterone, the hormone currently used in birth-control pills. The pill contains hormones that cause the body to mimic pregnancy. While a woman is taking the pill, being "chemically pregnant," she cannot become "actually pregnant" again for a while. It has very few side effects and became a component of the oral pill for birth control. On May 9, 1960, the FDA approved a pill as safe for birth-control use. It was still in use in the year 2000, but of risk for smokers.[175] When women stopped taking the pill, this resulted in an increase in unwanted pregnancies and abortions.[176]

In 1994 only one-half of large-group insurance plans provided any form of coverage for birth control.[177] Beginning in 2000, California health-care plans that cover prescription drugs had to include prescription contraceptives and devices, such as diaphragms. Moreover, by 2003, California, New Mexico, Washington, and Alaska allowed pharmacists to dispense an emergency contraceptive drug directly to women without the need for a prescription or doctor's visit.[178]

Despite the efficacy of the copper intrauterine devices (IUDs), there was much concern that IUD use may cause pelvic infection. Although previously used by 10 percent of women in the United States, its use dropped to less than 1 percent.[179] IUD, an inexpensive contraceptive, cannot protect against sexually transmitted diseases (STDs). Formerly, one type of IUD, which resulted in infection (by sometimes drawing bacteria from the vagina to the uterus), caused lawsuits against its manufacturer (Dalkon Shield) to go into

---

175. "Current use of oral contraceptives (predominantly first generation) together with heavy smoking increased the risk of myocardial infarction to 32 times that of nonsmokers who did not use oral contraceptives." N. ENG. J. MED., Dec. 20, 2001, Vol. 345.

176. Jain *et al.*, number needed to treat and relative risk reduction. ANNALS OF INTERNAL MEDICINE, 128, 72–73 (1998).

177. 1994 study by the Alan Guttmacher Institute.

178. ORANGE COUNTY REGISTER, May 19, 2003, at News 11. There are 20 states that require private-sector insurance coverage for prescription contraceptives: Arizona, California, Connecticut, Delaware, Iowa, Georgia, Hawaii, Maine, Maryland, Massachusetts, Missouri, Nevada, New Hampshire, New Mexico, New York, North Carolina, Rhode Island, Texas, Vermont, and Washington. ORANGE COUNTY REGISTER, Nov. 30, 2003, at News 30.

179. ORANGE COUNTY REGISTER, Nov. 1998, p. 1.

bankruptcy. However, since the dawn of the century, three different types of IUDs became available to be inserted into the uterus, which neither destroy fertilized eggs nor induce abortion.[180] In 2002, a study concluded that among women from 35 to 64 years of age, current or former oral-contraceptive use was not associated with a significantly increased risk of breast cancer.[181] As of 2003, most oral contraceptives in widespread use have been combinations of an estrogen and a progestrin. The doses of estrogen and progestrin were "rapidly reduced during the 1960s and 1970s because of concern about safety and because the reduction of the doses did not reduce the contraceptive effectiveness."[182] Demands for "emergency contraception" to prevent pregnancy after unprotected sex went up 70 percent in 1998 over the previous year.[183]

In 2001 the FDA approved the birth-control patch and by 2004 it was estimated that 800,000 American women were on the patch. Although there were no requirements for follow-up studies beyond routine FDA reviews of reports called in by customers and others, the risk of suffering a potentially fatal blood clot while using the patch was found to be about three times higher than while using birth-control pills.[184] Starting in 2003, California hospitals must offer emergency contraception. It became available for an

180. These include Paragard (Copper T IUD) (which lasts 10 years), Mirena (Levonorgestrel IUS) (to be replaced in five years), and Progestasert (which needs replacement every year): "The reliability rate of an IUD, in place over a long period, is similar to that of tubal ligation. The device costs $200 to $300 (less than a year's worth of oral contraceptives) and may be covered by insurance." UC Berkeley Wellness Letter, July 2004, pp. 4-5.

181. N. Eng. J. Med., 2002; 346:2025-32. "The importance of this finding for public health is enormous, because more than 75 percent of the women in the study used oral contraceptives." N. Eng. J. Med., June 27, 2002, p. 2078.

182. Petitti, Combination Estrogen-Progestin Oral Contraceptives, N. Eng. J. Med., Oct. 9, 2003, p. 1443. "The favorable risk-benefit ratio for healthy women applies to all oral contraceptives containing a low dose of estrogen (less than 50 g.). The risks associated with oral-contraceptives use outweigh the benefits for women with a history of stroke, ischemic heart disease, or venous thrombo-embolism." *Id.* p. 1449.

183. N. Eng. J. Med., Aug. 23, 2001, p. 608.

184. Orange County Register, July 17, 2005, at News 17. By contrast with the pill, the patch rate of nonfatal blood clots was about 12 out of 10,000 users during the clinical trials, while the rate of deaths appears to be 3 out of 200,000. Accordingly, an FDA reviewer said "the label should clearly reflect this reviewer's safety concern about a potential increased risk." *Id.*

estimated 8,000 rape victims who report the crime each year in California. It has been noted that "If the 'morning after' pill is taken within 72 hours of intercourse, it can prevent a fertilized egg from implanting itself in the uterus. An estimated 25,000 pregnancies resulted from sexual assaults in 1998, but as many as 22,000 could have been prevented by using emergency contraception."[185] However, only 10 percent of the public even has had knowledge of the existence of emergency contraception.

Of the methods of birth control introduced in the 1990s, Norplant, a hormonal implant, was engulfed in litigation. The female condom attracted a market in Africa and many parts of the developing world as an important weapon in the fight against AIDS.[186] Depo-Provera, an injectable hormone took 20 years from submission to the FDA to obtain its approval. As a result, many companies stopped such research.[187] Some people express fear that promulgation of the use of condoms would merely increase sexual promiscuity. However, a national campaign in Switzerland to promote use of condoms did not encourage people to have more sex.[188] At the dawn of the third millennium, late-night condom ads on TV in the United States began getting more favorable reviews than during the decade before. The steady sexual content on TV desensitized consumers to messages about birth control, becoming a $2 billion industry.[189] The advertising campaign by former

185. Orange County Register, Sept. 6, 2002, at News 22.

186. Orange County Register, July 24, 1997, at News 23.

187. Orange County Register, Dec. 27, 1995, at News 11. The Today Sponge contraceptive came back on sale in 2003. "Sponges priced at the U.S. equivalent of about $2.90 each will hit the shelves at 4,000 pharmacies. Roughly, 250 million polyurethane Today Sponges were sold from 1983 to 1995. It was taken off the market in 1995 after problems were found at the company's factory," http://member.compuserve.com/news/story.j . . . 3/4/03.

188. "The study published in the *American Journal of Public Health* found that a safe-sex program and education about the risks of contracting the AIDS virus led to dramatic increases in the use of condoms, but no increase in the number of sex partners or the rate of casual sex." Orange County Register, May 15, 1997, at News 10.

189. Orange County Register, April 5, 1999, p. 6. The World Watch Institute's 2003 *State of the World* reported that 350 million women lack access to contraception—a number that can be expected to grow as populations increase. "And an estimated 125 million women do not want to be pregnant but are not using any type of contraception." *Sierra,* November/December 2004, p. 40.

senator and presidential candidate Robert Dole on behalf of Viagra, together with the political impeachment trial of President Clinton for sexually related activity, also changed things for the media.

Despite the litigation, Norplant (levonorgestrel) was approved by the FDA in 1990 and is used by 1 million American women. This remains one of the most effective reversible methods of contraception. It is implantable and provides effective contraception for five years. In February of 1998, a mistrial was declared in the first Norplant suit to reach the courts.[190] The FDA earlier "reaffirmed its safety."[191] In 1996, a California judge ordered an abusive mother to use Norplant as a condition of being allowed probation, in lieu of a jail sentence, after she was convicted of severely beating her children. The decision was criticized and the judge was actually brought before the state Commission on Judicial Performance in San Francisco for making this seemingly reasonable decision.[192]

Although all state statutes cover prescription drugs for FDA-approved forms of contraception, the majority of these statutes also contain varying degrees of restriction. California exempts religious employers for whom "the inculcation of religious values is the purpose of the entity."[193]

---

190. L.A. Times, March 23, 1998, A19. *See* Broadman v. Commission on Judicial Performance, 959 P.2d 715 (Cal. 1998). Norplant presented a coercive use by judges as a condition of parole following a conviction for child abuse.

191. L.A. Times, Aug. 18, 1995, D3.

192. Orange County Register, June 27, 1996, at Metro 7. "The order was never implemented because she was jailed for violating another probation condition." Broadman v. Comm'n on Judicial Performance, 959 P. 24 715 (Cal. 1998).

193. Cal. Health & Safety Code §§ 1367.25, 10123.196 (West 2000 & Supp. 2002). However, in Catholic Charities of Sacramento, Inc., v. Superior Court of California, the court found that Catholic Charities were not exempt because the majority of its employees were non-Catholics and offered social services to people of varied faiths. 109 Cal. Rptr. 2d 176, 184, 2001.

good v. Good

Breast feeding in public is regarded as disgusting and even provocative. It should not be permitted, nor should any contraceptives, especially so-called "emergency contraceptives." Such uses are analogous to abortions, which kill babies.

v.

Our society is maturing and should recognize natural functions of women with infants. Breast feeding in public will encourage this, lengthen the interval between births, and thus partially reduce their number, as do all temporary contraceptives, including emergency ones.

Twenty-first Century Approach

1. Most state legislatures should enact laws (similar to that in California) allowing women to breast-feed in public in order to encourage more women to do so and significantly augment the interpregnancy interval.
2. Knowledge of the existence and availability of emergency contraceptives should be further promulgated.
3. State legislatures should consider enactment of a statute authorizing judges to order women convicted of serious child abuse to use drugs such as Norplant (which can prevent another pregnancy for as long as five years) and Depo-Provera (which is given to patients four times a year) as a condition of probation.

(iii) Unintended Pregnancies

In 1995, the Centers for Disease Control and Prevention reported that 57 percent of pregnancies in the United States were unintended and that 43 percent of births resulted from unintended pregnancies.[194] About half of all unintended pregnancies that result in live births have more maternal complications and poorer infant outcomes than intended pregnancies.[195] A Census Bureau study in 1999 found that more than 50 percent of first pregnancies

194. ORANGE COUNTY REGISTER, April 8, 1997, at News 2. *Dallas Morning News*, Oct. 12, 1995, 8A. Half of these pregnancies occur in women who are not using some contraceptive and half occur in those who sometimes do.

195. Dietz *et al.*, JAMA, Oct. 13, 1999, p. 1359.

occurred out of wedlock and that 89 percent of births to teenagers occurred before marriage in the 1990s, compared with less than 30 percent 60 years before.[196] Some studies found that only half of sexually active high school students used condoms the last time they were sexually active.[197] Although more than 10 percent of all females between ages 15 and 19 years of age were pregnant in 1995 according to a study by the nonprofit Alan Guttmacher Institute,[198] only one-third of these obtained abortions.[199] However, the teen birthrate fell from 73 in 1991 to 41.1 per 1,000 teens in 2002.[200] The United States had the highest birthrate in the world for unwed mothers.[201] Boyfriends who are considerably older than adolescent girls have been found to be responsible for a majority of teen pregnancies.[202] An important decision by a Canadian court in 1997 ordered a boyfriend to pay "espousal support" to cover his girlfriend's prenatal expenses.[203]

A study of 17,000 girls found that menstruation began at an average age of 12.16 years in blacks and 12.88 years in whites. This is earlier than commonly believed and suggests that sex education should begin sooner than it often does.[204] The rate of pregnancy in girls 14 and under did not drop at the

196. ORANGE COUNTY REGISTER, Nov. 9, 1999, at News 13.

197. Mellhaney Jr., INSIGHT, Sept. 29, 1997, p. 24. A study at the end of the 20th century estimated that "70 percent of the prisoners in the United States [are] the children of [former] teen mothers." *Education Digest*, 1999, p. 38.

198. ORANGE COUNTY REGISTER, Oct. 26, 1999.

199. ORANGE COUNTY REGISTER, Feb. 17, 1999.

200. ORANGE COUNTY REGISTER, May 10, 2004, at News 8. California is the only state that refused to accept federal sex-education grants which teach only sexual abstinence and avoid providing information on contraception. L.A. TIMES, July 16, 2004, A27.

201. The rates of unmarried women of childbearing age giving birth, broken down by ethnicity in a report by the federal Centers for Disease Control and Prevention, were: "For white women, the rate is 2.8 percent, for Hispanics, 9.3 percent, for black women, 7.6 percent (after peaking at 9.1 percent in 1989). The number of single parents tripled from 3.8 million in 1979 to 11.7 million in 1996, an increase from 1.9 percent of the population to 4.4 percent." ORANGE COUNTY REGISTER, Nov. 19, 1998, at News 27.

202. "Over half of all infants born to women younger than 18 years are fathered by adult men, with 40 percent of 15-year-old girls having infants with partners aged 20 years or older." JAMA, Aug. 19, 1998, p. 648.

203. Scott v. Miggs (1997), Carswell Sask 518, (Canada).

204. ORANGE COUNTY REGISTER, April 8, 1997, at News 2.

dawn of the twenty-first century; teenage girls who grow up without their fathers have sex earlier. A 15-year-old girl who has lived with her mother only is three times as likely to lose her virginity before her 16th birthday as compared to one who has lived with both parents. Further, their crime rates soar. A survey by the Justice Department found that more than half of 14,000 inmates had not lived with both parents while they were growing up.[205]

A 1995 Harris poll found 73 percent of women stated that men were "not responsible enough" to choose a birth-control method and 70 percent of the men either agreed or said they considered it "the female's responsibility." These attitudes accounted for 40 percent of unwanted births.[206] California spent $500,000 to publicize laws that allow parents to abandon newborns at hospitals without being prosecuted.[207]

In 1998, an appellate court upheld a California statute making it a crime for unemancipated minors to have sex with each other. The court stated, "[W]hile we do not ignore the reality that many California teenagers are sexually active, that fact alone does not establish that minors have a right of privacy to engage in sexual intercourse."[208] Although such a ruling had little effect,[209] it represents a step toward diminishing teenage pregnancy and unwanted births.

---

205. Smith, 1994, *The New Wave of Illegitimacy,* FORTUNE, April, pp. 81p94. Mellhaney Jr., INSIGHT, Sept. 29, 1997, p. 24. EDUCATION DIGEST, 1999, p. 38.

206. ORANGE COUNTY REGISTER, May 23, 1995, at News 3. Alfred Kinsey's 1953 volume *Sexual Behavior in the Human Female* stated that: "One in four wives committed adultery at some point, compared with about half of all husbands, and that half of all women had sexual relations before marriage." ORANGE COUNTY REGISTER, Jan. 25, 2003, at News 22.

207. This law became effective Jan. 1, 2001. It requires parents to leave babies with the hospital staff within 72 hours of birth. ORANGE COUNTY REGISTER, Feb. 19, 2002.

208. *See In re* T.A.J., 73 Cal. Rptr. 2d 331 (1998). An Alameda County 16-year-old boy, arrested for having sexual intercourse with a 14-year-old girl, challenged the so-called statutory rape law, relying on a recent state supreme court ruling that overturned a state law requiring parental consent for minors' abortions. *See* American Academy of Pediatrics v. Lungren, 940 P.2d 797 (Cal. 1997). *See also* Cal. Penal Code § 261.5, 1999 (unlawful sexual intercourse with person under 18; age of perpetrator; civil penalty).

209. Over three decades ago, anthropologist Margaret Mead concluded:

> Each new generation of young males learn the appropriate nurturing behavior and superimpose upon their biologically given maleness this

(iv) Emergency Contraception

In 1998, a number of physicians emphasized that if emergency contraception were available without a prescription, it would be used more often and could be expected to reduce the number of unwanted pregnancies. They urged, "As an interim measure, health professionals should educate women about emergency contraception and prescribe a supply of such contraception that can be kept at home."[210] For women lacking the desire to give birth to an infant, little is more important than avoiding the need to obtain an abortion rather than to proceed toward having an "unwanted child."

In 1997, the pill RU-486 was finally approved by the FDA as a possible way of avoiding pregnancy after intercourse and urged the makers of certain birth-control pills to formally seek approval for use. Such pills have been controversial because antiabortion groups consider it a form of nonsurgical abortion.[211] However, being made a prescription drug significantly restricted its usage. In the United States, contraception "has meant only anticipatory contraception," and "pregnancy begins with implantation, not just upon fertilization. Medical organizations and the federal government concur that fertilization is a necessary but insufficient step toward pregnancy." Accordingly, as noted above, "all women at risk for unintended pregnancy should keep a pack of oral contraceptives handy for emergency use."[212] The FDA recom-

---

> learned parental role. When the family breaks down . . . in periods of abrupt transition from one type of economy to *another,* this delicate line of transmission is broken. [However,] so far, in all known history, human societies have always reestablished the forms they temporarily lost.

MARGARET MEAD, MALE AND FEMALE: A STUDY OF THE SEXES IN A CHANGING WORLD, Dell, 1969, pp. 190-98.

210. Glasier and Bald, *The Effects of Self-Administering Emergency Contraception*, N. ENG. J. MED., July 2, 1998, p. 1; and STUBBLE FIELD, SELF-ADMINISTERED EMERGENCY CONTRACEPTION—A SECOND CHANCE.

211. ORANGE COUNTY REGISTER, Feb. 25, 1997, p. 3, and ORANGE COUNTY REGISTER, Sept. 19, 1996, p. 1. Testing of RU-486 in this country ceased in 1989 because the Bush Administration's Food and Drug Administration imposed an unwarranted "import alert" on the drug under pressure by the antiabortion right. See also L.A. TIMES, Dec. 8, 1994, M4.

212. *Emergency Contraception Expanding Opportunities for Primary Prevention*, N. ENG. J. MED., Oct. 9, 1997, pp. 1078-9.

mended approval of the morning-after pill to be used as soon as possible after unprotected sex because (a) it cuts a woman's chance of getting pregnant by up to 89 percent and (b) it works best in the first 24 hours.[213] Although the government rejected over-the-counter morning-after birth control in 2004, regulators left open the possibility that they will reconsider.[214] The FDA determined not to grant over-the-counter status to the emergency contraceptives, which preserved a major barrier to them despite the fact that:

> Six states have lowered that hurdle by allowing pharmacists to dispense emergency contraception without a prescription. In those states, patients can simply bypass physicians. But the FDA's decision means that patients cannot avoid pharmacists. Because emergency contraception remains behind the counter, pharmacists can block access to it. And some have done just that.[215]

It is currently estimated that half of the 3.5 million unintended pregnancies that occur each year in the United States could be averted if emergency contraception (EC) were easily accessible and used. Pregnancy rates were comparable in all groups as was the incidence of new sexually transmitted infections (STIs). Easier access to EC did not compromise regular contraceptive use or lead to an increase in risky sexual behavior. In 2005, a study found "clear evidence that neither pharmacy access nor advance provision compromises contraceptive or sexual behavior" and concluded "it seems unreasonable to restrict access to EC to clinics."[216]

---

213. ORANGE COUNTY REGISTER, Dec. 17, 2003, at News 26.

214. The FDA "was under intense political pressure from both sides on whether to lift the prescription requirement for emergency contraception." ORANGE COUNTY REGISTER, May 7, 2004, at News 25.

215. Canton and Baum, *The Limits of Conscientious Objection—May Pharmacists Refuse to Fill Prescriptions for Emergency Contraception?*, N. ENG. J. MED. 3, (Nov. 4, 2004) p. 2008.

216. Raine *et al., Direct Access to Emergency Contraception through Pharmacies and Effect on Unintended Pregnancy and STIs,* JAMA, Jan. 5, 2005, p. 54 and list at p. 93.

good v. Good

"Emergency contraceptives" (often considered to be a misnomer because it merely prevents implantation after conception) constitute early abortion and hence kills "infants." An attempt to outlaw intercourse between teenagers is to seek fantasy, as is the idea of seeking "spousal support" from their "boyfriends."

v.

The constitutional rights to contraception and abortion also apply to emergency contraception in order to prevent unwanted pregnancy. The equal protection clause of the Constitution may require "boyfriends" to pay "spousal support" to unwed females and laws prohibiting intercourse between minors would help reduce teenage pregnancies.

Twenty-first Century Approach

1. In view of the fact that more than half of 3.5 million unwanted pregnancies (and 43 percent of unwanted births) occur without use of contraceptives (and sometimes desspite their use), 44 state legislatures should consider the enactment of legislation that would make emergency contraceptives available without the requirement of a prescription.
2. Because it is unfair to place the entire burden of expenses on an unwed pregnant female, legislation is necessary to provide the equivalent of "spousal support" by the male responsible for impregnating her for the purpose of sharing the costs of their child's expenses.
3. In those states where teenage pregnancies and unwanted births are out of control, state legislatures should consider enacting statutes (similar to that in California) to provide for outlawing sexual intercourse between unemancipated minors.

8

Infertility, Impotence, and Cloning

1. *Fertility and Impotence*
   a) **Some Causes of Impotence**
   b) **Public Subsidizing of Reproductive Assistance and Mandatory Insurance Coverage**
   c) **Fertility Clinics, Lack of Regulation by States; Embryo Detraction**
   d) **Sperm Banks and Sperm Problems**
   e) **Egg Banks**
   f) **Frozen Embryos and Posthumous Reproduction**
   g) **Multiple Births Resulting from Fertility Pills**
   h) **Artificial Insemination of Single Mothers**
   i) **Surrogacy**
2. *Cloning in the Third Millennium*
   a) **Origins of Cloning**
   b) **Some Religious Reactions**
   c) **Some Political and Legal Statements on Human Cloning**
   d) **Some Views of Scientists and Physicians on Human Cloning**
   e) **Proposed limits on Cloning in Ethics and the Law**

---

1. Fertility and Impotence

a) Some Causes of Impotence

Infertility has been defined as "the failure to conceive after 12 months of unprotected intercourse; further, it apparently affects an estimated 10 percent of the population of reproductive age in the United States."[1] Healthy couples who do not achieve a pregnancy within one year of trying will conceive during the second year, according to a study by the European Society of Human Reproduction and Embryology. However, according to the U.S. National Institute of Environmental Health Sciences, many doctors would like to recommend techniques such as test-tube fertilization earlier than after one year of failure for older couples.[2] Although the Pregnancy Discrimination Act (PDA) prohibits discrimination on the basis of pregnancy and "related medical conditions," it was held not to extend to discrimination on the basis of infertility:

> Because reproductive capacity is common to both men and women, we do not read the PDA as introducing a completely new classification of prohibited discrimination based solely on reproductive capacity. [F]or a condition to fall under the PDA's inclusion of 'pregnancy. . . and related medical conditions' as sex-based characteristics, that condition must be unique to women.[3]

---

1. N. Eng. J. Med. 769 (2002). Unprotected copulation with animals is both unreproductive and prohibited by biblical rules. *See Exodus* 21:19; *Leviticus* 17:23.

2. Orange County Register, July 4, 2002, at News 15. Historically in Rome, during the sixth century B.C. "[w]hips snap[ped] as young women eagerly bare[d] their shoulders for a touch of the lash, believed to bestow fertility." Later priests "smeared goat's blood," which remained an important Roman festival until banned by the pope in A.D. 494. Nat'l Geographic, Jan. 2005, at 66–67.

3. Saks v. Franklin Covey Co., (2d Cir. 2003). Americans spend more than $1 billion a year to treat infertility. In about 40% of infertile couples, a female factor is solely responsible, but in 30% male infertility is the only culprit, and both partners have reproductive problems in another 30%." Harvey B. Simon, The Harvard Medical School Guide to Men's Health 279 (2002). "Obesity increases the risk of impotence. All in all, a man with a 42-inch waist is twice as likely to develop erectile dysfunction as a man with a 32-inch waist. *Id.* at 74.

An infertile person incapable of engaging in the activity of reproduction has been held "disabled" within the federal Americans with Disabilities Act.[4] "Nearly 40 percent of women in their 30s and 40s develop fibroids, non-cancerous growths of muscle fibers inside the uterus. No one knows what causes fibroids. More than 150,000 hysterectomies—surgical uterus removal—each year are due to fibroids. For women who still want children, options are limited."[5]

Some studies show that obese women have a higher risk for diabetes; higher maternal weight before pregnancy increases the risk of late fetal death.[6] Men who had undergone a radical prostatectomy after being diagnosed with prostate cancer became impotent, unless the surgery spared "one or two bundles of nerves."[7] The Male Aging Study reported impotence in 25 percent of 65-year-olds, 55 percent of 75-year-olds, and 65 percent of 80-year-olds.[8] Others claim that longer working hours may make it more difficult for a woman to find time or energy for coitus.[9] A study done at Prince of Songkla University in Thailand that found that women who worked more than 71

4. Bragdon v. Abbott, 107 F.3d 934 (1st Cir. 1997), *aff'd* 163 F.3d 87 (1st Cir. 1998); ORANGE COUNTY REGISTER, July 5, 1998 (discussing a study in which the men who smoked marijuana had only about half as many sperm as those who didn't and in which the women who smoked marijuana were also vulnerable to lower fertility).

5. http://member.compuserve.com/news/stor.../0001%2F20021112%2F085696253.htm&sc-150, Nov. 12, 2002. After attempting to conceive an infant without success for a year, a couple is considered infertile. GEOFFREY SHER ET AL., FROM INFERTILITY TO IN VITRO FERTILIZATION 21 (1988).

6. N. ENG. J. MED. (1998); *see infra* Chapter 12, Subsection 3(c) (discussing obesity). In 2003, the *British Medical Journal* published findings that pregnant women who drink eight or more cups of coffee a day could double their risk of stillbirth compared with pregnant women who do not drink coffee. ORANGE COUNTY REGISTER, Feb. 2, 2003, at News 18.

7. N. ENG. J. MED. (2000). In a recent study, 56% of those who had both bundles spared said they were impotent 18 months or more after the surgery. Viagra has been one of the drug industry's most successful products; it had $1.7 billion in sales during 2002, including a success rate with diabetes-related impotence. Diabetic men suffering from impotence had improved erections with other new drugs. WALL ST. J., Mar. 27, 2003, at D4.

8. SIMON, *supra* note 3, at 298.

9. ORANGE COUNTY REGISTER, Feb. 12, 1998, at News 16 (citing the journal *Occupational and Environmental Medicine*).

hours a week took longer to get pregnant. Another study was made of women recruited from union members who were 20 to 35 years old, were living with a partner, and had no children. It was concluded that psychological distress might be a risk factor for reduced fertility in women with long menstrual cycles.[10]

Timing is a factor; a 1995 study showed that nearly all pregnancies are a result of intercourse during a six-day period surrounding the day of ovulation.[11] It has even been claimed that adulterous matings have a greater chance of resulting in pregnancy than one "within marriage."[12] Some 10 percent of women are infected with chlamydia, which can lead to infertility secondary to scarring the fallopian tubes—particularly among promiscuous women.[13] Women suffering from epilepsy have lower fertility rates because the treatment impairs ovulation and decreases the rate of congenital anomalies.[14] Perhaps most important is a lack of interest in sex; a poll showed that about one-third of women had decreased libido and over a quarter of them did not

---

10. Hjollund et al., FERTILITY & STERILITY 47–53 (1999). The main finding of this study was substantially reduced fecundability among highly distressed women with long menstrual cycles. Distress did not affect women with a normal duration of menstrual cycles. *Id.* at 50.

11. 333 N. ENG. J. MED. 1517–21 (1995).

12. ROBIN BAKER, SPERM WARS: INFIDELITY, SEXUAL CONFLICT AND OTHER BEDROOM BATTLES (1996).

13. ORANGE COUNTY REGISTER, Sept. 10, 1998, at News 18. "A 50-state CDC survey found a 9 percent increase in the rate of gonorrhea—from 121.8 cases per 100,000 people in 1997 to 132.9 per 100,000 in 1998. The sexually transmitted disease can cause pelvic inflammation and infertility in women, and can also foster the transmission of the AIDS virus." ORANGE COUNTY REGISTER, June 23, 2000, at News 20; *see also* N. ENG. J. MED. 1748 (2000).

14. ORANGE COUNTY REGISTER, Dec. 18, 1998, at News 35. This difference was highly significant overall and in women aged 25 to 39 years. *Id.*

15. Research based on the 1992 National Health and Social Life Survey, a compilation of interviews with 1749 women and 1410 men. ORANGE COUNTY REGISTER, Feb. 1999, at 1 (citing the February 1999 issue of *JAMA*).

A May 2000 study showed that the drug Viagra did not help women with sexual dysfunctions, such as difficulty getting aroused. "Viagra did about as well as a placebo." ORANGE COUNTY REGISTER, May 23, 2000, at News 14 (reporting on a presentation at a meeting that day of the American College of Obstetrics & Gynecology in San Francisco).

achieve orgasms.[15] As one bioethicist wrote, "The difficulty of unraveling the ethics of reproductive technology may be due to our impoverished ability to recognize and appreciate what is normal about being human."[16]

It has been estimated that more than 30 million men over 40 years old suffer from impotence, most of which is caused by erection difficulties. In 2002 research disclosed that infertility in men has been inexplicable in two-thirds of cases.[17] These were addressed by Viagra for the first time in 1998.[18] Men in other countries also suffer from impotency.[19] The Food and Drug Administration's (FDA) approval of oral sildenafil citrate (Viagra) for the treatment of male erectile dysfunction followed research by three Americans—Ignarro, Furchgott, and Murad—who were awarded the 1998 Nobel Prize for elucidating the physiology of nitric oxide, which acts as a human

---

16. ALBERT R. JONSEN, THE BIRTH OF BIOETHICS (1998). Sexual pleasure may be undermined when each act of intercourse becomes a reminder of infertility. PAUL LAURITZEN, PURSUING PARENTHOOD: ETHICAL ISSUES IN ASSISTED REPRODUCTION (1993).

17. ORANGE COUNTY REGISTER, Sept. 6, 2002, at News 10 (reporting research published in the medical journal *Lancet* conducted by scientists at the U.S. Environmental Protection Agency, Wayne State University in Detroit, and Leeds University in England). Infertility in men may be linked to DNA that stops conception. FERTILITY & STERILITY 63; ORANGE COUNTY REGISTER, Aug. 14, 2002.

18. ORANGE COUNTY REGISTER, May 3, 1998, at News 6. About half of men ages 40 to 70 had difficulties obtaining or maintaining erections. ORANGE COUNTY REGISTER, Apr. 5, 1998, at News 32.

> A professor of urology at the Boston University School of Medicine said his research indicated that the problem for middle-aged women was the same as it was for middle-aged men: a paucity of blood flowing to the sex organs. And if Goldstein is correct, it makes at least theoretical sense that Viagra could be helpful for women.

Id. By 2003, three anti-impotence drugs were FDA approved: Viagra, Levitra, and Cialis. ORANGE COUNTY REGISTER, Nov. 22, 2003, at News 22.

19. The incidence of infertility treatment in Israel—measured by the number of clinics per capita—is the highest in the world, four times higher than in the United States, which excels in the provision of commercialized quality medical services. Carmel Shalev, SOC. SEC., Aug. 1998, at 75 (citing ISRAEL MINISTRY OF JUSTICE, REPORT OF THE PUBLIC-PROFESSIONAL COMMISSION IN THE MATTER OF IN VITRO FERTILIZATION ((1994)).

signal molecule.[20] Although the birth control pill remains banned in Japan at the end of the twentieth century, Viagra was approved there for men within a mere six months. This double standard was criticized by Japanese women, who remarked, "The drug that lets you get pregnant is approved, but not the one that would prevent pregnancy."[21] This is because the Japanese government's policy since the end of the twentieth century has been to increase its birthrate.

Japan's population is expected to peak at around 127 million as early as 2006 and fall rapidly over the next 50 years to roughly 100 million as the birthrate declines. Meanwhile, "the graying of Japan will accelerate, with people over 65 becoming 35.7 percent of the population by 2050, roughly double the 17.4 percent in 2000. . . . Japan has no large-scale immigration policy."[22]

A 1999 study of the use of Viagra by women found "that there was no significant change in intercourse satisfaction or in the degree of sexual desire after the patients had taken Viagra for 12 weeks."[23]

About 10 percent of all men who can produce little or no sperm suffer from a genetic defect in a part of the Y chromosome. They often use *in vitro* (Latin for "in glass") fertilization (IVF), a laboratory procedure to join an egg with sperm in a Petri dish and implant the resulting embryo into the uterus. When

---

20. JAMA 495 (1999). In ABRAHAM MORGENTALER, THE VIAGRA MYTH (2003), the author points out that "it continues to astound me that a pill that works on blood vessels can successfully treat a failing relationship. But it's true. This is the power of Viagra. This is the stuff that myths are made of." *Id.* at 156. It is the most commonly prescribed therapy for men after prostate cancer treatment, but it is frequently ineffective in these men, particularly after surgery. Viagra does not work at all for men when neither of the paired nerves has been spared, since Viagra acts by enhancing the nerve signal.

21. L.A. TIMES (1999). The AMA itself failed to disclose that Pfizer, the maker of Viagra, had paid the people who reviewed clinical trial data on it before the impotency drug was submitted for government approval. ORANGE COUNTY REGISTER, Feb. 11, 1999, at.

22. ORANGE COUNTY REGISTER, May 22, 2002, at News 32.

23. "Signaling an ongoing concern over Japan's plummeting birthrate and aging population, the Cabinet issued a report that urged people to work less and have more babies as part of a 'structural reform in life-style.'" Kaplan, Urology, ORANGE COUNTY REGISTER, Nov. 6, 1999, at News 10.

they breed using IVF, their sons tend to inherit the genetic defect.[24] While IVF may help otherwise infertile men produce sons, it will not necessarily ensure that there will be grandchildren. Male menopause (viropause or andropause) occurs in men after the age of 40 or 45. Their hormones and neuropeptides diminish and many develop an enlarged prostate;[25] unlike unaided females, the arrival of menopause permits them to continue to have children. Menopause abruptly ends the ability of women to have children in midlife when a woman's ovaries usually stop functioning and the menstrual cycle ceases.[26] Menopause also protects older women from the potentially fatal rigors of childbirth while ensuring that she will live long enough thereafter "to see that her last child survives safely into adulthood."[27] But the arrival of menopause has not stopped the appetite of some women to have children via fertility treatments such as artificial insemination and *in vitro* fertilization (IVF) via egg donation.[28] During the twenty-first century, a single sperm can be injected directly into a single ovum to overcome male-factor infertility; or an ovum from a young donor can be fertilized and implanted in a post-menopausal woman so she can carry the

---

24. The IVF study, published in the journal *Human Reproduction*, was the first to demonstrate that the technique of injecting a single sperm into an egg in a laboratory dish results in some men passing the genetic flaw responsible for their infertility on to their sons. Daughters are not affected, since females receive from their fathers an X chromosome instead of the Y chromosome. ORANGE COUNTY REGISTER, July 1, 1998, at News 18.

25. A study published in the April 2002 issue of the journal *Human Reproduction* showed that "[m]en's fertility starts dwindling after 35 . . . . At the age of 40, men were 40% less likely to get their partners pregnant in a month than they were at the age of 35." HUMAN REPRODUCTION, *available at* http://humrep.oupjournals.org. "On average, female fertility begins its meaningful slide at age 27. No decline in male fertility was seen before age 35, and the man's age only seemed to matter when the woman passed 35, the study found." *Id.*

26. Black women are three to four times more likely to die as a result of pregnancy than white women in the United States, the Centers for Disease Control and Prevention (CDC) reported. ORANGE COUNTY REGISTER, Feb 21, 2003, at News 21.

27. L.A. TIMES, Apr. 23, 1998, at A29 (citing research at the University of Minnesota and Brown, which was published in *Nature* that day). A 60-year-old woman in Wales has become the oldest woman in Britain to give birth without fertility drugs. ORANGE COUNTY REGISTER, Jan. 16, 1998.

28. ORANGE COUNTY REGISTER, Nov. 11, 1998 (discussing research to be published in the journal *Nature Medicine*).

fetus.[29] In 1997, research by Harvard scientists experimenting with mice found a way to keep ovaries from dying and blocking menopause. It has been remarked that childlessness has been more and more transformed into "medical problems" through the availability of these technologies.[30] These alleged medical problems have also given rise to bioethical matters—eight of which are discussed below.

b) Public Subsidizing of Reproductive Assistance and Mandatory Insurance Coverage

Over four million American women are unable to have children.[31] Infertile people claim that their medical condition deserves relief and will seek fertility treatments, typically *in vitro* fertilization.[32] To the extent that the infertility is genetically caused, insurance coverage may be denied if it is considered a preexisting condition,[33] or because it is not a disease. In Iowa, a

---

29. JAMA 1038 (2002). Postmenopausal hormone replacement therapy (HRT) is widely used for its benefits, including prevention of osteoporotic fractures and colorectal cancer. However, new evidence about its harm requires reconsideration of its use for the primary prevention of chronic conditions that include CHD, stroke, thromboembolic events, breast cancer with five or more years of use, and cholecystitis. JAMA 872 (2002).

30. Hast. Ctr. Rep., Sept.-Oct. 1995, at 16.

31. Mosher & Pratt, *Fecundity and Infertility in the United States: Incidence and Trends*, 56 FERTILITY & STERILITY 192–93 (1991).

32. As of 1998, over 58,000 *in vitro* fertilization cycles were performed annually, resulting in 17,943 pregnancies and 14,789 live births (or one out of four). Soc'y for Assisted Reproductive Tech. & Am. Soc'y of Reproductive Med., *Assisted Reproductive Technology in the United States: 1998 Results Generated from the American Society for Reproductive Medicine/Society for Assisted Reproductive Technology Registry*, 77 FERTILITY & STERILITY 18–31 (2002).

33. State Nat'l Life Ins. Co. v. Stamper, 312 S.W.2d 441 (Ark. 1958). The court held that in preexisting condition clauses, "the sickness should be deemed to have had its inception at the time it first manifested itself or became active, or when sufficient symptoms existed to allow a reasonably accurate diagnosis of the case, [not when] . . . the condition was latent, inactive, and perhaps not discovered." *Id.*

In *Bullwinkel v. New Eng. and Mut. Life Ins. Co.*, 18 F.3d 429, 432 (7th Cir. 1994), the court held that benefit coverage could be denied under the preexisting condition clause. The plaintiff found a breast lump that was examined for the first time days before the preexisting condition period expired. *Id.*; *see also* Neorette v. Principal Mut. Life Ins. Co., 56 F.3d 72 (9th Cir. 1995) (unpublished opinion) (hold-

couple was denied coverage for *in vitro* fertilization (IVF) because she did not suffer from a disease for which IVF was a medically necessary treatment. The magistrate found that although IVF allowed the woman to become pregnant, it did not treat the underlying disease. An appellate court reversed by somehow finding that the plan broadly defined "illness" such that infertility could be treated by IVF. But if a genetic defect is a disease, then surgery circumvents, rather than treats, the disease without changing an underlying bad gene.[34]

As one bioethicist has remarked, "There is no need or obligation for medicine to help each and every couple have a child."[35] Another has written, "We need not bring children into the world when this would contribute to a problem of overpopulation or of limited resources."[36] This has not been raised politically in this country during the past three decades. Ethics of obstetrics has been recognized a primacy of autonomy.[37] This raises the question whether assisted-reproduction technology should be made available to HIV-infected women. It has been pointed out that "[i]nfertile women with HIV infection have had to confront the possibility of mother-to-child transmission of HIV and the likelihood that the mother will die before her child reaches majority."[38]

---

ing that the coverage was properly excluded where the plaintiff received treatment for symptoms of undiagnosed cancer during the exclusion period), *cert denied*, 516 U.S. 1011 (1995). *But see* Holman v. Pac. Health & Life Ins. Co., 902 P.2d 106, 109 (Or. Ct. App. 1995) (holding that the jury reasonably found that a routine breast examination was not treatment for the later-diagnosed cancer; hence, the preexisting conditions clause did not apply).

34. Witcraft v. Sundstrand Health & Disability Group Benefit Plan, 420 N.W.2d 785 (Iowa 1988).

35. Daniel Callahan, False Hopes 153 (1998).

36. "The Interest in Existing Argument assumes that children with an interest in existing are waiting in a spectral world of nonexistence where their situation is less desirable than it would be were they released into this world." Hast. Ctr. Rep., Mar.-Apr. 1996, at 25. Overpopulation is discussed *infra* in Chapter 7.

37. Chervenak & McCullough, *What Is Obstetric Ethics?*, 35 Clinical Obstetrics & Gynecology 709–19 (1992); *see also In re* AC, 573 A.2d 1235 (D.C. Ct. App. 1990).

38. Minkoff & Santoro, N. Eng. J. Med. 1748 (2000).

At the turn of millennium, it was indicated that this should preclude her from becoming a candidate for reproduction.[39] However, it has been noted, "there are not many situations in which assisted reproduction would be considered an option knowing the mother will not likely survive longer than 10 years." *In vitro* fertilization is not included in Medicaid benefits and the California legislature has refused to mandate insurance coverage of IVF treatment;[40] some other states require insurance companies to offer[41] or to provide[42] such coverage. Mandated coverage has been criticized as "an unwarranted diversion of scarce societal resources," and fraud has been rampant by some doctors who trick insurance companies into paying for fertility treatments that are not covered by their policies.[43] The cost has been reported to be typically $10,000 per IVF with 25 percent chance of success, $40,000 per delivery;[44] of the live births from IVF, 36 percent involve multiple in-

---

39. In most cases, the maternal medical condition limiting survival, such as pulmonary hypertension, advanced cancer, or end-stage renal disease, would preclude her from being a candidate for assisted reproduction. One could argue that a woman with human immunodeficiency virus (HIV) infection has a limited life expectancy. However, because survival time after a diagnosis of HIV infection is increasing with current treatment, it is uncertain whether a woman with HIV could reproduce and care for her child into adulthood. Certainly, this is not the responsibility of health care professionals alone.

JAMA 1039 (2002).

40. CAL. HEALTH & SAFETY CODE § 1374.55(a) (West) covers only group health service plans. It does not apply to HMOs, *id.* § 1374.55(c), individual health plans, or public health insurance.

41. CONN. GEN. STAT. § 38a-536 (2000); TEX. INS. CODE ANN. art. 3.51-6, § 3A(a) (Vernon 1999).

42. ARK. CODE ANN. §§ 2-851-37, 23-86-118; HAW. REV. STAT. §§ 431:1OA-l 16.5, 432:1-604 (1999); 215 ILL. COMP. STAT. § 5.356m(a) (West 2000); MD. CODE ANN. INS. § 15-810 (West).

43. Dr. Neus Lauersen is accused of doing a common, unspoken practice among many doctors. He helped couples get pregnant by providing fertility treatments not covered by insurance, then submitted bills for various covered treatments. "This sort of thing is unfortunately very widespread," said Michael Diegel, spokesperson for the Coalition Against Insurance Fraud. He estimated that health insurance fraud cost insurers $53.9 billion in 1997. Much of that cost, he said, must be passed onto the consumer as higher rates. ORANGE COUNTY REGISTER, Feb. 16, 2000, at News 15.

44. N. ENG. J. MED. 687 (2002).

fants; with three or more infants, the costs increase to about to $340,000.[45] For older couples with more complicated conditions, the cost can run as high as $800,000. The Massachusetts Association of Health Maintenance Organizations said that its members pay $40 million more in premiums to cover infertility treatment for 2,000 couples.[46] It has been pointed out that by only requiring companies "to offer, would enable consumers to make rational choices when selecting insurance," as well as to "make rational and informed choices when deciding whether to attempt IVF and where to seek such services."[47] Insurers have also maintained that IVF is excluded from normal coverage as "experimental treatment" because of its limited success rate and otherwise mixed results.[48] The majority of insurance companies do not cover *in vitro* fertilization.[49] Nevertheless, it has been reported that:

> Each year, the Society for Assisted Reproductive Technology collects data on the outcomes of in vitro fertilization from its members,

---

45. Goldfarb et al., *Cost-Effectiveness of an In Vitro Fertilization*, 87 OBSTETRICS & GYNECOLOGY 18–21 (1996); Collins et al., *An Estimate of the Cost of In Vitro Fertilization Services in the United States in 1995*, 64 FERTILITY & STERILITY 538–45 (1995).

46. "The median cost of one IVF cycle is approximately $8,000; because success rates are low, many patients make several tries before having a baby or giving up. Using donor eggs raises the price still higher—$10,000 to $20,000—for each attempt." PARK RIDGE CENTER BULL., Aug.-Sept. 1998, at 13.

47. *In Vitro Fertilization: Insurance and Consumer Protection*, 109 HARV. L. REV. 1092, 2109 (1996).

48. The 50% success rate in 1990 was up from about 20% before 1960. JOHN YOVICH & GEDIS GRUDZINSKAS, THE MANAGEMENT OF INFERTILITY (1990); *see* Reilly v. Blue Cross & Blue Shield United of Wis., 846 F.2d 416 (7th Cir. 1988). The trial court granted summary judgment for the defendant. The appellate court concluded that disputed issues of fact remained, including "the reasonableness of employing a success ratio per se, particularly the 50% ratio." *Id.* at 423.

49. "As of November 2002, three states had laws mandating complete coverage (Illinois, Massachusetts, and Rhode Island), and five states had laws requiring partial coverage (Arkansas, Hawaii, Maryland, Ohio, and West Virginia). . . . On January 1, 2002, New Jersey become the fourth state to require complete insurance coverage for in vitro fertilization." N. ENG. J. MED. 666 (2002); American Society for Reproductive Medicine; State Infertility Insurance Laws, http://www.asrm.org/Patients/insur.html (last visited Aug. 6, 2002).

whose clinics perform more than 90 percent of all such procedures in the United States. Among 134,985 children conceived as a result of assisted reproduction technology between 1996 and 2000, 2,597 infants (1.9 percent) were reported to have a major birth defect. This rate is similar to the incidence of major abnormalities reported in general populations in both Europe and North America.[50]

good v. Good

In view of the primacy of autonomy, reproductive assistance should be offered to all infertile people. This should be at taxpayer expense to those who cannot afford it and insurance companies should be required to offer coverage for such assistance without extra charge.

v.

Such a financial incentive would not only aggravate the problem of overpopulation but also deplete our limited resources, and would appear to ignore the possible mother-to-child transmission of HIV and other unfortunate transmissions.

Twenty-first Century Approaches

1. Although unfortunate results are limited, there has been a lack of social need to mandate insurance coverage of expensive IVF treatment by state legislatures.
2. As an alternative, state legislatures should consider authorizing insurers to make IVF treatment available as an option (at an extra premium) for consumers who may evaluate their particular program's success together with their particular needs.
3. Insurance companies should authorize treatment of physical problems such as fibroids, scarring, tubal damage from infection, and ovulatory problems.

---

50. Steinkampf & Grifo, N. Eng. J. Med. 1449 (2002).

c) *Fertility Clinics and Preimplantation Genetic Diagnosis; Lack of Regulation by Many States; Embryo Detraction*

Large numbers of physicians also have little or no training in fertility and genetics. A Human Genome Project (HGP) study in 1998 found that 54 percent of the physicians surveyed had only one course in basic genetics. A shortage has existed of clinicians trained in genetic counseling and interpreting genetic tests,[51] and it was argued that because cures and treatments are not available for many genetic disorders, individuals could not be tested for such genetic diseases. However, much genetic screening helps individual decision making and contributes toward the discovery of treatments and cures.

A genetic counselor advises on risks for inherited disease or an abnormal pregnancy outcome and the chances of having children who may be affected. Such counseling is needed by women over 35 who are pregnant or planning to become pregnant.[52] It is also needed by couples who have a mentally retarded child or an inherited disorder where one of them has a job involving exposure to radiation, chemicals or drugs, or who belong to a particular ethnic group. For example, about 2.5 million pregnant women undergo screening for Down syndrome each year in the United States.[53] About 250,000 families in the United States include a person with Down syndrome, and about 4,000 Down syndrome babies are born annually. About 6,000 clinical

---

51. Hast. Ctr. Rep., July-Aug. 1998, at 17. A 2004 study by the University of California, San Diego, found that women seeking *in vitro* fertilization who were worried about the medical aspects of the procedure had 20% fewer eggs retrieved and eggs fertilized than women who were less inclined to worry about it. ORANGE COUNTY REGISTER, Apr. 22, 2004, at News 7 (citing the journal *Fertility and Sterility*).

52. SALK INST. SIGNALS, Summer 1998, at E. "If the mother is under age 35, the chance of a baby per cycle is probably greater than 40%, according to fertility scientists. But by age 45, that drops to less than 1%." ORANGE COUNTY REGISTER, July 25, 2003, at News 21.

53. 338 N. ENG. J. MED. 955–61 (1998). Thus an embryo that has an extra chromosome in pair 21 becomes affected with Down syndrome, referred to as trisomy 21, which is associated with that life-threatening or life-altering disease. Structures in cells that lined up in pairs as each cell undergoes division, with one copy of each structure transmitted to the new cell, were called "chromosomes," a Latin term meaning colored things, due to scientists' having stained the structures with colored dyes to observe their movement.

cycles of PGD were performed in 1990 and more than 1,000 children were born for these parents who decided which embryos to transfer into the woman's uterus.[54] Mental capacity for people with Down syndrome can range from severe retardation to intelligence that falls within the so-called normal range. Most people with Down syndrome live an average-length life. About 10 percent, however, are born with heart conditions or other physical problems that can shorten life expectancy.[55] Studies have shown that all fetuses were aborted by affected Ashkenazi Jews where prenatal screening also indicated Tay-Sachs disease.[56]

The Preimplantation Genetic Diagnosis (PGD) may be used to test a single-embryo cell (called a blastomere, which has been extracted from a three-day-old eight-celled embryo) for genetic diseases that are due to defective genes or abnormal chromosomes within days of fertilization and prior to the establishment of pregnancy. Although risky, it obviates the need for such couples to decide whether to abort a fetus at a later stage. Many people using *in vitro* fertilization (IVF) have become interested in the PGD technique, which can be used to determine a number of an embryo's hereditary diseases. A preconception technology called Microsort was used to help couples at risk for X-linked genetic disorders, like hemophilia and Duchenne's muscular dystrophy, and diseases most prevalent in males (see Chapter 15).

The technology was used by the Genetics and I.V.F. Institute to assure such a couple that they could select the infant's gender.[57] Their policy be-

---

54. Verlinsky et al., *Over a Decade of Experience with Preimplantation Genetic Diagnosis: A Multicenter Report*, 82 FERTILITY & STERILITY 292 (2004); Talbot, *Jack or Jill? The Era of Consumer-Driven Eugenics Has Begun*, ATLANTIC MONTHLY, Mar. 2002, at 25.

55. ORANGE COUNTY REGISTER, NOV. 18, 2002, at Local 4. An additional chromosome in pair 21 is often affected with Down syndrome. It is referred to as trisomy 21 (a third, or extra, chromosome in pair 21).

56. J. Tasca & Michael E. McClure, *The Emerging Technology and Application of Pre-implantation Genetic Diagnoses*, 26 J.L., MED. & ETHICS 7, 9 (1998). "For the many people who disapprove of abortion, PGD is a happy medium because healthy embryos can be selected when they are just three days old." WINSTON, SUPER HUMAN 231 (2000). In 2002, PGD was discovered to be able to screen for Alzheimer's disease. JAMA; L.A. TIMES, Feb. 27, 2002, at A18.

57. Hast. Ctr. Rep., Jan.-Feb. 2002, at 24. It has been concluded that the genetic basis for "sexual differentiation probably emerged with the appearance of eukaryotes

came one that limited the use of its technology Microsort to select a child of the "non-dominant" sex in their family. Microsort claims an average success rate of 90 percent for X-bearing (female) sperm and an average of 73 percent for Y-bearing (male) sperm.[58] At the turn of the twenty-first century the American Society of Reproductive Medicine (ASRM) originally issued a report that set forth reasonable methods of gender selection, indicating that "[p]hysicians should be free to offer [them] in clinical settings to couples who are seeking gender variety in their offspring if the couples met certain conditions."[59] However, this was significantly revised in 2002 when the ASRM concluded, "[i]t has not yet been clearly established that a couple's desire for gender variety in offspring is sufficient to outweigh the need to show special respect to embryos." This was based upon its revised moral position that "fertilized eggs and preimplantation embryos, although not persons or moral subjects in their own right, should not be treated like any other human tissue. Rather, because of the meanings associated with their potential to implant and bring forth a *new person,* they deserve 'special respect.'"[60] Its new posi-

---

at least 2 billion years ago; Y emerged from X perhaps 300 million years ago, and evolution has maintained the duet assiduously ever since." Levin, *The Puzzle of Aspirin and Sex*, N. Eng. J. Med. 1366 (2005); *see also* Ethics Comm. of the Am. Soc'y of Reproductive Med., *Sex Selection and Preimplantation Genetic Diagnosis*, 72 Fertility & Sterility 595 (1999).

58. "Microsort Current Result," http://www.microsort.net. China and India, the countries with the largest populations (the United States is the third most populous), have both had more significant preference for sons: The preference for boys, strong across most of Asia and particularly in China, where the official one-child policy has dramatically magnified the pressure for male children. In India's larger families, if the first-born is a daughter, she will normally be accepted. In the 2001 census count of children six or under, there were 927 girls for every 1,000 boys—down from 945 girls in 1991 and 962 in 1981. L.A. Times, Nov. 17, 2002, at A3.

59. Ethics Comm. of the Am. Soc'y for Reproductive Med., *Preconception Gender Selection for Non-medical Reasons*, 75 Fertility & Sterility 861, 863–64 (2001).

60. The ASRM concluded that "[i]t has not yet been clearly established that a couple's desire for gender variety in offspring is sufficient to outweigh the need to show special respect to embryos." Robertson, *Sex Selection for Gender Variety by Preimplantation Genetic Diagnosis*, 78 Fertility & Sterility 463 (2002). Sex selection is illegal in Austria, Chile, Germany, Ireland, and Switzerland. *See Preimplantation Genetic Diagnosis,* 81 Fertility & Sterility S38 (2004). Nevertheless, in the United States, sex selection remains largely unregulated.

tion is opposed to that expressed by the Supreme Court's rulings respecting a "person" in connection with legal abortions (see Chapter 9) and also represents the difference between the views of the U.S. presidents before and after 2001. The AMA *Code of Medical Ethics* states that "physicians should not participate in sex selection for reasons of gender preference" with the sole exception of "selection of sperm for the purposes of avoiding a sex-linked inheritable disease."[61] AMA policy constitutes no portion of the law in the United States; but sex selection is actually illegal in a number of European countries, including Austria, Germany, Ireland, and Switzerland.[62]

As of the turn of the third millennium, PGD began to be used to determine the presence of up to 30 genes that may lead to hereditary diseases.[63] No genetic technique has yet found a single gene that "determines IQ, height, or strength, intelligence, bearing and aggression. These matters will involve more complex discoveries."[64] Some people think of PGD as a eugenic (or "well produced" in Greek) product. They include "some advocates for the disabled and some disabled people themselves who fear that PGD is the beginning of genetic cleansing."[65] Following testing, some couples may de-

61. AMA CODE OF MEDICAL ETHICS § 2.04 (2003). After the turn of the twenty-first century, the GenSelect system began offering boy and girl kits on the Internet at $199 apiece. Their patents were approved in 2004 and "include a thermometer to help predict ovulation, special douches and 'gender-specific' mineral and herbal pills." ORANGE COUNTY REGISTER, Apr. 17, 2004, at News 15.

62. *See Preimplantation Genetic Diagnosis*, *supra* note 59, at S38. The British government's Human Fertilisation and Embryology Authority (HFEA) licenses clinics in accordance with its Code of Practice (2004), which states that they "should not use any information derived from tests on an embryo, or any material removed from it or from the gametes that produced it, to select embryos of a particular sex for social reasons."

63. *Id.* WINSTON, *supra* note 55, at 229. Although the Fertility Clinic Success Rate and Certification Act (FCSRCA) requires clinics to report pregnancy success rates, the medical diagnosis leading to IVF treatment, whether fresh or frozen embryos were used, and similar information, it does not require clinics to report the health status of the babies that are born.

64. WINSTON, *supra* note 55, at 241; *see infra* Chapter 15.

65. WINSTON, *supra* note 55, at 231.

sire to exercise the option of sex selection.[66] It has been argued that only grounds based upon sex-linked genetic disease "overrides concerns regarding sex selection."[67]

Reproductive technology has permitted sperm and oocytes to be joined (fertilized) outside the bodies of infertile would be parents. Embryos can be formed and inserted into the womb for implantation and later birth of one or more infants.[68] However, doctors in Western Australia found that infants born after Intracytoplasmic Sperm Injection (ICSI), when a single sperm is injected into the egg, were twice as likely to have a major birth defect and nearly 50 percent more likely to have a minor one.[69] Clients may also have their embryos screened in these clinics during *in vitro* fertilization for genetic disorders prior to implantation. The technique is used to prevent hereditary disorders such as cystic fibrosis (CF), Tay-Sachs disease, and as of 1999, sickle cell anemia as well.[70] This has turned infertility into a $2 billion

---

66. A method developed by the Genetics & IVF Institute in Fairfax, Va., involves sorting sperm by the amount of DNA they contain and then using them for artificial insemination. . . . Sperm with a Y chromosome have about 2.8% less genetic material. . . . Most couples wanted to choose their baby's sex for "family balancing." A few wanted sex selection because they were at risk of having babies with genetic diseases that afflict boys almost exclusively.

ORANGE COUNTY REGISTER, Sept. 9, 1998.

67. Damewood, JAMA 3144 (2001).
68. Reproductive technology includes in vitro fertilization (IVF) and embryo transfer, gamete intrafallopian transfer (GIFT), intrauterine insemination (IUI), zygote intrafallopian transfer (ZIFT), and intracytoplasmic sperm injection (ICSI). In 1996 a study of ICSI in human IVF (the leading method for treatment of male factor infertility) was conducted to determine occurrence of chromosomal abnormalities and rate of congenital malformation in neonates. It was concluded that they were within the range observed with standard in vitro fertilization.

JAMA 1893 (1996).

69. ORANGE COUNTY REGISTER, Nov. 14, 1997, at News 9.
70. Previously, parents who were sickle-cell carriers had to wait until 10 weeks into pregnancy for a test to indicate whether their fetus inherited the defect and [then consider whether or not to abort]. . . . One in 625 black children is afflicted.

ORANGE COUNTY REGISTER, May 12, 1999, at News 9.

a year business.[71] Nevertheless, it has been noted that by one year of age, "one or more major birth defects have been identified in 9 percent of babies conceived with assisted reproductive technology, as compared with 4.2 percent of those who were conceived naturally," slightly over twice as many.[72] In 1992, the Y required such clinics to annually report all their instances when oocytes and sperm are handled outside the body.[73] By 1998, 92 percent of 360 clinics held the data of 81,899 such instances, together with a total of 20,143 live-birth deliveries.[74] Their low use has been due to a high cost; the estimated mean 2002 cost of a single IVF cycle in the United States was $9,500.[75]

The fertility business is generally unregulated throughout the United States. Laws governing use of artificial insemination (AI) and IVF are largely nonexistent.[76] In 2004, regulation of assisted reproductive technology has

71. ORANGE COUNTY REGISTER, July 7, 1996, at News 18; Begley, *The Baby Myth*, NEWSWEEK, Sept. 4, 1995. "One out of six couples in the U.S. is infertile or fails to conceive within one year of deciding to have a child." H.R. 5397, 101st Cong. 2d Session 26 (1990).

72. N. ENG. J. MED. 769 (2002).

73. Pub. L. No. 102-493, 106 Stat. 3146 (1992) (codified at 42 U.S.C. §§ 263a-1.

74. JAMA 1522 (2002).

75. Smith et al., *Diagnosis and Management of Female Infertility*, JAMA (2003). "A conservative estimate places the costs of a successful pregnancy of this kind between $10,000 and $15,000. If we use the conservative figure of 500,000 for estimating the number of infertile women with blocked oviducts in the United Sates whose *only* hope of having children lies with *in vitro* fertilization, we reach a conservative estimate cost of $5 to $7.5 billion. Is it fiscally wise for the federal government to start down this road? Clearly not." LEON R. KASS, LIFE, LIBERTY, AND THE DEFENSE OF DIGNITY 111 (2002). In cases where the woman is over forty, the cost of in vitro fertilization (IVF) was estimated between $160,000 and $800,000. Neumann et al., N. ENG. J. MED. 239–44 (1994).

76. Wright, *Human in the Age of Mechanical Reproduction*, DISCOVER, May 1998, at 75, 76 (quoting one specialist who described the field as "the Wild West of medicine").

> Since 1978, the management of female infertility has been transformed by in vitro fertilization (IVF). Intracytoplasmic sperm injection (ICSI), first reported as an IVF laboratory technique in 1992, has similarly changed the management of male infertility.

JAMA 1767 (2003).

been called for by the President's Council on Bioethics. Currently, however, no U.S. laws require the licensing programs, nor is there much regulation governing private embryo research. The U.S. Fertility Clinic Success Rate and Certification Act of 1992 (FCSRCA)[77] requires infertility clinics to submit assisted reproductive technology success rate data that is reported on by the Centers for Disease Control and Prevention (CDC).[78] Nevertheless, with sperm provided by donors, 18 states "deregulate" by permitting AI children to obtain available information based on satisfaction of a "good cause" or similar standard.[79] Potential loss of anonymity might impact a donor's decision. In 2000, Supreme Court Justice Sandra Day O'Connor rejected an emergency request to delay Oregon's 1998 adoption records law from taking effect for adoptees over age 21. Birth mothers had argued that the new law violates the privacy of people like themselves who gave up their children for adoption and started new lives.[80] In May 2000 a California appellate court held that an anonymous sperm donor does not have an unlimited right to privacy and can be forced to testify in legal actions alleging that his donation resulted in genetic harm to a child he helped conceive. The donor in that case had sold about 320 deposits of semen to the bank over five years, collecting a total of $11,200. "The sperm bank's staff falsely represented to the Johnsons that the sperm they were buying had been tested and screened for infectious and genetic diseases. Cryobank failed to properly test and screen Donor No. 276 and [to] conduct further investigation or testing of the donor once they learned he had a family history of kidney disease."[81]

In 2002 Dr. Yury Verlinsky tested the embryos of a 30-year-old woman with a family history of Alzheimer's. Because she tested for the mutation, he and his colleagues used PGD to screen for the gene that causes early onset of the disease, and transferred only the unaffected embryos into the uterus. As a

---

77. 42 U.S.C. § 263a-1 to -7 (2000).

78. The CDC publishes annual success rates for such pregnancies and develops a model program for possible state legislation to license embryo laboratories.

79. Andrews & Elster, *Adoption, Reproductive Technologies, and Genetic Information*, 8 Health Matrix 125, 135 (1998).

80. *See* Does v. Oregon, 164 Or. App. 543 (1999); Orange County Register, May 31, 2000, at News 13.

81. The court said that the doctor-patient privilege is not relevant because the donor was not a patient. L.A. Times, May 20, 2000, at 1.

result, they reported how a child was delivered free of the predisposing gene.[82] In the same publication its editors, Dena Towner and Roberta Springer Loewy, pointed out that "[r]eproductive freedom is such a widely accepted norm in Western society that some even assume it to be an individual's absolute or inalienable right. But experience teaches that, like interests and goods, rights cannot be absolutely unqualified." Although Verlinsky and colleagues provided a woman with the opportunity to have a child free of an autosomal dominant form of Alzheimer's disease (AD) "she will likely manifest early symptoms of AD while this child is in the early, formative childhood." Analogously, "women who are dying of cancer while their children are still young grieve greatly because they will not be able to see their children grow up, participate in their lives, or help protect them from harm. It is early-onset AD." "Ultimately, patients and physicians are faced with the 'technology question': should a procedure be done simply because it can be done?—and the subsidiary ones: who is to decide, and when? Because societal values include equity and respect for persons, the goal is to represent the interests of all relevantly affected."[83]

The U.S. differs from most other countries that require sperm banks to make AI available only to married women.[84] It has been reported that children need "both a father and mother" and children are better off in a two-parent household.[85] The Fertility Clinic Success Rate and Certification Act was enacted in 1992, in order to require infertility clinics to report the preg-

82. Verlinsky et al., *Preimplantation Diagnosis for Early-Onset Alzheimer Disease Caused by V717L Mutation*, 287 JAMA 1018 (2002).

83. Towner & Loewy, *Ethics of Preimplantation Diagnosis for a Woman Destined to Develop Early-Onset Alzheimer Disease*, 287 JAMA 1038 (2002).

84. Nielsen, *Legal Consensus and Divergence in Europe in the Area of Assisted Conception—Room for Harmonization?*, *in* CREATING THE CHILD: THE ETHICS, LAW, AND PRACTICE OF ASSISTED PROCREATION (Donald Evans ed., 1996). The United Kingdom grants single, noncohabiting women access to sperm banks. DEREK MORGAN & ROBERT G. LEE, BLACKSTONE'S GUIDE TO THE HUMAN FERTILISATION & EMBRYOLOGY ACT, 1990: ABORTION & EMBRYO RESEARCH, THE NEW LAW (1991).

85. *The Formation of Families*, *in* FAMILY CHANGE AND FAMILY POLICIES IN GREAT BRITAIN, CANADA, NEW ZEALAND, AND THE UNITED STATES (Sheila B. Kamerman & Alfred J. Kahn eds., 1997). A number of states deny sperm donors all parental rights and responsibilities. Koehler, Comment, *Artificial Insemination: In the Child's Best Interest?*, 5 ALB. L.J. SCI. & TECH. 321, 332 & nn.79–80 (1996).

nancy success rates of their assisted reproductive technology programs and to certify embryo laboratories.[86] In 1995 the American Society of Reproductive Medicine (ASRM) planned to draft model state legislation to regulate the nation's fertility clinics.[87] In 1997, the CDC reported that, based upon information from 281 fertility clinics, almost 60,000 "cycles" (attempts) of assisted reproduction occurred resulting in 11,315 live births; this also resulted in almost 18 times as many multiple births as in the general public[88] (see Subsection (g) infra).

The need for some regulation was amply demonstrated by the experiences of clinics. For example, one that cost the taxpayers millions of dollars was the University of California–Irvine's (UCI) Center for Reproductive Health. From over 30 unconsenting donors, the clinic's three physicians took eggs and frozen embryos and transferred them to other women who believed the eggs were their own. These resulted in over seven live births to parents who were not genetically related to the children,[89] a serious violation of an ethical trust. The doctors attempted to have patients sign retroactive approval forms to coverup their improper conduct. In another case, due to an embryonic mixup at a New York fertility clinic, a woman gave birth to two boys: One, her biological child, was white; but the other was black, the son of other clients of this clinic.[90]

---

86. Fertility Clinic Success Rate and Certification Act of 1992, Pub. L. No. 102-493, 106 Stat. 3146 (codified at 42 U.S.C. § 263a-1).

87. Orange County Register, Oct. 10, 1995, at A1; Orange County Register, Oct. 21, 1995, at B4. The European Society for Human Reproduction and Embryology (ESHRE) is also dedicated to facilitating the study of all aspects of human reproduction and embryology.

88. L.A. Times, Dec. 19, 1997, at A23.

89. Karen T. Rogers, Comment, *Embryo Theft: The Misappropriation of Human Eggs at an Irvine Fertility Clinic Has Raised a Host of New Legal Concerns for Infertile Couples Using New Reproductive Technologies*, 26 Sw. U. L. Rev. 1133 (1997). The university acknowledged a failure to oversee the clinic and settled the suits on that basis. Meanwhile, "tanks containing sperm samples and hundreds of frozen embryos from former clinic patients hang in limbo at California Cryobank as officials try to locate the genetic parents, and await trial to force the university to take custody of the tanks." Orange County Register, June 23, 1999, at Metro 2; *see also* Dodge & Geis, Stealing Dreams: A Fertility Clinic Scandal (2003).

90. Orange County Register, July 17, 1999, at News 23. A New York judge awarded permanent custody of the black baby to the black couple, but gave the white

While donors normally retain undisputed rights over their eggs and sperm,[91] in the UCI case, they never intended to donate them or relinquish their rights to the children produced by their genes. Not having waived their rights, the unauthorized transfer of embryos violated their ownership right to choose when to procreate, a right that includes that of avoiding offspring.[92] Two of the three physicians liquidated their assets and fled the country. The third was convicted of defrauding insurance companies by listing his partners as assisting surgeons on procedures when they were not present. He was ordered held until sentencing so that he could not also flee the country like his former partners.[93] The university settled for some $20 million with the victims and their attorneys.[94] The whistleblowers (some of whom were fired) received hundreds of thousands in a settlement agreement that contained the questionable provision specifying a $100,000 penalty if they publicly revealed any information connected to the scandal.[95] As a result of this case,

---

couple visitation rights to the non-biologically-related infant. The first similar case in Britain occurred in 2002 when a white couple became the parents of black twins after a mistake by a fertility clinic during *in vitro* fertilization. The High Court forbade publicly identifying the clinic, the twins, or the parents. ORANGE COUNTY REGISTER, July 9, 2002, at News 9.

91. Maria R. Durant, Note, *Cryopreservation of Human Embryos: A Scientific Advance, a Judicial Dilemma*, 24 SUFFOLK U. L. REV. 707, 729, 742 n.114 (1990).

92. Perry & Schneider, *Cyropreserved Embryos: Who Shall Decide Their Fate?*, 13 J. LEGAL MED. 463, 472 (1992). The Tennessee Supreme Court, while referring to the IVF dilemma, described the right to avoid all parental links, even links involving no gestational burdens at all. Davis v. Davis, 842 S.W.2d, 588, 603 (Tenn. 1992).

93. ORANGE COUNTY REGISTER, Oct. 31, 1997, at 1.

94. ORANGE COUNTY REGISTER, Feb. 21, 1997 (discussing $2 million settlement); ORANGE COUNTY REGISTER, Nov. 4, 1997, at Metro 4 (discussing $695,000 settlement, and also noting "the taxpayer's tab now exceeds $20 million, including settlements, tentative settlements and legal fees"). "UCI also had other troubles, e.g., an $18 million malpractice judgment to cover a lifetime of treatment for a mother of two who emerged brain-damaged from hand surgery at the UCI Medical Center." ORANGE COUNTY REGISTER, Sept. 25, 1997, at News 14.

95. ORANGE COUNTY REGISTER, Mar. 26, 1998, at Metro 2; L.A. TIMES, Mar. 31, 1998, at B8. The university was seen as having sought to cover up the case. A lawyer who represented35 women during a UCI fertility scandal from 1993 to 1999 was convicted of defrauding ten clients of at least half a million dollars. She faces a maximum of 11 years in prison when sentenced. ORANGE COUNTY REGISTER, Sept. 8, 2004, at Local 15. She was actually sentenced to four years in prison. ORANGE COUNTY REGISTER, Oct. 30. 2004.

California's Governor Wilson signed a law making it a felony to steal human eggs or embryos and requiring written approval of donors before embryos, sperm, and reproductive materials can be harvested.[96] However, kidnapping statutes definitely do not apply, because California criminal statutes do not treat human eggs, or embryos, as humans. Penal Code provisions such as the law against homicide do not take effect until the second trimester of pregnancy, when the embryo has become a fetus.[97]

Genetically aberrant oocytes and waning ovarian functions are the main causes of the age-related decline in fecundity. However, it has been observed that women at age 40 can have implanted oocytes donated by younger women.[98] In 1997, a 63-year old woman who had lied about her age gave birth to a girl following in vitro fertilization at the University of Southern California's Program for Assisted Reproduction; she would not have qualified had she told the truth.[99] A question arises concerning the point at which nondiscriminatory age limits are valid. The United Nations declaration and convention of the rights of the child states children have the right to be raised by their parents. To exercise this right, it has been reasonably argued that the upper age of donor egg recipients should be set at 20 years younger than the average life expectancy of these women.[100] Studies of motherhood after meno-

96. CAL. PENAL CODE § 367(g) (West 1999). Even in the year 2000, this university kept paying the salary of one of the doctors who was convicted in 1997 of fraudulent billing practices in connection with the clinic. The state medical board had ruled that the conviction "is for a crime that includes dishonesty as one of its essential elements." ORANGE COUNTY REGISTER, Jan. 25, 2000, at News 9; *see also* 42 U.S.C. §§ 263a-1 to -7 (West 2000).

97. L.A. TIMES, May 29, 1995, at A3. "Pursuant to a Federal statute the Secretary of the Department of Health and Human Services, through the Centers for Disease Control (CDC), is directed to establish a model certification of embryo labs in each state."

98. Hull et al., *The Age-Related Decline in Female Fecundity: A Quantitative Controlled Study of Implanting Capacity and Survival of Individual Embryos After In Vitro Fertilization*, 65 FERTILITY & STERILITY 783–90 (1996).

99. ORANGE COUNTY REGISTER, Apr. 24, 1997, at 1. "Had the individual disclosed her actual age, she would not have qualified for treatment at USC, since the program uses an arbitrary upper age limit of 55 for women." *Id.*

100. Mori, *Egg Donation Should Be Limited to Women Below 60 Years of Age*, 12 J. ASSISTED REPRODUCTION & GENETICS 229–30 (1995).

pause have found no medical reason to prevent healthy women in their 50s from having babies with donated eggs.[101] These tests did not cover women in their 60s who received donated eggs with little or possibly no opportunity to bring up delivered children throughout the period of their youth.

Unlike the United States, France has the most comprehensive set of regulations governing artificial insemination, *in vitro* fertilization, etc. The French prohibit the insemination of single women, homosexuals, and the use of donated eggs or sperm of a person who is dead.[102] Some people in this country have quite different views. Every state has elaborate regulations for adoption, but few (e.g., Virginia and New Hampshire) comprehensively regulate assisted reproduction. It is difficult to understand why a child's safety should be better regulated if they are adopted than were they born through assisted reproduction by a donor in one of the nation's 280 fertility clinics—many of which have had documented problems.[103] Part of the reason lies in the absence of standards for this complex and fast-changing field.

good v. Good

No state legislation is needed to govern fertility clinics because it would only interfere with the autonomy needed by both sides in connection with agreements to provide assisted reproductive services. Settlement agreements are also private matters, as are the ages of people seeking these services.

v.

Abuses by some fertility clinics at the turn of the third millennium demand certain legislation that authorizes regulations to control those abuses.

Twenty-first Century Approach

1. Legislators in many states should consider enactment of laws requiring specific standards for donor consent agreements drawn up by

101. A study of 77 women who participated in the University of Southern California's assisted reproduction program found that there were no infant or mother deaths and no serious health problems in the babies. Although they were more likely to have Caesarean births and faced high rates of pregnancy-induced diabetes and high blood pressure, those conditions were temporary and treatable. JAMA 2320 (2002).

102. Arthur Caplan, Due Consideration 53 (1998).

103. These problems have been documented. *See* 281 Science 651–52 (1998).

fertility clinics. These would cover particular reproductive procedures and other matters dealing with possible IVF and abuses such as theft of eggs or embryos, or women who want a child but not the father, respectively. This legislation need not bc limitcd to fcderal standards being developed by the Department of Health and Human Services (HHS), which normally apply only where federal funds are involved (e.g., grants such as Medicaid, etc.). It should also minimize interference with procreational choices.

2. A statute should be considered that invalidates clauses in settlement agreements containing sanctions for revealing information concerning the unethical matters that are the subjects of such agreements.
3. In view of the fact that some older women now undergo IVF, consideration must be given to capping their qualifying age without discrimination.
4. More counseling needs to be provided by fertility clinics with regard to genetic abnormalities, financial burdens, and the emotional problems involved.
5. The growth in the number of diseases detectable by PGD should result in the authorization for clinics to normally provide these tests of embryos.
6. Gender selection criteria should be developed and used by fertility clinics as ethical standards for an accreditation program pending legislation thereon. Consideration might include portions of those developed by the American Society for Reproduction Medicine (ASRM) in the twenty-first century; namely, physicians should ascertain that the couples (a) are fully informed of the risks of failure, (b) are counseled about having unrealistic expectations concerning the behavior of children of the preferred gender, and (c) are offered the opportunity to participate in research to track and access the safety, efficacy, and demographics of preconception gender selection.

d) Sperm Banks and Sperm Problems

A 1997 study found that sperm counts in the United States have fallen an average of 1.5 percent a year since the 1930s.[104] As regards HIV-infected men, more than one in five semen samples contain live specimens of the virus.[105] Donor insemination with frozen sperm, first achieved in 1953, became common 20 years later; its success equals that of fresh sperm.[106] Except for the Catholic religion, few others consider such insemination as adultery.[107] Sperm banks have been proliferating across the country, primarily for use in the artificial insemination of women.[108] Sperm sorting is a preconception method, unlike Preimplantation Genetic Diagnosis (PGD), which involves discarded embryos, regarded as remaining actual lives by some religions.

At the turn of the third millennium, there was a substantial lack of racial

---

104. ORANGE COUNTY REGISTER, Nov. 24, 1997, at 7. Fertility refers to the ability to produce a baby, while fecundity refers to the ability to do it within a certain period of time. "The research, published today in the European journal *Human Reproduction*, provides the first clear evidence that the age of a man, as well as that of a woman, could be an important factor." ORANGE COUNTY REGISTER, Aug. 1, 2000, at 1.

105. ORANGE COUNTY REGISTER, Aug. 26, 1995, at News 17 (citing a report published in the September 1995 issue of the *Journal of Urology*).

106. The American Fertility Society issued guidelines in 1986 recommending that all donations be made with frozen sperm that has been quarantined for at least 180 days and then retested.

107.
> Some rabbis construe [donor insemination] as adultery. . . . [However,] the majority of recent authorities maintain that the legal category of adultery is incurred only when the penis of the man enters the vaginal cavity of the woman. That is clearly not the case when insemination takes place artificially.

ELLIOT N. DORFF, MATTERS OF LIFE AND DEATH 67 (1998).

108. SUSAN LEWIS COOPER & ELLEN SARASOHN GLAZER, BEYOND INFERTILITY: THE NEW PATHS TO PARENTHOOD 178–79 (1994). As of 2005:

> Sperm bank officials estimate the number of children born to donors at about 30,000 a year, but no one really knows. As half-siblings find one another, it is becoming clear that banks do not know how many children are born to each donor, or where they are. Men are paid about $65 to $100 per sample, and customers pay about $150 to $600 per vial, plus shipping.

ORANGE COUNTY REGISTER, Nov. 20, 2005, at News 33.

minority donors of eggs or sperm,[109] and only four states had sperm banks that tested for diseases other than HIV; about a dozen more states have laws requiring HIV antibody testing of semen donors.[110] Although guidelines have been issued on record keeping and donor screening, they have been voluntary.[111] However, sperm banks now must register with the Food and Drug Administration (FDA) to increase safety in this larger industry.[112] The largest sperm bank, the New England Cryogenic Center, subjected potential donors to questionnaires and examinations; each donor had to have at least a C average in college or graduate school, and every eight weeks, he was tested for all known communicable diseases.[113] In 2002 a study found the quality of semen significantly poorer in men from rural mid-Missouri than in males from urban areas, possibly caused by chemicals:

> Fertile men from mid-Missouri's Boone County were found to have a mean sperm count of about 59 million per milliliter, compared to 103 million for men in New York, 99 million in Minnesota, and 81 million in Los Angeles. Their sperm tended to be less vigorous.[114]

Physicians advise men about to undergo chemotherapy treatment to bank sperm because the procedure often leads to sterility. A Portland, Oregon,

---

109. ORANGE COUNTY REGISTER, July 25, 1999, at News 14.

110. Four of these states are Illinois, Indiana, New York, and Michigan. AMA CODE OF MEDICAL ETHICS § 2.05 states: "Frozen semen should be used for artificial insemination because it enables the donor to be tested for HIV infection at the same time of donation, and again after an interval before the original semen is used, thus increasing the likelihood that the semen is free of HIV infection."

111. Am. Fertility Soc., *Guidelines for Therapeutic Donor Insemination: Sperm,* 62 FERTILITY & STERILITY 101S (1994).

112. This new rule applies to all fertility clinics. ORANGE COUNTY REGISTER, Jan. 22, 2004, at News 6.

113. FERTILITY & STERILITY 1015–25 (1994).

> Each can offer sperm up to 8 times a month (at $70 per visit, for a monthly income of $560). The profiles list basics such as race, ethnic origin, blood type, height, weight, skin tone, eye and hair color, occupation, education, and interests. . . . Once the donor is chosen, it costs $170 to $195 per vial of semen.

ORANGE COUNTY REGISTER, May 26, 2002, at Accent 6.

114. ORANGE COUNTY REGISTER, Nov. 12, 2002, at News 9.

hospital banked the sperm of a man named Eubanks. When the hospital was unable to locate the sperm 10 years later, a suit arose. A jury awarded Eubanks $1.5 million for the loss of this sperm.[115] In 2000, for the first time in California, a sperm bank was closed that had stored tanks of human and dog sperm side by side, used dog names as codes for human donors, and handled specimens with dirty hands.[116] In the same year, the California Supreme Court unanimously upheld an appellate court decision demanding that a man, identified only as Donor 276, testify in a lawsuit against a Culver City sperm bank, thereby putting limits on a sperm donor's anonymity.[117] It has been reported that some military men were leaving deposits at sperm banks before heading overseas to fight in Iraq.[118]

Where a man has not only banked his sperm, but also indicated his intent regarding its use, the court will follow that intent. This was done by an author and businessman who had 12 vials of his sperm put in frozen storage before committing suicide and willing the vials to his lover. A California court gave her the vials. While confirming this ruling, the appellate court stated, "Seldom has this court reviewed a probate case where the decedent evidenced his or her intent so clearly."[119] The following year, sperm was extracted from another man's body 30 hours after arrival at a coroner's office in Los Angeles to fertilize eggs of his wife 15 months later.[120] However, where

115. Hast. Ctr. Rep., Jan.-Feb. 1998, at 47.

116. ORANGE COUNTY REGISTER, Apr. 19, 2000, at News 13.

117. The donor was allowed to give sperm, although he had a family history of kidney disease. He had received more than $11,000 for donations to California Cryobank since 1986. The Second District Court of Appeal in Los Angeles ruled that Donor 276 must testify in the case, and "his identity should remain undisclosed to the fullest extent possible." The court said the doctor-patient privilege does not apply because the donor was not truly a patient of the clinic. L.A. TIMES, Aug. 25, 2000, at A15.

118. http://msnbc.com/news/874893.asp (2003). By 2003, married men with cancer commonly stored their sperm prior to treatment. *See* LANCE ARMSTRONG, IT'S NOT ABOUT THE BIKE (2000); Tomlinson & Pacey, *Practical Aspects of Sperm Banking for Cancer Patients,* 6 HUMAN FERTILITY 100 (2003).

119. Hecht v. L.A. Superior Court, 59 Cal. Rptr. 2d 222 (Cal. Dist. Ct. App. 1996).

120. ORANGE COUNTY REGISTER, Oct. 16, 1998, at 1.

In 2004, a British woman gave birth to a baby boy using sperm from her husband that was frozen twenty-one years earlier. In 2004, the American

the intent is not clear, an opposite result may be reached. In 1995, while Stephen Blood was comatose and on a life-support machine, his wife asked doctors to take samples of his sperm. Later, a British court denied her use of the sperm "because she does not have his written permission." The ruling has been criticized as an "'unduly narrow approach' by requiring written consent."[121] However, the lack of a written living will by a person who has become incompetent does not prevent the introduction of evidence to permit end-of-life decisions where "clear evidence" of his intent is forthcoming from other sources. Accordingly, such evidence should also be available to authorize the wife of a man in a persistent or permanent vegetative state [PPVS] to undergo artificial insemination, or *in vitro* fertilization using her husband's sperm, should she desire to become pregnant.[122] Neither common law nor the Uniform Anatomical Gift Act gives them this right at present.[123] However, in 2004 the 9th U.S. Circuit Court of Appeals held that twins conceived

---

> Urological Association indicated that the use of frozen sperm eliminates the pressure of obtaining sperm on a specific day and unnecessary risk to the woman due to ovarian hyperstimulation.

ORANGE COUNTY REGISTER, May 13, 2004, at News 15.

121. ORANGE COUNTY REGISTER, Oct. 18, 1996, at News 22. In Israel, the attorney general published guidelines based on the assumption that:

> . . . a man who lived in a loving relationship with a woman would want her to have his genetic child after his death even if he never had the opportunity formally to express such a desire. Legal marriage is not perceived as a necessary condition for such a presumption. . . . Israeli culture tends to encourage genetic parenthood at almost all costs.

Hast. Ctr. Rep., Mar.-Apr. 2004, at 6.

122. *See* White, J.L. MED. & ETHICS 360 (1999).

> The AMA CODE OF MEDICAL ETHICS so provides; it also prohibits the sperm's donation to another person: If semen is frozen and the donor dies before it is used, the frozen semen should not be used or donated for purposes other than those originally intended by the donor. If the donor left no instructions, it is reasonable to allow the remaining partner to use the semen for artificial insemination but *not to donate it to someone else*.

AMA CODE OF MEDICAL ETHICS § 2.04. (Emphasis added.)

123. White, *supra* note 121, at 361. The Uniform Anatomical Gift Act (UAGA) only applies when the organs or tissues are used for "transplantation or therapy." UNIF. ANATOMICAL GIFT ACT, § 4, 8A U.L.A. (Supp. 1991).

from frozen sperm after their father died of cancer are eligible to collect Social Security benefits.[124]

good v. Good

Because the field of sperm banks has largely been developing at the turn of the third millennium, proposals for legislative controls are premature.

v.

Sperm banks have now developed to the point that regulation is needed for both testing and usage.

Twenty-first Century Approach

Where sperm is collected either before or after a man's death or when he is entering into a persistent or permanent vegetative state (PPVS) its use should be governed by his clearly expressed intent—whether or not this intent was solely evidenced by a writing.

---

e) Egg Banks

In 1996, the first egg bank was opened in Melbourne, Australia. Like dying men who may desire to preserve their sperm, women and girls undergoing cancer therapy that may compromise their fertility can have their eggs frozen. Unlike the United States, in 1996 the state of Victoria, Australia, made marriage a requirement of its assisted conception law before the eggs can be implanted, and the women must recover from their cancers first.[125] Doctors in China performed the world's first whole ovary transplant at the Zhejiang University Medical School in eastern China.[126] The first case in the

124. The court stated: "Developing reproductive technology has outpaced federal and state laws, which currently do not address directly the legal issues created by posthumous conception. The Social Security law should be construed liberally to ensure that children are provided for financially after the death of a parent." ORANGE COUNTY REGISTER, June 12, 2004, at News 25.

125. THE AGE (Austl.), Feb. 1, 1996, at A4.

126. A woman's ovaries and fallopian tubes were removed after she developed ovarian cancer that resulted in complications associated with early menopause. . . . Theoretically, Tang can conceive naturally in the future.

United States concerned a woman who gave birth after being implanted with eggs that had been frozen.[127] As with sperm banks, egg donor programs operate basically without federal regulation. The freezing of eggs or sperm does not raise the same ethical problems for those who consider frozen embryos to be babies just waiting to be born.

Obtaining eggs for donation differs from sperm donation; several weeks of hormone injections are needed to stimulate egg production before the donor undergoes outpatient surgery to retrieve the eggs. The donors usually received some $2,500 in the United States, whereas in Britain, such payments were capped at about $24, the same as for sperm donors there.[128]

A Harvard professor noted that "many of the women my student interviewed, even those who did not sell their eggs, saw nothing wrong with the process as such. A healthy human female produces between four hundred and five hundred eggs in a lifetime. She can bear only a tiny fraction of that number of children, so she is not losing any capacity to bear her own chil-

---

> However, because the eggs will be produced by the transplanted ovary, any child will genetically be her sister's, not hers.

http://www.reutershealth.com/archive/2002/04/11.

127. ORANGE COUNTY REGISTER, Oct. 17, 1997, at 1.

> Doctors thawed 23 eggs, 16 of which survived; 11 were injected with a single sperm from the woman's husband. Then four of the 11 that developed into embryos were implanted into the woman."

Id.

> It has been reported that "freezing is risky—the process can damage eggs, which have a higher moisture content than sperm or embryos and can form ice crystals. The procedure has a 3 percent pregnancy rate per thawed egg and has resulted in fewer than 100 births worldwide.

ORANGE COUNTY REGISTER, July 18, 2005, at News 6 (citing the *New England Journal of Medicine*).

128. L.A. TIMES, July 25, 1995.

> Women who donate their healthy eggs to infertile women are ethically entitled to be paid for their services, but should only rarely be paid more than $5,000 and never as much as $10,000, according to the American Society for Reproductive Medicine (ASRM), the nation's largest organization of fertility professionals.

FERTILITY & STERILITY (2000).

dren."[129] Our law has made the sale of certain organs illegal (see Chapter 11), but not the sale of sperm or eggs. People in Catholic and Muslim countries are also forbidden to sell the latter. The costs plus fees for egg retrieval has run as high as $20,000 and are generally not covered by insurance.[130] According to guidelines issued by the American Society for Reproductive Medicine, egg donors are to be paid for the "inconvenience, time, discomfort, and for the risk undertaken." Fees are justified for these painful extractions that are preceded by the injection of powerful hormones to boost egg production. The egg itself is "wasted" each month if not used. Some egg agencies will obtain them from donors fitting detailed requirements (IQ or SAT score, height, etc.) for a premium price,[131] although there is no guarantee that the egg purchased will result in having a child with such characteristics.

good v. Good

Women should not sell their "wasted eggs" and jeopardize their fertility.

v.

The number of such eggs is so many times higher than a woman's possible number of pregnancies that this alone is no ground for objection.

Twenty-first Century Approaches

In view of the large number of oocytes produced by virtually all women, their donation or sale should be authorized, provided they have been genetically tested prior to fertilization.

---

129. HARVEY COX, WHEN JESUS CAME TO HARVARD 60 (2004).

130. Mead, NEW YORKER, Aug. 9, 1999, at 54. "Schiller donates her eggs because she thinks that it's a worthy thing to do, and because it's a worthy thing to do for which she can be paid in sums that seem handsome to a heavily indebted student." *Id.* at 57.

131. A half-page ad in the *Stanford Daily* offered $50,000 for an "intelligent, athletic egg donor" who must be a least 5 feet 10 inches tall and "have a 1400-plus SAT score." ORANGE COUNTY REGISTER, Feb. 28, 1999, at News 7; *see also* MT. HERALD TRIB., Oct. 18, 1999, at 3.

f) *Frozen Eggs and Embryos and Agreements to Donate Them, Posthumous Reproduction*

Illinois was the first to find that state law did not prohibit the In Vitro Fertilization (IVF) procedure in which a couple sought to participate.[132] This benefits women who produce healthy eggs but cannot normally become pregnant. The physician Carl Wood and his Australian research team demonstrated that human embryos or eggs could be deep-frozen, thawed, fertilized (IVF), and then placed within the womb.

Human egg and sperm cells and even human embryos are routinely frozen during fertility treatments. The key to their survival seems to be freezing under carefully controlled conditions that minimize the damage to cells by promoting the formation of only very tiny ice crystals.[133] This cryopreservation spared women having to undergo multiple egg extractions.[134] A younger couple might choose to freeze embryos produced early in their marriage, delay having children for a number of years, and avoid risking the infertility problems that increase with age. Several other ethical and legal questions arise.

---

132. Smith v. Hartigan, 556 F. Supp. 157 (N.D. Ill. 1983). Three states currently regulate IVF: 720 Ill. Comp. Stat. Ann. § 51016 (West 2000); La. Rev Stat. Ann. §§ 9:121–9:133 (2000); 18 Pa. Cons. Stat. Ann. § 3213(e) (West 2000). Louisiana granted embryos the status of "judicial persons" with the right to sue and be sued.

133. Johnjoe McFadden, Quantum Evolution 24 (2001).

> In 2002, a professor of microbiology and immunology at the Indiana University School of Medicine disclosed how "human umbilical-cord blood [could be] frozen for 15 years. It was revived and able to grow and expand in laboratory mice." He then stated, "We could take the cells after defrosting and have them expand extremely well—as well as if we had used fresh cord blood."

Orange County Register, Dec. 31, 2002, at News 12.

134. Sci. Am., Apr. 1996, at 16.

> Cryopreservation is the process of freezing embryos in liquid nitrogen at the two, four, or eight cell stage of development for the purpose of storing them and preserving their viability for use at a future date either by the donor or by some third party.

To Be Genetically Tied or Not to Be: A Dilemma Posed by the Use of Frozen Embryos, 12 Women's Rts. L. Rev. 115 (1990).

An embryo is neither a sperm nor an egg. Rather, when these are united, they form a zygote, and after several subdivisions it becomes an embryo.[135] Technically this makes little difference inasmuch as all pre-embryos can be frozen.[136] A Swedish study in 1998 found that children born from frozen embryos are just as healthy as those conceived normally or through standards of *in vitro* fertilization.[137] When several embryos are implanted into the womb, they become potential human beings.[138] Many states allow sperm donors to give up their parental rights and be free of child support obligations; other states have no similar law specifically for embryo donors.[139]

A baby was born a "twin" of an embryo frozen in 1989 after the couple underwent infertility treatment and stored embryos in a hospital freezer until the second one was defrosted in 1997. He was called the "world's oldest" newborn. In 1998, Britain allowed embryos to be created from frozen eggs.[140] A spokesperson for the American Society for Reproductive Medicine stated, "The frozen embryos are probably good for 200 years. It doesn't seem to be big news that an embryo could be kept for several years and then used."[141] At the end of the twentieth century there were estimates

---

135. AMA Code of Medical Ethics § 2.14 (2003) actually also states that "any fertilized egg that has the potential for human life and that will be implanted in the uterus of a woman should not be subjected to laboratory research."

136. AMA Code of Medical Ethics § 2.141 (2003) notes:

> Cultural and legal traditions indicate that the logical persons to exercise control over a frozen pre-embryo are the woman and man who provided the gametes (the ovum and sperm). . . . The pre-embryos should not be available for use by either provider or changed from their frozen state without the consent of both providers.

137. The cryopreservation process does not adversely affect the growth and health of children during infancy and early childhood. Up to 2% of all births in Sweden are from IVF. The study was published in *The Lancet*. Orange County Register, Apr. 10, 1998.

138. In 2001, U.S. clinics made and transferred some 65,000 fresh embryos from women's own eggs. Of these, fewer than 22,000 resulted in a live birth, a failure rate of 69%. *See* Ctrs. for Disease Control & Prevention, 2001 Assisted Reproductive Technology Success Rates (2003); Jon Cohen, Coming to Term 261 (2005).

139. Orange County Register, May 2, 1998, at Metro 7.

140. Ian Wilmut et al., Second Creation 272 (2000).

141. Orange County Register, Feb. 18, 1998, at News 4.

that 100,000 frozen embryos were stored across the country, up to 20,000 of which may be objects of disputes. This is partly due to the absence of U.S. laws regulating the future of stored embryos. In March 1996 a Society's ethics committee issued guidelines concerning the disposition of "abandoned embryos" after five years:

> The Ethics Committee finds it is ethically acceptable for a program to consider *embryos* to have been *abandoned if more than five years* have passed since contact with a couple, diligent efforts have been made by telephone and registered mail to contact the couple at their last known address and no written instruction from the couple exists concerning disposition. The guidelines *allow for disposal of such abandoned* embryos *but forbid clinics to make them available to other couples or for use in research* unless the donors have given their consent.[142] (Emphasis added.)

In 1998, the New York State Court of Appeals held that the preconception contract between gamete donors regarding the disposition of embryos will be presumed valid. It may be enforced in the event of later disagreement between the parties following divorce. The court concluded that there was a mutual intention to donate any disputed embryos to the IVF program for research purposes.[143] The court also held that agreements between progenitors or gamete donors regarding disposition of their pre-zygotes should generally be presumed valid and binding, and enforced in any dispute between them. Disposition of these pre-zygotes does not implicate a woman's right of privacy or bodily integrity in the area of reproductive choice.[144] In a Tennessee case, ova were removed from a woman and fertilized with her husband's sperm in a petri dish. Some were transferred back into her uterus for implan-

---

142. ORANGE COUNTY REGISTER, Jan. 3, 1997, at News 12. "Germany and Spain prohibit the freezing of embryos. Under a 1990 British law, frozen embryos must be destroyed after five years unless consent is given allowing future storage." ORANGE COUNTY REGISTER, July 23, 1996.

143. Kass v. Kass, 696 N.E.2d 174 (N.Y. 1998).

144. *Id.* at 179; *Asexual Reproduction*, 8 ALB. L.J. SCI. & TECH. 1 (1990). "[L]iberty secured by the Constitution . . . does not import an absolute right in each person to be . . . wholly freed from restraint." (quoting Jacobson v. Massachusetts, 26 (1905)).

tation and the rest were frozen for subsequent implantation; but the initial efforts to conceive were unsuccessful. When the husband filed for divorce, she wanted to donate the embryos to a childless couple, and he wanted the frozen embryos to be destroyed. The Tennessee Supreme Court upheld the husband's desire, stating that "[o]rdinarily, the party wishing to avoid procreation should prevail, assuming that the other party has a reasonable possibility of achieving parenthood by [other] means."[145] Some states have enacted legislation requiring couples to execute written agreements that provide for embryo disposition in the event of divorce or death.[146] This legislation permits couples to express these desires free from the very different emotions that follow divorce proceedings. One state ruled to the contrary: In 2000 the Supreme Court of Massachusetts denied the former wife of a divorced couple the use of frozen embryos to have additional children even though the IVF consent form for freezing specified that in the event of separation that they "both agree[d]" to have the embryo(s) . . . return[ed] to [the] wife for implant." The former husband sought a permanent injunction to prevent his wife from using the remaining embryos. That court used judicial arguments in ruling that it "will not enforce contracts that violate *public policy*,"[147] (emphasis added) in order to indicate the creation of its own "policy." This decision was consistent with the AMA *Code of Medical Eth-*

---

145. Davis v. Davis, 842 S.W.2d 588, 604 (1992).

146. *E.g.*, FLA. STAT. ANN. § 742.17 (West 2000). In November 1997, the General Conference of the United Nations Educational, Scientific and Cultural Organization (UNESCO) adopted the Universal Declaration on the Human Genome and Human Rights, UNESCO Gen. Conf. Res. 29C/Res. 16 (Nov. 11, 1997). Its purpose is to protect human rights while advancing scientific research. But the United Nations has not yet developed binding standards.

147. UNIF. PREMARITAL AGREEMENT ACT § 3(a)(8), 9B U.L.A. 37 (1983). In England, the Human Fertilization and Embryology Act states that unless both parties consent to storage and use, the frozen embryos must be destroyed. In 2003, a High Court judge in London rejected a challenge by divorced women to the contrary, stating: "The act entitles a man in the position of these men to say that he does not want to become the father of a child by a woman from whom he has separated and with whom he now no longer has anything in common apart from the frozen embryos." Some 45,000 couples seek *in vitro* fertilization in Britain each year, and "27,000 couples receive the treatment, but mostly in the private sector." N.Y. TIMES, Oct. 2, 2003, at A5.

ics (2003), which recommends use of non-enforceable advance agreements for deciding the disposition of frozen pre-embryos in the event of divorce.[148] A French court ordered that a dead couple's bodies, which were frozen with the hope of future revival, actually be buried in accordance with its Civil Code.[149]

Dr. Vishvamath Karan, director of the *in vitro* fertilization program, Center for Human Reproduction, stated, "There are probably a lot of couples that are unable to decide what they want to do with their embryos, and just continue to pay the storage fees." He added that in July 2001 the Center had 5303 cryopreserved embryos, of which 359 were being donated to other couples, 529 had been designated for research and about 30 were destroyed each month at the couples' requests.[150] Some courts have allowed embryos to be destroyed or donated to research. For example, the New Jersey Supreme Court granted a divorced woman permission to dispose of seven frozen em-

---

148. The Code's Rule 2.141 states: "Advance agreements can help ensure that the gamete providers undergo IVF and pre-embryo freezing after a full contemplation of the consequences but should not be mandatory." AMA CODE OF MEDICAL ETHICS R. 2.141.

Fiestal, *A Solomonic Decision: What Will Be the Fate of Frozen Preembryos*?, 6 CARDOZO WOMEN'S L.J. 103, 110 (1999), concluded that, in legal disputes, the party opposing implantation ordinarily should prevail, and Haut, *Divorce and the Disposition of Frozen Embryos*, 28 HOFSTRA L. REV. 493, 519 (1999), concluded that, while a prior agreement should be binding, various contract defenses may apply where circumstances have substantially changed.

149. A biologist who once taught at France's most prestigious medical school, Dr. Raymond Martinot was fascinated by cryonics. He kept his dead wife's body in the refrigerated crypt of the family's chateau for 18 years with the hope of [her] being revived one day. However, a court ordered their son to bury them because the French civil code only envisages the burial or cremation of deceased persons *within a time limit of six days after death.*

ORANGE COUNTY REGISTER, Mar. 19, 2002, at News 19. (Emphasis added.)

150. Constable, http://www.dailyherald.com, July 29, 2001. In California, the transplanting of eggs without the donor's consent constitutes a felony. CAL. PENAL CODE § 367g (West). On the civil side, in 2004 a woman was awarded $1 million in damages to settle a malpractice lawsuit against a fertility specialist who accidentally implanted her with the wrong embryos. L.A. TIMES, Aug. 4, 2004, at B7.

bryos left over from the failed marriage.[151] When a child is born after one or more of its parents die, this may result from a "posthumous reproduction." For a woman, this occurs by extracting an egg from her *while she is alive,* fertilizing it with frozen sperm, and following her death, implanting it in a surrogate (see subsection i). For a man, his sperm is retrieved while he is alive; it is frozen and used for fertilization after his death.[152] Posthumous reproduction has often been favored on the ground that such "reproduction connects individuals with future generations and provides personal experiences of great moments in large part because persons reproducing see and have contact with offspring, or are at least aware that they exist."[153] However, it has been argued on the other side that two parents are needed and children are generally disadvantaged when reared without mothers.[154]

With regard to embryo research, pro-lifers would not liberate the frozen embryos beyond 14 days of development. In this connection it has been noted that "the NIH Embryo Research Panel in 1994, the National Bioethics Advisory Commission in 1999, and the pro-embryo research members of the President's Council on Bioethics have stated explicitly that we should not do research on embryos beyond 14 days of development."[155] President Bush also limited federal funding of embryonic stem cells to only the existing stem cell lines. However, no such limits apply to the private funding of research on embryos.

In 2004, the Centers for Disease Control (CDC) estimated that 400,000 frozen embryos are currently stored in fertility clinics; each has fewer than 100 cells, and about 60 percent of the clinics allowed them to be donated

---

151. The couple, who were married in 1992, began *in vitro* fertilization that produced 11 embryos. Four were implanted without result. They separated in 1996, and the woman filed a petition for permission to destroy the remaining embryos despite her former husband's objection. The ruling allowed the embryos to be either destroyed or used in research. http://www.reutershealth.com/archive/2001; *see also* J.B. v. M.B., 751 A.2d 613 (N.J. Super. Ct. App. Div. 2000).

152. *See* Hart v. Shalala, 94-3944 (E.D. La. 1993); Hecht v. Super. Court, 20 Cal. Rptr. 2d 275 (Cal. Ct. App. 1993).

153. Kristin Antall, *Who Is My Mother?: Why States Should Ban Posthumous Reproduction by Women,* 9 HEALTH MATRIX 203, 227 (1999).

154. *Id.* at 230; Ruth Landau, *Planned Orphanhood,* 49 SOC. SCI. MED. 185 (1999).

155. Cohen, NEW ATLANTIS, Summer 2003, at 15.

for research. Bioethicist Caplan, of the University of Pennsylvania, stated, "I don't think anyone who deals with these frozen embryos considers them to be persons. . . They see the potential for life in this material."[156] Only 34 percent of all embryo transfers will result in a live birth. At the end of the third millennium, Nightlight Christian Adoption of California started the only known American program that offers adoption services for embryos. It is a fertility treatment similar to sperm or egg donations:

> A couple that underwent in-vitro fertilization and wants to donate the embryos fills out an application detailing what would characterize ideal adoptive parents for their embryos. They may specify criteria including amount of contact between genetic parents and resulting children. They provide family information, including a three-generation health history report.[157]

Of course, successful frozen embryo donations do not result in legal "adoptions" of children and the parties to their contract do not transfer "parental rights" for embryos; they transfer rights to its "ownership."[158]

---

156. Orange County Register, Sept. 18, 2004, at News 25.

157. Orange County Register, Sept. 24, 2004, at News 5.

> Adoptive parents pay Nightlight up to $5,600 per match. If the embryos don't result in a pregnancy, they pay $1,000 for an additional match. Adoptive parents will receive at least six embryos per adoption, even if this requires more than one family match. They agree to thaw and transfer only the number of embryos that they would be willing to raise if each embryo resulted in a live birth. The donor couple pays nothing.

Id.

158. *Id.* Although only 62 babies have been born since this program began, the Christian organization believes it has had some effect against those who favor embryo stem cell research. *Id.*

good v. Good

The use of a frozen embryo years after conception violates God's laws demanding that they be allowed to proceed on to delivery without delay by humanity or used for research purposes.

v.

No scripture governs the use of frozen embryos and their use should be permitted subject to reasonable controls.

Twenty-first Century Approach

1. State legislatures should consider enactment of statutes:
   (a) authorizing the transfer of frozen embryos scheduled for destruction to institutions for research purposes for longer periods;
   (b) providing that embryo donors who give up parental rights be free of child support obligations in the same manner as is applicable to sperm and egg donors; and
   (c) authorizing competent couples to execute binding written agreements providing for the ethical voluntary disposition of frozen IVF embryos (including the events of the death of one or both parties).

When men are most sure and arrogant, they are commonly most mistaken.

David Hume

g) Multiple Births Resulting from Fertility Pills

Fertility pills currently tend to give rise to multiple births—some of whom may become unwanted children. Induction with intrauterine insemination is frequently used to treat infertility because among infertile couples such induction "is three times as likely to result in pregnancy as is intracervical [the canal of the cervix] insemination."[159] The Centers for Disease Control and Prevention has found that twin births alone climbed 42 percent since 1980 in

159. 340 N. Eng. J. Med. 177–83 (1999). "It is also twice as likely to result in pregnancy as is treatment with either superovulation and intracervical insemination or intrauterine insemination alone."

the United States due to such drugs.[160] With fertility drugs the chance of multiple pregnancies is many times higher than normal, and their use may produce other deleterious effects.[161] For example, due to the use of fertility drugs in IVF, 57 sets of quintuplets were born in the U.S. in 1995. Studies show that "mothers of multiple births experience a higher level of depression and anxiety than those who have a single baby, and divorce is more common."[162] On November 19, 1997, the McCaughey septuplets, a world record, were delivered in Carlisle, Iowa, to a woman taking fertility pills rich in follicle-stimulating hormone (FSH). All the infants were in critical condition—the smallest weighing 11 ounces and the largest one pound, 11 ounces. These births required an extensive team of 40 nurses, respiratory therapists, perinatologist, neonatologists (from the 20th week of gestation up to four weeks after birth), and anesthesiologists. All the septuplets were placed on ventilator support. These results did not have to occur; McCaughey's physicians could have removed some of her eggs and implanted a maximum (say, three) number of embryos. Critics of the current method viewed it as a "medical disaster."[163] In 1997 the Center for Disease Control (CDC) reported that of all the live births aided by fertility assistance, 37 percent were multiple births, compared with 2 percent in the general population.[164] Another woman in Michigan who used fertility drugs to become pregnant gave birth to sextuplets; the four boys and two girls were three months premature.[165] In the

160. L.A. Times, Aug. 20, 1998, at M2; Orange County Register, Oct. 8, 1998, at News 6 (reporting on research developed by the Stanford University Medical Center). "The ratio of triplet and higher-order multiple births has more than quadrupled and . . . a large proportion of this increase can be attributed to the use of ovulation-inducing drugs." JAMA 299 (2000).

161. According to the Centers for Disease Control and Prevention, twin births rose from 68,339 babies in 1980 to 97,064 in 1994. L.A. Times, Feb. 14, 1997, at A22. In 1998, multiple births (56%) were higher than the overall national average of 3%. The triplet and higher-order birthrate was 100 times higher than the national average of 2.16 in 1998. JAMA 1566 (2002).

162. http://www.independent!.co.uk.news.

163. Milhill & Boseley, *Multiple Births: When the Shine Wears Off a Miracle*, Guardian (Eng.), Nov. 21, 1997, at 17.

164. L.A. Times, Dec. 12, 1997, at A23.

165. Orange County Register, Jan. 18, 2004, at News 30.

United States, triplets or more accounted for 181 per 100,000 births.[166] It has been recommended that in future instances, some fetuses be aborted and that fewer be implanted. Bioethicists at the University of Pennsylvania's Center for Bioethics proposed that the government should either restrict hyperovulation drugs and allow couples to sue clinics for reproductive malpractice, or create insurance penalties or fines for couples who choose to have multiple births.[167]

As of the turn of the third millennium, only 15 states required insurance companies to provide coverage for assisted reproduction, which does not qualify for Medicaid.[168] However, the Supreme Court has held that since reproduction is a "major life activity," a substantial limitation on a person's ability to reproduce would actually fall within the definition of disability under the Americans with Disabilities Act (ADA).[169]

The average hospital bill for one baby—$5,700—soars to $90,000 for triplets. In 2005, it was shown that "New Jersey has the highest ratio of triplet births in the United States. Between the years 1998 and 2002 the state recorded 358 sets of triplets (and higher-order births) per 100,000 live births.

---

166. JOYCE A. MARTIN ET AL., BIRTHS: FINAL DATA FOR 2000, at 19 (Nat'l Vital Statistics Rep., Vol. 50, No. 5, 2002). As a result of reproductive technology, 62.9% were single births, 32.2% were twins, 4.7% were triplets, and .2% were higher order births. Soc. for Assisted Reproductive Tech. & Am. Soc'y for Reproductive Med., *Assisted Reproductive Technology in the United States: 1999 Results Generated from the American Society for Reproductive Medicine/Society for Assisted Reproductive Technology Registry*, 78 FERTILITY & STERILITY 918–31 (2002).

167. http://www.policy.com/news/dbrief/dbriefarc748.asp. In 2004, it was reported that the average number of embryos implanted per attempt dropped only from four to three between 1995 and 2001. However, the rate of twins held steady among *in vitro* pregnancies. N. ENG. J. MED. ; ORANGE COUNTY REGISTER, Apr. 15, 2004, at News 10.

168. Strong, J.L., MED. & ETHICS 278 (2003); *see also* Hughes & Giacomini, *Funding In Vitro Fertilization Treatment for Persistent Subfertility: The Pain and the Politics*, 76 FERTILITY & STERILITY 431–42 (2001).

169. Bragdon v. Abbot, 524 U.S. 624 (1988). *See also* Saks v. Franklin Covey Co., 316 F. 3rd 337 (2d Cir. 2003), in which the Court of Appeals for the Second Circuit held that the ADA (together with Title VII of Civil Rights Act of 1964 and New York state law, the Pregnancy Discrimination Act (PDA)) does not prohibit an employer's health plan from excluding surgical impregnation procedures from its coverage. The plan did not discriminate under the ADA because both males and females were equally disadvantaged by the exclusions. *Id.* at 348.

. . . Twins are not considered a significant health risk, but triplets are—frequently resulting in low-birth-weight babies and maternal health problems."[170] Although the rate of triplets has leveled off, the rate of twins has either increased or remained steady.[171]

Oregon's basic package of Medicaid services no longer covers infertility treatments, and some health maintenance organizations would eliminate similar coverage.[172] In a study of 13,206 expectant women entering Brigham and Women's Hospital in Boston from 1986 through 1991, the average total charges for a single baby were $9,845 (of which 15 percent required intensive care and about 2 percent died before discharge).[173] This is compared to $37,947 for twins and $109,765 for triplets (with a 78 percent intensive care rate for higher-order multiples, 13 percent of whom died before discharge).[174] Although Great Britain's regulations forbid the transfer of more than three embryos, there is no law limit to the number of embryo implantations in the United States. Some of our clinics make a limit of two embryos "to avoid the serious risks caused by triplets." Such limits are vital, and at the turn of the third millennium, the American Fertility Society issued a recommendation of transferring two or, at most, three embryos.[175] Evidence demonstrated that

170. NAT'L GEOGRAPHIC, Oct. 2005, at 120.

171. PRESIDENT'S COUNCIL ON BIOETHICS, REPRODUCTION AND RESPONSIBILITY: The REGULATION OF NEW BIOTECHNOLOGIES 23–87 (2004), *available at* http://www.bioethics.gov/reports/reproductionandresponsibility/chapter2.html.

172. L.A. TIMES, June 23, 1996, at M4 (stating that "[t]he technique and the increased use of more potent fertility drugs are the main reasons for the 34% increase in the number of triplets and higher order multiple births, according to the National Center for Health Statistics").

173. L.A. TIMES, May 14, 1996 (stating that "[t]he data also showed that if the 10 sets of triplets and 77 sets of twins produced by infertility treatments in 1991 had all been born as single babies, the savings would have topped $3 million in this one hospital alone").

174. *See* Gardner, *Improving Implantation Rates in In Vitro Fertilization*, 12 INFERTILITY & REPRODUCTIVE MED. CLINICS N. AM. 403–24 (2001); Karaki et al., *Blastocyst Culture and Transfer: A Step Toward Improved In Vitro Fertilization Outcome*, 77 FERTILITY & STERILITY 114–18 (2002).

175. Martine Nijs et al., *Preservation of Multiple Pregnancies in an In Vitro Fertilization Program*, 59 FERTILITY & STERILITY 1245–50 (1993).

> Older women undergoing IVF have far more abnormal eggs than younger women. The new technique permits viewing eggs and choosing the normal ones to be fertilized and placed in the uterus of a woman undergoing

"in terms of cumulative overall pregnancy rates, intrauterine insemination without ovarian stimulation is safer and cheaper than intrauterine insemination with stimulation and is just as effective," and single-embryo transfers have been effective.[176] A 2002 study reported that the increased risk of low birth weight associated with the use of assisted reproductive technology can be attributed largely to the higher rate of multiple gestations associated with such technology. It was found to account for a disproportionate number of very-low-birth-weight infants (see Chapter 10). This was in part because of increases in multiple gestations, but also in part due to higher rates of low birth weight among singleton infants conceived with this technology.[177] Nevertheless, in 2005 a surrogate mother in Arizona who had agreed to carry the couple's child for $15,000 actually delivered quintuplets, which she found to be very, very difficult to carry. To do that for someone else is extraordinary.[178]

> in vitro fertilization. It is experimental, but it could help reduce the high rate of multiple births after IVF procedures.

ORANGE COUNTY REGISTER, Nov. 5, 1996, at News 4.

176. *See* N. ENG. J. MED. 58–60 (2000). In 2004, a study in the United States found that since 1997 there have been some "decreases in both the number of embryos transferred per cycle and the percentage of pregnancies with three or more fetuses, as well as a consistent increase in the percentage of live births per cycle." Jain et al., N. ENG. J. MED. 1639 (2004). However, it was also pointed out that "fertility centers in Finland, Sweden, and Belgium have shown that the use of elective single-embryo transfers in patients reduces the rate of twin pregnancy without affecting overall pregnancy rates." *Id.* at 1644; *see also* Barlow, *The Debate on Single Embryo Transfer in IVF: How Will Today's Arguments Be Viewed from the Perspective of 2020?*, 20 HUMAN REPRODUCTION 1–3 (2005).

177. N. ENG. J. MED. 731 (2002). Premature births had accounted for 14% of twins and 41% of triplets. Alexander et al., *What Are the Fetal Growth Patterns of Singletons, Twins, and Triplets in the United States?*, 41 CLINICAL OBSTETRICS & GYNECOLOGY 115–25 (1998); *see infra* Chapter 10.

178. ORANGE COUNTY REGISTER, Apr. 27, 2005, at News 16. Five embryos were implanted to increase the chances that at least one of them would take hold. The babies were born by Caesarean section in the 33rd week of gestation "because the mother was experiencing elevated blood pressure and other complications."

good v. Good

In view of the fact that use of fertility pills (which give rise to multiple births) currently has a small effect on the increase in population, no efforts need be taken to curtail their use.

v.

Most prospective parents neither contemplate nor desire multiple births; when they occur, they have a higher level of depression and divorce rates increase.

Twenty-first Century Approach

With either intrauterine insemination without ovarian stimulation or with *in vitro* fertilization (and transfer), no more than two embryos at one time should be authorized.

---

h) Artificial Insemination of Single Mothers

Although most births by artificial insemination are to married couples, some 10 percent of all such births (amounting to several thousands per year) have been to unmarried women.[179] Half of all physicians who perform artificial inseminations do not screen prospective patients according to marital status.[180] This approach has been criticized by the bioethicist Daniel Callahan, who noted:

> Acceptance of the systematic downgrading of fatherhood brought about by the introduction of anonymous sperm donors, or perhaps it was the case that fatherhood had already sunk to such a low state, and male irresponsibility was already so accepted, that no one saw a problem. It is as if everyone argued: Look, males have always been fa-

179. U.S. CONG., OFFICE OF TECH. ASSESSMENT, ARTIFICIAL INSEMINATION: PRACTICE IN THE UNITED STATES 3–23 (1988).

180. *Id.* In 1991, doctors at the University of Michigan began offering donor insemination to unmarried women. RICHARD LOUV, FATHER LOVE 39 (1993). AMA CODE OF MEDICAL ETHICS § 2.04 (1993) states: "If the donor and recipient are not married, an appropriate legal rule would treat the situation as if the donor were anonymous: the recipient would be considered the sole parent of the child except in cases where both donor and recipient agree to recognize a paternity right."

thering children anonymously and irresponsibly; why not put this otherwise noxious trait to good use.[181]

Pregnancies of women who have reached the age of 42 result in miscarriages 50 percent of the time.[182] Egg donations from younger women are often made to older women. In 2000, 4,565 births resulted from transferred embryos to women ages 45 to 54.[183] Often the eggs are sold at prices up to $10,000—which has been declared inappropriate.[184] Studies show that many donors "are not providing their eggs for the money but largely for altruistic reasons."[185] Although some bioethicists have emphasized that a consumer's right "to make contracts with providers of gametes [should] not be prohibited or limited,"[186] others have pointed to adverse social consequences of accelerating the number of single parents in this manner.[187]

A legal victory for those who argue in favor of the need for a father in a family was obtained when a court ruled that the Minnesota Human Rights Act did not require physicians to provide artificial insemination to a lesbian couple. The court noted that the Minnesota parentage law on artificial insemination applied only to married couples, not single women, and the Minnesota law generally favored the institution of traditional marriage.[188]

Many have written that the decline of fatherhood is one of the most unprecedented and devastating developments of modern times. David Papenoe, professor at Rutgers, has demonstrated that the absence of a strong paternal presence at home is "a primary force behind the most disturbing problems that plague American society: crime and delinquency; premature sexuality

---

181. Callahan, *Bioethics and Fatherhood*, UTAH L. REV. 735, 741 (1992).

182. WINSTON, SUPER HUMAN (2000) at 214.

183. ORANGE COUNTY REGISTER, Feb. 10, 2002, at News 214.

184. *Id.* This statement was made by Dr. William Keye, president of the American Society for Reproductive Medicine.

185. *Id.*

186. *Alternative Reproductive Technologies: Implications for Children and Families: Hearing Before the House Select Comm. on Children, Youth, and Families*, 100th Cong. (May 21, 1987) (statement of Robertson).

187. Wasserman & Wachbroit, *Defining Families: The Impact of Reproductive Technology*, REP. FROM INST. FOR PHIL. & PUB. POL'Y, Summer 1993, at 4–6.

188. Heeney v. Erhard (Minn. 1995); *see also* Anna J. v. Mark C., 12 Cal. App. 4th 977 (Cal. Dist. Ct. App. 1991).

and out-of-wedlock births; poor educational achievement; addiction and alienation among adolescents."[189] In his book, *Life without Father*, Papenoe shows that "nearly 50 percent of American children may be going to sleep each evening without being able to say good night to their dads." Other scholars, including Jean Elshtain, Sara McLanahan, and Maggie Gallagher, all argue the same. Jean Elshtain, in *The Lost Children*, demonstrates that "teen parenting usually means parenting without a father." A study published in 2003 found that children growing up in single-parent families are twice as likely as their counterparts to develop serious psychiatric illnesses and addictions later in life.[190] Nevertheless, frozen sperm and embryos could result in inheritance by children born following the death of the father. The Massachusetts Supreme Judicial Court upheld the right of children, who resulted from the wife being artificially impregnated with the husband's sperm after he died, to inherit under Massachusetts' law of intestate succession.[191] The Uniform Parentage Act (UPA) specifically provides that if a spouse dies before placement of eggs, sperm, or embryo:

> The deceased spouse is not a parent of the resulting child unless the deceased spouse consented in a record that if assisted reproduction

---

189. Lost Fathers 4–7 (Cynthia R. Daniels ed., 1998).

190. It was reported in a study published in the medical journal *The Lancet* in January 2003.

> The scientists found that children with single parents were twice as likely as others to develop a psychiatric illness such as severe depression or schizophrenia, to kill themselves or attempt suicide, and to develop an alcohol-related disease. Girls were three times more likely to become drug addicts if they lived with a sole parent, and boys were four times more likely.

Orange County Register, Jan. 24, 2003, at News 18.

191. Woodward v. Comm'r of Soc. Sec., 760 N.E.2d 257 (Mass. 2002); Hast. Ctr. Rep., Nov.-Dec. 2002, at 8. The court stated: "[P]osthumously conceived children may not come into the world the way the majority of children do. But they are children nonetheless." The Massachusetts Supreme Judicial Court had previously refused to enforce a divorcing couple's embryo disposition agreement that would have awarded the embryos to the woman, A.Z. v. B.Z., 725 N.E.2d 1051 (Mass. 2000).

were to occur after death, the deceased spouse would be a parent of the child.[192]

good v. Good

A woman's autonomy clearly grants her the choice for artificial insemination without the burden of marriage or of living with the father.

v.

Such women are often self-centered and refuse to recognize a child's need for a father as well as the quite significant damages likely to follow to the children from his absence.

Twenty-first Century Approach

1. State legislatures will find a child's unique need for the presence of a father in a family and should limit artificial insemination to presently or formerly married couples (where one or both have died).
2. Many states should consider the enactment of the Uniform Parentage Act (UPA) that sets forth the conditions under which such posthumously conceived children should inherit under the intestacy laws.

---

i) Surrogacy

For purposes of equity between fertile and infertile couples, most states authorize agreements (contracts) between hopeful parents where the wife is infertile to use another woman as her surrogate (substitute) birth mother. Some birth mothers receive embryos for gestation. Others receive the husbands' sperm and contribute the egg. Only in the latter instance would the couple wish to examine her race, color, and intelligence and would she be more likely to use them in setting a price for her services.[193]

In 1991, the egg of a woman (Crispina) was combined in a laboratory petri dish with her husband's (Mark's) sperm. The resulting embryo was implanted in a gestational surrogate mother (Anna) who later gave birth to a healthy baby boy. Anna was clearly not genetically related to the child. As the court noted, "Historically there has never been an occasion when the

192. § 707, 9B U.L.A. 358 (2001).

193. "[H]eight, eye color, race, intelligence, and athletic ability" are, of course, monetized in the employment. ALAN HYDE, BODIES OF LAW 77 (1997).

'natural' mother was not unquestionably also the same woman who bore the child." The surrogacy agreement they all entered provided for the payment of $10,000 to Anna for her services. Disagreement arose when Mark heard that Anna had not disclosed that she had had several stillbirths and miscarriages. Anna was upset, feeling that she was abandoned upon the onset of premature labor, and indicated that she would not give up the child. The court began by noting that a "state, rather than the United States Supreme Court, is the final arbiter of the content of its own law" (unless the state's law conflicts with the U.S. Constitution). It concluded, "Mark and Crispina are the natural and legal parents of the child."[194]

Statutes in a half dozen states declared all such agreements void and nine other states prohibited only those agreements requiring the payment of broker's fees. For example, Arizona prohibits all surrogacy contracts "in which a woman agrees to implantation of an embryo not related to that woman" as well as contracts in which she "agrees to conceive a child through natural or artificial insemination and to voluntarily relinquish her parental rights to that child."[195] In 1997, for the first time, a fertilized egg left behind after a woman's death was implanted in a surrogate mother.[196]

In 1994, Julie Garber, a cancer patient, postponed sterilizing radiation therapy for her lymphoblastic leukemia for a month in order to have her still-healthy eggs harvested, fertilized by an anonymous sperm-bank donor, and frozen. Following her death in 1996, her parents sought help from the Center for Surrogate Parenting and Egg Donation in Beverly Hills, California, then the world's largest such center. The Center refused because they knew of no cases in which a dead woman's frozen eggs were carried

---

194. Johnson v. Calvert, 5 Cal. 4th 90 (1993).

195. ARIZ. REV. STAT. § 25-218(d) (LexisNexis); *see* Soos v. Superior Court of the State, 179 Ariz. Adv. Rep. 22 (Ct. App. 1994), *rev. denied*, CV-5-OOI-49-PR (Ariz. July 1, 1995).

196. ORANGE COUNTY REGISTER, Dec. 5, 1997, at Metro 1.

> In 1998, a Nevada woman had sperm removed from her nineteen-year-old son while he was on life support and had it frozen before he died; she intended to find an egg donor and a surrogate mother to become artificially inseminated and carry the fetus to term so that she could become a grandmother.

ORANGE COUNTY REGISTER, Oct. 2, 1998, at News 32.

to term by a surrogate. When her parents looked elsewhere, which included making appearances on TV talk shows, about 80 women contacted them volunteering to be surrogates.[197] The parents chose one of them and their daughter's embryos were successfully implanted. Because Garber had no living husband, her child will be brought up by its grandparents.

Where a woman gestates another couple's embryo through the *in vitro* fertilization, the resulting child is not biologically related to the surrogate birth mother. A woman's body will reject an organ transplant from an unrelated person, but her immune system does not affect the placenta. In California, John and Luane, a married couple, entered into a surrogacy agreement, pursuant to which a surrogate gave birth after being implanted with an embryo created from the egg and sperm of anonymous donors. A month before the birth, John filed for divorce, alleging that there were no children of the marriage. After the birth, both John and the surrogate disclaimed any responsibility for the child. The trial court agreed with him, noting that under California's version of the Uniform Parentage Act, his wife, Luane, was not the genetic mother, birth mother, or adoptive mother. It was therefore concluded that John was not the legal father and was thus not obligated to pay child support, a situation burdening taxpayers. An appellate court in 1998 reversed this ruling, holding that despite the absence of any biological tie to the child, the divorcing couple is its lawful parents because the husband was engaged in "procreative conduct."[198]

A Massachusetts statute also recognizes a surrogate father as a man who consents to the insemination of the fertile wife with the sperm of a donor. In 1998, the Massachusetts Supreme Judicial Court looked at the traditional surrogacy agreement in which a woman agrees to bear a child as a surrogate mother and surrender custody of the infant to the biological father. That court held the agreement to be invalid due to its provision for the payment of $10,000

---

197. ORANGE COUNTY REGISTER, May 23, 1997.

198. Buzzanca v. Buzzanca, 72 Cal. Rptr. 2d 280 (Cal. Ct. App. 1998). The court stated: "Just as a husband is deemed to be the lawful father of a child unrelated to him when his wife gives birth after artificial insemination, so should a husband and wife be deemed the lawful parents of a child after a surrogate bears a biologically unrelated child on their behalf."

for "services rendered in conceiving, carrying and giving birth to a child."[199] The AMA's *Code of Medical Ethics* (2003) pointed out that "Surrogate" motherhood "involves the artificial insemination of a woman who agrees, usually in return for payment." This is "permissible," and it adds that in surrogacy contracts "in which the surrogate mother has no genetic tie to the fetus, the justification for allowing the surrogate mother to void the contract becomes less clear. Gestational surrogacy contracts should be strictly enforceable (i.e., not voidable by either party)."[200] It is also noted that an Israeli Commission in 1994 recommended unanimously that the surrogate mother should be entitled to payment for her services, to cover the actual expenses of conception, pregnancy and birth, and in compensation for her time, suffering, loss of income or temporary loss of earning capacity.[201]

good v. Good

Surrogacy for pay is nothing but baby selling and should be prohibited by law.

v.

Becoming a birth mother (surrogate) to an infertile couple provides a great service to the couple for which the birth mother should have an enforceable contract and a right to a reasonable fee.

---

199. R.R. v. M.H., 426 Mass. 501 (1998). Certain states that permit surrogate parenting require that the intended mother be infertile or unable to carry the child without unreasonable risk. *See* FLA. STAT. ANN. § 742.15 (West 2002); N.H. REV. STAT. ANN. § 168-B:17 (West 2002); VA CODE ANN. § 20-160(8) ().

200. AMA CODE OF MEDICAL ETHICS R. 2.18 ()."Japan has refused to grant citizenship to a Japanese couple's twins because an American surrogate mother gave birth to them. The boys were born at a hospital in California which makes them U.S. citizens, a status Japan recognizes. However, the Japanese Health Ministry is opposed to surrogate births." L.A. TIMES, Oct. 24, 2003, at A .

201. ISRAEL MINISTRY OF JUSTICE, THE REPORT OF THE PUBLIC-PROFESSIONAL COMMISSION IN THE MATTER OF IN VITRO FERTILIZATION (1994); Shalev, *supra* note 19. As one female bioethicist has written, "the right to be a surrogate mother implies that we women, as human beings, are capable of exercising reason with respect to reproduction and of sharing our birth power with those less fortunate than we." CARMEL SHALEV, BIRTH POWER: THE CASE FOR SURROGACY 29–32 (1989). Although the National Center for Health Statistics reported that twelve children were born in 2002 to women ages 50 to 54 who carried triplets, no statistics for women 55 and over were maintained.

Twenty-first Century Approach

Surrogacy agreements should be made enforceable under state law and no qualified surrogate birth mother should be denied her contractual right to accept a reasonable fee for her services (including the return of the infant to its genetic parents).

If we have decided that a democratic, free society is what we want, it seems to follow that people's wishes should be obstructed only with good reason . . . the onus is on those who would ban it [cloning] to spell out the harm it would do, and to whom.

Richard Dawkins (1998), p. 66

A human being—born of clonal reproduction—most likely will appear on the earth in the next twenty to fifty years, and conceivably even sooner.

James D. Watson, Nobel laureate

The idea that men are created free and equal is both true and misleading: men are created different; they lose their social freedom and their individual autonomy in seeking to become like each other.

David Riesman, *The Lonely Crowd* (1950)

2. Cloning in the Third Millennium

a) Origins of Cloning

Cloning currently represents the deliberate generation of two or more animals that share virtually the same nuclear deoxyribonucleic acid (DNA) in their cells. One way to clone two cells is to remove the nuclei from a human or animal and fuse those nuclei with enucleated donor egg cells.[202] Under a method of cloning by twinning (blastomere separation), the protective covering of a

---

202. In 1944 Hans Spemann dreamed up the "fantastical" experiment of taking a nucleus out of a cell and transferring it into an egg cell that had lost its own nucleus. If an organism developed correctly from the genetic material provided by the transplanted nucleus, then the nucleus must contain all the information necessary to make the organism; sadly for Spemann, the techniques needed to express his vision didn't emerge until after his death. WILMUT ET AL., SECOND CREATION (2000) at 194. Blood cells are the only cells in the body that lack DNA, the key cloning ingredient.

cleaving embryo is dissolved, its cells separated, and the embryos are moved to a medium for cleavage and later transfer. Twinning is normally done to increase the number of the best offspring in animal biotechnology.[203] One of the parents will be the genetic parent, who will have contributed virtually all of the genetic material, and the offspring will have only a set of grandparents, each of whom contributed 50 percent of their genes. The splitting of embryos can occur naturally when an early embryo spontaneously divides in half, producing identical twins, or thirds or quarters early in development. Then, each piece can develop as an independent embryo.[204]

Private funding must be used for infertility development because of a current ban on federal funding of most embryo research and most cloning (an outgrowth of the abortion debate).[205] It has been written that in all non-federal ways "a couple should be free to choose cloning unless there are compelling reasons for thinking that this would create harm other procedures would not cause."[206]

Cloning is almost asexual reproduction, analogous to the normal formation of identical twins. However, such twins also contain identical mitochondrial genes—the "energy factories" of cells. Although Dolly the sheep, the first mammal to be cloned, received her nucleus from her mother, her egg cytoplasm (including the mitochondria) came from her surrogate, a different source. Dolly was made from cultured cells derived from an adult mammary gland cell. As noted by her creator:

> She was the first animal of any kind to be created from cultured, differentiated cells taken from an adult. Thus, she confutes once and for all the notion—virtual dogma for 100 years—that once cells are committed to the tasks of adulthood, they cannot again be totipotent.

---

203. Bonickson, *Procreation by Cloning*, J.L. MED. & ETHICS 273 (1997); Ian Wilmut et al., *Embryonic and Somatic Cell Cloning*, 10 REPRODUCTION FERTILITY & DEV. 639–43 (1998). The National Bioethics Advisory Committee's (NBAC's) report ambiguously observes in a footnote: "Any other technique to create a child genetically identical to an existing . . . individual would raise many, if not all of the same non-safety-related ethical concerns raised by . . . somatic cell nuclear transfer." p. iii, p. 1.

204. SCIENCE (2000); ORANGE COUNTY REGISTER, Jan. 14, 2000, at .

205. ORANGE COUNTY REGISTER, Oct. 10, 1998, at News 6; *see supra* subsection 1(f).

206. Robertson, N. ENG. J. MED. (1998) (citing Robertson, *Liberty, Identity, and Human Cloning*, 77 TEX. L. REV. 137–456 (1998)).

> The cell that created Dolly came from an adult ewe—indeed, the ewe that provided her genes was almost elderly—yet its ability to be *re-programmed* into totipotency was demonstrated.[207]

Like twins, clones are different from one another to some degree. They will vary because brains evolve within a given species that look much alike but actually vary at the microscopic level.[208] This is because the structure of the brain and function of the mind emerge from an interaction between maturation and experience that are never identical for any two people.[209]

Serious argument on cloning first began in the last third of the twentieth century. In 1967, Joshua Lederberg, a Nobel laureate, discussed the improvement of the genetic qualities of humankind.[210] Two years later, theologian Paul Ramsey wrote *Shall We Clone Man?* to oppose the cloning of humans. Virtually nothing seemed to happen in the debate on cloning for 29 more years. In 1982, President Reagan's commission to study ethical problems in medicine made the statement, "The technology to clone a human being does not—and may never—exist."[211]

In March 1997, after 277 attempts, a Scottish scientist, Ian Wilmut, working at the Roslin Institute, took a ewe's egg, removed its nucleus, injected the cell from the udder, then fused the two with electricity without the need for a sperm. He then let the new embryonic cell multiply and implanted the embryo into a third ewe, the surrogate mother ewe. Wilmut named the resulting newborn Dolly, after the U.S. country singer Dolly Parton (whom he claimed has been known for her mammaries). In 1999 Dolly gave birth to two males and one female lamb, thus demonstrating that she was able to breed normally and produce healthy offspring.[212] Statements were immediately pro-

---

207. WILMUT ET AL., *supra* note 138, at 209; *see* 385 NATURE 810–13 (1997); 394 NATURE 329–30 (1998) (showing that Dolly's DNA was identical with that in the cultured mammary cells that had given rise to her).

208. RICHARD M. RESTAK, BRAINSCAPES 94 (1995).

209. Eisenberg, *The Social Construction of the Human Brain*, 152 AM. J. PSYCHIATRY 1563–75 (1995).

210. LEGERBERG, AM. NATURALIST (1966).

211. PRESIDENT'S COMM'N FOR THE STUDY OF ETHICAL PROBLEMS IN MED. & BIOMEDICAL BEHAVIORAL RESEARCH, SPLICING LIFE: THE SOCIAL AND ETHICAL ISSUES OF GENETIC ENGINEERING WITH HUMAN BEINGS 10 n.6 (1982).

212. ORANGE COUNTY REGISTER, Apr. 2, 1999, at News 9.

claimed, including about "playing God," that formerly had greeted the introduction of *in vitro* fertilization (IVF). The former presidential press secretary George Stephanopoulos said, "What it creates is the possibility of immortality,"[213] a pejorative reference found to be politically correct near the dawn of the third millennium.[214]

For several months, no scientists could replicate Wilmut's feat, and some claimed that he might have selected a fetal cell, and not an adult cell, from which he cloned Dolly.[215] He failed to report that the adult sheep from which Dolly was cloned had died several years earlier, which prevented a direct comparison between Dolly and her donor. Wilmut himself decided not to replicate his experience, stating briskly, "I don't perceive a need. The principle is established. Repeating experience is boring and unimaginative."[216] The Roslin Institute announced its intention to patent the process. Although its deputy director stated this process would be for use on animals only, not humans, he added, "Our (patent) applications do apply to use in animals. But it is up to the relevant authority in each country to decide whether the term 'animals' should include humans." Asked if the institute intended to patent its process to include human cloning, he said, "That was the intention."[217] Others spoke somewhat differently:

---

213. Horgan, L.A. TIMES, Mar. 2, 1997.

214. In 1818, Mary Shelley warned in her novel *Frankenstein* that "supremely frightful would be the effect of any human endeavor to mock the stupendous mechanism of the Creator of the world." The censors took her words to heart in the final cut of James Whale's 1931 film version starring Boris Karloff. In the riveting laboratory scene when the monster is brought to life, Dr. Frankenstein roars, "It's alive." His lips keep moving but his voice disappears. The censors deleted the rest of the sentence—the forbidden words that have frightened cultures from ancient Greece to modern America—"Now I know what it feels like to be God."

215. In 1982 Professor August Weismann first proposed that germ cells (gametes, or eggs and sperm) and cells that comprise the eggs and sperm develop in the embryo quite separately from the somatic cells (body cells).

216. ORANGE COUNTY REGISTER, June 30, 1998, at News 10.

217. ORANGE COUNTY REGISTER, May 9, 1997. The applications have been submitted to the World Intellectual Property Organization, a U.N. agency based in Geneva. *Id.*; *see also* ORANGE COUNTY REGISTER, Dec. 9, 1998. Such cloning has also been referred to as genetically identical, different aged (GIDA).

> One reason to clone cattle would be to reproduce exact copies of animals that are superb producers of meat or milk. In fact . . . on a recent visit to Japan, [Roberts] saw a calf that other researchers asserted they had cloned from cells taken from the ear of a prize bull. The scientists who created that clone have not yet published their results. After cloning Germany's first cow in 1999, professor Eckhard Wolf of Munich's Ludwig Maximillian University stated the main practical benefit could be a new genetic transfer process, which . . . could be used to create important medicines for treating people.[218]

Nevertheless, in 2003 Dolly, the first mammal cloned from an adult cell, raised questions about the practicality of cloning. At the age of six, only about half the life expectancy of her breed she was put to death after premature aging and lung disease.[219]

Others have been actively cloning animals elsewhere. In March 1997, the Oregon Health Sciences Research Center stated that monkeys had been cloned, but this had not yet been reported in a scientific journal and only became confirmable in 2004.[220] In July of that year, Japanese scientists re-

---

218. L.A. TIMES, Jan. 7, 1999.

219. Richard Garner, a professor of zoology at Oxford University, said:

> We must await the results of the post-mortem on Dolly in order to assess whether her relatively premature death was in any way connected with the fact that she was a clone. If there is a link, it will provide further evidence of the dangers inherent in reproductive cloning and the irresponsibility of anybody who is trying to extend such work in humans.

ORANGE COUNTY REGISTER, Feb. 17, 2003, at News 9.

220. L.A. TIMES, Mar. 2, 1997, at A35.

> A team of University of Pittsburgh scientists has produced the world's first cloned monkey embryos with the help of the South Korean scientists who created the first mature human cloned embryos. Researchers created 30 blastocysts before they could isolate a stable line of embryonic stem cells.

ORANGE COUNTY REGISTER, Dec. 8, 2004, at News 13.

ported cloning calves from an adult cow,[221] as did a Wisconsin biotechnology company.[222] In 1998, Princeton University biologist announced that multiple mice cloning means that human cloning could occur within the next decade.[223] Researchers said their experiment demonstrates that cloning can be accomplished with the genetic material contained in the nucleus of somatic cells, or ordinary cells found throughout the body.[224] In 1999, Dr. Jerry Yang cloned the first mammal in the United States (a calf), and in 2003, the Food and Drug Administration (FDA) stated there was no evidence that meat or milk derived from healthy cloned farm animals can harm people.[225]

Dr. Richard Seed, a 69-year-old Harvard physicist and fertility researcher, dramatically announced, "It is my objective to set up a Human Clone Clinic in Greater Chicago; make it a profitable fertility clinic; and when it is profit-

---

221. Orange County Register, July 6, 1998, at News 15; Orange County Register, July 7, 1998, Cloning E. This work has continued in the United States into 2003.

> Advanced Cell Technology in Worcester, Mass., fused the banteng skin cells with 30 cow eggs that had their genetic material removed. Of the 16 resulting pregnancies, only two came to term last week—and one of the bantengs is 80 pounds, about twice as heavy as he should be. Even same-species cloning has a high failure rate, and survivors often are less healthy than naturally born animals.

Orange County Register, Apr. 8, 2003, at News 9.

222. Orange County Register, Aug. 17, 1997, at News 22; Orange County Register, Aug. 8, 1997, at News 12.

223. Dr. Lee Silver, a biologist at Princeton University, stated: "If we follow scientific protocol, it could take 5 to 10 years before in vitro fertilization clinics add human cloning to their repertoires." Orange County Register, July 23, 1998, at News 14. In 1992, scientists in Hawaii cloned a trio of identical mice using ordinary cells rather than DNA extracted from the female reproductive system. They used genetic material extracted from tail cells of adult male mice.

224. Orange County Register, June 5, 1999, at News 7. Also in 1998, the cloning of human cells from an infertile woman was reported by the Kyunghee University Hospital in Seoul, Korea. The resulting embryo could have grown into an infant; however, it was prevented from doing so because of the legal and ethical implications involved. Orange County Register, Dec. 17, 1998, at News 26. "The Korean government restricts the cloning of humans, though it does not ban it outright." *Id.*

225. Yang, Orange County Register, June 29, 2001, at Local; Orange County Register, Oct. 31, 2003, at News 20.

able, to duplicate it in 10 or 20 locations around the country, and maybe five or six internationally."[226] He told National Public Radio, "We are going to have as much knowledge and almost as much power as God." In 1998, Dr. Seed announced that his postmenopausal wife, Gloria, had agreed to carry an embryo that would be created by combining the nucleus of one of his cells with a donor egg. However, he refused to give any details on how he planned to accomplish this endeavor.[227]

Possible benefits of cloning include food production and the understanding and treatment of many diseases, and the cloning of some recently extinct mammals could become possible. A French-led expedition has chopped the fully preserved carcass of a 20,000-year-old woolly mammoth from the permafrost of Arctic Siberia, in the first successful salvage of an intact specimen of this ancient behemoth. The recovery will allow scientists an opportunity to analyze the fur, organs, and other soft tissues of an animal that has been extinct for 10,000 years. The team hoped to recover DNA or even sperm from the carcass, if it was a male, and either attempt to clone a mammoth or artificially inseminate a modern-day elephant.[228]

Little concern was provoked when cloning was done exclusively with animals. Major ethical discussion ensued when the procedure was first considered for human use.[229] The Roslin Institute in Edinburgh, where Dolly was cloned, has been working to create transgenic pigs as a source of organs for transplant to humans. But the institute backed off in August 2000 when its commercial arm, the U.S. biotech company Geron,[230] expressed concern about the risks that xenotransplantation might create viral diseases in humans.[231] Therapeutic cloning uses somatic cell nuclear transfer [SCNT], an

---

226. *Quoted in* Donnelly, Pres., Hast. Ctr. Rep., Feb. 1998.
227. L.A. Times, Sept. 7, 1998, at A16.
228. Int'l Herald Trib., Oct. 22, 1999, at 1.

Dinosaurs disappeared 65 million years ago, and no dinosaur flesh can likely ever be found. Animals cannot be re-created from DNA that does not exist.

229. Ramsey, Shall We Clone a Man? Fabricated Man (1970).
230. *See* 399 Nature 92 (1999).
231. However, others indicated that if Geron Bio-Med were to abandon xenotransplantation work, it would be purely on the grounds of refocusing research priorities, and not viral risk. *See* http://www.nature.com/cgi-taf/.

inefficient process in animals and people.[232] However, in October 2000, cloned cows began being auctioned at commercial sales in Madison, Wisconsin; the company only had a 5 percent success rate.

In attempting an analysis of the development of cloning in the twenty-fiirst century, mention must be made of some reactions of people in religion, politics, and science. Ethics and the law will then be approached.

Cloning, like all other technologies, is in and of itself morally neutral. Its moral valence depends upon how we use it.

Dorff, *Matters of Life and Health* (1998), p. 323

Not everything that can be done should be done.

Sir John Polinghorne, Anglican theologian

---

b) Some Religious Reactions

A President's Advisory Commission on cloning decided to consider religious traditions because they "would be instructive, even if not necessarily determinative for public policy." This approach may not be wholly consistent with the legal principle of separation of church and state; the religiously dominated National Right to Life Committee and other opponents of abortion are not only against most embryo research, but human cloning as well.[233]

232. In 2005, South Korean researchers cloned what scientists deemed the most difficult animal to duplicate, the dog. A total of 1,095 reconstructed embryos were transferred into 123 surrogates to create the two dogs—an efficiency rate of 1.6%. Their "research goal is to produce cloned dogs for studying disease models, not only for humans, but also for animals." ORANGE COUNTY REGISTER, Aug. 4, 2005, at News 3. An earlier study indicated that the efficiency of animal cloning has typically been about 1% to 2%, meaning that out of every100 embryos implanted in surrogate animals, 98 or 99 fail to produce live offspring. Autumn Fiester, *Creating Fido's Twin: Can Pet Cloning Be Ethically Justified?*, Hast. Ctr. Rep., July-Aug. 2005, at 35 (citing Coleman, *Somatic Cell Nuclear Transfer in Mammals: Progress and Application*, 1 CLONING 185–200 (1999));. *see also* Green, *The Ethical Considerations*, 286 SCI. AM. 48–50 (2002).

233. Op-Ed., N.Y. TIMES, Mar. 14, 1998.

The advent of the potential cloning of humans produced mixed reactions among religious communities. Ted Peters, a professor of systematic theology at Pacific Lutheran Theological Seminary in Berkeley, California, said, "The way I read the Bible, the status of that person before God would not be any different from anyone born the old-fashioned way."[234] In June 1997 a spokesperson for the pontifical Academy of Life, a panel set up by Pope John Paul II, said that the spiritual soul, "the constitutive kernel" of every human created by God, cannot be produced through cloning.[235] This "soulless" reaction was consistent with previous releases by the Vatican, which had forbidden any kind of artificial fertilization to create human life.

Theologian Sir John Polkinghorne (see epigraph), a member of the Human Genetics Advisory Commission in the United Kingdom, wrote, "Theology will not seek to stifle advances that could benefit humankind in acceptable ways; but it will insist that the means by which these desirable ends are achieved must themselves be of ethical integrity." Similarly, Rabbi Dorff (see epigraph) stated that cloning is "morally neutral." He recommended, "Human cloning should be regulated, not banned,"[236] contrary to other statements that "[i]f God is seen as Creator, and if we are created in the image of God, then might we not have a role to play as creators ourselves?"[237] Conservative Jews appear more liberal on cloning than Reform Jews. It has been reported how "conservative Rabbi Byron Sherwin observed that on the basis of their legal tradition, orthodox and conservative Jewish thinkers find it difficult to condemn human cloning per se, and that such opposition to human cloning, particularly in Reform Judaism, reflects categories outside Jewish law, such as human dignity."[238] Orthodox Jews do not favor human autonomy (dignity); most serious bioethical matters pertaining to life and death are reserved for rabbis to decide, not individual believers (see Chapter 14). Nevertheless, they find no biblical injunction against human cloning.

---

234. Professor Peters is the author of *Playing God?: Genetic Determinism and Human Freedom.*

235. Orange County Register, June 25, 1997, at News 12.

236. Dorff, *supra* note 106, at 322.

237. Cohen, *In God's Garden: Creation and Cloning in Jewish Thought*, Hast. Ctr. Rep., July-Aug. 1999, at 7–12.

238. Hast. Ctr. Rep., Sept.-Oct. 1997, at 11.

In condemning cloning as being man's abortive attempted step toward a sort of immortality, Catholic theologian Paul Ramsey had stated, "Religious people have never denied, indeed they affirm, that God means to kill us all in the end, and in the end He is going to succeed."[239] On the other side, theologian Ted Peters stated, "We cannot allow our ethics to derive from our fears of scientific advance. Rather, we need to construct ethical visions that take expanded choice into consideration. For cloned children as well as children born the old-fashioned way, we need to be reminded that God loves each of us regardless of our genetic makeup. And we should do likewise."[240]

In short, it may be concluded that there are a variety of religious views on cloning.

Moreover, to pretend that human cloning will not take place if it is banned in experiments funded with government money is simply unrealistic; it will happen with private funds in the United States and/ or abroad.

Dorff, *Matters of Life & Death* (1998), p. 33

c) Some Political and Legal Statements on Human Cloning and Stem Cell Research

At the end of the twentieth century, politicians in America, Europe, and China reacted adversely to the potential of human cloning for political reasons. In early March 1997, President Clinton imposed a ban on the use of federal money for cloning humans. Clinton attempted to justify the ban with the warning against "trying to play God"[241]—a statement refuted by Jewish commentators as noted above. In early 1998, he reiterated this "moral unacceptability."[242] Likewise, Harold Vorman, director of the National Insti-

239. RAMSEY, FABRICATED MAN: THE ETHICS OF GENETIC CONTROL (1970).

240. *Id.* In order to favor a form of immortality, the corporation, Clonaid, was created for the purpose of human cloning. *See* RAEL, THE TRUE FACE OF GOD (1998).

241. ORANGE COUNTY REGISTER, Mar. 5, 1997; ORANGE COUNTY REGISTER, Feb. 27, 1997, at News 18 ("it was out of his element").

242. ORANGE COUNTY REGISTER, Jan. 11, 1998. This occurred just a few weeks prior to announcing that he "never had sexual relations" with White House intern Monica Lewinsky, for which he would later undergo impeachment proceedings.

tute of Health, stepped out of his element when he echoed the president stating that "cloning human beings would be repugnant."[243]

At the state level, in 1997, California legislators, intending to ban human cloning, passed a bill that was characterized as having been "so sloppily worded that it prohibits a host of infertility treatments."[244] However, a legislative California Advisory Committee on Human Cloning unanimously reported on January 15, 2002, that it gave "a red light to reproductive cloning and a green one to cloning for medical purposes."[245] On March 24, 2003, Governor Huckabee of Arkansas signed into law a bill to ban the cloning of humans for any purpose.[246] But, on January 4, 2004, Governor McGreevy of New Jersey signed a law allowing researchers to clone an embryo, implant it into a woman, and gestate the clone all the way up to birth—as long as it is aborted. Legislation concerning destructive human embryo research has pended in Illinois and New York.[247]

In Europe, each country is governed through national legislation as well as the international European legislation that is passed by the European Union

---

243. Orange County Register, Mar. 2, 1997, at News 13.

> A 1997 poll sponsored by *Time* and CNN found that 64% of 18-to-29-year-olds favored therapeutic cloning—that is, cloning for replacement tissues and parts—compared to only 32% of those over 65 years.

Wolfson, *Biodemocracy in America*, Pub. Int., Winter 2002.

244. L.A. Times, Jan. 31, 1998, at B11.

> One such treatment would help women who have become infertile because their eggs have hardened with age. Infertility specialists can now get around this problem by removing the nucleus of a donor egg, replacing it with nuclei from the woman and her husband and implanting the egg in the woman's uterus. This technique, a far cry from cloning, has become illegal in California because the new law prohibits doctors from transferring any nucleus into a human egg.

Id.

245. So. Cal. Christian Times, Mar. 2002, at 9. The report stated: "We believe that use of this technology offers potential medical and scientific benefits while not raising many of the same concerns as human reproductive cloning."

246. http://washingtontimes.com/national/20030325-95699176.htm.

247. info@cbhd.org.

(EU).[248] In January 1998, 19 European nations signed a treaty declaring that cloning is "contrary to human dignity and thus constitutes a misuse of biology and medicine"; however, the protocol made no mention of sanctions against those who refuse to follow it.[249] Similarly, in 1997 the World Health Organization in Geneva stated that the use of cloning for the replication of human individuals is ethically unacceptable inasmuch as it would violate some of the basic principles that govern medically assisted procreation. The Council of Europe in Strasbourg followed suit with its Code of Bioethics: "Any intervention seeking to create a human being genetically identical to another human being, whether living or dead, is prohibited."[250] A draft of the European Patent Office excluded as "unpatentable" any inventions whose exploitation or publication would be contrary to public policy or morality, such as "human cloning."[251] In 2001 the British Parliament enacted a law making it a crime punishable by up to 10 years in prison for a person to place "in a woman a human embryo that has been created otherwise than by fertilization," while allowing research on therapeutic cloning.[252]

---

248. "These legislative and regulatory initiatives address two main ethical questions: First, does the production or use of human embryos in research threaten human dignity? And second, might therapeutic cloning lead to a commercialization of human eggs or embryos?" N. ENG. J. MED. 1579 (2002).

249. ORANGE COUNTY REGISTER, Jan. 13, 1998, at 1. "French President Jacques Chirac called for an international ban on human cloning, and two days after, President Clinton urged Congress to do the same." *Id.*

250. ROGER GOSDEN, DESIGNING BABIES 138 (1999).

251. *See European Parliament Approves Draft Biotech Patent Directive*, Pat. Trademark & Copyright L. Daily (BNA), Aug. 28, 1997, at D3; Draft of a Universal Declaration on the Human Genome and Human Rights, UNESCO, U.N. Doc. CIP/BIO/96/COMJUR 6/2 (Prov. 5) (1996).

252. Human Reproductive Cloning Act, 2001, c. 23 § 1 (Eng.); ORANGE COUNTY REGISTER, Feb. 12, 2004, at News 4. Yet, in 2004 Britain granted its first license for human cloning research. Japan's science council voted to adopt policy recommendations that would permit the limited cloning of human embryos for scientific research. Although South Korea created human embryos through cloning, their scientists said the technique was not designed to make babies. N. ENG. J. MED. (2004). France banned human cloning, making it punishable by 30 years in prison and a fine of approximately $9.3 million. It also banned stem cell research on human embryos for five years. ORANGE COUNTY REGISTER, July 10, 2004, at News 17. Similarly, on March 29, 2004, Canada passed the Assisted Human Reproduction Act (AHR), prohibiting

A Declaration of UNESCO in 1997 endorsed by the UN General Assembly in 1999 contains an Article 11, which prohibits the cloning of human beings as a practice contrary to human dignity;[253] however, UNESCO had been the most criticized and most "politicized" of all the specialized agencies.[254] It was for this reason that the United States withdrew from that part of the UN organization in 1985, and the U.K. withdrew in 1986.[255] The Chinese Academy of Sciences banned all research into human cloning, stating, "Banning the use of cloning to copy humans is absolutely necessary to maintain ethical morality that holds together today's human society."[256] At the same time, Chinese scientists were trying to preserve their endangered pandas through cloning.[257] The United Nation Economic and Social Council's non-binding Universal Declaration on Human Genome and Human Rights concluded, "Practices which are contrary to human dignity, such as reproductive cloning of human beings, shall not be permitted."[258]

---

human cloning or the creation of embryos for research purposes. But a license may be granted for research using human embryos remaining after infertility treatment. Hast. Ctr. Rep., May-June 2004, at 5.

253. Universal Declaration on the Human Genome and Human Rights, UNESCO Gen. Conf. Res. 29C/Res. 16 (Nov. 11, 1997), adopted by the General Assembly in 1999, G.A. Res. 152, U.N. Doc. A/RES/53/152 (March 10, 1999).

254. Clare Wells, The UN, UNESCO, and the Politics of Knowledge 1 (1987). The U.S. House Committee on Foreign Affairs defined politicization as the "introduction of highly charged, highly controversial, and extraneous issues into international debates that polarize parties to the debate and promote hostile, confrontational rhetoric that hardens positions."

255. Joel H. Rosenthal, The Withdrawal from UNESCO: International Organizations and the U.S. Role (Carnegie Council Case Study Series No. 10, 1990).

256. Orange County Register, May 13, 1997, at News 14.

257. Orange County Register, July 21, 1998, at News 17.

> By the year 2004, China said it had built the world's biggest panda-gene bank as a repository of cells for use in cloning the animal. However, James Harkness, the head of World Wildlife Fund China, said, "We don't believe it will do much to help panda conservation. The biggest threat to pandas is loss of habitat."

Orange County Register, Aug. 16, 2004, at News 6.

258. Universal Declaration on the Human Genome and Human Rights, Dec. 23, 1997; *see U.N. Weighs in on Cloning*, 278 Science 1407 (1997).

In the United States the June 1997 Report of a National Bioethics Advisory Commission, hurriedly appointed by President Clinton three months earlier, recommended a continuation of the moratorium on federal funding of any attempt to create a child by somatic cell nuclear transfer and that Congress make this the law. The commission also recommended that non-federally funded sectors comply voluntarily for three to five years.[259] On the other hand, Senator Tom Harkin, an Iowa Democrat who had lost two sisters to cancer, stated, "Human cloning will take place and it will take place in my lifetime. I think it is right and proper. It holds untold benefits for humankind in the future."[260] Congress showed no inclination to enact bills containing such recommendations.

On April 10, 2002, President Bush stated he believed that "all human cloning is wrong" and that therefore both reproductive cloning and research cloning "ought to be banned."[261] He gave three main reasons for this position: research cloning "would require the destruction of nascent human life"; anything other than a dual ban would result in "embryo farms" that would inevitably result in "the birth of cloned babies"; and the "benefits of research cloning are highly speculative."

Thus, President Bush even discouraged the process called "therapeutic cloning," which:

---

259. Hast. Ctr. Rep., Sept.-Oct. 1997, at 8–10.

260. Orange County Register, Mar. 13, 1997. Sen. Christopher S. Bond (R.-Mo.) introduced an emergency measure to ban research into human cloning, but congressional support collapsed.

261. N. Eng. J. Med. 1599 (2002).

> He gave three main reasons for this position: research cloning "would require the destruction of nascent human life"; anything other than a dual ban would result in "embryo farms" that would inevitably result in "the birth of cloned babies"; and the "benefits of research cloning are highly speculative." In 2003, "all 191 U.N. members agreed on the need for a treaty that would prohibit the cloning of human beings, but they are divided over whether to extend such a ban to stem cell and other research known as therapeutic cloning."

Orange County Register, Dec. 10, 2003.

> [I]n combination with the differentiation potential of embryonic stem cells, offers a valuable means of obtaining autologous cells for the treatment of a variety of diseases. It is important, therefore, to continue research aimed at improving our understanding of the molecular events that take place during nuclear reprogramming, in order to develop these potential new therapies.[262]

In July 2002, the majority of the President's Council on Bioethics (a 10-7 decision) endorsed a moratorium on research cloning for four years. The seven-person minority opinion held that research cloning should move forward and embryos be used for the purpose of saving human life. They stated that research using nuclear transfer–derived embryos "must be grounded in our judgment about the moral status of the embryos themselves, not the purpose of their creation. . . . Because the use of stem cells from cloned embryos may in the future provide treatment for serious human diseases, the creation of cloned embryos and their subsequent disaggregation to isolate stem cells can be justified."[263] A cell biologist, who was among those asked to leave the Council,

262. Hochedlinger & Jaenisch, N. ENG. J. MED. 295 (2003). In 2004, the National Institutes of Health (NIH) defended the controversial federal policy, indicating that the White House does not plan to back off the policy limiting federal funding of embryonic stem cell research. ORANGE COUNTY REGISTER, May 16, 2004, at News 27. However, Nancy Reagan, the wife of former President Reagan, made an impassioned plea for stem cell research, saying it could help cure illnesses such as Alzheimer's, which afflicted her husband. ORANGE COUNTY REGISTER, May 10, 2004, at News 7. She also supports the Juvenile Diabetes Research Foundation, whose money is part of the nearly $20 million that foundation is donating to advance stem cell research.

263. J.L., MED. & ETHICS 804–05 (2005); Lyon, Hast. Ctr. Rep., Sept.-Oct. 2002, at 7. Although Immanuel Kant (1724–1804) spoke of the "spurious principles of morality" in his *Fundamental Principles of the Metaphysics of Ethics* (par. 88), Leon Kass, the chair of George W. Bush's Council on Bioethics, indicated abandoned moral reasoning on cloning: "We are repelled by the prospect of cloning human beings not because of the strangeness or novelty of the undertaking, but because we intuit and feel, immediately and without argument, the violation of things that we rightfully hold dear. . . . Shallow are the souls that have forgotten how to shudder." STEVEN PINKER, THE BLANK SLATE 274 (2002). The Kass report of March 30, 2004, had stated, "Congress should prohibit attempts to conceive a child by any means other than a union of egg and sperm." N.Y. TIMES, Mar. 30, 2004, at D4.

noted that not one of the newly appointed members is a biomedical scientist.[264] Congress did not react; however, in September 2002 the state of California enacted a bill that made permanent its temporary ban on human cloning for reproductive purposes.[265] In November 2002, the United States was supported by 36 other nations to block an initiative by Germany and France for a worldwide ban on cloning to create human beings. It insisted that the ban should include all forms of human cloning, as distinguished from the contrary view expressed by the International Academy of Humanism.[266]

Nevertheless, the U.S. federal funding was opposed by the administration on both therapeutic cloning (the making of early-stage preimplanatation embryos for use as sources of stem cells) and reproductive cloning (the creation of cloned babies through the transfer of cloned embryos into a woman's uterus). But tissue may be rejected by the immune system. One potential solution is to replace nuclei with genetic material of the patient (therapeutic cloning).[267]

---

264. One, a pediatric neurosurgeon, had championed religious values in public life; another, a political philosopher, has publicly praised Kass's work; the third, a political scientist, has described as "evil" any research in which embryos are destroyed. . . . The best possible scientific information was not incorporated and communicated clearly in the council's report, suggesting that the presentation was biased.

Elizabeth Blackburn, *Bioethics and the Political Distortion of Biomedical Science*, N. Eng. J. Med. 1379 (2004).

265. Orange County Register, Sept. 23, 2002, at News 4.

266. Julia Preston, *U.S., Pushing for Broader Ban, Blocks U.N. Anti-Cloning Move*, N.Y. Times, Nov. 8, 2002, http://www.nytimes.com/2002/11/08/international/08CLON.html. "Views of human nature rooted in humanity's tribal past ought not to be our primary criterion for making moral decisions about cloning. . . . The potential benefits of cloning may be so immense that it would be a tragedy if ancient theological scruples should lead a Luddite rejection of cloning." Int'l Acad. of Humanism, Declaration in Defense of Cloning and the Integrity of Scientific Research (May 16, 1997), *reprinted in* Free Inquiry, Summer 1997, *available at* http://www.secularhumanism.org/index.php?section=library&page =cloning_declaration_17_3. In November 2003, the U.N. General Assembly's legal committee voted 80–79 to delay consideration of a cloning global treaty.

267. Andrea L. Bonnicksen, Crafting a Cloning Policy: From Dolly to Stem Cells (2002); Prentice, Stem Cells and Cloning (2003); *see also infra* Chapter 15.

A human being—born of clonal reproduction—most likely will appear on the earth in the next 20 to 50 years, and conceivably even sooner.

James Watson, Nobel laureate

There is inherent in nature's works no prudence, no artifices, no intelligence, but these only appear to our thinking to be there because we judge the divine things of Nature according to our special faculties and peculiar manners of thought.

William Harvey (1649)

d) Some Views of Scientists and Physicians on Human Cloning

In 1997, the International Academy of Humanists wrote an open letter calling cloning bans "the Luddite option." Its members included Nobel laureate Francis Crick (co-discoverer of the structure of DNA), Richard Dawkins (professor, Oxford University), Nobel laureate Herbert Hauptman (Chemistry), William V. Quine (professor of philosophy, Harvard University), Edward O. Wilson (also of Harvard), and Simone Veil (a former president of the European parliament). Their statement was: "The moral issues raised by cloning are neither larger nor more profound than the questions human beings have already faced in regard to such technologies as nuclear energy, recombinant DNA, and computer encryption. They are simply new."[268]

In December 1998, a British government advisory committee pointed to the lifesaving potential of cloning human tissue and even organs for therapeutic uses. Three years later, Britain became the first country in the world to legalize therapeutic cloning for purposes of extracting stem cells for medical research, though for therapeutic cloning in Britain one also needs a license from the government's Human Fertilisation and Embryology Authority (HFEA), which had not granted any as of April 2004.[269] Dr. Davor Solter,

268. FREE INQUIRY, Summer 1997. "Luddite refers to a band of 19th century English workmen who tried to prevent the use of labor-saving machinery by breaking it."

269. "Although such applications are still some years away, we believe that it would not be right at this stage to rule out limited research using such techniques, which could be of great benefit to seriously ill people." Solin Campbell, L.A. TIMES, Dec. 1998, at A3.

director, department of developmental biology, Max Planck Institute of Immunobiology in Freiburg, Germany, pointed out how cloning depends upon women. For example, a woman could use one of her own eggs to create an organ she needs. But a man would have to buy eggs from a woman.[270] It has been shown that, assuming the availability of a woman's eggs, altered DNA in embryonic cells could be returned to the mother for gestation. However, they lose their pluripotent capacity after several cell divisions, so that only a few cells could be manipulated. With cloning, the age and number of cells is unlimited. In addition, gene therapy might be conducted on cells from a parent; "A child cloned from those altered cells would be free of the genetic defect but in other ways a genetic duplicate of this donor parent."[271] A study with cows in 2002 showed that more than 80 percent of clones die during pregnancy or shortly after birth: "In clones, even though they can develop to full term, many abnormalities in gene expression exist, which may be partially responsible for the developmental abnormalities frequently observed, including death."[272] In 2002, the National Research Council wrote:

> Human reproductive cloning . . . is dangerous and likely to fail. The panel, therefore, unanimously supports the proposal that there should be a legally enforceable ban on the practice of human reproductive cloning. . . . The ban should be reconsidered only if at least two conditions are met: (1) a new scientific and medical review indicates that the procedures are likely to be safe and effective and (2) a broad national dialogue on the societal, religious, and ethical issues suggests that a reconsideration of the ban is warranted.[273]

Thus, although it may become safe in the future, it was considered currently unsafe to seek a human pregnancy with a cloned embryo. However, in 2003 the American Medical Association (AMA), the largest orga-

---

270. ORANGE COUNTY REGISTER, Feb. 22, 1998, at News 22. The going rate for a month's worth of such eggs today ranges from $2,500 to $5,000. *Id.*

271. Mirsky & Rennie, SCI. AM., June 1997, at 123.

272. ORANGE COUNTY REGISTER, May 27, 2002, at News 16 (reporting on an article in the journal *Nature Genetics*).

273. NAT'L RESEARCH COUNCIL, SCIENTIFIC AND MEDICAL ASPECTS OF HUMAN REPRODUCTIVE CLONING 1 (2002).

nization of doctors, endorsed cloning for ethical research purposes for the first time. This was contrary to the Bush administration and in favor of therapeutic cloning.[274]

We live in an age of the ethicist, a time when we argue about pragmatism and compromises in our quest to be morally right.

Kolata, *The Road to Dolly* (1998)

Look at what is happening within the working classes. Can you not see that their passions, from being political have become social? Can you not see that ideas are gradually spreading among them which are not only going to overthrow certain laws, but society itself, knocking it off the foundations on which it rests today?

Tocqueville, Speech to the Assembly (January 27, 1848)

e) Proposed Limits on Cloning in Ethics and the Law

All potential uses of cloning techniques require ethical evaluation. A philosopher who is a member of a medical ethics committee on which I served characterized cloning as a "mere extension of *in vitro* fertilization." This was an ethical endeavor expressed in the United States and other countries.

As stated in the epigraph (by Kolata), "We live in an age of the ethicist." The ethicist's job is to analyze moral positions and seek "pragmatism and compromises" by "balancing good against good." No ethicist should be swayed by media reports of national and international declarations that "distort ethical issues in science and medicine" and in the humanities.[275] As stated by the philosopher Roger L. Shinn: "No law from Sinai, no command from

---

274. ORANGE COUNTY REGISTER, June 18, 2003, at News 10. It added: "The proposal focused on a procedure designed to create human embryos for their stem cells, which are master cells that can potentially grow into any type of human tissue." Similarly, Professor Rudolf Jaenisch noted that both embryonic stem cells derived from a cloned embryo and embryos derived through have an identical potential to serve as a source for cells that may prove useful for research or therapy. *Human Cloning—The Science and Ethics of Nuclear Transplantation*, N. ENG. J. MED. 2787 (2004).

275. *See* Turner, *The Media and the Ethics of Cloning*, CHRON. HIGHER EDUC., Sept. 26, 1997, at B4.

a sacred Mount, nothing in the Buddhist eightfold path or the Muslim sharia decrees: Thou shalt or shalt not clone."[276]

Although ethicists must weigh arguments like those of Leon Kass that the cloning of human beings is unethical, immoral, and leads ultimately to the dehumanization of humanity, they also note that, like official Catholicism, Kass objects to any form of assisted reproduction that does not involve parental coitus.[277] Similarly, the bioethicist (and lawyer) George Annas describes cloning as "asexual child abuse" and would make it an international crime.[278] In 2002 he, together with two others, suggested language for a proposed international "Convention of the Preservation of the Human Species" that would outlaw all efforts to initiate a pregnancy by human replication

---

276. SHINN, HUMAN CLONING 106 (1997). Wesley Smith, an attorney for the International Anti-Euthanasia Task Force, argued that control of the fate of human evolution would undermine the intrinsic worth and equality of everybody. Wesley Smith, MINDSZENTY REP., Feb. 2001. However, as stated by Michael Sandel, Ph.D. Phil., at Harvard:

> [An] argument claims we must regard embryos as possessing the same inviolability as fully developed human beings. But this argument is flawed. The fact that every person began life as an embryo does not prove that embryos are persons. Consider an analogy: although every oak tree was once an acorn, it does not follow that acorns are oak trees, or that I should treat the loss of an acorn eaten by a squirrel in my front yard as the same kind of loss as the death of an oak tree felled by a storm. Despite their developmental continuity, acorns and oak trees are different kinds of things. So are human embryos and human beings.

N. ENG. J. MED. 207 (2004).

277. Leon R. Kass, *The Wisdom of Repugnance*, NEW REPUBLIC, June 2, 1997, at 17–26.

278. GEORGE J. ANNAS, SOME CHOICE: LAW, MEDICINE, AND THE MARKET (1998).

> Annas argues that to come into being by cloning is to be robbed of one of the most fundamental characteristics of being human—namely, to be reproduced by two human beings and not merely replicated by one. Cloning thus undermines the uniqueness of individuals.

Doughtery, *Some Choice: Law, Medicine, and the Market*, 340 N. ENG. J. MED. 1445 (1999) (book review). However, in 2002 he compromised to politically justify nuclear transplantation of human embryos as sources of stem cell research. *See* George J. Annas, *Cloning and the U.S. Congress*, 346 N. ENG. J. MED. 1599–1602 (2002).

cloning.[279] The proposed convention would note that "the increased power of genetic science has the power to diminish humanity fundamentally by reproducing a child through human cloning" and express their view of a need "to prevent the misuse of genetic science in ways that undermine human dignity and human rights." It would include:

> Article 2. Parties shall take all reasonable action, including the adoption of criminal laws, to prohibit anyone from utilizing somatic cell nuclear transfer or any other cloning technique for the purpose of initiating or attempting to initiate a human pregnancy or other form of gestation.
>
> Article 5. Reservations to this Convention are not permitted.
>
> Article 6. For the purpose of this Convention, the term "somatic cell nuclear transfer" shall mean transferring the nucleus of a human somatic cell into an ovum or oocyte. "Somatic cell" shall mean any cell of a human embryo, fetus, child or adult, other than a reproductive cell.[280]

But in 2003, another view argued that although medical research raises special moral problems, "these concerns should not ultimately lead us to oppose it because we have judged certain powerful considerations to outweigh the moral problems with it."[281] Although these points of view have qualifications, they are opposed by the majority of bioethicists.

Bioethicist Susan Wolf would "ban the ban" proposed by President Clinton, stating that his advisory commission was wrong to urge it in the first place. "Like any technology, cloning needs to be safe before used. But that counsels regulation, not a ban, which merely slows development of safe procedures."[282]

---

279. Annas & Andrews, *Protecting the Endangered Human: Toward an International Treaty Prohibiting Cloning and Inheritable Alterations*, 28 Am. J.L. & Med. 151–78 (2002).

280. *Id.*

281. Fitzpatrick, *Surplus Embryos, Nonreproductive Cloning, and the Intend/Foresee Distinction*, Hast. Ctr. Rep., May-June 2003, at 29–36.

282. Wolf, *Ban Cloning? Why NBAC Is Wrong*, Hast. Ctr. Rep., Sept.-Oct. 1997, at 12–15. "Instead of developing a legal response to cloning that addresses the core problems of private research and under-regulated reproductive technologies, NBAC simply called for a ban of cloning itself. That skirts the central problems, while adding new ones." *Id.*

A professor at University of California–Los Angeles has written, "If cloning is to occur, the central problem is to ensure that it be done only for two-parent families who want a child for their own benefit." Bioethicist John Robertson defines "procreative liberty" as "the freedom to reproduce or not to reproduce in the genetic sense."[283] "There is a right to have children who are genetically linked to us."[284] For a sterile man who is childless or whose son has died, "cloning would provide the only means of producing a biologically related son." Similarly, bioethicist Arthur Caplan has refuted the concept that cloning is unethical, stating, "There's nothing harmful about being a clone—it's not a public health concern. It will never be a mainstream way to make people, I'll still bet on sex."[285] A section of the AMA's *Code of Medical Ethics* (2003) states, "The medical profession should not undertake human cloning at this time, and pursue alternative approaches that raise fewer ethical concerns," and it adds that "physicians should help establish international guidelines governing cloning."[286] (This section was cited in an article concluding that cloning may benefit infertile couples and single people who want to have children without involving a third party in their procreational activities.)[287]

In the book *Second Creation* (2000), one of the creators of the cloned sheep Dolly argued "that he can envision no circumstances that could not be aided simply by adapting present techniques. . . . There is no circumstance in which a couple would absolutely need to produce a child 'by non-sexual means.'"[288] The Center for Bioethics and Human Dignity and Georgetown's

283. John A. Robertson, *Liberalism and the Limits of Procreative Liberty: A Response to My Critics*, 52 WASH. & LEE L. REV. 233, 236 (1995).

284. JOHN A. ROBERTSON, CHILDREN OF CHOICE (1994); *see also* John A. Robertson, *Genetic Selection of Offspring Characteristics*, 76 B.U. L. REV. 421, 438 (1996); John A. Robertson, *The Question of Human Cloning*, Hast. Ctr. Rep., Mar.-Apr. 1994, at 6–14.

285. UCI NEWS, May 26, 1999.

286. AMA CODE OF MEDICAL ETHICS § 2.147 (2003).

287. Orentlicher, *Cloning and the Preservation of Family Integrity*, 59 LA. L. REV. 1019, 1022 (1999) (referencing Opinion 2.147).

288. Ian Wilmut's statements regarding his colleague Keith Campbell included: "Of course, a child produced with the help of an egg donor is genetically related to only one of the prospective parents (unless the donor is related to the other parent)." WILMUT ET AL., *supra* note 138, at 288.

Center for Clinical Bioethics offer such material—not to hold back science but to encourage it to proceed in an ethical fashion.[289]

The National Advisory Board on Ethics in Reproduction, an independent privately funded group, has approved embryo twinning if done without freezing and if a maximum of four embryos were generated from the initial embryo.[290] As noted earlier, most bioethicists would not ban human cloning forever because the parents of the clonal child would love and care for their child as they would any other, and no evidence convincingly indicates that the clonal child will suffer more.[291] For such reasons, the various ethical concerns on the scientific developments do not materially differentiate future cloning from other modes of reproduction now in use.[292] In 2004 several bioethicists concluded that "over the past few years strong divergence among

---

289. Any serious consideration of the issues raised by research cloning must engage all basic ethical arguments against such a prospect, rather than merely dismiss them or regard them as secondary in importance.

N. Eng. J. Med. 1619 (2002).

290. J.L. Med. & Ethics 275 (1997).

291. In testimony before the National Bioethics Advisory Commission, Cloning Beings, on March 13–14, 1997, a bioethicist stated:

> The cloned individual would not otherwise have come into existence. He or she would have the presumed benefit to enjoy life and would, of course deserve all the legal protections of any human being brought into the world by any means, natural or with the aid of assisted reproductive technology.

Lawrence Wu, *Family Planning Through Human Cloning: Is There a Fundamental Right?*, 98 Colum. L. Rev. 1461, 1462, 1515 n.8 (1998).

292. As Lee N. Silver, a Princeton University biologist and geneticist, stated, "Americans will not be hindered by ethical uncertainty, state-specific injunctions or high costs in their drive to gain access" to new reproductive technologies. In the 21st century some scientists have indicated that "it might take as long as a century to prove that reproductive cloning is safe for humans. Certification [of its safety] 'might take 80 to 100 years.'" Hast. Ctr. Rep., May-June 2001; *see Issues Raised by Human Cloning Research: Hearing Before the Subcomm. on Oversight and Investigations of the H. Energy and Commerce Comm.*, 107th Cong. (2001).

countries and international organizations regarding policy approaches to human cloning and embryonic stem cell research has undermined efforts to develop any international regulatory framework." They noted the emergence of new actors and leaders from Latin America, Asia, and Africa in the quest for scientific and biotechnological developments.

Perhaps what is needed is to stop trying to find a utopian consensus and focus instead on fostering open, transparent, democratic, and meaningful public debates. In that way, coherent public policy could be adopted that reflects societal interests and concerns and balances public and scientific views.[293]

Nevertheless, bioethical individuals will have to wait several years of scientific development prior to becoming able to make ethical decisions on human clonal reproduction, as distinguished from its use for other purposes.

good v. Good

No medical advance is worth research on cloned embryos because it constitutes a direct affront to the sanctity of life. The door should be closed to all cloning, including reproductive cloning.

v.

In his *On Liberty*, John Stuart Mill wrote: "Mankind are greater gainers by suffering each other to live as seems good to themselves, than by compelling each to live as seems good to the rest." The constitutional separation of church and state activity declines the ability of the latter to permanently condemn all cloning. As proven by identical twins, they are autonomous individuals and are without identical brains.

Twenty-first Century Approach

As cloning humans becomes more feasible in the twenty-first century, it will be ethically accommodated in those instances where other methods of procreating are not possible for those therapeutic clones needed for medical reasons. Accordingly, an authorization should continue for all states that conduct research using nuclear transfer to obtain clones needed for medical reasons.

293. Isasi et al., *Legal and Ethical Approaches to Stem Cell and Cloning Research: A Comparative Analysis of Policies in Latin America, Asia, and Africa*, J.L., Med. & Ethics 636–37 (2004).

Week after week the importuning of women made me more and more distraught. I could not ignore their pleas for help. Despite the case pending against me, I reopened my clinic.

Ruth Barnett, abortionist at the Stewart Clinic in Oregon, who was sentenced to a county jail in 1952 and to state prison in 1968 (just five years prior to *Roe v. Wade*).

9

The Choice of Abortion

1. ***Legality, Privacy, and Substantive Due Process***
2. ***Restrictions on Public Funding***
3. ***Religion and Abortion***
4. ***Abortion Pills and Other Devices***
5. ***Unwanted Pregnancies, Unmarried Mothers, and Abortion Rates***
6. ***Spousal and Parental Consent for Abortions***
7. ***Harassing and Bombing Abortion Clinics***
8. ***Formerly Mandated C-Sections to "Save" a Viable Fetus***
9. ***Genetic Testing and Prenatal Counseling; Sex and Dwarf Selection***
10. ***Partial Birth and Post-viability Abortions***
11. ***Shortage of Physicians for Abortions***

This woman's life and work make a very strong case that it was not the practitioner who created the danger for women *before Roe v. Wade*, it was the law. *(Emphasis added.)*

Rickie Solinger, Barnett's biographer, in *The Abortionist* (1994).

"In my family, the worst thing that could be said about anybody was that he was an abortionist." (p. xi)

He is his own best friend, and takes delight in privacy, whereas the man of no virtue or ability is his own worst enemy and is afraid of solitude.

Aristotle

1. Legality, Privacy, and Substantive Due Process

It has been noted that "elective abortions were not unknown" even to colonial women. Not until the 1830s did the states begin to pass laws banning abortions as contrary to a "right to life." During the last third of the twentieth century, a pregnant woman having a single status asked her doctor to perform an abortion for her. He refused because Texas, the state where they lived, then had a law that made it a crime for him to do so. She obtained an attorney who filed a suit against Wade, the district attorney of Dallas County, and named her Jane Roe to protect her identity. Her lawyer argued in *Roe v. Wade* that the Texas law violated her right to an abortion under the Due Process Clause of the Fourteenth Amendment, which prohibited states from depriving her of that liberty. On January 22,1973, the Supreme Court with a 7-2 majority ruled that the Texas law was unconstitutional because of "the Fourteenth Amendment's concept of personal liberty," and is "broad enough to encompass a woman's decision whether or not to terminate her pregnancy."[2] The court, through Justice Harry Blackmun, looked at the three trimesters of a nine-month pregnancy; it denied the right to intervene with a woman's right to an abortion during the first two without "a compelling state interest" and medical evidence showing a need to protect her health and life.[3] However, a state could regulate

1. Zoila Acevedo, *Abortion in Early America*, 4 WOMEN & HEALTH 160–61 (1979).

2. Roe v. Wade, 410 U.S. 113 (1973). When President Lyndon B. Johnson died, Jane Roe's real name was Norma McCorvey. She got pregnant for the third time in 1969, and two previous children were given up for adoption. During the legal process, she gave birth to her third child. This decision constituted a change from the religious and historical culture under which no regard was given to her choice. Rebecca J. Cook & Bernard M. Dickens, *The Injustice of Unsafe Motherhood*, 2 DEVELOPING WORLD BIOETHICS 64, 73 (2002); *see also* Rebecca J. Cook et al., *International Developments in Abortion Law from 1988 to 1998*, 89 AM. J. PUB. HEALTH 579 (1999). Abortion had been legalized in New York in 1970. ORANGE COUNTY REGISTER, Apr. 29, 2005, at Com. 6.

3. Justice Lewis F. Powell, Jr., had a rather personal historical need:

or prohibit abortion during the third trimester. This was largely justified on the grounds that a partial-birth abortion usually takes place at or near the sixth month of a pregnancy. Abortion had actually been legalized in New York state in 1970, three years before it was legalized nationally by the Supreme Court. It should also be noted that "[a]fter natural sexual intercourse, an estimated 60 to 80 percent of all embryos generated through the union of egg and sperm spontaneously abort—many without our knowledge."[4] As of 2005, more than 25 million American women have had abortions since the Supreme Court decided *Roe v. Wade* in 1973.[5]

The legality of abortion at the close of the twentieth century had also been firmly settled by the Supreme Court's decision in *Planned Parenthood v. Casey* on June 29, 1992.[6] This opinion reaffirmed the "central holding" of *Roe v. Wade* (19 years before) but rejected some of its trimester approach; it

---

> When he was a lawyer in Richmond, Va., a young man came to him in despair. His pregnant girlfriend had tried to abort her fetus with his help, and she had bled to death. Powell went to the authorities to explain what happened. Thereafter, he was determined to see abortion made safe and legal.

L.A. TIMES, Sept. 14, 2005, at A16.

4. "So if we use IVF to create embryos and them implant only a select few, aren't we doing what nature does? We have simply replaced nature's techniques with modern scientific techniques for selecting the strongest embryos." MICHAEL GAZZANIGA, THE ETHICAL BRAIN 13–14 (2005). "Clearly, I believe that a fertilized egg, a clump of cells with no brain, is hardly deserving of the same moral status we confer on the newborn child or the functioning adult. Mere possession of the genetic material for a future human being does not make a human being." *Id.* at 17–18.

5. ORANGE COUNTY REGISTER, Sept. 18, 2005, at News 35. "More than half of all women who have abortions say they used a contraceptive method in the month they conceived, according to the Alan Guttmacher Institute, a research group that supports abortion rights." *Id.*

6. Planned Parenthood v. Casey, 505 U.S. 833 (1992). The majority opinion was written by Sandra Day O'Connor. Justice Ginsburg, who was appointed to the court after *Casey*, stated that the court should have relied upon the Equal Protection Clause rather than creating substantive law in this manner. In 1997 a political "scientist" wrote a book advocating self-defense as the reason, stating: "Even in a medically normal pregnancy, the fetus massively intrudes on a woman's body and expropriates her liberty. If the woman does not consent to this transformation and use of deadly force to stop it."

substituted the rule that state legislation thereon would be valid if it did not place an "undue burden," which was described as "a substantial obstacle in the path of a woman seeking an abortion of a nonviable fetus." Both *Roe* and *Casey* decisions relied upon the Court's previous insertion of "privacy" into the Constitution in the 1960s as a "penumbra"—a secondary shadow seen during solar eclipses. The "main shadow" (the Constitution itself) contains the term "Due Process" that most people equate with procedure; however, the Court of the 1960s allowed it to include an alleged substantive law that is not written into the Constitution called *Substantive Due Process*. Some courts have suggested a different approach to avoid this methodology; namely, that an abortion could be a person's right under the Equal Protection Clause of the Fourteenth Amendment.[7] The Supreme Court held that "the statute, viewed as a prohibition on contraception per se, violates the rights of single persons under the Equal Protection Clause of the Fourteenth Amendment."[8] A writer stated: A major justification in Casey (for upholding the central holding of *Roe v. Wade* after 19 years) was the principle of *stare decisis* or to "stand on decisions," which has been the law for some time. This principle goes back more than half a millennium under British law. As a legal historian put it, "Pragmatic analysis was replacing a priori reasoning."[9] The principle is unique to the common law in the judiciaries of the United States, England and certain English-speaking countries in the world. It does not apply to the civil law countries in Latin America, continental Europe or in most Asian coun-

7. *See* Eisenstadt v. Baird, 405 U.S. 438 (1972) (concerning the constitutionality of a Massachusetts law prohibiting the giving away of "any drug, medicine, instrument or article whatever for the prevention of conception").

8. A writer stated that it represented a good example of the confusion of due process with equal protection: If A and B both have a right to X and A is allowed X but B is not, B's primary complaint is a due process one, and his inequality vis-à-vis A is not his primary complaint. . . . The litmus test for distinguishing due process cases from equal protection cases is whether the complainants would be satisfied with the equality of lowering the comparison group to their level.

9. ALFRED KNIGHT, THE LIFE OF THE LAW 33 (1996). England's Chief Justice Priscot was quoted as stating to an attorney who wished to overturn a common-law precedent in 1454: "If we were to rule as you suggest, it would assuredly be a bad example to the young apprentices, for they would never have any confidence in their books if now we were to adjudge the contrary of what has been so often adjudged in the books." *Id.* at 36. Nevertheless, the U.S. Supreme Court has overturned its own precedent on rare occasions.

tries. As a result of this principle, the U.S. Supreme Court held steadfast to these decisions and in 1996 rejected Utah's law attempting to outlaw abortions after 20 weeks of pregnancy as an unconstitutional "undue burden" on pregnant women.[10] However, in 2001, South Carolina's Legislature enacted a measure offering the "Choose Life" license plate upon the payment of $4,000. Proceeds from the sale would go to health organizations not offering abortion services. The statute was declared unconstitutional in 2004.[11]

Abortion at this moment is the only way to limit certain major hereditary diseases, which are not eliminated through spontaneous abortions.[12] It may become important for pregnant women to have prenatal testing so that they may terminate the pregnancy when their fetus has cystic fibrosis (CF) and they are not just a carrier for CF.[13] Many states have laws specifically authorizing abortions for pregnant inmates of prisons and juvenile facilities.[14] In

10. *See* Leavitt v. Jane LORANGE COUNTY REG.., 518 U.S. 137 (1996).

11. Planned Parenthood of S.C. v. Rose, 373 F.3d 580 (Ct. App. 2004).

12. Women over 40 experience a fivefold increase in spontaneous abortion compared with women aged thirty-one to thirty-five. This increase in spontaneous abortion rates from 3.8% to 30% could be caused by problems with the oocytes, the embryo, or the uterus, or by other systemic factors. Smith & Buyalos, *The Profound Impact of Patient Age on Pregnancy Outcome After Early Detection of Fetal Cardiac Activity*, 65 FERTILITY & STERILITY 35–40 (1996).

13. Robinson, Hast. Ctr. Rep., Mar.-Apr. 2002, at 45.

14. *E.g.*, the California Code states:

> No condition or restriction upon the obtaining of an abortion by a female detained in any local juvenile facility pursuant to the Therapeutic Abortion Act. Females found to be pregnant and desiring abortions, shall be permitted to determine their eligibility for an abortion pursuant to law, and if determined to be eligible, shall be permitted to obtain an abortion.

CAL. WELF. & INST. CODE § 220 (West 2000).

> Some 28 states recognize a fetus as an independent murder victim. On April 1, 2004, President George W. Bush signed into law the Unborn Victims of Violence Act, which states that any "child in utero" is considered to be a legal victim if injured or killed during the commission of a federal crime of violence.

ORANGE COUNTY REGISTER, Nov. 13, 2004, at News 3. This was done in spite of the Supreme Court's ruling that the fetus is not a "person" under the Constitution (i.e., not an "unborn" under the Fourteenth Amendment). Roe v. Wade, 410 U.S. at 158.

the third millennium, there is a continued need to replace "back alley" abortions, which prevailed throughout the first three quarters of the twentieth century, an unfortunate part of the country's past.[15] By narrow vote in 1999, the Senate passed a resolution supporting *Roe v. Wade*.[16] In 1998 it was reported that Ohio Judge Patricia A. Cleary actually sentenced a woman to a term of jail in excess of state guidelines in order to confine the woman long enough so that she could not exercise her right to seek a legal abortion.[17] However, in 1999 a federal judge struck down South Carolina's 1996 abortion-clinic regulations as unconstitutional, stating that the state had "loaded these abortion clinics down with so many unnecessary requirements that this court has no choice but to conclude that the regulation unduly burdens a woman's fundamental right to undergo an abortion."[18]

Legality alone does not guarantee the complete competence of every physician performing an abortion or other operation or treatment. In New York, a doctor whose botched abortion also caused a patient to bleed to death

15. In some states, like Illinois, old abortion laws exempted "any person who procures or attempts to produce the miscarriage of any pregnant woman for bona fide medical or surgical purposes." 1867 ILL. LAWS 89. Physicians could only perform therapeutic abortions to save the life of the pregnant woman. This law is the same as that in Catholic Portugal.

16. It read, "It is the sense of the Congress that *Roe v. Wade* was an appropriate decision and secures an important constitutional right and such a decision should not be overturned." INT'L HERALD TRIB., Oct. 22, 1999, at 3. However, some judges and physicians have their own personal social agenda. A heart transplant candidate who was refused an abortion in Louisiana had to travel to Texas to have the procedure performed. ORANGE COUNTY REGISTER, Oct. 20, 1998.

17. Jane M.E. Peterson, *The Politicization and the Practice of Law*, FED. LAW., Dec. 15, 1998, at 5.

18. ORANGE COUNTY REGISTER, Feb. 6, 1999, at News 8; *see* Greenville Women's Clinic v. Bryant, 66 F. Supp. 2d 691, 695 (D.S.C. 1999). In March 2004, the U.S. Court of Appeals for the Seventh Circuit blocked the government from obtaining abortion records from a Chicago Hospital, stating: "If Northwestern Memorial Hospital cannot shield its abortion patients' records from disclosure in judicial proceedings, moreover, the hospital will lose the confidence of its patients, and persons with sensitive medical conditions may be inclined to turn elsewhere for treatment." http://www.cnn.com/2004/LAW/03/27/abortion.ruling.ap/index/html.

was convicted of murder—even though most such cases are settled through malpractice suits or disciplinary proceedings.[19]

There have been all kinds of stories concerning the health risks of abortions versus deliveries at term. Most of such risks have been shown to be of little merit. For example, it had been hypothesized that an interrupted pregnancy might increase a woman's risk of breast cancer. However, a detailed 1997 study of thousands of cases concluded that induced abortions have no overall effect on the risk of breast cancer.[20] Nevertheless, a federal appellate court upheld a public transportation agency's widespread display of a religious group's advertisement asserting that "Women Who Choose Abortion Suffer More & Deadlier Breast Cancer" as a First Amendment right to free speech.[21] In 2002, the National Cancer Institute stated that the evidence of association between abortion and breast cancer is inconclusive.[22] However, an important study was released in 2004 of 83,000 women worldwide showing that abortions do not increase a woman's risk of developing breast cancer later in life.[23]

Medical and ethical statutes in some states have prohibited tort claims based on the allegation that but for the defendant's negligence, the woman would have obtained an abortion.[24] These anti-abortion statutes are not to help care for the resulting children with genetic diseases or, in the case of a

19. Prosecutors charged that Benjamin left a three-inch rip in Negron's uterus and left her to die on an operating table covered in blood and vomit. Queens District Attorney Richard Brown said he knew of only one other murder conviction in 1989; namely, Dr. Milos Klvana was convicted of murder in Los Angeles in the stillbirth of one infant and the death of eight newborns. ORANGE COUNTY REGISTER, Aug. 9, 1995, at News 6.

20. "We identified 370,715 induced abortions among 280,965 women (2.7 million person-years) and 10,246 women with breast cancer. After adjustment for known risk factors, induced abortion was not associated with an increased risk of breast cancer." NEW ENG. J. MED. 81 (1997).

21. Christ's Bride Ministries Inc. v. Se. Pa. Transp. Auth., 148 F.3d 242 (3d Cir. 1998).

22. ORANGE COUNTY REGISTER, Dec. 27, 2002, at News 25.

23. The report of this study appeared in the journal *The Lancet* on March 6, 2004. Its authors, from the University of Oxford in England, indicated that it was the largest one to examine the issue. L.A. TIMES, Mar. 26, 2004, at 14.

24. MINN. STAT. ANN. § 145.424 (West 2000); S.D. CODIFIED LAWS § 21-55-2 (2000); UTAH CODE ANN. § 78-11-24 (2000).

botched tubal ligation, for the parents who would not have conceived the child if only the physician had not been negligent. But for these statutes, parents could prove negligence and causation to recover economic damages for the wrongful birth.[25]

Senator Dale Bumpers of Arkansas stated, "Think for a minute about the number one problem in the whole world: population. . . . It is an absolute unmitigated disaster in the making. But you never hear it talked about on the floor of the Senate, because somebody's afraid it's going to degenerate into a question of abortion." At this time in the third millennium, we should officially recognize that with respect to the unwanted children, abortion constitutes an important possible single step toward stopping the uncontrolled increase in population on our "shrinking planet" (see Chapter 7). In the final year of the twentieth century, "about 19 million over-the-counter test kits were sold in the United States."[26] Without such steps (and other steps), there exists a threat not only to our way of life in the twenty-first century but also to the lives of our descendants.[27] In mid-1999, the world population passed the 6 billion mark, according to the UN Population Fund (UNFPA). This rapid expansion "is a major factor in the rise of human diseases."[28] China, the country with the largest population, has taken affirmative action to control this problem:

> In China there was a rapid increase in population in the '50s and '60s resulting from lack of abortions. Finally, in 1980 the government announced efforts to achieve a goal of one child for each couple. Within two years a national goal established that 80 percent of all couples in the cities and 50 percent in the rural areas should have only one child (excluding, of course, those who already had more),

25. JAMA 1759 (2001).

26. Viccaro v. Milunsky, 551 N.E.2d 8 (Mass. 1990); Schroeder v. Perkel, 432 A.2d 834 (N.J. 1981); Becker v. Schwartz, 386 N.E.2d 807 (N.Y. 1978); Speck v. Finegold, 408 A.2d 496 (Pa. Super. 1979), *aff'd in part and rev'd in part*, 439 A.2d 110 (Pa. 1981).

27. Lester R. Brown et al., Saving the Planet 181 (1991).

28. "Without international cooperative efforts, disease prevalence will continue its rapid rise throughout the world and will diminish the quality of life for all humans." Pimentel et al., BioScience 824 (1998).

> using pledges, rewards and penalties; e.g., lack of subsidies, medical care, and ineligibility for three years for job promotions, wage increases, production awards, and participation in worker competitions. Multiple births are an exception to these rules. The rules apply to the "Han people"—95 percent of the Chinese.[29]

In 2004 the Bush administration indicated "the withholding of $34 million in dues from the United Nations Population Fund program for the third consecutive year because of its belief that the fund indirectly supports Chinese programs that force women to have abortions. However, Fund officials said a U.S. fact-finding team sent to China two years ago 'found no evidence that the fund had supported or taken part in management of Chinese programs of coercive abortion or involuntary sterilization.'"[30]

At the dawn of the third millennium, most Americans did not seem to understand the notion of "natural limits to growth" or the dire need to build an environmentally sustainable economy.[31] As stated by Jonathan Schell, "A society that systematically fails to take any steps to save itself cannot be called psychologically well."[32] This could be accomplished by amendment of the Universal Declaration of Human Rights to include abortion.[33] However, this is unlikely to occur early in the twenty-first century. For example, a review of the book *Life, Liberty, and the Defense of Dignity: The Challenge for Bioethics,* by Leon R. Kass (who became chair of President Bush's Council on Bioethics in 2003), stated:

> It seems hastily constructed and rushed to market. Possibly because abortion has been an intractable political issue for more than three decades, Kass does not devote a chapter to it. Instead, he expresses distaste for abortion rights in his descriptions and his asides. His central ethical tenet seems to be a product of his disgust with abortion

29. *See* BANTON ET AL., LAW IN RADICALLY DIFFERENT COUNTRIES 915–39 (1993).

30. L.A. TIMES, July 17, 2004, at A7. The anti-abortion group Concerned Women for America praised President Bush "for reaffirming his stand against coercive abortion." *Id.*

31. BROWN ET AL., *supra* note 27.

32. JONATHAN SCHELL, THE FATE OF THE EARTH 8 (2000).

33. *Universal Declaration of Human Rights*, 280 JAMA 469–70 (1998).

(which he believes is reasonable to consider "cruelty at best, and possibly murder)."[34]

It should be noted 32 states and the federal government recognize "fetal homicide" as applicable to an independent murder victim under a law that applies only to third parties. That is, a woman can choose to have an abortion, but if someone else causes the death of a fetus, it might result in an alleged murder victim as if it were a human being, according to a Pro-Life Council and those particular statutes.[35] In other states, the assault terminating the life of a fetus would be a crime other than murder of any degree.

In 2005 a study determined that prior to 29 weeks of gestation, a fetus has not developed enough to feel pain. In spite of this, in *Roe v. Wade,* the U.S. Supreme Court had authorized states to legislate abortion restrictions during a woman's second trimester.[36] Further (as discussed in Chapter 10), between 20- and 26-week deliveries, an overwhelming number of fetuses either fail to survive or suffer major defects. "By the thirty-second week the fetal brain is in control of breathing and body temperature."[37]

A change in the Supreme Court's methodology occurred in the late 1990s. The decision in *Roe* in the early 1970s held the Texas abortion statute uncon-

---

34. JAMA 2869 (2003).

35. National Conference of State Legislatures, Fetal Homicide, http://www.ncsl.org/programs/health/fethom.htm (last visited 2004). In California, that law applies to all fetuses beyond eight weeks' gestation. In April 2004, the California Supreme Court ruled (6–1) that the killing of a fetus constitutes second-degree murder, even if the accused had no idea the expectant mother was pregnant. People v. Taylor, No. 04 C.D.O.S. The Unborn Victims of Violence Act (H.R. 1997) was signed into law.

36. > Arkansas, Minnesota, and Georgia require physicians to tell women that 20-week-old fetuses can feel pain during the procedure unless they are anesthetized. A newly released review of the scientific evidence, however, suggests the premise of those laws is wrong [because] . . . until the third trimester, "the wiring at the point where you feel pain, such as the skin, doesn't reach the emotional part where you feel pain, in the brain."

DISCOVER, Dec. 2003, at 12; *see also* ORANGE COUNTY REGISTER, Aug. 24, 2005, at News 6 (citing the August 2005 issue of *JAMA* and noting that "in the United States, 1.4 percent of abortions are performed at or after 21 weeks of gestational age").

37. GAZZANIGA, *supra* note 4, at 6.

stitutional because it violated an aspect of the liberty protected by the Due Process Clause: As noted above, it was a demonstration of "substantive due process" asserting the right to privacy as being "broad enough" to encompass the concept that the "unborn have never been recognized in the law as persons in the whole sense." Towards the end of the twentieth century, the Supreme Court has generally refused to continue ruling on the basis of "substantive due process"; it would rather let state legislatures decide such matters. For example, this latter approach was taken on the decision to find no constitutional right to physician-assisted suicide (PAS) in 1997. Many people wrongly thought that this ended the matter; however, merely four months later, the Court let stand an Oregon statute specifically authorizing PAS for its residents[38] (see Chapter 14).

good v. Good

Abortions result in the unethical killings of innocent young "persons" and hence should be made illegal.

v.

The right to have an abortion is an expression of a woman's ethical autonomy over her own body and a young "fetus" is not initially regarded as a "person" by the courts. A pregnant woman is authorized to both diminish the number of unwanted children and to aid in fulfilling the current need to avoid depletion of the planet's limited resources by not helping to create too many humans.

Twenty-first Century Approach

1. Except with respect to constitutional matters in the First Amendment or elsewhere, courts should continue their current and traditional role with respect to procedural due process and leave most substantive bioethical matters to the legislatures of the states or initiatives voted upon by the people.

---

38. Washington v. Glucksberg, 521 U.S. 702 (1997). On the other hand, the European Court of Human Rights held that it could not rule on a case filed by a French woman who was forced to have an abortion after a doctor's mistake, stating that when the right to life begins was "a question to be decided at the national level . . . because the issue had not been decided within the majority of states." At the European level, there was "no consensus on the nature and status of the embryo and/or fetus." Orange County Register, July 9, 2004, at News 19.

2. State legislatures should repeal statutes prohibiting claims of parents alleging that but for the physician's negligence, a pregnant woman would have had an intended abortion.
3. Legislatures should also recognize that abortions for unwanted infants constitute steps toward ending the uncontrolled increase in the population and depletion of the resources on our "ever-shrinking planet."
4. International agreements should be sought to authorize abortions for unwanted children. For example, Art. 7 of the Universal Declaration of Human Rights declares "equal protection of the law" could be amplified to specifically include equal rights to abortions.
5. Due to (a) the large failure of survival and deliveries of severely defective fetuses between 20 and 26 weeks of gestation, (b) the fact that they begin to feel pain at 29 weeks, and (c) the brain controls of breathing and body temperature by the 32nd week, all states should authorize abortion at least to the 28th week of gestation.

---

2. Restrictions on Public Funding of Abortions

Over a million teenage girls have been getting pregnant in the United States annually, and about one-third of those pregnancies end in abortion. The staggering costs associated with teen pregnancy "range from $7 billion in actual costs . . . to $29 billion in combined social costs attributed to childbearing, caring for children and their adolescent teen parents."[39]

The use of federal funds for abortions has been barred (since 1977) by the "Hyde Amendment," except for those necessary to save the life of the pregnant female and (since 1994) those resulting from rape or incest.[40] The regulations issued under that statute by the Department of Health and Human Services (HHS), which concern the allocation of funds to projects providing counseling without referral for abortion, became known as the "gag

---

39. S. CAL. CHRISTIAN TIMES, May 2002, at 24. "According to the National Campaign to Prevent Teen Pregnancy's 'National Campaign Key Statistics,' if every birth to a teen mother in 1990 had been delayed until the mother was in her 20s, the federal government would have saved about $10 billion." *Id.*

40. *See also* Elizabeth Blackwell Health Ctr. for Women v. Knoll, 61 F.3d 170 (3d Cir. 1995).

rule."[41] However, the denial of such funding was not held to interfere with a woman's right to abortion.[42] However, in 1996 Congress also passed legislation known as the Coats Amendment, which prevents the federal, state or local government that receives federal financial assistance from discriminating against healthcare entities that refuse to perform, undergo training, or provide referrals for training in abortions.[43] In the international arena, however, President Clinton vetoed legislation to provide $1 billion in delinquent dues to the UN because it contained an unrelated anti-abortion restriction. The Hyde Amendment has been upheld, as has the refusal of state governments to subsidize abortions.[44] Some states attempted to bar the 1994 exceptions in the federal law. However, the courts have stopped 11 states from restricting public funding of abortion. In 1995, a federal appeals court struck down laws in Arkansas and Nebraska that sought to bar public funding for abortion in cases of rape or incest.[45] The Supreme Court overturned an addition to Colorado's state constitution (that paralleled the Hyde Amendment) barring use of state money for abortions except those to save poor women's lives. Colorado must pay for rape or incest abortions as part of its participation in Medicaid.[46]

In 1999, a unanimous decision by the Supreme Court of New Mexico held unconstitutional a state regulation that barred use of Medicaid funds for

---

41. *See* Memoranda of the President, 58 Fed. Reg. 7455 (1992). President Clinton recognized that the gag rule "endangered women's lives and health."

42. Rust v. Sullivan, 500 U.S. 173 (1991).

43. 42 U.S.C. § 238n (2003).

44. Harris v. McRay, 448 U.S. 297 (1980); *see also* Planned Parenthood of Se. Pa. v. Casey, 505 U.S. 833, 919 (1992).

> Nevertheless, in 2002 the 9th U.S. Circuit Court of Appeals denied an appeal by Federal government lawyers to stop the Navy from paying for an abortion for a woman carrying a fetus missing most of its brain—anencephaly—a severe defect in which a fetus lacks forebrain, cerebellum, and cranium.

Orange County Register, Aug. 18, 2002, at News 3d.

45. Little Rock Family Planning Servs. v. Dalton, 860 F. Supp. 609 (E.D. Ark. 1994); *see also* Planned Parenthood Aff. of Mich. v. Engler, 73 F.3d 634 (6th Cir. 1996).

46. Orange County Register, Dec. 5, 1995, at News 22 (citing Hern v. Beye, 57 F.3d 906 (10th Cir. 1995)).

abortion except when necessary to save a mother's life, to end ectopic pregnancy, or in cases of rape or incest.[47] The regulation violated the New Mexico Constitution's Equal Rights Amendment; it was "gender-linked" and "that does not apply the same standard of medical necessity to both men and women, and there is no compelling justification for treating men and women differently with respect to their medical needs in this instance." The National Crime Victimization Survey by the Justice Department's Bureau of Justice Statistics (1995) estimated that there were 500,000 sexual assaults on women annually, including 170,000 rapes and 140,000 attempted rapes.[48] A number of states—California and a dozen others—have broader protection for privacy under their state constitutions than under the federal Constitution. This may provide greater protection for women than does the federal government. In other states, courts have found such broader protections even when right of privacy is not expressed in the constitutions of those states.[49]

In 1987 Congress created the Prenatal Care Assistance Program to afford federal reimbursement to states providing prenatal care for women with household incomes exceeding the Medicaid eligibility standard. A New York prenatal care assistance program (which implemented the federal program but denied the use of the funds for abortions) was held not to violate New York law; it was found to fail to subsidize medically necessary abortions.[50]

---

47. N.M. Right to Choose/NARAL v. Johnson, M.M., 986 P.2d 450 (N.M. 1999).

48. SIGNAL , Aug. 17, 1995, at A2.

> Even Muslims in some countries have decided to allow rape victims to undergo abortions. In Algeria a panel of Muslim clerics has decided to allow women raped by Islamic militants to undergo abortion. . . . The High Islamic Council sent an order to the Solidarity Ministry authorizing "women raped by the terrorists to end their pregnancies."

L.A. TIMES, Apr. 14, 1998, at A8.

49. *E.g.*, Doe v. Maher, 515 A.2d 134, 146–47 (Conn. Super. Ct. 1986); Moe v. Sec'y of Admin. & Fin., 417 N.E.2d 387 (Mass. 1981); Right to Choose v. Byrne, 450 A.2d 925, 931–32 (N.J. 1982); Hope v. Perales, 571 N.Y.S.2d 972, 978 (N.Y. Sup. Ct. 1991); Planned Parenthood Ass'n v. Dep't of Human Res., 663 P.2d 1247, 1256–57 (Or. Ct. App. 1983).

50. Hope v. Perales, 611 N.Y.S.2d 811 (N.Y. 1994); FRED ROSNER, MODERN MEDICINE AND JEWISH LAW 45 (1972).

On December 14, 2004, "The Weldon Amendment" was enacted, which allows any "health care professional" hospital, HMO, health insurance plan, and "any other kind of health care facility, organization, or plan" to refuse to perform an abortion and to refuse to refer the patient to another doctor even in the case of rape or medical emergency. Although a state's violation has the potential to lose all or a substantial portion of federal funds for health services, much of the remainder of this statute appears to conflict with existing law. The National Family Planning and Reproductive Health Association and the California Attorney General filed suit in 2005 seeking a preliminary injunction to stop enforcement on the grounds that it infringes on state sovereignty and it violates women's constitutional right to emergency abortion care.

good v. Good

Since virtually all abortions are unethical, it follows that no public funds should be appropriated for such purposes

v.

The denial of funds for abortions results in discrimination against the poor, who cannot afford them, and increases the number of unwanted children to be cared for at public expense.

Twenty-first Century Approach

1. The federal government should consider the repeal of the Hyde and Weldon Amendments restricting use of federal funds for abortions, particularly those affecting military personnel and government employees who are stationed abroad.
2. State legislatures should appropriate sufficient funds for abortions for certain women who cannot afford to pay for them.

A union of government and religion tends to destroy government and to degrade religion.

Justice Hugo Black, *Engel v. Vitale*, 370 U.S. 421, 431 (1962)

It is general error to suppose the loudest complainers for the public to be the most anxious for their welfare.

Edmund Burke

The political machine triumphs because it is a united minority acting against a divided majority.

Will Durant

3. Religion and Abortion

Although the Bible is silent on abortion, Biblical interpreters have attempted to find interpretations against it by pointing to several specific instructions in the Old Testament to be "fruitful and multiply," beginning with those to Adam and Eve,[51] to Noah and his sons,[52] and Jacob.[53]

The Catholic equating of an abortion to murder (because of their position that a fetus is a "person") dates from modern times.[54] In 1869 Pope Pius IX decreed that a human life begins at conception. As of the end of the twentieth century the Catholic church has only allowed two types of abortive procedures; namely, (a) where a fetus is removed from a pregnant woman along with her cancerous womb, and (b) in the case of ectopic pregnancy, where the fetus is lodged in a fallopian tube which is removed along with the fetus.[55] This should not be confused with the double effect of painkilling drugs that may also cause a patient's demise where judged to be needed to

---

51. *Genesis* 11:28.

52. *Genesis* 9:1, 9:7.

53. *Genesis* 35:11. Today, for many Jews, abortions must be approved by a panel of doctors and rabbis:

> The panels typically approve abortions for unmarried woman [sic], [but] are more reticent to do so for married woman [sic] (who, in the view of the rabbis, should not squander the opportunity to bring more Jews into the world). Telling the panel, however, that the pregnancy is the product of an extramarital affair (better yet, an affair with a gentile) typically secures approval.

Noah J. Efron, Real Jews 228 (2003).

54. In 2002, the Bush administration said that states may classify a developing fetus as an "unborn child" eligible for government healthcare. . . . The plan will make a fetus eligible for healthcare (CHIP) under the state Children's Health Insurance Program. This was criticized as an attempt to establish the fetus as a person with legal standing, which could make it easier to "criminalize abortion."

http://member.compuserve.com/news/story.jsp. Jan. 31, 2002. As noted above, the fetus is not a "person" under the Constitution. *See supra* note 10.

55. Kathy Rudy, Beyond Pro-Life and Pro-Choice 24 (1996).

stop his pain (see Chapter 14, Right to Die). Abortions outside the two Catholic instances are considered illicit. It has been remarked how these restrictions apply to a pregnant woman who suffers from chronic hypertensive heart disease associated with severe renal insufficiency:

> If this woman does not receive an abortion, she will probably die from cardiac or renal dysfunction brought on directly by the pregnancy. However, under the Catholic conditions of double effect, even though this woman may die, an abortion is unacceptable . . . As long as she did not seek an abortion, she can rest assured that she did not sin. In the logic of double effect, it is better to allow one woman to die—or, if you are the pregnant woman, to die yourself—than to kill a fetus. Or as many moral theologians and church officials have articulated, two deaths are better than one murder.[56]

Under the Church-lobbied Amendment to the Public Health and Welfare Act, federally funded individuals and organizations that refuse to provide sterilization or abortion services are protected when they declare those services to be "contrary to [their] religious beliefs or moral convictions."[57] This was enacted shortly after *Roe v. Wade* (1973), discussed above. Similar statutes were enacted in 45 states so that their appropriate healthcare providers could decline to provide abortion services.[58]

Perhaps, at least partly as a consequence of such sharp rules, Catholics are 30 percent more likely than Protestants to have abortions.[59] In spite of this, one church official declared, "If they continue with this idea of practicing abortion, they know the penalty. They are excommunicated."[60] A court in a Catholic

56. *Id.* at 26, 32.

57. 42 U.S.C. § 300a-7.

58. National Abortion and Reproductive Rights Action League, Denial Clauses: Danger for Women's Health (Jan. 11, 2002). Alabama, Vermont, New Hampshire, West Virginia, and Mississippi did not enact such laws.

59. Rudy, *supra* note 55, at 159 (citing Barbara Ferraro & Patricia Hussey, No Turning Back: Two Nuns' Battle with the Vatican over Women's Right to Choose 252 (1999)).

60. Orange County Register, Dec. 14, 1996, at News 33. In Wisconsin, the bishop's decree denied Holy Communion to Roman Catholic lawmakers who support abortion. It stated that they are not to be admitted to Holy Communion, should they present themselves, until such time as they publicly renounce their support of these most unjust practices. Orange County Register, Jan. 10, 2004, at News 24.

country convicted a nurse performing illegal abortions and found one woman guilty of terminating her pregnancy. It had Europe's most restrictive abortion laws, and a woman convicted of having an illegal abortion was given a choice of four months in jail or a $108 fine.[61] Some Jews have experienced the same consequence. One rabbi has written, "By and large the Jewish tradition prohibits abortion"; and noted that "because of the high rate of abortion among Jews," many infertile couples "cannot find Jewish children to adopt."[62]

The Catholic Church sanctions have been applied abroad. A Roman Catholic archbishop in Brazil threatened to excommunicate a mother, doctors, and a judge who authorized the abortion of a fetus that had no brain (i.e., anencephalic); abortion is unacceptable for the church in cases to save a pregnant woman where the fetus itself could not survive.[63] In 1996 the predominantly Catholic Poland enacted legislation whereby women would be able to end pregnancies before the 12th week if they face financial or personal problems, but only after counseling and a three-day waiting period. The Vatican reacted immediately, saying, "Poland has chosen the path of death." Seven months later the Polish High Court overturned the law, stating that it violated the right to life contained in the Polish Constitution. A new law in 1997 again restricted abortions to those cases where giving birth would endanger a

61. ORANGE COUNTY REGISTER, Jan. 19, 2002, at News 3.

62. ELLIOT N. DORFF, MATTERS OF LIFE AND DEATH 109 (1998).

> Judaism does not see all abortion as murder, as Catholicism does, because biblical and rabbinic sources understand the process of gestation developmentally. Thus, the Torah stipulates that if a woman miscarries duc to an assault, the assailant is not held liable for murder but rather must pay only for the lost capital value of the fetus. That early law already indicates that the fetus is not to be viewed as a full-fledged human being but rather as part of one.

Id. at 128.

63. N.Y. TIMES, Oct. 25, 1996, at A6.

> There is no point in the life of a developing human being when you can say, "Yes, at that point it becomes a human being." There is no discrete event. . . . There is no "moment of conception." It is a continuum of biochemical events. The concept that there is a moment of conception is unintelligible. It cannot be made real when you reduce it to the level of molecular biology.

HOOKER, HOW TO THINK ABOUT ETHICS 5 (1992).

woman's health, the fetus is irreparably damaged, or the pregnancy resulted from rape or incest—which is essentially the same as the U.S. restriction on the use of federal funds for abortions.[64] In 1996, when the Pope appeared in France, a nation that is 82 percent Catholic, hundreds of outraged French Catholics demanded that their names be scratched from baptism rolls to protest the Pope's opposition to abortion and birth control.[65]

Under a law in Roman Catholic Ireland, women have the right to travel abroad to terminate a pregnancy. In 1997 a 13-year old rape victim was allowed to travel abroad for an abortion.[66] Until recently Roman Catholic Portugal permitted abortions only for strict medical reasons, such as a risk to the mother's life; however, in 1998 its parliament approved a law to allow abortions on request up to the 10th week of pregnancy.[67] In the Philippines the abortion rate is 16 for every 100 pregnancies. According to the U.S. Agency for International Development, that is higher than the rate in most Asian countries. Nevertheless, "this overwhelmingly Roman Catholic country has doubled its population in the past half-century, to 80 million, and could double again in less time than that."[68] In the United States, it has been ruled that although patient records must be treated as confidential, a regulation authorizing state health inspectors to review and photocopy all abortion clinic records without disclosing them to others does not violate patients' constitu-

64. L.A. TIMES, May 29, 1997, at A8; ORANGE COUNTY REGISTER, Dec. 24, 1997, at News 13.

65. ORANGE COUNTY REGISTER, Sept. 19, 1996, at News 10. "The Vatican, vociferous on the subject of abortion, joined forces with some Islamic countries concerned that the conference would become an international platform to promote the women's rights they opposed." HILLARY RODHAM CLINTON, LIVING HISTORY 299 (2003).

66. ORANGE COUNTY REGISTER, Dec. 2, 1997, at News 3. In 2002, Ireland rejected an amendment to overturn its 1992 Supreme Court ruling legalizing abortion for women whose pregnancies threatened their lives (including suicide). By then approximately 7,000 Irish women traveled to England each year for abortions. L.A. TIMES, Mar. 8, 2002, at A11.

67. ORANGE COUNTY REGISTER, Feb. 5, 1998, at News 19.

68. ORANGE COUNTY REGISTER, Mar. 21, 2003, at News 36.

> It is interesting to note that in the sixth century, Christian Emperor Justinian's code decreed that "those who the portions of the abortionist's, are subject to the full penalty of the law—both civil and ecclesiastical—for murder."

GEORGE GRANT, THIRD TIME AROUND 38 (1991).

tional right of privacy.[69] California had enacted the Women's Contraception Equity Act (WCEA) in 1999 to eliminate gender discrimination in healthcare benefits and required that certain healthcare plans cover prescription contraceptives.[70] In its lawsuit, Catholic Charities argued that the WCEA coerced a violation of religious beliefs. However, the Supreme Court of California held that it does not violate the California Constitution by discriminating against the Catholic Church because of its exemptions which do not impose a burden on any religious organizations that meet the criteria.[71]

In Russia, abortions declined from 4.6 million in 1998 to 1.7 million in 2002 following the collapse of the Soviet Union and the increased availability of contraceptives. Women could receive an abortion up to the 22nd week of their pregnancies by citing one of 13 special circumstances called "social indicators." However, in 2003, Russia restricted unlimited reasons for abortion up to 12 weeks (except for a medical need) and otherwise reduced abortion decisions to rape, imprisonment, the death or severe disability of the husband, or a court ruling stripping a woman of her parental rights. Nevertheless, for every 10 births in Russia, there are still nearly 10 abortions.[72]

---

69. Greenville Women's Clinic v. Comm'r, S.C. Dep't of Health and Envtl. Control, 306 F.3d 141 (4th Cir. 2002). In the year 2000, a federal statute exempted doctors with religious objections from being required to perform abortions. 42 U.S.C. § 300a-7 (2004).

70. Catholic Charities of Sacramento, Inc. v. Super. Court of Sacramento County, 85 P.3d 67 (Cal. 2004).

71. *Id.* As regards a "religious employer" the law states: "(A) The inculcation of religious values is the purpose of the entity[;] (B) The entity primarily employs persons who share the religious tenets of the entity[;] (C) The entity serves primarily persons who share the religious tenets of the entity[; and] (D) The entity is a nonprofit organization . . . ." CAL. HEALTH & SAFETY CODE § 1367.25 (West). Because Catholic Charities met these criteria, it was not compelled to offer coverage for contraceptives. 85 P.3d at 91–92.

72. N.Y. TIMES, Aug. 24, 2003, at 3.

> In the next 20 years, according to Goskomstat, the state statistics agency, the Russian National Security Council and the United Nations Population Division, Russia's population of 144 million could drop by a third. Russian women would have to have almost twice as many children (2.4) as they're having now (1.3) just to keep the population from declining, but *Russia has one of the world's highest abortion rates.* Some surveys suggest that there are more abortions than births.

ORANGE COUNTY REGISTER, Feb. 13, 2005, at News 13. (Emphasis added.)

In the United States, Protestants recognized:

> What pro-lifers regard as the legal promotion of a religious point of view seems like *the unconstitutional imposition of religious views upon government* to their opponents. But the event of these kinds of debates *is a result of the separation of church and state that Protestants helped bring about in early U.S. history.*[73] (Emphasis added.)

---

4. Abortion Pills and Other Devices

At the turn of the third millennium, millions of women do not want another child, either due to a lack of or refusal to use methods of family planning.[74] It has been pointed out that "a 15 percent increase in the use of contraceptives means, on average, about one fewer birth per woman."[75] The new Surgeon General's "Call for Action" pointed out that each year there are 1.4 million abortions annually, but that these are just 20 percent of all pregnancies.[76]

Contraceptive pills have been in legal use in the U.S. for over four decades. The latest research appears to show that there are few risks involved. A 1997 study demonstrated that ten years after going off the pill, women were at no higher risk of breast cancer than those who had not taken the

73. Mark A. Noll, Protestants in America 138 (2000).

74. An estimated 120 million couples in developing countries do not want another child soon but have no access to family-planning methods or have insufficient information on the topic. Consequently, pregnancy too often brings despair instead of joy. In many countries, laws create hurdles. Japanese women were until this past year [1999] forbidden access to the pill and so had to rely heavily on abortion.

Potts, Sci. Am., Jan. 2000, at 90.

75. *Id.* Indeed, the United States has cut its funding for international family-planning programs over the past few years. The Planned Parenthood Federation of America uses the abortion pill, including RU-486, at 222 of its 845 clinics.

76. Orange County Register, June 29, 2001, at News 7. He also stated that Americans account for some 40,000 new HIV infections each year and more than 100,000 are children victimized by sexual abuse annually.

pill.[77] The following year it was concluded that birth control pills appear to cut the chances of ovarian cancer in half among women who inherited a faulty gene that puts them at risk for the disease.[78] Other studies in Sweden indicated that the pill could strengthen women's bones and reduce the risk of hip fractures in older women.[79] A British study examining the pill and heart attacks concluded there is no increased risk associated with taking the oral contraceptive. Nevertheless, some pills or other contraceptive methods have been advertised without adequate warnings of their side effects. Courts are split on the adequacy of warnings such as "see your doctor" for protecting drug companies against liability in such instances.[80]

Unlike pills for contraception, some abortion pills prevent implantation in the uterus soon after conception or even later. The first and most famous of these was the French "abortion pill" designated RU486 (mifepristone), which was introduced in 1988 and then became the subject of many protests by Catholics and others in the United States and abroad. Opponents were particularly upset because this pill can be taken privately, that is, without the necessity of going to a clinic that could be picketed. It is sometimes called "the morning-after pill," because it does not prevent conception; it prevents a fertilized ovum from implanting, a process that occurs eight to 10 days after ovulation.[81] In 1999 the French government decided to make a similar pill,

---

77. The FDA approved the first oral contraceptive in 1960. L.A. TIMES, July 13, 2000, at B6. The 2003 study by England's Oxford University National Cancer Institute found that abortion did not increase the risk of breast cancer. ORANGE COUNTY REGISTER, Mar. 27, 2004, at News 24.

78. NEW ENG. J. MED. (1998); ORANGE COUNTY REGISTER, Aug. 13, 1998.

79. LANCET (1999); ORANGE COUNTY REGISTER, Apr. 3, 1999, at News 12.

80. ORANGE COUNTY REGISTER, Dec. 1, 1999, at News 17. In 1999, the U.S. Court of Appeals for the Fifth Circuit, based in New Orleans, ruled in *In re Norplant Contraceptive Products Liability Litigation*, 165 F.3d 374 (5th Cir. 1999), that the learned intermediary doctrine does not protect the drug maker. However, in *Perez v. Wyeth Laboratories Inc.*, 734 A.2d 1245 (N.J. 1999), New Jersey's high court rejected the doctrine and held that it is the manufacturers' responsibility to let physicians know of their duty as learned intermediaries to warn patients. A.B.A. J., Dec. 1999, at 36.

81. However, among women who conceive as a result of *in vitro* fertilization, the successful implantation of a conceptus may be detected as late as fourteen days after egg retrieval. NEW ENG. J. MED. 17969 (1999). In this connection it is noted that Justice Harry Blackmun's majority opinion in *Roe v. Wade* included the statement: "We need not solve the difficult question of when life begins. When those trained in

NorLevo, available for distribution to teenage girls. It is currently available without prescription. The Deputy Education Minister stated that teenage pregnancies are not just a family matter but also a public health problem that affects 10,000 girls under 18 each year.[82] In June 2000, the Council of State, France's highest administrative court, overruled this decision, the first of its kind, and required a prescription.[83] In 1999 Canada approved the "morning after pill"[84] and Japan, after a 34-year delay, also approved the pill.[85]

The first Bush administration slowed the testing of the pill RU-486 by withholding funds from certain World Health Organization (WHO) programs and forbidding the National Institutes of Health (NIH) from conducting research on abortion inducers.[86] In 1992 the Supreme Court upheld blocking of the importation of that pill.[87] Following a change of administrations, clinical trials began in 1994 when the nonprofit Population Council was given U.S. rights to seek FDA approval of RU-486. After an FDA panel approved the pill in 1996, the Vatican theologian Gin Concetti wrote "It is the pill of Cain: the monster that cynically kills its brothers."[88] It was planned to get the drug on the market in September 2000[89] after a study showed that 9 out of 10 women who took the pill said they would do it again and would recommend it to others.[90]

A study had recently found that this method for medical abortion was "acceptable to women and providers in the United States, including physi-

---

the respective disciplines of medicine, philosophy, and theology are unable to arrive at any consensus, the judiciary, at this point in the development of man's knowledge, is not in a position to speculate." Roe v. Wade, 410 U.S. 113, 159 (1973).

82. Orange County Register, Dec. 1, 1999, at News 17. In 1999, the U.S. Court of Appeals for the Fifth Circuit, based in New Orleans, ruled in *In re Norplant Contraceptive Products Liability Litigation*, 165 F.3d 374 (5th Cir. 1999), that the learned intermediary doctrine does not protect the drug maker.

83. Orange County Register, July 1, 2000, at News 30.

84. Orange County Register, Nov. 4, 1999, at News 24.

85. L.A. Times, June 3, 1999, at 1.

86. Cal. Law., May 1997, at 36.

87. Benten v. Kessler, 505 U.S. 1084 (1992).

88. Orange County Register, July 23, 1996.

89. Orange County Register, Apr. 30, 1998, at News 14.

90. The Sept. 30, 2000 date was reported in L.A. Times, June 18, 2000, at M5.

cians, nurses, and counselors."[91] On September 25, 2000, the FDA gave final approval for use of RU-486. However, the FDA required three meetings with a doctor and the signing of a written agreement by the pregnant woman that she will make these meetings. Such agreements are called "Med Guides."[92] Vivian Dickerson, president of the American College of Obstetricians and Gynecologists, said, "The overwhelming data is that it is safe, effective and usable across age groups." The American Medical Association (AMA) supported over-the-counter sales of morning-after birth control and indicated that the FDA was wrong to reject such sales. In 2001 the Health Minister of Canada noted that emergency contraceptives for women help prevent unwanted pregnancies without a prescription. Taking this pill represents a way for women to obtain non-surgical abortions from their own doctor without going to a clinic; however, some doctors (including gynecologists) are against abortions.[93]

As of 2003, a study of RU-486 estimated that more than 37,000 abortions were performed using the pills in the first six months of 2001. The study's author, Lawrence Finer, wrote, "The growing acceptance of

---

91. JAMA 1034 (1998). The regimen was highly acceptable. Nearly all women (95.7%) would recommend it to others; 91.2% would choose it again. Most providers and women thought that home use of misoprostol should be available for women who prefer it.

92. ORANGE COUNTY REGISTER, Oct. 18, 2000, at News 18. Some people have complained that these rules on RU-486 are too strict. ORANGE COUNTY REGISTER, Oct. 18, 1998, 2000, at News 18. Arkansas, Georgia, Idaho, Nebraska, Rhode Island, South Carolina, Tennessee, Texas, and Virginia have all indicated that the notification of at least one parent would be applied to RU-486. ORANGE COUNTY REGISTER, Oct. 11, 2000, at News 7.

93. ORANGE COUNTY REGISTER, Sept. 25, 2001, at News 14. However, the Women's Health Initiative, the largest women's health study on oral contraceptives, found that women taking the birth control pill had lower risks of heart disease and stroke, and no increased risk of breast cancer, contrary to previous studies. ORANGE COUNTY REGISTER, Oct. 21, 2004, at News 17.

"In the U.S., the fatality risk with mifepristone is slightly less than 1 per 100,000 cases, compared with 0.1 per 100,000 for surgical abortion at 8 weeks or less. Pregnancy itself carries a fatality risk of 11.8 per 100,000." *Abortion Options*, CONSUMER REP., Feb. 2005, at 38.

"Taken within 6 to 72 hours after mifepristone, misoprostol completely expels the pregnancy in more than 90 percent of users, usually within a day." *Id.*

mifepristone raises the possibility that the decrease in surgical abortion providers may be offset by an increase in the number of providers that offer medical abortion."[94]

Other methods were also being developed. A new device for women delivers contraceptives through the skin.[95] In 1995 a study demonstrated that the "combination of methotrexate and misoprostol represents a safe and effective alternative to invasive methods for the termination of early pregnancy."[96] The following year the FDA approved a Planned Parenthood study of the two-drug abortion method.[97]

5. Unwanted Pregnancies, Unmarried Mothers, and Abortion Rates

It has been shown that, on average, girls in households without a father reach menarche (the beginning of menstrual function) earlier than those with the father present; they engage in sex earlier and with more partners.[98] Since the turn of this twenty-first century, one in five students reported being physically and/or sexually abused by a dating partner. They constituted over 1.5 million women of which 25 percent were adolescents. Fur-

---

94. http://member.compuserve.com/news/story/0001%2F20030115%F104058379.htm&sc-150, Jan. 14, 2003. In January 2004, Mexico's Department of Health approved the use of the morning-after contraceptive pill as a family-planning method. "But the Catholic Church expressed outrage, and bishops threatened to excommunicate any woman who knowingly takes what the church calls an 'abortion pill.'" ORANGE COUNTY REGISTER, Jan. 29, 2004, at News 13.

95. The world's first birth control patch for women is a device that will deliver contraceptives through the skin for a week. The adhesive patch, which Johnson & Johnson calls Evra, is the size of a half-dollar and can be worn on an arm, abdomen, or buttocks. Evra is different from Norplant, which consists of capsules that are surgically inserted under the skin in the upper arm and last about five years. It would be the newest type of birth control to hit the U.S. market since the injectable drug Depo-Provera in 1993.

ORANGE COUNTY REGISTER, July 21, 1999, at News 13.

96. 333 NEW ENG. J. MED. 537–40 (1995).

97. ORANGE COUNTY REGISTER, Sept. 12, 1996.

98. JAMA 989 (1996). "Teen pregnancies are down worldwide, but the U.S. has by far the highest rate of any industrialized nation, according to a report from the Alan Guttmacher Institute." ORANGE COUNTY REGISTER, Feb. 3, 1997, at News 17.

ther, it was found that those women "are more likely to exhibit other serious health risk behaviors."[99]

One writer pointed out that people "who were pregnant as teenagers (girls), or caused a pregnancy as teenagers (boys), had a significantly lower average IQ percentile than those who were married as teenagers but not pregnant, or those who were neither married nor pregnant as teenagers."[100]

An analysis of pregnancy, abortion, and birth rates among adolescent girls in 1980, 1985, and 1990 found that despite efforts to reduce such activity in the United States, "pregnancy and birth rates for that group continue to be the highest among developed countries."[101] In 2000, the Centers for Disease Control and Prevention reported, "Each year in the United States, 800,000 to 900,000 adolescents aged 19 or less become pregnant." Further, this has been associated with adverse health and social consequences for young women and their children. In Mexico a 12-year-old girl, made pregnant by her father in 2001, had no abortion available because they are illegal at the location.[102] In 2005 in Florida, a judge ruled that a 13-year-old girl who had sex after running away from a state shelter and was 14 weeks pregnant could terminate her pregnancy.[103] In 1993, California had nearly 49,000 non-marital births to teenagers and about 23,000 other teens had

---

99. Silverman et al., JAMA 572 (2001). Laws require mandatory reporting of intimate partner violence to police. Because 55.7% favored mandatory reporting but 44.3% opposed it, it was recommended that "policymakers" should consider options that include consent of patients before wider implementation.

100. Having an illegitimate child was associated with a lower IQ, a shorter time in school, having children earlier, having more children, and coming from larger families than those who had legitimate children, regardless of when they were married. These observations indicate that having illegitimate children and getting married in the teenage years will select for the genes associated with a lower IQ and other genetically influenced behaviors in this group of individuals.

David E. Comings, The Gene Bomb 153 (1996).

101. The number of pregnancies was estimated as the sum of live births, legally induced abortions, and estimated fetal losses (i.e., spontaneous abortions and stillbirths) among females aged less than 19 years. JAMA 952 (2000).

102. Orange County Register, July 14, 2001, at News 27.

103. L.A. Times, May 4, 2005, at A10. Gov. Jeb Bush said Tuesday that the state would not appeal further.

publicly funded abortions.[104] The abortion rate remained unchanged when the Centers for Disease Control (CDC) reported 1,221,585 of them in 1998,[105] but increased to 1.31 million in 2000.[106] This was because almost half of all pregnancies in the United States are unintended ones.[107] Further, 31 percent of babies in the United States are born to unmarried mothers, and 66 percent of that group are born to single African-American women.[108]

The Centers for Disease Control and Prevention (CDC), which has been tracking abortions since 1972, reported that in 1993 there were 334 abortions for every 1000 live births,[109] and those numbers increased toward the end of the millennium. A survey of all known abortion providers in the United States revealed a total of 1,366,000 abortions in 1996.[110] Most important was the finding that 95 percent of these pregnancies were unintended and unwanted;[111] of the live births, 80 percent were to single women.[112] For example, a baby boy was found alive in a trash bin located in San Juan Capistrano, California. His umbilical cord was still attached and a bag was tied around his neck. He was brought to a hospital and put up for adoption.[113] The "Baby Moses Law" has been enacted in 44 states to discourage such mothers from leaving the unwanted babies to die. In California the mother has three days from the birth of the child to surrender it confidentially to any public or private hospital emergency room without the risk of arrest or pros-

---

104. "State and Federal costs for welfare programs associated with teen births are estimated at $5 billion to $7 billion." L.A. TIMES, Jan. 5, 1995, at B1.

105. L.A. TIMES, July 26, 1998, at A16; ORANGE COUNTY REGISTER, Aug. 8, 1996.

106. Finer & Henshaw, *Abortion Incidence and Services in the United States in 2000*, 35 PERSP. ON SEXUAL & REPRODUCTIVE HEALTH 6–15 (2003).

107. Henshaw, *Unintended Pregnancy in the United States*, 30 FAM. PLAN. PERSP. 24–29, 46 (1998).

108. L.A. TIMES, Aug. 21, 1996, at E5.

109. ORANGE COUNTY REGISTER, Dec. 5, 1995, at News 29.

110. JAMA 1170 (1999). "The latter figure is probably more complete than the Centers for Disease Control and Prevention total because of underreporting to state health departments." *Id.*

111. ORANGE COUNTY REGISTER, Apr. 3, 1996, at News 13.

112. ORANGE COUNTY REGISTER, Mar. 22, 1996, at 4.

113. ORANGE COUNTY REGISTER, Sept. 1, 2004, at Local 1. The case may be considered an attempted murder.

ecution for child abandonment.[114] The highest unplanned pregnancy rate is among 18- to 24-year-olds, mostly black or Hispanic. According to the Alan Guttmacher Institute, "Whether they end up in abortion or unplanned birth, unintended pregnancies come at a cost both to the people involved and to society."[115] At the turn of the third millennium, women were expected to have 1.4 unintended pregnancies by the time they reach age 45 years, and 43 percent will have had an induced abortion.[116]

A 1997 study revealed that 60 percent of all pregnancies in the United States are unintended.[117] A 1999 analysis of economists and criminologists concluded that unwanted children are most likely to commit crimes as adults. A Stanford University law professor stated that this analysis suggests that legalized abortion may account for as much as half of the overall crime-drop in the United States from 1991 to 1997.[118] The Supreme Court had stated, "Unwanted motherhood may be exceptionally burdensome for a minor.," while ruling that the constitutional right of privacy applied to mature minors.[119] Three youths who killed the fetus in a pregnant woman were charged with capital murder.[120] The CDC reported that California, the most populous state, continued to have the highest abortion rate in 1998.[121]

However, repeaters are also a major problem. The national statistics indicated:

---

114. Orange County Register, Mar. 6, 2004, at News 30. "The law also includes a 14-day 'cooling off' period, in which parents can reclaim children if they change their mind about giving them up. The hospital gives both parent and child an identification bracelet if the baby is dropped off anonymously." *Id.*

115. Orange County Register, Feb. 28, 1998, at News 21.

116. JAMA 1171 (1999).

117. Delbanco et al., *Public Knowledge and Perceptions About Unplanned Pregnancy and Contraception in Three Countries*, 29 Fam. Plan. Persp. 70 (1997).

118. The study found that states with high abortion rates in the 1970s experienced greater drops in crime in the 1990s than states with lower abortion rates, even when other factors that influence crime—income, racial composition and incarceration rates—were taken into account.

Orange County Register, Aug. 9, 1999, at News 8.

119. Bellotti v. Baird, 443 U.S. 622, 642–44 (1979). The court stated, "[T]here are few situations in which denying a minor the right to make an important decision will have consequences so grave and indelible." *Id.* at 642–44.

120. Orange County Register, Sept. 3, 1999, at News 21.

121. Orange County Register, July 3, 1998, at News 15.

> The prevalence of repeat adolescent pregnancies averaged 30 percent during the first and 40 percent to 50 percent during the second post-partum year. These statistics are of concern because the incidence of low birth weight and pre-term deliveries increases and the likelihood of completing high school, having a job, and of being self-supporting decreases with each additional birth during adolescence.[122]

These births generally result in unwanted children. A test was made of the hypothesis that a monetary incentive promotes peer support group participation and decreases repeat adolescent pregnancies. Unfortunately, it was concluded, "these discussions do not prevent repeat pregnancies."[123] The unwed birth rate peaked in 1994 at 32.6 percent and, unfortunately, it has remained about the same since then.[124] Men at the turn of the third millennium (particularly African-Americans) are generally unwilling to marry single women who become pregnant.[125] They either believe or become convinced that the fetus may not be theirs. This should be changed now that genetic tests for paternity are available for a few hundred dollars.[126]

Studies have shown that children are 100 times more likely to be abused by stepparents than by biological ones.[127] A survey of women seeking abortion at one urban clinic in 1996 showed that 57 percent of black women and 37 percent of white women suffered abuse.[128]

---

122. JAMA 977 (1997).

123. *Id.* In 1996 Congress authorized the distribution of funds to states in the form of welfare bonus money for reducing the rate of out-of-wedlock births—if the states could show abortions also declined. In 1999 a few states (California, Michigan, Alabama, Massachusetts) split the first pot of money. ORANGE COUNTY REGISTER, Aug. 1999, at News 6.

124. *Id.*

125. Black women who had conceived premaritally were far less likely to marry before giving birth (10%) than were white or Hispanic women (28%–29%). Only 16% of teenagers who conceived out of wedlock married before their infant was born, compared with 29% of women in their early twenties and 38% of those aged twenty-five to twenty-nine. Amara Bachu, *Trends in Marital Status of U.S. Women at First Birth: 1930 to 1994* (Population Div., U.S. Bureau of the Census, Working Paper No. 20, 1998).

126. ROBERT WRIGHT, THE MORAL ANIMAL: WHY WE ARE THE WAY WE ARE; THE NEW SCIENCE OF EVOLUTIONARY PSYCHOLOGY (1995).

127. MARTIN DALY & MARGO WILSON, HOMICIDE (1988).

128. Glander et al., *The Prevalence of Domestic Violence Among Women Seeking Abortion*, 91 OBSTETRICS & GYNECOLOGY 1002 (1998).

At the turn of the third millennium, actual access for women to legal abortion services often remained scarce; 86 percent of U.S. counties had no known abortion provider and 32 percent of U.S. women lived in these counties. Even in metropolitan areas 28 percent had no such provider;[129] the Alan Guttmacher Institute found that 65 percent of teen-age mothers from 15 to 19 had children by older men, 20 percent of whom were six years older or more. Some states are beginning to enforce statutory rape laws that prohibit sex between adults and minors.[130]

The mother of Nicolas Perruche had contracted German measles, which was harmful to her fetus. When only four weeks pregnant, she told her doctor she wanted an abortion. However, both the laboratory and doctor failed to diagnose it. Since his birth, Nicolas Perruche has been unable to speak, largely blind, and only able to be moved in a wheelchair. She began litigation that lasted 13 years. The highest court of France ordered that damages be paid not only to the Perruche family but also to Nicolas himself, because medical errors had allowed him to be born. This was the first high court ruling that a handicapped child has the right to recognition that his (or her) birth was a mistake; it could potentially be followed by similar suits or changed by the legislature.

---

129. To reduce the out-of-wedlock rate, the number of illegitimate births must fall more quickly than births to married women. *Id.*; JAMA 1170 (1999).

130. Orange County Register, May 19, 1996, at News 17. The reverse is also the case in some instances. In one case, a fifteen-year-old boy was seduced by a thirty-four-year-old woman, who was subsequently convicted of statutory rape for the event. After she gave birth to their daughter, she sued the boy and his family for reimbursement of the costs for Aid to Families with Dependent Children, and the court upheld the claim against the boy. County of San Luis Obispo v. Nathaniel J., 57 Cal. Rptr. 2d 843 (Ct. App. 1996).

good v. Good

Abortions are religiously or ethically wrong, and are especially wrong for teenagers and others. Except in order to preserve a woman's life, physicians should not diagnose any reason to consider an abortion.

v.

Because of the massive number of unmarried teenage pregnancies in the U.S., as well as the resulting dramatic increase in a population of unwanted children, attention is needed to provide more access for them to abortion clinics, which are in great shortage in many states early after the dawn of the third millennium.

Twenty-first Century Approach

1. Inasmuch as unwanted pregnancies and births—particularly among teenagers—are dangerous for the resulting children as well as for society, there is a need to strengthen the laws concerning the right and the ability of women to locate abortion clinics within a reasonable distance from where they live or be transported to a clinic at a longer distance.
2. In view of the fact that more than two out of three of U.S. women still live in counties with no known abortion providers, efforts must be undertaken early in the twenty-first century to increase their access.

---

6. Spousal and Parental Consent for Abortions

At the close of the twentieth century it was written, "Nearly every common law country in the world has rules that a man can neither block nor force a woman's decision to abort simply by virtue of his genetic connection to the embryo or fetus."[131] In a number of states, over 50 percent of minors seek abortion.[132] But automatic application of this rule of autonomy to mi-

131. Knowles, Hast. Ctr. Rep., Mar.-Apr. 1999, at 39. Even Catholic Italy's Parliament in 1988 affirmed a ruling by its Constitutional Court that a married woman has the right to abort without her husband's consent or even over his objection.

132. "Among the states with the highest rates of minors seeking abortions: New Jersey (where 58% of teenage pregnancies end in abortions), New York (56%), Massachusetts (53%), and Connecticut (50%)." N.E.H. Hull & Peter Charles Hoffer, Roe v. Wade 265 (2001).

nors seeking an abortion has met with more controversy. Teenage abortions have been inhibited by the laws of 27 states which require the notification or the consent of one or more parents.[133] This may delay the abortion procedure and increase the health risk and the expense of the procedure.[134] In March 1997 the Supreme Court upheld a Montana law that prohibits a physician from performing an abortion on an unemancipated minor unless the physician has notified one of the minor's parents or a court finds that notification "is not in the best interests of the minor." (This is the "judicial bypass exception").[135] On the other hand, it left standing the ruling by a circuit court of appeals that judicial bypass amendments (to Louisiana's parental consent abortion statute) violate the Fourteenth Amendment due process right of minors to seek an abortion.[136]

It acted similarly by refusing to hear South Dakota's appeal of a federal court order that struck down a 1993 law requiring pregnant girls generally to notify a parent before they obtain an abortion.[137] In August 2000, the New Jersey Supreme Court struck down a law requiring parental notice for unmarried girls under age 18. That law was held to be a violation of the equal protection guarantee in the state constitution by imposing a burden on minors seeking abortions that it did not impose on those who continue their

133. Corinne Schiff, *The Lawyer's Role in Restoring Adolescents' Abortion Rights*, FED. LAW., May 1997, at 44; *see also* Ohio v. Akron Ctr. for Reproductive Health, 497 U.S. 502 (1990) (upholding a parental notification statute). *Akron II* determined that the "not in the best interests of the minor" criterion for judicial bypass of parental notification was satisfied. *See* ORANGE COUNTY REGISTER, Apr. 18, 2002, at News 10.

134. O'Keefe & Jones, *Easing Restrictions on Minors' Abortion Rights*, ISSUES SCI. & TECH., Fall 1990, at 78.

135. Lambert v. Wicklund, 520 U.S. 292 (1997). Thirty-two states have laws requiring at least one parent be notified before a minor has an abortion. ORANGE COUNTY REGISTER, Oct. 11, 2000.

136. Ieyoub v. Causeway Medical Suite, 552 U.S. 943 (1997); *see also* Causeway Medical Suite v. Ieyoub, 123 F.3d 849 (5th Cir. 1997). The Supreme Court continued its five-year record of denying full review to any case dealing directly with abortion rights.

137. L.A. TIMES, Apr. 30, 1996, at A12.

pregnancies.[138] The AMA's 2002 *Code of Medical Ethics* is very flexible on this matter and would move physicians in alternative directions.[139] In 2000, a study found that over half of adolescent girls surveyed in Wisconsin would quit going to Planned Parenthood if their parents had to be told they wanted prescribed contraceptives; this indicated that parental notification could lead to more teen pregnancies, abortions, and the spread of sexually transmitted diseases. Accordingly, it was concluded, "Mandatory parental notification for prescribed contraceptives would impede girls' use of sexual healthcare services, potentially increasing teen pregnancies and the spread of [sexually transmitted diseases] STDs." [140]

In California, a plurality of its Supreme Court in 1997 declared unconstitutional a never-enforced statute requiring unemancipated minors to obtain parental consent to abortion or seek a judicial bypass,[141] even though the California Constitution has stronger privacy protections than the Federal

138. Planned Parenthood of Cent. N.J. v. Farmer, 762 A.2d 620 (N.J. 2000). "The court was more protective of abortion rights than the U.S. Supreme Court, which has upheld parental notification statutes as long as they provided for a waiver of the requirement by a judge. About 40 states have enacted laws requiring parental involvement." Orange County Register, Aug. 16, 2000, at News 5. "California currently has no parental notification statute. State legislators passed a law in 1987 that required girls under the age of 18 to get parental consent for abortions, but the law was never enforced and in August 1997, the California Supreme Court ruled in *American Academy of Pediatrics v. Lungren* that the statute violated the state Constitution's explicit right of privacy." Orange County Register, Apr. 28, 2005, at News 8.

139. "Physicians should ascertain the law in their state on parental involvement" and "strongly encourage minors to discuss their pregnancy with their parents." However, they should not "be compelled to require minors to involve their parents before deciding whether to undergo an abortion. Minors should ultimately be allowed to decide." Physicians "should encourage their minor patients to consult alternative sources if parents are not going to be involved in the abortion decision." Code 2.015 (issued in 1994) had no new changes.

140. JAMA 710 (2002).

141. American Academy of Pediatrics v. Lungren, 940 P.2d 797 (Cal. 1997). "A closely divided Supreme Court upheld a never enforced 1987 law requiring minors to receive the consent of a parent before obtaining an abortion." L.A. Times, Apr. 14, 1996, at E.

Constitution.[142] Justice Kennard's opinion contended that most minor girls are capable of making informed and mature decisions about abortion:

> Regardless of age restrictions . . . an adolescent's expectation of privacy is reasonable as to decisions with the following characteristics: (1) they are protected by the right of privacy for adults, (2) they are within the adolescent's competence, (3) they have serious and enduring consequences, and (4) they cannot be postponed to the age of legal majority.[143]

Since that decision, more courts that have ruled on the matter have done so in favor of the minors.[144] The Ninth U.S. Circuit Court of Appeals affirmed a district court's holding that Arizona's parental consent law is unconstitutional because it lacks a time limit for a judge to decide on an abortion for a minor who cannot consult her parents.[145] However, it upheld

142. "In California, the state must demonstrate a 'compelling' or 'extremely important and vital' reason for interfering in reproductive choices." *Id.*; *see also* American Life League, Inc. v. Reno, 47 F.3d 642 (4th Cir. 1995), *cert. denied*, 516 U.S. 809 (1995); United States v. Wilson, 73 F.3d 675 (7th Cir. 1995), *rev'g* 880 F. Supp. 621 (E.D. Wis. 1995); L. A. Times, Aug. 6, 1997, at A14.

143. *Lungren*, 940 P.2d at 840–41 (Kennard, J., concurring). This decision provoked an attempt to unseat the Chief Justice; however, he was easily reelected in 1998.

144. A judge issued an injunction barring enforcement of a Florida law that required doctors to notify the parents of girls under eighteen who are seeking abortions. Orange County Register, July 27, 1999, at News 12. In 2003, the Florida Supreme Court refused to enforce a state law requiring notification for minors requesting abortions. This decision was based on the grounds that an individual's privacy under Florida's constitution is a "fundamental right." N. Fla. Women's Health and Counseling Servs. Inc. v. Florida, 866 So. 2d 612 (Fla. 2003). The following year, an Idaho law requiring girls under age eighteen to get partial consent for abortions was declared unenforceable by a federal appeals court. The court stated that there was no reasonable explanation for limiting emergency abortions without consent to "sudden and unexpected" instances of physical complications. Orange County Register, July 17, 2004, at News 27.

145. Planned Parenthood of S. Ariz. v. Lawall, 193 F.3d 1042 (9th Cir. 1999).

a newer Arizona law requiring minors to get parental consent to have an abortion.[146] An Arizona statute permits a minor to use a fictitious name and bars public access to the records of such proceedings.[147] In 2002 the U.S. Court of Appeals for thc Ninth Circuit upheld the statute even though it did not exclude "public" state and federal employees who are authorized to copy court records.[148]

good v. Good

No teen-ager is mature enough to make a decision as momentous as abortion without parental guidance.

v.

Many, if not most, pregnant mature teen-agers are fearful about bringing their desire for an abortion to the attention of their parents. This has resulted in huge numbers of unwanted births where parental notification is required and is especially unfortunate in the large majority of counties in the United States, which have no abortion facilities.

146. Orange County Register, Oct. 10, 2002, at News 18. The U.S. Court of Appeals for the Fourth Circuit, however, upheld Virginia's law requiring that parents be notified when girls younger than eighteen seek abortions. Orange County Register, Aug. 21, 1998, at News 4. The U.S. Court of Appeals for the Seventh Circuit held that an unwed father has no constitutional right to notice from a public hospital before the mother obtains an abortion at the hospital. *See* Coe v. County of Cook, 162 F.3d 491 (7th Cir. 1998).

147. Ariz. Rev. Stat. § 36-2152(d) (LexisNexis). Mississippi and North Dakota require the consent of both parents before a minor can have an abortion. However, if the girl's father impregnated her, only the mother's consent is needed. Orange County Register, Dec. 28, 2004, at News 17.

148. Planned Parenthood of S. Ariz. v. Lawall, 307 F.3d 783 (9th Cir. 2002). The court stated that "complete anonymity is not critical to pass constitutional muster" and that the statute "provides young women with adequate protection to prevent unauthorized disclosure of personal information. . . . Moreover, public interest militates in favor of permitting authorized personnel to handle closed court records."

Twenty-first Century Approach

State legislation requiring parental consent to abortions by all minors who are mature enough to make an abortion decision on their own should be repealed.

---

7. Harassing and Bombing Abortion Clinics

The harassing and bombing of abortion clinics since the 1970s caused Congress to enact the Freedom of Access to Clinic Entrances (FACE) Act in 1994.[149] The term "reproductive health services" is expansively defined to include "medical, surgical, counseling or referral services . . . including services relating to . . . the termination of a pregnancy" whether those services are provided "in a hospital, clinic physician's office, or other facility."[150] It criminalized the use of force or physical obstruction to injure or interfere with any person either obtaining or providing reproductive health services. The act also provides for civil enforcement by the attorney generals of the states. An appellate court upheld this statute in 1995 as being within the Commerce Clause of the Constitution.[151] Although violence and harassment by abortion protesters across the nation dropped the next year,[152] it picked up again in the following years. In January 1997, two firebombs were thrown at an abortion clinic in Tulsa, Oklahoma; in February John Salvi was found guilty of murdering two abortion clinic workers in Brookline, Massachu-

---

149. 18 U.S.C. § 248. The statute provides longer prison terms and larger fines for repeat offenders, and shorter prison terms and smaller fines for offenses "involving exclusively a nonviolent physical obstruction." 18 U.S.C. § 248(b).

150. *Id.*

151. *See* United States v. Wilson, 73 F.3d 675 (7th Cir. 1995). The Freedom of Access to Clinic Entrances (FACE) Act carries a penalty of up to six months imprisonment and a $10,000 fine for a nonviolent physical obstruction of access to abortion clinics. It was held to be a petty offense for which the Sixth Amendment does not guarantee a right to a jury trial. United States v. Soderna, 82 F.3d 1370 (7th Cir. 1996).

152. L.A. Times, Apr. 4, 1996. "Death threats at Planned Parenthood clinics have dropped from 14 in 1995 to three this year [1998], bomb threats are down from 15 to seven; the number of protesters is down from 5034 to 1357, and the amount of damage is down from $15,581 to $8,130." *Id.*; *see also* Orange County Register, Sept. 27, 1996, at News 8

setts;[153] and in March a man wearing a crash helmet drove a truck with gasoline cans into an abortion clinic.[154]

In January 1998 a bomb blew up the New Woman All Women Healthcare Clinic, killing a man, just one week after the 25th anniversary of the *Roe v. Wade* decision; in February 1998 a man pleaded guilty of setting fires at seven western abortion clinics;[155] another man was sentenced to 15 years in a federal prison for a failed attempt to blow up an abortion clinic with gasoline;[156] and in July, acid was spilled at four abortion clinics in Houston, Texas, causing 10 people to be treated for breathing problems.[157] In 1998 James Kopp, another opponent of abortion, shot and killed an obstetrician-gynecologist. He fled the United States and was later extradited from France, after being sentenced to 25 years to life in 2003.[158] In 1994, Paul Hill, a 49-year-old minister, murdered abortionist Dr. John Britton and his escort, retired Air Force Lt. Col. James H. Barrett, and was sentenced to death in Florida. He did not appeal, and on the eve of his execution he stated, "The sooner I am executed . . . the sooner I am going to heaven. I expect a great reward in heaven. I am looking forward to glory. I don't feel remorse."[159]

One abortion clinic doctor in Colorado found that his name, address, and phone number began to appear on abortion-opposing fliers and a Web site listing of "baby butchers." He said, "I felt like a hunted animal, that I would be shot at any time."[160] The following month a federal jury in Oregon ruled against an anti-abortion web site that included the names and addresses of

---

153. L.A. Times, Jan. 2, 1997, at A18.

154. Orange County Register, Feb. 2, 1997, at News 22.

155. Orange County Register, Feb. 16, 1998, at News 4.

156. L.A. Times, Feb. 10, 1998, at A24. In 1997, Eric Rudolph set off a remote-controlled bomb at an abortion clinic in Alabama that killed an off-duty police officer and maimed a nurse. When he was sentenced to life in prison he simply declared that abortion must be fought with "deadly force." Orange County Register, July 19, 2005, at News 11.

157. L.A. Times, July 9, 1998, at A18. A man who died when he accidentally detonated a pipe bomb in his lap was planning to use the device to destroy an abortion clinic to keep an acquaintance from having an abortion. Orange County Register, Mar. 20, 1999, at News 13.

158. Orange County Register, May 10, 2003, at News 28.

159. Orange County Register, Sept. 3, 2003, at News 10.

160. Orange County Register, Jan. 11, 1999, at News 6.

doctors who perform abortions and featured photos of mangled fetuses and drawings of dripping blood.[161] It found that the site constituted a threat to abortion providers and awarded the plaintiffs $107 million in damages.[162] Abortion-rights groups in California, Maine, and New York said they plan to use advertisements showing that "six out of 10 women seeking abortions have experienced contraceptive failure."[163]

In 2001, the Ninth Circuit of Appeals held that information displayed on posters that results in attracting unfriendly attention to certain abortion providers but does not support violence is protected by the First Amendment.[164] The following year it ruled that abortion opponents' creation of posters and a Web site condemning abortion doctors can be held liable because this activity amounted to illegal threats, and not simply free speech.[165]

Legal remedies for and against abortion clinics that are being protested include the establishment of buffer zones and the seeking of injunctions to enforce them. In 1994, the Supreme Court upheld an injunction that created a free-speech buffer zone around a clinic and restricted the permissible level of noise from anti-abortion demonstrations.[166] These injunctions have been upheld in many other cases.[167] In June 2000, the Supreme Court upheld Colorado's so-called "bubble" law that prohibits people from counseling, distributing leaflets or displaying signs within eight feet of others without their consent whenever they are within 100 feet of a clinic entrance.[168]

---

161. Planned Parenthood of Columbia/Willamette, Inc. v. Am. Coal. of Life Activists, 23 F. Supp. 2d 1182 (D. Or. 1998).

162. ORANGE COUNTY REGISTER, Feb. 6, 1998, at News 9.

163. ORANGE COUNTY REGISTER, Nov. 4, 1998, at News 17.

164. Planned Parenthood of Columbia/Willamette, Inc. v. Am. Coal. of Life Activists, 244 F.3d 1007, 1019 (9th Cir. 2001).

165. ORANGE COUNTY REGISTER, May 17, 2002, at News 14. However, the appellate court ordered the lower court judge to reduce the $107 million in damages the jury awarded to four doctors who sued the 12 abortion foes.

166. Madsen v. Women's Health Ctr., Inc., 512 U.S. 753 (1994).

167. Schenck v. Pro-Choice Network of W. N.Y., 519 U.S. 357 (1997); Sabelko v. City of Phoenix, 68 F.3d 1169 (9th Cir. 1995), *vacated by* 519 U.S. 1144 (1997); Nat'l Org. for Women v. Operation Rescue, 37 F.3d 646, 648–49 (D.C. Cir. 1994); Horizon Health Ctr. v. Felicissimo, 638 A.2d 1260 (N.J. 1994).

168. The Court stated that Colorado's law simply empowers private citizens entering a healthcare facility to prevent a speaker, who is within eight feet and advancing, from communicating a message they do not wish to hear. ORANGE COUNTY REGISTER, June 29, 2000, at News 11.

A Seattle policeman was fired for attempting to prevent an abortion; he had stopped a car for speeding that contained a couple who were lost on their way to a Planned Parenthood office for a pre-abortion procedure. He coerced them to follow him to his church's anti-abortion counseling center. The couple arrived at Planned Parenthood much later and had their abortion. The officer was found to have violated the department's code of ethics and engaged in "unbecoming conduct."[169] On the other hand, a federal appellate court found that a Roman Catholic police officer who was assigned by the city of Chicago to protect abortion clinics could seek transfer to another police district because this was a reasonable accommodation. The officer was precluded from holding the city liable for religious discrimination under Title VII of 1964 Civil Rights Act.[170] A blow to abortion clinic objectors, called Operation Rescue, was the result of a class-action ruling under the Racketeer Influenced and Corrupt Organizations Act (RICO). A federal jury found two officers of the Pro-Life Action League to have engaged in 21 acts of extortion to shut down clinics; they, along with two groups that oppose abortion, were ordered to pay $258,000 in an action brought by the National Organization for Women.[171]

However, in 2003 the Supreme Court ruled that RICO was intended to be used against gangsters and organized crime, in cases of extortion. It found that the protesters of abortion did not extort money or valuables from the clinics when they tried to disrupt the business and interfered with clinic operations.[172] This view was retained by that Court in 2003 when it held "that petitioners did not commit extortion because they did not 'obtain' property from the National Organization for Women, Inc. (NOW), a 'national nonprofit organization that supports the legal availability of abortion.'"[173] The

169. L.A. Times, Nov. 21, 1995, at A20.

170. Rodriguez v. Chicago, 975 F. Supp. 1055 (7th Cir. 1998).

171. Orange County Register, Apr. 21, 1998, at 1; Orange County Register, Apr. 29, 1998, at News 5.

172. *Scheidler v. National Organization for Women, Inc.* and *Operation Rescue v. National Organization for Women*. It was a class-action suit filed by NOW and a group of abortion clinics against Pro-Life Action Network and other anti-abortion organizers. Chief Justice Rehnquist wrote: "But even when their acts of interference and disruption achieved their ultimate goal of 'shutting down' a clinic that performed abortions, such acts did not constitute extortion."

173. Scheidler v. Nat'l Org. for Women, Inc., 123 S. Ct. 1057, 1066 (2003).

lower court had issued a "permanent nationwide injunction prohibiting petitioners from obstructing access to the clinics, trespassing on clinic property, damaging clinic property, or using violence or threats of violence against the clinics, their employees, or their patients." The majority acknowledged that the petitioners "interfered with, disrupted, and in some instances completely deprived respondents of their ability to exercise their property rights." But the "petitioners neither pursued nor received 'something of value from' respondents that they could exercise, transfer, or sell."[174] Hence, the anti-racketeering laws were wrongly used to thwart Operation Rescue and other abortion opponents. However, other laws remain available to punish property damage, violence, and the like. At about the same time, Planned Parenthood was awarded $1 million in punitive damages from the Dallas-based Operation Rescue and Houston-based Rescue America for protests during the Republican National Convention.[175] One year after the turn of the third millennium, it was noted that some 200 of those facilities received Federal Express packages that contained poisonous anthrax threats.[176] A number of religious extremists continue bombing and harassing abortion clinics into the twenty-first century.

good v. Good

All abortion clinics must be protested until such deadly action is outlawed in each state.

v.

The Constitutional right to abortion must be maintained by the significant distances of buffer zones from clinics and more enforcement of existing laws against the perpetrators of violence.

Twenty-first Century Approach

Legal actions, including injunctions, the award of damages, limits with respect to the distance of "buffer zones" around abortion clinics, as well as other remedies, will have to be regulated. The twenty-first century fight against

174. *Id.* The Supreme Court had previously ruled that RICO could be applied to abortion protesters.

175. L.A. TIMES, Apr. 20, 1995, at A28; *see also* NEW ENG. J. MED. 81 (1997).

176. In the fall of 2001, the packages were traced to three locations in Virginia, Philadelphia, and Detroit. ORANGE COUNTY REGISTER, Nov. 9, 2001, at News 17.

this peculiar brand of terrorism within the United States can be eventually won by legal measures available under civil and criminal laws.

---

8. Formerly Mandated C-Sections to "Save" a Viable Fetus

Abortion rights are retained by pregnant females up to the point of the fetus's viability; thereafter, the state's interest may supersede rights that are, until then, solely those of such women. It was formerly argued that where physicians have determined that a cesarean was needed in the interests of a viable fetus, the woman should be compelled to undergo that operation. The considerably more expensive C-sections were more common until the practice came under criticism toward the last decade of the twentieth century. In 1989 it was reported by the Public Citizen Health Research Group, a consumer advocacy group, that "[p]hysicians overdiagnose complications in affluent childbearing women and then overprescribe cesareans, because the surgical procedure provides a bigger payoff than vaginal delivery does. Indeed, it was charged that avarice on the part of hospitals and physicians is largely to blame for the fourfold increase in the rate of cesareans in the U.S. since 1970."[177]

In 2002, a study found a retreat from the "natural childbirth" trend, and the cesarean section rate, which jumped 7 percent in 2001, reach 24.4 percent of all live births. The Centers for Disease Control (CDC) found that the percentage of women who chose a vaginal birth after having their previous child by C-section dropped from 23 percent to 15 percent between 1996 and 2000. The government set a goal for 2010 of 37 percent of women having vaginal births after having cesareans.[178] In 2002, it was found that "there is no hard evidence on relative risks and benefits of term elective cesarean delivery for non-medical or non-obstetric reasons compared with vaginal delivery,"[179] although the higher price of the cesarian has an advantage for phy-

177. Sci. Am., Oct. 1989, at 36.

178. Orange County Register, Nov. 8, 2002, at News 11. Cesarean section deliveries have been defined as "surgical incision of the walls of the abdomen and uterus for delivery of offspring." Merriam Webster's Collegiate Dictionary 188 (10th ed. 1993).

179. JAMA 2627 (2002).

sicians.[180] A case arose in Illinois where the state petitioned the court to order a pregnant woman to have a cesarean section after she had refused such an operation. A panel of three judges in an appellate court "suggested that an order compelling a pregnant woman to submit to an invasive procedure such as a cesarean section would violate her constitutional rights."[181] The court also held that the juvenile court did not have jurisdiction over a fetus.[182]

Despite opposition from counsel for the woman, the court granted the public guardian's request to be appointed guardian ad litem for the fetus.[183] In ruling in favor of the pregnant female, the court stated that it had "seen no case that suggests that a mother or any other competent person has an obligation or responsibility to provide medically for a fetus or another person for that matter."[184] It found no authority that "mandates balancing tests by which a court balances, as in this case, the right to life of a viable person versus the right of the mother to choose a medical procedure which may cause death or other injury." In holding that a competent woman could not be forced to undergo the procedure to save a viable fetus, it was stated that "Illinois courts should not engage in the balancing of the rights of the unborn viable fetus against the right of the competent woman to choose her own medical care and that "*a woman's competent choice* in refusing medical treatment as invasive as a cesarean section during her pregnancy *must be honored even in circumstances where the*

---

180. Such deliveries, as of 2002, cost approximately $10,000. Washington State Nat'l Org. for Women, *Contraceptive Equity* (Position Paper, 2000), *available at* http://www.wanow.org/pp/contraceptive_equity.pdf. On March 11, 2004, in Utah, a pregnant woman who allegedly ignored medical warnings to have a cesarean section to save her twins was charged with murder after one of the babies was stillborn. ORANGE COUNTY REGISTER, Mar. 12, 2004, at News 16.

181. *See In re* Baby Boy Doe, 632 N.E.2d 326 (Ill. App. Ct. 1994). In 2003, an infant was born with cerebral palsy due to loss of oxygen because of a uterine rupture, the possibility of which the parents were not told during labor. The mother had had two prior C-sections because of a cephalo-pelvic disproportion and claimed that Dr. Peggy Fletcher was negligent in ordering a vaginal birth. The parents were awarded $16.22 million by an Illinois jury. Foley v. Ingalls Health Ventures, No. 99-L-13945 (Ill. Cir. Ct.); NAT'L L.J., Nov. 3, 2003, at 4.

182. *In re* Baby Boy Doe, 632 N.E.2d 326.

183. *Id.* at 327.

184. *Id.* at 329.

choice may be harmful to her fetus."[185] (Emphasis added.) As if to underscore her right to avoid the performance of such an unwanted procedure, this woman successfully gave vaginal birth to a baby boy. The case set a precedent that was followed elsewhere in other states because a number of trial judges in several states had actually ordered pregnant women to undergo C-sections against their will. This Illinois case was cited in the British case that agreed so emphatically that it deserves to be quoted:

> The law is, in our judgment, clear that a competent woman who has the capacity to decide may, for religious reasons, other reasons, or for no reasons at all, choose not to have medical intervention, even though . . . the consequence may be the death or serious handicap of the child she bears or her own death. She may refuse to consent to the anesthesia injection in the full knowledge that her decision may significantly reduce the chance of her unborn child being born alive. *The fetus up to the moment of birth does not have any separate interest capable of being taken into account* when a court has to consider an application for a declaration in respect of a cesarean section operation.[186] (Emphasis added.)

Nevertheless, in March 2004 a woman in Utah was charged with murder for allegedly delaying a cesarean section that could have saved one of her twins. She was actually sentenced to 18 months' probation.[187]

---

185. *Id.* at 330 (citing Cruzan v. Dir., Mo. Dep't of Health, 497 U.S. 261, 277 (1990)). The court held that "the Due Process Clause of the Fourteenth Amendment confers a significant liberty interest in avoiding unwanted medical procedures" and that "the liberty guaranteed by the Due Process Clause must protect, if it protects anything, an individual's deeply personal decision to reject medical treatment." *Id.* at 331.

186. 38 BMLR 175.

> In 2003, out of 4.1 million births recorded in the United States, 1.13 million, or 27.6 percent, were Caesarean deliveries. The rate was up by a third since the 1996 Report from the National Center for Health Statistics.

ORANGE COUNTY REGISTER, Nov. 24, 2004, at News 27. (The report did not detail whether they were elective or medically required.)

187. The woman refused because she feared being scarred by the operation, and her baby was stillborn. She pleaded guilty and avoided prison. ORANGE COUNTY REGISTER, Apr. 30, 2004, at News 19.

Physicians prescribing C-sections may be interested in a study which found that for a woman with one prior cesarean delivery, the risk of uterine rupture is higher among those whose labor is induced. Although this particular risk is three times as high, the risk of uterine rupture (with vaginal birth after a cesarean) is normally less than .05 percent (or only 90 in 20,000).[188]

good v. Good

No woman ever has the right to condemn a "person" like her fetus to death when it might be saved by a C-section.

v.

Pursuant to the law of informed consent, all women have the right to refuse treatment, including invasive surgery such as a cesarean section for pregnant women.

Twenty-first Century Approach

In order to prevent some trial judges and some physicians from ordering cesarean sections over the objections of pregnant women, several more state legislatures should consider enacting specific legislation to prevent such an abuse of their autonomy during pregnancy.

Except in the case of man himself, hardly anyone is so ignorant as to allow his worst animals to breed.

Darwin, *The Descent of Man and Selection in Relation to Sex* (1871) Modern Lib. (1970), p. 501

---

188. New Eng. J. Med. 3 (2001). In 2003, a study of 15,307 women reported that the risk of urinary incontinence (a strain on the orifice [outlet] of the bladder) is higher among women who have had cesarean sections than among nulliparous women—those who have never borne a viable child. New Eng. J. Med. 900 (2003). The International Federation of Gynecology and Obstetrics concludes in its ethics statement that "performing cesarean deliveries for non-medical reasons is ethically unjustified," and maintains that "physicians have the responsibility to inform and counsel women in the matter." New Eng. J. Med. 946 (2003). Although a later study found the absolute risks are low, it confirmed that a trial of labor after prior cesarean delivery is associated with a greater perinatal risk than is elective repeated cesarean delivery without labor. Landon et al., *Maternal and Perinatal Outcomes Associated with a Trial of Labor After Prior Cesarean Delivery*, New Eng. J. Med. 2581 (2004) (asserting that the risk of problems "is fairly small, [and] many women will continue to opt for vaginal birth after cesarean").

9. Genetic Testing and Prenatal Counseling; Gender Selection

Rapid progress in gene discovery has dramatically increased diagnostic capabilities for carrier screening and prenatal testing of genetic diseases. It has been shown that education and genetic counseling increase understanding. The retention of genetic concepts and disease-related information will minimize test-related anxiety.[189] Most individuals are carriers of several lethal recessive genes.[190] Preventive strategies are likely to be introduced into the general population—particularly for multiple inherited diseases—through preconception or prenatal carrier screening. As we approached the end of the twentieth century, 89 percent of the public approved genetic screening for serious genetic defects[191] (see Chapter 15).

In 1997 the National Institutes of Health (NIH) issued a Consensus Statement that "over 99 percent of couples tested receive reassuring information regarding the improbability of having a child with CF" (cardiac failure) and that "most couples with no positive family history in this circumstance choose to terminate the pregnancy." The latter choice by virtually all couples is backed up by figures showing "the cost of DNA diagnosis testing for CF is between $50 and $150 per test." This is contrasted with the costs of care for children with cystic fibrosis that are estimated to "exceed $40,000 per year in direct medical costs and $9,000 per year in ancillary costs."[192] The cost savings from neonatal testing can be staggering; NIH studies showed that "the cost

---

189. "These findings emphasize the importance of genetic counseling for prenatal carrier testing and may improve understanding, acceptance, and informed decision making for prenatal carrier screening for multiple genetic diseases." 278 JAMA 1268 (1997). VICTOR A. MCKUSICK, MENDELIAN INHERITANCE IN MAN (1998), lists more than 5,000 inherited traits or disorders. Over 1,000 have already been mapped to specific chromosomal regions.

190. JAMA, *supra* note 189, at 1271; *see also* MOTULSKY, HUMAN GENETICS 559 (3d ed. 1996).

191. U.S. CONG., OFFICE OF TECH. ASSESSMENT, NEW DEVELOPMENTS IN BIOTECHNOLOGY: PUBLIC PERCEPTIONS OF BIOTECHNOLOGY 74–75 (1987).

192. Using a 3 percent discount rate, this implies a net present value of approximately $800,000 for direct and ancillary costs associated with a CF birth. *Genetic Testing for Cystic Fibrosis*, NIH CONSENSUS STATEMENT, Apr. 14–16, 1997, at 1–37, *available at* http://consensus.nih.gov/1997/1997GeneticTest CysticFibrosis106html.htm.

per identified cystic fibrosis fetus averted [avoided] ranged from $250,000 to $1,250,000 for a Caucasian population of Northern European ancestry."

Ashkenazi Jews (as distinguished from Sephardic Jews) constitute a group that includes 90 percent of the 6 million Jewish people in the United States. These Jews are descended from ancestors in Eastern and Central Europe. Many of them are known to inherit a number of diseases, including Down syndrome. An extra chromosome is the major cause of deafness, mental retardation, a flat facial profile and upward-slanting eyes, and half have heart defects and a lower life expectancy.[193] Fetuses can be tested and some parents will choose abortion rather than give birth to a child who cannot hear. A safer Down syndrome test combining blood and ultrasound tests from the first and second trimesters was found to be more accurate than standard screenings and would reduce the need for amniocentesis.[194] In 2005, a study of 38,167 patients for the presence of fetal Down syndrome incorporated measurements in both trimesters. It was concluded that "first-trimester combined screening at 11 weeks of gestation is better than second-trimester quadruple screening but at 13 weeks has results similar to second-trimester quadruple screening. Both stepwise sequential

---

193. NEW ENG. J. MED.(1998); ORANGE COUNTY REGISTER, NOV. 19, 1998, at News 6.

> An increased risk of fetal Down's syndrome (trisomy 21) is the most common reason for offering prenatal genetic diagnosis. Since the 1970s, an age of 35 years at delivery has been used as the cutoff for offering amniocentesis for this indication in the United States. After this age, the risk of Down's syndrome rises rapidly and is thought to balance the risks of amniocentesis. . . . We believe that offering second-trimester screening should continue to be the standard of care.

Mennuti & Driscoll, *Screening for Down's Syndrome—Too Many Choices?*, NEW ENG. J. MED. 1471 (2003).

194. NEW ENG. J. MED. (1999); ORANGE COUNTY REGISTER, Aug. 12, 1999, at News 20. About 1 in 700 babies has Down syndrome, which is marked by a broad, flat face with slanting eyes—and an early death. Two doctors in an editorial argue that Wald's method won't gain wide acceptance because it requires withholding results of first-trimester tests until they are combined with the later data. Women might prefer an earlier warning so they could terminate the pregnancy safely. *See also* Ohallan et al., *Methods to Increase the Percentage of Free Fetal DNA Recovered from the Maternal Circulation*, JAMA 1114 (2004).

screening and fully integrated screening have high rates of detection of Down syndrome, with low false-positive rates."[195]

However, even though (a) a pregnant couple underwent a series of tests in 2002 that indicated an 85 percent chance that the child would be born with Down syndrome, and (b) doctors in Utah led them to believe the tests had yielded a false positive and a baby girl arrived with Down syndrome, (c) the majority of the Supreme Court of Utah actually held that the doctors did not violate either Utah's Wrongful Life Act or the Due Process or Equal Protection Clauses of the Utah and U.S. Constitutions.[196] The statute prohibits any suit which claims "that but for the act or omission of another, a person would not have been permitted to have been born alive but would have been aborted." That court somehow indicated that the withholding of the test information was "too tenuous to hold that the statute has the effect of placing a substantial obstacle in the path of a woman who seeks an abortion."[197]

A cost-effectiveness study in England confirmed recommendations that universal, voluntary antenatal HIV screening should be implemented. Detection of HIV infection in pregnant women allows the risk of mother-to-

195. Malone et al., *First-Trimester or Second-Trimester Screening, or Both, for Down's Syndrome*, New Eng. J. Med. 2001 (2005). These findings were "likely to result in more early screening and lure insurance companies to cover the practice." Further, "earlier abortions are far safer than later ones." Orange County Register, Nov. 10, 2005.

196. Wood v. Univ. of Utah Med. Ctr., 67 P.3d 436 (Utah 2002). In New Jersey, an infant was born with a malformed brain and diagnosed with the condition known as holoprosencephaly (failure of the forebrain to develop properly). A sonogram taken at 16½ weeks showed the condition, but the doctor's and technician's failure to diagnose it essentially denied the option of terminating the pregnancy. The parents ultimately settled their wrongful birth suit against the obstetrician and the sonogram technician for $1.75 million. Battista v. Reinkraut, No. BER-L-007463-00 (Bergen Co., N.J., Super. Ct.); Nat'l L.J., Nov. 24, 2003.

197. Utah Code Ann. § 78-11-24 (2002). The dissent by the female Chief Justice pointed out that it was improper to so interpret that statute because it banned "all actions based on negligence where birth occurred instead of an abortion." *Id.* at 22. Further, although the U.S. Supreme Court under *Planned Parenthood of Southeastern Pennsylvania v. Casey*, 505 U.S. 833, 877 (1992), ruled that courts must determine if a statute has the "purpose or effect of placing a substantial obstacle in the path of a woman seeking an abortion," the Court normally denies certiorari to such cases. Kowitz, 61 Brook. L. Rev. 235, 257 n.105 (1995).

child transmission to be reduced. On November 8, 2002, the Food and Drug Administration (FDA) approved a test that can detect whether someone is infected with HIV, the virus that causes AIDS, in as little as 20 minutes. This in turn might slow the spread of the disease.[198] The Chinese Law on Maternal and Infant Healthcare provides that couples with unspecified genetic diseases "considered to be inappropriate for childbearing" be married only if both agree to practice long-term contraception or to be sterilized.[199]

On the other hand, religious and social prejudices weigh against prenatal screening and testing. In 1999 Pope John Paul II called upon Germany's Roman Catholic Church to end a morally "ambiguous" counseling program that provides permission slips for German women to obtain abortions.[200] Others try to claim that eugenics of any sort is somehow immoral. However, the United States has established what many believe to be an excellent form of eugenics by permitting early detection of abnormalities in a fetus that enables couples to terminate pregnancies instead of giving birth to defective children.[201] In his 1996 book *The Lives to Come*, Philip Kitcher, a philoso-

---

198. Postma et al., *Universal HIV Screening of Pregnant Women in England: Cost-Effectiveness Analysis*, 318 BRIT. MED. J. 1656–60 (1999); Sheryl Gay Stolberg, *Drug Agency Approves a Quick Test for H.I.V.*, N.Y. TIMES, Nov. 8, 2002, *available at* http://www.nytimes.com/2002/11/08/health/08IMMU.html. In 1992, the American Medical Association (AMA) began emphasizing HIV testing in its *Code of Medical Ethics*. AMA CODE OF MEDICAL ETHICS § 2.23 (1992). In 1996, New York became the first state to pass legislation mandating HIV testing of newborns (called "Baby AIDS"). *See* GERALD J. STINE, AIDS UPDATE 1999: AN ANNUAL OVERVIEW OF ACQUIRED IMMUNE DEFICIENCY SYNDROME (1999). However, it has been noted at one New York hospital that 30% of people seeking HIV testing in New York reported attempting suicide. Catalan & Pugh, *Suicidal Behavior and HIV Infection—Is There a Link?*, 7 AIDS CARE S117–21 (1995). Although blacks make up about 12% of the U.S. population, in 2002 they accounted for more than half of all HIV and AIDS cases. "AIDS is now the No. 1 killer of black women between the ages of 24 and 44." ORANGE COUNTY REGISTER, Mar. 14, 2004, at News 16.

199. SCI. AM., Mar. 1997, at 33.

200. Since 1995, German law has required women contemplating abortions to obtain certificates verifying that they have received counseling from one of 1,700 counseling centers operated by churches and other organizations such as the Red Cross. ORANGE COUNTY REGISTER, Sept. 22, 1999, at News 32.

201. *See* Plachot, Preimplantation Genetic Diagnosis: Technical Aspects, Third Symposium on Bioethics, CDBI (96) 14 (1996); *Friedman v. Glicksman,* 1996 (1) SA 1134 (W) (S. Afr.).

pher, stated, "Once we have left the garden of genetic innocence, some form of eugenics is inescapable, and our first task must be to discover where among the available options we can find the safest home."[202] He cautioned against "the tendency to try to transform the population in a particular direction, not to avoid suffering but to reflect a set of social values." His solution lies in individual choice by using "reliable genetic information in prenatal tests that would be equally available to all citizens." For Kitcher "there would be no societally imposed restrictions on reproductive choices—citizens would be educated but not coerced, there would be universally shared respect for difference coupled with a public commitment to realizing the potential of all those who are born." He himself called it "Utopian eugenics." However, this also presents a chance for "practical and responsible eugenics." The need for an authorizing individual choice as the solution was described at a 1997 meeting of the Society for Disability Studies. Some people with disabilities would use prenatal testing to selectively abort a fetus with the trait they themselves carry. Other people would not abort a fetus carrying their own disability, but they might abort a fetus if it carried a trait incompatible with their own understanding of a life they want for themselves and their child.

A two-year project concerning possible "discrimination against people with disabilities, increasing attention to the disability critique of prenatal testing, and the societal debate about abortion" ended in 1999. It found that "many who in general are against the right to abortion nonetheless approve of abortions performed on a fetus carrying a disabling trait."[203]

Some suggest testing for any traits that might be covered under the first part of the Americans with Disabilities Act (ADA), but consider it unreasonable to test for traits that are not covered. They might argue that a disability can only be tested if it constitutes a physical or mental impairment that substantially limits one or more of the major life activities of an individual. However, "drawing a line between traits covered under the ADA and those not covered would be entirely paternalistic; that is, doing so would be to

202. Kitcher, The Lives to Come: The Genetic Revolution and Human Possibilities 204 (1996).

203. Hast. Ctr. Rep., Sept.-Oct. 1999, at 510. "Virtually all the major work in the disability critique of prenatal testing emerges from those who are also committed to a pro-choice." *Id.* at 512.

make decisions for prospective parents that are rightly their own."[204] No consensus was reached about drawing only such lines. It was concluded that "even with the best information about the meaning of disability to various individuals and families, many (perhaps most) will choose to forego raising a child with a disability."[205] Thus we must recognize the fallacy of statements such as "I can cope with this, therefore you should do so." My sole object to this statement concerns the word "therefore." As Matt Ridley stated in his book *The Origins of Virtue*, "The roots of social order are in our heads, where we possess the instinctive capacities for creating not a perfectly harmonious and virtuous society, but a better one than we have at present."

Preference for male offspring in some countries received much publicity in the late twentieth century. An infant's sex, "La difference," begins at conception. Sex determination still remains a field full of speculation and with limited empirical evidence.[206] Without external interference, there are about 105 boys born for every 100 girls, a ratio that has persisted over millennia. Nevertheless, in the United States. and elsewhere, some people wish to take advantage of new technologies to facilitate gender identification (see Chapter 8). This is likely to continue rather than abate.[207]

---

204. *Id.* at 518.
205. *Id.* at 521.
206. The sex of a baby is fixed according to whether a male or female sperm fertilizes the egg. Half the sperm produced in each testis carry a Y chromosome, which is a rather short and mainly degenerate stretch of DNA carrying the genes needed to make the testes and produce sperm. The other half carries an X chromosome, like the eggs. If sperm and egg meet by chance there should be an equal number of XY, or male, and XX, or female, embryos. The type of sperm therefore fixes the sex of the embryo.

Roger Gosden, Designing Babies 162–63 (1999).

207. Doll, L.A. Times, Jan., 18, 1998, at M2.

    In 1999 the Genetics and IVF Institute in Virginia indicated that forty-six babies have been born in clinical trials for a patented procedure called Microsort, which sorts sperm so that couples can select the gender of their babies. In 2003 the Huntington Reproductive Center in Laguna Hills, California, became the second place in the United States that used this technology for gender selection.

Orange County Register, July 25, 2003, at News 3.

As noted above, at the end of the twentieth century, amniocentesis and other common prenatal procedures were used to test for fetal chromosomal problems and birth defects such as Down syndrome, Tay-Sachs, cystic fibrosis, and spina bifida. These tests are generally covered by health insurance—particularly where recommended by the woman's physician. If amniocentesis is performed at 11 to 14 weeks of gestation, this test may be associated with loss of amniotic fluid and vaginal bleeding. Ordinarily, however, it is given from the 15th to 18th week.[208] The usual blood screenings done in this country identify up to 75 percent of Down syndrome babies but yield results until about 20 weeks into pregnancy, when abortion may be somewhat more dangerous for women. However, in 2003, a new combination of blood tests and ultrasound can detect fetuses with Down syndrome and yield results at about 12 weeks.[209] Chromosomal analysis then reveals the sex of the fetus. In China, on the other hand, ultrasound is widely used to cull out girls, even though sex screening has been outlawed since 1989.[210] In order to reduce some of this practice, China's People's Congress enacted a new law easing the rules for adoption. The new law lowered the minimum age of Chinese adoptive parents from 35 to 30, dropped a requirement that they be childless, and allowed them to adopt more than one orphan. Parents from outside China annually adopted more than 4,000 Chinese babies, and 96 percent of the new parents were Americans.[211] Interpreters of Jewish law have indicated that screening for gender is "only acceptable when there is a family history of

208. *Id.*; Sci. Am., Dec. 1998, at 32.

209. Orange County Register, Oct. 9, 2003 (citing the *New England Journal of Medicine* of that date). "The absolute biggest advantage is this allows women to make private decisions before they are visibly pregnant." *Id.*

210. Orange County Register, July 2, 1996.

211. L.A. Times, Nov. 7, 1998, at A5. The INS and National Center for Health Statistics reported: "The cost of adopting a child born in China can range from $12,000 [to] $25,000. 4,206 children from China were adopted by American families in 1998." Orange County Register, June 25, 1999, at Accent 6. While Chinese origin of such children may be obvious, an appellate court has upheld a state statute authorizing disclosure of adoption records to adopted persons over the age of 21 and to a limited class of other eligible persons. It held that the statute does not violate federal constitutional privacy rights of birth parents, adoptive families, or entities involved in the adoption process. Doe v. Sundquist, 106 F.3d 702 (6th Cir. 1997).

gender-related diseases linked to the chromosome for the child's gender."[212] To the contrary, the Dutch Health Council concluded that sperm selection would involve very few people. "It would seem improbable that the position of women in our society would noticeably worsen as the result of availability of . . . sex-selective insemination."[213]

Some physicians have actually considered withholding this information from the woman along with her potential for delivering an infant with a genetic disease.[214] Any particular doctor may personally object to abortion "for such reasons." However, the woman should either receive the information or else be transferred to another physician who has no such inhibitions; patients are normally considered entitled to information about their own condition so that they, and not the physician, can make a decision on what to do about it. The AMA *Code of Medical Ethics* (2003) Rule 2.12, Genetic Counseling, states, "When counseling prospective parents, physicians should avoid the imposition of their personal moral values and the substitution of their own moral judgment for that of the prospective parents." A refusal by a physician to communicate his knowledge of such a test could lead to a malpractice suit (except under the Utah law, as noted above).[215]

A study based on a screening program in Wisconsin found that 70 percent of parents whose first child had cystic fibrosis eventually decided to have another child. They did so even though they realized each baby faces a one in four chance of also inheriting the incurable and fatal disorder.[216] In 2002, for the first time, a pre-implantation genetic diagnosis (PGD) was given

---

212. DORFF, MATTERS OF LIFE AND DEATH (1998).

213. GOSDEN, DESIGNING BABIES (1999).

214. Richard, *The Tailor-Made Child: Implications for Women and the State, in* EXPECTING TROUBLE: SURROGACY, FETAL ABUSE, AND NEW REPRODUCTIVE TECHNOLOGIES 14 (Patricia Boling ed., 1995).

215. Although some states, such as Illinois, have attempted to prohibit sex-selection abortions prior to viability, this has been held an "undue burden" on a woman's right to choose abortion (under *Casey* discussed *supra* notes 6, 197 and accompanying text) and would be held unconstitutional. Hence, with foresight the Illinois legislature added an express provision to its ban: "If the application of [the prohibition] to the period of pregnancy prior to viability is held invalid, then such invalidity shall not affect its application to the period of pregnancy subsequent to viability." 720 ILL. COMP. STAT. ANN. § 510/6-8 (West 2000).

216. ORANGE COUNTY REGISTER, July 7, 1998, at News 10.

to a woman in her 40s by screening her eggs in the laboratory for Alzheimer's; she then gave birth to a baby free of this defect. With genetic testing during pregnancy, they would be able to choose an abortion upon learning that by continuing on to delivery their infant would be likely to be afflicted with this disorder. It has been stated "genetic testing can immeasurably improve the quality of life for individuals, even entire families. To ignore the good it could do would be an act of immoral blindness and cowardice."[217]

good v. Good

Choosing to abort a fetus due to its sex, genetic or other disease thwarts God's will, which is that humanity should do everything possible to keep alive.

v.

The Constitutional right to choose abortion following genetic or other tests embraces the concept of autonomy of the pregnant woman to avoid the birth of unwanted children who will generally be a significant burden to society. Physicians should be held responsible to fully disclose the genetic test results.

Twenty-first Century Approach

1. Women should continue to remain free to choose abortion for reasons such as gender-related or other possible genetic disease of the fetus, in their discretion prior to the viability of their fetuses.
2. Upon request, physicians should (a) conduct tests indicating the gender-related and/or possible genetic diseases of a fetus and (b) either perform, or transfer the patient to another physician who will perform, an abortion particularly where a possible disease is shown to be the basis for a woman's choice of abortion.

---

10. Partial Birth and Post-viability Abortions

In both 1996 and 1997, President Clinton vetoed nearly identical bills that would have banned most second- and third-trimester abortions.[218] These

217. Sci. Am., June 1994, at 97.

218. In a 2003 Gallup Poll, 68% of Americans said abortion "should be generally illegal" in the second trimester; 84% said it should be barred in the third trimester. Orange County Register, Aug. 5, 2004, at Loc. 7.

bills would have prohibited so-called "partial-birth" or late-term abortions performed after 20 weeks of gestation—procedures used in less than 1 percent of the 1.5 million abortions performed annually. No such term as "partial-birth abortion" exists in medical terminology; the method is referred to by practitioners as "dilation and extraction" (D&X), generally performed in the second trimester.[219] In his 1997 veto message, the President stated, "Unfortunately [the bill] does not contain an exception to the measure's ban that will adequately protect the lives and health of the small group of women in tragic circumstances who need an abortion performed at a late stage of pregnancy to avert death or serious injury."[220] In such circumstances the Jewish *Mishnah* specifically and graphically states: "If a woman has [life-threatening] difficulty in childbirth, one dismembers the embryo in her, limb by limb, because her life takes precedence over its life."[221]

Twenty-two states have enacted laws patterned after these federal bills. In 2002 the Supreme Court held (5:4) that statutes in three states (Nebraska, Iowa, and Arkansas) violated the Fourteenth Amendment because they would apply to most prescribed methods and thus impose an undue burden on women's right to choose abortion. The majority stated that they could accept the claim that the Nebraska law "distinguishes between the overall 'abortion procedure' itself and the separate 'procedure' used to kill the unborn child."

---

219. It was developed by Dr. Martin Haskell, of Dayton, Ohio, and first described in 1992. *See* Women's Med. Prof'l Corp. v. Voinovich, 911 F. Supp. 1051 (S.D. Ohio 1995). In 2004, to defend the government against a lawsuit seeking to overturn a law passed by Congress outlawing what opponents call partial-birth abortions, the Justice Department demanded that hospitals turn over hundreds of patient records (but not the patients' names) on certain abortions. ORANGE COUNTY REGISTER, Feb. 12, 2004, at News 22.

220. N.Y. TIMES, Apr. 11, 1996. Opponents of the veto labeled the procedure "infanticide." L.A. TIMES, Jan. 17, 1997; *see infra* Chapter 10. "Republican Party leaders rejected a proposal to withhold support for candidates who fail to oppose a controversial late-term abortion procedure." *Id.*

221. DORFF, *supra* note 62, at 129. However, "once his head [or his "greater part"] has emerged, he may not be touched, for we do not set aside one life for another." DAVID BLEICH, BIOETHICAL DILEMMAS 275 (1998) states: "A fetus that cannot survive for a period of at least 30 days subsequent to birth is not considered to be a live creature for purposes of Jewish law and hence the prohibition against feticide does not apply."

Therefore, the law attempted to ban conventional abortions, in violation of the Court's *Planned Parenthood of Southwestern Pennsylvania v. Casey* decision of 1992.[222]

An appellate court struck down a Louisiana statute that created a tort cause of action against abortion providers for "any damage" caused to a mother or fetus by an abortion.[223] The court also found that the statute had the purpose and effect of driving abortion providers out of business and thus imposes an undue burden on a woman's right to seek abortion in violation of *Planned Parenthood of Southeastern Pennsylvania v. Casey.*[224] *Casey* prohibits any abortion regulation that "has the purpose or effect of placing a substantial obstacle in the path of a woman seeking an abortion of a nonviable fetus."[225]

Although the American Medical Association (AMA) first avoided taking a position in 1966,[226] it issued a report on May 17, 1997, recommending that such procedures not be used unless alternative procedures pose materially greater risk to women;[227] it switched and favored the ban two days later.[228] The following year a report stated that the AMA "ignored its own decision-making proce-

---

222. Carhart v. Stenberg, 530 U.S. 983, 938–39 (2000). The minority, including Chief Justice Rehnquist, agreed that the killing of the "partially born" fetus "closely borders on infanticide." Two of them quoted the AMA's statement that in a partial birth, unlike an abortion, the fetus was "killed outside of the womb," where it had "an autonomy which separates it from the right of the woman to choose treatments for her own body." *Id.* at 962–63, 979.

223. Okpalobi v. Foster, 190 F.3d 337 (5th Cir. 1999), *reh'g en banc granted*, 201 F.3d 353 (5th Cir. 2000).

224. Planned Parenthood of Se. Pa. v. Casey, 505 U.S. 833 (1992).

225. *Id.* In December 1999, Supreme Court Justice Stevens put in abeyance the laws of Illinois and Wisconsin banning a type of late-term abortion, pending Supreme Court review. ORANGE COUNTY REGISTER, Dec. 1, 1999, at News 25. An appellate court had upheld the Illinois and Wisconsin statutes. Hope Clinic v. Ryan, 195 F.3d 857 (7th Cir. 1999), *vacated by* 120 S. Ct. 2738 (2000); *see also* Ann McLean Massie, *So-Called "Partial-Birth Abortion" Bans: Bad Medicine? Maybe. Bad Law? Definitely!*, 59 U. PITT. L. REV. 301 (1998).

226. ORANGE COUNTY REGISTER, May 17, 1997, at News 15.

227. N.Y. TIMES, May 19, 1997, at A10.

228. ORANGE COUNTY REGISTER, Dec. 4, 1998, at News 20.

dures, got swept up in politics, and failed to protect patient welfare when it endorsed a GOP bill banning a late-term abortion procedure."[229]

State legislatures have also shown political views in this area. In 1997 the New Jersey legislature passed a partial-birth ban that was promptly vetoed by Governor Christie Whitman. Thereafter the state Senate overrode the veto.[230] Preliminary or permanent injunctions against enforcement of such state bans have been entered in a number of states. In 1997 the first court decision on the merits held that a Michigan ban was unconstitutionally vague and overbroad.[231] About the same time, a Michigan judge ruled that a 12-year-old girl who had been impregnated by her 17-year-old brother and was in her 28th week of gestation could travel to Kansas for a late-term abortion.[232]

Post-viability abortions are generally beyond the Fourteenth Amendment protection recognized in the 1992 decision of the Supreme Court.[233] However, at 24 weeks' gestation a fetus is generally below viability; although many have survived, most of them have several severe problems and a rather low quality of life. It takes about 28 weeks' gestation, which is more than three-fourths of full-term, before a fetus acquires the cognitive capacities

229. Orange County Register, Dec. 10, 1997, at News 4.

230. "It was also the first restriction on abortion to pass in the state since the U.S. Supreme Court legalized the procedure. Governor Whitman said she would not have the state attorney general defend it. That means the Legislature will have to hire its own attorney." *Id.*

231. Evans v. Kelley, 977 F. Supp. 1283 (E.D. Mich. 1997). The court specifically held that the statutory definition was "hopelessly ambiguous" such that "physicians looking to [its] language for direction as to what procedures are proscribed by the law simply cannot know with any degree of confidence what conduct may give rise to criminal prosecution and license revocation."

232. Orange County Register, July 25, 1998, at A15.

233. Planned Parenthood of Se. Pa. v. Casey, 505 U.S. 833 (1992).

> In October 2003, the President indicated that he would sign a bill that Congress enacted to make partial-birth abortion a crime without considering the health of the mother. A number of groups filed lawsuits to block that law from taking effect, namely, Planned Parenthood, the Center for Reproductive Rights, and the American Civil Liberties Union.

Orange County Register, Nov. 1, 2003, at News 22.

normally found after delivery.[234] An Ohio statute banned abortions after 24 weeks of pregnancy unless a physician determined "in good faith and in the exercise of reasonable medical judgment" that either the fetus is not viable, an abortion is necessary to prevent the woman's demise, or there is "a serious risk of the substantial and irreversible impairment of a major bodily function."[235] Similar laws have been enacted in 17 states. The first appellate court to rule on such a law, a federal circuit court, struck down the Ohio statutory ban.[236]

It noted that medical necessity emergency exceptions "contain subjective and objective elements in that a physician must believe that the abortion is necessary, his belief must be objectively reasonable to other physicians," and "physicians face liability even if they act in good faith according to their own best medical judgment."[237] In this connection the U.S. Court of Appeals for the Seventh Circuit held that the government was not entitled to abortion patient records due to the damage that their release might do to patients and hospitals *using the dilation and extraction method.*[238]

In some states, such as Kansas, *post-viability abortions* are permitted where the fetus has a "severe" or "life-threatening deformity or abnormality."[239] Post-viability abortions are also permitted in Israel. When the Su-

---

234. Flower, *Neuromaturation and the Moral Status of Human Fetal Life*, *in* ABORTION RIGHTS AND FETAL PERSONHOOD (Ed Doerr & James W. Prescott eds., 1989). Only a few abortions take place in the third trimester.

235. *See* OHIO REV. CODE ANN. § 2919.17 (West 2000); *Casey*, 505 U.S. at 878 (1992).

236. *See* Women's Med. Prof'l Corp. v. Voinovich, 130 F.3d 187 (6th Cir. 1997).

237. *Id.* at 204. Women's Med. Prof'l Corp. v. Voinovich, 130 F.3d 187 (1997). On March 23, 1998, the Supreme Court refused to reinstate the invalidated Ohio law, which stated:

> No person shall perform or induce an abortion when the fetus is viable unless such person is a physician and has a documented referral from another physician not financially associated with the physician performing or inducing the abortion and both physicians determine that (1) the abortion is necessary to preserve the life of the pregnant woman; or (2) the fetus is affected by a severe or life-threatening deformity or abnormality.

ORANGE COUNTY REGISTER, Mar. 24, 1998, at News 13.

238. Nw. Mem'l Hosp. v. Ashcroft, 362 F.3d 923 (2004).

239. KAN. STAT. ANN. § 65-6709 (1999).

preme Court reaffirmed the central holding in *Roe v. Wade* in 1992, it included the right of the state to restrict abortions after viability, stating:

> The concept of *viability*, as we noted in Roe, is the time at which there *is a realistic possibility of maintaining and nourishing a life* outside the womb, *so that the independent existence* of the second life can in reason and all fairness *be the object of the state protection that now overrides the rights of the woman.*[240]

The court adopted an evolving standard dependent upon the state of prenatal and postnatal technology. Because about 50 percent of infants survive with delivery at 24 weeks' gestation (without taking into consideration the greatly reduced quality of life of most of these that do survive), this might have been considered a mere presumptive point of viability near the end of the twentieth century. However, no such uncertain line should be used to delineate constitutional rights and state interests in the control of a woman over her body—particularly when it is accompanied by such devastating results with respect to those infants who are delivered alive (see Chapter 10).

At the dawn of the twenty-first century, five justices voted to overturn Nebraska's ban on partial-birth abortions; four dissenters disagreed.[241] Breyer's opinion for the Court held that "constitutional law must govern a society whose different members sincerely hold directly opposing views." Then he concluded, "This Court, in the course of a generation, has determined and then redetermined that the Constitution offers basic protection to the woman's right to choose. . . . We shall not revisit those legal principles. Rather, we apply them to the circumstances of this case." This view by the smallest majority of the Supreme Court was significantly in favor of such pregnant women at the dawn of the third millennium. President Bush signed

240. *Casey*, 50 U.S. at 846. Viability was at 28 weeks of gestation when *Roe* was decided in 1973. Roe v. Wade, 410 U.S. 113, 160 (1973).

241. Stenberg v. Carhart, 530 U.S. 914 (2000). The Nebraska law defines a partial-birth abortion as "a procedure in which the doctor delivers vaginally a living unborn child, before intentionally killing the child." It was a felony punishable by up to 20 years in prison, a fine of $25,000, and revocation of the doctor's license. The majority of the Supreme Court found the Nebraska statute vague and not interpretable as a ban only on the D & X procedure. The Court also vacated *Hope Clinic v. Ryan*, 195 F.3d 857 (7th Cir. 1999), which had upheld the Illinois and Wisconsin partial-birth abortion bans.

242. U.S. 2–8, 108th Cong., 1st Sess. (2003).

into law the Partial-Birth Abortion Ban Act of 2003,[242] which is intended to ban "a gruesome and inhumane procedure that is never medically necessary." It provides for a defendant accused of violating the law to "seek a hearing before the State Medical Board on whether the physician's conduct was necessary to save the life of the mother." The courts determined the unconstitutionality of the statute even though the number of partial-birth abortion procedures done in the United States is small.[243] The Supreme Court had ruled in favor of such abortions in 2000 (as noted above).

The best thing for the states would be to require any treatment over the pregnant woman's objection only where the fetus is capable of surviving outside the womb without artificial life support and would not be denied most of its significant and verifiable qualities of life.

good v. Good

All states should enact legislation specifically prohibiting abortions after 24 weeks' gestation or viability, whichever is sooner, in order to protect or preserve life of all fetuses to the maximum extent possible.

v.

No states should proscribe exercising a woman's desire for an abortion at or after 24 weeks' of gestation (using its definition of viability), where it is due to (a) the limited survivability of the fetus and (b) a very high proportion of defects likely to diminish or eliminate its quality of life.

Twenty-first Century Approach

1. State legislatures should not attempt to ban all abortions over the objection of pregnant women within the limits of safer viability (at or near 28 to 30 weeks' gestation) in view of current statistics indicating that at between 24 and 26 weeks, about half these infants will die and most of the remainder suffer severe physical or mental damage. States with such laws should repeal them.

---

243. The U.S. Court of Appeals for the Eighth Circuit in St. Louis upheld a ruling that the Federal Partial-Birth Abortion Ban Act is unconstitutional, stating: "[W]e believe when a lack of consensus exists in the medical community, the Constitution requires legislatures to err on the side of protecting women's health by including a health exception."

ORANGE COUNTY REGISTER, July 9, 2005, at News 26.

2. Those state legislatures desirous of protecting the life of a fetus at or near safer viability, and over the objection of the pregnant woman, should not limit any restrictions on abortions to the cases where it can be demonstrated that the fetus is quite defective and/or unlikely to be capable of surviving outside the womb without significant and/or continuous artificial life support.

---

11. Shortage of Physicians for Abortions

At the dawn of the twenty-first century, the United States faces a severe shortage of medical practitioners who perform legal abortions. In the 1990s, 84 percent of all counties and 94 percent of rural counties had no providers to permit women to exercise their constitutional right to an abortion.[244] In 1998 the Alan Guttmacher Institute, a nonprofit health policy research group, reported that "there are now nearly one-third fewer providers than the peak number in 1982 (2,908)." The Institute's study found a 14 percent drop in the number of clinics, hospitals, and doctors' offices offering abortion from 1992 to 1996.[245] Yet, these alone provide approximately 1.2 million abortions performed a year. At the dawn of the third millennium, direct abortion services were available in only 16 percent of the counties in the United States.

Accordingly, in 1994 the American College of Obstetricians and Gynecologists (ACOG) recommended that non-physicians be trained to perform low-risk first-trimester abortions.[246] It had earlier recommended "encouraging physicians and clinics to train and integrate mid-level clinicians into abortion service delivery."[247] Unfortunately, by the end of the twenti-

244. Schirmer, *Physician Assistant as Abortion Provider*, Nov. 1997, at 253.

245. Orange County Register, Dec. 12, 1998, at News 14.

246. Donald Judges, *Taking Care Seriously: Relational Feminism, Sexual Difference, and Abortion*, 73 N.C. L. Rev. 1323 (1995). In the 1990s, the Council of Resident Education in Obstetrics and Gynecology (CREOG) recommended that ob-gyn programs be required (by the Accreditation Council on Graduate Medical Education) to offer training in abortion and contraceptive services. However, "in the ensuing political outcry, the mandatory wording was dropped." Med. Econ., Aug. 10, 1998, at 140.

247. Nat'l Abortion Fed'n & Am. Coll. of Obstetrics and Gynecology, Who Will Provide Abortions? Ensuring the Availability of Qualified Practitioners: Recommendation from a National Symposium 24–25 (1990).

eth century, most states had laws discouraging or prohibiting such clinicians from doing so.[248]

Many health maintenance organizations (HMOs) and family-planning clinics depend upon nurse practitioners, certified nurse midwives, and physician assistants to provide many of the services once provided by physicians.[249] Statutes licensing midwives have been upheld by the courts in spite of objections by physicians to such competition. The U.S. Court of Appeals for the Third Circuit held that a New Jersey statute requiring licensing of midwives did not violate the due process rights of aspiring or unlicensed midwives, or the due process rights of parents desiring to use midwives in the birthing of their children. It found that the licensing scheme was rationally related to the legitimate state interest in protecting the health and welfare of mothers and children.[250] The Kansas Supreme Court held that the "traditional and time-honored techniques" of lay midwifery were not prohibited by Kansas's healing arts licensing laws. The court specifically found that "the fact that a person with medical training provides services in competition with someone with no medical degree does not transform the latter's practices into the practice of medicine."[251]

State laws limiting the abortion practice to physicians would appear to violate the 1992 decision of the Supreme Court in *Casey* by imposing an "undue burden" on a woman's decision to terminate her pregnancy. Such a burden was defined as "a state regulation [which] has the purpose or effect of placing a substantial obstacle in the path of a woman seeking an abortion of a nonviable fetus."[252] Under the Coats Amendment to the Omnibus Consoli-

---

248. Lieberman & Laiwani, *Physician-Only and Physician-Assisted Statutes: A Case of Perceived but Unfounded Conflict*, 49 J. AM. MED. WOMEN'S ASS'N 146–49 (1994).

249. BOSTON WOMEN'S HEALTH COLLECTIVE, THE NEW "OUR BODIES OURSELVES": A BOOK BY AND FOR WOMEN, 666, 671–75 (1992).

250. Sammon v. N.J. Bd. of Med. Exam'rs, 66 F.3d 639 (3d Cir. 1995). In *Lange-Kessler v. State Department of Education*, a federal appeals court upheld New York's Professional Midwifery Practice Act, a law that provided for the practice of midwifery under a written agreement with a licensed physician or hospital.

251. State Bd. of Nursing v. Ruebke, 913 P.2d 142 (Kan. 1996); *see also* Hast. Ctr. Rep., May-June 1997, at 28.

252. Planned Parenthood of Se. Pa. v. Casey, 505 U.S. 833, 846, 876 (1992).

dated Rescissions and Appropriations Act of 1996,[253] residency programs will be deemed accredited by the federal government to receive federal funds even if the residency program chooses not to provide abortion training to its students. However, state and local governments that receive federal funding must also treat these programs as accredited and cannot refuse them "legal status, . . . financial funding, or other benefits" if programs fail to comply with abortion training accreditation.[254] In 1997 that court upheld a Montana statute that prohibits physician assistants from performing abortions.[255] As a result, significant action by state legislatures is now required in order to ameliorate the current desperate situation of so many American women in need of an abortion. Court action may also be taken in those states whose constitutions contain privacy provisions that might invalidate laws inhibiting the obtaining of abortions.

There appears to be no reason why physician assistants, nurse practitioners, and certified nurse midwives could not perform early abortions. This is because they are currently performing procedures with several risks, such as cervical biopsies, endometrial biopsies, Norplant insertions and removal, pelvic exams, and testing and treatment for sexually transmitted diseases.[256]

A study of 2500 first-trimester abortions performed by physician assistants found no difference in the rate of complications for abortions performed

---

253. Pub. L. 104-134, Stat.

254. Foster, *Educational and Legislative Initiatives Affecting Residency Training in Abortion*, JAMA 1777 (2003). In 2002, California enacted a law (AB-2194) requiring either that abortion training be available at each of California's six public medical schools or that ob-gyn residents be given the opportunity to receive abortion training at other institutions. *Id.*

255. Mazurek v. Armstrong, 520 U.S. 968, 973 (1997) (quoting *Casey*, 505 U.S. at 885). The Court stated: "We emphasized that our cases reflect the fact that the Constitution gives the States broad latitude to decide that particular functions may be performed only by [physicians], even if an objective assessment might suggest that those same tasks could be performed by others."

256. *See* Schirmer, *supra* note 244, at 268–70. In 1998, it was found that "half of all Ob/Gyns who currently provide abortions are at least 50 years old" and "only 7 percent of abortions are performed in hospitals, and only 4 percent in physician's offices; the rest are done in clinics." MED. ECON., Aug. 10, 1998, at 144.

by them and those performed by physicians.[257] Abortion skills have been left out of the majority of obstetrical teaching programs, but this may be changing. In 2002, residents in New York City's public hospitals will be routinely trained to perform abortions, and this training will become mandatory in the four-year obstetrics/gynecology residency.

Inasmuch as the scarcity of abortion providers is severely burdening the rights of thousands of women seeking to terminate their pregnancies, such a practice by midlevel providers should be legalized throughout the United States.

good v. Good

Only qualified physicians can adequately perform abortions; permitting others to do so would return our society to the condition prior to *Roe v. Wade*, when they were performed in back alleys.

v.

Following the dawn of the third millennium, the severe shortage of physicians willing to abort babies in most parts of the United States justifies the training of physician assistants, nurses, and midwives to provide most types of such services—particularly in locations where no physicians are willing to do so.

Twenty-first Century Approach

1. Because of
   (a) the severe and increasing shortage of physicians willing to provide abortions in most counties of virtually all states, and
   (b) the dependence of private offices, family-planning clinics and HMOs upon various physician assistants to provide the bulk of services (once provided exclusively by physicians), physician assistants and midwives should be adequately trained and authorized to perform low-risk, first-trimester abortions in all states.
2. Laws inhibiting or prohibiting *all* abortions by various qualified physician assistants and others should be revised.

---

257. Schirmer, *supra* note 244, at 269. "Sixty-one percent endorse abortion in the first trimester, but that support plummets to 7 percent by the third trimester." MED. ECON., Aug. 10, 1998, at 143. Further, the vast majority of abortions are still performed in the first trimester. Most late-term abortions are performed because poor women haven't been able to get quick access to the procedure. *Id.* at 147.

10

Fetal Abuse and Severely Defective Newborns

1. *Current Liberty Interests of Most Pregnant Drug Users*
2. *Withholding Medical Treatment from Severely Defective Premature Infants with Low Birth Weight*
3. *Newborn Screening Requirements: Intersex Conditions*
4. *Attempts to Prevent Premature Infants with Very Low Birth Weight and Futile Results in Certain Prenatal Births*
5. *Poor* and *Expensive Outcomes at Neonatal Intensive Care Units (NICUs)*
6. *Proposal that Parents Share Some Expenses and Time*
7. *Infanticide and Abortions after Viability of Certain Types of Severely Defective Infants*

---

1. Current Liberty Interests of Most Pregnant Drug Users

A 1996 survey by the Department of Health and Human Services (HHS) found that more than 2.4 million youths between the ages of 12 and 17 admitted using an illicit drug at least once during the previous month.[1] This more than doubled the 1992 figures in a mere four years. When other users were also taken into account at the turn of the third millennium, some 3.6 million Americans had become chronically admitted to the drug cocaine alone,

---

1. L.A. TIMES, Aug. 21, 1996, at 1. The Centers for Disease Control and Prevention (CDC) noted that as of 2002 no state had reached the decade-end target of 4.5 infant deaths per 1,000 live births. ORANGE COUNTY REG., June 11, 2005, at News 22.

a major cause of crime. The virtual loss of the "war on drugs" can be gauged by the fact that in the year 2000 prices of cocaine and heroin had fallen to record lows and they remained widely available.[2] In this connection, it has been pointed out that the human species is distinguished by the fact that the action-releasing mechanisms of its central nervous system are "susceptible to the influence of imprintings from the society in which the individual grows up. For the human infant is born—biologically considered—some 10 or 12 years too soon."[3] Infant mortality in 2002 climbed for the first time in more than four decades, in part because older women are having multiple babies via fertility drugs.[4]

An analysis of reviews and studies confirm that prescribed drugs actually do not improve learning or academic performance.[5] In 2000 there was a sharp rise in the number of young children using Ritalin, Prozac and other powerful psychiatric drugs; thousands of these children die annually.[6] The brain is almost always impaired by psychiatric drugs.[7] Only at the turn of the

---

2. Orange County Reg., Mar. 23, 2000, at News 22.
3. It acquires its human character, upright stature, ability to speak, and the vocabulary of its thinking under the influence of a specific culture, the features of which are engraved, as it were, upon its nerves. The constitutional patternings, which in the animal world are biologically inherited, are in the human species matched largely by socially transmitted forms, imprinted during what have long been known as the "impressionable years," and rituals have been everywhere the recognized means of such imprinting.

Campbell, Myths to Live By 44 (1971).

4. Orange County Reg., Feb. 12, 2004, at News 15. The Centers for Disease Control and Prevention also reported that more babies are being born prematurely or at low birth weights because more doctors are inducing labor and using cesarean sections for delivery.

5. Int'l Herald Trib., Oct. 29, 1999, at 4.

6. Breggin, Talking Back to Ritalin: What Doctors Aren't Telling You about Stimulants for Children (1998); Parade, Jan. 6, 2000, at 5. In 2004, President Bush signed a law making it a separate crime to harm a fetus during the commission of a violent crime. The law defines an "unborn child" as a child carried in utero, which "means a member of the species homo sapiens, at any stage of development, who is carried in the womb." There are 29 states with an unborn-victims law. Orange County Reg., Mar., 2004, at News 15.

7. Breggin, Brain-Disabling Treatments in Psychiatry (1997).

third millennium are we becoming somewhat more cognizant of the extent of drug treatment for alleged "psychiatric disorders" and the fact that the propaganda for this remarkable perspective is financed by drug companies and spread by the media.[8] However, there are few public discussion soft politics of biopsychiatry, the lack of scientific confirmation, the claims being made about biological and genetic causes, and the efficacy of medications.[9] Nor are we always being well-informed. The notion that Prozac corrects biochemical imbalances has been speculation or propaganda from the biological psychiatric industry.[10] All of this activity adds fuel to the drug problem in the United States. Even in hospital patients, the incidence of serious and fatal adverse drug reactions was found to be extremely high at the end of the twentieth century. It is between the fourth and sixth leading cause of death,[11] and alcohol a leading cause of defective and drug-addicted infants.[12] At the same time women show variance in the effects of drugs both on themselves and on their fetuses.[13] For example, women with very severe acne, the kind that causes large cysts and leaves deep scars, when given the drug Accutane must (since 2004) be enrolled in a national registry. This is because if such a woman becomes pregnant, even within 30 days after stopping the drug, her baby can suffer severe brain and heart defects, mental retardation, and other abnormalities.[14]

---

8. BREGGIN & COHEN, YOUR DRUG MAY BE YOUR PROBLEM 4 (1999); *see* Ch. 6, § 3a.

9. *Id.*

10. *Id.* at 7. "Some people do turn out to have subtle, undetected biochemical imbalances. However, there is no reason to give them drugs like Prozac or Xanax that cause biochemical imbalances and disrupt brain function." *Id.*

11. Pomeranz & Corey, 279 JAMA 1200–05 (1998).

12. *A Review of the Neurobehavioral Deficits in Children with Fetal Alcohol Syndrome or Prenatal Exposure to Alcohol*, 22 ALCOHOLISM: CLINICAL & EXPERIMENTAL RES., 279–92 (1998); Coleman & J. Kay, *Biology of Addiction*, 25 OBSTETRICS & GYNECOLOGY CLINICS N. AM., 1–19 (1998).

13. Mattson & Riley, *supra* note 12, at 285–86; E. L. Abel & R. J. Sokol, *Incidence of Fetal Alcohol Syndrome and Economic Impact of FAS-Related Abnormalities*, 19 DRUG & ALCOHOL DEPENDENCY, 5170 (1987); Mattson et al., *Heavy Prenatal Alcohol Exposure With or Without Physical Features of Fetal Alcohol Syndrome Leads to IQ Deficits*, 131 J. PEDIATRICS, 718–21 (1997).

14. ORANGE COUNTY REG., Feb. 28, 2004, at News at 20. The FDA also requires drugstores that dispense it to register.

The recent doubling of drug use in a mere four years constitutes fetal abuse where the user is a pregnant woman. For example, methamphetamine (a central nervous system stimulant) constricts the woman's blood vessels, cutting off nutrients to the fetus and brings on constrictions that separate the placenta from the wall of the uterus, pushing it out weeks too soon, often with disastrous results.[15] All psychiatric drugs cross the placenta and enter into the fetal blood stream "readily, rapidly, and without limitation,"[16] and such drugs should be avoided during pregnancy. Women who smoke during pregnancy can increase their babies' risk of developing attention deficit disorder and learning difficulties.[17] Drug-related morbidity and mortality have been estimated to cost more than $136 billion a year and account for 140,000 deaths annually.[18] A 1997 study concluded that adverse drug events are associated with high numbers of infants born addicted to drugs, a significantly prolonged length of stay in hospitals, increased economic burden, and an almost two-fold increased risk of death.[19]

Cocaine is most addictive but speed lasts longer. In California criminal laws generally do not apply to fetuses.[20] A study in February of 2000 found that a single drinking binge by a pregnant woman could be enough to permanently damage the brain of her unborn child.[21] As one writer put it, "[T]hese

15. A study released by the National Center for Addiction & Substance Abuse, Columbia University, stated: "Eighth graders in rural America are 104 percent likelier than their counterparts in big cities to use amphetamines, including methamphetamines, and 50 percent likelier to use cocaine."

ORANGE COUNTY REG., Jan. 27, 2000, at News 16.

16. The FDA has not approved any psychiatric drugs for use during pregnancy or lactation. Stowe et al. (1998) (1992).

17. ORANGE COUNTY REG., Apr. 13, 1999.

18. JAMA 301–06 (1997).

19. In a single state (Washington) legislative researchers estimated that there are between 7,500 and 10,000 infants born each year who are addicted to drugs. Further, they are treated at a center where about 75 percent of infants had at least one older sibling who was treated there as well.

L.A. TIMES, Sept. 13, 1998, at B4.

20. *Id.* As noted above, a homicide law may apply to a third party who kills the fetus of a pregnant woman. *See supra* note 6.

21. In humans, this brain growth spurt starts in the sixth month of gestation and continues for two years. During the brain growth spurt, called synap-

infants are not given a choice as to whether or not they want to be born addicted, yet their mothers have direct control over their status of addiction or nonaddiction."[22]

Few pregnant women have been prosecuted for continuing to use drugs even when they have full knowledge of the probable consequences for their infant when and if delivered. Some of this reluctance may be due to political fears resulting from racial overtones because between 70 and 80 percent of pregnant women prosecuted have been black or Hispanic.[23] This presents ethical and legal crises. Consider this case that arose in Wisconsin: A 35-year-old intoxicated woman nearly nine months pregnant was wheeled into a local hospital for an emergency cesarean section the doctors deemed necessary. She then said, "I'm just going to go home and keep drinking and drink myself to death, and I'm going to kill this thing because I don't want it anyways." Later she gave birth to a girl whose blood-alcohol level was 0.199, nearly twice the threshold for a legal finding of intoxication in Wisconsin. The baby was smaller and her forehead was flatter than normal as with Fetal Alcohol Syndrome (FAS). The mother was actually charged with attempted murder, the first time when a fetus had not died that a prosecutor made such a charge. Like most states, Wisconsin had no statute prohibiting drinking while pregnant, no feticide statute, and it did not recognize a fetus as a human under its criminal law.[24] The judge stated that in his view:

> togenesis, brain cells must receive a balanced signal from two types of neurotransmitter chemicals, glutamate and GABA. If this signal is disrupted, the developing brain cells are programmed to commit suicide.

SCIENCE (2000); ORANGE COUNTY REG., Feb. 11, 2000.

22. Van Guusven, *Dilemmas of Providers Treating Pregnant Women Who Use Drugs: Patient Confidentiality versus the Duty to Report Drug Use*. J. HEALTH & HOSPITAL L. 243, 250. At a national meeting of the American Association for the Advancement of Science in February of 2004, it was indicated that in most women, two cocktails is enough to elevate blood alcohol levels to 0.07 percent. In unborn mice, this concentration is enough to kill developing brain cells., ORANGE COUNTY REG., Feb. 15, 1004.

23. ROBERTS, KILLING THE BLACK BODY 172–76 (1997).

24. Wisconsin's statute authorized "any necessary measure" to protect the fetus. WIS. STAT. ANN. § 48.13 (West 1997); LAFAVE & SCOTT, CRIMINAL LAW, § 7.1, at 607 (1986). "Being born alive required that the fetus be totally expelled from the mother and show a clear sign of independent vitality, such as respiration, although respiration was not strictly required." *Id.*

> There is no question that the young victim was born alive and qualifies as a human being under Wisconsin's homicide laws . . . the instrumentality of the attempted homicide in this case was not the shooting of a bullet or the plunging of a knife. Instead, it was the massive consumption of a potentially deadly quantity for intentional or reckless acts.[25]

She might have aborted the fetus (see Chapter 9). In this case, her "intentional and reckless act" resulted from her giving birth rather than not doing so. Nevertheless, on appeal, the Wisconsin Supreme Court overturned the lower court decision, ruling that she could not be charged with attempted murder because the fetus is not a human being.[26] It is noted that even in Canada, a fetus has no standing until birth and Canada's Supreme Court has ruled in a case involving a glue-sniffing mother that it will not expand existing laws to allow mothers to be ordered to protect fetuses.[27] State Supreme Courts in Florida, Kentucky, Nevada, Ohio, and Wisconsin had also struck down child-endangerment laws applied to fetuses. Most of them ruled, "a fetus was not a person" under their criminal laws.[28] However, in 2002 the federal DHHS reclassified developing fetuses as "unborn children." In this manner, it expanded prenatal care under the State Children's Health Insurance Program (SCHIP), including that of illegal aliens, to provide health

---

25. Courts have refused to let defendants escape criminal liability on the basis of voluntary intoxication. In *Montana v. Egelhoff*, 518 U.S. 37 (1996), for example, Egelhoff was convicted of murdering two strangers. He tried to argue that he was so intoxicated that he could not remember what had happened. In *Powell v. Texas,* 392 U.S. 514, 531 (1986), Powell argued that he was an alcoholic and was being punished for his "status" as an alcoholic. The court disagreed.

26. State *ex rel.* Angela M. W. v. Kruzicki, 561 N.W.2d 729 (Wis. 1997) *rev'g* 541 N.W.2d 482 (Wis. Ct. App. 1995). However, in 2005, a Chicago judge upheld a wrongful death lawsuit concerning the accidental destruction of a frozen embryo. He actually ruled that "a preembryo is a human being . . . whether or not it is implanted in its mother's womb." ORANGE COUNTY REG., Feb. 6, 2005, at News 24.

27. *"The common law does not clothe the courts with power to order the detention* of pregnant women for the purpose of preventing harm to her [sic] unborn child." L.A. TIMES, Nov. 1, 1997, at A3.

28. ORANGE COUNTY REG., Aug. 17, 1996, at News 5; *see In re* Dettrick, 263 N.W.2d 37 (Mich. Ct. App. 1977); *In re* Steven S., 126 Cal. App. 3d 23 (Cal. App. 1981).

insurance coverage for children in poor families. A "child" is defined as "an individual under the age of 19 including the period from conception to birth,"[29] which "has, therefore, been criticized as being more of a political maneuver to redefine the status of the fetus rather than a means to extend health care access and prenatal care to those in need."[30]

Near the end of the twentieth century, the South Carolina Supreme Court became the first state supreme court to uphold the criminal prosecution of pregnant women who used drugs under that state's law.[31] The court ruled that after a woman gave birth to a baby that tested positive for cocaine, the woman could be convicted of child neglect and sentenced for up to eight years in prison.[32] The court stated:

> The abuse or neglect of a child at anytime during childhood can exact a profound toll on the child herself as well as on society as a whole . . . However, the consequences of abuse or neglect that takes place after birth often pale in comparison to those resulting from abuse suffered by the viable fetus before birth. This policy of prevention supports a reading of the word "person" to include viable fetuses.[33]

The taking of cocaine during the third trimester of pregnancy was held under South Carolina's Supreme Court to constitute criminal child neglect.[34]

---

29. State Children's Health Insurance Program; Eligibility for Prenatal Care and Other Health Services for Unborn Children, 42 C.F.R. Pt. 457. This revision also appears to allow states to cover prenatal and delivery care for unborn children of illegal immigrant residents. *Id.*

30. Rogers, A.B.A. BIOETHICS BULL., Winter/Spring 2005, at 2.

31. *See* S.C. CODIFIED LAWS, §§ 34-23B-2, -20A-23 (Michie 1999).

> Any person having the legal custody of any child or helpless person who shall, without lawful excuse, refuse or neglect to provide . . . the proper care and attention for such a child or helpless person; shall be guilty of a misdemeanor.

32. Any woman who uses cocaine after the 24th week of her pregnancy is guilty of distributing a controlled substance to a minor. Whitner v. State, 492 S.E.2d 777, 780 (1997).

33. Whitner v. South Carolina, 1996 WL 393 164, at * (S.C. Oct. 27, 1997) (opinion withdrawn on grant of rehearing).

34. Whitner v. South Carolina, 492 S.E.2d 777 (1997), *cert. denied*, 523 U.S. 1145 (1998).

On the legislative front, in 1998 the Wisconsin and South Dakota state legislatures enacted laws authorizing their judges to confine pregnant women who abuse alcohol or drugs for the duration of their pregnancies. These statutes could become models in the twenty-first century, although they may also be improved. The Wisconsin statute defines "unborn child" as a "human being from the time of fertilization to the time of birth,"[35] and its provisions "apply throughout an expectant mother's pregnancy."[36] Their "need to be free from physical harm," must be protected when "an expectant mother of an unborn child suffers from a habitual lack of self-control" in the use of alcoholic beverages or controlled substances "to a severe degree." To accomplish this goal, a court may "determine that it is in the best interest of the unborn child for the expectant mother to be ordered to receive treatment, including patient treatment.[37] Officials who determine the unborn child is in immediate danger may "take any necessary action, including confinement of the pregnant women, to protect the unborn child."[38] A guardian ad litem can be appointed to represent the fetus, and after a hearing, a court can order mandatory commitment and treatment when the mother refuses voluntary treatment,[39] as long as it is necessary to protect the unborn child.[40]

Constitutional arguments continue to be raised. Just prior to the turn of the third millennium, a federal appellate court held that a South Carolina hospital policy of testing pregnant women for cocaine use without their consent did not violate the U.S. Constitution, the Civil Rights Act of 1964, or South Carolina common law.[41] However, on appeal after the turn of the third millennium, the Supreme Court reversed the Circuit Court and held that the

35. A better and more accurate term would have been "fetus."

36. Wis. Stat. § 48.02 (1998); (West 2000); S.D. Codified Laws §§ 34-20A-63 to -70 (2000). In South Dakota a pregnant woman can be placed in custody for up to nine months.

37. Wis. Stat. § 48.01 (1) (am).

38. *Id.* § 48.193.

39. *Id.* §§ 48.213, 48.235.

40. *Id.* §§ 48.205, 48.345, 48.347. This may take place in a private or public residential substance abuse treatment facility, or in a hospital. *Id.* § 48.207.

41. Ferguson v. Charleston, 186 F.3d 469 (4th Cir. 1999). The Supreme Court has noted that the state may have an interest in potential life at the previability stage. Planned Parenthood of Se. Pa. v. Casey, 505 U.S. 833, 870 (1992); *see also* Addington v. Texas, 441 U.S. 418 (1979).

Fourth Amendment was violated by subjecting pregnant women to nonconsensual drug tests and turning the results over to the police, without a warrant.[42] This means that there is no longer a reason for a woman to choose between abortion and treatment or incarceration and unfortunate loss especially for the poor.[43] Many medical associations condemn the prosecution of pregnant women in favor of counseling.[44] However, voluntary treatment programs have had a poor success rate,[45] whereas sanctions against pregnant drug users help steer them toward treatment in the same manner. This can be brought to the attention of the Court during the hearing phase in order to reasonably determine the "substantial risk" of fetal injury in a particular case.[46] This approach appeared to achieve the goals of "a proper balance between an individual's right of freedom and society's need to protect public health and safety."[47] Nothing precludes the woman from exercising her right to carry or not carry her fetus to term viability and appears to generally remain a non-legal issue in fetal abuse detention cases. Legislation to accomplish this goal provides Equal Protection because of the rational relationship between its purpose and the class being covered.[48]

42. 121 S. Ct. 1281 (2001). The Supreme Court held the state program within the "closely guarded category of constitutionally permissible suspicionless searches." The dissent (which included Chief Justice Rehnquist) noted the health - care professionals "ministering not just to the mothers but also to the children whom their cooperation with the police was meant to protect."

43. Frank et al., *Growth, Development, and Behavior in Early Childhood Following Prenatal Cocaine Exposure: A Systematic Review*, 285 JAMA 1613–25 (2001).

44. Lynn M. Paltrow, *Pregnant Drug Users, Fetal Persons, and the Threat to* Roe v. Wade, 62 ALB. L. REV. 999, 1044 (1999).

45. Shaver, *Prosecute the Mothers of Addiction*, CAL. LAW., Nov. 1989, at 72.

46. Handwerker, *Medical Risk: Implicating Poor Pregnant Women*, 8 SOC. SCI. MED. 665–75 (1994).

47. Wilton, *Compelled Hospitalization and Treatment During Pregnancy: Mental Health Statutes as Models for Legislation to Protect Children from Prenatal Drug and Alcohol Exposure*, 25 FAM. L.Q. 149, 150 (1991); *see also* NELSON & MARCHALL, ETHICAL AND LEGAL ANALYSES OF THREE COERCIVE POLICIES AIMED AT SUBSTANCE ABUSE BY PREGNANT WOMEN 95–111 (1997).

48. Rinaldi v. Yeager, 384 U.S. 305, 308–09 (1966). In *Rinaldi v. Yeager,* the court upheld a California disability insurance system excepting those resulting from normal pregnancy as not discriminatory and as serving a legitimate government interest in keeping costs low; *see also* General Electric Co. v. Gilbert, 429 U.S. 125 (1976).

good v. Good

The autonomy and liberty interests of pregnant drug users should be paramount to society's interest in the possible birth of nondefective, or drug-addicted "cocaine babies."

v.

Society's interest in avoiding and preventing such births must be paramount to the liberty interests of pregnant drug users who wish to continue such drug use throughout pregnancy.

Twenty-first Century Approach

Most states should consider enacting legislation authorizing mandatory counseling, certain treatments where necessary, and detention of pregnant drug users who plan to deliver in those instances where such use may endanger their fetuses that otherwise could become wards of the state and suffer a significant reduction in their quality of life.

> Pronouncements about the quality of life might be dangerous because they guide our lives by lights that are quite contrary to our own. Generations of recipients have responded to the efforts of well-intentioned do-gooders by protesting they "don't want to be done good to."
>
> Litcher, *The Lives to Come* (1996), p. 286

2. Withholding Medical Treatment from Severely Defective Premature Infants with Low Birth Weight

At the beginning of the third millennium, protestors against being "done good to" are being "done to" despite their protests. In the United States, 65 percent of infant deaths that occur among low birth weight infants accounted for 7.6 percent of all infants born at the turn of the third millennium.[49] The past quarter century has witnessed increased survival of premature and low

---

49. JAMA 195 (2002). As of 2004, the number of fetal deaths is now equal to the number of infant deaths in the United States. Furthermore, women whose first infant was severely small for its gestational age had significantly higher risk of late fetal death at 28 or more weeks of gestation in the subsequent pregnancy. Zhang & Klebanoff, N. Eng. J. Med. 754 (2004) (citing Surkan et al., 777–85).

birth weight infants,[50] but the price of surviving resulted in severe disabilities. For example, if they're born at 23 or 24 weeks gestation, up to 35 percent may survive; however, 50 to 70 percent of these survivors will have permanent disabilities, such as cerebral palsy, impaired hearing, and vision with major learning problems.[51] Consider these cases:

a) Karla Miller's daughter, 10-year-old *Sydney Miller, was born when* Karla was just *22 weeks pregnant*. She and her husband *decided to let the baby die* quietly in her arms. But a hospital official said it was a *policy to resuscitate all* babies who weigh at least a pound. A doctor they'd never seen took Sydney from them. "He put a breathing tube into her lungs and from that day forward, *she has been* tortured, *medically tortured*." So far, she has undergone *13 operations*. The Millers sued the hospital for medical expenses *and won $42 million*.[52]

b) When Ishmiel Azad Mohammad was born, he weighed about as much as a can of Coke and could fit in his doctor's hand. Doctors at Mission Hospital gave him less than a 5 percent chance of surviving. His twin brother, Jihad Amir, lived only an hour and a half. Their mother's pregnancy lasted only 24 weeks. They were delivered three months early by an emergency cesarean section.[53]

---

50. Grogaard et al., *Increased Survival Rate in Very Low Birth Weight Infants (1500 grams or less): No Association With Increased Incidence of Handicaps*, 117 J. PEDIATRICS 139–46 (1990).

51. Blackman, *Neonatal Intensive Care: Is It Worth It?*, 38 PEDIATRIC CLINICS N. AM. 1497–1511 (1991).

52. http://adbnews.go.dom/onair/CloserLook/wnt000705_preemie-feature.html. It has been noted that air sac development sufficient for gas exchange does not occur until the 23rd week of gestation or later. Beddis et al., *New Technique for Servo-Control of Arterial Oxygen Tension in Preterm Infants*, ARCHIVES DISEASE CHILDHOOD 278–80 (1979).

53. ORANGE COUNTY REG., Dec. 21, 2000. Very preterm birth and very low birth weight are principal risk factors for cerebral palsy, which affects motion and causes muscle movement problems due to brain abnormalities. Approximately 500,000 persons have cerebral palsy in the United States, and the condition cannot be cured. Crother et al., 2669–70 (2003).

The life of a person without any quality is not one desired; only he (or his surrogate) can decide what that quality is for him. As stated by the writer Kitcher, "Instead of focusing on pleasure or the absence of pain, I take the core of a minimally valuable life to be a person's chosen ideal of that life's direction."[54] As regards a standard, it has been remarked, "that a clear conceptual basis for quality-of-life measures is lacking, and the few attempts to develop models or operational definitions of quality of life have been woefully inadequate." Government attempts to do so have "often been based on arguments of authority rather than on rational debate."[55] As for physicians and counselors, such attempts are merely examples "of the medicalization of everyday life, that is, the unwarranted straying of medical practitioners into fields beyond their expertise."[56] Attention should be "shifted to means of assessing this perspective by way of methods capable of reflecting *individuals' concerns* when they become ill." In the case of healthcare, that person's choice should not necessarily be "the government's choice."

For example, the DHHS issued regulations that state child-protection programs must meet in order to qualify for participation in the state grant program.[57] In lieu of civil or criminal actions against doctors, hospitals, or parents for violation of the rules, they provide for withdrawal of federal assistance to state agencies.

The current Child Abuse Prevention and Treatment and Adoption act[58] sets forth federal rules on the treatment of severely handicapped infants:

> Withholding of medically indicated treatment means the failure to respond to the infant's life-threatening conditions by *providing* treatment (including appropriate *nutrition, hydration*, and medication), which, in the treating physician's or physicians' reasonable medical

54. KITCHER, THE LIVES TO COME, 287 (1996).

55. Le Plege & Hung, *The Problem of Quality of Life in Medicine*. 278 JAMA 47–50 (1997). Only at 28 weeks of gestation (or about 70 percent of development) "does the fetus acquire sufficient neocortical complexity to exhibit some kind of the cognitive capabilities typically found in newborns." SHERMER, THE SCIENCE OF GOOD AND EVIL (2004) (citing Flower, *Neuromaturation and the Moral Status of Human Fetal Life*, *in* ABORTION RIGHTS AND FETAL PERSONHOOD 1989)).

56. *Supra* note 55, at 50.

57. 45 C.F.R. § 1340.15(c)–(d) (2000).

58. 42 U.S.C.A. § 5106(g)(6) (West 2000).

judgment, will be most likely to be effective in amelioration or correcting all such conditions, *except* that the term does not include the *failure to provide treatment* (*other than* appropriate *nutrition, hydration*, or medication) to an infant *when,* in the treating physician's reasonable judgment:

(A) the infant is chronically and irreversibly comatose;
(B) the provision of *such treatment would* (i) merely prolong dying, (ii) not be effective in ameliorating or correcting all of the infant's life threatening conditions, or (iii) otherwise *be futile in terms of the survival of the infant*; or
(C) the provision of such *treatment would be virtually futile in terms of survival of the infant and* the treatment itself under such circumstance would be *inhumane*.[59] (Emphasis added.)

These rules appear to eliminate the well-founded rule that it is the parents' right to make medical decisions for their child.[60] The "substituted privacy right" is based upon the presumption that parents act in the best interest of their child.[61] An appellate court upheld a parent's decision as overriding any state interest where it was backed by "uncontroverted medical evidence as to the terminal condition of the child, and the illness is not reversible and not curable."[62] The AMA Code of Medical Ethics, Section 2.215, "Treatment Decisions for Seriously Ill Newborns," states:

> Physicians must provide full information to parents of seriously ill newborns regarding the nature of treatments, therapeutic options, and expected prognosis with and without therapy, so that parents can make informed decisions for their children about life-sustaining treatment. Ethics Committees should also be utilized to facilitate parental decision making.[63]

59. 42 U.S.C.A. § 5106(g)(8) (West 2000).
60. *In re* Guardianship of Barry, 445 So. 2d 365 (1981).
61. *In re* L.H.R., 321 S.E.2d 716, 722 (Ga. 1984).
62. *In re* Rosebush, 491 N.W.2d 633, 637 (Mich. Ct. App. 1992).
63. Sklansky, *Neonatal Euthanasia: Moral Considerations and Criminal Liability*, 27 J. MED. ETHICS 5, 8–9 (2001) (concluding that the moral and legal status of passive neonatal euthanasia is unclear and should be reviewed and clarified, and quoting Opinion 2.215.

Where federal funds are used, these considerations are replaced in the current law by listing the alternatives for lack of treatment that include "not be effective," "futile in terms of survival," and "virtually futile in terms of survival and the treatment itself . . . would be inhumane." These both limit nontreatment and leave room for interpretation by doctors and medical ethics committees. Nontreatment has often required that nutrition and hydration be given. Further, the use of the term "survival" eliminates "quality-of-life" considerations that may not be employed in the decision-making process.[64]

Many ethicists would consider this provision with conditions limited solely to those pertaining to survival as unethical where "quality of life cannot be excluded. It is irrational to condemn an infant to live without any, and when a denial of quality of life in decision making constitutes an absolute, it could lead to the virtual bankruptcy of the parents if they were required to care for such a severely defective child for decades without the possibility that its life could have any other meaning. It may constitute a religious (the sacredness of life) criterion or a cultural one, rather than another human concept of life. Many components of quality of life must be measured by addressing "each objective and subjective component (symptom, condition, or social role) that is important to members of the patient population and susceptible of being affected, positively or negatively, by intervention."[65] Thus, in England, Sir Stephen Brown, president of the High Court's family division, approved the doctors' plans to withdraw artificial respiration over the parents' objection. Doctors believed this would not be in her best interests because it would reduce her quality of life for the limited time left to the child:

> The case came to court because the parents oppose the action proposed by the doctors. But it is well-established British law that the courts will not compel doctors to carry out treatment against their wishes. Medical evidence was not in dispute, he said: The child was approaching death. The mother had stated, "Our religion is that life should always be preserved."[66]

---

64. 45 C.F.R. § 1340.15(b)(2)(ii) ( ).
65. N. Eng. J. Med. 836 (1996).
66. Orange County Reg., Nov. 20, 1997, at News 35.

Even though "the effects of medical treatments and programs on quality of life should not be ignored simply because such effects are difficult to measure,"[67] the DHHS has chosen to do so. This largely prevents physicians and patients (or their surrogates) from exercising their minds independently. Further, "[T]he estimated advantage of one treatment over another in cost-effectiveness can be substantially altered, and even reversed, by adjusting the primary measure of effectiveness for *quality of life*."[68] In 2003 a detailed study was made of school-age regional cohorts of children born in the 1990s with birth weights less than 1000 g (low birth weight) or earlier than 28 weeks' gestation (quite preterm). It was determined that at school age they "continue to display cognitive, educational, and behavioral impairments."[69] The largest study of early school-age children who had been born at 25 or fewer completed weeks of gestation in the United Kingdom and Ireland was published in 2005. It showed that "cognitive and neurologic impairment is common at school age":

> Among these extremely preterm infants, neonatal survival to discharge was low: 1 percent among those born at 22 weeks of gestation, 11

---

67. N. Eng. J. Med. 839 (1996). The lack of any objective measure for quality of life can be found in the attempted definition by the AMA in Section 2.17 (named "Quality of Life"), which states:

> Quality of life, as defined by the *patient's interest and values*, is a factor to be considered in determining what is best for the individual. It is permissible to consider quality of life when deciding about life-sustaining treatment in accordance with Opinion 2.215, "Treatment Decisions for Seriously Ill Newborn" and 2.22, "Do-Not-Resuscitate Orders" (DNR).

68. N. Eng. J. Med. 836 (1996).

69. Anderson & Doyle, *Neuro Behavioral Outcomes of School-Age Children Born of Extremely Low Birth Weight or Very Preterm in the 1990s*, JAMA 3264 (2003).

> Extensive neuropathologic abnormalities may be observed in preterm children; white matter injury is the most prominent cerebral abnormality associated with prematurity. In many instances, the trauma and problems associated with having a very premature infant are sufficient to result in short- and long-term consequences on the families' environment.

Id. at 3271.

percent at 23 weeks, 26 percent at 24 weeks, and 44 percent at 25 weeks. The rates of *survival with no disability* at 6 years of age were even *more troubling:* none among infants born at 22 weeks of gestation, 1 percent at 23 weeks, *3 percent at 24* weeks, and 3 percent at 25 weeks. These data are all *the more compelling because they are* the most up-to-date data available and are relevant to *current practices of obstetrical and neonatal intensive care.*

Some 92 percent of such children born at 25 weeks of gestation were shown to be disabled at 6 years of age (and thereafter) even after receiving neonatal intensive care.[70] The results of a 2005 study of very low birth weight infants weighing less than 1000 g showed how they fared substantially worse than the children born at term with a normal birth weight. They virtually exclusively included the most severe medical problems.[71]

good v. Good

"Quality of life" cannot be objectively defined and hence it cannot be a regulatory criterion in determining an infant's life-threatening condition, the provision of treatment, or as to its nutrition and hydration.

v.

"Quality of life" should be used as a criterion for a patient by the patient's parents (or surrogate).

Twenty-First Century Approach

1. An infant's parent is presumed to act in the child's best interest. A parent's informed decision (or the surrogate's) should be considered critical in connection with whether or not a treatment is futile.
2. The Department of Health and Human Services (DHHS) regulation should be changed to eliminate the criteria of "futility" solely in terms of "survivorship" and substitute "quality of life, rationally determined by competent parents" (or their surrogates).

---

70. Marlow et al., *Neurologic and Developmental Disability at Six Years of Age After Extremely Preterm Birth,* N. Eng. J. Med. 9 (2005); Vohr & Allen, *Extreme Prematurity—The Continuing Dilemma,* N. Eng. J. Med. 71 (2005).

71. Hack et al., *Chronic Conditions, Functional Limitations, and Special Health Care Needs of School-Aged Children Born with Extremely Low-Birth-Weight in the 1990s,* JAMA 318–25 (2005); *see also* Tyson & Suigal, *Outcomes for Extremely Low-Birth-Weight Infants Disappointing News,* JAMA 371 (2005).

3. Newborn Screening Requirements: Intersex Conditions

Although 36 states have established newborn screening advisory committees, it is not clear how often health departments ask bioethics committees to review or approve new tests. Two of these states (Maryland and Wyoming) required consumer participation; and the exclusion of consumer input has been condemned:

> Responses to our questions suggested that many experts in newborn screening perceive the public as unable to understand technical issues sufficiently to be involved in policy-making. In contrast, it is a fundamental assumption of the U.S. legal system that the general public can understand and utilize complex information to make significant decisions, such as when jurors consider highly technical forensic DNA data to reach verdicts.[72]

New genetic tests are marketed as services and were not regulated by the FDA because of the absence of federal legislation requiring newborn screening.[73] For example, the recent development of electrospray tandem mass spectrometry makes it possible to screen newborns for many rare inborn errors of metabolism. A 2003 study found that more cases of inborn errors of metabolism are diagnosed by screening with tandem mass spectrometry than are diagnosed clinically.[74] However, 48 states have such programs and 32 of them

72. Am. J. Pub. Health 1276 (1997). For example, in 2003 a study was made of 358 infants born at less than 30 weeks of gestation who remained dependant on supplemental oxygen at 32 weeks of postmenstrual age. It was concluded that a higher oxygen-saturation range in extremely preterm infants who were dependent on supplemental oxygen conferred no significant benefit with respect to growth and development and resulted in an increased burden on health services. Askie et al., N. Eng. J. Med. 959 (2003).

73. Am J. Pub. Health 1276 (1997). It is interesting to note that couples applying for a marriage license in New York receive information about genetic diseases along with data about rubella and HIV/AIDS. Hastings Ctr. Rep., Oct. 1996, at 42.

74. N. Eng. J. Med. 2304 ("It is not yet clear which patients with disorders diagnosed by such screening would have become symptomatic if screening had not been performed."). As regards previous increases in neonatal mortality for infants born on the weekend, a study in 2003 found "no evidence that the quality of perinatal care in California was compromised during the weekend." Gould et al., JAMA 2962 (2003).

require educational programs or materials for parents and/or the general public. Most states permit refusal by parents.[75]

The U.S. Supreme Court in 1986 ruled that Congress had not authorized federal agencies to regulate nontreatment decisions in hospitals and newborn nurseries.[76] Nevertheless, in Texas, after a woman who was 23 weeks into her pregnancy experienced contractions and possible rupture of membranes, a meeting was held to discuss a treatment plan. The father did not agree and refused to sign a consent form; however, the hospital's (HCA's) physician went ahead with it. But the child ended up with extreme mental and physical disabilities. When the parents sued the hospital, the jury found they had not consented to treatment, and granted them significant dollars in damages.[77] However, the Texas Supreme Court actually affirmed the reversal of that decision allowing the right of the physicians to go ahead with their treatments without a parent's consent. [78] A Texas bioethicist actually praised his state's high court ruling in 2004, while he was criticized by many others.[79]

---

75. AM J. PUB. HEALTH 1280–81 (1997). Maryland and Wyoming have been alone in requiring by law that all newborn screening be implemented only after obtaining a signed parental informed consent. *Id.* at 1286. The National Center for Human Genome Research has instructed the Ethical, Legal, and Social Implications Program (ELSI) to do research on "DNA data banking policies, DNA identification standards, legal definitions of 'genetic privacy,' and intrafamilial communications issues," BIOTECHNOLOGY, 1996, 199.

76. Bowen v. Am. Hosp. Ass'n, 476 U.S. 610 (1986). At the time of trial, she "could not walk, talk, feed herself, or sit up on her own . . .[,] was legally blind, suffered from severe mental retardation, cerebral palsy, seizures, and spastic quadriparesis in her limbs . . .[,] could not be toilet-trained, required a shunt in her brain to drain fluids, and needed care twenty-four hours a day." *Id.*

76. Miller v. HCA, 47 Tex. Sup. Ct. J. 12, 118 S.W.3d 758 (Tex. 2003). The court relied on its 83-year-old (1920) dicta that even though "there was an absolute necessity for a prompt operation," the situation was "not emergent in the sense that death would likely result immediately upon failure to perform it." Moss v. Rishworth, 222 S.W. 225 (Tex. App. 1920).

77. *Miller*, 118 S.W.3d 758. The court relied on its 83-year-old (1920) dicta that even though "there was an absolute necessity for a prompt operation," the situation was "not emergent in the sense that death would likely result immediately upon failure to perform it." Moss v. Rishworth, 222 S.W. 225 (Tex. App. 1920).

78. *Miller,* 118 S.W.3d 758.

79. *See* Robertson, *Extreme Prematurity and Parental Rights After Baby Doe*, Hast. Ctr. Rep. 2004, at 39. However, the following year he was criticized by eight

A physician will normally have a duty to warn the patient's immediate family members who may be subject to a genetically transmissible condition that the patient's relatives could be injured or endangered as a result of the physician not disclosing the information. Such courts will hold possible liability for damage to a third party (such as a member of his family).[80]

It has also been found that between 1.7 and 4 percent of the world population is born with intersex conditions.[81] Genital reconstruction surgery for children to conform to normal often results in an inability to achieve orgasm or even to reproduce through artificial insemination.[82] It has been held that a single such surgery will not necessarily permit a change in a child's birth certificate.[83] The Criminalization of Female Genital Mutilation Act of 1996

---

experts who were published in the *Hastings Center Report. See* Hastings Ctr. Rep., Jan.-Feb. 2005, at, 4–7. As one of them put it, with regard to such physicians: "We are now torturing babies to life, the way we used to torture them to death." *See also* Cole, *Extremely Preterm Birth—Defining the Limits of Hope*, 343 N. Eng. J. Med. 429–30 (2000). It has been pointed out how "unfortunate" the decision of the Texas Supreme Court was for the Millers and that were society to favor physicians, "public support may be morally obligatory." Avinas, *Extremely Preterm Birth and Parental Authority to Refuse Treatment—The Case of Sidney Miller*, N. Eng. J. Med. 2123 (2004).

80. Safer v. Pack, 677 A.2d 1188 (N.J. Super. Ct. App. Div. 1996).

> We see no impediment, legal or otherwise, to recognizing a physician's duty to warn those known to be at risk of avoidable harm from a genetically transmissible condition. In terms of foreseeability especially, there is no essential difference between the type of genetic threat at issue here and the menace of infection, contagion or a threat of physical harm . . . the individual or group at risk is easily identified, and substantial future harm may be averted or minimized by a timely and effective warning.

Id. at 1192.

81. Greenberg, *Defining Male and Female: Intersexuality and the Collision Between Law and Biology,* 41 Ariz. L. Rev. 265, 267 (1999).

82. Fausto-Sterling, Sexing the Body: Gender Politics and the Construction of Sexuality (2000).

83. *See In re* Estate of Gardiner, 22 P.3d 1086 (Kan. Ct. App. 2001). At least one court has held that where one whose certificate was changed from male to female married, that person was not authorized to sue for the wrongful death of the husband. Littleton v. Prange, 9 S.W.3d 223, 225–26 (Tex. App. 1999).

prohibits an operation on a female child to remove all or part of her genitals for other than reasons "necessary to the health of the person on whom it is performed."[84] However, it would appear those operations that are consistent with the patient's need for a normal distribution of X and Y chromosomes might be interpreted as being necessary for reasons of health. With regard to a surgical operation for reasons of health, the statute merely outlaws the congressional finding of the African "practice of female genital mutilation." The statute does this by stating that in determining exceptions for health:

> No account shall be taken of the effect on the person on whom the operation is to be performed of any belief on the part of that person, or any other person, that the operation is required as a matter of custom or ritual.

good v. Good

Once a physician discloses to his patient that carcinoma is a genetically transferable disease, he bears none of the responsibility to disclose this to his patient's family.

v.

The physician himself should disclose such information to those in the patient's family who might inherit the disease and need to be tested thereon.

Twenty-first Century Approach

1. Where a physician's patient has an inheritable disease, the physician must see that the members of the family are so informed.
2. Where a child is born with intersex conditions, its parents should be informed about possible genetical reconstruction surgery.
3. Because many state courts do not do so, state legislatures should seek to enact appropriate legislation to obtain this result.

---

84. 18 U.S.C. § 116.

4. Attempts to Prevent Premature Infants with Very Low Birth Weight and Futile Results in Certain Prenatal Births

In 2002 there were more than four million births in the United States, of which some 482,000 (or 12 percent) were premature. About 318,000 (7.9 percent) were listed as low birth weight, which was the highest level in more than 30 years.[85] Of the half a million babies born prematurely in the United States each year, 25,000 are extremely premature. A study concluded in the year 2000 of "children who were born at 25 or fewer completed weeks of gestation made at the time when they reached a median age of 30 months of age" demonstrated that "severe disability is common and remains a major challenge in this group of children."[86] In fact, "even mild and moderate preterm birth infants are at high relative risk for death during infancy and are responsible for an important fraction of infant deaths."[87] The prevention of premature birth—gestational age at birth less than 37 completed gestational weeks—is a very elusive goal in the United States.[88] In 1995 it was recorded that "The poor performance of the United States in the international ranking of infant mortality is mainly due to high rates of premature birth and associated low birth weight, which have changed little during the past 50 years and may

85. Orange County Reg., June 26, 2003, at News 18.

86. 343 N. Eng. J. Med. 378–84 (2000). Women who take more than a year to get pregnant have a slightly higher than normal chance of giving birth prematurely. "The rate of premature births in the United States has jumped in the last 20 years from about 9.4 percent of all births to almost 12 percent." Orange County Reg., Oct. 30, 2003, at News 12 (citing the journal Human Reproduction published at that time).

87. 284 JAMA 843–49 (2000); *see also* World Health Organization, The Prevention of Prenatal Mortality and Mortality and Morbidity, WHO Technical Report Series Rep. 457 (1970).

88. Volpe, 16 Clinics Perinatology 496 (1989). "Preterm delivery is more common among blacks, the poor, the unmarried, cigarette smokers, underweight women, women with multiple gestations, women with uterine anomalies, women with a history of previous preterm delivery, and women without prenatal care; our perinatal mortality rate is among the highest in the developed world." Greene, N. Eng. J. Med. 2453 (2003).

recently even have increased."[89] In 2002 it was reported that 65 percent of all infant deaths occur among low birth weight (LBW) infants (< 2500 g), and both environmental and genetic factors may play a role.[90]

Public health "experts" have tried to justify the expenditure of large amounts of the taxpayers' funds in prenatal care on the grounds that such prevention produces savings. However, studies have not shown that such savings occur:

> One cannot predict saving based simply on the costs of caring for an "average" low-birth-weight baby. Very small infants are extremely expensive, whereas moderately low-weight babies may require only a few extra days in the hospital after delivery. Although the greatest savings can be achieved by avoiding the smallest birth weights, these occurrences are in fact the hardest to prevent.[91]

In addition, it was pointed out in 2002 that the uncontrolled growth of neonatal intensive care specialty services "has less to do with the true need of communities for effective clinical services than with the financial incentives promoting specialization."[92] At the dawn of the third millennium, it was

---

89. N. Eng. J. Med. 1772 (1995).

> In 1996, there were 7.2 infant deaths per 1,000 live births, according to a summary of preliminary 1996 government statistics published in the December issue of Pediatrics, the journal of the American Academy of Pediatrics. That is 5 percent lower than in 1995 and the lowest ever recorded in the United States.

Orange County Reg., Dec. 2, 1997, at News 12.

90. JAMA 195 (2002). It has been estimated that as many as 15 percent of the most immature infants develop cerebral palsy and approximately half develop cognitive and behavioral deficits. A large cohort study suggested that extremely low birth weight infants' frequent neonatal infections are associated with poor neurodevelopmental and growth outcomes in early childhood. Additional studies are needed to elucidate the pathogenesis of brain injury in infants with infection so that novel interventions to improve these outcomes can be explored. Stoll et al., *Neurodevelopmental and Growth Impairment Among Extremely Low-Birth-Weight Infants With Neonatal Infection*. JAMA 2357 (2004).

91. Sci. Am., Apr. 1995, at 126.

92. N. Eng. J. Med. 1574 (2002). Summarizing Goodman et al. The relation between the availability of neonatal intensive care and neonatal mortality. 346 N. Eng. J. Med. 1538–44 (2002).

stated, "[T]he improvement in survival has not been accompanied by a substantial reduction, if any, of risk of prematurity-associated neurologic handicaps." Contrary to much media input, "making prenatal care available to more women or making more visits available to the small number of women has generally not reduced preterm births."[93] Further, a woman who gives birth prematurely is much more likely than other women to have a premature delivery the next time.[94] In 2003 a study was conducted in the form of a double-blind, placebo-controlled trial involving pregnant women with a documented history of spontaneous preterm delivery. It was found that "weekly injections of 17 alpha-hydroxyprogesterone caproate (17P) resulted in a substantial reduction in the rate of recurrent preterm delivery among women who were at particularly high risk for preterm delivery and reduced the likelihood of several complications in their infants."[95] However, not all trials reported positive results.[96]

Most interventions designed to prevent preterm birth do not work, and the few that do, "including treatment of urinary tract infection, cerclage, and treatment of bacterial vaginosis in high-risk women, are not universally effective and are applicable to only a small percentage of women at risk for preterm birth."[97] However, a Canadian study in 2002 found that recent ad-

---

93. Goldenberg & Rouse, N. Eng. J. Med. 313 (1998). It has been found that dental radiography during pregnancy is associated with low birth weight. Hujoel et al., *Antepartum Dental Radiography and Infant Low Birth Weight*, JAMA 1987 (2004).

94. N. Eng. J. Med. (1999); (Orange County Reg., Sept. 23, 1999, at News 6; *see also* Adams et al., 283 JAMA 1591–96 (2000).

> Women who have a hard time getting pregnant also have a higher risk of having a premature baby or of losing their child within the first week of life. Compared with women without fertility, those with untreated infertility were 3.3 times more likely to have prenatal death, and those with treated infertility were over 2.7 times more likely.

Id. Lancet (1999); Orange County Reg., May 21, 1999.

95. Meis et al., N. Eng. J. Med. 2374 (2003). "The mechanisms of action of 17P in prolonging gestation are not entirely known." *Id.* at 2383.

96. "Despite treatment, more than a third of the subjects studied by Meis et al. still delivered prematurely. The study does, however, demonstrate that at least some preterm deliveries can be prevented." N. Eng. J. Med. 2454 (2003).

97. *Id.* at 318. In this connection preeclampsia toxemia of late pregnancy will often force a premature delivery, SN May 10, 2003, 293; Feb. 14, 2004, 100. The

vances in prenatal diagnosis and selective termination of affected pregnancies (abortions) have led to declines in the birth prevalence of congenital anomalies and to decreases in infant deaths.[98] In Canada, infant deaths due to congenital anomalies constitute approximately 30 percent of all infant deaths.

> This decline [in infant deaths] was preceded and accompanied by a sharp increase in fetal deaths due to congenital anomalies and *termination of pregnancy at 20 to 21 weeks of gestation.* These data suggest that increases in prenatal diagnosis and pregnancy termination for congenital anomalies are related to decreases in overall infant mortality at the population level. Further *declines in infant deaths* due to congenital anomalies are *likely to occur* as prenatal diagnosis and *selective termination of affected pregnancies become more widely available.*[99] (Emphasis added.)

High teenage pregnancy is unique to the United States and a recent study found that teenagers are almost twice as likely as older women to deliver premature babies.[100] The youngest girls, those 13 to 17, were 90 percent more likely than the women in their early 20s to deliver prematurely.[101] The continuance of teenage sexual activity remains in the United States at an increased expense.

---

placental growth factor (PlFG) permits blood vessel growth in the placenta, so that the mother can nourish her fetus. Its "low levels at mid-gestation is strongly associated with subsequent early development of preeclampsia." Levine et al., JAMA 77 (2005).

98. Liu et al., JAMA 1561 (2002).

99. *Id.* at 1566. On Sept. 27, 2002, Health and Human Services issued rules allowing states to define a fetus as a child eligible for government-subsidized health care under the Children's Health Insurance Program (CHIP). It was criticized as an attempt to create precedent for viewing a fetus as a separate physical and legal entity with its own rights, thereby challenging the current legality of abortions. Hanna, *Aid to Fetuses with Dependent Mothers.* Hastings Ctr. Rep., 2002, at 8.

100. N. Eng. J. Med. (1998). The study was based on a "review of the cases of 134,088 white girls and women ages 13 to 24 who delivered their first babies from 1970 to 1990. Low birth weight is twice as common among blacks than whites. Blacks face a higher risk for conditions such as high blood pressure, diabetes, and ESRD," 160 Archives Internal Med., 1472–76 (2000).

101. The risk was still elevated, but less so for the older teenagers. Orange County Reg., Apr. 27, 1998.

good v. Good

Prenatal diagnosis and selective terminations and resulting abortions at or after 20 weeks of gestation are immoral actions and should be prohibited.

v.

Such actions should become widely available in order to decrease infant mortality of infants after birth where possible.

Twenty-first Century Approach

1. Early in the third millennium both prenatal diagnoses and the communication of selective termination of affected pregnancies should be made available at most hospitals; appropriate state legislation should be enacted to reduce infant mortality.
2. When there is high risk of significant damage resulting either with or without a physician's treatment to newborns, their cognitive parents should be given maximum control over whether or not such treatment should be performed.

---

5. Poor *and* Expensive Outcomes at Neonatal Intensive Care Units (NICUs)

Normally birth is a painful joy. But if the outcome is abnormal, it is often pain, followed by more pain. As might be expected, babies with birth defects have decreased survival as compared with those without birth defects, especially in the first years of life, and they have an increased risk of having children with the same defect. But low birth weight babies have more problems:[102]

> At any given gestational age, infants with low birth weight have relatively high morbidity and mortality. Each year in the United States, approximately 250,000 infants are born weighing less than 2,500 grams. These infants are classified as having low birth weight, although the majority of these infants are before term.[103]

102. 340 N. Eng. J. Med. 1057–62 (1999).

103. McIntire et al., 340 N. Eng. J. Med. 1234–38 (1999). A study in the *Journal of Epidemiology and Community Health* found that fetuses with shorter thighbones

Infants born before 26 weeks of gestation who survive stay in neonatal intensive care units (NICUs) for more than four months at a cost that averages $250,000 per infant.[104] A study of children admitted to intensive care units during the evening rather than during the day are slightly more likely to die in the first 48 hours of care.[105] The NICUs, created in 1960 by the late Dr. Louis Gluck, a professor I met at the University of California–Irvine, had mixed results. They are quite costly and have been said to be "tremendously stressful for preterm infants." They delay and possibly alter neurological development, resulting in longer hospital stays and negatively influencing long-term developmental outcomes.[106] The supply of neona-

---

at 24 weeks had higher blood pressure at the age of six than those with longer thighbones. "Scientists believe that when a fetus is undernourished, it diverts resources to areas it needs at the time, such as the brain, at the expense of organs it will need later in life. That may permanently change the baby's structure, functioning, and metabolism, experts believe," ORANGE COUNTY REG., Aug. 15, 2002, at News 12.

104. Cole, *supra* note 79, at 429.

> Newborns admitted to neonatal intensive care units (NICUs) undergo a variety of painful procedures and stressful events of a measurable analgesic effect. A study has found that "absence of a beneficial effect on poor neurologic outcome do not support the routine use of morphine infusions as a standard of care in preterm newborns who have received ventilatory support."

Simons et al., *Routine Morphine Infusion in Preterm Newborns Who Received Ventilatory Support*, JAMA 2419 (2003).

105. PEDIATRICS (2004); ORANGE COUNTY REG., June 8, 2004, at News 10. The mortality rate for children admitted during the day was 2.2 percent. But for children admitted during the evening (defined as 5 p.m. to 7 a.m.) the death rate increased to 2.8 %.

106. Brown & Heermann, APPLIED NURSING RES. 190–97 (1997). "Doctors and nurses are least likely to wash their hands in that part of the hospital where patients are often at their most vulnerable—the intensive care unit." ANNALS INTERNAL MED. (1999). A study at Duke University found that only 17 percent of physicians in intensive care units washed their hands appropriately. L.A. TIMES, Mar. 1, 2000, at E6. "Bacteria found under the long fingernails of two nurses may have contributed to the deaths of 16 sickly babies in 1997. All of the babies were newborns in the NICU." ORANGE COUNTY REG., Mar. 24, 2000, at News 27. "ICU-acquired infection is common and often associated with microbiological isolates of resistant organisms." Vincent et al., 274 JAMA 639–44 (1995).

tologists "appears to have grown beyond that needed solely for the care of ill newborns."[107]

Children at the upper end of the low birth weight range who require no intensive care have poorer outcomes than children with normal birth weight. It has been remarked: "Rates of neonatal mortality were approaching an irreducible minimum and that for the United States to lower its rates of infant mortality, preterm births and congenital malformations would need to be prevented."[108] Various "programs undertaken specifically to prevent preterm birth and low birth weight have been unsuccessful."[109] Consider these year 2000 statistics:

> The most common outcome of pregnancies that end between 20 and 25 weeks of gestation is stillbirth or death before admission to a neonatal intensive care unit (79 percent), the cheapest outcome, followed by death before discharge (12 percent) and survival to discharge (8 percent, with < 1 percent dying after discharge but before 2½ years of age). Among the survivors at 2½ years of age roughly half had some disability (about half of these had severe disability) and half had no disability.[110]

It has been pointed out that "parents understanding of an extremely premature infant's prognosis may be based more on their own background and beliefs than on valid statistics about the outcomes of similar infants."[111] How-

107. Many regions of the United States have "more neonatal intensive care resources than are needed to prevent the death of high-risk newborns," 346 N. ENG. J. MED. 1538–44 (2002). In addition it was recently discovered that "the experience of pain in early life may lead to long-term consequences" and found that "long-term behavioral changes can extend far beyond what would be considered the normal period of post-injury recovery. Chronic pain, including neuropathic pain, is far more common in children than was thought." Howard, *Current Status of Pain Management in Children.* JAMA 2464 (2003).

108. *Id.*

109. *Id.*

110. Wood et al., *Neurologic and Developmental Disability After Extremely Pre-Term Birth,* 343 N. ENG. J. MED. 378–84 (2000).

111. Cole, *supra* note 79, at 429 (citing McCormick, *Conceptualizing Child Health Status: Observations from Studies of Very Premature Infants*, 42 PERSPECTIVE VIOL MED., 372 (1999).

ever, physicians and their institutions tend to find reasons to justify a strongly interventionist therapy of life support, "even though statistics about longer-term survival, quality of life, or longer-term cognitive or physical deficits might not show such a favorable prognosis for these infants."[112] For example, physicians "may mask the level of uncertainty associated with the outcome of serious illness so as to encourage seriously ill patients to accept high-technology care provided by intensive care specialists."[113]

Others prefer to point out that in such cases, pediatric hospice (end-of-life) programs offer the ideal support structure for some palliative care.[114] Many serious problems face premature infants: intracranial hemorrhage (ICH) is a common serious complication in the very low birth weight preterm infant. Since 1985, the incidences of this were reported to be 13 percent to 48 percent. Studies of long-term outcomes in school-age children demonstrate that survivors of severe ICH remain abnormal:

> Cranial ultrasound has allowed accurate diagnosis of all grades of ICH, thus enabling physicians to correlate accurately ICH severity with outcome measures. Another example is periventricular-intraventricular hemorrhage (IVH), the most common neurologically important lesion of the premature infant. It has been noted, any premature newborn in an intensive care unit may be considered to be at risk in view of the very high incidence of the disorder in this population.

---

112. *See* Mullins et al., *Adequacy of Hospital Discharge Status as a Measure of Outcome Among Injured Patients* 279 JAMA 1727–31 (1998).

113. Culver et al., JAMA (2000) (citing Hefferman P. Hellig, *Giving "More Distress" a Voice: Ethical Concerns Among Neonatal Intensive Care Personnel,* 8 Cambridge Q. Healthcare Ethics 173–78 (1999), and Harvey, *Achieving the Indeterminate: Accomplishing Degrees of Certainty in Life and Death Situations*, 44 Soc. Rev. 78–98 (1996)). "We are aware that experienced neonatal caregivers may be reluctant to have their own ELBW infants resuscitated and treated; however, they were all too willing to force this care on our children against our wishes or without our informed consent." To do this they might even be relying in part on Section 2.17 of the *Code of Medical Ethics*, which states that a physician's "decisions for the [best] treatment of seriously disabled newborns" should not rest on "the avoidance of a burden to the family or to society."

114. Geromo & Fahner, *Symptoms and Suffering at the End of Life: Children with Cancer*, N. Eng. J. Med. (2000).

> Further, commonly used practices in newborn resuscitation may increase the likelihood of IVH in the premature infant.[115]

It is reported that about "*half of the surviving children* who weigh less than 750 g at birth experience moderate or severe disability, including blindness and cerebral palsy, and require special education. Many such infants have more than one disability. However, some of these children are educable and can function with their family unit."[116]

In 2002, an Institute of Medicine report recommended the improvement of training and methods to comfort terminally ill children and their families.[117] It has been urged that family members be informed of the range of morbidity and survival rates and of the rates of long-term disabilities that can be expected. This is because "decisions regarding obstetric management must be made by the parents and their physicians if the neonate's prognosis is uncertain."[118] The parents may also wish to be informed that studies have failed to document benefits of cesarean delivery for extremely premature infants, even those in the breech position.[119] Following the turn of the third millennium, unnecessary cesarian births again became prevalent; new guidelines stated that they should be reserved "for mothers and babies who truly need it."[120] A brief 1997 story concerns a one-year-old girl on a ventilator:

> Priscilla was born after a 17-year-old single mother's normal pregnancy. She had some malformations, a cleft palate, low-set ears, a small rib cage, and had some respiratory distress that led to her being put on a ventilator. After her cleft palate was repaired, she got off the

115. Ford et al., 143 AJDC, 1166 (1989).
116. 96 Pediatrics 974 (1995).
117. Orange County Reg., July 26, 2002, at News 24.
118. *Id.* at 975.
119. Nuffield Council on Bioethics, *Genetic Screening, Ethical Issues* (London 1993).
120. Now, with cesareans inching back up to 22 percent of U.S. births, the nation's leading obstetricians' group is issuing new guidelines to reduce unnecessary C-sections and reserve the surgery for mothers and babies who truly need it. C-sections have risen for three years, climbing 4 percent in 1999 to account for 22 percent of live births, the government reported this month. Those discrepancies suggest doctors' habits play a big role.

Orange County Reg., Aug. 29, 2002, at News 15.

> ventilator, but at four weeks of age, she had a respiratory arrest and an emergency tracheotomy and was put back on a ventilator. She couldn't get off. Over the next few months, she was very unstable. She tried to die a number of times. The doctors were of a mind to let her. They talked to Mom about do-no-resuscitate orders [DNR]. She refused to consider any limitation of treatment.[121]

The doctors recommended DNR. The mother's immediate refusal to consider it was a typical reaction of many Hispanics and Asians in California. In some instances, a medical ethics committee had been successful in effecting a change of mind on the part of some families. These cases raise the ethical issue of continuing expensive but futile treatment where doctors may withdraw, rather than continue, treatment (other than comfort care) that is not beneficial to the patient. However, when, as in Priscilla's case, the family insists upon its continuance, physicians generally do so despite opposing provisions contained in the current AMA *Code of Medical Ethics*.[122]

Some have suggested that two reasons may account for this: (1) fear of the possible malpractice lawsuit by the family if the physician were to stop treatment and (2) continuance of treatment results in the receipt of payment (e.g., through Medicaid or insurance). In such cases, they usually seek and find some medical reason for the treatment; however, one of them wrote in the *New England Journal of Medicine*, "Physicians tend to frame value judgments as medical decisions without being aware of it."[123] Even where the family agrees to stop treatment, they may "medicalize the decision in order to override the wishes of the patient, family, or other surrogates."[124] It has also been noted that "parents—who had been struggling with the issue for a long time—were more willing to stop treatment than some of the team members who were unfamiliar with the patient. In these cases the family often had to resist the request of a new and eager physician to 'try one more

121. LANTOS, DO WE STILL NEED DOCTORS? 323–33 (1997).

122. Rule 2.215, "Treatment Decisions for Seriously Ill Newborns"; 2.035, "Futile Care"; and 2.037, "Medical Futility in End of Life Care." These provisions are further discussed in Chapter 13.

123. Solomon, Letter, *Futility as a Criterion in Limiting Treatment*, 327 N. ENG. J. MED. 1239 (1992).

124. *See* Robert Veatch et al., *Medically Futile Care: The Role of the Physician in Setting Limits*. 18 AM. J. L. & MED. 15 (1992).

thing.'"[125] However, in 2003 it was written that: "French doctors are much more willing than their American counterparts unilaterally to discontinue treatment when babies are not doing well. They don't see this either as unjust or as rationing. Instead, they see it as a part of their professional responsibilities. Do the parents object, I asked? No, no. They accept it. They expect it." Further, the World Health Organization recently rated France's healthcare system the best in the world.[126]

In both instances, many such cases are not being referred to hospital ethics committees by the doctors or by the hospital departments involved; when pressed, some doctors tend to find that treatment is the "best interests" of the infant. As bioethicist Veatch has pointed out, "Anyone who has given his life on the field of professional specialization cannot be expected to value the contributions of that area in a reasonable way: musicians, artists, politicians . . . and doctors all over value the benefits of their field."[127] He proposed that the "best interests standard" for decision making should give way to a standard of reasonableness: "What would most persons who share the parents' deeply held values choose as most reasonable course of action?"[128] Either nontreatment or treatment may also be found in the "best interests" of the infant depending upon the subjective values of many nonmedical committee members who normally are dominated by physician members (see Chapter 6). The write-up of the committee's decision for nontreatment may hinge upon the interpretation of the term "futile" in the HHS regulation (see Chapter 13).

Some state courts have ruled that parents can make such decisions without intervention from the courts.[129] A Pennsylvania court upheld a parent's petition to terminate a child's life-sustaining treatment, while Texas ruled to the contrary.[130] Other courts seem to have thrown up their hands. A Michigan court

125. Orr & Perkin, *Clinical Ethics, Consultations with Children*, 5 J. CLINICAL ETHICS 323, 327 (1994).

126. Lantos, Hastings Ctr. Rep., May-June 2003, at 43–44.

127. 98 PEDIATRICS 1082 (1996).

128. Veatch, *Abandoning Informed Consent*. Hastings Ctr. Rep. (1995), at 25, S-12.

129. *In re* Rosebush, 491 N.W.2d 633 (Mich. Ct. App. 1992); *In re* Jobes, 529 A.2d 434 (N.J. 1987).

130. The court stated that such a decision is "fraught with pain and anxiety" that would only be compounded by "insensitive and unnecessary" court proceedings. *In re* Fiori, 652 A.2d 1350 (Pa. Super. Ct. 1995). In Texas, when a woman went into premature labor 33 weeks into her pregnancy, the parents refused to sign a consent

actually stated that making a life-sustaining decision involved an area "that neither law, medicine, nor philosophy [could] provide a wholly satisfactory answer to."[131] In such states, the legislature may feel it is not competent to act.

good v. Good

Only physicians should evaluate infants in neonatal intensive care units (NICUs) and determine their morbidity and survival chances and treatment choices.

v.

Competent parents both can and should be informed and evaluate their infant's future quality of life and should objectively determine treatment choices.

Twenty-first Century Approach

1. Competent parents should be reinstated by statute as objective decision makers for severely handicapped infants.
2. Where infants are in NICUs, family members must be informed about the range of morbidity and survival rates and the rates of long-term disabilities that can be expected in adolescence and adulthood.
3. Physicians cannot ethically use their subjective values to continue nonbeneficial (futile) treatment (other than comfort care) of infants where the family opposes it in an objective manner.
4. In those states where the courts have not upheld the parents' right to terminate treatment for severely defective infants, state legislatures should revise their laws in order to avoid federal regulations applicable to receipt of funds regarding the treatment of a futile case largely in terms of "Survivorship."

---

form allowing resuscitation on those conditions. But the obstetrician induced labor, and the baby had a brain hemorrhage as doctors had predicted. The woman was blind and severely mentally retarded, and had cerebral palsy, seizures, and spastic quadriplegia. Finding that the hospital performed the resuscitation without the parents' consent and that it was grossly negligent, the court awarded about $43 million in actual damages. Unfortunately, the Texas Supreme Court held the reverse and actually upheld the hospital's denial of parental consent. Miller v. HCA Inc., 118 S.W.3d 758 (Tex. 2003).

131. *In re* Martin, 538 N.W.2d, 399, 401 (Mich. 1995) (denying removal of life support from an incompetent adult due to lack of "clear and convincing" evidence regarding prior expressed wishes to terminate the treatment.). The court stated, "If we are to err, however, we must err in preserving life." *Id.* at 402.

6. Proposal that Parents Share Some Expenses and Time

Elsewhere I have suggested the need for parents of severely defective premature infants to be brought "closer to reality."[132] This might be accomplished by two methods: require (1) that they pay at least a fraction of the cost of their baby's care, and (2) that they personally participate by spending some amount of time in the care. Where parents are responsible for a minimum amount of the cost of and time for their infant's care, decisions about withholding or withdrawing treatment tend to be made more thoughtfully and expeditiously than when these are not being met by them. A proposal is to request such parents to spend an amount of time participating in their child's care, instead of coming in briefly (sometimes as infrequently as once a month). This was actually tried out with the following results:

> After only three days at their infant's bedside, their attitude toward continued treatment underwent a complete reversal. They concurred with the view of the staff that their baby was not only failing to thrive, but was actually being ravaged by an alleged well-intentioned intervention initially applied in the hope of saving its life. Very quickly, they consented to having their baby withdrawn from the breathing machine and medicated for pain. They then held him as he died hours later at peace.[133]

As regards moderate preterm infants, their parents might be informed about turn of the third millennium studies that demonstrate that children born more than five weeks premature are more likely to have reading difficulties and suffer from behavioral problems in adolescence.[134] Such children

132. KEYES, LIFE, DEATH, AND THE LAW 482–83 (1995).

133. YOUNG, ALPHA AND OMEGA 100.

134. According to Dr. Ann Stewart of University College London, studies show that "individuals born very prematurely show an excess of neuro-cognitive behavioral problems in adolescence, and more than half have abnormal MRI brain scans," LANCET; ORANGE COUNTY REG., May 14, 1999, at News 26. Black babies are almost twice as likely as white infants to be born prematurely. This rate was 160.7 in 1997 and only 83.7 for whites that year. ORANGE COUNTY REG., Sept. 22, 2000.

experience significant learning difficulties that persist into their teenage years[135] and into the adult years. A 26-year study publicized in February of 2000 showed that adults who were born small for gestational age had significant differences in academic achievement and professional attainment compared with adults who were born at normal birth weight.[136]

These should be more widely discussed and implemented to show that we can act more rationally in these dire circumstances. As one analyst has perceptively put it with regard to a mother: "So long as we elect not to take these measures, babies born to our babies will either die or be rescued at incalculable financial and emotional cost. For in rescuing them, we may well succeed only in adding to the swelling number of totally dependent people in our midst."[137]

The first parental extreme is about families that make "unreasonable demands" and insist on overtreatment of infants for whom the future higher quality of life is virtually nonexistent.[138] The second concerns dying children, those who cannot have any significant future quality of life and suffer needlessly because their parents (or their doctors) refuse to give up hope of curing them.[139] In this connection a Texas case was noted where the hospital's continued treatment of a 22-week gestation infant, against the parent's wishes, that initially resulted in compensatory and punitive damages of millions of dollars, was actually reversed on appeal.[140]

---

135. PEDIATRICS (2000); ORANGE COUNTY REG., Feb. 8, 2000, at News 13.

136. Stauss, 283 JAMA 625–32 (2000).

137. *Id.* at 106.

138. *See* SILVAMAN, WHERE'S THE EVIDENCE? DEBATES OF MODERN MEDICINE (1998).

139. A Harvard study of suffering among children with cancer showed that one-fourth of all children diagnosed with cancer will die of it. Their parents were told that of 103 such children, 92 had suffered "a great deal" or "a lot" from at least one symptom. "For most children with cancer, the primary goal of treatment is to achieve a cure. Considerations of the toxicity of the therapy, the quality of life and growth and development are usually secondary to this goal." Wolfe, N. ENG. J. MED. February 3, 2000.

140. *See* Miller v. Texas Women's Hospital/HCA, 118 S.W.3d 758 (Tex. 2003), at note 77. If a child is born with severe disabilities that threaten his or her life, however, heroic measures need not be employed to keep the infant alive. Here the same rules that govern the withholding and removal of life support systems from any human being apply to newborns. Specifically, until the child is 30 days old, she or he is not considered to be a person whose life is confirmed (bar Kayama), VITAL SIGNS, May 1999, at 2.

good v. Good

Severely defective infants should always receive continued treatment at taxpayer expense regardless of their condition and their parents should always be encouraged to support their survival together with those physicians and healthcare organizations who so believe and are willing to act in accordance with their belief.

v.

Because the very early births are severely defective infants who have low chances for survival and most of those that survive have very substantial defects, all parents should be encouraged to (a) consider the potential quality of life of such infants, particularly for those who will either have no such quality or a very limited one, rather than (b) limit their considerations to the mere survival of the infant regardless of its lack of quality.

Twenty-first Century Approach

Where treatment becomes nonbeneficial and the parents refuse to agree with respect to stopping such treatment, they should be requested to spend a *reasonable amount of time* at the hospital *observing (or possibly even participating) in the infant's care*, before making their final decision.

> Mentally defective or physically deformed or defective infants born from these matings will be put away immediately to die in some unknown place. Plato's bold treatment of genetic engineering and infanticide for the defective figures prominently in the current controversy on these issues.
>
> Lavine, *From Plato to Sartre* (1984), p. 62

7. Infanticide and Abortions after Viability of Certain Types of Severely Defective Infants

Although about 3,700 newborns die each year by the age of one month from smoking by the mother during pregnancy, none of these mothers have been prosecuted for homicide. As noted in Chapter 9 on abortion, a woman has a constitutional right to abort her fetus prior to its viability. At the dawn of the third millennium, only a few scientists such as Lynn Margolis and Peter Singer of Princeton University believe in and support "infanticide" for severely defective newborns.[141] They believe that families ought to have the right to terminate the life of their newborns, up to an age of one month, if the child's medical condition is so serious as to fundamentally compromise its own life or the family's long-term happiness. Margolis and her son wrote concerning them:

> While we certainly may not do anything actively to hasten death, we may, according to these authorities, do less to sustain the child's life than would be called upon to do with regard to people who had lived beyond 30 days.[142]

They noted that such activity in connection with a fetus, or defective fetus or newborn, is not morally equivalent to that concerning a "person" and is not wrong. In this connection two authors noted that "ancient Greece provides evidence that the ancient Greeks did not in fact prohibit infanticide."[143]

According to Hardy, "What we can be certain of is that infanticide (ranging from rare to common) has characterized human societies through history and prehistory."[144] Human neonates, "those 'exterogestate fetuses,' are particularly helpless by primate standards and require full-time carriage by a caretaker to survive." She noted that the "increased care and staggering parental and societal investment required for the exterogestate fetus to survive are carrying us into novel terrain. There are no precedents—emotional, conceptual, legal, or otherwise. This is totally uncharted territory."[145] The question of what determines a woman's reaction to her baby is often deemed

---

141. MARGOLIS & SAGEN, SLANTED TRUTHS (1997); J. FAMILY MED. at A17 (1995). The National Center for Health Statistics reported: "In 2002, the infant mortality rate was 7 deaths for every 1,000 live births, up from 6.8 in 2001. *Much* of the *increase* was attributed to *deaths in the first month* of life *especially* the first week," N.Y. TIMES, Dec. 3, 2004. (Emphasis added.)

142. *See also* DARFF, MATTER OF LIFE AND DEATH 273–74 (1998); ORANGE COUNTY REG., Mar. 26, 2000, at Accent 3; ORANGE COUNTY REG., Sept. 23, 1999, at News 19; ORANGE COUNTY REG., Oct. 2, 1999, at News 19.

143. Spielman & G. Agich, *The Future of Bioethics Testimony: Guidelines for Determining Qualifications, Reliability, and Helpfulness*, 36 SAN DIEGO L. REV. 1043, 1056 (1999). "Descriptive ethics would also include attention to what Greek laws and thinkers said about the practice of infanticide."

144. HARDY, MOTHER NATURE 480, 519 (1999).

145. *Id.* at 473.

too sensitive to discuss in Western society.[146] We are now into the third millennium and sometime therein perhaps we shall be able to chart a course in this area. It has been noted that medical doctors as well were calling for more open discussion of maternal ambivalence and infanticide.[147] The limit of abortion on "viability" has been challenged by Peter Singer,[148] along with a 2003 decision by the recent Texas Supreme Court by which outside of Texas "emergent circumstances" became an exception to the general rule that a physician commits a battery by providing medical treatment to a child without obtaining the parents' consent[149]—although informed consent is normally required in emergencies.

---

146. *Id.* at 474.

> In contemporary Western society, parents are respected and admired for caring for the same infants that in other societies mothers would be condemned by their neighbors for not disposing of. Some adopting parents in the West go out of their ways to select the neediest infants and commit themselves to years of therapy on behalf of children who will never repay that care in any material sense.

Id. at 460.

147. *Id.* at 296.

148. I can see how one could defend the view that fetuses may be "replaced" before birth, but newborn infants may not be. Nor is there any other point, such as viability, that does a better job of dividing the fetus from the infant . . . It may still be objected that to replace either a fetus or a newborn infant is wrong because it suggests to disabled people living today that their lives are less worth living than the lives of people who are not disabled. Yet it is surely flying in the face of reality to deny that, on average, this is so. The position taken here implies only that the parents of such infants should be able to make this decision.

SINGER, WRITINGS ON AN ETHICAL LIFE 191 (2000); *see also* SINGER, RETHINKING LIFE AND DEATH 215 (1994) ("In regarding a newborn as not having the same right to life as a person, the cultures that practiced infanticide were on solid ground. . . . In our book, *Should the Baby Live?,* my colleague Helga Kuhse and I suggested that a period of 28 days after birth might be allowed before an infant is accepted as having the same right to life as others.").

149. In *Miller v. HCA Inc.,* 18 S.W.3d 758 (Tex. 2003), the physician had advised the parents that if their child were revived, she would likely suffer permanent physical and mental injuries. "The parents had successfully sued Houston's The Woman's Hospital of Texas, a subsidiary of HCA, for

Infanticide is to be distinguished from the well-documented somewhat later killings of children by stepfathers. Daly and Wilson have noted that in North America, when the father of offspring under two years of age no longer lives in the home and an unrelated man or stepfather lives there instead, this rare event is 70 times more likely to occur.[150] In such cases "defense lawyers seek to exonerate their clients on the grounds that it was in their genes."[151] A Daly and Wilson study implied that this is in some way "natural," and if so it must be "right, or at least excusable." Others say that we should not admit, except under stress, that infanticide is "natural,"[152] even though examples exist with other primates.[153]

The not infrequent cases of the disposal of nondefective newborns must be distinguished; typical are cases of teenage girls who do so in order to hide the fact that they had been pregnant.[154] In the United States, infants born to a mother under 17 or being the second infant born to a mother under 19 are the highest-risk categories. Lacking kin support, the mother denies the pregnancy to herself, conceals it from others, and makes no preparation for the impending birth. An anthropologist found that throughout history women have killed their newborns in times of stress.[155] "Mothers do not automatically and unconditionally respond to giving birth in a nurturing way," which is contrary to our cultural concept that "in all cases" a mother's love is presumed to be unconditional.[156] Those under the age of 20 were more likely to

---

> negligence and battery in winning $43 million from a jury. That decision was reversed; it applies when there is no time to consult the parents or seek court intervention if the parents withhold consent."

NAT'L L.J., Oct. 13, 2003, at 4.

150. HARDY, *supra* note 144, at 296.

151. SCI. AM., Oct. 1995, at 180.

152. JOLLY, LUCY'S LEGACY 115 (1999).

153. "Male gorillas in the wild are highly infanticidal of other's infants. If the group silverback dies or is deposed, there is not much hope of raising his infant offspring." HARDY, *supra* note 144, at 117.

154. *Id.* at 573.

155. "In Bolivia, where researchers found that nearly every woman had killed a newborn of her own during a period of war and economic stress in the 1930s, when the prospects of raising a child with a suitable father were extremely poor, many of those women went on to become devoted mothers," L.A. TIMES, Dec. 26, 1999, at J58.

respond to poor circumstances by committing infanticide, while older mothers were far less likely to do so.[157]

Today state legislatures are resurrecting a modern equivalent of the medieval receptacle for handing unwanted newborns to certain places as an alternative to mothers leaving such infants to die behind bushes or inside Dumpsters without fear of prosecution. Texas became the first state to enact such a law in 1999, followed by California and 40 other states, which permit mothers to leave such babies in hospitals. However, in 2001, 38 babies were illegally abandoned, including 17 found dead.[158]

From the prospective of a "precivilized" mother, "abortion and infanticide are equivalent, except that in earlier times the former (abortion) was riskier to the mother, whereas today birth followed by infanticide is."[159] As

---

156. *See* Daly & Wilson, Homicide (1988); Fromm, The Art of Loving (1956); Hardy, *supra* note 144, at 297.

157. Daly & Wilson, *supra* note 156; N.Y. Times, Nov. 6, 2000, at 1. "The more we learn about the full range of material behaviors, the weaker become the grounds for ethnocentric moralizing, for making distinctions between 'civilized' and 'savage' people, between Christians and non-Christians, and so on." Hardy, *supra* note 144, at 299.

158. The laws' goal of protecting unwanted newborns enable women to avoid prosecution. They don't address the factors causing babies to be abandoned and they prevent abandoned children from ever learning their medical or genealogical histories. Most of the laws were enacted hastily and without the states sufficiently evaluating their effectiveness.

Orange County Reg., Mar. 10, 2002, at News 10. Illinois has a "safe haven" law under which "babies within three days of birth can be taken to any staffed fire station or hospital. You have no questions asked," Orange County Reg., Apr. 16, 2003, at Accent 2. However, in 2004, very sick seven-month-old conjoined twins were removed from their respirators and died with the consent of both their parents and physicians (who had decided they could not be separated).

159. Hardy, *supra* note 144, at 91.

> The risk of maternal death from "safe" abortion, as in developed countries, is 1 in 3,700 cases; from "unsafe" abortions it is 1 in 250. The World Health Organization estimates that globally 20 million unsafe abortions are performed each year, resulting in the death of 70,000 women. In other words, women often choose to abort, after taking into account a terrifying risk. This is not to say that deliberate abortion in any one case is right or wrong, only that, like both spontaneous abortion and infanticide, it is natural.

Id. at 118.

noted in Chapter 9, some methods of abortion were dangerous during the period when it was outlawed. But since becoming legal, it has been regarded as one of the safest medical procedures.[160]

At the turn of the third millennium, some people who are "pro-choice were against late-stage abortion; they argued that it appeared too much like infanticide. Nevertheless, almost all infanticide in traditional societies occurs right after birth, and is conceptually identical to late-stage abortion."[161] Today, it might be argued that because abortions are safe and legal, and information can be obtained regarding the condition of the baby, the choice must be made prior to viability (unlike Israel's choice of up to delivery). This is legal and is normally done whenever the mother knows that the fetus is severely defective prior to that time. Why does that standard not apply much later in those instances where she does not or could not become aware of severe defectiveness any earlier? In the year 2002 Congress enacted the Born-Alive Infants Protection Act, which granted every infant "born alive" full rights under federal law. It neither addresses the issue in which a baby is partially delivered from the mother before its demise[162] nor where the severely defective born child is an "unwanted one" and most likely to become a ward at the expense of states (such as Texas, California, and Illinois, as noted above).

China's policy of one child per family "has enhanced the well-being of wanted children and helped the country to catch up economically in spite of its possible abuses." A survey of "Methods of Abandonment of the Seriously

---

160. *Id.*

161. *Id.* at 314. In Holland, where a newborn would otherwise die shortly, physicians may euthanize the infant with drugs at its parents' request. Van der Maas et al., *Euthanasia and Other Medical Decisions Concerning the End of Life, translated in* 22 HEALTH POL'Y (1992).

162. The new law defines the term "born alive" as:

> The complete expulsion or extraction from his or her mother of [a human being], at any stage of development, who after such expulsion or extraction breathes or has a beating heart, pulsation of the umbilical cord, or definite movement of voluntary muscles, regardless of whether the umbilical cord has been cut, and regardless of whether the expulsion or extraction occurs as a result of natural or induced labor, cesarean section, or induced abortion.

S. CAL. CHRISTIAN TIMES, Sept. 2002, at 2.

Defective Infant" in China showed that three out of four preferred some form of euthanasia in such cases.[163] It was concluded that passive or active euthanasia of such infants "carried out by medical personnel does not violate humanitarian principles; on the contrary, it is morally indefensible to treat them at any cost or to leave them to the mercy of their largely unskilled families."[164] One study found that neonaticide mothers were single, with a mean age of 21, and only one of ten such mothers had previously suffered from a psychiatric illness.[165] In all the neonaticide cases studied, "the mothers made some attempt to conceal the death."[166] In England, they were found to differ from "filicide mothers who were over 25, married, psychotic, and suffering from serious depression."[167] Thus, the courts have found that "psychiatric disorder was thought to be responsible for the offense."[168] The mothers found to have killed their newborns have a greater than 50 percent chance of not even being indicted for the killing, and if indicted and convicted, the mother is likely to receive only probation under the British Infanticide Act.[169] There

---

163. Li Ben-fu, MED. LAW 554 (1998) (citing Shi Da-pu et al., *A Course of Lectures of Life Ethics in China*, 5 CHINESE MED. ETHICS 25–26 (1989)). The Netherlands has proposed guidelines for mercy killings of terminally ill newborns:

> Examples include extremely premature births, where children suffer brain damage from bleeding and convulsions; and diseases where a child could only survive on life support for the rest of their life, such as severe cases of spina bifida and epidermosis bullosa, a rare blistering illness . . . Experts acknowledge that doctors euthanize routinely in the United States and elsewhere but that the practice is hidden.

ORANGE COUNTY REG., Dec. 1, 2004, at News 27.

164. *Id.* at 562. An alternative proposed by Professor Qui Ren-zong was to establish ethical counseling committees that would consider the interests in each case of parents, doctors, hospitals, and society in general. Qui Ren-zong, *Ethics and Treatment of Defective Infants*, 5 MED. & PHIL. 25–26 (1986); *see also infra* Appendix B.

165. P. T. D'Orban, *Women Who Kill Their Children* 134 BRIT. J. PSYCHIATRY, 560 (1979).

166. *Id.* at 565.

167. Phillip J. Resnick, *Murder of the Newborn: A Psychiatric Review of Neonaticide*, 126 AM. J. PSYCHIATRY 1414 (1970).

168. D'Orban, *supra* note 165, at 560.

169. Marks & R. Kumar, *Infanticide in England and Wales*, 33 MED. SCI. L. (1993).

is no such statute in the United States; the only defense is insanity, even though a study in Iowa found that "in no case report was there an indication that the mothers were mentally ill" and there is no "neonaticide syndrome."[170] The New York Court of Appeals upheld a ruling that the defense had not established "neonaticide syndrome" as a generally accepted syndrome in the psychological community.[171] Dr. Barbara Kerwin who coined the phrase "designer defense" has written, "Syndromes are created to describe 'symptoms, traits and behaviors [that] are observed to occur together in an individual.'" She cited the examples of "battered woman syndrome, Vietnam syndrome, and sexual abuse syndrome." The presentation of these by alleged expert psychiatric witnesses "can be a far more simple, stylish, and comprehensible explanation for [a] horror than saying she simply chose to do evil. The jurors can thus justify their own willing suspension of disbelief."[172] Infanticides must normally be separated from mental illness or postpartum depression;[173] the AMA ethical principle is that "life-sustaining treatment may be withheld or withdrawn from a newborn when the pain and suffering expected to be endured by the child will overwhelm any potential for joy during his or her life." The Code states:

---

170. Edward Saunders, *Neonaticides Following "Secret" Pregnancies*, 104 PUB. HEALTH REP. 368 (1989).

171. People v. Wernick, 651 N.Y.S.2d 392, 393 (N.Y. 1996). "Perhaps even harder to accept, medical investigators in industrialized nations have shown that at least some cases of 'sudden infant death syndrome' (SIDS) are also, in reality, cases of infanticide." ELDREDGE, WHY DO IT 196 (2004) (citing HARDY, *supra* note 144).

172. BARBARA R. KIRWIN, THE MAD, THE BAD, AND THE INNOCENT: THE CRIMINAL MIND ON TRIAL 113–16 (1997).

173. > [P]ostpartum depression has been underrecognized ranging from "blues" to nonpsychotic major depression to depression with psychotic features . . . . [It] occurs in approximately 50 percent of women who have recently given birth. A high percentage of women with postpartum depression have ego-dystonic thoughts of harming their infants. . . . [W]hen a woman with severe postpartum depression becomes suicidal, she may also consider killing her infant and young children, not usually out of anger but stemming from a desire not to abandon her children.

Id. Miller, *Postpartum Depression*, JAMA 762 (2002).

Decisions about life-sustaining treatment should be made once the prognosis becomes more certain. It is not necessary to attain absolute or near absolute prognostic certainty before life-sustaining treatment is withdrawn, since this goal is often unattainable and risks unnecessarily prolonging the infant's suffering. Ethics committees or infant review committees should also be utilized to facilitate parental decision making.[174]

good v. Good

Infanticide (or neonaticide) of severely defective infants, like late-term abortion, is simply murder and neither it nor abortion after viability can be tolerated by a civilized society under any circumstances.

v.

Infanticide of severely defective infants shortly after birth is similar to the legal abortion of such fetuses, including abortions that occur after possible viability (which is justifiable in Israel).

Twenty-first Century Approach

1. Unlike in the past, technology of the third millennium often permits prospective parents to ascertain, during gestation and prior to viability, whether a newborn infant will be severely defective. In such cases, the pregnant woman has a constitutional right to determine whether to abort prior to viability. However, where a pregnant woman does not do so, or is unable to ascertain such severe defects in her fetus prior to viability, her right to seek an abortion should be extended beyond that time.
2. The live birth delivery of a severely defective infant should be subject to an objective position concerning its possible treatment or lack thereof.

174. AMA Code of Medical Ethics § 2.215 (2003).

11

Restrictions on the Sources and Allocation of Organ Transplants

1. *Restrictions on Sources for Organ Transplant and Embryo Transfer in the Twenty-First Century*
   a) The Need for Consent to Organ Transplants
   b) Religious and Cultural Restrictions on Transplants
      i) Embryo Transfer for Stem Cell Research
   c) Donations to or by Living Mature Minors and Other Legal Incompetents
   d) Sales of Organs
   e) Organs from Suicidal Donors and Executed Criminals (Murderers)
   f) Organs from Anencephalics
   g) Fetal Tissues and Mortality Rates in Hospitals; Age Limits
   h) Whole Brain Death (by Neurological Criteria) and Persistent/Permanent Vegetative State (PPVS)
   i) Xenotransplants: From Animals to Humans
2. *Allocation of Donated Organs*
   a) Allocation Priority to the "Sickest" Nationwide; HIV Patients
   b) Non-direct Donations by Living Individuals
   c) Directed Donations by Living Persons
   d) Transplants to Patients with End-Stage Renal Disease (ESRD)

No matter how impressive our technology, we still cannot create a human heart, a liver, a lung, a kidney. The power of our newest technology [transplantation] depends on our oldest values—love of life, love of neighbor.

Vice President Gore, December 16, 1997,
announcing an initiative for Americans to give the "gift of life"

1. Restrictions on Sources for Organ Transplant and Embryo Transfer in the Twenty-first Century

Consider this case: You are on vacation with your family driving a rented car on the island of Sicily, Italy. As you round a curve, a car passes you and someone inside shoots your 7-year-old son. You immediately get to a hospital for help. The doctors try for a day and a night to save your son. . . . In a true story about another family, the father continues in his words:

> We wanted only to go home, to take Nicholas with us, however badly injured, to help nurse him through whatever he faced, to hold his hand again, to put our arms around him. It had been the worst night of our lives.
>
> There had been no deterioration but no improvement either. "You know, there are miracles," said the man who had been appointed to act as our interpreter, but the doctors looked grave. In lives that only a few hours before had been full of warmth and laughter, there was now a gnawing emptiness. We asked how they knew his brain was truly dead, and they described their high-tech methods in clear, simple language. It helped. But more than that, it was the bond of trust that had been established from the beginning that left no doubt they would not have given up until all hope was gone.
>
> I remember the hushed room and the physicians standing in a small group, hesitant to ask crass questions about organ donation. As it happens, we were able to relieve them of the thankless task. We looked at each other. "Now that he's gone, shouldn't we give the organs?" one of us asked. "Yes," the other replied. And that was all there was to it.
>
> Within days our intensely personal experience erupted into a worldwide story. Newspapers and television told of the shooting at-

> tack by car bandits, Nicholas's death, and our decision to donate his organs. Since then streets, schools, scholarships, and hospitals all over Italy have been named for him. We have received honors previously reserved largely for kings and presidents. All this for a decision that seemed so obvious we've forgotten who among us suggested it.
>
> Yet we've been asked a hundred times: "How could you have done it?" And a hundred times we've searched for words to convey the sense of how clear and how right the choice seemed. Nicholas was dead. He no longer looked like a sleeping child. By giving his organs, we weren't hurting him but we were helping others.
>
> His toys are still here, including the flag on his log fort, which I put at half-staff when we returned home and that has stayed that way ever since. Donating his organs, then, wasn't a particularly magnanimous act. But *not to have given them would have seemed to us such an act of miserliness* that we don't believe we could have thought about it later without shame. The future of a radiant little creature had been taken away. It was important to us that someone else would have that future.
>
> It turned out to be seven people's future, most of them young, most very sick. The 60-pound 15-year-old who got Nicholas's heart had spent half his life in hospitals; now he's a relentless bundle of energy. One of the recipients, when told by his doctors to think of something nice as he was taken to the operating theater, said, "I am thinking of something nice. I'm thinking of Nicholas." Our joy in seeing so much eager life that would otherwise have been lost, and the relief on the families' faces, is so uplifting *that it has given us some recompense for what otherwise would have been just a sordid act of violence.*[1]

This donation was "so obvious," said Reg Green of Bodega Bay, California, Nicholas's father. What happened to the hate for the criminals who murdered Nicholas? From that, shouldn't the family have exercised its "property right" to their son's whole body when he was later buried in California? Were not such organs available from others? Didn't the Green family's culture or religion prevent them from donating those organs so quickly? Why

1. JAMA 1732 (1995) (emphasis added).

didn't they sell his organs? Did they inquire enough on whether the allocation process was fair in Italy? Why didn't they?

However, why don't we first ask: What can we learn about source restrictions on organ harvesting and allocation problems in the United States? Secondly, how can these processes be improved in the twenty-first century?

Permitting organs to remain alive outside the body was first achieved by a collaboration between the famous aviator Charles Lindberg and Nobel laureate Alexis Carrel in the mid-1930s:[2] "On April 5, 1935—using the Lindbergh Pump—a whole organ was successfully cultivated in vitro for the first time. Over the next two months, Carrel and Lindbergh performed more than two dozen experiments with the pump. They tested spleens, ovaries, kidneys, and hearts. In almost all instances, they succeeded in keeping them fully viable and free of infection."[3] However, immune-suppressive drugs were only found in the second half of the twentieth century. Near the close of that century, something had developed to the point of having conditional three-year grafts and patient survival rates of approximately 90 percent for organs other than lung and heart-lung.[4]

Donors, like the parents of Nicholas in the above case, turned out to be in the minority as of the dawn of the third millennium. The transplant waiting list for all organs increased by nearly 64 percent between 1988 and 1994, while available organs increased by only 33 percent.[5] By the beginning of 1994, the

2. Using Lindbergh's machine, Dr. Carrel performed surgeries that showed that circulation, even in such vital organs as the kidneys, could be interrupted for as long as two hours without causing permanent damage.

3. Berg, Lindberg 336 (1998); Lindberg, Autobiography of Values 133–38 (1977).

4. Lin et al., *1997 UNOS Report*, 280 JAMA 1153–60 (1998). In connection with surgeon selection for organ transplants, it is important to note that "patients can often improve their chances of survival substantially, even at high-volume hospitals, by selecting surgeons who perform the operations frequently." N. Eng. J. Med. 2117 (2003). Nevertheless, "some high-volume hospitals have relatively poor outcomes." *Id.* at 2159.

5. United Network for Organ Sharing (UNOS), gopher://info.med.yale.edu:70/00/Disciplines/Disease/Transplant/religion.txt. If an organ is transplanted but fails, there does not appear to be an obligation to restore it. "The common law has consistently held to a rule that provides that one human being is under no legal compulsion to give aid or rescue or to take action to save another. The rule is founded upon the very essence of our free society."

total number of people awaiting solid organ transplants was 64,423, an increase of about 300 percent since 1988. At the end of the twentieth century, 72,255 people were on transplant waiting lists, and 6,448 died while waiting for an organ.[6] As of July 2005 there were about 89,000 people on waiting lists, and the demand for organs remains far greater than the supply.[7] Some people actually believe that their doctors might not try as hard to save their lives if the doctors knew that they had agreed to be organ donors. Other people consent to donate an organ with the hope that they will partially live on while it is inside others; that is, they believe in the partial continued existence of their body in other persons, even after they have passed away. Thus, the United States has the lowest cremation rate in the world.[8] As will be noted below, there are very few religious objections to such donations.

Other possible objections arise. Although transplantation therapies have revolutionized care for patients with end-stage organs (the failure of their kidney, liver, heart, lung, and pancreatic B-cell), problems still persist with treatments designed to prevent graft rejection. The blood type is most important: After the death of a 17-year-old girl who received organs with the wrong blood type, the United Network for Organ Sharing (UNOS) concluded that none of the organizations involved in her transplant violated the network's policies in 2002. Accordingly, UNOS revised them to require that blood type be verified by at least four staff members, two at the organ recovery agency and two at the transplant hospital.[9] Anti-rejection drugs, which must be taken daily, are expensive, not always effective, and are associated with generally

6. JAMA (2001).

> Barriers at several steps are responsible for sociodemographic differences in access to cadaveric renal transplantation. Efforts to allocate kidneys equitably must address each step of the transplant process.

280 JAMA (1998).

7. N. ENG. J. MED. 441 (2005).

8. Siminoff & Chillag, *The Fallacy of the "Gift of Life,"* Hast. Ctr. Rep., 1999, at 34–41. "Minorities always tend to donate less," said Dr. David Imagawa, chief of transplantation at UCI Medical Center. ORANGE COUNTY REGISTER, Apr. 18, 1999.

9. ORANGE COUNTY REGISTER, June 28, 2002, at News 31. Also, transplant programs must check blood type when the organ arrives and compare it directly to the potential recipient's blood type.

well-known toxic effects.[10] Transplant patients normally must take anti-rejection drugs for the rest of their lives to keep the body from trying to destroy the foreign organ or tissue. Although the drugs tend to suppress the immune system, the potential gains of the procedure often justify the risks. In this connection, it is interesting to note that in 2004 Illinois became the first state to enact a law that would specifically allow HIV-infected people to donate organs to others who also had that virus.[11]

Nevertheless, by the turn of the third millennium, organ transplantation had become big business.[12] It has been noted that "although the use of grafts from living donors is standard practice in transplantation in children, their use in adults remains controversial." A recent survey found that adult-to-adult liver transplantation from a living donor is increasingly performed in the United States; it is concentrated in a few large-volume centers. Mortality among donors is low but complications in the donor are relatively common.[13] Skin cancers are the most common malignant conditions in transplant recipients and account for substantial morbidity and mortality in such patients.[14]

Approximately 30 percent to 50 percent of patients with squamous-cell carcinomas also have basal-cell carcinomas. According to a Scandinavian

---

10. JAMA 1076 (1999) (noting 21,000 of one type of organ).

11. However, before such donations can take place, changes are needed in the Department of Health and Services regulations. Currently they require that organs from HIV-infected patients be discarded to prevent them from being transplanted into uninfected patients and spreading the AIDS virus. ORANGE COUNTY REGISTER, July 16, 2004, at News 19.

12. ORANGE COUNTY REGISTER, Aug. 20, 1999, at News 13. In 1999, Congress enacted the Organ Donor Leave Act, H.R. 457, Pub. L. No. 106-56, Sept. 24, 1999, 113 Stat. 407, which provides a 30-day paid medical leave for all federal and state employees who donate an organ for transplantation.

13. "In summary, adult-to-adult liver transplantation from living donors currently accounts for approximately 5% of liver transplantations performed in adults. Overall mortality among donors is approximately 0.2%." Brown, Jr. et al., 348 N. ENG. J. MED. 818–25 (2003).

14. The mean interval between transplantation and diagnosis of a tumor (swelling) is eight years for patients who received transplants at approximately 40 years of age, but is only about three years for those who received transplants after the age of 60. The severity of these tumors is linked to their number. Euvrard et al., *Skin Cancers after Organ Transplantation*, N. ENG. J. MED. 1681 (2003).

study, 25 percent of patients with a first squamous-cell (scaly or platelike) carcinoma will have a second lesion discontinuity of tissue within 13 months, and 50 percent will have a second lesion within 3.5 years.[15]

The preservation time varies according to the organ matching of a donor organ to the recipient; these are critical in some, but not all organ transplants. Currently, the maximum preservation time is four hours for heart, lung, and heart-lung transplant, eight to 10 hours for pancreatic and liver transplants, and 36 hours for kidney transplants.[16]

Transplanted kidneys imperfectly matched to recipients fail more often in the first year when they have been shipped long distances. The researchers tied the greater risk to the longer time the organs spent in cold storage during shipment. A study showed a 17 percent higher failure risk in the first year for the organs that were shipped long distance, yet there was no difference in risk when transplanted kidneys were perfectly matched to the recipient's immune system. Hence, in some cases, it might be more important to wait for a better match even if the organ has to travel a long distance.[17] A close match between the genetic types of an organ donor and recipient reduces the risk of

---

15. *Id.*

16. *See also* Hauptman & O'Connor, *Medical Progress: Procurement and Allocation of Solid Organs for Transplantation*, N. Eng. J. Med. 422–31 (1997). Organ donations from the living reached a record high in 2001. Orange County Register, Apr. 23, 2002, at News 5. More than 6,400 people gave away a kidney or a piece of their liver. Yet, in 2001 there were 6,081 donor cadavers, and they often donate organs for three or four transplants. Multiple transplants are of many organs. "Domino use" refers to the placement of a transplant organ from one to another recipient. Praseedom et al., *Combined Transplantation of the Heart, Lung, and Liver*, 358 Lancet 812–13 (2002).

17. N. Eng. J. Med. (2001).

> However, a kidney is also affected by one's age. For example, between the ages of 40 and 80, the normal kidney loses some 20% of its weight and develops areas of scarring within its substance. Thickening of tiny blood vessels that are inside the kidney further decreases blood flow and results in destruction of the organ's filtering units, which are, of course, the crux of its ability to clear the urine of impurities. In time, some 50% of the filtering units will die.

Nuland, How We Die 54 (1995).

organ rejection.[18] The average transplant recipient lived another 20 years, while the average waiting-list patient, who instead remained on dialysis, lived another 10 years.[19] Lung transplantation for bronchioalveolar carcinoma has proven to be technically feasible, but recurrence of the original tumor within the donor lungs, up to four years after transplantation, was common.[20] As of 2001, the United Network for Organ Sharing (UNOS) found that 4,000 patients with end-stage lung disease are awaiting transplantation and that "most patients waiting for a lung graft will never receive one."

In 1998, a federal law required hospitals to report all deaths to organ collection agencies and have specially trained people approach families about the donation. Permission rates rise from around 60 percent to around 85 percent when the request comes from someone well versed at addressing common fears and questions about donation.[21]

Physicians have often been "uncomfortable in discussing the worst and discourage patients from honestly discussing their concerns," yet physicians can benefit by being honest with patients and helping them to consider making a living will, naming a healthcare proxy, preparing financial matters, or

---

18. Johns Hopkins Medical Test (2001), p. 68. Advances in immunosuppression for successful organ transplantation have eliminated the requirement of a genetic match. Terasaki, *The HLA Matching Effect in Different Cohorts of Kidney Transplant Recipients, in* Clinical Transplants 2000 (2001). Further, by 2001, the number of kidneys transplanted from living donors began to surpass the number from cadavers. United Network for Organ Sharing, 2001 Annual Report 32 (2002).

19. N. Eng. J. Med. (1999). Currently, there are 43,584 people in the United States on the waiting list for a kidney transplant, according to the United Network for Organ Sharing. Only 12,166 transplants were performed last year, and 2,295 people on the waiting list died. For 1997, the FDA approved Zenopox, the monoclonal antibody that blocks immune cells from attacking the new kidney during the first eight weeks after transplant, the riskiest period. Orange County Register, Dec. 12, 1997, at News 42.

20. Garner et al., 340 N. Eng. J. Med. 1071–74 (1993). There are currently 90 lung-transplantation centers in the United States and less than 20% of cadaveric donors have lungs suitable for harvest. Arca Soy & Kotloff, N. Eng. J. Med. 1081–88 (1999).

21. United Network for Organ Sharing, Critical Data: Facts About Transplantation (2001); JAMA 2721 (2001); *see also* http://www.boston.com/dailyglobe2/.../Cash_for_organs_Doctors_consider_a_loaded_idea+.shtm.

settling family affairs. It has been noted that one study indicated that "unarticulated concerns correlate with increased anxiety and depression" and "patients and families who prepare for a range of outcomes may be less likely to blame their physicians for the consequences of disease progression."[22]

a) *The Need for Consent to Organ Transplants*

According to UNOS, as of October 31, 1996, the number of registrations of potential and hopeful organ recipients was almost 50,000, "with the majority in the 18-to-49-year age range."[23] Only about 20,000 Americans received organ transplants in 1997, and some 4,000 died while waiting for a donor.[24] As for potential donors, an estimated 12,000 to 15,000 people died each year from whom organs could be used, but there were fewer than 5,500 donors in 1997[25] and 6,613 donors in 2002, according to UNOS.[26] In 1998 it was noted that well over 120,000 people in the United States could benefit from organ transplantation.[27] Why does this mismatch happen? Can it be changed? These are the questions to be answered in our twenty-first century.

When a patient has end-stage congestive heart failure and is waiting for a heart transplant, his doctor may indicate: (1) that a new heart will become available; (2) if not, and if the patient's heart became worse, the physician's need to ascertain the patient's desire for cardiopulmonary resuscitation (CPR) even though it usually becomes an ineffective therapy;[28] and (3) at some stage, should the patient express a wish to decline CPR, the doctor may ask if such a patient might be willing to become an organ donor, or had so expressed in an advanced directive, or so informed a designated surrogate. Section 2.157 of the American Medical Association's *Code of Medical Ethics* states that where "patients or their surrogate decision-makers request withdrawal of life support and choose to serve as organ donors, the organs can be preserved best by discontinuation of life support in the operating room so

22. Arnold & Quill, ANNALS INTERNAL MED. 440 (2003).
23. N. ENG. J. MED. 422 (1997).
24. ORANGE COUNTY REGISTER, Dec. 16, 1997.
25. ORANGE COUNTY REGISTER, June 18, 1998, at News 19.
26. ORANGE COUNTY REGISTER, Aug. 10, 2003, at News 15.
27. N. ENG. J. MED. 1780 (1998).
28. Arnold & Quill, *supra* note 22, at 441.

that organs can be removed two minutes following cardiac death." In other scenarios, patients who suffer unexpected cardiac death may be cannulated and perfused with cold preserving fluid (in situ preservation) to maintain organs.

Should there be any reason to suspect that the patient's consent may not be voluntary (or that surrogate decisions about life-sustaining treatment may be influenced by the prospect of organ donation), "a full ethics consultation should be required."[29] The dramatic shortage of kidney donors has triggered interest in other sources of organs, such as donors without a heartbeat. A study in 2002 found that "although the incidence of delayed graft function is significantly higher with kidneys from donors without a heartbeat than with kidneys from donors with a heartbeat [living donors], there is no difference in long-term outcome between the two types of graft."[30] In 2003, for the first time, six-way organ match became possible: "A woman from Miami, a woman from Pittsburgh and a teenager from Maryland, all of whom were on dialysis, had come to the Baltimore hospital with willing donors whose blood or tissue types didn't match their own—but did match one of the others."[31] In light of a limited number of donor kidneys for transplantation, the almost 100,000 American patients with end-stage renal disease must annually choose between hemodialysis (performed in an outpatient dialysis facility three times a week) and peritoneal dialysis (performed every day at home by the patient). A recent study showed that "After several weeks of initiating dialysis, patients receiving peritoneal dialysis rated their care higher than those receiving hemodialysis."[32]

---

29. AMA CODE OF MEDICAL ETHICS § 2.157 (1996). That section, titled "Organ Procurement Following Cardiac Death," provides: "Perfusion without either prior specific consent or general consent to organ donation violates requirements for informed consent for medical procedures and should not be permitted." *Id.* "Perfusion" is "the passage of a fluid through the vessels of a specific organ." DORLAND'S POCKET MEDICAL DICTIONARY.

30. N. ENG. J. MED. 248 (2002).

31. "The surgery lasted 11 hours, with two doctors, two nurses, and two anesthesiologists in each operating room." ORANGE COUNTY REGISTER, Aug. 2, 2003, at News 20.

32. Rubin et al., *Patient Ratings of Dialysis Care with Peritoneal Dialysis vs. Hemodialysis*, JAMA 697 (2004).

The widespread introduction of automated peritoneal dialysis, whereby

A study of the Harvard School of Public Health found that 27 percent of potential donors are either not identified or their families are not approached about donation.[33] This problem should be solved by the states. As we have seen in other areas of bioethics, some federal mandates can produce more problems than they solve. Property rights, including those valid under the Due Process Clause of the Fourteenth Amendment, are mostly a matter of state law. During one autopsy on Michigan decedents, their corneas were removed. Under Michigan law, next of kin have the right to control the disposal of their bodies and possession for burial. In lieu of suing in a state court, next of kin brought an action under federal law[34] for violation of their civil rights. A federal court of appeals held that Michigan provided the next of kin with a constitutionally protected property interest in the dead body of a relative.[35] Similarly, a court has held that a couple's embryo belonged to them even though it resided in a physician's laboratory. Thus, the doctor was the custodian of their tissue.[36] In order to obtain organs for transplant, it is normally necessary to obtain consent either from the deceased as an anatomical gift or from the next of kin. The AMA Code states:

> Full discussion of the proposed procedure with the donor and the recipient or their responsible relative or representatives is mandatory. The physician should ensure that consent to the procedure is fully informed and voluntary, in accordance with the Council's guidelines on informed consent. The physician's interest in advancing scientific

---

> dialysis fluid is exchanged frequently while the patient is asleep, improves ultrafiltration and reduces even further the time that the patient needs to devote to dialysis procedures, since daytime exchanges are often not required. . . . [Further,] a recently published randomized controlled trial found a significant survival advantage for patients who received peritoneal dialysis.

Heaf, *Underutilization of Peritoneal Dialysis*, JAMA 740 (2004).

33. *Supra* note 32. Public education is needed to modify attitudes about organ donation prior to a donation opportunity. 286 JAMA 71–77 (2001).

34. 42 U.S.C. § 1983. The Illinois Supreme Court in *Curran v. Bosze*, 566 N.E.2d 1319, 1325–36 (1990), refused to compel twin minors to donate bone marrow to a half-sibling.

35. Whaley v. County of Tuscola, 58 F.3d 1111 (6th Cir. 1995).

36. Ramirez, 1998 WL 345105, at *3.

> knowledge must always be secondary to his or her concern for the patient . . . no physician may assume a responsibility in organ transplantation unless the rights of both donor and recipient are equally protected.[37]

There is no common-law right of action to recover damages for negligent interference with a cadaveric organ donation. The most used model statute, the Uniform Anatomical Gift Act (UAGA), creates qualified immunity for good-faith compliance with its procedures.[38] If permission is to be sought from next of kin, time may be of the essence depending upon the organ being sought. This is because, as has been long known, various tissues and organs in the human body die at different times and rates.[39] A 2003 study showed the definite need to invest resources in order to increase the rate of organ recovery:

> We identified a total of 18,524 brain-dead potential organ donors during the study period. The predicted annual number of brain-dead potential organs lie between 10,500 and 13,800. The overall consent rate (the number of families agreeing to donate divided by the number of families asked to donate) for 1997 through 1999 was 54 percent, and the overall conversion rate (the number of actual donors divided by the number of potential donors) was 42 percent.

---

37. AMA Code of Medical Ethics R. 2.16(3) (2002–2003).

38. York v. Jones, 717 F. Supp. 421 (E.D. Va. 1989). Until the beginning of the third millennium, the most common reason for missed donation opportunities had been a denial of consent by the donor's family. Sade, *Cadaveric Organ Donation: Rethinking Donor Motivation*, Archives Internal Med. 428–42 (1999).

39. Orange County Register, Feb. 5, 1997. It has been argued that there is "no compelling ethical argument why viable solid organs should be treated differently from blood and some other tissues." Tamb et al., *Cadaveric Organ Donation: Encouraging the Study of Motivation*, 76 Transplantation 748 (2003). Yet the current means of organ donation themselves imply some notion of property right because "one cannot give away what one does not own any more than one can sell it." Council of Ethical and Judicial Affairs, American Medical Association, *Financial Incentives for Organ Procurement: Ethical Aspects of Future Contracts for Cadaveric Donors*, Archives Internal Med. 155 (1995).

Lack of consent to a request for donation was the primary cause of the gap between the number of potential donors and the number of actual donors. Since potential and actual donors are highly concentrated in larger hospitals, resources invested to improve the process of obtaining consent in larger hospitals should maximize the rate of organ recovery.[40]

The consent requirement is not worldwide. Since 1997 a Brazilian law allows federal authorities to use the organs of the deceased for transplants without the person's prior consent.[41] Physicians have argued for "mandated choice," requiring individuals to state whether they wished to donate their organs when they applied for licenses, filed their tax returns, or obtained official ID cards. Applications would not be accepted if the preference regarding the donation was not stated.[42]

However, in England, consent is presumed unless the deceased has otherwise set forth his desire not to donate his organs. This solution was characterized as an opting-out scheme in which willingness to donate is assumed unless there is a formal refusal on record. It is also similar to the law in France and some other countries.[43] For example, under Belgian law, the state

---

40. Sheely et al., *Estimating the Number of Potential Organ Donors in the United States*, N. Eng. J. Med. 667 (2003). The Uniform Anatomical Gift Act (UAGA) grants the next-of-kin *the right* to disposal of the body. Unif. Anatomical Gift Act § 3. The majority of states have adopted this law, which also indicates how persons can donate their bodies to medical schools and hospitals or permit members of their family to do so. *Id.* However, "The act does not compel organ donations nor does it establish a presumption that organs will be donated." Lyon v. United States, 343 F. Supp. 531, 536 (D. Minn. 1994).

41. Spital, Annals Internal Med. 125. It would also increase awareness about organ donation. *See* Klassen & Klassen, *Who Are the Donors in Organ Donation? The Family's Perspective in Mandated Choice*, 125 Annals Internal Med. 70–73 (1996) (describing the state of Virginia's program, where 31% registered as donors, 45% as non-donors, and 24% undecided).

42. Law of July 29, 1994. Finnish Law No. 355 of Apr. 26, 1985 [12, p. 195] provides that if the deceased consented to removal, "the procedure may be carried out despite the objection of the relatives." *Id.* § 4. At this annual conference in Belfast, the British Medical Association rewrote its 30-year policy on organ donation to back a nationwide program of "presumed consent" in which tissues could be removed for transplantation after an individual's death unless that person registers an objection. Orange County Register, July 7, 1999, at News 25.

43. Brahams, *Body Parts as Property,* 66 Med.-Legal J. 45 (1993).

makes the assumption of "presumed consent"—that people are willing donors of organs after death—unless they write down the contrary. After 10 years, it was a great success; only 2 percent of the people had declined in writing to donate; and this expression of the desires of donors could not be overcome by their families after they died. In Texas, a 1991 law provides that the wishes of organ donors 18 or older "shall be honored without obtaining the approval or consent of any other person."[44] Nevertheless, the AMA *Code of Medical Ethics* Section 2.155 still states that for "presumed consent to be ethically acceptable, effective mechanisms for documenting and honoring refusals to donate must be in place." However, in those states in which individuals are required to express their preferences regarding organ donation when reviewing their drivers' licenses, this becomes "an ethically appropriate strategy for encouraging donation and should be pursued."[45] The need for presumed consent is one of the problems to be solved in the twenty-first century.

---

b) Religious and Cultural Restrictions on Transplants

Regardless of the great benefits achieved by the transplantation of organs in the United States, religious and cultural restrictions have caused significant problems. For example, many Orthodox Jews maintain that because man was created in the image of God, he should be buried whole. Hence, if a limb has been removed from a person, it must be kept to be buried in the future when he dies. One of them carried a belief that at the end of time, the body itself will be resurrected.[46] When a funeral parlor in Florida lost an amputated leg of an Orthodox Jewish woman, it actually paid a $1.25 million

44. Hast. Ctr. Rep., Sept.-Oct. 1991, at 48.

45. Rule 2.155; *see Strategies for Cadaveric Organ Procurement: Mandated Choice and Presumed Consent*, 272 JAMA 809-12 (1994); Childress, *The Failure to Give: Reducing Barriers to Organ Donation*, 11 KENNEDY INST. ETHICS J. 1, 13–14 (2001).

46. Kohn v. United States, 591 F. Supp. 568, 573 (E.D.N.Y. 1984) (citing Fred Rosner, *Autopsy in Jewish Law and the Israeli Autopsy Controversy*, *in* JEWISH BIOETHICS 331–32, 335, 338 (1979)).

lawsuit settlement of a property right to her daughter in 1997.[47] Some Jews have also effectively blocked autopsies on the ground that the Orthodox tradition prohibits dissection of the body.[48]

Several of these matters have been debated by rabbis. According to Fred Rosner in his *Modern Medicine & Jewish Law* (1972):

> The prohibition of deriving benefit from the dead applies only to flesh (Bassar) or organs, but not to skin (Orr). The cornea of the eye, according to Rabbi Greenwald, "is considered as skin and not as flesh" (p. 163). Moreover, "Rabbi Greenwald, as most authorities, would only permit eye grafts for a person blind in both eyes. . . . Rabbi Glickman adds, however, that one may only perform a transplant if the donor gave permission prior to his death. Otherwise the donor is hindered from achieving atonement (Kaparah) for his sins through his death since one of his organs remains alive. If he gave permission, then he has voiced his acquiescence to delaying his atonement until his organ is later buried following the eventual death of the recipient." (p.164)

Rabbi Immanuel Jakobovitz has stated that a donor may endanger his own life or health to supply a "spare" organ to a recipient whose life would thereby be saved only if the probability of saving the recipient's life is substantially greater than the risk to the donor's life or health. (p. 167) "Where chances for success are minuscule and the recipient's life is probably shortened rather than lengthened by this procedure, heart transplantation must still be considered murder of the recipient." (per Rabbi Feldstein, p. 175)

Orthodox Jews are in the minority in the United States and other countries and even disagree among themselves. The former chief rabbi of the British Commonwealth and author of the first comprehensive book on Jewish medical ethics wrote: "Since the mortality risk to kidney donors is estimated to be only 0.24 percent and no greater than is involved in any amputa-

47. A woman was awarded $1.25 million as compensation in a suit against a funeral home for losing her mother's amputated leg. SUN-SENTINEL (Ft. Lauderdale, Fla.), May 16, 1997, at 1B.

48. Atkins v. Med. Exam'r, 418 N.Y.S.2d 839, 940–41.

tion, the generally prevailing view is to permit such donations as acts of supreme charity but not as an obligation."[49] An American rabbi has written, "Conservative and Reform rabbis have taken the same stance."[50]

Beliefs like these raise the serious question of whether people who refuse to donate should be entitled to receive scarce organs when they need them. Is it fair for them to be permitted to receive a life-saving organ from another when they would refuse to give up one of their own organs in like circumstances? The same question might apply to those traditional Korean-Americans who believe that donating bone marrow is viewed as "cutting off the link with your parents and ancestors." Of 3.1 million donors in 1997, only 5.8 percent were of Asian descent.[51] In Singapore, the Human Organ Transplant Act of 1987 provides for certain priorities, namely, "Any Muslim who has failed to opt in to posthumous donation, will be subordinated in priority to candidate recipients of transplantation who have not opted out, or who are Muslims who have opted in."[52] It has been noted that "blacks have lower rates of kidney transplants . . . even where no differences in insurance or ability to pay exists."[53] The Department of Surgery at Mount Sinai School of Medicine observed, "African-Americans have historically donated significantly fewer cadaver kidneys than they have received. They have benefited from Caucasian organ donation."[54]

49. Dorff, Matters of Life and Death 227 (1998).

50. *Id.* at 227, 330. Yet, he sadly noted that when asked to be donors, "about 60% of the general population and their families consent, but only about 5% of Orthodox Jews do so, and *the record of Conservative and Reform Jews is not much better*."

51. Orange County Register, Mar. 20, 1998, at Metro 4.

52. Dickens, *WHO Guidelines and the WHO Task Force*, *in* Organ Allocation 83–94 (Touraine et al. eds., 1998); Dickens et al., *Legislation on Organ and Tissue Donation*, *in* Organ and Tissue Donation for Transplantation 95–119 (Chapman et al. eds., 1997).

53. Smith, Healthcare Divided: Race and Healing a Nation 27 (1999). The United Network for Organ Sharing (UNOS) indicated that African-American patients decline the procedure more often than do whites. *Background Problems and Concerns in Equitable Organ Allocation Statement of Principles and Objectives App. D*, http://204.127.237.11/eg_bkgnd.htm.

54. Rozon-Solomon & Burrows, Tis Better to Receive Than to Give 274.

good v. Good

Body parts have been too personal to the dead to be distributed without the specific consent of the family or their owner while alive and competent.

v.

Unless otherwise indicated in writing or by the family, presumed consent by the dead would save the lives of thousands more people who die annually due to the lack of an available organ.

Twenty-first Century Approach

1. One solution to the lack of organs that should be pursued by most states in the twenty-first century is a "presumed consent" that authorizes the harvesting of organs from a deceased person for the benefit of others (and for therapeutic or research purposes). This may be possible where (a) the dying person's consent can be quickly ascertained while alive, or (b) after an initial reasonable attempt has been made to discover from others whether or not the person had previously made known his intent was to refuse such donations.
2. Consideration should be given to the lowering of the priority of organ recipients who themselves would refuse to consent to such donations.

The development of cell lines that may produce almost every tissue of the human body is an unprecendented scientific breakthrough. It is not too unrealistic to say that this research has the potential to revolutionize the practice of medicine and improve the quality and length of life.

Former NIH Director and Nobel Prize Winner Harold Varmus
http://www.hhs.gov/asl/testify/t981201a.html(1998)

i) Embryo Transfer for Stem Cell Research

At the end of the second millennium, anti-abortion activists also tried to block support of all embryonic cell and fetal stem cell research. These may become cells in the body used to treat various diseases.[55] With stem cells,

55. For example, transplantation of corneal epithelial stem cells can restore useful vision in some patients with severe ocular-surface disorders. 340 N. Eng. J. Med. 1697–1703 (1999). In 2004 researchers using embryonic stem cells at the University

human embryonic cells have been developed into functioning brain-transportable neural cells to treat Parkinson's disease and spinal cord injuries.[56] In 1995, President Clinton signed the Dickey Amendment, which was attached to the Health and Human Services authorization bill each year since 1995. It has denied federal funds to be used for both *the creation of human embryos for research purposes and "research in which a human embryo or embryos are destroyed, discarded, or knowingly subjected to risk of injury or death greater than that allowed for research on fetuses in utero."* This prohibition applies to "*any organism that is derived by fertilization, parthenogenesis, cloning, or any other means from one or more human gametes or human diploid cells.*"[57] However, James Thomson, professor at the Wisconsin Regional Primate Research Center at the University of Wisconsin, who developed the first line of human embryonic stem cells, declared, "These cell lines should be useful in human developmental biology, drug discovery, and *transplantation medicine*." (Emphasis added.)[58]

However, private funds could become available for such research. In 1999, the National Bioethics Advisory Commission (NBAC) recommended that private laboratories develop human stem cells from embryos while the government only backed researchers' use of discarded ones. The NBAC concluded that the potential benefits of stem cell research outweigh the disadvantages—provided the cells are drawn from embryos that would otherwise

---

of Wisconsin–Madison became the first to make human motor neurons, the spindly nerve cells that control nearly all movement in the body. However, more research may provide motor nerve cells that can be used to test new drugs intended to treat various nerve ailments. ORANGE COUNTY REGISTER, Jan. 31, 2004, at News 13.

56. ORANGE COUNTY REGISTER, May 31, 2002, at News 3. "Fetal tissue research is morally acceptable to many because it can be compared with cadaveric organ retrieval after homicide or suicide. Furthermore, in this case, the woman's voluntary choice to donate is protected by informed consent." FLETCHER, THE HUMAN EMBRYONIC STEM CELL DEBATE (2001).

57. *See* Balanced Budget Downpayment Act, Pub. L. No. 106-554, § 510, 110 Stat. 26 (1996); 45 C.F.R. § 46.111(a)(2). Section 46.208 applies to the Dickey Amendment and has been reenacted annually by Congress since 1996 and signed each time by the President.

58. Thomson et al., *Embryonic Stem Cell Lines Derived from Human Blastocysts*, SCIENCE 1145–47 (1998); *see also* Thomson et al., *Isolation of a Primate Embryonic Stem Cell Line*, PROC. NAT'L ACAD. SCI. U.S.A. 7844–48 (1995).

be discarded. The report *Stem Cells and the Future of Regenerative Medicine* (2001) for the National Academy of Sciences stated:

> In the last three years, it has become possible to remove these [human embryonic] stem cells from the blastocyst and maintain them in an undifferentiated state in cell culture lines in the laboratory. . . . Because of a misunderstanding of the state of knowledge, there may be an unwarranted impression that widespread clinical application of new therapies is certain and imminent. In fact, stem cell research is in its infancy, and there are substantial gaps in knowledge that pose obstacles to the realization of new therapies from either adult or embryo-derived stem cells.[59]

Rules that would allow the use of such privately produced cells were issued as guidelines by the Geron Corporation, a company that had succeeded in establishing culture lines of human embryonic stem cells. These include conditions that the blastocysts (cells prior to implantation) be donated by women who gave their informed consent for research and development approved by an independent ethics board and an Institutional Review Board (IRB). Under the applicable regulations, IRBs are also required to weigh as a *benefit* the "importance of the knowledge that may be expected to result" against the *risk* of "possible long-range effects of applying knowledge gained in the research (for example, the possible effects of the research on public policy)."[60] These donors also understand market implications. In 2002 a study found that adult blood stem cells were unable to transform themselves into other types of tissue. This supported the view that embryonic stem cells offer much promise for treating such conditions as heart disease, spinal injury, diabetes, and Parkinson's disease.[61] However, in 2005 it was noted that "[a]fter the fertilized egg, embry-

59. SCIENCE 502 (1999). In 2004, two members of the President's Council on Bioethics who had been outspoken in their support of research on human stem cells were not reappointed. N. ENG. J. MED. 1379–80, 1454 (2004). Yet federal advisory committees, under the Federal Advisory Committee Act, Pub. L. No. 92-463, _Stat. _ (1972), are to provide independent, expert, and objective advice on policy, the funding of research, and other issues.

60. *See* 21 C.F.R. § 56.111(a)(2); 45 C.F.R. § 46.111(a)(2).

61. Geron, Ethics Advisory Board, *Research with Human Embryonic Stem Cells: Ethical Considerations*, Hast. Ctr. Rep., 1999, at 31–36. "Researchers in many labs

onic stem cells are the most flexible stem cell type." Further, "[s]ome evidence suggests that umbilical cord blood and fetal stem cells are just as versatile as embryonic stem cells."[62] However, it has also been noted that miscarriages occur with 70 percent of conceptions.[63]

Some religious people argue that a person's embryo is a potential fetus, since it may later become embedded in a uterus and develop into a fetus, which they consider a "person" at several states. With this belief, the Roman Catholic Church contends that its use and destruction in research is "a gravely immoral act" and prohibited such action. As John T. Noonan wrote: "The positive argument for conception as the decisive moment of humanization is that at conception the new being receives the genetic code. . . . A being with the human genetic code is man."[64] Nevertheless, an August 2004 poll found

---

are now studying two basic types of stem cells. Somatic, or adult, stem cells come from mature tissue. Embryonic stem cells come from embryos that have been allowed to grow to a certain point and then are destroyed." ORANGE COUNTY REGISTER, Sept. 6, 2002, at News 7 (reporting on a study published in *Science*). In 2004, Harvard University proposed the cloning techniques to create embryonic stem cell lines expressing the genes for diabetes, and Parkinson's and Alzheimer's disease. SCIENCE 586 (2004). The California Institute for Projects plans to fund human cloning designed to create stem cells. ORANGE COUNTY REGISTER, Mar. 11, 2005, at News 16. In 2005, Brazilian legislators voted to legalize stem cell research using human embryos for research to find cures for ailments such as diabetes, Parkinson's disease, and spinal cord injuries. ORANGE COUNTY Register, Mar. 4, 2005, at News 11.

62. Brownlee, *Full Steam Ahead*, SCI. NEWS 218 (2005); *see also* 21 C.F.R. § 566.111(a). Embryonic stem cells come from the human embryo during the blastocyst stage, which is the period prior to implantation in the uterus. Embryonic stem cells can be induced to differentiate into various cell and tissue types. Goldstein, *Dipping into Uncle Sam's Pockets: Federal Funding of Stem Cell Research;Is It Legal?*, 11 B.U. PUB. INT. L.J. 229, 232 (2002).

63. STENCHEVER ET AL., COMPREHENSIVE GYNECOLOGY 414 (4th ed. 2001). Another study showed that women who were 18 to 35 years of age repeatedly suffered chromosomally normal miscarriages 64% of the time, and for women over 40 that percentage of miscarriages had *abnormal* chromosomes. Stephenson et al., *Cytogenetic Analysis of Miscarriages from Couples with Recurrent Miscarriage: A Case-Control Study*, 17 HUMAN REPRODUCTION 446–51 (2002).

64. Noonan, Jr., *An Almost Absolute Value in History*, *in* INTERVENTION AND REFLECTION 83–86 (Munson ed., 2001).

> When the embryo is four to seven days old, it's called a blastocyst, a ball of up to 150 cells. A researcher removes the inner cell mass from the

that 34 percent of those who reported high levels of religious commitment support stem cell research, while 66 percent of those with low levels of religious commitment support the research.[65] However, in a general poll in 2005 by the Coalition for the Advancement of Medical Research, 59 percent favored embryonic stem cell research, and 33 percent were against it.[66]

However, courts have denied the use and designation of the embryo as a "person" for such an objective reason as to prohibit its use for stem cell research that might save lives. Professor Farley at Yale University indicated that "[a] growing number of Catholic moral theologians, for example, do not consider the human embryo in its earliest stages . . . to constitute an individualized human entity," and Rabbi Elliot Dorff noted: "Genetic materials outside the uterus have no legal status in Jewish law, for they are not even a part of a human being until implanted in a woman's womb and even then, during the first 40 days of gestation, their status is 'as if they were water.' As a result, frozen embryos may be discarded or used for reasonable purposes, and so may stem cells be procured from them."[67]

In February 2003 it was noted that "some of the most exciting biomedical research of the twenty-first century isn't getting done. Research on stem cells from human embryos has become so entangled in politics and public

---

> blastocyst and places the cells in a dish filled with a broth of proteins and enzymes: the cells continue to multiply, but they become more specialized. They are called pluripotent and can generate cells for the body but cannot support fetal development. These are *embryonic stem cells.*

Stem Cell Research Foundation Home Page, http://www.stemcellresearchfoundation.org; ORANGE COUNTY REGISTER, Jan. 7, 2004, at News 16.

65. Pew Forum on Religion and Public Life, *GOP the Religion-Friendly Party, But Stem Cell Issue May Help Democrats*, http://pewforum.org/docs/index.php?DocID=51 (Aug. 24, 2004).

66. Coalition for the Advancement of Medical Research (CAMR) Home Page, http://www.camradvocacy.org.

67. 1 NAT'L BIOETHICS ADVISORY COMM'N, ETHICAL ISSUES IN HUMAN STEM CELL RESEARCH (1999), *available at* http://www.bioethics.gov/reports/past_commissions/index.html. In 2005 a Chicago judge actually ruled that "a test-tube embryo is a human" and where such an embryo is not preserved, the parents may sue for its wrongful death. Because such a ruling would appear to threaten *in vitro* fertilizations and abortions, it is expected to be reversed. ORANGE COUNTY REGISTER, Feb. 9, 2005, at News 15.

misunderstanding that researchers are worried about serious delays in understanding life-threatening diseases."[68] Some bioethicists immediately pointed out that "neither Jewish, Christian, or other theological literature help in any specific way with this problem, owing to the new nature of this research."[69] In 2000, the Vatican's "Declaration on the Production and the Scientific and Therapeutic Use of Human Embryonic Stem Cells" stated that the removal of a cell's mass from an embryo "is a gravely immoral act, and consequently is gravely illicit."[70] However, in the same year, Rabbi Elliot Dorff of the University of Judaism wrote: "[I]n light of our divine mandate to seek to maintain life and health, one might even argue that from a Jewish perspective, we have a duty to proceed with that research."[71]

In the final days of the twentieth century, it was noted that "[i]ndustry's share of total investment in biomedical research and development grew from approximately 32 percent in 1980 to 62 percent in 2000, while the Federal government's share fell."[72] The National Institutes of Health (NIH) had published guidelines for some federal funding. Embryonic cell lines could come from frozen "excess" embryos created during fertility treatments at private clinics. A Human Pluripotent Stem Cell Review Group would evaluate any newly derived cell lines to ensure they are in compliance with NIH rules.[73] Research and development of embryonic stem cells embarked with two sets

---

68. Bruce Agnew, *The Politics of Stem Cells*, Genome Network News, Feb. 21, 2003, http://www.genomenewsnetwork.org/articles/02_03/stem.shtml. "The moral principle that 'It's wrong to kill an innocent person' is violated . . . [However,] the aim of stem-cell research is definitely utilitarian, but it's an aim we can pursue without causing any significant moral harm." Munso, Raising the Dead.

69. Glenn McGee & Arthur L. Caplan, *What's in the Dish?*, Hast. Ctr. Rep., 1999, at 36–38. In Ronald Green, The Human Embryo Research Debates: Bioethics in the Vortex of Controversy, the author took the position denying that embryos have full moral status.

70. http://www.vatican.va/roman_curia/pontifical_academies/acdlife/documents/re_pa_acdlife_doc_20000824_cellulestaminali_en.html.

71. 3 Nat'l Bioethics Advisory Comm'n, Ethical Issues in Human Stem Cell Research C1 (2000), *available at* http://www.bioethics.gov/reports/pastcommissions/ index.html.

72. Bekelman et al., JAMA 454 (2003).

73. Science 2050 (1999).

of guidelines: one for private research and another more restricted one, using federal funds due to political and religious influences. President Bush initially allowed federal funds to be used for research using cells from only about 60 embryos that had already been somewhat destroyed at private laboratories, in order to discourage the further destruction of human embryos. In 2002, the President said, "Scientists could use taxpayer dollars only to study those self-sustaining colonies, or lines, of cells that had already been extracted from human embryos."[74] In November 2002, Australia gave a green light to controversial research on human embryos, but only after a marathon debate in the national parliament and a barrage of attempted legislative amendments.[75] An Australian report of December 19, 2005, on embryonic stem cell law recommended relaxing the country's current ban on therapeutic cloning and establishing a national stem cell bank. It also advised permitting the creation of cross-species (chimeras)—where human somatic cells are fused with non-human eggs—for research.[76] In 2004 strong support for embryonic stem cell research was also reported to include Israel, the Czech Republic, Singapore, Korea, and the United Kingdom.[77] However, President Bush's appointee as chairman of the official Bioethics Council pointed out that his "legitimacy was still being denied by many who opposed him and whose pro-life moral views and born-again Christian religious attachments are, let us be frank, strongly held against him in many scientific and intellectual circles. Moreover, our creation was linked to a highly charged single public issue, federal funding of embryonic stem cell research, about which feelings ran and still run enormously high."[78] Accordingly, the Council offered legis-

---

74. On Aug. 5, 2002, the President added: "This allows us to explore the promise and potential of stem cell research without crossing a fundamental moral line by providing taxpayer funding that would sanction or encourage further destruction of human embryos." ORANGE COUNTY REGISTER, Aug. 7, 2002, at News 18.

75. http://www.reutershealth.com/archive/2002/11/11/enline/links/20021111elin026.html.

76. NATURE 1062 (2005).

77. N. ENG. J. MED. 1789 (2004).

78. Leon Kass, *Reflections on Public Bioethics: A View from the Trenches*, KENNEDY INST. ETHICS J. 222 (2005).

lative proposals targeting allegedly "unethical or disquieting practices in human reproduction."[79]

Nevertheless, Leon Kass added that "busy legislators and government officials are not going to take the time to read lengthy reports and ponder. For this reason, I suspect that several of our projects have done little to advance public understanding in these narrow quarters."[80] In this connection, in 2005 Senate Republican leader Bill Frist, M.D., a heart-lung transplant surgeon, decided to support a bill to expand federal financing for embryonic stem cell research.[81] Efforts should be taken early in the twenty-first century to remove some of these restrictions. The limitation on use by federal researchers on cells that have to be taken from the embryos by privately funded researchers may only be viewed by some people as a metaphysical one from some ethical point of view. "The debates about embryo and human embry-

---

79.
1. Prohibit the transfer, for any purpose, of any human embryo (produced *ex vivo*) into the body of any member of a nonhuman species.
2. Prohibit the production of a hybrid human-animal embryo by fertilization of human egg by animal sperm or of animal egg by human sperm.
3. Prohibit the transfer of a human embryo (produced *ex vivo*) to a woman's uterus for any purpose other than to attempt to produce a live-born child.
4. Prohibit attempts to conceive a child by any means other than the union of egg and sperm, by using gametes obtained from a human fetus or derived from human embryonic stem cells, or by fusing blastomeres from two or more embryos.
5. Prohibit the use of human embryos in research beyond a designated stage in their development (between 10 and 14 days after fertilization).
6. Prohibit the buying and selling or patenting of human organisms at any stage of development.

Id. at 242–44.

80. *Id.* at 245.

81. Senator Frist stated: "While human embryonic stem cell research is still at a very early stage, the limitations put in place in 2001 will, over time, slow our ability to bring potential new treatments for certain diseases. Therefore, I believe the president's policy should be modified." L.A. TIMES, July 29, 2005, at 1; *see also* GENETICS & PUBLIC POL'Y CTR., PREIMPLANTATION GENETIC DIAGNOSIS: A DISCUSSION OF CHALLENGES, CONCERNS AND PRELIMINARY POLICY OPTIONS RELATED TO THE GENETIC TESTING OF HUMAN EMBRYOS (2004).

onic stem cell research reveal deeply divided judgments about the status of early forms of human life and are often occasions for polemics rather than respectful dialogue."[82] In 2002, a study indicated that:

> Stem cells have the unique capacity not only to give rise to more stem cells (self-renewal) but also to generate differentiated progeny. Human embryonic stem cells are now available and are at an early stage of validation. Verification of the presence of the critical properties of stem cells—self-renewal and differentiation—should be the gold standard for all such studies. Only the diseased cells have these genes. By contrast, embryonic stem-cell lines with the appropriate sets of inherited and acquired genes should prove invaluable for studying the cellular basis of many diseases.[83]

In 2005 South Korean scientists reported their discovery of how to more quickly produce human embryonic stem cells by the use of embryos that are clones of patients and thus won't be rejected by the immune system. This constitutes therapeutic cloning of cells that are genetic matches for injured or sick patients.[84]

---

82. Lustig, *Moral Pluralism and the Debate over Research on Embryonic Tissue*, Hast. Ctr. Rep., Sept.-Oct. 2002, at 41 (discussing GREEN, *supra* note 69, and HOLLAND-LEBAC-QZ-ZOLOTH, THE HUMAN EMBRYONIC STEM CELL DEBATE: SCIENCE, ETHICS, AND PUBLIC POLICY (2001)). "Green's optimism about the immediate importance of embryo (and by extension ES cell research) is shared by several contributors to the Holland-Lebac-Zoloth volume." *Id.* at 42.

83. N. ENG. J. MED. 1576 (2002).

> The most obvious, and most formidable, challenge to creating stem cell banks in the United States is the widespread disagreement about the moral status of early human life. Therefore, we believe that it is morally desirable to delay creation of the therapy band until there is solid evidence from early clinical trials that stem cell-based therapies will work.

Faden et al., *Public Stem Cell Banks: Considerations of Justice in Stem Cell Research and Therapy*, Hast. Ctr. Rep., Nov.-Dec. 2003, at 24.

84. ORANGE COUNTY REGISTER, May 31, 2005, at 1 (citing the journal *Science).* Leon Kass, chairman of the President's Council on Bioethics, complained that "whatever its technical merit, this research is morally troubling: It creates human embryos solely for research, makes it much easier to produce cloned babies, and exploits women as egg donors not for their benefit." *Id.*

The European Commission (EC) regulates research and clinical trials funded at European Union (EU) level. Member states have to address conflict of interest issues in the authorization process of research and clinical protocols of biomedicine under the control of ethics committees. Article 2(k) of the EU defines an Ethics Committee as "*an independent body* in a member state, consisting of healthcare specialists and non-medical members, whose responsibility it is to protect the rights, safety and well-being of human subjects involved in a trial." (See Chapter 6 regarding American ethics committees.) Directive 2001/20/EC states that prerequisites for the approval of clinical trials are (1) the independence of ethics committee members, and (2) the respect of clauses fixed in the Directive. Article 9 of the Directive states:

> The sponsor may not start a clinical trial until the Ethics Committee has issued a favorable opinion and inasmuch as the competent authority of the Member State concerned has not informed the sponsor of any grounds for non-acceptance. The procedures to reach these decisions can be run in parallel or not, depending on the sponsor.

Maurizio Salvi, European Commission, Research Directorate-General, Life Sciences–Bioethics, Belgium, has noted that these principles are implemented at national and local level. When the Ethics Committee is established, the Commission services check "whether members of such a panel have a 'direct' link with the projects they have to evaluate. Such experts are selected according to (1) the specific expertise needed for evaluating the proposals, (2) an interdisciplinary composition (lawyers, philosophers, scientists, sociologists, psychologists, experts of public policy and so on), and (3) a geographical distribution that also takes into account gender issues."[85]

---

85. Salvi, Science and Engineering Ethics 9, 101–08 (2003). However, in 2004, William Cheshire of the Mayo Clinic pointed out:

> The latest scientific advances have opened windows on the human embryo that challenge how society contemplates early human life. Ever since in vitro fertilization technology physically separated embryogenesis from the context of procreation, it has become possible to think of human embryos less as tiny offspring and more as products of the laboratory.

William Cheshire, *Human Embryo Research and the Language of Moral Uncertainty*, ajob, MIT (2004) at 1.

In April 2003 the Court of Appeal in England authorized couples to screen their test-tube embryos to ensure the baby's tissue provides a match to help cure a sick sibling.[86] It should be possible to develop sperm or eggs in view of the fact that embryonic stem cells in mice have given rise to all the cells in its body.[87] In 2003, Spain became the first Catholic country in Europe to authorize research on human embryos to obtain stem cells.[88] In November 2004, in spite of its then very high indebtedness and low credit rating, 59 percent of California's voters increased its controversial debts for such research. That state called for $3 billion in bonds (that would, with interest, cost a total of $6 billion from the general fund over 30 years) to establish a California Institute for Regenerative Medicine to regulate and fund stem cell research.[89] (Several of its universities and corporate laboratories were already into such work.) However, because of lawsuits challenging the ethics of embryonic research, no bonds have yet been sold to fund the program. On

---

86. ORANGE COUNTY REGISTER, Apr. 9, 2003, at News 26. In 2003, the British medical journal, *The Lancet*, reported that stem cells can help repair damaged hearts "only if they're given directions on how to get there. The SDF-1 molecules call the stem cells to the heart for repair work." ORANGE COUNTY REGISTER, Aug. 30, 2003, at News 22.

87. Giejsen et al., *Derivation of Embryonic Germ Cells and Male Gametes from Embryonic Stem Cells,* 427 NATURE 148 (2004). In 2004 Harvard University commenced plans to launch the largest privately funded American stem cell research project, and both New Jersey and California announced their plans to publicly fund such research. ORANGE COUNTY REGISTER, Mar. 1, 2004, at News 10; Hast. Ctr. Rep., Mar.-Apr. 2004.

88. Research will be allowed on human embryos stored at *in vitro* fertilization (IVF) clinics. http://www.biomedcentral.com/news/20030731/03. As noted in Chapter 8, Section 1(f), Nightlight Christian Adoptions, an adoption agency in California, started the first American fertility service for the donation of embryos under a contract to transfer its ownership. This was done in order to partially reduce the number of embryos used for stem cell research.

89. It was called Proposition 71, which created the California Stem Cell Research and Cures Initiative to locate these institutions over a ten-year period, at $350 million per year, using a competitive peer-reviewed process. While looking at all such bonds, the legislative accounting office stated that "the general fund deficit bonds will grow from $33 billion from May 1, 2004 to $50.75 billion by June 30, 2005; that's a 54% increase in 14 months." IRVINE REV., Nov. 2004, at 14. The institute will be guided by an "Independent Citizens' Oversight Committee."

April 26, 2005, the National Academy of Sciences (NAS) issued its *Guidelines For Human Embryonic Stem Cell Research* (Guidelines) "to advance the science in a responsible manner."[90] It concerns both ethical and religious viewpoints and even pointed out how stem cell research results in the destruction of the blastocysts, regarded by some people as human beings (NAS, 29). Further, it actually opposed all research (1) involving *in vitro* culture of any intact human embryo, regardless of derivation method, for longer than 14 days or until formation of the primitive streak begins, whichever occurs first; (2) in which human embryonic stem cells are introduced into non-human primate blastocysts or in which any ES cells are introduced into human blastocysts; and (3) involving animals into which human embryonic stem cells have been introduced at any stage of development; these animals should not be allowed to breed.[91] On May 20, 2005, Leon Kass, chairman of the President's Council on Bioethics, sent President Bush a "White Paper" entitled *Alternative Sources of Human Pluripotent Stem Cells*. Although it also summarized some arguments on possible research in compliance with the Dickey Amendment's prohibition on the use of federal funds, he had to admit that this inquiry was "no more than a preliminary hearing" thereon.

good v. Good

No person should be denied his belief that all stem cell research should be prohibited from the use of the embryos that constitute a person.

v.

Any couple willing to donate their stem cells for research should be encouraged to permit the use of their embryos since they are not "persons" under the law.

Twenty-first Century Approach

1. The embryos from which a cell is withdrawn for such research are not destroyed. Most people who have been asked normally give consent to donate their embryonic stem cells and consideration might be given to possibly *lowering the priority* of those who would otherwise refuse to consent to such donations.

90. *See* L.J. NEWSL. (L.S.N.), July 2004, at 7–11.
91. *Id.*

2. Efforts should be made to reduce (or remove) political and alleged religious restrictions on the research of embryonic stem cells to develop replacement organs, including the emphasis distinguishing the use of private versus federal funds.

The doctor explained that she had the same disease the boy had recovered from two years earlier. Her only chance of recovery was a transfusion from someone who had previously conquered the disease. Since the two children had the same rare blood type, the boy was an ideal donor.

"Would you give your blood to Mary?" the doctor asked.

Johnny hesitated. His lower lip started to tremble. Then he smiled and said, "Sure, for my sister."

Soon the two children were wheeled into the hospital room. Mary, pale and thin. Johnny, robust and healthy. Neither spoke, but when their eyes met, Johnny grinned.

As the nurse inserted the needle into his arm, Johnny's smile faded. He watched the blood flow through the tube.

With the ordeal almost over, Johnny's voice, slightly shaky, broke the silence.

"Doctor, when do I die?"

Only then did the doctor realize why Johnny had hesitated, why his lip had trembled when he agreed to donate his blood. He thought giving his blood to his sister would mean giving up his life. In that brief moment, he had made his great decision.

David C. Needham, from "Close to His Majesty"
in Gray, *Stories for a Teen's Heart*, p. 82

c) Donations to or by Living Mature Minors and Other Legal Incompetents

The American Academy of Pediatrics has noted that society "generally presumes that parents should exercise the right to refuse treatment when non-autonomous children cannot do so for themselves. The best interests standard serves as the basis for decisions for patients who have never achieved decision-making capacity, including infants." It points out how the use of the "best interests" standard involves weighing the benefits and burdens of life-sustaining medical treatment (LSMT) and that improved "quality of life" (which refers to the experience of life as viewed by the patient) after the LSMT has been applied (including reduction of pain or disability); and increased "physical pleasure, emotional enjoyment, and intellectual satisfaction. [But] the burdens of LSMT may include intractable pain; irremediable disability or helplessness; emotional suffering; invasive and/or inhumane interventions designed to sustain life; or other activities that severely detract from the patient's quality of life."[92]

Some state courts had held that children and the mentally retarded could not donate organs to siblings or family members on the grounds that they lack the capacity to give informed consent.[93] A 2000 study reported that recipients of cord-blood transplants from HLA-identical siblings have a lower incidence of acute and chronic graft-versus-host disease than recipients of bone marrow transplants from HLA-identical siblings.[94] Courts had begun to authorize the participation of incompetents to transplant an organ as a donor; however, as stated by an appellate court, "Nothing in this opinion is to be construed as being applicable to the situation where the proposed donee is not a parent or sibling of the incompetent."[95] In the United Kingdom, couples have been authorized by the Human Fertilization and Embryology Authority to both create a child for the purpose of donating its umbilical

92. Comm. on Bioethics, Am. Acad. of Pediatrics, *Guidelines on Forgoing Life-Sustaining Medical Treatment*, PEDIATRICS 532–33, 535 (1994).

93. Bleich, *Survey of Recent Halakhic Literature: Compelling Tissue Donations*, 27 TRADITION 59–89. In *Curran v. Bosze*, 566 N.E.2d 1319 (Ill. 1990), the Illinois Supreme Court denied a mother's request to tissue-type twins to see if they might be suitable marrow donors, on the ground that the minor twins had not themselves developed values or judgments.

94. Rocha et al., 342 N. ENG. J. MED. 1864–54 (2000); *see also* National Organ Transplant Act, Pub. L. No. 98-507, 98 Stat. _ (codified as amended at 42 U.S.C. §§ 273–274 (1988).

95. Little v. Little, 576 S.W.2d 493 (Tex. Civ. App. 1979); *see also* Strunk v. Strunk, 445 S.W.2d 145 (Ky. 1969); Annotation, 35 A.L.R.3d 683. It has been argued that surrogates making decisions regarding organ donation "have the right to take into account other factors, such as the demands for morality and the best interests of the family as a whole." Morley, *Proxy Consent to Organ Donation by Incompetents*, YALE L.J. 1215–49 (2002).

cord to a sick sibling and to select test-tube embryos whose tissue type matches that of the ailing child. Umbilical cord blood contains stem cells used in the treatment of leukemia and other immune diseases. The analysis of a cell taken from the embryo is made some three days after fertilization without destroying the embryo. They must obtain a license from the regulator. Although the British Medical Association welcomed this ruling, it pointed out how "[t]he BMA sees moral and practical differences between using umbilical cord blood for treatment—which involves no discomfort or risk to the child—and using the child him- or herself as a donor that would involve physical risk to the child."[96]

Ethicists have recommended that an exception be made if the organ is donated to another family member.[97] By equating the family's interest with that of the child, the latter "is not being treated solely as a means" even if the child is not yet legally of age to consent. According to a convention adopted by the Council of Europe in 1996, donation of bone marrow by minors will

---

96. Human Fertilization and Embryology Authority, http://www.hfea.gov.uk.

> Dear Miss Manners: A close friend of our 33-year-old son just completed testing to be a kidney donor for him. He is not a candidate for the gift. Would it be proper, as parents, to send him and his wife a gift certificate for a dinner out, to show our gratitude for his ultimate gift offer?
>
> Gentle Reader: He offered your son a kidney, and you are debating whether to offer him a steak? If you have a son in need of a kidney, you should not need Miss Manners to explain to you that this offer was priceless beyond compare. . . . As this gentleman is not actually able to be the donor, you need not look after him in the hospital as well as you do your son, which is the way you should treat the person chosen. But in addition to displaying your thanks and offering your eternal friendship, you should be endeavoring to discover anything serious that you might do for him.

Judith Martin, Miss Manners, ORANGE COUNTY REGISTER, July 1, 2003, at Accent 2.

97. Ross, 21 J.L. MED. & ETHICS 253 (1993). Of course, possible bias toward adult recipients must be avoided. For example, in 1999 a University of Pittsburgh study found that two-thirds of children's livers donated for transplant in the United States went to adults, while about 75 children a year died waiting for a new liver. ORANGE COUNTY REGISTER, May 18, 1999, at News 12.

be accepted if donor and recipient are siblings and if informed consent can be obtained from the donor.[98]

Where the risk is small, mature children may be allowed to volunteer their organs as donors for bone marrow and kidney donations to siblings in certain circumstances and become donors of blood during a severe blood shortage. However, it has been noted that "living transplantation in children with the use of grafts from living donors has been performed only by the most experienced liver-surgery programs in the world." Hence, "they are likely to vary among institutions according to the skill and experience of the team performing the surgery."[99] The AMA *Code of Medical Ethics* (2002-2003) states, "Minors should not be permitted to serve as source when there is a very serious risk of complications (e.g., partial liver or lung donation, which involve a substantial risk of serious immediate or long-term morbidity)." Even if "a child is capable of making his or her own medical treatment decisions," the Code states both that "physicians should not perform organ retrievals of serious risk without first obtaining court authorization" and courts ought to ensure that "ideally the minor should be the only possible source."[100] However, these criteria can be broadened.[101] In 1974, a girl at the age of 16

98. Convention for the Protection of Human Rights and Dignity of the Human Being with Regard to the Application of Biology and Medicine, Nov. 19, 1996, JAMA 1855–56 (1997). As stated by the British bioethicist Raanan Gillon: "Children too can 'be volunteered' by their proper proxies—normally their parents—both for medical research that will not benefit them but is hoped to benefit other, and as organ donors. Again the criterion of very small risk of harm is fundamental." RAANAN GILLON, BIRTH & DEATH 111 (1996).

99. "Centers that perform fewer than 20 transplantations involving cadaveric grafts per year have lower rates of survival among recipients than centers that perform 20 or more." N. ENG. J. MED. 1636 (2001).

100. Rule 2.117. Cases have upheld a parental decision for a minor to donate an organ, using the substituted judgment principle. *E.g.*, Hart v. Brown, A.2d 386 (1972). It has been argued that to reduce psychological problems, oocyte donors should be limited to those who are 21 or older. Am. Soc'y for Reproductive Med., *Guidelines for Oocyte Donation*, 70 FERTILITY & STERILITY 5S–6S (1998); 70 FERTILITY & STERILITY 9S (1998).

101. Other useful neurologic criteria can be established to inform equitable allocation decisions. Orr et al., *Should Children with Severe Cognitive Impairment Receive Solid Organ Transplants?*, 11 J. CLINICAL ETHICS 219, 222–23, 228 (2000); Chen, *Organ Allocation and the States: Can the States Restrict Broader Organ Sharing?*, 49 DUKE L.J. 261, 273 (1999) (discussing a federal rule mandating broader organ allocation).

years received a bone marrow transplant from her identical sister. In March 2001, 27 years later, the same sister donated her right liver lobe when the donee was a 42-year-old woman.[102] It is noted that only 2 percent of those who list their ancestry with the National Marrow Donor Program (NMDP) are multiracial. As of 2005 the NMDP began to study multiracial patients' medical records in order to ascertain what kind of marrow tissue they tend to inherit from their parents. The NMDP registry works with international registries that list 3.8 million individuals. Of the 9-plus million, nearly half are not identified by race."[103]

good v. Good

Because all minors lack capacity to consent, organs should never be harvested from them.

v.

Donations from consenting parents, or "consenting mature minors," to their siblings or others should be permitted in order to save lives.

Twenty-first Century Approach

In the twenty-first century we should rid ourselves of the legal problem that, solely because minors under the age of 18 have no right to vote (under Constitutional Amendment 26), no such child can be mature enough to legally consent to donate an organ. A child capable of making its own medical decisions need not be prohibited from acting as a possible source of organs.

---

102. Andreoni, *Liver Transplantation 27 Years After Bone Marrow Transplantation from the Same Living Donor*, N. ENG. J. MED. 2624 (2004). "As our understanding of the immune response improves, the potential for developing 'operational tolerance' may widen the recipient's benefits from living donor transplantation." *Id.*

103. Northern California is home to half of the 10 U.S. cities with the highest percentages of multiracial individuals, according to Census 2000 . . . . Whites in need of a bone-marrow transplant have about a 90 percent chance of finding a match . . . . Africans and their descendants globally have the most variation of any population in the world.

ORANGE COUNTY REGISTER, Jan. 31, 2005, at News 14.

Drugs are paid for, hospital stays are paid for, physicians are compensated. It is not so clear why ethical critics become so fastidious, so squeamish about paying for the life-giving organ itself.

Blumstein, *Birth & Death* (1996), p. 126

d) Sales of Organs

A person can make a testamentary gift of his organs; if he does not do so, his next of kin have a property interest under state law to the body of a deceased person.[104] However, ordinary sales of organs are still legally prohibited as being unethical.[105] The concept has been disputed by both physicians and ethicists.[106] Mark J. Cherry, author of a 2005 book on the sale of kidneys, stated that the view against sale of organs "fails to take adequate account of many of the issues central to the debate."[107] The price of sales "at cost has greatly escalated due to the law's allowance for recovery of so many costs as

104. *See* UNIF. ANATOMICAL GIFT ACT, § 3, 8A U.L.A. 40 (1993). Death means the transformation of the body into property. Rao, *Property, Privacy, and the Human Body*, 80 B.U. L. REV. (2000). "Nearly $6 billion is spent [annually] on a handful of solid organ transplant recipients." Evans, *How Dangerous Are Financial Incentives to Obtain Organs?*, 31 TRANSPLANTATION PROC. 1337–41 (1999).

105. National Organ Transplant Act (NOTA), 42 U.S.C. § 274e; ORANGE COUNTY REGISTER, Apr. 18, 2000, at News 11.

106. Similarly, in several states the transfer of blood is a "service" that can be provided only by a licensed professional. *See* ALA. CODE § 7-2-314(4) (1975); ILL. REV. STAT. ch. 91, para. 183, § 3; KY. REV. STAT. ANN. § 139.125 (West); OHIO REV. CODE ANN. § 2108.11 (West year); *see also* TRANSPLANTING HUMAN TISSUE: ETHICS, POLICY, AND PRACTICE 106, 123 (Younger et al. eds., 2004). NICHOLAS TILNEY, TRANSPLANT: FROM MYTH TO REALITY 274 (2003) notes that the black market "use of human tissues as market commodities has been described by one knowledgeable critic as 'neo-cannibalism.'" Yet, as the attorney Vanessa S. Perlman pointed out, "Tilney's association of tissue commodification with the sale of organs on the black market gives short shrift to tissue transplantation and demonstrates the public relations problem Younger et al. are trying to combat." Vanessa Perlman, *The Place of Altruism in a Raging Sea of Market Commerce*, J.L. MED. & ETHICS 166 (2005).

107. CHERRY, KIDNEY FOR SALE BY OWNER: HUMAN ORGANS, TRANSPLANTATION, AND THE MARKET (2005). He also wrote: "All organ procurement and distribution schemes commodify even donation. That is, on each ground one has specified a market in human organs, albeit a heavily regulated market, with carefully stimulated conditions for bearing the costs and benefits of procurement, distribution, and transplantation." Younger, JAMA 2366 (2005).

exceptions to the prohibition of sales for valuable consideration."[108] In the United States the National Organ Procurement and Transplantation Network matches body part donors with transplant patients. NOTA made it a federal offense to traffic in the sale of human body parts.[109] The AMA *Code of Medical Ethics* issued in 1984 (updated in 1994) states "it is not ethical to participate in a procedure to enable a living donor to receive payment, other than for the reimbursement of expenses necessarily incurred in connection with removal, for any of the donor's nonrenewable organs." Nevertheless, it added, "an adult" may agree to donate organs after death and "the donor's family or estate would receive some financial remuneration after the organs have been retrieved and judged medically suitable for transplantation." However, this "should be the lowest amount that can reasonably be expected to encourage organ donation."[110] This has undergone criticism for several reasons.[111]

---

108. Radcliff-Richards et al., *The Case for Allowing Kidney Sales*, 351 LANCET, 1950–52 (1998).

> It has been noted that the bone-graft industry alone was estimated to be worth $500 million in 2002, with most procedures costing $28,820. In 2004, average prices were listed as heart valves, $9,120; corneas, $4,800; sclera, $2,000; and fascia lata, $11,400 (fascia lata is the connective tissue covering the thigh muscles and typically yields 57 strips, 3mm x 15mm, worth $200 each).

ORANGE COUNTY REGISTER, Mar. 9 2004, at News 3.

109. However, NOTA then defines "valuable consideration" to include "[t]he reasonable payments associated with the removal, transportation, implantation, processing, preservation, quality control, and storage of a human organ or the expenses of travel, housing, and lost wages incurred by the donor of a human organ in connection with the donation of the organ." Further, "any person who violates subsection (a) of this section shall be fined not more than $500,000 or imprisoned no more than five years, or both." 42 U.S.C.A. § 274e (1988); *see* Wilson v. Adkins, 941 S.W.2d 440 (Ark. Ct. App. 1997).

110. Rule 2.15.

111. One article had opposed efforts to reverse existing prohibitions against use of financial incentives, Joralemon, *Shifting Ethics: Debating the Incentive Question in Organ Transplantation*, 27 J. MED. ETHICS 30, 31, 34 (2001), whereas another indicated a futures market system may offer a viable solution to the desperate shortage of organs and should be given more consideration, Jensen, *Organ Procurement: Various Legal Systems and Their Effectiveness*, 22 HOUS. J. INT'L L. 555, 578 (2000).

More than 200,000 patients with end-stage renal disease undergo dialysis in the United States each year, about two-thirds of them in for-profit centers. A low case study found for-profit ownership of dialysis facilities, as compared with not-for-profit ownership, is associated with increased mortality and decreased rates of placement on the waiting list for a renal transplant.[112] In the mid-1990s tissue banks were being mined in the Los Angeles Coroner's office for corneas. Buyers "paid between $335 and $215 for a set of corneas. They are then sold to transplant institutions for a current 'processing fee' of $3,400—among the highest in the nation. Seventy-two percent were homicide victims who were likely to have been young. . . . The average age was 27.7 years, much younger than the average age of all those autopsied."[113] The University of California–Berkeley opened a center to watch for and investigate illegal trafficking in human organs around the world.[114] At the turn of the third millennium, the director of the U.C.–Irvine body donation program was fired after selling body parts. Only schools with researchers nearby are allowed to use bodies.[115]

The Transplant Center at the University of Chicago and elsewhere makes use of a four-person organ exchange to encourage live-donor kidney transplantation. It applies to the situation in which a patient in need of an organ may have a relative or friend willing to donate that organ (as a living donor), but a blood incompatibility prevents that person from donating. The two sets

---

112. 341 N. Eng. J. Med. 1653–60 (1999). The Medicare End Stage Renal Disease (ESRD) program began in 1973. In 1997, Medicare spent approximately $11.76 billion on the ESRD program. *Id.* at 1691. Bids for a human kidney offered on the Internet auction site eBay reached $5.7 million before the company put a halt to the sale. The description read: "Fully functional kidney for sale. You can choose either kidney. Buyer pays all transplant and medical costs. Of course only one for sale, as I need the other one to live. Serious bids only." Orange County Register, Sept. 3, 1999, at News 29.

113. L.A. Times, Nov. 2, 997, at A35. California law states that corneas can be taken without family consent only in cases targeted for autopsy.

114. Orange County Register, Nov. 6, 1999, at News 24.

115. Orange County Register, June 16, 1998, at News 14. A government order made the offense punishable by two to seven years in prison. More than 500 Iranians have pledged to sell their kidneys to raise money for the slaying of author Salman Rushdie; *see also* Orange County Register, Dec. 29, 1999, at News 23.

of people are paired in order to allow simultaneous "swaps."[116] It has been remarked that criminalizing these "altruistic" transactions "would be a misinterpretation of the intent of the law, which was to prevent the exploitation of living persons who might be willing to sell their body parts for profit."[117] Beginning in the year 2000 the state of Pennsylvania began a sensible program of offering up to $300 directly to funeral homes in order to help families of organ donors cover their funeral expenses.[118] Some condemned the program as the evil selling of body parts, and argued that it would give Pittsburgh a competitive advantage in the transplant business.[119] It constitutes one step toward encouraging more people to be donors, which may be followed by other state legislatures.

It is not intrinsically unethical to sell organs at a reasonable price where allocation is adequately regulated. In the United States and most other countries, blood is procured by sales from persons willing to supply it for a price. This is also true of sperm and egg sales (see Chapter 8). Today, by not permitting organ sales, unless they are being made at "cost"—which may actually result in a price that may be unreasonable. Consider facts reported by a national medical journal:

> One is struck by the high level of organ procurement charges in view of the characterization of organ procurement as altruistic. Although the median organ procurement charges, documented by Evans, ranges from nearly $16,000 to nearly $21,000 (1991 dollars), there was not a penny for the accident victim's/organ donor's family. That some transplant hospitals routinely marked up charges they paid to organ procurement agencies by as much as 200 percent hardly seems consistent with altruism. . . . If everyone else involved in transplantation,

116. Menikoff, *Organ Swapping*, Hast. Ctr. Rep., 1999, at 28 (noting that under the tax law, "a cash sale of a building may cause a tax to be owed; a 'like kind' exchange of the building for a different building will defer taxation" (citing I.R.C. § 1031)).

117. Ross et al., Ethics of a Paired-Kidney-Exchange Program 1754.

118. N. Eng. J. Med. 1383 (2002).

119. Orange County Register, May 18, 1999, at Metro 7. "We don't see many doctors and hospitals donating their time and own resources to transplant organs for free." *Id.*

> the organ procurement organization, the surgeon, the nurses, and the hospital, receive compensation, why should the donor's family be excluded?[120]

In 2001, the AMA became upset with the congressional action in 1984, noted above, banning sales of organs for profit. Most of them must be volunteered. The United Network for Organ Sharing (UNOS) indicated that only 25 percent of 78,000 organ transplants needed occur in time to save a life. Most families of people who die unexpectedly decline to offer their deceased relatives' organs as donations. The AMA decided to propose a change in the law to authorize paying dying would-be donors and their families for needed vital organs,[121] and on April 29, 2004, Georgia's Governor Sonny Perdue signed a bill "and celebrated Donor Recognition Day at the State Capitol to recognize and honor donor family members and living organ donors who sacrificed to give others a second chance at life."[122]

In 2004 it was reported how 81 patients received an artificial-heart device that replaced both native cardiac ventricles and all cardiac valves. The survival rate was 79 percent, and it was concluded that the "device prevents death in critically ill patients who have irreversible biventricular failure and are candidates for cardiac transplantation." It is known as a "bridge" because with the device, "a considerable number of potential cardiac-transplant recipients who have no other suitable options may successfully await cardiac transplantation."[123]

---

120. JAMA 3155 (1999).

121. ORANGE COUNTY REGISTER, Dec. 3, 2001, at News 11.

122. TRANSPLANT CHRON. at 4. The new law allows living donors to receive a tax deduction up to $10,000 on costs incurred from organ donation, such as travel expenses, lodging expenses, and lost wages applicable to all taxable years beginning on or after January 1, 2005. Georgia and Wisconsin have enacted legislation to provide tax assistance for living donors. Legislation is pending in Illinois, Massachusetts, New Jersey, New York and Pennsylvania.

123. Copeland et al., *Cardiac Replacement with a Total Artificial Heart as a Bridge to Transplantation*, N. ENG. J. MED. 865 (2004). It has been noted that "'bridging' is not to be confused with mechanical circulatory support intended from the outset to be permanent treatment, known as 'destination therapy.'" Renland, N. ENG. J. MED. 849 (2004).

good v. Good

The sale of human organs for profit is not only considered unethical in itself, but it also results in profit making by those who prey upon the poor and the helpless.

v.

Current law authorizing organ sales "at cost" has often actually resulted in the payment of higher prices resulting from mark-ups by hospitals and others for alleged "costs without profit," while the family of the donor is excluded from benefiting in any manner.

Twenty-first Century Approach

1. In the twenty-first century, sales of organs should be authorized and regulated. The National Organ Transplant Act of 1984, which currently prohibits interstate sales of organs for profit but permits recovery of significantly inflatable costs of their procurement services without competition, should be amended along with the laws thereon in many states.
2. Other transplant centers should consider adopting a "four-person organ exchange" (as at the University of Chicago) to encourage live kidney transplantation.
3. Other states should consider enacting a statute (similar to that in Pennsylvania) authorizing the payment of a few hundred dollars to funeral parlors to offset the funeral expenses of the families of organ donors, as well as allowing living donors to receive tax deductions from organ donations (similar to that of Georgia).

e) Organs from Suicidal Donors and Executed Criminals (Murderers)

The so-called "dead donor rule" requires that donors not be killed in order to obtain their organs. The removal of nonvital organs prior to death would not necessarily violate the rule. (See Sections 1 (j) and 2, *infra.*) According to bioethicist John Robertson, this rule should prevent a person from committing suicide in order to provide organs to his family or others. He stated: "It prevents the killing of one person for organs that would save the three or more lives that can be saved by a single cadaveric donor."[124] However-

124. Robertson, Hast. Ctr. Rep., Nov.-Dec. 1999, at 6.

ever, suicide is not illegal in the United States, and while this approach may constitute a moral ideal, it is difficult to understand how the rule can prevent anyone from committing suicide in all circumstances in order to benefit a designated recipient. Under the Uniform Anatomical Gift Act (UAGA), adults of at least 18 years of age are permitted to donate their bodies upon their death.[125] There is no basis for arguing that the rule should prevent the designee from accepting the organ, if he so desires, provided that he had not conspired with or otherwise participated in the donor's demise.

Under the current UNOS policy, an individual's status as a prisoner is ignored with respect to an organ's allocation decisions.[126] The AMA *Code of Medical Ethics* sets forth a policy that "organ donation by condemned prisoners is permissible only if the decision to donate was made before the prisoner's conviction."[127] This policy would make it unethical for a physician to harvest organs from a convicted murderer who has agreed to donate his organs to save the life of another person. However, courts have ruled that the AMA's several policies in its attempt to prohibit physician participation in capital punishment matters in fact have no legal effect.[128] Most AMA objections come from those who oppose the death penalty in all instances, rather than discussion of any problem with the transplant itself.[129] It is difficult to

125. Under the UAGA, family members may posthumously donate a deceased's organs, provided that they lack actual notice that the deceased did not want the organs donated. But if a donor's intent is known, the donee's wishes take precedence over family objections.

126. *See* Douglas, *Prisoners Are Now Entitled to Organ Donation—So, Now What?*, 49 St. Louis U. L.J. 539 (2005); Hinkle, *Giving Until It Hurts: Prisoners Are Not the Answer to the National Organ Shortage*, 35 Ind. L. Rev. 593, 598–600 (2002).

127. Rule 2.06. This rule also requires that "the donated tissue [be] harvested after the prisoner has been pronounced dead and the body removed from the death chamber, and physicians do not provide advice on modifying the method of execution for any individual to facilitate donation," 365–68 (1993), updated June 1996; AMA policy statement E-2.06, Capital Punishment, reaffirmed June 2001.

128. Keyes, *Choice of Participation by Physicians in Capital Punishment*, Whittier L. Rev. 829 (2001).

129. Cameron & Helffenberg, Kidney Int'l 726, 730 (1999). Nevertheless, in 2004 a confessed serial killer was put away for the rest of his life (at taxpayer expense). By striking a deal with prosecutors he received immunity for 12 killings—11 in Texas and one in Michigan—without a death penalty. Orange County Register, Dec. 8, 2004, at News 18.

comprehend just how such a donation can be declared unethical solely because the defendant did not make the donation prior to his conviction for a heinous crime. In Alabama in 1996, "David Nelson's execution was halted by the state Supreme Court less than 24 hours before it was scheduled so Nelson could donate a kidney to his sick brother. But the brother was too ill for surgery and later died." However, "in Texas in 1998, Jonathan Nobles' request for a stay of execution to donate his kidney was rejected."[130] Why should he be deprived of the right possessed by virtually all others to volunteer their organs that are so desperately needed by society?[131]

Some maintain that it is unethical to refuse to accept organs from persons who are executed in a foreign country for political reasons and without their consent to donate. The Chinese government, for example, has claimed that condemned persons who donate organs actually gave their consent or that consent was obtained from their families.[132] The reliability of this statement from a communist country may be due to the absence of our concept of due process during their criminal trials.[133] In the United States, where "super due process" is required in trials for capital punishment,[134] the AMA is protesting such punishment by denying the physician's ethical right to enter the process in order to express a desire for a change in the law. The AMA should not attempt to control those physicians (a majority of whom may disagree)

---

130. Orange County Register, May 20, 2005, at News 30.

131. It was reported that Ronald Miller, M.D., "would allow transplantation of deceased prisoners' organs if, and only if, the society had a universal presumed-consent policy for all members of the society" except for those who chose to "opt out." *Id.* at 736. This would appear to possibly penalize U.S. citizens if a state refused to adopt the presumed-consent policy.

132. Orange County Register, Oct. 16, 1997, at News 15.

133. China's courts are staffed by poorly trained judges, many of whom do not have law degrees . . . . In China courts have no higher status than administrative agencies. Bureaucrats have the ultimate power to interpret their own rules, because in China the courts defer to the agencies on interpretations of their rules. The Chinese Communist Party remains superior to both. China already has agreed to publish laws before they take effect.

L.A. Times, Sept. 6, 1998, at M2.

134. Tex. Code Crim. Proc. Ann. art. 43.25 (Vernon 1998); Keyes, Life, Death, and the Law, chs. 16 and 26 (1995).

by making them appear to be acting in an unethical manner when harvesting and transplanting organs from executed prisoners.

It is acknowledged that the unclaimed bodies of executed inmates are routinely given to medical schools for anatomical study.[135] With respect to organ retrieval, a prisoner would request his method five to seven days before the execution date. This would not violate the so-called "dead donor rule" any more than other deaths.[136] The rule merely requires that human donors not be killed *solely* in order to harvest their organs; it does not apply to convicted criminals who are being executed for murder and who may wish to offer their organs to help others as a final act of repentance. Such an offer by the condemned should not be considered as a means to lower his sentence. However, there is no reason why state procedures cannot be modified to accommodate the retrieval of organs from prisoners who die after expressing a desire to donate their organs. In an article weighing both sides of the issue, the authors of a study concluded "that most of the revulsion that this practice attracts is based first on antipathy to the death penalty itself and, second, as a reaction to possible abuses of the conviction situation," such as in China. They ask "what the balance of harm may be between obtaining an organ from an individual who died from renal failure, and the obtaining of a kidney from an individual already dead by due legal process." They urge, "The use of executed prisoners' organs needs consideration."[137]

A taxpayer-supported heart transplant, costing $1 million, was given to a California prison inmate serving 14 years for robbery. This was done even

---

135. Statutes in five states provide that the corpse of a capital felon may be turned over to a medical center for research, if neither relatives nor friends request the corpse. MISS. CODE ANN. § 99-19-55(4) (1994); N.J. STAT. ANN. § 2C:49-9(a) (West 1995); N.Y. CORRECT. LAW § 662(1) (Gould 1996); 61 PA. STAT. ANN. § 2127 (West 1996); TEX. CRIM. PROC. CODE ANN. § 43.25 (Vernon 1979).

136. Bioethicist John Robertson has indicated the contrary, by leaving out the word "solely." Hast. Ctr. Rep., Nov.-Dec. 1999, at 9.

137. Cameron & Hoffenburg, KIDNEY INTERNAT'L 724–32 (1999):

> The fact that the executed prisoner's organs are going to save the life of another individual or relieve suffering may bring solace to a family making some amends for whatever wrongdoing he had committed and for his death. (p. 229)

though some 500 Californians waited for hearts.[138] The extent to which state legislators should change the law regarding such transplants now appears proper in view of the UNOS declaration that long-term liver failure, seen in alcoholics and drug addicts, remain well below the top of the list for liver transplants.[139]

good v. Good

The AMA restrictions on the harvesting of organs from any convicted murderers constitute a rational attempt to separate them from others.

v.

This separation does not always constitute a rational basis, and it aggravates the current critical lack of organs that cause more innocent people to die in the third millennium.

Twenty-first Century Approach

1. More reason must prevail in the twenty-first century regarding the use of organs of a felon executed after being duly convicted of murder. No valid distinction can normally be made between organs normally harvested from a suicidal donor (or an executed criminal) and those harvested from others.
2. Many AMA ethical rules are being adopted virtually verbatim in some states, and state medical associations should revise adverse policies on harvesting organs from criminals because its effect is to let legally innocent recipients of much-needed organs continue to suffer and die.
3. Convicted felons, however, should not receive organ transplants unless they are no longer needed by others who are innocent of any such crimes.

---

138. The California Department of Corrections cited both a 1976 Supreme Court ruling declaring it to be a "cruel and unusual punishment" to withhold necessary medical care from inmates and a 1995 federal court ruling ordering prison officials to give a kidney transplant to an inmate.

ORANGE COUNTY REGISTER, Jan. 26, 2002, at News 24.

139. United Network for Organ Sharing, http://www.unos.org.

f) Organs from Anencephalics

In 1994, the AMA declared it was ethically acceptable to take organs for transplant from anencephalics (infants born without a neocortex, the portion of the brain that allows thinking and speech). They have "no potential for future consciousness." Such infants are not legally "brain-dead" because of the current concept requiring death of the "whole brain." The AMA stated, "It is ethically permissible to consider the anencephalic as a potential organ donor, although still alive under the current definition of death."[140] This was a statement that such infants have no intrinsic value except with respect to their ability to donate organs to others. The woman who bore the infant may decline to offer its organs; however, because no treatment can possibly benefit the infant itself, it would appear that physicians should have no ethical duty to treat it.[141] In 1991 the Supreme Court of Florida refused to add "brain absent" as an exception to the brain-dead (or homicide) laws despite the request of parents who wished to donate their infant's organs. It stated it would defer to the state legislature. Baby Theresa in that state was diagnosed prenatally with anencephaly and her mother agreed to a cesarean section "with the express hope that the infant's organs would be less damaged and could be used for transplant in other sick children." In spite of its denying her the right to do so, the court recognized the pressing need for organ donors, and admitted: "[W]e have been deeply touched by the altruism and unquestioned motives of the parents of [Baby Theresa]. The parents have shown great humanity, compassion, and concern for others."[142]

140. Opinion 2, 162 (1994). A Florida case concerned a pregnant woman who was told by doctors that the baby she was carrying was missing most of its brain. (In fact, the child was later born without much of a brain and died after a few days.) On the advice of physicians, the woman agreed to continue the pregnancy to term and even agreed to a cesarean, with the expressed hope that the baby's organs could be used for transplants to other needful children. When the baby was born, the parents requested that it be declared legally dead for this purpose. The healthcare providers refused out of fear of civil or criminal liability. The Supreme Court of Florida actually found "no basis to expand the common law to equate anencephaly with death." *In re* T.A.C.P., 609 So. 2d 588 (Fla. 1992).

141. Sibbener, *Organ Wars*, Hast. Ctr. Rep., Nov.-Dec. 1999, at 8 (citing Child Abuse Amendments of 1984, 42 U.S.C.A §§ 5102, 5106, 5111–5113, 5115 (West 2000)).

142. *In re* T.A.C.P., 609 So. 2d at 594.

Nevertheless, in 1992 the AMA *Code of Medical Ethics* adopted a rule that "Retrieval and transplantation of the organs of anencephalic infants are ethically permissible only after such determination of death is made." It went so far as to rule that physicians "may provide anencephalic neonates with ventilator assistance and other medical therapies that are necessary to sustain organ perfusion and viability until such time as a determination of death can be made."[143] The Universal Determination of Death Act would more accurately define anencephalic infants as brain-dead, since their mental status is nonexistent.[144] The American Academy of Neurology asked the AMA to consider a change in its position. A neurologist at Johns Hopkins University pointed out that anencephalics have "zero potential for normal development."[145] Because the patient will never be conscious, treatment not only serves to prolong biological life without benefiting the patient as a person, but also may deprive others (who might benefit from its organs). Their condition is similar to that of persons in a vegetative state (see paragraph (h) below).

good v. Good

Although they retain virtually no brain and can never be cured, anencephalics are still alive, and because life is sacred, they must be treated medically solely to prolong their lives.

v.

Because anencephalics can never be cured, all medical treatment to prolong their lives is non-beneficial (futile) and constitutes a waste of taxpayers' funds that can be better used with the harvesting of their organs to benefit others who are in need.

Twenty-first Century Approach

Because the need for organs will greatly increase along with our increasingly larger population during the twenty-first century, a rational approach

143. Rule 2.162.

144. MUNSO, *supra* note 68, at 89.

145. DALLAS MORNING NEWS, Jan. 7, 1996, at 7A. Robertson, *The Dead Donor Rule*, Hast. Ctr. Rep., Nov.-Dec. 1999, at 6, 13 (discussing different proposals to amend the dead-donor rule because of shortages in available organs).

should be adopted that will cease expending funds to prolong the lives of anencephalics, except to the extent needed for providing their use as organ donors.

---

g) Fetal Tissue and Mortality Rates in Hospitals; Age Limits

During the last third of the twentieth century, the choice of abortion became the optional Constitutional right of pregnant women (see Chapter 9). It has been argued that use of fetal tissue in transplants should be prohibited because this practice might encourage abortions and lend moral legitimacy to them.

There is little or no evidence that using fetal tissue from elective abortions has had any impact on the rates of abortions; there has been no challenge that allowing research on fetal tissue transplants would necessarily increase the incidence of abortions.[146] Nevertheless, they somehow renew embroilment in continuing abortion debate.

As noted in section 1(d) above, under the National Organ Transplant Act (NOTA), Congress prohibited the interstate acquisition, receipt, or other transfer of human organ or fetal tissue for valuable consideration in human transplantation.[147] Beyond this, Congress neither regulates their interstate trade nor prevents donations of body parts across state lines.[148] At the state level, the Uniform Anatomical Gift Act (UAGA) applies to fetal tissue transplants.[149] Informed consent must be obtained from the mother before fetal tissue transplantation can occur. The AMA *Code of Medical Ethics* expressed its own ethical concern in the use of human fetal tissue for transplantation as being

---

146. Vawter & Caplan, *Strange Brew: The Ethics and Politics of Fetal Tissue.*

147. Organ Transplants Amendment Act of 1988, Pub. L. No. 100-607, § 407, 102 Stat. 3114, 3116 (1990) (codified at 42 U.S.C. § 274e(c)(1) (1994)).

148. *See* Jonathan Hersey, *Enigma of the Unborn Mother: Legal and Ethical Considerations of Aborted Fetal Ovarian Tissue and Ova Transplantations*, 43 UCLA L. Rev. 159, 172 (1995).

149. Under the Unif. Anatomical Gift Act § 1(1), 8A U.L.A. 1 (1987), a "part" of the body includes fetal tissue. *Id.* § (7). No distinction is made between a fetus donated from a miscarriage or one given through an elective abortion. The donation is prohibited if one of the parents objects. *Id.* § 2, *Transplant Research in the United States*, 120 J. Laboratory & Clinical Med. 30, 35.

"the degree to which the decision to have an abortion might be influenced by the decision to donate the fetal tissue." It requires that "a final decision regarding abortion be made before initiating a discussion of the transplantation use of fetal tissue" and that such tissue not be provided "in exchange for financial remuneration above that which is necessary to cover reasonable expenses."[150] Abortion opponents claim that because the fetus itself could not have controlled what is happening to its body parts, no transplantation should be authorized. However, fetal tissue has many important uses. It has been found that fetuses of women who miscarry during the second trimester of pregnancy have outstandingly effective stem cells for possible transplantation; they were 23 times more effective than adult marrow and eight times better than umbilical cord blood.[151] Experiments are being conducted with such cells; for example, in 1997 Florida researchers performed the nation's first transplant of fetal tissue into a person with a spinal cord disease to slow the progression of cord damage in a paralyzed man with a chronic disorder called syringomyelia.[152] A man incapacitated with Parkinson's disease became the first American to have dopamine-producing brain tissue taken from an aborted fetus and transplanted into his brain. After the operation he was again able to walk and returned to his hobby of woodworking in Denver.[153] Similarly:

> On September 11, 2001, two planes hit the World Trade Center towers, changing everyone's lives. One of the thousands of people in the first tower to be hit was a fifty-seven-year-old electrician name George Doeschner, who was working on the thirty-fourth floor. George has had Parkinson's disease since 1986. Most people with Parkinson's disease take L-dopa. He had been given him a fetal-cell transplant in January 1999, and a year later he was doing so well that he was able to stop taking the drug altogether. Now he had the test of his life . . .

150. Rule 2.161 (updated June 1996).

151. Orange County Register, May 4, 1997, at .

152. Orange County Register, July 11, 1997, at 1.

153. Pestack, Brainscanner 135 (1995); *see also* Joynt, *Neurology*, 263 JAMA 2660 (1990) (pointing out the relevance of fetal tissue transplantation to neurological research regarding Parkinson's disease). (This was the disease of Pope John Paul II, who died in 2005.)

His fetal dopamine cell transplant had kept him moving and had kept him alive.[154]

Thus, until recently, scientists had been excited at the prospect of transplanting fetal dopamine-producing cells to alleviate the symptoms of the disease. However, some of the long-term results of this procedure have been disappointing. Several patients who received these cells deteriorated irreversibly. More recently, transplanting human stem cells has been raised to only a possible therapeutic measure.[155]

Blood and its products have many uses. For example, blood taken from newborns' umbilical cords appears to offer a good source of lifesaving tissue for cancer victims.[156] At the dawn of the third millennium, just as the American Law Institute (ALI) was beginning to correct its direction on some products,[157] it could be going in the opposite direction in connection with blood products. This observation was set forth in section 402A of the ALI's *Restatement* (Second) ushered in the era of strict liability for defective products.[158] The *Restatement* (Third) rejects this approach in favor of one toward

154. FREED & LEVAY, HEALING THE BRAIN 259–60 (2002).

155. Use of a drug (in lieu of the fetal cell transplant) has been described by Dr. Rosenfeld as follows: "Levodopa (referred to as L-dopa) remains the basic treatment for millions of patients with Parkinson's disease. It does have side effects." ROSENFELD, POWER TO THE PATIENT 332 (2002). "Levodopa is useful and important medication but improves symptoms only partially. What's more, the longer you take it and the higher the dose, the greater are its debilitating side effects. That's why many doctors delay prescribing if for as long as possible." *Id.* at 34, 377.

156. ORANGE COUNTY REGISTER, Aug. 25, 1999 at . AMA CODE OF MEDICAL ETHICS R. 2.165 (2000-2003) states:

> Human umbilical cord blood has been identified as a viable source of hematopoietic stem cells that can be used as an alternative to bone marrow for transplantation. It is obtained by clamping the umbilical cord immediately after delivery . . . . [However] there is a risk that the infant donor will develop a need for his or her own cord blood later in life . . . . The possibility that an infant donor would be in need of his or her own umbilical cord blood is highly speculative.

157. N. ENG. J. MED. (1998). Eighty-one percent of the transplants in the study were considered successful, meaning the blood graft took hold within six weeks.

158. Keyes, *The Restatement Second: Its Misleading Quality, and a Proposal for Its Amelioration*, 13 PEPP. L. REV., 23 (1985).

whether or not the mission of an existing feasible alternative safer design was reasonable[159]—a return to the general negligence standard in lieu of the strict liability rule. Some manufacturers of drugs and medical devices are not held liable even if their products reasonably could have been made safer. The ALI rationales set forth in the *Restatement* (Third) [160] may not necessarily justify the bifurcation of products-liability law granted to makers of prescription drugs and other medical products. There are reasons why meaningful design-defect review is desirable for such products.[161]

At the very end of the twentieth century it was found that bone marrow transplants made it easier for the immune system to accept transplants. In August 1999, for the first time a patient received a kidney and bone marrow transplant from the patient's sister in a single operation. The organ rejection was prevented (after 73 days of anti-rejection drugs) by giving the transplant patients enough stem cells from the bone marrow.[162] It has also been found that transplanted hair cells somehow bypass the body's normal rejection mechanism for foreign tissue.[163] The first double-hand transplant operation was done in 1999 in Leon, France.[164] This recipient must take several anti-rejection drugs for the rest of his life. If they cannot control any potential rejection reaction, either or both of the new hands may have to be ampu-

159. Assessment of a product design in most instances requires a comparison between an alternative design and the product design that caused the injury, undertaken from the viewpoint of a reasonable person. That approach is also used in administering the traditional reasonableness standard in negligence. RESTATEMENT (THIRD) OF TORTS: PRODUCTS LIABILITY § 2 cmt. D (1998).

160. Manufacturers of medical products need not make a safer product if the existing product does more good than harm, *id.* § 6(c) and thus permits them to use less than reasonable care. It has been noted that the reports failed to explain why the cost or injuries from defectively designed drugs and medical devices should be borne by the injured rather than the designer and manufacturer. Ank, 109 YALE BUS. REV. 1087, 132 (2000).

161. Ank, *supra* note 160.

162. TRANSPLANTATION (1999); ORANGE COUNTY REGISTER, Aug. 25, 1999, at News 9. "Donor bone marrow treated ex vivo to induce energy to alloantigens from the recipient can reconstitute hematopoiesis in vivo with a relatively low risk of GVHD."

163. 340 N. ENG. J. MED. 1704–14 (1999); N. ENG. J. MED. (1999); ORANGE COUNTY REGISTER, June 3, 1999, at News 14.

164. ORANGE COUNTY REGISTER, Nov. 4, 1999, at News 22.

tated.[165] This raises ethical issues concerning whether the hazards of anti-rejection drugs are worth their risk.[166]

These have also raised a number of other *ethical issues concerning the age of the recipient*, the costs involved, and the volume of particular types of transplants being conducted at the institution doing them. Studies have demonstrated that as a group, liver-transplantation centers in the United States that perform 20 or less transplantations per year have mortality rates that are significantly higher than those at centers that perform more of them. It was recommended that such information be made widely available to the public.[167]

Unlike Canada, the United States has no age limit for potential transplant patients or organ donors.[168] In recognition of the somewhat poorer outcomes among older patients and the rigors of the more extensive surgical procedures, the following age limits have been recommended:

> Fifty-five years for candidates for heart-lung transplantation, 60 years for candidates for bilateral lung transplantation, and 65 years for candidates for single-lung transplantation.[169]

A 79-year-old Canadian, who was 14 years older than what previously was considered the maximum age for undergoing a heart transplant, was

---

165. ORANGE COUNTY REGISTER, Jan. 15, 2000, at 1. Federal regulations apply only to government-funded research, where funding is obtained through the Department of Health and Human Services. 45 C.F.R. § 46.201(a) (1995). Research is allowed only on nonviable or dead fetuses.

166. Dr. James H. Herndon, chief of the Partners Healthcare orthopedic service in Boston, argues that regaining the use of a hand doesn't justify subjecting patients to the risks of antirejection drugs, which can lead to infections, lymphoma, leukemia, and other complications. ORANGE COUNTY REGISTER, Aug. 17, 2000 at News.

167. Edwards et al., N. ENG. J. MED. 2049–53 (1999).
Thirteen centers had mortality rates that exceeded 40%, and the rate at one of these centers was 100%. It would be reasonable to assume that, given this information, patients would decide to undergo transplantation at a center with low mortality rates. Our findings suggest that the information available to patients and referring physicians is inadequate or that regional healthcare systems may be forcing patients to go to centers with poor results.

168. N. ENG. J. MED. 1083 (1999).

169. ORANGE COUNTY REGISTER, Jan. 8, 2000, at News 20.

brought before a transplant committee. In a "secret ballot" the committee voted to permit the transplant only if the patient received a heart that would otherwise go unused. When two independent cardiologists on the committee decided the heart was not appropriate for others on the eligibility list, the "70-plus-year-old received it."[170] This solution appears to have ameliorated the harsher rule that for some people exceeding a certain numerical age becomes an absolute bar to receiving a particular transplant.

Transplants are expensive and such *costs* should always be considered a relevant issue to the expedition of public funds. For example, liver transplantation can rescue about 90 percent of patients who undergo it from a terminal or pre-terminal state.[171] This has had an average cost of $203,434 for the operation and the first year's continuing care thereafter.[172] Should a transplant be considered at public expense for patients whose self-abuse caused alcoholic cirrhosis? It would appear that a negative answer would be correct except where the available liver was not appropriate for use by any others on the eligibility list (as in the heart case discussed above).

California law states, "Notwithstanding any other provision of law, a licensed embalmer, at the request of a licensed physician, may remove tissue from human remains for transplant, or therapeutic, or scientific purposes specified in, and pursuant to, the provisions of the Uniform Anatomical Gift Act."[173]

---

170. Schoenfeld et al., *Evolving Trends in Liver Transplantation: An Outcome and Charge Analysis*, 67 TRANSPLANTATION, 246–253 (1999).

171. JAMA, 1431 (1999).

172. Since no community's resources are limitless, each community must ensure that those who receive public assistance for healthcare deserve it. Thus if a person repeatedly endangers his or her health through practices—such as smoking, drug or alcohol abuse, or overeating—known to constitute major risks, the community may decide to impose a limit on the public resources that such a person can call upon to finance the curative procedures she or he needs as a consequence of these unhealthy habits. DORFF, *supra* note 49, at 304.

173. CAL. BUS. & PROF. CODE § 7634 (West 2000).

good v. Good

All hospitals should have the right to continue transplanting organs regardless of the fact that they cannot transfer as many of them as other hospitals do. Neither age nor self-abuse by recipients should result in their being given a lower priority to receive transplants. However, fetal tissue should not be used because it will encourage more women to seek abortions that kill fetuses.

v.

Limiting transplant hospitals to those that do many of them greatly increases their success rate and they deserve to be given a preference. Self-abusers of drugs and alcohol who often continue abusing and wasting organs that have been transplanted should receive lower preference. Fetal tissue is often a most effective tissue for transplantation processes.

Twenty-First Century Approach

1. Because of the very high costs and varied success with transplants, rules should be adopted that:
   a. Patients considering a transplant at a particular hospital be normally informed of the comparative mortality rates of that organization with those of other healthcare organizations;
   b. Institutions should normally plan to either have a fairly high concentration of transplants, or not be doing them.
2. Patients who (a) are rationally found to be in need of certain transplants, but above an age for receiving them, or (b) have a history of self-abuse (e.g., smoking, alcohol, or drug abuses) should receive transplants only to the extent that the available organ is considered to be less appropriate than for others on an eligibility list.
3. In the twenty-first century, state legislatures should consider enacting laws authorizing the use of fetal tissue for transplant and embryonic stem cells for research and development purposes from elective abortions as well as from those performed as a result of medical need.
4. State legislatures and the courts should consider holding manufacturers of drugs and medical devices to the same standards for negligence as manufacturers of other products.

Keeping necrotically dead bodies alive in a persistent vegetative state for year or even decades is simply unreasonable.

Devettere, *Practical Decision Making in Healthcare Ethics: Cases and Concepts* (1995) at 161

In a certain state it is indecent to live longer. To go on vegetating in cowardly dependence on physicians and machinations . . . ought to prompt a profound contempt in society.

Nietzsche, *The Twilight of the Idols*

Our brains are the most impressive achievement of evolution we know of. They are several million times more powerful than any computer we've created. Every one of our brains is so complicated that it would take thousands of our best experts many lifetimes to design and construct anything a fraction as powerful using our current technology. Indeed, the technology in our heads is millions of years ahead of us; we are still like cave dweller looking up at the sun in awe.

Bentley, *Digital Biology* (2002), p. 67
Queens College, Cambridge

h) Whole Brain Death (by Neurologic Criteria) and Persistent or Permanent Vegetative State (PPVS)

It has been written, "The greatest number of organs are procured from cadaveric (brain-dead) donors with intact circulation."[174] The concept of brain death was created in 1968 by a committee at the Harvard Medical School due to current advances in medical technologies. These included the ability to restart the heart and keep the lungs functioning artificially for an indefinite length of time, with consequent increased availability of organs for transplant.[175] It rendered the previous definitions of death (such as cessation of heartbeat and respiration) nonessential for purposes of human organ trans-

174. N. ENG. J. MED. 422 (1997).

175. FOX & SWAZEY, SPARE PARTS: ORGAN REPLACEMENT IN AMERICAN SOCIETY (1992); KIMBRELL, THE HUMAN BODY SHOP: THE ENGINEERING AND MARKETING OF LIFE (1993); Lock, *Death in Technological Time: Locating the End of Meaningful Life*, 10 MED. ANTHROP. 221 JAMA 48–53 (1972).

plantation. As Henry Beecher said in a 1971 address on the subject to the American Association for the Advancement of Science:

> The need is to choose an *irreversible state where the brain no longer functions*. It is best to choose a level where, although the brain is dead, usefulness of other organs is still present. This we have tried to make clear in what we have called the *new definition of death*.[176] (Emphasis added.)

Although developments such as this caused the 1990s to be designated the "Decade of the Brain," that definition actually excluded the "reptilian portion of the brain."

The Uniform Anatomical Gift Act, which was enacted in some form in all 50 states, requires that hospitals request organ donations from the family of everyone who has been, or soon may become, all but reptilian-brain-dead.[177] Beginning in 1998, Medicare required all participating hospitals to have a plan that such requests be made.[178]

The concept of "whole brain death" was a significant advance in the last quarter of the twentieth century. In the light of hindsight, it has become too narrow for use in many instances. Henry Beecher's reference to an "irreversible state where the brain no longer functions" was declared to be only the "whole brain" including the brain stem (or "reptilian brain"). This was the view incorporated into state laws that prevailed throughout the United States

---

176. Beecher, *The New Definition of Death, Some Opposing Viewpoints*, 5 INT'L J. OF CLINICAL PHARM. 120 (1971).

177. CAPLAN & COELHO, THE ETHICS OF ORGAN TRANSPLANTS 142–46; Caplan, *Ethical and Policy Issues in the Procurement of Cadaver Organs for Transplantation*, 311 N. ENG. J. MED. 981–83 (1984).

178. This is analogous to the requirements of the Patient Self-Termination Act of 1994 with respect to Advance Directives. *See infra* Chapter 13. "Even if organs for transplants were harvested from all the potential brain-dead donors, there wouldn't be enough to go around, the researchers concluded after reviewing hospital records from around the nation. There are 82,000 people on the nation's waiting list for transplants." N. ENG. J. MED. (2003); ORANGE COUNTY REGISTER, Aug. 14, 2003, at News 11.

at the turn of the third millennium.[179] For other reasons, thousands of bodies are being kept "alive" in a persistent or permanent vegetative state at an annual cost estimated in 1994 at up to $7 billion.[180] It has also been noted over thc past several decades that intensive care units have become increasingly sophisticated "surrogate brainstems," replacing respiratory and other functions.[181]

Some religious objections have remained. For example, the same year that the Harvard Committee adopted the new criteria, the Chief Rabbi of Israel stated that by Jewish law, one is dead when one has stopped breathing.[182] New Jersey, in 1991, enacted a statute stating that physicians cannot declare brain death if "such a declaration would violate the personal religious beliefs of the individual." In such cases, death is declared only when a patient's heart and lungs stop, and the latter event can long be delayed by a respirator at increased expense.[183] It has been stated that "eventually society

---

179. *E.g.*, CAL. HEALTH & SAFETY CODE § 7180 (West 2000). "The base of the brain, called the reptilian brain, is where the necessary command centers for living are located. These control sleep and waking, respiration, temperature regulation, and basic automatic movements and are way stations for sensory input." RATEY, ANSER'S GUIDE TO THE BRAIN 10 (2001).

180. Multi-Society Task Force on PVS, *Medical Aspects of the Persistent Vegetative State (Part I)*, 330 N. ENG. J. MED. 1499–1508 (1994) [hereinafter *Persistent Vegetative State Part I*]; Multi-Society Task Force on PVS, *Medical Aspects of the Persistent Vegetative State (Part II)*, 330 N. ENG. J. MED. 1572–79 (1994) [hereinafter *Persistent Vegetative State Part II*]. The term *vegetable* means "a person who is incapable of normal mental or physical activity, especially through brain damage; a person with a dull or inactive life." CONCISE OXFORD DICTIONARY.

181. GREEN & DANIEL WIKLER, BRAIN DEATH AND PERSONAL IDENTITY; Daniel Wikler, *Brain Death: A Durable Consensus?*, 7 BIOETHICS 239–46 (1993).

182. ROSNER, MODERN MEDICINE AND JEWISH LAW 130 (1972) (citing Talmudic and post-Talmudic stages, including the statement from the *Mishnah* as codified by Maimonides: "If upon examination no sign of breathing can be detected at the nose, the victim must be left where he is [until after the Sabbath] because he is already dead.").

183. New Jersey Declaration of Death Act, N.J. STAT. ANN. § 26.6A-8 (West 1991). It ignores the "Wall of Separation." *See supra* Chapter 3, § 3.

will come to accept that the body of a patient in a permanent vegetative state is simply that person's 'living remains.'"[184]

The major problem is the "whole brain" criterion itself. The destruction of the neocortex should become the criterion in those instances where this can be adequately demonstrated. We have seen above how the depth of irrationality regarding treatment was reached in the instance of anencephalics, infants born without most of their brain, and who should also be considered essentially clinically dead (except to the extent required by a donor of possible transplants). Kant's categorical imperative cannot apply to such infants because they are neither persons nor potential persons.[185] Brains cannot be transplanted. Nevertheless, in 1994 a court actually ordered a hospital to continue life-prolonging treatment to a non-organ-donating anencephalic.[186] The lack of most of the brain as well as the equally clear difference between such severe brain damage and brain death "means that life support is useless; brain death is the principal requisite for the donation of organs for transplantation."[187]

A male infant with severe heart disease was put on a ventilator and given drugs continuously to keep him alive. At the same time, in the adjacent bed in the intensive care unit, was another baby who had been well until a sudden and catastrophic collapse occurred. He had some abnormal blood vessels in his brain and these suddenly burst; there was massive bleeding into his brain, causing destruction of the whole of his cerebral cortex. However, his brain stem was partially functioning, and he had irregular gasping movements. These were not enough to enable him to survive off the ventilator, but they were evidence that he did not have death of his whole brain, so he was not currently legally dead:

184. Wikler, *supra* note 181. However, from an ethical standpoint, two decades ago the President's Commission stated: "Although undeniably disconcerting for many people, the confusion created in personal perception by a determination of 'brain death' does not . . . provide a basis for an ethical objection to discontinuing medical measures on these dead bodies any more than on other dead bodies." Hast. Ctr. Rep., Jan.-Feb. 1997, at 133 (citing PRESIDENT'S COMM'N, DEFINING DEATH 84; *see also infra* Chapter 14 (regarding PVS).

185. Kant, *Foundations of the Metaphysics of Morals*, *in* PHILOSOPHICAL WRITING 94 (1986).

186. *In re* T.A.C.P., 609 So. 2d 588 (Fla. 1992).

187. Wijdicks, N. ENG. J. MED. 1215 (2001).

> There was one child who was completely normal except for a dying heart in one bed, and in the next bed, a child with a dead brain but a normal heart. As it happened, the two children had the same blood group, so the heart of the child with no cerebral cortex could have been transplanted into the child with cardiomyopathy (heart disease).[188]

No other heart was available. But since the child whose cortex was destroyed was not yet legally dead, his heart could not be used under the "whole brain" rule to save the other's life. Both children died shortly thereafter (rather than only one). This type of situation should not continue throughout the twenty-first century. A bioethicist suggested that due to its internal inconsistencies, "the whole brain concept of death is undergoing 'impending collapse.'"[189] For example, with respect to one type of case it has been stated, "In this protocol, patients are taken to an operating room, where the ventilator is removed and, after a few minutes of asystole [cardiac arrest] the patient is declared dead and the organs are removed."[190] Another explains the "use of elective ventilation (EV), in which the person is dead according to brain-death criteria, to keep the organs perfused because of the patient's wish to be an organ donor."[191]

The neocortex, part of the brain involved with higher learning, reasoning, and perception, makes up about 85 percent of its total mass. The great preponderance of evidence is that the neocortex is the "locus" of consciousness, but that input from the reticular activating system of the brain stem is required to activate it.[192] The only rational objection to advancing the criterion to loss of function of the neocortical region involves the ability of physicians to accurately diagnose it.[193] Near the end of the twentieth century, it

188. Shann, *The cortically dead infant who breathes*, ANENCEPHALICS, INFANTS AND BRAIN DEATH TREATMENT OPTIONS, Op. 28, *cited in* SINGER, RETHINKING LIFE & DEATH 43 (1994); *see also* Unif. Determination of Death Act, 12 U.L.A. 589 (West Supp. 1997).

189. Veatch, Hast. Ctr. Rep., 1993, at 18.

190. YOUNGER ET AL., THE DEFINITION OF DEATH (1999).

191. GILLET, BIOETHICS IN THE CLINIC 199 (2004); *see also* Browne et al., *The Ethics of Elective (Non-therapeutic) Ventilation*, 14 BIOETHICS 42–57 (2000).

192. *Quoted in* FRANKLIN, ARTIFICIAL MINDS 218 (1995).

193. Hast. Ctr. Rep., Mar.-Apr. 1992, at 19.

was found that this could be done very accurately in many if not most cases, with machines such as the positron emission tomography (PET) scanning device.[194] The advance in PET diagnosis was termed a "'significant breakthrough' in demonstrating that permanently unconscious patients do not experience pain, and will make it increasingly difficult to justify continuing treatment."[195]

As stated in the epigraphs to this section, the nineteenth-century philosopher Nietzsche and the late-twentieth-century Devettere both condemn the practice of maintaining the life of a person in a permanent vegetative state. However, another twentieth-century writer would maintain the absolute basis of the "sanctity of human life" in spite of the legal "wall of separation" between the church and state. Although at the turn of the twenty-first century American law courts have sometimes attempted to allow both views to prevail, depending upon the patient's wishes, in 2004 a Florida circuit court judge struck down a law that empowered the governor to prolong the life of a severely brain-damaged woman against her husband's desires.[196] On the other hand, in the words of bioethicist Peter Singer, "the British courts ceased to give effect to the traditional principle of the sanctity of human life"[197] (see Chapter 13). Consider the case of a 17-year-old soccer fan who went to a stadium to see his team play a semifinal cup match:

> A fatal crush occurred, pushing hundreds of fans against some fencing that had been erected to stop them getting onto the playing field. Before order could be restored and the pressure relieved, 95 people

---

194. Dr. Michel Mathew Ter-Pogissian, who died in 1998, led the research on the PET scanner. It is now used in hospitals everywhere. N.Y. TIMES, June 21, 1996, at B13.

195. INTERNAL MED. NEWS, Dec. 15-31, 1987, at 3.

196. In an American case in which a woman remained unconscious for over 15 years, the court stated that the law "unjustifiably authorizes the governor to summarily deprive Florida citizens of their constitutional right to privacy." ORANGE COUNTY REGISTER, May 1, 2004, at News 1. Nevertheless, the case underwent much further litigation until February 2005, when the husband finally received a judge's permission to remove the brain-damaged woman's feeding tube. ORANGE COUNTY REGISTER, Feb. 26, 2005, at News 32.

197. SINGER, *supra* note 188, at 65.

> had died in the worst disaster in British sporting history. Tony Bland was not killed, but his lungs were crushed by the pressure of the crowd around him and his brain was deprived of oxygen. Later, in hospital, it was found that only his brain stem had survived; his cortex had been destroyed.[198] Bland was one of many persons who, due to advances of modern medicine, could be kept in a vegetative state for many years. However, his case was destined to make legal history in his country because he was incapable of consenting to the medical treatment needed "to keep him alive." The British courts held that with such a patient, doctors are under *no* legal duty to continue treatment. That is, mere biological life is no benefit to the patient. The case was appealed to the House of Lords.[199] One of the Lords stated, "To withdraw life-support was not only legally, but also ethically justified, since the continued treatment of *Anthony Bland can no longer* serve to *maintain* that combination of manifold *characteristics which we call a personality.*"[200] (Emphasis added.) That is, only a "subjective personality" was left in those who knew him in the past, not in Tony Bland. Another Lord stated that "the case for the universal sanctity of life assumes a life in the abstract and allows nothing for the reality of Mr. Bland's actual existence."[201]

Indeed, this decision went to the heart of the ethical problem of expensive non-beneficial (futile) treatment, which is still extant in the United States (see Chapter 13). The case of nonvoluntary termination of treatment for a patient in a permanently vegetative state (and consequent possible use of his organs for transplant) was faced in America by a Pennsylvania court in 1993. It concerned a patient who had been in a vegetative state at a nursing home since 1976. His mother and guardian requested his physicians to remove the

198. *Id.* at 58. This last statement that "his concept has been destroyed" clearly indicates a "permanent" vegetative state.

199. Airedale N.H.S. Trust v. Bland (1993), 2 W.L.R. 316, 368 (Eng.).

200. Singer, *supra* note 188, at 67.

201. *Id.* at 74. The director of a medical ethics center stated: "The Law Lords in the *Bland* case have in effect declared non-voluntary euthanasia lawful." Gormally, *Definitions of Personhood: Implications for the Care of PVS Patients*, 9 Ethics & Med. 48 (1993).

gastrostomy tube through which he received his medication and nourishment so that he might die. The nursing home refused to do so without a court order. The patient had never commented upon how he wished to be treated if he were in PPVS or was otherwise incompetent. The attorney general argued that without "clear and convincing evidence" of his wishes, his life should not be terminated. He cited a 1990 Supreme Court decision holding that life should be maintained where state law in Missouri required such evidence.[202] The court could not rely upon a famous 1976 decision of the New Jersey Supreme Court when it cut new ground by authorizing the withdrawal of a life-support ventilator from Karen Ann Quinlan, who was quoted by her family as saying that she never wanted to be kept alive by extraordinary means.[203] A New Jersey law provides that a person who did not believe in the concept of brain death cannot be declared dead using the current brain-death criteria:

> The death of an individual shall not be declared upon the basis of neurological criteria . . . when the licensed physician authorized to declare death, has reason to believe . . . that such a declaration would violate the personal religious beliefs of the individual. In these cases, death shall be declared, and the time of death fixed, solely upon the basis of cardio-respiratory criteria pursuant to section 2 of this act.[204]

Nevertheless, the Pennsylvania court had noted that the right to self-determination as to one's own medical treatment is "not absolute." It held that even without such evidence, the tube might be withdrawn if a family member or guardian receives written statements from two qualified doctors certifying that the patient does not have a reasonable possibility of recovery. The Pennsylvania court also found that state involvement through the courts was overly intrusive and violative of the individual's right to privacy; in the future, the decision to terminate treatment should be made only by such a family member or guardian.

A virtually identical approach has now been taken in Japan following that country's enactment of its Organ Transplant Law in 1997:

---

202. Cruzan v. Dir., Mo. Dep't of Health, 497 U.S. 261 (1990).
203. *In re* Quinlan, 355 A.2d 647 (N.J. 1976), *cert. denied*, 429 U.S. 922 (1976).
204. N.J. Rev. Stat. § 26:6A-5 (1999).

> The donor was a 44-year old woman who exhibited a subarachnoid hemorrhage and was hospitalized in February 1999. She subsequently became comatose and appeared to be brain-dead. Coordinators of the Japan Organ Transplant Network (JOTN) confirmed that her family had consented to organ removal and that her donor card included all the necessary information. Two doctors licensed to pronounce brain death carried out two rounds of tests at six-hour intervals for deep coma, dilated pupils, the loss of voluntary breathing, the loss of brain-stem response, and flat brain waves. On 28 February, the woman's heart, liver, kidneys, and corneas were removed and transplanted to recipients selected by JOTN from a computer network register. All operations were successful, and all the recipients were still alive as of 30 March 1999.[205]

This last part of the ruling is important for future terminations of treatment for PPVS and the possible harvesting of organs. The ruling also applies to patients who have living wills in those states that have statutes applying to nonterminal conditions such as "irreversible coma," patients who might otherwise be kept alive for many years.[206] Unfortunately, most states have yet to amend their living will statutes so that a person whose cerebral cortex can be shown to have been destroyed (as in the cases of Tony Bland and others) will not have their lives prolonged for no possible purpose. Many state governments continue to spend the millions of taxpayer dollars required to keep all permanently comatose bodies alive and prevent the donation of their organs to benefit others.

As noted above, sometimes physicians cannot predict that a person will or will not come out of a coma—a state of profound unconsciousness from which a patient cannot be aroused even by powerful stimuli.[207] For example, in 1998

---

205. Hast. Ctr. Rep., May-June 1999, at 48.

206. Haw. Rev. Stat. § 327-2(D); La. Rev. Stat. Ann. § 40:1299.58.10; *Id.* § 39-4503(3); N.M. Stat. Ann. § 24-7a-2B (West 1978); Tenn. Code Ann. § 32-11-103. Ohio and Arkansas extend the application of their living will statutes to patients in a permanent unconscious state. *See* Ark. Code Ann. § 20-17-201(7); Ohio Rev. Code Ann. § 2133.01(u) (West 2000).

207. Dorland's Medical Dictionary, *supra* note 29.

Gary Dockery, a Tennessee police officer aged 43, fell into a stupor after he was shot in the forehead by a drunken man who was angry at the police officer who reprimanded him for making noises that bothered his neighbors. Dockery fell into a comatose state where he remained for 7½ years. Suddenly and inexplicably on February 11, 1996, he stirred and started talking for several hours. Then he slipped back into a semi-comatose state and died.[208] Some people could say that those few hours awake were worth the many thousands of dollars spent to keep Dockery "alive"; others may disagree. In another case, an 18-year-old woman suffered severe head injuries in a car accident in 1987. After 15 months in a coma, she moved her eyes. Three years later, she was communicating with eye blinks. After five years she could say short phrases for several minutes. She was sent home wheelchair-bound. Her partial rehabilitation had cost well over a million dollars.[209] Another well-documented case of late improvement after permanent vegetative state had been reported.[210]

A Persistent or Permanent Vegetative State (PPVS) has normally been considered permanent if the coma lasts 12 or more months after traumatic injury. Irreversibility was established when the risk of prognostic error is "exceedingly small,"[211] and the diagnosis of irreversibility can be established with "a high degree of clinical certainty," that is, when the chance of regaining consciousness is exceedingly rare.[212] In one of the twentieth-century estimates for such patients, "life expectancy ranged from two to five years." The chance for survival of greater than 15 years was approximately 1/15,000 to 1/75,000.[213] A multi-society task force outlined guidelines and claimed

208. Straits Times (Sing.), Feb. 16, 1996, at 7.

209. Childs & Mercer, N. Eng. J. Med. 124 (1995); Orange County Register, Jan. 4, 1996, at News 5.

210. Patrick et al., *Unexpected Improvement After Prolonged Posttraumatic Vegetative State*, 48 J. Neurology, Neurosurgery, Psychiatry § 1200–03 (1985).

211. The Multi-Society Task Force on PVS, *Persistent Vegetative State Part I*, *supra* note 180, at 1499–1508. The acceptable risk of prognostic error was defined as 0.1 by the American Medical Association's Council on Scientific Affairs. Child & Mercer, N. Eng. J. Med. 24 (1996).

212. 45 Neurology 1015–18 (1995).

213. *Id.* at 1016. Further, feeding tubes have been found to be inappropriate for people with advanced dementia. Finucane et al., *Tube Feeding in Patients with Advanced Dementia: A Review of the Evidence*, JAMA 1365–70 (1999); Gillick, *Rethinking the Role of Tube Feeding in Patients with Advanced Dementia*, N. Eng. J. Med. 206–10 (2000).

"sufficient data are now available to make the diagnosis of permanent vegetative state in appropriate patients with a high degree of certainty."[214]

good v. Good

The "whole brain death" test initiated in the late 1960s is the only current criterion (other than cardiac arrest) that can guarantee a patient is really not alive for the purpose of harvesting his organs.

v.

At the turn of the third millennium, tests became available to determine the "actual permanency" of many patients who were in a persistent or permanent vegetative state (PPVS). Such tests should be utilized for the determination of those patients whose organs may be harvested to benefit others who need them.

Twenty-first Century Approach

1. The twentieth-century use of "whole brain death" as the sole criterion for brain death has outlived its unique usefulness in certain instances. It should be replaced by criteria based upon reliable evidence of destruction of the neocortex.
2. While some unconscious patients in a persistent or vegetative state (PPVS) may be diagnosed as more likely to have a provisional status, others can now be diagnosed as being in a permanent vegetative state.
3. Where a PPVS patient's condition has been shown to be permanent to a high degree of certainty from reliable techniques, the patient's organs should normally be "presumed" available for transplant to other patients who would die without these organs.

The creatures outside looked from pig to man, and from man to pig, and from pig again; but already it was impossible to say which was which.

George Orwell, *Animal Farm*

Consider your nature, you were not made to live as beasts, but to pursue virtue and knowledge.

Dante

---

214. The Multi-Society Task Force on PVS, *Persistent Vegetative State Part I*, *supra* note 180, at 1499.

i) Xenotransplants (from Animals to Humans)

For many years it has been required that experiments be conducted on animals where possible prior to experimentation on humans. Until recently, few questions have arisen about where animals were on the value scale; humans, being of "inestimable value," were at the top and other (or lower) animals were all nearer the other end of the scale. The value of animals has been for economics (such as food or for work as horses), education and entertainment (e.g., zoos), and for research (monkeys and especially rodents like mice or rats).[215]

The rights of animals have been of concern to some bioethicists (such as Peter Singer of Australia),[216] particularly of those animals on the Endangered Species List. The World Health Organization's Task Force on Organ Transplantation has not substantially concerned itself with whether xenotransplantation should or should not proceed. It largely looked at the screening of source animals, monitoring the recipients of organs and results of contacts with them.[217] With the advent of an awareness of environmental abuses, objections to the use of animals for these purposes have little diminished.

Nevertheless, xenotransplants (from the Greek word *xenos*, meaning "foreign" or "stranger") are widely regarded as "the earliest foreseeable means of alleviating the dire shortage of transplantable organs from human donors." A financial company estimated that "successful xenografting would produce a tenfold increase in transplants from 45,000 worldwide in 1994 to 455,000 in 2010."[218] Immanuel Kant's imperative against using humans as a means, but always as ends, does not apply to this use of animals. There are very few religious objectors to xenotransplantation. Buddhists and Hindus do not draw Kant's line between humans and animals (see Chapter 3). It has been written

215. Jane Goodall, *Foreword* to GREEK & GREEK, SACRED COWS AND GOLDEN GEESE (2000). "There are Nobel Prizes for alternative techniques."

216. SINGER, ANIMAL LIBERATION (1975). "An anencephalic baby clearly ranks lower on any possible scale of relevant characteristics than a chimpanzee. Yet as the law now stands, a surgeon could kill a chimpanzee in order to take her heart and transplant it into a human being." SINGER, WRITINGS ON AN ETHICAL SCALE 222 (2000).

217. DAAR, ETHICS OF XENOTRANSPLANTATION 981.

218. LANCET 1347 (1998).

that rationally, "those who believe in reincarnation should not object to organ donation; they are going to inhabit a new body anyway."[219]

Kant's dichotomy is close to the views of Christians and Jews who read in the Old Testament, "man was made in God's image and has dominion over all other creatures and all the earth."[220] Jews and Muslims consider that pigs are ritually unclean. However, this is "not a problem in Jewish law or ethics; if the use of animal parts can save a human life, we would have a moral and religious obligation to use them."[221] Each of these groups is living in the United States and they can be accommodated within the doctrine of informed consent as recipients of organs (see Chapter 5). When animals are not regarded as property, the issue of consent by the donor does not apply.

Another factor enters into the picture, namely, the possibility of introducing new infectious diseases into the human population. The closer two species are related, the more likely organs or grafts will be compatible. Yet the very closeness of primates (apes and humans) also makes for viruses to penetrate the immune barriers between species. For example, it is known that baboons carry many viruses, at least two of which might be transmitted into the human population and later mutate into dominant strains of HIV.[222] Until the twenty-first century, most doctors and insurers routinely rejected AIDS patients for normal transplants due to their lower life expectancies.[223] It has been noted in the twenty-first century:

> Although animal studies and the early unregulated human trials have not demonstrated whole organ xenograft survival rates high enough

219. Dorff, *supra* note 49, at 217.

220. *Genesis* 1:26. Similarly, those who believe in resurrection should also not object to organ donation. If resurrection is the blessing that most who believe in it hold it to be, God should surely be trusted to resurrect us in a better body than the one in which we died. Dorff, *supra* note 49, at 239.

221. Dorff, *id.*

222. Hast. Ctr. Rep., Sept.-Oct. 1995, at 5. Yet in 1997 naval researchers claimed that their experiments on monkeys with synthetically created antibodies showed that organ recipients may not need antirejection drugs for the rest of their lives. L.A. Times, Aug. 6, 1997, at A17.

223. Orange County Register, Aug. 3, 2001, at News 9.

to justify proceeding with clinical trials, active preclinical research on whole organ xenografting continues.[224]

A unique strain of genetically engineered pigs produces two human proteins that might help allow organs to be transplanted from one person to another. It has been predicted that pig hearts may be experimentally introduced into human patients in the twenty-first century.[225] Xenotransplantation is virtually not possible when the human immune system recognizes unmodified animal parts as foreign and rejects them. However, the "knocking out" of genes can be a vital step in producing pigs with organs that could be more safely transplanted into human patients. Cloned pigs have had one such gene knocked out (or inactivated).[226] Following these types of experiments, if the patient is found to harbor an infection that might prejudice public health, he may have to be quarantined with his prior or subsequent consent (or even without it).

In 1996 the Institute of Medicine's Council of Healthcare Technology issued a report concluding that the benefits of xenotransplantation were great enough to justify taking the risk of infections. The Institute recommended proceeding with clinical trials "when the scientific basis . . . is judged sufficient and the safeguards are in place." It was acknowledged that cellular xenotransplants were under way in the United States.[227] Researchers in England discovered that pigs have porcine endogenous viruses that can infect human cells in a manner as deadly as HIV, but remain dormant in pigs.[228] In

224. "Clinical trials and preclinical studies using xenotransplantation products have demonstrated that animal cells and tissues might have the potential to be used successfully for the treatment of various diseases and conditions. [Some recipients of certain] xenotransplantation products have survived for years after receiving them." JAMA 2304 (2001).

225. Orange County Register, May 1, 1995, at News 3.

226. The GGTA1 gene makes a sugar called alpha-1-galactose, which lines pigs' blood vessels. Because it is nearly identical to a bacterial sugar, the human immune system attacks it. As a result, pig organs transplanted into people would be destroyed almost instantly. Scientists announced that they have cloned piglets lacking both copies of the gene that makes the human immune system reject pig tissue. Orange County Register, Aug. 23, 2002, at News 29; *see infra* Chapter 15.

227. 21 World J. Surgery 980 (1997).

1998, despite concerns that pigs harbor a potentially transmissible retrovirus,[229] the Food and Drug Administration (FDA) approved operations in which humans were connected to pig livers via filtering lines. These procedures were to save dying patients while their liver organs recovered from medication poisoning.[230] In 2003 it was noted that:

> Several countries are performing or planning clinical xenotransplantation (e.g., pig islet-cell and Sertoli-cell transplantation in Mexico and goat hepatocyte transplantation in India), and some of these procedures are performed by commercial enterprises for cosmetic or medical treatment. . . . International cooperation is needed to develop universal procedures and standards for oversight, including ethical and monitoring guidelines.
>
> In addition to working toward international guidelines, we should encourage public health authorities in countries that do regulate xenotransplantation to develop ways to identify persons who enter such a country after undergoing xenotransplantation and to ensure monitoring.[231]

In 1999 a study of the safety of transplanting animal parts into humans found no evidence that people caught a worrisome pig virus.[232] But, the trans-

---

228. Weiss, *Xenotransplantation*, 317 Brit. Med. J. 931–34 (1998). *Xenotransplants and Retroviruses*, 285 Science (1999).

229. In 1996 the Public Health Service published the following notice in the Federal Register:

> Because transplantation bypasses most of the patient's usual protective physical and immunological barriers, transmission of known and/or unknown infectious agents through xenografts may be facilitated. . . . [S]ome agents, such as retroviruses and prions, may not produce clinically recognizable disease until many years after they enter the host, and some infectious agents may not be readily detected or identified in tissue samples by current techniques.

McCarthy, Hastings Ctr. Rep., Nov.-Dec. 1999, at 25.

230. L.A. Times, Mar. 20, 1998, at A19.

231. Sykes et al., N. Eng. J. Med. 1295 (2003).

232. Science (1999); Orange County Register, Aug. 21, 1999, at News 36. The study was concerned with the "porcine endogenous retrovirus" (PERV).

plant of larger organs has been shown to yield fatal results.[233] Nevertheless, in 1999 the FDA issued a document entitled "Guidance for Industry: Public Health Issues Posed by the Use of Non-human Primate Xenografts in Humans."[234] Sykes et al. indicated their belief that "xenotransplantation should be performed only with oversight by a governmental regulatory agency that has such guidelines (e.g., those of the U.S. Food and Drug Administration)."[235] Because of the fears expressed therein, it was interpreted as a request for a moratorium on requested approvals that would "not likely be lifted anytime soon." This approach was objected to by a member of the Food and Drug Administration (FDA) Subcommittee on Xenotransplantation.[236] Use of xenotransplants will dominate many ethical debates in the twenty-first century.[237] However, the director of FDA's division of cellular and gene therapies stated, "We do feel the potential benefits are great and that efforts can be made to make everyone responsible. There are ways to deal with problems should they arise."[238]

In 2001 the AMA *Code of Medical Ethics* issued Rule 2.169 based on the report "The Ethical Implications of Xenotransplantation," adopted by the

---

233. The recent history of fatal transplants is described in GREEK & GREEK, *supra* note 215. They set forth many examples. In a 1992 "pig-to-human transplant" at a Los Angeles hospital "the patient survived less than two days." "A pig heart xenotransplant was performed, with the same result." Chimpanzees were not useful in AIDS research.

234. The document provides guidance to industry concerning (1) the potential public health risks posed by non-human primate xenografts; (2) the need for further scientific research and evaluation of these risks, particularly infectious agents; and (3) the need for public discussion concerning these issues. Fed. Reg. 16,744 (Apr. 6, 1999).

235. N. ENG. J. MED. 1294 (2003) (citing GUIDANCE FOR INDUSTRY: SOURCE ANIMAL, PRODUCT, PRECLINICAL, AND CLINICAL ISSUES CONCERNING THE USE OF XENOTRANSPLANTATION PRODUCTS IN HUMANS (2001)).

236. "Instead of placing a moratorium on xenotransplantation research, we should strive for a balance between delimited infectious disease risks and the moral imperative of not turning away from patients 'whose suffering is both clearly visible to us and more clearly devastating in its impact on them.'" Harold Y. Vanderpool, Commentary, *A Critique of Clarit's Frightening Xenotransplantation Scenario*, 27 J.L. MED. & ETHICS 153, 155 (1999).

237. 351 LANCET 1347 (1998).

238. 21 WORLD J. SURGERY 180 (1997).

Council in December 2000. It offered eight guidelines for the medical and scientific communities, telling physicians to encourage public discussion of xenotransplantation because of the potential unique risks; support oversight for the development of clinical trial protocols, including disclosure of sexual contact; autopsy; and "a waiver of the traditional right to withdraw from a clinical trial until the risk of late xenozoonoses is reasonably known not to exist." Although there were several other guidelines,[239] the rule recognized that "research in this area may uncover physical and psychological conditions that require medical attention." With respect to preclinical cardiac xenotransplantations, it was noted that animal organs could satisfy the demand for solid organ transplants, which currently exceeds the limited human donor supply. The longest published median and individual survivals of transgenic pig hearts in a baboon has been 27 and 139 days. However, the 2004 "median survival of 76 days in a group of heterotopic porcine-to-baboon cardiac xenografts represents a major advance. . . . If further studies in the orthotopic position replicate these outcomes, criteria considered appropriate for clinical application of cardiac xenotransplantation would be approached."[240]

Federal approval was given to perform at least five experiments with the AbioCor, an artificial heart that weighs about four times the weight of a normal heart.[241] During the twentieth century, experience with artificial organs began to replace some of the current xenotransplant experiments. For the twenty-first century it has been argued that, with prior consent, conducting trials on bodies in a permanent vegetative status "is preferable to the use of human subjects without lack of brain function."[242]

239. Issued June 2001.

240. McGregor et al., *Cardiac Xenotransplantation: Progress Toward the Clinic*, 78 Transplantation 1569 (2004).

241. Orange County Register, Aug. 22, 2001, at News 4.

242. Ravelingien et al., *Proceeding with Clinical Trials of Animal to Human Organ Transplantation: A Way Out of the Dilemma*, 30 J. Med. Ethics 92–98 (2004).

good v. Good

Xenotransplantation research should not continue because this work is only experimental at the dawn of the twenty-first century and not covered by insurance.

v.

Limited types of research should continue because potential benefits have been shown to exist, even though fatal transplants of larger organs have occurred late in the twentieth century.

Twenty-first Century Approach

It would appear premature to issue regulations with respect to the potential for zoonosis (infections or transfer of viruses from animals to humans) in order to impede efforts needed early in the twenty-first century for possible clinical application of xenotransplantation. However, international and FDA guidelines are needed for ethical monitoring of those people who receive xenotransplantation abroad and return to their home country.

---

2. Allocation of Donated Organs

In the United States, the allocation of an individual's organs while he is living and upon his death is essentially a choice that can and should be made by that individual.[243] This is similar to choices made in connection with sterilization, abortion, and suicide, all of which are legal in all states. He may offer to donate to whomever he chooses provided that all medical requirements are met with respect to both the donor and the organ recipient.

Since the dawn of the third millennium, some artificial hearts have been devised and applied in the United States (as noted above). Although they

243. *See* Unif. Anatomical Gift Act § B, 8A U.L.A. 40. In 2004, 6,647 people became living kidney donors in the United States, according to the Organ Procurement and Transplantation Network. Ingelfinger, *Risks and Benefits to the Living Donor*, N. Eng. J. Med. 447 (2005); *see also* Delmonico, *A Report of the Amsterdam Forum on the Care of the Live Kidney Donor: Data and Medical Guidelines*, Transplantation S53–S66 (2005).

have their limits, some recipients have lived with the devices a long time.[244] If their success becomes more significant later in the twenty-first century, these devices will also partially replace the donors. For the first time in the twenty-first century, a 65-year-old man lived three years with an artificial heart (when no suitable donor was found), and he later underwent a successful implant of a living heart.[245]

a) Allocation Priority to the "Sickest" Nationwide; HIV Patients

Normally it is the potential recipient who seeks an organ from a qualified donor somewhere in the country, which can be difficult. In 1987 a private organization called the United Network for Organ Sharing (UNOS) made a contract with the government to allocate scarce transplantable organs to appropriate recipients.[246] Traditionally, an organ was offered to waiting local patients in medical need. If there was no match, it was offered to patients in the region. It was offered nationwide only if medical matches were not found locally or regionally. In 2000 it was noted that since 1982 there had been a substantial increase in short-term and long-term survival of kidney grafts obtained from both living and cadaveric donors: 12,166 transplants were performed and 2,295 people on the waiting list died.[247]

However, this was changed at the end of the second millennium so that the organs would go to the "sickest" person nationally regardless of where he was located.[248] There have been many requests that the position be recon-

244. Regarding one man with an artificial heart, it was stated, "He has lived with the device longer than anyone else in the United States." ORANGE COUNTY REGISTER, Aug. 11, 2001, at News 9.

245. ORANGE COUNTY REGISTER, Jan. 17, 2002, at News 18.

246. *See* National Organ Transplant Act of 1984 (NOTA), 42 U.S.C. § 274.

247. ORANGE COUNTY REGISTER, Mar. 2, 2000.

248. In 1998 a revised final rule stated that allocation should be distributed "over as broad a geographic area as feasible," and that "the stated principles underlying 'the Final Rule' include the creation of a 'level playing field' in organ allocation—that is, organs are allocated based on patients' medical need and less emphasis is placed on keeping organs in the local area where they are procured." However, due to much opposition, Congress suspended it temporarily. Sharon Mussong, *Administrative Developments: DHHS Issues Organ Allocation Final Rule*, 27 J.L. MED. & ETHICS 380 (1999). Under the rule, the secretary retained the right to approve or veto any allocation plans designed by the United Network for Organ Sharing (UNOS). ORANGE COUNTY REGISTER, Nov. 12, 1999.

sidered. In the third millennium patients in many centers complained that they may lose locally donated organs. They debated transfers to a region while not accepting transplants coast to coast.[249] Although each bank formerly would consider organs first locally, next regionally, then nationally,[250] the AMA *Code of Medical Ethics* has maintained since 1994 that "[o]rgans should be considered a national, rather than a local or regional resource." The rule also states:

> Geographical priorities in the allocation of organs should be prohibited except when transportation of organs would threaten their suitability for transplantation, [and] patients should not be placed on the waiting lists of multiple local transplant centers, but rather on a single waiting list for each type of organ.[251]

In April 2000, the House of Representatives voted 275-147 to return to a policy of distributing organs to medically eligible patients in a local area first, in order of need, before they are offered regionally and then nationally.[252] Some pointed out that the sickest people may be less likely to survive long with a transplant, and thus fewer patients with less-urgent conditions will get needed organs. A study of alcohol use by liver transplant recipients and others pointed out that "large numbers of patients do not comply with their prescribed treatment" and may waste the organ transplant.[253] In 1996, in spite of the fact that alcoholics represented 20 percent

---

249. Orange County Register, May 14, 2000, at News 12.

250. Orange County Register, Oct. 12, 2000, at News 16.

251. Rule 2.16. *But see* Chen, *Allocation and the States: Can the States Restrict Broader Organ Sharing?*, 49 Duke L.J. 261, 273 (1999) (raising constitutional questions, particularly with respect to state laws regulating allocation of organs).

252. L.A. Times, Apr. 5, 2000, at A16. A 1995 survey found that 62% of the American public agreed that "we have an obligation to take care of people in our local community who are in need, but not all the needy in the entire country." Schlesinger, *Paradigms Lost: The Persisting Search for Community in U.S. Health Policy*, 22 J. Health Pol., Pol'y & L. 996 (1997).

253. Orange County Register, Apr. 1, 1994, at 1. Until about 1988, "liver cells were not considered suitable targets for gene transfer for two reasons: (1) adult liver cells are not susceptible to infection with the types of viruses commonly used in gene transfer experiments and (2) liver cells could not be removed from and then reintroduced into the body." Nichols, Human Gene Therapy 107 (1988).

of all liver transplants, they were often on the top of the list for that organ, which cost between $180,000 and $300,000 each.[254] It has been recommended that patients with alcohol-related end-stage liver disease should be given lower priority as candidates for liver transplantation than those not responsible for their liver disease.[255] It has been pointed out that people who make "medically inappropriate lifestyle choices," such as smoking or alcohol, do not have the same right to healthcare as those who do not: "To smoke is to create an artificial and preventable health need. That is irresponsible, and to insist that smokers be treated like nonsmokers is to unjustly treat irresponsible people as responsible people."[256] Until the turn of the twenty-first century, patients infected with the human immunodeficiency virus (HIV) were excluded from organ-waiting lists because they were not expected to live much longer. However, because HIV patients are living longer with the drugs, some develop organ failure. They then argue that this alone makes them candidates for transplants of organs that would have gone to an otherwise healthy patient.[257]

In some locations they have successfully received liver transplants.[258] Until 2002, inequalities developed in the transplantation of livers based primarily on waiting time and subjective measures of procurement and allocation of deceased donors. A model for end-stage liver disease (MELD) score can serve as the basis to allocate livers of deceased donors according to the medical needs of patients. In 2004 a study pointed to "a significant disparity in MELD scores in liver transplant recipients in small vs. large" organ pro-

254. N. Eng. J. Med. 1469 (1999). "Under the new policy, livers must be offered first to any 'status 1,' suddenly ill, patients within the region. On May 15, 2000 it was reported that 132 of the less-sick patients—those classified as status 2b or 3—died while waiting for a transplant." http://www.cnn.com/2000/HEALTH/05/15/organ.transplants.ap/index.html "Tests showed those who become suddenly ill anywhere should receive the highest number for the 'sickest patients.'" Orange County Register, June 30, 2001, at News 8.

255. Moss & Seigler, *Should Alcoholics Compete?*; Yoger, *Individual Responsibility for Health*, Hast. Ctr. Rep., Mar.-Apr. 2002, at 26.

256. Kluge, *Drawing the Ethical Line*, Can. Med. Ass'n J. 746.

257. Orange County Register, Aug. 30, 2002, at News 23.

258. *Id.*

curement organizations.[259] In January 2002, a donor died after giving a portion of his liver to his brother at Mount Sinai Medical Center in New York. Subsequently, the number of liver transplantations involving living donors in the United States—which had increased from 395 in 2000 to 518 in 2001—decreased to 362 in 2002 and about 320 each in 2003 and 2004.[260] The Center for Medicare and Medicaid Service sets a minimum of 12 transplants a year to be certified for compensation under Medicare rules. The California Department of Health Services sets a minimum of 18 for MediCal certification.[261] It has been noted that the allocation of organs nationally based on "sickest first and widest distribution should not be applicable to all forms of organ transplantation." A Michigan professor stated:

> The philosophy of treating the sickest patient first assumes that the patients who are less sick will receive treatment later. This is not the case with liver transplantation. Need exceeds the annual supply of 4,000 donor organs by a factor of 5 to 10. If a liver is allocated to one patient, another patient may never receive one. In effect, most patients who ultimately undergo transplantation would have to endure years of pro-

259. "This disparity does not reflect the stated goals of the current allocation policy, which is to distribute livers according to a patient's medical need, with less emphasis on keeping organs in the local procurement area." JAMA 1871 (2004).

260. Steinbrook, *Public Solicitation of Organ Donors*, N. Eng. J. Med. 441 (2005).

261. Orange County Register, Dec. 5, 2005.

> The problem of the University of California Irvine (UCI) Medical Center's liver transplant program led the Center for Medicare and Medicaid Service to decertify the program.

Orange County Register, Nov. 16, 2005.

> At least 15 transplant candidates at UCI died while waiting for a liver. The one-year survival rate for UCI patients was 69%, well below the 85% rate for the rest of the state and Medicare's 77% requirement. The United Network for Organ Sharing (UNOS) gave UCI administrators a detailed picture of the problems that were contributing to high death rates.

Orange County Register, Dec. 23, 2005, at 1.

> At least 20 people have sued UCI for negligence in the liver program.

Orange County Register, Dec. 31, 2005, at Local 2.

gressive morbidity until critical care or at least hospitalization was required, while simultaneously risking death while waiting.[262]

The chairman of the UNOS Liver Committee had pointed out serious inequities resulting from this new policy.[263] UNOS exports up to 5 percent of organs when efforts to find a recipient in the United States have failed.[264]

In 2001, the head of HHS proposed methods to increase the volume of giving organs by promoting organ donation among business employees and even giving a national donor card to people who will serve as a donor. After noting that, we have not heard high officials moving toward this approach.[265] As one bioethicist stated: "*There can be no health reform without healthcare rationing. There can be no fair health reform without healthcare rationing for all.*"[266] (Emphasis added.)

good v. Good

A national policy for allocating the limited supply of organs in the U.S. to the "sickest" is the fairest choice for rationing.

v.

There is nothing necessarily fair about allocating scarce organs to those sickest patients, some of whom who may be smokers, drug abusers, or alcoholics, while forcing others to undergo years of increased suffering and/or death while waiting for organs.

Twenty-first Century Approach

A priority of the top allocations of organs should not normally be given to the "sickest" patients without taking an account their background or self-abuse, as well as other important matters.

---

262. Jeremiah G. Turcotte, N. Eng. J. Med. 963 (1998).
263. N. Eng. J. Med. 426 (1997).
264. *Id.*; *see* Rand, The Virtue of Selfishness 51 (1962).
265. Caplan, http://www.msnbc.com/news/560295.asp, May 12, 2001.
266. We cannot escape the need to make rationing decisions, especially the very painful decisions associated with last chance therapies. We can make these rationing decisions collectively in ways that we judge to be "just enough" and "caring enough," given that there are no perfectly just options available. Or we can allow bureaucrats and administrators and employers and stockholders to make these decisions for us.

Fleck, *Health Reform Requires Healthcare Rationing*, Hast. Ctr. Rep., 2002, at 35–36.

---

b) Non-Direct Donations by Living Individuals

Some people are not willing to have organs removed from their bodies prior to their being declared dead. However, more altruistic people exist who choose to do so.[267] In 1994, the President signed legislation that gives federal employees who make live-donor donations up to 30 days of paid leave.[268] However, White House officials acknowledged that they knew of no other employer who had adopted a similar policy.[269] There is no legal obligation to sacrifice one's kidney (or one's bone marrow or blood) to benefit others; but there is widespread agreement that any such action would be morally admirable, an example of altruism or action above and beyond the call of duty.[270]

The use of a living donor to provide a liver transplant for an adult was performed 100 times in the last two years of the twentieth century.[271] Liver transplantation is among the most costly of these medical services. A study of the recipients who were older, had alcoholic liver disease, or were se-

---

267. Sibberner, Hast. Ctr. Rep., Nov.-Dec. 1999, at 5.

268. ORANGE COUNTY REGISTER, Feb. 9, 2000.

269. GILLON, IN BIRTH AND DEATH 110 (1996). He added: "There would not be universal agreement about this—for some people and for some cultures the scope of one's obligation of beneficence might well be seen as requiring such self-sacrifice in the interests of, for example, one's child, sibling, spouse or friend." *Id.*

270. ORANGE COUNTY REGISTER, Aug. 2, 1999, at News 12; *see also* Goldman, *Liver Transplantation Using Living Donors: Preliminary Donor Psychiatric Outcomes*, 34 PSYCHOSOMATICS 235–40 (1993).

271. Shows et al., 281 JAMA 1381–86 (1999). "As of July 28, 13,519 people in the United States were awaiting liver transplants, according to the United Network for Organ Sharing. Last year, only 4,450 liver transplants were performed, and in 1997, 1,129 people died while on the waiting list." ORANGE COUNTY REGISTER, Aug. 2, 1999, at News 12.

272. When the University of Minnesota began its program to accept nondirected organ donations in August 1999, it decided to allow an interval of several weeks between the evaluations of the donor and the recipient, and their admission for surgery, "so that the donor would have sufficient time to reconsider." N. ENG. J. MED. 433–34 (2000). However, it restricted the honoring of requests by both donor and recipient to meet each other until after the transplantation. On the other side, that program provided limited financial assistance, when requested and needed, for both donors and recipients "to help defray the costs of travel, lodging, and meals."

verely ill found that they were the most expensive to treat; organ allocation criteria affect transplant costs.[272]

Medical tradition generally holds that physicians do no harm; they may only impose risk of harm on a patient if this is likely to produce net medical benefit for that patient. Where the risk of harm to the patient is smaller (as is the case of kidney transplant donor), then the medical benefit to another person may be taken into consideration as an exception to the traditional rule.

Consider the case of a "non-directed donation" brought to a hospital ethics committee on which I served in the late 1990s. A woman in her late thirties approached a local transplant surgeon to inquire about becoming a living donor for renal transplantation anonymously. She wanted to help somebody else less fortunate than herself, and her husband supported her in this desire. When interviewed by the hospital ethics committee, she appeared as a person fully competent to make such a donation. The risk of permanent harm from the operation was small. One doctor asked whether there should be a relationship between the donor and recipient; he said this was the "only issue" for consideration. The transplant surgeon also felt a need to contact UNOS to ascertain how this case fell within their transplant policies. The majority of the committee then agreed to withhold approval of the transplant until a qualified recipient was located, and request a policy input from UNOS.

For several committee members the very nature of such an altruistic offer raised a question in their minds concerning the donor's mental competence, and for some reason a majority of the committee actually insisted that she be examined by a psychiatrist. The woman was examined by a psychiatrist, who found her competent to make the offer (just as some of the other members of the committee had done prior to that examination). Next, UNOS was contacted and the chairman of its Ethics Committee replied that it had not yet heard of such a case before in the United States. He referred to this anonymous kidney donation as an "unused donor service" and stated that it was "not only ethically acceptable, but morally commendable." Accordingly, about one year after the time this altruistic donor first contacted the surgeon, the transplant was actually performed. These end results were not strange in view of the fact that in the same year, Carl Sagan published his final work, *Billions and Billions*. He wrote:

> There are more than 2 million Americans in the National Marrow Donor Program's volunteer registry, all willing to submit to a somewhat uncomfortable marrow extraction to benefit some unrelated perfect stranger. Millions more contribute blood to the American Red Cross and other blood donor institutions for no financial reward, not even a five-dollar bill, to save an unknown life. (p. 265)

Attending physicians themselves normally make the judgment concerning a patient's competency for surgery. It ought to have sufficed in this case because the committee found nothing about this woman to indicate a possible need for a psychiatric evaluation; her altruism should not have automatically called her competency into question. Similarly, as noted above, UNOS had no policy regarding such donors, and there was no need to contact that organization because the donor's autonomy prevailed in such matters.[273]

good v. Good

Any living person who plans to donate a kidney to another person should also be required to undergo a psychiatric examination prior to authorization of the transplant surgery.

v.

Every competent citizen has the right to determine whether he wishes to donate his body part to another. Physicians and surgeons normally determine the competency of their patients without referral to psychiatrists except for those whom they determine to have a need for special expertise.

Twenty-first Century Approach

1. State laws should not require that patients who are willing to donate their organs be referred to a psychiatrist unless the attending physician finds it likely that a lack of competence exists on their part.
2. Unused donor services should normally be considered ethically acceptable regardless of the lack of any direct relationship between the donor and recipient.

---

273. Kulynch, 49 STAN. L. REV. 1249, 1250 (1997); Mantell, *A Modest Proposal to Dress the Emperor: Psychiatric and Psychological Opinion in the Courts*, 4:331 WIDENER J. PUB. L. 60–61 (1994). It has been noted that unscientific testimony has been admitted by many courts on psychiatry "based upon notoriously unreliable clinical judgments that arguably fail to meet the minimal standard for admissibility."

c) Directed Donations by Living Persons

Cases arise where a person wishes to direct the donation of one of his organs only to a recipient who is a relative,[274] a member of his church (e.g., the Catholic Church), or a member of his community, (e.g., Vietnamese, Mexican, etc.). UNOS had actually debated the issue of possible discrimination in the late 1990s and reached no conclusion. Upon making inquiry, I was informed, "*UNOS does not have an allocation policy that regulates directed donations*. When a donor or a donor family directs the donation of anatomical gifts, the UNOS allocation policies become secondary to the directed donation."[275] At the state level, the Uniform Anatomical Gift Act (UAGA) allows the donor to designate the donee.[276]

Due to the existing nationwide shortage of organs donated for transplantation, it is fortunate that UNOS,[277] whose concept of directed donation clearly fulfills individual autonomy and enhances the principles of bioethics, has no policies regulating directed donations.[278]

Any altruistic donor exercising his autonomy as a private individual with respect to the most personal of his possessions (e.g., a kidney, bone marrow, etc.) may well have good reasons for limiting the benefit to a particular recipient or to one of several people falling within a class of designees. If so, that individual's desires should be honored whenever a medically qualified

---

274. 24 Transplantation Proc. 2234 (1992).

275. Letter, The Gans, Regional Administration, UNOS/Keyes, Aug. 4, 1998, with the article *General Principles for Allocating Human Organs and Tissues,* 24:5 Transplantation Proc. 2234 (October 1992). The University of Minnesota decided not to allow a donor to specify "the characteristics of the recipient (e.g., age or race)" to avoid the possibility that the university might "tacitly endorse such views by permitting this type of directed donation." N. Eng. J. Med. 435 (Aug. 10, 2000).

276. It also allows a donor to choose a hospital, physician or surgeon, or any other organization that will use the tissue for transplantation or research.

277. For "undirected donations" allocated by UNOS, UNOS policy states: "In prioritizing for transplantation, organ sharing policy does not discriminate based on political influence, race, gender relation, or financial or social status." Uniform Network for Organ Sharing, Policies & Bylaws, http:/www.unos.org/frame_Default.asp? Category=About (last visited Oct. 20, 1999).

278. *See* Sade, *Cadaveric Organ Donation: Rethinking Donor Motivation*, 159 Archives Internal Med. 438–42 (1999).

recipient within that designation is located. The argument that the recognition of one's autonomy in this manner might possibly result in a discriminatory choice cannot be used to override the need to receive an organ from such an altruistic donor and avoid a consequent dire result.

good v. Good

All discriminatory donations directed only to persons or classes of persons (races, churches, associations, etc.) violate public policy and should be banned.

v.

Such a regulation would both violate the autonomy of the individual donor over his most personal possessions and significantly diminish the already limited supply of organs available for transplant.

Twenty-first Century Approach

The United Network of Organ Sharing (UNOS) should continue to refrain from issuing any policy limiting the autonomy of donors who direct their donations to certain persons and classes of persons.

d) Transplants to Patients with end-stage renal disease (ESRD)

Worldwide at the end of the twentieth century, there were more than 20,000 kidney transplants being performed yearly and more than 150,000 patients alive with a functioning kidney transplant.[279] In the past few decades, the procedure evolved from a heroic one to a more routine clinical practice. A study has shown that kidney transplants from a spouse are slightly more likely to succeed than those from an unrelated cadaver.[280] Minorities represent 50 percent of the more than 40,000 patients seeking kidney transplants annually even though, as noted previously, they are less likely to become organ donors,[281] as is the case with blacks, who make up one-third of the

---

279. N. Eng. J. Med. 1779 (1998).

280. N. Eng. J. Med. 95 (1997); L.A. Times, Aug. 10, 1995, at News 12.

281. A new UNOS policy has increased the number of and percentage of HLA-matched transplants available for blacks and members of other minority groups. N. Eng. J. Med. 1468 (1999).

kidney transplant list.[282] Most of these transplants were for patients undergoing dialysis. It has been reported that for renal transplantation, significant numbers of patients are listed a second or third time because of failure of previous transplants:

> Transplantation of nonrenal organs can be associated with acute renal failure. . . . Hypertension is present in the majority of affected patients. . . . The predominant cause of these clinicopathological abnormalities is the long-term use of calcineurin inhibitors—either cyclosporine or tacrolimus. . . . Severe chronic kidney disease is relatively common after the transplantation of a nonrenal organ and is associated with increased mortality.[283]

End-stage renal disease occurred at a rate of 1.0 to 1.5 percent per year [and] may ultimately translate into a requirement for renal-replacement therapy (dialysis or renal transplantation) in many thousands of patients. The care of patients with ESRD is extremely expensive. For example, although such patients represent only 0.8 percent of the Medicare population, they account for almost 6 percent of Medicare expenditures.[284]

At the beginning of 1997, more than 34,000 patients undergoing dialysis were on the national waiting list for kidney transplants. They cannot cleanse their bodies of the wastes normally excreted by the kidneys in urine and hence they are kept alive by dialysis.[285] The treatment lasts about three hours and is performed three times each week. In 2005, it was noted that for "the

282. NEOPHROLOGY NEWS & ISSUES, Aug. 1997, at 31.

283. JAMA 1184 (1998).

284. Mageet Pascual, *The Growing Problem of Chronic Renal Failure after Transplantation of a Nonrenal Organ*, N. ENG. J. MED. 994 (2003). The National Institute of Diabetes and Kidney Diseases (NIDKD) reported that the rate for new cases of kidney failure in 2003 was 338 per 1 million individuals, about the same as the 340 cases per 1 million seen in 2002. JAMA 2563 (2005); http://www.usrds.org/atlas.htm.

285. During dialysis a patient's blood is conveyed from the arm through plastic tubing to an artificial kidney (dialyses), which removes wastes like salt and urea from the blood. The patient's blood is then returned to the body through a second tube.

62,500 patients awaiting kidneys, the expectation is that only about a quarter of them will receive a transplant within the next year."[286]

It has been demonstrated that transplantation improved longevity in all groups of recipients, including patients who were 60 to 74 years old at the time of transplantation.[287] Although only 8,600 cadaveric kidneys were transplanted that year, by the turn of the millennium over 42,000 patients would be on that list.[288] Of cadaveric kidneys, 99 percent came from brain-dead donors whose hearts were still beating. Use of kidneys from donors whose hearts have stopped could increase the supply by a factor of 2 to 4.5 even though they do not always work as well as those from the brain-dead.[289] A study has shown that transplantation of better-matched cadaveric kidneys could have substantial economic advantages.[290]

Unlike the treatment of other lethal diseases, since 1972 the federal government has paid most of the cost of dialysis for everyone who needs it. Some of these costs are covered by private insurance agreements. As a result, such therapy has become the *most expensive* single kidney or urologic disorder *billed to Medicare*. Its costs have been called "out of control" and inordinately high when compared with Europe and Japan.[291] The 24 percent mortality rate is the highest in the industrialized world. For example, in 1991 dialysis patients received 79 percent of all such Medicare expenditures; transplant patients got the rest.[292] With over 150,000 Americans kept alive by renal dialysis, the cost

---

286. N. Eng. J. Med. (2005).

> The average waiting time for kidney transplants is more than 1,000 days, and if the waiting list continues to grow at its current rate of 20% per year while the number of kidney transplants taking place remains below 10,000 per year, the average wait for a kidney will be 10 years by the year 2010.

Smith, *Organ Allocation at the Crossroads*, http://www.chfpatients.com/tx/txrules.htm.

287. N. Eng. J. Med. 1729 (1999).

288. Cho et al. 338 N. Eng. J. Med. 221–25 (1998).

289. N. Eng. J. Med. 48 (1998). Also, it should be noted that statutes in Florida, Virginia, and the District of Columbia do not allow *in situ* organ preservation while next of kin are being located. *Id.*

290. 341 N. Eng. J. Med. 1440–46 (1999).

291. JAMA 1118 (1996).

292. *Id.* at 1123.

exceeded $5 billion back in 1997.[293] Although Congress had been seeking a balanced budget by 2002, funding could become more limited for this and other diseases. In 2003, it was noted that decisions to withdraw dialysis now precede one to four deaths of patients who have end-stage renal disease; this indicates that the stopping of life-support treatment for patients with end-stage renal disease should be considered when the burdens outweigh its benefits.[294] During that year it was also reported that:

> The number of available cadaveric organs has remained essentially static over the past decade. As a result, the waiting list for renal transplants in the United States has grown to more than 52,000 patients, with average waiting times exceeding 1,000 days.[295]

good v. Good

Costs to the government must be considered a correct factor in providing dialysis for *all* those who need it.

v.

It is unfair to provide these huge amounts of taxpayer dollars for those in need of dialysis alone while so many others have equally critical medical needs, but often receive substantially less funding.

Twenty-first Century Approach

Rules currently used under UNOS standards for kidney and other transplants should also be considered for dialysis patients in order to approach a greater equalization of funding for the treatment of other lethal conditions.

---

293. Caplan, Am I My Brother's Keeper? 102 (1997).

294. Cohen et al., *Practical Considerations in Dialysis Withdrawal: To Have That Option Is a Blessing*, 289 JAMA 2113–19 (2003).

295. Langone & Helderman, *Disparity Between Solid-Organ Supply and Demand*, N. Eng. J. Med. 704 (2003) (citing Sheehy et al., N. Eng. J. Med. 667 (2003)).

Part III

The Right to Die with Dignity

"You said you are a healer, but you have some problems."

"I deal with life and death almost every day. Yet they say to me 'Physician heal thyself!'"

"Why do they ask?"

"You must both do those things that are within the goals of medicine and next look to others for things that are beyond."

"But can't you give me some guidance?"

"There are two goals:

- If your patient is among the few who are *within your tribe*, you do unto him *as you* would have *him* do unto *you*.
- If he is among the many who are *non-tribal*, you do unto him, *as he* would have *you* do onto *him*; it places your healing at different locations.

"Each of them retains their limitations."

"But, those limitations. How do I solve my problems there?"

"They are the next: they come from bioethical things, which often exceed the traditional goals of medicine." (Emphasis added.)

The Author

12

The Right to Dignity in the Dying Process

1. *The Need for End-of-Life-Care Professionals in the Dying Process*
2. *The Refusal to Acknowledge the Right to Die*
3. *Culture, Longevity, and Quality of Life Decisions; Monism and Dualism*
    a. Increasing Longevity
    b. Decreasing Longevity and Quality of Life
    c. Obesity
4. *Attempts to Change from the "Slow Course to Death" to "Varieties of Approaches to Death," Kubler-Ross*
5. *Pain and Its Partial Management*
    a. Pain
    b. Pain Management and Its Limitations; Social Security
6. *Costs Incurred Nearing End-of-Life Healthcare*
    a. Entitlements and Their Limitations
    b. Inequitable Allocations and Deceiving Healthcare Payers v. HMOs
    c. Long-term Home Care and Nursing Homes

Death, so call'd, is a thing which makes men weep:
And yet a third of life is pass'd in sleep.

Lord Byron, *Don Juan*

> *My research into the experience of over 700 widows and widowers across seven years has taught me that there is no norm, no set pattern for healthy* grieving *and recovery. These new findings should help a widow to see that her grieving is personal. (Emphasis added.)*
>
> Liverman, *Doors Close, Doors Open* (1996), p.1

> *Priests and ministers in many faiths bless the sick by anointing them with sacred oils, often touching them on the forehead, hands, or diseased part of the body. Once called "the last rites," Roman Catholics receive "the sacrament of the sick," in which communion, confession, and anointing occur. This sacrament, which is considered to be a healing one for the soul, can be received several times during an illness.*
>
> Lynn & Harrold, *Handbook for Mortals*, p.36

1. The Need for End-of-Life-Care Professionals in the Dying Process

When a patient is dying, he often wonders whether others may grieve for him after he is gone. Like the costs of medicine in the late 1980s and early 1990s, funeral prices were often rising three times faster than the cost of living at the turn of the third millennium.

It has been estimated that by 2020, 53.7 million people living in America will be 65 or over.[2] If the patient were well-known in his community, he might assume that his own obituary was being prepared in a manner similar to the speech that President Nixon had requested (but which was never delivered) for the first astronauts to land on the moon in the event they were unable to return.[3] As

1. "Funeral chains often raise prices soon after acquiring an established independent home, sometimes upping fees more than 100 percent. With a cemetery plot and marker, the typical American funeral now costs $8,000 or more." U.S. News & World Rep., Mar. 23, 1998. Health care costs prior to death are also rising. Jemal et al., JAMA 1255 (2005).

2. U.S. Census Bureau, Statistical Abstract of the United States: 2000.

3. This speech, which was drafted by William Safire, began with the words:

> Fate has ordained that the men who went to the moon to explore in peace will stay on the moon to rest in peace. These brave men, Neil Armstrong and Edwin Aldrin, know that there is no hope for their recovery. But they also know that there is hope for mankind in their sacrifice.

Orange County Reg., July 10, 1999, at News 17.

noted in the above epigraph from the *Handbook for Mortals,* if he were Catholic he might see a priest carrying the Viaticum, the last sacrament. If he were a Buddhist, he may be aware of *The Tibetan Book of Living and Dying,* about which it has been pointed out that "no one can die fearlessly and in complete security until they have truly realized the nature of mind. For only this realization, deepened over years of sustained practice, can keep the mind stable during the molten chaos of the process of death."[4] The hospital chaplain may have participated in clinical pastoral education programs approved by either the Association for Clinical Pastoral Education or the National Association of Catholic Chaplains.[5] The patient will be invited to consider making a living will or durable power of attorney, as provided by the law of his state.[6] Should he be in touch with certain bioethicists, he may wish to undertake discussions of a planned exit or an elective death. He knows that his wife's major fear may be less about her husband's death than whether she will be able to handle the final stages of his dying (or vice versa). This might be accomplished at home with hospice care. But many shy away from terms like "hospice" because they imply that death is imminent; they prefer euphemisms like "comfort care."[7]

Over 2 million persons die in the United States each year. Most of these deaths occur among elderly persons who have one or more disabling conditions. While "the absolute number of deaths and age at death continue to increase in the United States," these trends also "have major implications for health care" in an aging population. For example, although there has been "continuing progress in reducing the age-standardized death rate from heart

4. Lama Surya Das, Awakening the Buddha Within 45 (1997).

5. "The Association of Professional Chaplains (and other national and denominational groups), a respected national professional association, provides a process for board certification of chaplains." Lynn & Harrold, Handbook for Mortals 34 (1999).

6. *See infra* Chapter 13.

7. *See* Lo et al., Annals Internal Med. (1999). This article also sets forth in considerable detail some of the words that physicians should consider using during rather lengthy discussions about palliative care with a dying patient. It has been stated that "palliative care is a critical component of end-of-life therapy where treatment options for cure do not exist and relief from unnecessary suffering is all that can be offered." Oxford Textbook of Palliative Medicine 7 (Doyle et al. eds., 1993).

disease," the number of deaths continues to increase because of population growth and aging.[8] Although participation in traditional and organized religion has decreased following the turn of the twenty-first century, among persons who normally are not atheists (see Chapters 3 and 4), many more people participate in prayer and other religious practices.[9] In 1999, the Institute of Medicine released a report, "To Err Is Human," which at that time claimed that 44,000 people die each year because of medical errors. In 2001, coronary heart disease was reported as being "the most common cause of sudden death. Smoking, high blood pressure, and high levels of cholesterol in food put an individual at high risk."[10] Age-related macular degeneration (AMD) is the leading cause of blindness, which now affects more than 1.75 million aging individuals in the United States (i.e., older than 65 years).[11] In the last year of the twentieth century, the AMA published extensive and detailed procedures for physicians on end-of-life care.[12] For the first time, these went far beyond direct medical treatments. However, when reviewed at a series of meetings

---

8. Jeunal et al., *Trends in the Leading Causes of Death in the United States, 1970–2002*, JAMA 1255 (2005).

9. Lukoff et al., *Cultural Considerations in the Assessment and Treatment of Religious and Spiritual Problems*, Psych. Clinics N. Am. 467 (1995); Lukoff et al., *Towards a More Culturally Sensitive DSM-IV: Psychoreligious and Psychospiritual Problems*, J. Nervous & Mental Disease 673 (1992); Waldfogel & Wolpe, *Using Awareness of Religious Factors to Enhance Interventions in Consultant-Liaison Psychiatry*, Hosp. & Cmty. Psych. 473 (1993).

10. Guinness World Records, 2002.

11. Archives of Ophth. 254 (2004) (studying reports of the Eye Diseases Prevalence Research Group (EDPRG), which predicts that this number will increase to almost 3 million by 2020); *see also* Bressler, *Age-Related Macular Degeneration Is the Leading Cause of Blindness*, JAMA 1900 (2004).

12. Education for Physicians on End-of-Life Care (EPSC) Project, American Medical Association, 1999.

> Capacities that medicine now sometimes lacks for attunement to patients' individuality and sensitivity to emotional or cultural dimensions of care may be provided through a rigorous development of narrative skills. The potential costs of these new narrative practices, including the increased time it may take to meet with new patients and challenges to patient confidentiality, are being recognized and addressed.

Charon, N. Eng. J. Med. (2004); *see also* Charon, *Narrative Medicine, A Model for Empathy, Reflection, Profession, and Trust*, JAMA 1897 (2001).

beginning in 2000 at a large number of hospitals, most attending physicians agreed that (a) although all "physicians should be aware of them," (b) only "few physicians would actually take the time to follow them" (see Section 4). Thus, it appears that persons other than attending physicians must be trained to handle the large number of such procedures. However, the AMA published these as if virtually all of them were to be practiced by the physicians themselves. The AMA publication consists of about 10 explanatory modules. Knowledge of the difference between that which is clear medical work and other more "social" work is needed to treat such patients. However, it is often found to be too much time to be spent by many physicians[13] (see Chapter 13).

Most physicians in ICUs are unprepared in "giving bad news, defining end-of-life care, ensuring comfort and dignity both for the patients we were still trying to cure and for those who were obviously dying." This has "not been part of any curriculum," prior to the third millennium.[14] Making it a significant part of a medical school's curriculum may not significantly change the practice of a busy physician who is currently either ignoring or not allowing his time to be

---

13. For example, module 1, "Gaps in End-of-Life Care," begins by setting forth clear medical analysis, 1–6, though later it explores other issues, including financial pressures, place of the coming death, public expectations, barriers to end-of-life care, and legal issues, *id.* 13. It noted that it was "not intended to make every physician a palliative care expert." *Id.* In the third module, "Whole Patient Assessment," the first five pages deal with medical issues, and the next several pages discuss social assessment, financial, community, and spiritual issues. The importance of the many changes in the end-of-life care area hopefully will become of most importance in the new field for eschats in the third millennium.

14. Nelson, Annals Internal Med. 776 (1999). In their eight-volume series, Storey and Knight provide the "set of attitudes needed to support and palliate the patient-family unit through the complex interactions of physiological, psychological, social, and spiritual symptoms of terminal illness." Storey & Knight, Hospice/Palliative Care Training for Physicians: A Self-study Program (2d ed. 2003). However, Dr. Periyakoil, in reviewing this work, indicated that he was "disappointed to note that the booklet on nonpain symptoms is the shortest in the series, not doing full justice to the wide array of symptoms that may arise at the end of life." Periyakoil, JAMA 2485 (2003). Nevertheless, studies have shown that the presence of specially trained critical-care doctors in the intensive care unit (ICU) can reduce death rates by 30%. Orange County Reg., Nov. 17, 2004, at Bus. 3.

spent to the full extent demanded by these prescribed procedures.[15] It has been noted that most biogerontologists, those who study the biology of aging, "believe that our rapidly expanding scientific knowledge holds the promise that means may eventually be discovered to slow the rate of aging."[16] A different profession could be created to do most of this "end-of-life" care in hospitals. They might be called "eschats" because eschatology (from the Greek *eschata*) is the "study of last things." Some of these would include aspects of different religions (beyond most chaplains) plus many others that the AMA has prescribed. This work is not what some hospitals call a "bereavement coordinator," a person only trained in a standardized process for dealing with a patient's family *after* his death.[17] That is, attention must also be given to nonpatients; this is because "many physicians are uncertain about how to identify bereaved individuals who

---

15. In 2004, a 71-year-old Catholic patient named Anna, who was dying following six years of cancer, asked her physician "to pray to God." He immediately recalled that "prayer was far from the forefront of my mind," that "none of the training I received in medical school, residency, fellowship, or practice had taught me how to reply," and that "I consider my beliefs and prayers a private matter." Finally, he simply asked: "What is the prayer you want?" "Pray for God to give my doctors wisdom," Anna said. To that, I silently echoed, "Amen."

Groopman, N. Eng. J. Med. 1176–78 (2004).

16. Butler et al., *Is There an Anti-aging Medicine?*, 57A J. Gerontology: Biological Sci., B333–38 (2002). When 63-year-old atheist scientist and Nobel laureate Richard Feyman was dying of cancer, another physicist who met with him commented:

> The thought of actually talking to a dying person made me uncomfortable. Strangely, I would find that being one did not seem to have the same effect on him. I could see right away that there was still an energy about him, a gleam in his eye. He may have had terminal cancer, but his spirit still zig-zagged around the universe.

Mlodinow, Feyman's Rainbow 34 (2003).

17. Life Magazine, Mar. 2000, at 84 (describing process following the death of a newborn at Lutheran General Hospital, Park Ridge, Illinois). Eschatology (from the Greek uttermost, or last) deals with death (as hospices do) and with the "future state," and the like, in theology. "The world today is awash with competing eschatologies and, like older people, the students knew it, even though they were not familiar with the technical terms." Cox, When Jesus Came to Harvard 187 (2004).

need treatment.[18] People with such grief can be reliably identified by administering the Inventory of Complicated Grief (ICG) after the death of a loved one.[19] Such people would not be practicing medicine because they would largely be doing the "ultramedical." For example, New Jersey Health Decisions developed the Disability Ethics Network (DEN) using the model of hospital ethics committees to assemble an interdisciplinary team (social workers, physicians, nurses, advocates, and others) to analyze the medical, ethical, and legal issues relating to healthcare decisions for individual patients who are without decision-making capacity.[20]

That such workers are much needed prior to death has been demonstrated by a 1998 study of how the quality of physician communication that leads to the completion of written advance directives may influence their usefulness. The study showed that the majority of the physicians discussed scenarios involving uncertainty, often using vague language: "Meeting the needs of patients will require that physicians employ skills that are not traditionally taught in medical schools," even though "communicating with patients is a core skill of palliative medicine."[21] Patients' values were rarely explored in detail. It was

---

18. Prigerson & Jacobs, *Caring for Bereaved Patients: "All the Doctors Just Suddenly Go,"* JAMA 1369–76 (2001); Shear et al., *Treatment of Complicated Grief*, JAMA 2601 (2005)). Of the approximately 2.5 million people who die annually in the United States, it is estimated that over a million of them are expected to develop complicated grief. National Center for Health Statistics, Births, Marriages, Divorces, and Deaths, http://www.cdc.gov/nchs/data/nvsr52_22.pdf (last visited May 3, 2005).

19. Horowitz, *Diagnostic Criteria for Complicated Grief Disorder*, Am. J. Psychiatry, 904–10 (1997); Prigerson et al., *Consensus Criteria for Traumatic Grief: A Rationale and Preliminary Empirical Test*, Brit. J. Psychiatry 67–73 (1999); Prigerson et al., *Inventory of Complicated Grief: A Scale to Measure Maladaptive Symptoms of Loss*, Psychiatry Res. 65–79 (1995).

20. State Initiatives in End-of-Life Care 7 (2005). "DEN also offers training to social workers, medical care providers, and workers in the State Bureau of Guardianship Services. DEN has worked with the state to develop regulations that will provide for hospice and palliative care for people under guardianship who are expected to live less than a year."; *see* N.J. Admin. Code § 10:48B.

21. Morrison & Meier, *Palliative Care*, N. Eng. J. Med. 25823 (2004); Steinhauser et al., *Factors Considered Important at the End of Life by Patients, Family, Physicians, and Other Care Providers*, JAMA 2476 (2000); Sullivan et al., *The Status of Medical Education in End-of-Life Care: A National Report*, J. Gen. Internal Med. 685 (2003).

concluded that they did not address the topic "in a way that would be of substantial use in future decision making."[22] Many physicians and others have continued with their attempts to diminish the current law granting patients autonomy (and paramount legal authority) concerning commencement of treatment. One of them emphasized rules of courtesy over patient autonomy.[23] However, a prominent bioethicist has pointed out that this plea "for privileging physician etiquette over patient rights is not persuasive";[24] there may be a "gap between physicians' values and those of patients or their families." There is an ever-increasing significance of "the lack of reimbursement for time-consuming conversations with patients and families."[25] More than two million people have dementia in the United States and most of them are cared for at home by family members. A recent study found that:

> The National Institute on Aging contained a Strategic Plan for 2001-2005, declaring that "[t]he ultimate goal of this effort is to develop interventions to reduce or delay age-related degenerative processes in humans,"[26] which is significantly beyond the treatment of disease. In 2005 it was shown that "more Americans die in hospitals than anywhere

22. 129 Annals Internal Med. 441–50 (1998). "Although patients' values arose often with these discussions, physicians were unlikely to elicit or explore them." *Id.* at 446.

23. Schneider, The Practice of Autonomy: Patients, Doctors, and Medical Decisions (1998).

24. Annas, JAMA 931 (2000).

25. Gould et al., Conflicts Regarding Decisions to Limit Treatment. There are more than two million people with dementia in the United States, and most of them are cared for at home by family members. A recent study found that "end-of-life care for patients with dementia was extremely demanding of family caregivers. Intervention and support services were needed most before the patient's death. When death was preceded by a protracted and stressful period of caregiving, caregivers reported considerable relief at the death itself." Schulz et al., *End-of-Life Care and the Effects of Bereavement on Family Caregivers of Persons with Dementia*, N. Eng. J. Med. 1936 (2003).

26. Nat'l Inst. on Aging, Action Plan for Aging Research: Strategic Plan for Fiscal Years 2001–2005 (2001). Nurse understaffing has been "ranked by the public and physicians as one of the greatest threats to patient safety in U.S. hospitals." In 2003, in hospitals with higher proportions of nurses educated at the baccalaureate level or higher, surgical patients experienced lower mortality and failure-to-rescue rates. JAMA 1617 (2003).

else."[27] But, in the same year a cancer patient called his physicians "MIA doctors" (missing in action physicians).[28]

As stated in the epigraph to this chapter, these are most often personal rather than "tribal" matters. One person's reactions may follow the sequences of Elisabeth Kubler-Ross's famous stages of dying, which include "denial, anger, bargaining, depression, and acceptance."[29] These are quite similar to what have been called the "basic response stages to killing in combat"; namely, "concern about killing, the actual kill, exhilaration, remorse, rationalization and acceptance."[30] The difference is that they are being applied to himself rather than to another person. As in combat, "some individuals may skip certain stages, blend them, or pass through them so fleetingly that they did not even acknowledge their presence."[31]

---

27. KAUFMAN, AND A TIME TO DIE 25 (2005) (citing FLORY ET AL., PLACES OF DEATH: U.S. TRENDS SINCE 1980, 194–200). Walker and Jamozik evaluated one of these initiatives—the Keep Well At Home (KWAH) Project, which is a U.K. government policy mandating the introduction of "intermediate care services" to reduce emergency admissions to hospitals from the population aged 75 years or more. However, they concluded that "the KWAH Project has been ineffective in reducing emergency admissions among the elderly." *Effectiveness of Screening for Risk of Medical Emergencies in the Elderly*, 34 AGE & AGEING, 238–42 (2005).

28. ORANGE COUNTY REG., June 18, 2005, at News 23. She stated that "it feels like a desertion. . . . It's very painful. They feel abandoned." In the April 2005 issue of the *Annals of Internal Medicine*, Dr. Anthony Back of the Seattle Cancer Care Alliance wrote about the "MIA doctors" who go missing in action when their patients are about to die. He suggested a seven-step guide that "starts with choosing an appropriate time and place, and ends with advising them to reflect on their work with the patient." *Id.*

29. LIEBERMAN, DOORS CLOSE, DOORS OPEN (1996).

30. *Id.* Although the Department of Justice had estimated that more than 500,000 elderly people are maltreated annually, the actual number may be much higher, because few cases are reported or investigated. Axmaker, *Neglect and Abuse of the Elderly*, HEALTH PLUS 5.

31. ERICKSON, THE LIFE CYCLE COMPLETED 114 (1982).

> The leading causes of death in 2000 were tobacco (435,000 deaths; 18.1% of total U.S. deaths), poor diet and physical inactivity (400,000 deaths; 16.6%), and alcohol consumption (85,000 deaths; 3.5%). Other actual causes of death were microbial agents (75,000), toxic

Having spoken of the similarities between grieving over a recently deceased and over killing in combat, it is little wonder that they also exist in minds of living people who are contemplating death. After all, it has been noted, "Lacking a culturally viable ideal of old age, our civilization does not really harbor a concept of the whole of life." The modern writer Harold Brodkey wrote, "I can't remember ever wishing life and death had a perceptible, known, over-all meaning." The logician and atheist Bertrand Russell had a more positive outlook: "I should scorn to shrivel with terror at the thought of annihilation. Happiness is none the less true happiness because it must come to an end, nor do thought and love lose their value because they are not everlasting."[32]

The varieties of patients' desires during the process of dying are virtually without end. Former Harvard professor Timothy Leary, who pushed the use of the drug LSD during the 1960s, was reported to have reinvented the dying process years later. He recorded each stage of his cancerous decline with seeming delight. One evening may have been somewhat unique.[33] His artist friend Aileen Getty wrote:

> My favorite memory took place five days before Tim soared elsewhere . . . I changed in his bathroom. I slipped on, with the same assurance as a broad from the forties . . . a sheer gown, put a light shade of auburn on my lips and perfume . . . I stood up straight, shoulders lifting the air above, and walked slowly, wantonly down toward Timmy's bedroom. Timmy was elated [and] held me tightly, wandering within me. We kissed, both shaking, and Timmy quickly kissed me a couple more times as he ran his cold and trembling hands on my stomach and breasts, speaking of my beauty and his love for

---

> agents (55,000), motor vehicle crashes (43,000), incidents involving firearms (29,000), sexual behaviors (20,000), and illicit use of drugs (17,000).

Mokdad et al., *Actual Causes of Death in the United States, 2000*, JAMA 1238 (2004); *see also* Swagerty, Jr., *Elder Mistreatment*, AM. FAMILY PHYSICIAN 2804 (1999).

32. MOREHEAD, BERTRAND RUSSELL: A LIFE 551 (1993).

33. T. LEARY GRABOI, DESIGN FOR DEATH 202–03 (1997). In this connection it is of interest to note that among those who have exceeded one century before death in 2003 were the entertainer Bob Hope (100), Madame Chiang Kai-shek, widow of the Nationalist Chinese president of China (and later Taiwan) (105), and a Japanese woman believed to have been the world's oldest person (116).

> me. I felt him waning, tucked him in, switched off the light beside him and crawled in beside him. I listened to him breathe all night, frightened that he would stop.[34]

Leary had previously noted "most physicians are more in denial about death than their terminal patients." A somewhat less exciting but still "happy" death was described in the final thought of the writer Brodkey:

> The world still seems far away. And I hear each moment whisper as it slides along. And yet I am happy—even overexcited, quite foolish. But happy. It seems very strange to think one could enjoy one's death. Ellen has begun to laugh at this phenomenon. We know we are absurd, but what can we do? We are happy.
>
> Me, my literary reputation is mostly abroad, but I am anchored here in New York. I can't think of any other place I'd rather die than here. I would like to do it in bed looking out my window. The exasperation, discomfort, sheer physical and mental danger here are more interesting to me than the comfort anywhere else. I lie nested at the window, from which I can see midtown and its changing parade of towers and lights; birds flying past cast shadows on me, my face, my chest.
>
> I can't change the past, and I don't think I would. I don't expect to be understood. I like what I've written, the stories, and two novels. If I had to give up what I've written in order to be clear of this disease, I wouldn't do it.[35]

The satisfaction expressed by Brodkey in the last paragraph about not changing the past even if he could and about his writing being paramount with respect to his final disease may not fit everyone; they might only apply to similar persons. At the dawn of the third millennium, a medical study noted that a quality end-of-life care is increasingly recognized as an ethical obligation. It has been found to include "avoiding pain and inappropriate prolongation of dying loved ones."[36] Another writer wrote, "Each time I place myself in the hands of my well-intentioned doctors, nothing is more chilling than the possibility that the powers of modern medicine are out of touch with our high-

34. *Id.* at 213.
35. Brodkey, This Wild Darkness (1996).
36. 281 JAMA (1999).

est human values."[37] In 2003, the National Cancer Policy Board of the Institute of Medicine recommended that researchers study the care received by patients prior to their death in order to improve the quality of care that is provided to dying patients.[38] They found that "there is no substitute for actually observing care provided to patients who have been identified as dying" provided "the features that defined the patient as 'dying' are clearly specified from the outset."[39] Further, people considered alternative methods of dying; some will wish to make different choices. It is to this group that the remainder of this chapter is dedicated; as stated in a *Guide to Pre-Death Dreams and Visions*: "We do, however, believe an exclusively scientific *or* exclusively religious explanation of pre-death dreams is inadequate."[40]

---

37. Rozak, America the West 199 (1978). Palliative care aims to improve the quality of life when treatment aimed at curing illness and prolonging life is no longer a realistic objective. Bennot & Plum, Textbook of Their Families Medicine (1996); Mount, Care of Dying Patients (1996).

38. Nat'l Cancer Pol'y Bd., Describing Death in America: What We Need to Know (2003); Earle et al., *Identifying Potential Indicators of the Quality of End-of-Life Cancer Care from Administrative Data*, J. Clinical Oncology 1133–38 (2003).

39. Bach et al., *Resurrecting Treatment Histories of Dead Patients*, JAMA 2765–69 (2004).

40. Bulkley & Bulkley, Dreaming Beyond Death 4 (2005).

> The notion that God is the only one who has the right to terminate a life is nonsense; if that were the case we should never be able to perform an abortion, fight a war, execute a criminal, or even take a life in self-defense. If we extend the argument further and assume that God has the ultimate say in whether we live or die, would we not be equally wrong to circumvent his will by prolonging and preserving life as well? To remain logically consistent, we should not attempt to cure a child of leukemia or perform open-heart surgery, lest we be taking events out of his hands.

Danelek, Mystery of Reincarnation 262 (2005).

good v. Good

Every physician can handle all significant end-of-life processes by communication with dying patients, whether or not all their end-of-life care treatment is in accordance with the title "Educate Physicians on End-of-Life Care" (EPEC) that was listed therein by the American Medical Association.

v.

Because most physicians neither receive sufficient training in end-of-life matters nor take enough time to both train themselves (on all the matters listed by the AMA) and spend enough time with patients on such matters, other people should be trained and assigned to do so.

Twenty-first Century Approach

Until recently, most physicians (a) have not received adequate training in handling the ultra-medical end-of-life processes for their patients, or (b) cannot normally find the time necessary to communicate with their terminal patients in a manner "that would be of substantial use in future decision making." A new group of people (working as eschats) are needed who would be trained to both understand different religions and to handle ultra-medical end-of-life processes for terminal patients.

We don't know yet about life; how can we know about death?

Confucius (c. 551-479 B.C.)

Two generations ago money was not a subject for polite conversation, then sex became something that was generally discussable, and now it is the turn of death to lose its taboo status.

J. Med. Ethics, February 1996, Ltr., p.55

An old man once cut some wood and was walking
a long way carrying it. As he grew weary,
he put down his load and called Death to come.
When Death appeared and asked why
he had called for him, the old man said,
"To get you to take up my burden."

Aesop

Individuals regard man—every man—as an independent, sovereign entity who possesses an inalienable right to his own life, a right derived from his nature as a rational being.

Ayn Rand (1963)

Don't let yourself forget how many doctors have died after furrowing their brows over how many deathbeds. How many astrologers, after pompous forecasts about others' ends. How many philosophers, after endless disquisitions on death and immortality.

Marcus Aurelius, BV 4, 48

2. The Refusal to Acknowledge the Right to Die

In 1976 the New Jersey Supreme Court issued its landmark (7-0) ruling in the case of Karen Ann Quinlan, who was in a permanent vegetative state following an automobile accident.[41] The ruling that the constitutional right of privacy was broad enough to encompass the decision of a person or his surrogate to forgo certain life-sustaining medical treatment has been called the beginning of the right-to-die movement.[42] In 2002 it was noted that suicide ac-

---

41. *In re* Quinlan, 355 A.2d 647 (N.J. 1976). The Ad Hoc Comm. of the Harvard Med. Sch. to Examine the Definition of Brain Death, *A Definition of Irreversible Coma*, 205 JAMA 337 (1968), was followed by the Unif. Determination of Death Act, 12 U.L.A. 65 (1978). The Uniform Determination of Death Act (UDDA) provided a model for legislating the brain-cessation criterion as legal death, which was endorsed by the ABA, the AMA, and 44 states (including the District of Columbia). The UDDA states: "An individual who has sustained either (1) irreversible cessation of circulatory and respiratory functions, or (2) irreversible cessation of all functions of the entire brain, including the brain stem, is dead." Unif. Determination of Death Act § 1, 12A U.L.A. 593 (1980). It may be argued that the insertion of the term "brain stem" means that this definition of death does not apply to persistent or permanent vegetative state (PPVS).

42. *Id.* This movement may involve the disconnection of respirators, high dosages of pain relief medication, and other matters discussed below. The New Jersey Supreme Court upheld a father's petition that his unconscious daughter, if competent, would have the constitutional right to resist life-sustaining medical intervention.

counts for 1.5 percent of all annual deaths and became the eighth leading cause of death in the United States.[43]

In 1990, for the first time, a majority of the U.S. Supreme Court stated that a citizen has a right to die by withdrawal of medical treatment.[44] In that case Justice Scalia only concurred on the question of the lack of a dying patient's wishes and Missouri's law requiring the addition of "clear and convincing" evidence thereof.[45] Later, such evidence was produced, treatment was withdrawn, and the patient died.

However, in 1996 while two assisted suicide cases were before the Supreme Court, Justice Scalia publicly stated his religious views before a Catholic School of Philosophy: "It is absolutely plain that there is no right to die."[46] Some ethicists criticized him for discussing this issue. Philosophers and a number of theologians also had trouble with Scalia's view because as the philosopher David Hume pointed out a few centuries ago, medical efforts to postpone death would also appear to violate the prerogative of his deity. Man's increasing prerogatives have often prevailed in keeping with the more secular views of that Court's majority. In 1998, the scientist Lynn Margolis discussed cultural and religious restraints, calling them "trained incapacities," "thought collectives," and "social constructions of reality."[47] In 2002, one physician also pointed out:

> Medical anthropologists have long noted that health and illness are culturally and personally determined constructs. Medical decisions vary in the degree to which they are affected by such constructs . . . the most difficult choices are both medically complex and invested with the personhood of the patient, as in . . . most end-of-life decisions.[48]

---

43. Falsetti, *Fluoxetine-Induced Suicidal Ideation: An Examination of the Medical Literature, Case Law, and the Legal Liability of Drug Manufacturers*, 57 FOOD & DRUG L.J., 273 (2002).

44. *See* Cruzan v. Mo. Dep't of Health, 497 U.S. 261 (1990).

45. *Id.* at 293.

46. ORANGE COUNTY REG., Oct. 29, 1996, at News 141; DALLAS MORNING NEWS, May 29, 1996, at 10J. However, according to *Genesis* 3:19, God told Adam, "In the sweat of thy face shalt thou eat bread, till thou return unto the ground . . . for dust thou art, and unto dust thou shalt return."

47. MARGOLIS, SYMBIOTIC PLANET: A NEW VIEW OF EVOLUTION 2 (1998). She added, "These ideas are rejected as obsolete nonsense by the scientific worldview." *Id.*

48. 113 AM. J. MED. 74 (2002).

In this connection there has been a controversy as to whether certain antidepressants can increase suicidal tendencies in vulnerable individuals. A complaint was filed against a "drug manufacturer alleging that such information about safety and efficacy had been concealed. On August 26, 2004, the parties entered into a settlement whereby the manufacturer agreed to pay $2.5 million in damages and publicly disclose information on *all* clinical studies in the future."[49]

Humpty Dumpty sat on a wall.
Humpty Dumpty had a great fall.
And all the King's horses,
And all the King's men
Couldn't put Humpty together again.

Humpty Dumpty is the cosmic egg, the wall, the edge between transcendence and existence. As nothing breaks up into the world of things, the movement toward entropy becomes irreversible. Humpty Dumpty is the immortal soul before its fall into time, and neither God's animals nor His angels can put him back into the world beyond time. The Human condition is the fallen condition of time and fragmentation.

William Irwin Thompson, American philosopher

Secrets of evolution are death and time—the deaths of enormous numbers of life-forms that were imperfectly adapted to the environment; and time for a long succession of small mutations that were by accident adaptive, time for the slow accumulation of patterns of favorable mutations.

Carl Sagan, *Cosmos* (1985), p. 20

---

49. The People of the State of New York v. GlaxoSmithKline; *see* Falit, J.L. Med. & Ethics 174–78 (2005).

3. Culture, Longevity, and Quality of Life Decisions; Monism and Dualism

In the epigraph, Thompson did not just refer to the Second Law of Thermodynamics—which scientist-novelist C.P. Snow said should be known by all people who have ever heard of Shakespeare—i.e., irreversible "movement toward entropy."[50] He also stated, "The nursery rhyme is a memory of soul, a piece of an old cosmology from a lost culture lingering on in the rational world of science as a trivial piece of children's verse." The elderly are not children, and they may exist in either a "lost culture" or in a "rational world of science." As noted in Chapter 5, many, if not most, people appear to remain somewhat "schizophrenic" since, in imagination at least, they can enjoy both worlds. They wish to both (a) keep their life as long as there is any personal quality left in it, and (b) prepare for the time when the loss of any personal quality moves more closely to the irreversible loss of their energy (entropy). It was in this manner that the epigraph of Carl Sagan brought death into the process of evolution.

Over one and a half centuries ago, the 27-year-old Harvard graduate Henry Thoreau not only rebelled against many of society's questionable dictates, but also wrote about them:

> I went into the woods because I wished to live deliberately, to front only the essential facts of life, and see if I could not learn what it had to teach, and not, when I came to die, discover that I had not lived. I did not wish to live what was not life, living is so dear; nor did I wish to practice resignation, unless it was quite necessary. I wanted to live deep and suck out all the marrow of life.[51]

Henry David Thoreau's entry into the woods, building a small shack, and 2½-year stay next to Walden Pond was done in order to think and write about what was and was not life. Most people leading busy lives do not (or argue that they cannot) do much of this prior to either (a) having a premature entry into the dying process or at least (b) retiring from an income-producing life to become a senior citizen. When dying of tuberculosis at the age of 44, Thoreau was asked, "Have you made your peace with God?" His reply, "I am not aware

50. JAMA 1872 (1996). Entropy concerns the loss of energy.
51. *Id.*

that we ever quarreled," could probably fit the condition of many more people at the beginning of the third millennium than in Thoreau's time.

In a large study of longevity and belief published back in the 1970s, researchers claimed to have taken precautions to study the effect of religious belief and people's faith in lieu of their lifestyle habits, as well as nonreligious people with equally healthy lifestyles. Their weekly churchgoers were found to have "died at the age of 83 on average while those who never went died at age 75."[52] In the early 1990s, Harvard began holding annual conferences called "Spirituality and Healing." By the turn of the third millennium, nearly half the medical schools (126 of them) began offering such conferences (up from only three in 1994).[53]

If people wait until retirement to "live deliberately," most of them will have much more time to do so in the twenty-first century because of a general increase in longevity. There is a limited amount of time "to learn only the essential facts of life." As Dr. Joseph Connors and Peter Drucker respectively stated: "You can *either* live for the rest of your life, *or* die for the rest of your life."

Most of the clients of eschats will have some belief in a religion; however, a smaller but growing number may not retain it with respect to their souls for Heaven, Hell, or a temporary time in Purgatory. This may be largely due to their having become what a modern scientific writer called "monists" who believe that there is a oneness of brain and self, rather than "dualists," who "see the brain and at least some aspect of the mind as independent entities detached at death. This aspect is sometimes called the soul, spirit,"[54] or the Cartesian dualism that mind and body are distinct (from René Descartes). Although cell death is crucial for all sizes

52. Strengthen your Immune System 215 (2001).

53. *Id.* Aging effects, specific subregions: "As we age, nerve cells in some regions of our brain become less numerous or efficient, which we can visualize with PET as a negative correlation: the older we get, the lower the flow in some brain regions." Andreasen, Brave New World 153 (2001).

54. Thus, with all the discoveries about the brain flooding the technical journals, it is clear we cannot possess a soul. After death, there can be no reincarnation of self because the self is forever tethered to the body it was developed within after birth; it cannot go somewhere else and wait to enter a new body some time in the future.

Volk, What Is Death? 41–42 (2002). Both terms are also set forth in McFadden, Quantum Evolution 284 (2002).

of animals,[55] the brain's cells turn over hardly at all;[56] "this stability is probably crucial for the feeling of continuity we have as individuals, so intricately connected are the cells to each other in vast nerve networks."[57] As Tyler Volk remarked:

> For those of the dualist persuasion, gratitude is welcomed not only for its earthly benefits but because it can induce a state of thanks directed to the higher powers on the other plane of the afterlife. But gratitude is especially important to monists, like me, who are in more dire need in this life for a way to counteract the fear of death as final annihilation. In the case of monists, gratitude might be vital for well-being and even sanity in the face of mortality. As gratitude spreads from consciousness into the unconscious, and as it allays the fear of death, it changes the totality of the self to a state of desiring even more gratitude.[58]

Of course, no dualist member of a hospital ethics committee needs to acknowledge any commitment toward the monist viewpoint; however, he must be able to see the difference in the gratitude of such a terminal patient and communicate with a monist in an appropriate manner.[59]

So live, that when thy summons comes to join
The innumerable caravan, which moves
To that mysterious realm, where each shall take
His chamber in the silent halls of death,
Thou go not, like the quarry-slave at night,
Scourged to his dungeon, but, sustained and soothed

---

55. *Supra* note 54, at 183. "Every single second in the adult human body, one hundred thousand cells die, and approximately one hundred thousand cells are newborn each second. From these numbers we deduce that on average the body's cells turn over about once a year." David L. Vaux & Stanley J. Korsmeyer, *Cell Death in Development,* 96 CELL 245–54 (1999).

56. LeGrand, *An Adaptationist View of Apoptosis,* Q. REV. BIOLOGY 142.

57. VOLK, *supra* note 54, at 185.

58. *Id.* at 71.

59. CAMPBELL, MYTHS TO LIVE BY 226 (1971).

By an unfaltering trust, approach thy grave
Like one who wraps the drapery of his couch
About him, and lies down to pleasant dreams.

William Cullen Bryant

If any man is favored with long life, it is God that has lengthened his days.

Incrase Mather, Puritan clergyman, 1716

A free man thinks of all of death, and his wisdom is a meditation on life, not on death.

Spinoza, *Ethics*

Now that we had added "years to life," it was time to think about how we might add "life to years."

President John F. Kennedy (1961)

a) Increasing Longevity

Lars Tornstam coined the term "gero-transcendence" to describe a shift in old-age metaperspective "from a materialistic and rational vision to a more cosmic and transcendent one, normally followed by an increase in life satisfaction."[60] A child born in the United States in the third millennium can expect to live about 80 years (as compared with only 45 years of age a century ago).[61] In

60. Tornstam, *Gero-Transcendence: A Theoretical and Empirical Exploration*, *in* AGING AND THE RELIGIOUS DIMENSION 203 (Eugene Thomas & Susan Eisenhandler eds., 1994). An AARP report showed that "four of five women are widows by age 75." Nevertheless, "for those 75 and older with partners, about one in four have sex at least weekly," although about "one in four male respondents acknowledged being either 'completely' or 'moderately impotent.'" ORANGE COUNTY REG., Aug. 4, 1999, at News 5. Apparently parenthood provides longevity in all primate males, including chimpanzees and humans who actively participate in childrearing. AARP BULL., Apr. 1996, at 1.

61. NAT'L CTR. FOR HEALTH STATISTICS, Gov't Print. Off. 3000.7, DHHS No. PhS 2000-1232-.1. The International Longevity Center in New York predicted one million by 2050. ORANGE COUNTY REG., Oct. 31, 2001, at News 17. Much longevity was noted for prisoners sentenced to life without parole. ORANGE COUNTY REG., Oct. 7, 2001, at News 38. A person's peak exercise capacity as measured on a treadmill predicts a lifetime, according to Veterans Affairs Palo Alto Healthcare System, Stanford University. ORANGE COUNTY REG., Feb. 14, 2002, at News.

2003 it was noted that "currently 35 million persons aged 65 years or older live in the United States, accounting for nearly 13 percent of the population. These older adults are the fastest-growing segment of the population.[62] A recent analysis also showed "not only that persons in good health at 70 years of age can expect to live longer and to have more years of good health than those in poor health at age 70, but also that their total expected medical care expenses appear to be no greater than those for less healthy persons, even though healthier persons live longer."[63] Since 1990, most medical schools established identifiable academic geriatric medicine programs; they need resources to train faculty for roles as teachers and researchers to develop their size and scope equivalent to other academic disciplines.[64] During the entire course of human existence, the chance of surviving from 50 to 80 years has increased 21-fold, and the chance of surviving from 80 to 100 years has jumped 3,000-fold.[65] Yet in a 1999 sur-

---

62. JAMA 1659 (2003). "By 2030 it is projected that 70 million persons, or 1 in every 5, will be 65 years old or older." In part this is due to certain operations. For example, in December, 2004, the Saddleback Memorial Medical Center in California informed me about a woman who "was recovering from life-saving triple bypass surgery. Today, at age 86, she is feeling fine and looking forward to celebrating with her family."

63. N. ENG. J. MED. 1048 (2003).

> A person with no functional limitation at 70 years of age had a life expectancy of 14.3 years and expected cumulative health care expenditures of about $136,000 (in 1998 dollars); a person with a limitation in at least one activity of daily living had a life expectancy of 11.6 years and expected cumulative expenditures of about $145,000.

Id. at 1054.

64. JAMA 2313 (2000).

65. JAMA 1872 (1996). In spite of being the richest citizens on the globe, Americans do not enjoy the highest longevity in the world. Across 28 countries belonging to the Organization for Economic Cooperation and Development (OECD), American men ranked 22nd out of 28 OECD countries for life expectancy in 1996, while American women ranked 19th. KAWACHI & KENNEDY, THE HEALTH OF NATIONS 45–46 (2002). On Dec. 9, 2005, the National Center for Health Statistics reported that life expectancy in the United States hit a high of 77.6 years and that deaths from heart disease, cancer, and stroke continued to drop. ORANGE COUNTY REG., Dec. 9, 2005, at News 24.

vey, 63 percent of the responders stated that they do not wish to live to 100.[66] It has been noted, "The biblical prophets who expired satiated with life provided good examples of this."[67] (If in a terminal condition, this might cause them to seek approaches discussed in Chapter 14.) A 1998 study projected that American women who turned 80 in 1987 would live 9.1 more years while men were expected to live 7 more years.[68] A report by the United Nations (UN) indicated "the people aged 60 or older will increase from 1 in 10 now to 1 in 5 by 2050. In some developed countries that proportion will increase from 1 in 5 to 1 in 2 in 2050. The population 80 years or older is projected to increase from 11 percent of those older than 60 years now to 19 percent by 2050, and the number of centenarians is expected to increase 15-fold to 2.2 million."[69]

However, the gender gap narrowed in the late 1990s because "women are posting stubbornly high rates of such *smoking-related illnesses* as heart disease and cancer."[70] Nevertheless, as of the end of the twentieth century, for every 100 men 65 years of age and older, there were about 150 women. By the age of 85, this ratio became about 100 to 250.[71] By comparison, Russian

66. CLARK, A MEANS TO AN END 198 (1999); ORANGE COUNTY REG., May 26, 1999, at News 18. Yet there were 70,000 of them in 1999, half of whom were in nursing homes. ORANGE COUNTY REG., Jan.15, 2000, at News 24.

67. HEYNICK, JEWS AND MEDICINE 327 (2002).

> Studies about general attitudes toward old age show that younger respondents have more negative attitudes than do older respondents, and gerontological discourses also hypothesize a gendered ageism, with especially negative attitudes toward elderly women. The empirical study of embodied aging among 1,250 Swedes aged 20-85 years contradicts these hypotheses. The results show rather positive attitudes toward embodied old age.

56 J. INT'L AGING & HUMAN DEV. 133–53 (2003).

68. N. ENG. J. MED. 1232 (1993).

69. The rate of aging of the population is greatest in developing nations, a growing challenge for nations with few healthcare resources. By 2050, the ratio of people 65 or older to those ages 16 to 64 will double in developed nations and triple in developing nations. Winker, JAMA 1320 (2002).

70. *Gov't Ann. Study of Dept. of Health and Human Services*, L.A. TIMES, July 30, 1998, at A17.

71. N. ENG. J. MED. 794 (1996). "The Centers for Disease Control and Prevention said life expectancy increased by two-tenths of a year from 2000. Americans reached an all-time high of 77.2 years in 2001. For men it rose to 74.4 years in 2001. For women, it went to 79.8 years." ORANGE COUNTY REG., Mar. 15, 2003, at News 14.

life expectancy has fallen sharply in the 1990s to 68.8 years for men and 74.4 years for women.[72] In 2003, a Danish study revealed that earlier detection and wide use of hormone treatment have driven *death rates from prostate cancer* dramatically lower over the past 10 years in North America and Western Europe.[73] Nevertheless, prostate cancer strikes 220,000 U.S. men annually and kills almost 29,000. In 2002 more than 1,285,000 Americans had invasive cancer and at least 555,000 will die.[74] More than 75 percent of the decline in Russia was due to increased mortality rates for ages 25 to 64 years. These changes were due to "economic and social instability, high rates of tobacco and alcohol consumption, poor nutrition, depression, and deterioration of the healthcare system." In 1994, "a Russian man aged 20 would have only a one in two chance of surviving to 60 years, compared with a nine in ten chance for men born in the United States or Britain. In 1994, life expectancy at birth for men was 13 years less than for women." However, studies in Iceland appear to show that longevity is inherited.[75] China in 2020 will be the location of one quarter of the world's elderly due in part to a birth-control program, which has reduced overall population growth. Since 1949, "China has evolved from a high-birthrate and high-mortality country to a low-birthrate, low-mortality country,"[76] which is a need of most other countries as noted in Chapter 7.

---

72. JAMA 7901 (1998).

73. "Death rates from prostate cancer have dropped by one-third in North America and by 20 percent in Europe since 1990 among men ages 65 to 74." ORANGE COUNTY REG., Sept. 23, 2003, at News 11.

74. SIMON, HARVARD MEDICAL SCHOOL GUIDE 61 (2002); ORANGE COUNTY REG., Sept. 17, 2003, at News 18. A study reported in *JAMA* showed that a recurrence of cancer after a diseased prostate is removed could largely be stopped by radiation treatment (known as salvage radiation). ORANGE COUNTY REG., Mar. 17, 2004, at News 18. In 2005 it was reported that "the absolute reduction in the risk of death after 10 years is small, but the reductions in the risks of metastasis and local tumor progression are substantial." Bill-Axelson et al., *Radical Prostatectomy Versus Watchful Waiting in Early Prostate Cancer*, N. ENG. J. MED. 1977 (2005).

75. "The entire population had descended from a small band of ninth- and tenth-century Norse settlers mixed with a few early Irish slaves. The island had almost no immigration from then until the Second World War, so for more than a thousand years, Icelanders bred among themselves." Specter, *Decoding Iceland*, NEW YORKER, Jan. 18, 1999, at 1.

76. L.A. TIMES, July11, 1996, at 1.

A 1998 survey of people over 50 found that nearly half the unretired adults said they expect retirement will be "a chance for a new beginning." Over half of the actual retirees who were polled said that old age has been a continuation of life before "a step down." Many retirees were remaining quite healthy as of the end of the twentieth century. To be healthy, many of the elderly in Western society pursue some of the *sports and physical activities* of their youth. Others seek activities more conducive to a lessening of the capacity to continue with robust sports. A Harvard study has suggested that simply mixing with other people may offer as great a benefit as regular exercise.[77] These are ethical approaches toward finding personal meanings to life.

A 1998 study was also made of the association between walking and mortality in a cohort of retired men who were nonsmokers and physically capable of low-intensity activity on a daily basis. After 12 years, it was found that the *distance walked* remained inversely related to mortality.[78] Such physical activity has also reduced mortality in postmenopausal women.[79] Walking is the most common activity among older adults, and is associated with substantially lower risk of hip fracture in postmenopausal women.[80]

The brain stays too. In 1997, the Salk Institute for Biological Studies demonstrated in laboratory animals that the right kind of *mental gymnastics* could dramatically increase the number of cells in a key region of the adult brain.[81] The same year, a 16-year study at Washington University in St. Louis

77. ORANGE COUNTY REG., Apr. 5, 1998, at News 25. However, exercise-induced ventricular ectopy (displacement) during recovery is a better predictor of an increased risk of death than ventricular ectopy occurring only during exercise. N. ENG. J. MED. 781 (2003).

78. People over 65 who are disabled have dropped to 24.9%. ORANGE COUNTY REG., Mar. 18, 1997, at News 14. People without major risk factors for cardiovascular disease in middle age live longer than those with unfavorable risk-factor profiles. 339 N. ENG. J. MED. 1122–29 (1998).

79. Elderly people who like to eat out, play cards, go to movies, and take part in other social activities live an average of two and one-half years longer than more reclusive people. ORANGE COUNTY REG., Aug. 20, 1999, at News 29.

80. JAMA 2300 (2002); *see also* Alpeter & Marshall, *Making Aging 'Real' for Undergraduates*, 29 EDUC. GERONTOLOGY, 739–56 (2003) ("We developed a one-credit, undergraduate course, 'Introduction to Aging,' to sensitize students to the aging experience, the diversity of the aging population, and the value of intergenerational communication.").

81. N. ENG. J. MED. 94 (1998).

was completed of 25 people aged 71 to 95, comparing the brains of those with Alzheimer's disease to those who maintained mental clarity; if brain function becomes impaired, it is the result of disease, not only age.[82]

Alzheimer's disease is a fatal brain disorder that annihilates the elderly patient's mind during a period of years. It affects over 20 percent of people 75 years of age or more, and 40 percent over 85 years of age.[83] Another study showed that low idea density and low grammatical complexity in autobiographies written in early life were associated with low cognitive test scores as well as Alzheimer's disease in late life.[84] Some 4.6 million Americans have this disease.[85] In 2002, a study determined that the high dietary intake of vitamin C and vitamin E may lower the risk of Alzheimer's. However, there is no cure and clinical trials of certain drugs were suspended (amyloid beta) or failed testing (Neotrofin);[86] others are still undergoing tests.[87] For example, one study claimed

---

82. N. ENG. J. MED. 1046 (1997). In 2002, it was pointed out that nearly half of community-dwelling persons aged 60 years and older express concern about declining mental abilities. However, the effectiveness and durability of the cognitive training interventions have been found to improve cognitive abilities. "Training effects were a magnitude equivalent to the amount of decline expected in elderly persons without dementia over 7- to 14-year intervals." JAMA 2271 (2002).

83. TANZI & PARSON, DECODING DARKNESS xiv (2000). At 25 there may be as many people beginning with those who inherit this genetic disease. It is being attacked in American laboratories in the twenty-first century; *see also* 6 NATURE GENETICS 20.

84. L.A. TIMES, Apr. 3, 1997, at 1. "Most neurologists have argued that brain tissue declines markedly the older one gets and especially beyond middle age. But a study by researchers at Oregon Health Sciences University in Portland challenges that notion with the discovery that healthy 85-year-olds lose brain tissue no faster than healthy 65-year-olds." ORANGE COUNTY REG., Dec. 21, 1998, at News 24.

85. The number of Americans with Alzheimer's disease could more than triple to 16 million by 2050. Unsurprisingly, the biggest surge will be among people aged 85 and older. Economists estimate that, in the United States alone, it costs at least $100 billion a year to look after people with Alzheimer's.

ORANGE COUNTY REG., July 23, 2002, at News 6.

86. ORANGE COUNTY REG., Nov. 15, 2002; JAMA 3223 (2002).

87. Overstimulation of the N-methyl-d-aspartate (NMDA) receptor by glutamate is implicated in neurodegenerative disorders. Antiglutamatergic treatment reduced clinical deterioration in moderate to severe Alzheimer's disease,

that "common pain medications such as ibuprofen and naproxen may dissolve the lesions that clog the brains of Alzheimer's patients. The University of California, Los Angeles (UCLA) discovered the use of a new chemical marker called FDDNP. It goes straight for the damaging protein plaques that seemed to break up."[88] Another study in 2003 found that people 65 and older who ate fish—including tuna sandwiches, fish sticks, and shellfish—once a week had a 60 percent lower risk of Alzheimer's than those who never or rarely ate fish.[89] Although the number of cancer survivors has more than tripled to almost 10 million, the estimated 64 percent chance of surviving is for five more years.[90]

At age 50, the loss of a brain's weight is 2 percent a decade. The "motor of a frontal cortex loses between 20 percent and 50 percent of its neurons" and "the visual area in the back loses about 50 percent." However, the higher intellectual areas of the cerebral cortex have a significantly lower degree of cell disappearance." It has been shown that "reasoning and judgment are quite often unimpaired until late senescence . . . dendrites of many neurons continue to grow in healthy old people who don't have Alzheimer's disease."[91]

Nevertheless, longevity-increasing activities often require the assessment of time-trade-off techniques according to a patient's current state of health. Although patients vary, it was found that most were unwilling to trade much time for excellent health.[92] An early bioethicist at Harvard named Lawrence J. Henderson, who died in 1942, described the beginning of medical improvements in the twentieth century: "Somewhere between 1910 and 1912 in this country, a random patient with a random disease consulting a doctor chosen at random had, for the first time in the history of mankind, a better than 50-50 chance of profiting from the encounter."[93]

---

which is associated with distress for patients and creates a burden on caregivers, and for which other treatments are not available. 348 N. Eng. J. Med. 1333 (2003).

88. Orange County Reg., Mar. 13, 2003, at News 13.

89. Orange County Reg., July 22, 2003, at News 35.

90. Orange County Reg., June 25, 2004, at News 21.

91. Nuland, How We Die 55 (1995). The accuracy of PET scans for early detection of Alzheimer's is 90%. JAMA (2001). However, skeptics say that the evidence in favor of PET scans isn't strong enough in comparison with clinical evaluations and existing medications. Orange County Reg., Apr. 2, 2003, at Accent 1.

92. JAMA 526 (1996).

93. China Post (Taiwan), June 10, 1998, at 6.

A study has pointed out that 73 percent of all deaths in the United States occurred among persons 65 years and older and 23 percent among those 85 years and older. It was remarked that among continued increases in total life expectancy, "an extension of the period of disability prior to death may be an inevitable consequence of living to an advanced age"; U.S. residents may instead be living a longer life with worsening health.[94]

Some scientists discovered that the *reason people live to be near or even over 100* was found to have more to do with their parents; they started thinking about genetic origin and a need to study many genes. A study group, including 303 people with ages ranging from 91 to 109, began a laboratory movement in the twenty-first century.[95] In 2001, the Census Bureau estimated that the number of Americans aged 65 and older would be 70 million in 2030, about double the number today. The people aged 85 and older are expected to nearly double to about 8.5 million by 2030.[96]

Studies have also shown that eating less leads to longer life.[97] On November 29, 2002, the journal *Science* reported that studies with fruit flies, which have many genes similar to mammals, showed that an enzyme called Rpd3 histone deacetylase was a key to longevity.[98] Although *people who are older than 65 years* are the fastest-growing segment of the population that *account for the majority of cardiovascular disease (CVD)* morbidity and mortality, a

94. JAMA 371 (1996).

95. *Proceedings of the National Academy of Sciences*, ORANGE COUNTY REG., Aug. 28, 2001, at News 6.

96. http://www.reutersnealth.com/archive/2001/12/14/eline/links/200112114elin008.html "Following the turn of the third millennium, the world's oldest man died in Italy at the age of 112. He claimed that the secret of his longevity was a daily glass of red wine." ORANGE COUNTY REG., Jan. 5, 2002, at News 29. In 2005, Americans turning 65 can expect to live, on average, until they are 83, four and one-half years longer than the typical 65-year-old could expect in 1940. ORANGE COUNTY REG., June 12, 2005, at News 15.

97. A lifetime of low-calorie dieting would extend the human life span by about 7%. ORANGE COUNTY REG., Aug. 30, 2005, at News 7 ("Longevity is not a trait that exists in isolation; it evolves as part of a complex life history, with a wide range of underpinning physiological mechanisms involving, among other things, chronic disease processes." (quoting AGEING RES. REVS.)).

98. ORANGE COUNTY REG., Nov. 29, 2002, at News 25. A Harvard study found a compound called resveratrol in red wine, which may also contribute to a longer life. ORANGE COUNTY REG., July 16, 2004, at News 19.

study has shown that cereal fiber consumption late in life is associated with lower risk of incident cardiovascular disease (CVD).[99]

Other studies showed how *telomere loss has acted to regulate human longevity.* Telomerase, an enzyme, maintains the length of telomeres at the terminal ends of DNA strands in chromosomes.[100] Unless telomerase is active, telomeres become successively shorter. When telomeres disappear, cells may cease to replicate.[101] Some life has been extended by introducing telomerase into cultured human cells.[102] However, the reviewer of a volume on research regarding "the dream of life extension" in 2003 pointed out that "today, it is impossible to say whether a dramatic extension of human lifespan by cell therapy is a future reality or will remain mythical." The author noted that although the potential for the use of embryonic stem cells is huge, "messy details" will keep scientists busy for "a very long time."[103] Cancer cells, however, possess telemerase, which can make them "virtually immortal."

---

99. "Cardiovascular disease (CVD) is the leading cause of death and disability among these older adults, who also account for a disproportionately large share of the $200 billion annual U.S. healthcare expenditures for CVD." JAMA 1659 (2003). "[In 2003] in the U.S., for example, more than 550,000 new cases of heart failure were diagnosed, but only about 2,000 transplants were performed. For the remainder of patients, quality of life steadily erodes, and less than 40 percent will survive five years after the initial attack." Cohen & Lear, *Rebuilding Broken Hearts*, SCI. AM., Nov. 2004, at 45.

100. ATHERLY ET AL., SCIENCE OF GENETICS 302–03 (1999); FACTS FOR FAITH, Q. 1, 2001, at 22.

101. Andrea G. Bodnar et al., *Extension of Life-Span by Production of Telomerase into Normal Human Cells*, 279 SCIENCE 349–52 (1998).

102. Artandi et al., *Telomere Dysfunction Promotes Non-Reciprocal Translocations and Epithelial Concerns in Mice*, 406 NATURE 641–45 (2000); Elizabeth H. Blackburn, *Telomere States and Cell Fates*, 408 NATURE 53–56 (2000); Hanahan, *Benefits of Bad Telomeres*, 406 NATURE 573–74 (2000).

103. Hornsby, JAMA 1925 (reviewing STEPHEN HALL, MERCHANTS OF IMMORTALITY: CHASING THE DREAM OF HUMAN LIFE EXTENSION (2003)). *See also* the study, which noted that:

> Ideally, methods that can measure the telomere length in viable cells should be developed to further explore selection of cells on the basis of telomere length and proliferative potential. Alternatively, the telomere length in specific chromosome arms could possibly be measured in very small numbers of purified cells using a recently described polymerase chain reaction-based technique.

Death is still a fearful, frightening happening and the fear of death is a universal fear even if we think we have mastered it on several levels.

Kubler-Ross

Dying is personal. And it is profound. For many, the thought of an ignoble end, steeped in decay, is abhorrent. A quiet, proud death, bodily integrity intact, is a matter of extreme consequence.

Justice William Brennan, dissenting in *Cruzan v. Dir., Mo. Dept. of Health*, 497 U.S. 261, 310 (1990)

We seek the dead only to return to earth the body, of which no man is the owner, but only for a brief moment the guest must return to dust again.

Euripides, *Seven Against Thebes*

b) Decreasing Longevity and Quality of Life

Decreases in life expectancy are manifold. It has been written, "Evolution has not produced any kind of behavioral system that serves exclusively to help the old."[104] Royalty believed that use of a lightning rod was not God's will, according to those alive in the eighteenth century;[105] the view expressed

---

Van Ziffle et al., *Telomere Length in Subpopulations of Human Hematopoietic Cells*, 21 STEM CELLS 654–60 (2003), *available at* http://www.stemcells.com. Werner syndrome (WS) patients suffer from multiple signs of premature aging. Crabbe and others propose that WRN is necessary for efficient replication of G-rich telomeric DNA, preventing telomere dysfunction and consequent genomic instability. Crabbe et al., SCIENCE 1951 (2004).

104. ORANGE COUNTY REG., Apr. 9, 1999, at News 10.

105. "When Benjamin Franklin invented the lightning rod, the clergy, both in England and America, with enthusiastic support of George III, condemned it as an impious attempt to defeat the will of God. . . . Therefore, if God wants to strike anyone, Benjamin Franklin ought not to defeat His design." BERTRAND RUSSELL, UNPOPULAR ESSAYS 74 (1950). Russell added, "But God was equal, the occasion Massachusetts was shaken by earthquakes, which Dr. Price perceived to be due to God's wrath at the 'iron points invented by the sagacious Dr. Franklin.'" *Id.*; *see also* CALLAHAN, FALSE HOPES 82 (1999); N. ENG. J. MED. 1283 (1999); Liao et al., 283 JAMA 512–18 (2000).

by King George, from whose rule we revolted in 1776, is still being religiously and practically tested today. In 1994 it was remarked, "Like a number of geriatricians, I have come to believe that modern medicine does not work well for old people. Old patients serve as a mirror, reflecting the limitations and sometimes the absurdities of modern medicine."[106] A study published in July 2000 stated: "The fact is that the U.S. population does not have anywhere near the best health in the world. Of 13 countries in a recent comparison, the United States ranks an average of 12th (second from the bottom) for 16 available health indicators."[107] In developed countries, many infectious diseases do not remain the major killers they were prior to the twenty-first century, including smallpox, which was declared eradicated worldwide in 1980.[108] However, in non-industrial countries the infectious diseases of tuberculosis, pneumonia, malaria, and measles continue to kill one in four, which is more than 10 million every year.[109] In 2003, a study showed that for individuals born in 2000, *diabetes mellitus* remained one of the most prevalent and costly chronic diseases in the United States:

---

106. KAMMER, IN QUEST OF THE SACRED BABOON (1995).

107. 284 JAMA 48 (2000).The United States ranked "7th for life expectancy at 65 years for females, 7th for males, 3rd for life expectancy at 80 years for females, [and] 3rd for males."

108. Following the terrorist activity in the United States on September 11, 2001, the possibility of biologic warfare caused the U.S. government to reintroduce the vaccination against smallpox. However, "no statistical model can predict the thoughts of a terrorist." N. ENG. J. MED. 381, 460 (2003). Although 61% of adults surveyed in November 2001 favored getting the vaccination, this number dropped to 54% in February 2003. ORANGE COUNTY REG., Feb. 12, 2003, at News 20.

109. WORLD HEALTH ORGANIZATION, REMOVING OBSTACLES TO HEALTHY DEVELOPMENT (1999).

> The death of an infant during the first year of life in cases in which all identifiable causes of death can be ruled out by appropriate assessment is called the sudden infant death syndrome (SIDS). A recent study found there is "a direct association between second-trimester maternal serum alpha-fetoprotein levels and the risk of SIDS, which may be mediated in part through impaired fetal growth and preterm birth." Thus it has become one of the best biochemical predictors of the risk of unexplained stillbirth.

N. ENG. J. MED. 978 (2004).

> The estimated lifetime risk of developing diabetes for individuals born in 2000 is 32.8 percent for males and 38.5 percent for females. Females have a higher residual lifetime risk at all ages. The highest estimated lifetime risk for diabetes is among Hispanics (males, 45.4 percent and females, 52.5 percent). For example, we estimate that if an individual is diagnosed at age 40 years, men will lose 11.6 life-years and 18.6 quality-adjusted life-years and women will lose 14.3 life-years and 22.0 quality-adjusted life-years.[110]

According to the Center for Disease Control, about 5 million Americans have *heart failure*, 70 percent of whom are at least 60 years old.[111] Coronary heart disease (CHD) is the leading cause of death in the United States, and half of all deaths from CHD occur in women. "Half of heart failure patients die within five years of diagnostics, and for many patients, death is sudden." This should encourage physicians "to address personal treatment goals and advance directives and pursue palliative care alongside optimal medical therapy."[112] The estimated mortality of 9 million in 1990 from cardiovascular causes in developing countries is expected to increase to 19 million by 2020 before 70 years of age.[113] In the early twentieth century, it had also been the leading cause of death in the United States.[114] A study in 1999 showed that men who lose a parent to sudden cardiac death are twice as likely as other men to meet the same fate.[115] A

110. Narayan et al., *Lifetime Risk for Diabetes Mellitus in the United States*, JAMA 1884 (2003). In 2004, an estimated 55,000 Americans received a diagnosis of cutaneous melanoma, and 7,900 will die from the disease. It is also the most common cancer among women 20 to 29 years of age. *See* et al., *Management of Cutaneous Melanoma*, N. Eng. J. Med. 998 (2004). Approximately 1.3 million Americans were diagnosed with some type of cancer in 2004.

111. Orange County Reg., Aug. 7, 1998, at News 16.

112. JAMA 2243 (2004); *see also* Pantilat & Steimle, JAMA 2476, 2481 (2004).

113. Reddy, *Cardiovascular Disease in Non-Western Countries*, N. Eng. J. Med. 2438 (2004).

114. Thom et al., *Incidence, Prevalence and Mortality of Cardiovascular Disease in the United States*, *in* Hurst, The Heart, Arteries and Veins 4 (Alexander et al. eds., 1998); *Achievements in Public Health, 1990–1999: Decline in Deaths from Heart Disease and Stroke—United States, 1900–1999*, 48 Morbidity & Mortality Wkly Rep. 644–56 (1999).

study of a large cohort of apparently healthy persons indicates that "the heart-rate profile during exercise and recovery is a strong predictor of sudden death.[116] Further, a high risk comes after myocardial infarction—particularly during the first 30 days—among patients with left ventricular dysfunction, heart failure, or both."[117] In 2002 it was noted that coronary artery disease "receives little attention by either the Federal government or the states."[118] "Heart failure is the reason for at least 20 percent of all hospital admissions among persons older than 65. Over the past decade, the rate of hospitalizations for heart failure has increased by 159 percent."[119] Nevertheless, in 2002, a battery-powered pump long used to help heart-failure patients survive received FDA approval as a heart-beater for some 20,000 to 30,000 people who failed to qualify for a heart transplant.[120] On the other hand, in 2005 the first fully implantable artificial heart

---

115. ORANGE COUNTY REG., Apr. 20, 1999, at News 9 (citing CIRCULATION). "Medical studies have found that cardiopulmonary resuscitation is far more effective than the Heimlich maneuver in forcing air into the lungs of people who have stopped breathing. Practicing the Heimlich instead of CPR goes against protocol for Orange County lifeguards [who] were recertified in CPR." ORANGE COUNTY REG., June 15, 2004, at News 3.

116. "Since heart-rate responses to exercise are under the control of the autonomic nervous system, these data support the concept that abnormalities in autonomic balance may precede manifestations of cardiovascular disease and may contribute to the early identification of persons at high risk for sudden death." Jouven et al., N. ENG. J. MED. 1951 (2005).

117. Solomon et al., N. ENG. J. MED. 2587 (2005). "Thus, earlier implementation of strategies for preventing sudden death may be warranted in selected patients."

118. Purnet, *After September 11: Rethinking Public Health Federalism*, J.L. MED. & ETHICS 208 (2002). "Heart disease will strike 1 out of every 3. More than 500,000 women die in the U.S. each year of cardiovascular disease, making it, not breast cancer (40,000 deaths annually), their No. 1 killer." http://member.compuserve.com/news/4/23/03.

119. Gregg et al., N. ENG. J. MED. (2003). "Increasing and maintaining physical activity levels could lengthen life for older women but appears to provide less benefit for women aged at least 75 years and those with poor health status." JAMA 2379 (2003).

120. Called a "left ventricular assist device," the HeartMate is hooked up to the heart and takes over the job of pumping blood that the organ's diseased left side can no longer perform. The pump works only as half a heart—it can't take over the right side. A major study released

(the Abio-Cor artificial heart) was rejected by FDA, which had tested it in only 14 men. Two died from the operation, one never regained consciousness, and the rest survived only an average of five months.[121] Special pacemakers are designed to correct a defect that affects close to two million American heart failure patients.[122] Research has shown that those who drank at least three days a week had about one-third fewer heart attacks than did nondrinkers by raising the level of "good" cholesterol, thinning blood and warding off clots that cause heart attacks.[123] Blood pressure is "the most consistent predictor of stroke, such that hypertension is casually involved in nearly 70 percent of all stroke cases."[124] In 2003 it was noted that "corona artery by-pass graft (CABG) surgery may be the most frequently studied of all surgical procedures; however, continued attempts should be made to explore the potential of readmission as a supplement to mortality in assessing provider quality."[125] The choice of hospitals is also a relevant factor. A 1999 study found that patients with acute myocardial heart wall infarction who are admitted directly to hospitals that have more experience treating myocardial infarction, as reflected by their case volume, are more likely to survive than are patients admitted to low-volume hospitals:[126] In "the absence of other information about the quality of surgery at the hospitals near them,

---

> last year showed the HeartMate also could add months to the lives of heart-failure patients who can't get a transplant.

ORANGE COUNTY REG., Nov. 7, 2002, at News 13.

121. The panel voted 7–6 that the heart's probable benefit didn't outweigh the risks. If approved, the implant is expected to cost about $250,000. But it is unclear if insurance would cover it. "It is too large for most women; the company is developing a smaller version." ORANGE COUNTY REG., June 25, 2005, at News 20.

122. ORANGE COUNTY REG., Feb. 12, 2003, at News 18 (citing JAMA).

123. ORANGE COUNTY REG., Jan. 9, 2003, at News 11. "Studies conclude that 15% to 18% of lawyers are problem drinkers or alcoholics. The figure is only about 10% for the general population." A.B.A. J., Jan. 2003, at 18. Alcohol is also the leading drug of abuse by teenagers in the United States. A study showed data to suggest "that underage drinkers and adult excessive drinkers are responsible for 50.1% of alcohol consumption and 48.9% of consumer expenditure." JAMA 989 (2003).

124. Staessen et al., JAMA 2420 (2003).

125. Hannan, et al., JAMA 773 (2003).

126. 340 N. ENG. J. MED. 1640–48 (1999). A study published in March 2000 showed referral of patients to high-volume hospitals to reduce overall hospital

Medicare patients undergoing selected cardiovascular or cancer procedures can significantly reduce their risk of operative death by selecting a high-volume hospital."[127]

There was a 60 percent decline in the age-adjusted death rate from 1950 to 1999.[128] During the same period, the incidence of heart failure has declined among women but not among men, whereas survival after the onset of heart failure has improved in both sexes.[129] Among men, *consumption of alcohol* at least three to four days per week was inversely associated with the risk of myocardial infarction.[130] Substantial evidence indicates that diets using nonhydrogenated unsaturated fats as the predominant form of dietary fat, and an abundance of fruits and vegetables, can offer significant protection against coronary heart disease.[131] Early use of aspirin after coronary bypass surgery is associated with a reduced risk of death. However, ReoPro, known as a "super aspirin" (which helps keep particles called platelets from sticking together and forming a clot that can cause a heart attack), does not improve hospitalized heart attack patients' chances of surviving a year. Research published in 2003 also found that those taking both aspirin and another drug called ibuprofen were twice as

---

mortality in California for the conditions identified." JAMA 1159 (2000). A study published in January 2000 found that men who are losing hair on the crown of their heads have up to a 36% greater risk of experiencing heart problems. ORANGE COUNTY REG., Jan. 24, 2000, at 1 (citing ARCHIVES INTERNAL MED.). But often myocardial infarction in younger women (but not older women) results in higher rates of death during hospitalization than men of the same age. 341 J. MED. 217–25 (1999).

127. Birkmeyer et al., *Hospital Volume and Surgical Mortality in the United States*, N. ENG. J. MED. 1128 (2002). "Death rates for all cancers had inched down by about 1.4 percent a year through the mid-1990s. By 2000, that decline leveled. . . . An estimated 556,500 Americans will die of cancer this year, and 1.3 million will be diagnosed with it." ORANGE COUNTY REG., Sept. 10, 2003, at News 8.

128. *Supra* note 114, at 649–56. Modifiable factors include high cholesterol, hypertension, physical inactivity, obesity, smoking, and diabetes. Kaplan & J. E. Keil, *Socioeconomic Factors and Cardiovascular Disease: A Review of the Literature*, 88 CIRCULATION 1973–88 (1993); Otsuka et al., *Acute Effects of Passive Smoking on the Coronary Circulation in Healthy Young Adults*, 286 JAMA 436–41 (2001).

129. N. ENG. J. MED. 1397 (2002).

130. N. ENG. J. MED. 2 (2003).

131. JAMA 2569 (2002).

likely to die during the study period as those who were taking aspirin alone or with pain relievers.[132] Further, daily use of *aspirin* is associated with a significant reduction in the incidence of colorectal (10 inches of the bowel) adenomas in patients with previous colorectal cancer,[133] and low-dose aspirin has a moderate chemopreventive effect on adenomas (tumors) in the large bowel.[134] Of the 4.9 million Americans who have been diagnosed with the disease, "hospital discharges for heart failure increased by 155 percent during the last 20 years, and heart failure is the most frequent cause of hospitalization in patients aged 65 years or older."[135] Over a seven-year period, a low-dose aspirin was associated with a 35 percent reduction in cardiovascular deaths and a 36 percent drop in total mortality. The Physicians' Health Study was the first to show that aspirin can protect healthy men older than 50 against heart attacks, but it was not the first to show that aspirin can help the heart.[136]

Many large hospitals have stopped treating people with *clot-dissolving drugs* (resulting in 9 percent deaths) in favor of primary angioplasty within the first two or three hours of a heart attack. With angioplasty, a balloon opens

---

132. N. Eng. J. Med. 1309 (2002); JAMA (2002), available at http://jama.amaassn.ord; Lancet (2003).

133. N. Eng. J. Med. 883 (2003).

134. *Id.* at 891. Colorectal cancer is rare, and a study found that between 94% and 98% of patients with positive tests did not have colon cancer. Mandel, *Reducing Mortality from Colorectal Cancer by Screening for Fecal Occult Blood*, 328 N. Eng. J. Med. 1365–71 (1993).

135. In this connection a study reported:

> Within this context, heart failure constitutes a public health problem singled out as an emerging epidemic. Although the clinical and public health importance of heart failure is undisputed, this epidemic is not adequately understood. . . . Thus, little is known about temporal trends in the incidence of heart failure and on survival after its onset.

Roger et al., *Trends in Heart Failure Incidence and Survival in a Community-Based Population*, JAMA 344 (2004). It only takes low doses of aspirin, between 75 and 325 milligrams a day, to produce this major benefit. At present, up to 25% of American heart attack survivors fail to take aspirin. It is a shame, since if all the heart attack patients who could take aspirin did so, it could prevent another 20,000 deaths annually.

136. Simon, *supra* note 74, at 164–65.

blocked blood vessels of the heart (with the lesser amount of 7 percent deaths).[137] The Ischemia Research and Education Foundation estimated that giving a 5-cent aspirin without hours of bypass surgery could prevent about 7,000 deaths and 51,000 serious complications annually worldwide.[138]

The mortality rate from *infectious disease* is the third leading cause of death in the United States and still the leading cause globally.[139] The prospects for treatment of disease like myeloblastic leukemia (AML) for patients over 60 "remain dismal." It has been shown in the third millennium that *human immunodeficiency virus (HIV)* demographics are shifting toward nonwhites and intravenous drug users. However, half of them "are at risk of making end of life decisions without prior discussions with their healthcare practitioners."[140] The United Nations (UN) indicated that AIDS could kill nearly 70 million people world-

---

137. ORANGE COUNTY REG., Oct. 5, 2003, at News 24. "Doctors estimate that fewer than one in five hospitals can offer emergency angioplasty around the clock." *Id.*

138. ORANGE COUNTY REG., Oct. 24, 2002, at News 15.

139. Hast. Cent. Rep., June 1998, at 44. "Strokes constitute the 3rd cause of deaths in the United States and some 700,000 adults have strokes, at an annual cost of about $40 billion. Some 85% of strokes are ischemic—i.e., due to functional constriction or actual obstruction of a blood vessel." Steven Cramer, Univ. of Cal., Irvine (UCI) Dep't of Neurology, Address at the Oasis Conference (May 25, 2004).

140. "Blacks, Latinos, intravenous drip users, and less educated individuals need advance care planning interventions in clinical HIV programs. Recent studies have focused on HIV-infected patients' interest in, and some practitioners' willingness to provide, euthanasia." JAMA 2880 (2001).

> I first met Elizabeth at the Democratic convention in 1992, where she spoke movingly about having contracted HIV through a blood transfusion while giving birth to her daughter, Ariel, in 1981. Unaware that she was infected, Elizabeth passed the disease on to her daughter through her breast milk and later to her son, Jake, in the womb. Elizabeth was outraged that medications available to her were denied to her daughter and son because they hadn't been tested for safety and efficacy for children. She and her husband, Paul Glaser, watched helplessly as their daughter succumbed to AIDS at age seven.

HILLARY RODHAM CLINTON, LIVING HISTORY 385 (2003). (Elizabeth died in 1994.)

wide by 2002.[141] The United States is the most heavily affected country in the industrialized world, with almost 1 million persons living with HIV. However, there have been reductions in AIDS incidence and deaths since 1996 through use of highly active antiretroviral therapy (HAART),[142] providing that such patients receive adequate palliative as well as curative care:

> The emerging biomedical paradigm of highly active antiretroviral therapy (HAART) has become the cornerstone of treatment and has helped to transform HIV into a manageable chronic disease, yet at the same time has resulted in a more narrow focus and a de facto separation between disease-specific "curative" and symptom-specific "palliative" care for patients with HIV/AIDS. As patients survive longer in the latter stages of progressive HIV disease, they may in fact have increasing need for comprehensive symptom management, as well as wide-ranging need for psychosocial, family, and care planning support. In the HAART era, the false dichotomy of curative vs. palliative care for patients with HIV/AIDS must be supplanted by a more integrated model.[143]

---

141. In rich countries last year, 500,000 people received life-extending AIDS drugs and 25,0000 people died of AIDS, the report said. In sub-Saharan Africa, 30,000 people received the antiretroviral medication and 2.2 million people died. Fewer than 4% of those who need treatment get it, said the report by the Joint United Nations Program on HIV/AIDS, or UNAIDS. ORANGE COUNTY REG., July 3, 2002, at News 22. In sub-Saharan Africa 9% of all adults are HIV-infected, and babies born by 2010 may not live more than 13 years. STANECKI, THE AIDS PANDEMIC (2000).

142. JAMA 236 (2002). As a side effect it may cause cholesterol levels to climb. "Of note, in part because of the successes of HAART, there has been a recent resurgence of unsafe behavior among men who have sex with men." *See* Karon et al., *HIV in the United States at the Turn of the Century: An Epidemic in Transition*, AM. J. PUB. HEALTH 91 (2001); *supra* Chapter 7. The United States has three to four times more HIV-infected men than women. SIMON, *supra* note 74, at 290. "HIV diagnoses among gay and bisexual men rose 7.1 percent in 2002 in 25 states with long-standing HIV-reporting procedures." ORANGE COUNTY REG., July 29, 2003, at News 12 (citing statistics from the Centers for Disease Control and Prevention).

143. Selwyth & Forstein, JAMA 806 (2003). In July 2003 a large study in Paris, France, reported three new cases of HIV-infected people who initially were doing well without drugs but became sick years later after contracting a second strain of the AIDS virus. It was suggested that all new AIDS patients

However, the first AIDS vaccine tested in a large population of people at high risk for the disease has proven to be largely ineffective, according to data released by the vaccine's manufacturer, because HIV mutates rapidly.[144] It has been pointed out that "Still absent from the politicized discussions about HIV/AIDS is a full knowledge of how ameliorating this disease can significantly improve the personal, economic, and political security of families and countries at risk."[145]

In view of this condition, the AMA *Code of Medical Ethics* not only encourages "diagnosis and treatment of HIV infection or for medical conditions that may be affected by HIV," but also adds that "wider testing is imperative to ensure that individuals in need of treatment are identified and treated."[146] Yet, the National Institutes of Health (NIH) refused to override patents on the AIDS

---

should be tested to determine the drug resistance of the strains infecting them. ORANGE COUNTY REG., July 10, 2003, at News 9. Further, at least half of those infected with HIV-AIDS also have genital herpes, which can increase HIV transmission. ORANGE COUNTY REG., July 30, 2003. "In industrialized nations, the use of HAART in pregnant women along with an optimal package of interventions, including cesarean section and formula feeding, has reduced the rate of mother-to-child transmission of HIV-1 from about 25 percent to 1 to 2 percent." Coovadia, *Antiretroviral Agents—How Best to Protect Infants from HIV and Save Their Mothers from AIDS*, N. ENG. J. MED. 289 (2004).

144. The trial of the vaccine, called Aidsvax, took place mostly in the continental United States. ORANGE COUNTY REG., Feb. 24, 2003, at News 7. A study showed that "during the years 1995 through 2001, overall mortality among HIV-infected VA patients decreased by over 75%, while rates of hospital admission or death due to cardiovascular or cerebrovascular disease remained constant or showed a slight decline." N. ENG. J. MED. 679, 702 (2003).

145. Luo, JAMA 1649 (2002). It has been argued that "AIDS-related stigma is manifested in prejudice, discounting, discrediting, and discrimination directed at people perceived to have AIDS or HIV and at the individuals, groups, and communities with which they are associated." J.L. MED. & ETHICS (2002) (citing Herek et al., *AIDS and Stigma: A Conceptual Framework and Research Agenda*, 13 AIDS & PUB. POL'Y J. 36–47 (1998)).

146. Code Section 2.23 also states: "When a healthcare provider is at risk for HIV infection because of the occurrence of puncture injury or mucosal contact with potentially infected bodily fluids, it is acceptable to test the patient for HIV infection even if the patient refuses consent." *See also* Kenworthy, *The Austrian Psychotherapy Act: No Legal Duty to Warn*, 11 IND. INT'L & COMP. L. REV. 469 (2001), which concludes that the duty to warn of HIV seropositivity should involve mandatory disclosure to public health officials.

drug Norvir when Abbott Laboratories raised its price.[147] Intensive chemotherapy is not very effective and is also unacceptably toxic. Control and cure are highly elusive. For many older patients, no treatment is offered beyond supportive care.[148] Although mortality rates declined in the twentieth century, persons with fewer years of education and black persons still live approximately six fewer years than better-educated persons and whites, respectively.[149]

Many of these diseases will cause some elderly patients to enter intensive care units (ICUs). If so, they may have to confront the 1996 statistic showing that they benefit from close surveillance and cardiovascular monitoring. In ICUs patients have *nosocomial infection* rates that are as much as five to 10 times higher than those in the general wards. In addition to "the underlying medical conditions and nosocomial bacteremia, nosocomial pneumonia [i.e., originating in a hospital] independently contributes to ICU patient mortality."[150] An analysis of the outcome of ICU patients with acute renal failure who received loop diuretics (agents that promote urine secretion) was associated with a 68 percent increase in in hospital mortality.[151] Anywhere outside the ICU they must take notice of the fact that in-hospital mortality for pneumonia patients older than 65 was 10.7 deaths per 100 discharges in spite of more than $3.5 billion having been spent to keep them alive.[152]

The incidence of serious and fatal adverse drug reactions in hospital patients has been increasing.[153] Adverse drug events affect millions of patients each year

147. ORANGE COUNTY REG., Aug. 5, 2004. Although sales have totaled more than $1 billion since Norvir was introduced, sales fell from a high of $250 million in 1998 to $100 million in 2003. The company maintained that the higher price was necessary to counter falling sales.

148. N. ENG. J. MED. 1712 (1995). "Among Medicare beneficiaries, 82 percent of those 85 and older have at least two chronic conditions." OUTREACH, Oct. 2002, at 22.

149. N. ENG. J. MED. 1585.

150. JAMA 866 (1996).

151. JAMA 2547, 2599 (2002). Sepsis, the presence of toxins in the blood or other tissue, is the leading cause of death in critically ill patients in the United States. "Sepsis develops in 750,000 people annually, and more than 210,000 of them die. After numerous unsuccessful trials of anti-inflammatory agents in patients with sepsis, investigators doubted that mortality could be decreased." 348 N. ENG. J. MED. (2003).

152. JAMA 2080 (1997).

153. ORANGE COUNTY REG., Feb. 28, 1999 (citing LANCET). The head of this

and are responsible for hospital admissions. A study in the year 2000 found iatrogenic illness to be the leading cause of 300,000 deaths per year that come from reactions to prescription drugs.[154] Although drug interactions are a preventable cause of morbidity and mortality, "some adverse drug events have life-threatening consequences and may prompt the removal of popular medications from the marketplace."[155] Because "many of these interactions could have been avoided," ethics committees may wish to see if this is adequately being done in their hospital. One recent study also showed that "when nurses give drugs intravenously, they are making mistakes in almost half of the injections because they are poorly trained."[156]A pharmacy has been held to have a duty to warn a customer when the pharmacy is aware of both a customer's drug allergies and that the prescribed *drug is contraindicated*[157] for a person with those allergies.[158] Many adults take multiple agents, and a substantial overlap between use of prescription medications and herbals/supplements raises concern about unintended interactions. However, documentation of usage patterns can provide a basis for improving the safety of medication use.[159]

---

research stated that the increase in death rate attributable to medication mistakes is sharper than the increase for any cause of death other than AIDS. *Id.*

154. NULL ET AL., DEATH BY MEDICINE (2003). The Children's Mental Health Screening and Prevention Act, H.R. 3063, 108th Cong. (2003), *available at* http://www.theorator.com/bills108/hr3063.html, recommends increased screening for suicidality and mental illness. On that particular matter, a study about the potential harm of suicide screening found "no evidence of iatrogenic effects of suicide screening emerged" and that "screening in high schools is a safe component of youth suicide prevention efforts." Gould et al., *Evaluating Iatrogenic Risk of Youth Suicide Screening Programs*, JAMA 1635 (2005).

155. JAMA 1652 (2003). "Many hospital admissions of elderly patients for drug toxicity occur after administration of a drug known to cause drug interactions."

156. http://news.bbc.co.uk/1/hi/health/2891327stm 3/28/03.

157. "An indication, symptom, or condition that makes inadvisable a particular treatment or procedure." MERRIAM-WEBSTER (1993).

158. Happel v. Wal-Mart Stores, Inc., 766 N.E.2d 1118 (Ill. 2002). The Happel couple filed suit against Wal-Mart for negligence after a Wal-Mart pharmacist failed to warn either the Happels or one of their physicians. The majority of the Illinois Supreme Court asserted that the pharmacy had a duty to warn Happel that the drug was contraindicated. The court found that the failure to warn could result in injury or death and that Wal-Mart "need only pass along to the customer or the physician the information it already possesses about the contraindication for this specific customer." *Id.* at 1124.

In 2003 the *World Cancer Report* was the first comprehensive examination of cancer around the globe. About 10 million people are diagnosed with cancer every year and 6 million people die from it. The report projects that the annual number of diagnoses will reach 15 million by 2020, based on current trends in smoking, diet, and exercise.[159] The previous year it was reported that "30 years after President Richard Nixon declared war on cancer and made a major Federal commitment to the research needed to win that war, the cancer survival rate has almost doubled—from 33 percent in 1971 to 62 percent."[161] It was also noted that there is no clear treatment in the elderly: "Most cancer drug trials exclude patients older than 70, and doctors are subsequently reluctant to give the medications to older patients because they fear the side effects may be too harsh for them."[162]

Toward the end of the twentieth century, over 50 million American women beginning at age 40 underwent an annual screening as then recommended by the American Cancer Society and over 300,000 without breast cancer had a biopsy.[163] However, following the turn of the millennium, it was found that early detection does not amount to prevention and "does not increase life expectancy, but only the time the patient consciously has to live

159. JAMA 337 (2002). The Supreme Court has held that commercial speech can be regulated only with regard to unlawful, false, or misleading activity, or to the extent that it directly advances a substantial government interest. Cent. Hudson Gas & Elec. Corp. v. Pub. Serv. Comm'n, 447 U.S. 557 (1980); *see also* Lorillard Tobacco Co. v. Riley, 533 U.S. 525 (2001) (holding that a potentially hazardous product itself should be regulated rather than the advertising of it).

160. http://member.compuserve.com/news/story.j. (Apr. 4, 2003).

161. Orange County Reg., May 14, 2002, at News 4. The American Cancer Society reported that in 2002, for the first time, cancer surpassed heart disease as the top killer of Americans younger than 85. According to their statistics, 476,009 Americans under 85 died of cancer, compared with 450,637 who died of heart disease. Orange County Reg., Jan. 20, 2005, at News 12. "Those under 85 make up 98.4 percent of the population." *Id.*

162. Orange County Reg., Oct. 21, 2002, at News 10. "Only modest cancer-specific survival differences are evident for blacks and whites treated comparably for similar-stage cancer." 287 JAMA 2106–13 (2002).

163. Lerman et al., *Psychological Side Effects of Breast Cancer Screening*, 10 Health Psychol. 259–67 (1991).

with the cancer."[164] The mortality reduction from age 50 on is about four in 1,000 women.[165]

It has been pointed out that in the United States, prostate cancer is more frequent than breast cancer. "It also takes almost as many lives. But there is no anxiety among men about prostate cancer comparable to women's anxiety about breast cancer."[166] A study in 2002 concluded that 29 percent to 44 percent of the men were "over-diagnosed." Many men over 60 are receiving unnecessary surgery for prostate cancer.[167] Another study of men undergoing prostatectomy (excision of the prostate) concluded that the rates of postoperative and late urinary complications would be significantly reduced if the procedure were performed in a high-volume hospital and by a surgeon who performs a high number of such procedures.[168] This may also become a more frequent conclusion of respectable hospital ethics committees.

The great variances in death rates among hospitals must also be considered. For example, the book *Consumers Guide to Hospitals* noted that 20 out of 189 coronary artery bypass patients at one California hospital died, but only one of 144 patients getting the *same surgery* at another hospital in the

---

164. GIGCRENZER, CALCULATED RISKS 57 (2002). However, in 2005 a study by researchers at the University of Texas Health Science Center found that very obese men had about 30% lower levels of the prostate-specific antigen (PSA) compared with men of normal weight. It was believed that "obese men produce more estrogen, which drives down testosterone levels and could affect cells that produce the antigen used in the test." ORANGE COUNTY REG., Jan. 24, 2005, at News 5.

165. *Supra* note 71. "These women also believed that breast self-examination is even more beneficial than 10 years of annual mammograms. These results indicate a striking degree of misinformation among American women. This is not to say that women in other countries are better informed; American women are just better studied." *Id.* at 72 (citing Schwartz et al., *U.S. Women's Reactions to False Positive Mammography Results and Detection of Ductal Carcinoma In Situ: Cross-Sectional Survey*, 320 BRIT. MED. J. 1635, 1640 (2000)).

166. "Interestingly, unlike breast cancer, prostate cancer is presented in the media as an old person's disease. However, its incidence, mortality rate and the mean age at diagnosis are in fact very similar to those of breast cancer." *Id.* at 79.

167. "A study in the *Journal of the National Cancer Institute* examined the use of a blood test to find prostate cancer in a group of patients ages 60 to 84 over a 10-year period." ORANGE COUNTY REG., July 3, 2002.

168. N. ENG. J. MED. 1138 (2002).

same area died. Accordingly, members of hospital ethics committees may recommend that either a higher volume be sought by their hospital in a particular area, or that such patients be requested to seek another hospital. In late 1999 the *Institute of Medicine (IOM)* reported on death and disability in U.S. hospitals: "Preventable adverse events are a leading cause of death and at least 44,000 and perhaps as many as 98,000 Americans die in hospitals each year as a result of medical errors."[169]

Although some physicians claimed exaggeration,[170] Lucian L. Leape, M.D., of the Harvard School of Public Health, was able to show that "the reliance on information extracted from medical records most likely led to a substantial underestimate of the prevalence of injury."[171] Most importantly, "the IOM report has galvanized a national movement to improve patient safety. It is about time."[172] If one is hospitalized, he must hope its physicians have redesigned their medical work. Because these deaths are even greater than those resulting from highway accidents, the President called for a nationwide system of reporting medical errors.

In 2000, a "group conducted a survey of the 51 boards regulating medical doctors to determine the current state of Internet-accessible disciplinary information." This occurred near the turn of the third millennium. It recommended that all "states adopt minimum, uniform standards to ensure that sufficient information on a given action is provided, that all of the information is presented in a user-friendly format, and that the information is comprehensive, current, and

169. To Err Is Human: Building a Safer Health System (Kohn et al. eds., 1999).

170. McDonald et al., JAMA 93–40 (2000).

171. All patients who had major surgery, acute myocardial infarction, pneumonia, or stroke who had an uncomplicated course (and therefore did not meet screening criteria) were excluded, as were patients who were admitted for planned terminal care, had a do-not-resuscitate order, or were extremely ill. Even many intensive care unit patients did not meet any of the screening criteria.

Jeupe, JAMA 96 (2000).

172. "Although the initial impact of the IOM report is in part due to the shocking figures (which, unfortunately, are not exaggerated), its long-term impact will result from the validity of its message that errors can be prevented by redesigning medical work." *Id.* at 97. "The Harvard Alumni Study found that smoking is the most dangerous lifestyle risk factor, increasing a man's risk of dying during the study by 76%. That's the bad news. But the study also found some good news: Quitting reduced the death rate by 41%." Simon, *supra* note 74, at 212.

retroactive for 10 years. Although 260 million patients live in the 41 states that provide some disciplinary information on the Internet, 114 million of them (44 percent) are in states that provide" adequate data, according to the survey. Patients "should be able to access disciplinary data regardless of a physician's license status, so if a doctor attempted to practice without an active license, patients could quickly determine that the doctor was practicing illegally."[173]

good v. Good

To err is human and no special investigations are needed for infections originating in a hospital [nosocomial], or deaths that occur at American hospitals. Investigation by state disciplinary boards is adequate to protect patients (consumers) and their findings should not be published on the Internet (or other media).

v.

More investigations of hospitals are needed due to the huge numbers of preventable adverse events leading to deaths in the United States. Adverse events here required immediate attention early in the third millennium. Likewise, the identity of disciplined physicians needs to be brought to the attention of consumers because they should have the right to know how to make a reasonable judgment in determining whether or not to seek their services.

Twenty-first Century Approach

1. State legislatures should require mandatory reporting by healthcare facilities of nosocomial infections and deaths, with adequate explanatory information.
2. All medical boards should set standards for providing information concerning an offense committed by the physician and/or disciplinary information.
3. This information should include all such actions recognized or taken in a preceding decade.

Eat to live—don't live to eat!

Hillel the Elder, in Forst, *The Laws of Kashrus* (1999), p. 28

---

173. Levy et al., Survey of Doctor Disciplinary Information on State Medical Board Web site, http://www.citizen.org/hrg/publications/1506.html (Aug. 9, 2000).

c) Overweight, Obesity, and Earlier Deaths

Among preventable causes of death, obesity may be second only to smoking. It has been underreported, with only 38 percent of patients so classified being reported as obese by their physician.[174] After the turn of this millennium, former U.S. Surgeon General Dr. C. Everett Koop found "obesity is escalating to epidemic levels"—obesity-related diseases cost Americans over $110 billion annually or 10 percent of the U.S. healthcare expenditures.[175] A National Health and Nutrition Examination Survey (NHANES III) reported that 33.4 percent of Americans are overweight, representing an increase during the full decade of the 1990s.[176] More than 300,000 adults die of obesity causes. A 2001 study concluded, "Interventions are needed to improve physical activity and diet in communities nationwide."[177] It has been

174. Stafford et al., *National Patterns of Physician Activities Related to Obesity Management*, 9 Archives Fam. Med. 631–38 (2000). In 2005 the association between obesity and smoking and psoriasis (hereditary recurrent dermatosis, a skin disease) was confirmed. *See* Herron et al., *Impact of Obesity and Smoking on Psoriasis Presentation and Management*, Arch. Dermatology 1527–34 (2005); Herron et al., *Obesity, Smoking, and Psoriasis*, JAMA 208 (2006); McGowan et al., *The Skinny on Psoriasis and Obesity*, Arch. Dermatology 1601–02 (2005).

175. Orange County Reg., Jan. 12, 1999, at Accent 7. For a discussion about avoiding childhood obesity in public schools, see U.S. Dep't of Health & Human Servs. Surgeon General's Call to Action to Prevent and Decrease Overweight and Obesity (2001) [hereinafter Surgeon General's Call to Action], *available at* http://surgeongeneral.gov/topics/obesity/ calltoaction/ CalltoAction.pdf (last visited Oct. 6, 2002), and Orange County Reg., June 2000, at News 24.

176. JAMA (1999). In 1999 a study showed that 58% of obese adult patients had not received any advice to lose weight from a physician in the previous 12 months. Galuska et al., *Are Healthcare Professionals Advising Obese Patients to Lose Weight?*, 282 JAMA 1576–78 (1999).

177. JAMA 1195 (2001). Further, a recent large study has found that not getting enough rest could lead to increased risks of gaining weight. The study used information on some 18,000 adults participating in the federal government's National Health and Nutrition Examination Survey. Orange County Reg., Nov. 18, 2004, at News 24. A South Dakota man who weighed 1,072 pounds and was dying of heart failure lost more than 450 pounds before leaving a hospital. At the beginning of 2005, he weighed 610 pounds and hopes eventually to slim down to 240 pounds. Orange County Reg., Jan. 6, 2005, at News 2.

noted that if adult humans in most Western countries restricted their caloric intake on averageby 20 percent to 25 percent, in the context of present-day standards of lifestyle and healthcare, we could increase in both average and maximum lifespan.[178] However, in the twenty-first century, well over one in three American adults may become classified as overweight or obese.

Childhood obesity continues to increase rapidly in the United States in the twenty-first century, particularly among African-Americans and Hispanics.[179] A recent AMA study showed that more than 10 percent of children ages two to five are overweight.[180] In 2004, a study found that blood pressure has increased over the past decade among children and adolescents, which "is partially attributable to an increased prevalence of overweight."[181] It has been written, "If the rate of obesity in the general population continues increasing at the same pace it has for the past two decades, half of the adults in the United States (50 percent) will be obese by the year 2025."[182]

---

178. CLARK, *supra* note 66, at 202. The October 2004 issue of the reproductive society's journal, *Fertility & Sterility,* published a study of Danish men undergoing military examinations indicating the sperm quality in those whose body mass index (BMI) was higher or lower than recommended. ORANGE COUNTY REG., Oct. 22, 2004, at News 23.

179. POOL, FIGHTING THE OBESITY EVIDENCE 6 (2000). Research into a gene called leptin (Greek for "thin") was successfully tried on mice and experiments on humans. However, in humans, it did not consistently result in a reduction in weight. SCIENCE 593 (1998); NATURE 1683 (1998). "But barger contestant David White (UK), also known as 'Mad Maurice Vanderkirkoff,' had a belly circumference of 54.2 in. (137.7 cm.) when measured on March 12, 2001." GUINNESS WORLD RECORDS 16 (2002).

180. ORANGE COUNTY REG., Apr. 17, 2005, at News 6. The Heart Association recommended that "toddlers should eat five fruits and vegetables a day along with fiber-rich grains and should switch from drinking full-fat dairy to 1 percent or fat-free dairy products after age 2." *Id.*

181. Munter et al., *Trends in Blood Pressure Among Children and Adolescents*, JAMA 2107 (2004). It has been noted that obesity currently causes over 400,000 deaths a year. Dietz et al., *Building Healthy Communities: Policy Tools for the Childhood Obesity Epidemic*, 30 J.L. MED. & ETHICS 83, 85 (2002). Further, obesity costs the United States over $117 billion per year. SURGEON GENERAL'S CALL TO ACTION, *supra* note 174.

182. JAMA 2845 (2001). The March 24, 2004 report by Duke University researchers stated that "childhood obesity has risen to a point that it can be considered a modern-day epidemic." ORANGE COUNTY REG., Mar. 25, 2002, at News 18.

There is now a standard way to define overweight, obesity, and morbid obesity. The body mass index (BMI) is calculated based on a person's height and weight. The weight in kilograms (2.2 pounds per kilogram) is divided by the square of the height in meters (39.7 inches per meter). A BMI of 25 or more is considered overweight, 30 or more is considered obesity, and 40 or more, morbid obesity.[183] This measurement could be required in our hospitals and schools with respect to the applicable patients and students. The results for students should be sent to their parents.[184] A study found that "among American 15-year-olds, 15 percent of girls and nearly 14 percent of boys were obese, and 31 percent of girls and 28 percent of boys were more modestly overweight."[185]

---

183. JAMA 2918 (2002). In a large community-based sample, increased body mass index was associated with an increased risk of heart failure. "As compared with subjects with a normal body mass index, obese subjects had a doubling of the risk of heart failure."

> Obesity accounted for 11% of cases of heart failure in men and 14% of those in women.

N. Eng. J. Med. 305 (2002).

> A recent study found that "among different immigrant subgroups, number of years of residence in the United States is associated with higher BMI beginning after 10 years. The prevalence of obesity among immigrants living in the United States for at least 15 years approached that of U.S.-born adults."

Goel et al., *Obesity Among U.S. Immigrant Subgroups*, JAMA 2860 (2004).

184. As stated by a physical education program leader: "Parents who receive health and fitness report cards were almost twice as likely to know or acknowledge that their child was actually overweight than those parents who did not get a report card." Orange County Reg., Aug. 12, 2003, at News 5. However, a 2005 study on data from 27,098 people in Europe, Asia, Africa, and the Americas found that the hip-to-waist ratio is a better predictor of the risk of heart attack than the BMI. "The risk of heart attack rose progressively as the ratio of waist size increased in proportion to hip circumference. The 20 percent of the survey who had the highest ratio were 2.5 times more at risk than the 20 percent with the lowest ratio." Orange County Reg., Nov. 4, 2005, at News 10 (reporting on that day's issue of *The Lancet* medical journal).

185. Orange County Reg., Jan. 6, 2004 (reporting on a study by the National Institute of Public Health in Copenhagen, Denmark, published in the January issue of *Archives of Pediatrics and Adolescent Medicine*).
One of the students of Socrates stated:

Bariatrics is the field of medicine that specializes in treating obesity. Two weight-loss medications have been approved by the Food and Drug Administration for long-term use: "fen-phen" and dexfenfluramine. Prescriptions were written for both between 1995 and 1997. They were withdrawn from the market after being discovered to be associated with heart disease. Further, many of those who took these medications were not mostly at risk because of obesity.[186] Excess weight risks premature mortality, cardiovascular disease, Type 2 diabetes mellitus, osteoarthritis, and certain cancers.[187] Type 2 diabetes mellitus affects some 17 million people in the United States. It has increased rapidly in parallel to the obesity epidemic along with sugar-sweetened beverages.[188] In 2003, a study of more than 900,000 U.S. adults found that increased body weight was associated with increased death rates for all cancers combined and for cancers at multiple specific sites.[189] However, in 2005, the

---

> He disapproved of over-eating followed by violent exercise, but approved of taking enough exercise to work off the amount of food that the mind accepts with pleasure; he said that this was quite a healthy practice and did not hinder the cultivation of the mind. . . . He believed that those of his associates who accepted the principles which he himself approved would be good friends all their life long to himself and to one another.

XENAPHON, *Memoirs of Socrates, in* CONVERSATIONS OF SOCRATES (trans., 1990).

186. Yanovski, N. ENG. J. MED. 2187 (2005). Nevertheless, "overweight and obesity are significant predictors of death from any cause during 20 years of follow-up and as recently as calendar year 2002." N. ENG. J. MED. 2197 (2005). It is estimated that more than 1 billion adults worldwide are overweight.

187. JAMA 229 (2003). "Type 2 Diabetes is diagnosed in children as young as five." 30 J.L. MED. & ETHICS (2002).

188. Schulze et al., *Sugar-Sweetened Beverages, Weight Gain, and Incidence of Type 2 Diabetes in Young and Middle-Aged Women*, JAMA 927 (2004).

> It is through type 2 diabetes that obesity seems to pose the biggest threat to public health. Doctors have found biological connections between fat, insulin, and the high blood sugar levels that define the disease. The CDC estimates that 55 percent of adult diabetics are obese, significantly more than the 31 percent prevalence of obesity in the general population. And as obesity has become more common, so, too, has diabetes, suggesting that one may cause the other.

SCI. AM., June 2005, at 76.

189. Trends of increasing risk with higher body-mass-index values were observed for death from cancers of the stomach and prostate in men

Centers for Disease Control and Prevention found that although obesity (extreme overweight) is very lethal, mere overweight is considerably less so.[190]

The chairman of the Interdisciplinary Council on Lifestyle and obesity management said, "Without a comprehensive law that addresses the many issues that contribute to the increasing prevalence of this disease, we can only expect that number to grow in years to come."[191] Another study has found that obesity increases the likelihood of premature death.[192] California's high school students are growing alarmingly fat and lazy on a steady diet of potato chips and video games.[191] One-third of the state's two million teens could face "chronic and debilitating health problems," such as diabetes. The 18 million Americans who have diabetes also find it to be a leading cause of blindness, associated with obesity, kidney failure, and heart disease. Diabetes claims 180,000 lives, the sixth-leading annual cause of death in the United States.[194] "The rising prevalence and severity of obesity are already reducing life expectancy among the U.S. population. A failure to address the problem could impede the improvements in longevity that are otherwise in store."[195] For His-

---

> and for death from cancers of the breast, uterus, cervix, and ovary in women. On the basis of associations observed in this study, we estimate that current patterns of overweight and obesity in the United States could account for 14% of all deaths from cancer in men and 20% of those in women.

Calle et al., N. Eng. J. Med. 1625 (2003). Howe.

190. Gibbs, *Obesity: An Overblown Epidemic*, Sci. Am., June 2005, at 70; Orange County Reg., Apr. 20, 2004, at News 16. Overweight, which is well below obesity, drops from No. 2 to No.7 as a leading cause of death.

191. Orange County Reg., Nov. 21, 1997, at News 37; Orange County Reg., Feb. 22, 2000, at News 10.

192. Orange County Reg., Jan. 1, 1997.

193. "Among primary care patients with Type 2 Diabetes, inadequate health literacy is independently associated with worse glycemic control and higher rates of retinopathy. Inadequate health literacy may contribute to the disproportionate burden of diabetes-related problems among disadvantaged populations." JAMA 475 (2002). However, there has been no link between dietary fat and dementia, according to a study published in the journal *Neurology*. Orange County Reg., Dec. 24, 2002, at News 5.

194. Orange County Reg., Jan, 2004, at Health 7.

195. Preston, *Dead Weight?—The Influence of Obesity on Longevity*, N. Eng. J. Med. 1137 (2005).

panic and black teens, the chances of being overweight can be twice the rate of whites and Asian Americans.[196] As stated by one expert, "The United States is not just the heaviest society in the world but probably the heaviest society in the history of the world."[197] The lowest mortality rate among women was for those who weighed at least 15 percent less than the U.S. average for women of similar age. Further, it was specifically noted, "there is little evidence of harm in being very thin."[198]

Because mammals share many philological characteristics, I place considerable confidence in a discovery made at Cornell University back in 1948 when it was found that the *life spans of rats* were extended by 33 percent simply by placing them on *low-calorie diets*.[199] A similar study by the University of Wisconsin in 1999 found that with a drastically reduced diet, the genetic processes of aging slow to a crawl. A recent research reported infants that who are breast-fed exclusively for 6 months to a year were 43 percent less likely to be obese. *Breast-feeding* beyond the first birthday was giving babies a 72 percent lower chance of becoming obese children.[200] A diet with about one-third fewer calories than a normal diet can prolong life.[201] It has also been found that *reducing television* and video game use may be a promising population-based approach to preventing childhood obesity.[202] Counseling is used to identify and address these issues.[203]

---

196. Orange County Reg., Sept. 26, 2000, at News 9.

> In the United States, diabetes is the leading cause of blindness in people aged 20 to 70. The American Diabetes Association estimates that 12,000 to 24,000 people lose their sight each year because of diabetes.

Orange County Reg., Feb. 17, 2003, at News 29 (citing Nature Medicine).

197. Orange County Reg., Jan. 1, 1998, at News 7.

198. N. Eng. J. Med. 777–23 (1995). Obesity is higher among black women. JAMA (2000).

199. Sci. Am., Jan. 1996, at 46.

200. Orange County Reg., July 11, 1999, at 1; *see also* JAMA 2461 (2001).

201. Science (1999). "Veterinarians consider it the leading health problem, affecting an estimated 30% of cats and dogs. Their life span is short enough already, without making them even shorter by allowing obesity to occur." L.A. Times, Feb. 18, 2000, at A28.

202. Robinson, *Reducing Children's Television Viewing to Prevent Obesity: A Randomized Controlled Trial*, JAMA 1651–1567 (1999).

203. 282 JAMA 1576–78 (1999).

A study has found television "as the number one cause of childhood and adolescent obesity. . . . Most experts agree that the resulting lack of physical activity is the primary reason why children become overweight."[204] The estimated 60 percent to 65 percent of adults in the United States who are overweight could achieve "significant weight loss and improved cardiorespiratory fitness through the combination of exercise and diet during 12 months"; however, a 2003 study indicated that "no differences were found based on different exercise durations and intensities in their group of sedentary, overweight women."[205] Another study in 2005 distinguished changes in posture and movement that are associated with the routines of daily life called non-exercise activity thermogenesis (NEAT). It was found that "posture allocation did not change when the obese individuals lost weight or when lean individuals gained weight, suggesting that it is biologically determined."[206] Secondly, it has been noted that "the drive toward thinness is fueling a $33 billion weight-loss industry" and several drugs are medically and intermittently used to reduce one's appetite.[207]

---

204. AKERS, OBESITY 36 (2000).

> A 1996 survey revealed that one-quarter of all Americans do not engage in any physical activity whatsoever during their leisure time. Another survey showed that Americans watch 1,589 hours of television, 51 hours of videos, and play 21 hours of video games a year. That's close to 4.5 hours per day spent sitting in front of a television screen.

A study, published as Hu et al., JAMA 1785 (2003), emphasized the importance of reducing prolonged TV watching and other sedentary behaviors for preventing obesity and diabetes.

205. Jakicic et al., *Effect of Exercise Duration and Intensity on Weight Loss in Overweight, Sedentary Women*, JAMA 1323 (2003). It was also found in one study that "almost one-third of obese men reversed their ED after exercise more than losing weight." Saigal, *Obesity and Erectile Dysfunction*, JAMA 3012 (2004).

206. 307 SCIENCE 584 (2005).

207. *Commercial Weight Loss Products and Programs: What Consumers Stand to Gain and Lose,* Report of the Presiding Panel, A Public Conference on the Information Consumers Need to Evaluate Weight Loss Products and Programs (Oct. 16–17, 1997), *available at* http://www.ftc.gov/os/1998/9803/weightlo.rpt.htm. "Because successful weight loss . . . [is] difficult, easily obtained non-prescription diet pills are an appealing alternative to the increasingly overweight population. . . . [However,] no population-based studies have examined the relationship between use of overall non-prescription weight loss products and use of prescription weight loss pills." JAMA 930 (2001).

Bariatric surgery, a surgical subspecialty that performs operations to treat morbid obesity, has been around for a half-century. The National Institutes of Health (NIH) endorsed a procedure in 1991 whereby surgeons staple the stomach to create a small pouch that accommodates up to three tablespoons of food. The person cannot eat more without becoming nauseated or vomiting. The American Society for Bariatric Surgery indicated that the number of these operations grew from 23,100 in 1997 to 63,100 in 2002. Massachusetts' health insurers have been paying $8,000 to $15,000 per operation, plus the surgeon's fee.[208]

In 2002 the Surgeon General noted that although many people believe that dealing with overweight and obesity is a personal problem, it is also a *community responsibility*: "When school lunchrooms or office cafeterias do not provide healthy and appealing food choices, that is a *community responsibility*. When new or expectant mothers are not educated about the benefits of breast-feeding, that is a *community responsibility*."[209] A hospital ethics committee is an important representative of community responsibility that can and ought to be taking appropriate action on the national increase in overweight and obese patients.

In 2004, Medicare declared that obesity is a disease. The Secretary of Health and Human Services (HHS) said, "Obesity is a critical public health problem in our country that causes millions of Americans to suffer unnecessary health problems and to die prematurely." Hence, they can make medical claims for treatments such as stomach surgery and diet.[210] However, they do not always win. More than three-quarters of people polled said that individuals themselves are to blame for being overweight or obese.[211] Nevertheless, in 2004 a New Jersey

208. ORANGE COUNTY REG., Jan. 12, 2003, at News 18. "At Tufts-NEMC, which has one of the largest programs in the country, gastric-bypass surgery is now third after transplants and open-heart surgery in profitability among the most common surgeries. 'Many surgeons and hospitals got into this for the revenue,' said Dr. William Mackey, the hospital's chairman of surgery." *Id.* "A study tallied that $93 billion per year goes to treat health problems of the overweight." ORANGE COUNTY REG., May 14, 2003, at News 12.

209. SURGEON GENERAL'S CALL TO ACTION, *supra* note 174. It has been pointed out that "the number of gastrointestinal surgeries performed annually for severe obesity increased from about 16,000 in the early 1990s to about 103,000 in 2003. . . . The number of practicing surgeons who are members of the American Society for Bariatric Surgery (ASBS) increased from 258 in 1998 to 1,070 in 2003." Robert Steinbrook, *Surgery for Severe Obesity*, N. ENG. J. MED. 1075.

210. ORANGE COUNTY REG., July 17, 2004, at 1.

211. ORANGE COUNTY REG., May 29, 2004, at News 7.

high school student who sued her basketball coach for telling her to lose 10 pounds was awarded $3 million. However, that award was held invalid because she had not proven any permanent damages.[212] In another case, a class action alleged that McDonald's restaurants failed to adequately disclose that their foods were less healthful, and hence they had been a "substantial factor contributing to [plaintiffs'] obesity." A New York court dismissed the action on the grounds that the minor plaintiffs did not render sufficient information indicating that consumption of vendor's food caused their obesity.[213]

Finally, it is interesting to note that according to the International Obesity Task Force 2005 adviser to the European Union, seven European countries—Cyprus, the Czech Republic, Finland, Germany, Greece, Malta, and Slovakia—have a higher percentage of men who are obese or overweight than the estimated 67 percent of men in the United States.[214] Nevertheless, lawsuits alleging discrimination based on overweight are increasing. In 2002 the New Jersey law that bars discrimination against the overweight protected a 400-pound woman whose obesity was allegedly due to a genetic condition.[215] Similarly, the Second U.S. Circuit Court of Appeals in New York City actually held the federal Americans with Disabilities Act may cover obesity if an employer regards it as a physi-

---

212. Besler v. Bd. of Educ. of W. Windsor-Plainsboro Reg'l School Dist., No. MER-L-000236-98; Nat'l L. J., May 17, 2004.

213. Pelman v. McDonald's Corp. A two-year study found that a "Mediterranean-style diet might be effective in reducing the prevalence of the metabolic syndrome and its associated cardiovascular risk." Esposito et al., JAMA 1440 (2004).

> A study of dietary patterns and lifestyle factors associated with mortality from all causes, coronary heart disease, cardiovascular diseases, and cancer, and cause-specific mortality found that among individuals aged 70 to 90 years, adherence to a Mediterranean diet and healthful lifestyle is associated with a more than 50% lower rate of all-causes.

Knaops et al., JAMA 1433 (2004).

214. Orange County Reg., Mar. 16, 2005, at News 17. In June 2005, the European Congress on Obesity estimated that about half a million children in Europe are suffering classic middle-age health problems because they are too fat. Orange County Reg., June 2, 2005, at News 21.

215. Viscik v. Fowler Equip. Co., 800 A.2d 826 (2002).

ological impairment or the plaintiff is so morbidly obese that it affects normal life activities.[216]

good v. Good

Eating so much as to become significantly overweight or obese is a matter of personal taste and should not become subject to disfavorable comment or discrimination by other people.

v.

Obesity can develop in a child due to lack of breast-feeding (less than six months to one year), overwatching TV, and being overfed for several years on a high-caloric diet; such a child also then becomes subject to diabetes and other diseases, as well as a shorter lifetime.

Twenty-first Century Approach

1. Preventable causes of significant overweight or obesity should receive greater emphasis as a matter of public policy to stop it from also causing many diseases and diminishing life's duration.
2. Although initial discrimination concerning preventable obesity should not be grounds for a lawsuit, a standard body-mass index (BMI) should be prescribed based upon a patient's height and weight, and physicians should confidentially report those who exceed the standard.

The surrounding world of nature, which preceded us and will succeed us, offers us the spectacle of longevity and an endurance that are denied us. This spectacle can be a source of anguish or of reassurance, depending on the relation we maintain with ourselves. A great deal of the destructiveness in our dealings with nature arises, it seems, from a stubborn refusal to come to terms with our finitude, to accept our fundamental limitations.

Cronin, *Uncommon Ground*, 1995, p. 436

Probably at no period in no culture have the old been so completely rejected as in our own country, during the last generations.

Lewis Mumford

---

216. Francis v. City of Meriden, 129 F.3d 281 (1997). Because the firefighter was fired simply for failing to meet generally applicable weight criteria, his complaint should be dismissed, the court said. Tobo, *As Obesity Claims Increase, Experts Say the Issue Isn't Nature or Nurture—It's Bias*, A.B.A. J., June 2005, at 17.

4. Attempts to Change from the "Slow Course to Death" to "Varieties of Approaches to Death," Kubler-Ross

With so many people living longer lives as of the turn of the twenty-first century, some seniors are flocking to classes to learn how to die and deliver themselves from terminal suffering. As stated by Cronin in the epigraph, these people at least are not maintaining a "stubborn refusal to come to terms with our finitude." In 1969 Kubler-Ross noted:

> The more training a *physician had the less he was ready to become involved* in this type of work. . . . Other authors have studied the physician's attitude toward death and the dying patient. We have not studied the individual reasons for this resistance but have observed it many times.[217] (Emphasis added.)

In the past 30 years, there have been a rather significant number of changes. An AMA study concluded, "By any standard one chooses, medical schools in the United States fail to provide even adequate education in the care of the dying." In fact, the evidence reveals a "well-established pattern of neglect of medical education in the care of the dying."[218] Most physicians are trying to do their best; however, some of them lure people to higher expectations of increasing technological development in spite of warnings that their patients are "doing better and feeling worse."[219] At the end of the twentieth century it was stated, "*Successful* medical treatment regularly causes a *slow course to death*. Yet, modern medicine has largely failed to note how a patient lives during the now prolonged course toward dying."[220] (Emphasis added.) Some have talked about an idealized concept of a "compression of morbidity," the shortening of the period of illness prior to death.[221] However, it has not been supported by enough evidence to instill confidence.[222] A Swedish writer coined

217. ELISABETH KUBLER-ROSS, ON DEATH AND DYING 245 (1969).

218. Hill, *Treating the Dying Patient*, ARCH. INTERNAL MED. 23 (1995).

219. Knowles, *Introduction,* DAEDALUS, Winter 1997, at 1–7.

220. JAMA 474 (1996).

221. Fries, *Aging, Illness & Health Policy: Implications of the Compression of Morbidity*, PERSP. BIOLOGY & MED. 407–28 (1986).

222. Olshansky, *Trading Off Longer Life for Worsening Health: The Expansion of Morbidity Hypothesis*, J. AGING & HEALTH 194–216 (1991); Verbruge, *Survival Curves, Prevalence Rates, and Dark Matters Therein*, J. AGING & HEALTH, 217–361 (1991).

the term *gerotranscendence,* which he defined as "a shift in meta perspective, from a materialistic and rational vision to a more cosmic and transcendent one, normally followed by an increase in life satisfaction." After a certain age, "people should think transcendently, i.e., beyond most physical and temporal limits. The question is can they change at that point in their lives. If so, this is not despair." Aldous Huxley wrote in his volume of essays *Tomorrow and Tomorrow and Tomorrow*:

> Despair is only the penultimate word, never the last. *The last word is realism*—the *acceptance of facts* as they present themselves, *the facts of nature and of human nature*, and the primordial fact of that spirit which transcends them both and yet is in all things. (Emphasis added.)

His reference to acceptance is the last word anticipated in *On Death and Dying* (1969) by Elisabeth Kubler-Ross (1926-2004), who identified five stages many patients go through in confronting their own deaths and was behind the hospice system (discussed in Chapter 14).[223] In 2004 Mark Kuczewski pointed out that "Kubler-Ross did not believe there to be an appreciable difference in adjustment to dying between those who were religious and those who were not. . . . [She] fell from grace with most scholars interested in the care of the dying because she went 'over the top' in her interest and speculations concerning the afterlife. . . . Insisting on a fact of life after death can easily seem to be a form of the denial of death. Certainly it is not a reasonable goal for bioethics to teach the world anything one way or another about the afterlife. But, how shall we deal with such matters in a way that connects with patients but keeps them at the center of the circle of the relationship? Bioethics has only just begun to consider such questions."[224] These

---

> 233. She helped turn thanatology, the study of the physical, psychological, and social problems associated with dying, into an accepted medical discipline. . . . The center she built in California in the late 1970s burned, and the police suspected arson. She set up another center in 1984 to care for children with AIDS. That center also was burned in 1994, and arson was again suspected.

Noble, ORANGE COUNTY REG., Aug. 26, 2004, at News 10.

224. Mark G. Kuczewski, *Re-Reading* On Death & Dying*: What Elisabeth Kubler-Ross Can Teach Clinical Bioethics*, AM. J. BIOETHICS, 2005, at W18 (citing KUBLER-ROSS, *supra* note 215, at 266); *see also* Cohen et al., *Walking a Fine Line: Physician Inquiries into Patients' Religious and Spiritual Beliefs*, Hast. Ctr.

include: "How can we foster our emotional, spiritual, and moral development so that we may better perform our service? How can we be or become the kind of persons who look at the matters on which we are consulted and let them show themselves as what they are, however they present themselves? Perhaps this is the future agenda of clinical bioethics, an agenda that is suggested by the excesses as well as the insights of Elisabeth Kubler-Ross."[225]

A patient realizes that both physicians and families often ignore his wish to be allowed to die or that these people may refuse to help him do so with dignity. Dignity is the maximization of control over one's self. Hence, such patients may stockpile barbiturates so as to be able to end their lives on their own terms when the time is propitious. They may read Derek Humphrey's book, *Final Exit,* that explains in detail how to end life in this manner.[226] Their desire to avoid pain and a prolonged life without a personal quality meaningful to themselves may lead them to seek medical help different from traditional medicine (see Chapter 14). Since the late 1990s, more physicians are willing to help dying patients carry out such wishes.[227]

An approach to delivering a prognosis in advanced cancer has been to anchor the estimate in an average survival for similar patients, disclosing the median survival[228] and "the prognostically variable interquartile ranges (IQRs)."

Rep., 2001, at 29–39; Lo et al., *Discussing Religious and Spiritual Issues at the End of Life*, JAMA 749–54 (2002).

225. He cites her last volume, KUBLER-ROSS, THE WHEEL OF LIFE: A MEMOIR OF LIVING AND DYING (1997), stating: "Her preoccupation with out-of-body, near-death experiences, channeling spirits, seeing 'fairies,' [and] reincarnation and her insistence on proving that death does not exist, are just too much for most of us. I have struggled to articulate what exactly is the nature of this discomfort with her."

226. More than 300 seniors at Leisure World's Laguna Hills and Seal Beach, California, learned from an expert what it would take to deliver themselves from terminal suffering. ORANGE COUNTY REG., Feb. 4, 1996, at Metro 2.

227. The percentage of physicians willing to help patients with AIDS kill themselves has nearly doubled over the past five years, according to a San Francisco study that suggests a growing acceptance among doctors of assisted suicide. "And more than half the doctors surveyed, all of whom specialize in AIDS treatment, said they have already helped at least one terminally ill patient obtain the narcotics needed to commit suicide, even though the practice is illegal in California." ORANGE COUNTY REG., Feb. 6, 1997, at News 22.

228. *E.g.*, the use of the National Cancer Institute's Surveillance, Epidemiology, and End Results (SEER) statistical software, Version 4.2, *available at* http://www.seer.cancer.gov/seerstat.

While communicating to patients, comments can be made upon "how the patient's existing performance status, symptoms, and subsequent treatment response might modify the estimate." In 2002 it was noted that this will provide the patient with a survival "estimate" that is reasonable.[229] Discussions can draw on the expertise of several disciplines, and the creation of a new professional role specializing in this area might be considered.[230]

good v. Good

Toward the end of life, often little can or should be given for its prolongation; comfort care should be the role of the medical profession.

v.

Near the "end of life" is still "life"; in spite of morbidity, it may be possible to make a willing patient feel and act alive by increasing its quality.

Twenty-first Century Approach

More emphasis must be placed upon the manifold methods of compressing morbidity at the end of life in lieu of emphasis on the "slow course to death."

Through Job we learn that it is not suffering that destroys people, but suffering without meaning. To cut oneself off from the possibility of suffering is to cut oneself off from love, and to cut oneself off from love is to cut oneself off from life itself.

Gunderman, *Is Suffering the Enemy?,* Hastings Center Report 41

It is only through suffering that we achieve wisdom.

Aeschylus

---

229. Although physicians report that they have mixed feelings about using survival estimates and statistics in their discussions with patients about poor prognoses, the technique has the clear advantage of anchoring patients in a prognostic estimate that is reasonable. With further discussion of the IQR, patients may come to understand the prognostic range and can plan accordingly.

Lamont & Christakis, *Complexities in Prognostication in Advanced Cancer*, JAMA 101–02 (2003).

230. Lawson & Tobin, JAMA 1977 (2000).

> *So this is how a thoughtful person should await death: not with indifference, not with impatience, not with disdain, but simply viewing it as one of the things that happen to us.*
>
> *The Meditations* by Marcus Aurelius, Bk 9.3

5. Pain and Its Partial Management

a) Pain

It has been noted that *reflex responses do not necessarily signify pain.*[231] For example, a federal district court held that without medical testimony that the fetus "experiences a conscious awareness of pain," a state should not ban a reflex on D&X, or partial-birth abortion.[232] Estimates of the commonality of pain in dying people vary: up to three out of four cancer patients suffer severe pain.[233] In a national survey, family members reported that about half the recent deaths of conscious patients were marked by moderate to severe pain most or all

---

231. Lloyd-Thomas & Fitzgerald, 313 Brit. Med. J. 797 (1996). Reflexes more often have to do with "feel" rather than "pain."

> Low back pain is a leading cause of disability. It occurs in similar proportions in all cultures, interferes with quality of life and work performance, and is the most common reason for medical consultations. Few cases of back pain are due to specific causes; most cases are non-specific. . . . No single treatment is superior to others; patients prefer manipulative therapy, but studies have not demonstrated that it has any superiority over others.

81 Bull. World Health Org. 671–76 (2003).

232. Women's Med. Prof'l Corp. v. Voinovich, 911 F. Supp. 1051 (S.D. Ohio 1995).

> Whether the fetus feels pain, however, hinges not on its biological development but on its conscious development. Unless it can be shown that the fetus has a conscious appreciation of pain after 26 weeks, then the response to noxious stimulation must still essentially be reflex, exactly as before 26 weeks.

Derbyshire & Ann Furedi, *"Fetal Pain" is a Misnomer*, 313 Brit. Med. J. 795 (1996); *see also supra* Chapter 9, § 10.

233. Bonica, *Cancer Pain, in* 1 The Management of Pain 400–60 (Bonica ed., 2d ed. 1990); Coyle et al., *Character of Terminal Illness in the Advanced Cancer Patient: Pain and Other Symptoms During the Last Four Weeks of Life*, 5 J. Pain & Symptom Mgmt. 83–93 (1990).

of the time.[234] Although pain can be eliminated by anesthesia and profound sedation,[235] this elimination of the quality of living may be too high a price for many to pay.[236] A 1997 study showed that "most elderly and seriously ill patients died in an acute care hospital often with severe pain. More than one in three conscious patients had severe pain, even in disease categories in which severe pain might not have been expected."[237] It was stated that "no one can assure him that his dying will be emotionally painless."[236] Increasing doses of painkillers does not always shorten these lives. Patients receiving higher doses of drugs such as morphine often live as long as those who do not.[239] A former widely used description of pain is: "an unpleasant sensory and emotional experience associated with actual or potential tissue damage, or described in terms of such damage."[240] (Pain relief under Emergency Department (ED) physicians is covered in Chapter 5, Section 2f.)

---

234. JAMA 1640 (1997). As regards surgery:

> Since every surgical intervention necessitates the infliction of some pain and suffering upon the patient, only those interventions that result in improvements in symptoms are justifiable. Even though there are differences between curative and palliative surgery, the differences are minimized when we identify the goals of surgery to include those of the palliative care model.

DUNN & JOHNSON, SURGICAL PALLIATIVE CARE 37 (2004).

235. Schmitz et al., *The Care of the Dying Patient*, *in* HAZZARD ET AL., PRINCIPLES OF GERIATRIC MEDICINE AND GERONTOLOGY 383–90 (1994).

236. JAMA 475 (1996). Although 38% to 74% of cancer patients experience significant pain, Cowan, *The Dying Patient*, CURRENT ONCOL. REP. 331–37 (2000), it has been reported in a European journal that approximately 50% of them receive inadequate analgesia, Ripamonti et al., *Pain Experienced by Patients Hospitalized at the National Cancer Institute of Milan: Research Project "Towards a Pain-Free Hospital*," TUMORI 412–18 (2000) (Italy).

237. ANNALS INTERNAL MED. 103–04 (1997).

238. PECK, DENIAL OF THE SOUL 202 (1997); *see also* MCDOUGALL, THEATERS OF THE BODY 129; *infra* Chapter 14 (regarding physician-assisted suicide (PAS) and double effect).

239. ORANGE COUNTY REG., July 2, 2000, at News 9 (citing *The Lancet*). However, "long-lasting (chronic) severe pain is prevalent among patients in substance abuse, especially patients in methadone maintenance treatment programs (MMTPs)." Rosenblum et al., JAMA 2370 (2003).

240. Association for the Study of Pain, Subcommittee on Taxonomy, International 1979.

The realization of the dying process may make applicable the aphorism attributed to Henry Maudsley, a nineteenth-century anatomist: "The sorrow that has no vent in tears makes other organs weep." Surveys show that women have been found to sense pain more often than men.[241]

Though the rock of my last hope is shiver'd,
And its fragments are sunk in the wave,
Though I feel that my soul is deliver'd
To pain—it shall not be its slave.

Lord Byron, Stanza to Augusta

And in most cases what Epicurus said should help: that pain is neither unbearable nor unending, as long as you keep in mind its limits and don't magnify them in your imagination.

The Meditations by Marcus Aurelius, Bk. 64

b) Pain Management and Its Limitations

The Joint Commission on Accreditation of Healthcare Organizations (JCAHO) accredits more than 14,000 healthcare organizations in the United States[242] that are required to address care at the end of life, including managing pain aggressively and effectively with blood pressure, temperature and respiration as vital signs, and using a ten-point scale—where zero signifies no pain and ten is the pain of an amputation without anesthesia—to evaluate and record the experience.[243] Patients in pain associated with terminal illness have been recognized by state medical boards as not being at substantial risk of addiction, and they are likely to need large doses of pain medica-

241. Orange County Reg., Apr. 8, 1998.

242. The Fertility Clinic Success Rate and Certification Act calls for accreditation organizations to inspect and certify embryo laboratories. *See* 42 U.S.C. § 263a(e) (1996).

243. Rosenfeld, The Truth about Chronic Pain 94 (2003); Joint Comm'n on the Accreditation of Healthcare Org., Comprehensive Accreditation Manual for Hospitals R. 1127 (1996). They would be independently certified for Medicare participation. *See* 42 U.S.C. § 1395x(e). For hospital certification regulations regarding pharmaceuticals, see 42 C.F.R. § 482.25 (1998).

tion over a long period of time.[244] In addition to almost 5,000 hospitals, the JCAHO guidelines also apply to home-care agencies, nursing homes, behavioral health facilities, outpatient clinics, and health plans. Pain management is defined as "a comprehensive approach to the needs of patients, residents, clients, or other individuals served who experience problems associated with acute or chronic pain."[245] Compliance with these guidelines (standards) is necessary for hospitals to maintain their accreditation.[246] However, in some states an immunity provision may protect physicians from discipline.[247]

During the end of the twentieth century, many physicians had been doing poor pain management for their patients.[248] "[D]espite good intentions and genuine concern for patients' comfort on the part of physicians, repeated evaluations of the state of pain therapy over the past 20 years suggest that many

244. Hoffman & Tarzian, *Achieving the Right Balance in Oversight of Physician Opioid Prescribing for Pain: The Role of the State Medical Boards*, 31 J.L. Med. & Ethics 21–40 (2003).

245. Joint Comm'n on the Accreditation of Healthcare Org., Pain Assessment and Management: An Organizational Approach 3 (2000). This differs from the AMA *Code of Medical Ethics*' emphasis on "dignity" and "autonomy": Rule 2.2.0 states that "physicians have an obligation to relieve pain and suffering and to promote the *dignity* and *autonomy* of dying patients in their care." AMA Code of Medical Ethics R. 2.2.0. (Emphasis added.)

246. The JCAHO standards effective for accreditation as of 2001 include the patient's right to appropriate assessment, recording the results in a way that allows regular reassessment, staff competency in pain assessment, procedures to support appropriate prescription, and education of patients and families about effective pain management.

247. The Texas Intractable Pain Treatment Act grants immunity from disciplinary action by the medical board to physicians when they prescribe opioids for intractable pain. Addiction has been associated with prescription opioid abuse. Michna et al., *Predicting Aberrant Drug Behavior in Patients Treated for Chronic Pain: Importance of Abuse History*, 28 J. Pain & Symptom Mgmt. 250–58 (2004); Reid et al., *Use of Opioid Medications for Chronic Noncancer Pain Syndromes in Primary Care*, J. Gen. Internal Med. 173–79 (2002). California also requires that all patients have a consultation in order that the physician may qualify for immunity. Cal. Bus. & Prof. Code § (West 1990).

248. AMA Department of Young Physicians Services, Pain Management: Resources for Physicians, http://ww.texnet.net/paincare/acute.htm (last visited July 20, 2000).

patients receive inadequate pain relief."[249] It should also be noted that in most states, advanced practice nurses are also authorized by law to prescribe medications for conditions requiring pain management.[250] A former study of cancer patients found 86 percent of physicians reported that most patients with cancer were undermedicated for their pain. It also found that non-cancer patients receive even less adequate pain treatment than patients with cancer-related pain.[251] The most severely neglected group was the population over 85 years of age, a fast-growing segment of the population in the United States.[252] A 1998 study found that among patients 85 and older who were in pain, 30 percent got no pain medicines, compared with 21 percent of patients aged 65 to 74.[253] "The oldest old were one-third more likely than the youngest patients to receive morphine or other strong opiates (38 percent vs. 18 percent)."[254]

In 1990 for the first time, a healthcare provider was held liable for failure to treat pain adequately.[255] A jury awarded $15 million in damages to the

---

249. *See also* COMMITTEE ON CARE AT THE END OF LIFE, APPROACHING DEATH: IMPROVING CARE AT THE END OF LIFE 5 (M. J. Field & C. K. Cassel eds., 1997). Nevertheless, it has been stated that "the financial burden of chronic pain in lost wages, lost productivity, and lost time may amount to as much as $100 billion a year." ROSENFELD, *supra* note 241, at 287.

250. NAT'L COUNCIL OF STATE BDS. OF NURSING, PROFILES OF MEMBER BOARDS 2000, at 251 (2001).

251. N.Y. STATE TASK FORCE ON LIFE & THE LAW 43–44 (1994).

> About one in three individuals in the developed world will be diagnosed with cancer and half of those patients will die of progressive disease. More than 80% of patients with cancer develop pain before death. That is also the percentage of those whose "cancer pain can be controlled with simple treatments." The remaining 20% require use of a multidimensional approach.

Bruera & Kim, JAMA 2476 (2003).

252. J.L. MED. & ETHICS 307 (1997) (reviewing PAIN IN THE ELDERLY (Ferrell & Ferrell eds., 1996)).

253. ORANGE COUNTY REG., June 17, 1998, at News 13.

254. *Id.*

255. Estate of Henry James v. Hillhaven Corp. But it has been found that "chronic pain is independently related to low self-rated health in [the] general population." It is expensive "because of the resulting disability and absence from work." Mäntyselkä, *Chronic Pain and Poor Self-rated Health*, JAMA 2435 (2003).

family of Henry James, a cancer patient in a nursing home. Although his physician had prescribed 7.5 cc of oral morphine every three hours as needed, a nurse substituted a mild tranquilizer because she believed that James was becoming "addicted to morphine." The court found this to be "inhumane treatment" inflicted without regard to the consequences and without care as to whether or not the patient received analgesic (pain) relief and without care that the result and procedures were torture of the human flesh.[256]

In 1992 and 1994 respectively, the U.S. Department of Health and Human Services Agency for Healthcare Policy and Research issued Acute Pain Management Guidelines and Cancer Pain Management Guidelines.[257] These called for individualized pain control plans developed and agreed to by patients, their families, and providers. Not being regulations, neither of these guidelines have the effect of law.

Back in 1986 the World Health Organization (WHO) issued a set of guidelines under the title Cancer Pain Relief. One guideline, the "three-step analgesic ladder," was based on the premise that most patients throughout the world should have adequate pain relief with relatively inexpensive drugs. However, a 1995 study concluded that there are insufficient data to estimate *confidently* the effectiveness of the WHO analgesic ladder for the management of cancer pain.[258] As of the turn of the third millennium, U.N. drug-control experts decried the near absence of morphine in the developing world.

---

256. *See also* Angarola, *Inappropriate Pain Management Results in High Jury Award*, 6 J. PAIN & SYMPTOM MGMT., 407 (1991).

> In 2002 Dr. James Graves, Florida's prescriber of OxyContin, became the nation's first doctor to be tried and convicted of manslaughter, racketeering, and unlawful delivery of a controlled substance. The 1,000 patients at his pain-management office brought in $500,000 a year. . . . Reports from 20 U.S. metropolitan areas show increased OxyContin-related deaths and emergency-room episodes since OxyContin's 1996 introduction.

ORANGE COUNTY REG., Feb. 20, 2002, at News 11.

257. Public Health Service Act § 901(b), 42 U.S.C. § 299(b)1; § 912 (a)(1)-(2), 42 U.S.C. § 299b-1(a)-(2).

258. JAMA 1870 (1995). Pain is significantly experienced by 38% to 74% of cancer patients, 50% of whom receive inadequate analgesia (which may be accompanied with loss of consciousness). Cowan, *supra* note 234, at 331–37; Ripamonti et al., *supra* note 234, at 412–18.

Two reasons appeared to account for this, and neither primarily related to the cost of medicine. The first was that they are overwhelmed with fighting other basic medical problems, such as malnutrition. The second was that both governments and doctors are overly restrictive in distributing morphine because they fear it can be diverted to illegal markets.[259] Physicians handling patients who are in pain prior to referring them to surgeons aften think about the prescription of a drug to alleviate such suffering. However, almost 9 out of 10 surgeons indicate they should consult with the patients prior to any prescriptions of medication for pain.[260] This would appear to be a major exception in view of the fact that relief of pain has been called a "fundamental imperative for any clinician."[261] As regards post-operative pain data, it has been suggested that:

> Many patients continue to fear severe pain after surgery, and many post-operative patients continue to have significant pain. Further improvements in the quality of pain control will not occur unless it is recognized as a priority by healthcare and an institutional approach is taken to assure that high-quality analgesic care is consistently provided. [Further,] many healthcare institutions continue to lack any organized institutional approach for the management of acute pain.[262]

259. ORANGE COUNTY REG., Feb. 24, 2000, at News 15. Nevertheless, it was recently stated that, in NICUs, "in daily practice, a newborn in pain who receives ventilatory support needs to receive analgesic treatment, independent of any routine morphine administration. If an infant was in pain, morphine was given." JAMA 2425 (2003).

260. Graber et al., *Informed Consent and General Surgeons' Attitudes Toward the Use of Pain Medication in the Acute Abdomen*, AM. J. EMER. MED. 113–16 (1999). Analgesia should be administered "only with the knowledge and consent of the surgeon who assumes responsibility for decision-making." Nissman et al., *Critically Reappraising the Literature-Driven Practice of Analgesia Administration for Acute Abdominal Pain in the Emergency Room Prior to Surgical Evaluation*, AM. J. SURGERY 291–96 (2003).

261. Fishman et al., *Pain Management in the ED*, 22 AM. J. EMER. MED. 56 (2004). Further, even in emergency medicine, "as a guiding principle of medicine and core covenant with our patients, every EP [emergency physician] must embrace *providing* timely and *effective pain control as a fundamental duty*." *Id.* (Emphasis added.)

262. Blau et al., *Organization of Hospital-Based Acute Pain Management Programs*, 92 HOSPITAL-BASED PAIN MGMT. 466, 467 (1999).

Previous reviews also showed inadequate treatment of pain for infants and children. Studies suggest that neonates, infants, and children can receive analgesia and anesthesia safely, with proper age-related adjustments in clinical practice and dosing. Making the hospital environment a less terrifying place may reduce anxiety and fear, which can themselves exacerbate pain. Conversely, nonpharmacologic approaches should not be used as an excuse to withhold appropriate analgesics.[263]

By 1996, 10 states had enacted statutes to permit prescribing of controlled substances for intractable pain, and six of them offer physicians and surgeons protection from disciplinary action.[264] Maryland has provided an option on the advance directive form to allow the patient to direct that pain medication not be given if it would shorten the patient's life.[265] Each state enacted and produced regulations on the use of drugs by healthcare professionals, some of which may be more stringent than their federal counterpart. On the other hand, if a federal regulation is more stringent than the state regulation, the federal regulation will govern in most cases. However, one year after the turn of the third millennium, California enacted the first law of its kind: doctors are required to take courses on pain management—particularly at the end of life.[266] A recent study found that the number of Americans who begin misusing painkillers each year had almost quadrupled from 1990 to 2001.[267]

---

263. N. Eng. J. Med. 1094 (2002).

264. For example, the California statute states: "No physician and surgeon shall be subject to disciplinary action by the board for prescribing or administering controlled substances in the course of treatment of a person for intractable pain." Cal. Bus. & Prof. Code § 2241.5(c) (West); *see also* Nev. Rev. Stat. § 630.3066 (1999); Or. Rev. Stat. § 677.474 (1999). Only Oregon requires the patient's written consent to pain medication. Or. Rev. Stat. § 677.485 (1999).

265. Md. Code Ann., Health–Gen. § 5-603 (1999); *see also* Federal Controlled Substances Act of 1970, 21 U.S.C. § 801 (enacted as part of the Comprehensive Drug Abuse Prevention and Control Act of 1970, Pub. L. No. 91-513k, 84 Stat. 1242.

266. Orange County Reg., Nov. 1, 2001, at News 21. "That law also requires the state medical board to track complaints of doctors mishandling pain care and ensure that those complaints are reviewed by a pain specialist. . . . Physicians will have four years to complete the courses to renew their licenses." *Id.*

267. Orange County Reg., Oct. 17, 2003, at News 14.

Regardless of their effectiveness, painkillers are often given to terminal patients in conjunction with efforts to prolong their lives. As stated by Omar Mendez, in his *Death in America*: "Sometimes because of legal issues we are driven to the point of doing the inhumane by unnaturally prolonging suffering and pain when there is no hope for recovery, or artificially maintaining a body that has no cognitive functions despite the family's requests and even the previously expressed wishes of the patient."[268] This has been called "harmful," "useless," and "disrespectful."[269] It has been noted that "to respect competent patients in these circumstances, providers should not generally use the presence of pain itself as a reason to doubt the capacity for a dying patient's decision-making."[270]

In this connection, in 2005 it was noted that "the available drugs to treat neuropathic pain have incomplete efficacy and dose-limiting adverse effects." This has quite a broad application:

> Neuropathic pain is a complication of cancer, diabetes mellitus, degenerative spine disease, infection with the human immunodeficiency virus, the acquired immunodeficiency syndrome, and other infectious diseases, and it has a profound effect on quality of life and expenditures for health care.[271]

The AMA *Code of Medical Ethics* Section 2.20 titled "Withholding or Withdrawing Life-Sustaining Medical Treatment" contains a provision that prescribes the obligation of physicians "to relieve pain and suffering and to promote the dignity and autonomy of dying patients in their care. This includes providing effective palliative treatment *even though it may foreseeably hasten death*" (i.e., result in PAS or euthanasia; but see Chapter 14). The state of Michigan brought criminal action against a physician for the murder of a patient by lethal injection. The physician's conviction of second-degree murder was affirmed on appeal in reliance on the state Supreme Court's decision determining that there is no constitutional right to commit euthanasia so that

268. MENDEZ, DEATH IN AMERICA 614.

269. Hast. Ctr. Rep., Jan.-Feb. 1998, at 37.

270. *Id.* at 26.

271. Gilron et al., *Morphine, Gabapentin, or Their Combination for Neuropathic Pain*, N. ENG. J. MED. 1324 (2005).

an individual can be free from intolerable and irremediable suffering.[272] This type of ruling was also adopted by the *Code of Medical Ethics* Section 2.211 with respect to physician-assisted suicide (PAS), which it stated to be "fundamentally incompatible with the physician's role as healer, would be difficult or impossible to control, and would pose serious societal risks." It adds, "Patients near the end of life must continue to receive emotional support, comfort care, and adequate pain control" without overruling its condemnation of PAS. Section 2.21 also states, "Euthanasia is fundamentally incompatible with the physician's roles as a healer . . . [it] heightens the significance of its ethical prohibition" (see Chapter 14 on both of these).

In 2002 it was stated that "despite the millions of people in pain every day, our medical establishment does not view pain management as a priority. You may find this statement surprising. Then again, if you've spent years going from doctor to doctor, receiving conflicting diagnoses and advice, you may not."

Pain is different from most other conditions doctors are trained to treat.

> For one thing, *no tool exists that can measure pain.* Doctors can't locate it with an MRI scan or measure it with a blood test. In January 2001, JCAHO decreed that to receive its accreditation, hospitals must consider pain the fifth vital sign, as important to measure as blood pressure or pulse. Doctors who dismiss their patients' pain are now at risk for losing their licenses. The commission also ordained that pain treatment *in addition* to pharmaceuticals be available to patients in hospitals.[273]

However, it has recently been noted that "virtual reality immerses users in a three-dimensional computer-generated" headset and is "uniquely suited to distracting patients from their pain." It does this by reducing the amount of pain-related activity in the brain. The patient's attention is no longer focused on the wound and the pain but drawn into the virtual world (e.g., a "SuperSnow

---

272. People v. Kevorkian, 639 N.W.2d 291, 305 n.42 (Mich. Ct. App. 2001); *see also* Gauthier, *Active Voluntary Euthanasia, Terminal Sedation, and Assisted Suicide*, J. Clinical Ethics 43–44, 49 (2001). *See* ch. 14, section 5c.

273. Dillard, The Chronic Pain Solution xviii–xx (2002). Nevertheless, undertreatment of pain by physicians is considered to give rise to a threat of criminal prosecution or possible action for an illegal prescription practice. Bourguignon & Martyn, *Physician-Assisted Suicide: The Supreme Court's Wary Rejection*, U. Tol. L. Rev. 253–72 (2000).

World, which will feature lifelike human avatars that will interact with the patient"). Pain-related brain activity using functional magnetic resonance imaging (MRI) to scan the brain will show "a large increase in pain-related activity in five regions of the brain that are known to be involved in the perception of pain: the insula, the thalamus, the primary and secondary somatosensory cortex, and the affective division of the anterior cingulated cortex." The patients change with virtual-reality equipment showing them another world; pain-related activity in their brains decreases significantly (and they also reported large reductions of subjective pain), which result may be detected on the MRI. However, it has been noted that "more research is needed to determine whether virtual reality can enhance the treatment of post-traumatic stress disorder (PTSD)."[274]

Hypnosis has been accepted by the American Medical Association (but some people are more "hypnotizable" than others), as is acupressure, called Traditional Chinese Medicine (TCM) (which blocks nerve impulses from carrying pain messages to the brain and may be used unless the patient has "a tumor or serious infection").[275] A hypnotist in Dallas, Texas, informed me in 2004 that "hypnotists at this time are not licensed or registered." However, she holds two certifications: one from the American Board of Hypnotherapy (which was originally in California and has since moved to Hawaii) and the other from the International Medical and Dental Hypnotherapy Association (IMDHA), headquartered in Michigan.[276]

After the turn of the third millennium, it was found that a single gene can determine whether a person will be a "wimp or a stoic" when it comes to handling pain. A gene produces an enzyme called COMT that metabolizes the neurochemical dopamine, which acts as a signal messenger between brain cells.[277] Research is now under way to predict which drug would be best for the patient.

274. Hunter G. Hoffman, SCI. AM., Aug. 2004, at 60–65.

275. *Id.* at 148–56; *see also* LERNER, CHOICES IN HEALING: INTEGRATING THE BEST OF CONVENTIONAL AND COMPLEMENTARY APPROACHES TO CANCER (1998).

276. E-mail from Joanna Rodriguez (Aug. 6, 2004): "Both of the associations above have schools in Texas." They claim that hypnotherapy can help with weight loss, grief, etc.

277. "A quarter of the U.S. population carries the 'stoic' gene variation, while another quarter has the gene variant that makes them super-sensitive to pain." ORANGE COUNTY REG., Feb. 21, 2003, at News 18.

In 2003 a goal-directed delivery of sedative (allaying irritability) and an analgesic medication (relieving pain without causing loss of consciousness) was recommended as standard care in intensive care units (ICUs). The Richmond Agitation-Sedation Scale (RASS) became the first sedation scale to be validated for its ability to detect such changes in sedation status over consecutive days of ICU care.[278]

good v. Good

All pain should be relieved at all times by all physicians.

v.

Pain relief involves the choice of maintaining dignity for those who continue to have some, rather than to lose all their remaining quality of life.

Twenty-first Century Approach

1. More emphasis should be placed upon pain management due to the recognition that such management, including post-operative pain, has not been adequately enforced in many states. In some cases, the patient may lose his quality of life and his decision-making power.
2. State legislatures should consider authorizing physicians to use overdoses which cause a patient's death subject to the conditions currently authorized for such use in the state of Oregon (see Chapter 14).
3. As of the turn of the third millennium, third world (poorer) countries are severely deprived of pain-relieving drugs. Some effort should be made to provide such drugs to them independently of total consideration regarding economic relief.

---

278. The RASS takes less than 20 seconds to perform and requires minimal training. The driving unmet need for goal-directed sedation practice has been met—now an instrument has been shown to detect variations in level of consciousness over time. [RASS] should lead to better characterization of acute brain dysfunction as an organ failure, reductions in the random variation with which patients' sedatives are currently managed, and appropriate interventions aimed at prevention or reversal of acute brain dysfunction.

Ely et al., JAMA 2983 (2003).

I place economy among the first and most important of republican virtues, and public debt as the greatest of the dangers to be feared.

Thomas Jefferson

The important point is a lively recognition that, with aging populations and constantly developing technologies, the greatest future threat to a steady-state medicine will be the costs of healthcare for the elderly.

Callahan, *False Hopes* (1998), p. 256

"Keep 'em alive, boy! Keep 'em alive!" said an old physician to his young brother practitioner. "Dead men pay no bills."

Author unknown

6. Costs Incurred Nearing End-of-Life Healthcare

In 1997 Richard Epstein noted that the roots of the current problems with the American healthcare system lie in universal acceptance of a false premise concerning the existence of a fundamental right to healthcare and education.[279] However no such fundamental rights are set forth in the Constitution. Essentially, there are only the so-called "legislative entitlements." In 1976 the U.S. Supreme Court stated that costs "must be weighed with respect to any procedural safeguards required by the Due Process Clause":

> The Government's interest, and hence that of the public, in conserving scarce fiscal and administrative resources is a factor that must be weighed. At some point the benefit of an additional safeguard to the individual affected by the administrative action, and to society in terms of increased assurance that the action is just, may be outweighed by the cost. Significantly, the cost of protecting those whom the preliminary administrative process has identified as likely to be found undeserving may in the end come from the pockets of

---

279. RICHARD A. EPSTEIN, MORTAL PERIL: OUR INALIENABLE RIGHT TO HEALTHCARE? 1–4 (1997).

the deserving, since resources available for any particular program of social welfare are not unlimited.[280]

For example, the Balanced Budget Act of 1997 mandated both reductions in federal spending and tax cuts over the period from 1998 through 2002, part of which was to be secured by reducing growth in Medicare spending to all physicians and hospitals.[281]

In 2000, a unanimous Supreme Court denied recovery against a Health Maintenance Organization (HMO) for an alleged violation of a "fiduciary duty" by offering incentives to their doctors to hold down treatment costs. It held that since Congress "has promoted the formation of HMO practices" for 27 years, the courts have "no warrant to precipitate [such an] upheaval."[282] A citizen's rights do not include the publicly subsidized longevity of all these costs.[283] This point was expressed in a poll of HMOs taken in that year when 79 percent of the people polled gave high marks to their HMOs for timely treatment of illnesses or injuries.[284] Nevertheless, it was stated that "expenditures once lavished on tombs are now expended on terminal care."[285]

---

280. Mathews v. Eldridge, 424 U.S. 319, 348 (1976) (citing Friendly, *Some Kind of Hearing*, 123 U. PA. L. REV. 1267, 1303 (1975)); *see also* Goss v. Lopez, 419 U.S. 565, 583 (1975) (ruling that public schools could provide an extremely modest hearing to a student threatened with expulsion because "even truncated trial-type procedures might well overwhelm administrative facilities in many places and, by diverting resources, cost more than it would save in educational effectiveness").

281. *See* N. ENG. J. MED. 300 (1999). In 2005 it was reported by the Congressional Budget Office that the United States will accumulate a $5 trillion budget deficit by 2015 if Congress extends the tax cuts of President George W. Bush and diverts some Social Security money to private accounts. ORANGE COUNTY REG., Feb. 2, 2005, at News 8. His 2005 budget is $2.5 trillion.

282. Pegram v. Herdrich, 530 U.S. 211, 233 (2000). The statute was enacted "in order to federalize malpractice litigation in the name of fiduciary duty." 29 U.S.C. §§ 1001, 1002(21), 1009(d).

283. Hast. Ctr. Rep., Mar.-Apr. 1999, at 29; *see also* Harui Morreim, *Moral Justice and Legal Justice in Managed Care: The Ascent of Contributive Justice*, 23 J.L. MED. & ETHICS 247–65 (1995).

284. ORANGE COUNTY REG., Sept. 25, 2000, at 4.

285. HEINZ, THE RITE OF DEATH 17 (1999).

a) Entitlements and Their Limitations; Social Security

Near the turn of the twenty-first century, elderly persons had higher living standards than any generation over 65 in U.S. history. A national Center for Health Statistics report linked income to the state of a person's health.[286] Almost all such persons are receiving Social Security and Medicare or Medicaid benefits from the government. This is being accomplished at higher cost.[287] In 2004, Federal Reserve Chairman Alan Greenspan has said that the U.S. will face "abrupt and painful" choices in trimming the costs of these three benefits that have been promised to the 77 million of our people who were born within the first 20 years following World War II. They have been called "baby-boomers" and will begin turning 65 for retirement in 2012.

Social Security was created in 1935 as a payroll tax-based retirement, survivors, and disability benefits program. Old Age and Survivors insurance (OAS) provides a modest retirement supplement and payments to a surviving spouse. As of 1995 it became the single largest item in the federal budget; it accounts for 20 percent of all spending and all entitlements that "approach

---

286. "The median net worth of Americans 65 and over was $88,192 in 1991, compared with $36,623 for all households, according to the Census Bureau. For the elderly, the 1993 poverty level was $6,930 for a person living alone and $8,741 for a couple." ORANGE COUNTY REG., July 30, 1998, at News 6. In this connection, a Harvard University study found that about half of personal bankruptcies in 2001 were triggered by medical costs or illness. That year, there were 1.46 million personal bankruptcies in the United States. ORANGE COUNTY REG., Feb. 2, 2005, at Business 6.

287. In 2002 it was reported that three-fourths of a trillion dollars was spent annually on healthcare for chronic diseases, which are responsible for seven of every 10 deaths. U.S. DEP'T OF HEALTH & HUMAN SERVS., THE BURDEN OF CHRONIC DISEASE AND THEIR RISK FACTORS: NATIONAL AND STATE PERSPECTIVES, 2002. About 10% of patients account for 70% of costs. Abbo, N. ENG. J. MED. 425 (2005). In order to reduce rising healthcare costs, another physician proposed requirements that would limit the amount of medical care people could obtain. Ginsburg, *Controlling Health Care Costs*, N. ENG. J. MED. 1591–93 (2004). The federal administration is seeking to restrain growth in mandatory spending in the budget for 2006 by trimming costs in Medicaid, the joint program with states that pays the cost of poor people's healthcare. ORANGE COUNTY REG., Feb. 7, 2004, at 4. However, community health centers, heavily used by the poor, would increase almost 18%. *Id.* at 9.

two thirds of non-interest federal spending."[288] Over 70 percent of American families now pay more in Social Security taxes than they do in income taxes. Without any real asset to meet growing future liabilities, the only alternatives are higher taxes and borrowing or reduced benefits. The 1998 Annual Report on financial outlook for Social Security stated that after 2013, it will take $684 billion in new taxes and new borrowing each year to keep Social Security afloat. The original (1935) Act stated: "Every qualified individual . . . shall be entitled to receive, with respect to the period beginning on the date he attains the age of sixty-five . . . and ending on the date of his death, an old-age benefit." Social Security was designed to establish a floor to protect seniors; that is, it was always assumed that other retirement income would be necessary, and since 1984, 50 percent of those payments have been subject to taxation at a progressive rate for those whose income exceeds certain minimum amounts. The 2006 budget provided some estimates of the cost in increased government borrowing for the president's proposal to allow younger workers to set up private savings accounts as part of a change in Social Security.[289]

The Older Americans Act (OAA) of 1965, a statute designed to help certain seniors continue to live (and to pay for 200 million meals), was not renewed in 1995. The National Family Support Act of 2000 was renewed until 2005, giving states the option of charging sliding-scale fees based upon their income.[290]

As these entitlement programs for the elderly become larger, they are on a collision course with the congressional budget resolution. By the end of 1997 healthcare costs topped $1 trillion for the first time in a single year.[291] In 2000, a writer noted that "Everyone agrees that the United States now spends too

288. SAMUELSON, THE GOOD LIFE AND ITS DISCONTENTS 15 (1995).

289. ORANGE COUNTY REG., Feb. 7, 2005, at 4. "Social Security was never designed to be a retiree's sole source of income. And in the current debate about how to strengthen the system, we should not impose that burden upon it." RICHARD GEPHARDT, AN EVEN BETTER PLACE 78 (1999).

290. "The total budget for this was $1.0 billion, as distinguished from the $450 billion to the Department of Health and Human Services (HHS)." North, *Secure Retirement*, Jan.-Feb. 2002, at 22.

291. ORANGE COUNTY REG., Jan. 13, 1998, at News 8. "Unlike the politicians of both parties who ignore the coming baby-boomer Social Security and Medicare crunch, Fed Chairman Alan Greenspan states that the problem is enormous and must be addressed quickly." ORANGE COUNTY REG., Mar. 7, 2004, at Commentary 6.

much on healthcare."[292] Spending on prescription drugs grew more than any other category—it had climbed 15.4 percent in 1998 to $90.6 billion.[293] The following year the U.S. Attorney General named "healthcare fraud as the government's second highest priority after violent crime."[294] The General Accounting Office has estimated that as much as 10 percent of all government expenditures on healthcare are out of the system because of fraud or abuse.[295] (See Chapter 6, Subsections 6 and 7, and Appendix C.)

good v. Good

Because entitlements to Social Security, Medicare, and Medicaid are generally considered "fundamental rights" by the populace, they should be increased whenever the need arises.

v.

These entitlements cannot be considered "Constitutional" rights, and they are not enacted free from budgetary restraints.

Twenty-first Century Approach

A special effort is often needed in the twenty-first century to eliminate the widely held misconceptions that entitlements to healthcare and Social Security are either constitutional rights or remain free from budgetary restraints.

An honest, no-holds-barred look at healthcare rationing, with a plea for cost-effectiveness analysis. Who could argue? Current rationing is hardly rational.

Alfred I. Tauber, M.D., Boston University

---

292. Dwarkin, Sovereign Virtue 207 (2000). "Administrative expenses account for a significant part of hospital costs, and American doctors' salaries are extremely large by other nations' standards." *Id.* at 308.

293. Orange County Reg., Jan. 10, 2000, at News 5. The information was obtained from the Healthcare Financing Administration.

294. *See* L. Aussprung, *Fraud and Abuse*, J. Legal Med.; Friedman, *A Concerned Optimist: An "Exit Interview" with Bruce Vladeck, PhD*, 281 JAMA 757–61 (1999).

295. Office of Inspector Gen., U.S. Dep't of Health & Human Servs., Semiannual Report, October 1, 1998–March 31, 1999 (indicating that improper payments under Medicare fee-for-service alone were estimated at $12.6 billion during fiscal year 1998).

I can imagine a future version of myself at age 85, cognitively very debilitated as a result of a stroke. Suppose I also have life-threatening heart problems that can be corrected either with a totally implantable artificial heart or the left ventricular assist device at a cost of $160,000. I do not see that treatment as a reasonable use of limited healthcare resources—not for me, not for anyone in those circumstances. This is a rationing decision we have imposed on ourselves; it has not been imposed upon us by "bureaucratic others." We make this decision, in part, because we do not want to tax ourselves to pay for that intervention.

Leonard Fleck, Hastings Center Report,
September–October 2002, p. 4

b) Inequitable Allocations and Deceiving Healthcare Payers v. HMOs

Some writers have suggested that high federal expenditure on certain diseases such as HIV/AIDS are unethical, since they were made for political reasons in order to please special-interest groups. It is termed "exceptionalism" because so many other potentially lethal diseases are not receiving proper attention.[296] As regards children, the American Academy of Pediatrics stated that "components of palliative care are offered at diagnosis and continued throughout the course of illness, whether the outcome ends in cure or death."[297] About 50,000 children die and 500,000 of them undergo life-threatening conditions.[286] "Although it represents an ideal, the presence of a designated pediatric palliative care team in all health care facilities that serve life-threatened children is currently a luxury."[299]

Although costs for the elderly account for almost one-third of healthcare spending,[300] fewer Medicare dollars are spent on the final two years for a 90-

296. ANNALS INTERNAL MED. 756 (1998).

297. American Academy of Pediatrics, Committee on Bioethics and Committee on Hospital Care, *Palliative Care for Children*, 106 PEDIATRICS 351–57 (2000).

298. MacDorman et al., *Annual Summary of Vital Statistics—2002*, 112 PEDIATRICS 1215–30 (2003).

299. Himelstein et al., *Pediatric Palliative Care*, N. ENG. J. MED. 1752–62 (2004).

300. 332 N. ENG. J. MED. 999–1003 (1995).

year-old than for a 70-year-old. However, 20 percent of older U.S. adults have chronic disabilities,[301] and the cost of medical care for a disabled older person averages three times that for a non-disabled senior. Although a number of measures of old age disability and limitations have shown improvements in the decade preceding 2002, it has been concluded that "research into the causes of these improvements is needed to understand the implications for the future demand for medical care."[302] The *New England Journal of Medicine* had previously remarked, "Apparently physicians are less aggressive in hospitalizing their oldest patients and initiating tests and treatments for them."[303] One study indicated that the elderly spend one-fifth of their income on healthcare.[304] In part, this is because Medicare only pays 80 percent of the costs and they must pay premiums on private insurance to pick up most of the remainder.[305] It has been noted that:

> In 1997, Congress enacted a formula for what Medicare would pay for healthcare. If expenses go up in one year—and last year they were "significantly above the target rate"—then Medicare must pay less the next year. Medicare payments to doctors—5.4 percent this year, with a total decrease of 17 percent by 2005—apparently has prompted some doctors to refuse to take on new Medicare patients.[306]

It was suggested that in the twenty-first century, "the dying patient is no longer the priority for healthcare ethics."[307] At the turn of the third millennium

---

301. "Seven percent to eight percent have severe cognitive impairments, roughly one third have mobility limitations, 20% have vision problems, and 33% have hearing impairments. Women, minorities, and persons of low socioeconomic status are especially vulnerable." JAMA 3137 (2002).

302. *Id.*

303. N. Eng. J. Med. 1027–28 (1995).

304. Orange County Reg., Mar. 5, 1998, at News 10.

305. "In addition, Alzheimer's disease costs American businesses more than $33 billion per year in lost productivity and absenteeism because workers are caring for parents and spouses."

306. Orange County Reg., Sept. 10, 1998, at 1.

307. Gibson, Ethical Comments, Summer 1999, at 3.

it was indicated that the United States is not likely to develop a comprehensive national welfare state along Western-European lines.[308] Just two years before that turn, Congress enacted major welfare and balanced budget reforms that significantly reduced the funding for social programs. This followed the "Contract for America," which changed 40 years of representation in the House of Representatives by a single (and different) party. To accomplish this required the partial shutdown of the federal government in 1995-96, which constituted a turning point in American history. For example, billions of dollars were saved following the landmark welfare overhaul.

In his 1996 State of the Union address, the President declared, "The era of big government is over." The new millennium may find that America is *Still the New World*, as the title of Philip Fisher's 1999 volume states. He argued that this is because of what the economist Joseph Schumpeter called our habit of "creature destruction." That approach was first expressed in Emerson's "Circles" (1839), where in moral life every new virtue extinguishes another one "in the light of a better."[309] Throughout the twenty-first century, we shall continue to ask about the extent to which an end can be "better" in the fields of bioethics and the law.

The advent of HMOs in the 1980s diminished some of the excessive medical costs; however, less affected are fraud and abuse by many healthcare institutions, which are considered unethical under all standards. The three types of conduct that are generally prohibited by healthcare fraud laws are false claims, kickbacks, and self-referrals:

> False claims are subject to several criminal, civil, and administrative prohibitions, notably the Federal Civil False Claims Act. Kickbacks, or inducements with the intent to influence the purchase or sale of healthcare-related goods or services, are prohibited under the federal Anti-Kickback statute as well as by state laws. Finally, self-referrals—the referral of patients to an entity with which the referring

---

308. L.A. TIMES, Feb. 25, 2000, at A24.

309. RALPH WALDO EMERSON, *Circles, in* ESSAYS AND LECTURES 410–11 (Parte ed., 1983). "There is no virtue which is final, all are initial . . . every moment is new, the past is always swallowed and forgotten, the coming is only saved. Nothing is secure but life, transition, the energizing spirit." *Id.*

physician has a financial relationship—are outlawed by the Ethics in Patient Referral Act as well as numerous state statutes.[310]

It has been stated, "Where a physician always errs on the side of providing additional resources, this deliberate bending of the rules amounts to a deliberate misinterpretation of their spirit. If physicians bend too many rules, the healthcare system is bound to break."[311] A study published in 2000 demonstrated "that many physicians report regularly engaging in patient advocacy to the point of consciously deceiving payers." Of the responding physicians, 39 percent reported that they had "exaggerate[d] the severability" of a patient's condition, "change[d] a patient's official (billing) diagnosis," or "report[ed] signs or symptoms that a patient did not actually have." Such fraud has been on the increase at the turn of the millennium. Of these physicians, 54 percent reported "using deception of third-party payers to obtain needed benefits more often 'now' than 5 years before."[312] To their credit, an editorial in the *Journal of the American Medical Association* stated that such dishonest practices "merit moral condemnation." It urged, "Legal and ethics scholars, and public officials should give high priority to the task of defining the scope and limits of clinical advocacy."[313] Philosophers have pointed out that "deceit is immoral because it causes false beliefs, and false beliefs cause wrong acts."[314]

Some physicians on a medical ethics committee on which I served argued that physicians can be expected to "show compassion for their patients." That compassion was largely to be accomplished by the use of public funds or those of insurance companies.[315] Physicians who entered the new specialty of

310. JAMA 1163 (1999).

311. JAMA 1675 (1999).

312. Wynia et al., *Physician Manipulation of Reimbursement Rules for Patients: Between a Rock and a Hard Place*, 283 JAMA 1858–65 (2000).

313. BiPoche, *Fidelity and Deceit at the Bedside*, JAMA 1883 (2000).

314. Smith, *Two-Tier Moral Codes*, *in* FOUNDATIONS OF MORAL AND POLITICAL PHILOSOPHY 112–32 (Paul et al. eds., 1990).

315. There can be no justification on grounds of "civil disobedience" such as that demonstrated by Thoreau or Gandhi, who operated openly, using no deceit whatsoever, and were willing to accept the legal consequences of their actions. ORANGE COUNTY REG., Sept. 26, 1998, at News 8. Since the dawn of the third millennium, the number of people without health insurance has risen to over 43 million, and over half of these people have annual incomes of $75,000 or more.

becoming hospitalists—to care for and coordinate the safety of hospitalized patients—increased to 8,000[316] by the year 2001. About half of them are members of their Society of Hospital Medicine and they work in teaching hospitals as well as HMOs in a manner that appears to result in lower inpatient costs.[317] In spite of "concern about burnout among hospitalists,"[318] it has been pointed out "that the promise of the model is being realized—an all-too-rare example of evidence-based organizational change. Yet a staffing model is only as good as the people who fill its roles."[319]

At the turn of the third millennium, the media have led many people, including politicians, to believe that HMOs need tighter controls due to their financial policies. These policies resulted in an AMA-supported movement toward "doctor unions." The HMOs would have to increase their premiums, and consumers will be paying a price for insurance. Some independent practitioners, who account for six out of seven U.S. doctors, have been increasingly barred from becoming members of unions by anti-trust laws.[320] HMOs have been shielded by federal law from some lawsuits since 1973 when Congress enacted a federal statute to encourage their growth.[321] As noted above, a

316. Wachter & Goldman, *The Hospitalist Movement 5 Years Later*, 287 JAMA 487–94 (2002).

317. Hoff et al., *Characteristics and Work Experiences of Hospitalists in the United States*, ARCH. INTERNAL MED. 851–58 (2001).

318. The rates may be increasing owing to growing workloads and fragile institutional support. Some programs have seen rapid turnover in personnel; with such turnover, many of the model's promised benefits, such as leadership in quality improvement and patient safety, as well as strong relationships with administrators and with nurses who care for inpatients, may fail to materialize.

Wachter, *Hospitals in the United States—Mission Accomplished or Work in Progress?*, N. ENG. J. MED. 1936 (2004).

319. *Id.*

320. Nevertheless, "the number of unionized doctors is growing. Of the 600,000 doctors nationally, roughly 40,000 belong to unions, up from about 25,000 two years ago. Experts predict that the numbers will continue to grow because of the AMA vote." ORANGE COUNTY REG., June 26, 1999, at News 27.

321. 42 U.S.C. § 300e (1973).

unanimous Supreme Court upheld this law in June 2000.[322] Studies have noted that HMOs "do not provide much support for the view that cost containment has worsened the quality of care."[323]

Near the end of the twentieth century, more than 23.5 million people were enrolled in California's managed healthcare plans, but only a scant few—a mere nine-thousandths of 1 percent—have complained about treatment.[324] Although for-profit hospitals account for only 12 percent of U.S. hospitals, their Medicare spending was found to be greater than that of not-for-profit hospitals.[325] Some healthcare facilities have embodied what one study called "the worst outcome of critical care." This is not death itself; rather, it may be an extended death process in which a patient's, and his family's, suffering is prolonged by services that ultimately prove both impotent and costly. A study was made to ascertain if potentially ineffective care is delivered less often to Medicare patients enrolled in health maintenance organizations (HMOs) than those in traditional fee-for-service health plans. The HMOs came out ahead. They had a definite "lower risk of experiencing ineffective outcomes" and were "better at limiting or avoiding injudicious use of critical care near the end of life."[326] In the twenty-first century it can only be hoped that traditional health plans will change to achieve results similar to the HMOs in this particular respect. Many families of dying patients report that their loved ones not only

322. The opinion by Justice David Souter claimed: "No HMO organization could survive without some incentive connecting physician reward with treatment rationing; the profit incentive to ration is the very point of any HMO scheme." Pegram v. Herdich; ORANGE COUNTY REG., June 13, 2000; *see also infra* Chapter 6.

323. Orenlicher, *Medical Malpractice, Treating the Causes Instead of the Symptoms*, 38 MED. CARE 248–49 (2000); Miller & Luft, *Does Managed Care Lead to Better or Worse Quality of Care?*, HEALTH AFF., Sept.-Oct. 1997, at 7–25.

324. "California's official report of HMO complaints for 1998 was compiled by the Department of Corporations, which regulates health-care management organizations in California." ORANGE COUNTY REG., Aug. 30, 1999, at News 4.

325. 341 N. ENG. J. MED. 420–26 (1999). "Though the widespread conversion of nonprofit hospitals to for-profit status in the mid-1990s has slowed considerably, nonprofit hospitals in financial straits continue to be acquired by investor-owned chains." Kuttner, *The American Healthcare System*, N. ENG. J. MED. 667 (1999).

326. JAMA 1001 (1997).

experience moderate to severe pain during most or all of their last days, but also severe care-giving and financial burdens.[327]

The most thorough study was project SUPPORT which documented the financial consequences of those caring for dying members of a family. The results were stark:

> Only those who had illnesses severe enough to give them less than a 50 percent chance to live six or more months were included in this study. When these patients survived their initial hospitalization and were discharged, about one-third required considerable care-giving from their families; in 20 percent of cases a family member had to quit work or make some other major lifestyle change; almost one-third of these families lost all of their savings; and just under 30 percent lost a major source of income.[328]

good v. Good

The exaggeration (and even the deceit) used by one-third of physicians must be tolerated because it demonstrates their compassion for their patients; all cost-saving features of HMOs should be curbed because costs cannot be considered when treatment is needed.

v.

The current widespread practice of deceiving public health agencies (and insurance companies) cannot be condoned on the basis of compassion by a physician (who is paid by those companies) (see Appendix C); HMOs currently constitute a principal method of controlling the spiraling of medical costs.

Twenty-first Century Approach

1. In view of current evidence that significant numbers of physicians use exaggeration and even some deceit in dealing with public health agencies and insurance companies, many state legislatures should consider providing adequate sanctions for these practices.
2. The need to reduce rising medical costs and to balance the budget will require a change in public policy, with special attention being given to certain dying patients on the basis of cost-benefit analyses of care.

---

327. *Id.* at 1025; Covinsky et al., *The Impact of Serious Illness on Patients' Families*, 272 JAMA 1839–44 (1994).

328. Hast. Ctr. Rep., Mar.-Apr. 1997, at 30.

3. While correcting any abuses, HMOs may normally continue their rational approach as to controlling costs of critical care at the end of life.
4. Appropriate cost controls used by HMOs should be studied by fee-for-service and other facilities in order to control critical care costs at the end of life.

Somebody asked me, "Would you like to be younger?"
I said, "Yeah, I'd like to be seventy-one!"
"I wouldn't want to go back of that," he replied.

Joseph Campbell

There is a time for everything;
and a season for every activity under heaven;
A time to be born and a time to die . . .
A time to weep and a time to laugh.

Ecclesiastes 3:1-2, 4

c) Long-term Home Care and Nursing Homes

In the profession of nursing, Florence Nightingale wrote *Notes on Hospitals* (1859), which began the nineteenth century's contribution to "the hospital building's self-consciously hygienic design, its orderly and efficient administration."[329] At the turn of the third millennium, about 1.5 million people, largely of the World War II generation, have been living in nursing homes, a group that could grow to 5 million by 2040.[330] In 2002, it was pointed out that nursing is "an embattled profession." In hospitals, many nurses feel that they are "unable to provide good patient care." However, "As the population ages, the demand for nurses is expected to grow rapidly."[331] Further, both their quality and sufficiency have been recently criti-

---

329. Rosenberg, The Care of Strangers 124.

330. "Nursing-home care cost $78 billion in 1996, with more than half paid by the Federal and state Medicaid and Medicare systems." *Id.*; *see also* Orange County Reg., Aug. 20, 1999, at News 22.

331. N. Eng. J. Med. 1761 (2002). "Registered nurses represent the largest single healthcare profession in the United States. An estimated 95% of the nurses were women, 72% married, and 87% were white. Their average age was 45 years. . . . In 2000, the national average was 782 employed nurses per 100,000 population."

cized.[332] Medicare does not normally pay for long-term custodial care beyond 30 days. However, Americans over 65 run a 43 percent chance of going into a nursing home at least once in their lives and more than half of those will last at least one year.[333] That year can cost more than $100,000 in New York City; the national average is almost $50,000.[334] In 2002, the Department of Health and Human Services concluded that "it is not currently feasible" to require the government to achieve a minimum ratio of nursing staff as had been recommended.[335] As a result, unmarried elders are less likely to enter a nursing home.[336]

With escalating health costs, employers in 2003 were raising premiums and increasing co-payments for retirees; some may eliminate health coverage for future retirees. In a 2002 survey: "The majority of employers, over 95 percent, said they will continue offering health insurance to current retirees in the next three years. But future retirees may face an uncertainty."[337] Several million people remain uninsured. Many people are buying insurance policies to cover the need of long-term care in the twenty-first century for one to three years—the estimated duration of the need in most instances. The extra premiums they pay can significantly affect the quality of life they might otherwise maintain. The peace of mind comes from knowing that when this need arises

---

332. CTRS. FOR MEDICARE AND MEDICAID SERVS., APPROPRIATENESS OF MINIMUM NURSE STAFFING RATIOS IN NURSING HOMES, REPORT TO CONGRESS: PHASE II REPORT (2001); WALLACE ET AL., CHANGING THE U.S. HEALTHCARE SYSTEM: KEY ISSUES IN HEALTH SERVICES POLICY AND MANAGEMENT 205–17 (2d. ed. 2001).

333. A.B.A. J., July 2000. In 1999, annual nursing home costs averaged $50,000 in the United States, according to the Center for Long-Term Care Financing in Seattle. *Id.* at 62.

334. N.Y. TIMES, Feb. 16, 2000.

335. ORANGE COUNTY REG., Feb. 18, 2002, at News 24.

336. Freedman, 51B J. GERONTOLOGY: SOC. SCI. S61–S69 (1996).

337. Among those surveyed, 82% of employers said they plan to increase retiree premiums over the next three years; 85% of those plan to increase prescription drug co-payments, according to Drew Altman, president of the Kaiser Family Foundation, which conducted the survey. Current retirees are being asked to pay more for their health coverage, and current workers are less likely to get health benefits from their employer when they retire. http://member.compuserve.com (last visited Dec. 4, 2002).

they will not become a burden to others. This will also diminish the economic help that they might otherwise give to their children or grandchildren both during their lives and afterwards.[338]

A significant change near the end of the twentieth century was noted: namely, that life insurance benefits may also be paid to the insured while he is still alive. As of July 25, 1995, annuitants who had a life expectancy of 9 months or less and enrolled in the Federal Employees Group Life Insurance (FEGLI) program could elect to receive living benefits. If he receives them, his basic insurance amount (with certain reductions) will be paid to him in a lump sum before his death.[339] Others are within the "Viatical Settlement" in industry.

Viatical companies gamble on people's lives. They purchase the life insurance policy of a terminally or chronically ill patient, pay the premiums, and wait. In the meantime, the patient may use the cash proceeds from the transaction in any way. When the person dies, the viatical company receives the benefits of the policy. The sooner the patient dies, the greater the viatical company's profit. No national regulation exists, and less than 20 states require licensing of viatical settlement companies.[340]

In the United States alone, the projected figure for 2000 was $4 billion. It sometimes is called a "holistic" approach to considering illnesses. People in long-term care institutions have been characterized as suffering from "loneliness, helplessness, and boredom." They can be helped by being visited by

338. In a study conducted by the Commonwealth Fund, it was found that help from parents [for] children or grandchildren was nearly 15 times more common than the reverse. When grandparents care for grandchildren, for example, they report spending an average of almost 14 hours per week. The economic value of this help to the country has been calculated in the range of $17 billion to $29 billion per year.

JAMA 1872 (1997) (citing Bass & Caro, *The Economic Value of Grandparent Assistance*, GENERATIONS, Spring 1996, at 29–33).

339. If he elects living benefits, he may not assign his life insurance coverage. U.S. Off. Pers. Mgmt., Retirement Operation Center, Boyers, Pa.

340. 59 MONT. L. REV. 701, 701 (1998); *see also* 24 PEPP. L. REV. 99, 110 (1997).

pets; this may also reduce their health costs by decreasing their depression.[341] It has been noted that it is common to place a relative with dementia into a long-term care facility. A study found that the transition "is particularly difficult for spouses, almost half of whom visit the patient daily and continue to provide help with physical care during their visits." However, it also notes that "clinical investigations that better prepare the caregiver for a placement transition and treat their depression and anxiety following placement may be of great benefit to these individuals."[342]

The large increase in *nursing-home abuse* required significantly more attention. Some 5,283 were cited for abuse from 1999 to 2001. In the year 2000, some 16 percent were cited (a 10 percent increase since 1996).[343] In 2002, a study by the General Accounting Office, an investigative arm of Congress, reported that more than 30 percent of the nation's nursing homes have been cited by state inspectors for violations that harmed residents.[344]

good v. Good

No public funds should be spent on such frivolities as pets for older nursing home residents as long as other funds are available for them; life insurance is for the family of the unskilled after his death, not for while he is alive.

v.

Most medical costs for the elderly are covered through Medicare and insurance in hospitals (but generally not in nursing homes). Life insurance that allows payments to be made to annuitants should be encouraged during their last period of life expectancy.

341. ORANGE COUNTY REG., Oct. 12, 1998, at News 6.

342. Schultz et al., *Long-Term Care Placement of Dementia Patients and Caregiver Health and Well-Being*, JAMA 961 (2004).

343. ORANGE COUNTY REG., July 21, 2001, at News 8.

344. "About 1.6 million people live in 17,000 nursing homes nationwide. Medicaid and Medicare help pay for three-fourths of the patients and spent $58 billion for their care last year." ORANGE COUNTY REG., Mar. 3, 2002, at News 21.

Twenty-first Century Approach

1. Legislatures should consider the small funding of the introduction of inexpensive plants, pets, and appropriate interventions for people in nursing homes to decrease their depression (from loneliness, helplessness, and boredom), which has also been found to decrease some medical costs and burdens upon their families.
2. The use of life insurance policies that pay benefits while the terminally-ill insured is still alive (viatical settlements) should be encouraged, since it further reduces costs to taxpayers for their healthcare.

We, as a matter of course, reflect on death, voice hope and fear, only when a dear one is near death, or out of it. Why not speak of it while we're in the flower of good health? How can we envision our life, the one we now experience, unless we recognize that it is finite?

Terkel, *Will The Circle Be Unbroken?* (2000)

They could probably keep me alive with all kinds of devices, but the quality of life will be ridiculous, and I'm not interested in that.

Richard Feynman, Nobel laureate in physics, quoted in *The Beat of a Different Drummer* (1994), Mehra, p. 606

Do not go gentle into that good night,
Old age should burn and rage at close of day;
Rage, rage against the dying of the light.

Dylan Thomas, *New Directions* (1953)

It is as natural to die as to be born.

Bacon, "Of Death"

13

Improvements Needed in the Twenty-first Century Right to Die

1. *Training for End-of-Life Care*
   a) Advance Directives, Living Wills
2. *Surrogates, Durable Powers of Attorney, and Emergency Treatment (EMTLA)*
3. *Withholding and Withdrawal of Treatment; Limitations on Nonbeneficial (Futile) Treatment: DNR vs. CPR*
   a) Withholding Equals Withdrawal
   b) Withholding Nutrition and Hydration
   c) Patients in a Persistent or Permanent Vegetative State (PPVS) Without a Living Will or DPA; Subjective Personhood
4. *The Rights of Mature Minors to Die and Jehovah's Witnesses*
5. *The Choices of Do Not Resuscitate (DNR) vs. Cardiopulmonary Resuscitation (CPR): Legal and Ethical*
   a) Presumed Consent to Cardiopulmonary Resuscitation (CPR)
   b) Best Interests vs. Substituted Judgment
6. *Limitations on Nonbeneficial (Futile) Treatment*
   a) Treatment Beyond the "Goals of Medicine"
   b) Treatment Regardless of a Lack of Consensus on "Futility"

c) **The Unique Direction of Informed Consent**
d) **A Way to "Free" Physicians from Self-Interest or Self-Protection in Connection with Life Support; Anencephalics**
e) **The Authority and Responsibility of Physicians with ICUs and Authorized Executions**
f) **The Current Method of Stopping Futile Treatment of Dying Patients by the AMA and by Texas**

7. ***Slow Code***

---

1. Training for End-of-Life Care

As noted in Chapter 12, there is a lack of uniformity regarding medical care to extend a patient's life concerning both positive and negative effects on his "quality of life." As illustrated by the epigraphs of this chapter, people differ about this matter. It has been estimated that "almost 40 percent of all deaths in the United States take place following the withdrawal of life-sustaining treatments—often from a sedated or comatose patient and after protracted, agonizing indecision on the part of family members and physicians."[1] Each patient (or a surrogate) must make a personal decision regarding how much treatment he may desire. As stated by Elisabeth Kubler-Ross in her classic volume *On Death and Dying*:

> Though every man will attempt in his own way to postpone such questions and issues until he is forced to face them, he will only be able to change things if he can start to conceive of his own death. This cannot be done on a mass level. This cannot be done by computers. This has to be done by every human being alone. Each one of us has the need to avoid this issue, yet each one of us has to face it sooner or later. If all of us could make a start by contemplating the possibility of our own personal death, we may effect many things, most important of all the welfare of our patients, our families, and finally perhaps our nation.[2]

---

1. Sci. Am., May 1996, at 12.
2. Elisabeth Kubler-Ross, On Death and Dying 17 (1969).

Much of her excellent advice has often been ignored, not only by patients, but also by their physicians with respect to the welfare of their patients, and their families and the nation. It has been remarked near the end of the twentieth century that "many physicians find it easier to define success in terms of life and death than to try to determine what sort of existence is meaningful to an individual patient."[3] Just prior to the turn of the third millennium a study was published showing that "Medical education typically provides little training in care of the dying, almost no formal training in palliative care," and that "top-selling textbooks generally offered little helpful information on caring for patients at the end of life."

Most disease-oriented chapters had no or minimal end-of-care content. Specifically, textbooks with information about particular diseases often did not contain helpful information on caring for patients dying from those diseases.[4]

As noted above, most of this type of bioethical care is not generally medical care fully handled by physicians and, hence, other professionals should be trained in this specialty. Emerson's statement, "One man's justice is another's injustice," may apply to many physicians. Difficult analysis and reasoning is often mentally subordinated to mere subjective personal viewpoints about bioethics, medicine, and the law.

---

3. Gilligan & Raffin, *Whose Death Is It, Anyway?*, 125 ANNALS INTERNAL MED. 137–41 (1996). In 2002, the average cost of the last five days for a cancer patient aged 65 or older, prior to hospital death in Virginia, amounted to $12,319 for nonpalliative care and $5,313 for palliative care (which focuses on comfort, not cure). WALL ST. J., Mar. 10, 2004, at 1. Thus, palliative care is usually cheaper than standard care and can be found in programs in 844 community hospitals.

4. 283 JAMA 771–78 (2000).

> For example, during the past 20 years, both inpatient units and outpatient clinics have developed programs for geriatric evaluation and management. . . . A trial involving frail patients 65 years of age or older . . . were randomly assigned, according to a two-by-two factorial design, to receive either care in an inpatient geriatric unit . . . or usual outpatient care. [Such] care provided in inpatient geriatric units and outpatient geriatric clinics had no significant effects on survival.

Cohen et al., N. ENG. J. MED. 905 (2002).

Judaism cannot sanction a "living will" . . . Judaism denies man the right to make judgments with regard to quality of life.

Bleich, *Bioethical Dilemmas, A Jewish Perspective* (1998), p. 73

a) Advance Directives, Living Wills

All 50 states have statutes governing advance and end-of-life directives. Although their standards differ,[5] they acknowledge the right of a competent individual to decide aspects of his or her own healthcare, including the right to decline healthcare or to direct that current healthcare be discontinued—particularly after becoming terminal and incompetent.

In an attempt to give patients more control over their lives near their ending, advance directives (ADs) were designed "to give patients a safeguard against being overpowered by overzealous physicians or institutions" and by providing another way of protecting "the private sphere of self-determination."[6] Treatment directives (called "living wills") and "advance directives" were authorized by state statutes to give such patients procedural mechanisms to limit or increase treatment. They normally become effective only after the patient loses decision-making capacity.[7] They were reinforced by the federal Patient Self-Determination Act, which requires that all hospitals, nursing homes, and other Medicare and Medicaid providers ask patients on

---

5. Sabatino, *The Legal and Functional Status of the Medical Proxy: Suggestions for Statutory Reform*, 27 J.L. Med. & Ethics 52–68 (1999).

> In Europe, living will declarations are not considered mandatory. This position was supported in 1993 by the Standing Conference of European Medical Associations and also by Article 10 of the European Bioethics Convention, which was signed by twenty-one countries, in Spain, on April 4, 1997. (Britain was one of the few countries that failed to sign the declaration.) *See* Council of Europe: Convention on Human Rights and Biomedicine, Directorate of Legal Affairs, Strasbourg, 1996; Council of Europe: Explanatory Memorandum, Convention on Human Rights and Biomedicine, Directorate of Legal Affairs, Strasbourg, 1997; European Standing Conference of Medical Associations: Declaration of 1993.

6. N. Eng. J. Med. 744 (1996).

7. However, the Uniform Heath Care Decision Act permits a patient to designate a surrogate to make healthcare decisions even though the patient may retain decision-making capacity. Unif. Health Care Decision Act § 2(c) cmt. 9, part 1, U.L.A. 225.

admission whether they have executed an advance directive.[8] However, many hospitals ignored this federal legislation for years. Project SUPPORT (discussed below) demonstrated how most physicians continued to ignore the directives in the mid-1990s even after they have been specifically brought to their attention.[9] These statutes do not normally inform individual patients how to make their own specific medical decisions.[10] At that time, 44 states had statutes authorizing living wills and 33 states also permitted a patient to appoint an agent, or healthcare proxy, to make treatment decisions on his behalf when the patient becomes incompetent. The latter is called "durable power of attorney" (DPA) because under the common-law world other powers of attorney normally terminate upon the death of the principal.

These statutes vary. New York requires an advance directive in all cases for life-sustaining treatment to be forgone, but the directives may be oral if it can be proved by clear and convincing evidence. Missouri's statute is virtually the only one at the commencement of the third millennium containing a prohibition of the enforcement of an instruction to forgo artificial nutrition and hydration whether or not those things are necessary for the patient's comfort.[11] The California statute effective in 2000 specifically states that a health care institution "may not require or *prohibit* the execution or revocation of *an advance health care directive as a condition for providing health care*, admission to a facility, or furnishing insurance." [12] (Emphasis added.) Penn-

8. 42 U.S.C. §§ 1395, 1369 (1994). As the end of the twentieth century approached, only 9.8% of patients near the end of life had completed a living will. Handon & Rodgman, 156 ARCH. INTERNAL MED. 1018–23 (1996). "Those with ADs were significantly less likely to die in a hospital compared to subjects with no ADs." Hickman et al., Hast. Ctr. Rep., Sept.-Oct. 2004, at 5.

9. For a discussion of Project SUPPORT, *see supra* notes 78–81 and accompanying text.

10. KING, MAKING SENSE OF ADVANCE DIRECTIVES (1996); Kapp, J.L. MED. & ETHICS 153 (1996). In 2004, when the *Archives of Internal Medicine* doctors were asked to consider certain hypothetical medical cases, "65 percent of the decisions the doctors made were contrary to patients' advance directives." ORANGE COUNTY REG., Mar. 24, 2005, at News 3.

11. MO. ANN. STAT. § 459.010(3) (West 2000).

12. CALIF. PROB. CODE § 4650 (West). Further, a death resulting from withholding or withdrawing healthcare in accordance with a valid advanced directive "does not for any purpose constitute a suicide or homicide or legally impair or invalidate a policy of insurance providing a death benefit, notwithstanding any term of the policy or annuity to the contrary." *Id.* § 4656.

sylvania[13] provides options that were summarized as follows: (1) cardiac resuscitation; (2) mechanical respiration; (3) tube feeding or any other artificial or invasive form of nutrition (food) or hydration (water); (4) blood or blood products; (5) surgery or invasive diagnostic tests; (6) kidney dialysis; and (7) antibiotics if they are "in terminal condition or in a state of permanent unconsciousness." They may also state whether they want to make an anatomical gift of all or part of their body.[14] A study of the relationship between advance directives, patient discussion, and costs of care concluded that "an enormous cost-savings to society may be realized if such discussions take place."[15] Neither savings in costs nor in treatments were being fully realized at the close of the twentieth century.

Most people are reluctant to think about their own death and, prior to the twenty-first century, "only 10 to 20 percent of adult Americans have signed an advance directive."[16] Some doubted the utility or efficacy of advance directives.[17] Despite prior efforts, many hospitals had failed to achieve substantial completion rates. It has been recommended that these rates be significantly increased by a law linking the completion of advance directives to the time when health insurance is initiated or renewed. "This would relocate the time and locus of their completion from the emotional turmoil of hospital admission and acute illness to a more equanimous time when family and others can be consulted and involved."[18] In this connection it has been noted that family members authorizing the withdrawal of life support for dying loved ones, family stress was higher in the absence

13. 20 PA. CONS. STAT. § 5404 (2000).

14. Diane E. Hoffman et al., *The Dangers of Directives or the False Security of Forms*, 24 J.L. MED. & ETHICS 5, 6 (1996).

15. Chambers et al., *Relationship of Advance Directives to Hospital Charges in a Medicare Population*, 154 ARCH. INTERNAL MED. 541–47 (1994).

16. SCI. AM., May 1996, at 1.

17. McGee, *Paper Shields: Why Advance Directives Still Don't Work*, PRINCETON J. BIOETHICS, 42–56 (1998).

18. Eisner & Weiss, *The Underachieving Advance Directive: Recommendations for Increasing Advance Directive Completion*, AM. J. BIOETHICS, http://www.ajobonline.com/in_focus/in_focus_eiser-weiss.htm.The authors stated: "Amending the Patient Self-Determination Act to require providing advance directive forms and/or incentives could be more effective than the current arrangements." *Id.*

of advance directives, was lower when verbal advance directives guided the family, and was lowest when written advance directives guided the family.[19]

In the mid-1990s it was found that most physicians disregard advance directives even when they have been brought to their attention by nurses. As noted above, a large Study to Understand Prognoses and Preferences for Outcomes and Risks of Treatments (SUPPORT) involved more than 9000 patients at five hospitals during a four-year period. It revealed "substantial shortcomings in care for seriously ill hospitalized adults."[20] About half of those who were able to communicate in their last three days of life were in serious pain. For half the patients who had wanted CPR to be withheld, their physicians never wrote the Do Not Resuscitate (DNR) order. "Project 'SUPPORT' confirmed the worst fears of many people in this country, that efforts to prolong life too often merely prolonged dying."[21] The principal investigators reported that "an explicit effort to encourage use of outcome data and preferences in decision-making were completely ineffectual." Project SUPPORT concluded that "we are left with a troubling situation. The picture we describe of the care of seriously ill or dying persons is not attractive."[22] One bioethicist added, "Physicians simply have never taken the rights of hospitalized patients seriously."[23] He pointed to the 50 percent of physicians who were unconcerned that the pain their patients were predictably suffering amounted to systematic patient abuse.[24] A 1998 study of discussions with patients by "experienced physicians" about advance directives yielded some observations:

> On average, these conversations were short, with physicians talking two-thirds of the time. Physicians typically discussed extreme sce-

---

19. Tilden et al., *Family Decision-Making to Withdraw Life-Sustaining Treatments from Hospitalized Patients*, Nursing Res. 112 (2001).

20. *Id.* at 14.

21. 278 JAMA 968 (1997).

22. JAMA 1497 (1995).

23. George Annas, *How We Die*, Hast. Ctr. Rep., 1995, at 512. Hospitals have responded with "very passive and limited implementation strategies." Yates & Glick, *The Failed Patient Self-Determination Act and Policy Alternatives for the Right to Die*, J. Aging & Soc. Pol'y, 1997, at 29, 32.

24. Yates & Glick, *supra* note 23 at 13.

narios that elicited little variation in preferences. Discussions of uncertainty were vague, and physicians often described treatments, but rarely in any detail. Although patients' values arose often in these discussions, physicians were unlikely to elicit or explore them.[25]

Major challenges for the third millennium include finding methods of understanding why large numbers of physicians chose to ignore both the common law of informed consent and the written laws on living wills. When she heard the disturbing results of SUPPORT, former first lady Rosalynn Carter stated, "What we learned from that study was that the situation at the end of life was even more grim than we imagined. It is not going to be easy. It is not going to be immediate."[26]

The doctrine of informed consent, which only seriously began in the last third of the twentieth century, had not fully taken hold at the dawn of the twenty-first. Two millennia ago when most people lived in essentially tribal societies, the Golden Rule "Do unto others as you would have them do unto you" sufficed. In today's diverse society, it cannot do so adequately without a change in three pronouns. The third millennium will look toward the following new general rule for physicians: "Do unto others as *they* would have *you* do unto *them*."[27] A study published in 2001 indicated that, unlike the overwhelming number of physicians in Project SUPPORT, when doctors in an emergency room are given a patient's advanced directive they more accurately predict his desires than such doctors do without one.[28] Nevertheless, some people have claimed that the widely promoted policy of living wills has not sufficiently "produced results that patients' exercise of autonomy could extend beyond their span of competence."[29]

Although the federal Patient Self-Determination Act (PSDA)[30] became effective in 1991, it was significantly ignored by many hospitals (including

25. ANNALS INTERNAL MED. 446 (1998).

26. ORANGE COUNTY REG., Feb. 1, 1997, at Accent 1.

27. *See* KEYES, LIFE, DEATH, AND THE LAW (1995); *supra* Chapter 5, § 2.

28. Coppola et al., *Accuracy of Primary Care and Hospital-Based Physicians' Predictions of Elderly Outpatients' Treatment Preferences With and Without Advance Directives*, ARCH. INTERNAL MED. 431–40 (2001).

29. Fagerlin & Schneider, *Enough: The Failure of the Living Will*, Hast. Ctr. Rep., 2004, at 30–42.

30. Patient Self-Determination Act, Pub. L. No. 101-508, Title IV, §§ 4206, 4571, Nov. 5, 1990, 104 Stat. 1388-115, 1388-204.

one where I served as a community member on its ethics committee).[31] In 1995, the Healthcare Financing Administration issued regulations requiring healthcare facilities to document in each individual's record whether the individual had completed an advance directive to comply with respective state laws concerning them.[32] In 1998 Project SUPPORT, among other things, caused the AMA to make revisions in Rule 2.225 in the AMA *Code of Medical Ethics* to state, "There is need for better availability and tracking of advance directives, and more uniform adoption of form documents that can be honored in all states of the United States." It then added "improvement strategies" in order to "give physicians immunity from malpractice for following a patient's wishes." These include (a) discussions with the patient and the patient's proxy and (b) the "signing and recording the document in the medical record should not be delegated to a junior member of the healthcare team." By 2003 certain banks had stored their advance directive in person (not mailed) that allow healthcare providers to share information in some circumstances.[33]

A number of physicians have told me that they really did not comprehend the need for more communication during the process of dying as was demonstrated by Project SUPPORT. In 1999, a physician stated:

> I also became aware that despite four years of medical school and six years of residency and fellowship, I was almost completely unprepared for the tasks that occupied most of my time as an ICU attending physician—counseling patients and families in crisis, explaining in simple terms the complexities of intensive care, giving bad news, defining endoflife care, and enduring comfort and dignity both for the patients we were still trying to cure and for those who were obviously dying.[34]

---

31. A bioethics committee may make recommendations to improve performance, but it may be ignored or refused by the management. *Id.*

32. Advance Directives: Final Rule, 60 Fed. Reg. 33,262 (1995).

33. Without a central repository, many advance directives are difficult to find and some are never found. Montana providers can access directives for out-of-state healthcare providers, if they need these documents. A.B.A. J., July 2003, at 24.

34. ANNALS INTERNAL MED. 776 (1999).

Virtually nothing more serious was being done about it at the end of the second millennium.[35] There was little enforcement of these patients' rights against providers for intentionally failing to comply with patient preferences expressly stated in valid advance directives. Attorneys representing these providers often gave the advice "to play it safe and administer treatment rather than to permit a patient to die from withholding or withdrawing treatment."[36] In one case, $16 million was awarded in a medical battery case against a hospital that continued treatment for 38-year-old Brenda Young despite objections by her family that she would not want it.[37] If social workers were trained in advance directive completion, this would better assist patients. Prehospitalization patient educational mailing increased completion to 40 percent.[38]

Just prior to the turn of the twenty-first century, a physician who headed an obstetrics and gynecology department of a medical center feared strokes. He wrote an advance directive specifically denying the use of extraordinary care measures. When he fell into a persistent vegetative state, the center agreed with his family to follow the directive. But in spite of this, he was resuscitated in the center's emergency room. The family brought suit on the denial of his right to refuse medical treatment and lost in a federal court in 1999.[39] In 2004, a professor of both law and medicine at the University of Michigan cited an

---

35. *See* Lo et al., *Discussing Palliative Care with Patients*, Annals Internal Med. 744 (1999); *supra* note 34, at 772.

36. 2 Meisel, The Right to Die 352–53 (2d ed. 1995).

37. Osgood v. Genesee Reg'l Med. Ctr. (awarding $16 million); *see also* William Prip & Ana Moretti, *Compliance: The Missing Component of Patient Autonomy Laws*, Experience, Spring 1996, at 4. In *First Healthcare Corp. v. Rettinger*, 467 S.E.2d 243 (N.C. 1996), the North Carolina Supreme Court held that a patient with an advance directive remained financially liable until actual certification by two physicians that the patient was terminally ill. For some reason, that court did not consider as crucial the time the patient's surrogate had requested that life-sustaining treatment be terminated.

38. Rubin et al., 271 JAMA 209–12 (1994).

39. The court stated: "While Dr. Klavan's situation cries out for prompt and definitive judicial resolution, we nevertheless decline to exercise our discretion . . . precisely because of the gravity of his case." Klavan v. Crozer Chester Med. Ctr., 60 F. Supp. 2d 436 (E.D. Pa. 1999).

article reporting that *54 percent* of patients studied might be willing to have their surrogates override even a "perfect" living will.[40] Hence, he argued that "No court should want to enforce a document whose authors are so ambivalent and so likely not to want the document to be binding."[41] Nevertheless, he had also noted that "doctors seemed persistently and gruesomely to overtreat dying patients" and cited a writer who stated, "Nearly all living wills now written prohibit such behavior and attempt to prevent overtreatment."[42] For example, a California Statute on Advance Directives (effective in 2000) grants a hospital immunity from civil or criminal liability for actions under that will, which are:

> Complying with a health care decision on the part of a person that the healthcare provider or health care institution believes in good faith has the authority to make a health care decision for a patient, including a decision to withhold or withdraw healthcare.[43]

A hospital "may decline to comply with an individual's health care instruction or health care decision for reasons of conscience" only "if the policy was timely communicated to the patient or to a person then authorized to make health care decisions for the patient."[44] In such cases that hospital must immediately make all reasonable efforts to assist in the transfer of the patient to another health care provider or institution that is willing to comply with the instruction or decision.[45] Thus, advance directives may become enforceable in most of California's hospitals—a matter of particular importance to prevent such hospitals from giving treatment to patients in opposition to their directions.[46] A report issued in

---

40. Terry et al., *End of Life Decision-Making: When Patients and Surrogates Disagree*, 10 J. Clinical Ethics 286–93 (2000).

41. Schneider, *Liability for Life*, Hast. Ctr. Rep., July-Aug. 2004, at 11.

42. Tonelli, *Beyond Living Wills*, Bioethics Forum 13, 1997 at 6, 12 at 7.

43. Cal. Prob. Code § 4740 (West 2001).

44. *Id.* § 4734; *see also id.* § 4716.

45. *Id.* § 4736.

46. "A health care provider or health care institution that intentionally violates this part is subject to liability to the aggrieved individual for damages of two thousand five hundred dollars ($2,500) or actual damages resulting from the violation, whichever is greater, plus reasonable attorney's fees." *Id.* § 4742.

2000 estimated that more than 70 percent of patients preferred that their family make resuscitation decisions with their physician.[47]

A system of advance care planning called Physician Orders for Life-Sustaining Treatment (POLST) was first developed in Oregon.[48] It constitutes a document that sets forth patient treatment preferences with specific written medical orders. Patients' preferences are documented with individualized options integrated with advance directives (including DNR and other treatments at the end of life and for comfort measures). Only about 13 states have adapted versions of the POLST. They are also used by nursing homes and hospices[49] and by emergency medical services.[50]

good v. Good

If, regardless of their patient's advanced directives, physicians are doing what is medically appropriate in order to extend the lives of their dying patients, little more can or should be demanded of them.

v.

Both federal and state laws require that advanced directives be followed by physicians and healthcare organizations. Project SUPPORT, which demonstrated that these directives were being ignored by many physicians, indicated that major changes were needed in physician conduct with respect to the desires that their patients had expressed regarding continued treatment.

---

47. About 30% indicated their own preferences if they became physically unable to make a decision. Puchalski et al., *Patients Who Want Their Family and Physician to Make Resuscitation Decisions for Them: Observations from SUPPORT and HELP*, J. Am. Geriatrics Soc'y S84–S90 (2000).

48. Hickman et al., *Use of the POLST (Physician Orders for Life-Sustaining Treatment) Program in Oregon: Beyond Resuscitation Status*, 52 J. Am. Geriatrics Soc'y 1424–29 (2004). In 2004 a National POLST Paradigm Task Force was formed. Hickman et al., *Hope for the Future: Achieving the Original Intent of Advance Directives*, Hast. Ctr. Rep., Nov-Dec 2005, at 5.

49. Lee et al., *Physician Orders for Life-Sustaining Treatment (POLST): Outcomes in a PACE Program*, 48 J. Am. Geriatrics Soc'y 1219–25 (2002).

50. Schmidt et al., *The Physician Orders for Life-Sustaining Treatment (POLST) Program: Oregon Emergency Medical Technicians' Practical Experiences and Attitudes*, 52 J. Am. Geriatrics Soc'y 1430–34 (2004).

Twenty-first Century Approach

1. Increased emphasis should be placed upon the existence and value of Advance Directives beginning in the senior year at public high schools, as well as in colleges and universities.
2. Many state's laws pertaining to advance directives as well as the federal Patient Self-Determination Act (PSDA) should be amended to impose appropriate sanctions upon hospitals, physicians, or staff members who would ignore these directives either by intent or negligence.
3. Many state legislatures should also enact statutes regarding (a) the category of other professionals (eschats) who will actually deal with patients on end-of-life matters, and (b) the completion of advance directives, and the time when health insurance is initiated or renewed.
4. The American Medical Association (AMA) and state medical associations should adopt a rule for physicians in our diverse society (consistent with "informed consent"), namely, "Do unto others as *they* would have *you* do unto *them*," while recognizing that it does not authorize patients to require physicians to grant all types of requested treatments.

---

2. Surrogates, Durable Powers of Attorney, and Emergency Treatment (EMTLA)

At common law, adults can appoint others to make legal decisions for them (called a "power of attorney") that expires with the principal who made them. Statutes have authorized citizens to execute "durable powers of attorney" (DPAs) to designate others, called "surrogates," to make decisions for them that are effective after they (patients) become incompetent. As in a living will, the patient can also describe what medical treatment should or should not be used. The person receiving the durable power does not have to anticipate every possible medical situation that may arise. According to the Uniform Health-Care Decisions Act, which was approved by the American Bar Association (ABA) in 1994, a written power of attorney for healthcare need not be witnessed or acknowledged. The act would require an agent, or surrogate authorized to make health-care decisions for an individual, to make them in accordance with the instructions and known wishes of the individual. Otherwise, the agent or surrogate must make those decisions in accordance with the "best interest" of the individual in light of the individual's personal values known to the surrogate.[51]

These durable powers are to be distinguished from "standby guardians" who are individuals chosen by terminally ill parents to become their child's guardian when the parent is incapacitated or dies. A parent makes this permanent plan without ending his parental rights while living and competent.[52] Unlike a living will or a DPA, the standby guardianship process generally begins when either a parent or a legal guardian petitions the court. The parent must generally indicate whether he or she suffers from a "progressively chronic illness" or an "irreversibly fatal illness."[53]

Members of hospital ethics committees have found that in many instances a patient's family will insist upon resuscitation in spite of (a) the existence of an advance directive indicating that the contrary was the patient's desire or (b) the fact that treatment may be "nonbeneficial" (a problem discussed below).[54] As once stated by a bioethicist, "There is a world of difference between situations in which quality-of-life judgments are made by patients themselves and

---

51. This Uniform Act was drafted by the National Conference of Commissioners on Uniform State Laws, which recommended that all states adopt the Act. However, as of the end of the twentieth century, it has only been adopted by six states: Alabama, Delaware, Hawaii, Maine, Mississippi, and New Mexico.

52. *E.g.*, N.J. REV. STAT. § 3B12-68 (2000), which states:

> [T]here is an imperative need to create an expeditious manner of establishing a guardianship known as standby guardianship, in order to enable a custodial parent or legal custodian suffering from a progressive chronic condition or a fatal illness to make plans for the permanent future care or the interim care of a child without terminating parental or legal rights.

53. North Carolina law does not require identifying the illness. N.C. GEN. STAT. § 35A-1373(b), (c) (1996). In Illinois, the court appoints a guardian only if both parents consent. If one parent disagrees, then that parent must object to the request for a standby guardian at the hearing. 755 ILL. COMP. STAT. ANN. § 5/11-5.3 (West 2000). California's Health Care Directive consolidates a number of forms: the Durable Power of Attorney for Health Care, the Natural Death Act, and the directive to physicians.

54. Apparently, some emergency physicians actually view the withholding of treatment as more serious than later withdrawal because of their concept of society's expectations about emergency treatment. Byock, *A Slight Post Mortem Disagreement*, *in* ISERSON ET AL., ETHICS IN EMERGENCY MEDICINE 80–87 (2d ed. 1995).

those in which family or medical caregivers make the assessment."[55] This behavior is often justified on the grounds that emergency physicians and prehospital providers must act at a time when they may lack vital information about their patients' identities and wishes. Many patients entering the emergency departments of hospitals are in the condition that inquiry can be made with regard to possible advance directives prior to receiving extensive treatment for which hospitals may stand to gain.[56]

The Emergency Medical Treatment and Active Labor Act (EMTALA)[57] requires that emergency patients receive an "appropriate medical screening examination" and stabilizing treatment. In 2004, the revised regulations (guidelines issued by the Centers for Medicare and Medicaid Services) indicates that should an individual come to the hospital's emergency department and the request is not an emergency, the hospital is only to perform such screening as would be appropriate to determine that the individual does not have an emergency medical condition.[58] In a recent case, a woman brought her husband with a number of symptoms to a hospital. The doctor and nurses examined him, gave him some medicine, and discharged him. Later that day he was taken to another hospital where he was pronounced dead. His wife sued the first hospital for failing to diagnose a life-threatening medical condition that resulted in his death in violation of that statute. The U.S. Court of Appeals held in favor of the hospital because no evidence was submitted on the hospital's baseline of care or to show how the husband's screening was inadequate. That court held that the EMTALA only requires that a hospital's emergency room's procedures "be reasonably calculated to identify the patient's critical medical condition," but does not require that they be totally accurate.[59] In another case, that court stated

55. Macklin, Mortal Choices 53 (1987).

56. Hospitals also stand to gain because Medicare reimburses hospitals by the hour for emergency room visits at rates higher than the per-diem room rate on general medical floors. Medicare pays for tests and procedures performed even in the last few hours of life, although these are often futile and render useless data, and tests ordered in hospital emergency rooms often receive higher reimbursement than identical tests performed in a nursing home. Hast. Ctr. Rep., Nov.-Dec. 2002, at 47.

57. 42 U.S.C. § 1395dd (2003).

58. C.F.R. §§ 489.24, 489.20(l), (m), (q), (r), 72 U.S.L.W. 2724 (2004).

59. Del Carmen Guadalupe v. Agosto, 299 F.3d 15 (1st Cir. 2002). In an earlier case, that court had held that the statute only provides for "screening examination reasonably calculated to identify critical medical conditions . . . and provides that level of screening uniformly to all those who present substantially similar complaints." Correa v. Hosp. S.F., 69 F.3d 1184, 1992 (1st Cir. 1995).

that the plaintiffs ignored the distinction between a malpractice claim (a matter discussed in Chapter 1) and a claim under the EMTALA, the federal law that requires all hospital emergency rooms to receive patients suffering from emergency medical condition requirements.[60] In 2003 only about 5 percent of U.S. hospitals had written policies allowing family members' access when emergency procedures are performed:

> Traditionally, family members have been kept at a distance during medical emergencies—both because doctors and nurses felt that seeing them work on the patient would be upsetting. But studies in recent years have found that bedside visitation during emergency care has been rated an overwhelmingly positive experience by members and has drawn generally positive reactions from health professionals.[61]

Back in 1982, the Foote Hospital in Jackson, Michigan, established a program for family members to be allowed with the patient during controlled periods in the resuscitation room. More recently, at the Parkland Health and Hospital System in Dallas, Texas, all family members who were interviewed indicated how important it was to be present.[62] Physicians and nurses were generally reluctant to do so in spite of the fact that the Foote Hospital found "no apparent difference" in the success rate of resuscitation attempts according to whether family members were present or absent; at Parkland, 95 percent of family members present indicated that "everything possible was done to save the patient's life."[63] It has been recommended that most institutions should establish programs allowing family members to be present during re-

---

60. *Del Carmen Guadalupe*, 299 F.3d at 21.

61. "About a quarter of the nurses in emergency rooms and critical-care units said family presence was banned during cardiopulmonary resuscitation attempts and invasive procedures, even though their units had no formal policy prohibiting such access." ORANGE COUNTY REG., May 24, 2003, at News 18 (describing a study based on survey responses from 984 members of the American Association of Critical-Care Nurses). Hospital ethics committees may find it desirable to take action upon this policy matter.

62. N. ENG. J. MED.1019 (2002).

63. *Id.* at 1020 (citing Meyers et al., AM. J. NURSING (2000), at 32–43).

suscitation procedures.[64]

In view of the fact that bioethics and the law include ultramedical matters, many states have prohibited the appointment of physicians as proxies (surrogates) who are legally empowered to make healthcare decisions for their patients. For example, a New York statute was needed because a physician's professional ethos may overwhelm a patient's preferences.[65] A patient's choice of proxies may be necessary in order to prevent a patient's values being overturned by those of his physician. It was needed due to a possible paternalism by physicians who have no special expertise beyond medical diagnosis and treatment and "to prevent certain potential conflict of interest situations from arising."[66] These may be financial conflicts; however, they are frequently the result of accommodating the wishes of the family to avoid being sued for nontreatment. A physician's view of his obligation to serve the best interests of the patient may conflict with his actual "obligation as a proxy to advocate for the preferences of the patient." Many elderly people have no living relatives or friends who might be trusted to serve as proxies.[67] Laws authorizing the appointment of physicians should include a requirement that patients be advised of the potential risks inherent in such appointments.[68]

Previously, statutes and courts have actually authorized physicians to ignore the expressed wishes of both the patient and his surrogate regarding withdrawal

---

64.
> Recommendations for allowing the presence of family members during resuscitations, including guidelines for implementing formal programs, have been made by the Emergency Nurses Associations in the United States . . . . [T]he American Heart Association, in its 2000 guidelines for cardiopulmonary resuscitation, also recommends that this option be made available: "Parents or family members seldom ask if they can be present unless they have been encouraged to do so. Healthcare providers should offer the opportunity to family members whenever possible."

Id. at 1021.

65. N.Y. STATE TASK FORCE ON LIFE & THE LAW, LIFE-SUSTAINING TREATMENT: MAKING DECISIONS AND APPOINTING A HEALTHCARE AGENT (1987).

66. N.J. BIOETHICS COMM'N, PROBLEMS AND APPROACHES IN HEALTHCARE DECISION MAKING 148; N.Y. STATE TASK FORCE ON LIFE & THE LAW, *supra* note 65, at 110.

67. Rai et al., *The Physician as Healthcare Proxy*, Hast. Ctr. Rep., 1999, 14–19.

68. *See* Hamann, 38 VILL. L. REV. 103, 171 (1993); Wilbur & Reynolds, *Rethinking Alternatives to Guardianship*, 35 GERONTOLOGIST 248 (1995).

of treatment. For example, when Martha Duarte's neck was broken in an auto accident, she became comatose and was taken to a California community hospital where she was placed on a ventilator. She had told her family that "she would never want to be like some of the patients she had seen . . . just lying connected to a machine" in the hospital where she worked as a housekeeper. After 10 days, her husband and children requested that it be removed. The attending physician, Dr. Honzon Ou, refused and an agreement was worked out to do so providing that the hospital and physician were released from liability. However, Dr. Ou refused to sign it. Duarte finally died and the family sued the hospital and Dr. Ou for negligent treatment and intentional infliction of emotional distress. The family appealed from jury instructions disallowing damages for unconsented treatment.[69] Although a state statute then immunized physicians from civil liability for refusing to withdraw care, it was inapplicable because Mrs. Duarte had not executed a durable power of attorney. Nevertheless, the court denied the appeal.[70] This left the family with no remedy for the doctor's refusal to recognize their role as surrogate decision makers for their incapacitated next of kin. It has been pointed out as wrong to both allow the prolonging of treatment "over the objections of patients or their surrogates" and to deny "redress" to the family.[71]

good v. Good

Appointment of physicians as healthcare proxies should be allowed inasmuch as they will then adequately attend to all their patients' needs. Due to lack of time, etc., emergency physicians need not inquire about the existence of a patient's advance directives.

v.

The appointment of doctors as surrogates should not normally be allowed because many of them may have not evaluated advance directives (as shown by Project SUPPORT) and may only agree to the traditional Golden Rule (see Chapter 5 section 2(a)). Many patients in emergency departments arrive in a condition in which inquiry may be made concerning advance directives.

---

69. *See* Duarte v. Chino Cmty. Hosp., 85 Cal. Rptr. 2d 521 (1999).

70. However, as noted above, the statute was revised the following year. *See supra* note and accompanying text.

71. Capron, *Job in Court*, Hast. Ctr. Rep., Sept.-Oct. 1999, at 25.

Twenty-first Century Approach

1. In an increasingly mobile society where an advance health-care directive given in one state must frequently be implemented in another, there is a need for greater uniformity. State legislatures in approximately 40 states that have not enacted the Uniform Health-Care Decisions Act should consider such legislation (which has been approved by the ABA).
2. Hospital emergency departments should normally be required to seek information about a patient's advance directives prior to proceeding with treatments that might be contrary to such directives.
3. Because (a) traditional paternalism may cause a physician to allow his own values to supersede those of the patient, and/or (b) a physician's financial conflict of interests or his desire to accommodate family interests in conflict with the patient's expressed values, state legislatures should consider either to:
   a. generally ban the appointment of physicians as patient proxies (as is currently the law in several states) or
   b. require that patients be specifically warned that such appointments are generally deemed inadvisable except in those instances where no other persons are available who know the patient's values and desires.
4. Many state legislatures should consider revision of existing statutes in order to provide that where a physician wishes to decline to withdraw life-sustaining treatment from a reasonably determined permanently comatose patient (as appropriately requested by his surrogate), he must locate another physician who may be willing to do so, to avoid becoming subject to civil liability resulting from the expenses incurred by prolonging that patient's life.
5. In view of a significant absence of written policies thereon, hospital ethics committees should prepare appropriate procedures regarding family member access during emergencies.

Sure, we should fight to stay alive whenever possible; but when death is clearly happening, it becomes extremely important to stop fighting and be very calm. Use your intelligence to die well. You can be prepared for this from now on if you take care to live well.

Thurman, *Circling the Sacred Mountain* (1999), p. 319

The team needs to discover what it is that makes continued life so grievous . . . Reassurance and explanation about the likely nature of the final coming of death may well be needed if anxious fears are to be eased.

Saunders, "The Dying Patient," in Gillon and Lloyd
Principles of Health Care Ethics (1994)

3. Withholding and Withdrawal of Treatment; Limitations on Nonbeneficial (Futile) Treatment: DNR vs. CPR

Consider these cases about competent patients: Mr. McAfee was a quadriplegic on a ventilator because he was incapable of spontaneous respiration. He wanted to receive a sedative and have the respirator discontinued. When these were denied him, a state's Supreme Court held that he was entitled to both:

> Mr. McAfee's right to be free from pain at the time the ventilator is disconnected is inseparable from his right to refuse medical treatment. The record shows that Mr. McAfee has attempted to disconnect his ventilator in the past, but has been unable to do so due to the severe suffering he suffers when deprived of oxygen. His right to have a sedative (a medication that in no way causes or accelerates death) administered before the ventilator is disconnected is a part of his right to control his medical treatment.[72]

The result of such rulings indicates that healthcare providers may be liable for a refusal to follow a patient's instructions with regard to certain treatments. The next case is somewhat similar except that a mother, rather than a court, wrote about it:

72. State v. McAfee, 385 S.E.2d 651, 652 (Ga. 1989).

> Daniel was 21 years old. He was entering his senior year in college as a premed student when he had a tragic diving accident, rendering him a quadriplegic. Though critical care can keep him alive, doctors are unable to stop continuing deterioration of his condition. Slowly, day by day, Daniel loses more and more ground. He eventually learns that he will never be able to eat or drink, never be able to speak, never be off of the ventilator. With the help and support of his family, Daniel decides to remove the ventilator, thus quickly and effectively ending his life.
>
> Would he have any quality of life? Could he give such a life meaning? Would his values, goals, aspirations, be fulfillable in any way? Only Daniel had the right to make such a decision about his life. If I believed the things I subscribed to, taught, lived by, could I deny him his basic human rights? My answer was no. Accordingly, I asked that his mind-altering medications be stopped, so that his reasoning and comprehension could be at their fullest. We spoke with his pulmonologist, who agreed to abide by Daniel's decisions and to remove the ventilator if Daniel made that request.[73]

In a similar story, the prolific prize-winning writer James A. Michener, like thousands of other people, was on life-sustaining kidney dialysis. His many works argued for religious and racial tolerance, hard work, and self-reliance.[74] In October 1997, he decided that he did not want to go on "living like this," stopped the treatment, and died within a few days.

Some Catholic bishops supported withholding decisions to a coma victim, claiming stating, "The intention of discontinuing medical treatment . . . is to alleviate the burden and suffering of the patient and *not to cause her* death. It does not contradict Catholic moral theology."[75] But a Jewish writer discussed

---

73. ROTHMAN, SAYING GOODBYE TO DANIEL: WHEN DEATH IS THE BEST CHOICE 162–63 (1995); *see also* HALPERN, FROM DETACHED CONCERN TO EMPATHY: HUMANIZING MEDICAL PRACTICE (2001).

74. Michener's works included *Tales of the South Pacific*; *Sports in America* (1976); *Chesapeake* (1978); *Space* (1982); *Poland* (1983); *Alaska* (1987); *Journey* (1988); *Caribbean* (1989); *Mexico* (1992); *Literary Reflections* (1993); and *Recessional* (1994). In October 1997, he decided that he did not want to go on "living like this. He declined to continue on dialysis and soon died." ORANGE COUNTY REG., Oct. 13, 1997, at News 17.

75. KNOX, DEATH AND DYING 53 (2000); *see* § 3(c) (discussing PPVS).

"a person suffering from multiple incurable illnesses . . . who contracts pneumonia [which] would be treated with antibiotics." In such cases, "we may refrain from treating the pneumonia if that will *enable the patient to die* less painfully." This would be a decision that "judges according to the best interests of the patient."[76] Similarly, surrogates who lack an understanding of a patient's healthcare values normally base their decisions on the "best interest" of the patient.[77]

The AMA *Code of Medical Ethics* (2003) Section 2.20 states that a competent adult patient may in advance "provide a valid consent to the withholding or withdrawal of life-support systems in the event that injury or illness renders that individual incompetent to make such a decision." However, in 2000 it was noted that "most physicians do not discuss those preferences or often make inaccurate assumptions."[78] Further, a study of intensive-care patients found that a conflict that draws physicians into a negotiation occurred in almost half of the cases that involved the withdrawal of life support. However, these doctors should focus upon what the patient wants by discussion or by his advanced agreement.[79]

In 1997 the American Medical Association (AMA) issued guidelines to help doctors let people die with dignity. They listed "elements of quality care," including giving patients the opportunity to discuss their end-of-life care and managing

---

76. Dorff, Matters of Life and Death 207 (1998). In 2002 a 43-year-old British woman, who was paralyzed from the neck down and unable to breathe unaided, won the right to die. It was the first time in Britain that someone in control of their full mental faculties had asked doctors to switch off life support in that manner. http://member.compuserve.com/news/story.jsp 3/21/02.

77. Layde et al., *Surrogates' Predictions of Seriously Ill Patients' Resuscitation Preferences*, Archives Fam. Med. 518–23 (1995); Morris, *Surrogate Decision Making: Only One Piece of the Puzzle*, Arch. Fam. Med. 503–04 (1995).

78. Covinsky et al., *Communication and Decision-Making in Seriously Ill Patients: Findings of the SUPPORT Project*, 48 J. Am. Geriatrics Soc'y S44–S51 (2000).

79. Breen et al., *Conflict Associated with Decisions to Limit Life-Sustaining Treatment in Intensive Care Units*, 16 J. Gen. Internal Med. 283–89 (2001); Fettei et al., *Conflict Resolution at the End of Life*, 29 Critical Care Med. 921–25 (2001). Project SUPPORT found that nurse-led communication intervention had virtually no effect in the mid-1990s on the care that dying patients in intensive care received from physicians. 274 JAMA 1591–98 (1996); *see also* Lynn et al., *Perceptions by Family Members of the Dying Experience of Older and Seriously Ill Patients*, 126 Annals Internal Med. 126 (1997).

pain and other symptoms of terminal illness. Patients' preferences on life-sustaining measures constitute these guidelines:

> Whether the intervention be less complex (such as antibiotics or artificial nutrition or hydration) or complex and more invasive (such as dialysis or mechanical respiration) and whether the situation involves imminent or distant dying, *patients' preferences regarding withholding or withdrawing* intervention *should be honored* in accordance with the legally and ethically established right of patients.[80] (Emphasis added.)

good v. Good

The statement that dying patients' preferences regarding withholding or withdrawing intervention gives each doctor no discretion to ignore it.

v.

While only patients (or their surrogates) should make these decisions, one doctor, who disagrees, may also consider the patient's transfer to another doctor.

Twenty-first Century Approach

A hospital's guidelines should state that the discretion of physicians regarding withholding or withdrawing intervention treatment that conflicts with a patient's desires cannot normally be honored; however, the patient may be transferred to another doctor.

> *Though the right to be free of unwanted life-sustaining intervention is now established in case law, too many people still have insufficient opportunity to exercise their right, and the use of unwanted life-sustaining intervention has persisted.*[81]
>
> Emanuel L. Alpert

80. ORANGE COUNTY REG., June 23, 1997, at A1; *see also* AMA CODE OF MEDICAL ETHICS R. 2.20.

81. Emanuel L. Alpert, *Comparing Utilization of Life-Sustaining Treatments with Patient and Public Preferences*, 12 J. GEN. INTENAL MED. 175–81 (1998); SUPPORT, Principal Investigators, *A Controlled Clinical Trial to Improve Care for Seriously Ill Hospitalized Patients*, 275 JAMA 1232 (1996); 274 JAMA 1591–98 (1995).

a) Withholding Equals Withdrawal

As of the end of the twentieth century, both the law and bioethics deny any ethical differences between the withholding and withdrawal of medical treatment.[82] An attempt to distinguish these in terms of acts versus omissions is essentially a psychological or metaphysical preference;[83] if treatment is either declined by the patient or will not benefit him, it makes no difference whether it is not begun or stopped thereafter.[84] Only comfort care, including pain relief, may continue.

Despite legal and ethical principles favoring the right of patients or their surrogates to refuse life-prolonging therapy,[85] many dying patients are still

---

82. AMA Code of Medical Ethics § 2.20 (2003) states: "There is no ethical distinction between withdrawing and withholding life-sustaining treatment." *But see* Curtis et al., *Use of the Medical Futility Rationale in Do-Not-Attempt-Resuscitation Orders*, 273 JAMA 124–28 (1995); 160 Archives Internal Med. 1597–1601 (2000).

83. Cruzan v. Dir., Mo. Dep't of Health, 497 U.S. 261 (1990); Beauchamp & Childress, Principles of Biomedical Ethics 145–50 (3d ed. 1989); Fletcher, *Decisions to Forego Life-Sustaining Technologies When the Patient Is Incapacitated*, *in* Basic Clinical Ethics and Health Care Law (1990–1996).

84. Cohen et al., *Making Treatment Decisions for Permanently Unconscious Patients*, *in* Medical Ethics: A Guide for Health Professionals 188–204 (Monagle ed., 1988). In 1997 it was pointed out that a decision is made to withhold or withdraw life-sustaining treatment in 70% of hospital deaths, and, of course, in all hospice deaths. Field & Cassel, Approaching Death: Improving Care at the End of Life (1997).

85. The section of the AMA *Code of Medical Ethics* titled "Quality of Life" states:

> Rule 2.17. Quality of life, as defined by the patient's interest and values, is a factor to be considered in determining what is best for the individual. It is permissible to consider quality of life when deciding about life-sustaining treatment in accordance with Opinion 2.20, "Withholding or Withdrawing Life-Sustaining Medical Treatment."
>
> Rule 2.215. Care must be taken to evaluate the newborn's expected quality of life from the child's perspective. Life-sustaining treatment may be withheld or withdrawn from a newborn when the pain and suffering expected to be endured by the child will overwhelm any potential for joy during his or her life.

AMA Code of Medical Ethics R. 2.17, 2.215.

receiving unwanted treatment.[86] Some physicians may be less likely to withdraw treatment where adverse conditions in a patient result from treatment (iatrogenic conditions).[87]

good v. Good

Because withdrawal of treatment requested by a dying patient (or surrogate) is a more traumatic experience for some physicians and staff than withholding treatment, withdrawal can never be mandatory.

v.

Because no distinction can be made between them, they must become mandatory to preserve the autonomy of cognizant patients (or surrogates) and respect for the lack of consent to continued treatment.

Twenty-first Century Approach

Physicians must not refuse to withdraw prescribed treatment when so requested by cognizant dying patients (or their the valid surrogates).

---

b) Withholding Nutrition and Hydration

When a patient requests withdrawal or withholding of treatment, that includes all actions by healthcare staff intended to keep him alive; for some, this also includes food and water.[88] Recent studies have concluded that under certain circumstances tube feeding can cause rather than prevent suffering.[89] Although many professionals and staff remain unconvinced that withdrawal includes

---

86. N. Eng. J. Med. 852 (1997).

87. *Id.*

88. *Cruzan*, 497 U.S. 261. A comatose woman lived in Missouri for seven years in a vegetative state. Her parents petitioned to remove the gastrostomy feeding and hydration tubes. In the U.S. Supreme Court, Chief Justice William Rehnquist, writing for the majority, stated that "for the purposes of this case, we assume that the United States Constitution would grant a competent person a constitutionally protected right to refuse life-saving hydration and nutrition." *Id.* at 279; *see also* Gostin, *Decisions to Abate Life-Sustaining Treatment for Non-Autonomous Patients: Ethical Standards and Legal Liability for Physicians after* Cruzan, 264 JAMA 1846–53 (1990).

89. N. Eng. J. Med. 206 (2000).

nutrition and hydration,[90] near the end of the second millennium, a study involving more than 1,400 physicians and nurses found that a significant minority (34 percent of medical attending physicians and 45 percent of surgical attending physicians) actually believed that if all forms of life support (including mechanical ventilation and dialysis) are stopped, nutrition and hydration should be continued.[91] Despite this statement, an expert in at least one religion has stated that there appears "to be no grounds to suggest that Catholics should accept tube feeding."[92] Yet, in 2004, Pope John Paul II actually stated that "the administration of *water and food, even when provided by artificial means,* always *represents a natural means of preserving life*, not a medical act," and "if done knowingly and willingly," their removal is "euthanasia by omission."[93] His use of the word "nature" extends beyond its normal meaning. Further, it was pointed out how his approach "seems to represent a significant departure from the Roman Catholic bioethical tradition."[94] Similarly, Southern Baptists have been encouraging its followers to "consider the provision of nutrition and hydration by medical means to be compassionate and ordinary care."[95] Such statements have also been made with respect to some traditional Jews.[96]

In 2001 the California Supreme Court ruled against the authority to stop nutrition and hydration to a severely brain-damaged patient. The court found

---

90. Slomka, *What Do Apple Pie and Motherhood Have to Do with Feeding Tubes and Caring for the Patient?*, 155 ARCH. INTERNAL MED. 1258–63 (1995).

91. *Decisions Near the End of Life: Professional Views on Life-Sustaining Treatments*, 83 AM. J. PUB. HEALTH 14–23 (1993).

92. Brody, AM. J. MED. 740 (2000).

93. Papal Allocution to the International Congress on Life-Sustaining Treatment and Vegetative State: Scientific Advances and Ethical Dilemmas (Mar. 20, 2004). (Emphasis added.)

94. Shannon & Walter, *Implications of the Papal Allocution on Feeding Tubes*, Hast. Ctr. Rep. "While we certainly support every effort to prevent euthanasia, we do not support policies that require medical staff to provide unwanted medical treatment. Such policies might even drive people *toward* euthanasia, by making them feel that they have lost a traditional and sympathetic ally in their final journey." *Id.* at 20.

95. PAUL D. SIMMONS, RELIGIOUS BELIEFS AND HEALTHCARE DECISION: THE SOUTHERN BAPTIST TRADITION (2003). Nevertheless, the declarations of organizations within the Southern Baptist Convention are not binding on an individual's conscience.

96. Gillick, *Artificial Nutrition and Hydration in the Patient with Advanced Dementia: Is Withholding Treatment Compatible with Traditional Judaism?*, J. MED. ETHICS 12 (2001).

that he was able to engage in "clear, though inconsistent, interaction with his environment in response to simple commands." He was paralyzed, and dependent on a gastrostomy tube for nutrition and hydration;[97] but prior to his accident, he had expressed favor toward his life-sustaining treatment. The hospital ethics committee had supported his wife's request that she wanted his tube feeding stopped. He died of pneumonia prior to the rendered opinion.[98] It is one of the decisions demanding more justification to end nutrition and hydration for conscious, essentially incompetent patients who are not necessarily terminally ill. The court stipulated that in the case of a conscious patient, a conservator may withhold or withdraw life-sustaining treatment only if there is clear and convincing evidence that the patient "wished to refuse life-sustaining treatment or that to withhold such treatment would have been in his best interest."[99]

Not only ethically, but also factually, the medical staff may incorrectly believe that dying patients experience discomfort upon the withdrawal of tube feedings, parenteral nutrition, or intravenous hydration.[100] Further, there is no evidence to support the need to provide artificial (tube) to patients with advanced dementia.[101] Indeed, obliging terminally ill patients to take food may

---

97. N. Eng. J. Med., 207 (2000). Gastronomy tubes have not been shown to prolong life, ensure adequate nutrition, or prevent aspiration, and there is neither a secular nor a religious ethical imperative to use them. In addition, they are not necessary to prevent suffering. *Id.* at 208; *see also* Nat'l Conference of Catholic Bishops, Ethical and Religious Directives for Catholic Healthcare Services (1995).

98. Wendland v. Wendland, 110 Cal Rptr. 2d 412 (Cal. 2001).

99. The decision was also discussed in N. Eng. J. Med. 1489, 1491 (2002):

> [T]he California Supreme Court noted that if death occurred "despite the administration of life-sustaining treatment," it would be "unexpected." However, for many seriously ill patients, death within a year would not be unexpected. Clinically, the criterion that death would not be unexpected may be more useful than the criterion of terminal illness.

100. *See* 336 N. Eng. J. Med. 665 (1997) (citing Andrews et al., *Dehydration in Terminally Ill Patients: Is It Appropriate Palliative Care?*, Postgraduate Med. 1993, at 201–03, 206–08, and Printz, *Terminal Dehydration, a Compassionate Treatment*, 152 Arch. Internal Med. 697–700 (1992)).

101. JAMA 1365–70 (1999). Yet, "in New York the law requires documentation of patients' preferences about artificial hydration and nutrition for this treatment to be withdrawn without court approval." Cohen, 24 Preventive Med., 284–91 (1994).

cause them to suffer more and live longer in that state.[102]

As regards nursing homes, a memorandum of the New York Department of Health states that no part of its regulation requires a nursing home operator to provide a patient with adequate nutrients and fluids when there is clear and convincing evidence that the patient wishes to refuse such care and understands the consequences of such refusal.[103] In a similar manner, it was concluded that:

> [W]hen nursing home administrators and their employees, or the physicians who abide by the nursing home's policies, refuse to permit the forgoing of artificial nutrition and hydration, they unlawfully restrict the right of competent nursing home residents to refuse treatment and the well-accepted right of incompetent residents to have their surrogates decline treatment on their behalf.[104]

The AMA *Code of Medical Ethics* states, "Life-sustaining treatment may include . . . artificial nutrition and hydration."[105] Nevertheless, it also states, "Even if the patient is not terminally ill or permanently unconscious, it is not unethical to discontinue all means of life-sustaining medical treatment in ac-

---

102. *Supra* note 101 (citing McCann et al., *Comfort Care for Terminally Ill Patients: The Appropriate Use of Nutrition and Hydration*, 272 JAMA 1265–66 (1994)).

> It seems to be basic caring, but as death approaches, you will not "keep up your strength" by forcing yourself to eat when it makes you uncomfortable. *Not forcing someone to eat or drink is not letting him "starve to death."* There is no medical or clinical evidence that not using a feeding tube or IV leads to a more painful death. In fact, the research says just the opposite.

Lynn & Harrold, Handbook for Mortals 133 (1999). (Emphasis added).

103. State of N.Y. Dep't of Health, Memorandum No. 89-84; *In re* Westchester County Med. Ctr., 531 N.E.2d 607, 613–14 (N.Y. 1988); *see also* Conservatorship of Drabick, 245 Cal. Rptr. 840, 848 (Ct. App. 1988).

104. Am. J.L. & Med. 340 (1995). It noted that "nursing homes, unlike hospitals, are subject to a special set of Federal regulations (adopted under the authority of the Federal Medicaid statute)," *id.* (citing 42 U.S.C.A. §§ 1396a–1396u (West 1992 & Supp. 1995) and Requirements for Long Term Care Facilities, 42 C.F.R. § 483.25(i)–(j) (1994)), and asserted that those provisions had "unwittingly been transformed into an obligation to accept nutrition and hydration," *id.* at 346.

105. Section 2.20 (first paragraph).

cordance with a proper substituted judgment or best interests analysis."[106] In 2004 it was pointed out how many important medical articles had noted advanced dementia or other serious illnesses.[107] Although the percutaneous endoscopic gastrostomy (PEG) requires only two small incisions into the ab dominal wall, Dr. Michael Gauderer stated that "because of its simplicity and low complication rate, this minimally invasive procedure also lends itself to overutilization." Accordingly, he suggested that "much of our effort in the future needs to be directed toward the ethical aspects associated with long-term enteral feeding [because] . . . we as physicians must continuously strive to demonstrate that our interventions truly benefit the patient."[108]

good v. Good

The withholding of nutrition and hydration from dying patients indicates a lack of humanity and should be condemned.

v.

Without such withholding, the suffering of many dying patients may be unreasonably prolonged.

---

106. *Id.* (last paragraph); *see also* Fainsinger & Bruera, *The Management of Dehydration in Terminally Ill Patients*, J. Palliative Care, 1994, at 55–59. In 1996, an appellate court in New York upheld the parents' decision to decline to follow the recommendations of physicians to insert a feeding tube into their incompetent 28-year-old son. *In re* Matthews, 225 A.D.2d 142 (citing *In re* Storar, 420 N.E.2d 64 (N.Y. 1981) (holding that parents have the right to consent to medical treatment on behalf of their infants or incompetent adults in their care)). *But see In re* Storar, 420 N.E.2d at 73.

107. Orentlicher & Callahan, *Feeding Tubes, Slippery Slopes, and Physician-Assisted Suicide*, 25 J. Legal Med. 389–409 (2004). The authors cited Gillick, *Rethinking the Role of Tube Feeding in Patients with Advanced Dementia*, 342 N. Eng. J. Med. 206 (2000); Mitchell et al., *Clinical and Organizational Factors Associated with Feeding Tube Use Among Nursing Home Residents with Advanced Dementia: A Review of the Evidence*, 282 JAMA 1365 (1999); Callahan et al., *Outcomes of Percutaneous Endoscopic Gastrostomy Among Older Adults in a Community Setting*, 48 J. Am. Geriatrics Soc'y (2000); and Finucane & Christmas, *More Caution About Tube Feeding*, 48 J. Am. Geriatrics Soc'y 1167 (2000).

108. Gauderer, *Twenty Years of Percutaneous Endoscopic Gastrostomy: Origin and Evolution of a Concept and Its Expanded Applications*, 50 Gastrointestinal Endoscopy 879, 882 (1999).

Twenty-first Century Approach

1. Physicians and staff should be required to monitor dying patients in order to verify that nutrition and hydration are to be discontinued whenever such withdrawal of treatment is requested by a cognizant patient or should not be utilized or continued for other reasons.
2. Hospital policy statements should include express provisions with respect to the discontinuance of nutrition and hydration, which become applicable in such cases.

The lives of such [permanently unconscious people] are of no benefit to them and so doctors may lawfully stop feeding them to end their lives. With this decision, the law has ended its unthinking commitment to the preservation of human life that is a mere biological existence. In doing so, they have shifted the boundary between what is and what is not murder. Now, conduct intended to end life is lawful.

Peter Singer, *Rethinking Life and Death: The Collapse of Our Traditional Ethics* (1995), p. 80

A PVS patient becomes permanently vegetative when the diagnosis of irreversibility can be established with a high degree of clinical certainty, i.e., when the chance of regaining consciousness is exceedingly rare.

45 Neurology 1015-18 (1995)

c) Patients in a Persistent or Permanent Vegetative State (PPVS) Without a Living Will or DPA; Subjective Personhood

Many courts have held that a family member of a patient in a permanent (i.e., irreversible) vegetative state[109] may withdraw life-sustaining medical treatment without court approval and without an advance directive (living will) or other specific indication from the patient concerning his desires.[110] This status

---

109. The loss of all higher brain functions with brain stem autonomic functions is PVS. Consciousness after 12 months in that state is very unlikely. Weijer, *Cardiopulmonary Resuscitation for Patients in a Persistent Vegetative State: Futile or Acceptable?*, 158 Can. Med. Ass'n J. 491 (1998).

110. *See also In re* Fiori, 673 A.2d 905, 912 (Pa. 1996); The Multi-Society Task Force on PVS, *Medical Aspects of the Persistent Vegetative State (Part I)*, 330 N. Eng. J. Med. 1499–1508 (1994) [hereinafter *Persistent Vegetative State Part I*].

was also formerly estimated to consist of from 10,000 to 25,000 adults and from 4,000 to 10,000 children in the United States.[111]

First, it is recommended that because "PVS" has been referred to as consisting of a "persistent" or "permanent" vegetative state, consideration should be given the abbreviation to "PPVS." Second, there now exists equipment to determine the likelihood of its actual permanency. These decisions are consistent with the general rule that physicians should not give treatment that does not benefit the patient. Third, when such a clear and irrevocable determination is made, appropriate action should be taken toward extending the definition of death to include the permanent vegetative state.

In 1996, the Connecticut Supreme Court held that a patient whose life is in danger, but who could be cured by relatively routine treatment, has the right to refuse that treatment.[112] Nevertheless, some people objected to this even though suicide is not illegal in the United States. For example, during the same year, the AMA issued a rule of Medical Ethics, stating, "Physician-assisted suicide is fundamentally incompatible with the physician's role as a healer . . . and would pose serious societal risks"[113] (see Chapter 14). However, in 1990, the AMA's Council on Ethical and Judicial Affairs had stated, "Even if death is not imminent but a patient is beyond doubt permanently unconscious, and there are adequate safeguards to confirm the accuracy of the diagnosis, it is not unethical to discontinue all means of life-prolonging medical treatment."[114] This appears to be consistent with many religious views. For example, since the turn of the third millennium, it has been stated that "withdrawing medically assisted nutrition and hydration from patients who have been accurately diagnosed in a *persistent*

111. Hoffenberg et al., *Should Organs from Patients in a Permanent Vegetative State Be Used for Transplantation?*, *in* Caplan AL.

112. Stamford Hosp. v. Vego, 236 Conn. 646 (1996).

113. AMA Code of Medical Ethics R. 2.211 (2003).

114. Councils on Scientific Affairs and on Ethical and Judicial Affairs, *Persistent Vegetative State and the Decision to Withdraw or Withhold Life Support*, 263 JAMA 429 (1999). Without clarity regarding permanence, it has been suggested that such patients should be maintained on nutrition and hydration for at least some time—especially if they need no scarce resources—to guard against the possibility that they were misdiagnosed as being PVS patients and are instead in a reversible coma. This also will give family members time to get used to the reality of their loved one's impending death before removing the tubes. Dorff, *supra* note 76, at 216.

vegetative state is morally justified."[115] Despite these ethical statements, "lives" are often being maintained in a *permanent* vegetative state in the United States at the cost to the country's taxpayers. As was noted at the dawn of the twenty-first century:

> If providing treatment for the permanently unconscious is a violation of professional integrity—as research suggests is the common view among healthcare professionals—why does the U.S. have so many permanently unconscious patients maintained (perhaps 35,000).[116]

Many legal and medical experts have recommended that permanently vegetative patients be considered dead.[117] In 2003 the California Court of Appeal

---

115. Panicola, *Catholic Teaching on Prolonging Life: Setting the Record Straight*, Hast. Ctr. Rep., 2001, at 10.21. But the Pope had stated:

> Normally one is held to use only ordinary means—according to circumstances of persons, places, times and culture—that is to say, means that do not involve any grave burden for oneself or another. A more strict obligation would be too burdensome for most men and would render the attainment of the higher, morc important good too difficult.

Pius XII, *The Prolongation of Life*, *in* CRITICAL CHOICES AND CRITICAL CARE 192 (Wildes ed., 1995).

116. HEALTH CARE REFORM & ETHICS REV., Sept. 2000, at 4.

> It is interesting to note that a woman who received nutrition and hydration while she was comatose died after 25 years in a coma.

ORANGE COUNTY REG., Sept. 22, 1995, at News 25.

> A recent study concluded that simple physical examination maneuvers strongly predict death or poor outcome in comatose survivors of cardiac arrest. The most useful signs occur at 24 hours after cardiac arrest, and earlier prognosis should not be made by clinical examination alone. These data provide prognostic information, rather than treatment recommendations, which must be made on an individual basis incorporating many other variables.

Booth et al., *Is This Patient Dead, Vegetative, or Severely Neurologically Impaired?*, JAMA 870 (2004).

117. Schadfner et al., *Philosophical, Ethical, and Legal Aspects of Resuscitation Medicine*, 16 CRITICAL CARE MED. 1069–76 (1988). Such recommendations would also become useful to facilitate organ procurement from such people. *See supra* Chapter 11 (discussing organ donation).

upheld a juvenile court's decision to withdraw life-sustaining medical treatment for a then one-year-old dependent where physicians had testified to support the withdrawal of treatment or a DNR order.[118] The AMA *Code of Medical Ethics* was partly revised to set forth its policy in a more positive manner.[119] Patients without higher cortical function who are on a ventilator do not experience discomfort when the ventilator is withdrawn.[120] (Also see Chapter 11—Transplant, section A 8.)

The effect of religion and politics on such patients continued in Florida. "Terri Schiavo, 25, was found unconscious and gasping for air at home by her husband on the early morning of February 25, 1990 . . . Terri had a history of erratic eating habits, including probable bulimia, with a major weight loss several years before this event. In high school it has been reported that Terri went from more than 200 pounds down to 110. In November 1992 Michael Schiavo won a malpractice suit against Terri's physicians for failing to diagnose her health problems leading up to the cardiac arrest resulting directly from her eating disorder."[121] In the fall of 2003, the feeding tube of Terri, who had then spent 13 years in a coma, was ordered removed after the state's Department of Children and Families completed nine reports of abuse accusations made from 2001 to 2004, including neglect of hygiene, denial of dental care, poisoning, and physical harm.[122] The state legislature rushed through a

118. *In re* Christopher I, Cal. Rptr. 2d 122 (2003). Christopher I's biological father had physically abused the one-year-old child and rendered him comatose. His mother petitioned for a DNR order for Christopher and/or to have parens patriae power to make such determinations after applying the correct evidentiary and substantive standards. *Id.* at 145.

119. When a permanently unconscious patient was never competent or had not left any evidence of previous preferences or values, since there is no objective way to ascertain the best interests of the patient, the surrogate's decision should not be challenged as long as the decision is based on the decision-maker's true concern for what would be best for the patient. Rule 2.20 (updated June 1996).

120. The Multi-Society Task Force on PVS, Medical Aspects of the Persistent Vegetative State (Part II), 330 N. Eng. J. Med. 1572–79 (1994) [hereinafter *Persistent Vegetative State Part II*]; 333 N. Eng. J. Med. 130 (1995).

121. *In re* Guardianship of Schiavo, 780 So. 2d 176, 177 (Fla. Dist. Ct. App. 2001); *In re* Schiavo, No. 90-2908-GB-003 (Fla. Cir. Ct. Feb. 11, 2000); Cranford, *Facts, Lies, and Videotapes: The Permanent Vegetative State and the Sad Case of Terri Schiavo*, J.L. Med. & Ethics, 364 (2005).

122. Cary, N.Y. Times, Orange County Reg., Apr. 16, 2005, at News 11.

bill designed to keep her alive. Governor Bush (the brother of the president) quickly invoked the new law and ordered the feeding tube reinserted.[123] In September 2004 the Florida Supreme Court struck down that law as unconstitutional. Despite this, in March 2005 Congress enacted and the President signed a law that would allow the tube to be reconnected while the federal court reviewed the case. Following their being so requested, the constitutionally independent federal courts (including the Supreme Court) declined to review the case again. Gallup polls found that a majority of Americans supported the decisions by these courts, noting how the refusal to remove the feeding tube keeping the Florida woman alive was actually being driven by church attendance, party affiliation, and political ideology overall.[124]

A physician and a theologian claimed that "conflicts between religious convictions to continue treatment and clinical judgments to suspend treatment have been addressed infrequently in the literature." They then added: "In our view, the fact that the intervention is requested on religious grounds is not sufficient to trump this process, as long as clinicians act in good faith at each step."[125]

good v. Good

Family members of a patient who is found to be in a permanent vegetative state should always have the right to demand that the patient be maintained alive at taxpayer expense because of their religious and/or subjective feelings concerning such a patient.

v.

"Family members" are not "patients" and physicians have no duty to treat a patient either against his expressed desire or in a manner that can be shown to be of no benefit to the patient.

123. Orange County Reg., Oct. 23, 2003, at News 14.

124. Orange County Reg., Mar. 28, 2005, at News 4. Terri Schiavo died on March 30, 2005, nearly two weeks after her feeding tube was removed. The autopsy reported that at her death, she had a brain weight of 615 grams, half that of a normal brain. Her brain had been damaged by oxygen deprivation, etc. Fins & Schiff, Hast. Ctr. Rep., July-Aug. 2005, at 8.

125. Brett & Jersild, *'Inappropriate' Treatment Near the End of Life, Conflict Between Religious Convictions and Clinical Judgment*, Arch. Internal Med. 1645–46, 1648 (2003).

Twenty-first Century Approaches

1. Requests regarding treatment expressed by the family or friends of an adult patient (subjective personhood) should not be considered a basis for the expenditure of taxpayer funds beyond those needed for comfort care given to patients in a permanently vegetative state.
2. Families who are guardians of such patients on a ventilator should be informed that the patient will not suffer discomfort should it be withdrawn.
3. Physicians should be authorized under appropriate circumstances to properly withdraw treatment from patients reasonably determined to be in a permanently vegetative state without the need to seek a court order authorizing such action.
4. If for personal or other reasons a physician, in the face of such circumstances does not wish to do so, has a duty to locate another physician or staff member to carry out the request for withdrawal of treatment from a patient in a permanently vegetative state.
5. Consideration should be given by state legislators toward the enactment of an appropriate extension of the definition of death to those who have clearly been determined to be in a permanent vegetative state.

---

4. The Rights of Mature Minors to Die and Jehovah's Witnesses

Consider this case:

Physicians recommended to a 17-year-old mature leukemia patient that she receive a series of blood transfusions. She objected on the basis of her religious convictions as a Jehovah's Witness. Nevertheless, she was forcibly transfused pending the outcome of her court suit thereon.

During a portion of the past two decades, a number of appellate courts refused to hear appeals from adult Jehovah's Witnesses who had refused blood transfusions, on the grounds that the cases were "moot" in a sense that the patient had either died or else had survived forcibly inserted transfusions and the issue was no longer deemed extant. In the 1990s, some appellate courts finally accepted such cases and they now recognize that competent adult Jehovah's Witnesses can legally refuse to be transfused.[126] The Illinois Supreme Court became the first state supreme court to rule that a minor patient should be permitted to refuse medical treatment claimed to be necessary to save her life. The court stated:

> Although the age of majority in Illinois is 18 [particularly for voting purposes], that age is not an impenetrable barrier that magically precludes a minor from possessing and exercising certain rights normally associated with adulthood. *If* the evidence is clear and convincing that *the minor is mature enough to appreciate the consequences of her action . . . then* the mature minor doctrine affords her *the common law right* to consent to or *refuse medical treatment*.[127] (Emphasis added.)

Such evidence of maturity overcame the presumption to the contrary. Chief Justice Berger of the Supreme Court had stated: "The law's concept of the family rests on a presumption that parents possess what a child lacks in maturity, experience, and capacity, or judgment required for making life's difficult decisions."[128] The Supreme Court, referring to the common-law mature minor rule, had previously authorized healthcare providers to dispense contraceptives to minors without parental consent.[129] It maintained the decision not to reproduce constitutes "one of life's difficult decisions" that reflects the view that, even if not fully mature, female teenagers need protection against pregnancy. The Supreme Court had also permitted the states to require parental (or judicial) consent of one parent for minors seeking abortions.[130] Courts in some states, such as Alabama, have not permitted minors

---

126. *See* Keyes, *supra* note 27, at Chapter 14.

127. *In re* E.G., 549 N.E.2d 322, 325–28 (1989). "Traditionally, Jews become adults in Jewish law at the age of twelve-and-a-half for girls and thirteen for boys; at those ages, they become liable for all of the duties and prohibitions of Jewish law." Dorff, *supra* note 76, at 160.

128. Parham v. J.R., 442 U.S. 584, 602 (1979).

129. Carey v. Population Servs. Int'l, 431 U.S. 678 (1977).

130. Beliotti v. Baird, 443 U.S. 622, 643 (1979).

to obtain abortions without parental consent.[131] However, it has been remarked that such courts appear to be more interested in protecting "the independent rights of parents" than in those of minors.[132] There has been judicial ambivalence about the mature minor rule in some instances. Other courts have assumed that minors have a right to refuse treatment; however, they were not always able to "find" the minor was mature enough.[133] Some insert an analogy to their argument against the execution of 17-year-old murderers even though they may not demonstrate immaturity or mental retardation.[134] (In 2003, for the last time the Supreme Court overturned a stay and allowed Oklahoma to execute a man who killed two people when he was 17.[135]) Nevertheless, most states permit unemancipated teenagers to consent to care for sexually transmitted diseases,[136] substance abuse treatment,[137] and psychiatric care.[138]

The American Academy of Pediatrics recommended that "considerable weight" be given to the desires expressed by mature minors. In 2000, its Committee on Bioethics maintained that a dying adolescent's refusal to undergo life-sustaining medical treatment "ought to be respected," and that a minor's ad-

---

131. *In re* Anonymous, 597 So. 2d 709 (Ala. Civ. App. 1992).

132. Baird v. Belotti, 393 F. Supp. 847, 856 (D.C. Mass. 1975), *vacated by* 428 U.S. 132 (1976).

133. *See In re* Long Island Jewish Med. Ctr., 557 N.Y.S.2d 239 (1990).

134. Although mental retardation may be a defense against execution, in 2002 three Supreme Court Justices on the court's designated "liberal wing" sought delay in the death penalty. The National Coalition to Abolish the Death Penalty indicated that "16 of the 38 states that allow the death penalty prohibit executing those who were under 18 when they killed."

ORANGE COUNTY REG., Aug. 29, 2002, at News 9. See Section 6.f *infra.*

135. Patterson v. Texas, 153 L. Ed. 2d 887 (2002) (denying an execution stay for Patterson, who was sentenced to death for a crime he committed when he was 17 years old); Stanford v. Kentucky, 492 U.S. 361, 380 (1989) (holding that state statutes authorizing the death sentence for adolescents 16 or 17 years old at the time of commission of a capital offense do not offend the Eighth Amendment of the Constitution); ORANGE COUNTY REG., Apr. 4, 2003, at News 24.

136. Linda Sorenson Ewald, *Medical Decision-Making for Children: An Analysis of Competing Interests*, 25 ST. LOUIS U. L.J. 689 (1982).

137. 410 ILL. COMP. STAT. ANN. § 210/4 (West 2000).

138. MICH. COMP. LAWS § 330.1498(d)(e) (1999). However, the United Sates did not sign the U.N.'s 1989 Convention on the Rights of the Child, which allows certain children to veto parents' decisions on issues of health. U.N. Doc. A/44/49 (1989); *see also* Haas, AM. J.L. & MED. 66–68 (2004).

vance directive "should serve as strong evidence" of those wishes.[139] The law in The Netherlands now permits euthanasia when a 16-year-old requests it, with parental notification, and when a 12- to 15-year-old requests it and the parents agree.[140] Some have argued that maturity is a lifelong proposition and in some instances there is no precise way to measure it.[141] The Supreme Court of Appeals of West Virginia considered the decision-making ability of a 17-year-old patient who was afflicted with muscular dystrophy. When he was intubated and put on a respirator, his physicians spoke about his condition with his father, who consented to a Do Not Resuscitate (DNR) order. No one discussed the DNR order with his son, who died with the event of respiratory failure without a resuscitation. The court stated, "[I]t is difficult to imagine that a young person who is under the age of majority, yet, who has undergone medical treatment for a permanent or recurring illness over the course of a long period of time, may not be capable of taking part in decisions concerning that treatment." The court then remanded the case back to the trial court to ascertain whether the son had been a mature minor who should have decided if he would have consented to a DNR for prolonging his life.[142]

---

139. Am. Acad. of Pediatrics Comm. on Bioethics, *Palliative Care for Children*, 106 PEDIATRICS 351, 354 (2000).

140. Editorial, J. PEDIATRICS 584 (2005) (citing Vrakking et al., *Physicians' Willingness to Grant Requests for Assistance in Dying for Children: A Study of Hypothetical Cases*).The authors note that nearly a quarter of neonatal and infant deaths in their country (the Netherlands) occur after use of drugs that could shorten life when given for the purposes of symptom relief—a figure that might not differ much from what happens in the United States, though we lack comparable data—and 8% of the Dutch infant deaths follow "the use of lethal drugs". *Id.*

141. Hohrman et al., *Guidelines on Foregoing Life-Sustaining Medical Treatment*, 93 PEDIATRICS 535 (1994); Lantos & Miles, *Autonomy in Adolescent Decision Making*, 10 J. ADOLESCENT HEALTH CARE 460–66 (1989).

142. Belcher v. Charleston Area Med. Ctr., 422 S.E.2d 827.

> Doctors may not understand the need to consult the child regarding his or her opinion on further treatment. Yet, ask a child-life specialist or a pediatric social worker, and you'll find that the best course includes asking the child. Your child knows what hurts and what she feels is worth continuing to fight for. Deciding medical care for her without her input, even when she is very young, ignores the fact that sometimes continued attempts to cure becomes doing *To* instead of doing *For.*

LYNN & HARROLD, *supra* note 102, at 167.

This was consistent with several studies. For example, in 2000, it was found that both normal 14-year-olds and 21-year-olds will rationally make and voluntarily reach their decisions and that even younger ones may be informed about their terminal condition.[143] Similarly, the American Academy of Pediatrics (AAP) Committee on Bioethics had previously recommended that adolescents become involved in their health care decision-making.[144]

good v. Good

Teenagers cannot be considered mature enough to make significant medical decisions without parental oversight.

v.

"Mature minors" should not be denied authority to make their own health care decisions should they so choose.

Twenty-first Century Approaches

1. The common law's "mature minor rule" should continue to be used for permitting certain teenagers to exercise their autonomy in making life-sustaining decisions concerning themselves.
2. When applying this rule to rational teenage children, courts should not seek to create or sustain an additional "independent" right solely for their parents.

---

143. Schlam & Wood, *Informed Consent to the Medical Treatment of Minors: Law and Practice*, HEALTH MATRIX 141 (2000). A recent study showed that no parent who disclosed the condition "later regretted having talked with his or her child about death." Furthermore, because "many parents who had sensed that their child was aware of his or her imminent death later regretted not having talked about death emphasizes the clear responsibility of health care workers to help parents respond to the wants and needs of a terminally ill child." Andris Kreicbergs et al., *Talking about Death with Children Who Have Severe Malignant Disease*, N. ENG. J. MED. 1185 (2004).

144. Am. Acad. of Pediatrics Comm. on Bioethics, *Informed Consent, Parental Permission, and Assent in Pediatric Practice*, PEDIATRICS 1185 (1995); *see also* Mlyniec, *A Judge's Ethical Dilemma: Assessing a Child's Capacity to Choose*, FORDAM L. REV. 1873 (1996).

So shall thou rest, and what if thou withdraw
In silence from the living, and no friend
Take note of thy departure? All that breathe
Will share thy destiny. The gay will laugh
When thou art gone, the solemn brood of care
Plod on, and each one as before will chase
His favorite phantom; yet all these shall leave
Their mirth and their employments, and shall come
And make their bed with thee.

William Cullen Bryant

Finally, the intelligence. Think of it this way: You are an old man. Stop allowing your mind to be a slave, to be jerked about by selfish impulses, to kick against fate and the present, and to mistrust the future.

Marcus Aurelius, Bk 2.2

5. The Choices of Do Not Resuscitate (DNR) vs. Cardiopulmonary Resuscitation (CPR): Legal and Ethical

Consider the case of Dr. Stamler of Iowa:

When I got to the scene, I saw a nurse pumping the chest of an elderly, obese woman . . . We wrestled her onto a cart and wheeled her into the emergency room with barely a break in the rhythm of the chest compressions and ventilation. As one nurse pounded away on the patient's chest, another started an IV, someone hooked up the ECG leads and I prepared to intubate.

The patient presented every problem in the manual. Her blood pressure would sink, I would respond. Her heart rate would slow, I'd respond. Her heart would fibrillate, I'd respond. After the third defibrillation, though, I had run out of options. I was at the bottom of my list when the patient stabilized. I transferred her to the intensive care unit, where her family doctor took over.

I found the patient's son in the ICU hallway. I explained that his mother's heart had stopped several times, but we'd managed to bring her back each time.

> "Her heart has been weak for years," he said. "She can't walk. She can't travel. She can't sleep lying down. She can't even eat the food she likes. She's been miserable for years. You should have just let her go."
>
> He turned and walked away. I had done everything that could be done, but failed. And the family could tell.[145]

The doctor "failed" and knew it. But at the dawn of the twenty-first century, many physicians would actually say that they "succeeded" by prolonging such a "life." Would Dr. Stamler "fail" again the next time such a patient was brought to him? Should he be able to say that when the next such patient came in he would "succeed" by learning her true quality of life?

A study of the desires of 936 patients with heart disease found that physicians often misjudged those desires. In 24 percent of the cases, the doctors were wrong, mostly in cases where the patient did not want to be resuscitated.[146] John Dewey, the American pragmatic philosopher of the first half of the twentieth century, wrote, "action and opportunity justify themselves only to the degree in which they render life more reasonable in increasing its value." At the end of the twentieth century, it was shown that delayed DNR orders increased costs of care without improving morality rates.[147]

A large degree of ethical and legal anarchy prevails in the 21st century in this area. A study has shown that 34 percent of physicians continue life-sustaining treatments despite patient or surrogate wishes for discontinuation. Some of these decisions are made "without the knowledge or consent of the patient or surrogate, and some are made despite their objection."[148]

In his book, *The Moral Imagination*, Tivnan describes a cover story in a *Newsweek* magazine about a 69-year-old woman dying of emphysema and inoperable lung cancer:

---

145. MED. ECON., Apr. 18, 1996, at, 224–25.

146. ORANGE COUNTY REG., Aug. 18, 1998, at News 5.

147. De Jong et al., *The Timing of Do-Not-Resuscitate Orders and Hospital Costs*, J. GEN. INTERNAL MED. 190 (1999).

148. Mayo Clin. Proc. May 1996, p. 674 (citing Ash et al., *Decisions to Limit or Continue Life-Sustaining Treatment by Critical Care Physicians in the United States: Conflicts Between Physicians' Practices and Patients' Wishes*, AM. J. RESPIRATORY & CRITICAL CARE MED. 288–92 (1995)).

The doctor advised hooking her up to a life-support machine. The woman's daughter disagreed, and so finally did the patient. The doctor bristled, "If that were my mother, I'd do it." The family ignored him, and the next day, the woman died quietly. This is a doctor who is unable to face death, and therefore, some might be unable to say, "He is a good doctor."[149]

a) Presumed Consent to Cardiopulmonary Resuscitation (CPR)

Consider this case:

No sooner had the Committee on CPR been formed than the specter of lawsuits loomed; the committee, at its first meeting, heard about a Veterans Administration hospital nurse who, through application of closed-chest compression, saved a patient's life. That's the good news. The bad news is that he remained in a vegetative state. The family sued the nurse and the Veterans Administration. The *case was either dropped or settled* out of court *since no further information was contained in the committee minutes.*[150] (Emphasis added.)

The author claimed that as of that time "there has not been a successful suit against someone performing CPR." Healthcare facilities maintain a "presumed consent" policy that cardiopulmonary resuscitation (CPR) or code should be called on unless there are specific written directives to the contrary. The current AMA *Code of Medical Ethics* also indicates that efforts should be made to resuscitate patients except when circumstances indicate that CPR "would

149. TIVNAN, THE MORAL IMAGINATION 141 (1995).

If patient input results in that patient's dissenting to the DNR order, then individual physicians are not justified in overriding such dissent. To do so would give unjustifiable primacy to the values of the individual physician. Therefore patient consent is effectively required to write a DNR order.

Bieglar, *Should Patient Consent Be Required to Write a Do-Not-Resuscitate Order?*, 29 J. MED. ETHICS 359–63 (2003).

150. EISENBERG, LIFE IN THE BALANCE (1997).

be inappropriate or not in accord with the desires or best interests of the patient."[151] As stated in the volume *Handbook for Mortals*, "[W]hen you are seriously ill, you may find his procedure—called cardiopulmonary resuscitation—to be worthless or even deeply disturbing."[152]

More than 250,000 people experience sudden cardiac death outside of hospitals each year in the United States.[153] As a result, many ordinary citizens have been trained to some extent in CPR. However, a study of more than 2,000 cases of cardiac arrest in New York City found that in 17 percent of the cases where CPR was performed by a bystander, it was not done properly, but among those where it was done properly, only 4.6 percent survived to be sent home.[154]

Communication with regard to death and CPR is generally inadequate, and action should be taken early in the third millennium to change this practice. The AMA *Code of Medical Ethics*, Rule 2.22, states:

> Patients at risk of cardiac or respiratory failure should be encouraged to express in advance their preferences regarding the use of CPR, and this should be documented in the patient's medical record. These discussions should include a description of the procedures encompassed

---

151. AMA Code of Med. Ethics R. 2.22 (2003) states: "Efforts should be made to resuscitate patients who suffer cardiac or respiratory arrest except when circumstances indicate that cardiopulmonary resuscitation (CPR) would be inappropriate or not in accord with the desires or best interests of the patient."

152. "While you may be grateful for the time you have, or at least willing to endure whatever time you have, you might not want someone to disrupt your time of death with a flurry of activity that is doomed to have little effect." Lynn & Harrold, *supra* note 102, at 137.

> While the obligation to perform CPR may not exist, another obligation does. The patient's fears and worries (and perhaps those of the healthcare providers) ought to be acknowledged, examined, and then dealt with. Such a patient ought not to be simply told that since CPR is unlikely to succeed, it will not be performed. It would be worse than unkind not to help him deal with his suffering, his fear of death and/or fear of death by suffocation.

Robichaud, *Healing and Feeling: The Clinical Ontology of Emotion*, 17 Bioethics 59–68 (2003).

153. JAMA 1220 (1999).

154. Orange County Reg., Dec. 27, 1995, at News 5.

> by CPR and, when possible, should occur in an outpatient setting when general treatment preferences are discussed, or as early as possible during hospitalization.[155]

It has been written: "As an ethical issue, this problem is in part 'iatrogenic,' [adverse condition in a patient resulting from treatment by a physician] because it often results from physicians' failure to talk with their patients. The problem is compounded by physicians' fear of litigation."[156] In some states, doctors and healthcare providers are protected from liability for issuing written DNR orders that have also been signed by the patient.[157]

Bioethicists have often become concerned with cases where patients want to discuss spiritual and religious issues with physicians. This arises because many physicians feel unskilled and uncomfortable discussing them and some patients may not want to talk to the chaplain. Physicians ask a typical question: "If you were to have a heart attack, your heart stopped, and you died, would you want us to try CPR?"[158] Where the patient feels that the physician is pushing him to make a decision that he had no time to think through, both of

---

155. Teno et al., *Preferences for Cardiopulmonary Resuscitation Physician-Patient Agreements and Hospital Resources Use*, 10 J. Gen. Internal Med. 179–86 (1995); Tuklski et al., *How Do Medical Residents Discuss Resuscitation with Patients?*, 10 J. Gen. Internal Med. 436–42 (1995).

156. N. Eng. J. Med. 1909 (1997). In 1998 a physician was convicted of attempted murder of a terminally ill patient and intentional and malicious second-degree murder of another terminally ill patient for his failure to provide cardiopulmonary resuscitation. The court reversed the physician's convictions, stating that the jury could not disregard the testimony of several physicians who concurred with the defendant-physician's treatment of the deceased patients. State v. Naramore, 965 P.2d 211 (Kan. Ct. App.).

157. *E.g.*, a California statute amended the Probate Code to provide civil, criminal, professional, and administrative immunity for healthcare providers who honor DNR requests if there is a written document signed by the patient or surrogate and physician. In addition, it provides that in the absence of knowledge to the contrary, healthcare providers may presume that the written request to forgo resuscitation is valid and has not been revoked. Cal. Prob. Code § 4753 (West 2000).

158. Le et al., JAMA 749 (2002).

them will "feel frustrated." Many physicians consider that asking more questions would be "too psychologically oriented or forced."[159] It has been stated, "physicians should respect patients' religious and spiritual views and avoid expounding or imposing their own beliefs," because few of them have the training or expertise "to engage in theological discussion about the nature of God, sin, and punishment"; they should be "kept separate"[160] (see Chapter 12). Where a religious patient has almost no chance for improvement, is unable to participate in such discussions, and had agreed to the use of a surrogate (without being specific), it has been suggested that the physician (often a secular one) not tell the religious surrogate (or the family) that a miracle is rare. This is because he may dismiss religions that are based upon the "insistence on interventions." On the contrary, physicians must "maintain their integrity."[161]

Regarding hospital cardiac arrests among children, they also have poor prognosis; many survivors remain in a persistent or permanent vegetative state (PPVS).[162] As of the end of the second millennium, 42 states had adopted protocols permitting seriously ill persons to avoid unwanted resuscitation

---

159. *Id.* at 754. More questions were added to help the patient think through his preferences:

> "Let me be sure I've understood. It sounds like if you thought God was calling you, you wouldn't want us to try to revive you. But you're not sure when God is calling you. Is that right?" If the patient agrees, the physician might then say, "If I told you how likely it is that CPR would succeed in various situations, would that help you decide whether God was calling you?" or alternatively, "When you've known other people who have died, what do you think helped them to know when God was calling?" Through such discussions, the physician can help.

Id.

160. *Id.* at 51.

161. The family is "unlikely to be persuaded by someone who does not believe in miracles. Furthermore, using the family's religious terms to get them to agree with the physician's plan can be manipulative. As tension mounts, physicians and families may become polarized, and disagreements may escalate into conflicts." *Id.* AMA CODE OF MEDICAL ETHICS R. 2.22 states: "Physicians should not permit their personal value judgments about quality of life to obstruct the implementation of a patient's preferences regarding the use of CPR."

162. Fields et al., *Outcomes of Children in a Persistent Vegetative State*, CRITICAL CARE MED. (1995).

through the use of out-of-hospital DNR orders.[163] As of the turn of the third millennium, 15 states limit eligibility for EMS-DNR orders to patients who are diagnosed with a terminal condition; 6 more states required that the patient either be terminally ill or meet one or more other medical preconditions; in 21 states with protocols, no medical preconditions were imposed.[164] These protocols must be extended to also apply in-house where patients wish to request that a DNR order be issued. The Rights of Patients in a General Acute Care Hospital in California specifically grants patients the right to "refuse treatment" and to leave the hospital against the advice of physicians.[165] In some states, courts have held that parental consent is required to issue a DNR order for minors[166] (see Section 4, above).

Patient preferences not to receive CPR were associated with no difference in hospital survivals.[167] It has been noted, "In actuality, the success of emergency medical service programs in the United States and throughout the world is not very good." New York and Chicago have had survival rates after car-

---

163. Of these, 34 are so authorized by statutes, and almost all offer protection from civil and criminal liability and from disciplinary action. Charles P. Sabatino, *Survey of State EMS-DNR Laws and Protocols*, 27 J.L. MED. & ETHICS 297, 302 (1999). As regards the standard of care, a California statute states that the provider must have "no knowledge that the action or decision would be inconsistent with a healthcare decision that the individual signing the request would have made on his or her own behalf under like circumstances." CAL. PROB. CODE § 4753(a)(2) (West 2000).

164. J.L. MED. & ETHICS 299 (1999).

165. *See* CAL. CODE REGS. tit. 22, § 7070.

166. *E.g.*, *In re* Doe, 418 S.E.2d 3 (Ga. 1992); *see also* GA. CODE ANN. § 31-39-1 to -6. Twenty-six states permit parental consent for minors in their DNR protocols. Sabatino, *supra* note 163, at 302.

167. Some suggest that a resuscitation attempt is more likely when preferred by patients and when death is least expected. 282 JAMA 2333–39 (1999). The guidelines for cardiopulmonary resuscitation (CPR) were published over 30 years ago and last updated in 2000. They recommend target values "for compressions, ventilations, and CPR-free intervals allowed for rhythm analysis and defibrillation." However, in a recent study of CPR during out-of-hospital cardiac arrest, "chest compressions were not delivered half of the time, and most compressions were too shallow." Wik et al., *Cardiopulmonary Resuscitation in the Real World: When Will the Guidelines Get the Message?*, JAMA 299 (2005).

diac arrest of only 5 percent and 4 percent, respectively.[168] It has been stated that, "a hospital policy that permits CPR on demand even when not medically indicated is asking physicians to practice bad medicine."[169] It may cause physicians to go against their medical judgment and expend resources for nonbeneficial care. Conversely, it would appear that a patient's exercise of autonomy to refuse treatment means that a DNR order must be posted at or near his hospital bed.[170] As regards the entering of a DNR order, the AMA *Code of Medical Ethics* requires that "The physician also should be prepared to discuss appropriate alternatives, such as obtaining a second opinion (e.g., consulting a bioethics committee) or arranging for transfer of care to another physician."[171]

Some state statutes prohibit surrogates from basing their decisions on anything other than the patient's wishes.[172] A California statute, which became

---

168. Becker et al., *Outcomes of CPR in a Large Metropolitan Area: Where Are the Survivors?*, 22 Annals Emergency Med. 1652–58 (1993); Lombardi et al., *Outcome of Out-of-Hospital Cardiac Arrest in New York City*, 271 JAMA 678–83 (1994).

169. Health Progress, June-Feb. 1990, at 192. In 1991 the American Medical Association guidelines for the use of DNR orders regarding the offer of futile interventions stated: "A physician is not ethically obligated to make a specific diagnostic or therapeutic procedure available to a patient even on specific request, if the use of such a procedure would be futile." Council on Ethical and Judicial Affairs, *Guidelines for the Appropriate Use of Do-Not-Resuscitate Orders*, 265 JAMA 1868–1971 (1991).

170. *See* Sabatino, *supra* note 163, at 312 (1999).

171. Rule 2.22. The rule states: "DNR orders, as well as the basis for their implementation, should be entered by the attending physician in the patient's medical record. DNR orders only preclude resuscitative efforts in the event of cardiopulmonary arrest and should not influence other therapeutic interventions that may be appropriate for the patient." In the absence of written DNR orders, the use of inadequate resuscitation efforts ("slow codes") by physicians and nurses is unethical. American College of Physicians Ethics Manual (3d ed.); 117 Annals Internal Med. 947–60 (1992); *see infra* Section 7.

172. *E.g.*, Tex. Health & Safety Code Ann. § 313.004 (Vernon 2000); *see also* Tex. Health & Safety Code Ann. § 166.152 (Vernon 2000) (requiring surrogates to make healthcare decisions according to the agent's knowledge of the principal's wishes, including the principal's religious or moral beliefs, or in accordance with the principal's best interest, if the principal's wishes are not known).

effective in 2000, refers to a request to forgo resuscitative measures with the words "do not resuscitate" or the letters "DNR," a patient identification number, and a 24-hour toll-free telephone number, issued by a person pursuant to an agreement with the Emergency Medical Services Authority.[173] It established "a registry system through which a person who has executed a written advance healthcare directive may register in a central information center, information regarding the advance directive, making that information available upon request to any healthcare provider."[174] When a hospital or physician honors this request, they are no longer subject to "criminal prosecution, civil liability, [or] discipline for unprofessional conduct."[175]

b) *Best Interests v. Substituted Judgment*

Some states have provided mechanisms for resolving disputes regarding potential DNR orders where the patient is competent. This raises problems regarding his right to refuse treatment. For example, a New York state law provides up to 72 hours of mediation and requires that "persons participating in this process be informed of their right to judicial review."[176] Such review is not permitted with respect to a surrogate's refusal to authorize a DNR order for an incompetent patient unless the person challenging the refusal can show, by clear and convincing evidence, that the decision is contrary to the patient's wishes. In the absence of evidence of the patient's wishes, it might be argued that judicial review is needed when the decision is claimed to be contrary to the patient's best interests.[177]

The principle of seeking the best objective interest of an incompetent would normally appear to be to terminate treatment if he is found to be in a permanently vegetative state. Because the "best interest" test in such cases has been

---

173. CAL. PROB. CODE § 4780 (West).

174. *Id.* § 4800.

175. *Id.* § 4782.

176. N.Y. PUB. HEALTH LAW § 2973 (McKinney 2000) states:

> These participants include the patient, and attending physician, a parent, a non-custodial parent, or legal guardian of a minor patient, any person on the surrogate list, the hospital that is caring for the patient and, in disputes involving a patient who is in or is transferred from a mental hygiene or correctional facility, the facility director . . . .

177. *Id.*

seen as too paternalistic, the "substituted judgment" test has been used in the alternative to discern the patient's probable wishes. A problem would arise should the surrogate somehow show that the patient actually desired to be maintained indefinitely in such a state. This is because a physician might be asked to do something unethical, namely, to treat a patient who cannot be benefited by the treatment.

Where a surrogate lacks reliable information about a patient's preference, some courts have rejected the "best interests" test and ruled that providers should continue life-sustaining medical treatment.[178] The view of "futility" can often be decisive despite the characterization by the AMA that it "cannot be meaningfully defined."[179]

good v. Good

The presumed consent to a CPR rule, which is enforced in most hospitals in the absence of a DNR order, is an excellent way to avoid unjustified deaths, which are not in the patient's best interest.

v.

In many cases, such as a permanent vegetative state and others, the presumed CPR consent rule may result in unwanted and expensive prolongation of lives.

Twenty-first Century Approach

1. The presumed consent to CPR rule should be reexamined early in the twenty-first century due to its frequent failure to produce satisfactory results during the last quarter of the twentieth century.
2. Physicians should be legally (as they are now ethically) required to discuss the issuance of a DNR order with their patients (or surrogates) without fear of possible litigation on the basis of their own prognoses and without the necessity of prior judicial review.
3. These discussions should include (a) any possible inconsistencies with the advanced directive the patient may have executed, or with the wishes of his family and (b) appropriate steps to assume the patient's current view becomes paramount over such inconsistencies.

---

178. *See* MEISEL, *supra* note 36, at 397–400.
179. AMA CODE OF MEDICAL ETHICS § 2.035 (1994).

Whoever thinks that life support should be removed will often feel guilty in "abandoning" Mom or Dad, even if that is the only reasonable and the most beneficial thing to do in light of the medical diagnosis. Often, the fact, it is precisely the adult child who had the least good relationship with Mom and Dad who now wants to use every possible means to keep the parent alive, even when there is no chance of recovery and when continuing to use life support machines just prolongs the dying process. Relatives do that, not because they are looking out for the best interests of the patient, but rather so that they can tell themselves later on that they did every possible thing to keep Mom or Dad alive and therefore need not subsequently feel any guilt about his or her demise.

Doroff, *Matter of Life & Death* (1998), p. 171

But man sets out for his eternal abode,
With mourners all around in the street . . .
And the dust returns to the ground as it was,
And the life-breath returns to God
Who bestowed it.
Utter futility–said Kohelet–
All is futile!

Ecclesiastes 12:30

6. Limitations on Nonbeneficial (Futile) Treatment

Many courts have held that a certain patient (or his surrogate) may require that life-sustaining treatment be withdrawn without court approval.[180]

William Drabick, who was 44 years old, had been unconscious for two years as a result of an automobile accident. His brother, David, William's conservator, authorized the hospital to have cardiopulmonary resuscitation withheld from William.[181] However, a trial court refused and simply stated, "Continued feeding is in the best interest of a patient who is not brain dead." When

180. KEYES, *supra* note 27.

181. As noted in the previous section, most healthcare facilities maintain a "presumed consent" policy that cardiopulmonary resuscitation (CPR) or code should be called on unless there are specific instructions to the contrary (such as DNR).

David appealed, the appellate court reversed and held that the conservator had the authority to make medical decisions, which do not require court approval.[182]

Some studies have shown that 34 percent of physicians continue life-sustaining treatments despite patient or surrogate wishes for discontinuation. Further, some of these decisions are made "without the knowledge or consent of the patient or surrogate, and some are made despite their objection."[183]

Suppose in the above example, Drabick's physician and hospital had issued a Do Not Resuscitate (DNR) order and that the conservator had requested that it be withdrawn and not only insisted upon continued treatment, but also threatened to sue both of them if they did not do as he said. This arises in many healthcare facilities, frightens many physicians, and is often referred to a hospital's ethics committee.

The debate on futility is about the division of power between patients and their physicians. For patients, this means that it is about the limits of their autonomy (and that of their surrogates) to demand "futile treatment" which physicians agree cannot help cure the disease or condition. Many physicians may be willing to continue what they know to be futile treatments due to (a) their fear of possible or actual threats of malpractice or other liability, (b) the fact that this fear may be avoided (and the patient's families not left in despair) by their disregarding certain ethical objections (discussed below), and (c) their receipt of payment for providing such treatment. Two examples of actual cases follow:

> 1. I have in my practice an individual who has *cost* approximately *$2.5 million to maintain over the past six years.* During the time I have followed her, she has not moved, spoken or given any indication of consciousness. She is being supported by a tube in her windpipe attached to a respirator, by a tube in her stomach to continuously feed her and by around-the-clock nursing care. This is not the wish of her providers, who have repeatedly requested that she be allowed to die.

182. Conservatorship of Drabick, 245 Cal. Rptr. 840 (Ct. App. 1988).

183. Mayo Clin. Proc., May 1996, at 674 (citing Ash et al., *Decisions to Limit or Continue Life-Sustaining Treatment by Critical Care Physicians in the United States: Conflicts Between Physicians' Practices and Patients' Wishes*, 151 Am. J. Respiratory & Critical Care Med. 288–92 (1995)).

> *The family has insisted that all of this be done*, and in our *present environment there is no good way to stop this futility*. In other words, adherence to the wishes of patients is perceived as a path by some physicians *to avoid inquiries by risk managers or lawsuits* by disgruntled family members.[184] (Emphasis added.)
>
> 2. A 26-year-old black female developed end-stage renal disease (ESRD) in 1989, partly as a result of a long history of drug abuse. In 1995, she suffered cardiopulmonary arrest in the dialysis unit. Resuscitation was attempted and 45 minutes later her heart began beating again. She was intubated and transferred to the intensive care unit (ICU). *Her father and sisters* were approached about limiting aggressive therapy including continued hemodialysis. They *insisted that they fully expected a miracle* of healing and *threatened to sue the hospital* if there was any further indication that we would not do everything technologically possible to sustain her life. They also indicated *their belief* that *racial discrimination* was *involved*. The *hospital complied with their demands*. She died in 1996 after incurring *costs borne by Med-I-Cal and Medicare of $1.2 million* of public money without her ever regaining consciousness.[185] (Emphasis added.)

The author of the 1997 volume *Do We Need Doctors?* wrote, "The better we get at keeping sicker and sicker people alive, so that everybody gets longer and longer stays in the ICU at greater and greater cost, the worse off everyone will be."[186] At the dawn of the third millennium, a beginning has been made toward a more rational resolution of this problem with some ethical and legal guidelines.

---

184. Marshall B. Kapp, Commentary, *Anxieties as a Legal Impediment to the Doctor-Proxy Relationship*, 27 J.L. MED. & ETHICS 69 (1999).

185. Ernie W.D. Young, Symposium, *Living Longer: A Legal Response to Aging in America*, 9 STAN. L. & POL'Y REV. 267, 278–79 (1998).

186. LANTOS, DO WE STILL NEED DOCTORS? 28–29 (1997); TIVNAN, *supra* note 149, at 141.

> *But bodies which disease had penetrated through and through he would not attempt to cure. The art of medicine was not intended for their good.*
>
> Plato, Republic 3.406
>
> *To wrest from nature the secrets, which have perplexed philosophers in all ages, to track to their sources the causes of disease, to correlate the vast stores of knowledge, that they may be quickly available for the prevention and cure of disease—these are our ambitions.*
>
> Osler, "Chauvinism in Medicine," *Montreal Med. J.* 31:684-99 (1902)

a) *Treatment Beyond the "Goals of Medicine"*

As regards treatment, physicians have (a) technical skills (which also involve some judgment), (b) values consistent with the goals of medicine, and (c) "other personal values." When treating a patient exclusively with medical expertise, his personal values must be consistent with the goals of medicine:

> Contrary to the assertion that physiologic futility is value-free, we argue that it entails a value choice. Specifically, it assumes that the goals of medicine are to preserve organ function, body parts, and physiologic activity—an assumption that, in our estimation, departs dramatically from the patient-centered goals of medicine.[187]

Treatment maintained for values other than such goals fall under the exclusive expertise of the patient up to a point. Under the doctrine of informed consent, the patient (or surrogate) can refuse treatment of any sort, whether or not he is terminally ill.[188] One court noted that:

> A California Court of Appeals upheld the right of a non-terminally ill patient to disconnect his ventilator, stating, "In short, the law recognizes the individual interest in preserving 'the inviolability of the person'. . . The constitutional right of privacy guarantees to the indi-

187. Schneiderman et al., *Medical Futility, Response to Critiques*, ANNALS INTERNAL MED. 671 (1996). In 1996, a group launched by Daniel Callahan, founder of the Hastings Center, a bioethical institute, published a report stating four specific goals of medicine.

188. Bartling v. Super. Court, 209 Cal. Rptr. 220, 225 (1984).

vidual the freedom of choice to reject, or refuse to consent to, intrusions of his bodily integrity."[189]

Physicians are required to make the fullest reasonable disclosure of the risks of any proposed treatment,[190] but cannot insist upon a treatment under that doctrine, as will be shown below. In a given case, the physician's religion or "other personal values" may be opposed to those of the patient or his surrogate(s). For example, a physician who believes that all life is sacred may be opposed to performing abortions or to the withdrawal of treatment as demanded by the patient who is dying and suffering. He then has an ethical duty to transfer to another physician who may be willing to comply with such personal values of the patient.[191] A U.S. Court of Appeals has held that the ability to minimize the risk of transfer is measured by a hospital's own standard practices.[192]

---

189. Lane v. Candura, 376 N.E.2d 1232 (1978). In upholding the right of a patient to refuse amputation of a gangrenous leg, the court stated:

> The constitutional right of privacy, as we conceive it, is an expression of the sanctity of individual free choice and self-determination as fundamental constituents of life. The value of life as so perceived is lessened not by a decision to refuse treatment, but by the failure to allow a competent human being the right to choice.

Id. at 1233.

190. *See supra* Chapter 5.

> The very foundation of the doctrine [of informed consent] is every man's right to forego treatment or even cure if it entails what for him are intolerable consequences or risks, however warped or perverted his sense of values may be in the eyes of the medical profession, or even of the community, so long as any distortion falls short of what the law regards as incompetency.

F. Harper & F. James, The Law of Torts 61.

191. Conservatorship of Morrison, 253 Cal. Rptr. 530, 534 (1988). It should be noted that PVS patients (those without higher cortical function) who are on a ventilator do not experience discomfort when the ventilator is withdrawn. Multi-Society Task Force on PVS, *Persistent Vegetative State Part II, supra* note 120, at 1572–79; 333 N. Eng. J. Med. 130 (1995).

192. Ingram v. Muskogee Reg'l Med. Ctr., 235 F.3d 550 (10th Cir. 2000) A gunshot victim died after being transferred from the Medical Center. Her mother sued the center and physicians for wrongful death, alleging medical malpractice and violation of the Emergency Medical Treatment and Active Labor Act

At the turn of the third millennium, a new volume on Christian bioethics was written to advocate that doctors return to medical paternalism. It contains a statement against following all of the bioethical law by doctors stating, "Traditional Christians draw authority not from the consent [of the patient] but . . . from the revelation of the requirements of God."[193] A reviewer of this volume specifically noted that the author "does not explain why his definition of Christianity is correct or why the authorities he cites are the definition expositors of Christianity."[194]

> *When a patient cannot be returned to normalcy, when no treatment will do more than stabilize a fragile and failing life, the function of medicine is limited and its practitioners will see the performance of their craft as futile.*
>
> Koch, *The Limits of Principle* (1998), p. 122

b) Treatment Regardless of a Lack of Consensus on "Futility"

The determination of futility is often a difficult matter.[195] A physician's determination of futility, or the lack of the affectivity of any care to a critically ill patient, is often called a "prognostication."[196] It is usually a group process and may vary (like "terminal illness") according to physicians' specialty and clinical experience.[197] These professionals are also culturally constituted.[198] When discussing futility, we are concerned with the basic distinction between (a) instances where there is genuine debate concerning whether or not a treatment would in fact be futile and (b) cases where a general consensus exists

(EMTALA), 42 U.S.C. § 1395dd (1994). Because of a need for cardiovascular surgery and the fact that the Medical Center had no such surgeon, the patient was transferred to another hospital.

193. Engelhardt, Jr., The Foundation of Christian Bioethics 356 (2000).

194. Guium, Hast. Ctr. Rep., Nov.-Dec. 2001.

195. Helft et al., N. Eng. J. Med. 293 (2000).

196. Teno et al., *Prognosis-Based Futility Guidelines;* Schneiderman & Jecker, *Futility in Practice*, Arch. Internal Med. 437 (1993).

197. Christakis & Twashyina, *Attitude and Self-Reported Practice Regarding Prognostication in a National Sample of Internists*, Arch. Internal Med. 2389 (1998).

198. Sudnow, Passing On: The Social Organization of Dying (1967).

that the medical treatment is in fact futile. In this section, we are only concerned with the latter cases; "treatment" as used herein signifies something beyond comfort care (including pain relief).

Hippocrates stated when physicians ought not to treat: "Refuse to treat those who are overmastered by their diseases, realizing that in such cases medicine is powerless."[199] Modern medicine, encouraged by the media on occasion, has led many patients to expect miracles and to demand useless treatment. Professionals and societies have stated that physicians need not (and ought not) accede to such demands. One group wrote that qualitatively futile treatment is any treatment that "merely preserves permanent unconsciousness or that fails to end a patient's total dependence on intensive medical care.[200] Two bioethicists wrote that "in order to establish that care is futile, the clinician must claim that even though the care predictably will have some effect that changes the way that the patient dies, the effect is not beneficial on balance."[201] Several favored the physician's autonomy over that of patients.[202] In 1991 the American Thoracic Society stated: "A life-sustaining *intervention* may be withheld or *withdrawn from a patient without the consent of the patient* or surrogate *if* the intervention is *judged to be futile*."[203] (Emphasis added.) However, in its 1994 *Code of Medical Ethics*, the AMA's Council of Ethical and Judicial Affairs issued a different rule: "Denial of treatment should be

---

199. 2 Hippocrates 193 (Jones trans., Univ. Press 1967); *see also* Plato, Republic bk. 3, 407 c–e.

200. Schneiderman et al., *Medical Futility: Its Meaning and Ethical Implications*, 112 Annals Internal Med. 949–54 (1990).

201. Veatch & Spicer, *Medically Futile Care: The Role of the Physician in Setting Limits*, 18 Am. J.L. & Med., 15–36 (1992). Nevertheless, Veatch argued that physicians have no expertise in making decisions about such subjective matters as futility. Veatch, *Why Physicians Cannot Determine if Care Is Futile*, 42 J. Am. Geriatrics Soc'y 871–74 (1994); *see also* Brett & McCullough, *When Patients Request Specific Interventions: Defining the Limits of the Physician's Obligation*, 315 N. Eng. J. Med., 1347–51 (1986).

202. Daar, *Medical Futility and Implications for Physician Autonomy*, 21 Am. J.L. & Med. 221–40 (1995); et al., *Beyond Autonomy—Physicians' Refusal to Use Life-Prolonging Extracorporeal Membrane Oxygenation*, 329 N. Eng. J. Med. 354–57 (1993); Layson & McConnell, *Must Consent Always Be Obtained for a Do-Not-Resuscitate Order?*, 156 Arch. Internal Med. 2617–20 (1996).

justified by reliance on openly stated ethical principles and acceptable standards of care . . . not on the concept of 'futility,' which cannot be meaningfully defined."[204] This is not necessarily so in all instances. Further, the physician has an ethical obligation to honor the resuscitation preferences expressed by the patient. Physicians should not permit their personal value judgments about quality of life to obstruct the implementation of a patient's preferences regarding the use of CPR. The Council added:

> *If*, in the judgment of the attending physician, it would be *inappropriate to pursue CPR*, the *attending physician may enter a do-not-resuscitate order* (DNR) into the patient's record.[205] (Emphasis added.)

This final sentence appears consistent with DNR order only by (a) not merely recording its grounds as the "futility" of treatment, and (b) there being no patient request for CPR; these are current views of the AMA. An example of an ethical procedure was in *Healthcare Ethics Committee's: The Next Generation*, the American Hospital Association (AHA), which sets forth "The Parkland Approach to Demands for 'Futile' Treatment of the Ethics Committee," Parkland Memorial Hospital, Dallas County Hospital District. It states:

> If a competent patient or the family or surrogate of an incompetent patient requests or demands a "life-saving" or "life-preserving" treatment, and the attending physician is of the opinion that the treatment is useless and without benefit, the physician should "first" consult the

---

203. Am. Thoracic Soc'y, Am. Rev. Respiratory Disease 728 (1991).

204. Rule 2.035.

205. Rule 2.22; Council of Ethical and Judicial Affairs, *Code of Medical Ethics*, JAMA 52–53, 6–7 (1994). The Santa Monica Hospital Medical Center's Futile Care Guidelines state that in cases of futility, "Neither the doctor nor the hospital is required to provide care that is not medically indicated." Halevy & Brody, *A Multi-Institutional Collaborative Policy on Medical Futility*, JAMA 571–74 (1996). Some states like Texas have laws referring to "incurable" and "irreversible" conditions where death would result within a relatively short time without the application of life-sustaining procedures. Tex. Health & Safety Code Ann. § 672-002(6), (9) (Vernon).

> Institutional Ethics Committee," and then with "two (2) physicians from a relevant specialty to render an opinion about the usefulness of the disputed treatment." *If* the "Ethics Committee, attending physician, and both consultants are *unanimous in their opinion* of the uselessness or *futility* of the disputed treatment, the treatment should not be provided." If these conditions "have not or cannot be met, the disputed treatment should be administered until these conditions have been met."

A denial of disputed treatment is normally followed with comfort care or transfer to a hospice therefor.

> *We sit together, the mountain and me, until only the mountain remains.*
>
> J. Kabat-Zinn, *Wherever You Go, There You Are: Mindfulness Meditation in Everyday Life* (1994)

c) The Unique Direction of Informed Consent

The doctrine of informed consent generally prohibits physicians from treating patients without their consent or that of their surrogates. It does not imply that giving consent authorizes a patient to *demand treatment*.[206] There is no Constitutional (or "fundamental") right to overall healthcare;[207] only the legislature can create such a right by statute or regulation directing the methods of payment of the costs thereof.[208] It has been stated:

> In a just healthcare system, believers who reject the wider society's accepted standards of medical futility, definitions of death, and blood use have no claim on nonbelievers for a financial blank check.[209]

---

206. Those who extrapolate a right to access from a right to noninterference simply misunderstand the nature and scope of rights as well as the concept of respect for autonomy. Furthermore, other considerations, such as respect for professional integrity and commitments to beneficence and nonmalfeasance, can constrain the autonomy-based rights patients would otherwise have.
207. Maher v. Roe, 432 U.S. 464, 469 (1977). See Chapter 5.
208. Callahan, False Hopes A.13 (1998).
209. Post, Hast. Ctr. Rep., Sept.-Oct. 1995, at 28.

Many physicians speak of compassion for a patient's family as the sole ground for treatment of the patient himself. This arises out of what I have called "subjective personhood," which applies to thoughts and feelings about the patient by his family and others who knew him.[210] Dealings with a family will be with compassion; however, medical "treatment" is for patients alone and compassion for the family alone does not justify futile treatment of the patient. As stated by the American Medical Association's *Code of Medical Ethics*, "Patients [or surrogates] should not be given treatments simply because they demand them" and as regards the permanently unconscious, "it is not unethical to discontinue all means of life-prolonging medical treatment."[211] In the mid-1990s, the University of California at Los Angeles (UCLA) Medical Center issued the policy statement: "Physicians need not provide treatment that they consider medically inappropriate" and that "recognizing a distinction between medically appropriate and medically inappropriate treatment means that some patients or surrogates will not be offered treatment that they might want."[212] Similarly, insurers only reimburse healthcare providers according to a contract that normally limits coverage to services medically reasonable and "necessary" for the patient; it would not be expected to cover treatment of no benefit to the patient, solely in order to assuage the feelings of his family. Insurance policies could cover a patient's family problems, although this would not be expected according to standard clauses pertaining to the treatment of the patient.[213]

---

210. KEYES, *supra* note 27.

> Emotions are not simply a matter of feelings, although feelings are involved. Rather, they are cognitive; i.e., emotions are more like beliefs and desires than the feelings of hunger and itches. . . . Indeed, the cognitive content gives rise to the phenomenological aspect, e.g., without the judgment that someone is wonderful, thereby making their death a loss, there is no grief over losing them. . . . Those suffering emotionally are as much in need of support and help for this suffering as they are for their physical suffering.

Robichaud, *supra* note 152, at 59–68.

211. AMA COUNCIL ON ETHICAL AND JUDICIAL AFFAIRS R. 2.17, 2.20–2.21.

212. UCLA Med. Ctr., Life Sustaining Treatment Policy, No. 0027, 5-1-93.

213. *See* Rosenbaum et al., *Who Should Determine When Healthcare Is Medically Necessary?*, N. ENG. J. MED. 229 (1999).

Doctor,
If only you could see
How heaven pulls earth into its arms
And how infinitely the heart expands
To claim this world, blue vapor without end.

Lisel Mueller

d) A Way to "Free" Physicians from Self-Interest or Self-Protection in Connection with Life Support; Anencephalics

In 1998, the *Journal of Law, Medicine & Ethics* noted the California Court of Appeals "in a decision widely followed in other jurisdictions, [which] dismissed the criminal information because failure to continue unwanted or non-beneficial treatment is not unlawful failure to perform a legal duty, even when the physicians know that the patient will die as a result."[214] That court ruled "A physician has no duty to continue treatment, once it has proved to be ineffective . . . If the treating physicians have determined that continued use of a respirator is useless, then they may decide to discontinue it without fear of civil or criminal liability."[215] The court hoped to "ameliorate the professional problems."[216] Although it does not appear to have done so, throughout much of the country there have been exceptions. In Massachusetts, a terminally ill, comatose, 71-year-old woman on a ventilator was refused treatment by the hospital and its physicians. With the approval of the ethics committee chair, and despite the surrogate's statement that her mother told her that she wanted everything done to keep her alive, the physician wrote a DNR order and removed the ventilator over the family's objection. The family then sued the physician, the ethics committee chairperson, and the hospital for negligence. The jury denied any damages, having found

---

214. Ann Alpers, *Criminal Act or Palliative Care? Prosecutions Involving the Care of the Dying*, 26 J.L. Med. & Ethics 308, 319 (1998).

215. [The court offered] *a way to free physicians*, in the pursuit of their healing vocation, from possible contamination by *self-interest or self-protection* concerns that would inhibit their independent medical judgments for the well-being of their dying patients. We would hope that this option might be serviceable to some degree in ameliorating the professional problems under discussion. (Emphasis added.) *Id.*

216. *Id.*

that treatment would have been futile and the defendants were *not negligent*.[217] Similarly, a Texas court threw out a case involving a physician's decision to withdraw life-sustaining treatment from an adult without the patient's or the family's consent.[218]

Cases such as these stand in contrast to an opinion of an appellate court majority under the 1986 Emergency Medical Treatment and Active Labor Act (EMTALA), which was intended to prevent hospitals from "dumping" patients who were unable to pay for their care.[219] This statute provides that where an "emergency medical condition" exists, a hospital must either provide treatment sufficient to stabilize the patient's condition or transfer the patient to another medical facility. Violation may result in damages to the hospital, but not the physician.

Baby K, an anencephalic baby (born without almost all of what would be an expected brain) was put on a ventilator several times. Brains cannot be transplanted and treatment could serve no beneficial purpose whatsoever to Baby K, who had no possible quality of life. The physicians recommended the issuance of a "Do Not Resuscitate" order. However, the mother refused. The hospital filed a declaratory judgment action to determine whether the statute required treatment that the doctors believed was medically and ethically inappropriate. It failed to ask for a ruling under the state law, but requested a ruling on federal statutes. A majority of an appellate court actually

---

217. Gilgun v. Dec. Cassem and Mass. Gen. Hosp., Civ. A. No. 92-4820 (Mass. Dist. Ct. Apr. 21, 1995).

218. Duensing v. Sw. Tex. Methodist Hosp., No. SA 87, C.A. 1119 (W.D. Tex. Dec. 22, 1988). The court's ruling was based on the absence of expert testimony that termination of treatment violated the applicable medical standard of care. The plaintiff's own medical expert testified that he had discontinued treatment on a "hopelessly ill patient" without family consent as part of "good and humane medical practice." Nelson & Nelson, *Futility and Disproportionate Burden*, 20 Critical Care Med. 432 (1992).

219. Emergency Medical Treatment and Active Labor Act (EMTALA), 42 U.S.C. § 1395dd. In 1991, another case was discussed in the literature. Mrs. Helga Wanglie, an 87-year-old woman with preexisting lung disease, was in a persistent or permanent vegetative state (PPVS) and on a ventilator. When her family refused to authorize discontinuation of ventilatory support, a lawsuit was filed to resolve whether Mr. Wanglie was her guardian and could make decisions for her. Although the court found that he was her guardian, it did not rule upon whether or not to continue the ventilator. Rubin, When Doctors Say No 24–25 (1998).

required that the treatment be continued despite state law to the contrary (which it held preempted). As one expert stated: "Ventilating a patient like *Baby K* may stabilize the infant but cannot cure its condition. From the perspective of mechanism, the severely *anencephalic infant is not being* like you and me."[220] (Emphasis added.) The judge somehow equated the infant virtually without a brain to a sick adult who had one without making an actual analysis of the difference, stating: "Just as an AIDS patient seeking ear surgery is 'otherwise qualified' to receive treatment despite poor long-term prospects for living, Baby K is 'otherwise qualified' to receive ventilator treatment."[221] The difference is manifest.[222] A survey of 43 children's hospitals revealed, "Not one respondent endorsed the life-sustaining treatment of Baby K."[223] As pointed out by the dissent, the diagnosis of an anencephalic "should be regarded as a continuum, not as a series of discrete emergency medical conditions to be considered in isolation."[224] Other appellate courts have held quite differently concerning the meaning of the EMTALA. In the medical profession and others, the concept of confronting the "physiological futility" of the anencephalic Baby K by means of a ventilator could not overcome the fact that nonbeneficial (futile) treatment was continually being given that neither did, nor could, alleviate such a hopeless condition for which no cure was possible. Its treatment could

---

220. KOCH, THE LIMITS OF PRINCIPLE 122 (1998).

221. *In re* Baby K., 832 F. Supp. 1022, 1027 (E.D. Va. 1993); Strasser, *The Futility of Futility: On Life, Death, and Reasoned Public Policy*, 57 MD. L. REV. 505 (1998).

222. KOCH, THE DEVIL'S DETAILS: THE LIMITS OF PRINCIPLE 25 (1998). Anencephalic infants are by definition outside the protection because "it is only to persons, that is, those individuals who have the capacity to value their existence, that respect is owed." Harris, *The Value of Life*, 24 J.L. MED. & ETHICS 240 (1996).

223. Schneiderman & Manning, *The* Baby K *Case: A Search for the Elusive Standard of Medical Care*, 6 CAMBRIDGE Q. HEALTHCARE ETHICS 9–18 (1997). "Although the physicians interviewed claimed that maintaining life support for an anencephalic baby was not acceptable practice, neither they nor the hospitals were inclined to establish non-treatment as customary practice, or test their expert opinion in the courts." *Id.* at 13–14. For a discussion of the treatment of anencephalics for the transfer of organs, see *supra* Chapter 11.

224. *In re* Baby K, 18 F.3d 5909 (4th Cir. 1994). Unfortunately, the Supreme Court declined review. 513 U.S. 825 (1994).

only for a short time enhance a "subjective person" for its mother who refused to accept the fact of hopelessness.

The concept of "subjective personhood" grew out of my committee's experience with relatives of patients (whose conditions were clearly futile) who attempted the transfer of their expectations for a patient into a right as a nonpatient to expect that futile treatment be given to the patient. Although there exists the holistic view of compassion and need for discussion with a patient's relative, this does not automatically transpose to such a relative a right to obtain such treatment. Regarding children, parents have certain rights regarding whether and how to treat them. However, they have no right to demand that physicians give nonbeneficial (futile) treatment to them. The American Academy of Pediatrics issued a recommendation that when it becomes obvious a child will not recover, palliative care should replace medical treatment designed to cure. It argued for the use of "comprehensive hospice-like care." Taxpayer dollars are being spent in hospitals throughout much of the United States for futile treatment (that is not "comfort care" for the patient) solely to avoid upsetting the insistent relative and the risk of a possible malpractice suit.[225]

> Over the past three decades vastly advanced technology, third-party payment, and a near-exclusive focus on patient autonomy has transformed medicine into an *institution with no apparent limits on what could be demanded* and what would be attempted. *The furthest reach* of this out-of-control system is seen in *the now-infamous Baby K case*, where, at a mother's insistence, an anencephalic infant was resuscitated, ventilated, and maintained with intensive-care measures for some two and one-half years. (Emphasis added.)[226]

225. Marshal B. Kapp, Commentary, *Anxieties as a Legal Impediment to The Doctor-Proxy Relationship*, 27 J.L. MED. & ETHICS 69, 70 (1999); *see also* LO, RESOLVING ETHICAL DILEMMAS: GUIDE FOR CLINICIANS 130 (1995).

226. Paris & Post, *Managed Care, Cost Control, and the Common Good*, 9 CAMBRIDGE Q. HEALTHCARE ETHICS 183 (2000) (citing Paris et al., *Guidelines on the Care of Anencephalic Infants: A Response to* Baby K, 14 J. PERINATOLOGY, 318–24 (1995)).

This raises very important ethical, fiscal, legal, and other problems that remain in dire need of resolution.

In the late 1990s, the AMA's Council on Ethical and Judicial Affairs had stated, "Even if death is not imminent but a patient is beyond doubt permanently unconscious and there are adequate safeguards to confirm the accuracy of the diagnosis, it is not unethical to discontinue all means of life-prolonging medical treatment."[227] Nobody could be more permanently beyond hope of having any quality of life than that of an anencephalic baby and both legal and medical experts have recommended that such patients be considered dead.[228] It has been noted, "The facts of Baby K are remarkably distinct from the prototypical patient-dumping scenario involving an indigent or uninsured patient in need of emergency medical treatment."[229] Accordingly, because judicial interpretation has led EMTALA astray, diverting it from its narrow original mission, Congress should initiate revisions.

A nurse has written, "When asked to name their most common problem in dealing with patients, most nurses respond, 'Their families.'"[230] "Subjective life" is merely the image, feelings, and thoughts by the family or others about what the patient was prior to the terminal illness, becoming permanently unconscious, or in the case of an anencephalic, what a person might have become if a brain had actually been present. Although the family constitutes people with whom physicians normally have discussions because of their concern about the patient, the physicians and healthcare institutions often continue treatment (other than mere comfort care) at taxpayer expense.

227. Councils on Scientific Affairs and on Ethical and Judicial Affairs, *supra* note 114, at 429.

228. Schaffner et al., *Philosophical, Ethical and Legal Aspects of Resuscitation Medicine*, 16 Critical Care Med. 1969–76 (1988). For example, the New Mexico Uniform Heath Care Decisions Act of 1995 provides that *physicians are not required to deliver "medically ineffective treatment"*; that is, treatment that offers no significant benefit to the patient, as determined by the physician. N.M. Stat. Ann. §§ 24-7-1 to -18 (West).

229. Lynn Healey Scaduto, Comments, *The Emergency Medical Treatment and Active Labor Act Gone Astray: A Proposal to Reclaim EMTALA for Its Intended Beneficiaries*, 46 UCLA L. Rev. 943, 955 (1999).

230. Galann, Caring for Patients from Different Cultures: Case Studies from American Hospitals 55 (2d ed. 1997).

This is done despite the laws of states that specifically authorize healthcare providers to decline nonbeneficial (futile) treatment.[231]

good v. Good

While infants (such as anencephalics) remain alive, physicians should always apply treatments necessary to maintain its life as long as possible.

v.

Infants who are born without most of a brain cannot receive a transplant; however, virtually all its other organs could be transplanted to those persons in need of them.

Twenty-first Century Approach

1. A brief extension of the life of an anencephalic infant may only be justified where its parents have agreed to donate its other organs to the many transplant seekers.
2. Otherwise, the delivery of an anencephalic should be added to the definition of "death" or a "stillbirth."

---

e) The Authority and Responsibility of Physicians with ICUs and Authorized Executions

It has been indicated that the "Authority to render a judgment about quantitative futility rests squarely with physicians . . . It is physicians who are responsible for determining whether their patients meet the conditions."[232] Others have noted, "Physicians must be careful not to offer ICU care once there is no reasonable chance it will improve outcome."[233] ICU care has been exercised too infrequently and many patients were also sent to the ICU inappropriately. Bioethicist Howard Brody has written:

---

231. A physician's refusal can properly rest on a judgment that treatment is contrary to a child's best interests, or is futile, harmful, or disproportionately burdensome, but only after completing a process aimed at reaching an honest agreement and specifying the parents' and physician's values. Nelson & Nelson, *supra* note 218, at 427. Several examples are set forth. *See, e.g.*, King, The Law of Medical Malpractice 43–48 (1997).

232. Jecker & Perlman, *Medical Futility*, Arch. Internal Med. 1140 (1992).

233. Osborne, *Physician Decisions Regarding Life Support in the ICU*, 101 Chest 222–23 (1992).

> The physician must inform the patient or family when an intervention is being *withheld on the basis of futility*. This is different from seeking consent; it is simply a courteous disclosure of relevant information. The physician accepts full responsibility for the decision and in no way seeks to involve the patient or family in the decision-making process, but nevertheless he makes clear that the decision is being made without any attempt at concealment or misrepresentation.[234]

Medical practices for end-of-life care differ in many countries; although it moved from paternalism to one that "promotes autonomy in the United States," in Europe, "patient-physician relationships are still somewhat paternalistic." A study demonstrated that "end-of-life actions are routine in European ICUs. Life support was limited in 73 percent of study patients and 10 percent of ICU admissions. Both withholding and withdrawing of life support seem to be accepted by most European intensivists while shortening the dying process."[235]

In 2003, it was concluded that ethics consultations were useful in resolving conflicts that may have inappropriately prolonged nonbeneficial or unwanted treatments in the ICU. Among the 25 percent of patients in ICUs who chose ethics consultation, those who died spent three fewer days in the hospital, 1.4 fewer days in the ICU, and 1.7 fewer days receiving mechanical ventilation than nonsurvivors in the control group. Although there was a difference in overall mortality between patients offered the ethics consultation versus usual care, it was reported that "ethics consultation was associated with significantly fewer ICU days and life-sustaining treatments in patients who ultimately did not survive to discharge." Hence, these consultations (which were associated with reductions in hospital and ICU days), were found "to be useful in resolving conflicts that may be inappropriately prolonging nonbeneficial or unwanted treat-

234. *SUPPORT, A Controlled Trial to Improve Care for Seriously Ill Hospitalized Patients*, 274 JAMA 1597 (1995).

235. The northern region had more limitations, decreased CPR use, less time until a limitation of treatment was determined, and perhaps consequent shorter ICU stays. The present study demonstrates that physicians who were Protestant, Catholic, or with no religion more frequently used withdrawal of life-sustaining treatments than physicians who were Greek Orthodox, Jewish, or Moslem.

Spring et al., *End-of-Life Practices in European Intensive Care Units*, JAMA 790, 796 (2003).

ments at the end of life."[236] However, as Bernard Lo indicated, it was unfortunate that the study neglected to supply "more detail about what was actually done. Without such information, other institutions cannot judge whether the consultations in this study are similar to those at their own, or how they would replicate the intervention."[237]

A dying cancer patient may (a) desire "extraordinary treatment" in order to meet with her family one more time or (b) wish to avoid it because of the extent of her suffering. A physician indicated that this decision to call the treatment "extraordinary" would not depend on its "technologic complexity, or its potential for discomfort and injury," but rather how to "advance the patient's management goals." He concluded: "For any given patient, any treatment, however complex, can be considered ordinary, and any treatment, however simple, can be considered extraordinary. The difference lies in the purpose of the treatment, not in its cost or complexity."[238]

Back in 1980, the AMA Council on Ethics and Judicial Affairs adopted an opinion setting forth in detail the denial of a physician's right to choose any participation in connection with capital punishment.[239] This action by the AMA occurred before many important legal developments in bioethics transpired. It currently opposes the law in most of the 38 states (76 percent of the states) that authorize executions. Virtually all of these states now prescribe the use of lethal injection—a procedure invented in 1977 by Dr. Stanley Deutsch, chair of the Anesthesiology Department at Oklahoma University Medical School.[240]

---

236. Schneiderman et al., *Effect of Ethics Consultations in the Intensive Care Setting on Nonbeneficial Life-Sustaining Treatments*, JAMA 1166 (2003).

237. Future studies need to address other important questions. Ethics consultations vary in terms of who carries them out, how the patient or surrogate was involved in the consultation, what communication skills the consultants used to try to reach consensus, and how recommendations were communicated. The heated debate over whether ethics consultations are best carried out by individuals, teams, or committees ought to be answered with empirical data.

Lo, JAMA 1208 (2003).

238. Moroff, *Ethical and Legal Issues in the Medical Management of the Terminally Ill Patient*, Einstein Q.J. Biology & Med. 81 (1985).

239. AMA Council on Ethical and Judicial Affairs, *Physician Participation in Capital Punishment*, 270 JAMA 365 (1993); AMA Code of Medical Ethics R. 2.06 (2002-2003).

240. Fed. Law., Sept. 2002, at 34.

At the turn of the third millennium, the state medical associations were requesting copies of their policy statements on the AMA rule. From the responses received, approximately 54 percent of state medical associations had issued statements that either declared the participation unethical or else "discouraged" it.[241] Another AMA policy is to "inform state medical licensure boards and certification and recertification agencies that physician participation in supervising or administering lethal injections is a serious violation of ethical standards of the medical profession."[242]

The AMA rule outlawing such participation is rather similar to the case of physicians who, although personally favoring the constitutional right of women to choose abortion (see Chapter 9), nevertheless still carry the fear of being known as "abortionists."[243] In the last third of the twentieth century, many physicians fought courageously for the removal of legal sanctions connected with current rights to choose contraception, sterilization, abortion, withholding of life support, withdrawal of life support (see Chapter 13), donation of an organ (see Chapter 11), and also physician assisted suicide (PAS) (see Chapter 14). The Supreme Court has upheld the legality of such choices.[244] The AMA Rule also

241. AMA CODE OF MEDICAL ETHICS R. 2.06.

> Discusses policies that consider physician participation in capital punishment to be unethical. Observes that such policies are without merit. Argues that such policies attempt to override the will of the people, their elected representatives, and the administration of justice. Concludes that these policies should be rescinded by state and national medical associations. Quotes Opinion 2.06. Keyes, *The Choice of Participation by Physicians in Capital Punishment*, 22 WHITTIER L. REV. 809, 810–11, 838 (2001).

242. Sanctions may be imposed that include suspension or withdrawal of a license to practice and such codes may be introduced to show evidence of malpractice upon a showing of damages. Hence, the issue arises whether medical organizations, which consider participation in capital punishment to be immoral or unethical, may discipline their members in a manner that could actually require the suspension or revocation of a physician's license to practice in the medical profession for the act of participating in a legally authorized execution in his or her state.

Keyes, *supra* note 241, at 812.

243. *Id.* at 813.

244. *Id.* at 814.

constitutes a declaration that PAS is unethical for physicians who are correctly performing PAS in Oregon where the voters approved their right to do so; the U.S. Supreme Court denied hearing challenges to that law.[245] Further, many physicians commit either PAS or the type of euthanasia called "double effect" to perform a lethal injection where necessary to alleviate the pain of a terminal patient (see Chapter 14). Hence, the notion that physicians are solely engaged in healing and prolonging life has become a matter of severe dispute. Of the 38 states performing capital punishment, 28 of them require physicians' attendance at their executions.[246]

Courts have extended the duration of death penalty appeals by adding many new procedural requirements in the past few decades. As of the dawn of the third millennium, all such cases are subject to these "super due process" rules.[247] State and federal court appeals on procedural matters have resulted in extreme delays, which can go on for 15 or 20 years.[248] Comparisons with WWII enemy

245. AMA CODE OF MEDICAL ETHICS R. 2.211 (2002-2003); Keyes, *supra* note 241, at 515. In contrast to the AMA's position, a state panel decided in 1998 that Oregon tax money should help pay for terminally ill poor patients to have this "medical service" because it would be "most discriminatory" not to give this choice to the poor. Obviously, were the AMA policy made subject to sanction by court action in Oregon, its successful prosecution would be virtually nil. *Id.*

246. Keyes, *supra* note 241, at 817. "Besides the United States, 89 countries have the death penalty and use it at the turn of the 3rd millennium." *Id.* at 819. Many Christian churches continue the incorrect former translation of a Biblical commandment as "Thou shalt not kill," which has been corrected by more recent translations to read: "You shall not murder." *E.g.*, *Deuteronomy* 5, *in* FRIEDMAN, THE BIBLE WITH SOURCES REVEALED 319 (2003). But on March 25, 1995, Pope John Paul II signed his different encyclical, *Evangelium Vitae*, in which he ruled out capital punishment except in cases of "absolute necessity." A CONCISE HISTORY OF THE CATHOLIC CHURCH 493–94 (2004).

247. KEYES, *supra* note 241 at 823–24. "Justices ruled that Jesse James Andrews should die for a 1979 Los Angeles triple murder—23 years after the crime and 18 years after the conviction. California is home to the nation's most clogged death row, housing 616 condemned inmates. Juries add nearly two dozen inmates each year." ORANGE COUNTY REG., Sept. 17, 2002, at News 4.

248. KEYES, *supra* note 241 at 831.

> Unfortunately, even today some people attempt to argue that participation by physicians in capital punishment is somehow comparable to that of Nazi and Japanese doctors during World War II. However,

practices in Germany and Japan are "without merit" and the death penalty was prescribed 60 times in the Bible.[249] However, in 2005, the Colorado Supreme Court (3-2) actually threw out the death penalty in a rape-and-murder case because jurors had studied Bible verses such as an "eye for eye, tooth for tooth" during deliberations.[250]

There are some movements to preserve DNA evidence in order to allow inmates access to it for testing. Many convicted rapist-killers have been released from prison after receiving such tests. A couple of governors even suspended executions in their states pending studies of the fairness of capital punishments.[251] In the year 2002, the Supreme Court overturned its own 1990 decision to rule that only juries (and no longer just judges) can impose the

---

> as stated by a British barrister who has studied the matter in some detail, claims of any such comparison are utterly without merit.

Id. at 832.

> Others point to Biblical proscriptions favoring the punishment, such as the Sixth Commandment: "You shall not murder" and "Whoever shed the blood of man, by man shall his blood be shed." It has been noted that "in the five books of the Torah alone, the phrase 'put to death,' appears 60 times."

Id. at 833.

249. In California, after [a] death sentence is imposed, it takes four to six years for a lawyer to be appointed for the state-court appeal. Thereafter, the records range from 5,000 to 15,000 pages and can take two years to certify, and the brief-filing with the California Supreme [Court] and the process take an average of four years. The oral arguments are the scheduled about ater and the decision must be rendered within 90 days of the oral argument. The death-row inmate can begin the federal court appeal process, which can take four years or more.

ORANGE COUNTY REG., May 17, 2005, at News 3.

250. ORANGE COUNTY REG., Mar. 29, 2005, at News 12. The court stated that "at least one juror in this case could have been influenced by these authoritative passages to vote for the death penalty when he or she may otherwise have voted for a life sentence." *Id.*

251. FED. LAW., Sept. 2002, at 35; *see, e.g.*, Harvey v. Horan, 285 F.3d 298 (2002) (supporting inmate's access to post-conviction DNA testing). Lethal injection is prescribed by 37 states (although it is an alternative method to electrocution in 10 states, the gas chamber in five states, hanging in two states, and a firing squad in two states). Only one state requires electrocution. ORANGE COUNTY REG., Sept. 18, 2004, at News 30.

death penalty[252] and also ruled that the Eighth Amendment forbids states from executing the mentally retarded.[253] In 2003, that court authorized possible forced psychotropic medications to a mentally disturbed dentist charged with both medical fraud and attempted murder of an FBI agent. The court added the conditions that it be "substantially unlikely" that the drug will have effects that could render the trial unfair and that the drug be "medically appropriate, i.e., in situations in the patient's best medical interest in light of his medical condition,"[254] and neither side was defeated. Further, in 2005 a bitterly divided Supreme Court (5:4) actually ruled against the execution of killers who were under 18 years of age. This decision reversed its previous decisions that specifically allowed 16- and 17-year-olds to face such punishment for murders.[255] Section 4 of this chapter pointed to a 2003 Supreme Court decision allowing the execution of a man who killed two people when he was 17, as well as the right of "mature minors" to refuse life-saving treatment, have abortions, etc.

It has been pointed out how several American states offer condemned prisoners lethal injection *as an option* alongside other forms of execution. It was noted that such a prisoner "is consenting to the lethal injection in a morally significant fashion" comparable to "the autonomous choice of the termi-

---

252. Ring v. Arizona, 122 S. Ct. 2428 (2002); *see also* Kelly v. South Carolina, 534 U.S. 246 (2002) (ruling that jurors choosing between the death penalty and a life sentence must be told if a life sentence would include no chance of parole). A study of federal cases showed that the average cost for capital cases at the trial court level was nearly four times that for noncapital cases ($218,000 compared with $56,000). An Indiana study reflected that costs associated with the death penalty will exceed the total costs of sentences of life without parole by more than one third.

253. Atkins v. Virginia, 122 S. Ct. 2242, 2252 (2002). In 2003, the Supreme Court held that the failure to adequately investigate a client's past for evidence against a death sentence could amount to ineffective legal assistance. Wiggins v. Smith, 123 S. Ct. 2527 (2003). However, by 2005, Adkins' IQ scores rose significantly, which the defense claimed was due to his participation in years of litigation, a practice that drives scores higher. ORANGE COUNTY REG., Feb. 6, 2005, at News 24.

254. United States v. Sell, 539 U.S. 166 (2003); *see also* Riggins v. Nevada, 504 U.S. 127 (1992); Washington v. Harper, 494 U.S. 210 (1990) (upholding forced drug treatment that was "reasonably related to legitimate penological interests" in prison).

255. The ruling overturned 72 death sentences in 12 states and reversed a practice used in 19 states. ORANGE COUNTY REG., Mar. 2, 2005, at News 17.

nally ill patient for physician-assisted suicide." Although one's liberty is restricted, this need not diminish the moral significance of one's autonomous choices.[256] A capital prisoner claimed that his sentence to death by lethal injection *would constitute cruel and unusual punishment* due to his years of drug abuse. It was held to have been properly brought under the federal civil rights statute.[257] Finally, it is important to understand the case in which 13 physicians sued their State Department of Corrections, arguing that physician participation in the death penalty not only violates medical ethics standards, but also conflicts with a doctor's role as a healer. In September 1998 an appellate court held that physician participation in executions is not "unprofessional conduct" within that state's Medical Practice Act:

> We agree, as a general matter, that the ethical rules serve such a vital purpose. *We do not agree*, however, that *physician participation in executions is likely to erode trust between individual physicians and patients* who have not been sentenced to death for a capital crime, or *undermine public confidence in physicians* or the medical profession as a whole. Indeed, appellants acknowledge that physicians have long participated in and witnessed gas chamber executions in California, yet *they make no showing* or assertion *that such conduct has in any way affected the trusting quality of the physician-patient relationship* for the population at large.[258] (Emphasis added.)

256. "Silver, Prof. Philosophy, Univ. Det. Lethal Injection, Autonomy Blackwell Publishing Ltd. 2003 p. 207 (citing http://www.deathpenaltyinfo.org). In 2003 a serial sex killer and a man convicted of a jailhouse murder both selected the firing squad over lethal injection." ORANGE COUNTY REG., May 24, 2003, at News 34.

257. 42 U.S.C. § 1983; Nelson v. Campbell. The Court did not determine how to treat that lethal drug. However, it has been pointed out that many relatives of victims cannot properly begin to heal until the killer of their loved one has faced the ultimate penalty. For example, the president of a local chapter of Parents of Murdered Children stated: "Nothing can bring a loved one back, but executing the person who committed a heinous crime is a crucial way for many families to feel a sense of closure." ORANGE COUNTY REG., Dec. 14, 2004, at News 3.

258. Thornburn v. Dep't of Corr., 78 Cal. Rptr. 2d 584, 585–57 (1998).

good v. Good

Physician participation of any sort in capital punishment is both unethical and incompatible with the healing profession and erodes trust in the physician-patient relationship. They should be barred from any participation therein. Further, they should not initially certify the death of any executed murderer because this is too close to making him a participant. Hence, physicians who so participate should be made subject to sanctions for such actions.

v.

Such participation in capital punishments should not be barred to physicians for there is no demonstration of the erosion of the physician-quality relationship. Making certification for such a death only after someone else so declared presents a metaphysical distinction and it cannot be justified. Physicians should be able to continue to preserve their right to choose to render help in authorized punishments of convicted murderers in order to see that they are carried out as humanely as possible.

Twenty-first Century Approach

Current policies of the AMA (and of those state medical associations that follow them) deny the choice of any proper participation by physicians in capital punishment. These policies have been held to have a legal effect that neither the American Medical Association (AMA) nor medical associations should retain.

---

f) The Current Method of Stopping Futile Treatment of Dying Patients by the AMA and by Texas

In 1997 the AMA *Code of Medical Ethics* issued a rule stating, "All healthcare institutions, whether large or small, should adopt a policy on medical futility." It set forth a number of steps that "should be included in such a due process approach to declaring futility in specific cases."[259] This was followed up by a comparable publication on March 10, 1999, from the AMA Council on Ethical and Judicial Affairs. The Council's report proposed a

---

259. AMA CODE OF MEDICAL ETHICS R. 2.037 (1997). If transfer is not possible, the intervention need not be offered.

six-step "fair process" end of life in futile cases—the sixth step of which is to stop treatment.[260] It starts with a hearing process (which resembles arbitrations). These include both discussions used on a particular dying status by various medical speakers (with "discrepant values or goals of care"), and those used by a family in the community (including those who may be objecting for religious ones). The first four steps are deliberation by both sets of parties, which are less expensive than a court action. These are followed by two steps at securing alternatives.[261]

Step 1 would require the seeking of an "understanding" of acceptable limits "before the critical illness occurs." This may be impossible for patients other than those for whom it may be anticipated that a critical illness will occur. Step 2 requires "joint decision-making . . . at the bedside" using "outcome data" where available and establish "goals for treatment" with informed consent. It is assumed that such goals are consistent with the goals of medicine. The phrase "joint decision-making" should be revised to state that "sincere" attempts should be made toward a decision that could actually be a unilateral one. For example, when a legal contract is being negotiated but no agreement can be reached, normally *nothing is done* (beyond the preparation of a proposal) until they reach an agreement. In medicine, however, there has been the implication that *everything* (i.e., treatment) *will* continue to *be done* until an agreement is reached to do otherwise which raises an important issue about an ethical outcome. Steps 3 and 4 would use the services of a consultant to "facilitate discussions" and/or utilize a hospital's ethics committee's "ad hoc team" to resolve disagreements. It would appear that these types of services might often (or generally) delay or otherwise postpone the agreement. Step 5 would provide for transfer to "another physician within the institution," in the event that the first one "remains unpersuaded," or, in the alternative, consider a transfer to another institution. This step needs change to omit the qualification where the other physician or institution would recommend undertaking a treatment that may not be ethically performed.

260. JAMA 940 (1999).

261. *Id.* at 937. Such discussions with the patients were most unusual. It was pointed out that "the SUPPORT patients were all seriously ill, and their dying proved to be predictable, yet discussions and decisions substantially in advance of death were uncommon." JAMA 1505 (1995).

Finally, Step 6 states that if transfer is not possible, "the intervention in question *need not be provided.*" (Emphasis added.) The stopping of treatment in this manner appears to be in accordance with the law in some jurisdictions. But, the words "*need* not be provided" (emphasis added) may be interpreted to give an option to continue futile nonbeneficial treatment, and it needs further clarification. In 1999, Texas became the first state to enact a similar law together with some more precise conditions concerning the timing of the steps.[262]

Literature specifically indicating utilization of this process was published in October 1999.[263] One example concerned a 67-year-old man with metastatic pancreatic cancer who was admitted to the hospital for terminal care. He was competent and completed a form by writing: "I want every life-sustaining measure to be employed in the event of any untoward circumstance, which I might experience. I want to live as long as possible!" The statement was set forth on his Advance Directive (AD), which appointed a durable power of attorney. There was also agreement among physicians that stated, "There is no obligation to offer treatments that are futile and are not a community standard of care. . . . We support the decision to no longer provide treatments for [the patient] that are effective including continued antibiotics." The treatment was withdrawn and the patient died in the hospital two days later. The case report stated:

> [P]hysician-driven limitations to medical care are acceptable and ethical. . . . The ethics committee consultation provided the only documentation in the patient's record that physicians were under no obligation to provide futile or nonstandard care, notwithstanding requests made by the patient in the AD.

The American Bar Association is an organization similar to the American Medical Association in that it has no power to legally enforce its rules upon

262. TEX. HEALTH & SAFETY CODE § 166.046(a) (Vernon Supp. 2002). This law was combined with the Texas Advance Directives Act, which also provides that the attending physician may not participate as a member of the ethics committees and that patients and families should be informed of their rights, including the timing of their participation and their receipt of a written report, as well as the "moral safe harbor" of consultation with parties outside of the treatment team.

263. MED. GEN. MED., Oct. 12, 1999; *see also supra* notes 173, 174.

lawyers. The Model Rules of Professional Conduct show the balance of the client's and lawyer's interests. The lawyer may withdraw his representation either without "material adverse effect on the client . . . or other good cause for withdrawal exists."[264] The Supreme Court recognized that in withdrawing, the attorney is allowed to exercise his "professional judgment."[265] Although no court had yet ruled on the AMA recommended approach or the Texas statute, the experience of the Baylor University Medical Center (BUMC) demonstrated that improvement has taken place in Texas.[266]

The subject of nonbeneficial (futile) treatment at taxpayer expense has received inadequate attention in the literature. Such payments are largely being made by Medicare and Medicaid payments upon the basis of approval by a physician or by the healthcare organization in hospitals whether or not treatment has been beneficial. If not, the steps should be taken toward stopping such treatment as outlined above.

Twenty-first Century Approaches

1. Appropriate changes should be made in the goals of medicine that do not adequately include ceasing to continue a treatment to sustain the life of a patient which the physicians agree is completely "futile" (pursuant to a procedure similar to that prescribed by the AMA) and that no means should be used other than comfort care.
2. Physicians should not refuse to communicate with patients (or their surrogates) that the authority to render judgment on the futility of treatment largely rests upon physicians or ethic committees.

---

264. Model Rules of Prof'l Conduct R. 1:16(b), (c). However, Rule 1:16(c) allows a court to order him to continue representing him. *See* Bondreau v. Carlisle, 549 So. 2d 1074 (Fla. Dist. Ct. App. (1989)).

265. Jones v. Barnes, 463 U.S. 745, 751 (1993).

266. Consultations on general ethics increased by 39%; medical futility, 67%; and right-to-die, 100% two years after the enactment of the Texas Advanced Directive Act of 1999. Some of these cases were described as follows: "In our two-year experience at BUMC, no family member has chosen to go to court to seek to delay or overturn the judgment of the ethics consultation process." Fine & Maye, 138 Annals Internal Med. 745 (2003).

3. A current "presumed consent" to CPR policy applicable to all patients (in the absence of a written order to the contrary, such as a DNR order) should be modified to the extent the consent can be mutual or by following the "futile procedure."

---

7. Slow Code

Slow codes, also known as partial codes, are decisions not to "aggressively attempt" to recommence a patient's heart after cardiac arrest. Sometimes staff may only diminish the motions of CPR. It results from (a) the assumption that without a DNR order, there is a legal requirement for CPR and (b) the fact that some physicians refuse to discuss DNR with their patients or next of kin, or acknowledge the absence of ethical or legal requirements not to perform an intervention that would reasonably be expected to be of no medical benefit.[267] The most prevalent reason for doing so has been that slow code gives a patient's family a "ritualistic comforting pat on the shoulder of a grieving family member, rather than an aggressive, deceitful show" and that hence it "may alleviate some anguish on the part of a family."[268] Such actual action cannot be justified. There is virtually no data on slow code,[269] which has been declared unethical. Appropriate actions need to be taken toward more education on this matter and against those who refuse to act accordingly.

good v. Good

Because "slow code" provides a dying patient's family with the comfort that allegedly everything is being done to keep him alive and thus alleviate their guilt, it should be continued.

v.

Because slow code represents deceit and has been declared unethical, it should be discontinued.

---

267. 338 N. Eng. J. Med. 367–69 (1988).
268. 339 N. Eng. J. Med. 1921 (1998).
269. *Id.* at 467.

Twenty-first Century Approach

Measures should be instituted to see that physicians and staff of healthcare facilities do not diminish the motions of CPR when performing it, and:

(a) discuss CPR and DNR with patients (or surrogates) as a part of the informed consent process.

(b) seek to ensure that all physicians are fully cognizant that normally a duty may not exist to provide treatment (including CPR) to patients who cannot benefit by such interventions.

By 1961, at 61, Ernest Hemingway, author of The Old Man and the Sea *and winner of the Pulitzer Prize and the Nobel Prize for literature, suffered from depression, alcoholism, failing eyesight, diabetes, hepatitis, hypertension, and impotence. He could no longer write nor live up to his standards of manhood. On July 2, almost immediately after returning from a hospital for treatment of his mental illness,* he killed himself in exactly the manner his father had *32 years before. (Emphasis added.)*

Chetnik, *Father Loss* (2001), p. 213

Were the disposal of human life so much reserved as *the peculiar province* of the Almighty, *that it were an encroachment on his right for men to dispose of their own lives,* it would be equally criminal to act for the preservation of life *as for its destruction. (Emphasis added.)*

Hume, *Of Suicide*

The only purpose for which power can be rightfully exercised over any member of a civilised community, against his will, is to prevent harm to others. His own good, either physical or moral, is not a sufficient warrant.

John Stuart Mill, "On Liberty" (London: Dent, 1960), pp. 72-73

14

End-of-Life Choices of Terminal Patients

1. ***The Choice of Suicide***
   a) **Some Religious Views of Jews and Jehovah's Witnesses**
   b) **Various Reasons for Possible Suicide; Taking Risks (motorcycles)**
   c) **Limitations on the Duty to Deny Suicide**
   d) **The Option of Knowing That You Can**
2. ***The Goals of the Hospice Option***
   a) **Assistance and Lack of Hospice Assistance in Dying**
3. ***The Choice of Physician-Assisted Suicide (PAS); Double Effect; Costs of Capital Punishment Procedures***
   a) **Some Religious Views on PAS**
   b) **The Choice of Double Effect (Euthanasia) by the Use (or Abuse) of Lethal Doses of Pain Medication**
   c) **The Choice of PAS and/or Double Effect; Oregon**

4. Limitation in Model Statutes; Mandatory Psychiatric Examination

5. Progressive Nullification of Anti-PAS Statutes

a) **Civil Disobedience Actions by Physicians and Nurses**

b) **The Prohibition Analogy (Nullification)**

c) **The Example of Dr. Kevorkian**

6. Other Choices of Euthanasia

Most older people fear disability and the dependency and loss of dignity it brings more than they fear death.

Evans, "A Correct Compassion: The Medical Response to an Ageing Society," in *Increasing Longevity* (London, 1998)

1. The Choice of Suicide

a) Some Religious Views of Jews and Jehovah's Witnesses

In the seventeenth century, the poet and minister John Donne wrote *Biathanatos*, the first defense of suicide in English.[1] At the end of the nineteenth century, Durkheim's study, *Suicide* (1895), compared suicide rates in major countries of Europe. The figures were highest in Protestant countries and lowest in Catholic ones. Even though both groups thought suicide was wrong, Protestant societies such as that of David Hume, the eighteenth-century English writer of this chapter's epigraph, offered the individual greater freedom of thought and

---

1. Donne died in 1631. *Biothanatos* was published in 1646. At the time and for almost two centuries thereafter, the "consequence of suicide" was that stated by Lord Byron, whose father-in-law escaped it only through a claim of insanity:

> that a domestic Affliction would lay him in the Earth—with the meanest of Malefactors—in a Cross road with the Stake in his body— if the Verdict of Insanity did not redeem his ashes from the sentence of the Laws he had lived by interpreting or misinterpreting, and died in Violating.

2. *Quoted in* Daniel Pals, Seven Theories of Religion 97 (1996). The Islamic philosopher al-Farabi, who traveled among the Greeks in the 11th century A.D., particularly Plato and Aristotle, wrote that the end of life was happiness, in the Greek sense of *eudaimonia,* which consisted of achieving theoretical, not practical, perfection. Hillsdale College Press, 2002.

life. In contrast, Catholics belonged to a more integrated social community where priests were mediating between God and the believer. This difference led to the current statement, "The tighter the social ties, the lower the rate of suicide."[2]

Orthodox Jewish rabbis consider an adult who commits suicide to be a murderer because he is denying the divine mastery and ownership of his life and soul; the person who ends his life by suicide is not accorded full burial honors. Children, however, "are never regarded as deliberate suicides" by Orthodox Jews and are afforded such rites.[3]

Many Americans today feel that they are free from such religious restrictions. Although we may have annulled the so-called "nuclear double suicide" pact that characterized the Cold War, multiple suicide options remain on an individual basis.

The Jehovah's Witnesses have been held to have the right to die rather than accept blood transfusions,[4] and as that particular population ages, their need to exercise this right increases.[5] Toward the end of the twentieth century, appellate courts have upheld the right of this group to cling to their belief,[6] and a doctor who forces a blood transfusion can be liable for damages. For example, after a

---

3. Rosner, Modern Medicine and Jewish Law (1972), *citing* Rabbi Yechiel Michael Totzinski and others, noted that:

> Rabbinical rulings, going back at least to Judah ben Samuel in the 13th century and codified in the 16th century by Isserles, have explicitly condoned the active removal of obstacles to the departure of the soul of the dying, and at least one rabbinical authority specifically ruled that it is "forbidden to hinder the departure of the soul by the use of medicines."

Haynick, Jews and Medicine 555 (2002).

4. *Id.* at 193. *E.g., Genesis* 9:4 and *Deuteronomy* 12:16 are cited by Jehovah's Witnesses. "But Jews did not consider blood through tubes to be a case of 'eating.'" Douff, Matter of Life & Death 210 (1998).

5. Orange County Register, Dec. 19, 1998, at News 4.

6. Fosmire v. Nicoleau, 551 N.E.2d 77 (1990). *See also* Public Health Trust of Dade County v. Wons, 541 So. 2d 96 (Fla. 1989). In the U.S. case of *Bowen v American Hospital Association*, it was said, "The decision to provide or withhold medically indicated treatment is . . . made by the parents or legal guardian." Bowen v. Am. Hosp. Ass'n, 476 U.S. 610, 628 n.13 (1986); Parham v. J.R., 442 U.S. 584, 603 (1979); Malette v. Shulman, 72 O.R.2d 417, 424 (Ont. C.A. App. 1990); and *In re* Brown, 478 So. 2d 1033, 1041 (Miss. 1985).

7. Stamford Hospital v. Vega, 674 A.2d 821 (Conn. 1996).

woman in Connecticut delivered her baby, she began to bleed heavily but refused to be transfused. The physicians, believing it essential that she receive blood to survive, requested a court injunction to transfuse her. The trial court granted their request, stating that recognition of her autonomy over her body would constitute an abandonment of her baby. However, the Connecticut Supreme Court reversed and held that under the doctrine of informed consent, there was no right to thrust unwanted medical care on a patient who, having been informed of the consequences, competently and clearly declined that care.[7] The court cited a statement of the U.S. Supreme Court that: "The notion of bodily integrity has been embodied in the requirement that informed consent is generally required for medical treatment. . . . The logical corollary of the doctrine of informed consent is that the patient generally possesses the right not to consent, that is, to refuse treatment."[8]

Courts have also found that Jehovah's Witnesses may assume the risk of death, which, if preventable, may be viewed as a form of suicide. Thus, in one case, a Jehovah's Witness who had refused recommended blood transfusions died following a dilation and curettage; the physician had perforated her uterus. Having found the doctor negligent, the jury apportioned the damages: the woman was found to be 75 percent at fault because she refused a blood transfusion that would have prevented her death. The Washington Supreme Court upheld the jury's apportionment of fault.[9] The interests of the parties are balanced and the refusal of a Jehovah's Witness to accept certain treatment does not excuse a physician's negligence in a civil lawsuit. A criminal case is different; a drunken driver was convicted of manslaughter in the death of a Jehovah's Witness who died of her injuries after she refused blood transfusions. He was sentenced to 10 years in prison.[10]

The maternal-fetal rights of a Jehovah's Witness were considered in the case of a woman in her 34th week of pregnancy who suffered greater than anticipated blood loss from an elective surgery.[11] In spite of her having re-

8. Cruzan v. Director, Mo. Dep't of Health, 497 U.S. 261 (1990).

9. Shorter v. Drury, 695 P.2d 116 (Wash. 1985). Similarly, in *Corlett v. Caserta*, a witness experienced profuse bleeding after colorectal surgery, refused blood, and died. The physician was negligent but the damages recovered were reduced to "the extent that his refusal contributed to the witness's death." Corlett v. Caserta, 562 N.E.2d 257 (Ill. App. Ct. 1990).

10. Orange County Register, March 13, 1999, at News 6.

11. Some medical testimony indicated it was unlikely she would have survived even if she had gotten blood.

fused a transfusion based on her religious objection, she was transfused after being forcibly restrained and sedated. Prior case law specifically held that pregnant women, prior to viability, owe no duty to their fetuses,[12] and the trial court held that the state's interest in saving the life of a fetus overrides a competent patient's decision to refuse medical treatment. However, in 1997 the Illinois Supreme Court became the first state supreme court to establish that *pregnant women* have the *right to refuse blood transfusions* when the life of the fetus is endangered,[13] stating:

> The State may not override a pregnant woman's competent treatment decision, including refusal of recommended invasive medical procedures, to potentially save the life of the viable fetus. . . . A blood transfusion is "an invasive medical procedure that interrupts a competent adult's bodily integrity."[14]

Damages may be obtained from the physician, balanced against the risk of death assumed by the Jehovah's Witness, and subject to possible reduction to the extent that assumption of risk contributed to her death.

Attempts to find a substitute for blood continue to be plagued by lack of success.[15] Should such substitutes actually become available in the twenty-first century, Jehovah's Witnesses may be saved more often without having to refuse transfusion of human blood.[16]

12. *See, e.g.*, Stallman v. Youngquist, 531 N.E.2d 355 (Ill. 1988).
13. *In re* Brown, 689 N.E.2d 397 (Ill. App. Ct. 1997).
14. *Id.* at 405.
15. Scientific American, February 1998, at 77.
16. Apparently the rights of Jehovah's Witnesses continue to be violated in South America, a generally Catholic continent, as reported in this case: "Doctors [initially] refused to amputate the youth's cancerous leg because he and the grandparents who raised him are Jehovah's Witnesses and for religious reasons would not accept a blood transfusion. However, a court in Bogota, Colombia, authorized the operation." L.A. Times, Dec. 27, 1996, at A39.

good v. Good

Because all physicians have a duty to preserve their patients' lives, Jehovah's Witnesses have no right to refuse blood transfusions. In particular, pregnant Jehovah's Witnesses have no right to do so when this might concurrently result in the death of the "baby" they are carrying.

v.

All competent persons (including Jehovah's Witnesses) have the right to refuse treatment of any kind; when pregnant, such refusals also remain their right until their "fetus" becomes legally viable.

Twenty-first Century Approach

1. Physicians and healthcare facilities should recognize the right of all competent patients to refuse consent treatment (including blood transfusions), which might save their lives. This applies to pregnant females and to the life (or lives) of their fetuses, which they possess rights to abort prior to viability.
2. Where such persons (including Jehovah's Witnesses) die as a result of a negligent transfusion (carried out in spite of their objections to the procedure), damages may be apportioned (reduced) to the extent that the patient's risk assumption may have contributed to their demise.

Oh, East is East, and West is West, and never the twain shall meet.

Rudyard Kipling, *The Ballad of East and West*

She Cleopatra grew more fierce as she beheld her death
Bravely as if unmoved, she looked upon
The ruins of her palace; bravely reached out
And touched the poison snakes, and picked them up.

Ferry, *Odes of Horace* (1997), p. 97

b) Various Reasons for Possible Suicides; Taking Risks (Motorcycles)

No jurisdiction in the United States has a law making suicide attempts a crime, nor is it a crime in Europe.[17] Legal attempts at suicide prevention

17. Suicide is no longer a criminal offense in Europe. Attempted suicide was decriminalized in France in 1970 and in Prussia in 1751. *See* M.P. Batten, *Assisted Suicide: Can We Learn Lessons from Germany?*, Hast. Ctr. Rep., 1992, vol. 22, at 44-45.

have not normally proven successful. In 1998, there were 30,575 recorded suicides in the United States, leaving more in need of social support.[18] After a review of the literature, researchers "could find no evidence that suicide prevention legislation affected the suicide rate."[19] As previously noted, $17 million of taxpayer money has been spent during the 1990s in an attempt to change the suicide rate, without success.[20] In 2002, the Centers for Disease Control and Prevention (CDC) reported that 70 percent of the more than 250,000 self-inflicted injuries seen in U.S. emergency rooms each year appear to be unsuccessful suicide attempts.[21]

The intentional taking of life-threatening risks has demonstrated inherent conflicts. Bills requiring motorcycle riders in California to wear a safety helmet or face a $100 fine was stalled for several years. The riders showed that they have a right to risk their lives on motorcycles as with many other activities. It became law in 1992 only after it was demonstrated that in the majority of accidents, the rider did not die; rather he suffered serious injury, causing the taxpay-

18. *Survivors of Suicide Fact Sheet* (American Association of Suicidology, 1998).

19. Lester, *The Myth of Suicide Prevention*, 13 COMP. PSY. 555-60. At the turn of the third millennium, some 30,000 people commit suicide each year. They constitute 1.4% of all deaths in the United States. Suicide rates among teenagers rose from the ninth to the eighth leading cause of death.

20. KEYES, LIFE, DEATH, AND THE LAW, ch. 14B. In 2000, it was indicated that effective rates of suicide are highest among the very old, especially old white men who have seen a primary care physician in preceding months. Bruce et al., *Reducing Suicidal Ideation and Older Primary Care Patients' Depressive Symptoms*, JAMA 1081 (March 3, 2004). In the case of cancer, a study showed that a positive attitude does not improve the chances of surviving cancer, and doctors who encourage patients to keep up hope may be burdening them. ORANGE COUNTY REGISTER, Feb. 9, 2004, at News 5.

21. Almost 62% of the patients were white, 9% were black, and 7% were Hispanic. Poisoning accounted for 65% of self-inflicted injuries, the CDC found, with cuts from sharp instruments coming second at 25% and firearms accounting for only 1%. "Feeling suicidal in Beijing? Then be patient. Nine out of 10 Chinese calling into a suicide-prevention hotline were greeted by a busy signal." ORANGE COUNTY REGISTER, Jan. 2, 2005, at News 25.

22. "Roughly 70% of deaths to cyclists are due to head injury. One widely quoted study on the effectiveness of cycle helmets suggests that wearing a cycle helmet reduces the risk of head and brain injury by over 80%." 22 J. MED. ETHICS 41-45 (1996).

ers to pay large sums for his recovery.[22] It had also been remarked that "encouraging helmet use may encourage cyclists to take greater risks, a phenomenon known as 'risk compensation'"[23] or "an anti-moral hazard." Nevertheless, after many such riders became insured, they lobbied against the helmet law, noting that the taxpayer was not the one to pay for their injuries. For this reason, a bill in California to partly repeal the law for insured motorcyclists almost succeeded in 1998.[24] A more reasonable bill was introduced to the California Assembly that would have allowed motorcyclists age 21 and over to ride without helmets only if they carried with them proof of $1 million in health insurance. However, that bill was defeated by a close vote (34-32).[25] Some 19 states began to also require the use of bicycle helmets by children as of 2001[26] (or made laws effective in 2003).[27] Typically, insurance companies exclude coverage of suicides within two years of taking out the policy. However, in one case, an insurance company was denied access to the insured's records on the ground of their privileged status when the company was seeking to demonstrate that the plaintiff's death was suicide.[28]

In the mid-1990s, suicides were committed by 31,142 persons in the United States and constituted the ninth leading cause of death. This was almost the identical number of annual suicides in Japan "mostly over money problems."[29]

23. Simon v. Sargent, 346 F. Supp. 277, 279 (D. Mass. 1972).

24. Orange County Register, June 17, 1998, at 4.

25. Orange County Register, May 14, 2000, at News 10. In 2003, the National Highway Traffic System Administration (NHTSA) reported that motorcycle deaths increased over 50 percent in Kentucky and 100 percent in Louisiana after they repealed their mandatory helmet laws. Some 20 states still require motorcyclists to wear helmets. NHTSA: http://www.nhtsa.dot.gov.

26. LJN's Prod. Liab. Law & Strategy 10 (September 2002). In 2004 it was noted that many middle-aged Americans have been getting motorcycles to recapture their lost youth. According to the National Highway Traffic Safety Administration, the annual number of motorcycle fatalities among 40-plus riders tripled over the past decade. Orange County Register, Jan. 22, 2004, at News 27.

27. Orange County Register, Sept. 13, 2002, at News 16.

28. Comm'n v. Sanchez, Pa. Super. 416, 160, 174 (Pa. Super. 1994).

29. "The per capita suicide rate in Japan is roughly twice that of the United States. And a recent surge in debt-linked suicides is alarming this nation, [unlike the] West, where businesses are allowed to fail yet people make a fresh start on life." Orange County Register, March 16, 2003, at News 31.

Firearms were the leading method used throughout the country.[30] This conclusion was based on the assumption that the greatest reductions in fatal violence would be within states that were required to institute waiting periods and background checks for obtaining firearms, implementing the Brady Act.[31]

A Harvard study found that women who drink caffeinated coffee are less likely to commit suicide than those who do not.[32] A 1995 study showed that China was the only country in the world where the suicide rate for women exceeded the rate of men.[33] A study by the Centers for Disease Control (CDC) found that 9 percent of high school students had attempted suicide in 1996.[34] In 2000, a study showed that over a million Americans ages 12 to 17 tried to kill themselves.[35] Teenage girls are more likely to attempt it than boys, but they are

---

30. JAMA, Sept. 24, 1997, at 674. A 1999 study found that suicide was the leading cause of death among gun buyers in the first year after the weapon was purchased. "The first week proved to be a critical period. A person who buys a handgun was found to be 57 times more likely than a member of the general population to commit suicide within a week of purchasing the weapon." N. ENG. J. MED., Nov. 18, 1999.

31. Ludwig & Cook, JAMA 585 (August 2000).

32. L.A. TIMES, March 11, 1996. Also, approximately 1.5 million U.S. women experience intimate partner violence annually. Approximately 20% of these women obtain civil protection orders, "but the effectiveness of such orders in preventing future violence is unclear." Nevertheless, it has been found that "permanent, but not temporary, protection orders are associated with a significant decrease in risk of police-related violence against women by their male intimate partners." JAMA 589 (Aug. 7, 2002).

33. ORANGE COUNTY REGISTER, Nov. 12, 1999. The U.S. ratio is about four men per one woman. JAMA 2272 (June 23/30, 1999).

34. ORANGE COUNTY REGISTER, Sept. 27, 1996, at News 4.

35. *National Household Survey on Drug Abuse Report*, ORANGE COUNTY REGISTER, July 16, 2002, at News 7.

> Bipolar disorder—in which patients swing between deep depression and manic highs—affect[s] 1.3% to 1.5% of the population. As many as one in every five bipolar patients attempt suicide at some time in their lives. The new study, reported today in the *Journal of the American Medical Association*, "makes an argument that lithium should be reconsidered as the first-line treatment [for bipolar disorder. Patients] who take Depakote are 2.7 times as likely to commit suicide as those who take lithium."

L.A. TIMES, Sept. 17, 2003, at A14.

less likely to die.[36] Sometimes both die together, as they did in Shakespeare's *Romeo and Juliet.* It remains a theme without end for immature teenagers. In 1995, two eighth-grade sweethearts, Maryling Flores, 13, and Christian Davila, 14, who were forbidden by Flores' mother to see each other, drowned themselves in a canal, leaving suicide notes. Christian wrote:

> I can't go on living. I've lost Maryling. I'm escaping from the realm of reality into the darkness of the unknown. Because reality is, I can't be with Maryling.
>
> Maryling wrote to "Mom and Dad":
>
> You'll never be able to understand the love between me and Christian. You don't let me see him in this world, so we're going to another place. Please don't cry for me. This is what I want. I want to feel happy, because I'm going to a place where I can be with Christian.[37]

A study has also shown that suicide rates increase after severe earthquakes, floods, and hurricanes.[38] However, physical disabilities engender more suicides. In a review of 5,200 spinal cord injury patients, researchers found suicide to be two to six times higher than in the general population.[39] (Comparable rate increases have been found for alcohol and substance abuse, epilepsy, and Huntington's chorea.) With respect to suicide and depression, it has been stated, "Though we may profoundly disagree with a competent patient's wish to die, imperatives suggest bowing to the patient's determination of his/her best interests. The notions of informed consent and patient

---

36. Orange County Register, Aug. 28, 1995, at News 23. "Firearms are used in approximately half of all youth suicides." Further, Webster et al., JAMA 594 (Aug. 4, 2004), showed that "minimum age restrictions for the purchase and possession of firearms do not appear to reduce overall rates of suicide among the young."

37. Orange County Register, Nov. 9, 1995, at News 10. This tragedy is not unique. It was repeated by two high school sweethearts in May 2000. L.A. Times, July 30, 2000, at A 26.

38. "The immediate physical health consequences of disasters have been well described. Increased respiratory, gastrointestinal, and cardiovascular symptoms have also been reported for up to 5 years after a disaster." N. Eng. J. Med. 338 (1998).

39. Charlifue & Gerhart, *Behavioral and Demographic Predictors of Suicide after Traumatic Spinal Cord Injury*, 72 Arch. Physical Med. & Rehab. 488-92 (1991).

autonomy are meaningless without the patient's ability to reject treatment."[40] That is why "professionals cannot always succeed in helping patients—disabled or not—to reach a quality of life that they find acceptable. If we truly accept the idea of independence for the disabled, we must also accept this choice";[41] it is a profound lesson in law and bioethics in the twenty-first century that has to be learned by many professionals, as well as some members of the public.[42] Consider the case of a 67-year-old named Robert, who ended his life by holding a .357-caliber Magnum to his head and pulling the trigger while sitting on his porch at sunset.

> This happened three days after an office visit during which he agreed to additional abdominal surgery to mitigate the ongoing damage caused by severe diffuse atherosclerosis (cholesterol within arteries). Despite being a T120 paraplegic from a skiing accident at 35, Robert had maintained an active lifestyle. . . . We spoke frankly about the hurdles that lay ahead. But we also talked about how he had been a fighter his whole life, and that after another operation or two, things might settle down. When the country coroner called with news of the suicide, I was shocked. How could I have been so naïve? How could I have been so remiss? I should have been able to stop him from taking such a foolish step. . . . As time passed, I wondered whether it had truly been foolish, from Robert's point of view. He knew he was in for months or years of discomfort, and that he might never have been able to tend his ranch or fly again. There was no certainty, though some chance, he'd recover. Maybe there was nothing I could—or should—have done to stop him, even had I been more sensitive and intuitive.[43]

Should this doctor or his patient have acted differently? Neither psychotropic drugs nor hospitalization has clearly shown to effect a reduction in suicide rates.[44]

---

40. J. L. MED. & ETHICS 58 (Spring 1996).

41. *Id.* at 60.

42. In his 1995 novel, *Prizes*, Erich Segal has Adam, one of his heroes who is dying from Alzheimer's disease, commit suicide immediately after learning that he would win a Nobel Prize.

43. MED. ECON. 145 (March 24, 1997).

44. L.A. TIMES, Jan. 22, 1998, at 26, *citing* Melzer & Okayli, *Reduction of suicidality during clozaine treatment of neuroleptic-resistant schizophrenia; impact on risk-benefit assessment*, AM J. PSYCH. 152, 182-90 (1995).

good v. Good

Although suicide is legal in all states, these laws should be changed because suicide deprives God of his prerogative

v.

For some people, and under certain conditions, the autonomous act of suicide may be a natural act: These laws should not be changed.

Twenty-first Century Approach

While suicide may normally be discouraged, public policy should continue to recognize that suicide remains a legal right of choice which may be similar to the decision not to receive treatment that is possessed by all competent patients as free citizens in the United States.

Education is the ability to listen to almost anything without losing your temper or your self-confidence.

Robert Frost

c) Limitations on the Duty to Deny Suicide

In 1982, the U.S. Supreme Court considered the case of Nicholas Romeo, a profoundly retarded adult who suffered numerous injuries as a result of suicide attempts after being committed to a state facility. The court held that he enjoyed "constitutionally protected interests in conditions of reasonable care and safety." The right to personal security, they held, constitutes a "historic liberty interest" protected substantively by the Due Process Clause.[45]

Because several lower courts thereafter refused to distinguish between voluntary and involuntary patients, the U.S. Supreme Court in 1989 considered a claim alleging the state's failure to remove a child from his father's home as a violation of his liberty interest when faced with evidence of abuse by the father. The Court rejected this argument, stating, "The state's affirmative act of restraining the individual's freedom to act on his own behalf, through incarceration, institutionalization, or other similar restraint of personal liberty, is the 'deprivation of liberty' triggering the protections of the Due Process

45. Youngberg v. Romero, 457 U.S. 307, 315 (1982).

Clause."[46] Similarly, an appellate court held that a teenager who committed suicide while under treatment at a state psychiatric facility had a constitutionally protected liberty interest in a safe environment only if her status changed from a voluntary to an involuntary one during the course of her admission.[47] If a competent person attempts to commit suicide without endangering others, these attempts cannot normally be legally circumscribed by the state. Other instances may or may not be considered rational and ethical actions by the patient depending upon the circumstances. In 2002, the U.S. Court of Appeals for the Seventh Circuit held that a lawyer who allegedly caused a client to commit suicide was not guilty of legal malpractice. He was a tax attorney hired to represent a man named Cleveland who had become involved in a bitter dispute with the Internal Revenue Service. The attorney was warned by a therapist that Cleveland suffered from severe depression and had suicidal tendencies. Prior to a rescheduled audit, Cleveland killed himself, and his widow sued the attorney for precipitating the audit and causing her husband to suffer severe financial and psychological harm that led to his suicide. In dismissing the claim, the court stated: "By contrast, attorneys are medical lay people who cannot be reasonably expected to anticipate the mental health consequences of their legal advice. Attorneys cannot reasonably be expected to screen potential clients for suicidal tendencies."[48]

The American Bar Association's *Model Rules of Professional Conduct* normally authorizes a lawyer's relationship with such a client. It states:

a. When a client's capacity to make adequately considered decisions in connection with a representation is diminished, whether because of minority, mental impairment or for some other reason, the lawyer

46. DeShaney v. Winnebago County Dep't of Soc. Serv., 489 U.S. 189 (1989). The court also acknowledged that "in certain limited circumstances, the Constitution imposes upon the State affirmative duties of care and protection with respect to particular individuals." *Id.* at 198. *See* Karen M. Blum, *DeShaney: Custody, Creation of Danger, and Culpability*, 27 Loy. L.A. L. Rev. 435 (1994) (discussing post-*DeShaney* rights and duties).

47. Kennedy v. Schafer, 71 F.3d 292 (8th Cir. 1995). The 15-year-old had been identified as a suicide risk, which required a staff member to keep her in constant eyesight and interact with her at 15-minute intervals. She was found dead in her room more than 2 hours after her last contact with a staff member.

48. Cleveland v. Rotman, 7th Cir., No. 02-2488 (July 17, 2002).

shall, as far as reasonably possible, maintain a normal client-lawyer relationship with the client.

b. When the lawyer reasonably believes that the client has diminished capacity, is at risk of substantial physical, financial, or other harm unless action is taken and cannot adequately act in the client's own interest, the lawyer may take reasonably necessary protective action, including consulting with individuals or entities that have the ability to take action to protect the client and, in appropriate cases, seeking the appointment of a guardian ad litem, conservator, or guardian.[49]

In a 1999 study of dying patients, a will to live showed substantial fluctuation, with the explanation for these changes shifting while death approaches.[50] As stated by the chair of the Department of Psychiatry and Behavioral Sciences at Memorial Sloan-Kettering Cancer Center (with an authority cited for each of the associations that he listed): "Demoralization is associated with being elderly, disabled, disfigured, dependent, socially isolated, or alienated, concerned about being a burden, fearing a loss of dignity or control, and desiring death or becoming actively suicidal."[51]

good v. Good

Involuntary incarceration should make no difference with respect to the outlawing of suicides, all of which are unethical.

v.

Incarceration deprives a person of his freedom and reasonable attempts should be made to prevent suicide in those special cases where the state is fully aware of its likelihood.

Twenty-first Century Approach

1. Mere incarceration in jail or prison should not result in any specific duty to prevent suicide absent special circumstances. Where a suicide is committed outside of such limited circumstances, no liability to the state should result.

---

49. MODEL RULES OF PROF'L CONDUCT (2003 ed.), R. 1.14: Client with Diminished Capacity.

50. L.A. TIMES, Nov. 27, 1999, at B1.

51. Kissane, *The Contribution of Demoralization to End of Life Decisionmaking*, Hast. Ctr. Rep. (2004) at 24; Clark, *Autonomy, Rationality, and the Wish to Die*, 25 J. MED. ETHICS 457-62 (1999).

2. It should be recognized that reasonable legal restraints might also be applied to institutionally committed persons in those limited circumstances where suicide appears a quite reasonable probability.

Sometimes just having a means to suicide limits anxiety and allows the patient to enjoy life more fully. And making physician assistance legal might make it easier to monitor and regulate.

Lynn & Harold, *Handbook for Mortals* (1999), p. 146

The thought of suicide is a great consolation, by means of it one gets successfully through many a bad night.

Nietzsche, *Beyond Good and Evil*

Many patients on life supports, who are not depressed, judge the quality of their own life to be so poor that they do not want it prolonged further by artificial means. And those patients whose depression results from learning their diagnosis of terminal illness may wish to die sooner *rather than later. Such a wish might* then be entirely rational. *(Emphasis added.)*

Macklin, *Mortal Choices, Bioethics in Today's World* (1987) p. 67

d) The Option of Knowing That You Can

Beyond Nietzsche's epigraph, the knowledge of the availability of a means to commit suicide when life becomes intolerable can be consoling. As one physician put it, "Denying this option to persons who are physically unable to commit suicide smacks of discrimination."[52] A 2000 study showed that only one in 10 of such patients who had asked a doctor for a prescription to help them commit suicide actually did so.[53] Some psychotherapists strive to destroy illu-

---

52. Kirkland Ltr., N. Eng. J. Med. 1852 (Dec. 18, 1997).

53. Orange County Register, March 27, 1996, at News 15. A study in February 2000 indicated that in Oregon, which has had legalized physician-assisted suicide since 1997, "in six requests for a prescription for a lethal medication . . . one in 10 requests actually results in suicide. That is, palliative interventions lead some—but not all—patients to change their minds about assisted suicide." Further, "after two years of legalized assisted suicide in Oregon, we found little

sions. Yet, some such things have been acknowledged as "healthy illusions."[54]

We should ask why many psychiatrists in the process of earning a living have not learned this lesson and prescribe fewer drugs to our over-drugged society. Doctors who prescribe drugs to aid suicide by those who ask for them may or may not be acting rationally. The publication of Hemlock Society founder Derek Humphrey's book *Final Exit* promoted the "rational" right to die by terminally ill patients after they had ruled out other options. It contained an extensive list of the names and dosages of potentially lethal medications. Two years later, a study published in the *New England Journal of Medicine* in November 1993 found that the total number of suicides in New York City had not increased after the publication of *Final Exit*—even though more of those people who did commit suicide used methods described in that book. In the following year, books were actually repressed at libraries for presenting suicide as "a rational and sane alternative" in dire circumstances.[55] In February 2000, Humphrey narrated a video *of Final Exit.* It was shown on a public-access cable station over the objections of those who warned that it would encourage the desperate to act on their impulses. However, local police and others reported no increase in people trying to kill themselves that day or 24 hours after the broadcast.[56] In 2003 an Australian doctor announced his plans to build a new carbon monoxide "COGen" machine that could be used to assist in suicide after his prototype was seized as he left his native country. Humphrey said he looked forward to seeing the device.[57]

---

evidence that vulnerable groups have been given prescriptions for legal mediation in lieu of palliative care." 342 N. Eng. J. Med. 557-63 (2000).

54. Peck, In Search of Stones 99 (1995).

55. Roman, *Exit House* (1994), because it presented suicide as a rational alternative, was pulled from the Springdale, Ark., Public Library; *Ilund* (1994) was pulled from the Lexington, N.C., Middle School (1994) because of the way it addresses death.

56. Orange County Register, Feb. 4, 2000, at News 20.

57. Orange County Register, Jan. 18, 2003, at News 6.

good v. Good

Any preparation for a possible suicide (by storing drugs or otherwise) only indicates that a dying depressed person needs psychiatric help to prevent his carrying it out.

v.

Certain persons who are preparing to shorten the dying process when life becomes unbearable may be acting rationally. If so, and in spite of their depression, they may or may not change after the receipt of psychiatric counseling.

Twenty-first Century Approach

It should be publicly recognized that competent dying citizens may prepare themselves for exercising their legal option to shorten the dying process when, in their opinions, their quality of life diminishes below a minimum. (See Physician-Assisted Suicide [PAS] below.)

Be still prepared for death, and death or life shall thereby be sweeter.

Shakespeare

Who knows but life be that which men call death,
And death what men call life.

Euripides

As the days pass, you choose. Do you live in anticipation of death or face death in anticipation of life? The answer came to me in one sentence: "Cancer may kill me," and then I would add triumphantly, "but not today!" In that acknowledgment, I accepted mortality and life at the same time.

Sam Gulberson, *ABA Journal*, April 2004, p. 87

2. The Goals of the Hospice Option

Although many physicians are likely to intervene regardless of their Medicare patient's expected outcome,[58] the Medicare statute requires that a nursing home provide services "to attain or maintain the highest practicable physical, mental and psycho-social well-being of each resident. . . ."[59] Hence the

58. Tu et al., *Use of Cardiac Procedures and Outcomes in Elderly Patients with Myocardial Infarction in the United States and Canada*, 136 N. Eng. J. Med. 1500-05.

59. 42 U.S.C.S. §§ 1395i-3(a)-(h) and 1396r (a)-(h) (2003).

integration of curative and palliative care in hospitals and nursing homes could become beyond their normal function. That is, it could lead them toward "aggressive rehabilitation, medical treatment and maintenance of functional capacity"[60] and render to dying patients "unsatisfactory care at the end-of-life with pain and other distressing symptoms inadequately managed, and failure to address spiritual and emotional needs."[61] When a patient is discharged from a hospital to a nursing home, Medicare pays for the first 20 days of skilled care. However, during the next 90 days patients are required to pay a coinsurance amount equal to one-eighth of the Medicare inpatient hospital deductible.[62] In this connection, it has been noted that "overall, total government expenditures in the last month of life were significantly less for hospice versus non-hospice residents."[63]

The goal of a hospice, on the contrary, is solely for the palliation of terminal patients' physical and mental suffering—that is, to provide "comfort care" rather than "curative." Normally, enrollees must be diagnosed with only six months to live. About 89 percent of this care is delivered in their home.[64] In 1982, when the Medicare hospice benefit (MHB) began, there were 800 hospices; Medicare paid for 100 days in a nursing home. By 1994, there were over 2,500 hospice

---

60. Nat'l Hospice Org., *Nursing Home Task Force Report* 6 (1998).

61. Keay, *Palliative Care in the Nursing Home*, 23 GENERATIONS 96-98 (1999); Von Gunten et al., *Recommendations to Improve End-of-Life Care through Regulatory Change in U.S. Healthcare Financing*, 5 J. PALLIATIVE MED. 35-41 (2002). A study pointed out how it may become difficult for "nursing facility staff to switch between providing restorative rehabilitative care and palliative care." Gage et al., DHHS, *Important Questions for Hospice in the Next Century* (March 2000), *available at* http://www.aspe.hhs.gov/daltcp/reports/impques.pdf (last visited 2005).

62. 42 U.S.C.A. § 1395e(a)(3) (West 2003). In those instances a nursing home resident may become eligible for both the hospital and nursing home requirements.

63. Miller et al., *Government Expenditures at the End-of-Life for Short- and Long-Stay Nursing Home Residents: Differences by Hospice Enrollment Status*, 52 J. AM. GERIATRICS SOC'Y 1284-92 (2004).

64. Nevertheless, a 2001 survey found that 75% of Americans were not aware that hospice care can be provided in the home. National Hospice Foundation, *Public Opinion Research: Baby Boomers Fear Talking to Parents About Death* (2001), *available at* www.hospiceinfo.org; Silveira et al., *Patient Knowledge about Options at the End of Life: Ignorance in the Face of Death*, 284 JAMA 2483-88 (2000).

facilities serving over 300,000 beneficiaries at a cost of over $1.3 billion.[65] In the 1990s, Medicare paid a per diem to the hospice for home care of about $91 and inpatient care of $403. It also paid for a respite hospital stay of up to five days to give families a rest.[66] To meet Medicare's requirements, patients' life expectancy must be six months or less, and they will not receive any more treatment directed at the underlying disease. Hence, less than 30 percent of them receive hospice care. But as of 2004, many provide "hospice-like" services to patients who are continuing to receive some active treatment.[67] Some cost-based payments to hospitals and nursing homes have been lowered and none of them are specifically identified as "palliative care."[68]

The policy of one nearby inpatient hospice I visited is that no cardiopulmonary resuscitation (CPR) is given there. However, for in-home care, a patient may choose CPR or not. The reimbursement rates are almost the same as those for comprehensive long-term care in an institution.[69] Physicians must certify that patients are expected to survive less than six months.[70] The difficulty in predicting life expectancy often makes physicians unwilling to sign a hospice certification.[71] The average patient enters hospice one month

---

65. L.A. Times, July 18, 1996, at 172. "By electing the hospice benefit, beneficiaries waive rights to Medicare payment for curative treatment. They receive support services, which would not otherwise be covered."

66. Med. Econ., Sept. 25, 1995, at 81.

67. Quill, N. Eng. J. Med. 2030 (May 13, 2004). At the turn of the 3rd millennium, data collected from 21 hospices and 15,484 hospice patients found no "meaningful difference between the total average daily visit volumes in nursing homes versus non-nursing home settings." Miller, *Hospice Care in Nursing Homes: Is Site of Care Associated with Visitor Volume?*, 52 J. Am. Geriatrics Soc'y 1331-37 (2004).

68. Beeuwkes, et al., *What Is Known about the Economics of End-of-Life Care for Medicare Beneficiaries?,* 42 Gerontologist, Special Issue III 40-48 (2002), and Juskamp et al., *Providing Care at the End of Life: Do Medicare Rules Impede Good Care?*, 20 Health Affairs 204-11 (2001).

69. L.A. Times, Vol. 335, No. 3, at 201.

70. "A hospice physician and the patient's attending physician must certify a patient as 'terminally ill' 'in the first 90 day period' and a hospice physician must certify the patient as terminally ill 'in a subsequent 90- or 60-day period.'" 42 U.S.C.A. § 1395f(a)(7)(West Supp. 2004).

71. Fox et al., *Evaluation of Prognostic Criteria for Determining Hospice Eligibility in Patients with Advanced Lung, Heart, or Liver Disease*, 282 JAMA 1638-45 (1999); Christakis & Lamont, *Extent and Determination of Error in Physicians/Prognoses in Terminally Ill Patients: Prospective Cohort Study*, 320

before death, and 16 percent enter only one week before death.[72] Data has shown that hospice and advance directives could save between 25 percent and 40 percent of healthcare costs during the last month of life, with savings decreasing over the last six months of life.[73]

Only slightly over a quarter of hospices admit children. Cancer is the second leading cause of death in children, after accidents. A study published in February 2000 not only showed that children who die of cancer receive aggressive treatment at the end of life, but that many of them have substantial suffering in the last month of life. Because attempts to control their symptoms are often unsuccessful, greater attention must be paid to palliative care for children who are dying of cancer.[74]

Only the state of Oregon, with its Death with Dignity Act, has legalized physician-assisted suicide (PAS) for terminally ill persons who request it. Although physician-assisted suicide has often been viewed as an alternative to hospice care, 78 percent of persons in Oregon who died by assisted suicide between 1998 and 2001 were enrolled in hospice programs:[75]

BRIT. MED. J. 469-73 (2000); Nat'l Hospice Org., *Report of the Committee on the Medicare Hospice Benefit & End of Life Care: National Hospice Organization*, at 17.

72. ANNALS INT. MED., May 4, 1999, at 772. After finding that "no national study has adequately characterized the U.S. experience of dying," a study was made which concluded in 2004 that "family members of decedents who received care at home with hospice services were more likely to report a favorable dying experience." Teno et al., *Family Perspectives on End-of-Life Care at the Last Place of Care,* JAMA 88 (Jan. 7, 2004).

73. JAMA 1907 (June 26, 1996). A study commissioned by the National Hospice Association in 1995 somehow found that Medicare saves $1.52 for every dollar it spends on hospice care. MED. ECON., Sept. 25, 1995, at 74.

74. 342 N. ENG. J. MED. 326-33 (2000). "Physicians are more comfortable predicting death within six months for patients with cancer than for those with other illnesses." N. ENG. J. MED. 962 (Sept. 23, 1999).

75. N. ENG. J. MED. 582 (Aug. 22, 2002). In Oregon 0.13% and 0.14% of deaths took place by assisted suicide in 2002 and 2003, Ore. Dep't of Human Services, *Sixth Annual Report on Oregon's Death with Dignity Act* 11 (2004). Hedberg et al., *Five Years of Legal Physician-Assisted Suicide in Oregon*, 348 N. ENG. J. MED. 961 (2003): "In Oregon, many hospice nurses and social workers have provided care for a patient who requested assistance with suicide. They rated desire for control as a very important reason for these requests." *Id.* "In some instances they may need to continue providing care for the patient until

> Oregon nurses reported that some hospice patients choose to hasten death by stopping food and fluids, even though physician-assisted suicide is legal in Oregon, and that the quality of the process of dying for most of these patients is good. . . . However, patients who voluntarily chose to refuse food and fluids were, on average, almost a decade older and were more likely to have neurologic disease than were patients who chose physician-assisted suicide. . . . [It was concluded] on the basis of reports by nurses that patients in hospice care who voluntarily choose to refuse food and fluids are elderly, no longer find meaning in living, and usually die a "good" death within two weeks after stopping food and fluids.[76]

Physicians are asked to confirm that the patient is expected to die within six months if the disease were to follow its typical course. Because a significant proportion of the patients who elect the Medicare hospice benefit have outlived that prediction,[77] this has been an argument against the prohibition of long-term care coverage by Medicare. The U.S. Department of Health and Human Services, which oversees Medicare, found that hospices in Puerto Rico, Florida, and elsewhere were admitting patients who were not terminally ill and thus not eligible for Medicare. Some remained at hospice three or four years.[78]

Many individuals involved with care of the dying advocate expanding access to hospice care for persons with advanced lung, heart, or liver disease. However, data indicated that for such seriously ill hospitalized patients, recommended clinical prediction criteria are not effective in identifying a population with a survival prognosis of 6 months or less. It has been written that "70

---

arrangements are made for a different physician." HALEY, TASK FORCE TO IMPROVE THE CARE OF TERMINALLY ILL OREGONIANS, THE DEATH WITH DIGNITY ACT: A GUIDE BOOK FOR HEALTHCARE PROFESSIONALS (1998).

76. Ganzini, *Nurses' Experiences with Hospice Patients Who Refuse Food and Fluids to Hasten Death*, N. ENG. J. MED. 359, ch. 13, § 3.b (July 24, 2003).

77. CONDIFF, EUTHANASIA IS NOT THE ANSWER: A HOSPICE PHYSICIAN'S VIEW 62 (1992).

78. SCIENTIFIC AM. (May 1997) at 103. "Across the U.S., hospices were asked to turn over volumes of records, and many were told to reimburse the government for care given to patients who didn't meet guidelines. Particularly galling were audits in Puerto Rico, where hospices improperly billed Medicare to take care of patients suffering from chronic ailments such as obesity or arthritis." WALL ST. J., June 5, 2000, at 1 & A18.

percent of patients identified as terminal (in hospice terms, expected to die within six months) were still alive at six months."[79]

In Spanish-speaking countries, death is referred to as "the great equalizer." The years immediately preceding the end of the second millennium have seen some movements toward "equalization" for those who are approaching death within a short period. One of these is called a "compassionate release" program for dying (terminal) prison inmates in New York, California, and several less populous states. Implementation of these programs is proceeding cautiously. Under New York's Medical Parole Law promulgated in 1992, inmates who have been convicted of murder or other serious crimes are statutorily ineligible.[80] Although AIDS-related deaths alone had accounted for 55 percent of deaths during 1997 and 1998, only 13 inmates were approved annually (one release was granted for every 13 deaths).[81] At present, only a few

---

79. JAMA 1671 (Nov. 3, 1999). *See also* Parker-Oliver et al., *Hospice Nursing Home Residents,* 6 J. Palliative Med. 69-75 (2003). In discussing Project SUPPORT, it was noted that "many of these chronically ill patients never experience a time during which they are clearly dying of their disease."

> Many individuals involved with care of the dying advocate expanding access to hospice care for persons with advanced lung, heart, or liver disease. However, data indicated that for such seriously ill hospitalized patients, recommended clinical prediction criteria are not effective in identifying a population with a survival prognosis of 6 months or less.

JAMA 1638-45 (1999). *See also* Killick & Allan, Communication and the Care of People with Dementia (2001).

80. Excluded are those convicted of: "murder in the first degree, murder in the second degree, manslaughter in the first degree, any offense defined in article one hundred thirty of the penal law (sex crimes) or an attempt to commit any of these offenses." N.Y. Exec. Law § 259(1)(a).

81. Beck, *Compassionate Release from New York Prisons*, J.L. Med. & Ethics 216, at 219 (Fall 1999):

> HIV-infected patients are living longer; consequently, many are being released from prison at the completion of their sentences or through regular parole procedures before they become seriously ill. Because treatment regimens have slowed the course of the illness, many more HIV-infected inmates are surviving their prison terms, even if their health is declining.

Id. at 221.

hospices appear to be receiving terminal prison inmates who are being released on such grounds. For example, a Vermont law provides that a terminal inmate who is "unlikely to be physically capable of presenting a danger to society may be released on medical parole to a hospital, hospice, other licensed impatient facility or suitable housing accommodation as specified by the parole board."[82] Early in the third millennium, surveys may show the extent to which this new program is actually able to achieve its ethical goal of providing "humanitarian relief to many without exposing society to any significant risk of harm from these patients."[83]

a) Assistance and Lack of Hospice Assistance in Dying

Hospices generally offer only comfort care. Most, but not all, offer no assistance to die. It has been noted that professionals know that suffering occurs but choose not to listen when the subject of suicide is brought up. In 1997, however, a registered nurse (RN) and a certified hospice nurse in California set forth a new position, which could be adopted early in the twenty-first century; namely that "the philosophies of Hospice and the Hemlock Society can be united to give terminally ill patients what they want, need, and deserve: quality care, compassion, support, and the right to choose."[84]

Some hospices have come part way from a rigid "no assistance" position by having hospice staff call Hemlock to report that a patient is asking for its assistance. The literature is then sent directly to this person.[85] Most people

---

82. Vt. Stat. Ann. tit. 28, § 502a(d) (1999). *See also Code of Federal Regulations* (C.F.R.), whose standard of terminally ill is similar to those federal Social Security regulations that describe the eligibility for Medicare benefits for hospice care (e.g., "the individual has a medical prognosis that his or her life expectancy is six months or less if the illness runs its normal course." 42 C.F.R. § 418.3 (2000).

83. John Beck characterized the New York statute as having "imposed an unworkable standard for eligibility, with a review process that is protracted and unnecessary." But, he added, "With some modifications, however, the program could realize its original purpose." John A. Beck, *Compassionate Release from New York State Prisons: Why Are So Few Getting Out?*, 27 J.L. Med. & Ethics 216, 231 (1999).

84. Timelines (October/November/December 1997), at 5.

85. Larne, Playing God 42 (1996). "On the other hand, Hemlock members sometimes call headquarters and ask for the name of their nearest hospice. We stock informational materials in the hospice for these occasions." *Id.*

desire that hospices go further. A 1996 poll conducted for the National Hospice Organization itself indicated that 50 percent of the 1,007 adults surveyed said physician-assisted suicide should be legal and only 41 percent opposed it.[86] A 1998 study concluded, "Hospice care does not obviate their [resident's] desire for euthanasia or physician-assisted dying (PAS)."[87] An example of how a hospice has complied with such desires was shown by a report of the board of directors of the Hospice of Boulder County in Colorado. As of May 1995, the hospice staff have been permitted to be present when a patient self-administers lethal medication and during the dying process. The patient is to be transferred to another caregiver if there is conscientious objection to this practice.[88]

Prior to the decisive vote of November 4, 1997 (when Oregon voters decided by a 60 percent to 40 percent margin to oppose the repeal of the state's Death with Dignity Act), the Oregon Hospice Association itself had opposed such assistance. Thereafter, however, the decision was made that each hospice would make its own policy in response to the Act. As of 1998, a policy adopted by many hospices allows their employees the option of being at the bedside of a patient during the ingestion of a lethal medication. Other hospices (particularly Catholic ones) do not do so.

good v. Good

No hospice should authorize the Hemlock or any other society to distribute their material or authorize suicides because all such suicides are unethical and should be made unlawful.

v.

Hospices, as institutions providing only "comfort care" for those who are dying, should normally be considered suited to authorize their residents the comfort of "knowing that they can."

Twenty-first Century Approach

1. More children should be admitted to hospices and greater palliative care must be given to the thousands who die of cancer each year to alleviate their suffering.

---

86. Orange County Register, Oct. 4, 1996, at News 4.
87. JAMA 507 (Aug. 12, 1998).
88. Med. Ethics Advisor 43 (April 1996).

2. Steps should be taken to terminate the practice of many hospices of turning into long-term care institutions for patients.
3. Surveys of the compassionate "programs for terminally ill [normally those with less than six months to live] prison inmates [inside certain] hospices should be made early in the third millennium in order to determine the extent to which these programs have balanced humanitarian relief with protection of society."
4. Hospices should be authorized to provide certain information on dying (such as that distributed by the Hemlock Society) when so requested by a still competent patient.

In the anticipation that the great majority of people die peacefully in any event, treatment decisions are sometimes made near the end of life that propel a dying person willy-nilly into a series of worsening miseries from which there is no extrication - surgery of questionable benefit and high complication rate, chemotherapy with severe side effects and uncertain response, and prolonged periods of intensive care beyond the point of futility.

Nuland, *How We Die* (1993), p. 143

I will neither give a deadly drug to anybody if asked for it, nor will I make a suggestion to this effect.

From the Hippocratic Oath

The time is not far off when physician-assisted suicide in justifiable cases will be lawful in enlightened countries.

Derek Humphrey, *Final Exit*

3. The Choice of Physician-Assisted Suicide; Double Effect; Costs of Capital Punishment Procedures

One of the physicians on a hospital ethics committee I served possessed wonderful characteristics for handling his many patients who were dying. He often expressed to others the nuances of end-of-life care. A recent newspaper article quoted him as being "spiritual but not religious." Nevertheless, he has indicated, "We are knocking at death's door. God is beckoning. It is time to step back and recognize God's will." His manner of expressing himself is directly connected with religious ideals of most of his patients.[89] Although the AMA *Code of Medical Ethics* states, "Physician-assisted suicide is fundamentally incompatible with the physician's role as a healer," he was committed to the concept for the terminally ill as "a moral issue that pits him against the U.S. Attorney General, the Catholic Church and many Americans who view assisted suicide as euthanasia." It was noted that "[s]ociety should have very little impact on the dying person's choice. Death, at some point, should be an individual right. Ashcroft [the Attorney General] has 'no right' to override the nation's only law [the law in Oregon] allowing assisted suicide."[90]

The characterization of the legal right of an individual to avoid suffering as a "claim right" with incumbent duties was first developed by Wesley N. Hohfeld in 1913.[91] Just prior to the first century A.D., this had been justifiably claimed by Cleopatra, the Ptolemy Queen of Egypt. At the end of the twentieth century, a right was being claimed for physician-assisted suicide (PAS) by many people in the United States; but it is currently only legally recognized by a single state. The twenty-first century will decide whether this right will also be a characterization of many other states. It has been noted that because physician-assisted death is not available to most terminally ill patients, "some medical experts have suggested voluntary refusal of food and fluids as an alternative. Unlike physician-assisted suicide, the choice to stop eating and drinking is legal throughout the United States, available to competent patients, and does not necessarily require the participation of a physician."[92]

---

89. Orange County Register, Nov. 30, 2001, at Accent 1. On Dr. Ronald Koons, who is the chair of the University of California–Irvine's Medical Ethics Committee. Code of Medical Ethics § 2.211.

90. *Id.* As stated, "Often, the additions made to the Hippocratic Oath are as historically interesting as the deletions." Markel, *"I Swear by Apollo"—On Taking the Hippocratic Oath*, N. Eng. J. Med. 2028 (May 13, 2004).

91. Hohfeld, *Fundamental Legal Conception as Applied in Judicial Reasoning*, 23 Yale L.J. 16, 36-38 (1913).

92. N. Eng. J. Med. 360 (July 24, 2003):

> Some physicians assert that the moral basis for this choice is stronger than that for physician-assisted suicide or euthanasia. Other physicians challenge that assertion by asking whether this behavior is different from suicide, with or without a physician's assistance, and some physicians believe that collaboration with a patient who intends to hasten death is morally impermissible. *Id.*

Physician-assisted suicide (PAS) is sometimes referred to as physician-assisted death (PAD); it exists today in the form of building up a possible lethal dose of a medication at the request of a patient in order to alleviate his pain. Such assistance, euphemistically called "double effect," is discussed below. Dr. Nuland (see his epigraph) quoted a colleague: "When the ultimate end is as inevitable as it now appears to be, the individual has a right to ask his doctor to end it for him"—a single sentence that epitomizes the battle in which many are engaged.[93]

Although the Greek physician Hippocrates pledged the medical profession to the preservation of life (and not for personal gain),[94] at the turn of the third millennium we are living in an age more concerned with death costs, as well as the end desired by a larger number of people. When requested, PAS results in a reduction in future medical costs for the patient; but, when *capital punishment* is sought for a murderer, the dollar *costs* for these procedures rise above those spent for virtually any other crime;[95] this punishment is undergone both with the approval by portions of the Bible (*Gen.* 9:6) and the law in the overwhelming majority of states. In 2005, the Colorado Supreme Court invalidated a death sentence because a juror brought a Bible into the jury room during deliberations. A federal court of appeals reversed a district court's granting of habeas relief to a man whose sentencing judge quoted the Bible as an

93. NULAND, HOW WE DIE 153 (1995).

94. A portion of the Hippocratic Oath states: ". . . to live my life in partnership with him, and if he is in need of money to give him a share of mine."

95. The extra taxpayer expense growing out of the minimum standards for attorneys appointed to represent death penalty defendants in state trial courts has been demonstrated effective January 1, 2003, as part of Rule 4.117 of the California Rules of Court:

> They require that counsel must be appointed only if the court determines the lawyer has demonstrated the skill, knowledge, and proficiency to competently represent the defendant. The attorney also must meet detailed qualifications, including at least 10 years of litigation experience in the field of criminal law, prior experience as lead counsel in specified types of trials and completion of at least 15 hours of capital case defense training.

96. People v. Harlan, 109 P.3d 616. The Colorado Supreme Court allowed the jury to discuss "religious upbringing, education and beliefs." But the court

"additional source."[96] Similarly, the North Carolina Supreme Court refused to overturn a death sentence in which a prosecutor began his argument by anticipating that the defense attorneys would argue that "capital punishment may be somehow contrary to Christian ethics. . . . And they may quote such chapters from the Bible as 'thou shall not kill.'"[97]

Although in 1986 the Supreme Court held that the "Eighth Amendment ban on cruel and unusual punishment prohibits a State from carrying out a sentence of death upon a prisoner who is insane," it added that if he "is cured of his disease, the State is free to execute him."[98] In 2003, the Court held that under certain circumstances forcible psychotropic medication may be used to make a defendant competent to stand trial.[99] Presumably, prosecutors may normally find a physician to help administer such a medication. However, one writer has asked, "Do they obey the ethical obligation to treat him, or the ethical obligation not to be involved in the death penalty?"[100]

Nevertheless, the expense of the alternative of carrying out a life sentence in prison without parole for murderers normally amounts to a greater taxpayer expense,[101] even if he is over 50 years of age. In this connection, it is interesting to consider the assisted suicide that was recently performed by a nonphysician.

---

said "it was improper for a juror to bring the Bible into the jury room to share with other jurors the written . . . texts during deliberations; the texts had not been admitted into evidence or allowed pursuant to the trial court's instructions." However, a dissenting justice wrote, "It is without doubt that a juror may relate passages of Scripture from memory during deliberations, and that such recitation would not even be considered extraneous, much less prejudicial." *See also* Arnett v. Jackson, 393 F.3d 681 (6th Cir. 2005).

97. State v. Haselden, 577 S.E.2d 594 (2003); A.B.A. J., July 2005, at 15. This is a mistranslation of the commandment "Thou shall not murder."

98. Ford v. Wainwright, 477 U.S. 399 (1986), at 409-10 and 425 n.5.

99. Sell v. United States, 539 U.S. 166 (2003). The same year the Court declined to review a claim by a capital convict that he should not be forcibly medicated to be made competent for execution. Singleton v. Norris, 319 F.3d 1018 (Ct. App. 3 [Ark.] 2003), *cert. denied,* 24 S. Ct. 74 (2003).

100. Stone, *Condemned Prisoner Medicated and Executed*, 21:3 Psychiatric Times (March 2004).

101.
> Statistics estimate we'll spend close to $700,000 over the lifetime of a prisoner who is 50 or older. Entire geriatric units are being designed to house our aging inmates; they are essentially nursing homes with bars.

Crier, The Case Against Lawyers 115 (2002).

Huntington Williams, 74, cleaned a revolver that a friend, who was dying of prostate cancer, used to kill himself. He also told the man how to aim the weapon and was charged with second-degree manslaughter. However, he was merely given a special form of probation—accelerated rehabilitation—and will be allowed to have the charge erased from his record if he completes the conditions of probation. He was accused of aiding in the death, not causing it. As a result of this case, there is a bill in the Connecticut General Assembly to allow those who are accused of assisting in such a suicide to be eligible for this form of probation.[102]

A 1995 survey by the American Psychological Association found that 81 percent of the respondents believed in the concept of rational suicide in the presence of an unremittingly hopeless condition, with a decision made of free choice, and the presence of an informed decision-making process.[103] This is comparable to the testimony by Dr. Timothy Quill on April 29, 1996, before the House Committee of the Judiciary, who indicated that 1) the request must come from the patient, who must be mentally competent and fully informed about the alternatives; 2) the patient must be terminally ill, or suffering intolerably in ways that cannot be adequately relieved by palliative measures; 3) an independent second opinion, with expertise in palliative medicine, must verify that the patient meets agreed-upon criteria.[104] The first of these, "informed consent" (discussed in Chapter 5), is where "a substantial amount of pain can't be controlled without a pharmacologically induced sleep that socially approximates death."[105] Pain relief, without such consequences, has resulted in "the existing lack of trust in the ability of physicians to relieve their suffering."[106]

As noted in Chapter 3, we have long maintained a legal separation of church and state. Both a populace and elected representatives often follow traditions, causing them to demonstrate that they do not always adhere to this principle. As of the turn of the third millennium, 38 states have specific stat-

---

102. World Right-to-Die Newsletter, June 2005, at 8.

103. 25:2 Suicide and Life-Threatening Behavior 231 (Summer 1995).

104. Bull. of the N.Y. Acad. of Med. 117 (Summer 1997). At the close of the 20th century, bioethicists opposed to PAS appear to have represented a minority viewpoint among them, *e.g.*, Annas, Some Chronicle: Law, Medicine, and the Market (1998).

105. Jamison, Final Acts of Love 147 (1995).

106. *Id.* at 250.

utes against assisted suicide, eight forbid the practice by common law or by interpretation of the state homicide statute, and four remain silent thereon. As in the case of abortions, federal law also prohibits use of federal funds for PAS.[107] In similar manner, several European countries have made PAS a criminal offense;[108] only in the Netherlands has PAS been upheld by its courts under certain conditions.[109]

By contrast, in China, the most populous nation in the world, there has been no such dualism because for two and one-half millennia Chinese tradition has been more concerned with philosophy than religion. No God existed for Confucius, Tao (most of his time), or Buddha, and much Chinese philosophic tradition resembled secular law. A visit to China in 1998 together with a look at its most recent literature on physician-assisted suicide gave me some of such assurance as of the end of the 20th century. (A discussion of the reasons for this view is set forth in Appendix B.) Contrary to the edict of Rudyard Kipling at the end of the nineteenth century, it seems rather likely that "the twain might possibly meet" in bioethics and much of the law sometime in the twenty-first century.

The cost savings engendered by assisted suicide have never been described as a reasonable effect for its use. Like advance directives, substantial sums are being saved in states where nullification of the law against PAS is occurring. (Nullification is discussed in (b) below.) In 1997, the Supreme Court indicated, "If physician-assisted suicide were permitted, many might resort to it to spare their families the substantial financial burden of end-of-life healthcare costs."[110] A 1998 study estimated that "legalizing physician-assisted suicide and euthanasia would have saved more than $627 million in

107. Hast. Ctr. Rep., January-February 1999. On April 30, 1997, President Clinton signed the Federal Assisted Suicide Funding Restriction Act of 1997.

108. *See* Art. 20, French Penal Code; Act 580, Italian Penal Code; Act 135, Portuguese Penal Code; and Act 115, Swiss Penal Code.

109. Schwartz, *Euthanasia and Assisted Suicide in the Netherlands*, 4 CAMBRIDGE Q. 111-12 (1995).

110. Washington v. Glucksberg, 521 U.S. 702 (1997). As regards a constitutional "right to die," former Chief Justice Rehnquist observed, "Our opinion does not absolutely foreclose such a claim . . . [but it] would have to be quite different from the ones advanced [in this case]."

1995 dollars."[111] A study of elderly outpatients in North Carolina found that "African-Americans were almost three times as likely as white patients to want more treatment (42 percent v. 15 percent)."[112]

As noted in the introduction, a Death with Dignity initiative was enacted following the two-thirds majority of voters in Oregon, where PAS was authorized by the U.S. Supreme Court in 1997. During the next four years, more than 90 terminally ill resident patients (i.e., with less than six months to live) chose to utilize a physician to help hasten their deaths. The median age was 68 and it only applied for "about one to three people every month."[113] The experience of Colleen Rice, who died with the help of lethal drugs obtained legally from her physician, was described as follows:

> Rice initially thought she had worsening asthma. But CT scan revealed she had advanced lung cancer. The tumor was inoperable and was leaking fluid into a lung, making breathing nearly impossible. Rice loved to laugh; but doing that caused pain not even morphine could dull. She became very focused . . . . to make the use of the assisted suicide law . . . [Attorney General] Ashcroft's order to make it illegal for doctors to prescribe medicines to help people die terrified her.[114]

In another case:

> Jenny's mother succumbed at age 50 after a brief illness. Months later, Jenny attended the funeral of her friend's elderly mother. "I'm sorry,"

111. Emanual et al., *What Are the Potential Cost Savings from Legalizing Physician-Assisted Suicide?*, N. Eng. J. Med. 167 (July 16, 1998). Assuming that the number of people choosing such assistance was similar to that in Holland, it would be 2.7% of the deaths or 62,000 per year. Using the figures $10,118 for medical bills for the last month of life for those who die naturally, simple multiplication yields $627,316,000.

112. Similarly, African-American outpatients in Miami were much more likely to want life-prolonging therapy than white patients (37% v. 14%). McKinley et al. found that 37% of African-American cancer patients wanted cardiopulmonary resuscitation when terminally ill, while only 16% of white patients would desire such treatment.

113. Singer & Blacknall, *Negotiating Cross-Cultural Issues at End of Life*, JAMA 2995 (Dec. 19, 2001).

114. Orange County Register, Dec. 26, 2001, at News 25.

she said. "Losing your mother is almost more than a daughter can bear. I understand. I know exactly how you feel."

"No, you don't," her friend said. "Mother suffered all those years, in and out of hospitals. She was miserable, and I'm grateful to see her finally out of her misery. There's no reason to be sorry."[115]

The Australian bioethicist Helga Kuhse explained how two people could rationally reach very different conclusions depending upon their circumstances: *"One person may wish to undergo chemotherapy*, for example, to increase her chances of seeing her first grandchild being born; *another may wish to reject life-extending treatments*, to spare herself and her loved ones a slow and drawn-out dying process. In other words, I am suggesting that one of the best ways of caring for people is to respect their autonomy and their rights."

good v. Good

Anyone who helps another person to commit suicide should be prosecuted for manslaughter; but no doctor should ever administer psychotropic medications to an incompetent person who is subject to prosecution for murder.

v.

A person should only be prosecuted if the person he assists is neither in terminal condition (and undergoing significant physical suffering) nor has actually reasonably requested such help. Further, circumstances frequently arise when psychotropic medications should be administered to incompetent murderers so that they may become competent to stand trial.

Twenty-first Century Approaches

1. State legislatures should consider reduction or repeal of any law requiring the prosecution of individuals who reasonably assist in a suicide at the request of another person who is in terminal condition (within a period of less than 6 months) and undergoing continuous significant physical suffering.
2. State legislatures may wish to prescribe the circumstances under which psychotropic medications may be administered to a murderer who has become mentally incompetent to stand trial.

115. Moody & Arcangel, Life After Loss 83 (2001).

God blessed Noah and his sons and said to them: I will require a reckoning for human life; whoever sheds the blood of a human, by a human shall that person's blood be shed; for in his own image, God made humankind.

Genesis 9:6

a) Some Religious Views on PAS

As of the end of the twentieth century, all scholars agreed with the translation of the Commandment "Thou shalt not murder." (In the original King James Version of the Bible as "Thou shalt not kill" has been corrected in modern editions.)[116] Throughout the twentieth century, killing in self-defense, in wartime, and legal capital punishment have been found consistent with the translation of the Commandment. Most people of the Judeo-Christian religions do not consider the granting of help to end incurable physical suffering as wrong under that Biblical injunction.[117] In spite of the separation of church and state under the law, Roman Catholic bishops in Boston have fought against physician-assisted suicide (PAS) as much as they have continued to fight against the constitutional right to abortion.[118] The British have a similar history, which has been described, with reference their suffering from cancer, by Lord Bertrand Russell in the mid-twentieth century (when physician-assisted suicide was called euthanasia).[119] A 1998 survey showed

---

116. *E.g., see* NEW REV. STAND, *Exodus* 20:13.

117. Childress, *Life, Prolongation of*, WESTMINSTER DICTIONARY OF CHRISTIAN ETHICS 249-50 (Childress & Macquarrie eds., 1986); FLETCHER, MORALS AND MEDICINE 172-210 (1954); INGE, CHRISTIAN ETHICS AND MORAL PROBLEMS (1930); and RASHDALL, 7 THE THEORY OF GOOD AND EVIL: A TREATISE ON MORAL PHILOSOPHY 208-12 (1990). Karl Barth, a European Protestant theologian, stressed the reality of God and "impractical idealism of the Social Gospel," writing that "it is for God and God alone to make end of human life" and that God gives life to us "as an inalienable loan." *The Doctrine of Creation, Part 4, in* BARTH, CHURCH DOGMATICS, VOL. III, B.W. Bromily & T.F. Torrance eds., trans. A.T. Mackay et al. (1961), at 404, 425.

118. L.A. TIMES, March 23, 1996. These bishops budgeted $80,000 to file briefs in federal courts. *See also* KEYES, LIFE, DEATH AND THE LAW, ch. 9.

119. "Some years ago, in the English House of Lords, a bill was introduced to legalize euthanasia in cases of painful and incurable disease. The patient's consent was necessary, as well as several medical certificates." BERTRAND RUSSELL, UNPOPULAR ESSAYS 76 (1950).

"Catholic physicians were least likely, and Jewish physicians or those with no religious affiliation were most likely, to be willing to provide assistance or to have actually done so."[120]

In an unprecedented interfaith partnership, American Roman Catholic bishops and Muslim leaders met in Washington to condemn PAS and submitted a legal brief attacking it in 1996.[121] In 1997, Roman Catholic Cardinal Adam Muida of Detroit, Michigan, asked Catholics to resume observing meatless Fridays as penitence for assisted suicide.[122] In 1998, Anthony Cardinal Beilaqua, head of the Archdiocese of Philadelphia, asked the Senate to pass a bill called the "Lethal Drug Abuse Prevention Act," which attempted to give the Drug Enforcement Administration power to revoke the license of any doctor who intentionally prescribed lethal drugs.[123] In 2004, Pope John II spoke at an International Congress stating that "the moral principle is well known, according to which even the simple doubt of being in the presence of a living person already imposes the obligation of full respect and of abstaining from any act that aims at anticipating the person's death."[124]

However, the Synod of the United Church adopted a contrary statement that endorsed Living Wills, stating: "We believe there comes a time in the course of an irreversible terminal illness when, in the interest of love, mercy and compassion, those who are caring for the patient should say: 'Enough.' We do not believe simply the continuance of mere physical existence is either morally defensible, socially desirable or God's will." Similarly, in 1980, the General Council of the United Methodist Church adopted the following statement: "We recognize the agonizing personal and moral decisions faced by the dying, their physicians, their families, and their friends. Therefore, we assert the right of

---

120. N. Eng. J. Med. 1199 (Apr. 23, 1998). "Six percent have complied with such requests at least once. The prevalence of ever having acceded to a request for a prescription for a lethal dose of medication was . . . 13.5% among New England oncologists in 1994 and 18% among Michigan oncologists in 1993." *Id.*

121. Orange County Register, Nov. 12, 1996, at News 22.

122. Orange County Register, Nov. 15, 1997, at News 23.

123. It is unclear how a federal court could legally revoke a state license. *See also* the volume titled *The Case Against Assisted Suicide: For the Right to End-of-Life Care* (2002).

124. The International Congress on Life-Sustaining Treatments and Vegetative State: Scientific Advances and Ethical Dilemmas, March 20, 2004.

every person to die in dignity, with loving personal care and without efforts to prolong terminal illnesses merely because the technology is available to do so."

good v. Good

Physician-Assisted Suicide (PAS) is murder and hence it should never be legalized.

v.

Legal prohibition of such assistance to the terminally ill (whose suffering is based upon views of compassion) does not violate the concept of a separation of church and state when it is enacted into law by any state.

Twenty-first Century Approach

1. The laws that separate church from state should recognize that compassionate end-of-life physician assistance (PAS) to suffering terminal patients does not constitute murder.
2. Like the law in Oregon, other state legislatures should consider the enactment of laws authorizing PAS (upon the request of terminal patients) to those physicians who will take such action in accordance with the same conditions applicable in that state.

There are two highly controversial vows in the original Hippocratic Oath that we continue to ponder and struggle with as a profession: the pledges never to participate in euthanasia and abortion.

Markel, *New England Journal of Medicine*, May 3, 2004, p. 2026

b) *The Choice of Double Effect (or Euthanasia) by the Use (or Abuse) of Lethal Doses of Pain Medication*

When a competent patient requests more medication to reduce or eliminate his pain, the physician may decide to use (or prescribe) lethal doses. If he were to do this without divulging his knowledge concerning that effect, such treatment would violate the doctrine of informed consent.[125] However,

125. *See* KEYES, LIFE, DEATH AND THE LAW (1995), ch. 5. In 1997, it was determined that patients' families were becoming more willing take part in the medical decisions at the end of life. WEIR, PHYSICIAN ASSISTED SUICIDE. In 2003, this was expanded to show that "[y]ounger people, who frequently are decision makers for elderly relatives, place less emphasis on a patient's age itself than do older people." 29 J. MED. ETHICS 330-36 (2003).

a patient may make an ultimate decision with respect to the acceptance of pain medication that his physician knows will be likely to have such a double effect.

Assuming that the requirement of informed consent has been satisfied, the prescribed dosages of pain medication, up to but not exceeding the amount needed to alleviate the pain of a terminal patient, became a practice in many parts of the country near the dawn of the third millennium. Consider this statement claiming that it may largely satisfy the Catholic position:

> A patient with lung cancer and widespread metastases would be permitted to reject the extraordinary means represented by a mechanical respirator. But what then would the patient's doctor do to combat the patient's pain and anxiety in the face of impending respiratory failure? According to Roman Catholic teaching, the physician would be permitted to give the patient morphine, even though such *treatment could hasten the patient's death*. This is based on the principle of double effect, according to which it is permissible to take an action that has some bad effects, provided that a good effect is intended. That bad effect, however, cannot be used as a means to achieve the good. Thus, the patient with cancer could be given increasing doses of morphine to control the pain and anxiety, even though this treatment may ultimately shorten the patient's life. To start with, *a lethal dose would not be permitted, because* in that case *the intention would be to cause death rather than relieve suffering*. Thus, the Roman Catholic position allows physicians to use effective means to alleviate the suffering of dying patients, but does not sanction direct killing.[126] (Emphasis added.)

In 1997, the U.S. Supreme Court considered whether a New York statute violated the Equal Protection Clause of the Fourteenth Amendment by making it a crime to assist another person in committing suicide. The court attempted to make a distinction from assisted suicide resulting from pain medication, noting that "in some cases, painkilling drugs may hasten a patient's

126. Dr. Robert Misbin, Arguing Euthanasia: The Controversy over Mercy Killing, Assisted Suicide, and the Right to Die 129 (1995), *quoted in* Peck, Dignity of the Soul 214 (1997).

death, but the physician's purpose and intent is, or may be, only to ease his patient's pain."[127]

When a drug is administered with full knowledge that it will cause death, the physician's concurrent purpose to ease pain can no longer be actually distinguished from a concurrent intent to do so by causing death. Any statement to the contrary makes a "metaphysical" distinction, and rarely a a rational one:[128] the proposition that a prescription given with "death foreseen" cannot be turned into an "unintended death."[129] Nevertheless, the AMA *Code of Medical Ethics* (2003 Edition) issued Rule 2.20 stating that physicians have "an obligation to relieve pain and suffering" by "providing effective palliative treatment even though it may *foreseeably hasten death*" (emphasis added). As recently noted by Timothy Quill, although PAS is generally illegal outside of Oregon, "in many parts of the country, the secret practice is quietly tolerated (according to a don't ask, don't tell policy)."[130]

"Providing" such treatment may amount to either PAS or euthanasia: Under 2.211, PAS is "providing the necessary means and/or information to enable the patient to perform the life-ending act" which is "fundamentally incompatible with the physician's role as a healer." Under 2.21, "Euthanasia is the administration of a lethal agent by another person to a patient for the purpose of relieving the patient's intolerable and incurable suffering," which also "is fundamentally incompatible with the physician's role as a healer"

---

127. Vacco v. Quill, 521 U.S. 793, 808 n.12 (1997).

128. For Rabbi Reisner, who does not accept my "double-effect argument," it would be outright murder even if the intent were not to bring about death but the physician knew that the given amount of morphine would do that. . . . If the physician knowingly administers enough morphine to kill a person, then even if she or he does so with the primary intent to reduce pain, the physician would be liable for injuring the patient.

Miller, et al., 132 ANN. INTERNAL MED. 470-75 (2000).

129. DORFF, MATTERS OF LIFE & HEALTH 373 (1998). In 2002 Dr. Peter Singer introduced several guidelines in an attempt to separate double effect to protect physicians from suspicions of euthanasia or mercy killing. Nevertheless, one of these guidelines opposed setting any figure for maximum dosage to prevent a threshold that doctors could cross in trying to assist patients in severe pain.

130. Quill, *Dying and Decision Making—Evolution and End-of-Life Options*, N. ENG. J. MED. 2031 (May 13, 2004).

and heightens the significance of its ethical prohibition. These appear to possibly be inconsistent with 2.20, which includes the "providing effective, palliative treatment even though it may foreseeably hasten death."

On March 20, 2004, Pope John Paul II addressed the participants in the International Congress on Life-Sustaining Treatments and Vegetative State; Scientific and Ethical Dilemmas, in which he condemned PAS, Double Effect, and Euthanasia as moral principles:

> I recall what I wrote in the Encyclical *Evangelium Vitae,* making it clear that ". . . *by euthanasia in* the true and proper sense must be understood *an action or omission which by its very nature and intention brings about death, with the purpose of eliminating all pain";* such an act *is always* "a serious *violation of the law of God, since it is the deliberate and morally unacceptable killing of a human person."*

The opinion of Justice Stevens indicated that "the case for the slippery slope is fairly made out here," and that "legislatures have superior opportunities to obtain the facts necessary for a judgment about the present controversy." From these words the justices were dealing with a subject too difficult to rule upon at the time. The majority of members of the current majority of the Supreme Court are less likely to use "substantive due process" to add a constitutionally required duty to assist suicide in the same manner as was done by its previous majority in certain other cases.[131] Furthermore, a physician who pursues double effect without consideration of patient autonomy would be treating the patient in violation of his need to grant informed consent.[132] Nevertheless, *in dicta,* that is, an opinion which does not completely embody the court's determination, Justice O'Connor's concurring opinion went further than the majority and stated: "A patient who is suffering from a terminal illness and who is

131. Vacco v. Quill, 521 U.S. 793 (1997). Two years later it was noted that "[d]octors' fears of disciplinary action and criminal prosecution are justified." Bourguignon & Marlyn, *Physician-Assisted Suicide: The Supreme Court's Wary Rejection*, U. Toledo L. Rev. 22 (1999).

132. In *Cruzan v. Director, Mo. Dep't of Health*, 497 U.S. 261 (1990) at 270, Chief Justice Rehnquist for the majority concluded that "[t]he logical corollary of the doctrine of informed consent is that the patient generally possesses the right not to consent, that is, to refuse treatment."

experiencing great pain has no legal barriers to obtaining medication from qualified physicians to alleviate that suffering, even to the point of causing unconsciousness and hastening death."[133] To diminish consciousness to halt the experience of pain is also called "terminal sedation." A 2004 survey split the opinions of physicians: Almost half of them (47 percent) agreed "with terminal sedation but not with PAS"[134] (see Double-Effect below). Although they were "likely to draw an ethical line between terminal sedation and assisted suicide," they do not seem to understand how their personal knowledge that the administration of terminal sedation resulting in the patient's death would actually constitute euthanasia—which is currently illegal in all states including Oregon (the only state where PAS is currently legal).

The predictability of heart failure as a "terminal illness" was involved in a 2003 study.[135] The case of state initiatives in Oregon implied that they might also be used to approve a statute resulting from those approaches that could amount to euthanasia by the use of drugs directly administered by the physician.[136] Physicians cannot normally rely directly upon a defense of "double effect" against a statute outlawing PAS after intentionally prescribing a lethal dosage of pain-killing drugs, except in Oregon (discussed below, where the procedures in its initiative are followed).

---

133. *See* Glucksberg v. Quill, p. 2303; Washington v. Glucksberg, 521 U.S. 702, 752 (1997) (O'Connor, J., concurring); Arco v. Quill, 521 U.S. 793, at 808.

134. Kaldjian et al., *Internists' attitudes toward terminal sedation in end of life care*, 30 J. MED. ETHICS 499 (2004).

135. The study found that an "externally validated index may assist clinicians in estimating heart failure mortality risk and in providing quantitative guidance for decision making in heart failure care." Lee et al., *Prediction Mortality Among Patients Hospitalized for Heart Failure*, JAMA 2581 (Nov. 19, 2003). They stated:

> In contrast to anecdotal experience, the heart failure index is an objective stratification of mortality risk. The index could be used as a framework to discuss prognosis and provides evidence to support rational decision making about end-of-life care in heart failure patients. (p. 2586)

136. *See* KEYES, LIFE, DEATH, AND THE LAW ch. 9.

good v. Good

So called "terminal sedation" often results in having a "double effect" where physicians are killing their patients, and this should be specifically made illegal regardless of the alleged excuse that it is intended only for pain relief.

v.

The intentional giving of pain relief drugs while aware that the dosage will cause a dying patient's death with a high degree of certainty should not remain illegal where the terminal patient so consents and needs such a fatal dosage to alleviate pain.

Twenty-first Century Approach

The right of all terminal patients to seek injection of drugs sufficient to hasten the dying process with informed consent by means of known "double effect" pain therapy should be considered by the legislatures of other states under the same tests used by Oregon, which has utilized legally recognized physician-assisted suicide by terminal patients since 1998.

How long must I continue to suffer such intolerable pain? When will I be permitted to accept "sweet peace" or even ask for it?

Sigmund Freud

To take into the air my quiet breathe;
Now more than ever seems it rich to die,
To cease upon the midnight with no pain.

Keats

c) The Choice of PAS and/or Double Effect in Oregon

Pressures from many religious groups opposed to abortion caused Congress to pass the Hyde Amendment in the 1970s prohibiting use of federal funds for abortions.[137] These same groups were active in 1997, when Congress passed the Assisted Suicide Funding Restriction Act,[138] which prohibits the use of federal funds to pay for assisted suicides. Because this restriction did not affect the use of private funds, state funds, or funds derived from insurance companies, the right to physician-assisted suicide (PAS) had to be settled in the courts.

137. *Id.*
138. Pub. L. No. 105-12 (1997).

In July 1997, the Supreme Court in two opinions initially refused to extend a fundamental constitutional right to PAS.[139] The best explanations for these opinions is that the Court's current membership no longer seeks to change thc law through alleged "substantive due process" as it did during the 1960s to the 1980s (when such a process was considered to include substantive as well as procedural matters). For example, in 1965 the Court found a Connecticut statute that prohibited the sale of contraceptives to be unconstitutional by inserting "privacy" into the Constitution as a "penumbra" of the Due Process Clause.[140]

In 1973, the Court held a Texas statute prohibiting abortion unconstitutional with the same grounds[141] and reconfirmed this decision in 1992.[142] Near the end of the twentieth century, the Court generally refused to directly extend this right of privacy; rather it has specifically deferred to state legislatures as being the proper forum for an "extensive and serious evaluation of physician-assisted suicide and other related issues."[143] Four months after its 1997 decision, the Supreme Court did just that by denying certiorari (refusing an appeal) to a decision of the Ninth Circuit Court of Appeals, thus upholding the constitutionality of the Oregon Initiative authorizing PAS. After all, the Tenth Amendment in the Bill of Rights specifically states that "[t]he powers not delegated to the United States by the Constitution, nor prohibited by it to the States, are reserved to the States respectively, or to the people." The Oregon law states:

139. Washington v. Glucksberg, 521 U.S. 702 (1997); Vacco v. Quill, 521 U.S. 793 (1997). Chief Justice Rehnquist stated:

> The decision to commit suicide with the assistance of another may be just as personal and just as profound as the decision to refuse unwanted medical treatment, but it has never enjoyed similar legal protection. Indeed, the two acts are widely and reasonably regarded as quite distinct. (p. 725)

140. Griswold v. Connecticut, 381 U.S. 489 (1965).

141. Roe v. Wade, 410 U.S. 113 (1973) (The Court held a Texas statute prohibiting abortion unconstitutional.).

142. Planned Parenthood of Se. Pa. v. Casey, 505 U.S. 833, 851 (1992).

143. Washington v. Glucksberg, 521 U.S. 702 (1997). Concurring opinion by Justice O'Connor and others.

> An adult who is capable, a resident, and has been determined by the attending physician and consulting physician to be suffering from a terminal disease, and who has voluntarily expressed his or her wish to die, may make a written request for medication for the purpose of ending his or her life in a humane and dignified manner in accordance with [this act].[144]

Anticipating the possibility of such a ruling, some people in Oregon prepared another initiative to repeal the first one; they spent $4 million, including monies raised by the U.S. Catholic Conference, Pat Robertson's Christian Coalition, and the Church of Latter-Day Saints.[145] However, this repeal attempt was defeated by an initiative with a two-to-one margin in November 1997, permitting doctors for the first time to legally prescribe a fatal dose of barbiturates and other drugs to affirmatively consenting terminally ill patients.[146] In January 1998, a Department of Justice team assigned by U.S. Attorney General Janet Reno reviewed Oregon's Death with Dignity Act[147] and concluded that the statute does not cause physicians to violate federal law.[148] In March 1998, an elderly woman with breast cancer and unable to breathe easily became the first known person to utilize Oregon's landmark law.[149]

---

144. Or. Rev. Stat. § 127.805 (1999). *See* Alpers & Lo, *Physician-Assisted Suicide in Oregon: A Bold Experiment*, 274 JAMA 483-87 (1995).

145. Time Lines, October-December 1997, at 1 (Hemlock Society). This result follows the Tenth Amendment to the Constitution that the "powers not delegated to the United States by the Constitution, nor prohibited by it to the States, are reserved to the States [legislatures] respectively, or to the people. [by Initiatives]"

146. L.A. Times, Nov. 5, 1997.

147. Or. Rev. Stat. § 127.800-.897 (1996):

> The Oregon Death with Dignity Act specifies an elaborate procedure consistent with the most rigorous standard of voluntariness. Requirements include two oral requests for lethal medication separated by at least fifteen days, a written request witnessed by two people followed by a fifteen-day waiting period, a determination that the patient is capable of making healthcare decisions, and the opportunity to rescind the request at any time.

Hast. Ctr. Rep., January-February 2003, at 18.

148. J. L. Med. & Ethics 77 (1998).

149. L.A. Times, March 26, 1998. Oregon's physicians wrote 33 legal prescriptions of lethal doses in 1999 and 39 of them in 2000.

The right to PAS is not specifically made available for nonresidents of Oregon. This matter was brought up at a 2004 conference of the World Federation of Right to Die Societies. An Italian delegate requested those nations that have lawful assisted suicide to allow Italian people to travel there for an assisted death: "Belgium, Switzerland and the Netherlands should accept cases from other countries. It would put pressure on the Italian government to change the laws on this." But the president of the Swiss-French right to die society disagreed, noting that "if we accept dying patients from Italy, many others would come, probably one every day. As an organization we would be soon burned out. The cases we handle within our own members are well prepared, we know the people, and that is how we handle it." Nevertheless, he said that he had no problem with the other right-to-die organization in Switzerland, DIGNITAS, helping foreigners and local people because that was within Swiss law.[150]

Although 35 states have banned doctor-assisted suicide, several state legislatures continue to consider bills in Oregon's footsteps. Nevertheless, after the turn of the third millennium, Oregon remains the only state in the union where physicians can legally assist terminal patients who are residents in the state to die.[151] They have been doing so legally since 1998.[152]

Oregon was not the first in the world to do so. In 1995, the Northern Territory of Australia enacted a statute that labeled assisted suicide and euthanasia "medical treatment" (for the first time);[153] several people were assisted

---

150. World Right-to-Die Newsletter, January 2005. The incoming president of the Netherlands "told the conference that laws in the European Union specifically forbade people moving within the community to take advantage of non-criminal laws in other countries. 'If we in Holland accepted Italian people, for instance, we would soon be in trouble with the EU for allowing this.'"

151. Volker, *Methodological Issues Associated with Studying an Illegal Act: Assisted Dying*, 27:2 Advances in Nursing Science 117 (2004); Meisel & Cerminara, The Right to Die 12-35 (3d ed., Aspen Publishers, 2004).

152. Emanuel, *Euthanasia and Physician-Assisted Suicide: A Review of the Empirical Data from the United States*, Arch. Internal Med. (2002); yet few physicians in other states have been prosecuted for engaging in such conduct. Meisel & Cerminara, The Right to Die 12-39 (3d ed. 2004); Reid, Criminal Law 227-35 (6th ed. 2004).

153. N. Eng. J. Med. 326 (Feb. 1, 1996).

thereunder.[154] However, in March 1997, the Australian legislature's federal senators narrowly voted 38:33 to overturn the territorial law.[155]

In February 1998, the Oregon Health Service Commission followed Australia's lead by deciding that delivering lethal doses of prescription drugs should be covered as a "medical service" for the state's 270,000 low-income residents. As its chair noted, "The most discriminatory thing would be not to give this choice to the poor."[156] Most private insurers have stated that they are prepared to cover the costs of lethal prescriptions.[157] The American Medical Association announced its strong opposition to the new Oregon law. However, as one bioethicist put it, "The AMA's intervention backfired."[158]

The Oregon Medical Association (OMA) prepared a non-binding "compliance checklist" to assist physicians in the implementation of the Act.[159] In July 2000, the Oregon authorities reported that no data support the hypothetical view that "vulnerable groups" either request or receive such assistance disproportionately in place of "palliative care."[160] In a study among Oregon's physicians, 36 percent responded that they had been asked by a patient if they were potentially willing to prescribe a lethal medication.[161]

---

154. N.Y. TIMES, Sept. 26, 1996, at A4. The first assist (under a statute) worldwide was an Australian man with prostate cancer who died from a lethal injection administered by Dr. Phillip Nischike at his patient's home in Darwin.

155. ORANGE COUNTY REGISTER, March 25, 1997, at News 21. Thereafter I met with the territorial leader in Darwin, Australia and invited him to the U.S.

156. The current federal law banning federal money from being used to cover PAS does not apply because Oregon covers the service using only state money.

157. ORANGE COUNTY REGISTER, Feb. 27, 1998, at News 9.

158. COLUMBIA DAILY SPECTATOR, Nov. 6, 1997, at 6. The bioethicist was Alexander Capron, a University of Southern California professor of law.

159. Woolfrey, Hast. Ctr. Rep., May-June 1998, at 10. "It may be hoped, though not very confidently, that the more human attitude will in time come to prevail, but so far the omens are not very propitious." LORD BERTRAM RUSSELL, UNPOPULAR ESSAYS (1950).

160. Ganzini, et al., N. ENG. J. MED. 152 (July 13, 2000). In 2003, Ganzini noted that among the hospice nurses in Oregon who were surveyed, nearly twice as many had cared for patients who chose voluntary refusal of food and fluids to hasten death as had cared for patients who chose physician-assisted suicide. N. ENG. J. MED. 325, 359-65 (July 24, 2003).

161. JAMA 2363 (May 9, 2001). Of those willing to write a lethal prescription and who had received a request from a patient, 27% were not confident they

In view of decisions of the Supreme Court upholding the right to PAS when authorized by the vote of the people in any state, and in spite of the literal AMA stand against it, many physicians were urging a change. As one of them put it: "Indeed, if we view physicians fundamentally as relievers of discomfort or disease with health promotion as part of that role, then assistance with suicide is not only compatible with the physician's role but quite possibly an obligation inherent in it." According to this view, what breeds distrust of physicians is not that many of them will dispense lethal agents but the possibility that they will refuse to do so. Patients fear that when they are suffering intolerably, they will be denied the drugs necessary to end their suffering.[162] Studies have shown that some 50 percent of dying patients would like the option of physician-assisted suicide (PAS) to be available for possible future use.[163] A large number of physicians and others in healthcare, seeing such a patient suffering, agreed that this option would be a proper medical role for them.[164] In this connection, it has been noted that physicians die by suicide more frequently than non-physicians.[165]

---

could always determine when a patient had less than six months to live. *Id.* In Oregon, 42 physicians wrote a total of 67 prescriptions for lethal doses of medication in 2003; 39 of these prescription recipients died after ingesting it. During 2003, 37 patients (88%) used pentobarbital as their lethal medication, four patients (10%) used secobarbital, and one (2%) used secobarbital/amobarbital (Tuinal). The prescribing physicians of patients who used PAS during 2003 had been in practice a median of 21.5 years; their specialties included internal medicine (12%), oncology (38%), family medicine (45%), and other (5%). OR. DEP'T OF HUMAN SERVICES, 6TH REP. ON OREGON'S DEATH WITH DIGNITY ACT, March 10, 2004.

162. Orentlicher, N. ENG. J. MED. 664 (Aug. 29, 1996).

163. Bascom & Tolle, JAMA 91 (July 3, 2002).

164. Pratt, *Too Many Physicians: Physician-Assisted Suicide after Gluckberg / Quill*, 9 ALBANY J.L. SCI. & TECH. 161, 200-01 (1999).

165. Hampton, *Experts Address Risk of Physician Suicide*, JAMA 1189 (Sept. 14, 2005). "The chances of dying by suicide are about 70% higher for male physicians than for men in the general population (including other professionals) and between 250% and 400% higher for female physicians than other women. . . . In addition, physicians who attempt suicide (usually with drugs) are more likely than nonphysicians to succeed in killing themselves, which may partly account for the increased rated of death by suicide." *Id.* (citing Schernhammer, N. ENG. J. MED. 2473-76 (2005).

At present, it is sometimes not clear whether a claim of discrimination might be made under the Americans with Disabilities Act (ADA) of 1990[166] if a hospital or hospice in Oregon were to refuse PAS. The report of the House Judiciary Committee expressly stated, "Nothing in the ADA is intended to permit discriminatory treatment on the basis of disability, even when such treatment is rendered under the guise of providing an accommodation service, aid, or benefit to the individual with disability."[167]

The concept of *double effect*, which has become the practice by a large number of physicians—prescribing lethal dosages of medication *to alleviate pain* in terminal patients *upon their request*—is not completely secure. It may become an even more common practice during the twenty-first century. That is, double effect appears to be the prelude to the facilitation of self-administered lethal drugs in order to end a life of suffering by administering the "ultimate dosage." Some physicians often use double effect even though it is far from actually becoming utilized or accepted almost everywhere, assuming the patient has fully given his consent.[168] One such twentieth-century case was contained in scenes from the 1996 film *Schindler's List*, where a doctor and nurse administered lethal injections for a Jewish patient too ill to be moved before the location was destroyed by Nazi troops. For many terminally ill Americans who are suffering intractably but not dependent upon life-sustaining treatment, the denial of a right to help end their lives may constitute a demand that they continue with their suffering.[169] Another court on the West coast was even more emphatic:

---

166. A.D.A. § 501(d), 42 U.S.C. § 1220.

167. H.R. Rep. No. 101-485, pt. 4, at 71-72 (1990). *See also* Mikochok, *Assisted Suicide and Disabled People*, De Paul L. Rev. 947, *citing* 28 C.F.R., pt. 36, App. B at 613 (1996); *cf. id.* § 36.203(c)(2) ("Nothing in the Act or this part authorizes the representative of an individual with a disability to decline food, water, medical treatment, or medical services for that individual.").

168. *See* Boyle, *Medical Ethics and Double Effect: The Case of Terminal Sedation*, 25 Theoretical Med. 51-60 (2004); Quill, Dresser & Brock, *The Rule of Double Effect—A Critique of Its Role in End-of-Life Decision Making*, 337 N. Eng. J. Med. 1768-71 (1997).

169. An appellate court on the East Coast claimed that "the ending of life by [withdrawal of life support] is nothing more nor less than assisted suicide." Quill v. Vacco, 80 F.3d 716 (2d Cir. 1996), *rev'd,* Vacco v. Quill, 521 U.S. 793 (1997).

> That right is strongest when the patient is terminally ill and wishes to hasten death because his remaining days are an unmitigated torture. We see no ethical or constitutionally cognizable difference between a doctor's pulling the plug on a respirator and his prescribing drugs which will permit a terminally ill patient to end his own life.[170]

These statements stand in both reason and experience even though the Supreme Court later declared that, unlike abortion, in the absence of a state statute (such as Oregon's), there was no required constitutional right to demand PAS. They are countered by those who would limit a patient's autonomy by declaring the paramount right of the state (1) to preserve his life regardless of its quality; (2) to prevent suicide (irrespective of the reason therefor); and (3) maintenance of the ethical integrity of physicians "to only heal," which, they claim, overrides the right of the patient.[171] Such declarations may be seen from one point of view as "good." However, as herein discussed with respect to all difficult bioethical matters, the balancing of "good against good" cannot be avoided. Two bioethicists pointed out that "medicine is too complex to be orientated toward a single fundamental goal," healing versus helping. Helping patients achieve a peaceful and dignified death may overlap with healing and become the better among the limited options available.[172]

As of the turn of the third millennium, there had been a somewhat limited examination of the detail of PAS in the United States.[173] A study conducted in 1998 determined that almost 16 percent (over one out of six) oncologists described clearly defined cases of PAS or euthanasia, and half of these indicated that they felt comfort having helped a patient with it. Nevertheless, 40 percent of them feared possible prosecution; they reported that 7.4 percent of their patients were experiencing unremitting pain or had such poor physical func-

---

170. Compassion in Dying v. Washington, 79 F.3d 790 (9th Cir. 1996).

171. *See* Krischer v. McIver, 697 So. 2d 97 (Fla. 1997).

172. Hast. Ctr. Rep., May-June 1995. A 1996 survey showed that only 37% of patients and 44% of the general public thought discussions between patients and physicians on "end-of-life care that would include explicit mention of euthanasia or physician-assisted suicide would reduce patients' trust in the physician." THE LANCET 1808 (June 29, 1996).

173. Tulsky et al., *Univ. of Pa. Center for Bioethics Assisted Suicide Consensus Panel*, 132 ANN. INTERN. MED. 494-99 (2000).

tioning that they could not perform self-care. Because less than one-third of these physicians had the patient initiate repeated requests or consult with a colleague, it was concluded that "the illegal status of euthanasia and PAS may itself prevent adherence to the primary safeguards, and legalization may improve adherence to these safeguards."[174] This would appear to be a strong argument for change among state legislators who are wavering on the issue of legalization of PAS. In 2002, Hawaii was indicated as possibly becoming the second state to allow physician-assisted suicide.[175] Physicians in many states do not favor the provision expressed in the AMA's current *Code of Medical Ethics* originally issued in 1994 claiming that in all cases, "Physician-assisted suicide is fundamentally incompatible with the physician's role as a healer."[176]

The last U.S. Attorney General of the twentieth century, Janet Reno, gave no federal drug enforcement officials the right to prosecute doctors from following the Oregon law. Her successor, John Ashcroft, attempted to reverse that view in the twenty-first century, claiming it violated the Controlled Substances Act, stating, "Physician-assisted suicide typically involves the use of a lethal dose of a combination of drugs, including controlled substances. First, the patient is sedated using either a barbiturate (e.g., sodium pentothal) or an opiate (e.g., morphine). Then, one or more drugs are used to paralyze the muscles and/or to stop the heart. The sedatives involved in these procedures are controlled substances under the CSA [Controlled Substances Act]."[177]

---

174. JAMA 507 (Aug. 12, 1998).

175. The bill was nearly identical in wording to Oregon's, adopted as a voter initiative in 1997 (L.A. Times, May 2, 2002, at A16), as was the initiative introduced in California in the fall of 2005.

176. Rule 2.211, Physician-Assisted Suicide. However, the American Association of Neurological Surgeons (AANS) established a Code of Ethics to evaluate qualifications for membership which, inter alia, states:

> The neurological surgeon shall be the advocate of the terminally ill patient to allow dignity in dying while providing relief of pain and suffering and avoiding unnecessary financial burdens for both patient and family. The lawful wishes of the competent patient shall be respected.

AANS Bulletin 8 (Spring 2002).

177. Memorandum for the Attorney General: Whether physician-assisted suicide serves a "legitimate medical purpose" under the Drug Enforcement Administration's regulations implementing the Controlled Substances Act. Wash-

However, in 2002 a federal judge restrained implementation of the U.S. Attorney General's decision. He issued a permanent injunction stating:

> No provision . . . demonstrates or even suggests that Congress intended to delegate to the Attorney General . . . the authority to decide, as a matter of national policy, a question of such magnitude as whether physician-assisted suicide constitutes a legitimate medical purpose or practice.[178]

On May 26, 2004, the Ninth Circuit Court of Appeals affirmed this ruling;[179] however, the following year the U.S. Supreme Court accepted an appeal thereon. On January 17, 2006, the Supreme Court upheld Oregon's Death with Dignity Act (6:3). It was the first loss for the new Chief Justice John Roberts (who replaced the former Chief Justice Rehnquist in 2005), and he joined Antonin Scalia and Clarence Thomas in a long dissent. Justice Anthony M. Kennedy wrote for the majority, "Congress did not have this far-reaching intent to alter the federal-state balance," and that the "authority claimed by the attorney general is both beyond his expertise and incongruous with the statutory purposes and design."[180]

---

ington, D.C.: Department of Justice Office of Legal Counsel, June 27, 2001. N. Eng. J. Med. 1918 (June 13, 2002). A U.S. district court held that the Oregon law on PAS does not violate the Federal Controlled Substances Act. Oregon v. Ashcroft, No. 011647. The governor of Oregon stated, "Oregonians are satisfied that we can responsibly implement physician aid in dying, and this is an unprecedented federal intrusion on Oregon's ability to regulate the practice of medicine."

178. Orange County Register, April 18, 2002, at News 4.

179. State of Oregon v. Ashcroft, 368 F.3d 1118 (9th Cir. 2004). The court declared: "The attorney general's unilateral attempt to regulate general medical practices historically entrusted to state lawmakers interferes with the democratic debate about physician-assisted suicide." Orange County Register, May 27, 2004. "Since 1998, at least 171 people have used the law to end their lives. Most had cancer."

180. Orange County Register, Jan. 18, 2004, at News 8. Justice Kennedy continued in *Gonzales v. Oregon*, stating that the "law structure and operation presume and rely upon a functioning medical profession regulated under the states' police powers." . . . He concluded that "the law's prescription requirement does not authorize the attorney general to bar dispensing controlled substances

Finally, it was noted in these reports that "the decision [is] one of the biggest expected from the Court this year," and that "the 6-3 ruling could encourage other states to consider copying Oregon's law."

good v. Good

Although no suicides are illegal, inasmuch as all suicides are unethical, physician assistance thereon merely compounds this immorality and it should no longer remain legal in Oregon or elsewhere.

v.

Many dying people suffer unnecessarily and wish to terminate their lives and suffering but are often unable to do so without assistance from their physician. Maintaining illegality of their giving assistance for terminal patients may be considered by some as being the infliction of "cruel and inhuman punishment" as well as a denial of "equal protection of the law." In states where PAS is not yet legal, physicians must currently indicate the need for these patients to seek other choices for their termination.

Twenty-first Century Approach

1. The American Medical Association's negative approach toward terminally ill patients, including Physician-Assisted Suicide (PAS), should be changed by deleting general references to the alleged unethical nature of PAS, and substituting the PAS conditions under the law in the state of Oregon.
2. The Americans with Disabilities Act of 1990 (ADA) might be clarified to indicate that in those states where PAS becomes an authorized treatment (as is now true only for residents of Oregon), its denial there would constitute discrimination.
3. It should be recognized that where competent terminal patients request PAS in any state where PAS has been declared illegal, physicians should discuss this with them (along with the lack of legality with respect to "double effect"), even when the informed consent requirement has been met.

---

for assisted suicide in the face of a state medical regime permitting such conduct. . . . The text and structure of the [law] show that Congress did not have this far-reaching intent to alter the federal-state balance." L.A.Times, Jan. 18, 2006, at A14.

4. State legislators who enact a law authorizing PAS under certain conditions may also consider a provision allowing out-of-state residents who enter that state to receive PAS to do so subject to those conditions.

Now when the bardo of dying dawns upon me,
I will abandon all grasping, yearning, and attachment,
Enter undistracted into clear awareness of the teaching,
And eject my consciousness into the space of unborn Awareness.
As I leave this compound body of flesh and blood
I will know it to be a transitory illusion.

Padma Sambhava, *The Tibetan Book of the Dead*

4. Limitation in Model Statutes; Mandatory Psychiatric Examination

Many patients fear the loss of dignity (personal control) associated with the end stages of fatal illnesses more than death itself. A Washington study so indicated and added, "Notably, neither severe pain nor dyspnea [shortness of breath] was a common patient concern, suggesting that intolerable physical symptoms are not the reason most patients request physician-assisted suicide or euthanasia."[181] Another experienced physician stated: "Some people argue 'Well, if we just treated pain well enough, then there would be no demand for assisted suicide or active euthanasia' I believe *that is not true*; there are things besides pain that are much more difficult to treat adequately."[182] (Emphasis added.)

Model statutes are generally concerned with relief to terminally ill patients. A different approach might also include people who wish to hasten their death because of intractable and unbearable suffering, including people with advanced emphysema (a pathologic accumulation of air in tissues or organs), some forms of cancer, amyotrophic lateral sclerosis (degeneration of neurons), and multiple sclerosis (lesions of white matter in nervous system).[183]

---

181. L.A. Times, April 8, 1996, at B8, *citing* an article in JAMA of that date.

182. Cassel, Physician-Assisted Suicide: Progress or Peril? Birth to Death 211 (1996).

183. J. L. Med. & Ethics 74 (Spring 1998); and Rebecca C. Morgan & D. Dixon Sutherland, *Last Rights? Confronting Physician-Assisted Suicide in Law & Society: Legal Liturgies on Physician-Assisted Suicide*, Harv. J. Leg. 33, No. 1 (1996), at 11; Smith, Forced Exit: The Slippery Slope from Assisted Suicide to Legalized Murder 128 (1997).

One model concluded that such a statute is needed to encourage honesty and communication, and avoid the current deceitful practices.[184]

Other model statutes would subject a patient seeking PAS to a psychiatric examination even when the attending physicians find him competent. Clinical situations do not normally make such a requirement mandatory. The Oregon initiative requires a psychiatric consultation only in cases where the primary physician believes that the patient has a mental disorder affecting his judgment.[185]

However, one model state act would require a professional mental health provider "to evaluate the patient to determine that his decision is fully informed, free of undue influence, and not distorted by depression or any other form of mental illness."[186] The current Diagnostic System's code for psychiatrists specifically categorizes "suicidal ideas" as a criterion of mental disorder.[187] This could lead to a view that a competent suffering patient was actually incompetent because he entertained such ideas, in spite of the fact that many terminally ill patients consider suicide as a more rational choice possible under certain circumstances.[188] Such an intervention might base a

184. N. Eng. J. Med. 1383 (Nov. 5, 1992).

185. Oregon Death with Dignity Act (DWD), ballot measure 16, Nov. 8, 1994, general election:

> The DWD Act defines "incapable" as meaning that "in the opinion of a court or in the opinion of the patient's attending physician or consulting physician, a patient lacks the ability to make and communicate healthcare decisions to healthcare providers, including communication through persons familiar with the patient's manner of communicating if those persons are available." The law clearly gives the attending and consulting physicians the power to decide competency in virtually all cases.

186. Charles H. Baron, et al., *A Model State Act to Authorize and Regulate Physician-Assisted Suicide*, 33 Harv. J. Legis. 1, 1-34 (1996). A study of 819 participants between 66 and 96 years of age found that those who were undergoing the chronic stress of caregiving had a decrease in immunity, greater cardiovascular reactivity, and mortality risk 63% higher than the others did. Schulz & Beach, *Caregiving as a Risk Factor for Mortality*, 282 JAMA 2215-19 (1999).

187. DSM-IV, ICD-10.

188. Blendon et al., *Should Physicians Aid Their Patients in Dying? The Public Perspective*, 269 JAMA 590-91 (1995).

rational decision on a subjective evaluation concerning what a "reasonable" person finds to be inappropriate. This was demonstrated by former requirements for psychiatrists to evaluate women seeking an abortion. Such mandatory approval of psychiatrists was found to be clearly inappropriate.[189] Patients may rationally refuse life-sustaining treatment in most cases without first having a psychiatric evaluation. It has been stated that to require that psychiatrists "should be called in on all requests for physician-assisted suicide (PAS) is illogical."[190]

good v. Good

Any dying patient who requests the assistance of a physician to receive help in dying quickly is obviously in need of psychiatric counseling.

v.

Most dying patients have quite rational reasons for making their competence known to physicians when they request such treatment. Upon receipt of requests for assistance in dying, physicians should only be required to refer the patient to an outside mental expert when, in their judgment, such help is needed.

Twenty-first Century Approach

When state legislatures (other than Oregon's) consider the enactment of statutes providing for physician-assisted suicide (PAS), this should be based on a model of such legislation other than one that requires the patient to undergo a psychiatric evaluation (except in those instances where the attending physicians have reason to believe that the patient may have mental problems the physician is unable to determine).

---

189. Forcing women to get the approval of psychiatrists subjected them to an arbitrary, pseudoscientific, and potentially humiliating experience. Not surprisingly, health grounds were related to their moral views on abortion as well as other demographic factors, such as religion or the number of children the psychiatrists had. Sullivan et al., *Should Psychiatrists Serve as Gatekeepers for Physician-Assisted Suicide?* Hast. Ctr. Rep., July-August 1998, at 28.

190. "The view that death through assisted suicide is a greater harm to patients than death through treatment refusal reflects social ambivalence about suicide rather than clearly distinct clinical situations." *Id.* at 21.

Bad laws are the worst sort of tyranny.

Edmund Burke

All truth passes through three stages: First, it is ridiculed; Second, it is violently opposed; and Third, it is accepted as self-evident.

Arthur Schopenhauer, 19th-century German philosopher

5. Progressive Nullification of Anti-PAS Statutes

a) Civil Disobedience Actions by Physicians and Nurses

Much of the public, and many physicians, favor PAS as of the dawn of the twenty-first century. It has been found that: "52 percent of 1,004 persons in the United States said that they might consider assisted suicide or euthanasia. . . . Many physicians would consider helping a terminally ill patient commit suicide in selected cases."[191] A review of Charles McKhann's volume *A Time to Die: The Place for Physician Assistance*, notes:

> The leading reason a physician would not prescribe opiates in a terminal illness, according to surveys cited by McKhann, is fear of breaking the law or being perceived as breaking the law. In plethora [an excess of blood] polls that he cites, more than 50 percent of physicians were found to favor laws allowing assisted dying and 40 percent to 53 percent would participate if such acts were legal.[192]

As of 1992, it was estimated that some 6,000 deaths per day were in some way assisted.[193] In 1996, the first national survey of nursing practices in euthanasia or assisted suicide was conducted. It involved a national sample of 850 nurses who practiced exclusively in adult critical care units—16 percent of whom

191. ANN. INTERN. MED., Jan.14, 1997, *citing* Cohen et al., *Attitudes Toward Assisted Suicide and Euthanasia Among Physicians in Washington State*, 33 N. ENG. J. MED. 89-94 (1994); Bachman et al., *Attitudes of Michigan Physicians and Public Toward Legalizing Physician-Assisted Suicide and Voluntary Euthanasia*, 334 N. ENG. J. MED. 303-09 (1996); Lee et al., *Legalizing Assisted Suicide – Views of Physicians in Oregon*, 334 N. ENG. J. MED. 310-(1) (1996).

192. Sinnon & Houston, *Physician-Assisted Suicide*, JAMA 540-41 (Jan. 26, 2000).

193. Weir, *The Morality of Physician-Assisted Suicide*, 20 L. MED. & HEALTH CARE 116-26 (1992).

said they had carried out euthanasia at least once, and 35 nurses (4 percent) said they had hastened a patient's death by only pretending to carry out a life-sustaining treatment that had been ordered.[194] At the same time, nurses reported engaging in euthanasia at the request of attending physicians.[195]

Their activity might be regarded as being similar to that of Henry David Thoreau, an American of the ninteenth century, who at the age of 32 developed the principles of passive resistance to state coercion of questionable dictates. He did this as a result of his brief imprisonment for refusing to pay a tax apparently supporting slavery. His essay, "On the Duty of Civil Disobedience," became a classic that was acted upon in the twentieth century by both Mahatma Gandhi in India and Martin Luther King, Jr. in the United States. In his letter from Birmingham City Jail, King declared that the disobedient individual who breaks an unjust law as a form of protest and then "willingly accepts the penalty by staying in jail to arouse the conscience of the community" is "in reality expressing the very highest respect for law."

b) The Prohibition Analogy (Nullification)

When a large number of citizens disregard the law generally, it is called anarchy. However, it is quite different when large-scale disobedience is directed toward a particular law. This is termed "nullification." It occurred when Prohibition (the Volstead Act) became law on January 16, 1920. Thereafter, millions of gallons of alcohol were consumed illegally and as of the year 1927, there were over 27,000 "speakeasies" in the United States. By 1933, nearly 800 Chicago gangsters had been killed in bootleg-related shoot-outs, and Prohibition was repealed. President Coolidge had described it the "greatest social experiment of modern times," but it was almost a total failure. Walter Lippmann, the liberal

---

194. Recurring themes reported by nurses included concern about the overuse of life-sustaining technology, a profound sense of responsibility for the patient's welfare, a desire to relieve suffering, and a desire to overcome the perceived unresponsiveness of physicians toward that suffering.

N. Eng. J. Med. 1274 (May 23, 1996).

195. A nurse commented, "I've often felt that the sign over a critical care unit should read: 'Within Are Often Examples of Man's Inhumanity to His Fellow Man.'"

Id.

columnist whose career began in those years, denounced the "circle of impotence in which we outlaw intolerantly the satisfaction of certain persistent human desires, and then tolerate what we have prohibited."[196] A 2001-2002 study of 43,093 people by the National Institute on Alcohol Abuse and Alcoholism reported that 4.65 percent of the adult population reported alcohol abuse, up from 3.03 percent a decade earlier.[197] It has also been emphasized that the legalization of PAS could be of significance for those who do not actually choose to seek such assistance; the mere existence of this remedy will help many people to endure more disability and pain than they would have done otherwise.[198] As stated by Howard Brody and Frank Miller: "Today's physicians might [legitimately] still conclude that so many things have changed since the time of Hippocrates . . . [that the times] warrant a *reconstruction* of the internal morality so that assisted suicide in certain defined circumstances is permissible."[199]

c) The Example of Dr. Kevorkian

People at the end of the twentieth century have observed the practice of civil disobedience in bioethics as evidenced by the persistence of a doctor who practiced it openly, rather than covertly like so many others described above. The most significant individual example of nullification of laws against physician-assisted suicide (PAS) was that set by Dr. Jack Kevorkian, a retired pathologist.[200] He anticipated the horrendous result uncovered by project

---

196. *Quoted in* BEHR, PROHIBITION 238-39 (1996).

> One Christian surgeon wrote, "If the medical profession accepts a role in active euthanasia, as it has for abortion on demand, the whole medical establishment will suffer severely and eventually collapse." This attempted analogy proved to be inept due to the fact that no such collapse has occurred during the period of more than a third of a century since Supreme Court decision authorized abortions on demand.

SCHEMMER, BETWEEN LIFE AND DEATH 139 (1988).

197. ORANGE COUNTY REGISTER, June 12, 2004, at News 20.

198. CAPLAN, AM I MY BROTHER'S KEEPER? 86 (1997).

199. Brody and Miller, *The Internal Morality of Medicine: Explication and Application to Managed Care*, 23 J. MED. & PHILOS. 397 (1998).

200. When many doctors euphemistically say, "We did everything we could," they don't say, "We put this person through hell before he died." *See* Bok, HARV. ALUM. BULL., Winter 1997, at 19.

SUPPORT, discussed above.[201] He was aware that his record of civil disobedience activities led to some criticism that he may have been too enthusiastic to help certain people.[202] Except in the state of Oregon, a change in statutes has been absent during the period of his portion of the nullification of PAS.

good v. Good

All PAS is not only against traditional religions, but also unethical and illegal in all states except in Oregon; the current progressive nullification of their laws (including prohibiting certain types of "double effect" by euthanasia) should be stopped, and the laws against PAS should be otherwise vigorously enforced

v.

Legitimizing PAS would help many people recognize the current significant increase in nullification of current laws prohibiting PAS (as was done by the 21st Amendment, which repealed the 18th Amendment's prohibition of the sale of "intoxicating liquors").

Twenty-first Century Approach

By recognizing the continuing current partial nullification of the majority of those states' laws outlawing the practice of physician-assisted suicide (which includes "double effect" or euthanasia), when state legislators are legalizing PAS, they should also consider the enactment of the appropriate controls contained in the law of Oregon.

> 201. Physicians misunderstood patients' preferences regarding CPR in 80% of the cases. Furthermore, physicians did not implement patients' refusals of interventions. When patients wanted CPR withheld, a do not resuscitate (DNR) order was never written in about 50% of the cases.

Id. at 18.

202. *E.g.*, the assist was for Judith Curren, who apparently was overweight and depressed but showed no signs of chronic fatigue syndrome, which, according to the coroner, leaves no signs in the bodies of its victims. ORANGE COUNTY REGISTER, Aug. 20, 1996.

And art made tongue-tied by authority,
And folly, doctor-like, controlling skill,
And simple truth miscalled simplicity,
And captive good attending captain ill.
Tir'd with all these, from these would I be gone,
Save that to die, I leave my love alone.

Shakespeare, Sonnet 66

In many circles, the term "death with dignity" *has become synonymous with the* right to assisted suicide and euthanasia, *removing it from its place as a principle of bedside care for patients nearing death. (Emphasis added.)*

Chochnov, "Dignity-Conserving Care,"
JAMA, May 1, 2002, p. 2254

As to diseases, make a habit of two things—to help, or at least, to do no harm.

Hippocrates, *Epidemics*

6. Other Choices of Euthanasia

Euthanasia is derived from the Greek words eu (good) and thanatos (death). A verbal distinction is made between physician-assisted suicide (where the physician, upon request, gives the patient the means to terminate his own life) and euthanasia (where the physician's requested action results in the termination of life). Where PAS remains possible, no vital need may normally exist for a physician to agree to euthanize his patients in the majority of instances. However, where the patient's condition makes it impossible to act upon a request of PAS, euthanasia still remains illegal in the United States and Europe (except in the Netherlands and Belgium, where both became law in February and September 2002, respectively).[203] The debates against PAS simulta-

203. *See* http://www.minjust.nl.8080/c-actual/persbet/ph0715.hrm *and* USA TODAY, Sept. 24, 2002, at 4A. The French code, which was modified in 1995, and the Italian code, which was modified in 1988, have confirmed their prohibitions. Article 20 of the French *Code of Medical Ethics* states that the Euthanasia ruling was announced on May 22, 1997. ORANGE COUNTY REGISTER, at News 27.

neously argue against legalizing euthanasia. For example, in 2004, the late Pope John Paul II stated that "euthanasia *brings about death*, with the purpose of eliminating all pain; such an act is always a serious violation of the law of God, since it is the *deliberate and morally unacceptable killing of a human person.*"[204] The current AMA *Code of Medical Ethics* condemns euthanasia in the same manner it does as PAS; namely, "Euthanasia, regardless of its help, is fundamentally incompatible with the physician's role as a healer."[205] However, when making a correct comparison of the two, their merits are indistinguishable in morals, logic, and philosophy; thus it has been noted "that none of these arguments is sound. For all that, they have shown the case for legalizing active euthanasia is *morally indistinguishable* from the case for legalizing physician-assisted suicide. Fairness requires that if we legalize physician-assisted suicide, then we also make active euthanasia legally available to patients."[206] The difference between PAS and euthanasia was noted as:

> The moral argument in favor of permitting physician assistance in suicide is grounded in the conjunction of two principles: self-determination (or, as bioethicists put it, autonomy) and mercy (or the avoidance of suffering). . . . Because living one's life as one chooses must also include living the very end of one's life as one chooses, the matter of how to die is as fully protected by the principle of self-determination as any other part of one's life. Choosing how to die is part of choosing how to live.[207]

---

204. Pope John Paul II's March 20, 2004 address at the International Congress on Life-Sustaining Treatments and Vegetative State; Scientific Advances and Ethical Dilemmas. Of the four Catholics who are now members of the U.S. Supreme Court, two have generally followed certain things within Catholic rules.

205. Code § 2.21, Euthanasia.

206. Nicholas Dixon, *On the Difference between Physician-Assisted Suicide and Active Euthanasia*, 281:5 Hast. Ctr. Rep. 25-29 (1998).

207. Battin, *Is a Physician Ever Obligated to Help a Patient Die?*, *in* REGULATING HOW WE DIE (Emanuel ed., 1998). Physicians may not intentionally try to promote the death of the patient when alleviating suffering. Hast. Ctr. Rep., July-August 1999, at 47. In 1995, among all deaths reported, 2.5% were due to euthanasia. ROYAL DUTCH MEDICAL ASSOCIATION: VISION ON EUTHANASIA (Utrecht, 1995).

Like the United States, Turkey has a secular Constitution and remains 90 percent Muslim. While the United States is 75 percent Christian, it is also recognized for its substantial diversity of other religions. Euthanasia is regarded as homicide, even though one of the elements of the crime of murder, called "bad intention," may not exist in euthanasia. There has been no lawsuit about such euthanasia in Turkey: "Religious authorities did not approve of some of these procedures, nor totally reject. But society made use of these rights liberally. Euthanasia is considered as one of the subjects of that kind in Turkey. Undoubtedly, fundamentalists will react sharply against the idea of legal acceptance of euthanasia, but their effect will be limited by their political power."[208]

On the other hand, the United States has been quite different. In jury trials on PAS, Jack Kevorkian had always been found to be innocent when he used "good intention" in those cases. However, when he used this same intention in the late 1990s to perform euthanasia in Michigan, he was convicted of second-degree murder and received a sentence of 10 to 25 years in prison. The case involved one Thomas Youk, who was terminally ill with Lou Gehrig's disease, and had requested and received euthanasia by Kevorkian. At trial, Kevorkian showed that he had videotaped himself injecting Youk with lethal doses of potassium chloride and gave the tape to CBS's "60 Minutes." The Michigan Court of Appeals, in affirming his conviction of second-degree murder, stated:

> But for defendant's self-described zealotry, Thomas Youk's death would in all probability not have been the subject of national attention much less a murder trial. . . . Defendant, in what is now apparently something of an afterthought, asks us to conclude that euthanasia is legal and therefore, to reverse his conviction on constitutional grounds. We refuse. Such a holding would be the first step down a very steep and slippery slope.

---

208. Oguz, *Euthanasia in Turkey*, EUBIUS J. ASIAN & JUTEV. BIOETHICS 1770-71 (November 1996). It states:

> Under these conditions, it is important to keep the discussion healthy and alive. Any effort that tends to reach a consensus on this subject will help us in finding a way towards a rational solution for the euthanasia debate in Turkey.

However, as noted above, in 2002 both the Netherlands and Belgium legalized active euthanasia under certain criteria: In April of 2001, the Dutch Parliament, after two decades of debate, adopted the Termination of Life on Request and Assisted Suicide (Review Procedures) Act. The Criminal Code was amended to provide that those acts are not illegal "if committed by a physician who fulfills the due care criteria" requiring that:

a. the physician holds the conviction that the request by the patient was voluntary and well-considered;
b. the physician holds the conviction that the patient's suffering was lasting and unbearable.[209]

Although certain methods of Sigmund Freud, the originator of his psychoanalyis, have been discredited, he requested euthanasia. Having become addicted to nicotine after smoking cigars for 40 years, by 1939 he was suffering from great pain as a result of malignancy in his throat. A Dr. Schav had agreed to Freud's plea "not to forsake me when my time comes." He did not. In 1972, Dr. Schav wrote:

> I informed Anna (Freud's daughter) of our conversation, as Freud had asked. When he was again in agony, I gave him a hypodermic of two centigrams of morphine. He soon felt relief and fell into a peaceful sleep. The expression of pain and suffering was gone. I repeated this dose after about twelve hours. Freud was obviously so close to the end of his reserves that he lapsed into a coma and did not wake up again.[210]

Three things are clear: (1) Freud was both relieved of his pain and died from the morphine overdose and not of his cancer (like a double effect); (2) the overdose was *intended to end* both his suffering and *his life* (In the context of double effect, such treatment to end his life would not only be his alleged "subjective intent" to reduce his pain.); and (3) Dr. Schav's "treatment" given to end Freud's

209. Mendelson & Just, J. L. MED. & ETHICS 137 (Spring 2003); Kimsma & van Leeuwen, *The New Dutch Law on Legalizing Physician-Assisted Death*, 10 CAMBRIDGE Q. HEALTHCARE ETHICS 445-50 (2001).

210. SCHUR M. FREUD, LIVING AND DYING (1972), *quoted in* ARCH. INTERN. MED., July 26, 1999, at 1521.

life was in his patient's best interest. Nevertheless, some physicians continue to deny that anything called "treatment" can have the purpose of ending a patient's life, even when it has been requested by him and may be in his best interests. In the United States, Dr. Schav might remain vulnerable to the charge of having committed a homicide from which he might have been prosecuted under the current law in all states except Oregon.

It has been written, "Freud certainly had a good death. As ideally should be true of planned deaths, Freud's end was purposeful, brief, relatively devoid of suffering, consistent with ego ideals, and allowed for resolution and reconciliation."[211] In 2003, a book called *I Ask the Right to Die* was written by a 22-year-old French paraplegic, Vincent Humbert, calling for the legalization of euthanasia, which "has transfixed the nation."[212] The question arises "why are our terminal citizens being denied the right to call a similar case a 'good death?'"[213] A professor at Sydney Law School in Australia called "direct administration of lethal injections or infusions at the patient's request (active voluntary euthanasia, or AVE)." He also indicated that "medicine is shot through with clinical scenarios that require the exercise of a discretion: PAS/AVE is hardly novel."[214] He also concluded, "While my

---

211. Arch. Int. Med., July 26, 1999, at 1522; Phillips, Darwin's Worms 105 (1999).

212. He was partially paralyzed, mute and blind following an accident. Two days before an overdose of sedatives was put into his body intravenously, he wrote:

> Then, so that you understand me better, so that the debate about euthanasia finally reaches another level, so that this world and this act are no longer a taboo subject, so that we no longer let live lucid people like me who want to put an end to their own suffering, I wanted to write this book that I will never read.

Orange County Register, Sept. 27, 2003.

213. *See* http://news.excite.com/news/r/000303/18/health-Iwa, March 12, 2000. National Ethics is a government-sponsored organization whose members are drawn from a broad cross-section of society, including medics and churchmen. *Id.* at 1523.

214. Magnusson, *Underground Euthanasia and the Harm Minimization Debate*, J. L. Med. & Ethics 486 (Fall 2004). After referring to the "safeguards under the Dutch PAS/AVE regime," he adds, "Curiously, euthanasia opponents seem more concerned about statistics in the Netherlands than with the implications of underground euthanasia in their home countries." (p. 490).

own initial assessment leads me to favor legalization of PAS/AVE under a statutory protocol, it is clear that if we take the problem of underground euthanasia seriously, the harm-minimization debate presents many more opportunities for constructive engagement."[215] It has been written:

> People who want an early, peaceful death for themselves or their relatives are not rejecting or denigrating the sanctity of life; on the contrary, they believe that a quicker death shows more respect for life than a protracted one. Once again, both sides in the debate about euthanasia share a concern for life's sanctity; they are united by that value, and disagree only about how best to interpret and respect it.[216]

Others have indicated that "[t]he medical profession in the United States has reflected our society's unwillingness to accept death as part of life and to face it with some humility . . . how sterile and technological our profession has become."[217]

As noted above, sections of the AMA *Code of Medical Ethics* stated that physician-assisted suicide (2.211) and euthanasia (2.21) are "fundamentally incompatible with a physician's role as a healer." However, in spite of these statements, the code adds that "an attending physician may enter a do-not-resuscitate (DNR) order" if he of she "cannot be expected either to restore cardiac or respiratory function to the patient" (2.21) and "it is not unethical to discontinue all means of life-sustaining medical treatment in accordance with a proper substituted judgment in best interest analysis." (2.22) ("There is no

---

215. *Id.* at 493. He also cites Fitzgerald, who argues that:

> Lack of resources and supports, along with the isolation and exclusion many people with disability experience, compound the intensity of their experience of disability, so much so that euthanasia becomes the most attractive option available. All these things override the intrinsic will of a person with a disability to live.

Fitzgerald, *Bioethics, Disability and Death: Uncovering Cultural Bias in the Euthanasia Debate*, in DISABILITY, DIVERS-ABILITY, AND LEGAL CHANGE 267-81, at 272 (Jones & Marks eds., 1999).

216. WORKIN, LIFE'S DOMINION 238 (1995).

217. C. Cassel & D. Meier, *Morals and Moralism in the Debate over Euthanasia and Assisted Suicide*, 323 N. ENG. J. MED. 750, 751 (1990).

difference between withdrawing and withholding life-sustaining treatment.") Thus, totally non-healing actions are actually ethical for physicians to perform on millions on patients.

In 2004, R.S. Magnusson, of the University of Sydney, Australia, pointed to studies suggesting that from 4 percent to 10 percent of doctors have intentionally helped a patient to die. Accordingly, he argued that legalizing physician-assisted suicide and active euthanasia (PAS/AE) may be safer, and therefore a preferable policy alternative, to prohibition. [218]

good v. Good

Euthanasia, the direct ending of the lives of self-asserting dying patients by members of medical professionals who are known as "healers," should never be legalized

v.

Because PAS may become treatment "in the best interest" of dying patients who make the request (as is legal in Oregon), euthanasia should be recognized as a legal "alternate treatment" in those instances that qualify for all elements required for physician-assisted suicide, where it is also impractical or impossible for the patient to assist himself.

Twenty-first Century Approach

1. State legislators should also consider the conditions needed in an appropriate law respecting actions taken in both physician-assisted suicide and euthanasia for terminal patients.
2. While euthanasia statutes should have the same protections against abuse as those applicable for PAS, euthanasia should also contain one concerning the impracticability or impossibility of the patient's ability to carry out the final portion of the treatment by himself.

218. Magnusson, *Euthanasia: above ground, below ground*, J. MED. ETHICS 442-46 (2004).

Part IV

Bioethics and Future Somatic and Germline Gene Therapy

It is not the strongest of the species that survive, nor the most intelligent, but the one most responsive to change.

Charles Darwin

When you come right down to it, the reason that we did this job is because it was an organic necessity. If you are a scientist, you cannot stop such a thing. If you are a scientist, you believe that it is good to find out how the world works; that it is good to find what the realities are; that it is good to turn over to humanity at large the greatest possible power to control the world.

Robert Oppenheimer

Transcendence is the only real alternative to extinction.

Václav Havel, former president of Czechoslovakia, July 4, 1994.
http://www.worldtrans.org/whole/havelspeech.html

At the end of the next century the Human Genome Project will probably be seen as a second Manhattan Project (the project that developed the atomic bomb), and the scientists involved in it as meddling Dr. Frankensteins—or else the mapping of the human genetic heritage will be regarded as the greatest advance made in the history of our species since the first primates walked upright.

Daniel Cohan and J. Craig Venter 17 (2000).

15

Bioethics on Genetics Superseding the Human Genome Project

1. *The Human Genome Project (HGP)*
2. *Genetically Modified Crops*
3. *Genetic Approaches to Disease; Confidentiality, Non-disease and Privacy (HIPPA)*
4. *The Absence of Ethics and Morals in Patents on Genetically Engineered Living Things*
5. *The Possible Future of Inheritable Genetic Interventions and Changes in Homo Sapiens*
   a) Clinical Trials on Genetic Research
   b) Distinctions between Somatic and Germline Therapy, Ethical Limits
   c) Somatic Gene Therapy (SGT) and Some Needs for Mandatory Testing
   d) The Publicity Needed for Recombinant DNR Research by Public or Private Entities
   e) Germline Gene Therapy (GLGT)

New discoveries about rules governing how genes encode proteins have revealed nature's sophisticated "programming" for protecting life from catastrophic errors while accelerating evolution.

Freeland and Hurst, "Evolution Encoded,"
Scientific American (April 2004) p. 84

The structure of the genotype is perhaps the most challenging remaining problem of evolutionary biology.

Ernst May, *What Evolution Is* (2001)

Along with Bach's music, Shakespeare's sonnets and the Apollo Space Programme, the Human Genome Project is one of those achievements of the human spirit that makes me proud to be human.

Richard Dawkins, 2001

We are living in exciting times, for science is in the process of shattering old myths and rewriting a fundamental belief of human civilization.

Lipton, *Biology of Belief* (2005) p.17

1. The Human Genome Project

In his nineteenth-century volume *Origin of Species* (1859), Darwin wrote, "The laws governing inheritance are for the most part unknown. At whatever period of life a peculiarity first appears, it tends to reappear in the offspring at a corresponding age, though sometimes earlier." (p.31) In his *The Descent of Man* (1871), he referred to the "indelible stamp of his body origin." In the twenty-first century, this will be significantly modified as a result of knowledge attained with respect to genetics—very different matters. He was unaware of the reasons for such bioethical issues due to his lack of any contact with Gregor Mendel, the Augustinian monk of his century who founded the different field.[1] Although they were contemporaries, Mendel's

---

1. *See* Tudge, The Impact of the Gene 10 (2000). These different matters were commenced by the monk Gregor Mendel's work *Experiments in Plant Hybridization* (known as his laws of genetics), whose "Introductory Remarks" in 1865 state:

> Experience of artificial fertilization, such as is effected with ornamental plants in order to obtain new variations in color, has led to the experiments which will be here discussed. The striking regularity with which the same hybrid forms always reappeared whenever fertilization took place between the same species induced further experiments to be undertaken, the object of which was to follow up the developments of the hybrids in their progeny. . . . That, so far, no generally applicable law governing the formation and development of hybrids has been successfully formulated can hardly be won

research on heredity was not reconfirmed until the dawn of the twentieth century when it was combined with Darwin's concept of evolution by natural selection.

In 1944 Oswald Avery, the bacteriologist at the Rockefeller Institute in New York, and his colleagues showed that DNA could be transferred from one strain of bacteria to another to mark the identification of DNA as the genetic material. A Nobel Prize committee mistake was its failure to award him the prize.[2] Erwin Chargaff, fascinated by Avery's report, used "chromatography to analyze DNA from a variety of organisms, [and] showed that the four bases appeared in proportions that, although constant in all tissues of a given species, varied widely from one species to another."[3] On February 28, 1953, Watson and Crick determined the double-helix structure of DNA, and understated, "It has not escaped our notice that the specific pairing we have postulated immediately suggests a

---

> dered at by anyone who is acquainted with the extent of the task, and can appreciate the difficulties with which experiments of this class have to contend. A final decision can only be arrived at when we shall have before us the results of detailed experiments made on plants belonging to the most diverse orders. . . . It requires indeed some courage to undertake a labor of such far-reaching extent; this appears, however, to be *the only right way by which we can finally reach the solution of a question the importance* of which cannot be overestimated *in connection with the story of the evolution* of organic forms. (Emphasis added.)

Blumberg, *Table of Contents for Mendel's Paper*, at Mendel Web, http://www.mendelweb.org/Mwpaptoc.html

2. In two decades of experiments with pneumococcus, step by step he isolated and identified what was called "the transforming principle." This was a substance extracted from one true-breeding variety of pneumococcus, which, when added to a live culture of a second, separate, distinct variety, would produce in the next generation bacteria of the first type. In 1944, Avery and colleagues reported the magisterial, definitive proof, the summation of those years of experiments, that *the transforming principle is not protein but DNA*.

Judson, N. Eng. J. Med. 1712 (April 24, 2003).

3. Chargaff wrote in 1949 . . . that in all the DNAs he had analyzed, "[t]he molar ratios . . . of adenine to thymine and of guanine to cytosine were not far from 1." *Id.*

possible copying mechanism for the genetic material."[4] By the early 1960s when they received Nobel Prizes, Watson entered the field of indirect performance of scientific research. However, Crick continued work on the DNA code-making for genes. In 1958 he expressed the central dogma of molecular biology as being "the precise determination of sequence."[5]

It took several decades to ascertain the structure of ribonucleic acid (RNA), an acid found in all living cells[6] (see Chapter 2). Others went well beyond; the duration of genetics and heredity constitute the several billion years' history of our genome. Unlike books written in English using 26 letters with words of variable length, genomes are written using only four letters.[7] In the 1980s, Francois Jacob and Jacques Monod of France discovered regulatory genes that have the ability to switch other genes acting as "inducers" from outside the cell.[8]

In 1992, the World Medical Association's Declaration on The Human Genome Project reported, "One can state that the understanding of all human

---

4. WATSON, J.D., THE DOUBLE HELIX (1968). "Science seldom proceeds in the straightforward, logical manner . . . its steps are often very human events in which personalities and cultural traditions play major roles." In 1998 the U.N. General Assembly recognized the Universal Declaration on the Human Genome and Human Rights, setting forth a close relationship between science and the future of humanity. The representatives of science at a UNESCO meeting on "Bioethics: International Implications" in October 2001 agreed.

5. Judson, N. ENG. J. MED. 1712 (April 24, 2003).

6. MCETHENY, WATSON AND DNA 78 (2003). This world is dominated by three types of RNA: ribosomal, transfer, and messenger (p. 116). About this time, it was noted that eugenicist scientists were expected to promote the creation of a biologically perfect human organ. This has been called "playing God":

> Princeton theologian Paul Ramsey coined the phrase "playing God" to capture the threats posed to humanity by the scientists' work on human genetic engineering. Among others, Leon Kass, now chair of President Bush's Council on Bioethics, joined in criticism of the rapid pace of and lack of moral reflection on human genetic engineering.

Keranen & Parker's review of Evans, *Playing God? Human Genetic Engineering and Rationalization of Public Bioethical Debate* (2002), JAMA 1313 (March 12, 2003).

7. "A,C, G, and T (which stand for adenine, cytosine, guanine, and thymine) are written on long chains of sugar and phosphate called DNA molecules to which the bases are attached as side rungs. Each chromosome is one pair of (very) long DNA molecules." RIDLEY, GENOME.

8. MONOD, CHANCE AND NECESSITY (1997).

biology is enclosed in the identification of 50,000 to 100,000 genes in the human body's chromosome." It added that "the accessibility to personal genetic data should be allowed only with the patient's informed consent." However, in 2003, the count was estimated to be well under 30,000 genes.[9] It was then that the discovery of all human genes was finally achieved and the use of such items for bioethical pursuit commenced. Nevertheless, in 2005 it was emphasized that "we can no longer use genes to explain why humans are at the top of the evolutionary ladder. It turns out there is not much difference in the total number of genes found in humans and those found in primitive organisms."[10]

The Human Genome Project (HGP) officially started between 1988 and 1990 as a 15-year international research program to characterize the genomes of humans and other organisms by sequencing genes. The International Human Genome Sequencing Consortium (IHGSC) was composed of more than 2,000 scientists at 20 institutions in 6 countries; the first five U.S. centers received grants for producing the particular sequences.[11] James Watson, one of the two scientists who had discovered the structure of DNA, became chief of the $3 billion HGP in October 1988. He made it into an international project on the "infrastructure" side of the genome centers, rather than only on genetic diseases.[12] Watson later noted that "opposition to the Human Genome Program is seen as the most vis-

9. PENNISI, SCIENCE 1040-41 (2003). *See also* the former National Research Council, *Evaluating Human Genetic Diversity* (1997) (which reported that the various participants "had quite different perceptions of the intent of the project") *and* Pennisi, *NRC Oks Long-Delayed Survey of Human Genome Diversity*, 278 SCIENCE 568 (1997).

10. LIPTON, THE BIOLOGY OF BELIEF 64 (2005).

11. (1) Whitehead Institute for Biomedical Research, Center for Genome Research, Cambridge, Mass., U.S.A.; (2) The Sanger Centre, Cambridge, U.K.; (3) Washington University Genome Sequencing Center, St. Louis, Mo., U.S.A.; (4) U.S. Dep't of Energy Joint Genome Institute, Walnut Creek, Cal., U.S.A.; (5) Baylor College of Medicine Human Genome Sequencing Center, Houston, Texas, U.S.A.

DENNIS & GALLAGHER, THE HUMAN GENOME (2001). The HGP was not to work on any human germ-line engineering projects. MCGEE, THE PERFECT BABY: PARENTHOOD IN THE NEW WORLD OF CLONING AND GENETICS (2000).

12. M. ETHENY, WATSON AND DNA 258 (2003). Of the three co-discoverers, James Watson is the sole survivor. Francis Crick and Maurice Wilkins, who also received Nobel Prizes, both died in 2004.

ible symbol of the evolutionary biology/genetics-based approach to human existence." The opponents argue that "we would be making better use of our monies in trying to improve the economic and moral environments of humans as opposed to the finding of genes that they believe will only marginally affect our health and social behavior." Then Watson added:

> With time, however, the truth must emerge that monies so spent have effectively no chance of rolling back the fundamental tragedies that come from genetic disease. Therefore, I believe that over the next several decades *we shall witness an ever-growing consensus* that humans have *the right to terminate the* lives of *genetically unhealthy fetuses*[13] (see Chapter 10).

Scientists in two groups of nonprofit international enterprises (including France, Germany, and China) continued into the third millennium sequencing human and other genomes. In the United States, funds for the Project came to the nonprofit enterprises from the National Institutes of Health. On the other hand, the Celera Genomics Group, a for-profit enterprise, was run independently by Craig Venter, who stated, "We're selling information about the vast universe of molecular medicine." The White House administration in the year 2000 declared that it supported the patenting of "non-natural" genetic sequences.[14] Private research may be "guaranteed information" and "confidential business information."[15] In 2001 both a private and a public organization announced completion of their first analysis of the human genome.[16] Yet, that draft included

---

13. WATSON, ETHICAL IMPLICATIONS OF THE HUMAN GENOME PROJECTS 176 (1994).

14. FERRIS PRESTON, THE COST OF AMERICAN SCIENCE WRITING 64 (2001). Over 150 prisms, the fastest DNA sequences, were used by Callera. The 2000 White House Administration declared that it supported the patenting of genes. *Id.* at 86. But the U.S. Patent and Trademark Office receives an incentive to grant patents by collecting fees of applicants, which rose to $423 million in 1993. A patent examiner receives a salary bonus in accordance with the number of his patent allowances; his rejections may be challenged and delay a possible bonus. Merges, *As Many as Six Impossible Patents Before Breakfast: Property Rights for Business Concepts and Patent System Reform*, 14 BERKELEY TECH. L.J. 607-10 (1999).

15. Kaplan, Annotation, *What Constitutes "Trade Secrets and Commercial or Financial Information Obtained from Person and Privileged or Confidential," Exempt from Disclosure Under Freedom of Information Act (5 U.S.C.A. § 552 (b)(4))*, 139 A.L.R. Fed. 225 (1997). In order to apply, a manufacturer must stamp the documents "confidential" and not waive the privilege. 40 C.F.R. §§ 2.201-2.320 (2003).

16. RYAN, DARWIN: BLIND SPOT 123 (2002).

thousands of gaps in the long sequence of DNA base pairs. By 2003 all but 400 of those gaps had been closed among the 3 billion pairs of DNA chemicals within 24 chromosomes. The genes that control growth, functions, and aging are made of specific sequences, a small change in which can cause disease. In 2005 it was noted how "the old notion of one gene/one protein has gone by the board: It is now clear that many genes can make more than one protein."[17]

Some funds were also provided to address the Ethical, Legal and Social Implications (ELSI) portion of HGP that might represent the world's largest bioethics program.[18] It remains one of the foundations for biometrically minded people to understand the genome throughout the twenty-first century and beyond. However, virtually all applicable portions of the AMA *Code of Medical Ethics on Genetics* were issued in the 1980s and 1990s. In 2003, a study of 31 of 42 cystic fibrosis care centers by local Institutional Review Boards (IRBs) showed the existence of "uneven human subjects projection." Thus, a need for reform was "underscored by the ever-increasing rate of genetic discoveries facilitated by the Human Genome Project and the unprecedented opportunity to assess the relevance of genetic variation to public health."[19] However, even in 2004, it has been argued that relevant data to address the issue on the evolutionary differences between humans and chimpanzees "have only recently started to become available."[20]

---

17. "Regulatory proteins, RNA, noncoding bits of DNA, even chemical and structural alterations of the genome itself control how, where, and when genes are expressed. Figuring out how all these elements work together to choreograph gene expression is one of the central challenges facing biologists." SCIENCE 80 (July 1, 2005).

18. It received 3%-5% of the annual HGP budgets of the U.S. Department of Energy and the National Human Genome Research Institute. Information about ELSI can be obtained at www.nhgri.nih.gov or www.ornl.gov/hgmis. The Advisory Committee on Health Research prepared a report that dealt with genomics for world health, including intellectual property. WEATHERALL, BROCK & CHEE, GENOMICS AND WORLD HEALTH (2002).

19. McWilliams et al., *Problematic Variation in Local Institutional Review of a Multicenter Genetic Epidemiology Study*, JAMA 360 (July 16, 2003). "The need for reform appears necessary if we are to reap the full potential of the Human Genome Project." *Id.* at 365.

20. Khaitovich et al., *Parallel Patterns of Evolution in the Genomes and Transcriptomes of Humans and Chimpanzees*, SCIENCE, Sept. 16, 2005 (citing Preuss, Caceres, Oldham, Geschwind, NAT'L REV. GENET. 5, 850 (2004)).

Early in the third millennium, most scientists seek further understanding of the role genes play in the diagnosis and treatment of disease. Normally, counseling is to help the patient make informed decisions. The AMA *Code of Medical Ethics* states: "Physicians engaged in genetic counseling are ethically obligated to provide prospective parents with the basis for an informed decision for childbearing, which should include reasons for and against testing." As regards prospective parents, it specifies that "physicians should avoid the imposition of their personal moral values."[21] But in 1996 it was noted there are "professionals who deal with the emotional repercussions of genetic information" and that "physician-geneticists prefer to leave that messy sort of work to non-MD counselors."[22]

In 1883, the term "eugenics" was coined by Francis Galton to improve Homo sapiens by means of selective marriages.[23] In the twentieth century, eugenics (derived from the Greek "wellborn") was feared as a possibly undesired attempt to not only seek particular physical traits, but also to improve IQs or personalities and organ formation.[24] "These traits are controlled by hundreds—perhaps thousands—of genes interacting with multiple factors in the environment."[25] It is presently quite difficult to select genetic combinations needed for this purpose; however it could become technically possible and better viewed by bioethicists of future generations.[26] For example,

21. Section 2.12, *Genetic Counseling*: "Genetic selection refers to the abortion or discard of a fetus or pre-embryo with a genetic abnormality."

22. MURRAY, THE HUMAN GENOME PROJECT 212 (1996); BOSK, ALL GOD'S MISTAKES: GENETIC COUNSELING IN A PEDIATRIC HOSPITAL (1992).

23. SMITH, THE EUGENIC ASSAULT ON AMERICA (1993).

24. Horgan, *Eugenics Revisited*, SCIENTIFIC AM. 122 (June 1993); INSTITUTE OF MEDICINE, NATIONAL ACADEMY OF SCIENCES, ASSESSING GENETIC RISKS: IMPLICATIONS FOR HEALTH AND SOCIAL POLICY (1993). *See* Section 5(e)2 below.

25. NICHOLS, HUMAN GENE THERAPY 12 (1988). The study of the molecular mechanisms by which environment controls is called *epigenetics*. "In recent years, molecular biology has shown that the genome is far more fluid and responsive to the *environment* than previously supposed. It has also shown that information can be transmitted to descendants in ways other than through the base sequence of DNA." EVA JABLONKA (a philosopher) & MARION LAMB (a biologist), EPIGENETIC INHERITANCE AND EVOLUTION—THE LAMARCKIAN DIMENSION (1995).

26. Beyond doubt, some of the genes are more directly involved than others. Those that produce particular neurotransmitters or help organize particular synapses are clearly among the most closely involved.

in 2003, the role of DNA as the "master molecule" determining phenotypic expression was altered: "The discovery of RNA interference (RNAI) and a vast new world of tiny regulatory RNAs had profoundly changed the way we think about gene regulation in animals, plants and fungi."[27]

The HGP has been mapping human genes and the sequences of their bases by DNA. In 2001 Craig Venter noted that "the sequence is only the beginning."[28] This mapping, called genomics, could result in consideration of disease reduction for healthcare, as well as eugenics to increase human qualities. Toward the end of the twentieth century, the latter became available to some degree by the use of sperm banks (as described in Chapter 8). In lieu of being limited to sperm and egg banks and certain diseases, genetics may be able to indicate how to produce better athletes, musicians, and possibly how IQ could be enhanced. Most of these will require bioethical decisions to be made. It has been expressed that "we are products of a long and complex process of genetic and cultural evolution and gene-culture coevolution. Our past was complicated, so is our present, and so will be our future."[29] In 2003

> It would not be at all surprising, then, if in time biologists discovered particular alleles, or combinations of alleles, that are closely associated with high IQ and that seem to have a particularly positive effect on IQ. Such alleles surely will be identified as geneticists map and sequence the genomes of more and more people.

JUDGE, THE IMPACT OF THE GENE 284 (2001).

27. MATZKE & MATZKE, 301 SCIENCE 1060-61 (2003). As the late Paul Silverman, from the University of California-Irvine, informed me:

> We know now that the nucleotide sequence that is identified as a gene is an important but only one part of a complex process in genetic expression. From nuclear heterchromatin to cytoplasmic RNAs, epigenetic factors and micro RNAs regulate gene expression. The extra-nuclear regulatory factors are particularly susceptible to environmental influences at the cellular and tissue levels. Some preliminary research has revealed that diet and vitamins can alter the expression outcome. In fact, some studies have suggested that a single "gene" may be expressed in over 100,000 different variants depending on environmental factors.

28. With his computers, he was "able to sequence the human genome at 3 to 5 percent of the cost of the public project, and substantially faster as well." NAAM, MORE THAN HUMAN 140 (2005).

29. In 2020, we will know how to build a functioning cell capable of free-

one writer noted, "the ultimate in diagnostic technologies may soon be available to tell me about my future."[30]

That same year, the HGP *itself* was well on course to being "essentially complete." It was the fiftieth anniversary of the publication of the double helix and Dr. Watson felt justified in hailing "the possibilities of a marvelous new weapon in our fight against disease and, even more, a whole new era in our understanding of how organisms are put together and how they operate, and of what it is that sets us apart biologically from other species—what, in other words, makes us human."[31] Others have pointed out that "the darker parts of the ge-

---

> living existence . . . but there still will be unanswered questions about it. We will be virtually no closer than we are today to the mysteries of such true "emergent" properties as intelligence in complex multicellular organisms.

Drell & Adamson, *Fast-forward to 2020: what to expect in molecular medicine*, *at* http://www.ornl.gov/hgmis/medicine/tnty.html (2002). *Quoted in* JAMA 2477 (Nov. 20, 2002). *See also* Ehrlich, Human Natures 331 (2000).

30. This technology will speedily decipher my entire genome in only about 30 minutes. With only a pinprick of blood, GenEngine will take out some DNA and shoot it through the system to get a complete genome scan. At the moment, the instruments can read 200,000 base pairs of DNA, but by 2006, they will read three billion base pairs, says Eugene Chang, the inventor of GenEngine and the founder and director of U.S. Genomics, which is developing GenEngine. It should be able to decipher a genome in less than 10 minutes. Interestingly, this is the same length of time my physician, under my current health plan, is allotted for an entire physical exam. And the cost is expected to be under $1,000.

Nelkin, N.Y.U., Hast. Ctr. Rep. (January–February 2003) at 9 (speaking on GenEngine patent No. 6,355,429).

31. Watson, DNA, The Secret of Life 193 (2003).

> On April 14, 2003, scientists announced to the world that they had finished sequencing the human genome—logging the three billion pairs of DNA nucleotides that describe how to make a human being. But finding all the working genes amid the junk in the sequence remains a further challenge, as does gaining a better understanding of how and when genes are activated and how their instructions affect the behavior of the protein molecules they describe. So it is no wonder that Human Genome Project . . . called the accomplishment only "the end of the beginning."

Freeland & Hurst, *Evolution Encoded*, Scientific Am. 84 (April 2004).

nome are still perceived dimly"; Carmen Sapienza asserted that the HGP was just the beginning of the job. We now need to produce a similar description of "the epigenetic landscape,"[32] and David More stated: "Notwithstanding all of the hoopla attending the Human Genome Project, the finished product will leave us much closer to the beginning than to the end of our quest to understand the origins of our biological and psychological traits."[33] For example, alternative splicing, which involves the removal of introns around an RNA gene copy before it has been transplanted into protein, occurs in about one-half of all human genes.[34] Following the HGP, the National Institutes of Health (NIH) sponsored the cataloging of all the proteins in a human body.[35] NIH's Protein Structure

---

32. SCIENTIFIC AM. 115 (Nov. 2003). This new work is seen as "the primary means to make medical sense out of the map of the human genome, and bioinformatics capabilities continue to expand exponentially. Consequently, the demand for access to human biological samples and medical information has never been greater." J.D. Malinowski, J. L. MED. & ETHICS 54 (Spring 2005).

33. MOORE, THE DEPENDENT GENE 208 (2001). Francis S. Collins called HGP accomplishment only "the end of the beginning." SCIENTIFIC AM. 84 (April 2004). For example, it has been noted that the introns within genes and the long stretches of intergenic DNA between genes "were immediately assumed to be evolutionary junk," and that this "may well go on as one of the biggest mistakes in the history of molecular biology." "Scientists are discovering many noncoding genes that give rise to surprisingly active RNAs, including varieties that can silence or regulate conventional genes." SCIENTIFIC AM. 49-51 (Nov. 2003).

34. Modrek & Lee, *A Genomic View of Alternative Splicing*, NATURE GENETIC 13-19 (2002).

> In the case of the Down syndrome cell-adhesion molecule (the Dscam gene), it was found that if it were to be actually spliced into all possible combinations of exons, it could produce 38,016 different kinds of protein from that gene.

Schmucker, et al., *Drosophila Dscam is an Axon Guidance Receptor Exhibiting Extraordinary Molecular Diversity*, CELL 671-84 (2000).

35. Hopkin, *The Post-Genome Project*, SCIENTIFIC AM. 16 (2001); HUSI & GRANT, PROTEOMICS OF THE NERVOUS SYSTEM 259-66 (2001).

> All the work is making clear that buried in DNA sequence is a regulatory code akin to the genetic code "but infinitely more complicated," says Michael Eisen, a computational biologist at Lawrence Berkeley National Library in California.

Searching for the Genome's Second Code, 306 SCIENCE (Oct. 22, 2004).

Initiative (PSI) catalogue hopes to develop mass-production of structures applicable to biology and medicine. A three-dimensional protein shape is called a fold.[36] Ronald Dworkin has argued that with our future genetic engineering, we should not be afraid to "play God." This was because we should be able to challenge those who claim it would actually be mistaken traditional moral thinking, but we "have found that some of the most basic presuppositions of our values are mistaken."[37] Although "genetic engineering shifts the boundary between chance and choice, between what is simply given and what can be had through human endeavour," Dworkin rightly observes that "because of all the potential benefits of the new genetics, we should not allow this fear to stop us moving forward."[38] Many of the reasons for such benefits follow along with appropriate cautions during its progress.

The word "cause" is an altar to an unknown god.

William James

2. Genetically Modified Crops

Changes in crop breeding have been debated between (a) "the biotechnology industry, which is seeking new products and markets," and (b) "the consumer and environmental movements, which perceive only impending medical and ecological disasters." The major problem concerns "the proper balance between benefit and risk that we must find to safely exploit these new techniques."[39] In the four years preceding the year 2002, more than two-thirds of

36. 307 SCIENCE 1554 (March 11, 2005).

37. Richter, *The Fear of Playing God*, *in* ETHICAL ISSUES IN THE NEW GENETICS (2003), ch. 5, *citing* Dworkin's paper titled *Playing God*, published in PROSPECT magazine.

38. *Id.* at 48. Richter added, "I agree, but argue that we should do so slowly, for both obvious practical reasons and less obvious philosophical reasons."

39. WINSTON, TRAVELS IN THE GENETICALLY MODIFIED ZONE (2002), *Prologue*:

> From 1996 to 2001, the global acreage of genetically modified crops grew from 4 to 125 million acres, mostly in the United States (68 percent of all GM crops planted worldwide) but also in Argentina (17 percent), Canada (7 percent), and China (1%). . . . Biotechnologists emphasize the pragmatic benefits of genetically modified crops, but objections expressed by environmentalists accentuate the serious risks.

Id. at 5.

the global acreage of genetically modified crops (over two million acres) were planted in the United States, where up to 50 percent of American farm income comes from subsidies paid directly to farmers by the government.[40] Recombinant DNA (rDNA) technology constitutes the removal of DNA from one species and recombining it with the genes from another species. It has been pointed out that "almost all (rDNA) research today is conducted with few restrictions." The National Institutes of Health (the biosafety protocols) mandate for all (rDNA) research stated, "The hazards may be guessed at, speculated about, or voted upon, but they cannot be known absolutely."[41] However, in 2005 it was reported that "genetically modified (GM) crops most popular with growers—corn, soybeans, and cotton—are high-yielding and possess other good qualities. For peak performance, GM crops require heavy applications of synthetic fertilizers, insecticides, and weed killers—chemicals that can damage the environment."[42] A majority of Americans (up to 58 percent) say they "would prefer to buy GM-free food if they had the choice."[43]

Greenpeace, an international organization with headquarters in Amsterdam, has one quarter-million members in the United States. Kellogg's, one of the

---

40. *Id.* at 15, 30. In 2004, the British government approved the commercial cultivation of a type of genetically modified corn. However, its planting would be under strict rules. ORANGE COUNTY REGISTER, March 10, 2004, at News 13. In the United States, *see* 21 C.F.R. § 50, "Protection of Human Subjects."

41. *Id.* at 41.

> Because the regulatory genes that tend to be of most interest to genetic engineers for modifying physiology are deleterious unless regulated precisely, the record to date is largely one of impaired, rather than improved, plant performance (e.g., Chen & Murata, 2002). This suggests that genetic engineers will have to work extremely hard if they are to move agronomic traits further than breeders, now empowered by candidate gene and marker-aided selection, are able to do.

Strauss, *Regulating Biotechnology as though Gene Function Mattered*, BIOSCIENCE 454 (May 2003), *citing* CHEN & MURATA, 5 CURRENT OPINION IN PLANT BIOLOGY 250-57 (2002).

42. THE FUTURIST 39 (September-October 2005). "Farmers who want to help save the environment are disappointed by present-day GM crops, because they believe the crops' dependence on chemical treatments is a serious deficiency."

43. *Id.* at 41.

U.S. cereal giants, promised its European customers not to use ingredients from genetically modified crops, but continues to use such crops in the United States.[44] However, Greenpeace complained about Gerber's Mixed Cereal for Babies' unlabeled genetically altered products, so the company promised to stop using genetically engineered corn and soybeans.[45] Corn crops, which are notoriously hard to control and known for spontaneous cross-pollination, were shown to have been contaminated with StarLink corn. This was genetically engineered for animal feed to produce its own pesticide, which contained a potential allergen—a substance capable of inducing allergy or hypersensitivity. In 2003, Aventis CropScience, the North Carolina-based company that developed StarLink, paid $110 million to settle a class action with farmers who claimed the contamination caused lower corn prices.[46]

In November 2001, the United States initially refused to sign an international agreement governing the genetic code of plants. This treaty is intended to preserve plant diversity and food supplies by safeguarding the genetic materials and is administered under the auspices of the United Nations Food and Agriculture Organization (FAO) in Rome. One year later (November 2002) the United States changed and became the 76th country to sign this treaty—although Congress has not ratified it. In July 2000, the European Parliament had approved legislation requiring strict labels for food

---

44. *Id.* at 114-15, 240. As of the present time, "there is no evidence that Kellogg's products or other transgenic food have actually harmed anyone." "George W. Bush accused Europe on Wednesday of aggravating hunger in Africa with restrictive trade policies on genetically modified food. As a result, African nations have been reluctant to try growing genetically modified crops 'for fear their products will be shut out of European markets.' The European regulatory system for genetically modified foods complies with trade rules." ORANGE COUNTY REGISTER, May 22, 2003, at News 22.

45. PLANT BIOLOGY 251. "The Food and Drug Administration (FDA) cannot guarantee the safety of genetically modified foods because it is unable to obtain all scientific data from biotech companies." *See* http://new.yahoo.com/new?tmpl=story2&cid=571&ncid=571&e=4&u=/nm/20030108/hl_nm/food_biotech_dc (last visited Jan. 8, 2003).

46. McDonough, *Theory Crops Up in Cases Against Makers of Genetically Modified Seeds*, A.B.A. J. 16 (August 2003), *in re* StarLink Corn Products Liab. Litig., 212 F. Supp. 2d 828 (N.D. Ill. 2003).

and feed made with genetically altered ingredients—a move that was "pilloried by American farmers."[47] In December 2002, the European Union environment ministers agreed upon new labeling and traceability rules for food and livestock feed derived from genetically modified organisms.

The Monsanto Company developed and patented genetically altered soybean seeds. However, when Monsanto sold them, its technology agreement stated that buyers could use the seed "only for a single season" and could not "save any seed produced from this crop for replanting." Likewise, a purchase contract for the seeds stated that buyers can use the seed "only for a single season" and not "save any seed produced from this crop for replanting." In 1998, a farmer breached this clause following his payment of $24,000 for 1,000 bags of seeds, including a "technology fee" of $6.50 a bag. Monsanto sued him in a federal court in St. Louis and was awarded $780,000 for what was called the planting of saved seed piracy.[48]

In 2004 Associate Law Professor Rebecca Bratspies studied "bioharming—a process in which plants are genetically modified so that they endogenously produce specialty pharmaceutical or industrial proteins." She concluded: "The integrity of the U.S. food supply is at stake. With a clear likelihood of contamination and no evidence that these crops are safe to consume, even in low levels, permitting commercial development of these products cannot be

---

47. The Bush administration criticized the legislation Wednesday, saying it would be burdensome for food producers, could prejudice consumers against genetically modified food and become a barrier to free trade. Hayden Milberg, a trade specialist for the National Corn Growers Association, estimated that American corn farmers lose $300 million a year.

ORANGE COUNTY REGISTER, July 3, 2003, at News 32.

48. ORANGE COUNTY REGISTER (from N.Y. TIMES), Nov. 2, 2003, at News 19.

About 45 percent of the corn produced in the United States is genetically modified. Mexico imports about 5.6 million tons of American corn a year. Thirty percent to 50 percent of that corn is genetically modified. A study released Monday by the trilateral Commission for Environmental Cooperation suggested that Mexico mill biotech corn as soon as it crosses the border to ensure that it is used only in animal feed and not planted.

ORANGE COUNTY REGISTER, Nov. 11, 2004, at News 23.

justified as scientifically sound or as a reasonable assessment of the costs and benefits."[49]

On the other side, it was pointed out that Monsanto's genetically engineered crops, called Roundup Ready, is a widely used agricultural herbicide. In Trefil's *Human Nature* (2004), he stated:

> A useful gene from one species is inserted into the DNA of another, and the host acquires the advantages that natural selection has conferred on the donor (p. 166). . . . As of this writing, there is no firmly documented case of a gene from a genetically modified plant being transferred into the wild, but it's hard to imagine that it won't happen sooner or later (p. 168). . . . A field is sprayed with Roundup before planting to get rid of existing plants, then sprayed as needed to control *the weeds* while the crop is growing. The result: the soil is never *plowed* up and erosion is reduced to a minimum (p. 176). . . .

In this connection it has been pointed out that "the biggest use of genetic engineering in the United States today, however, is not to be found in medicine, but in agriculture."[50] To this was added that in traditional molecular biology, genes can be passed "from parent to offspring," but "not from one species to another." However, "there is now good evidence that what scientists call 'gene flow' occurs. In general, the flow is between cultivated plants and their wild relatives."[51] In 2004,

---

49. Bratspies, *Consuming (F)ears of Corn: Public Health and Biopharming*, AM J.L. & MED. 404 (2004). "The health risks from consuming these adulterated foods could be considerable. Nevertheless, industry and governmental regulators have failed to impose obvious biological controls that would greatly protect the public's safety, while still permitting exploitation of this technology." *Id.* at 175. *See also* Somerville & Bonetta, *Plants as Factories for Technical Materials*, PLANT PHYSIOLOGY (2001).

50. "Between 2001 and 2002, for example, the amount of genetically modified agricultural planting increased by 13 percent. Today, almost three-quarters of the soybeans, a third of the corn, and almost a quarter of the cotton grown in this country has been genetically modified." TREFIL, HUMAN NATURE 166 (2004).

51. *Id.* at 100. In 2004, the Environmental Protection Agency reported that bentgrass, modified to resist Roundup weedkiller, pollinated grasses as far as 13 miles away. "But the fear is that the modified crops can spread the bacterial gene to surrounding plants, creating hard-to-kill 'super-weeds,' or perhaps having unforeseen effects on other crops." ORANGE COUNTY REGISTER, Sept. 22, 2004, at News 25.

it was noted that "genetic variety in wild relatives of our modern crops is only beginning to be explored. In rice and tomatoes, an estimated 80 percent of each species' total allelic diversity remains untapped."[52] In 2005 it was pointed out how "a human gene that Japanese researchers have inserted into rice enables the plant to break down a portfolio of chemicals now used on farms to kill weeds."[53] The U.N. Food and Agriculture Organization (FAO) that year determined that certain "food plants currently on the market pose no risk to human health, although multiple-gene transformations now in development need further study."[54] However, *Science* magazine has described the "routes into Europe of the western corn rootworm (*Diabrotica virgifera virgifera,* WCR), the most destructive pest of corn in the United States."[55] Transgenes (genes artificially inserted into an organism's native genome) have been planted by farmers in some 65 million hectares, or 5 percent of total arable land area. "The United States and Argentina together account for nearly 90 percent of transgenic production, with Canada and China accounting for most of the remainder."[56]

Nevertheless, in 2005, it was also noted that "there is no central planning and policy-framing agency overseeing genetic engineering in the United States,

52. "So without the technology to use genes or chromosomal loci as molecular markers, scientists will find identifying some of these desirable traits or moving them into modern crops nearly impossible." Groff & Salmoron, plant geneticists who led a U.S. team in producing a draft sequence of the rice genome. SCIENTIFIC AM. 47 (August 2004).

53. SCIENCE NEWS, April 16, 2005, at 246.

54. *The Green Gene Revolution*, SCIENTIFIC AM. (August 2004) at 8. *See also* Watanabe, et al., *Japanese Controversies over Transgenic Crop Regulation*, SCIENCE 1572 (Sept. 10, 2004). As of 2004, 55 nations have ratified the Treaty on Plant Genetic Resources for Food and Agriculture, including Canada, the United Kingdom, Germany, Finland, Italy, India, Egypt, Spain, and Sweden. SCIENCE NEWS, July 17, 2004, at 45.

55. SCIENCE, Nov. 11, 2005, at 992. "Our finding that there have been at least three independent transatlantic introductions of WCR suggests that incursions from North America are chronic. Prevention of future WCR invasions will require action against multiple invasion routes, which have apparently been used repeatedly and are potentially predictable." *Id.*

56. Statistical databases, Food and Agricultural Organization, 2004. KLEIMAN, KINCHY & HANDESON, CONTROVERSIES IN SCIENCE & TECHNOLOGY 119 (2005). "Transgenes are contaminating nontransgenic seed stocks at a low but measurable level." *Id.* at 121.

although other nations have such agencies."[57] Although GM crops became an industry in 1996, in January 2006 it was reported that "6 out of 10 adults in the U.S. are unaware that [genetically modified] crops exist, while only 25 percent realize that GM foods have been on sale in the U.S. for the past 10 years. . . . When further informed about the pros and cons of GM foods, more than 6 out of 10 respondents said they would oppose the importation of GM crops into the U.S."[58] Accordingly, the suggestion has been made that "it is past time to establish a Genetic Science Commission" (GSC), similar in importance to the former Atomic Energy Commission (AEC).[59]

good v. Good

Without specific proof that genetically modified organisms (GMOs) are harmful, no labeling should be required for such a product.

v.

No customer should be denied the right to know of a genetically modified product by not so labeling it, and consideration should be given to establishing a Genetic Science Commission (GSC).

Twenty-first Century Approach

1. Transgenes (genetically modified organisms) should be required to maintain a label on distribution so that consideration may be given to whether proper choices are being made.
2. Congress may currently wish to consider whether to establish a Genetic Science Commission.

---

57. THE FUTURIST 40 (September-October 2005). Currently, three U.S. federal agencies share the responsibility for reviewing genetically modified organisms: the U.S. Department of Agriculture, the Environmental Protection Agency, and the Food and Drug Administration.

58. NEW SCIENTIST 10 (Jan. 21, 2006).

59. *Id.* In this connection, *see* 42 U.S.C. § 287 A-2, "Biomedical and beahvioral research facilities, (a) modernization and construction facilities." "The American biotech industry wants to expedite development, not delay it. While some geneticists think today's regulatory system is adequate, others see the need for tightening regulations. How extensive the tightening should be is an open question." (p. 42) The article is by Clifton E. Anderson, University of Idaho professor emeritus.

3. Genetic Approaches to Disease; Confidentiality; Non-Disease and Privacy

Comparison of the Manhattan Project (for atomic weapons) to the HGP did not cease in the twentieth century. The size of those multinational debates concerning nuclear reactors and nuclear weapons during that century are becoming matched by those on the bioethical matters during the third millennium. Nevertheless, there are relatively few legislative bioethical mandates,[60] even though they were considered and funded by HGP's component, to support discussion on ethical, legal, and social implications of new genetic knowledge.[61]

The 1984 congressional Office of Technology Assessment (OTA) report, which was called *Human Gene Therapy—A Background Paper,* concluded: "Civic, religious, scientific, and medical groups have all accepted, in principle, the appropriateness of gene therapy of somatic cells in humans for specific genetic diseases. . . . Whether somatic cell gene therapy will become a practical medical technology will thus depend on its safety and efficacy, and the major question is when to begin clinical trials, not whether to begin them at all."[62] Private genetic testing is not normally subject to

---

60. Happily, the proposal to control recombinant-DNA research through legislation never got close to enactment. And when anti-DNA doomsday scenarios failed to materialize, even the modestly restrictive governmental regulations began to wither away. In retrospect, recombinant DNA may rank as the safest revolutionary technology ever developed; to my knowledge, not one illness, much less fatality, has been caused by a genetically manipulated organism.

WATSON, VIEWPOINT: ALL FOR THE GOOD—WHY GENETIC ENGINEERING MUST SOLDIER ON (1999).

61. WATSON, GENES AND POLITICS 202 (1997). Watson added:

> The problems are with us now, independent of the genome program, but they will be associated with it. We should devote real money to discussing these issues. People are afraid of genetic knowledge instead of seeing it as an opportunity. We cannot make policy . . . all we can do is educate.

M. ETHENY, WATSON AND DNA 259 (2003).

62. *See* NICHOLS, HUMAN GENE THERAPY 12 (1988). "Public discussion about potential abuses of new technologies is an important part of the decision-making process in modern society." *Id.*

regulation by the Food and Drug Administration (FDA) at this time. In 2002 it was pointed out that "genomics will revolutionize healthcare," and "healthcare will be on the leading edge in the struggle to extract real human value from the discoveries to come."[63]

Parts of this struggle are in current research, and it has been indicated that Genomics will be to the twenty-first century what infectious disease was to the twentieth century for public health.[64] However, it has been emphasized that "differences in social structures, lifestyle, and environment account for much larger proportions of disease than genetic differences."[65] In this connection, the current AMA *Code of Medical Ethics* still continues the unique position it adopted in 1994; namely, "it would not be ethical to engage in [genetic] selection on the basis of non-disease-related characteristics or traits."[66] In the 1980s the initial publicized attempt at gene therapy led to fear.[67] The current AMA *Code of Medical Ethics* (since 1996) specifically states, "Efforts to enhance 'desirable' characteristics through the insertion of a modified or additional gene, or efforts to 'improve' complex human traits—the eugenic development of offspring—are contrary not only to the ethical tradition of medicine, but also to the egalitarian values of our society."[68] That same year the Code at-

63. Carlson & Stemeling, The Terrible Gift (2002), *Prologue*.

64. J.L. Med. & Ethics, Supp. to Vol. 30:3, Fall 2002. It has been noted that dogs share about 75 percent of their genes with humans. "Although all (dogs) are members of the same species, this selective breeding has resulted in amazing variations. There are some 400 breeds of dogs." Orange County Register, May 22, 2004, at News 20 (reporting on work at Fred Hutchinson Cancer Research Center in Seattle).

65. Holtzman & Marteau, *Will Genetics Revolutionize Medicine?* 343:2 N. Eng. J. Med. 141 (July 13, 2000). "In our rush to fit medicine with the genetic mantle, we are losing sight of other possibilities for improving the public health." *Id.*

66. Code § 2.12. *Also see* Mehlman, *The Law of Above Averages: Leveling the New Genetic Enhancement Playing Field*, 85 Iowa L. Rev. 517, 527-28 (2000).

67. M.D. Prchal, *Delivery on Demand—A New Era of Gene Therapy?*, N. Eng. J. Med. 1282 (March 27, 2003).

68. Because of the potential for abuse, genetic manipulation to affect non-disease traits *may never be acceptable* and *perhaps should never be pursued.* If it is ever allowed, at least three conditions would have to be met before it could be deemed ethically acceptable: (1) there would have to be a clear and meaningful benefit to the person, (2) there would have to be *no trade-off with other characteristics or traits*, and (3) *all citizens would have to have equal access* to the genetic technology,

tempted to place limits on genetic testing of children: "Genetic testing for carrier status should be deferred until either the child reaches maturity, the child needs to make reproductive decisions, or in the case of children too immature to make their own reproductive decisions," such testing "decisions need to be made for the child."[69] For example, based upon her Ashkenazi Jewish background, a woman underwent genetic disease tests, 10 tests, all of which were negative, before becoming pregnant. However, after the birth of her son she discovered he was deaf and learned that a simple blood test could have alerted her to that possibility before conception.[70]

In 2003 it was expressed that "substantial progress has been made in overcoming these obstacles." Although manipulation of the genome to improve it remains difficult, some religious people have stated that the gaining of control over our evolution was not "more troubling than discussing any other human capacity to alter the natural world."[71] However, in 2000, a committee of the

---

> irrespective of income or other socioeconomic characteristics. (Emphasis added.)

Code § 2.11.

69. (1) When a child is at risk for a genetic condition for which preventive or other therapeutic measures are available, genetic testing should be offered or, in some cases, required.
(2) When a child is at risk for a genetic condition with pediatric onset for which preventive or other therapeutic measures are not available, parents generally should have discretion to decide about genetic testing.
(3) When a child is at risk for a genetic condition with adult onset for which preventive or other therapeutic measures are not available, genetic testing of children generally should not be undertaken.

AMA Code § 2.138, Genetic Testing of Children.

70. ORANGE COUNTY REGISTER, July 24, 2004, at News 22. In Ashkenazi Jews, "80% to 90% of inherited deafness in their children is caused by mutations in the gene." *Id.*

71. Freundel, *Gene Modification Technology*, *in* ENGINEERING THE HUMAN GENE LINE: AN EXPLORATION OF THE SCIENCE AND ETHICS OF ALTERING THE GENES WE PASS TO OUR CHILDREN 121, Stock & Campbell eds. (2000). He added:

> I take this position as a result of Judaism's teaching that human beings are the most important part of God's created universe. . . . God has entrusted this world to humankind's hands, and the destiny of this world has always been our responsibility and our challenge. Whether or not we live up to that challenge is our calling and essential mission.

American Association of the Advancement of Science (AAAS) had actually sought to block "inheritable genetic modifications," including therapies for eggs or sperm, unless they were found to be "theologically acceptable."[72] In 1998, John Fletcher, chief of the bioethics program at the NIH, made a comment regarding how wrong it would be not to treat the gene pool "because of the belief that people had a right to an untampered genetic patrimony."[73] For example, some genetic variants are called "Ancestry Informative Markers" (AIMs) for people whose ancestors came from several different locations.[74] There are confidentiality limits regarding genetic testing and counseling.[75] Further, the management of clinical genetic family studies often constitutes a bioethical issue.[76]

In this connection, an editorial in the *New England Journal of Medicine* claimed that "'race' is biologically meaningless."[77] However, it has been

72. Frankel & Chapman, Human Inheritable Genetic Modifications: Assessing Scientific, Ethical, Religious, and Policy Issues (2000); Stock, Redesigning Humans 154 (2002).

73.
> Suppose we really knew how to treat cystic fibrosis or some other very burdensome disease and didn't do it because of the belief that people had a right to an untampered genetic patrimony. Then you met a person 25 years later and did the Golden Rule thing and said, "Well, you know, we could have treated you for this, but we wanted to respect your right to your untampered genetic patrimony. Sorry." It doesn't take a highfalutin' ethicist to realize that's just plain wrong. You violate one of the basic principles of morality, namely that you . . . treat a person as you would want to be treated.

Id. Stock, Redesigning Humans 132 (2002).

74. Leroi, Mutants 339 (2003); Collins-Schramm, *Ethnic-Difference Markers for Use in Mapping by Admixture Linkage Disequilibrium*, Am. J. Human Genetics (2002) at 737-50; and Shriver, et al., *Skin Pigmentation, Biogeographical Ancestry and Admixture Mapping*, Human Genetics 387-99 (2003).

75. Dugan et al., *Duty to Warn At-Risk Relatives for Genetic Disease: Genetic Counselors' Clinical Experience*, Am. J. Med. Genetics 27-34 (2003); White, *Making Responsible Decisions: An Interpretive Ethic for Genetic Decision-Making*, Hast. Ctr. Rep. 14-21 (1999).

76. Botkin, *Privacy and Confidentiality in the Publication of Pedigrees,* JAMA 1808-12 (1998); Friedman-Ross, *Disclosing Misidentified Paternity*, Bioethics 114-30 (1996).

77. Schwartz, *Racial profiling in medicine*, N. Eng. J. Med. 1392-93 (2001).

pointed out that "every race and even ethnic group within the races has its own collection of clinical priorities based on differing prevalence of diseases."[78] In this connection, this year Jonathan Pritchard, University of Chicago, headed a study showing that human beings are still evolving. His researchers detected some 700 regions of the human genome where genes appear to have been reshaped by natural selection; that also may help explain races.[79]

Genetic testing issues arise with respect to employers who may want to exclude workers with certain genetic risks from the workplace because these workers may become disabled prematurely and incur higher healthcare costs. Since 1991, the AMA *Code of Medical Ethics* has prescribed "minimum" conditions for genetic testing in the exclusion from the workplace of workers who have a genetic susceptibility to injury.[80] However, the Code also provides that "insurance companies or employers should not be permitted to discriminate against carriers of genetic disorders through policies that have the ultimate effect of influencing decisions about testing and reproduction."

---

78. Risch et al., *Categorization of humans in biomedical research: genes, race and disease*, GENOME BIOL. 11 (2002). *See also* SMEDLEY, STITH & NELSON, UNEQUAL TREATMENT: CONFRONTING RACIAL AND ETHNIC DISPARITIES IN HEALTH CARE (Inst. of Med., 2003) *and* Lee, AM. PUB. HEALTH 2133 (December 2005).

79. Jonathan Pritchard said, "There is ample evidence that selection has been a major driving point in our evolution during the last 10,000 years, and there is no reason to suppose that it has stopped." Nicholas Wade (of *The New York Times*) in the ORANGE COUNTY REGISTER, March 8, 2006, at News 10.

80. (a) [The] irreversible injury would occur before monitoring of either the worker's exposure to the toxic substance or the worker's health status could be effective in preventing the harm.

> (b) The genetic testing is highly accurate, with sufficient sensitivity and specificity to minimize the risk of false negative and false positive test results.
>
> (c) Empirical data demonstrate that the genetic abnormality results in an unusually elevated susceptibility to occupational injury.
>
> (d) It would require undue cost to protect susceptible employees by lowering the level of the toxic substance in the workplace.
>
> (e) Testing must not be performed without the informed consent of the employee or applicant for employment.

Code § 2.132, Genetic Testing by Employers. (Adopted in 1991)

This was called the AMA's "direction of future genetic screening tests" with respect to "social policy."[81] In 2003, it was found that in the U.S. genetical health disparities fall under social and environmental influences.[82] Current portions of the Code also offer guidelines on the future possibilities of multiplex genetic testing, where "tests are offered for several different medical conditions in a single session."[83] It has been noted that some testing has been "offered to populations who have a very low incidence of disease or for which the specificity of the test is unknown." Thus, "Asian-Americans have a one in 32,000 risk of having a child with cystic fibrosis (CF) with each pregnancy, about 10 percent of the risk faced by Caucasians." Although there is no current data on the percentage of carriers among Asian-Americans or about the reproductive risk such couples face if only one partner is found to be a carrier,[84] it was still recommended that (CF) carrier testing be offered to all pregnant couples and couples contemplating pregnancy, regardless of their race or ethnicity.[85] Yet, it has been noted, "This autosomal

---

81. *See* Hoffman, *Preplacement Examinations and Job-Relatedness: How to Enhance Privacy and Diminish Discrimination in the Workplace*, 49 U. Kan. L. Rev. 517 (2001). (Indicate that such testing must relate to the potential employee's ability to do the work to which he has been assigned.)

82. Institute of Medicine, Unequal Treatment: Confronting Racial and Ethnic Disparities in Healthcare (2003).

83. (1) Physicians should not routinely order tests for multiple genetic conditions.
(2) Tests for more than one genetic condition should be ordered only when clinically relevant and after the patient has had full counseling and has given informed consent for each test.
(3) Efforts should be made to educate clinicians and society about the uncertainty surrounding genetic testing.

Code § 2.139, Multiplex Genetic Testing. *See also* Fedder, *To Know or Not to Know: Legal Perspectives on Genetic Privacy and Disclosure of an Individual's Genetic Profile*, 21 J. Legal Med. 557, 563 (2000).

84. 30 J.L. Med. & Ethics 293 (Summer 2002), *citing* American College of Medical Genetics, Preconception and Prenatal Carrier Screening for Cystic Fibrosis: Clinical and Laboratory Guidelines (2001).

85. It is easier for the clinician simply to do a test, particularly when it involves drawing one more tube of blood or doing a cheek swab, than it is to counsel about what the test is for and the factors that weigh for and against testing and to document a refusal of testing in a way that protects the clinician in the event of an adverse outcome. *Id.*

recessive disease affects one out of every 2,500 babies born in the United States; more than 8 million Americans may be carriers of the gene. The population at risk of being carriers of cystic fibrosis are all persons of Caucasian descent."[86]

In 2000 the Equal Employment Opportunity Commission (EEOC), the agency authorized to enforce the Americans with Disabilities Act (ADA), issued a guidance including genetic discrimination connected with disabilities under the statute.[87] The same year one state enacted a law "to protect the results of genetic tests as private information," which prohibits employers and insurers from requiring genetic tests as an employment condition.[88] However, the 2001 survey by the American Management Association found employers' use of genetic monitoring was under 1 percent. An AMA study indicated 68 percent of patients feared that the results of their genetic tests would actually be used against them by their employers or their insurers.[89] Although a majority of states have enacted genetic nondiscrimination laws, following a conditional offer of employment, employers may require that such individuals sign a blanket authorization to release all of their medical records.

86. Jonsen, The Human Genome Project 14 (1996).

87. E.E.O.C. Notice 915.002, Enforcement Guidance: Disability-Related Inquiries and Medical Examinations of Employees under the American with Disabilities Act (ADA) (July 27, 2000). A 1995 issuance by the EEOC stated that when genetic predisposition to disease is used for discrimination against employment, that disease will be regarded as falling within the definition of disability under the ADA. EEOC Order 915.002, Definition of the Term Disability, at 902-45 (March 15, 1995).

88. An Act Relative to Insurance and Genetic Testing and Privacy, 2000 Mass. Acts 254, § 1. The term "genetic information" is the identifiable result of a genetic test as defined by this section or explanation of such a result or family history pertaining to the presence, absence, variation, alteration, or modification of a human gene or genes, and "genetic tests" are any tests of human DNA, RNA, mitochondrial DNA, chromosomes or proteins for the purpose of identifying genes or genetic abnormalities, or the presence or absence of inherited or acquired characteristics in genetic material. Mass. Gen. Laws ch. 151B, § 1.

89. Mitka, *Genetics Research Already Touching Your Practice*, Am. Med. News, April 6, 1998, at News 3. *Also see* Lapham et al., *Genetic Discrimination: Perspectives of Consumers*, 274 Science 621-24 (1996); Andrews & Jaeger, *Confidentiality of Genetic Information in the Workplace*, Am. J. L. & Med. (1991).

However, the laws of two states restrict employers' access to *any* non-job-related medical information.[90] Others have set forth particular jobs justifying an employer's use of genetic information.[91] An actual or potential employer may not test employees for "highly sensitive" medical and genetic information without the employee's consent,[92] but the employer should be authorized to require that consent be obtained for safety purposes for assignment to such particular jobs.

Under the Health Insurance Portability and Accountability Act (HIPAA) of 1996,[93] medical privacy regulations are issued by the Department of Health and Human Services (HHS). The privacy provision applies to health insurers, hospitals, doctors and pharmacists. Further, if based upon an individual's current health or "genetic information," HIPAA prohibits employer-sponsored group health plans from charging individuals higher rates or excluding certain medical conditions from coverage.[94] However, doctors can share information with the patient's spouse or family members, as well as other hospital doctors. The regulation generally provides that "A person with appropriate knowledge of and experience with generally accepted statistical and scientific principles and methods for rendering information not individually identifiable, . . . [a]pplying such principles and methods, determines that the risk is very small that the information could be used, alone or in combination with other reasonable available information, by an anticipated recipient to identify an individual who is a subject of the informa-

90. Cal. Gov't Code §§ 12926, 12940; Minn. Stat. Ann. § 363.01 02. *See also* Rothstein et al., *Protecting Genetic Privacy by Permitting Employer Access Only to Job-Related Employee Medical Information: Analysis of a Unique Minnesota Law*, 24 Am. J. L. & Med. 399-417 (1998).

91. Yesley, *Protecting Genetic Difference*, 13 Berkeley Tech. L.J. 653, 663 (1998).

92. Norman-Bloodsaw v. Lawrence Berkeley Lab., 135 F.3d 1260 (9th Cir. 1998).

93. Health Insurance Portability and Accountability Act of 1996, Pub. L. No. 104-191, 42 U.S.C. § 30ugg (2000). This statute allows patients access to their medical records while limiting the ability of others to obtain such information. Confidentiality is addressed in the so-called "Common Rule" concerning privacy research with human subjects. 45 C.F.R. § 46A and 45 C.F.R. §§ 160, 164.

94. 42 U.S.C. §§ 300gg(b)(1)(B), 300gg-1(a)(F), and 300gg-(1)F.

tion,"[95] together with a number of specific types of information. Hospitals can disclose names to clergy, but there are statutory limits on the information that emergency medics may disclose. Police and fire departments may release names and information about accident victims, homicides, and other matters.[96] Neither the statute nor regulation preempt state laws to the extent that they do not conflict; that is, some states may require even more protection of privacy.[97] Further, employers have not been prohibited from seeking and obtaining the results of genetic tests recorded in individual medical records by their employment laws,[98] and much genetic information is beyond the scope of the HIPAA, including employers, pharmaceutical companies, types of insurers, and researchers. Accordingly, geneticists Lin and Altman from the Department of Genetics at Stanford University School of Medicine, and Owen, of the Department of Statistics, recommended "clarifications to rules of legislation (such as HIPAA), so that they could explicitly protect genetic privacy and set strong penalties for violations."[99] That is certainly true for

---

95. 45 C.F.R. § 164.514(a) (2001); but the compliance date for most entities was April 14, 2003; 45 C.F.R. § 164.534 (2004).

> There are criminal penalties for disclosing or obtaining protected information, and imprisonment may result when it is used for commercial advantage. The first HIPAA violation conviction was on November 5, 2004. http://www.usdoj.gov/usao/waw/press_room/2004/nov/gibson.htm A man named Gibson obtained a cancer patient's name, birth date, and Social Security number, and used that information to obtain four credit cards in the patient's name. He then purchased more than $9,000 of video games, jewelry, home figurines, clothes, groceries, and other items.

JAMA 1767 (April 13, 2005).

96. USA Today, Oct. 17, 2003, at 2A.

97. HIPAA § 264(c)(2), Subtitle F of Title II. *See also* 45 C.F.R. §§ 160.202-160.203. Although the nondisclosure policy "individually identifiable health information" exceptions include cases where there is a "serious and imminent threat to the health or safety of a person of the public." 45 § 164.512(j).

98. Rothstein, *Genetic Privacy and Confidentiality: Why They Are So Hard to Protect,* J.L. Med. & Ethics 198-203 (1998).

99. Lin, et al., *Genomic Research and Human Subject Privacy*, Science 183 (July 9, 2004). The HIPAA and "Privacy Rules of 2003" generally forbid sharing identifiable data without patient consent. However, they do not specifically

most cases. However, appellate courts in several states have ruled that physicians have a duty to inform third parties in certain instances. For example, New York state's top court so ruled, stating: "When the service performed on behalf of the patient is necessarily implicated in protection of . . . other identified persons foreseeably at risk because of the relationship with the patient, who the doctor knows or should know may suffer harm by relying on prudent performance of the medical service."[100]

Similarly, a woman sued her mother's physician for his failure to warn her of hereditary thyroid cancer. The Supreme Court of Florida upheld "the standard of helping a patient's children overcome the relevancy of privacy."[101] The Minnesota Supreme Court also held that "a physician's duty regarding genetic testing and diagnosis extends beyond the patient to biological parents who foreseeably may be harmed by a breach of that duty."[102] In this

---

address use or disclosure policies for human genetic data. 45 C.F.R. Parts 160 and 164, Subparts A and E. As of April 14, 2003, this is enforced through HHS's Office for Civil Rights (OCR). Hodge, Jr., *Health Information Privacy and Public Health*, 31 J.L. MED. & ETHICS 663-71 (2004).

100. Tenuto v. Lederle Labs., 90 N.Y.2d 606 (N.Y. 1997). For challenges to privacy laws under the HIPAA Privacy Rule and many state laws regulating access to genetic databases by researchers and biotechnology companies, *see* Burnett et al., *The 'GeneTrustee': A Universal Identification System that Ensures Privacy and Confidentiality for Human Genetic Databases*, 10 J.L. & MED. 506-12 (2003), and the Patient Safety and Quality Improvement Act of 2005, Pub. L. No. 109-41, 42 U.S.C. § 299 et seq. (see ch. 13, sec. 3).

101. Pate v. Threlkel, 661 So. 2d 278 (Fla. 1995). The court stated that "in many circumstances in which the physician has a duty to warn of a genetically transferable disease, that duty will be satisfied by warning the patient."

102. Molloy v. Meier, 679 N.W.2d 711 (Minn.), May 20, 2004.

> In this case, which stems more from an alleged failure to perform a diagnostic test than from a failure to breach confidentiality to warn of a genetic disease, the mother and her second husband stated they would not have conceived another child if they had known of the diagnosis of fragile X syndrome in the mother's first child.

Offit et al., *The 'Duty to Warn' a Patient's Family Members About Hereditary Disease Risks*, JAMA 1471 (Sept. 22/29, 2004). *See also* Safer v. Estate of Pack, 677 A.2d 1188 (N.J. App.), *app. denied*, 683 A.2d 1163 (N.J. 1996).

connection, the Council on Ethical and Judicial Affairs of the AMA encouraged physicians to "make themselves available to assist patients in communicating with relatives to discuss opportunities for counseling and testing, as appropriate."[103] On the other side, a "double coding" has been suggested to allow researchers access only to anonymized information to go from the patient/subject to the researcher.[104] Some examples of current genetic difficulties are set forth in Appendix E.

good v. Good

While gene testing and therapy in parts of the genome may be of some help during the twenty-first century, this remains less important than social structures, life styles, and environmental matters. Furthermore, all such testing must also remain confidential to the patient at all times.

v.

During the twenty-first century gene testing will become important to enable people to know what to do under a number of circumstances. Although normally these results will remain confidential, in certain instances physicians should notify other people.

Twenty-first Century Approach

1. Gene test information should normally remain confidential except in those cases where there is serious and imminent threat toward the health or safety of others.
2. When such a threat exists, particularly with respect to the patient's family, relatives or acquaintances, the physician has a duty to warn them about the hereditary disease risks in those instances where the patient has refused to do so.

---

103. *Id. See* AMA Code of Medical Ethics § 2.139, Multiplex Genetic Testing. The American Society of Clinical Oncologists also advised oncologists to "remind patients of the importance of communicating test results to family members." Policy Statement of the American Society of Clinical Oncologists, *Update: Genetic Testing for Cancer Susceptibility*, 21 J. Clin. Oncol. 2397-2406 (2003).

104. Knoppers, *Biobanking: International Norms*, 33 J. L. Med. & Ethics 7-14 (2005). A "double coding" has been suggested to allow researchers access only to anonymized information to go from the patient/subject to the researcher. *Id.* at 7-14.

Then God said, "Let Us make man in Our image.

Gen. 1:26-27

The image of God is the "key to the Christian understanding of human nature."

Nigel M. de S. Cameron, *The New Medicine: The Revolution in Technology and Ethics* (1991), p. 172

All medicine is politics, and politics is nothing but medicine on a large scale.

Rudolf Virchow, German pathologist (1821-1902)

4. The Absence of Ethics and Morals in Patents on Genetically Engineered Living Things

Benjamin Franklin, an eighteenth-century participant in the preparation of our secular Constitution, was also a great inventor. However, he never sought to obtain any patents on what he invented, saying, "I never made, nor proposed to make, the least profit by any of them."[105] In 1790, the year he died, the first U.S. patent was issued together with putting in effect the "first to invent" principle,[106] meaning that a patent should be issued to the inventor who can prove he was the first to come up with the invention. During the nineteenth century Thomas Edison invented the stock ticker (in

---

105. American History (February 2006) at 42.

106. The controversial proposed Patent Reform Act, H.R. 2795, is now under consideration in Congress to abandon the first-to-invent principle in favor of the "first-to-file" approach that is used by the rest of the world. "The ABA jumped on the first-to-file bandwagon in February 2005, when the association's policy-making House of Delegates approved the Intellectual Property Law Section's recommendation that the association support 'legislation providing that the right to a patent shall belong to the inventor who first files an application for patent' containing adequate disclosure of information describing the invention. The House adopted the recommendation by a 277-121 vote." A.B.A. J. 52 (January 2006). *See also The First-to-Invent Rule in the U.S. Patent System Has Provided No Advantage to Small Entities*, J. Patent & Trademark Office Soc'y, June 2005, at 514; *Are the U.S. Patent Priority Rules Really Necessary?*, Hast. Ctr. Rep., July 2003, at 1299.

1869). Although Alexander Graham Bell received a patent for his telephone on March 7, 1876, Thomas Edison discovered the telephone's carbon transmitter (1876), the phonograph (1877), the electric light (1879), motion pictures (in the late 1880s), etc. Hence in 1899, the commissioner of U.S. patents asked that his office be abolished because "everything that can be invented has been invented."[107] But he completely lacked the ability to anticipate the continuance of our new ways of thinking—let alone those of his other contemporaries—such as Albert Einstein, who worked as an examiner at the Swiss Patent Office in Bern.

A patent examiner refused to grant a patent for the genetically engineered version of the bacterium *Pseudomonas,* which could break down components of crude oil and thus help control oil spills. He had stated that it was a "product of nature" and a living thing. In 1980 the Supreme Court, in rejecting these objections by the narrow majority (5:4), pointed out that although physical natural phenomena are not patentable, this was "not nature's handiwork." Although the Patent Office turned down a patent application from General Electric for a bacterium that had been genetically engineered to "eat" oil spills, that office was forced to change its policy by the Supreme Court in 1980, which held that:

> The laws of nature, physical phenomena, and assurance ideas have been held not patentable. Thus, a new mineral discovered in the earth or a new plant found in the wild is not patentable subject matter. Likewise, Einstein could not patent his celebrated law, $E=mc^2$; nor could Newton have patented the law of gravity. *Such discoveries are "manifestations of . . . nature, free to all men* and reserved *exclusively to none."* Judged in this light, respondent's micro-organism plainly qualifies as patentable subject matter. His claim is not to a hitherto unknown natural phenomenon, but to a nonnaturally occurring manufacture or composition of matter—a product of human ingenuity "having a distinctive name, character [and] use." [In contrast to previous cases, here] *the patentee has produced* a new bacterium with *markedly different characteristics from any found in nature* and

107. *See* GATES, THE ROAD AHEAD xiii (1995).

> one having the potential for significant utility. His discovery is not nature's handiwork, but his own; accordingly it is patentable subject matter under [the patent law].[108] (Emphasis added.)

That same year, the Bayh-Dole Act of 1980[109] was enacted, which authorized U.S. privately owned patents on federally funded research to be obtained by universities and researchers and thus increase the commercialization of medical research. Nonprofit and small business contractors may grant large corporations exclusive licenses to their government-funded inventions.[110] Further, NIH grantees have the right under the Act to elect to retain title to subject inventions and are free to choose to apply for patents should additional biological experiments reveal convincing evidence for utility.[111] The Bayh-Dole Act encouraged U.S. universities to patent the results of federally funded research, and income remaining after payment of royalties to the inventor and expenses must be used "for the support of scientific research or education." A survey by the Association of University Technology Managers (AUTM) of its members found that academic licenses brought in over $1 billion in 2003.[112]

In 1996, the National Center for Human Genome Research (NCHGR) pointed out how the HGP would "have a major impact on health care and disease prevention" and provide "enormous scientific and technological advances... having both basic and *commercial applications.*" As Brody noted, this would allow for the patenting of uses of the genes (or of sequences)

108. Diamond v. Chakrabarty, 447 U.S. 03 (1980).

109. Codified as amended at 35 U.S.C. §§ 200-212 (2004).

110. The government retains a nonexclusive, nontransferable, irrevocable license to practice that invention. Schofield, J. L. Med. & Ethics (Winter 2004) at 777. *E.g.,* a biotechnology corporation obtained a patent for the protein that regulates the production of red blood cells in human beings (the DNA sequence for erythropoietin). Amgen, Inc. v. Chugai Pharm. Co., 927 F.2d 1200 (Fed. Cir. 1991).

111. The National Center for Human Genome Research, Policy on Availability and Patenting of Human Genomic DNA Sequence Produced by NCHGR Pilot Projects (1997).

112. Josephine Johnston, Association for Ethics, Law, and Society (2005).

providing, of course, that it was the use and not just the gene (or sequence) that had been discovered.[113] The Patent and Trademark Office's (PTO) guidelines of 2001 require that the specification describe "[a] utility that is specific to the subject matter claimed [in] contrast [to] a general utility that would be applicable to the broad class of the invention."

It is interesting to note that although Harvard has already been awarded a patent on the "oncomouse" (a mouse genetically engineered to have a predisposition to cancer) in the United States, the Canadian Supreme Court ruled that Harvard did not meet the Canadian rules for patentability. The court held that the patenting of higher life forms was to be decided by its legislature. But Canada does not stand alone thereon;[114] that mouse was only partially protected by the European Patent Office in 2004 (i.e., that patent was limited to rodents and not to all mammals).

When Craig Venter was twice rejected for his proposal on the genome project, he left the NIH in order to go after brain genes and patent his results.[115] He used Expressed Sequence Tags (ESTs) (rather than unexpressed, then called "junk" DNA), which Watson stated was only "cream-skimming"

---

113. Brody, The Ethics of Biomedical Research 95 (1998). The U.S. Constitution's Article, Section 8 grants Congress power to "promote the Progress of Science and Useful Arts." The statute, 35 U.S.C. § 101, mandates that "[i]nventions be patentable" and "new and useful." *But see* Juicy Whip Inc. v. Orange Bang Inc., 185 F.3d 1364, 1366 (Fed. Cir. 1999) ("The threshold of utility is not high.").

114. Hast. Ctr. Rep., March–April 2003, at 6. In this connection, it is noted that a provision was drafted and presented at a September 21–23, 2001, conference at Boston University that stated: "No patents shall be granted on human genes, parts of human genes or unaltered products of human genes, nor on the genes of bacteria, viruses, or other infectious agents that cause disease in humans." Annas, Andrews & Isasi, *Protecting the Endangered Human*, Am. J. L. & Med. 156 (2002).

115. Adams et al., *Complementary DNA Sequencing: Expressed Sequence Tags and Human Genome Project*, Science 251 (June 21, 1991), 1651-56. In Article 4 of the Universal Declaration on the Human Genome and Human Rights, the human genome *in its natural state* shall not give rise to financial gains. Nevertheless, in 2001 some representatives at UNESCO believed that such concerns warrant further consideration.

without the "utility" required by patent law; Watson rejected patents as "interfering with scientific progress."[116] Others disagreed with him. In 1998 the AMA Council on Ethical and Judicial Affairs concluded that "granting patent protection should not hinder the goal of developing new beneficial technology"; its guidelines specify a lack of preference for patents on "genes or DNA sequences":

> *Patents on processes*—for example, processes used to isolate and purify gene sequences, genes and proteins, or vehicles of gene therapy—do not raise the same ethical problems as patents on the substances themselves and are thus preferable.
> *Substance patents*—on purified proteins present fewer ethical problems than patents on genes or DNA sequences and are thus preferable.
> *Patent descriptions*—should be carefully constructed to ensure that the patent holder does not limit the use of a naturally occurring form of the substance in question. This includes patents on proteins, genes, and genetic sequences.[117]

The head of the Biotechnology Examination Group of the U.S. Patent and Trademark Office denied that any ethical issue had ever been raised challenging an American patent.[118] Nevertheless, an expert on plant biotechnol-

116. ETHENY, WATSON AND DNA 265, 276 (2003). In order to be patented in the United States, the matter involved must be useful, novel, and nonobvious. 35 U.S.C. §§ 101-154. A written description, enablement, best mode, and definiteness are also needed. 35 U.S.C. § 112. Once it is issued, for 20 years (formerly it was only 17 years) the patent holder has the right to exclude others from making, using, selling, offering to sell, or importing the invention. 35 U.S.C. § 154.

117. AMA CODE OF MEDICAL ETHICS § 2.105, Patenting Human Genes (1998). It has been noted by the humanitarian group Doctors Without Borders that many developing countries grant protection even when it is not required under international law. ORANGE COUNTY REGISTER, May 23, 2003, at News 18.

118. In my twenty-six years *in the patent office*, I have never seen a morality rejection made. I don't know of any case law that is directly on point to morality in biotechnology in the United States. . . . We deal with statutes and *we generally don't get involved in ethics*. We almost always sidestep the issue, and a lot of people have tried to draw us into that discussion. Our point is that *we are an administrative agency; statutes* are passed by Congress; they *are interpreted by the judicial system*; we administer the law. (Emphasis added.)

ogy patent law also noted, "There are so many broad and fundamental patents that, in essence, every major actor may be violating a patent held by every other major actor."[119] However, in June 2005, the Supreme Court held (9-0) that big pharmaceutical manufacturers are free to use compounds patented by rivals in conducting research on new medicines.[120]

Stanford law professor Hank Greely and other ethicists and geneticists contributed to the Human Genome Diversity Project (HGDP) a "Model Ethical Protocal" as a guide, stating, "The HGDP requires that researchers . . . show that they have obtained the informed consent of the population, through its culturally appropriate authorities where such authorities exist."[121] The need for a guide has been argued because geneticists often present the information in such a way that only one decision is logically possible, and many people are hesitant to ask questions in such settings. Some ethicists ask whether informed consent is an unattainable ideal.[122] The Human Genome

---

MARK WINSTON, TRAVELS IN THE GENETICALLY MODIFIED ZONE 190 (2002). Professor Winston, of Simon Fraser University, British Columbia, is quoting John Doll of the U.S. Patent & Trademark Office.

119. Barton, *The Impact of Contemporary Patent Law on Plant Biotechnology Research*, *in* INTELLECTUAL PROPERTY RIGHTS III: GLOBAL GENETIC RESOURCES; ACCESS AND PROPERTY RIGHTS (Crop Science Soc'y of Am., 1998). It should be noted that prior to 1996, a jury had been assigned to determine the meaning of patent claims. However, in that year the U.S. Supreme Court affirmed the decision by a federal circuit court that patent claims constituted a matter of law to be decided by the trial judge, not a jury. Markman v. Westview Instruments, Inc., 152 F.3d 967 (Fed. Cir. 1995) (en banc), *aff'd*, 517 U.S. 370 (1996).

120. L.A.TIMES, June 14, 2005, at Bus.1. Justice Antonin Scalia wrote: "Scientific testing is a process of trial and error. In the vast majority of cases, neither the drug maker nor its scientists have any way of knowing whether an initially promising candidate will prove successful after a battery of experiments. That is the reason they conduct the experiments." *Id.*

121. OLSON, MAPPING HUMAN HISTORY 216 (2002); *Proposed Model Ethical Protocol for Collecting DNA Samples*, 33 HOUS. L. REV. 1431-73 (1997). In the 1970s, India's patent laws recognized processes only and did not necessitate clinicals. However, in 1995 India signed the World Trade Organization (WTO) agreement, and commencing in 2006 it will honor product patents. SCIENTIFIC AM. (August 2004) at 23.

122. Olson, *id.* at 218. *Beyond Consent: Are Ethical and Social Issues in Genetic Consulting Possible?,* 338 LANCET 998-1001 (1991). Immediately following the publication of the discovery of the hemachromatosis gene, many

Organization (HUGO) is an international organization of genetics researchers. In lieu of financial compensation, benefits are recommended (by making) "agreements with individuals, families, groups, communities or populations that foresee technology transfer, local training, joint ventures, provision of health care or of information infrastructures, reimbursement of costs, or the possible use of a percentage of any royalties for humanitarian purposes."[123] However, a court has gone further with respect to a supplier of blood and tissue specimens to a research institute for deposit in its biobank. Using it, the institute discovered and patented the gene and genetic for Canavan disease. The supplier's tort suit resulted in a recovery under an unjust enrichment theory.[124]

In 2005, people with the Illinois Institute of Technology, Chicago-Kent College of Law examined 1,167 claims contained in 74 relevant patents on human genetic material. They found that 38 percent of the claims were "problematic"; they indicated that "the conclusions of project personnel are not necessarily predictive of eventual validity determinations by the courts or the U.S. Patent & Trademark Office (USPTO)." As a result they noted, "Something needs to be done about the number of human gene patents being granted that arguably do not measure up to the federal patent law."[125]

The European Patent Convention permits denial of patents for the human body on the grounds of public interest, and where issuance would be contrary

---

U.S. laboratories started to test for its mutations. However, when a patent on the gene was granted, 30% of them discontinued testing for its disease. Merz, et al., *Diagnostic Testing Fails the Test*, 415 NATURE 577-78 (2002).

123. HUGO Ethical, Legal, and Social Issues Committee Report to HUGO Council, *Statement on the Principled Conduct of Genetics Research, Recommendation 9*, March 21, 1996, *at* http://www.gene.ucl.ac.uk/hugo/conduct.htm.

124. Greenbery v. Miami Childrens Hospital, 264 F. Supp. 2d 1064 (S.D. Fla. 2003). *See* Anderlik & Rothstein, Canavan *Decision Favors Researchers over Families*, J. L. MED. & ETHICS (2003) at 450-53, *and* Andrews, *Harnessing the Benefits of Biobanks*, J. L. MED. & ETHICS (2005) at 22-30.

125. Paradise, Andrews & Holbrook, *Patents on Human Genes: An Analyasis of Scope and Claims*, 307 SCIENCE, March 11, 2005, at 1566. They also noted, "Currently, patent examiners are encouraged with monetary bonuses to grant patent applications." *Id.* at 1567.

to *ordre public* or morality.[126] Thus, although the European Patent Office (EPO) grants patents for biotechnology inventions, "which are susceptible of industrial application," the EPO will not grant patents for inventions that are contrary to "order public" (public policy) or morality. The General Conference of UNESCO has stated that the human genome is part of the heritage of humanity and "in its natural state shall not give rise to financial gains."[127]

good v. Good

No ethical reason exists for tampering with the legal patenting of new living things in any manner whatsoever.

v.

All processes of patenting non-natural living objects should become more open to public awareness.

Twenty-first Century Approach

In the event the current voluntary disclosure method proves to be inadequate, the law should be modified in the twenty-first century to augment the require-

126. Art. 53. *See also* Cunningham, 19:2 ETHICS & MEDICINE (2003) at 85-98, *citing* Explanatory Report to the Convention for the Protection of the Human Rights and Dignity of the Human Being with Regard to the Application of Biology and Medicine: Convention on Human Rights and Biomedicine, Directorate of Legal Affairs (Strasbourg, May 1997) DIR/JUR (97)5, para. 7. *See also* Chin, *Unilateral Technology Suppression: Appropriate Antitrust and Patent Law Remedies*, 66 ANTITRUST L.J. 441, 450 (1998).

127. *Id.* Cunningham, *citing Universal Declaration on the Human Genome and Human Rights*, 1997, Art. 4 (accessed September 19, 2001) and Article 19(a)(iii). The World Trade Organization's dispute settlement system neglects health and human welfare. *See* Bloch & Jungman, 31 J. L. MED. & ETHICS (2003) at 529-45; Hilary, *The Wrong Model: GATS, Trade Liberalization and Children's Right to Health*, 46 SAVE THE CHILDREN (October 2001) (stating that WTO dispute settlement panels that consider cases involving member states' health regulations "rule . . . on the basis of trade considerations, not public health concerns"), *and* Gilles de Wildt et al., *Which Comes First—Health or Wealth?*, 357 LANCET 9262 (April 7, 2001), at 1123 (discussing the UN Subcommission on Human Rights' proclamation that the TRIPS agreement "does not adequately reflect the fundamental nature and indivisibility of all human rights," including the right to health). *Also see* Gitter, *International Conflicts Over Patenting Human DNA Sequences in the United States and the European Union: An Argument for Compulsory Licensing and a Fair-Use Exemption*, 76 N.Y.U. L. REV. (2001) at 1681.

ments regarding public discussion of the biological aspects of patentable non-natural living objects that are related to the human species.

To expect the unexpected shows a thoroughly modern intellect.

Oscar Wilde

There's a high probability that for Homo sapiens, the process of evolution as we currently think about it, as natural selection, is for all intents and purposes over. It is going to be replaced by our desire and capability to tinker.

Stuart Schreiber, Harvard geneticist

Scientific knowledge may enrich esthetic and moral perceptions, and illuminate the significance of life and the world, but these are matters outside science's realm.

Francisco Ayala, Evolutionist,
University of California, (1998) Vatican Observatory Publications

Already in our day man is turning round from his remade environment, and beginning to remake himself.

Durant, *The Greatest Minds and Ideas* (2002), p. 101

5. The Possible Future of Inheritable Genetic Interventions and Changes in *Homo Sapiens*

By the end of the twentieth century, evolutionists such as Francisco Ayla (above epigraph) and Richard Dawkins[128] pointed to evidence showing that "blind change" together with "natural selection" accounted for the creation of much life on earth over billions of years (see Chapter 2). Although professors of theology maintain "that the purely natural causes science is obliged to look for in explaining evolutionary process do not logically rule out the

128. Richard Dawkins, The Blind Watchmaker (1986), River Out of Eden (1995); Daniel C. Dennett, Darwin's Dangerous Idea: Evolution and the Meaning of Life (1995).

simultaneously effective action of God,"[129] a massive diminution in the time for continued human evolution may result from work of successors to the human genome project throughout the third millennium. As written by the professor of systematic theology at the Graduate Theological Union in Berkeley: "Perhaps if we could be confident that humans would play God as God does—with infinite love and compassion—the concern would be lessened."[130] It may not be possible to predict; however, a need is prescribed for us to draw some bioethical lines. The question has also been asked, "What general ideas and opinions in our culture might prevent us from drawing a line between the promise and the peril of biotechnology?"[131]

The United States has not yet established a comprehensive policy about the handling of genetic information, even though it invested huge sums to promote the Genome Project. Section 2.12 of the American Medical Association's *Code of Medical Ethics* (2003), called "Genetic Counseling," was updated in 1994 when the Genome Project was just beginning. Although "it is ethically permissible for physicians to participate in genetic selections to prevent, cure, or treat genetic disease," the Code specifically states, "It would not be ethical to engage in [genetic] selection on the basis of non-disease-related characteristics or traits." On the contrary, Fritz Allhoff in 2005 has stated, "Genetic enhancement would aim, *most fundamentally*, at the improvement of the human experience. What end could possibly be more noble than that?"[132] He added that:

---

129. HAUGHT, DARWIN'S GIFT TO THEOLOGY 393. But other writers have accessed bioethicists for their junction with scientists toward becoming more secularistic. EVANS, PLAYING GOD: HUMAN GENETIC ENGINEERING AND THE RATIONALIZATION OF PUBLIC BIOETHICAL DEBATE (2002).

130. TED PETERS, PLAYING GOD? GENETIC DETERMINISM AND HUMAN FREEDOM, 2d ed. (2003).

131. WOLFSON, THE PUBLIC INTEREST 334 (Winter 2002). "Scientific knowledge may enrich esthetic and moral perceptions, and illuminate the significance of life and the world, but these are matters outside science's realm." Ayala, University of California-Irvine, recipient of the 2002 National Medal of Science.

132. Allhoff, *Germ-Line Genetic Enhancement and Rawlsian Primary Goods*, 15 KENNEDY INST. OF ETHICS J. 48 (March 2005).

> To honor the Kantian principle, genetic intervention would be morally permissible only if *every* future generation would rationally consent to the genetic alterations made in the germ-line. Can this criterion be satisfied? Absolutely. More interestingly, however, the answer suggests which sorts of germ-line enhancements are morally permissible and which are not.[133]

This is a heuristic matter, one that serves to discover or to stimulate investigations, and some currently speak of the "posthuman species."[134] However, few, if any, of the inheritable changes would appear likely to fall into that category early in the third millennium, which might eventually affect all humans on the planet Earth. Nevertheless, in the twenty-first century we shall begin to supersede the speed of Darwin's slow natural selection for evolution. While referring to the greater and greater role in courts of law, a writer adds: "The great challenge will be integrating the future research on genetic predispositions with the law's notion of free will. This will demand a sophisticated understanding of the science and considerable creativity in the law."[135]

In 1998, Israel enacted a statute that outlawed inheritable interventions using reproductive cells that have intentional genetic modification for five years.[136] Only four years later, the three American authors who had actually expressed their view "that the time is ripe for a flat-out international ban on human cloning" (see Chapter 8) stated that they "also advocate a similar ban on inheritable genetic interventions."[137] They even suggested enactment of an international

133. *Id.* at 48. In this connection the Harvard philosopher John Rawls wrote: "[I]n the original position . . . the parties want to insure for their descendants the best genetic endowment. . . . The pursuit of reasonable policies in this regard is something that earlier generations owe to later ones." RAWLS, A THEORY OF JUSTICE (rev. ed. 1999).

134. *E.g.*, FRANCIS FUKUYAMA, OUR POSTHUMAN FUTURE: CONSEQUENCES OF THE BIOTECHNOLOGY REVOLUTION (2002).

135. FAIGMAN, LABORATORY OF JUSTICE 353 (2004).

136. Prohibition of Genetic Intervention Law No. 5759 § 3(2) (1998).

137. Annas, Andrews & Isasi, *Protecting the Endangered Human: Toward an International Treaty Prohibiting Cloning and Inheritable Alterations*, 28 J. L. MED. & ETHICS 172 (2002).

convention on the "Preservation of the Human Species."[138] They appear to claim that a need exists to ban changes, which, for biologists, are "reproductively isolated from other natural populations." Because all animal organisms differ genetically to some degree, geneticists have had some difficulties in identifying many species. Although this does not appear related to the genetic work being sought by researchers during this portion of the third millennium, the Convention would go much further by outlawing use of "embryos or reproductive cells that have undergone intentional inheritable genetic modification."[139] The desired absolutism used in those bans, and the need to obtain such results, appear to have been largely based upon traditional religious beliefs in the divine identification of species. Near the end of the twentieth century almost half of the state legislatures enacted statutes pertaining to practices associated with trait selection, including research on embryos or fetuses.[140]

---

138. It set forth their concern "that by altering fundamental human characteristics to the extent of possibly producing a new human species or subspecies, genetic science will cause the resulting persons to be treated unequally or deprived of their human rights." Their draft included:

> Article 1. Parties shall take all reasonable action, including the adoption of criminal laws, to prohibit anyone from initiating or attempting to initiate a human pregnancy or other form of gestation using embryos or reproductive cells that have undergone intentional inheritable genetic modification. . . .

Article 5. Reservations to this Convention are not permitted. *Id.* at 154-55.

139. *See also* objections reflecting theological concerns. David Danks, *Germ-Line Gene Therapy: No Place in Treatment of Genetic Disease*, 5 HUMAN GENE THERAPY 151-52 (1994). Ronald Dworkin stated, "Our genetic identity . . . has been a paradigm of nature's responsibility and not ours." . . . [The] sphere of our own responsibility would destabilize much of our conventional morality." Dworkin, *Playing God: Genes, Clones, and Luck*, *in* SOVEREIGN VIRTUE: THE THEORY AND PRACTICE OF EQUALITY 448 (2000).

140. *See* Coleman, *Playing God or Playing Scientists: A Constitutional Analysis of State Laws Banning Embryological Procedures,* 27 PAC. L.J. 1331, 1354 (1996), *and* ANDREWS, STATE REGULATION OF EMBRYO STEM CELL RESEARCH, ETHICAL ISSUES IN HUMAN STEM CELL RESEARCH (Nat'l Bioethics Advisory Comm'n, January 2000). The Florida law, for example, states: "No person shall use any live fetus or live premature infant for any type of scientific, research, laboratory, or other kind of experimentation either prior to or subsequent to any termination of pregnancy procedure except as necessary to protect or preserve the life and health of such fetus or premature infant." FLA. STAT. ANN. § 390.0111(6) (West 2004).

The childhood of the human race is about to come to an end.

Kevin Davies, science writer

If a nation expects to be both ignorant and free, it expects what never was and never will be.

Thomas Jefferson

a) Clinical Trials on Genetic Research

A clinical trial is called a double-blinded trial if it blinds both the subjects and the treating professionals to the assignment of individual patients to the control or the treatment group. It has been said that blinding the investigators "should be carried out where possible because it helps prevent their knowledge of the assignment influencing their clinical assessment of the outcomes."[141] Further, a good informed consent process is one of the conditions required for controlling the risk/benefit ratio in legitimate human subjects research, because losses to those in the control group might be regarded as illegitimate despite their consent.[142] However, for certain cases there are regulations that provide exceptions from informed consent requirements for "emergency research" when approved by the responsible IRB (with concurrences of a member "who is not otherwise participating in the investigation").[143] The FDA

---

141. Brody, Ethics of Biomedical Research 142 (1998). Normally this is approved as ethical in advance by an Independent Review Board. Randomized controlled trials "are largely recruited by physicians in the context of a patient-physician relationship within which patients seek care." Hast. Ctr. Rep. (Sept.-Oct. 2003) at 4.

142. *E.g., see Proposed Recommendations of the Task Force on Genetic Testing*, 62 Fed. Reg. 4534-47 (Jan 30, 1997). Further, institutional review boards must contain at least one member who is not otherwise affiliated with the institution and at least one member from a nonscientific background (a lawyer, ethicist, or clergyman), which are requirements adopted in the 1996 draft of the International Harmonization Guidelines for Good Clinical Practice. Although IRBs are to consist of others besides the researchers, they may be colleagues in the field.

143. For example, the IRB may decide that (a) obtaining informed consent is not feasible because the "human subjects are in a life-threatening situation, available treatments are unproven or unsatisfactory, and the collection of valid scientific evidence through randomized placebo-controlled investigations is neces-

must be contacted when the criteria are not met.[144] In 2003, it was concluded "that important technical barriers stand in the way of using physician clinical performance assessment for evaluating the competency of individual physicians."[145] Finally, it has recently been stated: "Gene transfer-studies with exciting prospects are under way, but with perhaps heightened risk for participants." It was noted that the current federal regulations only stipulate "that the consent process should include an explanation as to whether medical care or other compensation will be provided and, if so, an explanation of what kinds of care and compensation are available."[146] The NIH peer review process can be seen as a barrier to certain clinical investigations.

---

sary to determine the safety and effectiveness of particular interventions." Further, it is necessary that "appropriate animal preclinical studies support the potential for the intervention to provide a direct benefit to them," "clinical investigation could not practicably be carried out without the waiver," and that there will be "public disclosure to the community in which the clinical investigation will be conducted." 21 C.F.R. § 50.24 (4-1-03 edition).

144. (e) If an IRB determines that it cannot approve a clinical investigation because the investigation does not meet the criteria in the exception provided under paragraph (a) . . . The sponsor of the clinical investigation must promptly disclose this information to the FDA.

Id. As of 2002, about 20 million Americans have participated in clinical research trials. Noah, *Informed Consent and the Elusive Dichotomy Between Standard and Experimental Therapy*, AM. J. L. & MED. 361 (2002). Institutions have been disciplined for rules violations by the federal Office of Human Research Protections (OHRP). Rettig, *The Industrialization of Clinical Research*, 19:2 HEALTH AFFAIRS 130 (2000).

145. Landon et al., *Physician Clinical Performance Assessment*, JAMA 1183 (Sept. 3, 2003). The International Committee of Medical Journal Editors (ICMJE) proposed that all its 11 member journals require, as a condition of consideration for publication, registration in a public trials registry or before the onset of patient enrollment by September 13, 2005. N. ENG. J. MED. 1250 (Sept. 16, 2004).

146. 45 C.F.R. § 46.116 (2001) (setting forth the general requirements for informed consent). The requirements relating to injury, addressed in paragraph (a)(6), are for projects with more than minimal risk. Scott, *Research-Related Injury: Problems and Solutions*, J. L. MED. & ETHICS (2003), at 419; Snyderan, *The Clinical Researcher—An 'Emerging' Species*, JAMA 882 (Feb. 18, 2004).

Clinical research in the United States is currently at a crossroads from both government and private entities. The establishment public-private partnership, called the National Clinical Research Enterprise (NCRE), has been proposed for participatory discussion on standards to improve the safety and efficiency of clinical investigation, and a transfer of basic research into practice to improve future health.[147] The NCRE board should bring an integrity derived from its "neutrality, equitability, and objectivity" because:

> no participant organization has been able to solve the current problems. Participation in any cooperative organization uniformly involves some sacrifice of individual autonomy. [The] fostering collaborative establishment of national standards for clinical research vocabulary, databases, and information technologies in clinical research is currently not possible.[148]

The journals published by the International Committee of Medical Journal Editors (including the *Journal of the American Medical Association, the New England Journal of Medicine,* and *Lancet*) require, "as a condition of consideration for publication," registration of clinical trials prior to patient enrollment.[149] The most critical basis for this result appears to apply to the need for clinical research in accelerated genetical breakthroughs. The private sector can extend protections to all human subjects in such research, regardless of funding

---

147. Crowley et al., *Clinical Research in the United States at Crossroads*, JAMA 1120 (March 3, 2004). Limited data exists concerning how physicians use and understand genetic tests. Freund et al., *Natural Settings Trials—Improving the Introduction of Clinical Genetic Tests*, J. L. Med. & Ethics (2004) at 106-10, *citing* Giardiello et al., *The Use and Interpretation of Commercial APC Gene Testing for Familial Adenomatous Polyposis*, 336:12 N. Eng. J. Med. 823-27 (1997).

148. *Id.* Crowley at 1123. "Many physicians had significant and serious misunderstandings about the application and interpretation of the genetic test for familial adenomatous polyposis." Giardiello, *supra* note 81. The genetic test was administered to 30 patients who lacked conventional indications (out of 177 total patients). *GeneTests: Medical Genetics Information Resource*, University of Washington and Children's Health System, Seattle, 1993-2004, hhtp://www.genetests.org (last visited Feb. 8, 2004).

149. DeAngelis et al., N. Eng. J. Med. 1250 (2004).

source.[150] With respect to recipients of federal grants, as of May 2, 2005, NIH-funded investigators are asked to submit voluntarily to PubMed Central (www.pubmedcentral.nih.gov) an electronic version of the author's final manuscript when the article is accepted for publication.[151] "As the electronic article increasingly becomes the authoritative and most useful document for researchers, and as scientists are actually computing on the contents of these documents—the text itself as well as the associated data—the impermanence of the publishers' Web sites presents a substantial risk. Creating such an archive is a historical and necessary NIH responsibility."[152]

The FDA had encouraged the use of placebo-control trials, and certain critics questioned its "orthodoxy."[153] However, Miller and Brody pointed out that clinical trials should be appropriately governed, and, if so, the "use of placebo controls when proven effective treatment exists is ethically justifi-

---

150. Proceeding of a conference sponsored by Howard Hughes Medical Inst. et al., February 2000. *The Declaration of Helsinki* stipulates "ethical principles for medical research involving human subjects." World Medical Association, *Declaration of Helsinki: Ethical Principles for Medical Research Involving Human Subjects*, 284 JAMA 3043-45 (2000). Principle number 5 states: "In medical research on human subjects, considerations related to the well-being of the human subject should take precedence over the interests of science and society." This has been said to provide "sound guidance for the clinical research insofar as it directs investigators to avoid exploiting research subjects and seriously compromising their well-being for the sake of science." Miller, *Research Ethics and Misguided Moral Intuition*, J. L. MED. & ETHICS 111-16 (2004).

151. Steinbrook, *Public Access to NIH-Funded Research,* N. ENG. J. MED. 1739 (April 28, 2005). Department of Health and Human Services, National Institutes of Health, *Policy on enhancing public access to archived publications resulting from NIH-funded research*, 70(26) Fed. Reg. 6891-6900 (2005). The NIH funds 212,000 researchers worldwide, and 5,000 scientists are direct employees of the institutes. Each year, these researchers publish 60,000 to 65,000 articles, accounting for about 10 percent of the articles in the nearly 5,000 journals indexed by PubMed, according to the NIH.

152. National Institutes of Health, *Final NIH public access policy questions and answers*, Feb, 24, 2005, http://www.nih.gov/about/publicaccess/index.htm (accessed April 8, 2005).

153. Rothman & Michels, *The Continuing Unethical Use of Placebo Controls*, N. ENG. J. MED. 394-98 (1994). *See also* Fredman et al., *Placebo Orthodoxy in Clinical Research*, J L. MED. & ETHICS 243-59 (1996). *See also* HHS regulations, 45 C.F.R. § 46.

able."[154] This could be accomplished by meeting the seven requirements proposed by others for all clinical research, which was "built on the difference between research and therapy and on the core value of *protecting research participants from exploitation*"; namely, (1) scientific or social value; (2) scientific validity; (3) fair subject selection; (4) favorable risk-benefit ratio; (5) independent review; (6) informed consent; and (7) respect for enrolled research participants.[155]

Although American medical students have not been tested on their clinical skills as part of a licensing examination since 1964, on June 2004, a clinical skills examination was added to Step 2 of the U.S. Medical Licensing Examination, which has been co-sponsored by the National Board of Medical Examiners. Beginning with the medical school graduating class of 2005, students were required to pass it in order to be licensed.[156] Of course, should they enter research, they must be taught precisely how clinical research trials differ from routine clinical practice (RCT). For example, only the RCT might involve a placebo.[157] Although 59 of 94 medical schools require passage of the examination of clinical skills for graduation, it is not clear whether this will cover all the details of clinical research.[158] Nevertheless, a 2004 study of gene expression patterns with drugs showed that since 1970, "the rate of cure of acute lymphoblastic leukemia in children has increased dramatically, from less than 30 percent to approximately 80 percent." Most important was the conclusion that "this remarkable improvement has resulted from the marriage of laboratory and clinical science."[159]

---

154. Miller & Brody, *A Critique of Clinical Equipose: Therapeutic Misconception in the Ethics of Clinical Trials*, Hast. Ctr. Rep. (2003) at 25.

155. Emanuel et al., *What Makes Clinical Research Ethical?*, 283 JAMA 2701-11 (2000). New York state has also received demands to require pharmaceutical companies to publicly disclose the results of all clinical trials even though this is not a legal mandate. Marshall, *Buried Data Can Be Hazardous to a Company's Health*, 304 SCIENCE (2004) at 1576.

156. Maxine A. Papadakis, M.D., N. ENG. J. MED. 1703 (April 22, 2004).

157. BERG, ET AL., INFORMED CONSENT: LEGAL THEORY AND CLINICAL PRACTICE 280-83 (2d ed. 2001).

158. *See* JAMA 280-83 (Sept. 1, 2004).

159. Winick et al., *Childhood Leukemia – New Advances and Challenges*, N. ENG. J. MED. 601 (Aug, 5, 2004) (citing Holleman et al., *Gene-Expression Patterns in Drug-Resistant Acute Lymphoblastic Leukemia Cells and Response to Treatment*, 3 N. ENG. J. MED. (Aug. 5, 2004), *id.* at 533).

In the year 2000, the World Medical Association's Declaration of Helsinki revised its Article 1(2) to state:

> Good clinical practice is a set of internationally recognized ethical and scientific quality requirements that must be observed for designing, conducting, recording and reporting clinical trials that involve the participation of human subjects. Compliance with this good practice provides assurance that the rights, safety and well-being of trial subjects are protected, and that the results of the clinical trials are credible.

In accordance with Article 22, Member States were to adopt and publish the necessary laws and regulations and to actually apply the provisions no later than May 1, 2004. However, apparently no Member State met this deadline.[160] "It can confidently be predicted," as one lawyer had noted, "that for a variety of reasons, human subject research will continue on a relatively large scale."[161]

> Homo sapiens, *the first truly free species, is about to decommission natural selection, the force that made us . . . Soon we must look deep within ourselves and decide what we wish to become.*
>
> Edward O. Wilson (*Wilson*, 1998)

b) Distinctions Between Somatic and Germline Therapy

In 1988, the European Medical Research Council's *Gene Therapy in Man* noted that "the expression of specific genes after their introduction into laboratory animals has raised the possibility that certain genetic defects in man

---

160. Fluss, *The Evolution of Research Ethics: The Current International Configuration*, J. L. MED. & ETHICS (Winter 2004) at 599.

161. McLean, *Regulating Research and Experimentation: A View fom U.K.*, *id.* at 608 (citing MCLEAN & MASON, LEGAL AND ETHICAL ASPECTS OF HEALTHCARE [London: GMM, 2003]). The journals published by the International Committee of Medical Journal Editors (including the *New England Journal of Medicine* and *Lancet*) require, "as a condition of consideration for publication," registration of clinical trials prior to patient enrollment. DeAngelis et al., N. ENG. J. MED. 1250 (2004).

might be corrected by applying similar techniques." It described how "foreign genes may be inserted either into somatic cells (i.e., any body cell except a germ cell) or into germ cells or cells that give rise to germ cells (e.g., early embryonic cells)." These two methods were then divided as follows:

> Insertion of genetic material into somatic cells and their subsequent transplantation is not fundamentally different from any form of organ transplantation or blood transfusion. Nevertheless, on the one hand Erik Parens indicated, "To oppose such genetic interventions would be cruel, if not stupid, to suggest that we ought never to use genetic technology to heal the sick."[162] On the other hand, "The insertion of genes into fertilized eggs or very early embryos is fundamentally different because these genes would be passed on to the offspring in subsequent generations. *Germline gene therapy should not be contemplated.*"[163] (Emphasis added.)

In 1988 it was noted: "Fears that eugenic genetic engineering might be used by an amoral government to implement coercive social programs have fueled much of the anxiety about the development of gene therapy."[164] Fourteen years later, Leon Kass, bioethical advisor to the current President Bush, noted "the fear of superhumanization, that is, that man will be 'playing God.'" He stated this fear to be "a complaint too facilely dismissed by scientists and nonbelievers. They stand in judgment of each being's worthiness to live or die (genetic screening and abortion)—not on moral grounds, as is said of God's judgment, but somatic and genetic ones; they also hold out the promise of salvation from

162. Parens, *The Goodness of Fragility: On the Prospect of Genetic Technologies Aimed at the Enhancement of Human Capacities*, KENNEDY INST. OF ETHICS J., at 151.

163. These had formerly been emphasized by the philosopher Hans Jonas, who actually argued for a "right to ignorance." "The ethical command here entering the enlarged stage of our powers is: never to violate the right to that ignorance which is a condition for the possibility of authentic action; or: *to respect the right of each human life to find its own way and be a surprise to itself.*" Hans Jonas, *Biological Engineering—A Preview*, in his PHILOSPHICAL ESSAYS: FROM ANCIENT CREED TO TECHNOLOGICAL MAN 161 (1974). *But see* Bentley Glass, *Science: Endless Horizons or Golden Age?*, 171 SCIENCE (1971), at 28.

164. NICHOLS, HUMAN GENE THERAPY 6 (1988).

our genetic sins and defects (gene therapy and genetic engineering). . . .Consider only that if scientists are seen in this godlike role of creator, judge and savior, the rest of us must stand before them as supplicating, tainted creatures. Despite the hyperbolic speech, that is worry enough."[165]

However, another American had differently written:

> Bombarded with white-coated claims that "Genes-R-Us," grateful for the absolution which such claims offer for our shortcomings and sins, and attracted to the promise of using efficient, technological means to fulfill our aspirations, rather than the notoriously unreliable moral or political ones, the idea that we are essentially self-replicating machines, built by the evolutionary process, designed for survival and reproduction, and run by our genes continues to gain.[166]

> *One of the fruits of the "genetic revolution" in biomedicine has been the ability to splice functional genes into the DNA of human cells to produce stable changes in the bodies of patients. The use of these 'human gene transfer' techniques to treat genetic disease has been widely endorsed, as long as two lines are not crossed . . . .*
>
> Juengst, *Medical Ethics Journal* (Sp. 1999), p. 1

---

165. KASS, LIFE, LIBERTY AND THE DEFENSE OF DIGNITY 129 (2002). A reviewer of this 2002 volume by Kass found it "disappointing, indeed, quite a disturbing book. . . . Kass completely ignores the extensive scholarly literature on the concepts of nature, natural law, and dignity. . . . Kass's castigation of his fellow bioethicists lacks any credibility . . . in no sense is this book internationally informed." Campbell, *Excoriating technology*, Hast. Ctr. Rep., January-February 2004, at 44.

166. Howard Kaye, *Anxiety and Genetic Manipulation: A Sociological View*, PERSPECTIVES IN BIOLOGY AND MEDICINE (Summer 1998) at 488. Similarly, following the turn of the third millennium, George Q. Daley, at Children's Hospital, Boston, wrote: "In numerous public lectures and scientific articles, I have consistently expressed my conviction that fundamental knowledge and lifesaving cell-based therapies will emerge from research into both embryonic and adult stem cells and that neither avenue should be excluded in favor of the other. In my own laboratory, I maintain active research programs on both classes of stem cells." N. ENG. J. MED. 1798 (Oct. 21, 2004).

c) Somatic Gene Therapy and Some Needs for Mandatory Testing

The evidence in favor of somatic gene transfer appears to be analogous to the legal introduction of DNA evidence for the accurate identification of a person in civil and criminal cases[167] as used by the FBI,[168] as well as others. For example, in 2003, Nobel laureate James Watson wrote:

> A DNA sample taken for fingerprinting purposes can, in principle, be used for a lot more than merely proving identity: it can tell you a lot about me—whether I carry mutations for disorders like cystic fibrosis, sickle-cell disease, or Tay-Sachs disease. Some time in the not so distant future, it may even tell you whether I carry the genetic variations predisposing me to schizophrenia or alcoholism—or traits even more likely to disturb the peace.[169]

---

167. The more regions in which the crime scene DNA matches a suspect's, the greater the probability of a match, and the more remote the chances that the crime scene DNA could have come from anyone else. Under the FBI's system, a DNA fingerprint is produced from the analysis of 112 such regions, plus a marker that determines the sex of the individual from whom the DNA sample is derived.

Watson, *DNA* 271 (2003).

168. In 1990, the FBI established its DNA database, CODIS (Combined DNA Index System), and by June 2002 it contained 1,013,746 DNA fingerprints. Of these, 977,895 are from convicted offenders and 35,851 are forensic crime scene samples for unsolved cases. Since its inception, CODIS has been used to make some 4,500 identifications that would not otherwise have been made.

Id. at 275.

169. *Id.* at 273.

> Dr. Alec Jeffreys, a British scientist, recognized the forensic potential of special areas of the genetic material that differed in size amongst individuals. Jeffreys et al., 314 Nature 67-73 (1985) and 316 Nature 76-79 (1985). The genetic patterns, called "DNA fingerprints," were first applied to resolving questions of family relatedness in immigration matters in the United Kingdom. Jeffreys et al., 317 Nature 818-19 (1986). The technique [is] of choice in most modern forensic science laboratories and government DNA databanks.

Hagman, Prevett & Murray, *DNA Handbook* (2003), Preface.

In addition, Watson expressed his "belief that the potential of this [somatic gene therapy] technology to lift the curse or genetic disease is simply too great for medicine to turn away from it."[170] The example he gave concerned the success of a team in Paris, France, with severe combined immune deficiency (SCID) by using cells modified from the infants' bone marrow to carry an inserted gene.[171] In October 2002 the doctors found that one of the two original patients was suffering from a side effect; namely, leukemia, a cancer of the bone marrow in which certain types of cells are overproduced. The gene therapy cured the SCID, which was "worse than the one it has caused," leukemia, and the baby boy in question apparently responded "to the chemotherapy used to treat the leukemia."[172] The French reported in January 2005 that 17 children had been successfully treated for SCID using gene therapy. However, when a third of such children died of leukemia, the French trial was halted and the U.S. Food and Drug Administration (FDA) suspended three U.S. SCID trials.[173] Embryos produced by *in vitro* fertilization (IVF) can be tested through

---

170. *Id.* at 356. Nevertheless, back in 1998 it was stated: "It can be argued that a phenotypic screening test is always preferable to a [genetic] test, because phenotype is the clinical concern. In this view, the availability of a phenotypic test should lead to increased caution in the use of a DNA-based test." Burke et al., 280 JAMA 176 (1998). *See also* Chapter 10.

171. For example, "When the infant appeared, the father commented on how healthy he looked. But everyone present knew that the howling newborn, already on his way to the brightly lit isolation room upstairs, had a rare, life-threatening genetic disease, severe combined immune deficiency (SCID). The parents' previous child also had been born with SCID. Specialists in pediatric immunology had diagnosed his condition at the age of 14 months, but by then he was too ill to tolerate experimental medical procedures that might have prolonged his life. He died before his second birthday, worn out from intractable diarrhea and countless respiratory infections." NICHOLS, HUMAN GENE THERAPY 1 (1988).

172. *Id.* at 354-55.

> The team of scientists investigating why two children in France developed leukemia after undergoing gene therapy has discovered that the virus used to replace the defective gene with a working copy inadvertently switched on a cancer-causing gene in the patients' DNA.

302 *Science* (2003), at 415-19.

173. SCIENCE 1028 (Feb. 18, 2005). However, a trial in Britain continues.

pre-implantation genetic diagnoses (PGD) for a number of genetic diseases.

A 30-year-old woman with a family history of Alzheimer disease had her embryos tested for the presence of the "predisposing gene mutation." When only the unaffected embryos were implanted into the uterus, she gave birth to a child who was free of such a gene.[174] In addition, even gender selection may be completely ethical whenever the PGD of one cell (a blastomere) withdrawn from the embryo with several cells (usually at least eight cells) reveals sex-linked genetic disorders, including hemophilia A and B, Lesch-Nyhan syndrome, Duchenne-Becker muscular dystrophy, and Hunter syndrome. Further, although gender may be so determined, such a choice need not be based upon gender for its own sake, nor does the PGD analysis destroy the embryo. It only destroys the blastomere cell that must be repeatedly heated and cooled[175] (see also Chapter 8, Section 1 c).

Prior to the turn of the third millennium, almost 300 of the 5,700 known genetic disorders were detectable with genetic testing.[176] PGD and prenatal diagnosis (PND) are currently used to test for some diseases, but the practice is controversial.[177] In 2005 a plan was announced recommending that all new-

174. Verlinsky et al., *Preimplantation Diagnosis for Early-Onset Alzheimer Disease Caused by V717L Mutation*, 287 JAMA 1018 (2002).

175. *See* Verlinsky et al., *Over a Decade of Experience With Preimplantation Genetic Diagonosis: A Multicenter Report*, 82 FERTILITY & STERILITY 292 (2004).

176. CONNER & FERGUSON-SMITH, ESSENTIAL MEDICAL GENETICS 142 (4th ed. 2004).

> PGD of embryos created by in vitro fertilization (IVF) has become an increasingly available option for couples seeking to avoid the birth of a child with a genetic disorder.

Norman Fost, Editorial, JAMA 2125 (May 5, 2004).

> Britain's Human Fertilization and Embryology Authority began granting or denying licenses for reproductive PGD testing. A study showed that neither IVF nor embryo biopsy pose serious threats to embryos.

Verlinsky et al., *Over a Decade of Experience with Preimplantation Genetic Diagnosis: A Multicenter Report*, 82 FERTILITY & STERILITY 292-94 (2004); Marques et al., *Genomic Imprinting in Disruptive Spermatogenesis*, 363 LANCET (2004), at 1700-02; and Ram research associate at Dalhousie University.

177. PARENS & ASCH, PRENATAL TESTING AND DISABILITY RIGHTS (Georgetown Univ. Press, 2000); GENETICS AND PUBLIC POLICY CENTER, PREIMPLANTATION GENETIC DIAGNOSIS: RELATED TO THE GENETIC TESTING OF HUMAN EMBRYOS (2004).

borns be screened for 29 rare medical disorders. Sharon Terry, of the Genetic Alliance, an advocacy group for people with genetic disorders, said, "Giving parents the result, saying, 'Here's the mutation; we are not sure what the outcome will be,' is better than not telling,"[178] On the other hand, Dr. Lainie Friedman Ross, a medical ethicist at the University of Chicago, actually claimed that "reporting test data for which there are no systems in place for follow-up testing and treatment is not rejecting paternalism, but it is patient abandonment."[179]

On the Human Genome Project, the World Medical Association declared in 1995: "One should respect the will of persons screened and their right to decide about participation and about the use of the information obtained."[180] Although some future studies are expected to identify genes of small effect,[181] Professor B. Brody of Biomedical Ethics and director of the Ethics Program at the Methodist Hospital in Houston, Texas, argued that the new genetics has raised crucial issues about both freedom and responsibility that "the voluntarist approach fails to confront." These are "genetically determined diseases which result in the inevitable death of young children" and where "relatively cheap and painless genetic tests are available." Partners who are carriers should "procreate only by use of in vitro fertilization (IVF), preimplantation zygote screening, and disposal of affected zygotes"—"only with an understanding that all fetuses will be tested and that those that will develop the disease will be terminated."[182]

Huntington's disease, an autosomal (non-sex-determining chromosomal) -dominant disease, which affects certain nerve cells in the brain, is one of the

---

178. Orange County Register, Feb. 21, 2005, at News 30.

179. *Id.*

180. World Medical Ass'n, Handbook of Declarations 17 (Ferney-Voltaire, France, 1995).

181. Lohmueller et al., *Meta-Analysis of Genetic Association Studies Supports a Contribution of Common Variants to Susceptibility to Common Disease*, 33:2 Nature Genetics (2003), at 177-82.

182. Brody, *Freedom and Responsibility in Genetic Testing*, Bioethics 343 (2002). "Genetic tests are now available for a very wide range of disorders, including Huntington's disease, cystic fibrosis, neurofibromatosis, polycystic disease of the kidneys, Tay-Sachs disease, fragile-X syndrome, hemophilia, and many, many others, in countries such as the United States, where abortion is legal for any purpose." Coming, The Gene Bomb 265 (1996).

most genetic of fatal mental illnesses.[183] One study has shown that "women with Huntington's disease have more illegitimate children than their unaffected siblings."[184] It is due to a single genetic mutation, which has been located on the short arm of Chromosome 4:

> We now have highly accurate clinical tests that permit people who have a parent with Huntington's disease to determine whether or not they have inherited the gene. If they are in the unlucky 50 percent, they are guaranteed to become ill, and if they have already married and have children, each of their children has a 50/50 chance of developing the illness. If they did not inherit the gene, however, they have escaped the family curse, and they are also free to conceive children without risk of passing the illness on.[185]

Cystic fibrosis (CF), located on a gene in chromosome 7, which becomes nonfunctional, is another inherited disease that keeps the lungs and digestive system from functioning normally and results in morbidity and premature death to some one in 25 Americans who are its genetic carriers. Like Huntington's, those with CF may not live much beyond 30-plus years

---

> 183. In 1872, George Huntington, a New York physician, described the disorder from families in Long Island, New York. Among their ancestors was one Jeffrey Ferris who emigrated from Leicester, England, in 1634. He almost certainly had the disease, as do many of his descendants today.

Leroi, *Mutants* 298 (2003). *See also* Rubinsztein, *Lessons from the Animal Models of Huntington's Disease*, 11 Trends in Modern Genetics 202-09 (2002).

184. *Id.* at 303 and Reed & Neel, *Huntington's chorea in Michigan*, 11 Am. J. Human Genetics 107-635 (1959).

> Huntington's fatal disease usually affects *some 30,000 people in the United States* in their 30s; yet many, if not *most of them, refuse testing and start giving birth to children* without ascertaining the inheritance. Work is currently being done at the University of California-Irvine toward finding a clue as to what causes this incurable and fatal brain malady.

Orange County Register, April 4, 2004, at Local 6.

185. Anderson, Brave New Brain 268 (2001). Although there have been trials for genetic disorders such as cystic fibrosis, the results are not yet very encouraging. Cavazzana-Calvo et al., *The future of gene therapy*, Nature 779-81 (2004).

(unless women with positive tests undertake abortion).[186] In Louisiana, a couple who decided not to terminate the pregnancy were first informed by the HMO that had paid for the testing that such a preexisting condition would not be covered—although in that earlier case the HMO actually later agreed to cover their child.[187] During the 1990s the CF screening was limited to those whose families reflected it or it was suggested that such people decide to stop normal reproduction or to use gametes instead.[188] Also, pressure toward more testing has been due to the possible threat of a malpractice suit should the patient's child be delivered with a disabling genetic condition.[189] Opposite results have been reached by courts in different states. In New York, when a couple was seeking fertility treatment, they proceeded with the Ovum Donor Program, which screened donors for various diseases, including cystic fibrosis. After a donor was accepted, a girl was born who was diagnosed with cystic fibrosis (CF) and allegedly would have to "remain under a doctor's and/or hospital's care for the rest of her life." However, the court dismissed the couple's malpractice action; citing an earlier case, it actually stated that "whether it is better never to have been born at all than to have been born with even gross deficiencies, is a mystery more properly left

---

186. Office of Technology Assessment, U.S. Congress, Cystic Fibrosis and DNA Tests: Implications of Carrier Screening (1992); Wilfond & Fost, *The Cystic Fibrosis Gene: Medical and Social Implications for Heterozygote Detection*, JAMA 2777 (1990).

187. Billings et al., *Discrimination as a Consequence of Genetic Testing*, Am. J. Human Genetics 476 (1992). "No insurer—life or health—currently requires genetic tests. One simple and practical reason is cost. . . . Above all, it cannot be emphasized enough that insurers are not using genetic tests in risk assessment, nor are there any plans to do so." *Report of the ACLI-HIAA Task Force on Genetic Testing*, Am. Council of Life Insurers and Health Ins. Ass'n of Am., 1991, at 5-6.

188. Minn. Inst. of Med., Assessing Genetic Risks: Implications for Health and Social Policy. In 2004, consideration is being given to test CF patients in placebo-controlled clinical trials of the curcumin compound. Zeitlin, *Can Curcumin Cure Cystic Fibrosis?*, N. Eng. J. Med. 608 (Aug. 5, 2004).

189. Charo, *Legal and Regulatory Issues Surrounding Carrier Screening*, Clin. Obst. & Gynec. (1993), at 580-83.

to the philosophers and theologians."[190] In Minnesota a woman ordered genetic along with chromosomal tests when she found that her three-year-old daughter was developing. The doctor had mentioned Fragile X Syndrome (a well-known cause of mental retardation) but did not test for it. Yet, a doctor advised her that her daughter's problems were not genetic in origin. Following her delivery of a son two years later who had Fragile X disorder, she sued the doctors for malpractice and recovered. The Minnesota Supreme Court noted that "only a few other jurisdictions have addressed the question of whether a physician owes a legal duty to the family of a patient who received negligent care in the field of genetics. . . .The standard of care thus acknowledges that families rely on physicians to communicate a diagnosis of the genetic disorder to the patient's family."[191]

It has been noted that "the sequence specificity of RNAi provides a promising approach to silencing genes that cause dominantly inherited neurodegenerative diseases, such as Alzheimer disease, Huntington's disease, and spinocerebellar ataxia (SCA), *for which there are now no effective therapies.*"[192] (Emphasis added.) Brody argues, "John Stuart Mill's 'harm principle' affirms that a necessary condition for the legitimacy of the use of the coercive powers of the state to prevent certain behavior is that the behavior would produce a harm to which no one consented."[193] He even indicated the

---

190. Paretta v. Med. Offices for Human Reproduction, 195 Misc. 2d 568, 760 N.Y.S.2d 639 (2003).

> Soon after the discovery of abnormal chloride transport in cystic fibrosis, Collins, Riordan, Tsui, and colleagues identified the gene that is responsible for the disease with the use of linkage-based techniques, independently of any prior knowledge of the structure of the cystic fibrosis protein. With a measure of circumspection, they named the gene product CFTR. . . . Clinical advances directed at the correction of CFTR function predict an optimistic future for patients with cystic fibrosis and their families.

Rowe et al., *Cystic Fibrosis*, N. Eng. J. Med. 1992, 1999 (May 12, 2005).

191. Molloy v. Meier, 679 N.W.2d 711 (Minn. May 20, 2004).

192. JAMA 1372 (March 16, 2005). "The RNAi machinery, which is expressed in all eukaryotic cells, has also been found to regulate the expression of key genes involved in cell differentiation in plants and animals." Shanker et al., *The Prospect of Silencing Disease Using RNA Interference*, JAMA 1367 (March 16, 2005).

193. *Id.* at 348. *See* Mill, On Liberty 9 (Hackett Pub., 1978).

need for screening a teenager's driver's license ahead of seeking to obtain a marriage license.[194] While states often offer voluntary tests for carriers, he adds, "I know of no attempts to analyze the issues of coercion, exploitation, and commodification in work related to financial incentives for encouraging minimally responsible reproductive behavior."[195] Currently in China a law forbids a mentally disabled person from marrying unless he or she has been sterilized.[196]

It has been noted by Mark Rothstein, director of the Institute for Bioethics, Health Policy, and Law at the University of Louisville School of Medicine, that "if a law prohibiting genetic discrimination did not include family history, it would not prohibit discrimination against an individual known to have a 50 percent risk of an autosomal dominant disorder (such as Huntington's disease) that had afflicted one of the individual's parents."[197] However, he added that "definitions of genetic that include information derived from family histories

---

194. *Id.* at 350, "One cannot, then, invoke the moral right of reproductive freedom as the basis for a moral argument limiting state compulsion to help prevent the harms produced by people who fail to be screened." *Id.* at 351.

195. *Id.* at 352. It was noted that "many people with a family history of Huntington's disease have refused to be tested, preferring to live with uncertainty." *Id.* at 356, *citing* Higgins et al., *Ethical and Legal Dilemmas Arising during Predictive Testing for Adult-Onset Disease: The Experience of Huntington's Disease*, AM. J. HUMAN GENETICS 4 (1990).

196. THE HUMAN GENOME (Nature Pub. Group, 2001). Britain has constructed a potential Universal Genetic Database. Its National Health Service would gather and store the genetic profile of every child at birth as recommended in 2003 in a government white paper called "Our Inheritance, Our Future—Realizing the Potential of Genetics in the NHS." NEW ATLANTIS 103 (Fall 2003). Routine newborn screening began in the United States as "screening for a single biochemical genetic disorder, phenylketonuria, in the 1960s. Tandem mass spectrometry now allows newborn screening for more than 20 biochemical genetic disorders. A study in 2003 concluded that "expanded newborn screening may lead to improved health outcomes for affected children and lower stress for their parents." Waisbren et al., JAMA 2564 (Nov. 19, 2003). Tandem mass spectrometry is now mandated in 24 states. *Id.*, Holtzman, at 2606.

197. Rothstein, *Genetic Exceptionalism and Legislative Pragmatism*, Hast. Ctr. Rep. (2005) at 29, *citing* TEX. LAB. CODE ANN. §§ 21.401-21.402 (enacted 1997) (defining genetic information as the results of a DNA-based test).

are usually considered too broad,"[198] because the political issue "is simply that it is difficult to make a moral argument that discriminating against people on the basis of genetic information is impermissible, but that discriminating against them on the basis of other medical information is okay."[199]

A lawyer bioethicist, Nancy King, summarized distribution in the body of genetic material with measuring experiments by noting that "the stuff goes all over the place." She wrote: "In my view, though the possibility of inadvertent germline gene transfer (i.e., into embryos or reproductive cells) is acknowledged, it has not been adequately considered." She then asked, "How can we possibly contemplate germline effects of any kind without calling for what [another lawyer bioethicist] George Annas has described as 'species-level' decision-making about germline GT?"[200] However, a professor of pediatrics indicated that "studies are mandatory before gene transfer can be attempted in humans."[201] She stated, "The risk of detecting a birth defect caused by gene transfer appears to be less than one in one million," and hence "the small risk involved would seem to justify the potential benefit to society of developing new approaches to a wide range of diseases." Similarly, Wylie Burke, M.D. wrote,

198. *Id., citing* N.J. STAT. ANN. § 10:5-10:12 (enacted 1996) (defining genetic information to include "inherited characteristics that may derive from an individual or family member").

199. *Id., citing* Daniels, *The Functions of Insurance and the Fairness of Genetic Underwriting, in* GENETICS AND LIFE INSURANCE: MEDICAL UNDERWRITING AND SOCIAL POLICY, M.A. Rothstein ed. (2004), and Lemmens, *Selective Justice, Genetic Discrimination, and Insurance: Should We Single Out Genes in Our Laws?,* MCGILL L. REV. 45 (2000), at 347-412.

200. King, *Accident and Desire: Inadvertent Germline Effects in Clinical Research*, Hast. Ctr. Rep., March-April 2003, at 24, 26, and 29. However, it has been noted that "the groups of people considered to be of different races have allelic differences of at most 15 percent, too little to constitute subspecies." Brownlee, *Code of Many Colors: Can researchers see race in the genome?,* 167 SCIENCE NEWS, April 9, 2005, at 232.

201. HIGH, THE RISKS OF GERMLINE TRANSFER, *id.* at 3, without any direct citation concerning "mandatory studies," *but citing* Kay et al., *Safety and Efficacy Studies of AAV-Mediated, Liver-Directed Gene Transfer for Hemophilia B*, THROMBOSIS AND HAEMOSTASIS Suppl., 2001 Abstract #OC2491; V.R. Arruda et al., *Lack of Germline Transmission of Vector Sequences Following Systemic Administration of Recombinant AAV-2 Vector in Males*, 4:6 MOLECULAR THERAPY 586-92 (2001).

"The success of genomic research to date suggests that this ambitious research enterprise will also ultimately succeed."[202] The AMA *Code of Medical Ethics* (2003) contains section 2.11 (issued in 1998), which states "it is appropriate to *limit* genetic intervention *to somatic at this time*." (Emphasis added.)

good v. Good

Without their consent, no one should be required to undergo somatic gene therapy by injection into the body. This could also cause changes in reproductive cells.

v.

Individuals who are identified as being more likely to have inherited severe genetically determined diseases (such as CF and Huntington's disease), resulting in suffering and death, together with the future suffering and death of their children, should be required to undergo available genetic tests for screening such diseases. They should also be offered somatic gene therapy when it becomes available (or sterilization where it is not available).

Twenty-first Century Approach

1. Human clinical trials remain essential to prescribing somatic gene therapy in order to have such therapy approved and become generally available.
2. Where (a) tests for the most serious genetic diseases (such as Huntington's and cystic fibrosis) are available to determine the high probability of suffering and death of a person and of his children, and (b) this can be adequately screened to identify carriers of such genes, consideration should be given by state legislatures toward making such somatic gene tests mandatory.

---

202. Burke, *Genomics as a Probe for Disease Biology*, NEMJ, Sept. 4, 2003, at 969. Campbell and Ross, who explored the attitudes of parents and healthcare professionals toward behavioral genetic testing for violence (about which no genetic tests are currently available), actually concluded that such "professionals do not want to offer testing if there is no clear treatment, while parents may want this information to shape environmental influences." 30 J. MED. ETHICS 580-86 (2004).

Things don't turn up in this world until somebody turns them up.
James A. Garfield

d) The Publicity Needed for Recombinant DNA Research by Public and/or Private Entities

The NIH has issued guidelines relating to recombinant DNA research. This is done by the Recombinant DNA Research Advisory Committee (RAC), which serves its role in promoting medical, safety, and ethical issues associated with human gene transfer research.[203] One guideline required that "all serious adverse events" occurring in conjunction with human gene transfer trials be reported immediately to the Office of Biotechnology Activities (OBA), among others. The RAC has indicated that the public disclosure of adverse events "is essential to public understanding and evaluation of gene transfer in humans," and that "adverse event reports must not be designated as confidential."[204] A "serious adverse event" is defined as any event occurring at any dose that results in any of the following outcomes: death, a life-threatening event, in-patient hospitalization or prolongation of existing hospitalization, a persistent or significant disability/incapacity, or a congenital anomaly/birth defect.[205]

NIH is not a regulatory agency; however, it places conditions upon the public funds that it awards to private institutions, which include compliance with the NIH Guidelines.[206] The Data Access Act, enacted in 1999, requires

---

203. RAC consists of 15 voting members; eight are knowledgeable in certain scientific fields and at least four are knowledgeable in applicable law, standards of professional conduct, and practice. 66 C.F.R. §§ 64052-64054, Dec. 11, 2001.

204. 66 C.F.R. §§ 57970-57977, Nov. 19, 2001. It was stated that although the "NIH and FDA will continue to work closely together in analyzing gene transfer adverse events, NIH OBA cannot reply on disclosures from the FDA to achieve the objective of public disclosure of the scientific and safety issues."

205. *Id*; Section I-E-8 also adds that:

> Important medical events that may not result in death, be life-threatening, or require hospitalization also may be considered a serious adverse event when, upon the basis of appropriate medical judgment, they may jeopardize the human gene transfer research subject and may require medical or surgical intervention to prevent one of the outcomes listed in this definition.

206. *Id. See also* 42 C.F.R. § 52.8.

that all "data needed to validate a federally funded study" be made available to requesting parties under the Freedom of Information Act.[207] The Data Quality Act of 2001 requires federal bureaus to develop a formal complaint procedure "for ensuring and maximizing the quality, objectivity, utility, and integrity of information (including statistical information) disseminated by federal agencies."[208] Although most biotechnology companies do not receive such funding for recombinant DNA research, when a company conducts such research in collaboration with an institution that receives any NIH funding for recombinant DNA research, all such research conducted at or sponsored by that institution is subject to the NIH Guidelines and "many companies have chosen such voluntary compliance."[209] It has been noted that in 2003 only 24

---

207. Amendment, Pub. L. No. 105-277, 112 Stat. 268-495 (1998). *See also* OMB Circular A-110, 64 Fed. Reg. 54,929, which requires research "if they were 'produced under an award that [was] used by the Federal Government in developing an agency action that has the force and effect of law'"; and Shelby, *Accountability and Transparency: Public Access to Federally Funded Research Data,* 37 HARV. J. LEGIS. 369, 379 (2000).

208. Treasury and Gen. Gov't Appropriations Act for Fiscal Year 2001, Pub. L. No. 106-554, Stat. 5658 (2000). Science that is "disseminated" by an agency does not include "distribution limited to correspondence with individuals or persons, . . . public filings . . . or adjudicative processes." Guidelines for Ensuring and Maximizing the Quality, Objectivity, Utility, and Integrity of Information Disseminated by Federal Agencies, 67 Fed. Reg. 369, 377-78 (Jan. 3, 2003). *See also* 67 Fed. Reg. 369, 374, which states that the requirements for publicly available data does not "override" other compelling interests such as privacy, trade secrets, intellectual property, and other confidentiality protections.

209. 66 C.F.R. § 57974.

> For more than a decade, genetic tests have been on the way that would tell patients about genetic variations that might increase their susceptibility to disease. And all the while, bioethicists have been warning the U.S. government of the impending need to regulate such tests when they arrive. . . . Now, researchers at Baylor College of Medicine in Houston, Texas, are offering pregnant women a genome-scanning test that checks for abnormalities in developing fetuses. But because of the way this test was developed, it is not currently subject to FDA review—and neither are tests made by private companies that can do prenatal screening but haven't yet been used to do so.

percent of 18,733 institute applications were awarded grants. The scientific director of the National Institute of Child Health and Human Development stated that what he called a "major issue" became "the recent and dramatic decline in patient-oriented research, i.e., research that requires interaction between a physician-investigator and a patient."[210]

Pursuant to a regulation effective in April 2003 under the Health Insurance Portability and Accountability Act (HIPAA) were details about individual patients. The company would provide periodic overall updates without personally identifying information. Normally, patients should provide certain information because complete nondisclosure of important research is inappropriate.[211]

Nature, Dec. 8, 2005, at 711.

> If, on the one hand, particular diseases or diseases in general come to be seen as largely genetic, those who were reluctant to take care of their fellows in need could argue that society did not cause the person's illness, bad genes did, and that therefore society had no obligation to remedy a problem it did not cause. On the other hand, the more forewarned, and the more individuals can do to prevent disease, the more society may be inclined to declare that those individuals are responsible for their own misfortune and that the community has no obligation to step in with assistance. But there is nothing intrinsic about genetic causality of disease that assures that society will respond compassionately.

Murray, The Human Genome Project 213 (1992).

210. Inouye & Fiellin, *An Evidence-Based Guide to Writing Grant Proposals for Clinical Research*, 142 Annals of Internal Med. 274-82 (2005).

211. 45 C.F.R. Parts 160, 162, 164: Final Rule at 68 Fed. Reg. 3334 (Feb. 20, 2003). As noted in section 5(a) above, the International Committee of Medical Journal Editors (ICMJE) proposed that all its 11 member journals require, as a condition of consideration for publication, registration in a public trials registry. The editors stated their view:

> that enhanced public confidence in the research enterprise will compensate for the costs of full disclosure. Patients who volunteer to participate in clinical trials deserve to know that their contribution to improving human health will be available to inform health care decisions. The knowledge made possible by their collective altruism must be accessible to everyone.

N. Eng. J. Med. 1251 (Sept. 16, 2004).

No changes are made that would be passed on to the patient's children; and no attempts are made to exceed restoration of the normal, that is, to "genetically enhance" human traits like stamina, beauty, memory or intelligence . . . because it captures an important moral intuition and, at the same time, is at constant risk of collapsing under the weight of basic conceptual arguments.

Juengst, Genetic Enhancement, *Med. Ethics J.* (Sp. 1999), p. 2

With what other than condemnation is a person with any moral sense supposed to respond to a system in which the ultimate purpose in life is to be better than your neighbor at getting genes into future generations?

Williams, *Plan and Purpose in Nature* (Basic Books, 1997), p. 157

Then the Lord God said, "See the man has become like one of us, knowing good and evil; and now, he might reach out his hand and take also from the tree of life, and eat, and live forever"—therefore the Lord God sent him forth from the garden of Eden, to till the ground from which he was taken. He drove out the man; and at the east of the garden of Eden he placed the cherubim, and a sword flaming and turning to guard the way to the tree of life.

Genesis 3:22

A human being is "l'être dont l'être est de n'être pas," the being whose essence is in not having an essence.

Simone de Beauvoir

e) Germline Gene Therapy

Germline gene therapy normally means the insertion of DNA into a single cell (a sperm or an egg cell), or an embryo cell. In 1996, Article 13 of the Council of Europe's Convention on Human Rights and Biomedicine proclaimed that "an intervention seeking to modify the human genome" may only be undertaken "if its aim is not to introduce any modification in the genome of any descendants."

On the contrary, three writers in the special "Justice and Genetic Enhancement" issue of the March 2005 *Kennedy Institute of Ethics Journal* endorsed the view that human germ-line genetic enhancement should be permitted, while it also recognized Lindsay's expressed need "to take appropriate action to manage the consequences of genetic interventions,"[212] and that such modification needs particular attention. We must keep in mind the historic 2001 report of Venter et al. that "a single gene may give rise to multiple transcripts, and thus multiple distinct proteins with multiple functions by means of alternative splicing and alternative transcription initiation and termination sites."[213] Although the NIH's Recombinant DNA Advisory Committee (RAC) is not yet seeking to grant funds for germ-line therapy, it may do so when the needed scientific information becomes available (see Chapter 9, subsection 9).

James Watson suggested that "we should attempt human experimentation only after we have perfected methods to introduce functional genes into our close primate relatives"; namely, "monkeys and chimpanzees (an even closer match)."[214] In his view, "we should give serious consideration to germ-line gene therapy"; and "should not be intimidated by the inevitable criticism. Some of us already know the pain of being tarred with the brush once reserved for emergencies. But that is ultimately a small price to pay to redress genetic injustice. If such work be called eugenics, then I am a eugenicist."[215] He noted that al-

212. Lindsay, *Enhancements and Justice: Problems in Determining the Requirements of Justice in a Genetically Transformed Society*, at 32. *See also* Allhoff, *Germ-Line Genetic Enhancement and Rawlsian Primary Goods,* and Loftis, *Germ-Line Enhancements of Humans and Nonhumans*, both in the same issue.

213. J. Craig Venter et al., *The Sequence of the Human Genome*, 291 SCIENCE, at 1304-51 (2001). In recognition of the inadequacy of the term "gene," Venter et al. introduced the term "transcription unit" to cover the post-transcriptional expression of multiple variants.

214. To the contrary, the vegetarian professor and founder of the Great Ape Project, bioethicist Peter Singer of Princeton University, stated that he "would stop experimentation on chimpanzees," our nearest animal cousins. Further, he noted that "in Britain, the government has said that it will no longer allow great apes to be the subject of harmful scientific experimentation." SINGER, WRITINGS ON ETHICAL LIFE 82, 84 (2000).

215. WATSON, DNA, THE SECRET OF LIFE 401 (2003).

> Chapters of the Eugenics Society organized competitions at state fairs, giving awards to families apparently free from the taint of bad

though "the board of scientific directors of the Eugenics Record Office, at Johns Hopkins, wrote in 1928 that 'orthodox eugenicists are going contrary to the best established facts of genetical science,' the time had finally arrived for an assault on the greatest biological mystery of all: what is the chemical nature of the gene?"[216] Some bioethicists have claimed the contrary to be necessary for "the preservation of the human species"(Homo sapiens) and avoiding the development of a "posthuman" one.[217] However, scientists have spoken differently: In his *Wonderful Life: The Burgess Shale and the Nature of History* (1989, the late Harvard paleontologist Stephen Jay Gould wrote: "*Homo sapiens* may be the brainiest species of all, but we represent only a tiny twig, grown but yesterday on a single branch of the richly arborescent bush of life. . . . The world is not there for us. We are not the object of creation, but rather the product of random forces." In his *What Do YOU Care What Other People Think?* (1985), the late Nobel laureate in physics Richard Feynman stated: "We are at the very beginning of time for the human race. It is not unreasonable that we grapple with problems. But there are tens of thousands of years in the future. Our responsibility is to do what we can, learn what we can, improve the solutions, and pass them on. It is our responsibility to *leave the people of the future a free hand.*" (Emphasis added.)

Others have also accepted the concept of genes being natural resources under the control of their community, but argued that they could be either

---

genes. Effectively these were efforts to encourage positive eugenics—inducing the right kind of people to have children. Eugenics was even *de rigueur* in the nascent feminist movement. *Id.* at 28-29.

216. *Id.* at 32-33. "Racism is not implicit to eugenics—good genes, the ones eugenics seeks to promote, can in principle belong to people of any race." *Id.* at 28. "Races are not even a subspecies." Brownlee, 187 SCIENCE NEWS, April 9, 2005, at 232.

217. Cloning and inheritable genetic alterations can be seen as crimes against humanity of a unique sort: they are techniques that can alter the essence of humanity itself (and thus threaten the foundation of human rights) by taking human evolution into our own hands and directing it toward the development of a new species sometimes termed the "posthuman."

Annas, Andrews & Isasi, *Protecting the Endangered Human: Toward an International Treaty Prohibiting Cloning and Inheritable Alterations*, AM. J. L. & MED. 28 (2002) at 153.

backward or fruitless: Mehlman, professor of law, Case Western Reserve University, recently wrote on the backwardness of genetically enhanced individuals: "Equal opportunity will disappear . . . upward mobility for those without access to enhancement services will become a thing of the past. The result will be . . . a society divided between those with access to the genomen and a genetic underclass . . . permanently enslaved to their genetic endowments."[218] Back in 1992, Professor Murray of Howard University had claimed that changes will be largely fruitless: "To the extent that these humanly valued traits are genetic at all, the genetic links are complex, buried many layers beneath the environmental and developmental forces that shape them. Efforts to enhance such traits genetically will probably turn out to be fruitless, at least for the foreseeable future."[219]

The AMA *Code of Medical Ethics* (2003) contains section 2.11 (issued in 1998), which states, "It is appropriate *to limit genetic intervention to somatic at this time*," and to GIGT added:

> If it is ever allowed, at least three conditions would have to be met before it could be deemed ethically acceptable: (1) there would have to be a clear and meaningful benefit to the person, (2) there would have to *be no trade-off with other characteristics or traits*, and (3) all citizens would have to have equal access to the genetic technology, *irrespective of income or other socioeconomic characteristics*. (Emphasis added.)

The AMA may be partly restrained by Leon R. Kass, former professor in social thought at the University of Chicago, who was the chairman of President Bush's Council on Bioethics until September 30, 2005[220] (see Chapter

---

218. Mertz, McGee & Sankar, *Iceland, Inc.? On the Ethics of Commercial Population Genomics*, 58 SOCIAL SCIENCE & MED. (2004) at 1201-09.

219. MURRAY, THE GENOME PROJECT 220-21 (1992).

220. The 66-year-old Leon Kass was replaced last year by President Bush with Edmund Pellegrino, as an 85-year-old physician and bioethicist at Georgetown University who is a "Jesuit-trained" Catholic. SCIENCE, Sept. 16, 2005, at 1800. This council was described by the editor-in-chief of *The New Atlantic* as "one very much opposite of the National Bioethics Advisory Commission (NBAC) set up by President Clinton" because it was not only "putting a stop to embryonic research" (see Chapter 11) but also to "a number of other scientific and clinical projects objectionable to the far right wing of the Republican party, and in particular, they were Southern Baptists." McGee, Summer 2003, Vol. 3, p. vii.

3). In 2003, Kass, in *The New Atlantic* magazine, attempted to explain the use of biotechnical powers to pursue "perfection," both of body and of mind. In conclusion, he stated, "I believe the deepest source of public anxiety about biotechnology is represented in the concern about 'man playing God.'" He called such playing the insertion of "new genes into various parts of the adult body, and someday soon also into gametes and embryos to enhance muscle performance and endurance."[221] He noted that health providers and insurance companies "have for now bought into the distinction, paying for treatment of disease, but not for enhancements." He asked, "What could be wrong with efforts to improve upon human nature?" and replied that the reasons offered "do not get to the heart of the matter." He declared, "To generalize: no biological agent used for purposes of self-perfection will be entirely safe... It surely makes sense, as an ethical matter, that one should not risk basic health trying to make oneself 'better than well.'" He pointed to Aldous Huxley's 1932 book, *Brave New World* (which satirized the mechanical world of the future), stating how it "strongly suggests that 'biotechnical powers' used to produce contentment in accordance with democratic tastes threaten the character of human striving and diminish the possibility of human excellence." (p. 17) But, as noted in the above epigraph, "man has become like one of us, knowing good and evil; and now, he might reach out his hand and take also from the tree of life. . . . Mankind is now going well into the molecules of life's tree and its branches and turning away the sword of its guard."

Kass went on to state that a common man's reaction to these prospects is the complaint of men playing God: "There can be no such thing as the full escape from the grip of our own nature. To pretend otherwise is indeed a form of hu-

---

221. Kass, *Ageless Bodies, Happy Souls: Biotechnology and the Pursuit of Perfection*, New Atlantic 10-11 (September 2003). The NBAC (now called the President's Council on Bioethics) "was accused of bias, although in its case the claims were lodged by the pro-life movement." Weed, *Ethics, Regulation, and Biomedical Research*, 14 Kennedy Inst. Ethics J. (December 2004), at 361. The Federal Advisory Commissions Act of 1972 (FACA), under which NBAC was established as an advisory body, requires that commissions must have both men and women as members (Pub. L. 92-463, 86 Stat. 770 (1972)). But "they are not forced to write their reports with an eye toward the practical problems of the policies that they might propose." *Id.* at 362. "Their members are appointed simultaneously and serve at the whim of a single official." *Id.* at 366.

bristic and dangerous self-delusion." (p. 10) His key word, "full," could only be correct if all changes in "our nature" were made within the basic laws in the universe of physics and chemistry. In fact, a significant amount of manipulation can be accomplished without violation of laws in nature. However, he quoted a working paper prepared for the President's Council on Bioethics, which stated "its resonance reached beyond religion," (p. 19) and added, "Biotechnology may serve or threaten," since "nothing good comes easily. One major trouble with biotechnical (especially mental) 'improvers' is that they produce changes in us by disrupting normal character of being-at-work-in-the-world . . . . We will be at a loss to attest [whether] our bodies and our minds are, in the fullest sense, our own as human." Apparently for him, "an enhancement" would not be "human in the fullest sense." However, he would deny the philosopher John Rawls's position on his "justice as fairness" enabling our community "to insure their descendants the best genetic endowment. . . . The pursuit of reasonable policies in this regard is something that earlier generations owe to later ones."[222] In this connection, M.J. Sandel expressed his concern with enhancements he called "hyperagency—a Promethean aspiration to remake nature, including human nature, to serve our purposes and satisfy our desires. The problem is not the drift to mechanism but the drive to mastery. And what the drive to mastery misses and may even destroy is an appreciation of the gifted character of human powers and achievements."[223] Although the neuroscientist Steven Rose shared many of Sandel's concerns about enhancements, he believed that his fear about hyperagency was misplaced. "Hyperagency, in Sandel's sense, will happen in the modern world of science and discovery. We will always try to perfect our procedures, and misuse of those procedures will occur. But . . . as a species, we ultimately make decisions that assure our future. We may invent the atom bomb, and even use it, but then we go to great lengths to assure it is never used again [see Chapter 5, Section 2.f]. The human being is a curious creature and this is a good thing."[224]

222. RAWLS, A THEORY OF JUSTICE (rev. ed. 1999). His "reasonable policies" could also bring us within the Lord's statement in the epigraph that "man has become like one of us, knowing good and evil."

223. Sandel, *The Case Against Perfection: What's Wrong with Designer Children, Bionic Athletes, and Genetic Engineering?*, ATLANTIC MONTHLY (April 2004), at 58.

224. ROSE, THE FUTURE OF THE BRAIN 40 (2005).

Of man's first disobedience, and the fruit
Of that forbidden tree whose mortal taste
Brought death into the world, and all our woe,
With loss of Eden, till one greater Man
Restore us, and regain the blissful seat,
Sing, Heavenly Muse. . .

John Milton's *Paradise Lost* (1667)

As Hippocrates famously said, "Life, is short, the art long, opportunity fleeting, experience perilous, and the crisis difficult," but the legacy of medicine suggests that we are capable of fulfilling this noble charge.

Markel, "I Swear by Apollo"—On Taking the Hippocratic Oath, *New England Journal of Medicine* (May 13, 2004) p. 2029.

To everything, there is a season and a time to every purpose under heaven: a time to be born, and a time to die.

Ecclesiastes 3:1-8

The duration of our body does not depend upon its essence, nor upon the absolute nature of God. . . . But an adequate knowledge of the way in which things are constituted exists in God in so far as He possesses the ideas of all things, and not in so far as He possesses only the idea of the human body.

Spinoza, *Ethics* (1678)

i) Age Limits

Although more Americans have recently exceeded one century in age, it has been pointed out how "(sea turtles may live for more than two hundred years) life does thrive."[225] Although the U.S. Constitution requires presidential nominees to meet a minimum age of 35 years,[226] it does not identify any

225. McFadden, Quantum Evolution 23 (2001). *See* Chapter 12, § 3 (b).
226. U.S. Const. art. II, § 1.

maximum age—even though currently health risks largely increase with age.[227] Each mitochondrion provides a cell with energy mtDNA that becomes riddled with mutations as it ages. In a Swedish hospital, genetically engineered mice developed mtDNA mutations faster than normal. Their younger deaths demonstrated the mitochondrial theory of aging, which was declared "a fundamental advance in aging research."[228] But we have a longer way to go.[229] Mice are animals with no legal protection,[230] and most genes in the human brain are closely related to the genes in the brain of a mouse.[231] Further, unlike the number of cells in one's body, the brain's neurons modify their wiring largely without creating new neurons.[232] Although it was thought that damaged neurons were irreplaceable, researchers at Princeton University found that new neurons (brain cells) in monkeys could be grown. They concluded

---

227. National Bureau of Economic Relief, *How Health Declines with Age,* http://www.nber.org/aginghealth/summer03/w9821.html (2005). For example, the normal rule for retirement age is 60 for airline pilots, and "there are fifteen medical conditions that will disqualify a pilot from all classes of medical certifications" under Federal Aviation Administration (FAA) regulations. Wiernicki, *Pilot Medical Certification: Current Standards and Procedures,* J. AIR L. & COM. (1999), at 479-80.

228. SCIENCE NEWS (July 10, 2004), at 26. "The product is an energy-rich molecule, adenosine triphosphate (ATP), which cells use to drive myriad chemical reaction. Without ATP, cells would be powerless and die. . . . It is this DNA that some researchers suspect may hold the keys to aging." *Id. See also* Schriner et al., SCIENCE (2005), at 909; Miller, SCIENCE, June 24, 2005, at 1875. Schriner et al. reported that "overexpression of human catalase in the mitochondria of mice extends median and maximal life span by about 20%."

229. Douglas Wallace of the University of California-Irvine "plans to engineer mice to have mtDNA similar to the human versions that he argues promote longevity." *Id.* at 27.

230. Mukerjeo's review of Carbone, *What Animals Want: Expertise and Advocacy in Laboratory Animal Welfare Policy*, SCIENTIFIC AM. (August 2004), at 96. "About 90% of its estimated 25,000 and 30,000 genes have counterparts in humans and mice." ORANGE COUNTY REGISTER, April 1, 2004, at News 11, *citing* NATURE (May 10, 2004).

231. Tecott, *The Genes and Brains of Mice and Men*, AM. J. PSYCH. (2003), at 646-56; Waterston, et al., *Initial Sequencing and Comparative Analysis of the Mouse Genome*, NATURE (2002), at 520-62.

232. Gould, et al., *Neurogenesis in the Neocortex of Adult Primates,* SCIENCE (1999), at 548-52.

this may be possible in elderly humans with nutrients, such as Phosphatidyl Serine (PS).[233]

As noted below, genetic changes may be found that will develop life-extensions needed by future generations for travel aboard space ships. In the meantime, we may place considerable confidence in a discovery made at Cornell University in 1948 when it was found that the lifespans of rats were extended by 33 percent simply by placing them on low-calorie diets.[234] A diet with about one-third fewer calories than a normal diet can also prolong the lives of cats and dogs.[235] Combinations of genetic and dietary intervention have increased median and maximal lifespan of a given mammalian species by about 80 percent.[236] However, our brains "lose weight" as we age: 5 to 10 percent of our brain volume is lost[237] and we also suffer losses in memory with aging.[238]

Telomeres cap the ends of chromosomes that are extended by a special enzyme (telomerase) that is composed of both protein and RNA. Their length

---

233. "Researchers administered 300mg of PS a day to 149 patients over the age of 50. Participants were given a panel of memory and learning tests before and after the trial. After just 12 weeks, the participants' scores reflected those typical of someone 12 years younger, concluding the PS supplementation can reverse cognitive decline by at least a decade. . . . PS also may be particularly useful in early stages of Alzheimer's disease . . . no participants reported adverse effects caused by PS." Leigh Connealy, M.D., COAST, Aug. 2005, at 226.

234. SCIENTIFIC AM. (January 1996), at 46. A similar study by the University of Wisconsin in 1999 found that with a drastically reduced diet, the genetic processes of aging slow to a crawl.

235. SCIENCE (August 27, 1999), at News 29.

> Veterinarians consider it the leading health problem, affecting an estimated 30% of cats and dogs. Their lifespan is short enough already, without making them even shorter by allowing obesity to occur.

L.A. TIMES, Feb. 18, 2000, at A28.

236. Bartke et al., NATURE (2001), at 412.

237. Terry & Katzman, *Life Span and Synapses: Will There Be a Primary Senile Dementia?*, NEUROBIOLOGY OF AGING (2001), at 347.

238. Temporal memory is linked to prefrontal cortex. Gallagher & Rapp, *The Use of Animal Models to Study the Effects of Aging on Cognition*, ANNUAL REV. OF PSYCHOL. (1997), at 339-79. *See also* WHALLEY, THE AGING BRAIN (Weidenfeld & Nicolson) (2001).

becomes shortened until the cells can no longer reproduce. As the telomeres diminish in length, the cells also diminish in the duration of their lives. Only cancer cells currently possess an enzyme to make cells become "virtually immortal." Further, a December 2004 report of the *Proceedings of the National Academy of Sciences* concluded that "mothers of chronically ill children show more cellular aging, as evidenced by shortened chromosomal tips, than do mothers of healthy children. . . . The scientists say oxidative stress is the most likely mechanism for translating a harried life into shortened telomeres."[239] The effect upon our cells might be changed later in the twenty-first century. The year 2005 saw the "mapping of the first locus that determines mean telomere length in humans. Identification of the gene involved and elucidation of its mechanism of action could have important implications for our understanding of chromosomal assembly, telomere biology, and susceptibility to age-related diseases."[240] As stated by a Buddhist writer, "When we dedicate ourselves to a cause larger or longer-lasting than our own mortal selves, we edge in the direction of immortality."[241] During the third millennium, a portion of this direction may become more necessary in order to reach planets that are located light-years away, even though "any observer would genuinely perceive time to slow down somewhat when looking at a moving clock, a phenomenon known as *time dilation*" pursuant to Einstein's twentieth-century theory of special relativity.[242] Currently, how-

239. SCIENCE, Dec. 3, 2004, at 1666.

240. Vasa-Nicotera et al., *Mapping of a Major Locus that Determines Telomere Length in Humans*, 76 AM. J. HUMAN GENETICS 147-51 (2005).

241. LAMA SURYA DAO, AWAKENING BUDDHA WITHIN 67 (1997).

242. "The extent of the time dilation depends on the speed of the clock or object in question compared with the speed of light. In the above example the time dilation is significant because Alice's carriage is traveling at 80% of the speed of light, which is 240,000,000m/s. However, if the carriage were traveling at a more reasonable speed of 100m/s (360km/h), then Bob's perception of Alice's clock would be almost the same as her own. Plugging the appropriate numbers into Einstein's equation would show that the difference in their perception of time would be just one part in a trillion. In other words, it is impossible [at this time] for humans to detect the everyday effects of time dilation." SINGH, BIG BANG 113 (2004).

ever, our satellites now moving within our solar system normally travel at speeds essentially within those calculated using Isaac Newton's methods.[243]

Leon Kass proceeded to "argue that a concern with one's own improving agelessness is finally incompatible with accepting the need for procreation and human renewal: a world of longevity is increasingly a world hostile to children."

> Inasmuch as I no longer cling so hard to the good things of life when I begin to lose the use and pleasure of them, I come to view death with much less frightened eyes. This makes me hope that the farther I get from life and the nearer to death, the more easily I shall accept the exchange . . . we find no shock when youth dies within us.[244]

His statement that "the pursuit of an ageless body is finally a distraction and a deformation" is subject to much disagreement by those who refer to Methuselah, the Biblical grandfather of Noah, who lived for 969 years.[245] The words of Kass are similar to those also written in 2003 by Daniel Callahan, director of the international program at The Hastings Center: "If we cannot expect improved societies, nor any amelioration of the pathologies of civilized life, then why should we seek much longer lives? The fact that many of us want it does not seem to me a good reason."[246] In this connection, we should

---

243. However, "the Global Positioning System (GPS), which relies on satellites to pinpoint locations for devices such as car navigation systems, can function accurately only because it takes into account the effect of special relativity. These effects are significant because the GPS satellites travel at very high speeds and they make use of high-precision timings. . . . Spacetime consists of four dimensions, three of space and one of time, which is unimaginable for most mortals, so it is generally easier to consider just two dimensions of space." *Id.*

244. He quoted Montaigne (1533-1592), a mild religious skeptic.

245. Noah is also mentioned in the Muslim book *Q'uran.*

246. Hastings Ctr. Rep., July-August 2003, at 3. Leon Kass stated, "The finitude of human life is a blessing for every human individual, whether he knows it or not," in *L'Chaim and Its Limits: Why Not Immortality?*, First Things (May 2001), at 17-24.

take cognizance of the astronomers inspecting nearby sunlike stars who have found more than 100 planets that lie within 200 light-years of Earth (see subsection (3), infra). As Nobel laureate Feynman wrote in his *The Pleasure of Finding Things Out*:

> It is one of the most remarkable things that in all of the biological sciences there is no clue as to the necessity of death [and] there is nothing in biology yet found that indicates the inevitability of death. This suggests to me that it is not at all inevitable, and that it is only a matter of time before the biologists discover what it is that is causing us the trouble and that that terrible universal disease or temporariness of the human's body will be cured.[247]

> *Intelligence may come by birth, but good sense is a gift of the god, not wealth.*
>
> Euripides

> *Minds are like parachutes; they work best when open.*
>
> Lord Thomas Dewar

ii) Intelligence Quotient; Millionaires; Brains of Males and Females; and Sizes in Animals

Genetic enhancement of some capability or trait might include one's mathematical and other reasoning, musical abilities, creativity, intelligence or talent.[248] "One of the major insights from the Human Genome Project has been the discovery of the number of human genes: roughly 23,000, with perhaps 50 percent of these expressed in the brain."[249] It has been pointed out that "With its hundred billion nerve cells, with their hundred trillion interconnections, the human brain is the most complex phenomenon in the

247. FEYNMAN, THE PLEASURE OF FINDING THINGS OUT 100 (1999).

248. CLARK & GRUNSTEIN, ARE WE HARDWIRED?: THE ROLE OF GENES IN HUMAN BEHAVIOR 221-38 (2000); Daniels et al., *Molecular Genetic Studies of Cognitive Ability*, HUMAN BIOLOGY 281-96 (1998).

249. JAMA 2222 (Nov. 2, 2005) (citing Sandberg et al., 200097 PROC. NAT'L ACAD. SCI. USA 11,038-43).

known universe—always, of course, excepting the interaction of some six billion such brains and their owners within the socio-technological culture of our planetary ecosystem!"[250] A normal person's IQ is 100. For a child of 10 who completes expected tasks (such as simple arithmetic, memory of certain facts, etc.), his IQ will be 100. If his mental age were that of a 12-year-old, it would be 120. If his mental age were 8, his IQ would be 80. Learning to read increases children's activity in the left hemisphere and it decreases in the right hemisphere.[251] Patients who have suffered extensive damage to their right hemispheres are generally able to speak well.[252] However, a person's true potential in all areas may not always be determined by IQ tests. William Shockley, who had co-invented the transistor in 1947 at Bell Laboratories, started a research and development company in Palo Alto, California, in 1955. He was labeled "a low empathizer" because of what was called "his crude eugenic proposal of offering $100 per IQ point below 100 to individuals with such intelligence scores who volunteered to sterilize themselves."[253]

On the side of the people who proceeded from the naturalistic premise that when we speak of body and mind "we are in both cases speaking about the natural world . . . we believe that our experience emerges out of natural (as

250. ROSE, THE FUTURE OF THE BRAIN (2005), at 3. "After all, it only takes twenty divisions for a single cell to multiply to over a million." (p. 65).

251. Turkeltaub et al., *Development of neural mechanisms for reading*, NATURE NEUROSCIENCE (2003), at 767-73. Dr. Francis Collins, director of the National Human Genome Research Institute, said that whether genetic changes have anything to do with brain size or intelligence "is totally unproven and potentially dangerous territory to get into with such sketchy data." ORANGE COUNTY REGISTER, Sept. 10, 2005, at News 20.

252. GARDNER ET AL., COGNITIVE PROCESSES AND THE RIGHT HEMISPHERE (E. Perelman ed., 1983). In his new book, *The Ethical Brain,* it has been noted how Michael Gazzaniga, who currently serves on the President's Council on Bioethics, indicated that "testing the so-called 'split brain' patients revealed that their two brain hemispheres operated independently, each hemisphere acting almost like a distinct person." Patricia S. Churchland, *Brain-Based Values*, AM. SCIENTIST, July-August (2005), at 350.

253. BARON-COHEN, THE ESSENTIAL DIFFERENCE 164 (2003). He also distinguished the IQ difference between whites and blacks, but recognized how it was demonstrated by the statistics (*i.e.,* the higher-IQ'ed blacks were only insignificantly fewer numbers.)

opposed to extra-natural or meta-physical) processes. . . . [We include] not only natural scientists, but many social scientists, philosophers, theologians, and others."[254] For example, the mind in the brain also includes emotions and survival advantages that go beyond mere information processing.[255] There are difficulties in ascertaining whether genetic or environmental causes excluded one another, as well as such effects that govern social and economic behaviors.[256] In *The Bell Curve* (1994), Richard Herrnstein and Charles Murray concluded that intelligence is largely heritable, as did Euripides two millennia ago. Without actually stating that inferior IQ scores are caused by inferior genes for intelligence, they claimed that studies of heritability of intelligence show that variation in IQ scores is better "accounted for" by variation in genetic rather than environmental factors. Although a 2003 study showed the influence of socioeconomics on IQ scores,[257] others had noted how more than

---

254. Parens, *Genetic Differences and Human Identities: On Why Talking about Behavioral Genetics Is Important and Difficult*, Hast. Ctr. Rep., Special Supp. 34, no.1 (2004), at 24.

255. Rose, The Making of Memory (Bantam, 2d ed., 2003); LeDoux, The Emotional Brain: The Mysterious Underpinnings of Emotional Life (1996); Damasio, The Feeling of What Happens (1999). When commenting upon Damasio's book and Descartes' famous "cogito ergo sum"—"I think, therefore I am"—"To understand the evolution of brains and behavior, and the emergence of humanity, we need at the very least to insist on 'Emotion ergo sum.'" Rose, The Future of the Brain 54 (2005), *citing* H.A. Rose, *Changing constructions of consciousness*, J. Consciousness Studies (1999) at 249-56.

256. Block, Cognition 106.

> Genes survive if they build bodies that flourish in their normal environment. . . . Natural selection therefore favors those genes that cooperate harmoniously in the joint enterprise of building bodies within the species. I have called the genes "selfish cooperators." There turns out to be, after all, an affinity between the harmony of a body and the harmony of an ecosystem. There is an ecology of genes.

Dawkins, A Devil's Chaplain 227 (2003). Costello et al., *Relationships between Poverty and Psychopathology: A Natural Experiment*, 290 JAMA 2023-29 (2003); M. Rutter, *Poverty and Child Mental Health: Natural Experiments and Social Causation*, 290 JAMA 63-64 (2003).

257. Turkheimer et al., *Socioeconomic Status Modifies Heritability of IQ in Young Children*, 14:6 Psychological Science (2003), at 623-25. "The claim, for example, that heredity accounts for 60% of the variance in IQ does not mean

one gene may explain phenotypic differences with respect to intelligence or other behavior.[258] For example, a key to becoming a bioethicist is recognition of the fact that although it is in a multidisciplined field, you would seek to ascertain the results of other fields—including law, medicine, and of several disciplines in the field of science.[259] In 1995 Bill Gates, of Microsoft, was considering computers in the fields of education, regulation, and the balance between individual privacy and community security. He stated: "We've got a good number of years to observe the course of the coming revolution, and we should use that time to make intelligent rather than reflexive decisions."[260] A

---

that 60% of an individual's IQ score, or deviation from the mean IQ, is genetically determined (if such claims are even intelligible)." Wasserman, J. L. MED. & ETHICS (2004), at 24-25, *citing* Sober, *Separating Nature and Nurture, in* GENETICS AND CRIMINAL BEHAVIOR, D. Wasserman & R. Wochbroit eds. (Cambridge Univ. Press, 2001), 47-48, at 55-56.

258. Plomin, *Genetics, Genes, Genomics, and G,* 8:1 MOLECULAR PSYCHIATRY 1-5 (2003); PLOMIN ET AL., BEHAVIORAL GENETICS 65; NUFFIELD COUNCIL OF BIOETHICS, GENETICS AND HUMAN BEHAVIOR 57; Gottesman & T.D. Gould, *The Endophenotype Concept in Psychiatry: Etymology and Strategic Intentions,* 106:4 AM. J. PSYCHIATRY 636-45 (2003); and Rutter, *Nature, Nurture, and Development: From Evangelism through Science toward Policy and Practice,* 73:1 CHILD DEVELOPMENT 1-21, at 4 (2002). "Knowing that a trait is genetically influenced . . . is of zero use on its own in understanding causal mechanisms."

259. For example, like much of the IQ test, the LSAT of the Law School Admissions Council "doesn't ask you to repeat memorized facts or to apply learned formulas to specific problems. In fact, all you'll be asked to do on the LSAT is think—thoroughly, quickly, and strategically. The LSAT is designed to test the critical reading, data management, and analytical thinking skills that have been deemed necessary (by the governing body of law schools themselves) for success in the first year of law school." *See* Kaptest.com, LSAT 2006 Edition.

260. GATES, THE ROAD AHEAD 252 (1995). The following year, a study of genetic factors in intelligence found them to be "even more controversial than the study of the genetics of behavior in general. . . . The fact that families at the lower end of intelligence tend to have more children than families at the higher end has been a subject of considerable concern. Since all agree that genes play some role, this would tend to select for those genes contributing to a lower intelligence level. Whether this is actually happening has been a subject of much discussion. All factors considered suggest that in the twentieth century there has been a slight downward trend per generation in IQ and that this trend may accelerate as fertility rates in advanced countries continue to decline." COMINGS, THE GENE BOMB 48 (1996).

decade later (2005) John D. Loike, a bioethicist at Columbia University, said: "We're entering territory we're just beginning to understand." However, the National Academy of Sciences published guidelines advising scientists to avoid implanting human embryonic stem cells into the developing brains of nonhuman primates. To the contrary, Evan Snyder, program director and professor at the Burnham Institute in San Diego, injected monkeys with human brain cells as part of his research on pediatric brain diseases. He asked, "If we're going to make mistakes, isn't it better to discover them in animals rather than having them pop up in your child?"[261]

Back in 1987, a cognitive psychologist declared, "Psychologists cannot agree about what it is they're measuring. IQs are an important part of intelligence, but they are only a part. Problems, however, arise when we equate IQ with intelligence; in general, especially in adults . . . mental age levels off somewhere around twenty, while chronological age keeps increasing."[262]

Persons have autism when they showed abnormalities in communication (not speaking or poor language ability) and low IQs.[263] As of 2003, about one in 200 children had one of the autism spectrum conditions, which includes Asperger Syndrome (AS), a variant of autism with the same conditions, but having a normal or high IQ.[264] Their sex ratio is at least *10 males to every female*, which

261. Weintraub, BUS. WEEK (Jan. 16, 2006), at 75.

262. PETERS, PRACTICAL INTELLIGENCE 22 (1997). "To continue to regard the traditional intelligence tests as a general intelligence measure when applied after the age of 20 is pure illusion." Raymond Cattell, *Are IQ Tests Intelligent?*, *in* READINGS IN PSYCHOLOGY TODAY, James Mass ed. (1974).

> A revised version of the Scholastic Aptitude Test (SAT) college-entrance exam begins to be used on March 12, 2005. Its basic changes include new kinds of grammar questions, more advanced math, more reading comprehension, and less vocabulary.

ORANGE COUNTY REGISTER, Feb. 21, 2005, at News 31.

263. BARON-COHEN, THE ESSENTIAL DIFFERENCE 134 (2003); Am. Psychiatric Ass'n, DSM-IV DIAGNOSTIC AND STATISTICAL MANUAL OF MENTAL DISORDERS, 4th ed. (1994). For example, "dyslexia, a learning disorder that afflicts at least 5 percent of elementary school children, is characterized by difficulties in perceiving sounds within words, spelling and reading problems, and troubles with written and oral expression. . . . Genetic testing for susceptibility to dyslexia is a realistic possibility in the future." SCIENCE NEWS, Nov. 5, 2005, at 292.

264. *Id.* at 136. The number rose to 1 in 200 by the year 2000. IQ tests measure verbal comprehension, analytical skills, perceptual organization, working

suggests that autism is heritable and it is difficult to predict their feelings, thoughts, and behavior.[265] They have "a different kind of intelligence."[266]

Although IQs are often called "predictors of success," they have "serious shortcomings." That is, "the fact that something (IQ) is correlated with something else (success) doesn't [always] mean that the first thing *caused* the other."[267] In this connection, one author noted, "It appears that the apparent trend to higher IQs may represent some poorly understood aspects of test-taking rather than a true increase in intelligence and cognitive ability. Hernstein and Murray termed this the *Flynn effect*"—the suggestion that "the increases might be more related to improvements in test-taking skills than intelligence *per se*."[268] A study has found that "successful intelligence has three components, and analytical intelli-

---

memory, and processing speed. *Also see* Gardner, *Intelligence Reformed: Multiple Intelligences for the 21st Century* (2000), and Thompson et al., *Genetic Influences on Brain Structure*, NATURE NEUROSCIENCE (2001), at 1253-58 (showing that differences in intelligence scores vary with the amount of gray matter in the frontal lobes).

265. Baron-Cohen et al., *in* J. CHILD PSYCHOLOGY & PSYCHIATRY, at 241-52. In 2004, a study's epidemiological findings were sufficient to state that "reading disability is truly more frequent in boys than girls. There now needs to be research to determine the causal influences that underlie the sex difference." Rutter et al, *Sex Differences in Developmental Reading Disability*, JAMA 2007 (April 28, 2004).

266. BARON-COHEN & BOLTON, AUTISM: THE FACTS (1993); MYERS ET AL., THE EXACT MIND (2003). In a 2004 study, scientists expressed their belief that 5-10 genes produce autism. AM. J. PSYCHIATRY (April 2004); ORANGE COUNTY REGISTER, April 2, 2004, at News 9.

267. *Id.* PETERS, *supra* note 261, at 23-24. A worldwide study found that the United States had the highest percentage of citizens suffering from some form of clinically diagnosed mental disorder. Kessler et al., *Prevalence, Severity, and Unmet Need for Treatment of Mental Disorders in the World Health Organization World Mental Health Surveys*, JAMA 2581-90 (June 2, 2004).

268. COMINGS, THE GENE BOMB 45 (1996) (citing HERNSTEIN & MURRAY, THE BELL CURVE 845 (1994)); Flynn, *Massive IQ gains in 14 nations: What IQ tests really measure* (1987), PSYCHOLOGICAL BULL. 171-91 (1987). A study in 1985 found adults without children had slightly higher IQs than those with children, showing that a small decrease in the IQ rates has been occurring during most of the twentieth century. Van Court & Bean, *Intelligence and fertility in the United States*, INTELLIGENCE 23-32 (1985). That is, those with fewer or no children would have the higher IQs.

gence is just one. The others are creative intelligence and practical intelligence, or common sense."[269] Thus, it has been shown that "many millionaires did not have a high enough grade point average to be hired by major corporations. But they still wanted to become financially successful, so many chose to be self-employed. They hired themselves when other employers would not."[270] Unlike the lower ages for obtaining an IQ, millionaires indicated that "they were able to develop, nurture, and enhance their courage when they were in their forties and even fifties."[271] Someday in the twenty-first century, the enhancement of genes necessary to obtain such courage may be brought to the attention of those who would choose to have it. It has been argued that allowing a genetic enhancement for ourselves, or as parents, "we must make a choice; the natural lottery or rational choice. Where an enhancement is plausibly good for an individual, we should let that individual decide. And in the case of the next generation, we should let parents decide. To fail to allow them to make these choices is to consign the next generation to the ball and chain of our squeamishness and irrationality."[272]

Although in the past it was considered that the brains of men and women

---

269. STANLEY, THE MILLIONAIRE MIND 14 (2000), *citing* STERNBERG, SUCCESSFUL INTELLIGENCE (1996). "Most of us (93 percent) indicated that our school and college experience was influential in determining that hard work was more important than genetic high intellect in achieving by the results of standardized test scores what we were not: intellectually gifted, of law-school caliber, [or] medical-school material."

270. *Id.* STANLEY at 20.

> I'm a full professor with an endowed chair at Yale. I've won many awards, published over six hundred articles and books, and been awarded about $10 million in research grants and contracts. . . . I bombed IQ tests when I was a kid. Why was that so lucky? Because I learned in elementary school that if I was going to succeed, it wasn't going to be because of my IQ. And . . . just as low scores on tests of inert intelligence don't preclude success, neither do high scores guarantee it.

STERNBERG, SUCCESSFUL INTELLIGENCE 12 (1996).

271. *Id.* STANLEY, at 141. In his survey of millionaires, only one-third were "religious millionaires" who "pray when confronted with fear and worry. Only 8 percent of the other millionaires (OMs) employ prayer in a similar fashion. That is a ratio of more than nine to one." *Id.* at 174.

272. Professor Julian Savulescu, *New breeds of humans: the moral obligation to enhance*, Reproductive Biomedicine Online: www.rbmonline.com/

were similar except for aspects of reproduction, beginning near the turn of the third millennium, that view "has now been knocked aside by a surge of findings that highlight the influence of sex on many areas of cognition and behavior, including memory, emotion, vision, hearing, the processing of faces and the brain's response to stress hormones."[273] It has now been shown that "anatomical differences occur in every lobe of male and female brain. For instance, many regions are proportionally larger in females than in males but other areas are larger in males." Accordingly, "growing numbers now agree that going back to assuming we can evaluate one sex and learn equally about both is no longer an option."[274]

In 2001, the National Institutes of Health found insufficient basic and preclinical data to justify the conduct of *in utero* gene transfer clinical research; significant additional preclinical and relevant clinical studies are required along with further deliberations of the ethical issues. Hence, the NIH guideline concluded, "it is premature to undertake any *in utero* gene transfer clinical trial." However, as noted above, when prescribed criteria concerning the efficacy and risks of human *in utero* gene transfer are met, the NIH's Recombinant DNA Research Advisory Committee (RAC) "would be willing to consider well rationalized human *in utero* gene transfer clinical trials."[275] When this occurs, the members of that committee might be requested by the chairman of

---

Article/1643, on Dec. 9, 2004: "If we were really serious that embryos were people, we would force couples undergoing IVF to donate spare embryos to other infertile couples just as we force couples who do not or cannot care for their children to have them adopted by other couples. But of course, most people really do not believe embryos are children. More importantly, no one would object to the treatment of disability in a child, if it were possible. Why then not treat the embryo with genetic therapy if that intervention is safe? Even though not a child, it might later be a child. And better that child without disability than with disability. This is no more thwarting God's will than giving antibiotics is." *Id.*, and 1:1 J. ETHICS, L. & MORAL PHILOSOPHY OF REPRODUCTIVE BIOMEDICINE (March 2005).

273. Cahill, *His Brain, Her Brain*, SCIENTIFIC AMERICAN, May 2005, at 42.

274. *Id.* at 47.

275. 66 C.F.R. §§ 1146-1147, Jan. 5, 2001. In 1994, the Report of the Human Embryo Research Panel indicated that "National Institutes of Health refers to a fertilized ovum *in vitro* that has never been transferred to or implanted in a uterus. This includes '*ex utero* preimplantation embryo' or 'preimplantation embryo,' a fertilized ovum that has been flushed from a woman before implantation in the uterus." Its Ethical Considerations stated:

the President's Council on Bioethics "to act in accordance" with a procedure that recognizes a more religious than an agnostic point of view. But, as stated by the co-discoverer of its structure, "Our DNA, the instruction book of human creation, may well come to rival religious scripture as the keeper of the truth."[276] Such a viewpoint could be adopted some time during the third millennium by those people who embrace the current science of evolution (see Chapter 2). Near the middle of the twentieth century, the geneticist Theodosius Dobzhansky noted that "nothing in biology makes sense except in light of evolution." Since the turn of the third millennium, a professor in human biology noted that scientific changes in evolution have approached human life as "creatures of the earth."[277] Another has argued that there is a moral obligation to test for genetic contribution to non-disease states such as intelligence and memory. On the basis of "procreative beneficence," people should select, based on the relevant, available information of the children they could have, the one who is expected to have the best life, or at least as good a life as the other. Although that may be a valid autonomous goal, as noted above, more work is needed on intelligence as a genetic change.[278]

We must also consider the brains of animals that are either smaller or

---

> In recommending public policy, the Panel was not called upon to decide which of these views is correct. Rather, its task was to propose guidelines for preimplantation human embryo research that would be acceptable public policy based on reasoning that takes account of generally held public views regarding the beginning and development of human life in light of the best.

276. WATSON, DNA, THE SECRET OF LIFE 404 (2003).

277. Hulbert, *From Biology to Biography*, THE NEW ATLANTIC 50 (Fall 2003). "The root meaning of the word *human*—derived from the Latin for 'earth' or 'soil.'"

278. In 2000, for example, it was argued that:

> There must be multigenerational data showing that the modification or improvement of a specific genetically determined trait is stable and effective and does not interfere with the functioning of other genes. [And] it could take another sixty to eighty years to have any multigenerational data.

FRANKEL & CHAPMAN, HUMAN INHERITABLE GENETIC MODIFICATIONS: ASSESSING SCIENTIFIC, ETHICAL, RELIGIOUS, AND POLICY ISSUES 23 (American Association for the Advancement of Science, 2000).

larger than those in humans. In 2002, Donald Kennedy, the editor-in-chief of the journal *Science* stated: "As more and more is learned about the behavior of animals, it becomes for me at least more and more difficult to get closure on a set of properties that are uniquely and especially human, [and] can be defined unambiguously in that way. So, as we learn more and more about the neural and behavioral capacities of animals, I think the zone of what we think of as uniquely human is gradually shrinking. And as we learn more about how their brains work it may well change our attitudes about how different we are from them, thus reducing our sense of being all that special. . . .There's this awkward growth of knowledge."

Amazing to many people was Darwin's volume *The Descent of Man* (1871), in which he wrote about the ability of ants to communicate information to each other, and unite for the same work, etc., and then actually concluded that "the mental faculties of man and the lower animals do not differ in kind, though immensely in degree." People thought that Darwin should not have selected ants for this conclusion. However, Jeremy Narby, in reviewing these quotations along with others concerning bees, noted that "for scientists, the great advantage of the bee brain is that it can handle complex mental tasks with less than a million neurons."[279] In this connection he pointed out that "our species is very young. In comparison, octopuses have been around for several hundred million years, which has given them time to hone their skills. By comparison, we are just getting started."[280]

It has been noted that although "the brain of an elephant, for example, is about four times as large as a man's, a blue whale's almost six times as large," their "size, it turns out, isn't everything." It has been pointed out that the folds of the brain may be important because they "are more developed in man than in any of the beasts."[281]

In this regard, it was recently noted that for years "scientists assumed that only neurons specify the connections they make to other neurons. But evidence shows that glia can strongly influence how many synapses a neuron

279. NARBY, INTELLIGENCE IN NATURE 61 (2005) (citing Giurfa, Martin et al. (2001), *The concept of 'sameness' and 'difference' in an insect*, 410 NATURE at 930-33). Narby pointed out, "Though bees have brains the size of pinheads, they can master abstract rules." *Id.* at 56.

280. *Id.* at 82.

281. BURRELL, POSTCARDS FROM THE BRAIN MUSEUM 18 (2004).

forms and where it forms them":

> The recent book *Driving Mr. Albert* tells the true story of pathologist Thomas Harvey, who performed the autopsy of Albert Einstein in 1955. After finishing his task, Harvey irreverently took Einstein's brain home, where he kept it floating in a plastic container for the next 40 years. From time to time, Harvey doled out small brain slices to scientists and pseudoscientists around the world who probed the tissue for clues to Einstein's genius. Marian C. Diamond of the University of California . . . [discovered] a suprisingly large number of nonneuronal cells known as glia—a much greater concentration than that found in the average head. . . . Propped up by glia, neurons were free to communicate across tiny contact points called synapses and to establish a web of connections that allow us to think, remember and jump for joy. . . . Perhaps a higher concentration of glia, or a more potent type of glia, is what elevates certain humans to genius.[282]

However, others found reasons to criticize Diamond's findings.[283] More information on the importance of making such changes may be discovered in the twenty-first century. Jerry Fodor, professor of philosophy at Rutgers University, stated, "If neuroscience cannot start until psychology gets finished, neuroscience is likely to be in for a long wait."[284] For example, with respect to

---

282. Fields, SCIENTIFIC AMERICAN (April 2004), at 55, and Marian Diamond et al., *On the Brain of a Scientist: Albert Einstein*, 88 EXPERIMENTAL NEUROLOGY 198-200 (1985). *See also* ABRAHAM, POSSESSING GENIUS: THE BIZARRE ODYSSEY OF EINSTEIN'S BRAIN (2004).

283. *Id.* Burrell (p. 282), *citing* Kantha, *Albert Einstein's Dyslexia and the Significance of Brodmann Area 39 of His Left Cerebral Cortex*, 37 MEDICAL HYPOTHESES 119-22 (1992); and Terence Hines of Pace University noted that Diamond's study "is so seriously flawed that its conclusions should not be accepted." EXPERIMENTAL NEUROLOGY (1998).

284. On one hand, you can't really ask serious questions about how the brain works if you have no idea what line of work it is in. You need some psychology to prime the neurologist's pump. But on the other hand, *there is not much psychology around that you can rely on if* the aspects of mind *you wish to study* are *the "higher cognitive" processes* (such as, in particular, thinking). (Emphasis added.)

Fodor, *Making the Connection*, TIMES LITERARY SUPP., May 17, 2002, at 3.

functional magnetic resonance imaging (fMRI), a technique that first appeared in the early 1990s to explore complex brain functions involved in human motivation, reasoning, and social attitudes, it was noted that there has been little to no research done on how people understand fMRI results.[285] Yet, it has already been demonstrated how a fetus's neuron's synapses—points of contact—begin "to reach tentacles out to each other, establishing at a rate of two million a second." However, shortly before birth, the trend reversed. "Groups of neurons competed with each other to recruit other neurons into expanding circuits with specific functions. Those that lost died off in a pruning process scientists call 'neural Darwinism.'"[286]

good v. Good

Laws should be enacted to prohibit all modifications of the *Homo sapiens* species by the transfer of embryos or reproductive cells that have undergone inheritable genetic modification.

v.

Studies of germline gene therapy must include necessary clinical tests prior to authorization to conduct such therapy. Authorization will require that (a) the risks of side effects cannot be of great significance, and (b) such therapy will not constitute a change of *Homo sapiens* into another "biological species" for purposes of reproduction. As noted by Jeremy Narby: "The word *sapiens* means 'wise' in Latin. Whether this label truly corresponds to humans remains to be determined."[287]

---

285. Friedrich, *Neuroscience Becomes Image-Conscious as Brain Scans Raise Ethical Issues*, JAMA 781 (Aug. 17, 2005). Robert Klitzman, M.D., professor of clinical psychiatry and codirector of the Center for Bioethics at Columbia University, lists some of the questions that need to be addressed: "How are findings from brain scans interpreted? What do patients understand of what's been said? Is any sort of counseling being provided to help the patient cope with results?" *Id.*

286. National Geographic, March 2005, at 10. Arthur Toga, director of the Laboratory of Neuro Imaging at UCLA, stated, "In the old days, people said the brain is like a computer. I'd say no. Images get decomposed and then recomposed. It's very distributed, closer to the Internet." *Id.* Functional magnetic resonance imaging noninvasively monitors increases in blood flow to measure cognitive activity.

287. Narby, Intelligence in Nature 77 (2005).

Twenty-first Century Approach

1. Germline gene therapy should continue to be authorized by the FDA or NIH when a study demonstrates that the risk of side effects is remote and insignificant, and
2. Clinical tests of future Germline Gene Therapy must also be shown to entail significant efficacy when they are authorized to be conducted with respect to clinical transfers in humans.

The brain—is wider than the sky.

Emily Dickinson

It's only fair to warn you, right at the start, that this is a story with no ending.

Arthur C. Clark, from *The Other Side of the Sky* (1957)

More than 120 extrasolar planet have been discovered, mostly gas giants. But what everyone really wasn't to find are "exo-Earths" orbiting other stars. Any day now, say the astronomers.

Robert Zimmerman

iii) The Importance of Germline Gene Therapy for Space Travel

The space age began on October 4, 1957, when the Soviet Union launched its little Sputnik satellite. On January 21, 1958, the United States entered the age with its first successful launch into orbit of the satellite Explorer. This was followed between 1969 and 1973 by several manned Apollo missions to land on the moon; but no human has yet traveled much beyond Earth's orbit. The Commercial Space Act, which the first President Bush signed into law in 1992, made a finding that "activities of the private sector have substantially contributed to the strength of both the United States space program and the national economy."[288] They now include the means to place "a launch vehicle

288. 15 U.S.C.A. § 5801.

Under the Commercial Space Launch Amendments Act of 2004, the FAA must wait eight years to issue regulations to protect the safety of passengers and crew, unless a serious safety problem, injury or death occurs before then. Space entrepreneurs formed a group last month called the Industry Consensus Standards Organization to set their own safety standards.

Orange County Register, Feb. 11, 2005, at News 24.

and its payload in outer space."[289] The U.S. Olsen and an American-Russian crew hurtled toward the International Space Station . . . on a Soyuz craft. "He reportedly paid $20 million for a seat on the Expedition 12 flight."[290]

As noted in Chapter 2, visits by astronauts to the planct Mars (which is about half as big as Earth) will soon take place in the twenty-first century. I had prepared the contract to use nuclear power for NASA's twin Mars Viking rovers of the late 1970s. Viking I landed on July 20, 1976. Viking 2 carried television cameras, meteorological instruments, atmospheric analysis equipment, soil analysis equipment, a seismometer, and a small chemical laboratory designed to detect signs of life in the soil.[291] Gerald Soffen, project scientist for the Viking missions, stated, "All the signs suggest that life exists on Mars, but we can't find any bodies," and the biology team added that "the Viking results do not permit any final conclusion about the presence of life on Mars."[292] When these were duplicated in 2004 by NASA's satellites Spirit and Opportunity utilizing solar power, one of them found more evidence of previous water

289. *Id.* § 5802. A Commercial Space Achievement Award was created by Congress to help push such aerospace enterprises. *See* www.aerospace commission.gov. "With funding from Microsoft co-founder Paul Allen, Rutan and his Scaled Composites firm developed the spaceship and a launching aircraft that in October secured the Ansari X-Prize. It was a $10 million reward for flying the first private manned spacecraft to an altitude exceeding 328,000 feet twice within a 14-day period. By 2050, [Burt Rutan] predicted, spaceliners would become so common that passengers 'will be bored looking out of a suborbital space flight as we do on an airliner.'" ORANGE COUNTY REGISTER, Nov. 21, 2004, at News 8.

290. ORANGE COUNTY REGISTER, Oct. 2, 2005, at News 25. "U.S. law bars NASA from making such payments to Russia. The U.S. Senate has agreed to amend the measure and lift the ban on NASA purchases of Soyuz seats until 2012. However, the House had not yet acted on the issue." *Id.*

291. *Also see* KIDGER, ASTRONOMICAL ENIGMAS 115 (2005).

292. *Id.* at 117; THE NEW SOLAR SYSTEM 278 (3d ed., Beatty & Chaiken eds. (Sky Pub., 1990)). "On Mars, the temperature races down at sunset. Viking I measured a typical variation at Chryse, from -35°C at 6:00 P.M. to -50°C at 8:00 P.M., to -75°C at midnight. If such variations occurred on Earth, the average temperature would not be 15°C but would instead be well below the freezing point." KIDGER at 217. "At present, the Martian atmosphere raises the average temperature of the surface by only 5°C." *Id.* at 218.

on the surface as well as at the polar caps.[293] Mars's gravity is .38 that of the Earth (if you weighed 100 pounds on Earth, you would only weigh 38 pounds on Mars), has a much lower atmosphere surface pressure, and *no magnetic field.* It contains some resemblances to the Earth: although only half its size, Mars's rotation on its own axis is similar (24 hours and 37 minutes), but its orbit of the sun takes 687 days (almost twice as long as the earth). It has been noted how the story of Mars "carries a moral of a good planet gone bad, and thus a warning to residents of its vibrant neighbor. Exploring Mars, then, might help save the Earth."[294]

Even though "oxygen is the most common element in Martian soil, it amounts to a mere .13 percent of the Martian air. Oxygen has only itself to blame. Earth's air has oxygen only because the planet (below) has life. If all terrestrial life perished, atmospheric oxygen would vanish just 4 million years later."[295] Are all planets in the solar system lifeless except Earth? "The answer may reside in the rocks of Mars" because "if Mars independently gave birth to life, no matter how simple, then the chance of life arising on a good

293. Sky & Telescope 16 (October 2004) and 19 (August 2004). This was observed even though the solar-powered rovers needed to stay near the equator. Sky & Telescope 42, March 2005. NASA also then sent an unmanned orbiter of Saturn for exploration. It was named "Cassini" for an Italian-French astronomer who noticed in 1675 the two major sections of Saturn's rings, which bear his name. It released the Huyens lander, which made a perfect landing on the moon Titan on January 14, 2005.

294. Croswell, Magnificent Mars (2003). With regard to its fossils, the writer added: "Although the Earth's age is 4.6 billion years, the most ancient known terrestrial rocks, in the Northwest Territories of Canada, are 4.0 billion years old, and the oldest known fossils, in Western Australia, 3.5 billion years old. Scientists already have a rock from Mars older than both." (p.28) *The Real Mars* (2004), by Michael Hanlon, reveals how this solar planet "went from a world teeming with life to one of utter desolation, to the current mainstream view of a once-watery and habitable planet that has long since dried up." Naeye, Sky & Telescope 101 (May 2005).

295. *Id.* at 75. "Microbes such as bacteria are just one possible explanation for the methane, detected this year by three separate teams using space-based and Earth-based instruments." Orange County Register, Jan. 4, 2005, at News 7.

planet [would be] closer to 100 percent."[296] However, in 2004, President Bush indicated his desire to have NASA begin to prepare for sending Americans to Mars within a score of years. NASA's new administrator noted that our expansion into space is "about where human beings go and what they do when they get there and what that means to the future of the human race."[297] The European Space Agency is pushing a multidecade plan of exploration called Aurora, "its most ambitious incarnation."[298] That trip itself will take two to three years, and demands that procedures be taken to prevent an astronaut from being debilitated over much of that period.

Although the search for extraterrestrial intelligence (SETI) terminated as a NASA program in 1992, with the support of private donors, the SETI Institute continues research and education projects.[299] The most ambitious plan to date is to erect 350 radio antennas across about 150 acres of a region

---

296. *Id.* at 161. Astronomy Professor Steven Squyres of Cornell University, who is also the current principal investigator for the Mars Exploration Team, stated: "I'm a huge fan of sending robots to Mars—that's what I do for a living. But even I believe that the best exploration, the most comprehensive, the most inspiring exploration is going to be conducted by humans." *Father of Spirit and Opportunity*, SCIENTIFIC AMERICAN, October 2004, at 46.

297. On May 11, 1990, then President Bush had "set a target for landing humans on Mars in the year 2019, the half-century anniversary of the Apollo 11 landing on the moon." KLERHX, LOST IN SPACE 290 (2004). NASA chief Michael Griffin believes that a majority of people "want to make sure that as humankind expands into space, the United States is there in the forefront." ORANGE COUNTY REGISTER, June 2, 2005, at News 18.

298. *Id.* at 242-43. On January 14, 2005, a satellite was scheduled to arrive on Saturn's moon, Titan, after release from NASA's Cassini orbiter. ORANGE COUNTY REGISTER, Jan. 14, 2005, at News 17. It landed there later and reported information thereon.

299. *Id.* at 6-7.

> SETI pioneer Frank D. Drake (SETI Institute), Stone and about a half dozen other astronomers and students use the 40-inch Nickel reflector to watch for coherent flickers of finely tuned light – possible laser signals from technologically advanced civilizations on planets orbiting other stars.

SKY & TELESCOPE 36 (June 2004).

north of California's dormant Lassen Peak volcano.[300] It has been noted that the International Space Station (ISS), the largest man-made object ever to orbit the Earth, should be able to discover how humans survive for long periods of time in space and solutions to the sensation of weightlessness.[301] A physicist at the Institute for Advanced Study at Princeton, New Jersey, pointed to the necessary space activities in this century: "The dream of expanding the domain of life from Earth into the universe makes sense as a long-term goal but not as a short-term goal. . . . Any affordable program of manned exploration must be centered on biology."[302] In 1976 Herman Kahn and his co-author's book *The Next 200 Years* concluded that our "capability for self-supporting existence in space would make possible the continuation of earth's civilization and the resuscitation of human life on the planet following an irreversible tragedy in the case of thermonuclear war or ecological catastrophe." Because the Cold War ended in 1991, it would appear that an "irreversible tragedy" in the third millennium would more likely relate to ecological and other possible catastrophies (see Chapter 1, Section 2).

To a great extent the needed changes in our genetic conditions (including the extension of our ages) will depend upon those forthcoming in the field of germline gene therapy. This will present our confidence regarding "interstellar space travel in a light entirely different from that cast on it by a rapidly evolving society of short-lived individuals."[303] Although bioethics in space

300. This SETI project, called the Allen Telescope Array, will have a collecting area of more than two acres, making it one of the world's largest steerable radio telescopes. . . . A conservative estimate from the Drake equation is that around 10,000 civilizations are advanced enough for SETI observations to find, estimates SETI astronomer Seth Shostak. "The bottom line comes out that even if you take Drake's more pessimistic ideas, you'll trip across a civilization by the year 2025."

301. NASA's studies on mice and rats born in space imply that the absence of gravitational clues in infancy may have serious effects on adult animals. Much research remains to be done in this area before we celebrate the first space-born human astronaut. But once the first has been born, can the one millionth be far behind?

GOLDSMITH, VOYAGE TO THE MILKY WAY 35 (1999).

302. *Id.* at 119 (quoting from Freeman Dyson).

303. *Id.* at 230.

> Of all extrapolations into the future, one of the most straightforward assumes that humans will continue the habits of previous millennia by seeking new worlds to explore and to inhabit.

has hardly begun to come of age, NASA is jointly supporting a project with the American Association for the Advancement of Science to examine "space exploration that may transport microbial life beyond earth, or have an impact on extraterrestrial life." This is at least a commencement of NASA's "research roadmap for astrobiology."[304] Astrobiologists normally define a "habitable zone" around stars as the distance at which "an orbiting planet might sustain liquid water on its surface for extended durations." This generally may be like planets with the orbits of Venus and out to Mars, depending upon the star's mass and age.[305] It is also the Earth's overpopulation (see Chapter 7) that suggests space colonization will commence in 2085. Astrophysicist Professor Freeman Dyson at Princeton University said: "To give us room to explore the varieties of mind and body into which our genome can evolve, one planet is not enough."[306] That suggestion is bound to take place during the third millennium. Discoveries of extrasolar planets are coming quickly; over 170 were announced in 2006.[307] It was formerly claimed that most stars are multiple stars, and single stars are less common; however, in March 2006 it was noted that "over two-thirds of all stars are single. Star formation theories will have to be adjusted to cope with the new statistics. Since theoretically planets form

---

304. PLANETARY REPORT 11 (July/August 2004).

305. Davies, SKY & TELESCOPE 45 (June 2004). In this connection it is helpful to recall the late Carl Sagan's description of the Earth as "a lonely speck in the great enveloping cosmic dark." SAGAN, PALE BLUE DOT 7 (1994).

306. FREEMAN DYSON, THE SUN, THE GENOME AND THE INTERNET, TOOLS OF SCIENTIFIC REVOLUTIONS (1999). In 2005, scientists utilizing a new light sensor at the W.M. Keck Observatory in Hawaii discovered a new planet, dubbed a "super-Earth," orbiting a star 15 light-years away—"a milestone in the search for a world outside our solar system that could sustain life." L.A. TIMES, June 14, 2005, at 8.

307. "One of these objects is the third known planet orbiting its star, making it the fourth known triple-planet system around another Sun-like star." SKY & TELESCOPE 19 (May 2005).

> Our first port of call is a mere 4.4 light-years away from the Sun, the Alpha Centauri binary star system. Alpha Centauri A and B are two Sunlike stars separated on average by a little over two billion miles. . . . This is disappointing because Alpha Centauri is so close. It is reachable in a relatively short 44 years by a spaceship traveling at one-tenth the speed of light.

VILLARD & COOK, INFINITE WORLDS (2005).

more often about single stars, this new finding implies planets are more common than previously thought."[308]

It has also been noted that "an atomic engine would not be able to generate more than an extremely low thrust—perhaps only a tenth of a gravity." However, "after three weeks of accelerating," the speed would be "1,800 kilometers per second, or well over 150 times the escape velocity of the Earth. In that time the spacecraft would have traveled 1,700 million kilometers, which would put it in the orbit of Saturn. Besides greatly reducing the medical problems associated with zero gravity, such a ship could reach Mars in less than a month, even allowing for deceleration halfway."[309] Such a method would also appear to be applicable to our travels to extrasolar planets.[310] During the previ-

---

308. Don Lynn, *Astrospace Update gathered from NASA and other sources*, Sirius Astronomers 8 (March 2006).

309. Kidger, Astronomical Enigmas 237 (2005).

> In March 2006, the Jet Propulsion Laboratory in Pasadena, which operates the probe for the Cassini spacecraft, captured images of jets of water vapor exploding from underground through vents on Enceladus, a moon of Saturn. This moon may have chemicals needed for life to arise, according to a JPL planetary scientist who deals with Cassini.

Orange County Register, March 1, 2006, at News 6.

> It has also been said that Saturn "offers the attraction of the dense, hydrogen-rich atmosphere of its moon, Titan, which would make it an ideal place for refueling. Interplanetary travelers preparing to return home could stop there to refill their fuel tanks. Especially because Titan would probably be the solar system's one and only gas station, it is potentially a very valuable piece of real estate."

Kidger, Astronomical Enigmas 240 (2005).

310. The Nuclear Engine for Rocket Vehicle Application (NERVA) may be used as a nuclear reactor to heat liquid hydrogen. "There were plans to use this engine either in the third stage of an expanded Saturn V rocket for a manned mission to Mars or in the Nova booster that was initially the rival of, and later proposed as the successor to, the Saturn V. . . . The trouble was that, as originally conceived, spacecraft equipped with these engines would have been flown in the atmosphere, and this raised fears about accidents. In open space, though, nuclear propulsion is without doubt the most sensible way forward. In a test on the ground, a later version of NERVA ran nonstop for fifty hours, thereby demonstrating that rapid, continuous-thrust flight to the planets is far from a fantasy." Kidger, at 238. A spacecraft traveling at 17,000 kilometers per second would be going some 5% of the speed of light. *Id.* at 256.

ous year Voyager I passed the border between our solar system and interstellar space. By 1980 it had taken images of Jupiter, Saturn, and some of their moons, and took the next 25 years to reach the heliosheath—some nine billion miles (or 90 astronomical units) from the sun—a space region none of our spacecraft has ever visited before; further, it was flying under the guidance of controllers at the Jet Propulsion Laboratory in Pasadena, California. Although most ground-based telescopes cannot directly observe extrasolar planets, they have been detected around stars by their Doppler shift. The gravity of an orbiting planet tugs ever so slightly on the star. This causes a periodic shift in the star's spectrum. Roughly 5 percent of main-sequence stars appear to have planets around them.[311]

Some such planets may have been "seeded" from outer space because amino acids can be synthesized from water, methanol, ammonia, and hydrogen cyanide in interstellar space and may warm up and change upon arrival at a proper planet.[312] While traveling there, an astronaut might experience Space Adaptation Syndrome. Although it can cause headach2es and nausea to more severe illness such as vomiting, it usually clears up during the first few days of a space mission. Weightlessness, which has a more serious impact, is counteracted by undergoing regular periods of exercise, certain amounts of acceleration or deceleration, and/or by having the spacecraft rotate to create "centrifugal force to provide some or all of the gravitational pull experienced on Earth."[313]

Ray Kurzweil suggested that an "ultimate thinking machine" could produce "ways in which humans are intelligent, including musical and artistic aptitude, creativity, physically moving, and even responding to emotion" during the first half of the twenty-first century. [314] This would require considerable extension of the work by persons currently producing software programs at Microsoft. [315] As Harvard's Edward Wilson pointed out in 1998,

311. National Geographic Encyclopedia of Space (2005).

312. Francis Crick suggested such "panspermia" of seeding from outer space. *See* Crick, FHC, Life Itself: Its Origin and Nature (1981); Rose, The Future of the Brain 13 (2005).

313. National Geographic Encyclopedia of Space (2005).

314. Kurzweil, *The Evolution of the Mind in the Twenty-First Century, in* Are We Spiritual Machines? 12, 29, 44-45 (Richards ed., Discovery Institute, 2002).

315. Halpern, The Great Beyond 298 (2004).

modern evolutionists agreed that "conscious experience is a physical and not a supernatural phenomenon."[316] Although some bioethical criticisms have been made about such machines,[317] they might become useful in space trips to the planets when they are preliminary to humans traveling there. Its machine and software are still not identical to human beings. In 1995, Bill Gates, CEO of Microsoft, stated: "Another fear people express is that computers will be so 'smart' they will take over and do away with any need for the human mind. Although I believe that eventually there will be programs that will re-create some elements of human intelligence, it is very unlikely to happen in my lifetime. For decades computer scientists studying artificial intelligence have been trying to develop a computer with human understanding and common sense."[318] As Nobel Prize winner John Eccles said, there's "no evidence whatsoever for the statement made that, at an adequate level of complexity, computers also would achieve self-consciousness."[319] Although Feynman, another such laureate, noted that almost all "conventional computers work on a layout or an architecture invented by von Neumann" (1903-1957), he also stated that it is always better to make the machines smaller, and the question is, how much smaller is it still possible,

---

316. WILSON, CONSILIENCE 132 (1998), *quoted in* STROBEL, THE CASE FOR A CREATOR 248 (2004).

317. Critics included Professor John Searle, University of California, who wrote, "You can expand the power all you want, hooking up as many computers as you think you need, and they still won't be conscious, because all they'll ever do is shuffle symbols," and William Dembski of Baylor University said, "Kurzweil is peddling science fiction and bad philosophy." *Id.* at 248.

318. GATES, THE ROAD AHEAD 255 (1995). "Every prediction about major advances in artificial intelligence has proved to be overly optimistic. Today even simple learning tasks still go well beyond the world's most capable computer. When computers appear to be intelligent it is because they have been specially programmed to handle some task in a straightforward fashion—like trying out billions of chess moves in order to play master-level chess." *Id.*

319. *Id.* at 261 (citing a quotation in ROBERT AUGROS & GEORGE STANCIU, THE NEW STORY OF SCIENCE 170). In his volume *Radical Evolution* (2004), Joel Garrean stated, "Those who worship the idea that computers are becoming sufficiently smart to be a successor species to humans would have you believe that soon we will be morally obligated to bring silicon beings inside our circle of empathy." (p. 21) [However,] "[h]uman nature is the ultimate example of the immeasurable." (p. 197)

in principle, to make machines according to the laws of Nature?[320] These would also be useful in many ways (including medical analysis) when accompanied by humans in space.

The spacecraft named for Galileo collected data orbiting Jupiter and its moons for eight years before it was sent into the planet in September 2003. John Casani, NASA Original Project Manager for the Galileo Mission, has indicated, "If we find that there's any life or evidence of life on Jupiter's satellites, it will be because we will have gone and *looked* for it—inspired by what we've seen from Galileo."[321]

Another astrophysicist, Neil Tyson, recently wrote about "the efforts made by Sir John Marks Templeton, the wealthy founder of the Templeton investment fund, to find harmony and consilience between science and religion." Apparently, "Templeton's annual religion award, with a cash value rivaling that of the Nobel Prize, has recently been won by several prolific religion-friendly scientists." [322] However, Tyson also noted a recent survey of science professionals on their religious beliefs,[323] which showed that "although 65 percent of the mathematicians (the highest rate) declared themselves to be religious," this only amounts to "22 percent of the physicists and astronomers (the lowest rate)"[324] who expressed their belief in a traditional religion. All hospital bioethicists who appreciate their own multidisciplinary needs must

320. Feynman, The Pleasure of Finding Things Out 29 (1999).

321. Kristin Lentwyler, The Moons of Jupiter (2003). Afterword by James Casani. Casani's work noted that the pictures contained in this volume "suggest there's water on Europa, Ganymede and Callisto." He also quoted from Galileo's letter in July 1637, stating in part: "By my remarkable observations, the sky . . . the world . . . the universe . . . was opened a hundred or a thousand times wider than anything seen by the learned of all the past centuries." *Id.*

322. *Id.* at 185, *citing* Larson & Witham, 394 Nature (April 3, 1997), at 313.

323. Tyson, The Sky Is Not the Limit 182 (2004).

> The national average among scientists was around 40 percent and has remained largely unchanged over the past century. For reference, 90 percent of the American public claims to be religious (among the highest in Western society), so either nonreligious people are drawn to science or studying science makes you less religious.

324. *Id.* at 186.

both evaluate how to treat the overwhelming majority of patients with their various religions (see Chapter 3) and recognize the lack of direct relationship between science and traditional religions (see Chapters 2 and 4). It appears that the beliefs most of our future astronauts may continue to remain closer to professionals on nature (like physicists and astronomers) than they are to specialists on subjects that are largely independent of nature itself.

In *The Great Beyond* (2004), physics professor Paul Halpern took us from (a) the three-dimensional universe of historically normal humans to (b) the fourth dimension of Albert Einstein (with physical evidence thereof). Potentially, we may even go into (c) the string theory universe of some modern physicists concerning tiny strands of energy vibrating in 10 dimensions (currently without physical evidence), but they are hopeful for finding it through particle accelerator experiments at Europe's Large Hadron Collider beginning in 2007.[325]

As noted above, when a number of people heard about potential germline gene transfers, they actually stated, "No trespassing beyond existing Homo sapiens status." However, following completion of the Human Genome Project in 2002, more people who are considering the advantages of bioethical aspects are saying something like this:

> Perhaps there exists some aversion to the current limits of Homo sapiens. We want to know what may become available in nature for us to cross the frontiers of evolved human beings beyond the current knowledge of our entire current genome. We are disturbed to be told "No trespassing beyond the point of our current evolution without awaiting the thousands of years normally required for the acceleration of important changes." We recognize that many dimensions of such changes may take us into and well beyond the duration of the third millennium.

325. SKY & TELESCOPE 15 (November 2004). "Perhaps it is the human aversion to limits. We want to know what is just outside the frontiers of knowledge. It disturbs us to be told, 'No trespassing beyond this point.' If nature counts to three, we want to count to four, five, or more." HALPERN, THE GREAT BEYOND (2004).

good v. Good

We must continue the current limits on modifications of *Homo sapiens* and make none that are solely for those astronauts who travel to distant planets.

v.

Due to (a) prospective overpopulation on earth, (b) environmental problems, and (c) the needs and future peoples' desires, special attention must be given to Gene Therapy for both Space Travel (GTST) and those who remain on earth.

Twenty-first Century Approach

Proper research and finances must be given special attention for human needs in connection with Gene Therapy for Space Travel (GTST).

By the time you read the Epilogue you will know that our world is not doomed, it is not fatally wounded, but neither is it healthy.

Pimm, The World According to Pimm (2001), p.8.

I believe in God, only I spell it Nature.

Frank Lloyd Wright

In general, when physicians believe a law is unjust, they should work to change the law. In exceptional circumstances of unjust laws, ethical responsibilities should supersede legal obligations.

AMA Code of Medical Ethics, Section 1.02

As long as I have any choice in the matter, I will live only in a country where civil liberty, tolerance, and equality of all citizens before the law are the rule.

Albert Einstein (1933)

Our peculiar security is in the possession of a written Constitution. Let us not make it a blank paper by construction.

Thomas Jefferson

Epilogue

The epilogue summarizes what may be considered the principle bases for the Twenty-first Century Approaches to medicine and the law.

The three and one-half decades preceding the third millennium were marked for all citizens with initial bioethical changes, performed by black-robed judges, which significantly affected white-robed physicians. Several were concerned with how we regard ourselves and our loved ones as we enter and leave our lives. In his recent memoirs, Arthur M. Schlesinger, Jr., wrote that "history is forever haunted by the crises and preoccupations of the age in which it is written." While writing his chapter "A Revolution Begins," Bill Gates, the chairman of Microsoft Corporation, stated, "In short, just about everything will be done differently. I can hardly wait for this tomorrow, and I'm doing what I can to help make it happen."[1] During the first decade of this new millennium, we are beginning to place greater emphasis on how and why we should utilize our knowledge and experiences, which have become the goal of the recommended approaches set forth herein. Several centuries ago, Michael Montaigne wrote, "A certain amount of intelligence is required to learn what you do not know." His motto was "*Que sais-je?*"—What do I know? That was also Socrates's motto.

In the twenty-first century, major portions of education applicable to the multi-disciplined field of bioethics began entry into our public schools. Many of them are introducing the teaching of modern astronomical and biological aspects of science into their educational motto. Only by educating maturing youths more broadly can they be accorded true freedom to choose their medical treatments from certain limited bioethical options previously offered to them. Where do we begin?

I suggest we start with the early twentieth century's third decade, when Edwin Hubble scientifically confirmed Immanuel Kant's eighteenth-century suggestion that other galaxies exist beyond our Milky Way. Hubble also discovered the velocity-distance shown by the "red shift" in the spectra caused by the recession of galaxies. When it was calculated backwards in time, this soon led to the discovery of the "Big Bang" theory of the universe. Following the turn of the third millennium, astronomers were able to show that the creation of the universe occurred between 13 and 14 billions of years ago. This provided a partial explanation of why scientists agree that they remain unable to make any determination about nature prior to its creation. Many physicists who study the existing laws of nature also support the rationality toward their Creator's existence. Again, I quote Montaigne:

---

1. Gates, The Road Ahead 7 (1995). "I'm enthusiastic about what my generation, which came of age the same time the computer did, will be able to do. We'll be giving people tools to use to reach out in new ways. I'm someone who believes that because progress will come no matter what, we need to make the best of it. I'm still thrilled by the feeling that I'm squinting into the future and catching that first revealing hint of revolutionary possibilities. I feel incredibly lucky that I am getting the chance to play a part in the beginning of an epochal change for a second time." (p.11) "The information revolution is just beginning." (p.21)

> Whenever a new discovery is reported to the scientific world, they first say, "It is probably not true." Thereafter, when the truth of the new proposition has been demonstrated beyond question, they say, "Yes, it may be true, but it is not important." Finally, when sufficient time has elapsed fully to evidence its importance, they say, "Yes, surely it is important, but is no longer new."

At some point in the twenty-first century, sufficient time will have elapsed that most of our high school graduates may more fully appreciate scientific truths so our society will have the ability to use them in as rational a manner as feasible. Prior to the collapse of the Soviet empire in 1991, its dissident writer Aleksandr Solzhenitsyn said in his Nobel speech: "One word of truth outweighs the entire world."

Of course, any prelude to sources of bioethical standards in this our twenty-first century must also consider those inherited from traditional cultures, and religions from the preceding ones. A number of these have resulted in conflicts with significant scientific and rational developments, which have led us toward the current greater emphasis thereon.[2] We have also found the need for the maximization of a person's autonomy during the last third of the twentieth century, together with a responsibility toward others. Hence more effort will be needed to combine autonomy with a deeper understanding of what this may mean to a society governed by ethics and the law. Important findings resulted in the emergence of new and rational approaches to bioethical issues connected with the beginning and ending of lives, as well as a number of actions necessary to preserve the earth's environment for those who continue to live on it.[3]

At the dawn of the third millennium, people are again comparing the differences between the "republic" we created over two centuries ago and the purely "democratic" government created in Athens, Greece, four centuries prior to the

---

2. Many religious values remain consistent with bioethics and the law. However, it has also been noted that to choose traditional "religious or ideological dogmatism in the name of freedom is as foolish as for a jailed man to exercise his right to remain in prison. To choose bondage may be the familiar choice, the traditional choice, but it is fatal all the same." Carey, The Third Millennium 40 (1995).

3. "For you inhabiting the last days of the human species' infancy, awakening is proceeding by stages, subsequent layers of illusion falling away one after another like the peeling layers of an onion." *Id.* at 145.

first year A.D. The latter's history caused new and significant thinking by the drafters of our Constitution—"knowing how badly Athens had gone wrong, [they] were firmly opposed to any attempt to replicate Athenian democracy."[4] Their system lasted for nearly 200 years, whereas ours has already lasted for 218 years. (The United States Constitution went into effect with voting for our officials and representatives in 1788.) Socrates, who lived in what was then a pre-scientific age, would openly recognize his lack of many aspects of knowledge; he would talk with the common man using logical terms to resolve differing views expressed by the people in their society, using arguments of "good v. Good." Pericles, their frequently elected leader of three decades (from 461 to 429 B.C.), got the people to help construct the Parthenon, which was dedicated to Athena, "the goddess of wisdom." He increased Athens' trade by sea, like England did on a much greater scale over two millennia later. But when at war with nearby Sparta, he died during a plague that cost the lives of almost one-third of Athens' citizens. That democracy then attempted the conquest of Syracuse, Sicily, a disaster costing thousands of Athenian lives (which has recently been compared with the cost to the United States in the ungained war of Vietnam).[5] Finally, when some Persians joined the Spartans, the remaining Athenians lost their preeminence as a democracy in the Middle East. In 2005 it was pointed out how "the chief founders of the United States, knowing how badly Athens had gone wrong, were firmly opposed to any attempt to replicate the Athenian form of democracy."[6] They accomplished this by creating the three

---

4. WOODRUFF, FIRST DEMOCRACY 123 (2005).

5. *See* BOK, FALSIFYING THE EVENTS IN THE GULF OF TONKIN 180-85 (1979). "After the defeat in Sicily (413) and the rebellion of much of the empire (412), Athens was reeling." WOODRUFF, *id.* at 3. "Democracy is hard. They did not get it entirely right in Athens 2,500 years ago." (p.4) "Understanding the ancient debate, we can clear the clouds and cobwebs away from the ideas we are trying to express in democracy today." (p.7) "The Ancient Athenians who invented democracy learned this lesson the hard way." (p.3) Pericles said: "You see, our empire is really like a tyranny—although it may have been thought unjust to seize, it is now unsafe to surrender." *Id.* WOODRUFF, *supra* note 4, at 124 (citing *The Eydides,* HISTORY 2.63).

6. *Id.* at 12. "From a modern point of view, the Athenian laws did not recognize human rights. Rights, as a concept, had not evolved at this time. Even so, the Athenians had the conceptual tool they needed to justify protecting what we would call rights—a *theory of natural equality. If they had followed* their own principles, *they would not have violated the ones we call 'rights.'*" *Id.* at 124.

complementary, separate but equal, elements of our republic—the executive, legislative, and judicial branches. Accordingly, it was noted that "the Constitution is a magnificent solution to the problems faced by the founders, but it is not democratic in itself, and the virtually scriptural authority that it now enjoys is a drag on the volution of democratic processes in the United States."[7]

Following the defeat of Athens, Socrates was convicted and executed for what he had been doing, as Jesus would over four centuries later.[8] This occurred to each of them while they were in the presence of their followers. Plato, who was one of Socrates' youthful disciples, became the greatest among their initial philosophers and he wrote in terms of discussions utilized by Socrates. His successor, Aristotle, initiated writing about the science of biology and on our classical fields of logic and ethics—the teachings of which ended shortly thereafter.

At the dawn of the second millennium, the largest centers of learning were initially located in both the Middle East and China. Many knowledgeable prognosticators of that epoch thought that those locations would most significantly develop the sciences and arts. However, within a few centuries such developments were accelerated by the struggling and competitive nations of the West; these nations would culminate their struggle using Immanuel Kant's eighteenth century slogan "Dare to know."[9] Many people had not accepted this challenge by the turn of the third millennium. Nevertheless, larger numbers appear to have now modified it to read: "The quest for knowledge is by far the noblest and most meaningful of all human activities."[10] Kant's *Critique*

---

7. *Id.* at 4.

8. There were no professional judges or prosecutors; "501 men served on the panel that convicted Socrates, who was tried on charges of impiety." *Id.* at 32. He "was an intellectual at a time when intellectuals as a class seemed to be a threat to religion and public order. That was enough to damn Socrates in the eyes of many citizens." (p.55) "A democratic jury had him killed." (p.100) The outcome of the trial, in any case, has been held against Athenian democracy ever since. (p.55)

9.
> Enlightenment is humanity's departure from its self-imposed immaturity. This immaturity is self-imposed when its cause is not lack of intelligence but failure of courage to think without someone else's guidance. Dare to know! That is the slogan of Enlightenment.

Kant (1783).

10. Hogan, The End of Science 5 (1997).

of Pure Reason acknowledged reason as an objective moral law; his major moral laws included truths that could be known only by experience. Although he termed "individual autonomy" as "the supreme principle of morality," his "categorical imperative" is to "so act that you treat humanity, both in your own person and in the person of every other human being, never merely as a means, but always at the same time as an end."[11] This largely corresponds with the belief of modern bioethicists in an autonomy that would treat every action as a right "if it allows each person's freedom of choice to coexist with the freedom of everyone in accordance with a universal law."[12] As of the turn of the twenty-first century, many civilized countries began to recognize that a determination should be made of the proper balance of autonomy with the individual's responsibility to others and the essential need to resolve those issues in a more rational manner. It was then written: "Neither unhealthy nor repressive truth is a vital requirement not only for individuals who would live a good life, but for free societies that would remain free."[13]

The emphasis on truth and reason have been historically tied to the English, American, and French revolutions. The words of the philosopher John Locke in 1664 were that reason "teaches all mankind who will but consult it, that being all equal and independent, no one ought to harm another in his life, health, liberty or possessions." These words justified the "bloodless English revolution" of 1688, which finally ended the absolutism of the British monarchy and placed power in the people and their representatives in Parliament. A historian stated their application to our American nation:

---

11. Now I saw that man, and in general every rational being, exists as end in itself, not merely as means to be used by this or that will as it pleases. Such a being must, in all his actions, whether they are directed to himself or to other rational beings as well, always be regarded at the same time as an end.

KANT, GROUNDWORK OF THE METAPHYSICS OF MORALS 425-28 (1785).

12. Thus, if my action, or in general my situation, can coexist with the freedom of everyone in accordance with a universal law, anyone who hinders me in these respects does wrong, for this hindrance (this resistance) cannot coexist with freedom in accordance with universal laws.

Id. at 230-31.

13. O.S. GUINESS, TIME FOR TRUTH 13 (2000).

> Locke's words affirming a law of nature and natural right for all human beings were soon used to justify the American Revolution of 1776 and the French Revolution of 1789. And they entered loud and clear into the Declaration of Independence of the United States in the language of Thomas Jefferson: "We hold these *truths* to be *self-evident*, that *all men are created equal*, that they are *endowed by their Creator* with certain inalienable rights, that among these are life, liberty and the pursuit of happiness.[14] (Emphasis added.)

Although Jefferson was raised in the Anglican Church, he soon became a part of the eighteenth century's "Age of Enlightenment." He joined those who became *deists*—"persons who asserted God exists, and that He created the world, but that He has no present relation to the world."[15] They made this assertion to harmonize science and rejected "revelation as the test of religious truth, accepting reason instead."[16] After the mid-twentieth century, Ernst Mayr (1904-2005) wrote *Animal Species and Evolution* (1963) and founded *The Journal of the History of Biology* in 1967. His *Growth of Biological Thought* (a 1982 history of organismal biology) has been called—"required reading for evolutionists." [17] This was very near to the turn of the third millennium statement of Nobel laureate Feynmann: "I don't believe that a real conflict with science will arise in the ethical aspect, because I believe that moral questions are outside of the scientific realm."[18] This was also similar to that contained in Professor Francisco Ayala's "Darwin's Devolution: Design with Designor,"

---

14. LAVINE, FROM PLATO TO SARTRE 137 (1989). In 2005 it was pointed out how John Locke had "argued that behavioral development was determined by unseen factors (of course, the gene hadn't been discovered yet)." Nelson & Gottesmann, *A Piece of a Neuroscientist's Mind*, SCIENCE, Feb. 2005, at 1204.

15. THE WORLD BOOK ENCYCLOPEDIA (1964). This view became consistent with that of current science and evolution (see Chapter 2, Sec. C).

16. *Id.* "Who remembers now that the man from Nazareth was once construed as both the exemplary paradigm of a pared-down ethical deism by Thomas Jefferson." COX, WHEN JESUS CAME TO HARVARD 301 (2004).

17. Coyne, SCIENCE, Feb. 25, 2005, at 1212.

18. FEYNMAN, THE PLEASURE OF FINDING THINGS OUT 254 (1999). "I actually believe that more than half of the scientists really disbelieve in the father's God; that is, they don't believe in a God in a conventional sense." (p.246) "What I mean is the kind of personal God, characteristic of the Western religions, to whom you pray." (p.247)

for which approach he received the highest American award in science from President Bush in 2002. Therein he had concluded that scientific knowledge "is satisfying and useful. But once science has had its say, there remains much about reality that is of interest, questions of value and meaning that are forever beyond science's scope."[19] In this connection the editor of the November 2004 issue of the *National Geographic* commented upon an article on evolution, noting that it "explains how scientists, by using techniques unheard of in Darwin's time, are finding more evidence than ever of the evolutionary links among all living things."

While Jefferson was succeeding Benjamin Franklin as our minister to France in 1787, Madison sent him a draft of the new Constitution. Although Jefferson approved what it then constituted, he strongly urged the addition of "a bill of rights." Jefferson and Franklin were largely responsible for the initial ten amendments to the Constitution, which became effective on December 15, 1771.[20] The first of these states, "Congress shall make no law respecting an establishment of religion, or prohibiting the free exercise thereof; or abridging the freedom of speech." These enabled the Supreme Court to make several bioethical rulings prior to the end of the twentieth century (see Chapter 3, Section 3) that fall within the words in the Constitution prohibiting the enactment of laws solely to establish their consistency with those set forth in the Bible, the Koran, or any other religious works. Thus every holder of a traditional religion may follow its rules up to the point where they would be found to embrace an illegal action. However, they may not require any other persons to limit their autonomous actions solely because they constitute a violation of a particular religious rule. As expressed by a recent physicist:

> This is not a new idea; this is the idea of the age of reason. This is the philosophy that guided the men who made the democracy that we live

19. Ayala, Evolutionary and Molecular Biology; Scientific Perspectives on Divine Action (Vatican Observatory Pubs., Berkeley, Cal.). In 2003 Ayala wrote *Intelligent Design: The Original Version*, in which he concluded that "[s]cientific knowledge may enrich esthetic and moral perception, and illuminate the significance of life and the world; but these are matters outside science's realm." (See Chapter 3.)

20. Theology and Science 31 (2003).

> under. . . . This method was a result of the fact that science was already showing itself to be a successful venture at the end of the eighteenth century. Even then it was clear to socially minded people that the openness of the possibilities was an opportunity, and that doubt and discussion were essential to progress into the unknown.[21]

When Thomas Jefferson was elected president of the United States in 1800 by his party (which had been called "Republican" since 1792), he asserted the separation church and state. Members of the other "Federalist" party of that time were said to be "supporters of established churches." Unlike the "democracy of ancient Greece," the United States had become a "republic" and Jefferson said his election was tantamount to "the revolution of 1800." James Madison, who was secretary of state during Jefferson's two terms, succeeded him as president for the next two-term period.[22]

Just as the balance between autonomy and responsibility is equated to ethical living, so must the proper balance between autonomy and diversity in education become a goal in the twenty-first century. However, it is of interest to note that the middle of the twentieth century Jean Paul Sartre, a nonscientistic existentialist philospher, which had lead him to prevent a universe containing the being who is "endlessly acting upon projects to achieve valued goals from which it will always be separated by a gap-goals wish which free nihilating consciousness will never coincide."[23] Although existentialism and tragedy are

---

21. FEYNMAN, THE PLEASURE OF FINDING THINGS OUT 148 (1999).

22. FERLONG, ADAMS VS. JEFFERSON 148 (2004). In 1825 the political group of (later) President John Quincy Adams adopted the title *National Republican.* The current name *Republican* was made in 1854 and Abraham Lincoln became the first Republican president; *see also id.* at 208.

23. He "finds it extremely embarrassing that God does not exist, for there disappears with Him all possibility of finding values in an intelligible heaven. . . . It is nowhere written that 'the good' exists, that one must be honest or just not lie, since we are now where there are only men." JEAN PAUL SARTRE, EXISTENTIALISM IS A HUMANISM (1947):

> Thus we have neither behind us, nor before us, in a luminous realm of values, any means of justification or excuse. We are left alone, without excuse. That is what I mean when I say that *man is condemned to be free*. (Emphasis added.)

profoundly un-Americn ways of looking at the place of the self in the world, in the 1950s they were in perfect step with the younger faculty's view that we were teaching students to have inquiring minds and not to accept without question established and obvious views.[24] This approach has been critized for having "made ethics impossible by rejecting any general principles or ideals as the foundation for moral choice." If its definition of "authenticity has not provided any justification for a choosing A rather than B, then my choice is arbitrary; it is simply my choice, but it has no foundation."[25] As noted below, Sartre's view also conflicted with a third millennium writer who was studying evolutionary biology and experimental psychology, and indicated that when he "turned to science and philosophy for moral answers and began to try different ethical systems" he found existentialism appealing. However, he soon changed his mind after discovering that "[m]ost existentialists believe that life is 'absurd' because we exist in a meaningless, irrational universe—any attempt to find ultimate meaning can only end in absurdity. . . ."[26] "I may be free from God, but the god of nature holds me to her temple of judgement no less than her other creations."[27]

Although a minority of scientists continue to hold memberships in traditional churches, neither the majority of current scientists nor future ones may rationally be able to become pure atheists (see Chapter 4). As stated by the late Harvard University professor Stephen Gould: "*Science* simply *cannot adjudi-*

Quoted in Lavine, From Plato to Sartre 370-71 (1989). While a student in France, I attended his plays. He remained a bachelor but often lived with many women and sought for changes in, or the disregard of, many laws. He became a communist in 1952, claiming to follow Hegel and Marx (see his *Critique of Dialectical Reason*) and was awarded the Nobel Prize for literature in 1964 (which he refused to accept). *Id.* at 386 and 392-93.

24. Boylan, Portraits of Guilt 99 (2000). "Western literature has been existentialist from Job to *Waiting for Godot,* and existentialism therefore made literature appear not simply another art but the definitive image of the human condition."

25. *Id.* Lavine, at 372-73 and 393. Similarly, the Oxford University evolutionary biologist Richard Dawkins wrote that "the universe we observe has precisely the properties we should expect if there is, at bottom, no design, no purpose, no evil and no good, nothing but blind, pitiless indifference." Dawkins, River Out of Eden (1995).

26. Shermer, The Science of Good & Evil 162 (2004).

27. *Id.* at 22.

cate the issue of God's possible *superintendence* of *nature*."[28] (Emphasis added.) A classical philosopher throughout most of the twentieth century explained the major change following the turn of the third millennium: "I now say my goal to be philosophy of organic matter or a philosophical biology. . . . The natural sciences bring to light relevant for the concept of being, which is, after all, philosophy's concern." It was biology that "revealed to me the wonders of life, its evolution, its abundance of forms, modes of functioning, etc."

> This awareness, along with that of many others, finally forced me to turn from the theoretical to practical reason—that is, to *ethics,* which became the *central interest and theme* of the last stage of my intellectual journey. Practical philosophy was my final personal experience in the course of my intellectual career.[29]

This will become a major change in philosophy applicable to bioethics. During the launch of "secular humanism" views, moral theory has many limitations with respect to the supernatural or the metaphysical characteristics of Middle Eastern and Western religions. The modern philosopher Rawls stated that a "carefully considered view [constitutes] a sufficient moral basis in the political convictions of the community."[30] The Supreme Court in 1961 wrote an opinion on Eastern religions (Buddhism, Taoism, Ethical Culture, Secular Humanism, and others) as those that neither retain nor worship the same currently active God[31] who has been the central concept of the Judeo-Christian-Muslim view of religion.[32] Furthermore, since President Jefferson's time, the U.S. courts under our Constitution have declined to support the use in public schools of textbooks promoting the establishment of any religion.[33]

Bioethicists must also be able to examine and evaluate one "good" against another "good" in a rational bioethical setting. During the twenty-first cen-

---

28. Gould, *Impeaching a Self-Appointed Judge*, SCIENTIFIC AM., July 1992.

29. Jonas, Hast. Ctr. Rep., July-August 2002, at 32-34.

30. RAWLS, A THEORY OF JUSTICE 367.

31. Torcaso v. Watkins, 367 U.S. 488, 495 n.11 (1961). *See also* Smith v. Bd. of School Comm'rs of Mobile County, 655 F. Supp. 939 (S.D. Ala. 1987).

32. Melnik, *Secularism in the Law: The Religion of Secular Humanism*, OHIO N.U. L. REV. 329, 329-57 (11th Cir. 1987).

33. *E.g., see* Smith v. Bd. of School Comm'rs of Mobile County, 827 F.2d 684, 693 (11th Cir. 1987), and Santa Fe Ind. School Dist. v. Doe, Docket No. 99-62, *aff'd*, 5th Cir., June 19, 2000.

tury, the medical profession must begin to recognize that although broad principles promulgated on bioethics near the end of the twentieth century constitute its background (values and inspirational material), they alone may not normally resolve difficult ethical, medical, or legal cases. Specific bioethical rationales must be found and used in order to make such decisions, as well as to educate medical and other professionals who may become members of a hospital's bioethics committees. Ethical actions will generally involve choices between alternatives. Buddha, the secular humanist, said, "I have shown you the path of liberation. Now liberation depends on you." Thus, most of the twenty-first century approaches set forth herein grant options to those who desire to exercise one action or another with the same autonomy allowed for individuals to pick a particular religion (or retain an agnostic position). Among U.S. citizens, 25 percent are Catholics, almost 50 percent are Protestants, and the remainder are members of other religions or are agnostics or alleged atheists. Until shortly after the dawn of the twenty-first century, few states still maintained laws prohibiting the teaching of the scientific version of evolution in their public schools. However, this has now undergone a major change regarding the legality of such teaching (although some school districts still make attempts to prohibit their teachers from doing so). Following completion of such teaching for another generation or two in all public high schools, this could modify "significant numbers" of graduates to express a desire to exercise bioethical options in favor of the recommended "twenty-first century approaches." These would then add the tools of public health toward the goals of the majority of our citizens.

But do the judges of the Supreme Court of the United States really care about the heart cry of the American people? Do they care enough to stop the moral disintegration of our society? Quite simply, a five-member majority of the nine judges thumb their noses at tens of millions of American citizens and their elected representatives. Well, to that majority of black-robed tyrants we say, "We care very deeply. You have usurped power that was never given to you. Your ill-conceived decisions, more than any other cultural phenomenon, have shredded the moral fabric of this nation. The blood of millions of innocent unborn babies is on your hands. And the time has come for you unjust judges to step down, so that truth and justice may at last be restored to this land! *(Emphasis added.)*

Courting Disaster (p.xxiv) by Pat Robertson, a religious broadcaster who, in 1988 was a candidate for the Republican nomination for the presidency of the United States, and holds a juris doctor degree from Yale University Law School and a master of divinity degree from New York Theological.

This Constitution . . . shall be the supreme Law of the Land. . . (Article VI, clause 2);

"The Constitution is what the judges say it is."

Chief Justice Charles Evans Hughes:
Speech at Elmyra, New York (May 3, 1907)
(quoted in Bork, *The Tempting of America* [1990], p.176)

Learning without thought is labor lost; thought without learning is perilous.

Confucius

It is a terrible thing that science has grown so distant from the rest of our intellectual life, for it did not start out that way. The writings of Aristotle, for example, despite their notorious inaccuracies, are beautifully clear, purposeful, and accessible. So is Darwin's Origin of Species. *The opacity of modern science is an unfortunate side effect of professionalism, and something for which we scientists are often pilloried—and deservedly so.*

Laughlin, *A Different Universe* (2005), p.xi.

I. Twenty-first Century Bioethical Decisions within Traditional Cultural, Religious, or Nontraditional Categories by Federal and State Courts

Following the arrival of the third millennium, the "crisis of values" continues to exist between two different and important groups of citizens. The current significantly large majority of the first category consists of those who are members of traditional churches. They maintain their autonomy (liberty) to follow the duties and restrictions expressed in the Testaments of the Bible or other religious volumes. The second category has a sufficient scientific background to choose differently in order to maximize their autonomy. That is, like the eighteenth-century deists Franklin, Paine, Presidents Adams and Jefferson, and others who were among the important drafters of the Constitution, many of those who now have a scientific background also consider God as the Creator of the Universe who set the laws of nature. They would normally favor the adoption of the bioethical approaches approved by the courts during the final third of the twentieth century as well as most of those approaches that are reasonably recommended for the second category in the twenty-first century.

Under the first amendment to the Constitution, those who fall within the first category no law can be made "respecting an establishment" of all their traditional religion-based duties and restrictions upon those who are within the second category. Further, although all people under Category 1 possess a constitutional right to practice most of their religiously prescribed morals, pursuant to the First Amendment, Congress cannot make any law "prohibiting the free exercise" of any religious persons falling within Category 2, who believe that God only created the laws of nature in the universe. The latter category not only includes the deists among our founders and presidents in the eighteenth and early nineteenth centuries (as noted above), but also twenty-first century appreciators of science, such as Professor Stephen Hawking, the physicist, along with many others.[34] In 2005 he stated: "Actually, the idea that God might want to change His mind is an example of the fallacy, pointed out by St. Augustine, of imagining God as a being existing in time. Time is a property only of the universe that God created. Presumably, He knew what He intended when He set it up!"[35] Over a half century earlier, Albert Einstein

---

34. STEPHEN HAWKING, THE UNIVERSE IN A NUTSHELL (2002). In his volume *A Brief History of Time* (1988), physicist Stephen Hawking concluded that with the discovery of a complete theory, "then we shall all, philosophers, scientists, and just ordinary people, be able to take part in the discussion of the question of *why it is that we and the universe exist.* If *we find the answer* to that, it would be the ultimate triumph of human reason—for *then we would know the mind of God.*" (p.175) (Emphasis added.)

35. HAWKING, A BRIEF HISTORY OF TIME 134 (2005). In 2005 the provost of a great modern western private university noted that "[i]n the last century, many of *America's* great [private eastern] *universities* such as Harvard, Princeton, and Yale *yielded to decidedly secular ideas* and abandoned their religious heritage.

had pointed out that "religion without science is blind" and "science without religion is lame."

Michael Foucault, a famous modern philosopher at the prestigious *College de France*, described how the "self" and the "care of self" were conceived during the period of antiquity, beginning with Socrates in connection with contemporary moral thought. He added "The search for a form of morality acceptable to everybody in the sense that everyone should submit to it, strikes me as catastrophic."[36] As expressed by Spinoza in his *Ethics* (1675); "A desire which springs from reason can never be in excess."[37] Accordingly, the options, which are exercised by the autonomous people in the second category, are neither based upon traditional religious views of morals, nor upon science; they are based upon the combined background of science, plus experience and logic. In their view, this should normally permit them to obtain the courts' authorization for current "deists," agnostics, and atheists to exercise their bioethical liberty (autonomy) under the Constitution regardless of

---

Today, elegant chapels remain on these campuses as reminders of the faith that once bound the academy to the church. . . . Any quest for knowledge without a commitment to ethical or moral foundations does not lead to progress or a secular paradise, but to chaos. A *society that emphasizes knowledge but ignores the search for the eternally good is in trouble.*" (Emphasis added). Provost Darryl Tippens, PEPPERDINE PEOPLE (Fall 2005) at 26. However, we now realize the deists who were our founders together with today's scientists actually are among those who both searched and who continue "to search for the eternally good" by increasing our knowledge concerning how to do so.

36. FOUCAULT, THE HUMAN ENTICS [Interpretation] OF THE SUBJECT 532 (2005). However, in connection with certain forms of introspection that he saw on the West Coast of the United States. he states: "Not only do I not identify this ancient culture of the self with what you might call the Californian cult of the self, I think they are diametrically opposed." *Id.* at 533.

37. SPINOZA, ETHICS 233 (Hafner, 1949), Prop. LXL. In addition, he stated:

> It is impossible that a man should not be a part of Nature and follow her common order; if he be placed amongst individuals who agree with his nature, his power of action will by that very fact be assisted and supported. But if, on the contrary, he be placed amongst individuals who do not in the least agree with his nature, he will scarcely be able without great change on his part to accommodate himself to them."

Id. at 241, App. VII.

objections raised by those who fall within the first ccategory. Most people who fall within the second category might also consider the first category's objections to be quite analogous to those that were made against Galileo for publishing his *A Dialogue on the Two Principal Systems of the World*, a masterpiece published in 1632. Those objections resulted in his being called before an Inquisition at the Catholic hierarchy in Rome and sentenced to spend the remainder of his life in one place.[38] Although the Inquisition itself was abolished in 1834, the condemnation of his work by Rome remains effective. (In 1992 the late Pope John Paul II created a commission to redeem Galileo's condemnation; however, it apparently failed to "rehabilitate" him.[39])

Recommended Twenty-first Century Approaches are many. This Epilogue emphasizes that although some laws affecting bioethics have been improved, additional efforts must be made not only by government officials, but also by hospitals and their ethics committees. The American Hospital Association (AHA) stated: "The processes of making changes have often inadvertently been neglected. It is the job of ethics committees—working in their institution, within networks, and with their community—now and in the future to remind their institutions what those values are and why they are still important."[40] Accordingly, twenty-first century recommendations include the current need for a modification of the American Medical Association position that "a majority of the [hospital ethics] committees consist of physicians, nurses, and other health care providers."[41] The current dominance on such committees of physicians who have a limited bioethical education should

38. In his review of Finocchiaro, *Retrying Galileo* (2005), on March 15, 1990, Machamer wrote how then Cardinal Joseph Ratzinger (now Pope Benedict XVI) stated: "At the time of Galileo the Church remained much more faithful to reason than Galileo himself. The process against Galileo was reasonable and just." Perhaps this portends the next story in the grand saga of the Galileo affair. "Galileo had verified the solar system (using the circulation of the four Jupiter's moons visible to his telescope) as well as the mathematical calculations in Copernicus's volume *De revolutionibus* on May 24, 1543, as he lay dying in Frauenburg." *See* MANCHESTER, A WORLD LIT ONLY BY FIRE 295 (1993).

39. SCIENCE, July 2005, at 58.

40. ROSS ET AL., HEALTHCARE ETHICS COMMITTEES: THE NEXT GENERATION 188 (1993).

41. AMA CODE OF MEDICAL ETHICS § 9.11.

be changed to other professionals with expertise in "ultra-medical disciplines."[42] Until the third millennium, medical school classes in ethics hardly ever went beyond teaching future physicians how to avoid malpractice. Such a change will also increase public confidence on bioethical matters decided by the members of hospital ethics committees.[43]

Charles Darwin commenced his nineteenth-century writing on evolution with his original volume *The Voyage of the Beagle* (1835), concerning his study of differences in the birds on the equator at the Galápagos islands in the Pacific Ocean off South America. In his *Descent of Man* (1871) he stated, "I fully subscribe to the judgment of those writers who maintain that of all the differences between man and the lower animals, the moral sense of conscience is by far the most important. . . . The following proposition seems to me in a high degree probable—namely, that any animal whatever, endowed with well-marked social instincts, the parental and filial affections being here included, would inevitably acquire a moral sense of conscience." The twentieth century ended with its confirmations by the volume *What Evolution Is,* by Ernst Mayr, and and the writings of Steven Jay Gould and Francisco Ayala. Nevertheless, these facts were actually denied in biology textbooks in Cobb County, Georgia, during 2005, which carried a warning label reading "Evolution is a theory, not a fact."[44] Einstein himself had noted in the mid-twentieth century that "the

---

42. The largely interdisciplinary ultramedical aspects of this volume concern the extent that they may need consideration by a hospital's patients as well as the members of its ethics committees and others. Examples of the chapter numbers pertaining to such aspects include: humans as a part of nature (see Chapter 2), religion (3), science and philosophy (4), autonomy and responsibility (5), committee responsibilities (6), controls of the beginnings of life (7-10), transplants (11), death with dignity (12-14), and the future of genetics (15). These are all related to ethical aspects of law and medicine.

43. As stated by the co-discoverer of its structure, "Our DNA, the instruction book of human creation, may well come to rival religious scripture as the keeper of the truth." WATSON, DNA, THE SECRET OF LIFE 404 (2003).

44. Evolution is a fact, just as gravity is a fact. Darwin's theory of evolution by natural selection is still around for the same reason that Einstein's general theory of relativity is still around: it successfully explains what we observe in nature, even as new discoveries come along. [But] each new discovery [is] hardly satisfying for people of faith.

Feinberg (ed.), SKY & TELESCOPE, April 2005.

theory must not contradict empirical facts," and that its "external confirmation" is concerned "with the 'inner perfection' of the theory."[45]

In 2005 it was noted that "the epilogue of physics, already written by some physicists, was consigned to the dustbin of history."[46] Thus, early in the twentieth century, the delay in Einstein's receipt of a Nobel Prize represented an example of previous opposition to his theories on relativity. Carl Wilhelm Oseen knew that committee members were emotionally opposed to relativity, so he developed a strategy. Although Einstein's proposed particle theory of light was widely rejected by physicists and could not be the basis of the prize, Oseen argued that Einstein's other initial theory of the photoelectric effect, which also appeared in his March 1905 paper and had been verified by Robert Millikan, was worthy of the prize. Oseen's strategy included promoting Niels Bohr for the prize at the same time. When the committee voted, they agreed to give the 1921 physics prize to Einstein for the photoelectric effect and the 1922 prize to Bohr.[47] It was noted how "physicists and nonphysicists alike get comfortable with our common-sense ideas and hate to give them up."[48] Nevertheless, by the turn of the twenty-first century a famous astrophysicist could state: "The equations of general relativity are his best epitaph and memorial. They should last as long as the universe."[49]

Nobel prizes are now being awarded to scientists in several areas (including physics, chemistry, and physiology or medicine) as well as nonscientists (for

---

45. *Autobiographical Notes, in* ALBERT EINSTEIN: PHILOSOPHER-SCIENTIST, vol. 1, Paul Arthur Schilpp ed. (1959), at 21-23.

46. The decade immediately preceding 1905 was a lively time in the physics profession. Recent, totally unexpected discoveries, including X-rays (1895), radioactivity (1896), the electron (1897), and the quantum (1900), had ripped the covers off the book of physics and had given physicists notice that Newton's mechanics and Maxwell's electromagnetism were not the final chapters.

RIGDEN, EINSTEIN 1905 77 (2005).

47. *Id.* at 100.

48. *Id.* at 103.

49. *Id.* at 149, *citing* Stephen Hawking, *A Brief History of Relativity*, IN TIME, vol. 154, no. 27, Dec. 31, 1999, at 81. In his book *A Brief History of Time,* Hawking asked "why it is that we and the universe exist," and concluded, "If we find the answer to that, it would be the ultimate triumph of human reason—for then we would know the mind of God."

their work as peacemakers, writers of literature, or economists). However, just as no such awards have been made posthumously in the fields of mathematics,[50] evolution, or bioethics, all retain key aspects of essential leads toward the future of humankind. Our bioethical future has nothing to do with obtaining such a prize. It constitutes a multidisciplinary area whose key is entirely based upon the use of objective reasoning for application to environment, medicine, and the law. As noted by Thomas Paine, who supported the American Revolution and helped lead the eighteenth-century movement toward rationalism: "When men yield up the privilege of thinking, the last shadow of liberty quits the horizon."[51] In the foreword to a 2003 reprint of his *Common Sense, Rights of Man, and other Essential Writings*, Jack Fruchtman, Jr. pointed out that:

> Paine was the most colorful and successful pamphleteer in the age of the American and French Revolutions. Arguably America's first "bestseller," *Common Sense* sold as many as 150,000 copies and the *Rights of Man* well into the hundreds of thousands. . . . [He] gave birth to several familiar slogans that we associate with his turbulent era: "we have it in our power to begin the world over again," "the birthday of a new world is at hand," and "these are the times that try men's souls." Most vividly, he gave life to the phrase that we typically associate with the eighteenth-century Enlightenment: "the Age of Reason."

---

50. Einstein remarked, "Insofar as the statements of mathematics refer to reality, they are not certain, and insofar as they are certain, they do not refer to reality." ALBERT EINSTEIN, GEOMETRIE UND ERFAHRUNG (Berlin: Julius Springer, 1921). Aristotle had previously pointed out why there are no sacred numbers. On the contrary, only people can project meaning upon numbers. Even though this may also be due in part to Gödel's theorem of 1930, which made mathematicians agree that some of their questions cannot be decided, the Oxford mathematician Roger Penrose called it "one of the greatest achievements of the twentieth century." NATURE'S IMAGINATION 23, John Cornwell ed. (1995). In 2006 it was noted that mathematics "is somehow independent of the universe. Results and theorems, such as the properties of the integers and real numbers, do not depend in any way on the particular nature of reality in which we find ourselves. Mathematical truths would be true in any universe." Chailin, *The Limits of Reason*, SCIENTIFIC AM., March 9, 2006, at 79.

51. In 1792 Thomas Paine wrote, "What Athens was in miniature, America will be in magnitude." *See* ROBERTS, ATHENS ON TRIAL: THE ANTIDEMOCRATIC TRADITION ON WESTERN THOUGHT 180 (Princeton Univ. Press 1994).

Over 200 years later it was pointed out how, genetically speaking, "we are just an intermediate step on one branch of the tree of life. But from this point on, we can choose the directions in which we grow and change. We can choose new states that benefit us and benefit our children, rather than benefiting our genes. . . . We are, if we choose to be, the seed from which wondrous new kinds of life can grow. We are the prospective parents of new and unimaginable creatures."[52] Another writer added: "If bioengineering made the myth of the 'self-made man' come true, it would be difficult to view our talents as gifts for which we are indebted rather than as achievements for which we are responsible. This would transform three day features of our moral landscape humility, responsibility, and solidarity."[53]

Although a certain amount of progress has been made in bioethics as of the dawn of the third millennium, significantly greater needs remain to be accomplished that are consistent with our eighteenth century Constitution, together with Amendments made thereto in that century and the succeeding ones.[54] In this connection, particular emphasis is placed upon their provisions respecting our religions,[55] liberty, due process of law, and equal protection of the laws.[56] Further, international agreements, which may constitu-

52. NAAM, MORE THAN HUMAN 233-34 (2005).

53. Michael Sandel's statement, *quoted in* Erik Parens's article, *Authenticity and Ambivalence*, Hast. Ctr. Rep., May-June 2005; Sandel, *The Case Against Perfection*, ATLANTIC MONTHLY (April 2004) at 51-62. Parens commented thereon as follows: "Leaving aside whether Sandel's predictions are accurate, we can discern in his words the sort of view symbolized by Jacob's questions [in the *Book of Genesis*] if we forget that life is a gift—albeit from an unknown giver; we will make a mistake about the sort of creatures we really are and the way the world really is." Although we may never be able to eliminate the questions raised by atheists and agnostics with regard to the "Creator" of the universe, I believe we cannot forever avoid our future responsibility for undertaking the bioengineering of the "self-made man" mentioned by Sandel.

54. In particular, *see* Chapter 2, § 2 and Chapter 3, § 3.

55. *See* Amendment I (1791).

56. *See* Amendments V, prohibiting being "deprived of life, liberty or property without due process of law"; IX, which states, "The enumeration in the Constitution of certain rights shall not be construed to deny or disparage others retained by the people"; and X, which states that "powers not delegated to the United States" are "reserved to the States respectively or to the people." *See also* the 1868 Amendment XIV, § 1, which states, "No State shall make or enforce

tionally influence the rights of our people, are beginning to enter the field of bioethics at this time.[57] As stated by Professor Arthur Caplan: "Bioethics has taken a road from which there is no return."[58] It is also true that there is not yet any easy route for our politicians. An example occurred 2,300 years ago by Euclid, the Greek who founded the first school of mathematics in Alexandria. When he completed his most useful book on elementary geometry, King Ptolemy I asked him if there were not an easier way to learn it. Euclid replied, "There is no royal road to geometry."[59]

Article VI specifies that all treaties "made under authority of the United States, shall be the supreme Law of the Land; and the Judges in every State shall be bound thereby, anything in the Constitution or Laws of any State to the contrary notwithstanding." This Article was contemporary with eighteenth-century Immanuel Kant's principle: "I ought never to act except in such a way that I can also will that my maxim should become a universal law."[60] As was

---

any law which shall abridge the privileges or immunities of citizens of the United States; nor shall any State deprive any person of life, liberty, or property, without due process of law; nor deny to any person within its jurisdiction the equal protection of the laws." As Demosthenes observed in ancient Greece, "Each man shares equally in the system of government, but not in wealth."

57. *See* Introduction, § 9.

58. Caplan is professor of bioethics and chairs the Department of Medical Ethics at the University of Pennsylvania. His article is *A Foreshadowing of Today's Idealogical Disputes in Bioethics*, Hast. Ctr. Rep., May-June 2005, at 12-13.

59. *See* Jones, THE WORLD BOOK ENCYCLOPEDIA 303 (1963).

60. KANT, THE MORAL LAW: KANT'S GROUNDWORK OF THE METAPHYSICS OF MORALS 70 (H.J. Patton trans. 1961).

61. Sharing power is a much more difficult process when dealing with groups that are outside one's orbit of shared humanity and reality. In an interdependent world, however, all groups must share in decisionmaking. To make this possible, power must be used to support reflective institutions that enable collective decisionmaking. Investing in international bodies and relinquishing power after using it to create these reflective representative bodies will take time and will require greater respect for each others' national institutions than now exists. (*Id.* at 438) Although it sounds utopian, ironically, it's a necessary reality in an interdependent world. (p. 339) In other words, in an interdependent world, all human survival must be the goal of all groups and become part of a global effort. (p. 441)

GREENSPAN & SHANKER, THE FIRST IDEA 429 (2004)

explained in the twenty-first century "Individuals in a group that responds adaptively experience the immediacy with each other that is needed to develop a shared sense of reality and humanity."[61] The authors of America's Constitution pointed out how "valid federal statutes and treaties would become part of the 'supreme Law of the Land,' with priority over any inconsistent state-law norm, even one that appeared in a state constitution."[62]

Consistent with the words of Thomas Jefferson in 1776 during the Age of Enlightenment,[63] Abraham Lincoln stated that the words "all men are created equal" represent a "*promise* embodied in the Declaration of Independence that in due time the weights would be lifted from the shoulders of all men and that all should have an equal chance."[64] Today we continue working toward the completion of that promise in this manner;[65] namely that:

> Two hundred and thirty years ago, a new nation was conceived in liberty and dedicated to the proposition that normal humans are engaged in testing whether that nation, or any nation so conceived and

---

62. AMAR, AMERICA'S CONSTITUTION 300 (2005). They also noted "that a duly enacted treaty cannot itself authorize a new expenditure, impose a new internal tax, create a new federal crime, raise a new army, or declare a war. (p.304) (citing JOHN C. YOO, *Globalism and the Constitution: Treaties, Non-Self-execution, the Original Understanding, and Treaties as "Supreme Law of the Land"* (p. 2095), and *Rejoinder* (p. 2018); and RESTATEMENT (THIRD) OF FOREIGN RELATIONS LAW OF THE UNITED STATES (1987), § 111, cmt. I & rep.'s note 6.)

63. In the first sentence of "The unanimous Declaration of the thirteen United States of America," Thomas Jefferson wrote, "In the Course of human events, it becomes necessary to assume among the powers of the earth, the separate and equal station to which *the Law of Nature and of Nature's God* entitle them." (Emphasis added.) *See also* Chapter 3, Section 3. The "Age of Reason" was also the movement called "Age of Rationalism." While the Constitution was being drafted, "Thomas Paine (who wrote the famous volume *Common Sense* and had supported our Revolution) was also one of those who led that movement." WORLD BOOK ENCYCLOPEDIA.

64. *See* WILLS, INVENTING AMERICA: JEFFERSON'S DECLARATION OF INDEPENDENCE xix. "We feel as connected to the past and future as we do to the present. Seeing the connectedness of everything opens life up and brings tremendous well-being to the phenomenal world." KONGTRIL, IT'S UP TO YOU 127 (2005).

65. Derived from his Gettysburg Address, Nov. 19, 1863, and speech in Springfield, Ill., on June 16, 1858.

> so dedicated, can and will endure altogether fitting and proper bioethical rights—particularly for those who no longer consecrate in the traditional manner. A house divided against itself cannot stand. We can never forget the unfinished work thus far so nobly advanced. It is rather for us to be dedicated to the great task remaining before us. We should resolve to grant a newer birth of freedom and responsibility; and that government of, by, and for the people shall neither perish from the earth nor elsewhere in space.

I have also concluded in the manner Bill Gates did at the end of his *The Road Ahead*. After indicating that "this is a wonderful time to be alive. There have never been so many opportunities to do things that were impossible before," he added: "I was writing this book to help get a dialogue started and to call attention to a number of the opportunities and issues that individuals, companies, and nations will face. My hope is that after reading this book you will share some of my optimism, and will join the discussion about how we should be shaping the future."[66]

I should also finalize this book by noting, as did Mark Twain in one of his writings, "I'm sorry this letter is so long, but did not have time to make it shorter."

---

66. Unlike this volume, other works have actually been funded in the field of bioethics. For example, nanotechnology in the United States is funded by the National Science Foundation, global health by the Gates Foundation, and the $100-million "Allen Brain Atlas" by the Allen Institute of Brain Science in Seattle, set up by Paul Allen, who, along with Gates, founded Microsoft. *See* NATURE, Feb. 23, 2006, at 1026.

Glossary

Biological, legal, and medical terms applicable to bioethics.

ADVANCE DIRECTIVE—A document in which a person gives directions about medical care or designates who should make medical decisions (a Surrogate) when the designator loses decision-making capacity, or both. There are advance directives for treatment itself and help guide (The Surrogate).

AGNOSTIC—A term derived from the Greek *agnostos* (unknowable) coined in 1869 by Thomas Henry Huxley. One who holds that the ultimate cause (God) and the essential nature of things are unknowable, or that human knowledge is limited to experience (*New Century Dictionary*). Although they have no traditional religion, they cannot be defined as atheists.

ALLELE—Each variant or alternative version of any one gene (an abbreviation of Bateson's proposed term, "allelomorph"). Humans carry two sets of most of them. We have about 34,000 loci each, and the genes that are positioned on many of those loci do occur in two or more alleles, some of them in many different forms. TUDGE, THE IMPACT OF GENES (2001), p. 110.

ALZHEIMER'S DISEASE—A progressive, irreversible brain disorder whose incidence rises sharply with advancing age. Symptoms of the disease include memory loss, confusion, impaired judgment, personality changes, and loss of language skills.

AMICUS CURIAE—A friend of the court. Also a person who has no right to appear in a suit but is allowed to introduce argument, authority, or evidence to protect his interests (*Black's Law Dictionary*).

AMINO ACID—Twenty molecules that are combined to form proteins in living things.

AMNIOCENTESIS—Serves to identify genetic defects, chromosomal anomalies, and the sex and blood type of the fetus. It constitutes surgical transabdominal or transcervical penetration of the uterus for aspiration of amniotic fluid.

ASSISTED REPRODUCTIVE TECHNOLOGIES (ART)—The various methods that utilize cryopreservation of embryos and sperm to assist reproduction. Typically, such technology is used in conjunction with In Vitro Fertilization (IVF). For a discussion of various types of reproductive technologies, see Ethics Committee of the American Fertility Society, *Ethical Considerations of Assisted Reproductive Technologies*, 62 FERT. & STER. 35S (Supp. 1, November 1994).

ATTENDING PHYSICIAN—The doctor on record at a hospital who is in charge of a particular patient. To coordinate care, the patient should receive the attending physician's name, telephone number, and pager number.

AUTONOMY—From the Greek *autos* (self) and *nomos* (rule or law). Its limits were stated by J. S. Mill: "As soon as any part of a person's conduct affects prejudicially the interests of others, society has jurisdiction over it." It is not tied to any particular religion or philosophy, but it reflects the statement of the slave Epictetus: "No man is free who is not master of himself." See FREEDOM OF ACCESS TO CLINICAL ENTRANCES (FACE).

AUTOSOME—A chromosome not involved in sex determination. The diploid human genome contains 22 pairs of autosomes and one pair of sex chromosomes.

BARIATRICS—The medical field specializing in obesity.

BEST INTERESTS—Refers to the principle of seeking the best objective interests of an incompetent. Because it has sometimes been seen as too paternalistic, the Substituted Judgment test has been used as an alternative. Courts determine, from an allegedly neutral perspective, if the treatment is in the objective best interest of the incompetent. *E.g.*, Curran v. Bosze, 566 N.E.2d 1319 (Ill. 1990); *In re* Guardianship of Pescinski, 226 N.W.2d 180 (Wis. 1975); Little v. Little, 576 S.W.2d 493 (Texas 1979).

BIG BANG—A term invented by the astronomer Fred Hoyle signifying the initial explosion marking the origin of time and space for the universe and of its expansion in accordance with Hubble Law. It appears to coincide with a portion of the Biblical story of the Creation.

BIOETHICS—A term derived from the Greek *bios* (life) and *ethike* (ethics). Study of ethical, religious, moral, philosophical, and legal problems that often largely relate to activities of the health professions. It focuses on principles such as autonomy, beneficence, and justice. Bioethics is an area of interdisciplinary studies, especially "Concrete Ethical and Legal Problems," and social issues, such as those associated with public health, occupational health, international health, and the ethics of population control. In the 1970s, the word *bioethics* was begotten (by the oncologist Van Rensselaer Potter). ENCYCLOPEDIA OF BIOETHICS, *Introduction* XIX and XXI (see ETHICS).

BIOPHYSICAL THESIS—A theory holding that all life processes, including mental and physical behavior, and physical attributes are or will be explained in physical terms and possibly someday may be subject to human control.

BIOTECHOLOGY—"Techniques that use living organisms to make or modify products, improve plants or animals, and develop micro-organisms for specific purposes." National Research Council, 1994.

BIOTERRORISM—The intentional release of viruses, bacteria, or toxins for the purpose of harming or killing civilians. CENTERS FOR DISEASE CONTROL AND PREVENTION, U.S. DEP'T OF HEALTH AND HUMAN SERVICES, THE PUBLIC HEALTH RESPONSE TO BIOLOGICAL AND CHEMICAL TERRORISM, INTERIM PLANNING GUIDANCE FOR STATE PUBLIC HEALTH OFFICIALS (July 2001).

BLASTOCYST—A hollow ball of a number of cells surrounding an inner cell mass of an embryo (or blastocyst stage) between four and six days old. However, some authors use the term to describe any cells of an embryo before it cavitates to become a blastocyst. *Alternative Sources of Human Pluripotent Stem Cells*, President's Council on Bioethics (2005).

BRAIN DEATH—In order to increase the supply of organs for donation, criteria were developed designating legal death upon diagnosis of irreversible cessation of all functions of the entire brain, including the brain stem. There is no voluntary or involuntary movement (except spinal reflexes) and no brain stem reflexes. This would not embrace Persistent Vegetative State (PVS), in which the brain stem operates to permit the patient to breathe.

CANCER—Many diseases, each with its own distinctive characteristics; what is common is that some cell undergoes changes that allow it to grow in an abnormal fashion, multiplying uncontrollably. The continuing growth of these deviant cells leads to the development of a mass or growth called a tumor. These cells may travel to a distant part of the body where they begin the development of another tumor, called a metastasis. The most common site of metastatic spread is the bones, lungs, liver, brain, and central nervous system. LYM & HAROLD, HANDBOOK FOR MORTALS (1979), p. 100.

CAPACITY—A patient's ability to understand the nature and consequences of proposed healthcare, including its significant benefits, risks, and alternatives, and to make and communicate a healthcare decision. California Probate Code 4609 (2000): There shall exist a rebuttable presumption affecting the burden of proof that all persons have the capacity to make decisions and to be responsible for their acts or decisions. *Id.*, E 810. A determination that a person is of unsound mind to make medical decisions shall be supported by evidence of a deficit. However, most ethicists tend not to talk about capacity as a global matter, but rather about "decision-specific capacity": "Does this patient have enough ability at this time, given these circumstances, to make this decision?" Dugel, Ethics on Call (1992), p. 128. See COMPETENCE.

CARDIOPULMONARY RESUSCITATION (CPR)—A technique (also called Code) to prevent death as a result of sudden cardiac arrest. Introduced in the 1960s and approved by the American Heart Association in 1974, it remains the only such medical intervention routinely performed by many nonphysicians without a physician's order, unless a Do Not Resuscitate (DNR) order or similar notice has been documented.

CENTRAL DOGMA OF BIOLOGY—The result of the discovery in the latter part of the twentieth century that the information defining the sequences in DNA could go from nucleic acids to a particular protein, but never in the reverse direction; that is, never from protein to RNA or DNA or to other proteins. This would rule out biological rebirth or resurrection of the body as stated in the Apostle's Creed of the Catholic and other denominations. See TWO CULTURES.

CERTIORARI—"To be informed." A party seeking review by the Supreme Court submits a petition to the Court for a writ of *certiorari*. Only if four justices vote in favor of "cert." is it granted and the case comes to the Court for its decision. Each year approximately 5,000 petitions are sent to the Court seeking a writ of *certiorari*. Less than 5% are "cert." If *certiorari* is denied, the decision of the lower court is sustained. However, a denial is not evidence of the Supreme Court's opinion on the issue in the case.

CHROMOSOME—A structure inside the nucleus of a cell that has packed strands of DNA. The diploid human genome has 23 pairs of chromosomes, 46 in 22 pairs of autosomes, and 2 sex chromosomes. See AUTOSOME.

CIVIL SENTENCE—Forcing a completely innocent invalid to suffer for long periods of time by refusing to authorize withdrawal of treatment (Double Effect or Assisted Suicide or Euthanasia). Not directly prohibited by the

Constitution's bar on "cruel and unusual punishment," which is applicable to the sentencing of a guilty criminal.

CLINICAL EQUIPOISE—Research conducted in a Randomized Clinical Trial (RCT) that normally refers to the equality of weight given to the tests of uncertain difference between two or more types of treatment. The report of the National Commission for the Protection of Human Subjects of Biomedical and Behavioral Research, *The Belmont Report* (1979), p. 3, held that Clinical Research differs fundamentally from medical practice, and others would rule out Placebo-controlled trials in many cases. *The Logic of Clinical Purpose*, IRB 12. [1] 6 (1990):1 æ #æ6.

CLONE—A precise copy of biological material, such as a DNA segment.

CLONING—Generating precise copies of a particular piece of DNA to allow it to be sequenced.

CODE—The use of all available means to resuscitate a failing patient. See DO NOT RESUSCITATE and CARDIOPULMONARY RESUSCITATION.

COMA—A pathologic loss of consciousness or awareness and responsiveness to self and the environment, which may or may not be permanent. However, sleep is not coma but a depression of consciousness from which a person can be aroused. See PERSISTENT VEGETATIVE STATE (PVS).

COMMON LAW—The law contained in decisions of courts in countries with an Anglo-American tradition, which is a practical combination of logic and experience, with the latter in predominance.

COMPASSION—To "suffer pity or distress on the misfortunes of another." WEBSTER'S THIRD NEW INTERNATIONAL DICTIONARY, p. 462.

COMPETENCE—Mental ability to comprehend measured by the clinical assessment of a patient's mental capacity or ability to give Informed Consent to treatment. It is not generally judged by the standard tests for "serious mental illness"; rather it varies with the complexity of the decision to be made and the risks involved. There is no single generally accepted test. See CAPACITY.

CONCUSSION—A temporary loss of brain function.

CONTUSION—Bruising of the brain.

CONFIDENTIALITY—A legal obligation owed by a professional such as a physician or attorney not to disclose information given by a patient or client without their consent, which would diminish their right to privacy. It is subject to many exceptions. See PRIVACY.

CREATIONISM ACT—Creationists believe the world is very young, and that all its life forms were created suddenly by God, and no new species have appeared since. A number of states enacted laws that introduced creationism into public school curricula. A trial on the merits in federal court in Arkansas in 1982 held that while it would "never criticize or discredit any person's testimony based on his or her religious beliefs, creation science has no scientific merit or educational value as science," and "the only real effect of [the] Act is the advancement of religion," which meant that it failed to pass constitutional muster for public schools.

CRO-MAGNON—Early variety of modern human, *Homo sapiens.*

CRYOPRESERVATION—A process whereby "excess" pre-embryos fertilized during the IVF procedure are frozen and stored for future implantation. Women normally produce only one egg each month; fertility drugs increase the number of eggs available during each cycle and the success rate of the IVF procedure, and improve the chances of achieving pregnancy. See PRE-EMBRYO.

CULTURAL COMPETENCE—The change from knowledge about individuals and groups of people into standards, policies, practices, and attitudes that can be used in appropriate cultural settings to increase the quality of services and produce better outcomes. See Davis, Exploring the Intersection between Cultural Competency and Managed Behavioral Health Care Policy of County Mental Health Agencies. Alexandria, Va.: National Technical Assistance Center for State Mental Health Planning (1997).

DEAD-DONOR RULE—Vital organs can only be taken from dead patients and living patients must not be killed by organ retrieval. Youngner & Arnold, *Ethical, Psychological, and Public Policy Implications of Procuring Organs from Non-Heart-Beating Cadaver Donors*, 269 JAMA 2769-74, at 2771 (1993). It is generally assumed that a violation even with consent would constitute euthanasia.

DEATH—An individual who has sustained either (1) irreversible cessation of circulatory and respiratory functions, or (2) irreversible cessation of all functions of the entire brain, including the brain stem, is dead. A determination of death must be made in accordance with accepted medical standards. Uniform

Determination of Death Act. See BRAIN DEATH, and PHYSICIAN-ASSISTED SUICIDE (PAS), and EUTHANASIA.

DEFENDANT—The party opposing the basis of a lawsuit filed by a PLAINTIFF.

DEVELOPMENT DISABILITY—a disability that is not an illness.

DEONTOLOGY—A term that is derived from the Greek *deon* (duty) and *logos* (a rational principle); used in a philosophy alleging that many actions are morally obligatory regardless of their consequences. It is used in contrast to the Utilitarian theory of considering several alternatives in order to produce the greatest happiness for the greatest number of people to be affected. See BEST INTERESTS, SUBSTITUTED JUDGMENT.

DEGENERATIVE EVOLUTION—Reversals, as in blind cave fish, where eyes degenerate because they are not used.

DIPLOID—A set of genetic paired chromosomes, one from each parental set. Most cells except the gametes have a diploid set of chromosomes. Compare HAPLOID.

DISCOVERY—Discovery is a pretrial device in which each party to a lawsuit seeks information from the other party as well as from non-parties believed to have knowledge relevant to the issues in the case. ABA Div. Pub. Edu.

DISEASE—A term generally set by its usual role with healthcare. See STATES OF DISEASE.

DIVERSITY—This term is used whenever a federal court has jurisdiction over a case that does not involve a question of federal law. While there are several types of diversity jurisdiction, the most common type has two requirements: (1) the plaintiff and the defendant are residents of different states and (2) the dollar amount of the dispute between the parties is at least $75,000 (exclusive of interest and costs). ABA Div. Pub. Edu.

DNA (deoxyribonucleic acid)—A chain-like molecule that encodes genetic information linked head-to-tail with smaller molecules that form DNA chains containing information needed to direct the production of proteins in cells.

DNA NUCLEOTIDES—Bases contained in DNA consisting of adenine (A), guanine (G), cytosine (C) and thymine (T); a double helix has bonds between base pairs of nucleotides where A pairs with T and G with C.

DO NOT RESUSCITATE (DNR)—A directive that Cardiopulmonary Resuscitation (CPR) should not be used to attempt the restoration of breathing and heartbeat to a particular patient. Patients or their surrogates who agree to the entry of DNR in medical records usually show that CPR would be useless or burdensome.

DOUBLE EFFECT—Treatment intended to relieve pain and suffering that may also knowingly shorten life or cause a patient's death.

DUE PROCESS—Limitations set forth in the U.S. Constitution's Fourth, Fifth, and Fourteenth Amendments. They have been classified as either procedural or substantive.

DURABLE POWER OF ATTORNEY—An Advance Directive authorized by state statutes permitting an individual to designate another to make treatment decisions when the patient (the principal) becomes incompetent. See LIVING WILL, LIFE-SUSTAINING TREATMENT.

ENTROPY—The measure of that part of the heat or energy of a system not available to perform work; it increases in all natural (spontaneous and irreversible) processes (*Orland Medical Dictionary*) and measures the amount of disorder in a physical system. High entropy means that rearrangements of the ingredients making up the system would go unnoticed. Low entropy means few rearrangements would go unnoticed, and the system is highly ordered.

EPIGENETIC—A heritable change in the pattern of gene expression that is mediated by mechanisms other than alterations in the primary nucleotide sequence of a gene. See N. ENG. J. MED., Nov. 20, 2003, p. 2042.

EPILEPSY—A chronic neurological condition characterized by recurrent, unprovoked seizures.

EPISTEMOLOGY—The theory, knowledge, or science of the method and grounds of knowledge with reference to its limits and validity. WEBSTER'S COLLEGIATE DICTIONARY.

ETHICS—A term derived from the Greek *ethos* (character). It has been connected with morals and varies with cultures; however, some people think that morals reflect more on character than does ethics. Its special concern is with the human power to make judgments and choose between alternative values, the motives that prompt them, and the consequences to which they give rise. It may also include etiquette. See BIOETHICS (and Codes of Ethics in the Introduction).

ETHICS COMMITTEES—Committees concerned with Ethics in almost 5,000 hospitals in the United States that attempt to (a) educate staff members, (b) draft policies, (c) undertake consultations with staff, patients, and family members, and (d) review individual patient cases and make recommendations. As of 1992, the Joint Committee for Accreditation of Healthcare Organizations (JCAHO) has required a "mechanism for the consideration of ethical issues arising in the care of patients and to provide education to caregivers and patients on ethical issues in healthcare." See ETHICS, BIOETHICS.

EUGENICS—A term coined in 1883 by Francis Galton that is derived from the Greek meaning "good in birth" and defined by him as "the study of agencies under social control which may improve or impair the racial qualities of future generations physically or mentally." According to the *American Heritage Desk Dictionary* (1981), it is "[t]he study of hereditary improvement of a breed or race, especially of human beings, by genetic control." Tudge, in *The Impact of Genes* (2001), p. 279, wrote:.

> I perceive it as the attempt to improve human beings by adjusting the human gene pool as a whole, as well as the genomes of an individual. People will also differ, of course, in what they mean by "improvement." The general approach, however, must be to reduce the frequency of alleles that are seen to detract from the intended goal, and to increase the frequency of alleles that seem to lead in the required direction. (See ALLELES.)

EUKARYOTE—Means "good" or "well" in Greek, and *Karyon* is for "kernel." Hence, it covers cells that include a nucleus, animals, plants, and fungi. Compare PROKARYOTES.

EXON—The protein-coding DNA sequence of a gene. Compare with intron.

EUTHANASIA—"The good death," "a gentle death," or "mercy killing" following suffering. It is generally restricted to administration of a drug in order to accelerate death or directly terminate a patient's life of suffering. The former terms "active" and "passive" euthanasia have been generally disregarded for lack of a logical distinction between them. See DOUBLE EFFECT.

EXPERT WITNESS—An expert who can testify regarding his opinion. Rule 702 of the Federal Rules of Evidence authorizes testimony of one who has "scientific, technical, or other specialized knowledge that will assist the trier of fact to understand the evidence or determine a fact in issue."

FALSIFIABLE—A scientific method derived from the writings of Sir Karl Popper that although nothing can be completely verified to be always true, working hypotheses must be tested by systematic attempts to refute them using a rational hypothesis. BRYAN MAGEE, KARL POPPER 32 (1973). It concerns rules appraising solutions, which is somewhat analogous to decisional law. See SCIENTIFIC EVIDENCE, TWO CULTURES.

FECUNDITY—Refers to the ability to produce a baby within a certain period of time.

FERTILITY—Refers to the ability to produce a baby.

FETAL ALCOHOL SYNDROME (FAS)—A term describing damage to some fetuses by pregnant women who drink alcohol during pregnancy such as to cause stunted growth, disfigurement, and mental retardation. FAS particularly affects fetuses during the first trimester, as bones and organs are forming.

FREEDOM OF ACCESS TO CLINIC ENTRANCES (FACE)—A statute enacted in 1994 making it a crime to interfere "by force or threat of force or by physical obstruction" with anyone who is seeking or performing an abortion or other reproductive health services. It has been held not to violate the First Amendment's free speech or religious clauses. See AUTONOMY.

FUNDAMENTAL RIGHT—A right, the restriction of which is subject to "strict scrutiny" by the courts and the modification of which requires "a compelling state interest." An example is the right of a competent adult to refuse medical treatment of any kind. It is also a "liberty interest" under the Fourteenth Amendment, which constitutes the general freedom of people in a free society; although fundamental rights are subject to strict scrutiny, liberty interest may be subject to state regulation provided that there is a legitimate state interest, public health or safety. Bop & Cleson, *Webster and the Future of Substantive Due Process*, 28 DUQ. L. REV. 271, 280 (1990).

FUTILITY—Nontreatment because treatment would not benefit the patient. It has also been defined as being "either predictably or empirically so unlikely that its exact probability is often incalculable" or a treatment that benefits one organ system but fails to benefit the person as a whole. Schneiderman, Jecker & Jonsen, *Medical futility—its meaning and ethical implications*, 112 ANN. INT. MED. 949 (1990). The U.S. Department of Health and Human Services requires medical treatment by healthcare providers receiving federal funds except where it would be "virtually futile in terms of the survival of the infant and the treatment itself under such circumstances would be inhuman." It thus excludes nontreatment in terms of morbidity or the quality of life of the infant, which has been described as a largely "illusionary" concept but may survive due to its ambiguity. The 2003 edition of the AMA's *Code of Medical Ethics* issued in 1994 still states that "[d]enial of treatment should be justified by reliance on openly stated ethical principles . . . , not on the concept of 'futility,' which cannot be meaningfully defined." Section 2.035.

GAIA—Derived from the Greek for Goddess of Earth. A current hypothesis that the earth itself is evolving as a large living organism; it is "the health of the planet that matters, not that of some individual species or organism" and "this homeostasis is maintained by active feedback process operated unconsciously by the biota." Its health is being threatened by major changes in our ecosystems. In 1926 Vladimir Vernadsky coined the term "biosphere" to indicate that the Earth constitutes an integrated living system. The hypothesis of Gaia was introduced in the 1970s by Lovelock.

GAMETE—A mature germ cell (either sperm or ovum) possessing a haploid chromosome set and capable of initiating the formation of a new individual by fusion with another gamete. An Ovum is a female gamete. A Sperm is a male gamete. WEBSTER'S THIRD NEW INTERNATIONAL DICTIONARY 726.

GAMETE INTRA-FALLOPIAN TRANSFER (GIFT)—The process whereby a woman receives the same hormonal treatment as in IVF to stimulate ovulation. A laparoscopy is done to harvest oocytes (eggs). Fresh semen is then mixed with the oocytes and transferred to one or both fallopian tubes during the laparoscopy. Fertilization occurs in the tube and the fertilized egg moves down into the uterus and implants as in a regular conception cycle. See IN VITRO FERTILIZATION.

GENE—"Any portion of chromosomal material . . . which is small enough to last for a large number of generations." RICHARD DAWKINS, THE SELFISH GENE (1976), p. 30.

GENE POOL—The set of all genes in all the genomes of a sexually breeding population. See DAWKINS, THE ANCESTOR: TALE (2004), p. 437.

GENETIC CODE—Nucleotides, coded in triplets (codons) along the m RNA, that determine the sequences of amino acids in protein synthesis. The DNA sequence of a gene can predict the m RNA sequence, and the genetic code can predict the amino-acid sequence.

GENETIC ENGINEERING—Strives to insert proper genetic information and remove genes causing problems. It adds the missing gene for producing a crucial enzyme or determines how genes interact.

GENOME—The DNA makeup of an organism including genes and non-coding DNA; a set of instructions embedded in a cell's nucleus. Genome is the set of genes within one individual.

GENOMICS—The attempt to find out which bits of DNA in an organism correspond to which genes. It has three goals: first, to map the positions of all the genes on their various chromosomes; second, to work out the sequences of bases within the functional genes, and within all the bits in between; and third, to show what each gene actually does, whether making proteins, transfer RNA, or ribosomal RNA. JUDGE, THE IMPACT OF THE GENE (2001), p. 259.

GENOTYPE—The entire genetic constitution of an individual. The genes an individual carries; also refers to the pair of Alleles that a person has at a given region of the genome.

GOLDEN RULE FOR PHYSICIANS—Disclose unto patients as they would have you disclose unto them. See INFORMED CONSENT and THERAPEUTIC PRIVILEGE.

HABEAS CORPUS—Under the federal habeas corpus statute, 28 U.S.C. § 2254 (1994), a person held in state/local custody who believes that his or her custody violates federal law—typically, the Constitution—may challenge that custody by filing a petition for a writ (i.e., an order) of habeas corpus in federal district court. If the petitioner wins, he or she must be released or retried, at the option of the prosecuting authority. ABA Div. Pub. Edu.

HAPLOID—Half the full set of genetic material present in the egg and sperm cells of animals and in the egg and pollen cells of plants.

HEALTH—"A state of complete physical, mental, and social well-being and not merely the absence of disease or infirmity." Definition from the Constitution of the World Health Organization.

HEALTH MAINTENANCE ORGANIZATION (HMO)—A Medicare managed-care plan; a group of doctors, hospitals, and other healthcare providers agree to give healthcare services to enrolled people for a fixed amount; it is both a health insurer and healthcare delivery system. A federally qualified HMO must apply to the federal government for qualification and administration by the Office of Prepaid Healthcare of the Healthcare Financing Administration (HCFA).

HEDONIC DAMAGES—Damages sometimes awarded in wrongful death actions for the lost pleasure of living. The hedonic value of life is separate from the economic or productive value of the individual.

HEMATOMA—A blood clot.

HEMODIALYSIS—Filtering the blood by passing through an artificial kidney machine.

HIGH BLOOD PRESSURE (HBP)—The pressure in one's arteries when the heart contracts (systolic) is consistently 140 mm Hg or higher and the pressure in arteries between beats (diastolic) is consistently 90 mm Hg or higher. Studies showed that a high systolic reading is a more serious warning sign of potential health risks in adults 65 or older. It is also called hypertension (high tension in the arteries [not nervous tension]) HBP is inheritable: about 25% from one parent and 60% from both. One in four adults in the United States has HBP; although there is no cure, it is treatable. SHEPS, MAYO CLINIC ON HIGH BLOOD PRESSURE (2002), and National Heart, Lung, and Blood Institute (NHLBI).

HIPPOCRATIC OATH—An oath taken by medical students upon graduation from medical school. It was named after an ancient Greek. It includes the statement, "I will not give to a woman an instrument to produce abortion." However, after the *Roe v. Wade* decision by the Supreme Court in 1973, this portion of the oath was deleted.

HOMO SAPIENS—The subspecies of the genus of primates.

HOSPICE—A term derived from the Latin word for both *host* and *guest*, constituting a program of palliative and supportive services (providing physical, psychological, social, and spiritual care) for dying persons and their families. Hospice and palliative care are used interchangeably. In the Middle Ages it described places where weary pilgrims could meet. Hospices include outpatient and in-patient care but do not engage in medical efforts to prolong life; it was first applied to specialized care for dying patients in 1967, at St. Christopher's Hospice in a London, England, suburb. See PALLIATIVE CARE.

HOSPICE SERVICES—Care for persons in the last phase of incurable disease to provide that the dying live fully and comfortably. It neither hastens nor postpones death.

HOSPICE TEAM—An interdisciplinary group of professionals, with expertise in palliative pain and symptom control, who attend to the psychosocial needs of both the patient and the family. 24 J. L. MED. & ETHICS 365-68 (1996).

HOSPITALISTS—Physicians who provide inpatient care in place of primary care physicians or academic one-month-per-year attending. Since 1996 the hospitalist model of inpatient care experienced hospitalist workforce growth to 1,900 in 2002 (comparable in size to cardiology). "Hospitalists are fundamentally generalist physicians who provide and coordinate inpatient care, often aided by myriad subspecialists. Academic hospitalists are emerging as core teachers of inpatient medicine." J 23/30 JAMA 487 (2002).

HUMANIST MANIFESTO II—A document that favors the cultural relativity of concepts of disease because there is no canonical interpretation of nature that exists as the result of physical processes. Kurtz (ed.), HUMANIST MANIFESTOS I AND II THEUS (1973), p. 25.

HYDE AMENDMENT—An amendment to a bill in Congress that has prohibited the use of federal funds for abortion except when the woman's life would be endangered if the fetus were carried to term. State legislation ("little hydes") was enacted in 14 states, restricting funding for the cost of abortion for poor women on welfare.

HUBBLE'S LAW—The linear relationship between the recession velocity and galactic distances; the velocity of recession is equal to the distance multiplied by a quantity (called Hubble's constant). By computing this expansion backward, as if the universe were imploding, one can ascertain the age at which it existed as a condensate of exceedingly high density.

HUMAN GENOME PROJECT (HGP)—An international project involving a number of laboratories to complete DNA sequence of the human genome and characterizing the human genome's functional organization.

HYPOTHERMIA—Lowered body temperature.

IATROGENIC—The adverse condition of a patient resulting from a treatment by a physician or surgeon.

IN BANC—*In banc* (sometimes spelled *en banc*) literally means "full bench." As a general rule, when an appellate court sits *in banc*, all active judges sit. However, in the federal system, some circuit courts of appeals have so many active judges, e.g., the Ninth Circuit with 28 judges, that sitting literally *in banc* is not feasible. Thus, for those circuits with 15 or more active judges, the size of an *in banc* court is set by circuit rule, with the exact number varying by circuit. ABA Div. Pub. Edu.

INCOMPETENT CAPACITY—A patient may lack the legal Capacity to give or withhold Informed Consent, if it is determined that the patient's lack is documented evidence of a substantial deficiency in a mental function.

INFERTILITY—The inability to conceive after one year of trying or the inability to carry a pregnancy to term without a serious threat of physical harm. One in six couples in the United States has infertility, or over 5 million couples. In some 90% of the cases, infertility has an identifiable medical basis. See IN VITRO, IN VIVO.

INFORMED CONSENT—A term first used in 1957 by a California court and expanded upon by a Kansas court in 1960 to significantly increase the autonomy of a patient's right to control treatment (the liberty interest in refusing unwanted medical treatment) and requiring physicians to adequately disclose reasonable risks to him. Liability may be incurred where the patient would not have consented to the treatment had he been informed of those risks. It is the physician's responsibility to set forth the alternative and not rely upon the patient's duty to ask. See GOLDEN RULE FOR PHYSICIANS and THERAPEUTIC PRIVILEGE.

INTRACYTOPLASMIC SPERM INJECTION (ICSI)—A reproductive technology whereby a single, washed sperm is inserted into a woman's extracted ovum, which would then be implanted.

INTRAVENOUS (IV)—A method to insert medication or nutrition through a tube (catheter) or by injection into a vein.

IN VITRO FERTILIZATION (IVF)—A technique in which several eggs are taken from a woman, fertilized in a petri (glass) dish, and returned to the prospective mother's womb. In one state, a human embryo is defined as "an in vitro fertilized human ovum, with certain rights granted by law, composed of one or more living human cells and human genetic material so unified and organized that it will develop in utero into an unborn child." LA. REV. STAT. ANN. 9:121 (West 1991).

IN VIVO—A method of conception in which the ovum of a woman is artificially inseminated by the sperm of a man; the fertilized ovum is then flushed from her and transplanted into the uterus of another woman who may be able to bear a child. See INFERTILITY.

JUDICIAL SUPREMACY—The result of a decision by the Supreme Court in *Marbury v. Madison* in 1803 which established that the Constitution is the supreme law of the land, and "it is emphatically the province and duty of the judicial department to say what the law is." This right of the judiciary was not expressed in the Constitution itself. See PRIVACY.

KANT'S CATEGORICAL IMPERATIVE—To act as if we were legislating for all, to treat others as ends and never as means, and to democratically accept others as being competent to make certain legislative decisions. It is a variant of the classical Golden Rule. See GOLDEN RULE FOR PHYSICIANS.

KARYOTYPING—A technique by which chromosomes are located through microscopic examination, photographed, and systematically ordered on the basis of physical characteristics.

KIDNEY TRANSPLANT—Obtaining a biologically compatible kidney from a living donor or someone who recently died and surgically implanting it into another person.

LAW—Society's making of a certain amount of overflowing life (AUTONOMY) by balancing it with responsibility through the use of reason.

LIFE—The temporary fight against equilibrium, or increased Entropy. It is also a concept that has no single definition but has several generally accepted characteristics, such as a high degree of organization, acquisition and use of energy, reproduction, growth and development, response to stimuli, adoption to the environment, and other characteristics. See WHARTON, LIFE AT THE LIMITS (2002), p. 221. A biological definition includes evolutionary components. The Declaration of Independence speaks of an inalienable right to "life," but this is not mentioned in the U.S. Constitution, except for the Fifth Amendment, which states that no person shall be "deprived of life" by the state "without due process of law." There is a growing movement toward a right to have one's life terminated under certain circumstances. See DOUBLE EFFECT, PHYSICIAN-ASSISTED SUICIDE, EUTHANASIA, METABOLIC, PHYSIOLOGICAL, and THERMODYNAMIC.

LIFE-SUSTAINING TREATMENT—Any medical intervention, procedure, or medication including surgery, use of a ventilator, respirator, dialysis, CPR, or artificial nutrition and hydration that is administered to prolong the process of dying. Uniform Rights of the Terminally Ill Act (URTIA) 1 (3), 9B U.L.A. 611. This has been applied "whether or not the treatment is intended to affect the underlying life-threatening disease(s) or biologic processes." Hastings

Center Guidelines; Hafemeister, *Guidelines for State Court Decision Making in Life-Sustaining Medical Treatment Cases*, 7 Issues L. & Med. 443, 445 (1992). See also Section 2.20, AMA Code of Medical Ethics (2003 ed.).

LIVING WILL—An Advance Directive authorized by state statutes giving an individual the right to forego Life-Sustaining Treatment under certain circumstances in the event he becomes terminally ill and incompetent. Such directives are intended to be additions to the common law of the state. Failure to observe such a directive may result in liability. See DURABLE POWER OF ATTORNEY.

MAGNETIC RESONANCE IMAGING (MRI)—Tests that use magnetic energy to produce pictures of brain tissue, bones, and other structures.

MANAGED CARE—Refers to prepaid healthcare organizations such as HMOs for care provided under a fixed budget and costs capable of being "managed." The term is being used to include forms of indemnity insurance coverage.

METABOLIC—Considers an organism, such as a membrane, which separates it from its non-living environment. Exchange of materials with its surroundings enables the organism to consume energy. It could also apply to a non-living entity such as a candle flame. Warton, Life at the Limits (2002), p. 220. See LIFE.

METASTASIS—The spread of cancer cells from the original site of the tumor to other areas of the body.

MICROCHIMERISM—A small population of cells or DNA in one individual that derives from another genetically distinct individual. *E.g.,* fetal microchimerism (or maternal microchimerism) constitutes cell traffic between mother and fetus during pregnancy (or from twin to twin transfer *in utero*).

MITOCHONDRIA—Organelles inside cells (other than the nucleus) that "produce" energy for use by the cell.

MOOT—Seeks to determine an abstract question that does not arise upon existing facts or rights. Blacks Law Dictionary.

MORULA—An early stage of embryonic development (roughly 16-64 cells) at which the embryo is a solid spherical mass of cells, resulting from the early cleavage divisions of the zygote; so called because of its resemblance to a "little mulberry" (in Latin, *morula*). The President's Council on Bioethics, *Alternative Sources of Human Pluripotent Stem Cells* (Washington, D.C.: President's Council on Bioethics, 2005).

MULTIPLE EMBRYOS—In In Vitro Fertilization (IVF), use of more embryos for implant to ensure reproduction. Splitting the embryos into twins, triplets, or quad triplets makes it more probable that a woman could become pregnant. Ordinary identical twins, triplets, or more are sometimes called clones. Additional embryos are often frozen and implanted years later.

MYOCARDIAL INFARCTION—A heart attack due to the sudden blockage of a major blood vessel supplying the heart's left ventricle by an obstruction, such as a clot. Some of the cardiac muscle (myocardium) becomes deprived of blood (and hence of oxygen), which kills the heart's contractile muscle cells (cardiomyocytes).

NATURAL LAW—A belief that law is rooted in something bigger than the people who hand it down. Thomas Aquinas and St. Augustine praised it. To some extent, the law constitutes a melding of the thoughts of people who hold such beliefs with those who do not. Lon Fuller defined its "fundamental tenet as an affirmation of the role of human reason in the design and operation of legal institutions." See NEUTRAL PRINCIPLES.

NEUTRAL PRINCIPLES—An attempt by some analysts of the Supreme Court's constitutional law decisions to avoid the conclusion of nihilists that the court is merely a "naked power organ." Because of the difficulty arising from asking "whose neutral principles" are to be applied in deciding difficult cases, a modern view is that strict neutrality is unattainable and that non-neutral valuation is an inescapable component of such decisions. Although the aphorism "we are a government of laws and not of men," or Natural Law, is promulgated, the only check upon judicial power in that court is the exercise of self-restraint. See NATURAL LAW.

NORPLANT—Synthetic hormones that are contained in and gradually released from six capsules implanted beneath the skin of a woman's upper arm to prevent pregnancy for five years. It is more than 99% effective in preventing conception. A California judge ordered its use as a condition of probation by a woman convicted of child abuse. See DO NOT RESUSCITATE (DNR).

NULLIFICATION—Large-scale actions inconsistent with the law. This was illustrated by the so-called Prohibition Era in the twentieth century resulting from the anti-alcohol law that led to the repeal of the Eighteenth Amendment of the Constitution by the Twenty-first Amendment. Nullification is occurring in the twenty-first century to some extent with respect to state laws prohibiting ASSISTED SUICIDE or EUTHANASIA and by those that are restricting abortion.

OBESITY—The state of overweight that increases the risk of development of high blood pressure. The more the body mass, the more blood needed to supply oxygen and nutrients to tissues. The volume of blood being circulated through blood vessels increases, adding force on the artery walls.

OCCAM'S RAZOR—If there are two competing theories or explanations, then all other things being equal, the simpler one is more likely to be correct. William Occam put it thus: *pluralitas non est ponenda sine necessitate* ("plurality should not be posited without necessity"). William of Occam was a fourteenth century English Franciscan theologian who argued that religious orders should not own property or wealth. He was run out of Oxford University and moved to Avignon, France, where he accused Pope John XII of heresy and was excommunicated. He succumbed to the Black Death in 1349. SINGH, BIG BANG (2004), p.45.

OFFICE OF HEALTH MAINTENANCE ORGANIZATIONS (OHMO)—Division of Department of Health and Human Services (DHHS) responsible for overseeing federal activity concerning HMOs.

ONCOGENES—Genes for cancer-causing proteins or tumor-inducing viruses.

PAIN— "An unpleasant sensory and emotional experience associated with actual or potential tissue damage, or described in terms of such damage, or both." The International Association for the Study of Pain. The Intractable Pain Treatment Act (IPTA) requires consultation from another physician for every chronic pain patient. However, patients with pain are excluded if they are past or current substance abusers. Texas was the first state to subsequently amend its IPTA.

PALLIATIVE CARE—From Latin, to cloak or to cover. Without the goal of prolonging life, make medical and surgical interventions to alleviate suffering, discomfort, and dysfunctions for other purposes than cure. It has been reported that "between 10% and 50% of patients in programs devoted to palliative care still report significant pain one week before death." See Quill, *Physician-Assisted Suicide*, JAMA, Dec. 17, 1997, p. 2099. The World Health Organization defines Palliative Care as follows:

> The active total care of patients whose disease is not responsive to curative treatment. Control of pain, of other symptoms, and of psychological, social, and spiritual problems, is paramount. The goal of palliative care is achievement of the best quality of life for patients and their families.

World Health Organization, *Cancer Pain Relief and Palliative Care* (Geneva: World Health Organization, Technical Report Series, 1990). See HOSPICE.

PANSPERMIA—The concept that life on planets such as Earth and Mars was seeded from Space. Also, it has been argued that collisions of the planet with comets and asteroids could lead to the dispersal of terrestrial life to other portions of the galaxy.

PATERNALISM—The assumption that when properly informed one shares his personal approach of what he thinks is best; the preferences of others are denied. It is also what is received or inherited from the father.

PARKINSON'S DISEASE—The degeneration of nerve cells that produce a neurotransmitter called dopamine, which is needed to control muscle activity.

PEER REVIEW—"The process by which physicians judge the competence of their fellow professionals and recommend disciplinary actions for the incompetent." Manion v. Evans, 986 F.2d 1036, 1037 (6th Cir. 1993), *cert. denied*, 114 S. Ct. 71 (1993).

PER CURIAM OPINION—This term literally means "the opinion of the court," the Supreme Court or any appellate court. Because the opinion is the court's opinion, there is no indication of which justice/judge wrote it. ABA Div. Pub. Edu.

PERINATAL PERIOD—Viability of a fetus from about 24 weeks of gestation until birth and the first hours or days thereafter. See PERSON.

PERITONEAL DIALYSIS—Filtering blood by passing a solution into the abdomen, withdrawing wastes and excess water from the blood through the peritoneal membrane (lining the abdomen), which acts as an artificial kidney.

PERSISTENT or PERMANENT VEGETATIVE STATE (PPVS)—Designates a state of Coma (unconsciousness) of the patient, who retains respiration and a few primitive reflexes, but is unresponsive to meaningful stimuli over a prolonged period. If the coma is permanent, it is often referred to as "wakefulness without awareness." Unlike BRAIN DEATH, it includes situations where the brain stem is functioning but the neocortex is not; 85% of the public indicated that if they were in PVS they would prefer being taken off food and water to being maintained in that state.

THE TIMES, June 26, 1990, A-18. Thousands of people with PVS are being artificially kept alive at great expense.

PERSON—A human fetus after viability for purposes of the Fourteenth Amendment and laws concerning abortion. In 1973, the U.S. Supreme Court declared that an early fetus is not a "person," and made abortion a constitutional right; the right to privacy was confirmed in 1992, and the potential to become a person is not legally equivalent to being a person. See AUTONOMY and PERINATAL PERIOD.

PHARMACOGENETICS—The relationship of genetic factors with a particular patient's response to drugs.

PHENOTYPE—The properties and physical characteristics of an organism. "The entire physical, biochemical and physiological makeup of an individual as determined both genetically and environmentally." DORLAND'S MEDICAL DICTIONARY, 23rd Edition.

PHYSICAL DEPENDENCE—A physiological phenomenon of patients who have the capacity for an abstinence syndrome (addiction) after days of repeated drug doses. AMERICAN PSYCHIATRIC ASSOCIATION, DIAGNOSTIC AND STATISTICAL MANUAL OF MENTAL DISORDERS.

PHYSIOLOGICAL—Life that includes functions such as feeding, growing, metabolizing, excreting, reproducing, moving, and responding to stimuli.

PLACEBO—An inactive substance given to satisfy the patient's need for drug therapy, used in controlled studies in which a medical substance is also given to others in order to determine the efficacy of medicinal substance. See CLINICAL EQUIPOISE.

PLAINTIFF—The one who files a lawsuit.

PLURALITY OPINION—This term denotes an opinion of the U.S. Supreme Court in which there is no majority opinion; that is, fewer than a bare majority of five justices were able to agree on the legal basis for the Court's action in affirming, reversing, or vacating a lower court decision. A partial plurality opinion is one in which at least one part of the opinion represents the views of four or fewer Justices. ABA Div. Pub. Edu.

POSTHUMOUS REPRODUCTION—Either extracting an egg from a dying woman *while she is still alive,* fertilizing it with frozen sperm, and implanting it in a surrogate after the genetic contributor's death, or retrieving sperm from a dying man *while still alive*, freezing it, and using it for fertilization after the man has died. SPERLING, MATERNAL BRAIN DEATH (2004), p. 2495 (citing *Hecht v. Superior Court,* 20 Cal. Rptr. 2d 275 [Cal. Ct. App. 1993] and *Hart v. Shalala*, No. 4-3944 [E.D. La. 1993]).

PREEMPTION—Under the Supremacy Clause, U.S. CONST. art. VI, § 2, federal law—whether based on the Constitution, a statute, or a treaty—takes precedence over state or local law on the same matter. In other words, if federal law addresses a matter, either expressly or by implication, it trumps and renders unenforceable any state or local law on the matter. ABA Div. Pub. Edu.

PRESUMED CONSENT—A concept that would do away with the necessity of actual notice to next-of-kin (following attempts to contact them) prior to the harvesting of organs from a recently deceased person. A step in the direction of "presumed consent" occurred with a Texas law which provided that the wishes of organ donors 18 or older "shall be honored without obtaining the approval or consent of any other person." In some states individuals who renew, reinstate, or replace their driver's license or personal identification card may decide whether they wish to become donors.

PRIVACY—A right the Supreme Court added to the Constitution in 1965; it voided a Connecticut statute that prohibited the use of contraceptives by married couples. The Court stated, "The First Amendment has a penumbra [shadow or partial illumination] where privacy is protected from governmental intrusion." In 1973 it was held to envelop the right to abortion, and later the Court stated that "both the common law and the literal understanding of privacy encompass the individual's control of information concerning his or her person." In 1989 the Court indicated that the right is not absolute.

PROGNOSTICATION—A physician's determination of the lack of effectiveness of any care (or FUTILITY) to a critically ill patient.

PRE-EMBRYO—The first 14 days after conception have been designated as an outer limit for research thereon produced by in vitro fertilization; an embryo does not come into being until two weeks after conception; the initial fertilized entity is a pre-embryonic conceptus. Near the 15th day after conception, fetal cells begin to arrange themselves into the "primitive streak," a dark, longitudinal band out of which various systems of the body develop, including the nervous system. TABER'S CYCLOPEDIC MEDICAL DICTIONARY 128, 13th Edition.

PROKARYOTES—Cellular organisms without a nucleus or nuclear membrane. Bacteria and blue-green algae are prokaryotes, whereas other cellular organisms, plant and animal, are EUKARYOTES.

PUBLIC HEALTH LAW—"The legal powers and duties of the state to assure the conditions for people to be healthy in the population and the limitations on the power of the state to constrain the autonomy, privacy, liberty, proprietary, or other legally protected interests of individuals for the protection or promotion of community health." GOSTINE, PUBLIC HEALTH LAW: POWER, DUTY, RESTRAINT (2000).

QUALITY OF LIFE—A criterion advocated by a Presidential commission in the early 1980s but use of which is denied by the Department of Health and Human Services (HHS) and subject to sanction by the withdrawal of federal funds from healthcare facilities if it used in connection with severely defective neonatals.

QUICKENING—The stage of pregnancy when the fetus can be felt to move in the uterus. St. Thomas Aquinas stated in his *Summa Theologica* that animus (life or soul) entered the body of the unborn infant when it moved in the womb. Aristotle and Gelan had indicated that movement occurred 40 days after conception; however, today others have stated that pregnant women will usually not feel the child's movements until four months after conception.

RANDOMIZED CLINICAL TRIALS—See EQUIPOISE and PLACEBO.

RESEARCH—Defined by the Common Rule of the Federal Regulations as a systematic investigation including research development, testing, and evaluation designed to develop or contribute to "generalizable knowledge." 45 C.F.R Section 46.102(d) (emphasis added). The Centers for Disease Control and Prevention (CDC) requires Common Rule review for "research" but not practice activities. Guidelines for Defining Public Health Research and Public Health, Non-Research; state law provides little or no "Common Rules" for public health practice.

RU–486—A French antiprogesterone "abortion pill" that was found to be almost 100% effective when taken within 49 days of the patient's last menstrual period. Progesterone, secreted by the "corpus ovulation," is needed at all times during the pregnancy, and if removed at any time, the fetus will be lost. Antiprogesterones, such as RU–486, are progesterone receptor blockers that bind to the receptor and make implantation impossible, and are apparently free from side effects.

SCHIZOPHRENIA—From the Greek words for "splitting of the mind"; a group of "reaction" disorders based on organic, psychological, and sociocultural factors:

> The basic flaw in the brains of many schizophrenics is that cells migrate to the wrong areas when the brain is first taking shape, leaving small regions of the brain permanently out of place. Brain misconnections might develop when the mother catches a virus early in pregnancy.

ORANGE COUNTY REGISTER, June 1, 1996, at News 16, citing the Schizophrenia Research Branch at the National Institute of Mental Health in Bethesda, Md. Currently, no known cure exists. However, treatment can reduce its impact using pharmacotherapy, behavior therapy, and individual psychotherapy. NEW HARVARD GUIDE TO PSYCHIATRY (1978), p. 272.

SCIENTIFIC EVIDENCE—Evidence used by judges and juries in order to become acquainted with scientific principles applicable to a case. A rule that developed, the so-called Frye test, is that such principles must "have gained general acceptance in the scientific community" and it is applied in most states, including California. In 1993 it was modified somewhat by the Supreme Court with regard to the evidence introduced into federal courts by the addition of theses factors: namely, whether the expert's technique has been tested, whether it has been subjected to peer review and publication, and what the known or potential rate of error is for any test of the technique that has been employed. The Court stated, "By retaining the general acceptance test, trial judges will look to scientists rather than deciding themselves the question of reliability." (See Chapter 1, Sec. 3a.) Courts may take judicial notice of generally accepted scientific facts (as has been done in many courts with respect to blood types and DNA), thus relieving a party from the burden of establishing a fact by producing sworn witnesses, authenticated documents, or other evidence. See FALSIFIABLE and TWO CULTURES.

SECULAR HUMANISM—The word "secular" is derived from the Latin "saeculum" (age). It means long-lasting in contrast to contemporary or quick-changing. Its use in connection with Humanism derives from two Manifestos in 1933 and 1973. The first maintained that "all associations and institutions exist for the fulfillment of human life." The second states, "Science affirms that the human species is an emergence from natural evolutionary forces," and that "moral values derive their source from human experience." As Dr. Francis Schaeffer put it, the culture now pursues only two objectives: that of personal peace, by which is meant "You have your life to live and I have mine; you leave me alone and I'll leave you alone," and affluence, by which is meant the pursuit and acquisition of more and more material things. See THOMAS, THE DEATH OF ETHICS IN AMERICA (1987), p. 1156.

SLIPPERY SLOPE—The refusal to draw a line (rule) because of the argument that future decision-makers are either immature or will be unable to judge and distinguish another case which, although similar in some respects, may differ in crucial ones. Although an instant case may be innocuous in itself, others may later extend that line in an inappropriate manner. It has been stated that "without empirical [scientific] evidence of such systematic skewing, the slippery slope argument has nothing on which to stand."

SENESCENE—Aging with the loss of physiological functions needed for survival and reproduction.

SEX CHROMOSOME—The X or Y chromosome in humans. Virtually all females have two X chromosomes in diploid cells; males have X and Y chromosomes. They comprise the 23rd chromosome pair of a human genome.

SLOW CODE—Less than Code. Doctors and nurses go through the motions of cardiopulmonary resuscitation but do so in a way that attempts to minimize the chances that resuscitation will be successful. The approach seems to defy all moral notions of honesty, truth telling, and medical standards of care. LANTOS, DO WE STILL NEED DOCTORS? (1997), p. 43.

SPECIES—"Groups of interbreeding natural populations that are reproductively isolated from other such groups . . . bacteria and blue-green algae do not reproduce sexually, but by fission. The definition of species applies only to organisms able to interbreed." AYALA, EVOLUTION AND MOLECULAR BIOLOGY (1998), p. 43. A multiplicity of new environments giving rise to several different lineages and species from a single ancestral lineage is called adaptive radiation.

STARE DECISIS—(Lat.) To abide by or adhere to decided cases; the policy of courts to stand by precedent and not to disturb a settled point. It binds courts of equal or lower rank.

STASIS—Periods when there is no evolution at all, progressive or otherwise.

STATES OF DISEASE—Affairs that are (1) difficult (or are likely to make it difficult) for individuals to perform functions considered to be appropriate for individuals of a particular age and sex, (2) involve pain or distress which are considered to be inappropriate under the circumstances, (3) cause disfigurements or a lack of grace, or (4) cause premature death. ENGLEHAVIT, BIOETHICS AND SECULAR HUMANISM (1991), p. 113.

STEM CELLS—Cells that often can divide for indefinite periods and give rise to specialized cells.

STEM CELL LINE—Stem cells that have been cultured under *in vitro* conditions that allow proliferation without differentiation for months to years. (NIH)

STENTS—Tiny stainless steel wire-mesh tubes that open arteries to help restore blood flow in non-emergency procedures for people with heart attacks. 48 AARP May/June 2002.

STRICT SCRUTINY— A searching level of judicial review applied to governmental actions—federal, state, and local—challenged as unconstitutional. Strict scrutiny requires the governmental actor to show that it had compelling reason to take the challenged action and that the action taken goes no further than necessary—is narrowly tailored—to advance the cited compelling reason. ABA Div. Pub. Edu.

SUBJECTIVE PERSONHOOD—The use (or abuse) of the thoughts and feelings about a patient by another who knew him in the past to justify maintaining the life of one who has permanently lost consciousness, and hence his personality. For example, a relative (child, spouse, or parent) of a patient in a Persistent or Permanent Vegetative State (PPVS) may request that the patient be maintained by a healthcare facility for an indeterminate period of years, and at significant expense to the taxpayer.

SUBSTANTIVE DUE PROCESS—A doctrine that focuses on the law that has been violated (or under which one has been convicted and punished) rather than upon the procedure by which this was done. Although the assertions in the Fifth and Fourteenth Amendments that no person shall be deprived of life, liberty, or property "without due process of law" was originally construed by the Supreme Court to mean only procedural due process rights, that Court later began to interpret these as a limitation on some of the more substantive regulatory powers of the states. See ABORTION and PRIVACY.

SUBSTITUTED JUDGMENT—A doctrine under which a court (or a Surrogate) makes a decision by seeking to ascertain what would be in accordance with the interest of an incompetent patient, as a person with his characteristics, preferences and prospects. Its use was intended to permit the refusal of treatments in certain cases that physicians viewed as being in the patient's Best Interest—a test that was criticized as more paternalistic because it was traditionally almost solely concerned with treatment necessary to preserve life. Luce, 263 JAMA 696, 697 (1990).

SUFFERING—It is often broader than pain and physical symptoms and arises from anguish, fear, distress, hopelessness, indignity, and meaninglessness. People experience suffering throughout illness, and also when close to death.

SUMMARY JUDGMENT—This is the name of a procedural device available to either party to a civil lawsuit that enables one or the other party to win without a trial. A party seeking summary judgment is entitled to a judgment in its favor if there is no genuine dispute about the pertinent facts, and, based on those undisputed facts, the law compels a judgment for the party who asked for a favorable ruling. ABA Div. Pub. Edu.

SUPER DUE PROCESS—Numerous procedural requirements have been added for capital punishment during the final third of the twentieth century. They have added immeasurably to both (a) the delays currently connected with carrying out the punishment and (b) the excessive costs to the community in which the case is tried.

SUPERNOVA—A star that suddenly bursts into great brilliance; it is brighter than a nova.

SURROGATE—One who is acting for an incompetent patient but is normally not a public guardian. Typically, it is a family member who has concern for the patient's welfare or appointment by means of a durable power of attorney. See IN VITRO, IN VIVO, and SURROGATE MOTHER.

SURROGATE MOTHER—Normally a woman capable of either (a) conceiving a child under a contract with a married man desirous of having a child biologically related to him by a woman who agrees to *in vitro* fertilization of his sperm, or (b) the implantation of a couple's embryo. In both instances, she also agrees to her termination of all parental rights to a child produced as a result of the agreement.

SURVIVAL STATUTES—Statutes providing that a cause of action held by a decedent immediately before or at death is transferred to his personal representative, who can then seek to recover the damages to which the decedent would have been entitled at death.

TELEOLOGICAL—A view that the world and life was created for a purpose. It was termed a metaphysical argument by Francis Bacon, a founder of the scientific method; but Charles Darwin stated, "The old argument from design to nature, . . . which formerly seemed to me so conclusive, fails, now that the law of Natural Selection has been discovered." See SECULAR HUMANISM, TWO CULTURES.

TELOMERE—The end of a chromosome involved in the replication. Telomeres maintain the stability of chromosomes. Their loss results in the lack of stability of linear DNA molecules.

> Researchers recently have learned how to produce an enzyme, telomerase, which resets this "death clock" by restoring the telomeres. In time, study of the enzyme may allow us to rejuvenate human cells at will and thus change a large number of genes.

Carlson & Stimelry, The Terrible Gift (2002), p. 25.

TERMINAL—Normally a condition that is incurable and irreversible and will cause the patient's death within a short period not to exceed one year without the administration of Life-Sustaining Treatment. Use of the word *terminal* has been criticized as confusing decisions about life-sustaining treatment. The Oregon Death with Dignity Act defines terminal disease as "an incurable and irreversible disease that has been medically confirmed and will, within reasonable medical judgment, produce death within six months." The model statute defines terminal illness as "a bodily disorder likely to cause the patient's death within six months." See also ADVANCED DIRECTIVE, LIVING WILL, DURABLE POWER OF ATTORNEY, BRAIN DEATH, PHYSICIAN-ASSISTED SUICIDE, and EUTHANASIA.

THANATOLOGY—The study of physical, psychological, and social problems associated with dying, into an accepted medical discipline. (Noble N.Y. Times and Orange County Register, Aug. 26, 2004, at News 10. See also Derek Humphrey's book *Final Exit*, which explains in detail how to end life.)

THERAPEUTIC PRIVILEGE—The waning doctrine that would permit physicians to conceal information from competent patients that in their opinion would upset the patient and hence be counter-therapeutic; the privilege claim of a Hippocratic physician. It has been found to have been overused as an excuse for not informing patients of facts they are entitled to know. The President's Commission stated, "Little documentation exists for claims that informing patients is more dangerous to their health than not informing them, particularly when the informing is done in a sensitive and tactful fashion." See INFORMED CONSENT.

THERAPEUTIC CLONING—Nuclear transplantation to produce stem cells.

THERMODYNAMIC—Life sees organisms as being in contradiction of the second law of thermodynamics. It states that the amount of disorder (entropy) is always increasing, with the universe (as a whole) moving toward a state of disorder and randomness. Organisms, at first sight, defy this by maintaining order. However, they maintain their structure by consuming energy from their surroundings, including sunlight, and undergo a gradual increase in entropy. WARTON, LIFE AT THE LIMITS (2002), p. 220. See LIFE.

TOLERANCE—A pharmacological property defined by the need for increasing doses to maintain effects. The term indicates that exposure to the drug is a driving force for change in response.

TRANSGENIC ANIMAL—One or more genes from another species have been inserted into its egg cells. When foreign genes are incorporated into the genome, they may be "express." When new genes have been planted into germ cells, the traits they carry are passed on from generation to generation.

TWO CULTURES—A major distinction in society between (a) those with a significant scientific background and (b) their lack of communication with most others who are without it. This was brought to the fore in a 1961 lecture by C.P. Snow at Cambridge University when he stated:

> They still like to pretend that the traditional culture is the whole of "culture," as though the natural order didn't exist. . . . So the great edifice of modern physics goes up, and the majority of the cleverest people in the western world have about as much insight into it as their Neolithic ancestors would have had.

Many believe that this gulf has increased into the third millennium since that time. See TELEOLOGICAL, BIG BANG, CENTRAL DOGMA OF BIOLOGY, CREATIONISM, SECULAR HUMANISM, and SCIENTIFIC EVIDENCE.

UTILITARIAN—One of the doctrines used by bioethicists that, *inter alia*, holds as a utility goal the method to bring "the greatest happiness to the greatest number." In the nineteenth century it was developed by Jeremy Bentham and his disciple, John Stuart Mill, who formed the Utilitarian Society in 1823 in order to bring about many social reforms.

VASECTOMY—A simple operation that divides the vas deferens, interrupting the flow of sperm. Since vasectomy has no effect on hormone production or the nervous system, the operation will not influence masculinity, libido, sexual performance, or erotic sensations. Vasectomy will not even reduce sperm production. SIMON, HARVARD MEDICAL SCHOOL GUIDE TO MEN'S HEALTH (2002), p. 283.

VIABILITY—The capacity of a fetus to survive outside the womb. After viability, the Supreme Court indicated that the state "may, if it chooses, regulate and even proscribe abortion, except where it is necessary, in appropriate medical judgment, for the preservation of the life or health of the mother." About one-half of the states have attempted to further restrict abortions near viability. See BEST INTEREST, SUBSTITUTED JUDGMENT.

VIAGRA—A drug quite safe among the 30 million American men plagued by impotence, which should be used only to correct impotence. Viagra starts to work in about 30 minutes; its effects can persist for up to four hours; it should not be used more than once a day. Other options are available for men who cannot use it or fail to respond to it. About 3% of Viagra users have visual disturbances, and the FAA has prohibited pilots from using Viagra within six hours of flying. SIMON, HARVARD MEDICAL SCHOOL GUIDE TO MEN'S HEALTH, p. 307.

VIATICAL COMPANIES—They purchase the life insurance policy of a terminally or chronically ill patient, pay the premiums, and wait. In the meantime, the patient may use the cash proceeds from the transaction in any way. When the person dies, the viatical company receives the benefits of the policy. The sooner the patient dies, the greater the viatical company's profit. No national regulation exists, and fewer than 20 states require licensing of viatical settlement companies. 59 MONT. L. REV. 701 (1998), p. 701; 24 PEPP. L. REV. 99, 110 (1997).

ZYGOTE—It derives from the Greek *zygous*, meaning *yoked*. It is the diploid cell that results from the fertilization of an egg cell by a sperm cell. The zygote constitutes the earliest form of the embryo, but nearly half of them die at this stage. See NATIONAL ACADEMIES OF SCIENCE, SCIENTIFIC AND MEDICAL ASPECTS OF HUMAN CLONING 2-5. To form a zygote, the sperm tears a hole in the egg's outer coating, or zona pellucida, and works its way through to the egg's plasma membrane, and then fuses with the egg.

Author's note: Legal terms may be located in *Black's Law Dictionary* and the *New Century Dictionary* (2 volumes). In space see *Dictionary of Astronomical Terms*. Medical terms may also be located on computers from the following resources: *Medical Library Association*, mlanet.org/resources/medspeak/; *National Library of Medicine*, www.medlineplus.gov; *Merck Manual of Health Information*, www.merck.com/mrkshared/mmanual_home2/home.jsp; and *American Association for Clinical Chemistry*, www.labtestsonline.org; *Consumer Reports on Health* (November 2004), p.5.

Appendix A
Environmental Crises at the Turn of the Third Millennium

Adam and Eve are put into the Garden of Eden
"to work it and to preserve it."

Genesis 2:15

The World has been converted in an instant of time from a wild natural one to one in which humans, one of an estimated 10 million or more species, are consuming, wasting, or diverting an estimated 45 percent of the total net biological productivity on land and using more than half of the renewable fresh water. The scale of changes in Earth's systems . . . is so different from before that we cannot predict the future.

Peter Raven, President of the American Association for the Advancement of Science, *Science*, (August 9, 2002) p. 955

We are polluting ourselves into a corner. All of our natural systems are in decline. We are growing at a population rate that the earth may not be able to support. We act as though our resources will last forever, instead of replenishing them and building a sustainable future.

Kent M. Keith, The Paradoxical Commandments, (2000), p. 9

Each man has two countries, his own and the planet Earth.

Ward and Dubos, *Only One Earth* (1972), Introduction, p. 17

1. *Life's Ecological Needs and Purposes to Existential Philosophers; Environment Crises*
2. *Biomass and the Lack of Accounting for the Loss in its "Capital"*
3. *A Sustainable Environment; Forests, Insects and Other Animals*
4. *Social Structures of Human and Other Primates*
5. *The Oceans and Their Fish and Whales*
6. *The Human Need for Protection Against Release of Chemicals into the Atmosphere*
7. *Ozone Depletion*
   8. **Global Warming and the Loss of Biological Diversity**
   a. **Solar, Moon, and Wind Power vs. Coal, Gas, and Oil**
   b. **The Hydrogen Engine vs. Electric Cars**

It is a wholesome and necessary thing for us to turn again to the earth and in the contemplations of her beauties to know of wonder and humility.

Rachel Carson, *Silent Spring*

1. Life's Ecological Needs and Purposes According to Existential Philosophers; Environmental Crises

For many people, ecology has become the key branch of biological science. This is due to the current necessity to deal with the relations between living things and their environments; namely, plants and animals. These are divided into the marine and terrestrial ecologies. One of the most important parts of ecology constitutes studies on human destruction of significant populations of plants and animals. In 1962, this became better known to the public with the publication of Rachel Carson's *Silent Spring*, a volume that was immediately criticized by established economic interests. The National Environmental Policy Act of 1969 established a national policy for the environment and the U.S. Environmental Protection Agency (EPA) consolidated existing and new federal environmental programs.[1]

1. Pub. L. 91-190, 83 Stat. 852 (1970) at 42 U.S.C. §§ 4321-4375.

In 1972 *Only One Earth,* by Ward and Dubos, was published by the United Nations Conference on the Human Environment. Comments from 58 countries were received. Among those commenting were scientists, humanists, engineers, and other leaders in their fields. Ward and Dubostried "to define what should be done to maintain the earth as a place suitable for human life not only now, but also for future generations." They noted "so locked are we within our tribal units, so possessive over national rights, that we may fail to sense the need for dedicated and committed action over the whole field of planetary necessities." Because "no sovereignty can hold sway over the single, interconnected global ocean system that is nature's ultimate sink and man's favorite sewer," the authors concluded, "It is no small undertaking, but quite possibly the very minimum required in defense of the future of the human race" (p. 217).

In 1989 Bill McKibben noted that the necessary mental adjustments. . ."offers at least a shred of hope for a living, eternal, meaningful world."[2] He quoted Walter Truett Anderson with regard to life's purpose according to existential philosophers:

> The existential philosophers—particularly Sartre—used to lament that man lacked an essential purpose." "We find now that the human predicament is not quite so devoid of inherent purpose after all. To be caretakers of a planet, custodians of all its life forms and shapers of the future is certainly purpose enough.[3]

---

2. McKibben, The End of Nature 1121 (1989).

> Western industrial society, of course, with its massive scale and hugely centralized economy, can hardly be seen in relation to any particular landscape or ecosystem; the more-than-human ecology with which it is directly engaged is the biosphere itself. (p. 1113) If it took ten thousand years to get where we are, it will take a few generations to climb back down.

3. *Id.* at 1128.

> But, our reason should also keep us from following blindly the biological imperatives toward endless growth in numbers and territory. Our reason allows us to conceive of our species as a species, and to recognize the danger that our growth poses to it, and to feel something for the other species we threaten.

Id. See also Chapter 7, regarding our population, and McKebben's The Age of Missing Information (1993) and In Hope Human and Wild (1995).

In 1997, the UN Environment Program reported that Latin America had inflicted "terrible destruction to the environment."[4] In 1993, Darly Allves da Silva, the rancher who had ordered the slaying of Chico Mendes (a land reformer who was trying to preserve portions of the Amazon forest), was convicted of first-degree murder in the first trial in Brazil of its kind. It has been noted that nearly half the region's grazing areas have lost their crops in the past decade and about half the Latin America's mangrove swamps have been polluted by agricultural pesticides. The report, part of UINEP's annual report on the environment, warned that from 100,000 to 450,000 species of plants and animals could disappear during the next 40 years in Latin America alone.[5]

Thomas Lovejoy, an ecological consultant to the Smithsonian Institute on the forest fires in Indonesia and the Amazon stated "More of the world was on fire in 1997 than ever in recorded history."[6]

In 2000, the UN indicated that it would not consider "third-party" (academia, industry or public interest groups) generated data in its regulatory process until ethical issues were resolved. The following year, the Department of Health and Human Services (DHHS) issued ethical standards for the protection of human subjects in research conducted or sponsored by all federal agencies under its Common Rule.[7] In 2004, the National Research Council (NRC) recommended that all research used by the EPA be reviewed by an Institutional Review Board (IRB).[8] However, writers from a public health school and industry disagreed, suggesting:

> EPA requires that private entities obtain review under the Common Rule or its foreign equivalent before undertaking a study and provide documentation of this review in order to submit their data for regulatory purposes. By requiring studies to follow the Common Rule or a foreign equivalent, EPA can strongly discourage the practice of con-

---

4. VAN RENNSSELAER, BIOETHICS: BRIDGE TO THE FUTURE (1971). Da Silva escaped from a state penitentiary and was rearrested in 1996.

5. ORANGE COUNTY REGISTER, Feb. 22, 1997.

6. ORANGE COUNTY REGISTER, Dec. 4, 1997.

7. 40 C.F.R. pt. 26 (2001).

8. THE COUNCIL FOR INTERNATIONAL ORGANIZATIONS OF MEDICAL SCIENCES (CIOMS), INTERNATIONAL ETHICAL GUIDELINES FOR BIOMEDICAL RESEARCH INVOLVING HUMAN SUBJECTS (2002).

ducting human-subjects research and clinical trials outside the United States, to avoid federal scrutiny.[9]

Nevertheless, many types of human-subject ecological portions of bioethics research (discussed below) stated the rates, scales, and kinds of charges occurring in all the earth's ecosystems are "fundamentally different from those at any other time in history."[10] Of course, it is recognized that certain actions by the movement of tectonic plates causing earthquakes under the land and tsunamis under the sea (such as the magnitude 9.0 on December 26, 2004, off Indonesia, which killed over 150,000 people) take virtually no account of the plans of homo sapiens.

And God said, Let us make man in our image, after our likeness. So God created man in his own image, in the image of God created he him; male and female created he them. And God blessed them, and God said unto them, Be fruitful, and multiply, and replenish the earth, and subdue it; and have dominion over the fish of the sea and over the fowl of the air, and over every living thing that moveth upon the earth.

Genesis 1:26, 28

Man has long forgotten that the earth was given to him alone, not for consumption, still less for profligate waste.

George Perkins Marsh, *Destructiveness of Man*

2. Biomass and the "Lack in Accounting" for the Loss in its "Capital"

Late in the 1980s, an important article in the scientific journal *Bioscience* explored the production of biomass—the quantity of biological accounts in the earth's products of photosynthesis and how much that biological capital

---

9. Silbergeld, Lerman & Hushka, *Human Health Research Ethics*, SCIENCE (Aug. 13, 2004).

10. Jane Lubchenko at Oregon State University, *quoted by* Vandevelder in *Audubon*, July-August 2004, at 25. *Also see* ORANGE COUNTY REGISTER, Dec. 29, 2004, at Local 6.

and the interest both decrease each year.[11] The physician, musician, and philosopher, Albert Schweitzer, founded a large hospital at Lamarene (in the then equatorial Africa (now Gabon)). His ethics covered "all will to live" as the "fundamental principle of morality."[12] However, by the turn of the third millennium, the earth has been shown to be suffering from huge and unmistakable human impact. Some (the loss of species and the loss of tropical forest) are becoming irreversible. It has been asked, "What does this mean for our future as human beings? What will Earth be like as our numbers double?"[13]

Environmental costs have not been included in cost accounting by corporations, even where they constitute a major cost to the environment. In 1997, it was noted, "The market place only values a tree for its pulp . . . The tree gets no credit for locking up carbon dioxide and freeing oxygen, for

---

11. Vitousek et al., *Human Appropriation of the Products of Photosynthesis*, 36 BIOSCIENCE 368-73 (1986).

> From cookbooks and food labels, I will calculate how much one person eats per day. Multiply this number by 365 days and 6 billion people. Than add in what cows, sheep, pigs, and the rest of our domestic livestock eat and the food we consume for heating and building. This gives us only the biomass that we consume directly.

PIMM, THE WORLD ACCORDING TO PIMM 11 (2001).

12. Ethics thus consists in this, that I experience the necessity of practicing the same reverence for life toward all will-to-live, as toward my own. Therein I have already the needed fundamental principle of morality. It is *good* to maintain and cherish life; it is *evil* to destroy and to check life.
SCHWEITZER, CIVILIZATION AND ETHICS 246-47 (part II of *The Philosophy of Civilization)* (C.T. Campion trans., 2d ed. 1929).

13. Pimm at 233.

> Datable extinctions can be plotted to show the rates of extinction of birds, mammals and some other groups in historical time. The current rate of extinction of birds is 1.75 per year (about 1% of extant birds lost since 1600). If this rate of loss is extrapolated to all 20-100 million living species, then the current rate of extinction is 5,000-25,000 per year, or 13.7-68.5 per day. With 20-100 million species on Earth, that means that all of life, including presumably *Homo sapiens,* will be extinct in 800-20,000 years.

BENTON, WHEN LIFE NEARLY DIED 289-90 (2003).

keeping soil in place . . . for being a habitat for wildlife, for beauty." An ecological economist of the University of Maryland stated that the dollar value of the world's ecosystem services average $33 trillion per year. This constitutes 24 times the current amount spent per year by the United States on healthcare and more than half the output of human-made goods and services planet-wide, which was only $18 trillion.[14] In his volume, *The Diversity of Life,* E.O. Wilson stated:

> The rule of thumb, to make the result immediately clear, is that when an area is reduced to one tenth of its original size, the number of species eventually drops to one-half.[15]

Currently, corporate accounting does not include those costs.[16] Conservation has been a matter of politics since it began in the United States at the turn of the twentieth century with Gifford Pinchot's fight for the national forests. It has been recorded that President Taft's firing of Pinchot led to the split of the Republican Party and Theodore Roosevelt's Third Party, the Bull Moose campaign in 1912.[17] In 1996, President Clinton declared 1.7 million acres of southern Utah's red-rock cliffs and canyons a national monument. He used the 90-year-old Antiquities Act and stood in the same spot where Theodore Roosevelt stood to protect the Grand Canyon from development in 1908. The newspapers reported that "with just five electoral votes in Utah, there was not much

---

14. NEWSWEEK, May 26, 1997, at 73. *See also* NATURE'S SERVICES (Island Press).

> David Pimental and eight graduate students at Cornell University's College of Agriculture Land Life Sciences recently figured the tab for services we get free from the planet's plants, animals, and microorganisms. The total came to $319 billion for the U.S. and $2.29 trillion for the world.

SCIENTIFIC AMERICAN, February 1999, at 22.

15. SCHNEIDER, LABORATORY EARTH 103 (1997).

16. KEYES, *Environment, Conservation and Occupational Safety* (Ch. 23), *in* GOVERNMENT AND CONTRACTS UNDER THE FEDERAL ACQUISITION REGULATION (3d ed. 2003).

17. GINGRICH, TO RENEW AMERICA 194 (1995).

political risk for Clinton in offending the state's political establishment."[18] Further, it was in an area where, in 1997, for the first time since its founding a century before, the U.S. Forest Service admitted that its "commercial logging operations lost money."[19] Also for the first time, the government agreed to pay a timber company $475,000 not to cut trees in an Oregon national forest where dozens of antilogging protesters had been arrested. This resulted from a circuit court of appeals ruling that the logging was not subject to legal challenge because of the "salvage timer rider" Congress had enacted.[20] Nevertheless, a 2002 study indicated reducing pollution could cut medical spending as well.[21]

God has cared for these trees, saved them from drought, disease, avalanches, and a thousand straining, leveling tempests and floods; but he cannot save them from fools.

John Muir

The urgency is driven by the pending loss of so much of our world's natural areas and biodiversity in the first half of this century if we do not act immediately. There is much more that we need to know, but we clearly know enough to act. Our world is a spectacularly beautiful, interesting, and diverse place. Only by attending to its problems will it remain so.

Pimm, *The World According to Pimm* (2001) p. 249

---

18. Orange County Register, Sept. 19, 1996.

19. L.A. Times, Dec. 28, 1997, at A-12. In 2004, the World Wildlife Fund said in its regular *Living Planet Report* that humans consume 20 percent more natural resources than the Earth can produce. Orange County Register, Oct. 22, 2004, at News 25.

20. Posner, Overcoming Law 129 (1995).

21. Hospital admissions for respiratory problems were, on average, 19% higher in the 37 areas with the highest air pollution compared with the 37 areas with the least amount of pollution. With medical care spending exceeding $1 trillion per year, even a reduction of only a few percentage points would save society tens of billions of dollars annually.

Funch, Stanford University economist and lead author of the study, Nov. 11, 2002.

> *Our species, at more than six billion strong and heading toward nine billion by mid-century, has become a geophysical force more destructive than storms and droughts. Half the world's forests are gone. Tropical forests in particular, where most of Earth's plant and animal species live, are being clear-cut at the rate of perhaps one percent a year.*
>
> E.O. Wilson, *National Geographic,* January 2002, p. 86

3. A Sustainable Environment in Forest and for Animals

In the nineteenth century, the American Indian Chief Sealth said "the earth does not belong to us. We belong to the earth." Thus, it has been pointed out that man constitutes a part of the environment and his life and death are affected by what he does to the planet on which he lives.[22] As Mander has reported, "We have developed here, on this planet, and we are adapted to life here . . . we are not fit to live anywhere else."[23] Without plants to free the oxygen in the biosphere, we could not survive. Hence, the only course for survival is to preserve the plants.[24] What bioethical goals could be a more important than this one?

---

22. MARY MIDGLEY, BEAST AND MAN: THE ROOTS OF HUMAN NATURE 194-95 (Cornell 1978). Maimonides, a Jew who tried to bring some of Aristotle's reason into theology in the Middle Ages, wrote it should not be believed that all beings exist for the sake of man. On the contrary, all other beings too have been intended for their own sakes, and not for the sake of something else.

23. MANDER, FACING THE RISING TIDE 990 (1996).

> The idea, promoted in corporate circles, that first we must make countries wealthy through development and then take care of the environment, is high cynicism, since development does not produce wealth, save for a few people; the wealth that is produced is rarely spent on environmental programs; and anyway, by the time the theoretical wealth is generated, life will be unlivable. . . . Even for the biggest "winner," it will be like "winning at poker on the Titanic."

Id. at 12. *See also* GOLDSMITH, THE CASE AGAINST GLOBAL ECONOMY 78 (1996). Due to massive subsidies, farmers in 1992 paid less than 1% for a cubic meter of water. The city of Phoenix paid about 25 cents. WORLD BANK, WORLD DEVELOPMENT REP. (1992)

24. PIMM, THE WORLD ACCORDING TO PIMM 41 (2001). In 50% of tropical forests, the habitat of two-thirds of species have been logged or burned to clear land and 1 million square kilometers disappear every five to ten years. Pimm & Raven, *Extinction by Numbers*, NATURE 403, 943-44.

In 1967 came a breakthrough in agricultural productivity called the Green Revolution. Research on hybrid strains of rice and wheat made them two or three times more productive, provided they receive enough water, fertilizers, and pesticides. But in 1972, it was found that "the Green Revolution, for all its promise, will not necessarily lead to a balanced and ecologically satisfactory use of human and natural resources."[25] In 1972, the UN pointed out how the population growth left us with "the threat to humanity in developing lands." Further, "their populations are growing almost twice as fast as those of wealthier countries and twice as fast as did industrializing states."[26]

In 1987, the UN's World Commission on Environment issued its report "Our Common Future" to invoke the concept of "sustainable development" by which economic and social progress could be achieved without compromising the environment. However, in the 1990s that phrase appeared nowhere in the subsequently negotiated Uruguay Round of General Agreement on Tariffs and Trade (GATT) or its successor World Trade Organization (WTO). Hence, models of sustainability must be developed during the third millennium.

The eastern United States once had roughly 2.5 million square kilometers of forest (about a quarter of the current extent of the world's tropical forest). By 1920 only about 4 percent of the forests, previously uncut, remained. By 1990, virtually all the alleged protected parks that contained Costa Rica's original forests had been cut down. Studies later considered joining some of these parks due to a minimal critical size of a forest for each species to continue to exist:

> Newmark in 1985 investigated eight parks and park assemblages and found that even the largest reserve was 6 times too small to support minimum viable population of species such as grizzly bear, mountain lion, black bear, wolverine, and gray wolf. A study by Salwasser, et al., (1987) looked beyond park boundaries and included adjacent public lands as part of conservation networks. The results were the same. Only the largest area (81,000 square kilometers) was sufficient to protect large

25. The inputs—of improved seed, of fertilizer, of controlled water supplies—all require an increased investment of capital. If cooperative funds or public investment are lacking and the capital infrastructure of the new farming is left almost solely to individual initiative, the risk arises that the gains from the new agriculture will be concentrated in too few hands.
Ward & Dubos, Only One Earth 157, 167 (1972).

26. *Id.* at 204, 209.

> vertebrate species over the long term . . . Virtually every study of this type has reached a similar conclusion: No park in the coterminous U.S. is capable of supporting minimum viable populations of large mammals over the long terms, and the situation is worsening.[27]

As a result, some experts conclude "the prime step [is] to permit no development of any more virgin lands . . . whatever remaining relatively undisturbed land exists that supports a biotic community of any significance should be set aside and fiercely defended against encroachment."[28]

In 2001 a study revealed that Mexico has been losing forest coverage almost twice as fast as previously estimated. According to the Environmental Department, the country has the second highest deforestation rate.[29] Brazil is the number one country in forest loss. A joint commission of the Brazilian Congress approved a measure that would more than double the amount of the Amazon jungle that ranchers, loggers, and miners would be permitted to raze. Many environmentalists "were shocked by the deforestation in 2002, an increase of almost 40 percent vs. the previous year." Satellite photos and data "showed that 9,169 square miles of rain forest was cut down in the 12 months ending in August 2003," which was more than in the same period in the previous year.[30]

Instead of the 20 percent life for development, a new forestry code would enable landowners to use at least 50 percent of their holdings for "productive purposes."[31] China, on the contrary, "has embarked on a $10 billion, 10-year program to plant 170,000 square miles of trees, an area larger than California."[32]

---

27. GRUMBINE, ECOSYSTEM MANAGEMENT 46.

28. RODERICK NASH, THE RIGHTS OF NATURE 168-69; Paul Ehrlich, *Comments,* DEFENDERS OF WILDLIFE, November/December 1995; ANNE & PAUL EHRLICH, EARTH 242 (1987).

29. ORANGE COUNTY REGISTER, Dec. 4, 2001, at News 21.

> Mexico was losing about 1.5 million acres of forest a year to logging, fires, and the expansion of farms and ranches. But according to a study of satellite images taken between 1993 and 2000, forest loss in those years averaged about 2.78 million acres. *Id.*

30. ORANGE COUNTY REGISTER, April 8, 2004, at News 30.

31. ORANGE COUNTY REGISTER, Sept. 23, 2001, at News 42.

32. ORANGE COUNTY REGISTER, May 18, 2002, at News 5. Previously "decades of logging had left large swaths of the country looking like a desert wasteland." *Id.*

It has been pointed out that the Amazon jungle is sometimes called the world's "lung" because its billions of trees produce oxygen and absorb carbon dioxide from the atmosphere.[33] To the dismay of environmental groups in 2005 the Brazilian government has restored logging licenses that were suspended the previous year in an effort to impede deforestation in areas of the Amazon where the jungle is rapidly vanishing.[34]

E.O. Wilson at Harvard stated, "So important are insects and other land dwelling arthropods that if all were to disappear, humanity probably could not last more than a few months."[35] In regard to animals, ecologists Robert MacArthur and E.O. Wilson showed that as habitat area increases, species numbers also increase, and as area decreases, species number fall in a predictable way. The number of species present on islands is always fewer than on a similarly sized mainland or continental area if the habitats are otherwise identical. Furthermore, creating an "island" by cutting a forest into patches will greatly increase the rate of extinction. This is called "the equilibrium theory of island biogeography."[36] Size, which will vary according to species, will always merit consideration when determining the size limits or our

---

33. ORANGE COUNTY REGISTER, Feb. 19, 2004, at News 39.

34. "Throughout the Amazon jungle, lands owned by the federal government have been illegally occupied and subdivided and then repeatedly sold without proper land titles, often by and to logging groups. ORANGE COUNTY REGISTER, Feb. 13, 2005, News 17.

35. WILSON, THE DIVERSITY OF LIFE 125 (1992).

> So balanced is the biosphere that if those cycles were interrupted—for example, if the microbes in the soil or all the insects in the biosphere died out overnight—all higher plants and animal life, including humanity, would perish. We humans are as dependent on and intrinsically woven into the ecological fabric of our planet as any other life form.

RYAN, DONIVIN'S BLIND SPOT 234 (2002).

36. In 1967, Martin published his theory in great detail. He noted that the first humans known to have settled North America, the Clovis people, did so between 11,000 and 12,000 years ago. He also noted that by about 1,000 years after this initial colonization, most of the extinctions among large North American land mammals had been completed. Martin proposed that the Clovis people rapidly hunted many species to extinction.

WARD, THE CALL OF DISTANT MAMMOTHS 128 (1997).

wild habitats during the third millennium. In 2002, the National Audubon Society estimated more than one in four U.S. bird species was declining in numbers.[37]

Larger animals have long been hunted and extinguished by mankind. As stated by Paul Martin, "Large mammals disappeared too because they lost their food supply."[38] In 2000, a 12-nation conference estimated that populations of chimpanzees, gorillas, and orangutans are far lower than they were even a year or two ago, with some species down to a few thousand, or even a few hundred.[39]

The Endangered Species Act (ESA) of 1973[40] was enacted to protect animals that were on the brink of extinction[41] or perilously close to that brink.[42] The United States remained the only country to enact such a law. It now covers more than 1,000 species. Although the statute did not specifically define "harm," the secretary of the U.S. Department of the Interior did

---

37. The bird-conservation group estimates that 201 species in the continental United States, Hawaii, and Alaska are menaced by habitat destruction, pollution, diseases, and other threats. ORANGE COUNTY REGISTER, Oct. 24, 2002, at News 6. In England, "the population of 201 bird species, tracked to 1991, declined about 54 percent." ORANGE COUNTY REGISTER, March 19, 2004, at News 12.

38. *Id.*

39. In Indonesia, for example, orangutans are disappearing at a rate of more than 1,000 per year, with fewer than 15,000 remaining. They are the slowest breeding of the great apes, with females bearing one infant every eight years. The orangutans' habitat shrank by 50% in the 1990s alone, as illegal logging quadrupled. ORANGE COUNTY REGISTER, May 13, 2000, at News 23.

40. Endangered Species Act of 1973, 16 U.S.C. §§ 1531-1544.

41. *See id.* § 1533(a)(2)(A)(i).

42. *See id.* § 1533(a)(2)(A)(ii). Professor Peter Singer of Princeton University is a moral vegetarian and utilitarian who would save the lives of animals. Nevertheless, he stated:

> I have never said that I think all animal experimentation should stop immediately. If you can show that that is the only way of achieving a goal like curing a major disease, then I would say we should seek alternative ways of getting to the same goal, but in the meanwhile, I would not campaign to stop those particular experiments.

SINGER, WRITINGS ON ETHICS, LIFE 327 (2000).

so in a rather broad manner[43]—a definition of such animals was upheld by the Supreme Court in 1995.[44] For a time, this stopped loggers who wished to clear-cut timber in the Oregon forests where the red cockaded woodpecker and the northern spotted owl barely existed. Alteration of critical habitat became a prohibited "taking" and extended liability to certain individuals, corporations, and government entities.[45] However, Congress must continually reauthorize the EPA; hence it could modify this court decision by catering to those constituents who are seeking jobs and property rights.[46] With regard to the total number of plants at risk, a 1998 survey of 34,000 plant species by the World Conservation Union ranked the United States first among the nations of the world with regard to the total number of plants at risk. That represents 29 percent of the country's 16,108 known plant species.[47] As E.O. Wilson stated, "The destruction of biodiversity is the folly our descendants are least likely to forgive us."

---

43. The following definition was promulgated by the Secretary of Interior, Bruce Babbitt, through the Director of Fish and Wildlife Service:

> Harm in the definition of "take" in the Act means an act that actually kills or injures wildlife. Such act may include significant habitat modification or degradation where it actually kills or injures wildlife by significantly impairing essential behavioral patterns, including breeding, feeding, or sheltering.

See 50 C.F.R. § 17.3.

44. Babbitt v. Sweet Home Chapter of Communities for a Great Oregon, 515 U.S. 687, 696-708 (1995).

45. 16 U.S.C. § 1538 (1999). Defines "take" as "to harass, harm, pursue, hunt, shoot, wound, kill, trap, capture, or collect, or to attempt to engage in any such conduct." *See id.* § 1532 (19).

46. Crying "no more extinction" produces a noble sound, but it does nothing to stop extinction. And this has the potential to worsen the plight of biodiversity, because demanding the perfect can prevent us from obtaining the merely good. To do better, we will have to accept the responsibility that comes with being human at this time in history.

MANN & PLULMER, NOAH'S CHOICE: THE FUTURE OF ENDANGERED SPECIES 215 (1995).

47. Thus, it is not surprising to see that when new environmental challenges arise, species are able to adapt to them. More than 200 insect and rodent species, for example, developed resistance to the pesticide DDT in different parts of the world where spraying has been intense. Although the insects never before encountered this synthetic

The plummeting of birds and animals was due to the use of DDT during the 1940s through the 1950s. It was finally banned in 1972. Two years thereafter the decline in peregrine populations landed the bird on the endangered species list. Then it began its recovery and the federal government was finally able to take the peregrine off the list in 1999.[48] Because the Mississippi gushes 4.5 million gallons of water into the Gulf of Mexico per second, together with the nitrate fertilizer and other pollutants, this creates a coastal dead zone. The National Oceanic and Atmospheric Administration (NOAA) found that parts of 31 states and two Canadian provinces have been cut nearly in half from current amounts of nitrate releases to significantly shrink the annual Gulf dead zone. "Fertilizer applied to crops is the greatest contributor to the pollution that has created the Gulf of Mexico dead zone."[49]

The famous anthropologist Richard Leakey, who spent much of his life fighting to save Africa's natural treasures, wrote *Wildlife Wars* (2001), in which he noted that during the past century:

> The number of plant and animal species faced with extinction has shot up dramatically, largely because of the spread and indifference of humankind. I understood these things intellectually, as I expect most people do who care about the natural world around them. Ironically,

---

> compound, they adapted rapidly by means of mutations that allowed them to survive in its presence.

AN EVOLVING DIALOGUE: THEOLOGICAL & SCIENTIFIC PERSPECTIVES ON EVOLUTION 32 (2001).

48. ORANGE COUNTY REGISTER, June 20, 2002, at News 8. Their characters as predators were highly unusual:

> For thousand of years, the birds had prevailed in a narrow niche as the top gun of the skies. Diving at speeds approaching 200 mph, peregrines snatch ducks, songbirds and pigeons in midair, or knock them out and capture them before they hit the ground. *Id.*

In 2005 a U.S. court in Grants Pass, Oregon, rescinded a rule under the Endangered Species Act when it relaxed protections and allowed ranchers to shoot wolves on sight if they were attacking livestock. The coordinator for the Fish and Wildlife Service said no wolf had been killed by a private citizen since the new rule went into effect. ORANGE COUNTY REGISTER, Feb. 2, 2005, at News 13.

49. Raloff, *How to Curb River Pollution and Save the Gulf of Mexico*, SCIENCE NEWS (June 12, 2004) at 378.

> while I was working so hard to build a life around museums and fossils, I was also dying from a kidney disease—about which I'll say more later. Suffice it to say that I survived thanks only to a kidney transplant from my younger brother, Philip, in 1979. The transplant marked the beginning of a new life for me.[50]

The appropriate action he took was most effective, although the press seldom hesitated to criticize him.[51] In 1989, President Bush and Britain's Prime Minister Thatcher announced that ivory could no longer be imported; jewelry stores in both countries stopped selling it. The whole world knew about the African elephant crisis, and Kenya had taken the lead. In August of 1989, the European Economic Community formally banned further ivory imports; piano manufactures there and in Japan announced that they were switching from ivory to plastic keys.[52] Nevertheless, in 2002 the U.N. Environment Pro-

---

50. LEAKEY, WILDLIFE WARS 37 (2001):

> He spoke Swahili, Kenya's national language, and was an atheist (p. 257), an appointed director of the Wildlife Department in 1989. There were 300 employees at the headquarters. And some 4,000 others in the numerous parks and reserves spread out across the country. I had heard enough from Fiona Alexander and others to know that they were a demoralized and not entirely honest lot. My first job would be to mold them into a team ready to fight to preserve Kenya's wildlife. And the first step in building that team was to follow through on my own promises (p. 21). He became a member of Kenya's Parliament in 1997, the Director of KWS in 1998, and head of Kenya's Public Service in 1999.

51. Another reporter asked if I could arrange for the press to photograph dead elephants. "No, because there aren't going to be any more dead elephants. Soon the press will not be asking permission to film dead elephants but to film dead poachers." This was simply a bluff and bravado, but Kenyan newspapers carried headlines bearing my strong statements—and you could almost sense a new national pride.

Id.

52. *Id.* at 93.

gram said the earth faced more rapid change over the next three decades; namely, a quarter of the world's mammal species could face extinction.[53]

As he learns to observe their interdependence and their fragility, their variety and their complexity, he may remember that he, too, is a part of this single web and that if he breaks down too thoroughly the biological rhythms and needs of the natural universe, he may find he has destroyed the ultimate source of his own being. This may be too hard a lesson for him to learn anywhere.

Ward and Dubos, *Only One Earth* (1972), p. 114

4. Social Structures of Humans and Two other Primates

What some animals who are closest to humans can teach us has been noted in a 1995 volume which concluded with a remarkable statement; on the human characteristic's in two other primates:

> The fascinating thing about the patriarchal hamadryads males is their double role as husbands and as members of male clan. In most animal species, a permanent marriage prevents close cooperative relation among males, so that families live apart from one another. The male hamadryads baboon have managed to integrate permanent marriage into a cooperative male society, despite sharp competition for the females: Surrounded by their harem, they live side by side with their rivals. This social structure is found in only two other primate species: The gelada baboon of the high mountains in Ethiopia, and the human.[54]

This experience in humans and two other primates may no longer be possible throughout the next millennium because man's closest genetic relatives are facing pressures that could drive them to extinction in the wild.

---

53. The report stated that the world's biodiversity is under threat, with 1,130 of the more than 4,000 mammal species and 1,183 of the 10,000 birds regarded as globally threatened. ORANGE COUNTY REGISTER, May 22, 2002, at News 19.

54. KANMER, IN QUEST OF THE SACRED BABOON (1995).

One of the largest threats to the great ape is loss of habitat. Unsustainable logging, agricultural expansion, oil exploration, mining, and human migration into ape habitats are all causing the animals "forest" home to shrink as never before.[55]

The oceans are in trouble; the coasts are in trouble; our marine resources are in trouble. These are not challenges we can sweep aside.
James Watkins

5. The Ocean and Their Fish and Whales

About half the oxygen we breathe is produced by the photosynthesis that takes place in the oceans by plants called phytoplankton. The satellite research of National Aeronautics and Space Administration (NASA) has found that concentrations of ocean plants have declined in parts of the northern Pacific and Atlantic oceans by as much as 30 percent during the past two decades.[56]

In 2002, an estimated 24 cubic miles of ice were found to be disappearing annually from Alaskan glaciers. The following year, NASA satellites measured the melting of glaciers in the Patagonian ice fields of southern Argentina and Chile and found that the rate doubled from 1995 to 2000.[57] The U.S.

---

55. Orange County Register, April 11, 1997, at News 13.

> We need another and a wiser and perhaps a more mystical concept of animals. Remote from universal nature, and living by complicated artifice, man's civilization surveys the creature through the glass of his knowledge and sees thereby a feather magnified and whose image in distortion. . . . In a world older and more complete than ours, they move finished and complete, gifted with extensions of the senses we have lost or never attained, living by voices we shall never hear. They are not brethren, they are not underlings, they are other nations, caught with ourselves in the net of life and time, fellow prisoners of the splendor and travail of the earth.

Beston, The Outermost House 19 (1971).

56. Orange County Register, Aug. 30, 2003, at News 24.

57. Researchers estimate that the glaciers are losing the equivalent of 10 cubic miles of ice every year now. Orange County Register, Oct. 17, 2003, News 12.

Commission on Ocean Policy found that ocean pollution (largely from farmland and urban runoff) and fish stocks continue to be depleted (and the advice of scientists often is ignored).[58]

Concerning the great benefits of one species, in 2002 it was written: "Even people that don't like salmon know by now that it contains omega-3 fatty acids, which are believed to protect against cancer and cardiovascular disease."[59]

The U.S. Commission on Ocean Policy found that feeding and nursery areas for three fourths of U.S. commercial fish catches are disappearing each year. Of the fully assessed U.S. fish stocks, 40 percent are depleted or are being over fished, the commission said in an interim report. Also, 12 billion tons of ballast water from ships are spreading invasive alien species around the world.[60]

In his article "Swimming with Sharks," the author wrote, "Somewhere between 40 million to 70 million sharks were killed in 1994. The International Shark Attack File estimates that for every human being killed by a shark, 10 million sharks are killed by human beings."[61]

In 1999 Iceland and Norway resumed whaling despite the fact that the International Whaling Commission decided in 1986 to ban whaling amid concern that many whale species were endangered.[62] In spite of the ban,

---

58. The contribution from Alaska's glaciers is greater than all the other glacier fields put together, excluding fields in Greenland and Antarctica. Experts have attributed sea level rise to two primary effects: runoff from the melting of ice fields and an ocean expansion due to warming.

ORANGE COUNTY REGISTER, July 19, 2002, at News 10.

59. Adler, *The Great Salmon Debate*, 140:18 NEWSWEEK 54-56 (2002).

60. ORANGE COUNTY REGISTER, Sept. 23, 2002, at News 5. A cruise ship can dump up to 30,000 gallons of sewage a day, as well as 255,000 gallons of "gray water" from laundries, showers, sinks, and dishwashers.

61. Peter Benchley, AUDUBON 52-57 (1998).

62. ORANGE COUNTY REGISTER, March 13, 1999, at News 20.

> In 2003 the World Wildlife Fund released a study reporting that some 900 whales, dolphins, and porpoises drown every day after becoming tangled in fishing nets and other equipment. This first global estimate of the problem indicated that 308,000 of the marine mammals die unintentionally in fishermen's hauls.

ORANGE COUNTY REGISTER, June 16, 2003, at News 9.

Japan and Norway kill about a thousand whales a year, although "a recent poll found that 60 percent of the Japanese had never eaten whale meat, while a scant 1 percent eat it regularly."[63] But in 2005, Japan permitted its fleet to catch whales—400 Antarctic minkes—alleging that it was for research because scientists have published papers about the catch. The meat typically ends up in stores and critics protested the research, charging that "it mostly amounts to an excuse to go on whaling."[64]

In 2002, a study was released that found that the United States may have no stream left free from chemical contamination, and about one-fifth of animal species and one-sixth of plant types are at risk of extinction.[65] The United Nations has indicated that 1.1 million people worldwide live without access to safe drinking water and 2.5 billion of them lack proper sanitation. In order to have clean water by 2025, $100 billion per year must be spent for the world's poor.[66] In the United States, the EPA estimated it would cost $10 billion per year to add necessary sewer capacity and to repair and renovate existing sewer systems.[67] In November 2002, the EPA was required by environmental groups to regulate Clean Air Act violations to prevent 15,000 premature deaths, 350,000 cases of asthma, and one million cases of diminished lung function in children.[68] In 2005, new EPA rules were issued to limit emissions from coal-burning power plants, which produce smog and soot that are the byproducts of generating electricity.[69] However, the U.S. Court of Appeals for the District of Columbia Circuit upheld many of the Bush administration's revisions of

63. Foster, review of *Dominion,* by Matthew Scully, in THE FEDERAL LAWYER, February 2004, at 49. In 2004, the 9th U.S. Circuit Court of Appeals ruled that California gray whales and other marine mammals have no standing to sue to stop the U.S. Navy from using sonar under the Endangered Species Act, the Marine Mammal Protection Act or the National Environmental Policy Act. ORANGE COUNTY REGISTER, Oct. 27, 2004, at News 14.

64. SCIENCE NEWS, Feb. 26, 2005, at 131.

65. A five-year study released by the H. John Heinz III Center for Science, Economics and the Environment, Sept. 27, 2002.

66. ORANGE COUNTY REGISTER, March 21, 2003, at News 27.

67. USA TODAY, Aug. 20, 2002, at 3A.

68. L.A. TIMES, Nov. 14, 2002, at A16.

69. USA TODAY, Mar. 11-13, 2005, at 1. "By 2015, the EPA expects several hundred power plants to cut their emissions of smog-forming gases by 61% below 2003 levels. The plants, in 28 states and the District of Columbia, also are expected to trim their emissions of a type of soot by 73% from 2020 to 2025."

the Clean Air Act to allow plant operators to modernize without installing expensive new pollution-control equipment.[70]

> *They are geniuses when it comes to the scientific questions . . . but we are primitive when it comes to dealing with many of the social political, and economic issues.*
>
> Dr. Arnold Relman

6. The Human Need for Protection Against Release of Chemicals into the Atmosphere

In her 1959 landmark book, *Silent Spring*, biologist Rachel Carson stated, "We have put poisonous and biologically potent chemicals indiscriminately into the hands of persons wholly ignorant of their potentials for harm. We have subjected enormous numbers of people to contact with these poisons without their consent and often without their knowledge." For this, future generations will be least kind in judging the one living at the end of the second millennium. In both the United States and Canada, such releases have resulted in the increase of cancer, as the second highest cause of death. Such deaths normally are preceded by great pain, as discussed in Chapters 11, 12, and 13. Carson received numerous awards throughout the world for this writing. Nevertheless, she was ridiculed by the chemical industry for her views on abuse of chemicals, just as Charles Darwin was ridiculed 100 years earlier following his publication of *Origin of Species by Natural Selection*.

The incidence of cancer rose some 50 percent in the second half of the twentieth century—a rate that almost exactly matches the rate of increase of pesticides during that period.[71] In 1951, DDT was found to be a contaminant of

---

70. Orange County Register, June 25, 2005, at News 9. Eliot Spitzer, the New York state attorney general, said: "Anybody who cares about the quality of air can view the case a victory for enforcement and continued aggressive action to limit the violations of the Clean Air Act by power companies."

71. Steingraber, Living Downstream 13-14 (1997).

> A study under the Environmental Protection Agency (EPA) found that Americans now have a cancer risk from toxic chemicals in the air that is at least 10 times above an acceptable level. As a result, 12 million people experience risks 100 times higher, as of the year 2002. They are produced mainly by vehicles and industry.

Orange County Register, April 19, 2002, at News 28.

human breast milk. Although Carson had pointed to the dangers of DDT in 1959, its registration was not revoked in this country by the EPA until 1973:

> The Federal Insecticide, Fungicide, and Rodenticide Act (FIFRA) requires reevaluation of old, untested pesticides approved before the current requirements for scientific testing were put into place. Initially scheduled to be completed in 1976, this deregistration process has been repeatedly delayed and is now scheduled for completion in the year 2010. Until then, the old untested pesticides can be sold and used. As one critic has noted, it is as if the bureau of motor vehicles issued everyone a driver's license but did not get around to giving road tests until decades later. According to the National Research Council, only 10 percent of pesticides in common use have been adequately assessed for hazards; for 38 percent nothing useful is known; the remaining 52 percent fall somewhere in between.[72]

In 2000, under the terms of a United Nations Environmental Program Treaty, DDT was virtually outlawed worldwide. It remains manufactured only in China and India, solely for use in antimalarial programs. Similar damage caused by certain types of radiation from pesticides is incrementally cumulative and difficult to measure.[73] American women with breast cancer lose an average of 20 years of life (amounting to one million life-years each year).[74] There were also health problems in workers who handled the chemical and liver cancer in exposed fish.

It is interesting to contemplate the example of four steps of alleged "unexamined progress" in the twentieth century:

1. Man puts DDT molecules together into long-lasting poisons in order to fight noxious plants and insects.
2. Its inventor is awarded the Nobel Prize.

---

72. Spielman, Mosquito (2001), Preface & 100.

73. At the end of the twentieth century, DDT still endangers species on Santa Catalina, an island 28 miles from my home in Corona del Mar, California. In 1997, only one of the eggs laid by the 17 eagles on the island hatched in five years due to the continuing effects of DDT.

74. Steingraber, Living Downstream (1997).

3. The use of DDT is later forbidden and declared subject to punishment in some countries but not others.
4. The effects of past and current usage of DDT continue well into the third millennium.

A similar history has run such a course with the very popular pesticide 2, 4, D, called agent orange, which was linked to miscarriages and was also eventually outlawed.[75] By 1990, the EPA listed 32,645 sites of past hazardous chemical waste dumping. Its National Priorities List of 1996 counted 1,420 Superfund Sites—the most notorious areas where people suffered significantly high mortality from cancers of the lung, bladder, esophagus, colon, and stomach. Such residues were found in 35 percent of foods consumed in the United States.[76] One writer referred to the pain filled deaths resulting from environmentally caused cancers as a "form of homicide."[77] Most disturbing perhaps is the fact that as of 1998, pesticide sprayings are not subject to public disclosure, except in two states—New York and California.[78]

During the twentieth century, Congress refused to address the problem of nonhazardous solid waste, which more than doubled from 1960 to 1990. It was determined that by continuing such a rate, more than 80 percent of landfills with such waste would be overloaded by 2005.[79] Further, unlike many other areas of human bioethics discussed in this volume, state courts have

---

75. *Id.* at 52.

> However, researchers reported that atmospheric levels of methane, the second most important greenhouse gas, have stopped increasing. Temporarily, at least, it has stabilized between 1999 and 2002.

ORANGE COUNTY REGISTER, Nov. 23, 2003, at News 13.

76. *Id.* at 165.

77. *Id.* at 269.

78. *Id.* at 275. In the preface to *Clinical Environmental Health and Toxic Exposures* (2d ed. 2001), Sullivan and Krieger state that "the environmental regulatory process is inextricably intertwined with politics, economics, and the legal system of a given country" and "there are trade-offs and compromises that are not driven by above-the-fray scientific rationalism." (p.xvii)

79. Harpring, Comment, *Out Like Yesterday's Garbage: Municipal Solid Waste and the Need for Congressional Action*, 40 CATH. U. L. REV. 851-52 n.2 (1991) [quoting Office of Solid Waste, U.S. Environmental Protection Agency, Pub. No. EPA/530-SW-90-042A, CHARACTERIZATION OF MUNICIPAL SOLID WASTE IN THE UNITED STATES: 1990 UPDATE, EXECUTIVE SUMMARY, ES-9, ES-13 (1990)].

often show an insensitivity toward this problem. For example, Wisconsin passed a comprehensive solid waste management program in 1989. It prohibited the disposal of recyclable material in its landfills unless the waste was generated in a municipality that had a mandatory recycling program and applied to any community in an adjacent state that utilized Wisconsin landfills.[80] In 1995, the Seventh Circuit Court of Appeals struck down the statute as a violation of the Commerce Clause and the Supreme Court refused to review that decision.[81] A decision by the Supreme Court could conceivably be overturned early in the third millennium before effective recycling requirements are virtually stifled and almost 100 percent of landfills become overloaded. It was estimated that $1.4 trillion in worldwide governmental subsidies promote environmental pollution, which is twice global military spending.[82]

Under the Clean Air Act, the EPA is authorized to stop construction or modification of a major emitting facility if the plans do not conform to the requirements of the agency's prevention of significant deterioration program. In 2004, the Supreme Court upheld the EPA's ruling that blocked construction of a pollutant emitting facility, which had been approved for construction by the state of Alaska.[83]

---

80. National Solid Wastes Mgmt. Ass'n v. Meyer, 63 F.3d 652 (7th Cir. 1995), *cert. denied,* 517 U.S. 1119 (1996).

81. National Solid Wastes Mgmt. Ass'n v. Meyer, 63 F. 3d 652 (7th Cir. 1995), *cert. denied*, Motor v. National Solid Wastes Mgmt. Ass'n, 517 U.S. 1119 (1996).

82. UNDP (1998) Human Development Report 1998 (Oxford Univ. Press 1998). In 2002 the Supreme Court rendered one victory for environmentalists by drawing a "fundamental distinction" between a government seizure of private property and regulation that limits an owner's use of it, by rejecting a taking claim brought on behalf of 449 families who were denied permits to build on their lots near Lake Tahoe. Tahoe Sierra Preservation Council v. Tahoe Regional Planning Agency, No. 00-1167. A.B.A. J., June 2003, at 32.

83. Alaska Dep't of Envtl. Conservation v. EPA, No. 02-658, Jan. 21, 2004. The EPA reported that emissions of sulfur dioxide, which causes acid rain, rose 4% in 2003, but that other pollutants (carbon monoxide, nitrogen oxides, particulate matter or soot, volatile organic compounds, and lead) dropped nearly 2%. ORANGE COUNTY REGISTER, Sept. 23, 2004, at News 10. But the following year that agency reported that from 1990 to 2002, U.S. greenhouse gases increased 13.1% while Russian greenhouse gases decreased 38.5%. *Id.* Nov. 7, 2004, at News 22.

In December 2004, the National Commission on Energy Policy, a privately funded group, issued a report concluding unanimously that voluntary emissions reductions favored by the Bush administration will fall short of doing the job:

> The commission called for a mandatory system of permits that would limit how much emissions a plant could release, saying it would slow the growth of carbon dioxide emissions by 500 million to 1 billion tons a year by 2020. Emissions growth rate would be reduced 2.4 percent a year beginning in 2010. Commission co-chair and former Environmental Protection Agency chief William Reilly said the mandatory program would be "a meaningful first step" to address climate change.[84]

> *On our planet itself all dividing horizons have been shattered. We can no longer hold our loves at home and project our aggressions "elsewhere" any more. And no mythology that continues to speak or to teach of "elsewhere's" and "outsiders" meets the requirement of this hour.*
>
> Joseph Campbell, *Myths to Live By*, p. 275

7. Ozone Depletion

Increased cancers and cataracts are caused by the release of chemicals into the air, resulting in the depletion of ozone, a pale blue gas between nine and 19 miles high in the earth's upper atmosphere. (It has been determined that

---

84. The White House Council for Environmental Quality said the administration opposes mandatory regulation of greenhouse gases. "We believe that will do more harm than good to the nation's energy policy and to the nation's economy." ORANGE COUNTY REGISTER, Dec. 9, 2004, at News 9.
No American would want to live in a village where a house with only 4 percent of the village's population belched out 25 percent of the smoke from its large chimneys, with much of the soot falling on the rest of the village. But this is precisely the relationship between America and the world on global environmental issues. In per-capita terms, the American household is by far the largest polluter in the global village. . . . Today, there is a virtual consensus among the global scientific community that the earth's atmosphere does not have a limitless amount of space to absorb pollution. *Mabubani, Beyond the Age of Innocence* 171 (2005).

asthmatic children using maintenance medication are particularly vulnerable to ozone at levels below EPA standards.)[85]

Ten years after the discovery of a hole in the ozone layer in the stratosphere and the troposphere over the Antarctic in 1985, the UN's World Meteorological Organization (WMO) found the hole began to expand earlier than expected, reaching a maximum of 20 million square kilometers (8 million square miles).[86] This expansion will continue despite a reduction in ozone destroying chemicals because the chemicals and their effects last for 60 to 100 years.[87] In 1996, the WMO reported that ozone levels plummeted to record lows over Great Britain and Scandinavia, exposing inhabitants to some of the worst levels of ultraviolet radiation ever recorded in those areas.[88] In 1998, the WMO predicted that the Earth's protective ozone layer

---

85. Gent et al., *Association of Low-Level Ozone and Fine Particles with Respiratory Symptoms in Children with Asthma*, JAMA, Oct. 8, 2003, at 1859. Signs and symptoms of ozone toxicity have been demonstrated to occur in exercising adults at ozone concentrations as low as the revised federal ambient air quality standard (0.08ppm). CLINICAL ENVIRONMENTAL HEALTH AND TOXIC EXPOSURES 814 (2d ed. 2001). In 2004, a study investigated whether short-term (daily and weekly) exposure to ambient ozone is associated with mortality in the United States. It found "a statistically significant association between short-term changes in ozone and mortality on average for 95 large U.S. urban communities, which include about 40% of the total US population." JAMA, Nov. 17, 2004, at 2372.

86. ORANGE COUNTY REGISTER, Sept. 13, 1995, at News 14. After reaching a record 10.8 million square miles earlier in the year, a seasonal "ozone hole" over the South Pole refills with surrounding ozone-rich air as temperatures rise. ORANGE COUNTY REGISTER, Nov. 21, 2003, at News 16. The Bush administration is seeking extensive exemptions from the phaseout for what it contends are "critical uses" of the pesticide methyl bromide. But scientists say the proposed exemptions would reverse steady progress in healing the ozone layer. N.Y. TIMES, National, Nov. 10, 2003.

87. ORANGE COUNTY REGISTER, Nov. 24, 1995, at News 12. A confirmation of a hypothesis of solar-caused variability remains to be proven since solar-induced mechanisms of climate change are still difficult to verify. Baliunas et al., *Time Scale Trends*, Harv.-Smith No. 4528, April 29, 1997, at 4.

88. Regional ozone levels during January, February, and the early part of March averaged about 30% below the average levels recorded between 1975 and 1978, before man-made chemicals started destroying the ozone. ORANGE

would hit its all-time thinnest by 2000 or 2001. In 2001, the Japanese Meteorological Agency announced that a large hole in the ozone layer has been forming over Antarctica. The largest hole ever observed, grew to cover 11,226,460 square miles.[89] In 2003 the WMO's expert said, "The ozone hole is getting larger and deeper and is lasting longer."[90] The following year, NASA launched its Aura satellite into orbit 438 miles above the surface on a six-year mission to provide detailed measurements of the earth's ozone layer, pollution levels, and changing climate.[91]

An Idaho family lived next door to a county dump with appliances leaking chlorofluorocarbon chemicals into the atmosphere, which constitute a degrader of the ozone layer. Their lawsuit under the Clean Air Act was upheld by the Ninth Circuit Court of Appeals.[92]

The EPA abused its discretion in refusing to consider and weigh the effect of the proposed waiver on particulate matter pollution along with its effect on ozone levels. In this connection, it has been noted that:

County Register, April 25, 1996, at A20. It was reported that "Chlorofluorocarbons (CFCs) were triggering anywhere from a 2.7% to 7% decrease in global ozone each decade." Effective Jan. 14, 1998, the entire continent of Antarctica was set aside as a global wilderness preserve under an international agreement. Orange County Register, Jan. 14, 1998, at News 10.

89. Orange County Register, Aug. 24, 2001, at News 9.

> However, chlorofluorocarbons (CFCs) are falling and an Australian government-funded study stated that the hole in the ozone layer over Antarctic should close within 50 years.

Orange County Register, Sept. 19, 2002, at 1.

90. Orange County Register, Oct. 4, 2003, at News 24.

91. L.A. Times, July 16, 2004, at A16. The cost of the satellite was $785 million.

92. Covington v. Jefferson County, No. 04 C.D.O.S. 1067 (CCA-9). That court also overturned the EPA's denial of California's request for a waiver of a Clean Air Act requirement that gasoline contain the oxygenate additive ethanol. The appellate court rejected the EPA's assertion that the state had not shown how a waiver would have a beneficial effect on ozone pollution, stating, "EPA abused its discretion in refusing to consider and weigh the effect of the proposed waiver on particulate matter pollution along with its effect on ozone levels." Davis v. EPA, 9th Cir., No. 01-71356, July 17, 2003.

American innovation and ingenuity have served us well in the past, and the future should be no different. Our markets allow for voids in technology to fill in quickly with new inventions and innovations. When faced with a goal, we have always found a new way to meet it. When the country decided to ban chlorofluorocarbons because of its destruction to the ozone layer and to lower the coincidence of skin cancers, we did not have an alternative refrigerant gas. However, once the ban was set in Congress, a cost-effective replacement was quickly available.[93]

The measurement of a 10 percent decline in snow coverage at high altitudes during the past 40 years has increased the earth's reflectivity and helped make it warmer.

World Resources 1998–1999, p. 174

Several studies have tracked human activity as a major cause of global warming during the last decades of the twentieth century from 3° C to 6° C (from 5° F to 10° F).

U.S. Department of Commerce,
National Oceans and Atmosphere Administration,
National Climate Assessment, June 2000

No one says flatly, "To hell with nature." On the other hand, no one says, "Let's give it all back to nature." Rather, when invoking the social contract by which we all live, the typical people-first ethicist think about the environment short-term and the typical environmental ethicist thinks about it long-term. Both are sincere and have something true and important to say. The people-first thinker says we need to take a little cut here and there; the environmentalist [thinker] says

---

93. Browner, *Environmental Policy: Principles for the Next Generation of Protection*, Am. J. L. & Med. 116 (2004). A 2003 study by the University of California Irvine (UCI) found high levels of ethane, butane and propane in a 1,000-mile-wide area roughly centered around Oklahoima City, Okla. "Such gases aid in the formation of ground-level ozone, which can damage people's lungs and slow crop growth." Orange County Register, Oct. 7, 2003, at News 10.

nature is dying the death of a thousand cuts. Paul II has affirmed that "the ecological crisis is a moral issue," [as he] Patriarch Bartholomew I, spiritual leader of the world's 250 million Orthodox Christians.
E.O. Wilson, The Future of Life (2002) p. 152

8. Global Warming and the Loss of Biological Diversity

A major effect of humans on the biosphere concerns global warming. In 1985, British meteorologists predicted a future mean warming of 1.54 degrees per decade for greenhouse gases or 1.35 degrees Fahrenheit per decade with sulfate aerosols included.[94] The higher the concentration of the greenhouse gas, the warmer the planet gets. Some scientists disagreed with this conclusion.[95] Nevertheless, a Global Climate Treaty at Rio de Janeiro presented ath the Earth Summit in 1992 concluded that nations should stabilize these gases by requiring that fuel use be cut by 60 to 80 percent worldwide. This was renewed at Kyoto, Japan, in 1997. The U.S. produces 21 percent of these gas emissions, with motor vehicles accounting for almost one third of the country's total carbon dioxide (CO_2) output.[96] In 1995, scientists and government leaders from 100 nations met at the Inter-Governmental Panel on Climate Change (IPCC) in Madrid. They concluded, "The balance of evidence suggests a discernible human influence on global climate."[97] In July 1996, the Clinton administration announced its support for legally binding emission limits on greenhouse gases in lieu of mere voluntary ones. But in 1997, the Senate unanimously resolved (95-0) that the United States should reject the treaty as harmful to the economy unless it were to force large developing countries, such as China, Mexico, and South Korea, to join the industrialized nations in fighting heat-trapping pollution. It was noted that the resolution did not call for developing countries to make the same pollution reductions as the industrialized ones. Hence, at the end of the twentieth century we are at a stalemate on

94. Sulphates, produced mostly by power stations that burn fossil fuels such as coal and oil, tend to cool the earth by reflecting sunlight.

95. ORANGE COUNTY REGISTER, July 14, 1996, at Com. 2.

96. CONSUMER REPORTS, September 1996, at 42. Ralph Cicerone, the new head of the National Academy of Sciences (NAS), said, "Carbon dioxide in the atmosphere is now at its highest level in 400,000 years, and it continues to rise." ORANGE COUNTY REGISTER, July 22, 2005, at News 14.

97. ORANGE COUNTY REGISTER, Dec. 3, 1995, at News 9.

this key issue. Further, the proponents of the Gaia hypothesis have not been able to devise models on any feedback loops strong enough to prevent a warming climate from growing ever hotter.[98] Thus, we are again reminded of H.G. Wells' warning early in the century that "Human history becomes more and more a race between education and catastrophe." An April 2000 poll showed that most Americans actually did not consider the environment to be a top political priority.[99] In 2002, it was noted that we are pumping carbon dioxide into the atmosphere at truly heedless rates, despite overwhelming evidence that CO_2 concentrations have been a critical factor in tipping our planet's climate periodically toward disaster.[100] Prior to the extensive use of fossil fuels, the concentration of carbon dioxide in the atmosphere was about 280 parts per million. At the Mauna Loa Observatory in Hawaii, levels were measured at 379 per million in 2004 and 376 per million the previous year.[101]

All of the European Union nations ratified the Kyoto protocol against global warming by June 1, 2002.[102] Four days later, the Bush administration, for the first time, declared that gases humans release into the atmosphere are

---

98. Lovelock & Kump, *Failure of Climate Regulation in a Geophysological Model,* 369 Nature 732-35 (1994). Yet in 1998, one of the hottest years of the century, a Duke University study showed that global warming could yield drier conditions on the northern Great Plains, triggering drastic ecological challenges during the next century. Orange County Register, Aug. 5, 1998, at News 7. *See also* Chapple, Let the Mountains Speak, Let the Rivers Run (1997).

99. The poll indicated that one in six people are active with the environmental movement, while just over half say they're sympathetic to the movement though not active. But it ranks in the middle with such issues.

100. Ward & Brownlee, The Life and Death of Planet Earth 211 (2002).

101. Orange County Register, March 21, 2004, at News 9.

102. Orange County Register, June 1, 2002, at News 22.

> The Kyoto pact was aimed at cutting emissions of the greenhouse gases blamed for rising global temperatures. Negotiators from nearly 185 countries worked to hammer out the details of implementing a landmark treaty aimed at reducing greenhouse gases that lead to global warming. But since the United States has withdrawn from the Kyoto Protocol, many countries say the accord's impact has been diluted.

Orange County Register, Oct. 25, 2002, at News 22.

chiefly responsible for global warming. Yet, it did not push for a reduction in greenhouse gases.[103]

The WMO found that the average temperature in 2001 rose a fraction of a degree to 57.2 degrees, marking the twenty-third consecutive year that temperature has been above the 1961-90 mean, which is used as a baseline.[104] The United States—the world's largest emitter of greenhouse gases—announced in March 2001 that it was withdrawing from the Kyoto treaty. It claimed that it would "harm its economy."[105] It also argued that the unfairness of the treaty [is] "because it excuses developing countries like India and China from obligations."[106]

After the turn of the third millennium, an ocean trip to the North Pole, "to the surprise of everyone aboard the ship found only open water to the north. The ship's captain said he had never seen open water there."[107] In 2003, the Senate rejected a plan to curb global warming, but both supporters and opponents of the measure favored attempts to reduce emissions of carbon dioxide and other greenhouse gases from industrial smokestacks. By this time, the senators only voted 55-43 to defeat a bill.[108] Russia's ratification of the Kyoto

---

103. Orange County Register, June 4, 2002, at 1.

> A U.N. panel has predicted that average global temperature could rise as much as 10.5 degrees over the next century as heat-trapping gases from human industry accumulate in the atmosphere.

Orange County Register, Jan. 2, 2003, at News 13.

104. L.A. Times, Dec. 24, 2001, at A14.

105. Orange County Register, Aug. 3, 2001, at News 17.

106. Orange County Register, Dec. 19, 2001, at News 23. It was also noted that our courts have not allowed environmentalists to challenge the federal government's rules on the effects on global warming and that programs under their authority harmed plaintiffs' programmatic activities in disseminating information about the greenhouse effect to the public. Foundation on Economic Trends v. Watkins, 794 F. Supp. 395 (D.D.C. 1992).

107. Segre, A Matter of Degrees 80 (2002).

> In the summer of 2005, the floating sea ice on the Arctic Ocean shrank to its smallest size in a century. The increased open water absorbed solar energy that would be reflected back into space by bright white ice.

Orange County Register, Sept. 29, 2005, at News 13.

108. Orange County Register, Oct. 31, 2003, at News 25.

Protocol in November 2004 fulfilled the condition that countries responsible for more than 55 percent of the world's greenhouse gases have done so. Hence, it became effective on February 16, 2005, for 126 countries. This treaty commits ratifying countries to reduce their greenhouse-gas emissions, by 2012, to 5 percent below their 1990 emissions.[109] In 2004, a report indicated that heat-related deaths can be expected to increase up to 180 percent annually by 2050 and as much as 500 percent annually by the end of the century if the atmosphere continues to warm.[110]

---

a) Solar, Moon, and Wind Power vs. Coal, Gas, and Oil

In June 2002, the EPA relaxed air pollution rules to make it easier for utilities to extend their coal-burning power plants. This also gave that industry greater flexibility in expanding electricity production without having to install additional emissions controls. However, environmentalists indicate that this will produce millions of tons of additional pollution and amounts to rolling back the Clean Air Act.[111]

The International Energy Agency predicted a two-thirds increase in global-energy demand by 2030, particularly by the Middle East, which sits on 65 percent of known oil reserves. Without government address of the long-term need to end a reliance on foreign oil, a major portion of the country's economy is at

---

109. SCIENCE NEWS, Jan. 1, 2005, at 3.

> The United Nations Environment Program (UNEP) released figures showing that 2004 was the costliest year on record for worldwide hurricane and other weather-related damage. The price tag was $90 billion for the first 10 months of 2004, compared with $65 billion for all of 2003.

Id.

110. The report was by ATMOS Research and Consulting. ORANGE COUNTY REGISTER, September 2002, at News 8.

111. ORANGE COUNTY REGISTER, June 15, 2002, at News 20.

> Carbon dioxide from the burning of fossil fuels is the most prevalent of the so-called greenhouse gases, whose growing concentration in the atmosphere is believed to be warming the Earth.

ORANGE COUNTY REGISTER, June 30, 2001, at News 8.

risk.[112] At that time congress gave some consideration to the development of cars using alternative fuel sources by automakers. However, in 2002 the Senate rejected a measure to require that 20 percent of the nation's electricity be produced from renewable sources such as solar and wind power by 2020. This dealt environmentalists a defeat on energy policy.[113] In 2005, a bill in California proposed a "million solar roofs initiative," which would cut $50 to $75 a month from an electric bill. But utility companies opposed having to pay for electricity put back into the grid by solar-powered homes and businesses—a process known as net metering—and the construction unions wanted guaranteed high wages for solar installers. As a result, the initiative bill was defeated.[114]

On the other hand, a proposal commissioned by the "seven rich industrial nations" plus Russia, would commit them to get their power from renewable energy sources, like wind, water, the moon, and the sun.[115] Further, it has been shown that the adoption of alternatives to fossil fuels is compatible with economic growth.[116] For example, in 2003 homes in Norway began obtaining

112. USA TODAY, Sept. 25, 2002, at 12A. The U.S. Department of Energy estimated that foreign oil imports will rise to 62 percent by 2020 and that half of this will be derived from the Middle East. Coon & Phillips, *Strengthening National Energy Security by Reducing Dependence on Foreign Oil,* BACKGROUNDER, No. 1540 (April 24, 2002).

113. L.A. TIMES, March 15, 2002, at A27.

114. ORANGE COUNTY REGISTER, Sept. 11, 2005, at News 16.

115. ORANGE COUNTY REGISTER, July 14, 2001, at News 25.

The White House attributes opposition to the proposal based on a desire to let the marketplace, rather than government, decide how quickly renewable energy sources are adopted worldwide. *Id.* Nevertheless, in 2002 a California law required that 20% of utilities' electricity be produced from renewable sources, such as solar, wind, and geothermal, by 2017.

ORANGE COUNTY REGISTER, Sept. 13, 2002, at News 16.

116. Barrett et al., *Clean Energy and Jobs: A Comprehensive Approach to Climate Change and Energy Policy*, Economic Policy Institute (2002).

> The objective of the United Nations Framework Convention on Climate Change (UNFCC); "Stabilize greenhouse gas concentrations in the atmosphere at a level that would prevent dangerous anthropogenic interference with the climate system," although this level is not specified. Given the likely inevitability of substantial climate change over the next few centuries, strategies to adapt to its potential impacts on human health are critical.

JAMA, May 1, 2002, at 2287.

electricity from the orbit of the moon from a sub-sea power station driven by the rise and fall of the tide.[117]

Nevertheless, Sallie Baliunas, an astrophysicist at the Harvard-Smithsonian Center for Astrophysics and deputy director of Mount Wilson Observatory, argued that much of the earth's possible warming aspects are due to those made by the sun:

> Twentieth century temperature changes show a strong correlation with the sun's changing energy output. Although the causes of the sun's changing particle, magnetic and energy outputs are uncertain. The correlation is pronounced. It explains the early twentieth century temperature increase, which, as we have seen, could not have had much human contribution.[118]

One writer noted that, "on one side are those who argue that human-caused changes in climate will make our lives so difficult that millions of us may die. . . . On the other side are those who warn that effects to avoid such a hypothetical fate may cause us to commit economic suicide and trigger the decay that we fear."[119] However, he added:

> Serious discussion is needed on human response to these changes, which may lead to human actions. This may require a slightly different view of the problem than used in traditional economic analyses.

b) The Hydrogen Engine vs. Electric Cars

America uses fossils as the greatest producer of CO_2 gases, which are causing the warming of the earth and also producing a worldwide asthma epidemic now affecting more than five million children. Conversely, hydrogen, the makeup of about 90 percent of the universe, can be used to produce water (H_2O) and greatly lower our production of CO_2. The chair of the board of Ford Motor Company stated, "Hydrogen will put an end to the 100-year reign of the internal engine." Along with many other nations, our automakers

---

117. ORANGE COUNTY REGISTER, Sept. 21, 2003, at News 12.
118. 31:3 IMPRIMUS (March 2002).
119. ALLEY, THE TWO MILE MACHINE 2 (2000).

are beginning to develop fuel cell vehicles to run on hydrogen as one of the means needed to further protect the world.

Electrolysis of hydrogen creates no CO_2 in fossil fuel. First, using the ocean, whose plants are its consumer, reduces pollution of the air. It could then replace the current gas station for users of hydrogen fuel cell vehicles. A second method is to use a vehicle's gasoline engine (or natural gas) only to produce the hydrogen.

> The type of cell used in cars is called a proton exchange membrane (PEM) . . . Hydrogen is separated from oxygen by a cellophane-like membrane coated with a platinum catalyst. As the hydrogen tries to get to the oxygen, attracted as if magnetically, the membrane allows the hydrogen proton to pass through but blocks the electron . . . The electron goes around the membrane. In the process it is channeled into a circuit . . . a steady stream of electrons flows through the circuit which could be . . . powering an electric car motor. [The hydrogen] would end up on the other side of the membrane where they would become part of the water molecule.[120]

The amount needed is less than one-half of the CO_2 currently being produced by vehicles that are powered by gasoline. The hydrogen would then replace it for all of the vehicle's power. For example, in 2002 a vehicle called the Santa Fe has a "75 kilowatt motor, equivalent to about 110 horsepower. It has a top speed of 77 miles per hour and a range of 100 miles."[121] Adoption of this process would greatly reduce excessive gasoline imports.

In this regard, I am reminded of a famous physician who set up a clinic in Guatemala in the early 1960s. A decade later, he concluded that "curing the ailing from clinics and hospitals in jungles, savannas, and mountains was something like trying to empty the Atlantic Ocean with a teaspoon. It made the toiler feel active and useful and caused everyone to exclaim: 'My, what a beautiful teaspoon.'" He then completely revised his priorities to "reflect the opinions and feelings of the people we serve." His new priorities were:

---

120. Steve Thomas, ORANGE COUNTY METRO, Aug. 8, 2002, at 33.

121. *Id.* at 36. The California Air Resources Board states, "Improved electricity-gasoline hybrid vehicles have made some progress in hydrogen fuel cell vehicles." ORANGE COUNTY REGISTER, April 28, 2003, at Local 6.

1. social and economic injustice,
2. land tenure,
3. agricultural production and marketing,
4. population control,
5. malnutrition,
6. health training, and
7. curative medicine.

By correctly placing curative medicine as the last priority, he succeeded even though he remained outside the health bureaucracies of the world. After another decade, he was able to state that "with careful nurturing and persistence, an impact for wider change can be made from the most humble and inexpensive start. I think we have proved it in Guatemala."[122] A list could be drawn up of priorities needed in order to achieve a healthy community. It has been stated in a Report Commissioned by the Secretary General of the U.N. Conference on the Human Environment:

> We can begin with knowledge. The first step toward devising a strategy for planet earth is for the nations to accept a collective responsibility for discovering more—much more—about the natural system and how it is affected by man's activities and vice versa.[123]

122. Dr. Carroll Behrhorst, Health in the Guatemalan Highlands (1982).
123. Ward & Dubos, Only One Earth 213 (1972).

Appendix B
Some Views on Medicine in the Two Chinas and Tibet at the Turn of the Millennium

We must have some symbol to represent China for us—China, so gigantic in size that it calls itself "All Under Heaven," and so old that it records the doings of its kings for the last four thousand years.

Will Durant, *The Greatest Minds and Ideas* (2002), p. 109

1. ***Documentary Video "Beyond the Clouds" and an Interview with an Experienced Doctor***
2. ***Birth Control for Han Chinese (Majority) v. Minorities***
   a) **The Critical Appraisal of Dr. Ruiping Fan**
   b) **Some Remarks on Articles Contained in "Chinese v. International Philosophy of Medicine" (1998)**
3. ***Adoptions of Abandoned Girls by Foreigners in Beijing***
4. ***Reactions to a Presentation on American End-of-Life Laws in Beijing***
5. ***Reactions by Lawyers in the "Other China" (Taiwan)***
6. ***Tibet***
   a) **The Herbal Medicine Hospital in the Capital (Lhasa) a Teaching Hospital (and "Living Museum")**

On February 19, 1997, Deng Xiaoping, the last of China's major Communist revolutionaries, died. He had just first approved some capitalistic growth in the south. In 1999, China amended its constitution for the first time since 1949 to declare private business an "important component" of the economy and on March 19, 2004, to declare that "private property obtained legally shall not be violated." (*Orange County Register*, March 15, 2004, News 11.) Beginning in 2003, new marriage rules no longer require couples to get physicals to marry.[1]

In September 2002, almost a quarter of China's family planning programs became effective; its one-child policy was codified into law. There is pressure on parents to ensure that their one and only child is of "good quality." Now, there is a "growing interest in bioethics in China."[2]

Knowing of my volume *Life, Death, and the Law: Autonomy & Responsibility in Medical Ethics*, Qin Renzong, professor of philosophy and bioethics at the Institute of Philosophy, National Academy of Social Sciences, invited me to speak in Beijing. The topic was "The American Revolution in Medical Ethics and the Law." During the final quarter of the twentieth century, I also spent some time in Tibet. I report on my actual experiences in these countries.

1. Documentary Video "Beyond The Clouds" and an Interview with an Experienced Doctor

Lijiang is a city at an altitude of almost 6,000. Most of it is preserved as it was when Marco Polo visited about 700 years before. At the hotel we were shown a video prepared by the Chinese government in conjunction with BBC. It started by showing a girl complaining to a Dr. Tang about a headache. The English subtitle showed her father stating that she always gets sick when she has an exam at school. The father, who made furniture, needed treatment for a pain in the neck. While treating him, Dr. Tang jokingly said to the man's adolescent daughter that he (Dr. Tang) was cruel because he stuck needles into people. The daughter replied that he (the doctor) really helped people by doing that. However, the English subtitle showed the father later complaining that the acupuncture he received did not cure his neck problem. (Acupuncture, belonging to the traditional Chinese medicine, has also been accepted as

1. ORANGE COUNTY REGISTER, Oct. 2, 2003, at News 15.
2. THE NEW ATLANTIS, Spring 2003, at 139.

a method of healing by the World Health Organization, the National Institutes of Health, and the British Medical Association.)[3]

The video next showed the family looking at convicted murderers led to execution by two men wearing sunglasses. The daughter asked her father why they wore them. He replied, "It's because they want to remain unknown." (I then recalled that in Europe up to the nineteenth century, all executioners wore masks.) The video then showed an old woman receiving good glasses. She said, "Now, all I have to look forward to is death. I really don't look for that." Thereafter, it showed Dr. Tang's daughter, who had been apprenticed to him and then went to college. He advised her by saying:

> Never treat an illness. You must treat the whole person. You must be loved and respected. You must treat each patient's life as if it were your own. That is the way I've lived my life.

That afternoon, I visited the 70-year old Dr. Tang at his home. He was trained in the army as an orthopedic surgeon and learned acupuncture in 1953. I arranged for a private interview with a very interested local interpreter.

First, I explained our concept of "informed consent" (see Chapter 5). He stated that no such concept applied in China.

Second, I asked him if he might be prosecuted for withdrawal of treatment from a dying patient at the family's request (see Chapter 13). He replied that withdrawal was possible in China. He even noted a case where a "body" (in PVS) was being "kept alive" in Lijiang for 17 years.

Third, I asked about a newborn who was both physically and mentally defective and whose parents want him to cease treatment (see Chapter 10). He said he could not answer those particular questions.

---

3. 18 FETAL DIAG. THER. 418-21 (2003). "Most statutes require a license to practice acupuncture, but regulations regarding other [alternative] techniques vary considerably." ALTERNATIVE MEDICINE, Harv. Med. School (2003). Acupuncture is widely used to prevent migraine attacks, but the available evidence of its benefit is scarce. In 2005 it was concluded that "acupuncture was no more effective than sham acupuncture in reducing migraine headaches, although both interventions were more effective than waiting list control." Linde et al., *Acupuncture for Patients With Migraine*, JAMA, May 4, 2005, at 2118.

Fourth, a question was asked about the right of a family to abort a child found severely defective in the final weeks of pregnancy (see Chapter 9). He explained the Chinese "live at birth rule," which appears to be much more liberal than the current law in the U.S. The woman's unfettered right to choose abortion is constitutionally protected only up to the time of "viability." Our law limits the choice to several months earlier than abortion is permitted in both China and Israeli law (see Chapter 9).

Fifth, was a question about the use of "painkillers," such as dosage rates for morphine up to an amount that may prove fatal. This is the doctrine of double effect (see Chapter 14). The doctor simply replied that it is normally a doctor's duty to keep a patient alive.

Sixth, I finally posed a question on the issue of individual rights of patients in China. It concerned whether a hypothetical patient, who could live for six months with treatment (including surgery), but only one week without it. I asked whether he could refuse such treatment. Dr. Tang struggled for awhile and then, for the first time, stated that if the patient continually persists in his refusal, he will prevail in such a case, which resembles our law on informed consent (see Chapter 5).

I then terminated the discussion by asking about his daughter in college and alleviated his struggle with such direct questions.

2. Birth Control for Han Chinese Majority vs. Minority

Only the Han Chinese, who constitute some 95 percent of the population of China, are bound by the *one child per family rule*. Exceptions include divorce and remarriage. However, I was told that upon a third marriage, no child is permitted.

In 2003, it was reported how a strict enforcement of China's one-child policy is sometimes followed differently in certain places. In Guangdong, families who want more than their allotted quota ignore the limits without penalty or pay the government's modest fines. However, in Guangxi, population control is taken seriously.

Family planning boards tracking women's fertility are still a feature in many villages, and a roadside marcher carrying banners and signs painted with slogans exhort local farmers to "Have Fewer, Better Children to Create Prosperity for the Next Generation." Each subsequent birth brings a $3,500 fine, the equivalent of two decades worth of local farm income. Eighty percent of trafficked babies are girls, sociologist Yu said. The rest are boys with a health problem or deformity.[4]

"Since 1979, when the one-child policy was implemented [in China], it has pursued the most comprehensive birth control policy in the world."[5] Like the millions of illegal Mexicans in the U.S., each year a number of migrants into Chinese cities are sent back to their families in the countryside. Given the widening gulf between rural and urban China, this redistributive function was welcome.[6] That is, migrants can be deported to their village anytime.[7]

In 2005 it was reported that "25 percent of women of reproductive age having had at least one abortion, as compared with 43 percent in the United States." Chinese "women who proceed with an unapproved pregnancy are know to be reluctant to use antenatal and obstetric services because they fear they will face pressure to have an abortion or to have to pay fines for violating the one-child policy . . . When the one-child policy was introduced, the government set a target population of 1.2 billion by the year 2000. The census of 2000 put the population at 1.27 billion, although some demographers regard this number as an underestimate . . . In addition, many countries have had substantial declines in fertility during the past 25 years, and China's neighbors in East Asia have some of the lowest total fertility rates in the world: 1.04 in Singapore, 1.38 in Japan, and 0.91 in the Hong Kong Special Administrative Region."[8]

a) The "Critical Appraisal" of Dr. Ruiping Fan

In 2003 Dr. Ruiping Fan, a scholar in Hong Kong, described how contemporary Chinese scholars still debate about whether traditional Chinese medicine is a science. Many traditional Chinese medical physicians in China perform *double therapy*—"they prescribe to the same patient both Chinese herbal medicines based on the traditional Chinese medical diagnosis and modern chemical drugs based on the scientific medical diagnosis."[9]

4. ORANGE COUNTY REGISTER, July 20, 2003, at News 9.
5. HUTCHINS, MODERN CHINA 341 (2000).
6. *Id.* at 343.
7. KCET, "Wide Angle: China's Entry into the World Trade Organization Affects Its Citizens," July 31, 2003.
8. N. ENG. J. MED. 1172, Sept. 15, 2005.
9. 31 J. L. MED. & ETHICS 213–21 (2003).

"Modern scientific medical physicians are no longer pushed to learn and practice traditional Chinese medicine. However, the monostandard of science has been strengthened even more than before in evaluating. This political ethos has caused a series of problems for traditional Chinese medicine." They actually call it a "monostandard":

> The real explanation for the practice of double diagnosis and double therapy has much to do with the monostandard of the integrated Chinese healthcare system.
>
> Worse yet, many young traditional Chinese medical physicians (especially those practicing in big hospitals) have accepted the complementary role of traditional Chinese medicine under the monostandard. Some double therapy may assert a better effect than the use of only modern scientific medicine or traditional Chinese drugs. Some traditional Chinese medical physicians use both types of drugs according to their different efficacious characters for different stages of treatment. All this makes sense.
>
> What is necessary, however, is that when a traditional Chinese medical physician administers a double therapy, he must have a reason for it. It is misleading to simply offer double therapy to every problem or every patient he is treating without a clear reason. The result is that when the therapy has a good effect, he has no idea whether it was due to the modern scientific drug, the traditional Chinese drug, or a combined effect of both drugs.[10]

b) Some Remarks on Articles Contained in the Volume "Chinese and International Philosophy of Medicine" (1998)

In his paper, "Confucian Values of Life and Death and Euthanasia," Ping-cheung Lo, Ph.D. of Hong Kong defines euthanasia to include assisted suicide. He states that in limited circumstances "if the premise 'Euthanasia is the only way or best way to eliminate pain in the dying process' is empirically true, one can infer that euthanasia can be justified by Confucian ethics of ren" (benevolence or compassion). He stated "Confucianism . . . places heavy emphasis on the quality of life . . . However, the limit of the Confucian echo is

10. *Id.* at 219.

that Confucianism cares largely [for] the moral quality of life, and cares very little about the biological quality of life." p. 179. In Confucian values, individual autonomy has never been a cherished value . . . That one can decide on the time and circumstances of one's death is only implied." p. 180. He asks "who is right, and who is wrong" and states that this is something "every member of the global village to ponder about."

Professors Ping Dong and Xiaoyan Wang of Capital University of Medicine, Beijing, discussed the Daoist perspective. This allows the terminally ill patient naturally to accept death by *foregoing aggressive medical procedures* when such procedures cannot do more benefit than harm to the patient. This would appear to condone American practices with respect to withholding and withdrawal of treatment for such dying patients at their request. However, it is against the Dao to "destroy the natural mechanism and process of human life." Thus, Dao would be against assisted suicide.

Quingziu Guo, et al., noted a survey showing that the Chinese are deeply divided on the matter. Over one quarter (26 percent) of the terminally ill would accept euthanasia (undefined) by consent of the "patient himself," and over one-half (55 percent) "believe that the matter should be decided by the family and the physician." They noted that "the Western reader might be shocked by this outcome." p. 186 (This is not the Western result, however, due to the Western idea of the patient's choice, which accommodates both viewpoints.) Even Guo states that such a choice would be accepted if "the member clearly demands it." p. 184.

Instructor Lin Yu of Malaysia and Professor Dapo Shi, Xi'an of China looked at different surveys. They found that "most surveys showed that more than 50 percent of the Chinese people advocated euthanasia." p. 189. They also stated that "all [such] defendants have been sentenced only in the lightest sense of the crime of 'killing'" [Article 132 of the current Criminal Law]. This is because of the "good motivation" and the "mild harm to society" of their behavior. Hence, they advocate a "new particular statute that both legalizes and regulates such cases so that they are no longer considered in terms of the crime of murder." p. 190. Most interesting is their advocacy of choice. "Opponents of euthanasia should not use the law to prohibit advocators of euthanasia from accepting or performing euthanasia." *Id.*

From these writers it would appear ripe for a much larger discussion of the divided views of Chinese on assisted suicide and related matters. This might be accomplished by the formation of joint bioethical committees of lawyers, physicians, and philosophers not only in Beijing, but in various cities throughout the country. These committees would then be asked to consider the possibility of granting choices to the terminally ill, which would accommodate the different views noted above, as well as other bioethical/legal matters of grave concern to the people of China. Such committees may or may not decide to follow the pro-choice American view; however, it would be a great thing for them to consider how best to accommodate the views of the very large minority with respect to their own physical being near the end of life.

3. Adoptions of Abandoned Girls by Foreigners in Beijing

At the Hua Du Hotel, I was at the restaurant that served Western-style breakfasts. Almost every morning I met couples from abroad with Chinese babies. I met two couples from Toronto, Canada, each of whom had adopted a baby girl who had been abandoned. They had received health certificates, but nothing else; they received no genetic information. Their application took as long as a year and a-half to reach a successful conclusion. These couples were very happy about the adoptions—particularly the males who were continually insisting that it was their turn to hold the infant. They pointed out to me that, unlike some other countries and in spite of the year or longer wait, once the machinery was set in motion for the adoption, a successful result seemed assured in China.

4. Reactions to a Presentation on American End-of-Life Laws in Beijing

My presentation to the doctors at the Institute of Medicine, National Academy of Social Sciences, in Beijing commenced with a discussion of the differences between Western law in the United States and views of law in the Far East. I promised to speak slowly and apologized for my inadequate ability to speak Chinese. Professor Qin Renzong translated my words. I was not fully convinced that my explanations were always understood.

The Chinese physicians present would not distinguish between withdrawal of treatment, withholding of treatment, double effect, assisted suicide, and euthanasia. Even though I clearly explained the distinctions we draw between each of these, they were all considered to be euthanasia because they resulted in death of the patient. I also described the opinions of the Ninth Circuit Court of Appeals on a Washington statute prohibiting physician assisted suicide (PAS) and of the Second Circuit Court (on a similar New York statute)—both of which were held unconstitutional by the Supreme Court in July 1997 (see Chapter 14). However, when I explained that the same Supreme Court denied certiorari of the second Oregon initiative that overwhelmingly voted to authorize PAS (thus upholding it), they appeared astounded.

Their questions demonstrated the cultural divide between our two countries. I tried to help them by pointing out that Chinese-Americans may often choose to follow Chinese culture in the United States as a result of our concepts of individual liberty and privacy. I suggested that this occurs only after foreigners have been in the U.S. for several years; otherwise they generally tend to exercise their choice in a manner similar to their forbearers. Some of the attendant Chinese stated that it may take China 50 years to thoroughly debate these new concepts.

In 2005, it was pointed out that Chinese life expectancy has risen to 70 years. In major cities, it's even higher.

A law entitled "Protecting the Rights and Interests of the Elderly," says support for the elderly "shall be provided for mainly by their families," shifting responsibility away from the state. The law also requires spouses to "assist in meeting the obligation for their in-laws." Experts say the law is one of a kind in the world.

5. Reactions by Lawyers in the "Other China" (Taiwan)

On June 3, 1989, Chinese troops (from outside of the city) began their sweep of Beijing, shooting hundreds of student-led pro-democracy demonstrators. Thereafter, Taiwan became a full democracy. In 1997 Hong Kong was returned by the British to Chinese rule under a "one country, two systems" formula that promised wide autonomy. But the city's leader is now chosen by a pro-Beijing committee, and only half the legislators are directly elected. China has snubbed pro-democracy figures and barred diehards from visiting the mainland. However, in 2005 China actually invited all the lawmakers, even Leung Kwok-hung, famous for shouting democracy slogans, to

Gudong—one of China's two biggest manufacturing bases, powered by investment from Hong Kong companies.[11]

Mainland China and Taiwan had each claimed sovereignty over the other nation. In 1991, Taiwan renounced sovereignty over the mainland. I was invited to Taiwan by Lee Sheng-Long, chairman of the Bioethics Committee of the Taipei Bar Association, to discuss both end of life and cloning in American law. All members of this audience spoke some English and appeared to have more understanding than they did in mainland China of the difference in the concept of law. Taiwan is a civil law country whose codes were borrowed from both the French Code of 1803 and the German Code of 1900 (much like Japan). Hence, a true understanding of the American concept of the common law was as difficult for them as the code system was for me before I studied law in Paris, France, where, for example, I discovered that their courts can only decide upon the unconstitutionality of a piece of pending legislation. The People's Republic of China had succeeded in blocking Taiwan's appeal for membership in the World Health Organization (WHO) and Taiwan only applied as "health entity," which excluded it from the WHO's information-sharing procedures. See *The New Atlantic*, (Fall 2003), p. 88.

My recommendation to Lee Sheng-Long was that on behalf of the Taipei Bar Association, he ought to consider the funding of a joint committee on bioethics with the Taipei Medical Association. This recommendation was combined with that of suggesting to him that the resulting joint committee not seek a physician domination of such a committee. This recommendation is consistent with that contained in the volume of the American Hospital Association *Healthcare Ethics Committees, the Next Generation* (1993).

Ping Wu, MD, author of *Asian Longevity Secrets* (2003), was grateful "for the opportunity to have visited and worked with some of the first-rate scientists in China."

---

11. ORANGE COUNTY REGISTER, Sept. 26, 2005, at News 11. The Chinese government also signed an agreement with the United Nations human rights agency to collaborate on reforming Beijing's legal system in preparation for adopting a key U.N. treaty on civil and political rights. However, no target date was given for China's ratification of the pact. L.A.TIMES, Sept. 1, 2005, at A10.

6. Tibet

Apparently, Tibetans do not practice birth control. This is because of their Buddhist belief that a child's spirit comes from a recently deceased person or an animal. There is a period of three months into pregnancy when this spirit can enter the fetus. Birth control would interrupt the Karma cycle. If it is used, then some other woman will have twins or triplets in order to accommodate those spirits seeking reincarnation.

Traditional Tibetans have no concept of genetics. However, some believe that if they overbreed, the China Han policy will be applied to such minorities some time in the twenty-first century.

a) *The Herbal Medicine Hospital in Lhasa, Tibet, a Teaching Hospital*

The memorial room of this teaching hospital contains treatises (or copies) going back to the eighth century. The first one was by a doctor born in 1086 who was said to have lived for 125 years.

The teaching hospital program lasts five years and includes training in acupuncture. This hospital was founded in 1916 by the thirteenth (the penultimate) Dalai Lama. It has 500 patients and uses only 300 beds in which one can only stay a month. A doctor who wrote extensively on pharmacology died there in 1962. The hospital is currently being used by business and government people.

The Tanka Room is the teaching room for young doctors. Some of the charts there were 300 years old while others were only 20. Some charts showed the use of water clocks and the lunar month. Traditional Tibetan acupuncture of the head is practiced. In the twelfth century, some traditional Chinese medicine was added. There were 200 doctors and researchers who would respond to questions from many foreign countries. The stronger ultraviolet rays in Tibet affect plants differently than at lower altitudes. We were informed that the U.S. government awarded the college $600,000 in 1989. In 1971, machines were added that produced 100,000 drugs by 1990. Some of their books are now being made available on computers. The study of medicine includes the subject of ethics. However, I was not able to delve into the quality or extent of such teaching. This hospital also appears to be one that could be called a "living museum."

In the United States, complementary and alternative medicine (CAM) has become widespread since the creation of the National Center for Complementary and Alternative Medicine in 1998. However, it has been argued that research on CAM should adhere to the same ethical requirements for al clinical research, and randomized, placebo-controlled clinical trials should be used for assessing the efficacy of CAM treatments whenever feasible and ethically justifiable.

Appendix C
False Claims Act Settlements and Rulings in Healthcare Organizations

The False Claims Act (FCA)[1] has been called the government's "primary litigative tool for combating fraud."[2] The qui tam action under the FCA is derived from the phrase qui tam pro domino rege pro sic ipso in hoc parte sequitur ("he who sues on behalf of the king as well as himself.")[3] Such an action may be brought by a private individual (known as whistleblowers a relators), with or without concurrent action by the attorney general. In FCA actions between 1986 and 2003, the U.S. Department of Health and Human Services (HHS) recovered more than $5 billion.[4] However, a state has been held not to be a "person"—which constitute the class of defendants subject to FCA liability.[5] Hence, state-operated healthcare facilities are immune from FCA liability,[6] along with their laboratories and corporate physician groups.[7]

Cases from 1998 to 2004 are listed below to illustrate the payments in millions of dollars to the government by healthcare organizations. Many of these were based upon fraudulent actions reported by "relaters" who have often been employees of healthcare organizations discharged following their direct and independent disclosure of falsified records; they receive substantial reward under the False Claims Act—a "whistleblower statute," and the Sarbanes-Oxley Act that covers fraud against publicly traded companies.

---

1. 31 U.S.C. § 3729, et seq.
2. United States *ex rel.* Kelly v. Boeing Co., 9 F.3d 743, 745 (9th Cir. 1993).
3. Black's Law Dictionary.
4. www.tat.org/statistics.htm
5. 31 U.S.C. § 3729(a).
6. Vermont Agency of Natural Res. v. Stevens, 529 U.S. 765 (2000), and Donald v. Univ. of Cal. Bd. of Regents, 329 F.3d 1040 (9th Cir. 2003).
7. Adrian v. Regents of the Univ. of Cal., 363 F.3d 398 at 402 (5th Cir. 2004).

1. *Wellmark Inc.*, Des Moines, Iowa, agreed to pay nearly $6.9 million to health plan participants and sponsors for failing to pass on negotiated discounts with hospitals, thus overcharging for service provided by Blue Cross/Blue Shield of Iowa and South Dakota, the Labor Department announced February 26, 1999. *Herman v. Wellmark Inc.*, S.D. Iowa, No. 499-CV-10110, settlement February 24, 1999. 71 Fed. Cont. Rep. (FCR) 290 (1999).
2. *Yale University* agreed to pay $5.6 million to settle a whistleblower lawsuit against the Yale School of Medicine alleging the improper handling of credit balances resulting from medical services billings by the school. *United States ex rel. Jackson v. Yale Univ. Sch. Of Med.*, D. Conn., No. 3:97CV02023 AWT, settlement September 8, 1998. 70 FCR 246 (1998).
3. *Quest Diagnostics* agreed to pay $6.8 to settle a whistleblower's charges that six of its regional labs billed Medicare and other federal health programs without obtaining the required physician's authorization. 69 FCR 441, 448 (1998). The whistleblower will receive $1.1 million as her share of the recovery. The following year *Quest Diagnostics Inc.* and two dialysis companies agreed to pay $15 million to settle for allegedly conspiring to defraud Medicare. 70 FCR 209 (1998).
4. *HMO Cigna HealthCare* and the physicians who treated its patients settled a class action for $1.3 billion along with eight other HMOs. *In re Managed Care Litig.*, No. MDL-1334, 2003 U.S. Dist. Lexis 22066 (S.D. Fla. Dec. 8, 2003).
5. An Illinois hospital has agreed to pay the government $228,500 to settle allegations that it failed to refund Medicare overpayments (*United States ex rel. Richmond v. St. Anthony's Memorial Hospital*, DC SI11, No. 95-4160, settlement April 9, 1998). The whistleblower, Dirk Richmond, received $45,700 as his share of the government's recovery, 60 FCR 400 (1998).
6. *Highmark Inc. of Camp Hill, Pa.,* agreed to pay $38.5 million to settle three whistleblower lawsuits that alleged that its corporate predecessor, Pennsylvania Blue Shield, lied to federal regulators, obstructed federal audits, failed to process certain Medicare claims, and failed to recover Medicare overpayments over a roughly eight-year period. *United States ex rel. Bultena v. Medical Servs. Assn.*, E.D. Pa., No.

96-CV-4430, settlement September 3, 1998; *United States ex rel. Howell v. Pennsylvania Blue Shield*, E.D. Pa., No. 97-CV-0518, settlement September 3, 1998. Former Blue Shield employees, the Hickses will receive $4.52 million, Bultena $2.88 million, and Howell $576,000. 70 FCR 247 (1998).

7. *Blue Cross Blue Shield of Illinois* pled guilty to eight felony counts and agreed to pay $144 million in criminal and civil fines to the government stemming from allegations by a whistleblower. 70 FCR 87 (1998).
8. *Charter Behavioral Health Systems*, which operates a psychiatric hospital in the Orlando, Florida, area, agreed to pay $4.75 million to settle a whistleblower lawsuit alleging that it fraudulently admitted and extended the hospital stays of hundred of mentally incapacitated Medicare beneficiaries. *United States ex rel. Mettevellis v. Chapter Hosp. Inc.,* M.D. Fla., No. 94-1170-ORL-22, settlement August 19, 1998. 70 FCR 254 (1998).
9. Companies that provide emergency physician staffing were to pay the United States and 11 states $3.1 million to settle a whistleblower case alleging that they submitted false claims through their billing company and received overpayments based on the claims. *United States ex rel. Semtner v. Medical Consultants Inc.*, W.D. Okla., No. CIV-94-617, settlement September 29, 1998. 70 FCR 369 (1998).
10. *The Healthcare Co. (formerly Columbia/HCA Healthcare Corp.)* agreed to pay a record $840 million to settle charges of Medicare fraud. The Department of Justice announced that it was intervening in eight other qui tam actions under the False Claims Act against the hospital chain. The lawsuits allege (1) payments of kickbacks to physicians to increase the number of government-insured patients, (2) inflation of hospital cost reported by including unallowable charges, and (3) payment of kickbacks and unallowable expenses on cost reports for wound care services. 75 FCR 318 (2001).
11. A claim that *St. Luk's Episcopal Hospital* "knowingly" made "a false or fraudulent claim" with Medicare and the Civil Health and Medical Program of the Uniformed Services for services that were either medically unnecessary or were rendered by an unlicensed physician was upheld by a circuit court (reversing a district court's dismissal of the

qui tam suit). *United States ex rel. Riley v. St. Luke's Episcopal Hospital,* 5th Cir., No. 02-20825, 1/9/04.

12. In 2003, *HCA, Inc.* signed an agreement to pay the government a total of $1.7 billion, the largest for-profit amount in civil fines and criminal penalties. HCA operates 200 hospitals and 70 outpatient surgery centers in 24 states, England, and Switzerland. *Orange County Register*, June 27, 2003, Business 6, *In re: Columbia/HCA Healthcare Corp.*, D.C., No. 01-MS-50 (RCL), settlement 6/26/03, 79 FCR 782.
13. *Community Health Systems, Inc.*, a healthcare provider based in Brentwood, Tenn., agreed to pay the government $31 million for having submitted false claims to Medicare, Medicaid, and TRICARE. 73 FCR 564 (2000).
14. The former vice president of a Toledo, Ohio, medical supply firm was sentenced to 10 years in federal prison for her role in a $42 million Medicare fraud scheme *United States v. Browning*, N.D. Ohio, No. C:98CCR819, sentencing March 8, 2000. 73 FCR 338 (2000).
15. *University Medical Center (UMC) of Tucson*, Arizona, agreed to pay $329,000 to settle allegations that it submitted false claims to federal and state healthcare programs for outpatient clinical laboratory services. 73 FCR 285 (2000).
16. An appellate court reversed a $52 million award to the whistleblowers involved in a $325 million False Claims Act settlement by *Smith-Kline Beecham Clinical Laboratories Inc.* in 1997, remanding the case to the district court to determine whether the relators were the "original source" of the allegations. *United States ex rel. Merena v. SmithKline Beecham Corp.*, 205 F.3d 97 (3d Cir. 2000).
17. *Fresenius Medical Care North America* agreed to pay a record $485.9 million in criminal and civil penalties for healthcare fraud by three subsidiaries of its kidney dialysis business, National Medical Care Inc. *United States v. NMC Homecare, Inc.*, D. Mass., number unavailable, criminal information filed January 19, 2000. 73 FCR 117 (2000); See also *United States v. NMC Homecare Inc.,* D. Mass., No. 00-CR-10017, guilty plea February 2, 2000. 73 FCR 160 (2000).
18. *The University of Chicago* agreed to settle for at least $10 million due to allegations by a whistleblower that its hospitals overbilled Medicare and Medicaid for several years. *United States ex rel. Reppine v. the*

University of Chicago, N.D. Ill., No. 96C8273, settlement approved February 10, 2000. 73 FCR 202 (2000).

19. *The Century Health Services (CHS)*, a holding company for a number of home healthcare agencies, established an employee stock ownership plan (ESOP) and charged Medicare for reimbursements for anticipated contributions to the plan. The CHS withdrew almost all the amounts into its corporate accounts, transferred most of its operating assets to another entity, and transferred shares to ESOP based upon their value at earlier years. The U.S. Court of Appeals for the Sixth Circuit held CHS liable for $7.62 million in treble damages and $100,000 in civil penalties related to its submission of certified cost reports seeking reimbursement from Medicare. *United States ex rel. Augustine v. century Health services Inc.*, 6th Cir., No. 01-5019, May 7, 2002, 77 FCR May 14, 2002 p. 573.
20. In 2000, the *Tenet Healthcare Corp.* paid $54 million to the U.S. to settle government allegations that two doctors at its hospital in Northern California performed numerous unnecessary heart surgeries. *Los Angeles Times*, August 7, 2003, Bus. 1. The largest was in a case involving "medical necessity fraud." See also *Orange County Register*, August 7, 2003, Bus. 1; *United States ex rel. Ayers v. Tenet Healthcare Corp.*, S.D. Fla., No. 97-CV-2507, July 17, 2002. 78 FCR 125, July 23, 2003. After discovering that unnecessary heart operations were performed there, Blue Cross of California severed a Modesto hospital run by Tenet Healthcare Corp., *Orange County Register*, November 4, 2003. Tenet Healthcare Corp., the second biggest hospital chain in the U.S., indicated that it was selling 27 hospitals. *Orange County Register,* March 25, 2004. In December 2004, Tenet agreed "in principle" to pay $395 million to settle lawsuits claiming patients at Redding Medical Center, one of its former hospitals, received unnecessary heart procedures. The settlement was distributed among at least 750 patients. *Orange County Register,* Dec. 22, 2004, Bus. 3.
21. An action against the *Erie County Medical Center* for allegedly submitting for reimbursement $121,916 in fraudulent Medicare claims was dismissed because a municipal entity is not a "person" under the False Claims Act. *United States v. Erie County Med. Ctr.*, W.D. N.Y., No. 02-CV-03005E, October 30, 2002. 78 FCR 125, July 23, 2002.

22. The U.S. Court of Appeals for the Eighth Circuit held that an association of nurse anesthetists may proceed with a FCA suit against anesthesiologists who allegedly made false claims to the government by mischaracterizing the services they provided to Medicare patients, notwithstanding the fact that the allegations had previously been disclosed in the association's antitrust suit. *United States ex ret. Minnesota Assoc. of Nurse Anesthetists v. Allina Health Sys. Corp.*, 8th Cir., No 9902356, January 17, 2002. 77 CRD 112.
23. A laboratory conducting tests to detect and determine the stage of various types of cancer, agreed to pay the government $4.8 million to resolve allegations that it violated the FCA by improperly charging Medicare and other federal healthcare programs for certain medical tests it performed. *United States ex rel. Worner v. Dianon Systems Inc.*, D. Conn., 78 FCR 693, December 17, 2002.
24. The U.S. government intervened in whistleblower lawsuits alleging that 27 hospitals violated the FCA by improperly charging Medicare for experimental cardiac devices. *United States ex rel. Cosens v. Stanford Hospital and Clinics*, N.D. Cal., C99-4121 MJJ. 78 FCR 726, December 24, 2002.
25. Seven hospitals from five states will pay the United States more than $6 million for having violated the FCA by unlawfully charging Medicare and other Federal healthcare programs for nonreimbursable cardiac devices. *United States ex rel. Cosens v. Scripps*, S.D. California, No. 990CV-1264, settlement announced 6/4/02. 77 FCR 712.
26. *Quorum Health Group, Inc.* agreed to pay $85 million and enter a corporate integrity agreement to settle a whistleblower's allegations that the company defrauded Medicare by submitting false statements for reimbursement. *U.S. ex rel. Alderson v. Quorum Health Group, Inc.*, M.D. Fla. No. 99-413-CIV.T-23B, Settlement April 23, 2001. FCR May 1, 2001. Quorum owns and manages more than 2,000 hospitals around the country. The government and Alderson identified more than 2,000 inflated cost items from 450 cost reports prepared and submitted by Quorum from 1985 to 1999. FCR May 1, 2001.
27. *Health South Corp.*, a healthcare provider based in Birmingham, Ala., agreed to pay $7.9 million to resolve a whistleblower lawsuit alleging that Medicare and TRICARE programs were overcharged for equip-

ment and supplies. *U.S. ex rel Madrid v. HealthSouth Corp.* N.D. Ala. No. CV-97-H-3206-S, settlement May 22, 2001. FCR April 16, 2002, p. 336. In 2003 Health Couth Corp. founder Richard Scrushy was indicted on accounting fraud that has nearly bankrupted the company. He had ordered top executives to create $2.7 billion in fictitious income to deceive shareholders and became the first chief executive prosecuted under the Sarbanes-Oxley Act. *Orange County Register,* November 5, 2003, Business 1.

28. *TAP Pharmaceutical Products, Inc.*, agreed to pay the government $875 million and comply with a far-reaching corporate integrity agreement to resolve criminal and civil liabilities arising from charges of fraudulent drug pricing and marketing of its prostate cancer drug Lupron. As part of the settlement, TAP pleaded guilty to conspiring to violate the Prescription Drug Marketing Act. It will pay a $200 million criminal fine, the largest criminal fine ever in a healthcare fraud prosecution. To settle civil FCA liabilities, TAP agreed to pay $560 million for fraudulent claims it filed. FCR October 30, 2001, p. 467.
29. *PacifiCare Health Systems* agreed to pay the U.S. $87 million to settle a whistleblower's allegations that it and predecessor companies violated the FCA in claims submitted for federal employee health benefits. The agreement is the largest civil settlement involving contracts with the Office of Personnel Management to provide benefits to Federal workers under the Federal Employees Health Benefits Program (FEHBP). FCR April 16, 2002, p. 446.
30. *Catholic Healthcare West (CHW)* and Mercy Healthcare Sacramento agreed to pay $10.2 million to settle a whistleblower lawsuit alleging that two of their clinics submitted false claims to Medicare and other healthcare programs. *U.S. rel Baca v. Catholic Healthcare West*, E.D. Calif. No. CIV-S-98-0569DFL Settlement May 16, 2001. FCR May 29, 2001.
31. *Catholic Healthcare West* and its affiliate Mercy Healthcare Sacramento agreed to pay $8.5 million to settle allegations of defrauding Medicare and other federal health programs by filing false cost reports at 13 hospitals, *United States v. Mercy Healthcare Sacramento*, E.D. Cal., No. CIV-S-99-292 LKK-JFM, settlement 6/4/02.

32. There was a $76 million settlement of a FCA whistleblower lawsuit against an insurer regarding allegations it falsified error rates as a claims processor in the Medicare program. *United States ex rel. Riggs v. General American Life Insurance Co.*, E.D. Mo., No. 4:99CV00608RWS, settlement announced 6/25/02.
33. The *State of California and Los Angeles County* agreed to pay the federal government $73.3 million to resolve a whistleblower's allegations that they violated the FCA by submitting claims to Medicaid for services to individuals who did not qualify, the Justice Department announced June 20. FCR ISSN 0014-9063.
34. A *Michigan emergency physician* group will pay the Federal government and Michigan $1.6 million to resolve charges of filing false claims with Medicare and other government health programs. $1.4 million will go to the United States and $200,000 will go to the state of Michigan. 77 FCR 608 (May 23, 2002).
35. The nation's largest ambulance company has agreed to pay $20 million to settle allegations that it submitted fraudulent claims to Medicare throughout the 1990s. *United States ex rel. Rau v. American Medical Response Inc.*, D. Mass., 98-CV-12050RGS, complaint unsealed 6/6/02, 77 FCR 745.
36. A doctor convicted of Medicare fraud and sentenced to five years in prison was held not precluded by a criminal sentencing agreement from agreeing to "move for reinstatement" to Medicare programs upon release from five-year incarceration. *Sternberg* v. Secretary of Health and Human Services, 10th Cir., No. 01-3185, August 13, 2002. 71 LW 1130, August 27, 2002.
37. *Parke-Davis Laboratories* (part of Pfizer) did not pay rebates owed to the state and federal governments under the national drug Medicaid rebate program of Lipitor, a cholesterol-lowering drug. Pfizer Corp. agreed to pay $49 million to resolve allegations that a subsidiary violated the FCA. *United States ex rel. Foster v. Pfizer Inc.*, Warner-Lamber Corp., E.D. Texas, 78 FCR 537, November 5, 2002.
38. *Columbia University* in New York agreed to pay $5.1 million to settle a whistleblower case alleging that an obstetrical facility it operates had overbilled the Medicaid program by claiming that doctors had provided services that actually were rendered by midwives or ineligible

providers. *United States v. Columbia University*, S.D.NY., No. 00-8798, FCR January 7, 2003.

39. In 2003, *AstraZeneca Pharmaceuticals LP* settled for $355 million criminal and civil charges that it gave physicians free samples of its prostate cancer drug, Zoladex (goserelin acetate implant), knowing that the doctors would then bill the Medicare program for the average wholesale price of the drug. 80 FCR 21, 7-8-03.
40. *Bayer Corp.* and *GlaxoSmithKline* agreed to pay $344 million to settle allegations they illegally repackaged drugs and sold them at prices below those paid by the federal government and state Medicaid programs. 79 FCR 496, 4-22-03.
41. *Pfizer, Inc.* plead guilty to criminal charges for illegally marketing an epilepsy drug for unapproved uses such as migraines and pain. Its Warner-Lambert division promoted the drug, Neurontin, for uses it had no scientific evidence to support. *O.C.Reg.* May 14, 2004, Bus.2.
42. The former cief executive of *Rite Aid Corp.,* Martin L. Grass, was sentenced to eight years in prison and fined $500,000 for his role in a billion-dollar accounting fraud in SEC filing. *Orange County Register,* May 28, 2004, Bus. 4.

As of 2004, the FCA has generated $12 billion for the federal treasury and more than $1 billion for hundreds of whistleblowers. *range County Register,* Nov. 27, 2004, News 9 The Patient Referrals Act, 42 U.S.C. § 1395nn, prohibits a physician from referring patients to those with which the physician has a financial relationship for certain health services that are reimbursable by Medicare and from billing Medicare for a prohibited referral. HHS published August 14, 1995, 60 Fed. Reg. 41,914. See also 66 Fed. Reg. 856 (January 4, 2001) and 66 Fed. Reg. 60,154 (December 3, 2001).

Only the portion of the Federal Acquisition Regulation (FAR) pertaining to debarment and suspension rules also apply to "guarantees." These rules have been adopted by 28 Federal agencies, including the HHS. 45 C.F.R. Part 76; and 67 Fed. Reg. 3265 (January 23, 2002). See also 31 U.S.C. § 6101 et seq. that states:

> No Agency shall allow a party to participate in any procurement or nonprocurement activity if any agency has debarred, suspended, or

otherwise excluded (to the extent specified in the exclusion agreement) that party from participation in a procurement or nonprocurement activity.

A state agency audit report on a health services company's noncompliance with Medicaid reimbursement rules was held to be a public disclosure for purposes of the FCA, *United States ex rel. Hays v. Hoffman*, 8th Cir., No. 01-3888, 4/9/03. 79 FCR 518, 4-29-03. See Keyes, *Government Contracts under the Federal Acquisition Regulation*, 3rd Edition 2003.

As regards global price-fixing litigation on the prices of vitamins, the Swiss-based F. Hoffman-La Roche Ltd., the world's largest vitamin manufacturer, plead guilty at a $500 million fine—the largest the Justice Department has ever obtained in a criminal case. The German company BASF paid a $225 million penalty for its role on the antitrust list. *ABA Journal*, April 2004, p. 85.

Appendix D
Millennia

2nd B.C.	10th century B.C.E.	The Temple is constructed in Jerusalem.
1st B.C.	6th century B.C.E.	The Temple is rebuilt after the Babylonian conquest.
	c.4 B.C.E.	Jesus is born, perhaps in Bethlehem.
	c.30 C.E.	Jesus is crucified under Pontius Pilate.
	c.50s	A first written document, Q, now lost, compiles the sayings of Jesus.
	c.50-c. 60	The earliest (1 Thessalonians (c.51)). . .
	70	Romans attack Jerusalem and destroy the Temple.
2nd A.D.	1939	Pacelli becomes Pope Pius XII. He cancels Pius XI's in-progress encyclical condemning anti-Semitism.
	1943	When Jews are rounded up at the foot of Vatican Hill, Pius XII does not openly protest.
	1948	The state of Israel is established.
	1965	Vatican II issues Nostra Aetate, deploring anti-Semitism, rejecting the idea that all of the Jews can be charged with the death of Jesus.
	1978	Karl Wojtyla becomes Pope John Paul II and immediately sets out to heal the break between Catholics and Jews. . .
	1993	The Vatican recognizes the state of Israel. . .
3rd A.D.	2000	John Paul II issues an apology for the historic sins of members of the Church.

From James Carroll, *Constantine's Sword, the Church and the Jews* (2001)

Appendix E
Genetic Research and Economic Advantage of Abbreviated New Drug Applications

In 2000, it was pointed out that "some 5,000 patients have now undergone different forms of gene therapy for a variety of very serious diseases, all of them genuinely life-threatening and mostly after all other possible treatments have failed." One youth, whose gene therapy was unneeded, died.[1]

The following offers a some examples of genetic research and the economic advantage of abbreviated new drug applications (ANDAs).

1. *Severe combined immune deficiency (SCID)* caused by a defective gene on the X chromosome requires a child to be kept in sterile environment (a sterile bubble) in order to stay alive. In 2000, French physicians cured two such infants using a retrovirus, "to bring the therapeutic genes into some of the babies' defective progenitor cells in vitro,

---

1. Winston, Superhuman 247 (2000).

   The recent death of teenager Jesse Gelsinger is a kind of watershed. When this 18-year-old died in the year 2000—his death was the first mortality attributed to gene therapy itself rather than to the underlying disease process—the field was halted. Moreover, many people doubted that his gene therapy was really needed. In most people, usually infants, this disease caused sickness, coma, and death, but in some individuals, it is controlled by a low protein diet and drugs. In Jesse Gelsinger's case, he was an active, athletic young man in no danger from his disease provided he prudently took his pills. (p. 239).

then cultured and injected them."[2] Although six more were cured, the report that one of them came down with a sickness resembling leukemia caused the U.S. Food and Drug Administration (FDA) in 2002 to halt some trials in order to evaluate the chance of it happening again.[3]

2. *Retinitis pigmentosa (RP)* is a hereditary disease marked by progressive loss and contraction of the visual field. A man with RP stated that "going blind over many years has not been easy for me, sometimes causing panic and depression, still I would be shocked if anyone aborted solely for reasons of retinitis pigmentosa." Nevertheless, he added, "most of the people I know would abort a disabled fetus in spite acting and taking an enlightened belief in diversity. It's not at all clear what they're going to cure or how."[4]
3. In connection with the brain science of *schizophrenia* in 2001, it was noted that the responsible genes are so broadly dispersed and obscure that concerted work by many experts has failed to identify any gene as being definitely tied to the disease. However, during the following years Icelandic and Scottish patients were found to have the same variant pattern in the gene in a form that appears to double the risk of schizophrenia. Although some who carry the variant are normal, both populations inherit the variant from of the gene.[5]
4. More than 25 genes have been associated with hearing loss and recently it has been demonstrated to be a prominent manifestation of *mitochondrial DNA mutation*—a mutation in sites of ATP (adenosine triphosphate) syntheses to supply high energy. In 2003, it was declared that "efforts at developing gene therapy for hearing loss are underway."[6]

---

2. *Id.* at 26, reporting on Dr. Alain Fischer and his colleagues at Paris's Hospital Necker.

3. L.A. Times, No. 12, 2002, at A18.

4. Potok, A Matter of Dignity 176 (2002). Hobson & Leonard, Out of Its Mind; Psychiatry in Crisis 199 (2001).

5. Orange County Register, Dec. 13, 2002, News 26, *citing* The American Journal of Human Genetics.

6. JAMA 1561 (March 26, 2003). *See also* Chen et al., *Presymptomatic Diagnosis of Nonsyndromic Hearing Loss by Genotyping*, 124 Arch. Otolaryng. Head & Neck Surg. 20, 23 (1998).

5. Scientists say they've identified a flawed gene that appears to promote *manic depression* or *bipolar disorder*. Manic depression affects about 2.3 million American adults. The association is only statistical and researchers are looking for evidence that this variant of the gene acts abnormally.[7]
6. *Gehrig's disease*, is a fatal nerve disorder (amyotrophic lateral or ALS) that afflicts 30,000 Americans. In 2003, Massachusetts General Hospital found gene therapy to be the most effective treatment in experiments on lab mice. They now believe ALS is so terrible that human trials are justified.[8] In 1939 baseball great Lou Gehrig announced he had ALS and retired from the New York Yankees. He died two years later.[9]
7. A gene was found that is believed responsible for *age-related macular degeneration (AMD)*, a complex disease that typically affects the macula part of the retina in people 65 and older. The macula allows for vision of fine details needed for activities such as reading, sewing. and driving and it is estimated that the disease afflicts about 6 million Americans.[10]
9. Scientists have found an unexpected genetic link among three common autoimmune diseases. *Psoriasis* affects 2 percent of Americans; *rheumatoid arthritis*, up to 1 percent; and *systemic lupus*—which attacks the joints and can cause severe inflammation—one-twentieth of 1 percent.[11]
9. A gene linked to *osteoporosis* (characterized by brittle bones) has been identified by DeCode Genetics, an Icelandic company. The weaken-

7. ORANGE COUNTY REGISTER, June 16, 2003, at News 11. John Kelsoe, SAN DIEGO J. MOLECULAR PSYCHIATRY (June 16, 2003). Statistical evidence was found tying a particular variant of the GRK3 gene to the disease.

8. ORANGE COUNTY REGISTER, Aug. 9, 2003, at News 17. "Gene therapy remains controversial and whether treatment will work in humans can only be determined by doing human experiments." *Id.*

9. SALK, SALKS 5 (Fall 2003).

10. ORANGE COUNTY REGISTER, Oct. 22, 2003, at News 24. AMD sufferers are predicted to double by the year 2030 as the population increases.

11. N.Y. TIMES, Nov. 10, 2003; J. NATURE GENETICS (same date).

12. ORANGE COUNTY REGISTER, Nov. 3, 2003, at News 8, *citing* the online journal PUBLIC LIBRARY OF SCIENCE.

ing of the bones is measured by the bone mineral density test. The DeCode team identified a gene on chromosome 20 called BMP-2, for one morphogenetic protein-2.[12]

10. U.S. and Japanese researchers found a genetic mutation that causes *obsessive-compulsive disorder,* which is one of the top 10 leading causes of disability worldwide. The gene is called the human serotonin transporter gene, hSERT, and helps control how the body uses serotonin, which is linked to moods.[13]
11. *Spinal muscular atrophy* is a disease that strikes about 1 in 6,000 newborn Americans each year. In 1995, researchers in France identified a gene called survivor motor neuron. The SMN 2 genes are backup ones that determine the severity of the disorder. Several studies aimed at manipulating the gene are underway in the United States and Europe. It is estimated that getting a drug on the market may take more than a decade.[14]
12. In connection with cancers, in 2003, a study looked for mutations of the HRPT2 gene, which encodes the parafibromin protein, in *sporadic parathyroid carcinoma*. This was because germ-line inactivating HRPT2 mutations have been found in hyperparathyroidism-jaw tumor (HPT-JT) syndrome, which carries an increased risk of parathyroid cancer. No specific gene has been established as a direct contributor to the pathogenesis of the disease, but several important molecular clues have been uncovered.[15]

In 2003, "the Congressional Budget Office has estimated that all spending for prescription drugs by and on behalf of Medicare's eligible population will total $1.8 trillion during the period from 2004 through[16] the same year new initiatives to improve drug reviews were made with respect to generic drugs. It was noted that "generics have the same quality, safety, and strength as branded medicines. But for an average brand-name drug that costs $72, the

---

13. ORANGE COUNTY REGISTER, Oct. 24, 2003, at News 19.

14. N.Y. TIMES, Oct. 28, 2003, at 3.

15. Shattuch et al., *Somatic and Germ-Line Mutations of the HRPT2 Gene in Sporadic Parathyroid Carcinoma*, NEMJ, Oct. 30, 2003, at 1722.

16. *See* Iglehart, *Prescription-Drug Coverage for Medicare Beneficiaries*, NEMJ, Sept. 4, 2003, at 923.

generic version costs about $17 and bring consumers significant savings."[17]

Without fair compensation from meaningful patent protection, drug research and development would slow or stop.[18]

The generic companies submit abbreviated new drug applications (ANDAs) "showing that a generic drug has the same bioavailability as the brand-name drug, generic companies must prove that their products have the same active ingredient, follow the same quality manufacturing standards, and have similar labeling.[19]

---

17. A 2002 study by the Schneider Institute for Health Policy at Brandeis University in Waltham, Mass., concluded that if Medicare increased the rate of generic usage to that of similar high-performing private sector plans, its 40 million beneficiaries could see potential savings of $14 billion in 2003. Meadows, FDA CONSUMER/September-October 2003, at 13.

18. *Id.*

19. *See FDA Requirements for Generic Drugs*, FDA CONSUMER/September-October 2003 at 15.

Authors of Books Referenced in Footnotes

Abate, Tom, *The Biotech Invester, How to Profit from the Coming Boom in Biotechnology* 4

Adamson, *Forever Free* 191

Akers, Charlene, *Obesity* 741

Allison, Graham, *Nuclear Terrorism, The Ultimate Preventable Catastrophe* 243, 251

Almond, Brenda, and Michael Parker, *Ethical Issues in the New Genetics, Are Genes Us?* 934

Amar, Akhil Reed, *America's Constitution, A Biography* 1042

Anderson, *Brave New Brain* 976

Andre, Judith, *Bioethics as Practice* 164

Andreasen, Nancy C., *Brave New Brain, Conquering Mental Illness in the Era of the Genome* 227

Attenborough, David, *Life on Earth, A Natural History* 55

Baker, Robin, *Sperm Wars* 424

Balmer, Randall, and Lauren F. Winner, *Protestantism in America* 155

Barber, Nigel, *Kindness in a Cruel World, The Evolution of Altruism* 108

Baron-Cohen, Simon, *The Essential Difference, The Truth About the Male and Female Brain* 997

Barr, Stephen M., *Modern Physics and Ancient Faith* 36

Becker, *The Heavenly City of the Eighteenth Century* 194

Beiner, *What's the Matter with Liberals?* 195

Bennett, Deborah J., *Logic Made Easy, How to Know When Language Deceives You* 39

Bennett, William, *Book of Virtues* 290

Bennett & Plum, *Textbook of Medicine* 702

Benton, Michael J., *When Life Nearly Died, The Greatest Mass Extinction of All Time* 381

Berumen, Michael E., *Do No Evil, Ethics with Applications to Economic Theory and Business* 206

Bhagwati, Jagdish, *In Defence of Globalization* xlvi

Bleich, David J., *Bioethical Dilemmas, A Jewish Perspective* 550

Block, Ned, *Cognition* 998

Bobrick, Benson, *The Fated Sky* 44

Boorstin, Daniel J., *The Creators* 100

Boylan, *Portraits of Guilt* 1030

Breggin, *Brain Disabling Treatment in Psychiatry* 562

Breggin and Cohen, *Your Drug May Be Your Problem* 562

Brody, Baruch A., *The Ethics of Biomedical Research, An International Perspective* 955

Brown, Lester, *Saving the Planet* 505

Bulkely, Kelly and Bulkley, Patricia, *Dreaming Beyond Death* 703

Burrell, *Postcards from the Brain Museum, The Improbable Search for Meaning in the Matter of Famous Minds* 1005

Buxton, Richard, *The Complete World of Greek Mythology* 366

Cantor, Norman F., *Antiquity* (2003), Harper Collins Publishers 91

Carey, Ken, *The Third Millennium* 40

Cassel, *Physician-Assisted Suicide: Progress or Peril? Birth or Death* 907

Cherry, Reginald, *Healing Prayer, God's Divine Intervention in Medicine, Faith, and Prayer* 107

Clinton, Hillary Rodham, *Living History* 385, 408, 726

Cohen, Joel, *How Many People Can the Earth Support?* 387, 393

Cohen, Jon, *Coming to Term, Uncovering the Truth About Miscarriage* 454

Comings, David E., *The Gene Bomb* 523

A Concise History of the Catholic Church 87

Cox, Harvey, *When Jesus Came to Harvard, Making Moral Choices Today* 162, 452, 696

Crocker, *Triumph* 98

Croswell, Ken, *Magnificent Mars* 1010

Daar, *Ethics of Xenotransplantation* 669

Daly and Wilson, *Homicide* 525

Daniels, *Lost Fathers* 467

Dante, *The Inferno of Dante* 99

Darwin, Charles, *The Origin of Species* 46, 48

Dawkins, *The Selfish Gene* 61

Deckman, Melissa M., *School Board Battles, The Christian Right in Local Politics* 69

Dershowitz, Alan, *Rights from Wrongs, A Secular Theory of the Origins of Rights* 133

Dillard, *The Chronic Pain Solution* 758

Dodge, Mary, and Gilbert Geis, *Stealing Dreams, A Fertility Clinic Scandal* 441

Dorff, Elliot, *Matters of Life and Death* 175, 405, 411, 515, 549

Dorland, *Medical Dictionary* 614, 665

Dworkin, Ronald, *Sovereign Virtue, The Theory and Practice of Equality* 765

Eldredge, Miles, *Why We Do It, Rethinking Sex and the Selfish Gene* 602

Erickson, *The Life Cycle Completed* 698

Faigman, David L., *Laboratory of Justice, The Supreme Court's 200-Year Struggle to Integrate Science and the Law* 20, 85, 131

Ferlong, John, *Adams vs. Jefferson, The Tumultuous Election of 1800* 1029

Ficarra, Bernard J., *Bioethics' Rise, Decline, and Fall* 74

Fisher, John, *The Enduring Paradox, Exploratory Essays in Messianic Judaism* 103

Fletcher, Joann, *The Search for Nefertiti, The True Story of an Amazing Discovery* 76

Foucault, Michel, *The Hermeneutics of the Subject, Lectures at the College De France, 1981-82* 1035

Fox, Richard Wightman, *Jesus in America, Personal Savior, Cultural Hero, National Obsession* 92

Freeman, Dyson, *The Sun, the Genome and the Internet, Tools of Scientific Revolutions* 1012

Freud, Schur M., *Living and Dying* 917

Friedman, Richard Elliott, *The Bible with Sources Revealed, A New View into the Five Books of Moses* 102

Galton, *Inquiries into Human Faculty and its Development* 211

Garrean, Joel, *Radical Evolution, The Promise and Peril of Enhancing Our Minds, Our Bodies—and What It Means to Be Human* 1016

Gates, Bill, *The Road Ahead* 1043

Gazzaniga, Michael S., *The Ethical Brain* 499

Ghamari-Tabrizi, Sharon, *The Worlds of Herman Kahn, The Intuitive Science of Thermonuclear War* 251

Gigerenzer, Gerd, *Calculated Risks, How to Know When Numbers Deceive You* 223

Gillett, Grant, *Bioethics in the Clinic: Hippocratic Reflections* 661

Gingerich, Owen, *The Book Nobody Read, Chasing the Revolutions of Nicolaus Copernicus* 74

Goldsmith, Donald, *Voyage to the Milky Way, The Future of Space Exploration* 1012

Gordin, Michael D., *A Well-Ordered Thing, Dmitrii Mendeleev and the Shadow of the Periodic Table* 36

Gosden, *Designing Babies* 483

Gould et al., *Conflicts Regarding Decisions to Limit Treatment* 698

Gould, Stephen Jay, *Wonderful Life, The Burgess Shale and the Nature of History* 67

Graboi, T.L., *Design for Death* 700

Greene, Brian, *The Fabric of the Cosmos; Space, Time, and the Texture of Reality* 49

Greenspan, M.D. and Shanker, D. Phil, *The First Ide* 1041

Guillemin, Jeanne, *Biological Weapons, From the Invention of State-Sponsored Programs to Contemporary Bioterrorism* 253

Hagman, Cecilia, Derrill Prevett and Wayne Murray, *DNA Handbook* 972

Halpern, Paul, *The Great Beyond, Higher Dimensions, Parallel Universes, and the Extraordinary Search for a Theory of Everything* 1018

Hamer, Dean, *The God Gene, How Faith Is Hardwired into Our Genes* 105

Hamilton, Edith, *The Greek Way* 877

Hamilton, Nigel, *Bill Clinton, an American Journey, Great Expectations* 57

Hanson and Heath, *Who Killed Homer?* 325

Hawking, Stephen W., *A Brief History of Time, From the Big Bang to Black Holes* 1034

Heinz, *The Rite of Death* 762

Heynick, *Jews and Medicine* 329

Hill, Napoleon and the Napoleon Hill Foundation, *The Law of Success, Revised and Updated, Vol., IV* 153

Hoffer, Eric, *True Believer* 197

Humphrey, Derek, *Final Exit* 747

Jenkins, Philip, *The New Anti-Catholicism, The Last Acceptable Prejudice* 119

Johnston, Josephine, *Associate for Ethics, Law, and Society* 955

Jonsen, *The Birth of Bioethics* 425

Judge, *The Impact of the Gene* 931

Kammer, *The Quest of the Sacred Baboon* 730

Kass, Leon, *Life, Liberty, and the Defense of Dignity* 505

Kaufman, Sharon R., *And A Time To Die, How American Hospitals Shape the End of Life* 698

Kawachi, Ichiro and Kennedy, Bruce P., *The Health of Nations, Why Inequality Is Harmful to Your Health* 282

Kennedy, D. James and Jerry Newcombe, *What If Jesus Had Never Been Born? The Positive Impact of Christianity in History* 167

Keyes, Daniel C., *Medical Response to Terrorism, Preparedness and Clinical Practice* 248

Keyes, W. Noel, *Government Contracts Under the Federal Acquisition Regulation, 3rd ed.* 19, 359, 1083

Keyes, W. Noel, *Life, Death & the Law* 28, 786

Kidger, Mark, *Astronomical Enigmas* 1009

Kimball, Charles, *When Religion Becomes Evil* 98

Kirwin, *The Mad, the Bad, and the Innocent* 602

Kitshnar, Robert P., *The Extravagant Universe* 35

Klinghoffer, David, *Why the Jews Rejected Jesus, The Turning Point in Western History* 103

Knight, *The Life of the Law* 500

Knox, *Death and Dying* 799

Koch, *The Devil's Details: The Limits of Principle?* 840

Komans, *China, Hong Kong, Taiwan, Inc.* 397

Kongtril, *It's Up to You, The Practice of Self-Reflection on the Buddhist Path* 1042

Lama Surya Das, *Awakening the Buddha Within* 111

Lantos, *Do We Still Need Doctors?* 830

Laughlin, Robert B., *A Different Universe (Reinventing Physics from the Bottom Down)* 44

Lavine, T.Z., *From Plato to Sartre, The Philosophic Quest* 1027

Leeming, David, *Jealous Gods and Chosen People, The Mythology of the Middle East* 117

Lentwyler, Kristin, *The Moons of Jupiter* 1017

Leonard, Karen Isaksen, *Muslims in the United States, The State of Research* 119

Leroi, Armand Marie, *Mutants, On Genetic Variety and the Human Body* 944

Levin, Christoph, *The Old Testament, A Brief Introduction* 76, 89

Lieberman, *Doors Close, Doors Open* 698

Lincoln, Don, *Understanding the Universe from Quarks to the Cosmos* 81

Lindsay, *Enhancements and Justice: Problems in Determining the Requirements of Justice in a Genetically Transformed Society* 986

Lipton, Bruce H., Ph.D., *The Biology of Belief; Unleashing the Power of Consciousness, Matter and Miracles* 907

Lynn, Joanne, M.D., Joan Harrold, M.D. and The Center to Improve Care of the Dying, *Handbook for Mortals, Guidance for People Facing Serious Illness* 1046

Macklin, Ruth, *Mortal Choices, Bioethics in Today's World* 231

Manchester, *A World Lit by Fire* 367

Marks, Jonathan, *What It Means to Be 98% Chimpanzee, Apes, People, and Their Genes* 374

McFadden, Johnjoe, *Quantum Evolution* 393

McKibben, Bill, *Maybe One* 397

Mead, *Male and Female* 417

Meisel, *The Right to Die* 697, 729

Mendez, *Death in America* 757

Mlodinow, Leonard, *Feynman's Rainbow, A Search for Beauty in Physics and in Life* 151

Morgentaler, Abraham, *The Viagra Myth, The Surprising Impact on Love and Relationships* 426

Munso, *Raising the Dead* 626

Murray, Charles, *Human Accomplishment, The Pursuit of Excellence in the Arts and Sciences, 800 B.C. to 1950* 129, 139

Murray, Thomas H., Mark A. Rothstein, and Robert F. Murray, Jr., *The Human Genome Project and the Future of Health Care* 984

Naam, Ramez, *More Than Human (Embracing the Promise of Biological Enhancement)* 931

Narby, Jeremy, *Intelligence in Nature, An Inquiry into Knowledge* 1005

Nichols, Eve K., *Human Gene Therapy* 930

Niebuhr, *The Responsible Self* 196

Novak, Michael, *The Universal Hunger for Liberty, Why the Clash of Civilizations Is Not Inevitable* 126

Nuland, Sherwin B., *How We Die* 716

Null, *Death by Medicine* 730

O'Connor, Sandra Day, *The Majesty of the Law, Reflections of a Supreme Court Justice* 32, 209

Oldstone-Moore, Jennifer, *Taoism (Origins, Beliefs, Practices, Holy Texts, Sacred Places)* 125

Oppenheimer, Mark, *Knocking On Heaven's Door, American Religion in the Age of Counterculture* 109

Park, *The Fire Within the Eve* 220

Partridge, Christopher, *The Modern Western World* 79, 124

Peck, *In Search of Stones* 872

Pelikan, Jaroslav, *Interpreting the Bible & the Constitution* 119

Pestack, *Brainscanner* 651

Peters, F.E., *The Monotheists* 76

Peters, Roger, *Practical Intelligence* 1000

Peters, Ted, *Playing God? Genetic Determinism and Human Freedom* 480

Pinker, Steven, *The Blank Slate, The Modern Denial of Human Nature* 486

Pool, *Figuring the Obesity Evidence* 736

Prentice, *Stem Cells and Cloning* 487

Preston, Ferris, *The Best of American Science Writing* 141

Prothero, Stephen, *American Jesus, How the Son of God Became a National Icon* 92

Ramsey, *Fabricated Man: The Ethics of Genetic Control* 481

Rand, *Atlas Shrugged* 196

Restak, *Brainscapes* (1995) 474

Restatement of the Law Governing Lawyers 222

Rigden, John S., *Einstein 1905, The Standard of Greatness* 1038

Ridley, Matt, *Genome* 926

Roberts, *Killing the Black Body* 565

Rose, Steven, *The Future of the Brain, The Promise and Perils of Tomorrow's Neuroscience* 300

Rosenbaum, Thane, *The Myth of Moral Justice, Why Our Legal System Fails to Do What's Right* 380

Rosenfeld, Arthur, *The Truth About Chronic Pain, Patients and Professionals on How to Face It, Understand It, Overcome It* 751

Ross, Rubin, Siegler, et al., *Ethics of a Paired-Kidney-Exchange Program* 641

Rothman, *Saying Goodbye to Daniel: When Death Is the Best Choice* 799

Rozak, *America the West* 702

Rudy, Kathy, *Beyond Pro-Life and Pro-Choice* 513

Ryan, Darwin; *The Blind Spot* 928

Salk, *Signals* 393

Salvi, *Science and Engineering Ethics* 630

Samuelson, *The Good Life and Its Discontents* 764

Sarna, Jonathan D., *American Judaism, A History* 96

Schell, Jonathan, *Fate of the Earth* 505

Schirmer, *Physician Assistant as Abortion Provider* 556

Schweizer, Peter, *Reagan's War, The Epic Story of His Forty-Year Struggle and Final Triumph Over Communism* 126

Segal, Alan F., *Life After Death, A History of the Afterlife in the Religions of the West* 158

Shermer, Michael, *The Science of Good and Evil, Why People Cheat, Gossip, Care, Share, and Follow the Golden Rule* 1030

Shinn, *Human Cloning* 491

Silver, *Remaking Eden* 212

Simon, Harvey B., M.D., *The Harvard Medical School Guide to Men's Health* 392, 648

Singer, *Animal Liberation* 669

Singer, *Rethinking Life and Death* 597

Singh, Simon, *Big Bang, The Origin of the Universe* 39

Smil, Vaclav, *The Earth's Biosphere: Evolution, Dynamics, and Change* 393

Spinoza, *Ethics* 1035

Spoto, Donald, *In Silence, Why We Pray* 154

Stanley, Thomas J., Ph.D., *The Millionaire Mind* 1002

Stein, Stephen J., *Communities of Dissent, A History of Alternative Religions in America* 115, 122

Stenchever et al., *Comprehensive Gynecology* 624

Stern, David H., *Complete Jewish Bible* 103

Strobel, Lee, *The Case for a Creator, A Journalist Investigates Scientific Evidence That Points Toward God* 49

Sullivan, John B., M.D. and Gary R. Krieger, M.D., M.P.H., D.A.B.T., *Clinical Environmental Health and Toxic Exposures* 1099

Tilich, Paul, *Systematic Theology* 208

Trefil, James, *Human Nature, A Blueprint for Managing the Earth – by People, for People* 938

Tyson, Neil deGrasse, *The Sky is Not the Limit, Adventures of an Urban Astrophysicist* 817

Vawter and Caplan, *Strange Brew: The Ethics and Politics of Fetal Tissue* 650

Veatch, Robert M., *The Basics of Bioethics, second edition* 353

Villard, Ray, and Lynette R. Cook, *Infinite Worlds* 378

Villary, *Thomas Jefferson, Genius of Liberty* 129

Wachter, Rovert M., M.D. and Kaveh G. Shojania, M.D., *Internal Bleeding, The Truth Behind America's Terrifying Epidemic of Medical Mistakes* 317

Ward, Peter D., and Donald Brownlee, *The Life and Death of Planet Earth* 36

Watson, *Ethical Implications of the Human Genome Project* 932

Wikler, *Brain Death: A Durable Consensus* 659

Wilbur, Sibyl, *The Life of Mary Baker Eddy* 106

Wills, Garry, *Inventing America: Jefferson's Declaration of Independence* 1042

Wilmut, Campbell, Tudge, *Second Creation* 454

Wilson, Edward O., *Concilience, The Unity of Knowledge* 201

Winston, Robert, *Human* 434

Woodruff, Paul, *First Democracy, The Challenge of an Ancient Idea* 1024

Wright, *The Moral Animal: Why Are We?* 525

Wynne, Clive D.L., *Do Animals Think?* 61

Xenophon, *Conversations of Socrates* 738

Zakaria, Fareed, *The Future of Freedom* 18

Author Index

Below is a lisitng of authors quoted in the text and footnotes of this volume.

Abate, Tom 4
Abelard and Heloise 366
Acton 312
Adamson, Joy 191
Aeschylus 748
Aesop 703
Akers, Charlene 741
Alighieri, Dante xv, 667
Allison, Graham 243, 251
Almond, Brenda 934
Alpert, Emanuel L. 801
Alpher, Ralph 39
Alverez 31
Amar, Akhil Reed 1042
Anderson 976
Andre, Judith 164
Andreasen, Nancy C. 227
Angell, Marcia 296
Aquinas, Thomas 178, 225, 312
Aristotle 289, 391, 498
Arnold, Matthew 193
Assman, Jan 76, 78
Attenborough, David 55
Aurelius, Marcus 176, 242, 704, 749, 751. 818
Ayala, Francisco 63, 65, 157
Bacon xxxix, 148, 778
Bagiella, E. 141
Baker, Robin 424
Baliunas, Sallie 1110
Balmer, Randall 155
Barber, Nigel 108
Baron-Cohen, Simon 997
Barr, Stephen M., 36
Bastiaat, Frederic 347
Beauchamp and Childress 181
Becker 194
Beckstrom, John 68
Behe, Michael 48
Beiner, Ronald 195
Bennett, Deborah xxxix, 39
Bennett, Claude. 702
Bennett, William 290
Benton, Michael J. 381
Berlin, Isaiah 149, 220, 285
Berumen, Michael E., 206
Berwick, Don 318
Black, Hugo 511
Bleich, David J. 550
Block, Ned, 998
Bloomfield, Arthur I. 322
Boas, Franz 291
Bobrick, Benson 44
Bok & Bok 34
Boorstin, Daniel J. 100
Boylan 1030
Brahe, Tycho 37
Brandeis, Louis 214
Branden, Nathaniel 207
Bray, Justice Absalom F. 223
Breggin 562
Breggin and Cohen 562
Brennan, William 719
Brennon, Trogen 216, 315
Brodkey, Harold 700
Brody, Baruch A. xxxvii, 955
Bronowski, Jacob 70
Brown, Lester 505
Brownlee, Donald 36
Bryan, William Jennings 84
Bryant, William Cullen 710, 818
Buddha 176
Bulkely, Kelly 703
Bulkely, Patricia 703
Burbank, Luther 152
Burke, Edmund 511, 910
Burrell 1005
Buxton, Richard 366
Byron 691, 751
Callahan, Daniel 282, 338, 465, 761

Callahan, Sidney xxi
Campbell, Joseph 111, 117, 147, 773, 1101
Cantor, Norman F. 91
Carey, Ken 40
Carlson, Richard 164
Carrel, Alexis 608
Carson, Rachel 1078
Carter, Jimmy 98
Cassandra xxxi
Cassel 907
Cherry, Reginald 107
Churchill, Winston 68
Cicero 194
Clarke, Arthur C. 43
Clinton, Hillary Rodham 385, 408, 726
Cohen, Joel 387, 393
Cohen, Jon 454
Coleridge, John Duke 280
Comings, David E. 523
Compte, August 182
Confucius xxvii, 227, 703, 1033
Conrad, Joseph 80
Cook, Lynette R. 378
Copernicus, Nicolaus 30, 93
Cowley, Robert xxxiii
Cox, George W. 377
Cox, Harvey 162, 452, 696
Coyne, George 74
Crane, Roger 179
Crick, Francis 46
Crocker 98
Croswell, Ken 1010
Daar 669
Dalai Lama 162, 200
Daly and Wilson 525
Daniels 467
Dante 99
Darwin, Charles 33, 45-70, 73, 269, 193, 540, 1037
Dawkins, Richard 61, 472
de Tocqueville, Alexis 137, 490
Deckman, Melissa M. 69
Dennett, Mary Ware 364, 404
Dershowitz, Alan 133, 135
Dewey, John xxxii, 177, 194, 208, 215, 220, 819
Diamond, Jared 61
Dillard 758
Djerassi, Carl 411
Dobzhansky, Theodosius 53, 68
Dodge, Mary 441
Donne, John 145, 146, 858
Dorff, Elliot 4, 175, 405, 411, 479-481, 515, 549, 625, 828
Dorland 614, 665
Doyle, Michael xviii
Durand, Will xxvii, 27, 124m 141, 512
Durkheim, Emile 236
Dworkin, Ronald 765
Ehrlich, Paul D. 54, 73, 364
Einstein, Albert 35, 73, 80, 110, 114, 124, 147, 155, 164, 165, 218, 331, 1021, 1034, 1038
Eisenhower, Dwight 161
Eldredge, Miles 602
Elliot, David 68
Emerson, Ralph Waldo 169, 292
Englehardt, James 209
Epictetus xliii, 225, 261
Epstein, Richard 216
Erickson 698
Euripides 719, 873
Faigman, David L. 20, 85, 131
Fan, Ruiping 175
Ferlong, John 1029
Fermi, Enrico 190
Feynman, Richard 44, 124, 151, 226, 778, 1027
Ficarra, Bernard J. 74
Fisher, John 103
Fisher, Philip 768
Fleck, Leonard 766
Fletcher, Joann 76
Flexner, Abraham 328
Fossey, Dian 59
Foucault, Michel 1035
Fox, Richard Wightman 92
Franklin, Benjamin 1, 178, 1028, 1034
Freeman, Dyson 1012
Freud, Schur M. 917
Freud, Sigmund 76, 896, 917

Friedman, Richard Elliott 102
Frost, Robert 195, 868
Fuller, Lon 291
Galton, Francis 211
Gamow, George 39
Garrean, Joel 1016
Gates, Bill 1043
Gazzaniga, Michael S. 499
Geis, Gilbert 441
Ghamari-Tabrizi, Sharon 251
Gigerenzer, Gerd 223
Gillett, Grant 661
Gingerich, Owen 74
Goldikas, Biruto 59
Goldsmith, Donald 1012
Goodall, Jane 59, 191
Gordin, Michael D. 36
Gore, Al 606
Gosden, Roger 483
Gould, Stephen Jay 50, 67, 698, 1030
Graboi, T.L. 700
Greene, Brian 49
Greenspan 1041
Guillemin, Jeanne 253
Hagman, Cecilia 972
Halpern, Paul 1018
Hamer, Dean 105
Hamilton, Edith 877
Hamilton, Nigel 57
Hanson and Heath 325
Harrold, Joan 1046
Harvey, William 488
Haught, John 63
Hawking, Stephen 32, 40, 43, 1034
Hayes, F.A. 218
Hazlitt, William 275
Hegel, Georg 203
Heinz 762
Heynick 329
Heraclitus 207, 291
Herman, Robert 39
Hill, Napoleon 153
Hillel the Elder 734
Himmelfarb, Gertrude xliii
Hippocrates xxxv, 883, 914
Hobbes, Thomas 74
Hoffman, Paul 52
Hohfeld, Wesley N. 882
Holmes, Oliver Wendell xxxvii, 68, 178
Hooper, Finley 176
Hughes, Charles Evans 1033
Humbert, Vincent 918
Hume, David 182, 460, 705, 858
Humphrey, Derek 747, 872
Hunter, John 335
Huxley, Aldous xxiv, 746
Huxley, Thomas 53
Jackson, Robert H. 200
James, William xxxii, 123, 125
Jefferson, Thomas xxiv, 32, 151, 279, 761, 1021, 1027, 1028, 1029, 1031, 1034, 1042
Jenkins, Philip 119
Johnston, Josephine 955
Jonas, Hans 185
Jonsen, Albert R. xix, 22, 425
Judge 931
Kabat-Zinn, J. 28, 836
Kammer 730
Kant, Immanuel xviii, xxxii, xli. 28, 150, 177, 198, 199, 200, 1022, 1025
Kass, Leon R. 505
Kaufman, Sharon R. 698
Kawachi, Ichiro 282
Keats 896
Keith, Kent M. 1077
Kennedy, Bruce P. 282
Kennedy, D. James 167
Kennedy, John F. xv, xviii, 710
Kepler, Johannes 28
Keyes, Daniel 248
Keyes, W. Noel xix, 19, 28, 359, 786, 1083
Kidder, Alfred xli, 201
Kidder, Rushworth M. 331
Kidger, Mark 1009
Kimball, Charles 98
Kipling, Rudyard 862, 886
Kitshnar, Robert P. 35
Klinghoffer, David 103
Knight 500

Knox 799
Koch 840
Komans 397
Komesaroff, Paul xx
Kongtril 1042
Konopka, Gisela 145
Krieger, Gary R. 1099
Kriwaczek 79
Kronman, Anthony 202
Krutch, Joseph Wood 49
Kubler-Ross, Elisabeth 719, 745–749, 789
Kuhn, Thomas xxxii, 149
Lachieze-Rey 35
Lama Surya Das 111
Lantos 830
Laszlo, Erwin 176
Laughlin, Robert 44, 1033
Lavine, T.Z. 1027
Leaky, Louis 59
Leary, Timothy 700
Lederberg, Joshua
Leeming, David 117
Lentwyler, Kristin 1017
Leonard, Karen Isaksen 119
Leroi, Armand Marie 944
Levin, Christoph 76, 89
Lieberman 698
Lincoln, Abraham 81, 178
Lincoln, Don 81
Lindberg, Charles 6
Lindsay 790
Lipton, Bruce H. 907
Locke, John 1026
Luther, Martin 97, 366
Lynn & Harrold 692, 871
Lynn, Joanne 1046
Machiavelli 24
Macklin, Ruth 22, 231, 311, 871
Madison, James 207
Mae-Wan Ho 70
Maimonides 33, 130, 182
Malraux, Andre 212
Manchester 367
Mansfield, Edward xviii
Marcuse, Herbert 165
Margulis, Lynn 51, 595, 705
Marks, Jonathan 374
Marsh, George Perkins 1081
Marx, Karl 68
Mather, Increase 710
Mayer, Ernst 13, 47, 167, 1027
McFadden, Johnjoe 393
McKhann, Charles 910
McKibben, Bill 397
Mead, Margaret 410, 417
Meisel 697, 729
Mendez 757
Michener, James A. 799
Midgley, xxxi
Mill, John Stuart xlii, 137, 185, 221, 225, 396
Miller, Kenneth 49
Mlodinow, Leonard 151
Montaigne, Michael de xli, 1, 1022
Montesquieu xxxv, 208
Morgentaler, Abraham 426
Mueller, Lisel 838
Muhammad 115
Muir, John 6, 1084
Müller, Herman 46
Mumford, Lewis 744
Munso 626
Murray, Charles 129, 139
Murray, Robert F., Jr. 984
Murray, Thomas H. 984
Murray, Wayne 972
Naam, Ramez 931
Narby, Jeremy 1005
Needham, David C. 633
Newcombe, Jerry 167
Nichols, Eve K. 930
Niebuhr, Richard 195, 196
Nietzsche, Friedrich 41, 657, 871
Noonan, John T. 624
Novak, Michael 126
Nuland, Sherwin B. 716
Null 730
O'Connor, Sandra Day 32, 209
Oldstone-Moore, Jennifer 125
Oppenheimer, Mark 109
Orwell, George 209, 667

Oseen, Carl Wilhelm 1038
Osler, William 39, 831
Pagels, Elaine 88
Paine, Thomas 127, 1034, 1039
Papenoe 466
Park 220
Parker, Michael 934
Partridge, Christopher 79, 124
Pascal, Blaise xv, xix
Patterson, Gareth 190
Pauling, Linus 46
Peck 872
Pelikan, Jaroslav 119
Pericles 1024
Pestack 651
Peters, F.E. 76
Peters, Roger 1000
Peters, Ted 480
Pimm, Stuart 1021, 1084
Pinker, Steven 486
Planck, Max 51, 70
Plato 214, 831
Plum, Fred, 702
Plutarch 123
Polinghorne, John 479
Pool 736
Pope, Alexander 221
Popper, Karl 148, 170
Posner, Richard xxxii, 180, 283
Prentice 487
Preston, Ferris 141
Prevett, Derrill 972
Prothero, Stephen 92
Proust, Marcel 28
Quill, Timothy 885, 893
Ramsey, Paul 474, 481
Rand, Ayn 196, 704
Raven, Peter 1077
Rawls, John 82, 291
Reagan, Ronald 199
Relman, Arnold 73, 174, 1097
Restak 474
Richard, Mathieu 113
Riesman, David 472
Rigden, John S. 1038
Ridley, Matt 926
Rilke, Rainer Maria 10
Roberts 565
Robertson, Pat 1033
Rodrigues-Telles, 65
Roosevelt, Theodore 179
Rose, Steven 300
Rosenbaum, Thane 380
Rosenfeld, Arthur 751
Rosner, Fred 619
Ross, Rubin, Siegler, et al. 641
Rostow, Walt 262
Rothman 799
Rothstein, Mark A. 984
Rousseau, Jean-Jacques 189, 190, 194
Rozak 702
Rudy, Kathy 513
Russell, Bertrand 177, 179, 889
Ryan, Darwin w928
Sagan, Carl 32, 160, 395, 681, 706
Salk, Jonas 393
Salvi, Maurizio 630
Sambhava, Padma 907
Samuelson 764
Sarna, Jonathan D. 96
Schell, Jonathan 505
Schirmer 556
Schlesinger, Jr., Arthur M. 1021
Schlessinger, Laura 198
Schopenhauer, Andre 910
Schweitzer, Albert 221
Schweizer, Peter 126
Segal, Alan F. 158
Shakespeare, William 53, 873, 914
Shanker, D. Phil 1041
Shapley, Harlow 38
Shaw, George Bernard xxxvii, 207
Shermer, Michael 1030
Shinn, Roger L. 490, 491
Shojania, Kaveh G. 317
Silver 212
Simon, Harvey B. 392, 648
Singer, Peter 60, 595, 597, 662, 669, 808
Singh, Simon 35, 39
Smil, Vaclav 393
Smith, Adam 355
Smoot, George 40

Snow, Charles 12
Snyder, Jack xviii
Socrates xl, 127, 146, 184, 293, 312, 1022, 1024, 1025, 1035
Solinger, Rickie 497
Solzhenitsyn, Aleksandr 127, 1023
Spinoza, Baruch 225, 330, 366, 710, 1035
Spoto, Donald 154
Stanley, Thomas J. 1002
Stein, Stephen J. 115, 122
Stenchever et al. 624
Stern, David H. 103
Stevenson, Robert Louis 218
Strauss, David Friedrich 83
Strobel, Lee 49
Sullivan, John B. 1099
Tauber, Alfred I. 765
Temple, William 152
Terkel, Studs 778
Thomas, Dylan 778
Thompson, William Irvin 706
Thoreau, Henry David 6, 707, 911
Tilich, Paul 208
Tolstoy, Leo 45
Tornstam, Lars 710
Trefil, James 938
Tuchman, Barbara 146
Tyson, Neil deGrasse 817
Varmus, Harold 621
Vawter and Caplan 650
Veatch & Spicer 327
Veatch, Robert M. 353
Villard, Ray 378
Villary 129
Virgil 22, 192
Wachter, Robert M. 317
Wall , Franz de 289
Ward and Dubos 1077, 1093
Ward, Peter D. 36
Watson, James D. 46, 472, 488, 932, 1094
Wechsler, Herbert 180
Weinberg, Steven 131, 148
Wells, H.G. xxx
Whitehead, Alfred North 166, 176
Whitman, Walt 27
Whittington, Harry 66
Wikler 659
Wilbur, Sibyl 106, 659
Wills, Garry 1042
Wilmut, Campbell, Tudge 454
Wilson, E.O 201, 284, 1085, 1105
Wilson, Woodrow 189
Winner, Lauren F. 155
Winston, Robert 434
Wolpe, Paul Root 208
Woodruff, Paul 1024
Worster, Donald 214
Wright, Frank Lloyd 525, 1021
Wynne, Clive D.L. 61
Xenophon xl, 738
Zakaria, Fareed 18

for a lesbian 466
Ashkenazi Jew genetic disease tests 943
aspirin after coronary bypass surgery 724
assisted suicide by a nonphysician 884
Assisted Suicide Funding Restriction Act 896
Association for Clinical Pastoral Education 693
Association of American Medical Colleges 302
asteroid impact 29
astrobiologist definition of habitable zone 1013
astrobiologists 41
astrobiology 44
atheistic 59, 124
Atomic Energy Commission (AEC) 41
attempt to change the suicide rate without success 863
Aurelius, Marcus 176
autonomy 190–211, 1023, 1026, 1029–35
autonomy and responsibility 190–211
Avery, Oswald 925
showed DNA to be a genetic material 925
Ayala, Francisco 63, 157

B

babies illegally abandoned 599
Baby Moses Law 523
Bacon, Francis 148
Bagiella, E. 141
Baptists 97
Southern Baptist Convention 97
basis for quality-of-life measures is lacking 572
Bayh-Dole Act 954
Beauchamp and Childress 181
Beckstrom, John 68
beginning of the right-to-die movement 704
Behe, Michael 48
Beiner, Ronald 195
being kept "alive" in a persistent or permanent vegetative state 659
Berlin, Isaiah 149, 220, 285
best interest 792, 800, 805, 826–31
Big Bang 32–41, 1022
Bill of Rights xviii, xix, xl
bioethical revolution xviii–xxi, xix
bioethical rulings 1028
bioethics xvii–xxxv
biogerontologists 696
Biological Weapons Convention 253
biology's "Big Bang" 52
biosafety protocols for all (rDNA) research 935
birth-control patch 412
births outside of marriages 368
blood pressure, predictor of stroke 723
Boas, Franz 291
body mass index 737
Bohr, Niels 1038
bone marrow transplants 634
Born-Alive Infants Protection Act 600
Brady Act 865
Brahe, Tycho 37
brain death 657–67
brain-dead potential organ donors 616
Bray, Justice Absalom F. 223
breast-feed in public 415
breast-feeding beyond the first birthday 740
Brennan, Troyen 216
Brodkey, Harold 700
Bryan, William Jennings 84
Buddhism 110–13, 125
Dalai Lama 113
Buddhists 163
Burbank, Luther 152

C

California criminal statutes do not treat human eggs or embryos 443
California Due Process in Competence Determination 271
California Institute for Regenerative Medicine 631
California law, abandonment of newborns at hospitals 417
Callahan, Daniel 282, 338, 465

Cambria explosion 66
Campbell, Joseph 117, 147
cancer cells possess telemerase 718
cancer patients undermedicated for pain 753
capital punishment costs 881–89
Carrel, Alexis 608
Carter, Jimmy 98
Cassandra xxxiii
Catholic conditions of double effect 513
Catholics more likely than Protestants to have abortions 513
Centers for Disease Control and Prevention 255
cereal fiber and lower risk of cardiovascular disease 718
cesarean 537
  delivery for extremely premature infants 589
  overprescribed 537
  woman's competent choice to refuse 538
chemical castration 403
chemotherapy treatment to bank sperm 447
Child Abuse Prevention and Treatment and Adoption Act 572
childhood obesity 736
children born more than five weeks premature 593
chimpanzees 60
China
  population control policy 380
  screening for gender 547
Chinese Academy of Sciences 484
Chinese standards 175
Christian Science Church 105
  Eddy, Mary Baker 105
chronic obstructive pulmonary disease 694
Church of Jesus Christ of Latter-day Saints 108
  Mormons 135
Churchill, Winston 68
Cicero 194
civil marriage 367, 375
Clark, Arthur 43
class-action ruling under the Racketeer Influenced and Corrupt Organizations Act 535
clinical equipoise 171
cloning 472–95
  ban on the use of federal money for cloning humans 481
  by twinning (blastomere separation) 472
  extension of in vitro fertilization 490
  for ethical research 490
  gestate the clone 482
  law allowing 482
  therapeutic cloning 483–90
Coats Amendment 557
Commandment "Thou shalt not murder" 889
Commercial Space Act 1008
Competency Interview Schedule 271
competent patient's autonomy 204
complementary and alternative medicine 173
Comprehensive Accreditation Manual for Home Health Care xxvii
Compte, August 182
computerized clinical decision support systems 172
Comstock Act 404
confine pregnant women who abuse alcohol or drugs 568
Confucianism 125
Confucius 227
congenital anomalies 584
consent to Do Not Resuscitate 816
contraceptives 409, 412, 418
  morning-after birth control 419
  no higher risk of breast cancer 503
  RU-486 418
  some pills prevent implantation 420
control and elimination of chemical and biological weapons 254
Copernicus, Nicolaus 30, 93
coronary artery bypass graft (CABG) surgery 723

coronary heart disease, leading cause of death 694
cosmic background explorer 39
cosmic background radiation 33
cosmic evolution 28–45
Council for International Organizations of Medical Sciences 240
Council of Europe 985
  Convention on Human Rights and Biomedicine 996
Council of Europe in Strasbourg, Code of Bioethics 483
Council on Ethical and Judicial Affairs 351
Coyne, George 74
Crick, Francis 46, 925
crime for unemancipated minors to have sex with each other 417
Criminalization of Female Genital Mutilation Act 579
cryopreservation, human embryos 453
cultural evolution 54
cultural inheritance 65
cystic fibrosis carrier testing 946

D

Dalai Lama 162, 200
Darwin, Charles 33, 47, 50–56, 59, 62, 64, 68–70, 1037
Data Quality Act of 2001 983
de Tocqueville, Alexis 137
dead donor rule 643
death penalty appeals 847
  super due process rules 847
death penalty prescribed in the Bible 848
death rates from prostate cancer 713
Death with Dignity Act 876, 880, 898, 905
  Oregon 876
deaths occur among low birth weight (LBW) infants 582
debate on futility 829
decision to withhold or withdraw healthcare 789
decisions to withdraw dialysis 687
Declaration of Independence 28, 32, 81, 85
  Jefferson, Thomas 32
Declaration of UNESCO 484
declining social and moral values 289
decreased libido 424
Defense Against Weapons of Mass Destruction Act 253
Defense of Marriage Act 371
deism 81, 111
denial of quality of life in decision making 574
Dennett, Mary Ware 404
denying that marriage was a sacrament 367
Department of Health and Human Services 300, 346, 359, 561, 576
  health insurance benefits 566
  impose penalties 346
Depo-Provera 413, 415
Descent of Man 924, 1005, 1037
determination of futility 833
Dewey, John 177, 194, 208, 215, 220, 819
diagnostic genetic disease tests 943
diagnostic system code for psychiatrists 908
dialysis in the United States 640
dialysis, most expensive kidney disorder billed to Medicare 686
Diamond, Jared 61
Dickey Amendment 622, 632
difference between a grant and a cooperative agreement 358
difficulty in predicting life expectancy 875
disability under the Americans with Disabilities Act (ADA) 462
Djerassi, Carl 411
DNA sample taken for fingerprinting purposes 972
Dobzhansky, Theodosius 1004
doctor unions 770
doctrine of informed consent 786, 831, 836
  does not authorize patient to demand treatment 836

Dolly, the first mammal cloned 476
domestic-partner rights 375
Donne, John 146, 858
"Don't Ask, Don't Tell" policy for gays 373
Dorff, Elliot 481, 625
double effect 881–89
double-blinded trial 964
Down syndrome test 433
drug-related morbidity and mortality 564
drug use among pregnant women 561–603
dual MD/JD degree program 21
durable powers of attorney 791–98

E

eating less leads to longer life 717
Eccles, John 1016
ecology 8
education 280–85
egg agencies 452
egg "wasted" each month 452
eggs in a lifetime 451
Ehrlich, Paul D. 54
Einstein, Albert 35, 80, 110, 114, 124, 147, 155, 164–65, 195–96, 1034, 1038
Eisenhower, Dwight 161
Electronic Event Reporting System 318
Elliot, David 68
Elshtain 467
embryo research 473, 479, 482
 cryopreservation 453
 implanted become potential human beings 454
embryo twinning 494
emergency contraception 418, 419
emergency contraceptive drug 411
Emergency Medical Treatment and Active Labor Act 242, 793, 839, 840, 842
Emergency Preparedness and Response Triad 250
emergency treatment for illegal immigrants 391
Emerson, Ralph Waldo 169
Employee Retirement Income Security Act of 1974 349–55, 352
employer-sponsored health plans 353
Encyclopedia of Bioethics 5
end-stage renal disease 684–87
Englehardt, James 209
entropy 706–07
epilepsy, lower fertility rates 424
epistemology 62
Epstein, Richard 216
Equal Access Act 131
Equal Employment Opportunity Commission 210, 309, 947
escalating health costs 774
eschatology 696
espousal support for prenatal expenses 416
ethical questions arising out of nondisclosure 314
eugenics 211–22, 930, 987
Europe is a post-Christian society 123
European Commission 630
 Directive 2001/20/EC 630
European Economic Community xlvi
European Parliament requiring labels for food 936
European Patent Office 959
European Union treaty 482
European Union Convention on Human Rights 9–11
euthanasia 880–920
evolution 64, 66–68, 68
evolutionists 163
examining life 293
executing the mentally retarded 849
execution of killers under 18 849
existentialists 1030
exorcism 92
expert witness malpractice 14–22
Expressed Sequence Tags 955

F

failure to disclose 314, 329
Faith Tabernacle church 219
Fan, Ruiping 175
Federal Acquisition Regulation 357, 360

Federal Employees Group Life Insurance 775
federal employees who make live-donor donations 680
Federal Grant and Cooperative Agreement Act 358
Federal Insecticide, Fungicide, and Rodenticide Act 250
Federal Patient Self-Determination Act 258
federal Peer Review Improvement Act 308
Federal Rules of Evidence 20
Federal Sentencing Guidelines 348
female circumcision 387
female genital mutilation 579
Fertility Clinic Success Rate and Certification Act 439–40
fertility pills 461, 465
  multiple births 441, 460, 465
Fetal Alcohol Syndrome (FAS) 565
fetal homicide 506
fetal tissue transplants 650
Feynman, Richard 44, 124, 151. 987, 996, 1016, 1027
First Amendment 286–87
Fisher, John 103
Flexner, Abraham 328
Fodor, Jerry 1006
follicle-stimulating hormone 461
Food & Drug Administration
  over-the-counter morning-after birth control 419
foreign-born population in U.S. 390
Fossey, Dian 59
Foucault, Michael 1035
four-person organ exchange 640
Franklin, Benjamin 178, 1028, 1034
Freedom of Access to Clinic Entrances (FACE) Act 502
French Code 223
French Revolution 84
Freud, Sigmund 917
Frost, Robert 195
Fuller, Lon 291

G

Galilei, Galileo 93
Gallup poll 100, 139
Galton, Francis 211, 930
Gamow, George 39
Gates, Bill 999, 1016, 1043
gene flow 938
gene transfer studies 965
General Conference of UNESCO 959
genetic disorders, detectable with genetic testing 974
genetic information beyond scope of HIPAA 949
Genetic Science Commission 940
genetic screening 433
  screening for Down syndrome 433
  Tay-Sachs disease 434, 437
genetic tests not regulated by the FDA 577
genetically engineered crops 935, 938
genomes, using four letters 926
geriatricians 720
gerotranscendence 746
gestational surrogacy contracts, not voidable 471
Goethe, Johann Wolfgang 123
Golden Rule 124, 786, 796
Golden Rule prohibition 226
Goldikas, Biruto 59
Goodall, Jane 59, 191
Gorbachev, Mikhail 100
Gospel according to Thomas 89
Gould, Stephen Jay 50
graft-versus-host disease 634
grants xxi, xxvi
gravity's speed 42
Gray, Thomas 138
great ape family 61
great equalizer 878
Green Revolution 392
Greenpeace 935
guidelines, the disposition of abandoned embryos 455

H

habitable zone 42
Han Chinese 163
handicapped child, right to recognition that birth was a mistake 526
harvesting and allocation problems in the United States 608
Haught, John 63
Hawking, Stephen 32, 40–43, 1034
Health and Human Services 241
  Privacy Rule 241
  Agency for Healthcare Policy Research (AHCPR) guidelines 754
healthcare expenditures 217
Health Care Quality Improvement Act 265, 309
Health Insurance Portability and Accountability Act 267, 345, 948, 984
  privacy rule 948
Health Maintenance Organization (HMO) 4, 349–55, 762
healthcare 280–85
healthcare costs topped $1 trillion 764
healthcare rationing 679
Hegel, Georg 203
Heidegger, Martin 331
Heraclitus 291
Herman, Robert 39
Herrnstein, Richard 998
heuristics 339
Hinduism 121
  dharma 121
  law of karma 121
  Untouchables 121
Hippocrates 883
Hippocratic oaths 287
Hispanic babies, California 394
Hobbes, Thomas 74
Hoffer, Eric 197
Hoffman, Paul 52
Hohfeld, Wesley N. 882
Holmes, Oliver Wendell 68, 114, 178
Hospice and the Hemlock Society 879
hospice facilities 874
  admit children 876
  admitting patients who were not terminally ill 877
  Medicare requirements 877
hospital's ethics committee 14–22
  physicians 14
Howard Hughes Medical Institute (HHMI) 46
Hubble, Edwin 31, 34, 1022
  galaxies began to intensify
human embryonic stem cells 628, 632
human gene inserted into rice 939
Human Genome Diversity Project 957
Human Genome Organization 957
Human Genome Project (HGP) 433, 924–34, 961, 975, 1018
  Ethical, Legal and Social Implications portion of HGP 929
human immunodeficiency virus (HIV) 726
human papillomavirus, (HPV) 409
humanities 180–85
Humbert, Vincent 918
Hume, David 182, 705, 858
Humphrey, Derek 747, 872
Huntington's disease 975, 978
Huxley, Aldous 746
Hyde Amendment 508
hypnosis 759
hysterectomies 423

I

iatrogenic injuries 263
iatrogenic treatment 803, 822
illegal status of euthanasia 904
implantable artificial heart (the Abio-Cor artificial heart) 722
in vitro fertilization 427–34. 453, 460, 475
  cost per IVF 430
  egg donation 469
  not included in Medicaid 430
inborn errors of metabolism 577
Inbreeding or incest 370
independently of management 343
industry-sponsored studies 297
infanticide for severely defective newborns 595
infant's gender 434
infertile women with HIV 429
infertility 422, 437, 453, 460, 482
  and breast-feeding 410

informed consent 190, 230–77
by minors 277
Institute of Medicine's Council of Healthcare Technology 670
institutional review boards 171, 234, 338
under federal research regulations 338
waiver of informed consent 239
insurance for assisted reproduction 462
insurer refusal to cover doctor-recommended medical treatment 353
intensive care unit 234–37, 270
Intergovernmental Bioethics Committee 11
International Academy of Humanism 487
International Atomic Energy Agency 251
International Committee of Medical Journal Editors 304
International Human Genome Sequencing Consortium 927
International Law Commission xlvii
International Medical Tribunal xlix, l
International Space Station 1012
Internet-accessible disciplinary information 733
interquartile ranges 747
intersex conditions 577–80
intestate succession 467
intimate partner violence 235
intracranial hemorrhage in low birth weight pre-term infants 588
intracytoplasmic sperm injection 437
intrauterine devices (IUDs) 411
intrauterine insemination without ovarian stimulation 464
involuntary sterilization and castration of rapists 398
irreversible coma 665

J

Jacob, Francois 926
James, William 123, 125
Jefferson, Thomas 32, 81–83, 109, 128, 1027–34, 1042
Jehovah's Witnesses 114, 219, 813–18, 858–62
maternal-fetal rights 860
right to die rather than accept blood transfusion 861
Jewish tradition 100
Job 43
Joint Commission on Accreditation of Healthcare Organizations (JCAHO) 249, 261, 327, 331–32, 348, 751
guidelines 752
Jonas, Hans 185
Justinian Code 90

K

Kant, Immanuel 177, 198, 200, 1022, 1025
Kantian principle 288
Kennedy, John F. xx
Kevorkian, Jack 912–20
intention to perform euthanasia 913
Kidder, Alfred 201
kidney transplants from a spouse 684
killings of children by stepfathers 598
King James translation 133
Kipling, Rudyard 886
Kitcher, Philip 544
Kronman, Anthony 202
Krutch, Joseph Wood 49
Kubler-Ross, Elisabeth 745–49, 780
Kuhn, Thomas 149
Kurzweil, Ray 1015

L

late-term abortions 550
Laughlin, Robert 44
Laws Governing Lawyers xxv
laws of nature, not patentable 953
Leaky, Louis 59
Leary, Timothy 700
Lederberg, Joshua 474
lethal injection as an option 849
life spans extended by low-calorie diets 740
life-threatening situations 237–42
Lincoln, Abraham 178
at Gettysburg 161
Lindberg, Charles 6, 608
liver transplantation 610, 654–55

living donor to provide a liver transplant 680
Locke, John 1026
longer life with worsening health 717
loss of brain weight 716
low birth weight babies 585
low-birthrate, low-mortality country 713
Luther, Martin 97, 366

M

Macklin, Ruth 311
Madison, James 207
Maimonides 182
Maimonides, Moses 33
malpractice insurance 15
malpractice suits or disciplinary proceedings 503
Malraux, Andre 212
mandatory testing 972–82
Manhattan Project compared to Human Genome Project 941
manufacturers of antidepressants 297
Marcuse, Herbert 165
Margolis, Lynn 51, 595, 705
Mars 41
Mars, American flight to 1009
Marx, Karl 68
mature minor rule 814
maxims of the law 326
Mayer, Ernst 13, 47, 167, 1027
McKhann, Charles 910
Mead, Margaret 410
mechanism for considering ethical issues 332
medical errors not reported 312–22
medical ethics xxi, xxxi, xxxv
medical malpractice lawsuit 14
medical schools 21
Medicare 242, 248, 317, 322, 330, 356–61
Medicare and Medicaid funds 107
  to Christian Science nonmedical "healers" 107
Medicare declared that obesity is a disease 742
Medicare payments to doctors 767
Megan's Law 399
Mendez, Omar 757
menopause 427
mentally handicapped women 402
messianic Jews 101, 103
Michener, James A. 799
Mill, John Stuart 396
Miller, Kenneth 49
millionaires of average intelligence not hired 1002
minorities less likely to become organ donors 684
minute of silence in public schools 132
misusing painkillers 756
model statutes, terminally ill patients 907
"monkey" trial 56
monkeys cloned 476
Monod, Jacques 926
monotheism 76–80
monotheists 163
Montaigne, Michael 1022
Montesquieu 208
morality 288–90
moratorium on federal funding 485
Mormon 286
Muir, John 6
Müller, Herman 46
multitheists 163
Murray, Charles 998
Muslim 103, 115–21, 139
  Allah 117
  Ayatollah Khomeini 120
  Muhammad 115–17
  Qur'an 117–20
myeloblastic leukemia 726

N

National Academy of Science 131, 623
  Guidelines for Human Embryonic Stem Cell Research 632
National Advisory Mental Health Council 272
National Association of Catholic Chaplains 693
National Association of Insurance Commissioners 283

National Bioethics Advisory Commission 171, 458, 485, 622
National Clinical Research Enterprise 966
National Family Support Act of 2000 764
national goal, adequate family planning services 383
National Health and Nutrition Examination Survey 735
National Hospice Organization 880
National Institute of Allergy and Infectious Diseases 255
National Institutes of Health 303, 626, 928–35, 1003
  funds from 928
  guidelines for some federal funding 626
  Recombinant DNA Advisory Committee 982, 1003
National Marrow Donor Program 637
National Organ Transplant Act 650
National Organization for Women, Inc. 535
National Practitioner Data Bank 262–65
National Research Council 254
National Right to Life Committee 479
nation's healthcare budget 283
natural selection 46–56, 63–68
need for rationing 284
neonatal intensive care units 585–93
neonaticide mothers 601
neonaticide syndrome 602
new definition of death 658
Niebuhr, Richard 195
Nightingale, Florence 773
no biblical injunction against human cloning 480
non-directed donation 681
non-exercise activity thermogenesis 741
nonsurgical method of sterilizing women 405
Noonan, John T. 624
Norplant 413–15
nosocomial infection rates 729
not forcing someone to eat or drink 806
Nuclear Nonproliferation Treaty 251
nuclear power for Mars Viking 1009
nullification 58
Nuremberg Code 237
nursing-home abuse 776

O

obesity 735–49
Older Americans Act (OAA) 764
organs from anencephalics 648–650
Orthodox Jewish rabbis 859
  suicide as denial of divine mastery 859
Orwell, George 209
Oseen, Carl Wilhelm 1038
Osler, William 39
out-of-hospital DNR orders 824

P

Pagels, Elaine 88
pain, the fifth vital sign 758
Paine, Thomas 1034, 1039
Papenoe 466
parent decisions on withholding, withdrawing treatment 593
parental consent for abortions 527–32
parental notice for unmarried girls under 18 528
Parkinson's disease 651
Parkland approach for futile treatment 835
partial-birth abortions 549–56, 749
Partial-Birth Abortion Ban Act of 2003 555
Patent and Trademark Office 955
patenting nonnatural genetic sequences 960
paternalism by physicians 205, 795
patient autonomy 222
patient, no right to inspect psychotherapy writing 346
Patient Self-Determination Act 782, 786
patients in ICUs who chose ethics consultation 844
patient's management goals 845
patients undergoing dialysis 685
patients with cancer were undermedicated for their pain
Patterson, Gareth 190

Pauling, Linus 46
peer review committees, hospitals 17, 308
Pericles 1024
permanently vegetative patients considered dead 810
persistent or permanent vegetative state 823
Pharaoh Akhenaten 77
Philippines 515
philosophical biology 1031
photosynthesis 46
physical activity in postmenopausal women 714
physical disabilities engender suicides 866
physician participation in capital punishment 203, 845
Physician Profile Act, Massachusetts 263
physician-assisted suicide 21, 881–912
  federal funds prohibited for 886, 896
  for nonresidents of Oregon 899
  Northern Territory of Australia 899
physicians 293–325
  accepted by physicians, problem 343
  are not scientists 147
  as proxies (surrogates) 795
  paternalistic 342
  programs for addicted physicians 308
  reducing dominance of 341
  suicide by 901
pill RU-486 418
placebo 293–325
  control trials 967
  controlled investigations 237–42
  effect against pain 298
Planck, Max 51
planets seeded from outer space 1015
Pledge of Allegiance 85, 129, 287
Pontius Pilate inscription 100
Pope John II 890
  Encyclical Evangelicum Vitae 894
Pope John Paul II 153, 384
Pope Pius XII 35, 153
Popper, Karl 148, 170
porcine endogenous 670
positron emission tomography 662
Posner, Richard 180, 283
post-viability abortion 549–56
  in Israel 553
posthumous reproduction 453, 458
potential germline gene transfers 1018
practical reason to ethics 1031
prayers at high school football games 132
pre-implantation genetic diagnosis 548
pregnancy and birth rates highest among developed countries 522
Pregnancy Discrimination Act 422
prelingually deafened children 209
prescribed drugs do not improve learning 562
preservation time 611
President Richard Nixon, war on cancer 731
President's Council on Bioethics, moratorium on research cloning 486
presumed consent 618
procreative conduct 470
principle of *stare decisis* 500
prisoners 281
privacy 24
privacy regulations 345
Project SUPPORT 772, 783–96
prostate cancer 713, 732
prostate cancer surgery 423
protestant churches 97
Public Health and Welfare Act 513
Public Health Security and Bioterrorism Preparedness Act 254
publicly funded education vouchers at religious schools 136

Q

quality improvement organizations 330
Quill, Timothy 885, 893

R

Rabbi Dorff 480
radiation dispersal devices 252
Ramsey, Paul 474
Rand, Ayn 196

Rawls, John 82, 291, 990
Reagan, Ronald 100
Recombinant DNA (rDNA) technology 935
recommended DNR 590
rediction in AIDS through highly active antiretroviral therapy (HAART) 727
registered nurses 229
registration of clinical trials 966
religion 74–86
Religious Freedom Restoration Act 134
religious views that there is no right to die 704–06
Relman, Arnold 174
reproductive assistance and insurance 428
reproductive health services 532
republic from purely democratic government 1023
Resolution on Terrorism 1540 251
responsibility 1029, 1040
Restatement: Law Governing Lawyers 222
revelation of the requirements of God 833
rhythm method 409
Richard, Mathieu 113
right of physicians to proceed with treatments without a parent's consent 578
right to choose 209
Rights of Patients in a General Acute Care Hospital 824
Rilke, Rainer Maria 10
rising drug costs 296
risk of breast cancer 503
Roman Catholicism 90, 94, 140
  American Roman Catholics 92
  Catholic Church 90
  divorced and remarried Roman Catholics 94
  John Paul II 95
  Law, Cardinal Bernard 96
  Pope Benedict XVI 94
  Pope Paul VI 94
  Ratzinger, Cardinal Joseph 94
Romeo and Juliet 866
  immature teenagers 866
Roosevelt, Theodore 179
Roslin Institute, Edinburgh 474, 478
Rosner, Fred 619
Rostow, Walt 262
Rousseau, Jean-Jacques 190, 194
Russell, Bertrand 177, 889
Russian abortions 516
Rwanda, highest growth rate in the world 394

S

safety helmets 863
  bicycle helmets for children 864
  motorcycle riders, risks of 863
Sagan, Carl 32, 138, 160, 395, 681, 707
sale of sperm or eggs 452
sales of drugs by manufacturers 295
Salk Institute for Biological Studies 714
Salvi, Maurizio 630
same-sex marriages 371
Schiavo, Terri 811
schizophrenia 273
Schlessinger, Laura 141
Schrag, Dan 52
scientific creationism 57, 58
scientific creationism bill 130
  statute's term 130
scientific integrity, ranking 293
Scopes, John 56–58
screening examination 242
screening for genetic defects 541
search for extraterrestrial intelligence 1011
section 402A of the ALI's Restatement (Second) 652
secular humanism 1031
secular, official in France 140
selling of body parts 641
Sells, Michael 119
severe combined immune deficiency (SCID) 973
sex determination 546
sex-linked genetic disease 437
sexual promiscuity 413
sexually transmitted diseases (STDs) 411
sexually transmitted infections (STIs) 419
Shapley, Harlow 38

Shinn, Roger L. 490
shortened telomeres 994
sickest person nationally 675
Singer, Peter 60, 595, 662
single children have higher IQs 397
six-step "fair process" end of life in futile case 852
skin cancers 610
Sloan, R.P. 141
slow codes 855
  after cardiac arrest 855
  diminish the motions of CPR 855
smokers 217
smoking-related illnesses 712
Smoot, George 40
Snow, Charles 12
social science not a science 184
Social Security taxes 764
Society of Hospital Medicine 770
Socrates 293, 312, 1022–25, 1035
  characterization of virtue 292
Solzhenitsyn, Aleksandr 1023
somatic cell nuclear transfer 478
Somatic Gene Therapy 972–82
Southern Baptist Convention 369
Space Adaptation Syndrome 1015
special pacemakers 723
speed of natural selection 962
sperm banks 446–50
  New England Cryogenic Center 447
sperm put in frozen storage before committing suicide 448
sperm sorting 446
Spirituality and Healing 708
State Children's Health Insurance Program 566
state Emergency Medical Assistance Compact 247
state genetic nondiscrimination laws 947
state legislation, valid if not undue burden 500
statutes licensing midwives 557
step toward stopping uncontrolled increase in population 504
sterilization 382, 398–405
Strauss, David Friedrich 83
subjective personhood 808–20
Substantive Due Process xviii
substitute for blood 861
substituted judgment 826–31
suicides
  after publication of *Final Exit* 872
  ninth leading cause of death 864
supernova
  type I 38
  type II 38
surgical castration and reduction of recidivism rate 400
surrogate (substitute) birth mother 468

T

Talmud 101
Taoism 125
Task Force on Standards for Bioethics Consultation 336
tax exemption limitation 382
  birth of third child 382
teens likely to deliver premature babies 584
television as cause of adolescent obesity 740
telemerase 718
telomeres 994
Ten Commandments 133
  large park 134
  three school buildings 133
terminal patients with six months to live 874
Termination of Life on Request and Assisted Suicide Act 917
terrorist attacks 242–50
testamentary gift of organs 638
textbooks promoting religion 1031
therapeutic cloning of cells 629
Thoreau, Henry David 6, 707, 911
Tornstam, Lars 710
Toxic Substances Control Act 250
transplant waiting list 608
transplanted kidneys 611
transplanting organs from executed prisoners 646
treatment alternatives 230

tribolic 55
tribolites 55
tubal ligation 405
Tuchman, Barbara 146

U

U.S. Constitution 24, 80–143, 280–85, 1034
  antiestablishment clause 82
  Article VI 82
  First Amendment xxxvi, 84, 286, 287
  secular purpose 186
U.S. cremation rate 609
U.S. Medical Licensing Examination 968
ultramedical 16
ultraphysicians 25
UNESCO 11
Uniform Anatomical Gift Act (UAGA) 449, 616, 644, 650, 655, 658
  state 683
Uniform Health-Care Decisions Act 791, 797
Uniform Parentage Act 467
unintended pregnancies 415–17, 524
Unitarian Universalist Association 109–10
  Unitarian church 110
United Methodist Church 99
  right to die in dignity 890
United Nations Food and Agriculture Organization 936
United Nations Fund for Population Control 398, 504
United Nations Security Council xlv
United Network for Organ Sharing 609, 612, 642, 675
  allocation secondary to direct 683
  exports up to 5 percent of organs 679
Universal Declaration of Human Rights 9, 196
University of California–Irvine's Center for Reproductive Health 441
upper age of donor egg recipients 443
upper limit for sustained population levels 379
Urantia 110
using good v. Good 1024
Utopian eugenics 545

V

vagueness, ground to declare statute invalid 325
variation in IQ scores 998
Venter, Craig 928, 931, 955, 986
Viagra 414
  use by women 426
viatical settlement in life insurance 775
Volstead Act 911
  greatest social experiment of modern times 911
voluntary sterilization 382

W

waivers 238
Wall, Franz de 289
wall of separation 127–43
war on drugs 562
Washington, George 86
Watson and Crick determined the double-helix structure 925
Watson, James 46, 925–32, 955, 972, 986
Web site condemning abortion doctors 534
Wechsler, Herbert 180
Weinberg, Steven 131, 148
Weldon Amendment 511
Wells, H.G. xxxii
Westermarck Effect 370
when telomeres disappear, cells may cease to replicate 718
whistleblowers under the False Claims Act 356
Whitehead, Alfred North 166
Whittington, Harry 66
whole brain death 657–67
withdrawal of life-sustaining treatments 780
withdrawing medically assisted nutrition and hydration 809
withholding equals withdrawal 802–03
  withholding and withdrawing of life support 844

Wolf, Congressman Frank R. 112
Wolpe, Paul Root 208
Women's Contraception Equity Act 516
World Cancer Report 731
World Health Organization (WHO) 483, 754
  guidelines 754
  rating of France's healthcare system 591
World Medical Association
  Declaration of Helsinki 969
World Trade Organization (WTO) xlviii

X

xenotransplants 668–75